Isotope Geochemistry

T0256801

Isotope Geochemistry

SECOND EDITION

William M. White

Department of Earth and Atmospheric Sciences
Cornell University
Ithaca, NY, USA

This second edition first published 2023
© 2023 John Wiley & Sons Ltd

Edition History
John Wiley & Sons Ltd (1e, 2015)

All rights reserved. No part of this publication may be reproduced, stored in a retrieval system, or transmitted, in any form or by any means, electronic, mechanical, photocopying, recording, or otherwise, except as permitted by law. Advice on how to obtain permission to reuse material from this title is available at http://www.wiley.com/go/permissions.

The right of William M. White to be identified as the author of this work has been asserted in accordance with law.

Registered Offices
John Wiley & Sons, Inc., 111 River Street, Hoboken, NJ 07030, USA
John Wiley & Sons Ltd, The Atrium, Southern Gate, Chichester, West Sussex, PO19 8SQ, UK

For details of our global editorial offices, customer services, and more information about Wiley products, visit us at www.wiley.com.

Wiley also publishes its books in a variety of electronic formats and by print-on-demand. Some content that appears in standard print versions of this book may not be available in other formats.

Trademarks: Wiley and the Wiley logo are trademarks or registered trademarks of John Wiley & Sons, Inc. and/or its affiliates in the United States and other countries and may not be used without written permission. All other trademarks are the property of their respective owners. John Wiley & Sons, Inc. is not associated with any product or vendor mentioned in this book.

Limit of Liability/Disclaimer of Warranty
While the publisher and authors have used their best efforts in preparing this work, they make no representations or warranties with respect to the accuracy or completeness of the contents of this work and specifically disclaim all warranties, including without limitation any implied warranties of merchantability or fitness for a particular purpose. No warranty may be created or extended by sales representatives, written sales materials or promotional statements for this work. This work is sold with the understanding that the publisher is not engaged in rendering professional services. The advice and strategies contained herein may not be suitable for your situation. You should consult with a specialist where appropriate. The fact that an organization, website, or product is referred to in this work as a citation and/or potential source of further information does not mean that the publisher and authors endorse the information or services the organization, website, or product may provide or recommendations it may make. Further, readers should be aware that websites listed in this work may have changed or disappeared between when this work was written and when it is read. Neither the publisher nor authors shall be liable for any loss of profit or any other commercial damages, including but not limited to special, incidental, consequential, or other damages.

Library of Congress Cataloging-in-Publication Data
Name: White, William M., 1948– author.
Title: Isotope geochemistry / William M. White, Department of Earth &
  Atmospheric Sciences, Cornell University, Ithica, New York, USA.
Description: 2nd edition. | Hoboken, NJ : Wiley, 2023. | Includes
  bibliographical references and index.
Identifiers: LCCN 2022033471 (print) | LCCN 2022033472 (ebook) | ISBN
  9781119729938 (paperback) | ISBN 9781119729921 (adobe pdf) | ISBN
  9781119729945 (epub)
Subjects: LCSH: Isotope geology. | Geochemistry. | Earth sciences.
Classification: LCC QE501.4.N9 W55 2023 (print) | LCC QE501.4.N9 (ebook)
  | DDC 551.9–dc23/eng20230106
LC record available at https://lccn.loc.gov/2022033471
LC ebook record available at https://lccn.loc.gov/2022033472

Cover Design: Wiley, based on a design by Kasia White
Cover Image: © USGS Hawaii Volcano Observatory; courtesy of William White; © NASA, © L. Sue Baugh/Wild Stone Arts (2002);
© José Braga & Didier Descouens, used under CC BY-SA 4.0; © L. Sue Baugh/Wild Stone Arts (2002)

Set in 11/12pt SabonLTStd by Straive, Pondicherry, India
Printed and bound by CPI Group (UK) Ltd, Croydon, CR0 4YY

C9781119729938_060223

Contents

Preface

In the preface to the first edition of this book, I noted the vast impact that isotope geochemistry has had on the earth sciences in the previous half century. That impact has only grown over the last decade. I stated that "nearly every earth scientist needs some exposure to, if not fluency in, isotope geochemistry." That is even truer today.

Not only has the impact of isotope geochemistry increased, but the field itself has also vastly expanded, particularly in stable isotope geochemistry, which now encompasses essentially every element having more than a single isotope. A decade ago, isotope geochemists were merely exploring the variations in the isotopic compositions of what were called "non-traditional isotopes." The old joke about a geochemist being someone with a technique in search of a problem seemed to apply. However, that exploration has paid off enormously and led to many new and important insights into how the Earth works and how it has evolved over time. It seems to me that this is particularly true in three areas: the nature of and processes within the nebula of gas and dust from which the solar system formed; the extent to which geochemical cycles encompass entire outer 3000 km of the planet, and possibly even the core as well; and how the atmosphere and climate have evolved over Earth's history.

This expansion of the field has necessitated an expansion of the book and a reorganization of it as well. It is now organized more around topics rather than groups of isotopes. The first four chapters cover the essentials of radiogenic isotope geochemistry and geochronology. Because stable isotope geochemistry is so important to understanding the solid Earth and Solar System it formed within, the fundamentals of staple isotope fractionations are now presented in Chapter 5. Chapter 6 then covers Solar System formation from a perspective of both stable and radiogenic isotope geochemistry. Chapters 7 and 8 cover the mantle and crust, mostly from a radiogenic isotope geochemistry perspective; however, stable isotopes are brought into the picture as well. Chapter 9 then examines the stable isotope geochemistry of the solid Earth more fully. In the first edition, the so-called "non-traditional isotopes," which were treated in a separate chapter, have become more traditional, so to speak, and are now integrated with the light stable isotopes in this chapter. Chapters 10 and 11 cover processes at the Earth's surface, the *exogene,* with the light stable isotopes covered in Chapter 10 and radiogenic and heavier stable isotopes covered in Chapter 11. Having laid out the operation of the modern exogene, Chapter 12 is then devoted to the evolution of the exogene, particularly climate and atmospheric history. Chapter 13 provides a brief look at how isotope geochemistry is being applied to biological evolution, paleoecology, and archeology. Chapter 14 is then devoted to noble gas geochemistry.

This book is intended not only primarily as a textbook for advanced undergraduate and graduate students, but also as a professional reference. As a textbook, there is far more material than can be presented in a single-semester course. This is intended, giving instructors the opportunity to pick and choose. For example, a course focused on the solid Earth might include Chapters 1 through 3, Chapters 5 through 9, and Chapter 14. A course focused on surficial processes might include just Chapters 1, 4, and 5 and Chapters 10 through 13.

I will conclude by pointing out that some of what was written in the first edition of this book has turned out to be wrong. And no doubt, some of what is written in this edition will also be shown by future studies to be wrong also. That is the nature of science: nothing is written in stone. Then there is the issue of the inevitable "typos" and other errors that I have inevitably introduced. I have found through experience that I can never catch and correct all of them and for this I apologize in advance.

Finally, I would like to thank all those who have commented (and pointed out some of my errors) on this manuscript and the first edition. And again, I would like to extend my thanks to the fine people at Wiley who have worked to transform this typescript to an actual book.

Ithaca, NY
July, 2022

About the companion website

This book is accompanied by a companion website.

www.wiley.com/go/white/isotopegeochem2

This website includes:

- Figures from the book as PowerPoint slides
- Tables from the book as PDFs
- Solutions to Problems

Chapter 1

Atoms and Nuclei: Their Physics and Origins

1.1 INTRODUCTION

Isotope geochemistry has grown over the last 60 years to become one of the most important fields in the earth sciences. It has two broad subdivisions, namely, radiogenic isotope geochemistry and stable isotope geochemistry. These subdivisions reflect the two primary reasons why the relative abundances of isotopes of elements vary in nature, which are radioactive decay and chemical fractionation; in this context, "fractionation" is any process in which the isotopes of the same element behave differently. One might recognize a third subdivision, i.e., cosmogenic isotope geochemistry, where interactions with high-energy cosmic rays produce nuclear changes.

The growth in the importance of isotope geochemistry reflects its remarkable success in attacking fundamental problems of earth science, as well as problems in astrophysics, physics, and biology (including medicine). Isotope geochemistry has played an important role in transforming geology from a qualitative, observational science into a modern quantitative one. To appreciate the point, consider the Ice Ages, a phenomenon that has fascinated the geologist and the layman alike for more than 150 years. The idea that much of the Northern Hemisphere was once covered by glaciers was first advanced by Swiss zoologist Louis Agassiz in 1837. His theory was based on observations of geomorphology and modern glaciers in the Alps. Over the next 100 years, this theory advanced very little, other than the discovery that there had been more than one ice advance

and that it was a global phenomenon. No one knew exactly when these ice advances had occurred, how long they lasted, or why they occurred. Stable and radiogenic isotopic studies in the last 60 years have determined the exact times of these ice ages and the exact extent of temperature change (i.e., about 3°C or so in temperate latitudes, and more at the poles). Knowing the timing of these glaciations has allowed us to conclude that variations in the Earth's orbital parameters (the Milankovitch parameters) and resulting changes in insolation have been the pacemaker of these ice ages. Comparing isotopically determined temperatures with carbon dioxide (CO_2) concentrations in bubbles in carefully dated ice cores leads to the conclusion that changes in atmospheric CO_2 concentration in response to these changes in insolation were the immediate cause of these climate swings. Careful uranium–thorium (U–Th) dating of corals has also revealed the detailed timing of the melting of the ice sheet and the consequent rise in sea level. Comparing this with stable isotope geothermometry shows that melting lagged warming (not too surprisingly). Other isotopic studies revealed changes in the ocean circulation system as the last ice age ended. In turn, changes in ocean circulation were likely the principal driver of changes in atmospheric CO_2. Forty years ago, all this seemed very interesting, but not very "relevant." Today, it provides us with critical insights into how the planet's climate system works. With the current concern over potential global warming and greenhouse gases, this information seems extremely "relevant."

Isotope Geochemistry, Second Edition. William M. White.
© 2023 John Wiley & Sons Ltd. Published 2023 by John Wiley & Sons Ltd.
Companion Website: www.wiley.com/go/white/isotopegeochem2

Some isotope geochemistry even seeps into public consciousness through its application to archeology and forensics. For example, a *National Geographic* television documentary described how carbon-14 dating of 54 beheaded skeletons in a mass grave in Dorset, England revealed these skeletons were from the tenth century and how strontium and oxygen isotope ratios revealed they were those of Vikings executed by Anglo-Saxons and not vice versa, as originally suspected. And the story of how isotopic analysis of the remains of English King Richard III, which were found under a parking lot in Leicester, revealed details of his life was widely reported in newspapers. Forensic isotopic analysis gets occasional mention both in shows, such as *CSI: Crime Scene Investigation*, and in newspaper reporting of real crime investigations.

Other examples of the impact of isotope geochemistry would include diverse topics, such as ore genesis, mantle dynamics, hydrology and hydrocarbon migration, monitors of the cosmic ray flux, crustal evolution, the origin of life, plate tectonics, volcanology, oceanic circulation, atmospheric evolution, environmental protection and monitoring, and paleontology. Indeed, there are few, if any, areas of geological inquiry where isotopic studies have not had a significant impact.

One of the first applications of isotope geochemistry remains one of the most important, namely geochronology, the determination of the timing of events in the history of the Earth and the Solar System. The first "date" was obtained in 1907 by Bertram Boltwood, a Yale University chemist, who determined the age of uranium ore samples by measuring the amount of the radiogenic daughter of U, namely, lead (Pb), present. Other early applications include determining the natural abundance of isotopes carried out by Alfred Nier in the first half of the twentieth century, providing key constraints on models of the nucleus and the origin of the elements (nucleosynthesis). Work on the latter problem still proceeds. The origins of stable isotope geochemistry date to the work of Harold Urey and his colleagues in the late 1940s. Paleothermometry was one of the first applications of stable isotope geochemistry as it was Urey who

recognized the potential of stable isotope geochemistry to solving the riddle of the Ice Ages.

This book touches on many, though not all, of these applications. Chapter 1 begins with a brief consideration of the physics of the nucleus and then that of the process that created the elements, called nucleosynthesis. Chapters 2 through 4 explore geochronology, and the myriad ways in which isotopes can be used as clocks. We will then introduce the fundamental principles underlying stable isotope geochemistry. With a basic understanding of both stable geochemistry and radiogenic geochemistry, we can then explore how they are used to understand the origin of the Solar System and the Earth. The following three chapters (i.e., Chapters 7–9) explore the isotope geochemistry of the mantle and crust, with radiogenic isotope geochemistry in Chapters 7 and 8 and stable isotope geochemistry in Chapter 9. Chapters 10 and 11 focus on the stable isotope geochemistry of the interactions occurring in the oceans, atmosphere, and land surface, called the "exogene." Chapter 13 elaborates the use of isotopes in the history of the life and archeology. Finally, Chapter 14 explores the isotope geochemistry of those elements at the first right of the periodic table, the noble gases, whose isotopic variations are due to both nuclear and chemical processes and provide special insights into the origins and behavior of the Earth.

1.2 PHYSICS OF THE NUCLEUS

1.2.1 Early development of atomic and nuclear theory

John Dalton, an English schoolteacher, first proposed in 1806 that all matter consists of atoms. In 1815, William Prout found that atomic weights were integral multiples of the mass of hydrogen (H). This observation was strong support for the atomic theory, though it was subsequently shown to be only approximate. Joseph John Thomson of the Cavendish Laboratory in Cambridge developed the first mass spectrograph and in 1912, his analysis of neon[1] showed why: those elements not having integer weights had several *isotopes*, each of which had mass that was an integral

---

[1] Thomson identified neon-20 and neon-22. Decades would pass before the much less abundant isotope neon-21 was discovered.

multiple of the mass of H. In the meantime, Ernest Rutherford, who was also from the Cavendish Laboratory, had made another important observation: that atoms consisted mostly of empty space. This led to Niels Bohr's model of the atom proposed in 1910; the model held that the atoms consisted of a nucleus, which contained most of the mass, and electrons in orbit about it.

In 1913, Frederick Soddy, then at the University of Glasgow, formulated the concept of isotopes: that certain elements exist in two or more forms, which have different atomic weights but are indistinguishable chemically. It was nevertheless unclear why some atoms had different atomic weights than other atoms of the same element. The answer to this question finally provided in 1932 by James Chadwick, who had been a student of Rutherford, was "the neutron". Chadwick won the 1935 Nobel Prize in physics for his discovery, although Walther Bothe and Herbert Becker from Germany had earlier discovered the particle but mistook it for gamma radiation. Various other experiments showed the neutron could be emitted and absorbed by nuclei; hence, it became clear that differing numbers of neutrons caused some atoms of an element to be heavier than others. This bit of history leads to our first basic observation about the nucleus that it consists of protons and neutrons.

1.2.2   Some definitions and units

Before we consider the nucleus in more detail, let us set out some definitions – N: the number of neutrons; Z: the number of protons (same as atomic number since the number of protons dictates the chemical properties of the atom); A: mass number ($N + Z$); M: atomic mass; and I: neutron excess number ($I = N - Z$). *Isotopes* have the same number of protons but different number of neutrons; *isobars* have the same mass number ($N + Z$); and *isotones* have the same number of neutrons but different number of protons.

The basic unit of nuclear mass is the unified atomic mass unit (*u*) or dalton

(Da)[2], as one-twelfth the mass of ^{12}C; that is, the mass of ^{12}C is 12 unified atomic mass units. 1 u = $1.66053906660 \times 10^{-27}$ kg = 931.49410242 MeV.

The masses of atomic particles are:

Proton: 1.007276467 u = 1.67262178
$\times 10^{-27}$ kg = 938.2720 MeV/c^2
Neutron: 1.008664916 u
Electron: 0.0005485799 u = 9.10938291
$\times 10^{-31}$ kg = 0.5109989 MeV/c^2

1.2.3   Nucleons, nuclei, and nuclear forces

Figure 1.1 is a plot of N versus Z showing which nuclides are stable. A key observation in understanding the nucleus is that not all combinations of N and Z result in stable nuclides. In other words, we cannot simply throw protons and neutrons (collectively termed *nucleons*) together randomly and expect them to form a nucleus. For some combinations of N and Z, a nucleus forms but is unstable, with half-lives from >10^{15} yr to <10^{-12} s. A relative few combinations of N and Z result in stable nuclei. Interestingly, these stable nuclei generally have $N \approx Z$, as shown in Figure 1.1. Also, notice that for small A, $N = Z$, and for large A, $N > Z$. This is another important observation that will lead us to the first model of the nucleus.

Roughly half of the nucleus consist of protons, which obviously tend to repel each other by coulombic (electrostatic) force. From the observation that nuclei exist at all, it is apparent that another force must exist that is stronger than coulomb repulsion at short distances. It must be negligible at larger distances; otherwise, all matter would collapse into a single nucleus. This force, called the *strong nuclear force*, is one of the fundamental forces of nature (or a manifestation of the single force in nature if you prefer unifying theories). If this force is assigned a strength of 1, then the strengths of other forces are as follows: electromagnetic force: 10^{-2}; weak force (which we will discuss later): 10^{-5}; and gravitional force: 10^{-39}. Just as electromagnetic

---

[2] A previous unit, the atomic mass unit or *amu*, was defined as 1/16th the mass of ^{16}O. Physicists tend to use "u" while biochemists tend to use "Da". Both are accepted as SI units by the General Conference on Weights and Measures, which is the international authority responsible for such matters. *amu* continues to be used among geochemists, as the difference between *u* and *amu* is negligible.

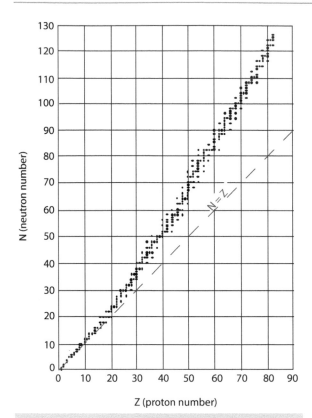

Figure 1.1   Neutron number versus proton number for stable nuclides.

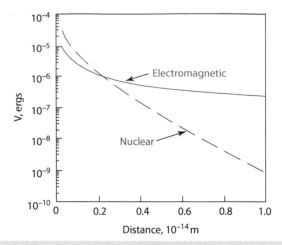

Figure 1.2   The nuclear and electromagnetic potential of a proton as a function of distance from the proton.

forces are mediated by a particle, the photon, the nuclear force is mediated by the *pion*. The photon carries one quantum of electromagnetic force field and the pion carries one quantum of nuclear force field. A comparison of the relative strengths of the nuclear and electromagnetic forces as a function of distance is shown in Figure 1.2.

1.2.4   Atomic masses and binding energies

The carbon-12 atom consists of six neutrons, six protons, and six electrons. However, using the masses listed above, we find that the masses of these 18 particles sum to more than 12 u, the mass of ^{12}C atom. There is no mistake; they do not add up. What has happened to the extra mass? The mass has been converted into the energy binding the nucleons. It is a general physical principle that the lowest energy configuration is the most stable. We would expect that if ^{4}He is stable relative to two free neutrons and two free protons, ^{4}He must be a lower-energy state compared to the free

particles. If this is the case, then we can predict from Einstein's mass–energy equivalence:

$$E = mc^2 \qquad (1.1)$$

that the mass of the helium nucleus is less than the sum of its constituents. We define the *mass decrement* of an atom as:

$$\delta = W - M \qquad (1.2)$$

where W is the sum of the mass of the constituent particles and M is the actual mass of the atom. For example, W for ^{4}He is $W = 2m_p + 2m_n + 2m_e = 4.03298$ u. The mass of ^{4}He is 4.002603 u, so $\delta = 0.0306767$ u. Converting this into energy using Equation (1.1) yields 28.28 MeV. This energy is known as the *binding energy*. Up on divided by A, the mass number, or the number of nucleons, gives the *binding energy per nucleon*, E_b:

$$E_b = \left[\frac{W - M}{A}\right] c^2 \qquad (1.3)$$

This is a measure of nuclear stability: those nuclei with the largest binding energy per nucleon are the most stable. Figure 1.3 shows E_b as a function of mass. Note that the nucleons of intermediate mass tend to be the most stable. This distribution of binding energy is important to the life history of stars, the abundances of the elements, and radioactive decay, as we shall see.

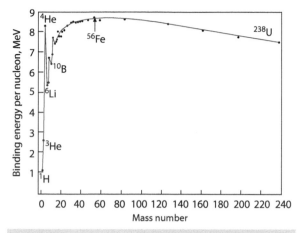

Figure 1.3 Binding energy per nucleon versus mass number.

Some indication of the relative strength of the nuclear binding force can be obtained by comparing the mass decrement associated with it to that associated with binding an electron to a proton in a hydrogen atom. The mass decrement we calculated previously for He is of the order of 1%, or one part in 10^2. The mass decrement associated with binding an electron to a nucleus is of the order of one part in 10^8. Therefore, bonds between nucleons are about 10^6 times stronger than bonds between electrons and nuclei.

Pions and the nuclear force

As we noted, we can make an *a priori* guess as to two of the properties of the nuclear force; namely, it must be very strong and it must have a very short range. Since neutrons and protons are subject to the nuclear force, we may also conclude that it is not electromagnetic in nature. What inferences can we make on the nature of the force and the particle that mediates it? Will this particle have a mass, or be massless like the photon?

All particles, whether they have mass or not, can be described as waves according to quantum theory. The relationship between the wave properties and the particle properties is given by the *de Broglie equation*:

$$\lambda = \frac{h}{p} \tag{1.4}$$

where h is Planck's constant, λ is the wavelength, called the *de Broglie wavelength*, and p is momentum. Equation (1.4) can be rewritten as:

$$\lambda = \frac{h}{mv} \tag{1.5}$$

where m is mass (relativistic mass, not rest mass) and v is velocity. From this relation, we see that mass and de Broglie wavelength are inversely related: massive particles will have very short wavelengths.

The wavefunction associated with the particle may be written as:

$$\frac{1}{c^2} \frac{\partial^2 \psi}{\partial t^2} - \nabla^2 \psi(x, t) = -\left(\frac{mc}{\hbar}\right)^2 \psi(x, t) \tag{1.6}$$

where ∇^2 is simply the Laplace operator:

$$\nabla^2 \equiv \frac{\partial^2}{\partial x^2} + \frac{\partial^2}{\partial y^2} + \frac{\partial^2}{\partial z^2}$$

The square of the wavefunction, ψ^2, describes the probability of the particle being found at some point in space x and some time t. In the case of the pion, the wave equation also describes the strength of the nuclear force associated with it.

Let us consider the particularly simple case of a time-independent, spherically symmetric solution to Equation (1.6) that could describe the pion field outside a nucleon located at the origin. The solution will be a potential function $V(r)$, where r is radial distance from the origin and V is the strength of the field. The condition of time independence means that the first term on the left in Equation (1.6) will be 0, so the equation assumes the form:

$$\nabla^2 V(r) = -\left(\frac{mc}{\hbar}\right)^2 V(r) \tag{1.7}$$

where r is related to x, y, and z as:

$$r = \sqrt{x^2 + y^2 + z^2}$$

and

$$\frac{\partial r}{\partial x} = \frac{x}{r}$$

Using this relationship and a little mathematical manipulation, the Laplace operator in Equation (1.7) becomes:

$$\nabla^2 V(r) = \frac{1}{r^2}\frac{\mathrm{d}}{\mathrm{d}r}\left(r^2 \frac{\mathrm{d}V(r)}{\mathrm{d}r}\right) \tag{1.8}$$

and Equation (1.7) becomes:

$$\frac{1}{r^2}\frac{\mathrm{d}}{\mathrm{d}r}\left(r^2 \frac{\mathrm{d}V(r)}{\mathrm{d}r}\right) = -\left(\frac{mc}{\hbar}\right)^2 V(r)$$

Two possible solutions to this equation are:

$$\frac{1}{r}\exp\left(-r\frac{mc}{\hbar}\right)$$

and

$$\frac{1}{r}\exp\left(+r\frac{mc}{\hbar}\right)$$

The second solution corresponds to a force increasing to infinity at infinite distance from the source, which is physically unreasonable; thus, only the first solution is physically meaningful. Our solution, therefore, for the nuclear force is:

$$V(r) = \frac{C}{r}\exp\left(-r\frac{mc}{\hbar}\right) \tag{1.9}$$

where C is a constant related to the strength of the force. The term $mc/\hbar$ has units of length^{-1}. It is a constant that describes the effective range of the force. This effective range is about 1.4×10^{-15} m. This implies a mass of the pion of about 0.15 u. It is interesting to note that for a massless particle, Equation (1.7) reduces to:

$$V(r) = \frac{C}{r} \tag{1.10}$$

which is just the form of the potential field for the electromagnetic force. Thus, both the nuclear force and the electromagnetic force satisfy the same general Equation (1.9). Because pion has mass while the photon does not, the nuclear force has a very much shorter range than the electromagnetic force.

A simple calculation shows how the nuclear potential and the electromagnetic potential will vary with distance. The magnitude for the nuclear potential constant C is about 10^{-27} Jm. The constant C in Equation (1.10) for the electromagnetic force is e^2 (where e is the charge on the electron) and has a value of 2.3 × 10^{-28} Jm. Using these values, we can calculate how each potential will vary with distance. This is just how Figure 1.2 was produced.

1.2.5 The Liquid-drop model

Why are some combinations of N and Z more stable than others? The answer has to do with the forces between nucleons and how nucleons are organized within the nucleus. The structure and organization of the nucleus are questions still being actively researched in physics, and full treatment is certainly beyond the scope of this text, but we can gain some valuable insight into nuclear stability by considering two of the simplest models of nuclear structure. The simplest model of the nucleus is the *liquid-drop model*, which was proposed by Niels Bohr in 1936. This model assumes all nucleons in a nucleus have equivalent states. As its name suggests, the model treats the binding between nucleons in a way similar to the binding between molecules in a liquid drop. According to the liquid-drop model, the total binding energy of nucleons is influenced by four effects: a volume energy, a surface energy, an excess neutron energy, and a coulomb energy. The variation of three of these forces with mass number and their total effect are shown in Figure 1.4.

In the liquid-drop model, the binding energy is given as a function of mass number, A, and neutron excess number, I (= N - Z), as:

$$B(A, I) = a_1 A - a_2 A^{2/3} - a_3 I^2/4A - a_4 Z^2/A^{1/3} + \delta \tag{1.11}$$

where

a_1: heat of condensation (volume energy $\propto A$) = 14 MeV
a_2: surface tension energy = 13 MeV
a_3: excess neutron energy = 18.1 MeV
a_4: coulomb energy = 0.58 MeV
δ: even–odd fudge factor, with binding energy greatest for the even–even factor and smallest for the odd–odd factor

Some of the nuclear stability rules here can be deduced from Equation (1.11). Solutions for Equation (1.11) at constant A, i.e., for isobars, result in a hyperbolic function of I as illustrated in Figure 1.5. In other words, for a given number of protons, there are an optimal number of neutrons: either too many or too few results in a higher-energy state. For odd A, one nucleus will lie at or near the bottom of this function (energy well). For even A, two curves result: one for odd–odd and the other for even–even. The even–even curve will be the one with the lower (more stable) one. Nuclei with either too many or too few neutrons tend to decay to nuclide with the optimal number of neutrons. From the perspective isotope geochemistry, the importance of these double curves is *branched decay*, in which an unstable nucleus can decay to two possible daughters: ^{40}K can decay to either ^{40}Ca or ^{40}Ar, and ^{138}La can decay to either ^{138}Ba or ^{138}Ce.

1.2.6 The Shell Model of the nucleus

1.2.6.1 Odd–even effects, magic numbers, and shells

Something that we have alluded to and the liquid-drop model does not explain is the even–odd effect, which is illustrated in

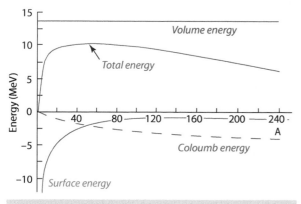

Figure 1.4 Variation of surface, coulomb, and volume energy per nucleon and their total versus mass number.

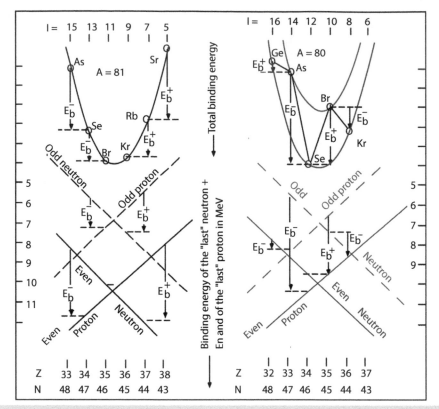

Figure 1.5 Graphical illustration of total binding energies of the isobars of mass number $A = 81$ (left) and $A = 80$ (right). Energy values lie on parabolas – a single parabola for odd A and two parabolas for even A. Binding energies of the "last" proton and the "last" neutron are approximated by the straight lines in the lower part of the figure. Source: Adapted from Suess, H. E. (1987).

Table 1.1 Numbers of stable nuclei for odd and even Z and N.

Z	N	A	Number of stable nuclei	Number of very long-lived nuclei
		$(Z + N)$		
Odd	Odd	Even	4	5
Odd	Even	Odd	50	3
Even	Odd	Odd	55	3
Even	Even	Even	165	11

Table 1.1. Clearly, even combinations of nuclides are much more likely to be stable than odd ones. This is the first indication that the liquid-drop model does not provide a complete description of nuclear stability. Another observation that is not explained by the liquid-drop model includes the so-called *magic numbers*. The *magic numbers* are 2, 8, 20, 28, 50, 82, and 126. Some observations about magic numbers are as follows:

1. Isotopes and isotones with magic numbers are unusually common (i.e., there are a lot of different nuclides in cases where N or Z equals a magic number).
2. Magic number nuclides are unusually abundant in nature (high concentration of the nuclides).
3. Delayed neutron emission occurs in fission product nuclei containing $N^* + 1$ neutrons (where N^* denotes a magic number).
4. The heaviest stable nuclide occurs at $N = 126$ (and $Z = 83$)[3].

---

[3] de Marcillac et al. (2003) demonstrated that ^{209}Bi was not entirely stable, undergoing α-decay with a half-life of 1.9×10^{19} yr. The weak energy of the α-particle emitted had evaded earlier detection. ^{208}Pb, with $N = 126$ and $Z = 82$, thus inherits the title of the heaviest stable nucleus.

5. Binding energy of last neutron or proton drops for $N^* + 1$.
6. Neutron-capture cross-sections for magic numbers are anomalously low.
7. Nuclear properties (i.e., spin, magnetic moment, electrical quadrupole moment, and metastable isomeric states) change when a magic number is reached.

The electromagnetic spectra emitted by electrons are the principal means of investigating the electronic structure of the atom. By analogy, we would expect that the electromagnetic spectra of the nucleus should yield clues to its structure, and indeed it does. However, the γ spectra of nuclei are so complex that not much progress has been made in interpreting them. Observations of magnetic moment and spin of the nucleus have been more useful (nuclear magnetic moment is also the basis for the nuclear magnetic resonance [NMR] technique used to investigate relations between atoms in lattices and the medical diagnostic technique nuclear magnetic imaging).

Nuclei with magic numbers of protons or neutrons are particularly stable or "unreactive." This is clearly analogous to chemical properties of atoms: atoms with filled electronic shells, the noble gases, are particularly unreactive. In addition, just as the chemical properties of an atom are largely dictated by the "last" valence electron, properties, such as the nucleus' angular momentum and magnetic moment, can often be accounted for primarily by the "last" nucleon. These observations suggest the nucleus may have a shell structure similar to the electronic shell structure of atoms and leads to the shell model of the nucleus.

In the shell model of the nucleus, the same general principles apply as to the shell model of the atom: possible states for particles are given by solutions to the Schrödinger equation. Solutions to this equation, together with the Pauli exclusion principle that states that no two particles can have exactly the same set of quantum numbers, determine how many nucleons may occur in each shell. In the shell model, there are separate systems of shells for neutrons and protons. As do electrons, protons and neutrons have intrinsic angular momentum, called *spin*, which is equal to $^{1}/_{2}\hbar$ ($\hbar = h/2\pi$, where h is Planck's constant and has units of momentum, $h = 6.626 \times 10^{-34}$ J s). The total

Table 1.2 Nuclear spin and odd–even nuclides.

Number of nucleons	Nuclear spin
Even–Even	0
Even–Odd	1/2, 3/2, 5/2, 7/2 ...
Odd–Odd	1, 3

nuclear angular momentum, somewhat misleadingly called the nuclear spin, is the sum of (1) the intrinsic angular momentum of protons, (2) the intrinsic angular momentum of neutrons, and (3) the orbital angular momentum of nucleons arising from their motion in the nucleus. Possible values for orbital angular momentum are given by ℓ, the orbital quantum number, which has integral values. The total angular momentum of a nucleon in the nucleus is thus the sum of its orbital angular momentum plus its intrinsic angular momentum or spin: $j = \ell \pm {}^{1}/_{2}$. The plus or minus results because the spin angular momentum vector can be either in the same direction or in the opposite direction of the orbital angular momentum vector. Thus, nuclear spin is related to the constituent nucleons in the manner shown in Table 1.2.

Let us now return to the magic numbers and see how they relate to the shell model. The magic numbers belong to two different arithmetic series:

$$n = 2, 8, 20, 40, 70, 112 \ldots$$
$$n = 2, 6, 14, 28, 50, 82, 126 \ldots$$

The lower magic numbers are part of the first series and the higher ones are part of the second series. The numbers in each series are related by their third differences (the differences between the differences between the differences). For example, for the first series:

	2	8	20	40	70	112
Difference	6	12	20	30	42	
Difference		6	8	10	12	
Difference			2	2	2	

This series turns out to be solutions to the Schrödinger equation for a three-dimensional harmonic oscillator (see Table 1.3). (This solution is different from that for particles in an isotropic coulomb field, which describes electron shells.)

Table 1.3   Particles in a three-dimensional harmonic oscillator (solution to the Schrödinger equation).

N	1	2		3			4			
ℓ	0	1		0	2		1	3		
j	1/2	1/2	3/2	1/2	3/2	5/2	1/2	3/2	5/2	7/2
State	s$^+$	p$^-$	p$^+$	s$^+$	d$^-$	d$^+$	p$^-$	p$^+$	f$^-$	f$^+$
No.	2	2	4	2	4	6	2	4	6	8
Σ	2	6		12			20			
Total	(2)	(8)		(20)			(40)			

N is the shell number; No. gives the number of particles in the orbit, which is equal to $2j +1$; Σ gives the number of particles in the shell or state, and "Total" denotes the total of particles in all shells filled. Magic numbers fail to follow the progression of the first series because only the f state is available in the fourth shell.

1.2.6.2   Magnetic moment

A rotating charged particle produces a magnetic field. A magnetic field also arises from the orbital motion of charged particles. Thus, electrons in orbit around the nucleus, and also spinning about an internal axis, produce magnetic fields, much as a bar magnet. The strength of a bar magnet may be measured by its magnetic moment, which is defined as the energy needed to turn the magnet from a position parallel to an external magnetic field to a perpendicular position. For the electron, the spin magnetic moment is equal to 1 Bohr magneton $(\mu_e) = 5.8 \times 10^{-9}$ eV/G. The spin magnetic moment of the proton is 2.79 nuclear magnetons, which is about three orders of magnitude less than the Bohr magneton (hence, nuclear magnetic fields do not contribute significantly to atomic ones). Surprisingly, in 1936, the neutron was also found to have an intrinsic magnetic moment, equal to −1.91 nuclear magnetons. Because magnetism always involves motion of charges, this result suggested there is a non-uniform distribution of charge on the neutron, which was an early hint that neutrons and protons were composite particles rather than elementary ones.

Total angular momentum and magnetic moment of pairs of protons cancel because the vectors of each member of the pair are aligned in opposite directions. The same holds true for neutrons. Hence, even–even nuclei have 0 angular momentum and magnetic moment. Angular momentum, or nuclear spin, of odd–even nuclides can have values of 1/2, 3/2, and 5/2, and non-zero magnetic moment (see Table 1.2). Odd–odd nuclei have an integer value of angular momentum or "nuclear spin." From this we can see that the angular momentum and magnetic moment of a nucleus are determined by the last nucleon added to the nucleus. For example, ^{18}O has 8 protons and 10 neutrons, and hence zero angular momentum and magnetic moment. Adding one proton to this nucleus transforms it to ^{19}F, which has an angular momentum of 1/2 and a magnetic moment of $\approx$2.79. For this reason, the shell model is also sometimes called the single-particle model since the structure can be recognized from the quantum-mechanical state of the "last" particle (usually). We will find in Chapter 5 that for heavy elements, nuclear spin can affect electrons in inner orbits about the nucleus and consequently produce very slight differences in the chemical behavior of isotopes of an element.

Aside: nuclear magnetic resonance

NMR has no application in isotope geochemistry (it is, however, used in mineralogy), but it has become such an important and successful medical technique that, as long as we are on the subject of nuclear spin, a brief examination of the basics of the technique seems worthwhile. In brief, some nuclei can be excited into higher nuclei spin energy states by radio frequency (RF) radiation – the absorption of this radiation can be detected by an appropriate RF receiver and the frequency of this absorbed radiation provides information about the environment of that nucleus on the molecular level.

In more detail, it works like this. A sample, a crystal or perhaps a person, is placed in a constant magnetic field of ≤20 T. As we have seen, even–odd and odd–odd nuclei have a nuclear spin.

A nucleus of spin j will have $2j + 1$ possible orientations. For example, ^{13}C has a nuclear spin of ½ and two possible orientations in space of the spin vector. In the absence of a magnetic field, all orientations have equal energies. In the presence of the applied magnetic field, however, energy levels split and those spin orientations aligned with the magnetic field have lower-energy levels (actually, spin vectors precess around the field vector) than others and consequently more nuclei will populate the lower-energy states by aligning themselves with the field (this is known as the Zeeman effect). A pulse of RF energy of appropriate frequency (typically sixty to several hundred megahertz, similar to very high frequency [VHF] television frequencies, where the value depends on the strength of the magnetic field and the isotope of interest) is then applied, producing a weak oscillating magnetic field that perturbs the static one causing the nuclei to reorient themselves and precess around the main field in response. Nuclei then relax to their initial state in the static field by emitting photons with energies being similar to, but differing from, those of the pulse (hence the term resonance). These photons are detected by a sensitive RF detector.

The precise energy difference between spin states, and hence the precise RF frequency that must be absorbed for the transition to occur, depends on the strength of the applied magnetic field, the nature of the nucleus, and also the atomic environment in which that nucleus is located. The latter is a consequence of magnetic fields of electrons in the vicinity of the nucleus. Although this effect is quite small, it is this slight shift in energy that makes NMR particularly valuable in probing the molecular environments of atoms. In addition, it is a non-destructive technique, which is an advantage for many analytical problems, particularly, as you can easily imagine, when the sample is a person! In geochemistry, NMR spectroscopy is being used to study reaction rates in solution at elevated pressures within the crust and, in conjunction with diamond-anvil cells, crystal structure and transitions at mantle pressures.

The three-dimensional harmonic oscillator solution explains only the first three magic numbers; magic numbers above that belong to another series. This difference may be explained by assuming there is a strong spin–orbit interaction, resulting from the orbital magnetic field acting upon the spin magnetic moment. This effect is called the Mayer–Jensen coupling. The concept is that the energy state of the nucleon depends strongly on the orientation of the spin of the particle relative to the orbit, and that parallel spin–orbit orientations are energetically favored; that is, states with higher values of j tend to be the lowest energy states. This leads to filling of the orbits in a somewhat different order; that is, such that high spin values are energetically favored. Spin–orbit interaction also occurs in the electron structure, but it is less important.

1.2.6.3 Pairing effects

In the liquid-drop model, it was necessary to add a "fudge factor" – the term δ – to account for the even–odd effect. The even–odd effect arises from a "pairing energy" that exists between two nucleons of the same kind. The result is the greater stability and consequently greater abundance of even–even nuclides (see Table 1.1).

1.2.6.4 Capture cross-sections

Information about the structure and stability of nuclei can also be obtained from observations of the probability that a nucleus will capture an additional nucleon. This probability is known as the *capture cross-section* and has units of area. Neutron-capture cross-sections are generally of greater use than proton-capture cross-sections, mainly because they are much larger. The reason for this is simply that a proton must overcome the repulsive coulomb forces to be captured, whereas a neutron, being neutral, is unaffected by electrostatic forces. Neutron-capture cross-sections are measured in barns[4], which have units of 10^{-28} m^2, and are denoted by σ. The physical cross-section of a typical nucleus (e.g., Ca) is of the order of 5×10^{-29} m^2, and increases

[4] The name for this unit has its origins in the need for secrecy of the Manhattan Project and humor: physicists choose the name of something very large to describe something very small.

somewhat with mass number (more precisely, $R = r_0 A^{1/3}$, where A is mass number and r_0 is the nuclear force radius, 1.4×10^{-15} m). While many neutron-capture cross-sections are of the order of 1 b, they vary from 0 (for ^{4}He) to 10^5 for ^{157}Gd and are not simple functions of nuclear mass (or size). They depend on nuclear structure, being, e.g., generally low at magic numbers of N and much higher for nuclides with one less neutron than a magic number. This also reveals the particular stability of magic numbers. Capture cross-sections are also dependent on the energy of the neutron, with the dependence varying from nuclide to nuclide.

1.2.7   Collective model

A slightly more complex model is called the collective model. It is intermediate between the liquid-drop model and the shell model. It emphasizes the collective motion of nuclear matter, particularly the vibrations and rotations, both quantized in energy, in which large groups of nucleons can participate. Even–even nuclides with Z or N being close to magic numbers are particularly stable with nearly perfect spherical symmetry. Spherical nuclides cannot rotate because of a dictum of quantum mechanics that a rotation about an axis of symmetry is undetectable and, in a sphere, every axis is a symmetry axis. The excitation of such nuclei (i.e., when their energy rises to some quantum level above the ground state) may be ascribed to the vibration of the nucleus as a whole. On the other hand, even–even nuclides far from magic numbers depart substantially from spherical symmetry and the excitation energies of their excited states may be ascribed to rotation of the nucleus as a whole.

1.3   RADIOACTIVE DECAY

As we have seen, some combinations of protons and neutrons form nuclei that are only "metastable." These ultimately transform to stable nuclei through the process of radioactive decay. This involves emission of a particle or particles and is usually accompanied by emission of a photon as well. In some cases, the photon emission is delayed and the daughter nuclide is left in an excited state. Just as an atom can exist in any one of a number of

excited states, so can a nucleus also have a set of discrete, quantized, excited nuclear states. The behavior of nuclei in transforming to more stable states is somewhat similar to atomic transformation from excited to more stable states, but there are some important differences. First, energy level spacing is much greater; second, the time an unstable nucleus spends in an excited state can range from 10^{-14} s to 10^{11} yr, whereas atomic lifetimes are usually about 10^{-8} s. Like atomic transitions, nuclear reactions must obey general physical laws, conservation of momentum, mass-energy, spin, and so on, and conservation of baryon number (effectively, the total number of nucleons).

Nuclear decay takes place at a rate that follows the law of radioactive decay. There are two extremely interesting and important aspects of radioactive decay. First, the decay rate is dependent only on the nature and energy state of the particular nuclide; it is independent of the history of the nucleus, essentially of external influences, such as temperature, pressure, and so on. It is this property that makes radioactive decay so useful as a chronometer. Second, it is completely impossible to predict when a given nucleus will decay. We can, however, predict the probability of its decay in a given time interval. The probability of decay in some infinitesimally small time interval, dt is λdt, where λ is the *decay constant*. Therefore, the rate of decay among some number, N, of nuclides is:

$$\frac{dN}{dt} = -\lambda N \qquad (1.12)$$

The minus sign simply indicates N decreases. Equation (1.12) is a first-order rate law known as the *basic equation of radioactive decay*. Essentially, all the significant equations of radiogenic isotope geochemistry and geochronology can be derived from this simple expression.

1.3.1   Gamma decay

A gamma ray is simply a high-energy photon (i.e., electromagnetic radiation). Just as an atom can be excited into a higher-energy state when it absorbs photon, nuclei can also be excited into higher-energy states by absorption of a much higher-energy photon. Both excited atoms and nuclei subsequently decay to their ground states by emission of a photon. Photons involved in atomic excitation and decay have energies ranging from the visible to the X-ray part of the electromagnetic spectrum (roughly 1 eV to 100 keV); gamma rays involved in nuclear transitions typically have energies greater than several hundred kiloelectron volts. Although nuclei, like atoms, generally decay promptly from excited states, in some cases, nuclei can persist in metastable excited states characterized by higher nuclear spin for considerable lengths of time.

The gamma-ray frequency is related to the energy difference by:

$$\hbar\nu = E_u - E_l \qquad (1.13)$$

where E_u and E_l are simply the energies of the upper (excited) and lower (ground) states, respectively, and $\hbar$ is the reduced Planck's constant ($h/2\pi$). The nuclear reaction is written as:

$$^A Z^* \rightarrow {}^A Z + \gamma \qquad (1.14)$$

Gamma emission usually, but not invariably, accompanies α-decay and β-decay as a consequence of the daughter being left in an excited state; this generally occurs within 10^{-12} s of the decay, but, as noted here, can be delayed if the daughter persists in a metastable state.

1.3.2 Alpha decay

An α-particle is simply a helium nucleus. Since the helium nucleus is particularly stable, it is

not surprising that such a group of particles might exist within the parent nucleus before α-decay. Emission of an α-particle decreases the mass of the daughter nucleus by the mass of the α-particle plus the kinetic energy of the α-particle and the daughter nucleus (because of the conservation of momentum, the remaining nucleus recoils from the decay reaction). The daughter is often in an excited state, from which it will decay by γ-decay. In some cases, the α-particle may leave the nucleus with any of several discrete kinetic energy levels, as is illustrated in Figure 1.6, but the sum of the kinetic energy of the α and the energy of the γ is constant.

The escape of the α-particle is a bit of a problem, because it must overcome a very substantial energy barrier, a combination of the strong force and the coulomb repulsion, to get out. For example, α-particles fired at in ^{238}U with energies below 8 MeV are scattered from the nucleus. However, during α-decay of ^{238}U, the α-particle emerges with an energy of only about 4 MeV. This is an example of a quantum effect, called *tunneling*, and can be understood as follows. We can never know exactly where the α-particle is (or any other particle, or you or I for that matter); we only know the probability of its being in a particular place. This probability is given by the particle's wavefunction, $\psi(r)$. The wave is strongly attenuated through the potential energy barrier; however, it has a small but finite amplitude outside the nucleus, and hence a small but finite probability of its being located there.

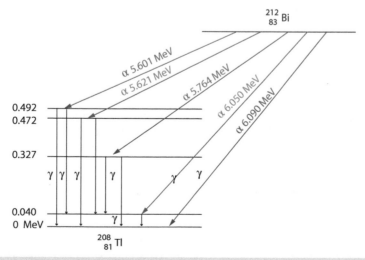

Figure 1.6 Nuclear energy-level diagram showing decay of bismuth-212 by α-emission to the ground and excited states of thallium-208.

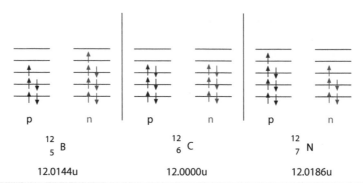

Figure 1.7 Proton and neutron occupation levels of boron 12, carbon 12, and nitrogen 12.

α-decay occurs in nuclei with mass above the maximum in the binding energy curve at ^{56}Fe (see Figure 1.3). Possibly, all such nuclei are unstable relative to α-decay, but most of their half-lives are immeasurably long.

1.3.3 Beta-decay

β-decay is a process in which the charge of a nucleus changes, but the number of nucleons remains the same. If we plotted Figure 1.1 with a third dimension, namely, energy of the nucleus, we would see that the stability region forms an energy valley. α-decay moves a nucleus down the valley axis; β-decay moves a nucleus down the valley walls toward the axis. β-decay results in the emission of an electron or positron, depending on which side of the valley the parent lies. Consider the three nuclei in Figure 1.7 (these are isobars, since they all have 12 nucleons). Of these, ^{12}C has an even–even nucleus and has the least mass. We can conclude that it is the most stable. ^{12}B decays to ^{12}C by the creation and emission of a β^--particle and the conversion of a neutron to a proton. ^{12}N decays by emission of a β^+ and conversion of a proton to a neutron.

Physicists encountered a problem in β-decay. The ^{12}C nucleus has integral spin as do ^{12}B and ^{12}N. However, the β-particle has 1/2 quantum spin units. An additional problem is that rather than having discrete kinetic energies, β-particles exhibit a spectrum of kinetic energies, although there is a well-defined maximum energy. Thus, β-decay appeared to violate both conservation of angular momentum and energy. The solution, proposed by Enrico Fermi[5] following a suggestion by Wolfgang Pauli, was the existence of another, nearly massless particle, called the *neutrino*, having ½ spin and variable kinetic energy. Thus, in β-decay, a neutrino is also released and the sum of the kinetic energy of the β and neutrino, plus the energy of any gamma, is constant.

β-decay involves the weak force, or weak interaction. The weak interaction transforms one flavor of quark into another and thereby a charged particle (e.g., a proton) into a neutral one (e.g., a neutron) and vice versa. Both the weak and electromagnetic forces are thought to be simply a manifestation of one force, called electroweak, which accounts for all interactions involving charge (in the same sense that electric and magnetic forces are manifestations of electromagnetism). In β^+-decay, e.g., a proton is converted to a neutron, giving up its +1 charge to a neutrino, which is converted to a positron. This process occurs through the intermediacy of the W+ particle in the same way that electromagnetic processes are mediated by photons. The

[5] Enrico Fermi (1901–1954) made many contributions to modern physics and particularly to the physics of the nucleus, both experimentally and theoretically. Born in Rome and educated at the University of Pisa, he was appointed professor at the University of Rome (also known as Sapienza) in 1926. He won the 1938 Nobel Prize in physics for his work on induced radioactivity. But rather than returning to Italy after receiving the prize in Stockholm, Fermi and his wife, Laura Capon who was Jewish, traveled directly to America to escape Mussolini's new racial laws. He quickly accepted a professorship at Columbia University (and later at the University of Chicago). He produced the first sustained nuclear chain reaction in 1942 and went on to play a key role in the Manhattan Project.

photon, pion, and W particles are members of a class of particles called bosons that mediate forces between the basic constituents of matter. However, the W particles differ from photons in having a very substantial mass (around 80 GeV or almost two orders of magnitude greater mass than the proton). Interestingly, *Nature* rejected the paper in which Fermi proposed the theory of β-decay involving the neutrino and the weak force in 1934!

1.3.4 Electron capture

Another type of reaction is electron capture. This is sort of the reverse of β-decay and has the same effect, more or less, as β^+-decay. Interestingly, this is a process in which an electron is added to a nucleus to produce a nucleus with less mass than the parent! The missing mass is carried off as energy by an escaping neutrino and, in some cases, by a γ. In some cases, a nucleus can decay due to either electron capture, β^-, or β^+ emission. An example is the decay of ^{40}K, which decays to ^{40}Ca by β^- and to ^{40}Ar by electron capture or β^+. We should point out that electron capture is an exception to the environmental independence of nuclear decay reactions in that it shows a very slight dependence on pressure. Pressure forces inner electrons closer to the nucleus, increasing the probability of capture.

A change in charge of the nucleus necessitates a rearrangement of electrons in their orbits. This is particularly true in electron capture, where an inner electron is lost. As electrons jump down to lower orbits to occupy the orbital freed by the captured electron, they give off electromagnetic energy. This produces X-rays from electrons in the inner orbits.

1.3.5 Spontaneous fission

Fission is a process in which a nucleus splits into two or more fairly heavy daughter nuclei. In nature, this is a very rare process, occurring only in the heaviest nuclei, namely, ^{238}U, ^{235}U, and ^{232}Th (however, spontaneous fission rates are significant only for ^{238}U). It also occurs in ^{244}Pu, an extinct radionuclide (we use the term "extinct radionuclide" to refer to nuclides that once existed in the Solar System, but now have subsequently decayed away entirely). The liquid-drop model perhaps better explains this particular phenomenon than the shell model.

Recall that in the liquid-drop model, there are four contributions to total binding energy: volume energy, surface tension, excess neutron energy, and coulomb energy. The surface tension tends to minimize the surface area while the repulsive coulomb energy tends to increase it. We can visualize these nuclei as oscillating between various shapes. It may very rarely become so distorted by the repulsive force of 90 or more protons that the surface tension cannot restore the shape. Surface tension is instead minimized by splitting the nucleus entirely.

Since there is a tendency for N/Z to increase with A for stable nuclei, the parent is much richer in neutrons than the daughters produced by fission. Thus, fission generally also produces some free neutrons in addition to two nuclear fragments (the daughters). The daughters are typically of unequal size, with the most common fission products forming two clusters – one with mass numbers between 85 and 110 and the other with mass numbers between 130 and 150. The average mass ratio of the high to low mass fragment is about 1.45. Even though some free neutrons are created, the daughters tend to be too neutron-rich to be stable. As a result, they decay by β^- to stable daughters. It is this decay of the daughters that results in radioactive fallout in bombs and radioactive waste in reactors (a secondary source of radioactivity is production of unstable nuclides by capture of the neutrons released).

Some non-stable heavy nuclei and excited heavy nuclei are particularly unstable with respect to fission; an important example is ^{236}U. Imagine a material rich in U. When ^{238}U undergoes fission, one of the released neutrons can be captured by ^{235}U nuclei, producing ^{236}U in an excited state. This ^{236}U then fissions producing more neutrons, and so on. This is the basis for nuclear reactors and bombs (the latter can also be based on Pu). The concentration of U is not usually high enough in nature for this sort of thing to happen. However, it apparently did at least once, 1.7 billion years ago in the Oklo U deposit in Africa. This deposit was found to have an anomalously high $^{238}U:^{235}U$ ratio (227 versus 137.82), indicating some of the ^{235}U had been "burned" in a nuclear chain reaction. Could such a natural nuclear reactor happen again? Probably not, because there is a lot less ^{235}U around now than there was 1.7 billion years ago. With equations, which we will introduce in the following two chapters, you should be able to calculate just how much less.

Individual natural fission reactions are less rare (although still rare, only 1 in 5×10^{-5} ^{238}U atoms will fission rather than undergoing α-decay). When fission occurs, there is a fair amount of kinetic energy produced (maximum about 200 MeV), with the nuclear fragments literally flying apart. These fragments damage the crystal structure through which they pass, producing "tracks," whose visibility can be enhanced by etching. This is the basis for fission-track dating, which we will describe in Chapter 4.

Natural fission can also produce variations in the isotopic abundance of elements among the natural, ultimate product. Xenon is an important product, as we will learn in Chapter 14. Indeed, the critical evidence showing that a nuclear chain reaction had indeed occurred in the Oklo deposit was the discovery that fission product elements, such as Nd and Ru, have anomalous isotopic compositions. Analysis of the isotopic composition of another fission product, Sm, fed a controversy over whether the fine-structure constant, α, has changed over time. α characterizes the electromagnetic interaction between elementary particles and is related to other fundamental constants as:

$$\alpha = \frac{1}{4\pi\varepsilon_0} \frac{e^2}{\hbar c} \qquad (1.15)$$

where e is the charge of the electron and ε_0 is the permittivity of a vacuum (and, as usual, $\hbar$ is the reduced Planck's constant and c is the speed of light). Thus, a change in the fine-structure constant raises the possibility of a change in c. The change, if it occurs, is quite small, i.e., less than 1 part in 10^7, and could be consistent with some observations about quasars and the early universe; however, thus far, experiments have failed to show any evidence that it varies.

1.4   NUCLEOSYNTHESIS

A reasonable starting point for isotope geochemistry is a determination of the abundances of the naturally occurring nuclides.

Indeed, this was the first task of isotope geochemists (although those engaged in this work would have referred to themselves as physicists). This began with Thomson, who built the first mass spectrometer and discovered that Ne consisted of two isotopes (as we noted earlier Thomson's instrument did not have the sensitivity to detect ^{21}Ne). Much of the initial subsequent work was carried on by American physicist Alfred Nier at the University of Minnesota. Having determined the abundances of nuclides, it is natural to ask what accounts for this distribution and, even more fundamentally, what process or processes produced the elements. The process is known as nucleosynthesis.

Physicists, like all scientists, are attracted to simple theories. Not surprisingly then, the first ideas about nucleosynthesis attempted to explain the origin of the elements by single processes. Generally, these were thought to occur at the time of the Big Bang. None of these theories was successful. It was really the astronomers, accustomed to dealing with more complex phenomena than physicists, who successfully produced a theory of nucleosynthesis that involved a number of processes. Today, isotope geochemists continue to be involved in refining these ideas by examining and attempting to explain isotopic variations occurring in meteorites and in presolar grains some of them contain – a topic we will explore in Chapter 6.

The origin of the elements is an astronomical question, and perhaps even more a cosmological one. To understand how the elements formed we need to understand a few astronomical observations and concepts. The universe began about 13.8 Ga[6] ago with the Big Bang. Since then, the universe has been expanding, cooling, and evolving. This hypothesis follows from two observations: the relationship between redshift and distance and the cosmic background radiation, particularly the former. This cosmology provides two possibilities for formation of the elements: (1) they were formed in the Big Bang itself or (2) they were subsequently produced. As we shall see, the answer is both.

Our present understanding of nucleosynthesis comes from three sorts of observations: (1)

---

[6] Ga is an abbreviation for giga-annum or 10^9 yr. Other such abbreviations used in this book include – Ma: 10^6 yr; ka: 10^3 yr.

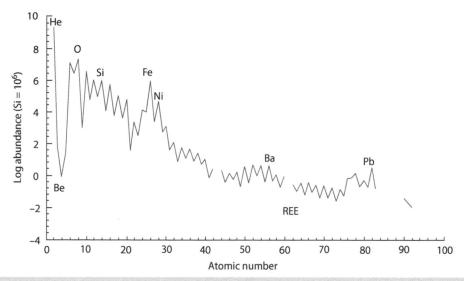

Figure 1.8   Solar System abundance of the elements relative to silicon as a function of atomic number.

the abundance of isotopes and elements in the Earth, the Solar System, as shown in Figure 1.8, and cosmos (spectral observations of stars); (2) experiments on nuclear reactions that determine what reactions are possible (or probable) under given conditions; and (3) inferences about possible sites of nucleosynthesis and about the conditions that would prevail in those sites.

Various hints came from all three of these observations. For example, it was noted that the most abundant nuclide of a given set of stable isobars tended to be the most neutron-rich one. We now understand this to be a result of shielding from β-decay (see the discussion about the r-process). Another example is the discovery of presolar grains in meteorites. These grains show wide variations in isotopic abundances reflecting their formation in a variety of environments. They allow astrophysicists to calibrate models of the physics of stellar interiors.

Another key piece of evidence regarding formation of the elements comes from looking back into the history of the cosmos. Astronomy is a bit like geology in that just as we learn about the evolution of the Earth by examining old rocks, we can learn about the evolution of the cosmos by looking at old stars. It turns out that old stars (such old stars are particularly abundant in the globular clusters outside the main disk of the Milky Way) are considerably poorer in heavy elements than are young stars. This suggests much of the heavy element inventory of the galaxy has been produced since these stars formed some 10 Ga or more ago. On the other hand, those stars seem to have about the same He : H ratio as young stars. Indeed, ^{4}He seems to have an abundance of 24–28 % in all stars. Another key observation was the identification of technetium emissions in the spectra of some stars. Since the most stable isotope of this element has a half-life of about 100 000 yr and does not exist in the Earth (with the exception of quite small amounts produced by ^{238}U fission), it must have been synthesized in those stars. Thus, the observational evidence suggests that (1) H and He are everywhere uniform implying their creation and fixing of the He : H ratio in the Big Bang and (2) subsequent creation of heavier elements (heavier than Li, as we shall see) by subsequent processes occurring in stars.

As we mentioned, early attempts ($\approx$1930–1950) to understand nucleosynthesis focused on single mechanisms. Failure to find a single mechanism that could explain the observed abundance of nuclides, even under varying conditions, led to the present view that relies on a number of mechanisms operating in different environments and at different times for creation of the elements in their observed abundances. This view, often called the polygenetic hypothesis, is based mainly on the work of Burbidge[7], Burbidge, Fowler and Hoyle. Their classic paper summarizing the theory, "Synthesis of the elements in stars" was published in *Reviews of Modern Physics* in 1957 and is widely known as B$_2$FH. Interestingly, the abundance of trace elements and their isotopic compositions were perhaps the most critical observations in development of the theory. An objection to this polygenetic scenario was the apparent uniformity of the isotopic composition of the elements. However, variations in the isotopic composition have now been demonstrated for many elements in meteorites. Furthermore, there are quite significant compositional variations in heavier elements among stars (e.g., Burbidge and Burbidge, 1956). These observations provide strong support for this theory.

B$_2$HF laid out the basic theory of nucleosynthesis but, as we shall see further in the text, it has undergone considerable refinement and revision over subsequent decades and this process continues to this day. To briefly summarize it, the polygenetic hypothesis proposes four phases of nucleosynthesis. *Cosmological nucleosynthesis* occurred shortly (really, really shortly!) after the universe began and is responsible for the cosmic inventory of H and He, and a bit of Li. Helium is the main product of nucleosynthesis in the interiors of normal or "main-sequence" stars. The lighter elements, up to and including Si, but excluding Li and Be, and a fraction of the heavier elements may be synthesized in the interiors of larger stars during the final stages of their evolution (*stellar nucleosynthesis*). The synthesis of the remaining elements occurs as large stars exhaust the nuclear fuel in their interiors and explode in nature's grandest spectacle, the supernova, and in neutron star mergers (*explosive nucleosynthesis*). Finally, Li and Be are continually produced in interstellar space by interaction of cosmic rays with matter (*galactic nucleosynthesis*). In the following sections, we examine these nucleosynthetic processes as presently understood.

1.4.1   Cosmological nucleosynthesis

Immediately after the Big Bang, the universe was too hot for any matter to exist – there was only energy. Some 10^{-11} s later, the universe had expanded and cooled to the point where quarks and anti-quarks could condense from the energy. The quarks and anti-quarks, however, would also collide and annihilate each other. Therefore, a sort of thermal equilibrium existed between matter and energy. As things continued to cool, this equilibrium progressively favored matter over energy. Initially, there was an equal abundance of quarks and anti-quarks, but as time passed, the symmetry was broken and quarks came to dominate. The current theory is that the hyperweak force was responsible for an imbalance favoring matter over anti-matter. After 10^{-4} s, things were cool enough for quarks to begin to associate with one another and form nucleons: protons and neutrons. After 10^{-2} s, the universe had cooled to 10^{11} K. Electrons and positrons were in equilibrium with photons and neutrinos; antineutrinos were in equilibrium with photons; antineutrinos combined with protons to form positrons and neutrons; and neutrinos combined with neutrons to form electrons and protons:

$$p + \bar{\nu} \rightarrow e^{+} + n$$

and

$$n + \nu \rightarrow e^{-} + p$$

---

[7] Margaret Burbidge (1919–2020), born Eleanor Margret Peachy in Davenport, England, received her PhD from University College London in 1943. In 1945, she was turned down for a fellowship at the Carnegie Mount Wilson Observatory in California as only men were allowed to use the telescope. She married Geoffrey Burbidge (1925–2010) in 1948 and in 1953 the pair, who had published numerous papers on stellar compositions, were invited to work with Fred Hoyle (1915–2001) and William Fowler (1911–1995) at the University of Cambridge on Hoyle's ideas about nucleosynthesis in stars. When Fowler returned to Caltech in 1955, the Burbidges followed. They joined the faculty of the University of California at San Diego in 1962, where, except for Margaret's brief tenure as director of the Royal Greenwich Observatory, they remained for the rest of their careers, making many additional seminal contributions to astrophysics. Margaret was a campaigner for women's rights and turned down the American Astronomical Society's prize for women, stating, "If my strong feeling is against any kind of discrimination, I have to stretch that to include discrimination *for* women too." She was later awarded the American Astronomical Society's highest honor. When Fowler won the 1983 Nobel Prize in physics for his work on nucleosynthesis, he expressed shock that the Burbidges and Hoyle were overlooked.

where ν designates a neutrino and the overbar indicates an antineutrino. This equilibrium produced about an equal number of protons and neutrons. However, the neutron is unstable outside the nucleus and decays to a proton with a half-life of about 10 min. Therefore, as time continued to pass, protons became more abundant than neutrons.

After a second or so, the universe had cooled to 10^{10} K, which shut down the reactions above. Consequently, neutrons were no longer being created, but they were being destroyed as they decayed to protons. At this point, protons were about three times as abundant as neutrons.

It took another 3 min to for the universe to cool to 10^9 K, which is cool enough for ^{2}H, created by

$$p + n \rightarrow {}^2\text{H} + \gamma$$

to be stable. At about the same time, the following reactions could also occur:

$$^2\text{H} + {}^1n \longrightarrow {}^3\text{H} + \gamma; {}^2\text{H} + {}^1\text{H} \longrightarrow {}^3\text{H} + \gamma$$
$$^2\text{H} + {}^1\text{H} \longrightarrow {}^3\text{He} + \beta^+ + \gamma; {}^3\text{He} + n \longrightarrow {}^4\text{He} + \gamma$$

and

$$^3\text{He} + {}^4\text{He} \longrightarrow {}^7\text{Be} + \gamma; {}^7\text{Be} + e^- \longrightarrow {}^7\text{Li} + \gamma$$

One significant aspect of this event is that it began to lock up neutrons in nuclei where they could no longer decay to protons. The timing of this event fixes the ratio of protons to neutrons at about 7:1. Because of this dominance of protons, hydrogen is the dominant element in the universe. About 24% of the mass of the universe was converted to ^{4}He in these reactions; less than 0.01% was converted to ^{2}H, ^{3}He, and ^{7}Li (and there is good agreement between theory and observation). Formation of elements heavier than Li was inhibited by the instability of nuclei of masses 5 and 8. Shortly thereafter, the universe cooled below 10^9 K and nuclear reactions were no longer possible.

Thus, the Big Bang created H, He, and a bit of Li (^{7}Li/H $< 10^{-9}$). Some 300 000 years or so later, the universe had cooled to about 3000 K,

cool enough for electrons to be bound to nuclei, forming atoms. It was at this time, called the "recombination era," that the universe first became transparent to radiation. Prior to that, photons were scattered by the free electrons, making the universe opaque. It is the radiation emitted during this recombination that makes up the cosmic microwave background radiation that we can still detect today. Discovery of this cosmic microwave background radiation, which has the exact spectra predicted by the Big Bang model, represents a major triumph for the model and is not easily explained in any other way.

1.4.2   Stellar nucleosynthesis

1.4.2.1   Astronomical background

Before discussing nucleosynthesis in stars, it is useful to review a few basics of astronomy. Stars shine because of exothermic nuclear reactions occurring in their cores. The energy released by these processes results in thermal expansion that, in general, closely balances gravitational collapse. Surface temperatures are very much cooler than temperatures in stellar cores. For example, the Sun has a surface temperature of 5700 K and a core temperature thought to be 14 000 000 K.

On a plot of luminosity versus wavelength of their principal emissions (i.e., color), called the Hertzsprung–Russell diagram (Figure 1.9), most stars fall along an array, known as the *main sequence*[8] defining an inverse correlation between these two properties. Stars on the main sequence are classified based on their color (and spectral absorption lines), which in turn is related to their temperature. From hot to cold, the classification is: O, B, F, G, K, and M (the mnemonic is "*O Be a Fine Girl, Kiss Me!*"), with subclasses designated by numbers, e.g., F5. The Sun is class G. Stars are also divided into populations. Population I stars are second or later generation stars and have greater heavy element contents than Population II stars. Population I stars are located in the main disk of the galaxy, whereas the old first-generation stars

---

[8] It was originally believed that stars evolved from hot and bright to cold and dark across the diagram, hence the term "main sequence." This proved not to be the case. Stars do, however, evolve somewhat to hotter and brighter during the main sequence part of these lives. The Sun is now about 30% brighter than it was when it first reached the main sequence; this, however, is small compared to the orders of magnitude range in luminosity.

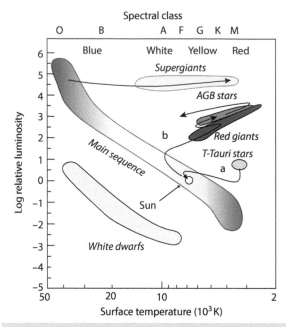

Figure 1.9   The Hertzsprung–Russell diagram of the relationship between luminosity and surface temperature. Arrows show evolutionary path for a star with the size of the Sun in (a) pre- and (b) post-main-sequence phases. Stars brighten somewhat during the main-sequence lives.

of Population II predominate in globular clusters that circle the main disk.

Since wavelength is inversely related to temperature, this correlation means simply that hot stars are more luminous and give off more energy than cooler stars. Mass and radius are exponentially related to temperature and luminosity in main-sequence stars; hot stars are big, whereas small stars are cooler and dimmer and release less energy. A star with a mass six times that of the Sun is almost two orders of magnitude brighter than one with a mass only twice that of the Sun. Thus, O and B stars are large, luminous, and hot; K and M stars are small, cool, and (comparatively speaking) dark. Stars on the main sequence produce energy by *hydrogen burning*, fusing hydrogen to produce helium. Since the rate at which these reactions occur depends exponentially on temperature and density, there is an inverse relationship between the lifetime of a star and its mass. A star with a mass of $2M_{\odot}$ (where $\odot$ is the astronomical symbol for the Sun; hence, $M_{\odot}$ is the solar mass) will remain on the main sequence for 10^9 yr while one with a mass of $12M_{\odot}$ will be on the main sequence for only

53 million years. The most massive stars, with a mass up to $100M_{\odot}$, have main-sequence life expectancies of only about 10^6 yr or so, whereas small stars, with a mass as small as $0.01M_{\odot}$, remain on the main sequence for more than 10^{12} yr.

Although the Sun is often referred to as an average star, it is more massive than nearly 80% of all stars and half of all stars have masses less than 25% of the Sun's mass. Truly massive stars are rare. B-type stars, which have masses more than twice that of the Sun, make up less than 1% of all stars, and O stars, those more than 15 times as massive as the Sun, are far rarer. Despite their rarity, large stars are responsible for almost all heavy element production in the universe because elements heavier than He are almost exclusively produced in the late stages of stellar evolution – after they have evolved off the main sequence. Large stars have short main-sequence lives and they enter these late phases and begin synthesizing heavy elements quickly, while even the oldest of the smallest stars have yet to enter this phase. Although rare, these large stars make up most of the ones you see in the night sky, small stars are generally too dim to be seen with the naked eye; however, they are responsible for the milky appearance of the Milky Way.

The two most important exceptions to the main-sequence stars – the *giant stars* and the white dwarfs – represent stars that have exhausted the H fuel in their cores and have moved on in the evolutionary sequence. When the H in the core is converted to He, it cannot be replenished because the density difference prevents convection between the core and outer H-rich layers. In the absence of energy generation, interior part of the core collapses under gravity, raising temperatures and pressures to the point where helium burning can begin. Giant stars are ones that have entered these late stages of stellar evolution where fusion of heavier elements, beginning with helium, provides energy to resist gravitational collapse. The collapse causes the star to heat up and it becomes opaque to ultraviolet radiation; as a result, the outer layers of the star expand and cool, making them appear red, while their luminosity increases. The star becomes a *red giant* or, in the case of massive stars, a *supergiant*, both of which are overluminous relative to main-sequence stars of the same color (see Figure 1.9). When the Sun reaches this phase, in perhaps another 5 Ga, it will expand to the Earth's orbit. During this time, the reduced gravitational force experienced by the expanded outer layers and combined with radiation pressure results in a greatly enhanced stellar wind, of the order of 10^{-6} to 10^{-7}, or even 10^{-4}, $M_\odot$ per year. For comparison, the present solar wind is 10^{-14} $M_\odot$ per year; thus, in its entire main-sequence lifetime, the Sun will blow off 1/10 000 of its mass through solar wind. Giant phase stars can lose as much as half of their mass due to this strong stellar wind. The ejected outer shells can form so-called "planetary nebulae" consisting of gas and dust surrounding the star. Betelgeuse, one of the brightest stars in the Northern Hemisphere winter sky, is a good example of a red supergiant star. It is estimated to be 10 million years old and have a mass 10–20 times that of the Sun. It dimmed dramatically in late fall and early winter of 2019–2020 before brightening again. The leading theory to explain this event is ejection of a large cloud of gas and dust from its surface, apparently from its poles. The stellar winds of giant phase stars account for most of the dust in galaxies and this is the material that will seed the next generation of stars.

The further evolution of stars depends on their mass. The stars with lowest mass, those with masses $M \lesssim 0.5M_\odot$, undergo helium burning in the central part of the core in the red giant phase before exhausting their He fuel and then contract, with their exteriors heating up as they do so, to become *white dwarfs*. Without an energy source to resist gravity, they collapse, heating up as they do so, and become *white dwarfs*. They continue to radiate energy from that produced previously by nuclear reactions, as well as gravitational potential energy released as the stars slowly contract. They are underluminous relative to stars of similar color on the main sequence. They can be thought of as little more than glowing embers.

In stars with $M \gtrsim 0.5M_\odot$, contraction increases pressure and temperature to the point where H burning in the layer immediately above the core, now consisting of He and some C and O produced by successive proton captures in the CNO (carbon-nitrogen-oxygen) cycle (explained in more detail in the following section), will ignite, providing energy to reverse the cooling and stabilize the star. Simultaneously, the outer convective envelope deepens leading to mixing of the outer envelope with regions that have experienced some fusion and mixing products of H burning to the surface. This is called the "first dredge-up." In stars in this mass range, the core becomes *electron degenerate*, which happens when density reaches to a point where all electron energies' levels are filled (the Pauli exclusion principle, which states that particles cannot share the same energy state). When that happens, electrons cannot give up energy by moving to lower-energy states; no thermal energy can be extracted and adding heat does not increase the speed of most of the electrons, because they are stuck in fully occupied quantum states. Consequently, temperature becomes independent of pressure and the core becomes supported by *degeneracy pressure* and cannot compress further.

In stars with $M \lesssim 2M_\odot$, He burning ignites explosively producing a *core He flash*. This also extinguishes the hydrogen burning shell. When the He-burning shell nears the base of the surrounding former H-burning shell, the heat reignites H burning. The trend toward lower surface temperature and higher luminosity initially reverses before resuming along the *asymptotic giant branch* (AGB) track, which is parallel to the red giant track but at higher luminosity (see Figure 1.9). The structure of stars at this point is illustrated in Figure 1.10. Stars with $M \gtrsim 2M_\odot$ make the transition to He burning and to the AGB track smoothly without the core helium flash. They can experience a second dredge-up bringing material that has experienced complete H burning to the surface in the early AGB phase. Karakas and Lattanzio (2014) provide an excellent summary of the evolution of these stars.

As He burning continues, the core, consisting of the carbon and oxygen ashes of He burning, expands and the He shell thins as the star evolves up the AGB track (see Figure 1.9). Eventually, the He-burning regions extend outward to the point where temperatures and pressures are very low and He burning extinguishes. The resulting cooling and contraction reignites the H-burning shell and the star enters a period of quiescent H-shell burning. Eventually, He burning can reignite explosively at the base of the intershell region (Figure 1.10), which is composed of material exposed to previous He-shell flashes plus the accumulated ashes of H-shell burning. This thermal pulse can happen repeatedly in this thermally pulsing

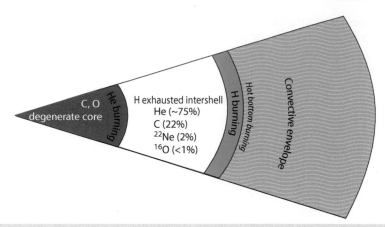

Figure 1.10   Cross-section of an AGB star (Adapted from Karakas and Lattanzio, 2014). Layers are not to scale; in particular, the thickness of the H-burning layer is exaggerated.

AGU phase. The pulse again initiates deep convection, again dredging up material from deep in the star (this, along with subsequent thermal-pulse-driven dredge-ups, is known as the third dredge-up).

These repeated dredge-ups from the deep burning shells enrich the outer part of the star in newly synthesized material, particularly ^{12}C and elements made by the s-process (which will be discussed later). At the same time, this outer envelope is being lost through massive solar winds. Although most stars start out with more O than C, these dredge-up events gradually increase the C : O ratio until it exceeds 1. This has a profound effect on dust chemistry: in stars with C/O < 1, infrared emission measurements indicate the presence of O-rich phases, such as silicates and oxides in the outflow, whereas in AGB stars with C/O > 1, these observations indicate the presence of phases, such as SiC and elemental C. As we will find in Chapter 6, grains of these minerals are found in primitive meteorites.

For stars with $M < 6M_\odot$, this pattern eventually ends as temperatures and pressure can never reach the threshold for heavier element burning and, like the stars with lowest mass, the star simply collapses and radiates it energy away as a white dwarf. More massive stars can continue to derive energy from fusion of successively heavier elements.

Very massive stars (those with $M \gtrsim 20M_\odot$) that began life on the main sequence as O-class stars can shed their outer hydrogen shells and as much as half of their initial mass exposing the hot He-rich layers below to become *Wolf–Rayet* stars. With their deep layers exposed,

Wolf–Rayet stars are extremely luminous and extremely hot, with surface temperatures of many tens of thousands of kelvins, reaching over 100 000 K in some cases. In addition to He, their optical spectra show emission from nitrogen, carbon, or oxygen produced in prior CNO cycle or He-burning phases, and they are subdivided based on the dominant element in their spectra (e.g., WO, if oxygen is dominant). Hydrogen is weak or absent in the spectra in all cases. They have massive stellar winds, blowing off mass at high velocities at rates of as much as 10^{-5} $M_\odot$ per year; some are surrounded by planetary nebulae as a consequence. Rapid rotation in Wolf–Rayet stars can result in nuclei newly synthesized by the s-process, which is discussed further in the text, being mixed to the surface where they can be incorporated in these stellar winds. This is significant because, as we will see in Chapter 6, such newly synthesized nuclei were present in the giant molecular cloud in which our Solar System formed.

The ultimate fate of stars with $M \gtrsim 6$ to $8M_\odot$ – once fusion reactions can no longer be sustained – is to gravitationally collapse. In most cases, this releases enormous energy and also blasts newly synthesized elements into the surrounding interstellar medium as Type Ib/c and Type II supernovae. The exception may be the most massive stars with $M \gtrsim 40M_\odot$ where gravity is so strong that neither energy nor mass is released as they collapse into black holes and the star simply disappears. Collapse occurs when the electron degeneracy pressure in the core can no longer support the star.

For stars with $6M_\odot \gtrsim M \lesssim 10M_\odot$, this can occur through electron capture by magnesium in a degenerate O/Ne/Mg core causing gravitational collapse followed by explosive oxygen fusion; these are known as *electron-capture supernovae*. In larger stars, collapse occurs when the core has been entirely converted to iron and fusion reactions and can no longer supply thermal energy to resist collapse. Depending on the initial mass and *metallicity*[9] and the evolutionary history of the star, the product of these events is either a neutron star or a black hole, as well as newly synthesized elements that eventually contribute to the gas and dust of the interstellar medium. White dwarfs can also explode as supernovae if they merge with or accrete mass from a companion star that ignites explosive fusion reactions (these are Type Ia supernovae).

In the following sections, we describe in more detail the nuclear reactions powering stars as they pass through this evolution and elements synthesized in the process.

1.4.2.2 Hydrogen, helium, and carbon burning in main sequence and giant stars

For quite some time after the Big Bang, the universe was a more or less homogeneous, hot gas. However, small quantum fluctuations in the initial stages of the Big Bang ultimately led to small inhomogeneities in the distribution of matter. These inhomogeneities enlarged in a sort of runaway process of gravitational attraction and collapse. Thus were formed protogalaxies, thought to date to about 0.5–1.0 Ga after the Big Bang. Instabilities within the protogalaxies collapsed into stars. Once this collapse proceeds to the point where temperatures reach 1 million K, deuterium burning can begin:

$$^2\text{H} + {^1\text{H}} \longrightarrow {^3\text{He}} + \gamma$$

This occurs while pre-main stars are still accreting mass and growing and temporarily stabilizes the star against further collapse. This may continue for several million years in smaller stars, such as the Sun. (In low-mass objects that will never reach temperatures

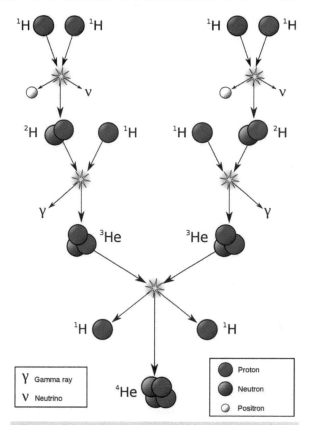

Figure 1.11 Schematic of the *pp* I chain in which protons fuse to produce ^4He in main-sequence stars. Source: Sarang, Wikimedia Commons, public domain.

and pressures for hydrogen burning to initiate, which are known as brown dwarfs, deuterium burning can occur and continue for hundreds of millions of years before the deuterium is exhausted. This requires a mass at least 13 times that of Jupiter to occur.)

When deuterium is exhausted and the stellar core reaches a density of 6×10^4 kg/m^2 and a temperature of 10–20 million K, *hydrogen burning*, or the *pp process* (Figure 1.11), begins and continues through the main-sequence life of the star. There are three variants, or branches: *pp* I:

$$^1\text{H} + {^1\text{H}} \longrightarrow {^2\text{H}} + \beta^+ + \nu;$$
$$^2\text{H} + {^1\text{H}} \longrightarrow {^3\text{He}} + \gamma;$$
$$\text{and } {^3\text{He}} + {^3\text{He}} \longrightarrow {^4\text{He}} + 2{^1\text{H}} + \gamma$$

---

[9] In astronomer speak, all elements heavier than He are referred to as "metals"; hence, the total content of these metals is known as *metallicity* and is represented as Z. Average metallicity of the galaxy is about 0.14. It is often expressed relative to solar metallicity as $Z_\odot$. The ratio [Fe]/[H] is often used as a proxy for metallicity.

pp II:

$$^3\text{He} + {}^4\text{He} \longrightarrow {}^7\text{Be}; \quad {}^7\text{Be} \longrightarrow \beta^- + {}^7\text{Li} + \nu;$$

$$^7\text{Li} + {}^1\text{H} \longrightarrow 2\,{}^4\text{He}$$

pp III:

$$^7\text{Be} + {}^1\text{H} \longrightarrow {}^8\text{B} + \gamma;$$

$$^8\text{B} \longrightarrow \beta^+ + {}^8\text{Be} + \nu; \quad {}^8\text{Be} \longrightarrow 2\,{}^4\text{He}$$

Which of these reactions dominates depends on temperature, but the net result of all is the production of ^{4}He and the consumption of H (and Li). All main-sequence stars produce He, yet over the history of the cosmos, this has had little impact on the H : He ratio of the universe. This in part reflects the observation that for stars with small mass, the He produced remains hidden in their interiors or their white dwarf remnants, and for stars with large mass, the later reactions consume the He produced during the main-sequence stage.

Once some carbon had been produced by the first-generation stars and supernovae, second- and subsequent-generation stars could synthesize He by another process as well – the *CNO cycle*:

$$^{12}\text{C}(p,\gamma)^{13}\text{N}(\beta^+,\gamma)^{13}\text{C}(p,\gamma)^{14}\text{N}(p,\gamma)^{15}$$

$$\text{O}(\beta^+,\nu)^{15}\text{N}(p,\alpha)^{12}\text{C}$$

(Here we are using a notation commonly used in nuclear physics. The reaction: ^{12}C $(p,\gamma)^{13}$N is equivalent to: $^{12}\text{C} + p \rightarrow {}^{13}\text{N} + \gamma$.)

It was subsequently realized that this reaction cycle is just part of a larger reaction cycle, which is illustrated in Figure 1.12. Since the process is cyclic, the net effect is consumption of four protons and two positrons to produce a neutrino, some energy, and a ^{4}He nucleus. Thus, to a first approximation, carbon acts as a kind of nuclear catalyst in this cycle: it is

neither produced nor consumed. When we consider these reactions in more detail, not all of them operate at the same rate, resulting in some production and some consumption of these heavier nuclides. The net production of a nuclide can be expressed as:

$$\frac{dN}{dt} = (\text{creation rate} - \text{destruction rate})$$

(1.16)

Reaction rates are such that some nuclides in this cycle are created more rapidly than they are consumed, while for others, the opposite is true. The slowest of the reactions in Cycle I is $^{14}\text{N}(p,\gamma)^{15}$O. As a result, there is a production of ^{14}N in the cycle and net consumption of C and O. Because of these rate imbalances, the CNO cycle may be the principal source of nitrogen in the universe. The CNO cycle will also tend to leave remaining carbon in the ratio of ^{13}C : ^{12}C of 0.25. This is quite different than the Solar System (and terrestrial) abundance ratio of about 0.01.

The CNO cycle and the *pp* chains are competing fusion reactions in main-sequence stars. Which of the reactions dominates depends on temperature, which in turn depends on mass, as well as on the metallicity, since some pre-existing carbon is required to initiate the cycle. Recent measurements of the energy spectrum of solar neutrinos have revealed that the CNO cycle (whose neutrino energies differ from that of the *pp* chain) accounts for about 1% of the Sun's He and energy production with the *pp* reactions accounting for the remainder (Agostini et al., 2020). However, if the Sun were only ≈30% more massive (and consequently a few million kelvin hotter), the CNO cycle would dominate energy production.

When H is exhausted in the core, fusion ceases, and the core collapses in the absence

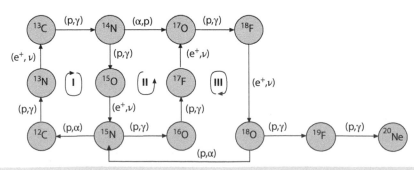

Figure 1.12 Illustration of the CNO cycle, which operates in larger second- and later-generation stars.

of thermal energy required to resist gravity. As described in the previous section, the outer part of the star expands as the star enters the giant phase. A star will remain in the giant phase for of the order of 10^5–10^8 yr, depending on mass. Temperatures in the He core eventually reach 3×10^8 K and density of 10^8 kg/m^3 igniting *He burning*:

$$^4\text{He} + {}^4\text{He} \longrightarrow {}^8\text{Be} + \gamma$$

and

$$^8\text{Be} + {}^4\text{He} \longrightarrow {}^{12}\text{C} + \gamma$$

The half-life of ^{8}Be is only 7×10^{-16} s, so three He nuclei must collide nearly simultaneously (this reaction is sometimes called the triple-alpha process for this reason); hence, densities must be very high. Once ^{12}C becomes sufficiently abundant, ^{16}O begins to build up through the reaction:

$$^{12}\text{C} + {}^4\text{He} \longrightarrow {}^{16}\text{O} + \gamma$$

^{14}N created by the CNO cycle in second-generation stars can be converted to ^{22}Ne; there can also be some production of ^{20}Ne and ^{24}Mg.

As we noted in the previous section, there is an H-burning shell surrounding the He-burning shell, which is in turn surrounded by an H-rich envelope (Figure 1.10). Successive thermal pulses resulting from explosive He-burning initiation can convectively mix He-burning products into overlying layers. In stars with $M \gtrsim 5M_\odot$, temperatures in a thin layer at the base of the convective envelope can reach 5×10^8 K, which can initiate *hot bottom burning*, initiating the Ne–Na and Mg–Al chains. In a manner analogous to the CNO cycle, these chains involve a series of proton-capture reactions beginning with ^{19}F synthesizing ^{4}He, but with significant production of nuclear byproducts, particularly ^{23}Na and ^{27}Al. This process is also an important source of the short-lived (700 000 yr) radionuclide ^{26}Al that was abundant in the early solar system.

Note that Li, Be, and B have been skipped: they are not synthesized in these phases of stellar evolution. Indeed, they are actually

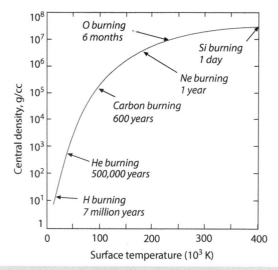

Figure 1.13 Evolutionary path of the core of star of 25 solar masses (after Bethe and Brown, 1985). Note that the period spent in each phase depends on the mass and composition of the star: massive stars evolve more rapidly.

consumed in stars, in reactions, such as *pp* II and *pp* III.

Because the densities and temperatures necessary to initiate further nuclear reactions cannot be achieved, evolution for stars with $M < 6M_\odot$ ends when helium in the core is exhausted, but stars with greater masses undergo further core contraction. Evolution now proceeds at an exponentially increasing pace (Figure 1.13), and these phases are poorly understood. However, if temperatures reach 600 million K and densities 3×10^9 kg/m^3, *carbon burning* becomes possible:

$$^{12}\text{C} + {}^{12}\text{C} \longrightarrow {}^{20}\text{Ne} + {}^4\text{He} + \gamma$$

Some intermediate-mass stars, particularly those with $\lesssim 8M_\odot$, can be catastrophically disrupted by the ignition of carbon burning. The outer envelope of the star is ejected, leaving an O–Ne–Mg white dwarf. For stars in the mass range of 8–$10M_\odot$, carbon burning continues until their cores consist of O, Ne, and Mg. Depending on a number of factors, such as metallicity, rotation speed, and mass loss rates, the core can reach the *Chandrasekhar limit*[10],

---

[10] Named for Subrahmanyan Chandrasekhar (1910–1995) who calculated it taking account of relativistic variation of mass with the velocities of electrons in extremely dense matter while completing his PhD at Cambridge University. Born in Lahore, British India (now Pakistan), he was educated at the University of Madras (now Chennai) and joined the faculty of the University of Chicago in 1934 and remained there for the rest of his life, making many additional fundamental contributions to physics and astrophysics. He was awarded the Nobel Prize in physics in 1983.

corresponding to a core mass of $\approx 1.4 M_\odot$. This is the point where the gravitational pressure exceeds the *electron degeneracy pressure* and electrons are captured by ^{24}Mg and ^{20}Ne nuclei. When this happens, the core collapses producing what is known as an *electron-capture core-collapse supernova*.

In stars with masses greater than $\approx 10 M_\odot$, the sequence of production of heavier and heavier nuclei continues. After carbon burning, there is an episode, called Ne burning, in which ^{20}Ne "photodisintegrates" by a (γ, α) reaction. The α's produced are consumed by those nuclei present, including ^{20}Ne, creating heavier elements, notably ^{24}Mg. The next phase is oxygen burning, which involves reactions, such as:

$$^{16}O + {}^{16}O \longrightarrow {}^{28}Si + {}^{4}He + \gamma$$

and

$$^{12}C + {}^{16}O \longrightarrow {}^{24}Mg + {}^{4}He + \gamma$$

A number of other less abundant nuclei, including Na, Al, P, S, and K, are also synthesized at this time.

During the final stages of evolution of massive stars, a significant fraction of the energy released is carried off by neutrinos created by electron–positron annihilations in the core of the star. If the star is sufficiently oxygen-poor that its outer shells are reasonably transparent, the outer shell of the red giant may collapse during last few 10^4 yr of evolution to form a *blue supergiant*.

1.4.2.3 The e-process

Eventually, a new core consisting mainly of ^{28}Si is produced. At temperatures above 10^9 K and densities above 10^{11} kg/m^2, a process known as *silicon burning*, or the *e-process* (for equilibrium), begins and lasts for only day or so, again depending on the mass of the star. These are reactions of the type:

$$^{28}Si + \gamma \rightleftharpoons {}^{24}Ne + {}^{4}He$$
$$^{28}Si + {}^{4}He \rightleftharpoons {}^{32}S + \gamma$$
$$^{32}S + {}^{4}He \rightleftharpoons {}^{36}Ar + \gamma$$

While these reactions can go either direction, there is some tendency for the buildup of heavier nuclei with masses 32, 36, 40, 44, 48, 52, and 56. Partly as a result of the e-process, these nuclei are unusually abundant in nature. In addition, because of a variety of

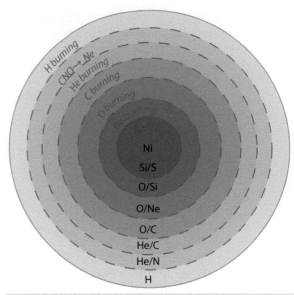

Figure 1.14 Schematic diagram of stellar structure just prior to core collapse. Nuclear burning processes (blue italics) and dominant composition are listed for each shell. Shells are not to scale.

nuclei produced during C and Si burning phases, other reactions are possible, synthesizing a number of minor nuclei. The star is now a cosmic onion of sorts (Figure 1.14), consisting of a series of shells of successively heavier nuclei and a core of Fe. Though temperature increases toward the interior of the star, the structure is stabilized somewhat, although not entirely, with respect to convective mixing between shells because each shell is denser than the one overlying it.

Fe-group elements may also be synthesized by the e-process in Type Ia supernovae. Type Ia supernovae occur when white dwarfs of intermediate-mass (3–$10 M_\odot$) stars in binary systems accrete material from their companion. When their cores reach the Chandrasekhar limit, C burning is initiated and the star explodes. This theoretical scenario has been confirmed in recent years by space-based optical, gamma-ray, and X-ray observations of supernovae, such as the image of supernova remnant Cassiopeia A, which exploded around 1681, shown in Figure 1.15.

1.4.2.4 The s-process

In second- and later-generation stars containing heavy elements, yet another nucleosynthetic process can operate. This is the slow

Figure 1.15   A false-color composite image of supernova remnant Cassiopeia A: red is infrared from the Spitzer Space Telescope, gold is visible from the Hubble Space Telescope, and blue and green are X-rays from the Chandra X-ray Observatory. The small, bright, baby-blue dot just off-center is the remnant of the star's core. Source: NASA.

neutron capture or *s-process*. It is so-called because the neutron density is generally low, i.e., $\lesssim 10^8/cm^3$, although it can reach $10^{15}/cm^3$ for brief periods, i.e., ≈ 10 yr, during thermal pulses of larger AGU stars (nucleosynthesis in these episodes of high neutron density, but nevertheless orders of magnitude lower than for the r-process, is sometimes referred to as the *intermediate neutron-capture process or i-process*). Consequently, the rate of capture of neutrons is slow, compared to the *r-process,* in which neutron densities can reach $10^{20}/cm^3$. We discuss the r-process in the following section.

Spectral observations reveal that the s-process operates in the He-burning AGB stage of low-to-moderate-mass stars with neutrons being primarily produced through the reactions:

$$^{14}N(\alpha, \gamma)^{18}F(\beta^+, \nu)^{18}O(\alpha, \gamma)^{22}Ne(\alpha, n)^{25}Mg$$

$$^{12}C(p, \gamma)^{13}N(\beta^+, \nu)^{13}C(\alpha, n)^{16}O$$

The latter mechanism dominates lower-mass stars and operates, particularly during thermal pulses, primarily in the intershell region (Figure 1.10) where both ^{4}He and ^{1}H are present. The former requires higher temperatures and operates primarily in stars with $M \geq 4M_\odot$. In massive stars, it can operate in the H-burning phase in stars of sufficient metallicity as well as in the He- and C-burning shells of Wolf–Rayet stars where neutrons are also produced by the above reactions and:

$$^{17}O(\alpha, n)^{20}Ne$$

Detection of gamma-rays produced by decay of ^{99}Tc and observed enhanced abundances of several s-process elements in these stars confirms that the s-process is occurring.

These neutrons are captured by nuclei to produce successively heavier elements. The principal difference between the r-process and the s-process (discussed in the following text) is the rate of capture relative to the decay of unstable isotopes. In the s-process, nuclei capture neutrons only infrequently. If the newly produced nucleus is unstable with a half-life shorter than the capture frequency, it will β^--decay to the next heavier element before another neutron is captured. In such cases, instabilities cannot be bridged as they can in the r-process discussed next. In the s-process, the rate of formation of stable species is given by:

$$\frac{d[A]}{dt} = f[A-1]\sigma_{A-1} \qquad (1.17)$$

where $[A]$ is the abundance of a nuclide with mass number A, f is a function of neutron flux and neutron energies, and σ is the neutron-capture cross-section. A nuclide with one less proton might contribute to this buildup of nuclide A, provided that the isobar of A with one more neutron is not stable. The rate of consumption by neutron capture is:

$$\frac{d[A]}{dt} = -f[A]\sigma_A \qquad (1.18)$$

From these relations, we can deduce that the creation ratio of two nuclides with mass numbers A and $A-1$ will be proportional to the ratio of their capture cross-sections:

$$\frac{[A]}{[A-1]} = \frac{\sigma_{A-1}}{\sigma_A} \qquad (1.19)$$

Nuclides with odd mass numbers tend to have larger capture cross-sections than even-numbered nuclides. Hence, the s-process will lead to the observed odd–even differences in abundance. The s-process also explains why magic number nuclides are particularly abundant. This is because they tend to have small capture cross-sections and hence are less likely to be consumed in the s-process. The r-process leads to a general enrichment in nuclides with N up to 6–8 greater than a magic number, but not to a buildup of nuclides with magic numbers.

The s-process yields depend on neutron density, which depends on mass and metallicity of the star as do the relative abundances of the nuclides produced. For example, ^{85}Kr has a half-life of 11 years. At low neutron densities and capture frequencies, it will decay to ^{86}Rb before it captures another neutron. ^{86}Rb has a half-life of 18 days and will then decay to ^{86}Sr at low and intermediate capture frequencies. At higher capture rates, ^{85}Kr will capture another neutron to become stable ^{86}Kr (as we will find in Chapter 14, this led to some variation in the ^{86}Kr : ^{84}Kr ratio in solar system materials).

1.4.3   Explosive nucleosynthesis

1.4.3.1   Supernovae and the r-process

The e-process stops at mass 56. Figure 1.3 shows that ^{56}Fe has the highest binding energy per nucleon, which means it is the most stable nucleus. Fusion can release energy only up to mass 56; beyond this, the reactions become endothermic, meaning that they absorb energy. Thus, once the stellar core has been largely converted to Fe, a critical phase is reached: the balance between thermal expansion and gravitation collapse is broken. The stage is now set for the most spectacular of all natural phenomena: a supernova explosion, the ultimate fate of large stars that have survived earlier evolutionary stages. The energy released in the supernova is astounding as supernovae can emit more energy than an entire galaxy (the recent supernova SN2011fe in the Pinwheel Galaxy, M101, provides an example; at its peak brightness, the supernova was visible with a small telescope, even though the galaxy was not). Stars that explode without having shed their outer layers to become Wolf–Rayet stars form Type IIa supernovae; exploding Wolf–Rayet stars are also the progenitors of Type Ib supernovae, distinguished by a lack of hydrogen in their spectra, and Type Ic supernovae, distinguished by a lack of both hydrogen and helium spectral emission.

When the mass of the iron core reaches the Chandrasekhar mass ($1.4M_\odot$), gravitational force exceeds the electron degeneracy pressure and the core collapses. The supernova begins with the collapse of this stellar core, which would have a radius similar to that of the Earth's before collapse, to a radius of 30 km or so. This occurs in a few tenths of a second, with the inner iron core collapsing at 25% of the speed of light. Forty percent of matter in the center of the core is compressed to the density of nuclear matter (3×10^{18} kg/m^2). If the core mass is $\lesssim 15M_\odot$, collapse is stopped by neutron degeneracy pressure and it rebounds colliding with the outer part of the core, which is still collapsing, sending a massive shock wave back out in less than a second after the collapse begins. The central core then forms a neutron star.

As the shock wave travels outward through the core, the increase in temperature resulting from the compression produces a breakdown of nuclei by photodisintegration, e.g.:

$$^{56}\text{Fe} + \gamma \rightarrow 13\,^3\text{He} + 4\,^1n;$$

$$^4\text{He} + \gamma \rightarrow 2\,^1\text{H} + 2\,^1n$$

This results in the production of a large number of free neutrons (and protons). Neutron densities can reach 10^{20} neutrons/cm^3 and are rapidly captured by those nuclei that manage to survive this hell. A supernova remnant having the mass of the Sun would form a neutron star of only 15 km radius.

In the case of Wolf–Rayet stars with masses greater than $25M_\odot$, neutron degeneracy pressure cannot stop the collapse and it continues, forming a singularity of infinite density from which light cannot escape: a black hole. In this case, there is no core rebound and little in the way of electromagnetic emission as well, with the star simply disappearing into the night. The fate of massive stars, in particular whether they explode as one of the various types of supernovae or simply collapse into a black hole, also depends on their metallicity and the amount of mass lost in earlier evolutionary stages as well as their initial mass.

Another important effect is the creation of huge numbers of neutrinos by positron–electron annihilations, which in turn had "condensed" as pairs from gamma rays. The energy carried away by neutrinos leaving the supernova, which can comprise 10% of the star's rest mass, exceeds the kinetic energy of the explosion by a factor of several hundred, and exceeds the visible radiation by a factor of some 30 000. Even though neutrinos interact very weakly with matter, in some supernova models it is the neutrino flux from the core that drives the rebound. The neutrinos leave the core at nearly the speed of light. While their departure is delayed slightly by interaction with matter, they travel faster than the shock wave and are delayed less than electromagnetic radiation. Thus, neutrinos from the 1987A supernova arrived at Earth (some 160 000 years after the event) a few hours before the supernova became visible.

The shock wave eventually reaches the surface of the core, and the outer part of the star is blown apart in an explosion of unimaginable violence. Amid the destruction, new nucleosynthetic processes are occurring.

This first of these is the *r-process* (rapid neutron capture) and is the principal mechanism for building up the heavier nuclei. In the r-process, the rate at which nuclei with mass number $A + 1$ are created by capture of a neutron by nuclei with mass number A can be expressed simply as:

$$\frac{dN_{A+1}}{dt} = fN_A\sigma_A \qquad (1.20)$$

where N_A is the number of nuclei with mass number A, σ is the neutron-capture cross-section, and f is the neutron flux. If the product nuclide is unstable, it will decay at a rate given by λN_{A+1}. It will also capture neutrons itself, so the total destruction rate is given by:

$$\frac{dN_{A+1}}{dt} = -fN_{A+1}\sigma_{A+1} - \lambda N_{A+1} \qquad (1.21)$$

An equilibrium distribution occurs when nuclei are created at the same rate they are destroyed at, i.e.:

$$N_A\sigma_A f = \lambda N_{A+1} + N_{A+1}\sigma_{A+1}f \qquad (1.22)$$

Thus, the equilibrium ratio of two nuclides A and $A + 1$ is:

$$N_A/N_{A+1} = (\lambda + \sigma_{A+1}f)/\sigma_A f \qquad (1.23)$$

Eventually, some nuclei capture enough neutrons that they are not stable even for short periods (in terms of Equation (1.23), λ becomes large, and hence N_A/N_{A+1} becomes large). They β^--decay to new elements, which are more stable and capable of capturing more neutrons. This process reaches a limit when nuclei beyond $Z = 90$ are reached. These nuclei fission into several lighter fragments. The r-process is thought to have a duration of 100 s during the peak of the supernova explosion. Figure 1.16 illustrates this process.

During the r-process, the neutron density is so great that all nuclei will likely capture a number of neutrons. And in the extreme temperatures, all nuclei are in excited states, and relatively little systematic difference is expected in the capture cross-sections of odd and even nuclei. Thus, there is no reason why the r-process should lead to different abundances of stable odd and even nuclides.

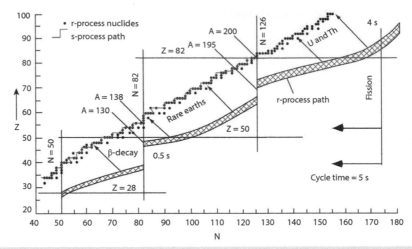

Figure 1.16   Diagram of the r-process and the s-process on a Z versus N diagram. Dashed region is the r-process path; the solid line through stable isotopes shows the s-process path.

On 23 February 1987, a blue supergiant star became the closest supernova (Figure 1.17) observed since the time of Johannes Kepler appeared in the Large Magellanic Cloud – a small satellite galaxy of the Milky Way visible in the Southern Hemisphere. This provided the first real test of models of supernovae as the spectrum could be analyzed in detail. Overall, the model presented earlier was reassuringly confirmed. The very strong radiation from ^{56}Co, the daughter of ^{56}Ni and parent of ^{56}Fe, was particularly strong confirmation of the supernova model. Observation of γ-rays from these short-lived radionuclides confirmed that the r-process occurs in supernovae. Nevertheless, there were some differences between prediction and observation, such as an overabundance of Ba. These differences provided the basis for model refinement – a process that continues with each new observation.

1.4.3.2   Neutron star mergers: kilonovae

Theoretical models and computer simulations of supernovae had not been entirely successful in reproducing the observed abundances of r-process nuclides (e.g., Freiburghaus et al., 1999a; Thielemann et al., 2011). Consequently, astronomers searched for another high-energy, neutron-rich environment where the r-process might occur. Lattimer and

Schram (1974) had suggested that tidal disruption of a neutron star by a neighboring a black hole might be an appropriate environment. Subsequent research suggested neutron star mergers, events known as "kilonovae," might be suitable environments (e.g., Li and Paczynski, 1998; Freiburghaus et al., 1999b).

On 17 August 2017, gravitational waves from an event that occurred a long time ago in a galaxy far, far away and shook the universe finally arrived at Earth and provided new insights into explosive nucleosynthesis. The US Laser Interferometer Gravitational-Wave Observatory (LIGO), which had been operational for less than two years, and the Italian Virgo observatory, which had been operational for only weeks, detected gravitational waves coming from an area of roughly 28 deg^2 of the sky, an event labeled by its date, GW170817. LIGO had previously detected several gravitational waves produced by black hole mergers[11], which by their nature produced no electromagnetic signal, but this event was different. In this same area of sky, a gamma-ray burst was detected 1 min and 1.7 s later in the galaxy NGC4993 located roughly 100 million light years away by NASA's Fermi Gamma-ray Space Telescope. These features were consistent with the predicted results of merger of two neutron stars (Figure 1.18) in the mass range of 1.17–1.60 $M_\odot$ (Li and

---

[11] Among other things, this confirmed the prediction Einstein's general theory of relativity that such waves existed. Rainer Weiss, Kip Thorne, and Barry C. Barish won the 2017 Nobel Prize in physics for the successful development of gravitational wave detectors.

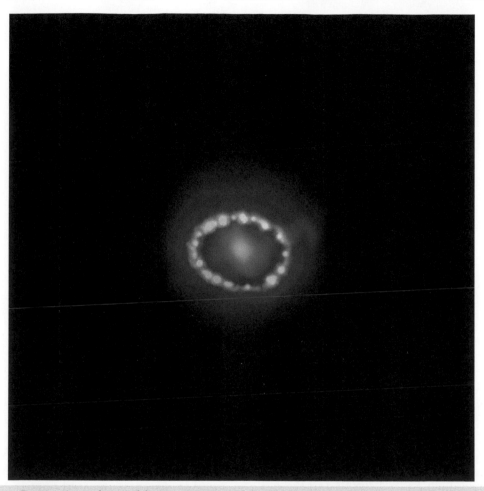

Figure 1.17 Composite radio-visible-X-ray image of the SN1987A supernova remnant. Red: radio image from the Atacama Large Millimeter Array (ALMA); green: visible Hubble image; and blue: Chandra X-ray image. The bright "string of pearls" results from the shock waves impacting a nebula expelled earlier from the star. Source: NASA.

Paczynski, 1998, LIGO Scientific Collaboration and Virgo Collaboration, 2017). In the days and weeks following this event, telescopes around the work were trained on this object, analyzing the entire electromagnetic spectrum. Although optically bright, 10^8 times more luminous than the Sun, it was less bright and dimmed more rapidly than a supernova, but remained luminous in the infrared. Analysis of the spectrum confirmed an ejected mass of 0.03–0.05 $M_\odot$ expanding at a rate close to a fifth of the speed of light. The infrared spectrum revealed this outflow was rich in heavy elements, such as the rare earths, platinum and gold (Pian et al., 2017). That, together with the gamma-ray burst, confirmed that r-process nucleosynthesis had occurred. Some of this is relatively "cold" neutron capture as

tidal forces fling material from the approaching stars and synthesize mainly heavy nuclides ($A > 140$). As the stars merge, additional matter is squeezed out of polar regions by shock heating, synthesizing lighter nuclides ($A < 140$). The timescales for these processes are a several seconds (Kasen et al., 2017).

Although neutron star mergers might seem unlikely, most stars are binaries, so a pair of large stars will eventually produce a pair of neutron stars. As they revolve around their common center of gravity, tidal forces and emission of gravitational waves steal angular momentum and they slowly spin inward over hundreds of millions of years until they finally merge. Kilonovae are now thought to be the source of mysterious short gamma-ray bursts long recognized by astronomers (long

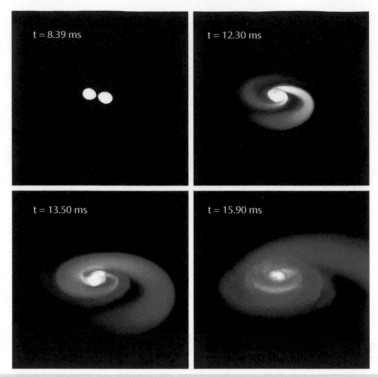

Figure 1.18 Illustration of the merger of a 1.4 and a $1.5M_\odot$ neutron star. Source: Rosswog et al. (2019), Springer Nature.

gamma-ray bursts are produced primarily by supernovae). Based on the amount heavy elements ejected during GW170817 and the estimate rate of neutron star mergers in our galaxy (about one every 2–60 million years), Kasen et al. (2017) concluded, "such mergers are a dominant mode of r-process production in the universe." Subsequent work suggests as much as $0.07\ M_\odot$ was ejected and this, combined with estimates of the rate of such events, produces an "uncomfortably large" amount of heavy elements in the universe (Rosswog, 2019). Future observations by gravitational wave observatories should refine these estimates and produce a better understanding of r-process nucleosynthesis in these events.

The r-process produces broad peaks in elemental abundance around the neutron magic numbers (50, 82, and 126, corresponding to the elements Sr and Zr, Ba, and Pb, respectively). This occurs because nuclei with magic numbers of neutrons are particularly stable and have very low cross-sections for capture of neutrons, and because nuclei just short of magic numbers have particularly high capture cross-sections. Thus, the magic number nuclei are both created more rapidly and destroyed

more slowly than other nuclei. When they decay, the sharp abundance peak at the magic number becomes smeared out.

1.4.3.3   The p-process

The r-process tends to form the heavier isotopes of a given element. A few nuclides, however, sit to the low N side of the s-process path (Figure 1.16) and are also shielded from r-process production; they must be produced by some other process. These tend to be the lightest isotopes of an element and are produced by the *p-process* and are known as p-nuclei or p-only nuclei. Burbudige et al. (1957) suggested that these nuclides were produced by proton capture in supernovae (hence termed the p-process). The probability of proton capture is much less likely than neutron capture and contributes negligibly to the production of most nuclides. The reason should be obvious: to be captured, the proton must have sufficient energy to overcome the coulomb repulsion and approach to within 10^{-16} m of the nucleus where the strong nuclear force dominates over the electromagnetic one. Astrophysicists have come to realize

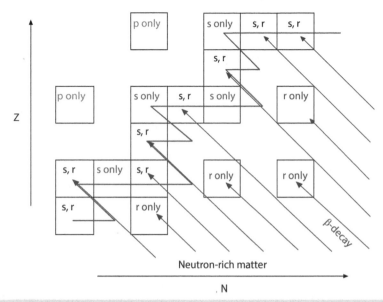

Figure 1.19 Z versus N diagram showing production of isotopes by the r-, s-, and p-processes. Squares are stable nuclei; blue lines are β-decay path of neutron-rich isotopes produced by the r-process; the red line through stable isotopes shows the s-process path.

that in environments where energies are high enough for protons to overcome this coulomb barrier, photodisintegration and (γ, p) reactions resulting in a decrease in proton number can proceed at higher rates than proton capture unless proton abundances are very high. In such environments, photodissociation reactions, such as (γ, n) and (γ, α), may be more likely sources of these p-only nuclides than proton capture. In general, a series of such reactions would be required to produce them, hindering development of robust theories. An inverse correlation between p-nuclide abundance with their photodisintegration rates, however, suggests such reactions are indeed important. α-capture reactions, such as:

$$^{190}\text{Pt}\,(\alpha, \gamma)^{194}\text{Hg}$$

may also be involved. Both Type I and Type II supernovae as well as neutron star mergers may be appropriate astrophysical sites for such reactions. In addition, in the massive neutrino outflows from Type II supernova, neutrino-induced reactions, such as:

$$^{85}\text{Rb}\,(\nu_{e}, \beta^{-}, n)^{84}\text{Sr}$$

and

$$^{181}\text{Hf}\,(\nu_{e}, \beta^{-}, n)^{180}\text{Ta}$$

(where ν_{e} is the electron neutrino) may also contribute to p-process nuclides (Rauscher et al., 2013).

The origin of p-nuclides remains incompletely understood and an area of active research in astrophysics. Since isotopic abundances cannot be determined spectroscopically in stars (except for the most abundant elements, such as O and C), the abundances of these nuclides in meteorites, including extinct ones with short half-lives, such as ^{146}Sm and ^{92}Nb, are important constraints on such models.

Figure 1.19 illustrates how the s-, r-, and p-processes create different nuclei. Note also the shielding effect. If nuclide X has an isobar (nuclide with the same mass number) with a greater number of neutrons, that isobar will "shield" X from production by the r-process. The most abundant isotopes will be those created by all processes; the least abundant will be those created by only one, particularly by only the p-process.

1.4.4 Nucleosynthesis in interstellar space

Except for production of ^{7}Li in the Big Bang, Li, Be, and B are not produced in any of the earlier situations. One clue to the creation of these elements is their abundance in

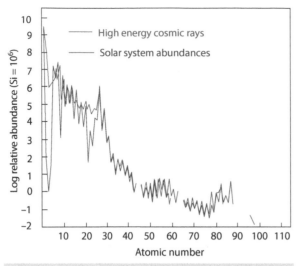

Figure 1.20 Comparison of relative abundances in cosmic rays and the solar system.

galactic cosmic rays: they are over abundant by a factor of 10^6, as is illustrated in Figure 1.20. They are believed to be formed by interactions of cosmic rays with interstellar gas and dust, primarily reactions of ^1H and ^4He with carbon, nitrogen, and oxygen nuclei. These reactions involve extremely high energies, but occur at low temperatures where the Li, B, and Be can survive. These products can themselves be accelerated to high energies as cosmic rays.

1.5 SUMMARY

Figure 1.21 is a part of the Z versus N plot showing the abundance of the isotopes of elements 56 through 62. It is a useful region of the chart of the nuclides for illustrating how the various nucleosynthetic processes have combined to produce the observed abundances. First, we notice that even-numbered elements tend to have more stable nuclei than odd-numbered ones – a result of the greater stability of nuclides with even Z, and, as we have noted, a signature of the s-process. We also notice that nuclides "shielded" from β^--decay of neutron-rich nuclides during the r-process by an isobar of lower Z are underabundant. For example, ^{147}Sm and ^{49}Sm are more abundant than ^{148}Sm, even though the former have odd mass numbers and the latter has an even mass number. ^{138}La and ^{144}Sm are rare because they are "p-process-only" nuclides: they are shielded from the r-process and also not produced by the s-process. ^{148}Nd and ^{150}Nd are less abundant than ^{146}Nd because the former are r-process-only nuclides while the latter is by both s-process and p-process. During the s-process, the flux of neutrons is sufficiently low that any ^{147}Nd produced decays to ^{147}Sm before it can capture a neutron and become a stable ^{148}Nd. Promethium has no stable nuclides and does not exist in the Earth except for extremely small amounts produced by ^{238}U fission.

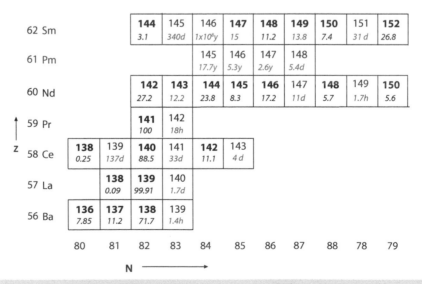

Figure 1.21 View of part of chart of the nuclides showing the light rare earths. Mass numbers of stable nuclides are shown in bold with gray background; their isotopic abundance is shown in italics as percent. Mass numbers of short-lived nuclides are shown in plain text with their half-lives given in red.

REFERENCES

Agostini, M. and the Borexino Collaboration. 2020. Experimental evidence of neutrinos produced in the CNO fusion cycle in the Sun. *Nature* **587**: 577–82. doi: 10.1038/s41586-020-2934-0.

Burbidge, E.M. and Burbidge, G. 1956. The Chemical compositions of five stars which show some of the characteristics of Population II. *The Astrophysical Journal* **124**: 116.

Burbidge, E.M., Burbidge, G.R., Fowler, W.A., et al. 1957. Synthesis of the elements in stars. *Reviews of Modern Physics* **29**: 547–650.

De Marcillac, P., Coron, N., Dambier, G., et al. 2003. Experimental detection of α-particles from the radioactive decay of natural bismuth. *Nature* **422**: 876–8. doi: 10.1038/nature01541.

Freiburghaus, C., Rembges, J.F., Rauscher, T., et al. 1999a. The Astrophysical r-process: a comparison of calculations following adiabatic expansion with classical calculations based on neutron densities and temperatures. *The Astrophysical Journal* **516**: 381–98.

Freiburghaus, C., Rosswog, S. and Thielemann, F.K. 1999b. R-process in neutron star mergers. *The Astrophysical Journal Letters* **525**: L121–4.

Karakas, A.I. and Lattanzio, J.C. 2014. The Dawes Review 2: Nucleosynthesis and stellar yields of low- and intermediate-mass single stars. *Publications of the Astronomical Society of Australia* **31**: e030 (62 pages). doi:10.1017/pasa.2014.21.

Kasen, D., Metzger, B., Barnes, J., et al. 2017. Origin of the heavy elements in binary neutron-star mergers from a gravitational-wave event. *Nature* **551**: 80–4. doi: 10.1038/nature24453.

Lattimer, J. M. and Schramm, D. N. 1974. Black-hole-neutron-star collisions. *The Astrophysical Journal* **192**: L145–7.

Li, L.-X. and Paczyński, B. 1998. Transient events from neutron star mergers. *The Astrophysical Journal Letters* **507**: L59–62. doi: 10.1086/311680.

LIGO Scientific Collaboration and Virgo Collaboration, Abbott, B. P., Abbott, R., Abbott, T. D. et al. 2017. GW170817: Observation of gravitational waves from a binary neutron star inspiral. *Physical Review Letters* **119**: 161101. doi: 10.1103/PhysRevLett.119.161101

Rosswog, S. 2019. Neutron star mergers as r-process sources, in Formicola, A., Junker, M., Gialanella, L. and Imbriani, G. (eds). *Nuclei in the Cosmos XV, 219*. Cham: Springer International Publishing, 105–10 pp.

Suess, H. E. 1987. *Chemistry of the Solar System*. New York: John Wiley & Sons, Inc.

Thielemann, F. K., Arcones, A., Käppeli, R., et al. 2011. What are the astrophysical sites for the r-process and the production of heavy elements? *Progress in Particle and Nuclear Physics* **66**: 346–53. doi: 10.1016/j.ppnp.2011.01.032.

PROBLEMS

Useful facts for these problems:

Avogadro's number is 6.02252×10^{23} $(\text{mol})^{-1}$

Speed of light: $c = 2.997925 \times 10^8$ m/s

$1 \text{ eV} = 1.60218 \times 10^{-19}$ J; $1 \text{ J} = 1 \text{ kg m}^2/\text{s}^2$; $1 \text{ J} = 1 \text{ VC}$

Electron charge: $q = 1.6021 \times 10^{-19}$ C (coulomb)

$1 \text{ u} = 1.660541 \times 10^{-27}$ kg

$1 \text{ G} = 2.997925 \times 10^4$ J/mC

1. (a) How many moles of Nd ($AW = 144.24$ u; u is unified atomic mass units) are there in 50 g of Nd_2O_3 (AW of oxygen is 15.999 u)?
   (b) How many atoms of Nd in this?

2. Given an electron and positron of equal energy, how much more energy is the positron capable of depositing in a detector?

3. What are the binding energies per nucleon of ^{87}Sr (mass = 86.908879 u) and ^{143}Nd (mass = 86.90918053 u)?

4. What is the total energy released when ^{87}Rb (mass = 86.909183 u) decays to ^{87}Sr (mass = 86.908879 u)?

5. Using the equation and values for the liquid-drop model, predict the binding energy per nucleon for ^{4}He, ^{56}Fe, and ^{238}U (ignore even–odd effects).

6. How many stable nuclides are there with $N = 82$? List them. How many stable nuclides are there with $N = 83$? List them too. Why the difference?

7. Calculate the maximum β^- energy in the decay of ^{187}Re to ^{187}Os. The mass of ^{187}Re is 186.9557508 u; the mass of ^{187}Os is 186.9557479 u.

8. Twenty-eight percent of ^{228}Th atoms decay to ^{224}Ra by emitting an α of 5.338 MeV. What is the recoil (kinetic) energy of the ^{224}Ra atom? Is the ^{224}Ra in its ground state? If not, what is nuclear energy in excess of the ground state? (Mass of ^{228}Th is 228.0288 u; mass of ^{224}Ra is 224.0202 u; and mass of α is 4.002603 u.)

9. A section of the chart of the nuclides is shown next. Mass numbers of stable isotopes are shown in bold; unstable nuclides shown in plain typeface can be assumed to be short-lived. The chart shows all nuclides relevant to the following questions.
   (a) Show the s-process path beginning with ^{134}Ba.
   (b) Identify all nuclides created, in part or in whole, by the r-process.
   (c) Identify all nuclides created *only* by the p-process.
   (d) Which of the stable nuclides shown should be least abundant and why?
   (e) Which of the cerium (Ce) isotopes shown would you expect to be most abundant and why? (Your answer *may* include more than one nuclide in [d] and [e].)

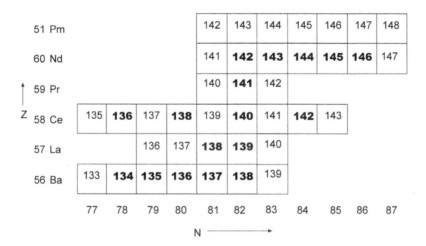

   Detailed information on the nuclides can be found on a web version of the Chart of the Nuclides maintained by the Brookhaven National Laboratory website (https://www.nndc.bnl.gov/).

10. Both ^{122}Te and ^{123}Te are created only in the s-process. ^{122}Te constitutes 2.55% of Te and ^{123}Te constitutes 0.89% of Te. What was the ratio of capture cross-sections of these nuclides during the s-process? (Your answer might differ from capture cross-sections listed in tables as the probability of neutron capture varies with neutron energy.)

11. What is the recoil energy of ^{208}Tl (mass = 207.9820187 u) in the 5.601 MeV α-decay illustrated in Figure 1.7?

12. A certain radionuclide emits radiation at the rate of 15.0 μW at one instant of time and at that of 1.0 μW one hour latter. What is its half-life?

Chapter 2

Decay Systems and Geochronology I

2.1 BASICS OF RADIOGENIC ISOTOPE GEOCHEMISTRY

2.1.1 Historical background

We can broadly define two principal applications of radiogenic isotope geochemistry. The first is geochronology, which makes use of the constancy of the rate of radioactive decay to measure time. Since a radioactive nuclide decays to its daughter at a rate independent of everything, we can determine a time simply by determining how much of the nuclide has decayed. We will discuss the significance of this time at a later point. Geochronology is fundamental to our understanding of nature and its results pervade many fields of science. Through it, we know the age of the Sun, the Earth, and our Solar System, which provides a calibration point for stellar evolution and cosmology. Geochronology also allows us to trace the origins of culture, agriculture, and civilization back beyond the 5000 years of recorded history; the origin of our species to some 350 000 years; the origins of humans, the genus *Homo*, to more than 2 million years; and the origin of life to at least 3.5 billion years. Most other methods of determining time, such as so-called molecular clocks, are valid only because they have been calibrated against radiometric ages.

The history of geochronology begins with physicist Ernest Rutherford (a New Zealander then working at McGill University in Montreal) and chemist Bertram Boltwood (an American pioneer of radiochemistry then working at Yale University). In 1896, Henri Becquerel at the École Polytechnique of Paris had discovered that uranium (U) was radioactive and by 1898, Marie and Pierre Curie had shown that thorium is also radioactive and had discovered some of the daughter products of uranium, including polonium and radium. Across the Atlantic, Rutherford quickly became interested in radioactivity and in 1899 demonstrated it produced two kinds of rays, which he called alpha and beta (gamma radiation was discovered the following year). Rutherford also recognized that helium was produced by radioactive decay, although it was not until 1907 that he could prove his suspicion that alpha particles were helium nuclei. In 1903, Rutherford and Frederick Soddy published their *Law of Radioactive Change*, stating that the rate of decay of radioactive substances was constant. Rutherford quickly recognized the significance of this for determining geologic time; he measured the amount of helium and uranium in several U-rich minerals and concluded that they must be at least 500 million years old (Rutherford, 1906).

Meanwhile, Boltwood, who was corresponded frequently with Rutherford, had deduced that lead was the ultimate decay product of uranium. In 1907, he analyzed a series of uranium-rich minerals, determining their U and Pb contents, and found that the Pb/U ratio in these minerals was the same when minerals were from the same geologic period and increased with apparent geologic age (Boltwood, 1907) (the *relative* geologic

Isotope Geochemistry, Second Edition. William M. White.
© 2023 John Wiley & Sons Ltd. Published 2023 by John Wiley & Sons Ltd.
Companion Website: www.wiley.com/go/white/isotopegeochem2

timescale, more or less as we know it today, had been worked out by the mid-nineteenth century). From laboratory experiments, he had calculated that 1 g of Pb was produced per year for every 10 billion grams of uranium. Based on this, he calculated ages for his samples ranging from 410 to 2200 million years. Rutherford's and Boltwood's ages meant that the Earth was far older than what Lord Kelvin (William Thomson) had estimated a decade earlier from cooling of the Earth, which was about 20–40 million years.

Boltwood's ages turned out to be too old by almost a factor of 2. There are a variety of reasons for this. For one thing, he was unaware that there were two isotopes of uranium decaying to two isotopes of lead at different rates. Indeed, isotopes had not yet been discovered: it would be five more years before Joseph John Thomson at Cambridge University would demonstrate their existence and it would not be until 1939 that Alfred Nier[1] was able to accurately measure the $^{235}U/^{238}U$ ratio (Nier, 1939). However, the main reason why Boltwood's ages were too old was probably his failure to account for thorogenic Pb; neither Boltwood nor anyone else knew that Pb was also the ultimate decay product of Th.

Physicists at first resisted the idea of a much older Earth, but geologists felt vindicated as they had believed that Kelvin's age was too young. As it turns out, radioactivity also helps explain why Kelvin's age was wrong: the radioactive decay of U, Th, and K heats the Earth, as Rutherford recognized in 1906. The great age of the Earth also required some new, as yet unknown, energy source for the Sun and the stars, which Hans Bethe, following an earlier suggestion of Arthur Eddington, later deduced to be nuclear fusion.

It was not until the 1950s that Aldrich et al. (1956) demonstrated that $^{40}K-^{40}Ar$ and $^{87}Rb-^{87}Sr$ decay systems could also provide useful geochronological tools. Another landmark was reached when Gast (1960) inferred

from $^{87}Sr/^{86}Sr$ ratios of basalts that the Earth's mantle was depleted in alkali elements relative to chondritic meteorites, demonstrating that radiogenic isotope ratios had utility beyond geochronology. This was followed by a study of Sr and Pb isotope ratios in oceanic island basalts (OIBs) by Gast et al. (1964), which, they argued, demonstrated that the Earth's mantle was chemically heterogeneous. Subsequent work added the $^{147}Sm-^{144}Nd$, $^{176}Lu-^{176}Hf$, $^{187}Re-^{187}Os$, and $^{138}La-^{138}Ce$ decay systems to the radiogenic geochemistry tool set. Radiogenic isotope ratios have been applied to problems extending far beyond measuring time, ranging from forensics to understanding the formation and evolution of continents, ocean circulation, climate evolution, and planetology. These applications will be the subject of Chapters 6 thorough 8 and will appear again in Chapters 10, 13, and 14. Those applications depend on the principles and equations we will introduce in this chapter and the following two chapters.

2.1.2 The basic equations

Table 2.1 lists the principal decay systems used in geology; these are also illustrated in Figure 2.1. As is suggested by the various footnotes, decay constants continue to be a field of active research (and also note that in this text, we will be using some recently determined values that have not yet become the official ones). All of these decay systems obey the *basic equation of radioactive decay*, which is:

$$\frac{dN}{dt} = -\lambda N \qquad (2.1)$$

where λ is the decay constant, which we define as the probability that a given atom would decay in some time dt. It has units of time^{-1}. Each radioactive nuclide has a unique decay constant governing its probability of decay. With the minor exception that we noted in Chapter 1 of electron capture, decay constants

---

[1] Alfred Nier (1911-1994) received his PhD from the University of Minnesota in 1936 and became an assistant professor there in 1938, where he remained for the rest of his life with the exception of 2 years he spent working in the Manhattan Project. His developed the magnetic sector mass spectrometer, which remains the workhorse of isotope ratio measurements to this day. With this instrument he was able to make the first radiometric U-Pb *isotope* age determinations. He also made the first determinations of the isotopic composition and isotope masses of many elements and discovered the radioisotope ^{40}K. He was able to isolate a pure sample of ^{235}U, which led directly to induced fission, the basis of bombs and nuclear reactors. Later in his career, he designed miniature mass spectrometers which flew on NASA missions to Mars and Venus to measure the isotopic compositions of those atmospheres.

Table 2.1   Geologically useful long-lived radioactive decay systems.

Parent	Decay mode	λ	Half-life	Daughter	Ratio
^{40}K	β^-, e.c, β^+	5.531×10^{-10} yr^{-1}*	1.253×10^9 yr	^{40}Ar, ^{40}Ca	^{40}Ar/^{36}Ar
^{87}Rb	β^-	1.42×10^{-11} yr^{-1}£	48.8×10^9 yr	^{87}Sr	^{87}Sr/^{86}Sr
^{138}La	β^-	2.67×10^{-12} yr^{-1}	2.59×10^{11} yr	^{138}Ce, ^{138}Ba	^{138}Ce/^{142}Ce, ^{138}Ce/^{136}Ce
^{147}Sm	α	6.54×10^{-12} yr^{-1}	1.06×10^{11} yr	^{143}Nd	^{143}Nd/^{144}Nd
^{176}Lu	β^-	$1.867^\dagger \times 10^{-11}$ yr^{-1}	3.6×10^{10} yr	^{176}Hf	^{176}Hf/^{177}Hf
^{187}Re	β^-	1.67×10^{-11} yr^{-1}	4.16×10^{10} yr	^{187}Os	^{187}Os/^{188}Os, (^{187}Os/^{186}Os)
^{190}Pt	α	1.54×10^{-12} yr^{-1}	4.50×10^{11} yr	^{186}Os	^{186}Os/^{188}Os
^{232}Th	α	4.948×10^{-11} yr^{-1}	1.4×10^{10} yr	^{208}Pb, ^{4}He	^{208}Pb/^{204}Pb, ^{3}He/^{4}He
^{235}U	α	9.8571×10^{-10} yr^{-1}‡	7.07×10^8 yr	^{207}Pb, ^{4}He	^{207}Pb/^{204}Pb, ^{3}He/^{4}He
^{238}U	α	1.55125×10^{-10} yr^{-1}	4.47×10^9 yr	^{206}Pb, ^{4}He	^{206}Pb/^{204}Pb, ^{3}He/^{4}He

Note: ^{147}Sm and ^{190}Pt also produce ^{4}He, but a small amount compared to U and Th.
*This is the value suggested by Renne et al. (2011) who also calculated a branching ratio, i.e., ratios of decays to ^{40}Ar to total decays of ^{40}K of 0.1041. The conventionally accepted decay constant and branching ratios (Steiger and Jager, 1977) are 5.543×10^{-10} yr^{-1} and 0.117, respectively. Min et al. (2000) calculated these values as 5.463×10^{-10} yr^{-1} and 0.1061, respectively.
£The officially accepted decay constant for ^{87}Rb is that shown here. However, recent determinations of this constant range from 1.421×10^{-11} yr^{-1} by Rotenberg (2005) to 1.393×10^{-11} yr^{-1} by Nebel et al. (2011).
†This is the value recommended by Söderlund et al. (2004).
‡Value suggested by Mattinson (2010). The conventional value is 9.8485×10^{-10} yr^{-1}.

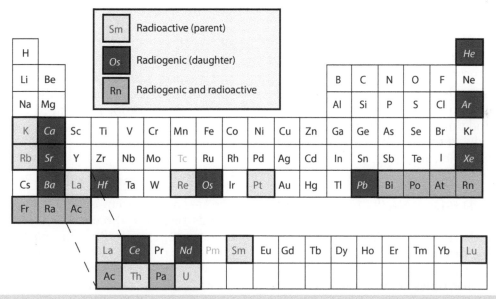

Figure 2.1   Periodic table showing the elements having naturally occurring radioactive isotopes and the elements produced by their decay.

are true constants: their value is the same throughout all space and all time. The other equations that we will introduce in this chapter and the subsequent five chapters are derived from this one equation.

Let us rearrange Equation (2.1) and integrate:

$$\int_{N_0}^{N} \frac{dN}{N} = \int_0^t -\lambda dt \qquad (2.2)$$

where N_0 is the number of atoms of the radioactive, or parent, isotope present at time $t = 0$. On integrating, we obtain:

$$\ln \frac{N}{N_0} = \lambda t \qquad (2.3)$$

This can be expressed as:

$$\frac{N}{N_0} = e^{-\lambda t} \text{ or } N = N_0 e^{-\lambda t} \qquad (2.4)$$

Suppose, we want to know the amount of time for the number of parent atoms to decrease to half the original number, i.e., t when $N/N_0 = 1/2$. Setting N/N_0 to $1/2$, we can rearrange Equation (2.3) to obtain:

$$\ln 1/2 = -\lambda t_{1/2} \quad \text{or} \quad \ln 2 = \lambda t_{1/2}$$

and finally:

$$t_{1/2} = \frac{\ln 2}{\lambda} \qquad (2.5)$$

This is the definition of the *half-life*, $t_{1/2}$.

Now, the decay of the parent produces a daughter, or *radiogenic*, nuclide. The number of daughter atoms produced, $D*$, is simply the difference between the initial number of parents and the number remaining after time t:

$$D^* = N_0 - N \qquad (2.6)$$

Rearranging Equation (2.4) to eliminate N_0 and substituting that into Equation (2.6), we obtain:

$$D^* = Ne^{\lambda t} - N = N\left(e^{\lambda t} - 1\right) \qquad (2.7)$$

This tells us that the number of daughters produced is a function of the present number of parents and time. Since, in general, there will be some atoms of the daughter nuclide around to begin with, i.e., at $t = 0$, a more general expression is:

$$D = D_0 + N\left(e^{\lambda t} - 1\right) \qquad (2.8)$$

where D is the present number of daughters and D_0 is the original number of daughters.

There is a simple linear approximation of this function for times short compared to the inverse of the decay constant. An exponential function can be expressed as a Taylor series expansion:

$$e^{\lambda t} = 1 + \lambda t + \frac{(\lambda t)^2}{2!} + \frac{(\lambda t)^3}{3!} + \ldots \qquad (2.9)$$

Provided $\lambda t \ll 1$, the higher-order terms become very small and can be ignored; hence, for times that are short compared to the decay constant inverse (i.e., for $t \ll 1/\lambda$), Equation (2.8) can be written as:

$$D \cong D_0 + N\lambda t \qquad (2.10)$$

Let us now write Equation (2.8) using a concrete example, such as the decay of ^{87}Rb to ^{87}Sr:

$$^{87}\text{Sr} = {}^{87}\text{Sr}_0 + {}^{87}\text{Rb}\left(e^{\lambda t} - 1\right) \qquad (2.11)$$

As it turns out, it is generally much easier, and usually more meaningful, to measure the ratio of two isotopes than the absolute abundance of one isotope. We therefore measure the ratio of ^{87}Sr to a non-radiogenic isotope, which by convention is ^{86}Sr. Thus, the useful form of Equation (2.11) is:

$$\frac{^{87}\text{Sr}}{^{86}\text{Sr}} = \left(\frac{^{87}\text{Sr}}{^{86}\text{Sr}}\right)_0 + \frac{^{87}\text{Rb}}{^{86}\text{Sr}}\left(e^{\lambda t} - 1\right) \qquad (2.12)$$

Similar expressions can be written for other decay systems. In some cases, such as the K–Ar system in young volcanic rocks or the Re–Os system in molybdenites, the abundance of the stable isotope of the daughter element, ^{36}Ar or ^{188}Os, is so low that it cannot be accurately analyzed, in which case Equation (2.11) is preferred.

It must be emphasized that ^{87}Rb/^{86}Sr ratio in Equation (2.12), which we will call the "parent–daughter ratio," is the ratio at time t, i.e., *present* ratio. If we need to know this ratio at some other time, we need to calculate it using Equation (2.4).

2.1.3   A special case: The U–Th–Pb system

The U–Th–Pb system is somewhat of a special case as there are three decay schemes producing isotopes of Pb. In particular, two U isotopes decay to two Pb isotopes, and since the two parents and two daughters are chemically identical, combining the two provides a particularly powerful tool.

Let us explore the mathematics of this. First some terminology. The ^{238}U/^{204}Pb ratio is called μ (mu) and the ^{232}Th/^{238}U is called κ (kappa). The ratio ^{238}U/^{235}U is constant, or nearly so (we discuss the consequences of small variations in the ratio in subsequent chapters), at any given time in the Earth and today is 137.82. We can write two versions of Equation (2.8):

$$^{207}\text{Pb}/^{204}\text{Pb} = \left(^{207}\text{Pb}/^{204}\text{Pb}\right)_0 + \frac{\mu}{137.82}\left(e^{\lambda_{235}t} - 1\right)$$
$$(2.13)$$

and $^{206}\text{Pb}/^{204}\text{Pb} = \left(^{206}\text{Pb}/^{204}\text{Pb}\right)_0 + \mu\left(e^{\lambda_{238}t} - 1\right)$

$$(2.14)$$

These can be rearranged by subtracting the initial ratio from both sides and calling the difference between the initial and the present ratio Δ. For example, Equation (2.13) becomes:

$$\Delta^{207}\text{Pb}/^{204}\text{Pb} = \frac{\mu}{137.82}\left(e^{\lambda_{235}t} - 1\right) \quad (2.15)$$

Dividing by the equivalent equation for $^{238}\text{U}-^{206}\text{Pb}$ yields:

$$\frac{\Delta^{207}\text{Pb}/^{204}\text{Pb}}{\Delta^{206}\text{Pb}/^{204}\text{Pb}} = \frac{\left(e^{\lambda_{235}t} - 1\right)}{137.82\left(e^{\lambda_{238}t} - 1\right)} \quad (2.16)$$

Notice the absence of the μ term. The equation holds for any present-day ratio of $^{207}\text{Pb}/^{204}\text{Pb}$ and $^{206}\text{Pb}/^{204}\text{Pb}$ we measure and thus for all pairs of ratios. The left-hand side is simply the slope of a series of data points from rocks or minerals formed at the same time (and remaining closed systems since time t) on a plot of $^{207}\text{Pb}/^{204}\text{Pb}$ versus $^{206}\text{Pb}/^{204}\text{Pb}$. This means we can determine the age of a system without knowing the parent–daughter ratio. The bad news is that Equation (2.16) cannot be solved for t. However, we can guess a value of t, plug it into the equation, calculate the slope, compared the calculated slope with the observed one, revise our guess of t, calculate again, and so on. This is pretty laborious, but making "educated guesses" of t and using a computer, this is pretty easy. In fact, using simple minimization algorithms, we can generally converge to a high degree of accuracy after a few iterations.

2.1.4   Caveat: isotope fractionation

The elements in Table 2.1 all experience chemical isotopic fractionation to some extent. That is to say, their isotopic ratios vary because of slight differences in their chemical behavior. We will discuss the underlying physics and chemistry of these isotopic fractionations in Chapter 5. We can ignore these effects in this chapter for several reasons. First, natural chemical fractionations for the elements of interest are generally quite small. For example, Krabbenhöft et al. (2010) documented variations in the $^{86}\text{Sr}/^{88}\text{Sr}$ ratio of less than ½ per mil in rivers, carbonate rocks, and the ocean. Second, Ar, Sr, Nd, Hf, and Os isotope ratios

are routinely corrected for such fractionation by measuring the extent to which the ratio of a pair of non-radiogenic isotopes (e.g., ^{86}Sr and ^{88}Sr or ^{146}Nd and ^{144}Nd) differs from an accepted value and by applying a correction for fractionation, assuming the magnitude of the fractionation depends on the masses of the isotopes of interest. The details of this correction are explained in Section 2.2.3. This correction is essential because fractionations occurring in the mass spectrometer can be much larger than natural ones (1% or more in the case of the $^{86}\text{Sr}/^{88}\text{Sr}$ ratio). Correction for fractionation in this manner is not possible for Pb isotopes because there is only one non-radiogenic isotope. However, there are several techniques that have now come into wide use to correct Pb isotope ratios for fractionation during analysis, leading to much greater precision than in the past. We will discuss those in Section 2.2.3.

2.2   FUNDAMENTALS OF GEOCHRONOLOGY

2.2.1   Isochron dating

Let us rewrite Equation (2.12) in more general terms:

$$R = R_0 + R_{\text{P/D}}\left(e^{\lambda t} - 1\right) \quad (2.17)$$

where R_0 is the initial ratio and $R_{\text{P/D}}$ is the parent/daughter ratio at time t. Measurement of geologic time is most often based on this equation, or on various derivatives of it. We will refer to it as the *isochron equation*. First, let us consider the general case. Given a measurement of an isotope ratio, R, and a parent–daughter ratio, $R_{\text{P/D}}$, two unknowns remain in Equation (2.17), namely, t and the initial ratio. In general, we can calculate neither of them from a single pair of measurements. In the special case where the initial concentration of the daughter is very small, as is sometimes true in the K–Ar or Re–Os systems, we can neglect R_0 or, if $R \gg R_0$, simple assumptions about R_0 may suffice. However, in the general case, we must measure R and $R_{\text{P/D}}$ on a second sample for which we believe t and R_0 are the same. Then, we have two equations and two unknowns, and subtracting the two equations yields:

$$\Delta R = \Delta R_{\text{P/D}}\left(e^{\lambda t} - 1\right) \quad (2.18)$$

which eliminates R_0 from the equation and allows us to solve for t. This can be rearranged as:

$$\frac{\Delta R}{\Delta R_{P/D}} = e^{\lambda t} - 1 \qquad (2.19)$$

In practice, one measures many pairs and solving for $\Delta R/\Delta R_{P/D}$ by regression (indeed, geochronologists would not generally accept an age based on only two measurements), t may then be solved for as:

$$t = \frac{\ln\left(\Delta R\big/\Delta P_{P/D} + 1\right)}{\lambda} \qquad (2.20)$$

For a given value of t, Equation (2.17) has the form $y = a + bx$, where y is R, a is R_0, b is $e^{\lambda t} - 1$, and x is $R_{P/D}$. This is, of course, an equation for a straight line on a plot of R versus $R_{P/D}$ with slope $b = e^{\lambda t} - 1$, and intercept $a = R_0$. Thus, on such a plot, the slope of the line depends only on t (since λ is a constant for any given decay system). A line whose slope depends only on t is known as an *isochron*. Note that on a plot of $^{207}Pb/^{204}Pb$ versus $^{206}Pb/^{204}Pb$, a line may also be an isochron, since its slope depends only on t.

Regression is simply a statistical method of calculating the slope of a line. Regression treatment yields both a slope and an intercept. The latter is simply the initial ratio since, as may be seen from Equation (2.17), $R = R_0$ when $R_{P/D} = 0$. The geochronological information is contained in the slope, since it depends on t; however, important information can also be obtained from the value of the intercept, the initial ratio, since it gives some information about the history prior to time $t = 0$ of the system being investigated.

There are two important assumptions, or conditions, built into the use of Equation (2.20):

1. The system of interest was at isotopic equilibrium at time $t = 0$. Isotopic equilibrium in this case means the system had a homogeneous, uniform value of R_0.
2. The system as a whole and its each analyzed part were closed between $t = 0$ and time t (usually the present time). By "closed" we mean there has been no transfer of the parent or the daughter element into or out of the system.

Violation of these conditions is the principal source of error in geochronology. Other errors arise from errors or uncertainties associated with the analysis. If the range in variation in measured R and $R_{P/D}$ is small, these analytical errors can be the limiting factor in the determination of an age. Note that both R and $R_{P/D}$ must be known accurately.

Finally, of course, we must also know λ accurately. Decay constants are not fundamental constants that can somehow be deduced from the fundamental laws; instead, each must be measured and there are limits to the accuracy with which they have been measured. As technology advances and analytical precision increases, the accuracy of radiometric ages is increasingly limited by how well the decay constants are known. Decay constants can be determined in three ways, which we will refer to as counting, accumulation, and calibration. In counting, a known amount of the nuclide of interest is placed in a detector of known efficiency and the numbers of α, β, or γ rays emitted in a fixed time are counted. In accumulation, a known mass of highly purified parent nuclide is allowed to sit for a fixed amount of time (decades in some cases), after which the daughter nuclide is extracted and its mass determined. In the calibration approach, isotope ratios and parent–daughter ratios of two systems, e.g., Lu–Hf and U–Pb, are determined in rocks or minerals that are known to meet the above two conditions. The age is determined using the system whose decay constant is well known, and then Equation (2.19) is solved for λ for the second system, using t determined in the first system. Decay constants for U, Th, and K are now known within an uncertainty of considerably better than 1% (but even at this level, uncertainty in decay constants can limit the precision of age determinations and revisions to the ^{235}U decay constant have recently been suggested). Decay constants for Rb, La, Lu, and Re are less well known and continue to be active research topics, and there have been a number of recent suggested revisions to these values, as indicated in Table 2.1. These nuclides emit relatively low-energy βs and weak or no γs, so the counting approach has proved problematic. Indeed, two recent attempts to determine the ^{87}Rb decay constant by accumulation and calibration disagree by 1.5%. The situation for ^{176}Lu was even worse in the early part of this century, with values varying by 6%. However, the most recent determinations by counting and calibration agree within 1%.

The requirement of a closed and initially homogeneous system described here suggests a meaning for the nature of the *event* dated by radiogenic isotope geochemistry, and a meaning for *time* in the first paragraph of Section 2.1.1 in this chapter. In general, the *event* is *the last time the system was open to complete exchange of the parent and daughter elements between the various subsystems* we sample and analyze; that is, *the last point in time that the system had a homogeneous, uniform value of R*. Since the rate at which diffusion and chemical reactions occur increases exponentially with temperature, *this event is generally a thermal one*, i.e., the last time the system was hot enough for such exchange between subsystems to occur. Exactly what temperature is implied can vary widely, depending on the nature of our samples and the particular decay system we are using. Minerals, such as biotite and hornblende, will lose Ar at temperatures of a few hundred degrees. On the other hand, minerals, such as pyroxene, can remain closed to Sm and Nd exchange up to nearly 1000°C. The "closure" temperatures of various isotope systems in various minerals can be used to advantage: in some cases, an analysis of a variety of decay systems on a variety of sample types has recorded entire cooling histories.

The process accomplishing isotopic homogenization of a "system" usually involves diffusion, the rate of which, like other reaction rates, increases exponentially with temperature. Diffusion rates will also vary depending on the element and the properties of the material through which the element diffuses. We can nevertheless make the general observation that the greater the length scale, the greater will be the time (or the higher the temperature required) for isotopic homogenization to be achieved. For the same temperature and duration of a thermal event, diffusion will more readily achieve isotopic homogenization on a small scale than on a large one. Thus, if our samples or subsystems are "whole rocks (WRs)" collected meters or perhaps kilometers apart, the event dated will generally be a higher-temperature one than an event dated by analysis of individual minerals from a rock specimen whose scale is only a few centimeters. We will discuss diffusion and closure

temperatures in more detail in Section 2.3.1 in the context of K–Ar geochronology.

2.2.2 Calculating isochrons

The idea of *least-squares regression* is to minimize the *squares* of the deviations from the function relating one variable to another (i.e., deviations from a line on a graph of the two variables). In the simplest case, the relationship is assumed to be linear, as it is in the isochron equation. The quantity to be minimized is the sum of the squares of deviations:

$$\sum_{i=1}^{n} e^2 = \sum_{i=1}^{n} (y - a - bx)^2 \tag{2.21}$$

where y is the observed value, $a + bx$ is the predicted value, n is the number of observations, and e is the difference between the observed and the predicted value, i.e., the *deviation*.

The use of the squares of the deviations means that large deviations will affect the calculated slope more than small deviations. By differentiating Equation (2.21), it can be shown that the minimum value for the left side occurs when the slope is:

$$b = \sum_{i} \frac{(x_i - \bar{x})(y_i - \bar{y})}{(x_i - \bar{x})^2} \tag{2.22}$$

where $\bar{x}$ and $\bar{y}$ are the means of x and y, respectively, and x_i and y_i are the ith pair of observations of x and y, respectively. We can see from Equation (2.22) that the regression slope is the cross product of the deviations of x and y from the means divided by the square of the deviations of x from the mean of x. A more convenient computational form of Equation (2.22) is:

$$b = \frac{\sum_{i}^{n} x_i y_i - \dfrac{\sum_{i}^{n} y_i \sum_{i}^{n} x_i}{n}}{\sum_{i}^{n} x_i^2 - \bar{x}^2 n} \tag{2.23}$$

The intercept is then given by:

$$a = \bar{y} - b\bar{x} \tag{2.24}$$

The error on the slope is:

$$\sigma_b = \sqrt{\left[\sum y_i^2 - \bar{y}^2 n - \frac{(\sum(x_i y_i) - \bar{y}\bar{x}n)^2}{\sum x_i^2 - \bar{x}^2 n}\right]\left[\frac{1}{(n-2)(\sum x_i^2 - \bar{x}^2 n)}\right]} \tag{2.25}$$

The error on the intercept is:

$$\sigma_a = \sqrt{\left[\sum y_i^2 - \overline{y}^2 n - \frac{\left(\sum(x_iy_i) - \overline{yx}n\right)^2}{\sum x_i^2 - \overline{x}^2 n}\right]\left[\frac{1}{n} + \frac{\overline{x}^2}{\left(\sum x_i^2 - \overline{x}^2 n\right)}\right]\left[\frac{1}{n-2}\right]}$$

$$(2.26)$$

Statistics books generally give an equation for linear least-squares regression assuming one dependent and one independent variable where the independent variable is assumed to be known absolutely. While it is true that in the isochron equation, R is a function of $R_{P/D}$ in a geologic sense and hence may be considered the dependent variable, in practice both R and $R_{P/D}$ are measured quantities, and neither is known absolutely: both have errors of measurement associated with them. These must be taken into account for a proper estimate of the slope and the errors associated with it. In some cases, the errors in measurement of x and y can be correlated, and this must also be taken into account. The so-called *two-error regression* algorithm takes account of these errors. This calculation is, however, considerably more complex than that mentioned previously. The approach is to weight each observation according to the measurement error (the weighting factor will be inversely proportional to the estimated analytical error so that observations with larger errors are less important than those with small ones). A solution was published by York (1969), among others. The regression slope is:

$$b = \frac{\sum\left(Z_i^2(y_i - \overline{y})\left[\frac{x_i - \overline{x}}{\omega(y_i)} + \frac{b(y_i - \overline{y})}{\omega(x_i)} + \frac{r_i(y_i - \overline{y})}{\alpha_i}\right]\right)}{\sum\left(Z_i^2(x_i - \overline{x})\left[\frac{x_i - \overline{x}}{\omega(y_i)} + \frac{b(y_i - \overline{y})}{\omega(x_i)} + \frac{br_i(y_i - \overline{y})}{\alpha_i}\right]\right)}$$

$$(2.27)$$

where $\omega(x_i)$ is the weighting factor for x_i (generally taken as the inverse of the square of the analytical error), $\omega(y_i)$ is the weighting factor for y_i, r_i is the correlation between the error of measurement of x_i and y_i, $\alpha = \sqrt{\omega(x_i)\omega(y_i)}$, $x = \sum Z_ix_i/\sum Z_i$, $y = \sum Z_iy_i/\sum Z_i$ (weighted means), and Z_i is:

$$Z_i = \frac{\alpha_i^2}{\omega(y_i) + \omega(x_i) - 2br\alpha_i}$$

Note that the expression for b (Equation 2.27) contains b. This requires an iterative solution: not something you want to do in your head, but reasonably easy with a computer. For example, the first estimate of b could be made using Equation (2.23). The difference between this method and the standard one is not great, so convergence is generally quick. The intercept is calculated as in Equation (2.24). Calculating the errors associated with a and b is fairly complex, but approximate solutions are given by:

$$\sigma_b = \sqrt{\frac{1}{\sum Z_i(x_i - \overline{x})^2}} \qquad (2.28)$$

$$\sigma_a = \sqrt{\frac{\sigma_b}{\sum Z_i}} \qquad (2.29)$$

From the error on the slope, the error on the age can be derived by simple algebra. The error so derived, however, does not include uncertainty in the value of the decay constant, which may or may not be significant.

A useful measure of the fit of the data to the regression line (isochron) is the mean squared weighted deviation (MSWD). If, as is usual, the weight factors are taken as the inverse square of the estimated analytical errors and assuming errors are uncorrelated, the MSWD is calculated as (Wendt and Carl, 1991):

$$\text{MSWD} = \frac{\sum\limits_{i}^{N} \frac{(y_i - bx_i - a)^2}{\left(b^2\sigma_{x_i}^2 + \sigma_{y_i}^2\right)}}{N - 2} \qquad (2.30)$$

where σ_{xi} and σ_{yi} are the errors on x_i and y_i, respectively. An MSWD value less than or equal to 1 indicates the deviations from prefect linear correlation between x and y are less than or equal to those attributable to associated analytical errors; an MSWD greater than 1 indicates other (geological) factors have contributed to the deviations from linearity and suggests that conditions (1) and (2) mentioned earlier in the text have been violated. This does not mean the isochron result must be discarded, particularly if it is not far from 1, but it does signal that some caution should be taken in interpreting the age.

Today, there are programs available that implement these equations, so there is no need to code them anew. One that has been in use for three decades and is still widely used is *Isoplot*, a Visual Basic Add-in for Microsoft's Excel® written by Ken Ludwig of the Berkeley Geochronology Center (BGC) and obtainable at https://www.bgc.org/isoplot. This software is useful for a wide variety of other geochronological problems that we will explore in this chapter and subsequent chapters, including concordia diagrams and $^{40}Ar/^{39}Ar$ dating. However, this software is not compatible with the most recent versions of Excel® and is no longer supported. A very similar program, *IsoplotR*, has been written by Pieter Vermeesch and is available online, and as offline and command line versions at https://www.ucl. ac.uk/~ucfbpve/isoplotr/home/index.html, and described in Vermeesch (2018). This program is highly recommended for completing some of the problems in this book.

2.2.3 Correcting mass fractionation

One of the most important sources of error in mass spectrometry results from slightly different behavior of isotopes of an element during analysis. The fractionations in mass spectrometry are essentially kinetic, and not equilibrium, and the form of mass dependence can vary slightly depending on the details of the process. Fractionation can also occur in preparative chemistry, including purification of the element of interest (generally ion exchange for non-gases) or reaction to the desired gaseous form (e.g., CO_2 for oxygen and carbon isotopic analysis), as well as in the instrument. In thermal ionization mass spectrometry (TIMS) and secondary-ion mass spectrometry (SIMS), the principal cause is the tendency of the lighter isotopes of an element to evaporate and ionize more readily than the heavier isotopes. In gas source mass spectrometry, lighter isotopes can be more readily ionized. In inductively coupled plasma mass spectrometry (ICP-MS) and SIMS, coulomb repulsion in the plasma or ion beam preferentially pushes light isotopes to the margin of the beam making them more likely to be excluded from entry into the mass filter of the instrument; fortunately, corrections can be made. The detailed physics of these processes is incompletely understood; therefore, mass fractionation corrections tend to be empirical rather than theoretical.

A common approach is to measure the ratio of two isotopes that are not radiogenic. For example, for Sr, we measure the ratio of $^{86}Sr/^{88}Sr$. By convention, we assume that the value of this ratio is equal to 0.11940. Any deviation from the value is assumed to result from mass fractionation in the mass spectrometer. The simplest assumption about mass fractionation is that it is linearly dependent on the difference in mass of the isotopes we are measuring. In other words, the fractionation between ^{87}Sr and ^{86}Sr should be half of that between ^{88}Sr and ^{86}Sr. Therefore, if we know how much the $^{86}Sr/^{88}Sr$ has fractionated from the "true" ratio, we can calculate the amount of fractionation between ^{87}Sr and ^{86}Sr. Formally, we can write the *linear mass fractionation law* as:

$$\alpha(u, v) = \left[\frac{R^N_{uv}}{R^M_{uv}} \right] \bigg/ \Delta m_{uv} \qquad (2.31)$$

where $\alpha_{u,v}$ is the fractionation factor between two isotopes u and v, Δm is the mass difference between u and v (e.g., 2 for 86 and 88), R^N is the assumed "true" or "natural" u/v isotope ratio (e.g., 0.11940 for 86/88), and R^M is the measured ratio. The correction to the ratio of another isotope pair i,j (e.g., $^{87}Sr/^{86}Sr$) is then calculated as:

$$R^C_{ij} = R^M_{ij} \left(1 + \alpha(i, j) \Delta m_{ij} \right) \qquad (2.32)$$

where R^C is the corrected ratio and R^M is the measured ratio of i to j and

$$\alpha(i, j) = \frac{\alpha(u, v)}{1 - \alpha(u, v) \Delta m_{vj}} \qquad (2.33)$$

If we choose isotopes v and j to be the same (e.g., to both be ^{86}Sr), then $\Delta m_{vj} = 0$ and $\alpha(i,j) = \alpha(u,v)$. (A convention that is unfortunate in terms of the above equations; however, it is that we speak of the 86/88 ratio, when we should speak of the 88/86 ratio [= 8.37521]). Using the 88/86 ratio, the "normalization" equation for Sr becomes:

$$\left(\frac{^{87}Sr}{^{86}Sr} \right)^C = \left(\frac{^{87}Sr}{^{86}Sr} \right)^M$$
$$\left[1 + \left\{ \frac{8.37521}{\left(^{88}Sr/^{86}Sr \right)^M} - 1 \right\} \bigg/ 2 \right] \qquad (2.34)$$

An alternative description of mass fractionation is the *power law*. The fractionation factor is:

$$\alpha = \left[\frac{R^N_{uv}}{R^M_{uv}} \right]^{1/\Delta m_{uv}} - 1 \qquad (2.35)$$

The corrected ratio can be computed as:

$$R_{i,j}^C = R_{i,j}^M [1 + \alpha]^{\Delta m_{i,j}} \qquad (2.36)$$

which can be approximated as:

$$R_{i,j}^C = R_{i,j}^M \left(\frac{m_i}{m_j}\right)^{\alpha m_j}$$

$$= R_{i,j}^M \left[1 + \alpha \Delta m_{ij} + \alpha^2 \frac{\Delta m_{ij}^2 (\Delta m_{ij} - 10)}{2} + \ldots \right]$$

$$(2.37)$$

Finally, the actual fractionation may be best described by an *exponential law*, from which the fractionation factor may be computed as:

$$\alpha = \frac{\ln \left(R_{uv}^N / R_{uv}^M\right)}{m_j \ln \left(m_u/m_v\right)} \qquad (2.38)$$

and the correction is:

$$R_{i,j}^C = R_{i,j}^M \left(\frac{m_i}{m_j}\right)^{\alpha m_j}$$

$$= R_{i,j}^M \left[1 + \alpha \Delta m_{ij} - \alpha \frac{\Delta m_{ij}}{2 m_j} + \alpha^2 \frac{\Delta m_{ij}^2}{2} + \ldots \right]$$

$$(2.39a)$$

Higher terms in Equations (2.37) and (2.39a) can be neglected. The exponential law appears to provide the most accurate correction for mass fractionation in at least some cases. However, all the above laws are empirical rather than theoretical. Fractionation processes are complex and difficult to treat theoretically. A more detailed discussion of these laws can be found in Wasserburg et al. (1981).

For stable isotopes, where the objective is to discover the extent of natural fractionation, as well as that of Pb, which has only one non-radiogenic isotope, this approach obviously cannot be used. Gas source mass spectrometers used for light stable isotopes are designed to minimize instrumental fractionation; furthermore, they can quickly switch from sample to standard and back. A correction for any instrumental mass fractionation can be made based on the difference between the measured and "true" isotopic composition of the standard. This approach can also be used in ICP-MS analysis, but it assumes fractionation of the standard is the same as that of the sample, which is not necessarily the case if purification of the sample is not complete. Rapid switching is not practical for TIMS instruments and,

traditionally, standards have been analyzed intermittently and a correction has been applied to samples based on the observed mass fractionation in the standards. For light elements, such as Li or B, analysis of a polyatomic ion containing a much heavier element minimizes mass fractionation. Two other alternatives are available.

In the first of these, for elements with three or more isotopes, a "spike" consisting of two or more isotopes of the element in proportions very different from the natural ones can be added to the analyte. Ideally, this spike consists of isotopes that do not exist in nature, such as ^{203}Pb and ^{205}Pb. Knowing the precise composition of the spike and measuring the isotope ratio of the spike, the fractionation factor is readily determined. One downside of that approach is that such spikes are radioactive, although the amounts are quite small and the hazard is minimum. A bigger downside is that they must be artificially produced and are very expensive if they are available at all. More often, the spike consists of two naturally occurring isotopes, generally, the least abundant ones, e.g., ^{204}Pb and ^{207}Pb for Pb analysis. In this case, two separate analyses are required – the sample without spike and the sample with spike. In the first case, isotope ratio deviates from the true ratio due to fractionation, and in the second case, it deviates due to both fractionation and the addition of spike. By solving a series of simultaneous equations both the ratios of sample to spike and the true isotopic composition of the sample can be found, provided that the resulting variations due to the addition of spike and fractionation do not produce correlated isotope ratio variations. The details are given by Gale (1970). It is used quite successfully for Pb (e.g., Galer, 1999) as well as for stable isotopes (Krabbenhöft et al., 2010), and can be used in both TIMS and ICP-MS analyses. Fractionation in the chemical purification process is also corrected for if the spike is added beforehand. An additional benefit of the spiking technique is the concentration of the element can be determined from amount of spike added. Indeed, adding an isotopic spike is perhaps more commonly used to determine concentrations in a technique known as *isotope dilution*.

The last approach is limited to ICP-MS analysis. A "spike" consisting of an element with masses similar to that of the element of interest and whose isotopic composition is known is added. For example, Maréchal et al. (1999) used Cu (with isotopes ^{63}Cu and ^{65}Cu) to correct for isotopic fractionation during analysis of Zn (isotopes ^{64}Zn, ^{66}Zn, ^{67}Zn, ^{68}Zn, and ^{70}Zn) and Zn to correct for isotopic fractionation in analysis of Cu. For Pb, thallium (Tl), with isotopes ^{203}Th and ^{205}Th, can be used. However, when White et al. (2000) used both the standard-sample and Tl spike techniques in Pb ICP-MS isotope analysis, they found that the value of the fractionation coefficient, f, differed between Tl and Pb, although the ratio f_{Pb}/f_{Tl} was constant and the two approaches could be combined.

2.3 THE K–Ar–Ca SYSTEM

We have now introduced the basic aspects of radiogenic isotope geochronology and we will now consider the various decay systems separately. Many of these have special aspects, but all share a common foundation based on Equation (2.1) – *the basic equation of radioactive decay*. We begin with the K–Ar–Ca system. The ^{40}K–^{40}Ar decay was one of the first to be used for geochronology (Aldrich et al., 1956), following the early U–Pb studies.

Two aspects of the K–Ar–Ca system make it special. First, it is a branched decay: a ^{40}K nucleus (an odd–odd nuclide) may decay to either a ^{40}Ca by β^- or to a ^{40}Ar atom by electron capture (or much more rarely by positron emission — which is just as well for us). It is impossible to predict how a particular ^{40}K atom will decay, just as it is impossible to predict when it will decay. We can predict quite accurately what proportion of a large number of ^{40}K atoms will decay in each way, however, in a given amount of time. The ratio of electron captures to β-decays is called the *branching ratio* and is defined as:

$$R_b = \frac{\lambda_{ec}}{\lambda_\beta} \qquad (2.39b)$$

where the two lambdas are the decay constants (i.e., the probability of decay) for each mode. According to recent work by Renne et al. (2010), the branching ratio is 0.1037, $\lambda_{ec} = 0.5755 \times 10^{-10}$ yr^{-1}, $\lambda_\beta = 4.9737 \times 10^{-10}$ yr^{-1}. The total decay constant for ^{40}K is:

$$\lambda = \lambda_\beta + \lambda_{ec} = 5.5492 \times 10^{-10} \text{yr}^{-1} \qquad (2.40)$$

We need to take account of this branched decay in our equation, because while a K atom decaying to Ca does not produce radiogenic Ar, it is no longer available for ^{40}Ar production. Thus, our equation for radiogenic daughter production (i.e., Equation [2.7]) becomes:

$$^{40}\text{Ar}^* = \frac{\lambda_e}{\lambda}{}^{40}K\left(e^{\lambda t} - 1\right) \qquad (2.41)$$

where the asterisk indicates radiogenic ^{40}Ar (designating the *radiogenic* atoms of an element in this manner is a widely used convention and we will follow it in this book). Note we can write a similar equation for ^{40}Ca* by substituting λ_β for λ_e.

Most, although not all, of the work on the K-Ca-Ar system has focused on Ar because the ^{40}K/^{40}Ca ratio is usually small. ^{40}K is the least abundant of the K isotopes (0.01167%), whereas ^{40}Ca is the most abundant ^{40}Ca isotope (96.92%), and Ca is a more abundant element than K (^{40}Ca is even–even and ^{40}K is odd–odd). As a result, variations in the ^{40}Ca/^{42}Ca ratio resulting from radioactive decay are quite small and difficult to measure (indeed, there is usually more variation in this ratio due to other causes, which we will discuss in Chapters 9–12). Only in very favorable circumstances, such as evaporite deposits, geochronology is practical.

As one might expect, particularly in view of the previous discussion, one of the most important criteria for a useful radiometric chronometer is that the variations in the radiogenic isotope be large relative to the precision and accuracy with which they can be measured. In this respect, a short half-life is advantageous, and K has one of the shortest half-lives of the long-lived radioactive nuclides. Because of the volatility of Ar, the Earth either lost much of its Ar during its formation or never acquired much, giving the Earth a rather high K/Ar ratio. Furthermore, much of the Ar the Earth retained is now in the atmosphere (as we will learn in Chapter 14). As a result, ^{40}K/^{40}Ar ratios in the solid Earth tend to be quite high. Because of the high ^{40}K/^{40}Ar ratios and the relatively short half-life of ^{40}K, the K–Ar system is often the one of choice when the task at hand is to date very young events. Meaningful ages (meaning the uncertainty is small relative to the age) of less than 30 000 years have been determined in favorable circumstances. Much of what we know of the timing of the human evolution is based on $^{40}K/^{40}Ar$ dating (including $^{40}Ar/^{39}Ar$ dating that we will discuss shortly).

Much of what is special about K–Ar derives from Ar being a noble gas and its resulting refusal to be chemically bound in crystal lattices. Ar in rocks and minerals is simply trapped there. It has difficulty escaping because the atoms of the lattice block its escape path, but it does not form chemical bonds with other atoms in the lattice. Thus, when a mineral crystallizes from lava, it will generally do so with very little Ar. Pillow basalts formed on the seafloor, however, can trap substantial amounts of Ar, which can prove quite useful in understanding atmospheric and mantle evolution, as we will find in Chapter 14. Similarly, minerals crystallizing from magma at depth within the Earth (plutonic rocks) may also retain Ar.

In favorable circumstances, essentially no Ar will be trapped in a mineral crystallizing from lava. The great advantage of this, from a geochronological viewpoint, is we have only one unknown, namely, t, and we can use Equation (2.33) to solve for it by measuring the ^{40}K and ^{40}Ar in one sample. Actually, one need not assume that no "initial" Ar whatsoever is present. Indeed, in detail, this would seem a poor assumption since a mineral crystallizing in contact with the atmosphere can be expected to absorb a small but finite amount of atmospheric Ar. This atmospheric Ar is readily corrected for since the atmosphere has a uniform $^{40}Ar/^{36}Ar$ ratio of 296.16[2]. By measuring the amount of ^{36}Ar present, we can deduce the amount of atmospheric ^{40}Ar initially present. Our age equation (i.e., Equation [2.17]) becomes simply:

$$\frac{^{40}Ar}{^{36}Ar} = 296.2 + \frac{\lambda_e}{\lambda}\frac{^{40}K}{^{36}Ar}\left(e^{\lambda t} - 1\right) \qquad (2.42)$$

If we suspect that the composition of "initial" Ar differs significantly from atmospheric Ar, it is then necessary to employ the isochron approach, measuring K and Ar in a number of cogenetic samples and solving simultaneously for t and the initial $^{40}Ar/^{36}Ar$ ratio.

2.3.1 Diffusion, cooling rates, and closure temperatures

Because Ar is not chemically bound in lattices, the K–Ar clock will generally be reset more readily than other systems. We concluded earlier that an event that "resets" a radiometric clock is generally a thermal one. In the case of K–Ar, we might guess that the system would be reset whenever temperatures are high enough to allow Ar to diffuse out of the rock or mineral of interest. It is worth considering this on a slightly more quantitative level.

It can be shown both theoretically and experimentally that the rate at which a species will diffuse through a medium is related exponentially to temperature as:

$$D = D_0 e^{-E_A/RT} \qquad (2.43)$$

where D is the diffusion coefficient, D_0 is the "frequency factor," E_A is the activation energy, R is the gas constant, and T is thermodynamic, or absolute, temperature (i.e., kelvin). The diffusion "flux" is then related to the concentration gradient by Fick's first law:

$$J = -D\left(\frac{\partial C}{\partial x}\right) \qquad (2.44)$$

where C is the concentration and x is distance. The distribution of a diffusing species in time and space is given by Fick's second law:

$$\left(\frac{dC}{dt}\right) = D\left(\frac{\partial^2 C}{\partial x^2}\right) \qquad (2.45)$$

Figure 2.2 shows a plot of experimentally determined values of D for Ar in biotite plotted against the inverse of temperature. The point to be made here is that relatively small increases in temperature result in large increases in the diffusion coefficient. For example, increasing the temperature from 600 to 700°C results in a two-order-of-magnitude increase of the diffusion coefficient, and, for a given concentration gradient, of the Ar diffusion flux. Using the values of E_A and D_0 given in the figure, we can calculate the diffusion coefficient for temperatures not shown in the graph. The value of R is 8.314 J/kelvin-mole (1.987 cal/kelvin-mole). For a temperature of 300 K (27°C), D would be 4×10^{-36} cm^2/s. For any reasonable concentration gradient,

---

[2] This is the more recent value reported by Mark et al. (2011). The older, conventional value is 295.5.

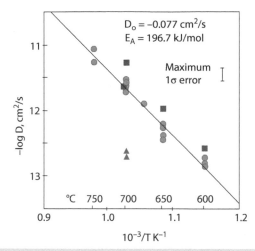

Figure 2.2 Log of the diffusion coefficient for argon in biotite against the inverse of thermodynamic temperature. Circles and squares indicate different-sized fractions of biotite used in 1 kbar experiments. Triangles are 14 kbar experiments. Source: Adapted from Harrison et al. (1985).

there would be no significant Ar loss from biotite, even over extremely long times. At 600 K (327°C), we obtain $D = 6 \times 10^{-19}$ cm²/s, which implies a slow, but significant, diffusion rate. At 700°C, however, loss of Ar would be quite rapid: over 1/3 of the Ar would be lost from a biotite crystal of 97 μ radius in two to three weeks (you can understand then why the experiments were done at these temperatures and not at lower ones).

How rapidly Ar will be lost from a mineral grain can be determined by solving Fick's second law. Rather than having a single solution, however, there are many possible solutions depending on boundary conditions. Consequently, the solution will depend on geometry and size, as well as the diffusion coefficient. In the case of a sphere of radius a, the solution gives the fraction of Ar remaining after time t as:

$$f = 1 - \frac{6}{\pi^2} \sum_{n=1}^{\infty} \frac{1}{n^2} e^{-\left(n^2 \pi^2 \frac{Dt}{a^2}\right)} \qquad (2.46)$$

(McDougall and Harrison, 1988). Unfortunately, this equation does not readily converge, particularly at small values of Dt/a^2, so two approximations are useful:

$$f \approx 6 \left(\frac{Dt}{a^2}\right)^{1/2} - 3 \frac{Dt}{a^2} \quad \text{for } f \leq 0.85 \qquad (2.46a)$$

$$f \approx 1 - \frac{6}{\pi^2} e^{-\left(\pi^2 \frac{Dt}{a^2}\right)} \quad \text{for } f \geq 0.85 \qquad (2.46b)$$

McDougall and Harrison (1988) list similar equations (which also do not readily converge) for a plain sheet of infinite dimension and a cylinder of infinite length (these equations are also listed in Reiners et al., 2017). Figure 2.3 compares values of f for a sphere and infinite length cylinder with a diameter of 150 μ and a plain infinite sheet of 150 μ thickness computed in MATLAB to many iterations using these equations. As we might expect, the sphere loses Ar more readily, but since plane or cylindrical mineral grains do not have infinite dimension, this computation exaggerates the differences due to geometry. Regardless, the rate of diffusive loss is greater for small crystals than for large ones. Solutions to Fick's second law for many boundary conditions can be found in Crank (1975).

Let us consider the geological implications of this diagram. Imagine a body of rock, either igneous or metamorphic, cooling from a temperature that is high enough so that all Ar is lost. Let us pick up the story when the body is still at 400°C and is cooling at a rate of 100°C/Ma. At this temperature, a sheet-like biotite grain would be just beginning to retain radiogenic Ar; that is, it is not being lost quite as fast as it is being created. After the first additional million years, it would have cooled to 300°C, and biotite would be retaining most of its radiogenic Ar (a loss rate of about 10%/Ma). If cooling continues at this rate for another million years (in the real world, it is unlikely that cooling rates would be so constant), biotite would be losing Ar at a rate of only a tenth of a percent per Ma, a fairly insignificant rate. If the body then cooled completely, and if we sampled biotite for K–Ar dating some 100 Ma later, assuming the biotite was not reheated, the "age" we would calculate would refer to that 2 Ma period when the biotite cooled from 400 to 200°C, and probably closer to the time it passed from 400 to 300°C. We say the biotite "closed" at that time, and we can estimate the closure temperature to be around 300°C.

Suppose cooling was slower, say 10°C/Ma. In this case, 10 Ma would be required to cool from 400 to 300°C, and 20 Ma to cool to 200°C. A much smaller fraction of the radiogenic Ar produced while the biotite was in the 200–400°C range would have been

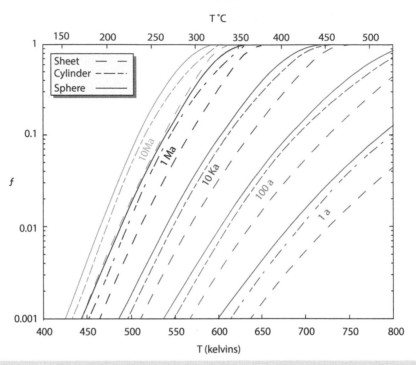

Figure 2.3 Calculated fraction of Ar lost from a sphere and an infinite cylinder of 150 μ radii and an infinite sheet of 150 μ thickness based on the frequency factor and activation energy in Figure 2.2 as a function of temperature for various heating times. All Ar is lost in 10 Ma at 320–340°C, or in 10 ka at ≈450°C.

retained. The "age" we would calculate using Equation (2.41) would be younger than in the previous example. It would thus seem that under these circumstances, the *closure temperature* would depend on the cooling rate. This is indeed the case.

Dodson (1973) derived an equation for closure temperature (also sometimes called blocking temperature) as a function of diffusion parameters, grain size and shape, and cooling rate:

$$T_c = \frac{E_A}{R \ln \left(-\frac{A R T_c^2 D_0}{a^2 E_A \tau} \right)} \quad (2.47)$$

where E_A and D_0 are the previously defined activation energy and frequency factor of diffusion, τ is the cooling rate, dT/dt (for cooling, this term will be negative), a is the characteristic diffusion dimension (e.g., radius of a spherical grain), and A is a geometric factor (equal to 55 for a sphere, 27 for a cylinder, and 9 for a sheet) and temperatures are in kelvin. Unfortunately, this is not directly solvable

since T_c occurs both in and out of the log, but it can be solved by indirect methods.[3]

There are several important notions we can come away with. First, a closure temperature is a useful concept, but a mineral will not suddenly stop losing Ar, or any other radiogenic component, at its closure temperature. Closure temperature reflects a trade-off between loss and creation of the radiogenic component. Second, there are some ultimate geological limitations on the meaning of an age of a slowly cooled rock, such as a large intrusion or regionally metamorphosed body of rock. We might also expect the age we obtain will depend on the mineral we use for dating (since the diffusion coefficient will vary), and perhaps on its composition (there is in fact some compositional dependence of the Ar diffusion coefficient on the Fe/Mg ratio in biotite, but apparently none in hornblende). Third, closure temperature depends on the diffusive properties of the element of interest. Consequently, we can expect that different decay systems will

[3] The Solver, an add-in tool available for Microsoft Excel™, can be used to solve problems, such as this. Programs, such as Mathematica™ and MATLAB™, also have tools for indirect solution built-in.

have different closure temperatures and we would want to take this into consideration in deciding which decay system to use in any particular geochronological problem. The U–Pb system in zircon has quite a high closure temperature, as do Lu–Hf and Sm–Nd in minerals such as garnet and clinopyroxene. In contrast, closure temperatures for the K–Ar system can be quite low, i.e. a few hundred degree Celsius. Rather than posing a challenge, this can be useful: by combining use of several decay systems, we can trace the cooling of buried rocks; we will explore this topic, *thermochronology*, in Section 4.3. Finally, we get the sense that it might also be rather easy for the K–Ar system to be *partially* reset. This is certainly the case. We discuss next a technique that can allow us to identify sources of errors in the K–Ar system and, in favorable cases, nevertheless provide a reasonable estimate of an "age."

2.3.2 ^{40}Ar–^{39}Ar Dating

If you look at a table of isotopes, you will see ^{39}Ar has a half-life of 269 years and does not occur naturally. You might justifiably wonder how it could be used for dating. The so-called *40–39 Method* is actually a version of ^{40}K–^{40}Ar dating that employs a somewhat different analytical technique for potassium that was first described by Merrihue and Turner (1966). The key is the production of ^{39}Ar by a nuclear reaction on ^{39}K, the most abundant of potassium's three isotopes:

$$^{39}\text{K}\,(n,p)^{39}\text{Ar}$$

The reaction is produced by irradiating a sample with fast neutrons in a reactor. It is important to distinguish this reaction from simple neutron capture, but we can nevertheless define a reaction cross-section. The amount of ^{39}Ar produced is then a function of the amount of ^{39}K present, the reaction cross-section, the neutron flux, and the irradiation time. Since the ^{40}K/^{39}K ratio in the Earth is constant or nearly so (at any given time), the amount of ^{40}K can be calculated from ^{39}Ar. In practice, the situation is more complex because the reaction cross-section is a function of neutron energy and there typically is a spectrum of neutron energies present. The production of ^{39}Ar from ^{29}K can be expressed as:

$$^{39}\text{Ar} = {}^{39}\text{K}\tau \int \phi(\varepsilon)\sigma(\varepsilon)\mathrm{d}e \qquad (2.48)$$

where ε is the neutron energy, $\phi(\varepsilon)$ is the flux of neutrons with energy ε, $\sigma(\varepsilon)$ is the capture cross-section for that energy, and τ is irradiation time. The ^{40}Ar*/^{39}Ar is then:

$$\frac{^{40}\text{Ar}^*}{^{39}\text{Ar}} = \frac{\lambda_e}{\lambda}\,\frac{{}^{40}\text{K}\left(e^{\lambda t}-1\right)}{{}^{39}\text{K}\tau \int \phi(\varepsilon)\sigma(\varepsilon)\mathrm{d}e} \qquad (2.49)$$

In practice, the analysis is performed by simultaneously irradiating and analyzing a standard of known age. The flux, capture cross-section, decay constant terms, and the ^{40}K/^{39}K ratio will be the same for the standard as for the unknown sample. We can combine them into a single term, J, as:

$$J = \frac{^{39}\text{K}}{^{40}\text{K}}\,\frac{\lambda_e}{\lambda}\tau \int \phi(\varepsilon)\sigma(\varepsilon)\mathrm{d}e \qquad (2.50)$$

and hence Equation (2.49) becomes:

$$\frac{^{40}\text{Ar}^*}{^{39}\text{Ar}} = \frac{\left(e^{\lambda t}-1\right)}{J} \qquad (2.51)$$

The value of J can be determined from analysis of the standard whose age is known. In practice, a number of standards are irradiated along with samples in typical analysis because the neutron fluence can vary even on the centimeter scale in the reactor. An additional problem is the production of both ^{39}Ar and ^{40}Ar, as well as ^{36}Ar and ^{38}Ar, by other reactions that include ^{40}K$(n,p)^{40}$Ar, ^{40}Ca$(n,n\alpha)^{36}$Ar, and ^{42}Ca$(n,\alpha)^{39}$Ar. Corrections for these must be made. These undesirable reactions can be somewhat reduced with the use of cadmium shielding, which adsorbs low-energy neutrons. Correction for these effects can be made from measurement of ^{37}Ar, produced by ^{40}Ca $(n,\alpha)^{37}$Ar. After correction for these effects, the age of a sample is calculated as:

$$t_u = \frac{1}{\lambda}\ln\left[1 + \left(\frac{^{39}\text{Ar}_K}{^{40}\text{Ar}^*}\right)_s \left(e^{\lambda t_s}-1\right)\left(\frac{^{39}\text{Ar}_K}{^{40}\text{Ar}^*}\right)_u\right] \qquad (2.52)$$

where the subscripts u and s denote the unknown (i.e., the sample) and standard, respectively, and the subscript K denotes the measured ^{39}Ar after correction for ^{39}Ar produced by reactions other than

^{39}K(n,p)^{39}Ar. On defining an intercalibration factor, R_s^u, as:

$$R_u^s = \frac{\left(e^{\lambda t_u} - 1\right)}{\left(e^{\lambda t_s} - 1\right)} = \frac{\left(\frac{^{39}Ar_K}{^{40}Ar^*}\right)_u}{\left(\frac{^{39}Ar_K}{^{40}Ar^*}\right)_s} \qquad (2.53)$$

Equation (2.52) becomes:

$$t_u = \frac{1}{\lambda} \ln\left[1 + \left(e^{\lambda t_s} - 1\right)R_s^u\right] \qquad (2.54)$$

2.3.2.1 Standards, calibration, and astronomical tuning

^{40}Ar–^{39}Ar dating depends on precisely knowing the age of the co-irradiated standards; consequently, considerable effort has gone into determining these and cross-calibrating between standards. Primary standards are those, such as *GA1550*, a biotite from Dromedary igneous complex of southeast Australia, which have been precisely dated by conventional ^{40}Ar/^{40}K (97.8 ± 0.9 Ma and later revised to 98.5 Ma). So-called secondary standards are those whose age has been determined by ^{40}Ar–^{39}Ar analysis using primary standards or by another method, usually U–Pb zircon dating (Section 3.3). A particularly commonly used standard is "Fish Canyon sanidine (FCs)" based on sanidine from the Fish Canyon Tuff, an enormous dacitic ignimbrite erupted 28 million years ago from the La Garita Caldera in the San Juan volcanic field of southwestern Colorado, which has been repeatedly dated by ^{40}Ar–^{29}Ar against primary standards. The age is constrained by the number of U–Pb analyses of zircons from the same tuff that have yielded ages from 27.5 to 28.5 Ma, although the most recent determinations range only from 28.18 to 28.20 Ma.

An important technique used to improve the precision of standards is astronomical or orbital tuning. The Earth's orbit around the Sun and the angle of its rotation axis with the orbital plane vary cyclically. For example, the eccentricity of the Earth's orbit varies between 0 and 0.07 with principal periods of 95, 99, 124, 131, 405, and 2260 ka. The obliquity of the rotational axis has a principal period of 41 ka and several shorter periods, and precession (timing of perihelion in relation to eccentricity) has a principal period of 24 ka and several shorter ones. Some of these cycles are subject to geophysical

(movement of continents) and astronomical disturbances (tidal dissipation, asteroid impacts, and passing stars), but others are quite stable because they are regulated by gravitational interaction of the Earth with other planets and, in the case of obliquity, the Moon. For example, the 405 ka eccentricity cycle is thought to be stable on timescales of 100 million years (Laskar et al., 2004; Hinnov and Hilgren, 2012).

These variations in orbit and rotation are sometimes known as Milankovitch cycles and are the pacemaker of Pleistocene glacial–interglacial cycles; we will discuss these in detail in subsequent chapters. Even when they do not produce ice ages, however, they influence climate and, consequently, can produce clearly apparent cyclicity in sedimentary sequences. Sedimentary rocks usually cannot be radiometrically dated, but volcanic tephra layers interbedded in them can be. An example occurs in the Miocene sediments of the Melilla Basin of Morocco where silicic tephras interbedded with an astronomically driven alternating sequence of marl and diatomite sediments are exposed. Kuiper et al. (2008) performed over 60 ^{40}Ar–^{39}Ar analyses of single-sanidine crystals from these tephras using the *FCs* as a standard. Ages ranged from 6.9 to 6.3 Ma. These ages were then refined to fit the astronomical timescale, which has an uncertainty of ±10 ka. Then, using these samples as the standards and *FCs* as the unknown, they used Equation (2.45) to calculate an age of 28.201 ± 0.023 Ma for the *FCs* standard. Using the revised decay constant and branching ratio of Renne et al. (2010), this age becomes 28.394 ± 0.036 Ma.

Another important standard is *ACs* sanidine from the Alder Creek rhyolite of the Napa–Sonoma volcanic province in California. It has an age of 1.1848 ± 0.0006 Ma when astronomically tuned to sanidine from the same Moroccan sequence (Niespolo et al., 2017). This becomes 1.1891 ± 0.0008 Ma using the decay constant and branching ratio of Renne et al. (2010). These ages are roughly 10 000 years younger than the U–Pb age of zircon from the Alder Creek rhyolite. This is consistent with a range of studies indicating that zircons can crystallize from silicic magmas below their closure temperature and reside in magma chamber for periods of as much as 10^5 yr before eruption.

2.3.2.2   Release spectra and their interpretation

In conventional K–Ar dating, Ar is released from samples by fusing in vacuum. However, we might guess from our knowledge of diffusion that a sample will begin to lose Ar before it reaches its melting temperature. If the ratio of radiogenic ^{40}Ar to K (and therefore to ^{39}Ar) were distributed uniformly throughout the sample, a sample of gas taken before the sample fully melted would produce the same age as for total fusion. We might guess, however, that some parts of a crystal will preferentially lose Ar through diffusion during the initial cooling of the crystal, or perhaps during some subsequent reheating event. Since the diffusion rate is proportional to the concentration gradient, we can anticipate that diffusive loss will be faster from the rims of crystals where the concentration gradient is higher than in the interior of crystals. Therefore, we might expect crystal rims to experience Ar loss at lower temperatures than crystal interiors. The rims would then record younger ages. As we heat the sample, we would also expect rims to start to give up their Ar at the lowest temperatures because the Ar has less distance to go to get out. The lower ^{40}Ar/^{39}Ar of the gas in the rim would be seen as a lower age (which may or may not have significance). As we increased the temperature, the more retentive parts of the crystal would release their gas, and we could expect the ^{40}Ar/^{39}Ar and the apparent age to increase. If some parts of the crystals have lost no gas, their ^{40}Ar/^{39}Ar ratios would record the "correct" age, even though the crystal as a whole has suffered some loss. Figure 2.4 is an Ar release diagram for a basalt exhibiting this sort of behavior, with the first 5% or so of Ar released giving a lower age than the remaining Ar. Conventional K–Ar dating would have produced an age intermediate between the "correct" age and the apparent young age of those parts of crystal that have suffered loss of radiogenic ^{40}Ar. Thus, the combination of the ^{40}Ar/^{39}Ar method with step heating provides a means of retrieving useful geochronological information from samples whose value would have otherwise been compromised because of diffusional loss. In a certain sense, we are relaxing our requirement that the system must have remained closed: with ^{40}Ar/^{39}Ar dating, we require only that some parts of the system have remained closed.

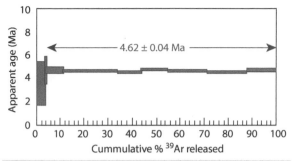

Figure 2.4 ^{40}Ar/^{39}Ar age spectrum produced by step heating of groundmass from basaltic lava sample GDA161 from Grenada. Filled areas represent analytical uncertainty in each step. (Analysis by P. Copland at the University of Houston; from White et al., 2017).

Many Ar release spectra are not so simple as that in Figure 2.4 – indeed, this spectrum is particularly simple and provides a very well constrained age. Figure 2.5 shows Ar release spectra for a series of hornblende samples taken at varying distances from the contact with an intrusive granodiorite. All show significant Ar loss as a result of heating from the intrusion. None retain, even at the highest release temperature, the true age of 367 Ma.

Many spectra are more complex. For example, some samples that have been reheated show false plateaus that correspond to ages intermediate between the crystallization age and the reheating age. An additional problem in interpreting such spectra is that samples that have not been subjected to reheating events but cooled slowly originally can show release spectra that mimic those of reheated samples in Figure 2.5. During a thermal event, ^{40}Ar diffusing out of some minerals may be taken up by other minerals. Since this ^{40}Ar is diffusing into the mineral grain, its concentration will be highest in the exterior of grains and thus will tend to be released at the lowest temperatures. An example is shown in Figure 2.6 where radiogenic argon from adjacent mineral grains has diffused into hornblende.

Recoil of ^{39}Ar produced by the ^{39}K$(n,p)^{39}$Ar reaction during irradiation can also cause problems. The recoil results in loss of ^{39}Ar from sites near the mineral surface. For large grains, this is generally insignificant, but for small grains, this can lead to significant ^{39}Ar loss, leading to erroneously old apparent ages.

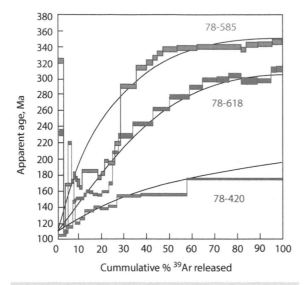

Figure 2.5   Ar release spectra for hornblendes taken from varying distances from a 114 million year old intrusion. The crystallization age of the intrusion is 367 Ma. Curves show calculated release spectra expected for samples that lost 31, 57, and 78% of their argon. Source: Adapted from Harrison and McDougall (1980).

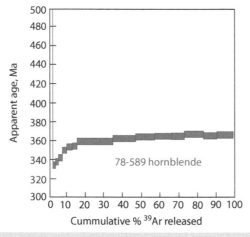

Figure 2.6   Ar release spectrum of a hornblende in a Paleozoic gabbro reheated in the Cretaceous by the intrusion of a granite. Anomalously old apparent ages in the lowest temperature release fraction results from diffusion of radiogenic Ar into the hornblende during the Cretaceous reheating. Source: Adapted from Harrison and McDougall (1980).

In most cases, the Ar present in a sample will not be pure radiogenic Ar. Non-radiogenic argon is often called *excess* Ar. $^{40}Ar/^{39}Ar$ ratios

used to calculate ages in release spectra are typically corrected for the presence of atmospheric Ar by measuring the $^{40}Ar/^{36}Ar$ ratio. Atmospheric argon has a constant $^{40}Ar/^{36}Ar$ ratio of 296.16. Only ^{40}Ar present in excess of this ratio is considered radiogenic and used to calculate the $^{40}Ar/^{39}Ar$ ratio. Nevertheless, some samples can have "initial" $^{40}Ar/^{36}Ar$ ratios greater than the atmospheric ratio; this will lead to too old an age if not properly accounted for. It is this "excess" argon that is of greatest concern.

Excess Ar can have two sources. First, it can arise when minerals crystallize under a finite partial pressure of Ar. For example, mantle-derived submarine basalts have been shown in some cases to have initial $^{40}Ar/^{36}Ar$ ratios of up to 40 000. The high $^{40}Ar/^{36}Ar$ ratio reflects production of ^{40}Ar by decay of ^{40}K within the mantle and crystallization at sufficiently high pressures on the seafloor that Ar cannot escape. Minerals crystallizing and glass freezing in the presence of this gas will trap some of this ^{40}Ar, which will result in an anomalously old age upon analysis. This is referred to as *inherited* Ar.

When excess Ar is held in more than one crystallographic site, for example, different minerals in the analyzed sample, release spectra can reveal a saddle shape. An example is shown in Figure 2.7. This sample is a calcic plagioclase from Broken Hill in Australia. The true metamorphic age is approximately 1600 Ma. Even the minimum values in the bottom of the saddle are too old. Electron microscopy of the plagioclase revealed that it had exsolved into a Ca-rich and Na-rich plagioclase. The saddle shape results from the fact that Ar in one of the phases diffuses readily and is thus released at low temperature, and diffuses more slowly in the other, resulting in release at high temperature.

A new technique, developed only in the last 25 years, involves releasing Ar from small areas of a sample through laser heating (after irradiation). This allows release of Ar from areas with diameters less than a millimeter and provides the possibility of spatial resolution of Ar diffusional loss.

2.3.2.3 $^{40}Ar/^{39}Ar$ isochrons

The data from various temperature release steps are essentially independent observations

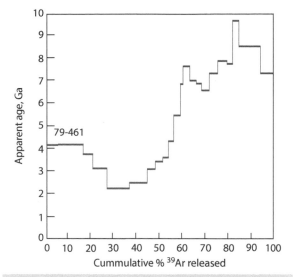

Figure 2.7 Ar release spectrum from a calcic plagioclase from Broken Hill, Australia. Low-temperature and high-temperature fractions both show erroneously old ages. This peculiar saddle-shaped pattern, which is common in samples containing excess Ar, results from the excess Ar being held in two different lattice sites. Source: Adapted from Harrison and McDougall (1981).

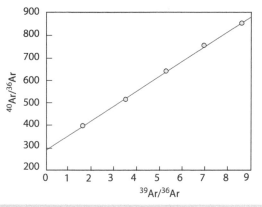

Figure 2.8 Hypothetical ^{40}Ar–^{39}Ar isochron diagram. The slope is proportional to the age and the intercept gives the initial $^{40}Ar/^{36}Ar$ ratio, which is commonly atmospheric as illustrated here.

of Ar isotopic composition. Because of this, they can be treated much the same as in conventional isochron treatment. The isochron equation, written for the K–Ar system, is:

$$\frac{^{40}Ar}{^{36}Ar} = \left(\frac{^{40}Ar}{^{36}Ar}\right)_0 + \frac{^{40}K}{^{36}Ar}\left(e^{\lambda t} - 1\right) \quad (2.55)$$

When $^{40}Ar/^{36}Ar$ data from a series of samples are plotted against $^{40}K/^{36}Ar$, the slope of the resulting line will be proportional to age, and the intercept gives the initial $^{40}Ar/^{36}Ar$ ratio. Since for all release fractions of a sample, the efficiency of production of ^{39}Ar from ^{39}K is the same and $^{40}K/^{39}K$ ratios are constant, we may substitute $^{39}Ar \times C$ for ^{40}K:

$$\frac{^{40}Ar}{^{36}Ar} = \left(\frac{^{40}Ar}{^{36}Ar}\right)_0 + \frac{^{39}Ar}{^{36}Ar}C\left(e^{\lambda t} - 1\right) \quad (2.56)$$

where C is a constant that depends on the efficiency of ^{39}Ar production during irradiation. Thus, when $^{40}Ar/^{36}Ar$ ratios from a series of release fractions are plotted against $^{39}Ar/^{40}Ar$, the slope of the resulting line will be proportional to the age of the sample, as is illustrated in Figure 2.8.

The use of the isochron diagram can help to identify excess Ar and its nature (atmospheric, inherited, etc.). It also provides a crucial test of whether ages obtained in release spectra are meaningful or not. A drawback of this diagram is that ^{36}Ar, which is the denominator in both the ordinate and abscissa, is often present in only trace amounts and is difficult to measure precisely. Because of this, errors in its measurements can produce correlations that imitate isochrons.

An alternative is to use a plot of $^{36}Ar/^{40}Ar$ against $^{39}Ar/^{40}Ar$ (Figure 2.9), often called an *inverse isochron plot*. We can think of the Ar in a sample as a mixture of a trapped, or inherited, component and a radiogenic component. As such, the data for various release fractions should plot as a straight line on such a plot. The radiogenic component has a $^{36}Ar/^{40}Ar$ ratio of 0 (because ^{36}Ar is not produced by radioactive decay), whereas the trapped, non-radiogenic component can be found by extrapolating to a $^{39}Ar/^{40}Ar$ ratio of 0 (corresponding to a $^{39}K/^{40}Ar$ ratio of 0; since ^{39}K is proportional to ^{40}K, this also corresponds to a $^{40}K/^{40}Ar$ ratio of 0). Thus, the age may be computed from the $^{39}Ar/^{40}Ar$ ratio obtained by extrapolating the correlation line to $^{36}Ar/^{40}Ar$ to 0, and the composition of the trapped component by extrapolating to $^{39}Ar/^{40}Ar$ of 0.

Figure 2.10 provides an example of how the inverse isochron plot may be used to identify trapped components. The original release data

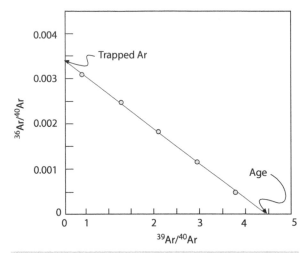

Figure 2.9 Plot of $^{36}Ar/^{40}Ar$ versus $^{39}Ar/^{40}Ar$, also called an inverse isochron diagram. Age is obtained from the value of $^{39}Ar/^{40}Ar$ corresponding to $^{36}Ar/^{40}Ar = 0$.

McDougall and Harrison (1999) and Reiner et al. (2017) provide greater detail on $^{40}Ar/^{39}Ar$ geochronology.

2.4 THE Rb–Sr SYSTEM

The K–Ar system is exceptional in that we can sometimes ignore or readily correct for initial Ar. In the systems we will discuss in the remainder of this chapter, both the initial ratio and the age are almost always unknown, meaning we must solve for both simultaneously through the isochron method. Consequently, this is an opportune time to briefly review and summarize the conditions that must be met to obtain a meaningful isochron age.

showed a disturbed pattern and lacked a plateau (not shown). The inverse isochron plot (Figure 2.10a) revealed two correlations suggesting the presence of two distinct trapped components. The lower intercept yielded an age of 149.1 Ma. When the data were corrected for the trapped component and replotted on a release spectrum, they produced a plateau corresponding to the same age as the isochron age (Figure 2.10b). The books by

1. *The ratio of parent to daughter should be large.* When this is the case, the amount of radiogenic daughter will be large relative to our ability to measure it. Under the best of circumstances, isotope ratios can be measured with a precision of a few parts per million. If the total amount of radioactively produced daughter is small relative to the amount present initially, for example, if the proportion of radiogenically produced daughter is only a few tens of parts per million or less of the total amount of daughter, accuracy of "ages" will be compromised.

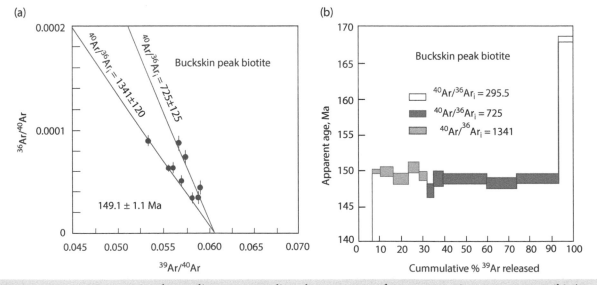

Figure 2.10 (a) Inverse isochron diagram revealing the presence of two excess Ar components. (b) Ar release spectrum for the same sample after correction for non-atmospheric excess Ar. Source: Adapted from Heizler and Harrison (1988).

2. *The parent/daughter should have a large range.* A large range in the parent–daughter ratio leads to a large range in isotope ratios in the daughter. The error on the regression slope, and ultimately the age, is a function of the range of values used in the computation. Therefore, given similar analytical precisions, we will obtain a more precise date with a decay system where the variations in the parent–daughter ratio are larger than with one where these variations are smaller.

3. *Deviations from closed system behavior must be minimal* subsequent to the event we are attempting to date. This should be considered when choosing both the decay system and the samples we plan to analyze. As we shall see, some elements tend to be more mobile than others, and some minerals are less reactive than others. Size also plays some role. A large sample is more likely to meet the closed system requirement compared to a small one (the elements have further to diffuse). Often, metamorphism will disturb a system on the scale of mineral grains, but not on a scale of "whole-rock" samples taken hundreds of meters apart (however, WRs will generally show less variation in parent/daughter ratios than minerals). One must also bear in mind that an atom created by radioactive decay will generally be a misfit in the lattice site it occupies (since the site was originally occupied by the parent). Furthermore, the decay process may damage the site. Such damage is more likely in the case of alpha decay than beta decay or electron capture because of the high energy of the alpha (typically 4 MeV), and the kinetic recoil energy of the daughter nucleus. These factors all lead to higher mobility of the daughter.

4. *The isotopic composition of the daughter must have been homogeneous* at the time of the event we wish to date. On a small scale, homogenization takes place through diffusion, which, as we have seen, is highly temperature dependent. The higher the temperatures obtained during the "event," the more rapidly and completely the system will be homogenized. On scales larger than 10 m or so, homogenization can only be achieved through convective-driven advective transport. This effectively means homogenization requires the presence of a fluid. This might be a magma or a hydrous fluid circulating through rocks undergoing metamorphism. In any case, both convection and diffusion will be more efficient at higher temperatures, so homogenization is more likely to be achieved at high temperatures than at low ones. Finally, the larger the range in parent/daughter ratios, and hence isotopic composition at the time we measure them, the less important will be any initial variations in isotopic composition.

We will now continue with our consideration of the various decay systems. However, Rb–Sr geochronology does not differ in principle from Sm–Nd geochronology or Re–Os geochronology. Thus, much of our discussion will focus on the geochemistry of these elements and the behavior of these systems with reference to the four points listed previously.

2.4.1 Rb–Sr chemistry and geochronology

Both Rb and Sr are trace elements in the Earth: their concentrations are generally measured in parts per million. Rb is an alkali element (Group 1) with a valence of +1. Like other alkalis, it is generally quite soluble in water and hydrous fluids. As a result, it is among the more mobile elements. Rb has an ionic radius of 148 pm. This large ionic radius means it is excluded from many minerals: it is simply too large to fit in the sites available. Elements that are not readily accommodated in common minerals, particularly those making up the Earth's upper mantle, are referred to as *incompatible* elements (and conversely, elements readily accommodated in mantle minerals are known as *compatible* elements). Upon melting of the mantle, these elements partition into the melt and are consequently enriched in the Earth's crust. Rb is one of the most incompatible elements and is strongly concentrated in the Earth's crust and depleted in its mantle. However, its radius is sufficiently similar to that of potassium (133 pm) that it substitutes readily for K in K-bearing minerals, such as mica and K-feldspar. As a result, no Rb minerals occur in nature; that is, it is not a stoichiometric component of any mineral.

Sr is an alkaline earth element (Group 2) with a valence of +2. The alkaline earths are also reasonably soluble in water and hydrous fluids, but not as soluble as the alkalis. Sr is therefore a moderately mobile element. Its ionic radius is 113 pm and sufficiently large for it to be excluded from many minerals; consequently, it is also an incompatible element, but not as highly incompatible as Rb. It substitutes for Ca (ionic radius 99 pm) to varying degrees. It is quite comfortable in the Ca site in plagioclase, with the solid/liquid partition[4] coefficient being about 2. It seems to be considerably less comfortable in the Ca site in clinopyroxene, with the Sr partition coefficient being only about 0.1. Thus, in most igneous and high-grade metamorphic rocks, most Sr will be in plagioclase (which typically constitutes about 50% of mafic igneous rocks). Sr can also substitute for Ca in other minerals, such as calcite, apatite, gypsum, titanite ($CaTiSiO_5$, also known as sphene), and so on. Sr is also concentrated in the crust relative to the mantle, but not to the degree that Rb is.

The Rb/Sr in the Earth as a whole is in the range of 0.021–0.029; we do not know this ratio exactly[5]. The ratio is lower in the mantle, and much higher in the crust. Mantle-derived rocks, such as basalts, also have low Rb/Sr ratios. Low ratios, such as these, violate condition 1 above; as a result, it is often difficult to obtain good Rb/Sr ages on mafic[6] and ultramafic rocks. However, igneous differentiation tends to increase the Rb/Sr ratio because Sr is removed by fractional crystallization of plagioclase while Rb remains in the melt. In felsic or silicic igneous rocks, the Rb/Sr ratio often exceeds 1 (a Rb/Sr ratio of 1 corresponds approximately to an $^{87}Rb/^{86}Sr$ ratio of 2.9,

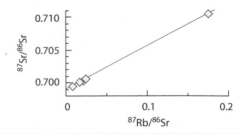

Figure 2.11 An Rb–Sr isochron. Five analyses from a clast in the Bholghati meteorite fall on an isochron, whose slope is related to the age of the system. The age in this case is 4.54 Ga. Data from Nyquist et al. (1990).

depending on the $^{87}Sr/^{86}Sr$ ratio). As a result, Rb/Sr dating can often be applied successfully to felsic igneous rocks. A large range in Rb/Sr ratio is also reasonably common. It may occur in whole-rock samples when the WRs represent various members of a comagmatic differentiation suite, or in mineral samples when both K- and Ca-bearing minerals are present. Rb–Sr geochronology can also be applied to metamorphic rocks, provided K-bearing, Rb-rich minerals are present, as they typically are. Figure 2.11 shows an example isochron of a meteorite.

A serious disadvantage of the Rb–Sr system is the mobility of these elements, particularly Rb. Because of their solubilities, Rb and Sr are readily transported by fluids, and may be moved into or out of the system. Furthermore, some K-bearing minerals, such as micas, are comparatively reactive, in the sense that some or much of the Rb may be present in exchangeable sites. These minerals are also subject to metamorphic resetting or partial resetting at relatively low temperatures. Thus, *Rb–Sr is a good system for dating silicic igneous rocks, where no intervening*

---

[4] The solid–liquid partition (or distribution) coefficient is a useful parameter in igneous trace element geochemistry. It is defined simply as the equilibrium ratio of the concentration of the element in the solid phase (e.g., a mineral, such as plagioclase) to the concentration in the magma. The partition coefficient provides a means of quantifying the term "incompatibility": the lower the partition coefficient, the higher the incompatibility.

[5] A reasonable compositional model for the Earth is that of chondritic meteorites, which we consider representative of the concentrations of non-gaseous elements in the Solar System. However, the Earth is demonstrably depleted in the more volatile of the non-gaseous elements so that this model of the Earth is valid only for the more refractory elements, such as Sr. The alkalis, including Rb, are among the volatile elements, for which this model is not valid.

[6] Mafic rocks are those rich in magnesium and iron (the term mafic comes from "MAgnesium and Ferric or Ferrous [fer being the Latin root meaning iron]). Ultramafic rocks are very rich in magnesium and iron. Basalt, the composition typical of many lavas, is mafic. The Earth's mantle is composed of peridotite, an ultramafic rock.

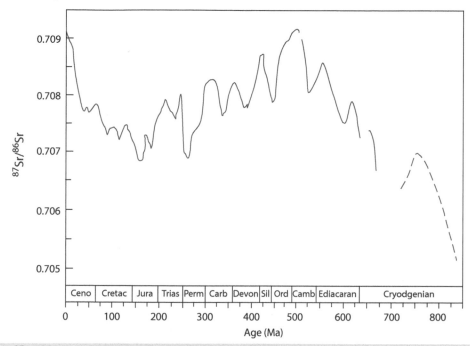

Figure 2.12 $^{87}Sr/^{86}Sr$ in seawater through Phanerozoic time determined from the analysis of phosphate and carbonate fossils. Source: McArthur et al. (2012)/With permission from Elsevier.

metamorphism or alteration has occurred, and for metamorphic rocks.

Rb–Sr dating can in special instances be applied to sedimentary rocks. Sedimentary rocks are generally difficult to date by any method because of the lack of high temperatures necessary for homogenization of initial Sr isotope ratios. However, minerals crystallizing from a homogeneous solution, such as seawater, will all have identical initial ratios. Thus, dates have been obtained using minerals, such as authigenic celadonite (a K- and Rb-rich mineral) and calcite (an Rb-poor, Sr-rich mineral). An additional advantage is that the evolution of $^{87}Sr/^{86}Sr$ in seawater is known. Thus, a reasonable assumption about the initial $^{87}Sr/^{86}Sr$ ratio may be made if the approximate age is known. To reiterate, however, successful dates of sediments are certainly rare.

2.4.2 Sr Isotope chronostratigraphy

Sr isotope ratios can, however, be used to date sediments in another way. Sr has a long residence time in the oceans, a consequence, in part, of its relatively high solubility. Consequently, it is uniformly mixed in the open ocean. As a result of that, its concentration and isotopic composition in the modern ocean are uniform.

Its isotopic composition has, however, changed over geologic time. The change over the Phanerozoic is illustrated in Figure 2.12.

Since seawater $^{87}Sr/^{86}Sr$ is geographically uniform at any time yet varies through time, the $^{87}Sr/^{86}Sr$ of minerals precipitated from seawater will be a function of time. Sr is concentrated in calcite and aragonite, in which many organisms, most notably mollusks, corals, and foraminifera, precipitate to form shells. By comparing the $^{87}Sr/^{86}Sr$ of a particular shell with the seawater curve provided in Figure 2.12, we can determine its age. This dating technique is called *Sr isotope chronostratigraphy*. There are, however, several caveats. First, $^{87}Sr/^{86}Sr$ is uniform only in the open ocean; it can vary in coastal areas due to continental inputs. Thus, for example, oyster shells would not be useful because oysters grow in brackish water. Second, while the ratio in a pristine shell should reflect the composition of the water it precipitates from, this ratio may change as a result of interaction with pore water. Finally, a given value of $^{87}Sr/^{86}Sr$ does not necessarily correspond to a unique age. For example, the value of 0.7080 occurred during Ordovician, Devonian, Mississippian, Permian, and Cenozoic times. Thus, the age of the fossil being dated needs to be approximately known before Sr isotope chronostratigraphy can be applied usefully. Finally, the accuracy of this technique depends on how accurately the $^{87}Sr/^{86}Sr$ value of seawater is known for any given time. For much of the Cenozoic, particularly the late Cenozoic, these values are very well known. Consequently, Sr isotope chronostratigraphy provides useful and accurate ages for these times. Values are less well known for the Paleozoic and poorly known for the Precambrian.

The change in seawater $^{87}Sr/^{86}Sr$ shown in Figure 2.12 has been very non-linear. Indeed, there have been times, such as the Permian and the Jurassic, when $^{87}Sr/^{86}Sr$ has actually decreased in seawater. This is perhaps initially surprising since the decay of ^{87}Rb to ^{87}Sr occurs at a constant rate. The variation in $^{87}Sr/^{86}Sr$ reflects the *open system* nature of the oceans and not simply radioactive decay. Salts are continuously added and removed from seawater; consequently, the oceans inventory of Sr is constantly, albeit slowly, renewed. Thus, the isotopic composition of seawater Sr reflects the isotopic composition of Sr added to seawater, i.e., the isotopic composition of the *sources* of Sr in seawater. We can broadly divide these sources into "continental" and "mantle." The continental source is dominantly the riverine input, and secondarily wind-blown and glacially derived particles that dissolve or partly dissolve when they reach the sea. The isotopic composition of the continental source will vary with the nature of continental material undergoing erosion at any time, as well as with the rate of erosion, but will generally have high $^{87}Sr/^{87}Sr$. The "mantle" source consists primarily of hydrothermal fluids of mid-ocean ridge hydrothermal systems. Secondary mantle sources include erosion and weathering (both subaerial and submarine) of young, mantle-derived basalts and will have a low $^{87}Sr/^{86}Sr$. Thus, the Sr isotopic composition of seawater primarily reflects the balance between continental and mantle inputs; we will discuss this further in Chapter 11.

2.5   RARE-EARTH DECAY SYSTEMS

Normally an electron is added to the outermost shell when atomic number increases. The rare-earth elements do not fit into this pattern and are therefore placed below the main part of the periodic table in its usual form (Figure 2.13). Five of these elements, namely, La, Sm, Lu, Th, and U, have long-lived radioactive isotopes. The lanthanoid[7] (or lanthanide)

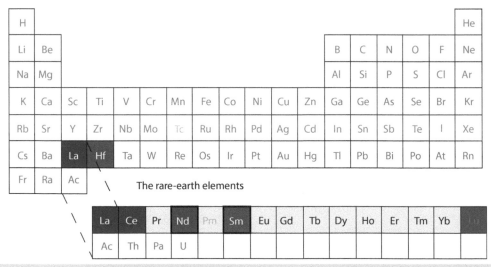

Figure 2.13   Periodic table highlighting the lanthanide rare earths and Nd–Sm, La–Ce, and Lu–Hf decay systems.

[7] The International Union of Pure and Applied Chemistry (IUPAC) recommends the terms lanthanoid and actinoid in place of lanthanide and actinide because the suffix -ide generally refers to negative ions or compounds of them (e.g., chloride and hydride).

rare earths, which include radioisotopes [138]La, [176]Lu, and [147]Sm, are part of period 6 and all have two electrons in their outer 6s orbital as neutral atoms, but the 4f and 5d electron shells, which were previously skipped, are progressively filled before filling the 6p orbitals progressing from La to Lu. In the following row, which includes actinoid (or actinide) rare earths, these are the 5f orbitals that are being filled. However, only two of the actinoid rare earths, Th and U, have long-lived nuclides. They are, of course, of great interest in radiogenic isotope geochemistry, but we will treat them separately. Henceforth, we will use the term *rare-earth elements* (REEs) to refer exclusively to the lanthanoids.

As the outer electron shells dictate the chemical behavior of elements and their configurations are identical in the rare earths, we would expect them to behave quite similarly. This is indeed the case. The rare earths generally have a +3 valence, with the most important exceptions being Eu, which is +2 under reducing conditions, and Ce, which is +4 under oxidizing conditions. The primary chemical difference between the rare-earth elements is the ionic radius, which shrinks systematically from 115 pm for La ($A = 57$) to 93 pm for Lu ($A = 71$). Since the rare earths form predominately ionic bonds with oxygen in the solid Earth, their ionic radius is a key factor in their geochemical behavior. Thus, there is a systematic variation

in their abundances in rocks, minerals, and solutions (see Box 2.1 on rare-earth plots). The ionic radii of Sm and Nd, which are separated by Pm (an element that has no stable or long-lived isotope), differ by only 4 pm (Nd = 108; Sm = 104 pm). The large ionic radii and relatively high charge of the rare earths, particularly the lighter ones, make them fairly unwelcome in many mineral lattices: they can be considered moderately incompatible, with Nd being slightly more incompatible than Sm. Ce is generally the most abundant rare earth and forms its own phase in rare instances (usually with considerable substitution by other light rare earths). The heavier rare earths are more readily accommodated in lattice structures of common minerals and become strongly compatible in some; for example, the partition coefficient of Lu in garnet is in the range of 4–10 (depending on the composition of the rock and the garnet). In mafic minerals, the lighter rare earths, which have the largest ionic radii, tend to be excluded more than the heavy ones, but in plagioclase, the heavy REEs are the most excluded (though partition coefficients generally are not less than 0.1). The high valence state of the rare earths results in relatively strong bonds. This, together with their tendency to hydrolyze (i.e., surround themselves with OH^- radicals), results in relatively low solubilities and low mobilities.

Rare-earth plots

The systematic contraction in the ionic radii of the rare-earth elements leads to systematic variation in their behavior. This is best illustrated by viewing their abundances on rare earth, or Masuda–Coryell, plots. The plots are constructed by first "normalizing" the concentration of the rare earth, i.e., dividing but the concentration of the element in a standard. Generally, this standard is the abundance in chondritic meteorites, but other values are also used (for example, rare earths in sediments and seawater are often normalized to average shale). This normalizing process removes the sawtooth pattern that results from odd–even nuclear effects, and also the decreasing concentration with atomic number. Those concentration variations, illustrated in Figure 2.14, reflect differences in nuclear stability and the nucleosynthetic process discussed in the previous chapter, and therefore affect the abundances of rare earths in all matter. Removing these effects by normalization highlights differences in concentration due to geochemical processes.

After normalizing, the log of the abundance of each element is plotted against atomic number, as is illustrated in Figure 2.15. The larger ionic radii of the light rare earths make them incompatible in mantle minerals and consequently they concentrated in melts of the mantle that ultimately make up the continental crust. The residual mantle has thus become depleted in light rare earths and this depletion is reflected in the rare-earth patterns of mid-ocean ridge basalts (MORBs).

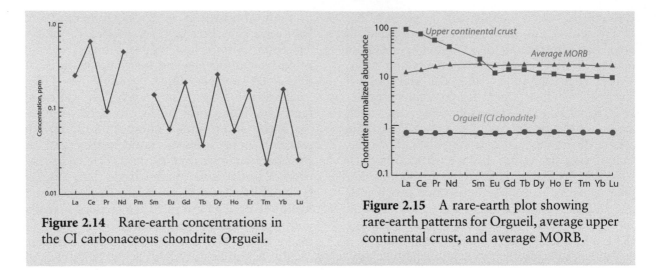

Figure 2.14 Rare-earth concentrations in the CI carbonaceous chondrite Orgueil.

Figure 2.15 A rare-earth plot showing rare-earth patterns for Orgueil, average upper continental crust, and average MORB.

2.5.1 Epsilon and mu notations

The rare earths have very nearly the same *relative* abundances (i.e., ratios to each other) in all classes of chondritic meteorites and they all exhibit flat patterns on rare-earth plots, such as Figure 2.15, although absolute concentrations vary significantly (see Box 2.1). In Figure 2.15, concentrations have been normalized to average ordinary chondrites. Rare earth concentrations in carbonaceous chondrites such as Orgueil are lower because they are diluted by carbonaceous material and other volatiles compared to ordinary chondrites. Chondrites are aggregations of nebular dust and hence preserve the composition of the condensable component of the solar nebula (although somewhat modified by subsequent processes on their asteroidal parent bodies). The uniformity of relative rare earth abundances indicates that the rare earths were not fractionated in the solar nebula, the cloud of gas and dust from which Solar System bodies, including the parent bodies of meteorites and the Earth, formed. This is also true of other refractory elements[8], such as Sr, Hf, Th, and U, but it is not true of elements that condense at lower temperatures. Much of the chemical variation among chondritic meteorites relates to volatility, and hence is apparently due to elements evaporating or condensing from nebular gas. That the relative abundance of refractory elements does not vary suggests nebular temperatures were generally not hot enough for significant fractions of these elements to evaporate.

Since the relative concentrations of rare earths in chondrite are uniform and identical (within analytical error) to those in the Sun, we expect that the nebular dust from which the Earth formed should also have had those same relative concentrations of rare earths. Thus, it has been widely assumed that the Earth also has chondritic relative abundances of the rare-earth elements as well as Hf, and that the $^{147}Sm/^{144}Nd$, $^{176}Lu/^{177}Hf$, and $^{138}La/^{142}Ce$ ratios of the Earth are equal to the mean chondritic values of these ratios. Assuming further that the solar nebula was isotopically homogeneous, we could also infer that the initial Nd, Hf, and Ce isotope ratios of the Earth should be identical to those in chondrites. Thus, for example, if the initial $^{143}Nd/^{144}Nd$ ratio and the Sm/Nd ratio of Earth are the same as chondrites, then the present $^{143}Nd/^{144}Nd$ should also be the same as the present ratio in chondrites.

These observations and assumptions led to a useful notation, namely, ε_{Nd} (epsilon-Nd), which is the relative deviation from the

---

[8] Here we define a refractory element as one that condenses from a gas phase at high temperature or forms compounds (usually oxides) that condense at high temperature. Elements that condense at temperatures above that of Mg, Fe, and Si, which are the most abundant condensable elements in the Solar System, are considered refractory; those condensing below this temperature are considered volatile.

chondritic ^{143}Nd/^{144}Nd value (DePaolo and Wasserburg, 1976). Since the deviations are small, they are expressed in parts in 10 000. Thus, ε_{Nd} is defined as follows:

$$\varepsilon_{Nd} = \left[\frac{^{143}\text{Nd}/^{144}\text{Nd}_{sample} - ^{143}\text{Nd}/^{144}\text{Nd}_{CHUR}}{^{143}\text{Nd}/^{144}\text{Nd}_{CHUR}}\right] \times 10\ 000$$

(2.57)

where ^{143}Nd/^{144}Nd$_{CHUR}$ is the value of the ratio in chondrites and the acronym *CHUR* stands for *chondritic uniform reservoir*. The present value of ^{143}Nd/^{144}Nd$_{CHUR}$ is 0.512630[9] when ^{146}Nd/^{144}Nd = 0.7219[10]. We can also calculate an ε_{Nd} value for any point in time using the ^{143}Nd/^{144}Nd ratios of the sample and chondrites at that time; the latter can be calculated from present chondritic ^{143}Nd/^{144}Nd$_{CHUR}$ and ^{147}Sm/^{144}Nd$_{CHUR}$ ratios, where ^{147}Sm/^{144}Nd$_{CHUR}$ = 0.1960 (corresponding to a Sm/Nd of about 0.325) (Bouvier et al., 2008). One advantage of this notation is that ε_{Nd} are small numbers of only two or three significant digits, with the range in ε_{Nd} among most terrestrial rocks being +14 to –20. This same range corresponds to ^{143}Nd/^{144}Nd from 0.5116 to 0.5132.

Originally introduced to describe Nd isotope variations, the same notation is used for Ce and Hf isotopic variations in an exactly analogous manner. It is also sometimes used to describe non-radiogenic isotope variations, including those due to chemical fractionation and nucleosynthetic effects.

As analytical precision has improved over the decades, isotope variations at the parts per million level can now be measured precisely and an additional notation has been introduced for such small variations, the μ notation. This, for example, is widely used to denote variations in the ^{142}Nd/^{144}Nd resulting from decay of the extinct radionuclide ^{146}Sm

(which is discussed below and in subsequent chapters):

$$\mu_{^{142}Nd} = \left[\frac{^{142}\text{Nd}/^{144}\text{Nd}_{sample} - ^{142}\text{Nd}/^{144}\text{Nd}_{Std}}{^{142}\text{Nd}/^{144}\text{Nd}_{Std}}\right] \times 10^6$$

(2.58)

Thus, the μ notation denotes parts per million variations. There is another important difference between this and the epsilon notation: *whereas the normalizing value in the epsilon notation is chondritic meteorites, it is a terrestrial laboratory standard in the case of the mu notation.*

2.5.2 The Sm–Nd decay system

The earliest developed and most widely applied of the rare-earth decay systems is Sm–Nd. ^{147}Sm decays to ^{143}Nd by alpha decay with a half-life of 106 Ga ($\lambda = 6.54 \times 10^{-12}$ y^{-1}). Sm and Nd are both intermediate rare-earth elements and behave similarly so that the Sm/Nd ratio shows only limited variation. Because of this and long half-life of ^{147}Sm, the resulting variations in Nd isotopic composition are small and require precise measurement. A second Sm isotope, ^{146}Sm, is also radioactive and undergoes α-decay to ^{142}Nd, but with a half-life of only 103 million years, it is no longer extant. Variations in the ^{142}Nd/^{144}Nd ratio in meteorites and some Archean rocks provide clear evidence that it was present in the early Solar System, including the Earth. The terrestrial ^{142}Nd/^{144}Nd ratio lies at the extreme of the range observed in meteorites, which places the assumption that the Sm/Nd ratio of the Earth is chondritic in question. The difference in ^{142}Nd/^{144}Nd ratios between meteorites and the Earth, roughly 10–30 ppm, implies the ^{147}Sm/^{144}Nd ratio of the Earth is about 6–8% higher than the chondritic one, corresponding to an ε_{Nd} of about +7

---

[9] This is the value of Bouvier et al. (2008); the original values of Jacobsen and Wasserburg (1984) were ^{143}Nd/^{144}Nd$_{CHUR}$ = 0.512638 and ^{147}Sm/^{144}Nd$_{CHUR}$ = 0.1966, and a considerable amount of the epsilon values in the literature has been calculated relative to these values.

[10] As we pointed out in Section 2.2.3, the isotope ratios of Sr, Nd, Hf, and Os are always corrected for mass fractionation occurring during analysis by "normalizing" the ratio of interest to an assumed "true" ratio of two non-radiogenic isotopes of the element of interest, for example, Sr isotope ratios are always corrected to ^{86}Sr/^{88}Sr = 0.11940. Unfortunately, two normalization schemes evolved for Nd. The "CalTech" normalization was to ^{146}Nd/^{142}Nd = 0.636151. Using this scheme, the present-day ^{143}Nd/^{144}Nd chondritic value is 0.511847. This normalization is now uncommon, in part due to the demonstration of radiogenic ^{142}Nd, but there is still considerable data in the literature based on it. The most common normalization is to ^{146}Nd/^{144}Nd = 0.7219. The value of ε_{Nd} for a given rock should be the same, however, regardless of normalization.

(Boyet and Carlson, 2005), assuming this difference is due to decay of ^{146}Sm. However, there are also small variations in ^{142}Nd/^{144}Nd between classes of chondritic meteorites. These are unrelated to variations in the Sm/Nd ratio and instead correlate with the abundance of non-radiogenic Nd isotopes (Burkhardt et al., 2016), implying they are so due to incomplete mixing of material produced in different nucleosynthetic environments, such as red giants, supernovas, and kilonovas. How much of the difference in ^{143}Nd/^{144}Nd between the Earth and chondrites is nucleosynthetic and how much is radiogenic is debated. Thus, while the epsilon notation and CHUR concepts remain useful, we need to be aware of the possibility that the Sm/Nd ratio, and consequently the ^{143}Nd/^{144}Nd ratio of the Earth, may differ slightly, a few epsilon units, from chondritic. We will discuss this in more detail in subsequent chapters.

Figure 2.16a illustrates how ^{143}Nd/^{144}Nd has evolved in the Earth. If the Earth indeed has chondritic Sm/Nd, it evolves along the line labeled "CHUR." If the ^{147}Sm/^{144}Nd is higher, it evolves along a steeper trajectory in the gray area labeled "bulk (observable) silicate Earth." In either case, differentiation of the Earth into a light rare-earth-enriched crust and light rare-earth-depleted mantle results in the mantle evolving along a steeper (high Sm/Nd) trajectory, and crust evolving along a less steep one (low Sm/Nd). On converting ^{143}Nd/^{144}Nd to ε_{Nd} (Figure 2.16b), the CHUR value remains constant at $\varepsilon_{Nd} = 0$ while the observable Earth evolves toward a higher ε_{Nd} value in the gray area +7. The mantle evolves toward more positive ε_{Nd} while the crust evolves toward negative ε_{Nd}.

Perhaps the greatest advantage of Sm/Nd is the lack of mobility of these elements. *The Sm–Nd chronometer is therefore relatively robust with respect to alteration and low-grade metamorphism.* Thus, the Sm–Nd system is often the system of choice for mafic rocks and for rocks that have experienced metamorphism or alteration. An additional advantage is relatively high closure temperatures for this system making it useful for dating peak metamorphism.

Because of the low solubility and particle-reactive nature of the rare earths, the Nd isotope ratio of seawater is not uniform. Consequently, Nd isotope ratios in seawater-

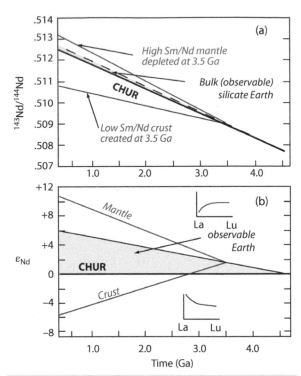

Figure 2.16 (a) Nd isotope evolution of CHUR, the chondritic uniform reservoir (bold line), the bulk observable Earth within the gray area, and residual mantle (red) resulting from continental crust (blue) produced at 3.5 Ga. (b) Evolution of bulk (observable) silicate Earth (gray area), crust (blue), and mantle (red) when ^{143}Nd/^{144}Nd is transformed to ε_{Nd}. Also shown are cartoons of the corresponding rare-earth patterns.

derived materials cannot be used as a geochronological tool. However, they can be used to investigate how ocean circulation has varied in time, a topic we will take up in Chapter 11.

There are also several drawbacks to the use of the Sm–Nd system in geochronology. First, the half-life of ^{147}Sm is relatively long, leading to relatively small variation in ^{143}Nd/^{144}Nd and imprecise ages, particularly for young rocks. The second is the limited variation in Sm/Nd. As things turn out, however, Sm–Nd complements Rb–Sr nicely. Sm/Nd variations tend to be largest in mafic and ultramafic rocks and smallest in acid rocks, exactly the opposite of Rb/Sr. One application that has been highly successful is the use of the Sm–Nd isotope system to date garnet-bearing rocks. Garnets incorporate heavy rare earths (HREE) over light rare earths (LREE) as they grow, which lead to very high Sm/Nd ratio in these minerals.

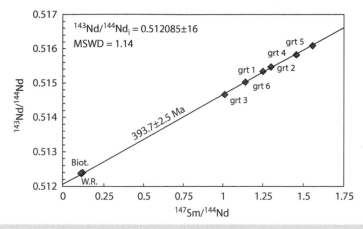

Figure 2.17 Sm–Nd isochron based on six garnet fractions, two WRs, and a biotite fraction from the Qinling metamorphic belt in China. Errors are smaller than the symbols. Source: Adapted from Cheng et al. (2011).

Therefore, garnet-bearing rocks, such as eclogites, can be dated to relatively high precision with the Sm–Nd system. Figure 2.17 is an example of a Sm–Nd isochron of a garnet-bearing granite from the Qinling–Tongbai–Dabie–Sulu ultrahigh-pressure metamorphic belt in China (Cheng et al., 2011). The 394 Ma age is younger than the 434 ± 7 Ma zircon age of the granite. The garnets likely formed from incongruent breakdown of biotite. The age reflects time the granite cooled through the closure temperature of Sm–Nd, in this case, some 40 million years after crystallization, suggesting it remained deeply buried.

2.5.2.1 Sm–Nd model ages and crustal residence times

A general assumption about the Earth is that the continental crust has been created from the mantle by magmatism (and radiogenic isotope geochemistry provides compelling evidence to support this assumption). When a piece of crust is first created, it will have the $^{143}Nd/^{144}Nd$ ratio of the mantle, although its Sm/Nd ratio will be lower than that of the mantle (a consequence of Nd being more incompatible and partitioning more into the melt than Sm). Let us make the simplistic assumption that the mantle has the same Nd isotopic history as CHUR. This means a piece of crust will have the same $^{143}Nd/^{144}Nd$ as the mantle and as CHUR when it is created, i.e., $\varepsilon_{Nd} = 0$. Based on that assumption and the additional assumption that its Sm/Nd ratio has been undisturbed,

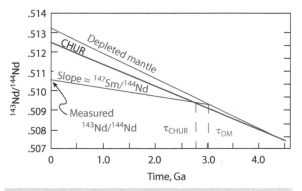

Figure 2.18 Sm–Nd model ages. The $^{143}Nd/^{144}Nd$ is extrapolated backward (slope depending on Sm/Nd) until it intersects a mantle or chondritic growth curve.

we can estimate the age of this piece of crust by measuring its present-day Sm/Nd and $^{143}Nd/^{144}Nd$ ratios.

Figure 2.18 illustrates how this is done graphically. We can see in Equation (2.10) that on a plot of an isotope ratio versus time, the slope of a line is proportional to the parent/daughter ratio. Thus, to find the "CHUR model age," we extend a line from the present $^{143}Nd/^{144}Nd$ ratio with a slope corresponding to the present $^{147}Sm/^{144}Nd$ until it intersects the CHUR evolution line. That intersection occurs at the CHUR model age, τ_{CHUR}.

Let us now see how this is done mathematically. What we want to find is the intersection of line describing the evolution of the sample and that describing the evolution of the mantle. To do so, we simply need to subtract one equation from the other. The closed system isotopic evolution of any sample can be expressed as:

$$^{143}Nd/^{144}Nd_{sam} = {}^{143}Nd/^{144}Nd_0$$
$$+ {}^{147}Sm/^{144}Nd_{sam}\left(e^{\lambda t} - 1\right) \qquad (2.59)$$

The chondritic evolution line is:

$$^{143}Nd/^{144}Nd_{CHUR} = {}^{143}Nd/^{144}Nd_0$$
$$+ {}^{147}Sm/^{144}Nd_{CHUR}\left(e^{\lambda t} - 1\right) \qquad (2.60)$$

The CHUR model age of a system is the time elapsed, $t = \tau$, since it had a chondritic $^{143}Nd/^{144}Nd$ ratio, assuming the system has remained closed. We can find τ by subtracting Equation (2.60) from Equation (2.59), which yields:

$$^{143}Nd/^{144}Nd_{sam} - {}^{143}Nd/^{144}Nd_{CHUR}$$
$$= \left\{{}^{147}Sm/^{144}Nd_{sam} - {}^{147}Sm/^{144}Nd_{CHUR}\right\}\left(e^{\lambda t} - 1\right) \qquad (2.61)$$

Solving Equation (2.61) for τ, we obtain:

$$\tau_{CHUR} = \frac{1}{\lambda} \ln\left(\frac{{}^{143}Nd/^{144}Nd_{sam} - {}^{143}Nd/^{144}Nd_{CHUR}}{{}^{147}Sm/^{144}Nd_{sam} - {}^{147}Sm/^{144}Nd_{CHUR}} + 1\right) \qquad (2.62)$$

An age obtained in this way is called an *Nd model age* (the model is that of chondritic evolution of the mantle), or a *crustal residence age*, because it provides an estimate of how long this sample of Nd has been in the crust. Note that we explicitly assume the sample has remained a closed system, in the sense of no migration in or out of Sm or Nd. Because of the immobility of these elements, the assumption often holds, although generally only approximately.

We can obtain somewhat better model ages by making a more sophisticated assumption about the Nd evolution of the mantle. Since the crust is enriched in Nd relative to Sm, the mantle must be depleted in Nd relative to Sm (analyses of mantle-derived rocks confirm this) and the mantle should evolve along a line steeper than chondritic. Once we decide on Sm/Nd and present-day $^{143}Nd/^{143}Nd$ ratios for this "depleted mantle" (the latter can be estimated from the $^{143}Nd/^{143}Nd$ of MORB), we can calculate a model age relative to the depleted mantle by substituting the depleted-mantle terms

for the CHUR terms in Equations (2.60) and (2.61).

To calculate the depleted-mantle model age, τ_{DM}, we use the same approach, but this time we want the intersection of the sample evolution line and the depleted-mantle evolution line. Therefore, Equation (2.53) becomes:

$$\tau_{DM} = \frac{1}{\lambda} \ln\left(\frac{{}^{143}Nd/^{144}Nd_{sam} - {}^{143}Nd/^{144}Nd_{DM}}{{}^{147}Sm/^{144}Nd_{sam} - {}^{147}Sm/^{144}Nd_{DM}} + 1\right) \qquad (2.63)$$

The depleted mantle (as sampled by MORBs) has an average ε_{Nd} of about 9, or $^{143}Nd/^{144}Nd = 0.51310$. The simplest possible evolution path, and the one we shall use, would be a closed system evolution since the formation of the Earth, i.e., 4.55 Ga ago (i.e., a straight line on a $^{143}Nd/^{144}Nd$ versus time plot). This evolution implies a $^{147}Sm/^{144}Nd$ of 0.213.

Because the Sm/Nd ratio is so little affected by weathering, and because these elements are so insoluble, Sm/Nd ratios in fine-grained sediments do not generally differ much from the ratio in the precursor crystalline rock. Thus, the system has some power to "see through" even the process of making sediment from crystalline rock. The result is we can even compute crustal residence times from Nd isotope ratio and Sm/Nd measurements of fine-grained sediments. This generally does not work for coarse-grained sediments though because they contain accessory minerals whose Sm/Nd ratios can be quite different from that of the WR. Nevertheless, we should not confuse these "model ages" with actual radiometric ages. For one thing, our assumption that Sm/Nd ratios do not change does not hold in the exact. For another thing, crustal segments and sediments derived from them are produced over a range of time, and, as we will see in subsequent chapters, production of continental crust often involves incorporation of older crust as well as new material from the mantle, in which case the model age will be older than the time of actual crust production.

2.5.3 The La–Ce system

^{138}La is an odd–odd nucleus and decays to both ^{138}Ba by β^+ and electron capture and to ^{138}Ce by β^- (34.5%) and is also quite rare (0.09% of La). ^{138}Ba is among the most abundant heavy isotopes (because it has

82 neutrons, a magic number); consequently, only a small fraction of ^{138}Ba is radiogenic, so attention has focused on ^{138}Ce. The rarity of ^{138}La, its slow decay ($t_{1/2}$ = 449 Ga), and most of this decay (65.2%) going to ^{138}Ba means variations in the relative abundance of ^{138}Ce are small, requiring high analytical precision to extract useful information. In addition, ^{138}Ce is also a rare isotope of Ce (0.251%) and that together the presence of isobaric interferences from neighboring elements and the presence of very abundant ^{140}Ce requires extremely careful preparative chemistry. Some early data on Ce isotopic variations were reported by Tanaka and Masuda (1982), Tanaka et al. (1987), and Dickin (1987), but because of the analytical challenges, relatively little subsequent work was done for several decades. Advances in analytical instrumentation have spurred new interest in the La–Ce system. While the long half-life and small La/Ce ratios have thus far precluded its use as a geochronometer, and a number of new studies published in the last decade show the La–Ce system can be a useful geochemical tool.

Most studies have reported the ^{138}Ce/^{142}Ce ratio, but because the abundance of ^{138}Ce is ≈50 times lower than that of ^{142}Ce, Willig and Stracke (2019) have measured the ^{138}Ce/^{136}Ce ratio instead. ^{138}Ce/^{136}Ce can be converted to ^{138}Ce/^{142}Ce by multiplying with ^{136}Ce/^{142}Ce = 0.01688 if the same fractionation normalization is used. Like Nd, Ce is a rare-earth element and it is reasonable to assume that the Earth has a chondritic ^{138}Ce/^{142}Ce ratio or nearly so. We therefore define an ε_{Ce} in a manner analogous to ε_{Nd}:

$$\varepsilon_{Ce} = \left[\frac{\left(^{138}Ce/^{142}Ce\right)_{sample} - \left(^{138}Ce/^{142}Ce\right)_{chon}}{\left(^{138}Ce/^{142}Ce\right)_{chon}} \right]$$
$$\times 10\,000 \qquad (2.64a)$$

$$\text{or}: \varepsilon_{Ce} = \left[\frac{\left(^{138}Ce/^{136}Ce\right)_{sample} - \left(^{138}Ce/^{136}Ce\right)_{chon}}{\left(^{138}Ce/^{136}Ce\right)_{chon}} \right]$$
$$\times 10\,000 \qquad (2.64b)$$

where $(^{138}Ce/^{142}Ce)_{chon}$ = 0.02256577 (Israel et al, 2020) and $(^{138}Ce/^{136}Ce)_{chon}$ = 1.336897 (Willig and Stracke, 2019).

As Figure 2.19 shows, there is an overall strong negative correlation between ε_{Ce} and

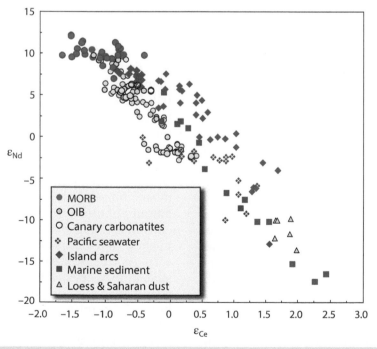

Figure 2.19   High-precision Ce and Nd isotope data on terrestrial materials (Doucelance et al., 2014; Bellot et al., 2015, 2018; Boyet et al., 2019; Willig and Stracke, 2019; and Israel et al., 2020) show an inverse correlation expected from the relative behaviors and parents and daughters.

ε_{Nd}. This is what we expect as all elements involved are rare earths and the parent, La, is more incompatible than the daughter, Ce, but the daughter of the Nd–Sm system is more incompatible than the parent. Young mantle-derived basalts, including MORBs and OIBs, have generally negative ε_{Ce} and positive ε_{Nd}, while continent-derived materials, including marine sediments, glacial loess, and desert dust, have positive ε_{Ce} and negative ε_{Nd}, consistent with light rare-earth depletion of the former and enrichment of the latter.

As we noted earlier, unlike the other rare earths, Ce can be present in the 4^+ valence state under oxidizing conditions. Tetravalent Ce is even more insoluble and more readily absorbed on surfaces than trivalent Ce, which results in a negative Ce anomaly in seawater and positive and negative anomalies in marine sediments. Almost uniquely among mantle-derived igneous rocks, some island arc lavas also exhibit negative Ce anomalies. Some of the recent Ce isotopic work was undertaken to assess whether these Ce anomalies are inherited from subducted marine sediments. Island arc magmas do appear to be offset to higher ε_{Ce} in Figure 2.19 compared to MORBs and OIBs, and Bellot et al. (2018) concluded that this is indeed the case.

2.5.4 The Lu–Hf system

Lu is the heaviest rare-earth element, with a valence of +3 and an ionic radius of 93 pm. It has two isotopes, ^{175}Lu (97.4%) and ^{176}Lu (2.6%). Like ^{138}La, ^{176}Lu is an odd–odd nucleus and doubly unstable. It decays to ^{176}Hf through β^- emission with a half-life of 37.1 billion years and may also decay to ^{176}Yb through positron emission, but there is uncertainty as to the branching ratio; decay to ^{176}Yb certainly constitutes less than 3% of the decay and probably much less. Hf is a member of the Group IVB elements, which includes Ti and Zr and consequently is chemically similar to them. This similarity is quite strong in the case of Zr and Hf because both occur naturally only in the 4^+ valence and their ionic radii are nearly identical: 71 pm in six-fold coordination (83 in eight-fold coordination) for Hf; 72 pm in six-fold coordination (84 pm in eight-fold coordination) for Zr. The radius of Ti^{4+} is much smaller, i.e., 61 pm, and Ti also exists in several valence states. Lu can be considered a slightly-to-moderately incompatible element; Hf is moderately incompatible (its incompatibility is very similar to that of Sm).

The Lu–Hf system shares many of the advantages of the Sm–Nd system: both are relatively insoluble and immobile elements and both are refractory, and hence we have reason to believe that the Lu/Hf ratio of the Earth should be the same as in chondrites or nearly so. We can define an ε_{Hf} value in a manner exactly analogous to ε_{Nd}:

$$\varepsilon_{Hf} = \left[\frac{\left(^{176}Hf/^{177}Hf \right)_{sample} - \left(^{176}Hf/^{177}Hf \right)_{CHUR}}{\left(^{176}Hf/^{177}Hf \right)_{CHUR}} \right] \times 10\,000 \qquad (2.65)$$

where $^{176}Hf/^{177}Hf_{CHUR} = 0.282793$ for zero-age samples. Calculation of initial $\varepsilon_{Hf}(t)$ is done using $^{176}Lu/^{177}Lu_{CHUR} = 0.0338$ (Blichert-Toft, 2018).

Isotopic analysis of Hf is made difficult by its extremely limited aqueous solubility and its nearly identical chemical behavior to Zr. An additional problem is that the temperatures required for ionization are extremely high; as a result, the ionization efficiency by thermal ionization is low, making analysis difficult by this method. As a consequence, the first survey of Hf isotopic variations was not carried out until the work of Patchett and Tatsumoto (1980). This problem has been overcome with the development of multiple-collector magnetic sector ICP-MS (MC-ICP-MS), in which the analyte is introduced into an Ar plasma with a temperature of ≈6000 K and completely ionized.

As the analytical problems with the Lu–Hf system were overcome, other problems emerged that needed to be resolved. Perhaps most importantly, there was a worrisome amount of uncertainty of the value of the decay constant. Determining the decay rate of a nuclide that decays as slowly as ^{176}Lu is not easy. As we noted earlier, there are several possible approaches. These approaches produced results that did not agree as well as one would hope. Counting experiments performed since 1975 has yielded a range of decay constants ranging from 1.70×10^{-11} a^{-1} to 1.93×10^{-11} a^{-1}, a 14% range. The "calibration" approach has also produced a range of values, ranging from $1.865 \times 10^{-11} \pm 0.015$

determined by Scherer et al. (2001) on U–Pb dated terrestrial samples in age ranging from 0.91–2.06 Ga while Bizzarro et al. (2003) calculated a value of 1.983×10^{-11} a^{-1} from an isochron on chondritic and eucritic meteorites and an assumed age of 4.56 Ga. Use of the smaller value of Scherer et al. (2001) tended to produce ages of meteorites older than the known age of the Solar System. At one point, it seemed that one value of the decay constant applied to meteorites and another to terrestrial samples, which hardly seemed likely.

Subsequent studies of meteorites appear to have resolved this issue in favor of the "terrestrial" decay constant of Söderlund et al. (2004), and 1.867×10^{11} yr^{-1} is the currently accepted value. Amelin (2005) carried out a calibration study using phosphates (such as apatite) in the Acapulco and Richardton meteorites and found decay constants in good agreement with this value. Subsequent work by Bouvier et al. (2008) found that the least thermally metamorphosed chondrites (petrologic classes 1–3) showed much less scatter on a Lu–Hf isochron plot than metamorphosed or "equilibrated" chondrites (Figure 2.20). Using

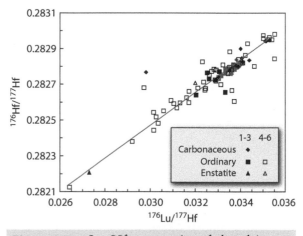

Figure 2.20 Lu–Hf systematics of chondrite meteorites. Data from Bouvier et al. (2008) in red, and data from Blichert-Toft and Albarede (1997), Bizzarro et al. (2003), and Patchett et al. (2004) in black. Petrologic classes 4–6 (open symbols) are considered "equilibrated" while classes 1–3 (closed symbols) are considered "unequilibrated." The red line is the isochron of Bouvier et al. (2008). The scatter is likely due to thermal and shock metamorphism and Lu mobility in phosphate compounds on the meteorite parent bodies.

only the most precise data and assuming an age of 4568.5 Ma, they calculated a ^{176}Lu decay constant of $1.884 (\pm 0.060) \times 10^{-11}$ yr^{-1}, which was in good agreement with the value obtained in studies of terrestrial rocks. Lu–Hf isochrons of meteorites that correspond to ages in excess of 4.568 Ga are a result of either subsequent metamorphic disturbance or terrestrial contamination.

In addition to questions about the decay constant, there were also questions about the ^{167}Lu/^{177}Hf ratio and the present and initial ^{176}Hf/^{177}Hf ratio of chondrites. Part of the problem is that there is a 28% variation in the ^{167}Lu/^{177}Hf ratio in chondrites (compared to only 3% variation in the Sm/Nd ratio). Bouvier et al. (2008) found that when only the least thermally metamorphosed chondrites are considered, the scatter in ^{167}Lu/^{177}Hf ratio reduces to only 3%, which is comparable to that observed for Sm/Nd. They argued the problem likely relates to the presence of phosphate phases in meteorites. These can have very high ^{167}Lu/^{177}Hf ratios and can be easily mobilized and recrystallized during thermal metamorphism. Using only the least thermally metamorphosed chondrites, Bouvier et al. (2008) calculate a mean ^{176}Lu/^{177}Hf ratio of 0.0336 ± 1 and a mean ^{176}Hf/^{177}Hf = 0.282785 ± 11, corresponding to an initial ^{176}Hf/^{177}Hf = 0.279794 ± 0.000023. However, the question of the initial ^{176}Hf/^{177}Hf of the Solar System persists. For example, Bizzarro et al. (2012) reported a ^{176}Lu–^{176}Hf internal isochron age of 4869 ± 34 Ma for a pristine achondritic meteorite (the angrite SAH99555), which had been precisely dated by U–Pb to 4564.58 ± 0.14 Ma. Thus, the Lu–Hf age is roughly 300 Ma, which is too old.

Subsequently, Iizuka et al. (2015) analyzed measured ^{176}Hf/^{177}Hf in a zircon from the achondrite *Agoult* (a eucrite, which, as we will find in Chapter 6, comes from the asteroid *4 Vesta*. The very low Lu/Hf in this zircon meant that there was very little radiogenic ingrowth of ^{176}Hf, allowing for a precise determination of the initial ratio as 0.279781 ± 0.000018, well within error of Bouvier, et al. (2008)'s value, but a slightly higher ^{176}Lu/^{177}Hf ratio of 0.0338 ± 0.0001. Bast et al. (2017) reported a similar initial ^{176}Hf/^{177}Hf, 0.279796 ± 0.000011, for the achondrite *Almahata Sitta* (an unusual uralite that fell in Sudan in 2008), within error of the value of Iizuka et al. (2015). Thus, the issues of decay constant and initial Solar System value for the Lu–Hf system now seem to be resolved and the values of Iizuka et al. (2015) are taken as the chondritic $^{176}Hf/^{177}Hf$ and $^{176}Lu/^{177}Hf$ ratios for calculation of ε_{Hf}.

The Lu–Hf system has several advantages, in principle at least, over the Sm–Nd system. First, because the half-life of ^{176}Lu is shorter than that of ^{147}Sm (37 versus 106 Ga) and the range of Lu/Hf ratios in common rocks and minerals is greater than that of Sm/Nd, the variations in $^{176}Hf/^{177}Hf$ and ε_{Hf} are larger than that of $^{143}Nd/^{144}Nd$ and ε_{Nd}. Second, because of the chemical similarity of Hf to Zr, Hf is concentrated in zircon, a very robust mineral that also concentrates U and can be dated using the U–Pb system.

The general similarity between the Lu–Hf system and the Sm–Nd system is demonstrated by Figure 2.21, which shows that ε_{Hf} and ε_{Nd} are well correlated in crustal rocks of all ages. It also shows that the variations in ε_{Hf} are about half again as large as those of ε_{Nd}. The correlation holds for "terrigenous sediments," which are marine continent-derived sediments, but breaks down for "hydrogenous" sediments that contain a significant component derived

from seawater, which defines a shallower "seawater array." This results from two effects. The first is the "zircon effect" (White et al., 1986; Patchett et al., 2004). When continental rocks are weathered, the rare earths, including Sm, Nd, and Lu, concentrate in clays, but a significant fraction of unradiogenic Hf remains concentrated in zircons, which resist both chemical and mechanical weathering. The zircons tend to remain in coarser-grained sediments on continental shelves; thus, the Hf flux to the oceans from the continents is more radiogenic than the continents themselves. Secondly, Hf is extremely insoluble in stream and ocean water, but less so in hydrothermal solutions where it forms soluble complexes with fluorine (Bau and Koschinsky, 2006). As a result, the flux of radiogenic Hf to seawater from the oceanic crust is larger than the hydrothermal Nd flux (White et al., 1986). One consequence of the different behaviors of Hf and Nd during weathering is that there would be little point to calculating Lu–Hf crustal residence times analogous to Sm–Nd residence times from sedimentary rocks. However, for igneous and meta-igneous rocks, Lu–Hf model ages, both τ_{CHUR} and τ_{DM}, can be calculated in a manner exactly analogous to Sm–Nd

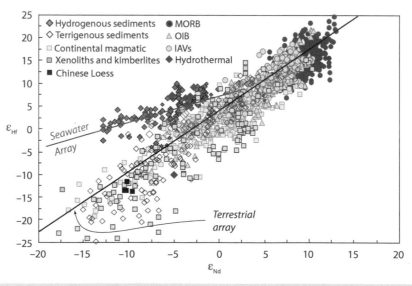

Figure 2.21 ε_{Hf} and ε_{Nd} in the Earth. The ε_{Hf} and ε_{Nd} in magmatic rocks of the continental and oceanic crust define the "terrestrial array" ($\varepsilon_{Hf} = 1.21 + 1.55\ \varepsilon_{Nd}$; Vervoort and Blichert-Toft, 1999; Vervoort et al., 2011). Terrigenous sediments also lie on this array. Hydrothermal and hydrogenous sediments, including Mn–Fe crusts and nodules, define a shallower array, which is called the *seawater array* by Albarède et al. (1998), described by equation $\varepsilon_{Hf} = 7.1 + 0.55\ \varepsilon_{Nd}$ (Vervoort et al., 2011). Data from the database kindly provided by J. D. Vervoort.

model ages. For τ_{DM}, the relevant depleted mantle evolution parameters are present, $^{176}Hf/^{177}Hf = 0.282320$ ($\varepsilon_{Hf} = 14.3$) and $^{176}Lu/^{177}Hf = 0.0389$.

The tendency for Hf to concentrate in zircon and that of Lu to be excluded from it can be exploited to help us understand continental evolution. Zircons, as we will find in the next chapter, also concentrate in U and exclude Pb, and hence their ages can be determined with high precision. The high Hf concentration in zircon allows for precise measurement of $^{176}Hf/^{177}Hf$ ratios. With their very low Lu/Hf ratios, only a small correction is required to calculate initial $^{176}Hf/^{177}Hf$ ratios using the U–Pb age. This can be applied not only to igneous rocks, but also from detrital zircons found in coarser-grained ancient sedimentary rocks. Since the mantle evolves toward negative ε_{Hf} and continental rocks toward positive ε_{Hf}, the initial ε_{Hf} can help us distinguish new additions to continental crust from internally recycled continental material, just as ε_{Nd} values do. We will return to these topics in Chapter 8.

One important difference between Lu–Hf and Sm–Nd is that whereas Sm/Nd ratios usually do not change much in the weathering of a crystalline rock to form a sediment, Lu/Hf ratios do. In both cases, the elements are reasonably insoluble, and little is carried away by solution. Most of the rare earths end up in clays, but much of the Hf in felsic crystalline rocks of the continental crust is in zircon ($ZrSiO_4$), which, as we have already noted, is very resistant to both chemical and mechanical weathering. The clays are, of course, quite fine and can be carried great distances from their source. Zircon remains in the coarse (and hence less mobile) sand fraction. As a result, there are large differences between Lu/Hf ratios in fine and coarse sediments.

The Lu–Hf system has become an increasingly important geochronological tool, particularly for silicic igneous and metamorphic rocks for several reasons. First is the difference in partitioning behavior of Lu and Hf, where Hf partitions strongly into some minerals, such as zircon, while Lu partitions strongly into others, such as garnet. Second, both elements are relatively immobile and relatively insensitive to weathering effects. Third, the Lu–Hf system appears to have a comparatively high closure temperature.

A recent study by Johnson et al. (2018) of orthogneisses of the Blue Ridge Province of Virginia (in the US) illustrated how the use of the coupled Lu–Hf and Sm–Nd chronometers can constrain the conditions of metamorphism during orogenesis. The basement rocks in this region consist of metamorphosed granites (orthogneisses) emplaced between ≈ 1.2 and 1.0 Ga in the Grenville Orogeny that resulted from the collision of Amazonia with the southeastern margin of Laurentia during the assembly of the supercontinent Rodinia. U–Pb ages of zircon cores, interpreted as igneous crystallization ages, range from 1199 ± 19 to 1140 ± 22 Ma; zircon rims are typically 100–200 Ma younger. Johnson et al. (2018) analyzed Lu–Hf and Sm–Nd in garnet separates and WRs in four of these rocks. Lu–Hf ages ranged from 1019 ± 3 to 1032 ± 3 Ma while Sm–Nd ages are 69–79 Ma younger, ranging from 929 ± 2 to 962 ± 6 Ma. All eight ages had MSWD values less than 1; hence, the difference is unlikely to be an analytical artifact. Figure 2.22 illustrates the isochrons obtained from one of these samples. After ruling out several other possibilities, Johnson et al. (2018) concluded the age differences resulted from higher closure temperatures of the Lu–Hf system than the Sm–Nd system. As we found in Section 2.3.1, closure temperatures are complex functions of grain size and cooling rate in addition to diffusion coefficients. In this study, the same grains were analyzed for both systems, so the difference would reflect only diffusion rates. Higher closure temperatures for the Lu–Hf system is consistent with a variety of early studies that have indicated that Lu–Hf closure temperatures are as much as 100°C higher than those of the Sm–Nd system. The average 75 Ma difference between the Lu–Hf and Sm–Nd ages suggests quite slow cooling of ≤2°C/Ma following peak metamorphic conditions. That in turn is consistent with some tectonic reconstructions of the Grenville orogeny.

Blueschist-facies metamorphic rocks provide another example of the utility of Lu–Hf geochronology. These rocks form in the high-pressure, low-temperature regime of subduction zones; as a result of the latter, they have proven difficult to date by methods other than K–Ar. Among their characteristic minerals are glaucophane ($Na_2Mg_3Al_2Si_8O_{22}(OH)_2$), which gives blueschist-facies rocks their characteristic blue color, and lawsonite,

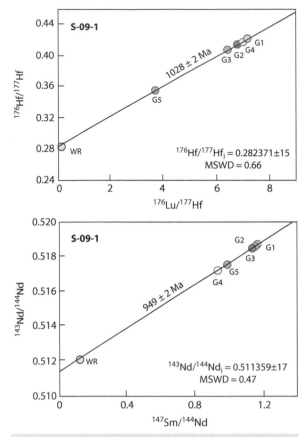

Figure 2.22 Lu–Hf and Sm–Nd isochrons from orthogneiss sample S-09-1 of the Blue Ridge Province of Virginia formed during the Grenville orogeny. Analyses were performed on the same WR and five garnet separates (G1 through G5). Both systems yield excellent isochrons with MSWD < 1, but the Sm–Nd age is 79 Ma younger. Three other analyzed orthogneisses yielded similar Lu–Hf and Sm–Nd ages. The difference likely reflects the lower closure temperature of the Sm–Nd system and slow cooling from peak metamorphic conditions in the Grenville orogeny. Source: Johnson et al. (2018)/ With permission of Elsevier.

$(CaAl_2Si_2O_7(OH)_2 \cdot H_2O)$. The latter can, in some cases, incorporate heavy rare earths, such as Lu, while glaucophane preferentially incorporates Hf. The ages reflect closure of these to diffusion; the closure temperatures are not known but are probably high relative to the conditions typical of blueschist-facies metamorphism. The Franciscan Formation, which outcrops along hundreds of miles of coastal California (including in San Francisco, whence its name), is a classic subduction complex that

includes blueschist-facies metamorphic rocks. Mulcahy et al. (2009) analyzed lawsonite and glaucophane separates and WR samples from the Ring Mountain on the Tiburon Peninsula (which projects into San Francisco Bay) and obtained the result shown in Figure 2.23. In this case, the MSWD is rather high, indicating the misfit to the isochron is greater than expected from estimated analytical errors alone. This may be a consequence of open-system behavior – the lawsonite formed under retrograde conditions while the rocks were being exhumed – or, in this case, inclusion of Hf-rich phases within the separates, such as titanite. The age nevertheless agrees well with other ages from the Tiburon Peninsula, which range from 157 to 141 Ma.

2.6 THE Re–Os–Pt SYSTEM

After early efforts by Hirt et al. (1963), the Re–Os system was largely ignored due to the analytical challenges it presented. The problems are two-fold: (1) Os is an extremely rare element, rarely present at concentrations above a part per billion (and often much lower), and (2) Os metal is extremely refractory, evaporating and ionizing only at extremely high temperatures. This has been overcome through analysis of the negative ion of OsO_3^-, which, in contrast to the metal, evaporates and ionizes at quite low temperature. This technique has proved to be extremely sensitive, making it possible to determine Os isotope ratios on extremely small amounts of Os (Creaser et al., 1991). Subsequently MC-ICP-MS has provided an additional technique. As a result, the Re–Os system has become a useful geochronological tool over the past two decades, although the applications are somewhat limited.

The decay systems we have discussed up to this point have involved *lithophile* (derived from Greek words for "rock" and "love") elements (the exception is Ar, which is an atmophile element). Lithophile means simply that, given the choice, the element partitions preferentially into a silicate or oxide phase (in fact a better term would be oxyphile) over a sulfide or metal phase. *Chalcophile* elements partition preferentially into the sulfide phases and *siderophile* elements partition into metal phases given the opportunity (*atmophile* elements, as

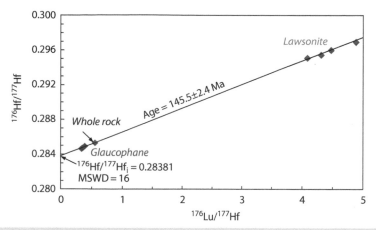

Figure 2.23 Lu–Hf age for a blueschist from the Franciscan of California. The high MSWD indicates the isochron fits the data poorly and suggests either open-system behavior or the effect of zircon inclusions. Source: Adapted from Mulcahy et al. (2009).

their name implies, are elements preferring a gas phase). Re and Os are both *siderophile* elements, though both, Re in particular, also have some *chalcophile* tendencies. Their siderophile nature accounts for their low concentrations in the crust and mantle: most of the Earth's Re and Os inventory is in the core. Os is one of the platinoid, or platinum-group, elements (the second and third transition series members of Group VIII elements) that include Ru, Rh, Pd, and Ir, as well as Os and Pt. Like the rare earths, the platinum-group elements (PGEs) behave coherently, although their valences and ionic radii differ. The equilibrium valence state of Os in equilibrium with the Earth's atmosphere is +4; its ionic radius is 0.69. The valence of Re is also +4 and its ionic radius is 0.63. However, in the mantle and magmas, these elements are likely in +1 or 0 valence states. Re is a moderately incompatible element whereas Os is highly compatible one: it partitions into a silicate melt only very sparingly. Hence, the crust has a much lower Os concentration than the mantle (the core, of course, should have a higher concentration than both). Together with Re and Au and several other metals, these elements are also referred to as *highly siderophile metals* or HSMs.

The older convention for Os isotope ratios, established by Hirt et al. (1963), reported the isotope ratio as $^{187}Os/^{186}Os$ (normalized for fractionation to $^{192}Os/^{188}Os$ of 3.08271). The difficulty with this normalization is that ^{186}Os is itself radiogenic, being the product of α-decay of ^{190}Pt. ^{109}Pt is sufficiently rare

and its half-life is sufficiently long (450 billion years) that in most cases the amount of radiogenic ^{186}Os is insignificant. However, measurable amounts of radiogenic ^{186}Os have been observed, as discussed below. This discovery prompted a shift in the convention and now all laboratories report Os isotope analyses as $^{187}Os/^{188}Os$. $^{187}Os/^{186}Os$ ratios may be converted to $^{187}Os/^{188}Os$ ratios by multiplying by 0.12035.

Figure 2.24 illustrates the evolution of Os isotope ratios in the crust and mantle. As expected from the difference in compatibilities of Re and Os, much higher $^{187}Os/^{188}Os$ ratios are found in the crust than those in the mantle. Puecker-Ehrenbrink and Jahn (2001) estimate the average $^{187}Os/^{188}Os$ of continental crust to be 1.4, implying a $^{187}Re/^{186}Os$ ratio of 46, about two orders of magnitude greater than the estimated primitive mantle ratio of 0.42. The average $^{187}Os/^{188}Os$ of the depleted upper mantle (DMM) is 0.1247 ± 0.0075 based on abyssal peridotites (Lassiter et al., 2014).

Unlike the rare-earth decay systems we discussed above, chondritic meteorites have variable Re/Os ratios that have resulted in an approximately 10% variation in their $^{187}Os/^{188}Os$ ratios, ranging from ≈0.120 to 0.132 with a poorly defined mean value of ≈0.127, reflecting quite variable Re/Os ratios (e.g., Horan et al. (2003) found that Re/Os ratios are 8% lower in ordinary and enstatite chondrites than carbonaceous ones). Interestingly, the $^{187}Os/^{188}Os$ ratio of the mantle, as well as by implication presumably the Re/Os

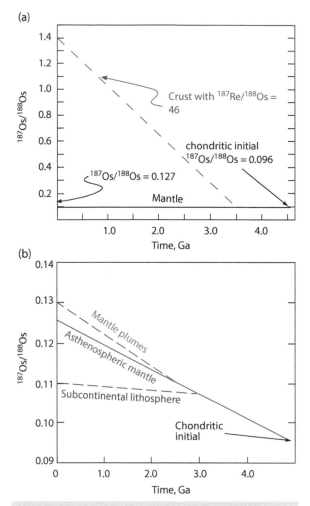

Figure 2.24 (a) Schematic evolution of Os isotope ratios in the mantle and crust. (b) $^{187}Os/^{188}Os$ evolution in the mantle (note the difference in scale). The mantle root of continents (lithospheric mantle) appears to have been particularly strongly depleted in Re by melt extraction.

ratio of the silicate Earth, falls within this range (Allègre and Luck, 1980), but unlike the Sm–Nd and Lu–Hf systems, there is no theoretical expectation that it should be. Indeed, this is a bit surprising if most of Earth's Re and Os have been extracted to the core. If the core and mantle are in equilibrium, then mantle concentrations will be determined by metal–silicate partition coefficients, which are large for both elements. Since the partition coefficients of these two elements are different, the ratio of the two in the mantle should be quite different from chondritic. The same is true of other PGEs: their relative mantle abundances are

close to chondritic even though their concentrations are two orders of magnitude lower than chondritic. The explanation proposed by Chou (1978) is that an additional 1% or less of chondritic material, a "late accretionary veneer," was added to the Earth after core formation had completed. In this model, the highly siderophile elements, including Re and Os, are nearly quantitatively extracted from the silicate Earth by core formation. The inventory of highly siderophile elements in the crust and mantle comes from that last percent of material accreted to the Earth.

2.6.1 Re depletion ages and Re–Os model ages

Since the silicate Earth appears to have a near-chondritic $^{187}Os/^{188}Os$ ratio, it is useful to define a parameter analogous to ε_{Nd} and ε_{Hf}. Walker et al. (1989) defined γ_{Os} as:

$$\gamma_{Os} = \frac{\left(^{187}Os/^{188}Os\right)_{sample} - \left(^{187}Os/^{188}Os\right)_{PUM}}{\left(^{187}Os/^{188}Os\right)_{PUM}} \times 100$$

(2.66)

where PUM denotes *primitive upper mantle* whose value is taken as 0.1296 (Meisel et al., 2001) corresponding to a $^{187}Re/^{188}Os$ of 0.4353 assuming an initial $^{187}Os/^{188}Os$ of 0.09517 deduced from iron meteorites. Assuming the Earth was initially isotopically homogenous, a default assumption in the absence of evidence to the contrary, this is the same as bulk silicate Earth. Thus, the gamma parameter is analogous, although not identical to the epsilon one, but where the latter denotes deviations in parts per 10 000 and the former denotes percentage deviations.

Since the mantle $^{187}Os/^{188}Os$ evolution curve is known to a first approximation, an estimate of age, or model age, analogous to Sm–Nd model ages, for Re-poor samples can be obtained simply by comparing the measured $^{187}Os/^{188}Os$ ratio with the mantle evolution curve. The association of platinoid metal deposits with mantle-derived ultramafic rocks would be one example of where such model ages can be obtained. The PGMs occur as very fine (down to a micrometer or so) metal alloys and sulfides. Os occurs in these principally as osmiridium (OsIr) and laurite ($Ru[Os,Ir]S_2$). These minerals have Re/Os close to 0. As a result, the $^{187}Os/^{188}Os$ ratio ceases to change once these minerals form. Assuming a Re/Os ratio of 0, we can define a *rhenium depletion age* as:

$$T_{RD} = 1/\lambda_{187}$$

$$\times \ln\left\{ \frac{\left(^{187}Os/^{188}Os\right)_{PUM} - \left(^{187}Os/^{188}Os\right)_{sample}}{\left(^{187}Re/^{188}Os\right)_{PUM}} + 1 \right\}$$

$$(2.67)$$

As we will find in subsequent chapters, peridotite xenoliths from the subcontinental lithosphere also have very low Re/Os ratios as a consequence of earlier melt removal. In many cases, what little Re is present in these xenoliths appears to have been introduced during eruption of the host. In that case, the $^{187}Os/^{188}Os$ is corrected for ^{187}Re decay to the eruption time and Equation (2.58) becomes:

$$T_{RD} = 1/\lambda_{187}$$

$$\times \ln\left\{ \frac{\left(^{187}Os/^{188}Os\right)_{PUM} - \left(^{187}Os/^{188}Os\right)_{sampleEA}}{\left(^{187}Re/^{188}Os\right)_{PUM}} + 1 \right\}$$

$$(2.67a)$$

where sampleEA denotes the ratio at the eruption age. We can also define a model age analogous to Sm–Nd and Lu–Hf model ages as:

$$T_{MA} = 1/\lambda_{187}$$

$$\times \ln\left\{ \frac{\left(^{187}Os/^{188}Os\right)_{chon} - \left(^{187}Os/^{188}Os\right)_{sample}}{\left(^{187}Re/^{188}Os\right)_{chon} - \left(^{187}Re/^{188}Os\right)_{sample}} + 1 \right\}$$

$$(2.68)$$

While Sm–Nd and Lu–Hf model ages have been useful in understanding continental evolution, rhenium depletion and Re–Os model ages have proved useful in understanding the evolution of the mantle lithosphere underlying continents.

The evolution of the Os isotope composition of seawater has also proved quite informative. The $^{187}Os/^{188}Os$ ratio of modern seawater is about 1.07. Like $^{87}Sr/^{86}Sr$, $^{187}Os/^{188}Os$ depends on the balance of continental fluxes (e.g., rivers, with $^{187}Os/^{188}Os \approx 1.4$) and oceanic crustal fluxes (e.g., hydrothermal activity, with $^{187}Os/^{188}Os \approx 0.13$). In addition,

however, cosmic fluxes ($^{187}Os/^{188}Os \approx 0.13$), which include both cosmic dust, which continually settles through the atmosphere into the oceans, and large meteorite impacts, may be significant for Os. Variations in the proportions of these fluxes have resulted in systematic changes in the $^{187}Os/^{188}Os$ ratio through time. We will return to this topic and discuss these changes in Chapter 11.

2.6.2 Re–Os geochronology

Given the rarity of these elements in the mantle and crust, geochronological applications are limited, but in specialized circumstances, such as iron meteorites, sulfide ore deposits, hydrocarbons, and some ultramafic rocks, such as komatiites, Re–Os geochronology has proved quite valuable. Because of the differences in compatibility, Re/Os variations are huge, at least by comparison to the other systems we have considered, and consequently accurate ages can be calculated even though analytical precision is often poorer than that for other systems.

One of the earliest applications of Re–Os geochronology was its use in determining the ages of iron meteorites. Indeed, since other radioactive elements are lithophile, this is the only tool available (however, some ages have been determined on silicate inclusions in iron meteorites). Figure 2.25 shows an example of an Re–Os isochron derived from 14 Group IVA iron

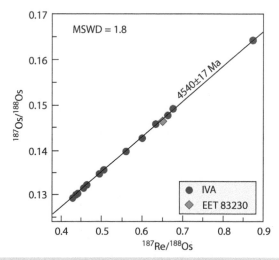

Figure 2.25 Re–Os isochron on IVA iron meteorites. Source: McCoy et al. (2011)/With permission of Elsevier.

meteorites and the Antarctic meteorite Elephant Moraine (EET) 83230, which, while ungrouped, is similar to IVA irons. Group IVA meteorites are "magmatic," meaning that they formed as the cores of asteroidal bodies. IVA irons appear to have cooled particularly rapidly, suggesting their parent body had been stripped off its silicate mantle by an impact. The age in this case indicates the parent body formed and cooled within 28 million years of the formation of the Solar System at 4567 Ma. As we will find in Chapter 6, chronology based on the short-lived ^{182}Hf–^{182}W decay system indicates most irons, including those of Group IVA, formed much more rapidly, within a few million years of the start of the Solar System.

2.6.2.1 Re-Os dating of diamonds

One novel application of the Re–Os system is its use in dating diamonds by analyzing sulfide inclusions within them. Since Re and Os are somewhat chalcophile, they can be concentrated in mantle sulfide minerals, such as pyrrhotite (FeS), chalcopyrite ($CuFeS_2$), and pentlandite (($Fe,Ni)_9S_8$). These minerals are sometimes found as small inclusions in diamond (Figure 2.26). Encapsulated in diamond, individual inclusions become closed systems and accumulate ^{187}Os in proportion to the amount of ^{187}Re they contain. Figure 2.27

shows a 2.9 Ga Re–Os isochron derived from inclusions in diamonds from the Kimberley Mine in South Africa. Studies, such as these, have shown that diamond formation in the mantle occurs in discrete events, perhaps related to subduction (Shirey et al., 2004).

2.6.2.2 Re-Os dating of ore deposits

Rhenium has strong chemical similarities to molybdenum and is concentrated in molybdenite (MoS_2), a mineral common in some types of sulfide ore deposits, while Os is excluded. Indeed, initial or "common" Os concentrations can be so low that they are below detection limits while Re concentrations can be hundreds of parts per million. Furthermore, molybdenite resists chemical exchange with other minerals and closure temperatures appear to be high. Porphyry copper deposits are the world's primary source of copper. The sulfide ores precipitate in veins from magmatic hydrothermal fluids associated with subduction-related magmatism. Typically in this type of deposit, a series of hydrothermal veins occur, which are labeled A through D in presumed chronological order with temperature decreasing from $\gtrsim 600°C$ for A-veins to $350°$–$250°C$ for the D-veins with the main mineralization occurring in association with B-veins at $\approx 350°$–$550°C$.

Figure 2.26 A sulfide inclusion in diamond. Fractures result from the fact that the sulfide expands more than the diamond as it is decompressed during ascent in the kimberlite eruption. Source: Reproduced with permission from J. W. Harris.

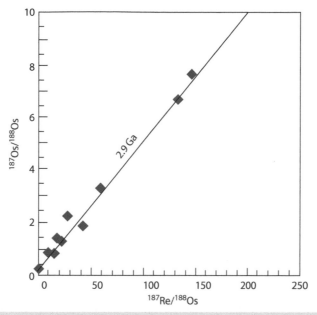

Figure 2.27 The 2.9 Ga Re–Os isochron from high-Os sulfide inclusions in diamonds from the Kimberley Mine, South Africa. Source: Adapted from Richardson et al. (2001).

The Los Pelambres porphyry copper deposit in the Chilean Andes is one of the largest porphyry copper deposits in the world with estimated reserves of 4.9 billion tons of copper ore. Stein (2014) determined Re–Os model ages on molybdenites in individual veins in this deposit. He found that many B-veins, which ranged in age from 10.56 to 10.95 Ma, were actually older than A-veins, which had ages of 10.63 and 10.64 Ma. Furthermore, molybdenites in many of the D-veins, which ranged in age from 11.94 to 10.15 Ma, were older than dated A- and B-veins. The likely explanation is that ore deposition occurred in multiple pulses with temperatures repeatedly cycling from high to low.

The Haigou gold deposit in northeastern China provides another example of the utility of Re–Os geochronology in resolving the timing of ore deposition. The gold occurs in quartz veins, which also contain pyrite and molybdenites, and these veins are associated with Carboniferous and Cretaceous granites that intrude Proterozoic schists. Zhai et al. (2019) determined U–Pb zircon ages of the veins of 318, 311, and 134 Ma. Re–Os model ages of 19 individual molybdenite samples, however, ranged from 467 to 155 Ma. This large range was defined only by some coarse-grained molybdenites, which Zhai et al. (2019)

attributed to Re spatial heterogeneity and small sample sizes. Nine fine-grained samples, however, defined a 309 ± 8 Ma isochron (Figure 2.28), indicating mineralization was associated exclusively with cooling of the Carboniferous granitic stock, which was dated at 323 and 320 Ma. In this case, the isochron is based only on ^{187}Re and ^{187}Os concentrations rather than isotope ratios as ^{188}Os concentrations are so low that poor precision in their measurement would considerably increase scatter.

Rhenium can be concentrated in other sulfides, so Re/Os dating is not restricted to molybdenites. One such sulfide is carrolite ($CuCo_2S_4$), which is an important ore of cobalt, a metal critical to modern technologies, such as Li-ion batteries. The Katanga Copperbelt, which straddles several central African countries, is the largest known stratiform ore deposit in the world and supplies over half the world's cobalt. The ores were deposited by saline hydrothermal fluids and are hosted by an ≈880–727 Ma sedimentary sequence that includes siliciclastic and carbonate rocks, as well as four thick evaporite layers, and is overlain by Cryogenian glacial diamictites. These were subsequently deformed during the Pan-African Lufilian orogeny around 600–500 Ma. There has been debate for nearly

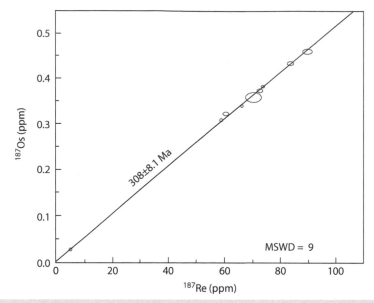

Figure 2.28 Re–Os isochron of fine-grained molybdenites from auriferous quartz veins of the Haigou gold deposit. Ellipses represent analytical uncertainties. ^{187}Re and ^{187}Os concentrations are used because ^{188}Os concentrations were too low to measure reliably. Source: Adapted from Zhai et al. (2019).

a century as to when the ore was deposited, with proposals ranging from syn- and post-depositional to post-orogenic. Saintilan et al. (2018) used Re–Os dating of carrolites to demonstrate that the first episode of ore formation was the precipitation of carrolite (CuCo$_2$S$_4$) replacing evaporite minerals in breccias and occurred at 609 ± 5 Ma (Figure 2.29). The main episode of Co–Cu mineralization occurred between 540 and 490 Ma during compressional tectonics when the collision between the Congo and Kalahari cratons drove fluid flow. Initial ^{187}Os/^{188}Os ratios > 3 indicate a crustal source for the metals, inferred to be Mesoproterozoic basement. A final late stage of mineralization in which bornite (Cu$_5$FeS$_4$) can be seen replacing carrolite occurred around 473 ± 3 Ma based on Re–Os dating of bornite. The bornite has lower ^{187}Os/^{188}Os of 0.4, indicative of a juvenile igneous source, likely the adjacent Hook Batholith that would have also provided heat to drive fluid flow.

2.6.2.3 Re–Os dating of hydrocarbons

There is, perhaps, no geologic material that has so transformed society since metals replaced stone as hydrocarbons, in particular petroleum. Hydrocarbons consist almost

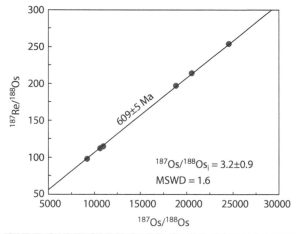

Figure 2.29 Re–Os isochron for carrolites replacing evaporite minerals in the Katanga copper belt dates the first episode of ore deposition. The initial ^{187}Os/^{188}Os indicates a crustal source for the ore-forming fluids. Data from Santilian et al. (2018).

exclusively of C and H, making them a challenge to date. Economic petroleum deposits can form if a series of conditions are met: an environment in which relatively high concentrations of organic matter are preserved in sediment as kerogen during deposition of the *source rock*; subsequent burial and conversion of the kerogen to hydrocarbons through gentle heating; and migration and concentration of the petroleum thus produced in a sufficiently porous *reservoir rock*. Through a variety of methods (some of which involve isotopic analysis), it is often possible to identify the source rock, and standard geologic correlation techniques can determine the age of source rock deposition. It has been much harder to determine the age of other steps in the process, particularly the timing of generation and migration. Researchers have recently had some success in dating hydrocarbon migration using Re–Os.

Perhaps surprisingly, noble metals, and Re and Os in particular, can be present in *relatively* high concentrations in petroleum. Work by Selby et al. (2007) found that Re and Os are primarily present in *asphaltene* fraction of petroleum. Asphaltenes are heavy (with molecular masses of around 750 u) polycyclic aromatic hydrocarbons. Although detailed studies have not been conducted, it seems likely that Re and Os are bound in porphyrins in the asphaltene fraction. Porphyrins are rings of pyrroline and pyrrole groups (five-sided hydrocarbon rings containing N or NH) that typically contain a metal ion in the center of the ring. These organometallic complexes make up part of the chlorophyll porphyrin – where Mg occupies the metal site – and the hemoglobin molecule – where Fe occupies this site. During diagenesis, the original metal is often replaced by a transition metal and consequently porphyrins in hydrocarbons can have high concentrations (up to a part per million or so) of transition metals, such as Ni, V, and Mo. Re concentrations in petroleum can be as high as 50 ppb, and it is more strongly concentrated than Os whose concentration reaches only 300 ppt. Consequently, $^{187}Rb/^{188}Os$ ratios can exceed 1000, and that in turn results in high $^{187}Os/^{188}Os$ ratios.

Selby and Creaser (2005) used Re–Os to date petroleum of the Western Canada Sedimentary Basin (WCSB). As may be seen in Figure 2.30, *apparent* isochrons yield ages of 111–121 Ma, depending on which samples are included. We say "apparent" isochrons because the MSWD (see Section 2.2.2) is greater than 1, indicating that not all deviations from regression can be accounted for by analytical error. Selby and Creaser (2005) argue that, given the regional scale involved, this is more likely due to incomplete homogenization of the initial isotope ratio than due to subsequent disturbances of Re/Os ratios. Precisely what event is being dated here is, however, somewhat unclear. Hydrocarbons in the WCSB are thought to have been derived from source rocks on the western side of the basin and then migrated east. Selby and Creaser (2005) noted the high initial $^{187}Os/^{188}Os$ compared to $^{187}Os/^{188}Os$ in Mesozoic seawater suggests that the source rocks are of Paleozoic age.

Whereas the ages calculated by Selby and Creaser (2005) were based on petroleum samples taken over a wide region, subsequent studies have obtained Re–Os isochrons by separating the various fractions from single-petroleum samples. For example, Georgiev et al. (2016) analyzed Re and Os in asphaltene and maltene fractions from three oils from the Gela oil field of southern Sicily. The oils occur in the Triassic–Jurassic Streppenosa, Noto, and Sciacca Formations. They obtained an age of 27.5 ± 4.6 (MSWD = 1.6) on asphaltene fractions separated from Noto and Sciacca Formations oils. A much older age of 200 ± 5.2 Ma (MSWD = 0.53) was obtained from maltene fractions separated from Streppenosa oil. The data thus provide evidence of multiple episodes of petroleum generation. Georgiev et al. (2016) speculate that the Streppenosa petroleum, which is heavily biodegraded and somewhat thermally immature, was generated shortly after Late Triassic deposition of the source rock by magmatic heating. The much younger ages of fully thermally mature oils from the underlying Noto and Sciacca Formations indicate that they were generated at the onset of regional collisional tectonics in southern Sicily.

2.6.3   The ^{190}Pt–^{186}Os decay system

As we noted above, ^{186}Os is the decay product of ^{190}Pt. Chondrites have variable Pt/Os ratios and consequently variable $^{186}Os/^{188}Os$ ratios. $^{186}Os/^{188}Os$ is available only for H ordinary and some enstatite chondrites, which have a mean $^{186}Os/^{188}Os$ ratio of 0.1198398 ± 16 (Brandon et al., 2006). This is the assumed Solar System value. $^{186}Os/^{188}Os$ in most terrestrial materials varies little from this value. Day et al. (2017) estimated the $^{186}Os/^{188}Os$ of primitive upper mantle to be 0.1198388 ± 29, which was indistinguishable from the chondritic value. They estimated the $^{186}Os/^{188}Os$ of the DMM to be 0.1198356 ±

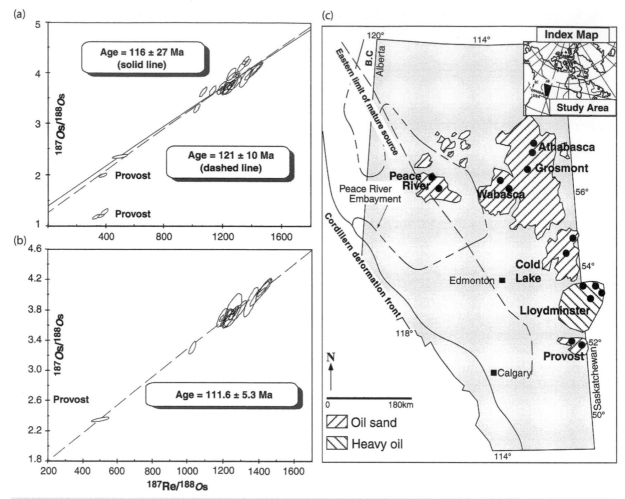

Figure 2.30 (a) Re–Os isochron diagram for 24 samples of oil from the WCSB. The slope when all data are included corresponds to an age of 116 ± 27 Ma. A slightly older, but more precise, age is obtained when the Prevost data are excluded. (b) Including just those samples whose calculated $^{187}Os/^{188}Os$ at 110 Ma is 1.4–1.5 yields a yet more precise age of 111.6 ± 5 Ma. (c) Location map for samples. Source: Adapted from Selby and Creaser (2005). Reproduced with permission of AAAS.

21, which was less than 30 ppm lower than the primitive upper mantle value. This lack of variation reflects the long half-life of ^{190}Pt, low $^{190}Pt/^{188}Os$ ratios (≈ 0.002), and limited variability in Pt/Os ratios in most materials.

Platinum, however, is strongly siderophile and can be concentrated in iron and copper sulfides while Os, which is only weakly chalcophile, is largely excluded from them. Walker et al. (1991) measured a $^{186}Os/^{188}Os$ ratio of greater than 0.3 in sulfides of the Sudbury (Ontario) ore deposit. Subsequent studies have found significant variations in $^{186}Os/^{188}Os$ in other sulfide ore deposits and in chromitites and ultramafic rocks. Large variations in Pt/Os ratios occur in

PGE placers derived from an ophiolite in Borneo; Re/Os ratios showed only limited variation in them. Coggon et al. (2011) were able to calculate a ^{190}Pt–^{186}Os isochron age of 197.8 ± 8.1 Ma (MSWD = 0.88). No other radiometric ages for the ophiolite are available for comparison, but the age is consistent with plate reconstructions. Coggon et al. (2012) were also able to construct a 1995 ± 50 Ma (MSWD = 1.16) ^{190}Pt–^{186}Os isochron for the Bushveld Complex, which is the world's largest PGE ore deposit. The age is slightly younger than the most precise U–Pb zircon age of 2054 ± 1.3 Ma, and Coggon et al. (2012) speculate it reflects late, low-temperature hydrothermal event.

Although most mantle materials have $^{186}Os/^{188}Os$ ratios indistinguishable from the chondritic value, Brandon et al. (1998) found $^{186}Os/^{188}Os$ ratios ranging from 0.119841 to 0.119853 in Hawaiian basalts, which were 20–120 ppm higher than the primitive mantle value. Brandon et al. (2003) found $^{186}Os/^{188}Os$ as high as ≈ 70 ppm above the PUM value in Cretaceous komatiites from Gorgona Island, Colombia. The Gorgona komatiites were produced in an early stage of the Galapagos mantle plume. These observations prompted speculation that these mantle plumes might contain a small component derived from the Earth's outer core. Brandon et al. (2003) argued that as the inner core crystallizes, Os would partition into the solid metal in preference to Pt, leading to high Pt/Os ratios in the outer core. This, however, requires a quite old age for the inner core to generate the high $^{186}Os/^{188}Os$ ratios. Furthermore, elevated $^{186}Os/^{188}Os$ ratios have not been found

in other mantle-plume-derived magmas and the $^{86}Os/^{188}Os$ is not correlated with PGE abundances as might be expected. Luguet et al. (2008) observed high Pt/Os ratios in pyroxenites and in sulfide pyroxenites, and they suggested that "involvement in the mantle source of either bulk pyroxenite or, more likely, metasomatic sulfides derived from either pyroxenite or peridotite melts can explain the ^{186}Os-^{187}Os signatures of oceanic basalts." Day et al. (2019) observed high $^{186}Os/^{188}Os$ in large layered intrusions, including the Archean Stillwater Complex in Montana, the Proterozoic Muskox intrusion of Canada, and the Cenozoic Rum Complex of Scotland, which they concluded as being resulted from assimilation of surrounding continental crust. They argued that recycled continental crust in some mantle plumes could explain the elevated $^{186}Os/^{188}Os$ observed. This debate continues and there is no consensus as to the cause of these high $^{186}Os/^{188}Os$ in Hawaii and Gorgona.

REFERENCES

Albarède, F., Simonetti, A., Vervoort, J. D., et al. 1998. A Hf-Nd isotopic correlation in ferromanganese nodules. *Geophysical Research Letters* **25**: 3895–8, doi: 10.1029/1998gl900008.

Aldrich, L.T., Davis, G.L., Tilton, G.R. et al. 1956. Radioactive ages of minerals from the Brown Derby Mine and the Quartz Creek Granite Near Gunnison, Colorado. *Journal of Geophysical Research* **61**: 215–32. doi: 10.1029/JZ061i002p00215.

Allégre, C.J., and Luck, J.M. 1980. Osmium isotopes as petrogenic and geologic tracers. *Earth and Planetary Science Letters* **48**: 148–54.

Amelin, Y. 2005. Meteorite phosphates show constant ^{176}Lu decay rate since 4557 million years ago. *Science* **310**: 839–41.

Bast, R., Scherer, E. and Bischoff, A. 2017. The ^{176}Lu-^{176}Hf systematics of ALM-A: a sample of the recent Almahata Sitta meteorite fall. *Geochemical Perspectives Letters* **3**: 45–54. doi: 10.7185/geochemlet.1705.

Bau, M. and Koschinsky, A. 2006. Hafnium and neodymium isotopes in seawater and in ferromanganese crusts: the "element perspective." *Earth and Planetary Science Letters* **241**: 952–61. doi: 10.1016/j.epsl.2005.09.067.

Bellot, N., Boyet, M., Doucelance, R., et al. 2018. Origin of negative cerium anomalies in subduction-related volcanic samples: constraints from Ce and Nd isotopes. *Chemical Geology* **500**: 46–63. doi: 10.1016/j.chemgeo.2018.09.006.

Bellot, N., Boyet, M., Doucelance, R., et al. 2015. Ce isotope systematics of island arc lavas from the Lesser Antilles. *Geochimica et Cosmochimica Acta* **168**: 261–79. doi: 10.1016/j.gca.2015.07.002.

Bizzarro, M., Baker, J.A., Haack, H., et al. 2003. Early history of Earth's crust-mantle system inferred from hafnium isotopes in chondrites. *Nature* **421**: 931–3.

Bizzarro, M., Connelly, J.N., Thrane, K., et al. 2012. Excess hafnium-176 in meteorites and the early Earth zircon record. *Geochemistry, Geophysics, Geosystems* **13**: Q03002. doi: 10.1029/2011gc004003.

Blichert-Toft, J. 2018. Hafnium isotopes. In: White W. M. (ed). *Encyclopedia of Geochemistry*. Cham: Springer International Publishing, pp. 631–636. doi: 10.1007/978-3-319-39312-4_224.

Blichert-Toft, J. and Albarède, F. 1997. The Lu-Hf isotope geochemistry of chondrites and the evolution of the mantle-crust system. *Earth and Planetary Science Letters* **148**: 243–58. doi: 10.1016/S0012–821X(97)00040-X.

Boltwood, B.B. 1907. Ultimate disintegration products of the radioactive elements; Part II, Disintegration products of uranium. *American Journal of Science Series* 4 **23**: 78–88. doi:10.2475/ajs.s4–23.134.78.

Bouvier, A., Vervoort, J.D. and Patchett, P.J. 2008. The Lu-Hf and Sm-Nd isotopic composition of CHUR; constraints from unequilibrated chondrites and implications for the bulk composition of terrestrial planets. *Earth and Planetary Science Letters* **273**: 48–57. doi: 10.1016/j.epsl.2008.06.010.

Boyet, M. and Carlson, R. L. 2005. ^{142}Nd evidence for early (>4.3 Ga) global differentiation of the silicate Earth. *Science* **309**: 576–81.

Boyet, M., Doucelance, R., Israel, C., et al. 2019. New Constraints on the Origin of the EM-1 Component Revealed by the Measurement of the La-Ce Isotope Systematics in Gough Island Lavas. *Geochemistry, Geophysics, Geosystems* **20**: 2484–2498. doi: 10.1029/2019gc008228.

Brandon, A.D., Walker, R.J. and Puchtel, I.S. 2006. Platinum–osmium isotope evolution of the Earth's mantle: constraints from chondrites and Os-rich alloys. *Geochimica et Cosmochimica Acta* **70**: 2093–103. doi: 10.1016/j.gca.2006.01.005.

Brandon, A.D., Walker, R.J., Morgan, J.W., et al. 1998. Coupled ^{186}Os and ^{187}Os evidence for core-mantle interaction. *Science* **280**: 1570–3. doi: 10.1126/science.280.5369.1570.

Brandon, A.D., Walker, R.J., Puchtel, I.S., et al. 2003. ^{186}Os–^{187}Os systematics of Gorgona Island komatiites: implications for early growth of the inner core. *Earth and Planetary Science Letters* **206**: 411–26. doi: 10.1016/S0012-821X (02)01101-9.

Burkhardt, C., Borg, L.E., Brennecka, G.A., et al. 2016. A nucleosynthetic origin for the Earth's anomalous ^{142}Nd composition. *Nature* **537**: 394–8. doi:10.1038/nature18956.

Cheng, H., Zhang, C., Vervoort, J.D., et al. 2011. Geochronology of the transition of eclogite to amphibolite facies metamorphism in the North Qinling orogen of central China. *Lithos* **125**: 969–83, doi:10.1016/j.lithos.2011.05.010.

Chou, C.-L. 1978. Fractionation of siderophile elements in the Earth's upper mantle and lunar samples. *Proceedings of the Lunar and Planetary Science Conference* **9**: 163–5.

Coggon, J.A., Nowell, G.M., Pearson, D.G., et al. 2011. Application of the ^{190}Pt-^{186}Os Isotope System to Dating Platinum Mineralization and Ophiolite Formation: an Example from the Meratus Mountains, Borneo. *Economic Geology* **106**: 93–117. doi: 10.2113/econgeo.106.1.93.

Coggon, J.A., Nowell, G.M., Pearson, D.G., et al. 2012. The ^{190}Pt–^{186}Os decay system applied to dating platinum-group element mineralization of the Bushveld Complex, South Africa. *Chemical Geology* **302–3**: 48–60. doi: 10.1016/j.chemgeo.2011.10.015.

Crank, J. 1975. *The Mathematics of Diffusion.* 2nd ed. Oxford University Press, pp. 414.

Creaser, R.A., Papanastassiou, D.A. and Wasserburg, G.J. 1991. Negative thermal ion mass spectrometry of osmium, rhenium, and iridium. *Geochimica et Cosmochimica Acta* **55**: 397–401. doi: 10.1016/0016-7037(91)90427-7

Day, J.M.D. and O'Driscoll, B. 2019. Ancient high Pt/Os crustal contaminants can explain radiogenic ^{186}Os in some intraplate magmas. *Earth and Planetary Science Letters* **519**: 101–8. doi: 10.1016/j.epsl.2019.04.039.

Day, J.M.D., Walker, R.J. and Warren, J.M. 2017. ^{186}Os–^{187}Os and highly siderophile element abundance systematics of the mantle revealed by abyssal peridotites and Os-rich alloys. *Geochimica et Cosmochimica Acta* **200**: 232–54. doi: 10.1016/j.gca.2016.12.013.

DePaolo, D. and Wasserburg, G. 1976. Inferences about magma sources and mantle structure from variations of ^{143}Nd/^{144}Nd. *Geophysical Research Letters* **3**: 743–6. doi: 10.1029/GL003i005p00249.

Dickin, A.P. 1987. Cerium isotope geochemistry of oceanic island basalts. *Nature* **326**: 283–4.

Dodson, M.H. 1973. Closure temperature in cooling geochronological and petrological systems. *Contributions to Mineralogy and Petrology* **40**: 259–74.

Doucelance, R., Bellot, N., Boyet, M., et al. 2014. What coupled cerium and neodymium isotopes tell us about the deep source of oceanic carbonatites. *Earth and Planetary Science Letters* **407L**: 175–86. doi: 10.1016/j.epsl.2014.09.042.

Galer, S.J.G. 1999. Optimal double and triple spiking for high precision lead isotopic measurement. *Chemical Geology* **157**: 255–74.

Gast, P.W. 1960. Limitations on the composition of the upper mantle. *Journal of Geophysical Research* **65**: 1287–97.

Gast, P.W., Tilton, G.R. and Hedge, C. 1964. Isotopic composition of lead and strontium from Ascension and Gough Islands. *Science* **145**: 1181–5. doi: 10.2307/1714243.

Georgiev, S.V., Stein, H.J., Hannah, J.L., et al. 2016. Re–Os dating of maltenes and asphaltenes within single samples of crude oil. *Geochimica et Cosmochimica Acta* **179**: 53–75. doi: 10.1016/j.gca.2016.01.016.

Harrison, T.M. and McDougall, I. 1980. Investigations of an intrusive contact, northwest Nelson, New Zealand–II. Diffusion of radiogenic and excess ^{40}Ar in hornblende revealed by ^{40}Ar^{39}Ar age spectrum analysis. *Geochimica et Cosmochimica Acta* **44**: 2005–20. doi: 10.1016/0016-7037(80)90199-4.

Harrison, T.M. and McDougall, I. 1981. Excess ^{40}Ar in metamorphic rocks from Broken Hill, New South Wales: implications for ^{40}Ar/^{39}Ar age spectra and the thermal history of the region. *Earth and Planetary Science Letters* **55**: 123–49. doi: 10.1016/0012-821X(81)90092-3.

Harrison, T.M., Duncan, I. and McDougall, I. 1985. Diffusion of ^{40}Ar in biotite: temperature, pressure and compositional effects. *Geochimica et Cosmochimica Acta* **49**: 2461–8.

Heizler, M.T. and Harrison, T.M. 1988. Multiple trapped argon isotope components revealed by ^{40}Ar^{39}Ar isochron analysis. *Geochimica et Cosmochimica Acta* **52**: 1295–303. doi10.1016/0016-7037(88)90283–9.

Hinnov, L.A. and Hilgen, F.J. 2012. Chapter 4 – Cyclostratigraphy and astrochronology, in Gradstein, F. M., Ogg, J. G., Schmitz, M. D. and Ogg, G. M. (eds). *The Geologic Time Scale.* Boston: Elsevier.

Hirt, B., Herr, W. and Hoffmester, W. 1963. Age determinations by the rhenium-osmium method, in *Radioactive Dating*. Vienna: International Atom Energy Agency, pp. 35–44.

Horan, M.F., Walker, R.J., Morgan, J.W., et al. 2003. Highly siderophile elements in chondrites. *Chemical Geology* **196**: 27–42. doi: 10.1016/S0009-2541(02)00405-9.

Iizuka, T., Yamaguchi, T., Hibiya, Y. et al. 2015. Meteorite zircon constraints on the bulk Lu–Hf isotope composition and early differentiation of the Earth. *Proceedings of the National Academy of Sciences* **112**: 5331–6. doi: 10.1073/pnas.1501658112.

Israel, C., Boyet, M., Doucelance, R., et al. 2020. Formation of the Ce-Nd mantle array: crustal extraction vs. recycling by subduction. *Earth and Planetary Science Letters* **530**: 115941. doi: 10.1016/j.epsl.2019.115941.

Jacobsen, S.B. and Wasserburg, G.J. 1984. Sm-Nd isotopic evolution of chondrites and achondrites, II. *Earth and Planetary Science Letters* **67**: 137–50. doi: 10.1016/0012-821X(84)90109-2.

Johnson, T.A., Vervoort, J.D., Ramsey, M.J., et al. 2018. Constraints on the timing and duration of orogenic events by combined Lu–Hf and Sm–Nd geochronology: an example from the Grenville orogeny. *Earth and Planetary Science Letters* **501**: 152–64. doi: 10.1016/j.epsl.2018.08.030.

Krabbenhöft, A., Eisenhauer, A., Böhm, F., et al. 2010. Constraining the marine strontium budget with natural strontium isotope fractionations ($^{87}Sr/^{86}Sr^*$, $\delta^{88/86}Sr$) of carbonates, hydrothermal solutions and river waters. *Geochimica et Cosmochimica Acta* **74**: 4097–109. doi: 10.1016/j.gca.2010.04.009.

Kuiper, K.F., Deino, A., Hilgen, F.J., et al. 2008. Synchronizing rock clocks of Earth history. *Science* **320**: 500–4. doi: 10.1126/science.1154339.

Laskar, J., Robutel, P., Joutel, F., et al. 2004. A long-term numerical solution for the insolation quantities of the Earth. *Astronomy and Astrophysics* **428**: 261–85. doi: 10.1051/0004-6361:20041335.

Lassiter, J.C., Byerly, B.L., Snow, J.E. et al. 2014. Constraints from Os-isotope variations on the origin of Lena Trough abyssal peridotites and implications for the composition and evolution of the depleted upper mantle. *Earth and Planetary Science Letters* **403**: 178–87. doi: 10.1016/j.epsl.2014.05.033.

Mark, D.F., Stuart, F.M. and de Podesta, M. 2011. New high-precision measurements of the isotopic composition of atmospheric argon. *Geochimica et Cosmochimica Acta* **75**: 7494–501. doi: 10.1016/j.gca.2011.09.042.

Mattinson, J.M. 2010. Analysis of the relative decay constants of ^{235}U and ^{238}U by multi-step CA-TIMS measurements of closed-system natural zircon samples. *Chemical Geology* **275**: 186–98. doi: 10.1016/j.chemgeo.2010.05.007.

McArthur, J.M., Howarth, R.J. and Shields, G.A. 2012. Strontium isotope stratigraphy, in Gradstein, F.M., Ogg, J.G., Schmitz, M.D. and Ogg, G.M. (eds.). *The Geologic Time Scale*. Boston: Elsevier.

McCoy, T.J., Walker, R.J., Goldstein, J.I., et al. 2011. Group IVA irons: new constraints on the crystallization and cooling history of an asteroidal core with a complex history. *Geochimica et Cosmochimica Acta* **75**: 6821–43. doi: 10.1016/j.gca.2011.09.006.

McDougall, I. and Harrison, T.M. 1999. *Geochronology and Thermochronology by the $^{40}Ar/^{39}Ar$ Method*. 2nd ed. New York: Oxford University Press, p. 269.

Meisel, T., Walker, R.J., Irving, A.J. et al. 2001. Osmium isotopic compositions of mantle xenoliths: a global perspective. *Geochimica et Cosmochimica Acta* **65**: 1311–23. doi: 10.1016/S0016-7037(00)00566-4.

Merrihue, C. and Turner, G. 1966. Potassium-argon dating by activation with fast neutrons. *Journal of Geophysical Research* **71**: 2852–7. doi: 10.1029/JZ071i011p02852.

Mulcahy, S.R., King, R.L. and Vervoort, J.D. 2009. Lawsonite Lu-Hf geochronology: a new geochronometer for subduction zone processes. *Geolog* **37**: 987–90. doi:10.1130/g30292a.1.

Nebel, O., Scherer, E.E. and Mezger, K. 2011. Evaluation of the ^{87}Rb decay constant by age comparison against the U–Pb system. *Earth and Planetary Science Letters* **301**: 1–8. doi: 10.1016/j.epsl.2010.11.004.

Nier, A.O. 1939. The isotopic constitution of uranium and the half-lives of the uranium isotopes. I. *Physical Review* **55**: 150. doi: 10.1103/PhysRev.55.150.

Niespolo, E.M., Rutte, D., Deino, A.L. et al. 2017. Intercalibration and age of the Alder Creek sanidine 40Ar/39Ar standard. *Quaternary Geochronology* **39**: 205–13. doi: 10.1016/j.quageo.2016.09.004.

Nyquist, L.E., Bogard, D.D., Wiesmann, H., et al. 1990. Age of a eucrite clast from the Bholghati howardite. *Geochimica et Cosmochimica Acta.* **54**: 2195–206.

Patchett, P.J. and Tastumoto, M. 1980. Hafnium isotope variations in oceanic basalts. *Geophysical Research Letters* **7**: 1077–80.

Patchett, P.J., Vervoort, J.D., Söderlund, U. et al. 2004. Lu-Hf and Sm-Nd isotopic systematics in chondrites and their constraints on the Lu-Hf properties of the Earth. *Earth and Planetary Science Letters* **222**: 29–41. doi: 10.1016/j.epsl.2004.02.030.

Peucker-Ehrenbrink, B. and Jahn, B.-M., 2001. Rhenium-osmium isotope systematics and platinum group element concentrations: loess and the upper continental crust. *Geochemistry, Geophysics, Geosystems* **2**(10): 1061. doi: 10.1029/2001gc000172.

Reiners, P.W., Carlson, R.W., Renne, P.R., et al. 2017. *Geochronology and Thermochronology.* Oxford: John Wiley & Sons.

Renne, P.R., Mundil, R., Balco, G., et al. 2010. Joint determination of ^{40}K decay constants and ^{40}Ar*/^{40}K for the Fish Canyon sanidine standard, and improved accuracy for ^{40}Ar/^{39}Ar geochronology. *Geochimica et Cosmochimica Acta* **74**: 5349–67. doi:10.1016/j.gca.2010.06.017.

Richardson, S.H., Shirey, S.B., Harris, J.W. et al. 2001. Archean subduction recorded by Re-Os isotopes in eclogitic sulfide inclusions in Kimberley diamonds. *Earth and Planetary Science Letters* **191**: 257–66.

Rotenberg, E., Davis, D.W. and Amelin, Y. 2005. Determination of the ^{87}Rb decay constant by ^{87}Sr accumulation. *Geochimica et Cosmochimica Acta* **69**(suppl): A326.

Saintilan, N.J., Selby, D., Creaser, R.A. et al. 2018. Sulphide Re-Os geochronology links orogenesis, salt and Cu-Co ores in the Central African Copperbelt. *Scientific Reports* **8**: 14946. doi: 10.1038/s41598-018-33399-7.

Scherer, E., Munker, C. and Mezger, K. 2001. Calibration of the Lutetium-Hafnium clock. *Science* **293**: 683–6.

Selby, D. and Creaser, R.A. 2005. Direct radiometric dating of hydrocarbon deposits using rhenium-osmium isotopes. *Science* **308**: 1293–5. doi: 10.1126/science.1111081.

Selby, D., Creaser, R.A. and Fowler, M.G. 2007. Re-Os elemental and isotopic systematics in crude oils. *Geochimica Cosmochimica Acta* **71**: 378–6. doi: 10.1016/j.gca.2006.09.005.

Shirey, S.B., Richardson, S.H. and Harris, J.W. 2004. Integrated models of diamond formation and craton evolution. *Lithos* **77**: 923–44.

Söderlund, U., Patchett, P.J., Vervoort, J.D. et al. 2004. The ^{176}Lu decay constant determined by Lu–Hf and U–Pb isotope systematics of Precambrian mafic intrusions. *Earth and Planetary Science Letters* **219**: 311–24.

Steiger, R.H. and Jäger, E. 1977. Subcommission on geochronology: convention on the use of decay constants in geo- and cosmochronology. *Earth and Planetary Science Letters* **36**: 359–62. doi: 10.1016/0012-821X(77)90060-7.

Stein, H.J. 2014. Dating and tracing the history of ore formation, in Holland, H.D. and Turekian, K.K. (eds). *Treatise on Geochemistry*, 2nd ed. Oxford: Elsevier. doi: 10.1016/B978-0-08-095975-7.01104-9.

Tanaka, T. and Masuda, A. 1982. The La–Ce geochronometer: a new dating method. *Nature* **300**: 515–8. doi: 10.1038/300515a0.

Tanaka, T., Shimizu, H., Kawata, Y., et al. 1987. Combined La–Ce and Sm–Nd isotope systematics in petrogenetic studies. *Nature* **327**: 113–7. doi: 10.1038/327113a0.

Vermeesch, P. 2018. IsoplotR: A free and open toolbox for geochronology. *Geoscience Frontiers* **9**: 1479–93. doi: 10.1016/j.gsf.2018.04.001.

Vervoort, J.D. and Blichert-Toft, J. 1999. Evolution of the depleted mantle; Hf isotope evidence from juvenile rocks through time. *Geochimica et Cosmochimica Acta* **63**: 533–56.

Vervoort, J.D., Plank, T. and Prytulak, J. 2011. The Hf–Nd isotopic composition of marine sediments. *Geochimica et Cosmochimica Acta* **75**: 5903–26. doi:10.1016/j.gca.2011.07.046.

Walker, R.J., Carlson, R.W., Shirey, S.B., et al. 1989. Os, Sr, Nd, and Pb isotope systematics of southern African peridotite xenoliths: implications for the chemical evolution of the subcontinental mantle. *Geochimica et Cosmochimica Acta* **53**: 1583–95.

Walker, R.J., Morgan, J.W., Naldrett, A.J., et al. 1991. Re-Os isotope systematics of Ni-Cu sulfide ores, Sudbury igneous complex, evidence for a major crustal component. *Earth and Planetary Science Letters* **105**: 416–29.

Wasserburg, G.J., Jacobsen, S.B., Depaolo, D.J., et al. 1981. Precise determination of SmNd ratios, Sm and Nd isotopic abundances in standard solutions. *Geochimica et Cosmochimica Acta* **45**: 2311–23. doi: https://doi.org/10.1016/0016-7037(81)90085-5.

Wendt, I. and Carl, C. 1991. The statistical distribution of the mean squared weighted deviation. *Chemical Geology: Isotope Geoscience Section* **86**: 275–85. doi: 10.1016/0168–9622(91)90010-T.

White, W.M., Copeland, P., Gravatt, D.R., et al. 2017. Geochemistry and geochronology of Grenada and Union islands, Lesser Antilles: The case for mixing between two magma series generated from distinct sources. *Geosphere* **5**: 1359–1391. doi: 10.1130/GES01414.1.

White, W.M., Patchett, P.J. and Ben Othman, D. 1986. Hf isotope ratios of marine sediments and Mn nodules: evidence for a mantle source of Hf in seawater. *Earth and Planetary Science Letters* **79**: 46–54.

White, W.M., Albarède, F. and Telouk, P. 2000. High-precision analysis of Pb isotope ratios by multi-collector ICP-MS. *Chemical Geology* **167**: 257–70. doi: 10.1016/S0009-2541(99)00182-5.

Willig, M. and Stracke, A. 2019. Earth's chondritic light rare earth element composition: evidence from the Ce–Nd isotope systematics of chondrites and oceanic basalts. *Earth and Planetary Science Letters* **509**: 55–65. doi: 10.1016/j.epsl.2018.12.004.

York, D. 1969. Least squares fitting of a straight line with correlated errors. *Earth and Planetary Science Letters* **5**: 320–4.

Zhai, D., Williams-Jones, A.E., Liu, J., et al. 2019. Evaluating the use of the molybdenite Re-Os chronometer in dating gold mineralization: Evidence from the Haigou Deposit, Northeastern China. *Economic Geology* **114**: 897–915. doi: 10.5382/econgeo.2019.4667.

PROBLEMS

1. As noted, ^{40}Ca is an even–even nuclide and is therefore more abundant than ^{40}K. What other factor might account for the high abundance of ^{40}Ca?

2. Use Dodson's equation (Equation [2.37]) to calculate the closure temperatures of biotite for the cases of a slowly cooled intrusion discussed in Section 2.3.1, namely, at 10°/Ma and 100°/Ma. Use the data given in Figure 2.1, which correspond to $E_A = 196.8$ kJ/mol and $D_0 = -0.00077$ m^2/s. Assume $a = 140$ µm and $A = 27$. The value of R is 8.314 J/K-mol. If we were to do K–Ar dating on these biotites long after they cooled (say 100 Ma later), how much different would the two ages be assuming the intrusion cooled at these rates from an initial temperate of 600°C? *(Hint, you can easily do this in Excel, either using the solver or iterating manually following an initial guess of the closure temperature – you can base that on the discussion in the text. Be careful to use consistent units.)*

3. You measure the following K$_2$O and ^{40}Ar on minerals from a small pluton. Calculate the age for each. What do you think the ages mean? Use the following:
   Branching ratio is 0.1157, $\lambda_e = 0.58755 \times 10^{-10}$ yr^{-1}, $\lambda_{total} = 5.5492 \times 10^{-10}$ yr^{-1}. (These are newly recommended values.)

   ^{40}K/K = 0.0001167, atomic weight of K is 39.03983.

	K$_2$O (wt.%)	Radiogenic ^{40}Ar, mole/g
Biotite	8.45	6.016×10^{-10}
Hornblende	0.6078	0.4642×10^{-10}

   Are the ages the same? If not, speculate on why not?

4. Use following data to answer this question:
   λ_{Rb}: 1.42×10^{-11} yr; ^{86}Sr/^{88}Sr: 0.11940; ^{84}Sr/^{88}Sr: 0.006756; ^{85}Rb/^{87}Rb = 2.59265, atomic weight of Rb: 85.46776
   Atomic masses of Sr:
   ^{88}Sr: 87.9056
   ^{87}Sr: 86.9088
   ^{86}Sr: 85.9092
   ^{84}Sr: 83.9134

   Calculate the abundances of the isotopes and atomic weight of Sr given that ^{87}Sr/^{86}Sr = 0.7045.

5. The following ^{40}Ar*/^{39}Ar ratios were measured in step heating of lunar Basalt 15555 from Hadley Rile. The flux monitor had an age of 1.062×10^9 yr and its ^{40}Ar*/^{39}Ar ratio after irradiation was 29.33. The ^{40}K/^{39}K ratio is 0.000125137. Calculate the age for each step and plot the ages versus percentage of release. From this release spectrum, estimate the age of the sample.

Cumulative % Ar released	^{40}Ar*/^{39}Ar
3	58.14
10	61.34
27	72.77
61	80.15
79	83.32
100	79.80

6. The following data were obtained on three minerals from a pegmatite. Calculate the age of the rock using the isochron method (you may use conventional regression for this problem). The data and approach used in Problem 4 will prove useful.

	Rb (ppm)	Sr (ppm)	$^{87}Sr/^{86}Sr$
Muscovite	238.4	1.80	1.4125
Biotite	1080.9	12.8	1.1400
K-feldspar	121.9	75.5	0.7502

7. The following data were measured on phlogopites (P) and phlogopite leaches (LP) from a kimberlite from Rankin Inlet area of the Hudson Bay, Northwest Territories, Canada. What is (1) the age of the rock, (2) the uncertainty on the age, (3) the initial $^{87}Sr/^{86}Sr$ ratio, and (4) the uncertainty on the initial ratio? The relative uncertainty on the $^{87}Sr/^{86}Sr$ is 0.005% and that of the $^{87}Rb/^{86}Sr$ is 1%. (*Hint: this is best accomplished using IsoplotR.*)

Sample	$^{87}Rb/^{86}Sr$	$^{87}Sr/^{86}Sr$
P1	46.77	0.848455
P2	40.41	0.828490
P3	34.73	0.810753
P4	33.78	0.807993
P5	0.1829	0.706272
P6	0.1373	0.705616
P7	1.742	0.710498

8. The following were measured on a coarse-grained metagabbro from the Cana Brava Complex in central Brazil. Plot the data on an isochron diagram, and calculate the age, errors on the age, and the initial ε_{Nd} and the error on the initial. The errors on the $^{147}Sm/^{144}Nd$ are all 0.0001 (2 sigma absolute). Two sigma errors on the $^{143}Nd/^{144}Nd$ shown below are in the fifth digit.

$^{147}Sm/^{144}Nd$	$^{143}Nd/^{144}Nd$	
Pyroxene	0.1819	0.51234 ± 2
Plagioclase	0.0763	0.51183 ± 4
WR	0.1678	0.51227 ± 4
Plagioclase	0.0605	0.51173 ± 4
Biotite	0.1773	0.51232 ± 4

9. The following data apply to WRs and separated minerals of the Baltimore Gneiss. Interpret these data by means of suitable isochron diagrams. Determine dates and initial $^{87}Sr/^{86}Sr$ ratios and errors on both using simple linear regression. Speculate on the geologic history of these rocks and minerals.

	$^{87}Rb/^{86}Sr$	$^{87}Sr/^{86}Sr$
Rock 1	2.244	0.7380
Rock 2	3.642	0.7612
Rock 3	6.59	0.7992
Biotite	289.7	1.969
K-feldspar	5.60	0.8010
Plagioclase	0.528	0.7767
Rock 4	0.2313	0.7074
Rock 5	3.628	0.7573
Biotite	116.4	1.2146
K-feldspar	3.794	0.7633
Plagioclase	0.2965	0.7461

10. The following data were obtained on an Egyptian diorite:

	$^{87}Rb/^{86}Sr$	$^{87}Sr/^{86}Sr$
Plagioclase	0.05124	0.705505
Amphibole	0.13912	0.706270
Biotite	0.95322	0.713847
Alkali feldspar	0.58489	0.710418
WR	0.33975	0.708154

Assume that the analytical error on the $^{87}Sr/^{86}Sr$ ratio was 0.006% and that the analytical error on the $^{87}Rb/^{86}Sr$ ratio was 0.1% in each case, and that these errors are uncorrelated. Use the two-error regression method to calculate the age and initial ratio and the errors on both.

(*Hint: This is best accomplished using IsoplotR.*)

11. A sample of granite has $^{143}Nd/^{144}Nd$ and $^{147}Sm/^{144}Nd$ of 0.51196 and 0.12990, respectively. The present chondritic $^{143}Nd/^{144}Nd$ and $^{147}Sm/^{144}Nd$ are 0.512638 and 0.1967, respectively. The decay constant of ^{147}Sm is 6.54×10^{-12} a^{-1}. Calculate the τ_{CHUR}, i.e., crustal residence time relative to a chondritic mantle, for this granite.

12. The following data were obtained on sulfide inclusions in diamonds from the Koffiefontein Mine in South Africa. Calculate the age of the diamonds assuming all analyzed samples are cogenetic.

	$^{187}Re/^{188}Os$	$^{187}Os/^{188}Os$
K310	104	2.19
K309	5.24	0.346
K308	116	2.28
K307	6.31	0.411
K305	80.6	1.78

Chapter 3

Decay systems and geochronology II: U and Th

3.1 INTRODUCTION

The U–Th–Pb system is certainly the most powerful tool in the geochronologist's tool chest. While we can use the three decay systems independently, the real power comes in using them in combination, particularly the ^{235}U–^{207}Pb and ^{238}U–^{238}Pb systems, as it allows a check of the fidelity of the age calculated and, in some circumstances, to obtain accurate ages despite disturbances to the system that violate the conditions we have already discussed in Chapter 2. We will begin by discussing U–Th–Pb dating, which is useful on a wide range of timescales, from hundreds of thousands to billions of years. Indeed, as we will see in this chapter and subsequent chapters, U–Pb dating provides the definitive ages of the Solar System and the oldest rocks on the Earth. Without question, it is the "gold standard" of geochronology.

Rather than decaying directly to lead, uranium and thorium decay through a chain of intermediate daughters, which have half-lives ranging up to hundreds of thousands of years. Measuring the ratios of intermediate radioactive parents and daughters enables an entirely new set of geochronological tools for use on those timescales. We will devote the second half of this chapter to uranium decay series geochronology.

3.1.1 Chemistry of U, Th, and Pb

U and Th are rare-earth elements and belong to the actinoid series rather than the lanthanoid series as noted and discussed in the previous chapter. As in the lanthanoids, an inner electron shell is being filled as atomic number increases in the actinoids. Both U and Th generally have a valence of +4, but under oxidizing conditions, such as at the surface of the Earth, U has a valence of +6. In sixfold coordination, U^{4+} has an ionic radius of 89 pm[1] (100 pm = 1 Å); U^{6+} has an ionic radius of 73 pm in sixfold coordination and that of 86 pm in eightfold coordination. Th^{4+} has an ionic radius of 94 pm in sixfold coordination. These radii are not particularly large, but the combination of a somewhat large radius and high charge is not readily accommodated in crystal lattices of most common rock-forming minerals, so both U and Th are highly incompatible elements. Th is relatively immobile under most circumstances. In its reduced form, U^{4+} is insoluble and therefore fairly immobile, but in the U^{6+} form, which is stable under a wide range of conditions at the surface of the Earth, U forms the soluble oxyanion complex, UO_4^{2-}. As a result, U can be quite mobile. U and Th can form their own phases in sedimentary rocks, uranite and thorite, but they

---

[1] In eightfold coordination, the effective ionic radius of U^{4+} is 100 pm. In zircon, a mineral with high concentrations of U, U is in eightfold coordination. This is probably a pretty good indication that eightfold coordination is the preferred configuration. The figure for sixfold coordination is given for comparison to other radii, which have been for sixfold coordination. Th has a radius of 105 pm in eightfold coordination.

Isotope Geochemistry, Second Edition. William M. White.
© 2023 John Wiley & Sons Ltd. Published 2023 by John Wiley & Sons Ltd.
Companion Website: www.wiley.com/go/white/isotopegeochem2

are quite rare. In igneous and metamorphic rocks, U and Th are either dispersed as trace elements in major phases or concentrated in accessory minerals (when they are present), such as zircon ($ZrSiO_4$) that concentrates U more than Th, and monazite ([La,Ce,Th]PO_4) that concentrates Th more than U. These elements may also be concentrated in other accessory phases, such as apatite ($Ca_5(PO_4)_3(OH)$), xenotime (YPO_4), and titanite (or sphene, $CaTiSiO_5$). However, zircon is far and away most important from a geochronological perspective.

U and Th are refractory elements, and we can therefore expect the Th/U ratio of the Earth to be the same as chondrites or nearly so. There is, however, some debate about the exact terrestrial Th/U ratio, a good estimate of which is 3.9 ± 0.1 (molar) (e.g., Wipperfurth et al. 2018). The ratio is 3.8 in the CI chondrite Orgueil, but it may be low due to mobility of U in hydrous fluid in the CI parent body[2].

The geochemical behavior of Pb is more complex than that of the elements we have discussed so far and, consequently, less well understood. It is a relatively volatile element, so its relative concentration in the Earth is lower than in chondrites. It is also a *chalcophile* element. If the core contains substantial amounts of S, it is possible that a significant fraction of the Earth's Pb is in the core. It is, however, difficult to distinguish loss of Pb from the Earth due to its volatility from loss of Pb from the silicate portion of the Earth due to extraction into the core; both are likely involved. Pb can exist in two valence states, namely, Pb^{2+} and Pb^{4+}. Pb^{2+} is by far the most common state; the Pb^{4+} state is rare and restricted to highly alkaline or oxidizing solutions. The ionic radius of Pb^{2+} is 119 pm in sixfold coordination and 129 pm in eightfold coordination. As a result of its large ionic size, Pb is an incompatible element, though not as incompatible as U and Th (incompatibility seems to be comparable to the light rare earths; the ratio of Pb to Ce is approximately constant in most mantle-derived rocks, but not in crustal rocks). The most common Pb mineral is

galena (PbS). In silicates, Pb substitutes readily for K (with an ionic radius of 133 pm) in potassium feldspar, but less so in other K minerals, such as biotite. Most naturally occurring compounds of Pb are highly insoluble under most conditions. As a result, Pb is usually reasonably immobile. However, under conditions of low pH and high temperature, Pb forms stable soluble chloride and sulfide complexes so that Pb can sometimes be readily transported in hydrothermal solutions.

Although Pb is less incompatible than U and Th, these three elements have been extracted from the mantle and concentrated in the crust to *approximately* the same degree; we will discuss this in more detail in Chapters 7 and 8.

3.1.2 The $^{238}U/^{235}U$ ratio and uranium decay constants

Up until the last two decades, it had been assumed that the $^{238}U/^{235}U$ ratio was constant. In this case, the $^{207}Pb*/^{206}Pb*$[3] ratio is a function only of time (and the decay constants). The conventionally accepted value of this ratio was 138.88 (Jaffey et al., 1971; Steiger and Jäger, 1977). However, as precision in isotopic measurements has improved, it became apparent (1) that this value varied somewhat and (2) that mean value in terrestrial materials is actually a little lower (e.g., Stirling et al., 2007; Weyer et al., 2008; Amelin et al., 2010; Mattinson, 2010; Hiess et al., 2012). Stirling et al. (2007) found a range of about 4 per mil in this ratio in natural terrestrial materials, while Weyer et al. (2008) found a range of about 1.4 per mil. Hiess et al. (2012) demonstrated a variation of 5 per mil in uranium-bearing minerals (zircon, apatite, monazite, xenotime, baddeleyite, and titanite) commonly analyzed in geochronological work, but this range is defined by relatively few "outliers" and almost all zircons fell within a much smaller range of 137.77–137.91, a 0.1 per mil variation. The outcomes of these variations are slight differences in bond strength and diffusivity that result from the mass differences

---

[2] Variability in chondrites reflects the mobility of U. The CI carbonaceous chondrites experienced mild alteration in hydrous conditions on the parent body. U was mobilized under these conditions and thus the U/Th ratio varies in these meteorites. For this reason, they cannot be used to precisely determine the U/Th ratio of the Solar System and the Earth.

[3] As we did in Chapter 2, we use the asterisk to designate the radiogenic component of an isotope. Thus $^{207}Pb*/^{206}Pb*$ is the ratio of radiogenic ^{207}Pb to radiogenic ^{206}Pb. We will use this notation throughout this chapte.

of the two U isotopes. We will postpone the discussion of the causes of isotopic variations resulting from chemical effects, such as these, until Chapter 9, where we will discuss them at length. Somewhat greater variations occur in meteorites as a consequence of slight chemical and isotopic heterogeneity in the solar nebula and the decay of the short-lived, and now extinct, radionuclide ^{247}Cm (more details on it will be provided in Chapter 6).

Amelin et al. (2010) estimated the mean terrestrial $^{238}U/^{235}U$ to be 137.821 ± 0.014; Hiess et al. (2012) estimated it to be 137.818 ± 0.045. There is excellent agreement between these two values, but both differ from the conventional value. Goldmann et al. (2013) have proposed a slightly lower value of 137.79 ± 0.03 based on measurements of meteorites. Herein, we will adopt a value of 137.82 for the $^{238}U/^{235}U$ and assume this value to be constant for geochronological purposes. However, one should be aware that, at least as of this writing, a value of 137.88 remains the "official" value (the one recommended by the International Union of Geological Sciences (IUGC) Subcommission on Geochronology) and that almost all the ages in the literature are based on that value. Furthermore, the highest precision geochronology may require the analysis of the $^{238}U/^{235}U$ and Pb isotope ratios. Finally, this value of 137.82 is the *present-day value*; it changes through time as a result of the two isotopes decaying at different rates. If we need to know the ratio at some other time (e.g., Problem 1), we need to calculate it based on Equation (2.4).

As Mattinson (2010) noted, a change in the accepted value of the $^{238}U/^{235}U$ ratio will require a re-evaluation of U half-lives, particularly that of ^{235}U; hence, further refinement of these values can be expected in the future.

3.2 Pb–Pb AGES AND ISOCHRONS

Table 3.1 summarizes this decay system. If the "Mad Men" of Madison Avenue were given the task of selling the U–Th–Pb system,

they would probably say that you get four dating methods for the price of one. We can calculate three ages using the conventional isochron approach, one for each of the $^{238}U–^{206}Pb$, $^{235}U/^{207}Pb$, and $^{232}Th/^{208}Pb$ systems. This proceeds exactly as for the decay systems we discussed in Chapter 2. However, we can also combine the ^{238}U and ^{235}U decays to calculate an additional age, known as a *Pb–Pb age*. If you bought the Madison Avenue sales pitch, you would probably discover that the conventional isochron approach for the first two systems mentioned was not particularly powerful, at least in comparison to either the Pb–Pb technique or when several approaches are used in combination. The reason for the power is simply that there are three parents decaying to three isotopes of Pb and, in particular, there are two isotopes of U that decay to Pb with very different half-lives. This is important because chemical processes will generally not change the ratio of the two U isotopes to each other significantly and will not change the ratio of the two Pb daughter isotopes to each other. The point is best illustrated as follows. First, we write the decay equation for each of the two U decay systems:

$$^{207}Pb^* = {}^{235}U\left(e^{\lambda_{235}t} - 1\right) \qquad (3.1)$$

$$^{206}Pb^* = {}^{238}U\left(e^{\lambda_{238}t} - 1\right) \qquad (3.2)$$

where λ_{235} and λ_{238} are the decay constants for ^{235}U and ^{238}U, respectively. If we divide Equation 3.1 by Equation 3.2, we obtain:

$$\frac{^{207}Pb^*}{^{206}Pb^*} = \frac{^{235}U\left(e^{\lambda_{235}t} - 1\right)}{^{238}U\left(e^{\lambda_{238}t} - 1\right)} \qquad (3.3)$$

Let us consider this in more detail. Assuming the present-day $^{238}U/^{235}U$ ratio is indeed constant, Equation 3.3 can be written as:

$$\frac{^{207}Pb^*}{^{206}Pb^*} = \frac{\left(e^{\lambda_{235}t} - 1\right)}{137.82\left(e^{\lambda_{238}t} - 1\right)} \qquad (3.4)$$

Table 3.1 Parameters of the U–Th–Pb system.

Parent	Decay mode	λ	Half-life	Daughter	Ratio
^{232}Th	α, β	4.948×10^{-11} yr^{-1}	1.4×10^{10} yr	^{208}Pb, 8 ^4He	$^{208}Pb/^{204}Pb$, $^3He/^4He$
^{235}U	α, β	9.8571×10^{-10} yr^{-1*}	7.07×10^{8} yr	^{207}Pb, 7 ^4He	$^{207}Pb/^{204}Pb$, $^3He/^4He$
^{238}U	α, β	1.55125×10^{-10} yr^{-1}	4.47×10^{9} yr	^{206}Pb, 6 ^4He	$^{206}Pb/^{204}Pb$, $^3He/^4He$

The good thing about Equation 3.4 is that the only variable on the right-hand side is time; in other words, $^{207}Pb*/^{206}Pb*$ is a function only of time.

In practice, this means that the age is independent of the parent/daughter ratio; that is, we need not measure the parent/daughter ratio. We shall see that this property actually allows us to relax our requirement that the system remain closed in some circumstances. We can also see that although we could write an equation similar to Equation 3.3 using ^{232}Th and ^{208}Pb instead of ^{235}U and ^{207}Pb, there would be little advantage to doing so because Th and U are different elements and could well be lost or gained in different proportions.

The Pb–Pb method, as it is called, can be quite useful when applied independently, particularly where there is reason to believe that there has been some recent change in the parent/daughter ratio.

The slope on a plot of $^{207}Pb/^{204}Pb$ versus $^{206}Pb/^{204}Pb$ is proportional to age since:

$$\frac{\Delta\left(^{207}Pb/^{204}Pb\right)}{\Delta\left(^{206}Pb/^{204}Pb\right)} = \frac{\left(e^{\lambda_{235}t} - 1\right)}{137.82\left(e^{\lambda_{238}t} - 1\right)} \quad (3.5)$$

Equation 3.5 is very similar to Equation 3.4. We would use Equation 3.4 when either the initial Pb is insignificant or the amount of initial Pb is sufficiently small that we can make a reasonable estimate of its isotopic composition and make a correction for it. We would use Equation 3.5 when initial Pb is present in significant quantities and has an unknown composition. Figure 3.1 shows an example of a Pb–Pb isochron that yielded a reasonably precise age based on Equation 3.5. Unlike a conventional isochron, the intercept in the Pb–Pb isochron has no significance and the initial isotopic composition cannot be determined without some additional information about parent/daughter ratios. As in the isochron approaches we discussed in Chapter 2, the slope is determined by regression.

There are a couple of reasons why we might suspect U/Pb ratios have changed and hence might prefer the Pb–Pb approach over a conventional U–Pb isochron approach. First, the solubility of U under oxidizing conditions often leads to mobility (open-system behavior) in the zone of weathering. It has often been

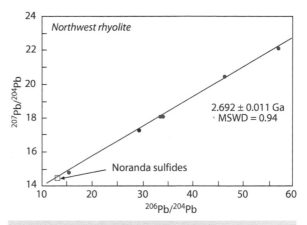

Figure 3.1 A Pb–Pb isochron obtained on volcanic rocks hosting the Noranda (Quebec) Cu-Zn sulfide deposit. Data from Vervoort et al. (1993).

found that U–Pb ages are spurious, yet Pb–Pb ages seem correct. This circumstance appears to result from recent U mobility as erosion brings a rock into the weathering zone. A second situation where parent/daughter ratios would have experienced recent change is in magma generation. When melting occurs, the U and Pb isotope ratios in the magma will be identical to those in the source (because the isotopes of an element are chemically identical), but the U/Pb ratio (and Th/Pb ratio) will change, as the chemical behaviors of U and Pb differ. Therefore, conventional dating schemes cannot generally provide useful geochronological information about sources of magmas. However, the Pb–Pb dating method can, at least in principle, provide useful information, because the Pb isotope ratios of a magma are representative of the source and the method does not depend on parent/daughter ratios. Essentially, what we are doing is allowing volcanism to "sample" the source, generally the mantle, but sometimes the lower continental crust. The sample is representative of the isotopic composition of the source, but not representative of the elemental chemistry of the source. The relationship between Pb isotope ratios in mantle-derived magmas has led to the conclusion that heterogeneities in the mantle must have existed for times on the order of 1–2 Ga. This is an important constraint not only on the chemical evolution of the mantle, but also on its dynamics.

3.2.1 Total U–Pb isochrons

The U–Pb system achieves its greatest power when we use the ^{238}U–^{206}Pb, ^{235}U–^{207}Pb, and ^{207}Pb–^{206}Pb methods in combination. In the ideal case where the system was isotopically homogeneous at time 0 and has remained closed since, the ^{238}U–^{206}Pb, ^{235}U–^{207}Pb, and ^{207}Pb–^{206}Pb ages should agree. In this case, the age obtained is said to be "concordant." Even when the three ages do not agree, it can be possible to "see through" open-system behavior and obtain an age of initial crystallization. In this section, we will consider an approach that is useful when the amount of initial Pb, often referred to as "common lead," is significant. Since ^{204}Pb is non-radiogenic, this is the case where significant amounts of ^{204}Pb are present and ^{206}Pb/^{204}Pb and ^{207}Pb/^{204}Pb ratios are relatively low. We will consider the case where nearly all the Pb is radiogenic in the subsequent section on zircon geochronology.

Tera and Wasserburg (1972), working with lunar samples, developed a graphical approach to evaluate the degree to which ^{238}U–^{206}Pb and ^{207}Pb/^{206}Pb ages agree; that is, they are concordant. On a *Tera–Wasserburg diagram*, or as Ludwig (1998) calls it, a *semi-total Pb isochron diagram*, measured ^{207}Pb/^{206}Pb ratios are plotted against ^{238}U/^{206}Pb ratios. Ratios are corrected for any contribution from analytical blank, but not for initial, or common, Pb.

On such a diagram (Figure 3.2), purely radiogenic Pb will have unique ^{207}Pb*/^{206}Pb* and ^{238}U/^{206}Pb* ratios at any given time and hence are defined by the blue "concordia" curve in Figure 3.2. If we measure a series of samples with different U/Pb values containing common Pb and if those samples meet the conditions of (1) isotopic homogeneity at time 0 and (2) no disturbance since, they will be plotted along a straight line that intercepts the "concordia" curve at a point where ^{207}Pb/^{206}Pb and ^{238}U/^{206}Pb ages are equal. The intercept of the regression line is the initial ^{207}Pb/^{206}Pb (because a sample with a ^{238}U/^{206}Pb ratio of 0 will retain its initial Pb isotopic composition).

The equation for the concordia line can readily be derived from Equations 3.1 through 3.4 (see Problem 2) and is:

$$\frac{^{207}\text{Pb*}}{^{206}\text{Pb*}} = \frac{\left(e^{\lambda_{235}t} - 1\right)\left(^{238}\text{U}/^{206}\text{Pb*}\right)}{137.82} \quad (3.6)$$

The slope of the line through the data, which in practice would be determined by regression, is:

$$b = \frac{d\left(^{207}\text{Pb}/^{206}\text{Pb}\right)}{d\left(^{238}\text{U}/^{206}\text{Pb}\right)} = \frac{^{235}\text{U}}{^{238}\text{U}}\left(e^{\lambda_{235}t} - 1\right)$$
$$- \left(\frac{^{207}\text{Pb}}{^{206}\text{Pb}}\right)_i \left(e^{\lambda_{238}t} - 1\right) \quad (3.7)$$

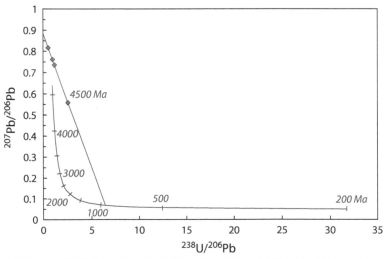

Figure 3.2 The *Tera–Wasserburg* or *semi-total Pb isochron diagram*. The blue line is the concordia curve where the ^{207}Pb*/^{206}Pb* age equals the ^{238}U/^{206}Pb* age, with ticks indicating the ages. A hypothetical data set with an age of 935 Ma and $(^{207}\text{Pb}/^{206}\text{Pb})_i = 0.8857$ is illustrated. The initial ^{207}Pb/^{206}Pb ratio is given by the *y*-intercept while the age is given by the intersection of the slope and the concordia curve and by the slope of the line (see Equation [3.7]).

Thus, this line is an isochron since its slope depends on t. The goodness of fit of the data to a straight line, of which the mean squared weighted deviation (MSWD) is a measure, depends on the degree to which conditions (1) and (2) here have been met. An analogous and more complete approach involves computing $^{207}Pb/^{206}Pb$, $^{238}U/^{206}Pb$, and $^{235}U/^{207}Pb$ isochrons simultaneously, which Ludwig (1998) calls *total Pb/U isochrons*. It is, however, difficult to represent this graphically, both because it is three-dimensional and because the relationship between $^{207}Pb/^{206}Pb$ and $^{235}U/^{207}Pb$ is non-linear (however, the $^{206}Pb/^{207}Pb$–$^{235}U/^{207}Pb$ relationship is linear). Ludwig (1998) explains that approach in detail.

3.2.2   Th/U ratios

Provided Th/U ratios are constant and known in a set of samples we wish to date, we can calculate ages from $^{208}Pb/^{204}Pb$–$^{206}Pb/^{204}Pb$ isochrons just as we can using ^{207}Pb and ^{206}Pb. However, although U and Th are geochemically similar and the Th/U ratio is not likely to vary much, it would not be prudent to assume the ratio is actually constant for geochronological purposes. Furthermore, there is little reason to do so, since we can already compute the age using ^{207}Pb and ^{206}Pb. However, it may be useful in some circumstances to turn the problem around and compute the Th/U ratio from the age and the slope of the data on a plot of $^{208}Pb/^{204}Pb$ versus $^{206}Pb/^{204}Pb$. The basis of this is as follows. We write the usual growth equations for ^{206}Pb and ^{208}Pb as:

$$^{206}Pb/^{204}Pb = \left(^{206}Pb/^{204}Pb\right)_0$$
$$+\ ^{238}U/^{204}Pb\left(e^{\lambda_{238}t} - 1\right) \qquad (3.8)$$

$$^{208}Pb/^{204}Pb = \left(^{208}Pb/^{204}Pb\right)_0$$
$$+\ ^{232}Th/^{204}Pb\left(e^{\lambda_{232}t} - 1\right) \qquad (3.9)$$

Subtracting the initial ratio from each side of each equation and dividing Equation 3.9 by Equation 3.8, we obtain:

$$\frac{\Delta\left(^{208}Pb/^{204}Pb\right)}{\Delta\left(^{206}Pb/^{204}Pb\right)} = \frac{^{232}Th/^{204}Pb\left(e^{\lambda_{232}t} - 1\right)}{^{238}U/^{204}Pb\left(e^{\lambda_{238}t} - 1\right)}$$
$$(3.10)$$

or

$$\frac{\Delta\left(^{208}Pb/^{204}Pb\right)}{\Delta\left(^{206}Pb/^{204}Pb\right)} = \frac{\kappa\left(e^{\lambda_{232}t} - 1\right)}{\left(e^{\lambda_{238}t} - 1\right)} \qquad (3.11)$$

where κ is used to designate the $^{232}Th/^{238}U$ ratio. Using μ to designate the $^{238}U/^{204}Pb$ ratio, the parent–daughter ratio of the Th–Pb system is the product $\mu\kappa$.

Equation 3.11 tells us that the slope of a line on a plot of $^{208}Pb/^{204}Pb$ versus $^{206}Pb/^{204}Pb$ is proportional to time and κ, provided that κ does not vary among the analyzed samples. If we can calculate t from the corresponding $^{207}Pb/^{204}Pb$–$^{206}Pb/^{204}Pb$ slope, we can solve Equation 3.11 for κ. If, however, κ varies linearly with μ, a straight line will still result on the $^{208}Pb/^{204}Pb$ versus $^{206}Pb/^{204}Pb$ plot and our estimate of κ will be incorrect.

3.3   ZIRCON DATING

Zircon ($ZrSiO_4$) is a mineral with a number of properties that make it extremely useful for geochronologists (Figure 3.3). First of all, it is very hard (with a hardness of 7½), which means it is extremely resistant to mechanical weathering. Second, it is extremely resistant to chemical weathering and metamorphism. For geochronological purposes, these properties mean it is likely to remain a closed system. Third, it concentrates U (and Th to a lesser extent) and excludes Pb, resulting in typically very high $^{238}U/^{204}Pb$ ratios. As we will see, however, the high U content can result in radiation damage and open-system behavior. Nevertheless, it is quite possibly nature's best clock. Finally, it is reasonably common as an accessory phase in a variety of intermediate to siliceous igneous and metamorphic rocks.

The very high $^{238}U/^{204}Pb$ ratios in zircon (and similar high-μ minerals, such as titanite and apatite) provide some special geochronological opportunities and a special diagram, called the *concordia diagram*, first introduced by Wetherill (1956), was developed to take advantage of them. The discussion that follows can be applied to any other minerals with extremely high $^{238}U/^{204}Pb$ ratios; however, in practice, zircons constitute the most common target for Pb geochronologists.

A concordia diagram is simply a plot of $^{206}Pb*/^{238}U$ versus $^{207}Pb*/^{235}U$, i.e., the ratios of the number of atoms of radiogenic

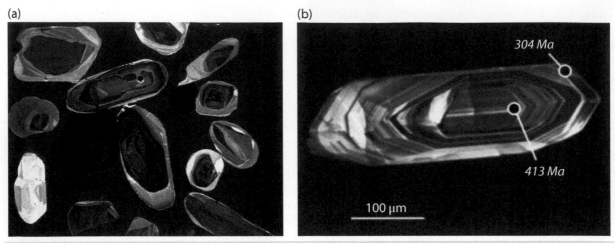

Figure 3.3   (a) Zircon grains viewed under cathode luminescence, which reveals zoning. (b) Spot analyses of a zircon revealing an old core surrounded by young overgrowths. Source: Reproduced with permission from J. D. Vervoort.

daughters produced to the number of atoms of radioactive parents. You should satisfy yourself that both of these ratios are proportional to time. In essence, the concordia diagram is a plot of the ^{238}U–^{206}Pb age against the ^{235}U–^{207}Pb age. The concordia curve on such a diagram that is the locus of points where the ^{238}U–^{206}Pb age equals the ^{235}U–^{207}Pb age. Such ages are said to be *concordant*. Figure 3.4 is an example of a concordia diagram.

One way to think about evolution of Pb/U ratios on the concordia diagram is to imagine that the diagram along with its axes grows over

down and to the left over time, while the actual data point stays fixed (not strictly true, of course). Let us take a 4.0 Ga old zircon as an example. When it first formed, or "closed," it would have plotted at the origin, because had anyone been around to analyze it, they would have found the $^{207}Pb*/^{235}U$ and $^{206}Pb*/^{238}U$ ratios to be 0. Initially, $^{207}Pb*/^{235}U$ would have increased rapidly, while the $^{206}Pb*/^{238}U$ would have been increasing only slowly. This is because 4.0 Ga ago there was a lot of ^{235}U around (recall that ^{235}U has a short half-life). As time passed, the increase in $^{207}Pb*/^{235}U$ would have slowed

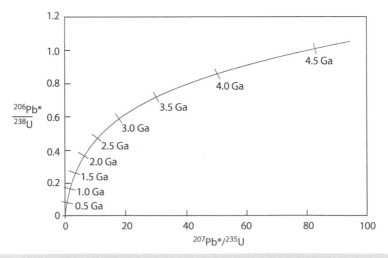

Figure 3.4   The concordia diagram. As usual, the asterisk is used to denote *radiogenic* Pb. In many concordia plots, the asterisk is not used, but it is nevertheless only radiogenic Pb in these ratios.

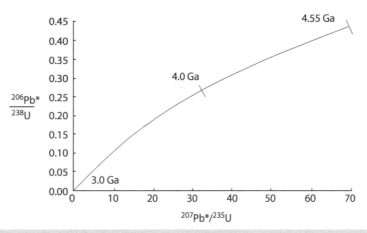

Figure 3.5 A concordia diagram as it would have been drawn at 3.0 Ga.

as the ^{235}U decayed away. We can imagine that the diagram initially "grows" or "expands" to the left, expanding downward only slowly. Had someone been around 3.0 Ga ago to determine "zircon" ages, they would have drawn it as it appears in Figure 3.5 (of course, they would have labeled the 3.0 Ga point as 0, the 4.0 Ga point as 1.0, etc.).

In a zircon that has remained as a completely closed system since its crystallization, the ^{206}Pb*/^{238}U and ^{207}Pb*/^{235}U will change as a function of age in such a way that it will always plot on the concordia line. What happens when a zircon gains or loses U or Pb? Let us take the case of Pb loss, since that is the most likely type of open-system behavior in zircons. The zircon must lose ^{207}Pb and ^{206}Pb exactly in the proportions they exist in the zircon because the two are chemically identical. In other words, a zircon will not lose ^{206}Pb in preference to ^{207}Pb or vice versa.

Let us take the specific case of a 4.0 Ga zircon that experienced some Pb loss during a metamorphic event at 3.0 Ga. If the loss were complete, the zircon would be reset and would be plotted at the origin in Figure 3.5. We could not distinguish it from one that formed 3.0 Ga. Now, suppose that zircon had lost only half its Pb at 3.0 Ga. Because ^{206}Pb and ^{207}Pb would have been lost in the same proportions in which they were present in the zircon, both the ^{206}Pb/^{238}U and ^{207}Pb/^{235}U would have decreased by half. Consequently, the point would have migrated halfway along a straight line between its original position and the origin. At 3.0 Ga, therefore, it would have plotted on a "cord," that is, a straight line, between its

initial position on the concordia curve, the 4.0 Ga point, and the origin at 3.0 Ga. Had other zircons lost some other amount of Pb, say 30% or 80%, they would have been plotted on the same cord, but further or nearer the origin and our concordia plot would appear as it does in Figure 3.6(a). The line is straight because the loss of ^{207}Pb is always directly proportional to the loss of ^{206}Pb. The origin in Figure 3.6(a) corresponds to the 3.0 Ga point on the concordia in Figure 3.6(b), which shows how those zircons would plot today assuming they had remained closed subsequent to the Pb loss at 3 Ga. Therefore, in Figure 3.6(b), the zircons would lay on a cord between the 4.0 Ga point and the 3.0 Ga point. We would say these are "discordant" zircons.

The intercepts of this cord with the concordia give the ages of initial crystallization (4.0 Ga) and metamorphism (3.0 Ga). Therefore, if we can determine the cord on which this discordant zircon lies, we can determine the ages of both events from the intercepts of that cord with the concordia. Unfortunately, if our only data point is a single zircon, we can draw an infinite number of cords passing through this point, so the ages of crystallization and metamorphism are indeterminate. However, we can draw only one line through two points. Therefore, by measuring two zircons (or populations of zircons) that have the same crystallization ages and metamorphism ages, but have lost different amounts of Pb, and hence plot on different points on the same cord, the cord can be determined. The closure age and partial resetting ages can then be determined from the intercepts. As usual in geochronology,

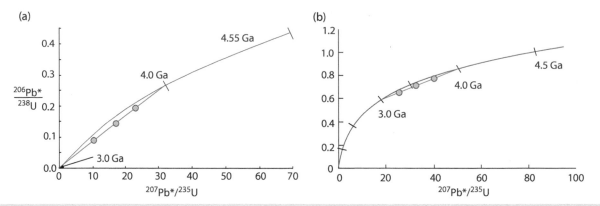

Figure 3.6   (a) Concordia diagram as it would have appeared at 3.0 Ga. Three zircons that experience variable amounts of Pb loss move from the 4.0 Ga point on the concordia curve (their crystallization age) toward the origin. (b) The same three zircons as they would plot at present. The three define a cord between 3.0 and 4.0 Ga. A possible interpretation of this result would be that 4.0 Ga is the crystallization age and 3.0 Ga is the metamorphic age.

however, we are reluctant to draw a line through only two points since any two points define a line; therefore, at least three measurements are generally made. In practice, different zircon populations are selected based on size, appearance, magnetic properties, color, etc. While zircon is generally a trace mineral, only very small quantities, i.e., a few milligrams, are needed for a measurement. Indeed, it is possible to analyze single zircons and even parts of zircons.

U gain, should it occur, would affect the position of zircons on the concordia diagram in the same manner as Pb loss; the two processes are essentially indistinguishable on the concordia diagram. U loss, on the other hand, moves the points away from the origin at the time of the loss (Figure 3.7). In this case, the zircons lie

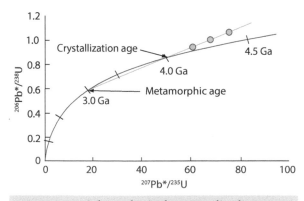

Figure 3.7   A hypothetical concordia diagram showing zircons that crystallized at 4.0 Ga and lost U during metamorphism at 3.0 Ga.

on an extension of a cord above the concordia. As is the case for Pb loss, the upper intercept of the cord gives the initial age and the lower intercept gives the age of U loss. However, such U loss is far less common than Pb loss. This is true for two reasons. First, U is compatible in the zircon, while Pb is not. Second, Pb occupies a site damaged by the alpha-decay process, particularly the recoil of the atom as it undergoes alpha decay, making diffusion out of this site easier. Radiation damage to the crystal lattice is a significant problem in zircon geochronology, and one of the main reasons ages can be imprecise. U-rich zircons are particularly subject to radiation damage. Heavily damaged crystals are easily recognized under the microscope and are termed *metamict*.

Pb gain in zircons does not have predictable effects on the concordia diagram because the isotopic composition of the Pb gained need not be the same as the composition of the Pb in the zircon. Thus, Pb gain would destroy any age relationships. However, Pb gain is far less likely than other open-system behaviors, of which Pb loss is by far the most common.

Zircons that have suffered multiple episodes of open-system behavior will have U–Pb systematics that are difficult to interpret and could be incorrectly interpreted. For example, zircons lying on a cord between 4.0 and 3.0 Ga that subsequently lose Pb and move on a second cord toward the 2.0 Ga could be interpreted as having a metamorphic age of 2.0 Ga and a crystallization age of between 4.0 and 3.0 Ga.

Continuous Pb loss from zircons can also complicate the task of interpretation. The reason is that in continuous Pb loss, zircons do not define a straight line cord, but rather a slightly curved one. Again imagining that the concordia diagram grows with time, a zircon losing Pb will always move toward the origin. However, the position of the origin relative to the position of the zircon moves with time in a non-linear fashion. The result is a non-linear evolution of the isotopic composition of the zircon.

Krogh (1982) showed that metamict regions of zircons could be removed abrasively using small air abrasion chambers designed and built expressly for this purpose. He demonstrated that abraded zircons were typically much more concordant than unabraded ones. Consequently, age uncertainties were considerably reduced. In addition to physical abrasion, numerous attempts have been made to "chemically abrade," or leach, zircons, with the idea being to remove the radiation-damaged regions of the crystal. The most successful of these methods has been that of Mattinson (2005), which involves first annealing the crystals at 800–1000°C for 48 hours (this repairs the radiation damage) before stepwise partial dissolution in acid at progressively higher temperatures. This stepwise dissolution allows for an approach similar to stepwise heating in ^{40}Ar–^{39}Ar dating, as illustrated in Figure 3.8. The uncertainty in this age is less than 0.1%, a level of accuracy otherwise unattainable. Figure 3.9 shows another example of how this technique improves accurate determination of crystallization ages. Three different fractionations of untreated zircons from a Finnish tonalite are discordant, but define a cord with an upper concordia intercept of about 1870 Ma. The chemically abraded zircon fraction is nearly concordant at this age, allowing a much more precise determination. In this diagram, the analyses are plotted by ellipses in order to represent analytical errors. The analytical errors in the ^{207}Pb/^{235}U and ^{206}Pb/^{238}U ratios are highly correlated, hence the elliptical shape. This "chemical abrasion" technique is now widely used in conventional thermal ionization analysis of zircons.

A variety of analytical methods have evolved for zircon analysis. The oldest is thermal ionization mass spectrometry (TIMS), in which U and Pb concentrations are determined by isotope dilution along with their isotope ratios.

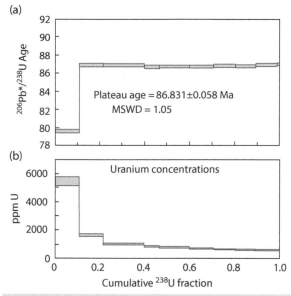

Figure 3.8 (a) Ages calculated from stepwise dissolution of zircons from the Sierra Nevada Batholith that had been abraded and then annealed at 850°C for 48 hours. (b) Corresponding U concentrations in the stepwise dissolution. The outer zones of the zircons are the most U-rich and consequently the most radiation-damaged and discordant. Shaded areas show the uncertainty in each step. Source: Mattinson (2005). Reproduced with permission of Elsevier.

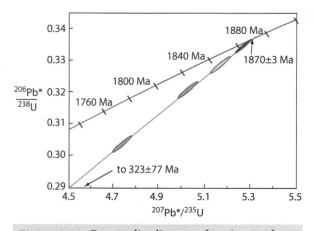

Figure 3.9 Concordia diagram for zircons from Tiirismaa tonalite in Finland. Untreated zircon analyses are shown as light-colored error ellipses, while the chemically abraded zircon analysis is shown as a dark-colored error ellipse. The elliptical shape of the error representation results from the correlation in analytical errors of ^{207}Pb/^{235}U and ^{206}Pb/^{238}U ratios. Source: After Lahtinen and Nironen (2010). Reproduced with permission of Elsevier.

Chemical abrasion is routinely done in connection with this method. This produces the most precise analyses. However, the technique lacks spatial resolution, so ages can be less precise if the zircons are zoned (although it is possible to analyze grain fragments of zircon crystals). Several techniques have been developed for in situ isotopic analysis and these have become increasingly important in geochronology.

The oldest of these is secondary ionization mass spectrometry (SIMS) in which an ion beam is focused at the polished surface of a sample. Atoms on the surface are ionized and swept into a double-focusing mass spectrometer for isotopic analysis. Both U and Pb can be analyzed simultaneously on spots ≈ 30 μm in diameter and 2–3 μm deep. The first such instrument employed for zircon dating was the SHRIMP instrument developed by a team at the Australian National University led by Bill Compston. A significant subsequent advance was the addition of multiple collectors allowing simultaneous measurement of several isotopes (MC-SIMS). Isotope ratios are measured less precisely than with TIMS, but ages can be more precise because a single zone of a crystal and areas free of radiation damage can be analyzed.

The newest technique is laser ablation inductively coupled plasma mass spectrometry (LA-ICP-MS), in which a laser is fired at a small spot on the zircon, ablating the surface. The ablated material is then swept into a mass spectrometer by a stream of Ar gas and analyzed. LA-ICP-MS has somewhat less spatial resolution, but is faster and cheaper than SIMS instruments and provides reasonable precision. The availability of these instruments has resulted in an explosion of zircon dating in the past decade.

U–Pb analysis has improved to the point where uncertainty in decay constants now constitutes a major part of the uncertainty in zircon ages. Because it is far less abundant, the decay constant for ^{235}U is the most uncertain. Based on highly concordant zircon analyses obtained through the chemical abrasion technique, Mattinson (2010) proposed that the ^{235}U decay constant should be revised from the conventionally accepted value of $0.98485 \pm 0.00135 \times 10^{-9}$ yr^{-1} to $0.98571 \pm 0.00012 \times 10^{-9}$ yr^{-1}.

To date, the oldest terrestrial rocks are the Acasta Gneisses of the Slave Province (Northwest Territories, Canada), which are as old as 4.03 Ga (Bowring et al., 1989). These ages were determined using an improved version of the SHRIMP instrument mentioned above to date the cores of zircon crystals extracted from these gneisses. The next oldest rocks are those of the Isua Gneisses in Greenland. These are roughly 3850 Ma old. However, individual zircon crystals have been found that are even older. Given its mechanical and chemical stability, it is not surprising that the oldest terrestrial material yet identified is zircon.

Zircons older than 4 Ga were identified in the Yilgarn Craton in Western Australia, first in the quartzites at Mt. Narryer by Froude (1983) and subsequently in conglomerates of the Jack Hills terrane by Compston and Pidgeon (1986). The area is geologically complex and has experienced multiple episodes of metamorphism. Outcrops include zircon-bearing sandstones that have been metamorphosed to quartzites. The >4 Ga ages were determined using the SHRIMP instrument and were at first quite controversial, but as confidence in this new technique grew through replication by conventional analysis, they were ultimately accepted. As can be seen in Figure 3.3, zoning in zircon is not uncommon and this zoning often reflects multiple episodes of growth. These zircons have had complex histories suffering multiple metamorphic events between 4260 and 2600 Ma, the effect of which was the growth of rims of new material on the older cores. Conventional analysis of these zircons would not have recognized the older ages. The cores of these zircons, however, proved to be nearly concordant at the older ages. Subsequent analyses using successor instruments to SHRIMP revealed ages as old as 4.4 Ga (Figure 3.10; Wilde, et al., 2001). Hadean[4] (i.e., >4 Ga) zircons have subsequently been discovered elsewhere; we will discuss their significance in Chapter 8. Thus, the oldest known terrestrial materials are approaching the oldest ages from other planetary bodies, including the Moon, Mars, and asteroids (as represented by meteorites). They remain, however, significantly younger than

---

[4] The Hadean Eon is defined as the time before the oldest rocks, which are the Acasta Gneisses with an age of 4 Ga.

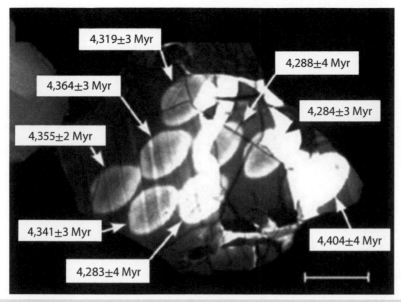

4,319±3 Myr

4,288±4 Myr

4,364±3 Myr

4,284±3 Myr

4,355±2 Myr

4,341±3 Myr

4,404±4 Myr

4,283±4 Myr

Figure 3.10 Photomicrograph of the oldest zircon known. Light areas are regions of ion probe analysis. Ages determined on these areas are shown. Source: Adapted from Wilde et al. (2001). Reproduced with permission of Nature Publishing Group.

the 4.567 Ga age of the Solar System. Nevertheless, these very old ages seem to demonstrate that it is zircons, and not diamonds, that "are forever."

3.4 U-DECAY SERIES DATING

Up to now, we have been discussing decay schemes that are based on measuring the amount of a stable daughter nuclide relative to the amount of the radioactive parent. Since the decay of the parent takes place at an invariant rate, this ratio of daughter to parent is proportional to time (in a closed system). In addition to Pb, decay of U and Th also produces ^4He, and thus the accumulation of ^4He in crystals can also be used for dating, as is discussed in Chapter 4, and a number of intermediate ephemeral radionuclides. In the remainder of this chapter, we will consider how the ratios of these intermediate decay products of U to their parents can be used as geochronological tools. U-decay series dating differs in a very fundamental way from the conventional techniques we have been

discussing. It does, however, share two features. First, the time we are measuring is the time since the system closed, and second, an accurate date requires the system to have remained closed. U-decay series are also useful in inferring the rate and extent of melting in the mantle; we will examine that application in Chapter 7 (Section 7.6). Far more details on U-series dating can be found in the book by Bourdon et al. (2003).

3.4.1 Basic principles

The fundamental principle involved in U-decay series dating is that in a closed system the ratio of parent to daughter will tend toward an equilibrium state in which the rate of decay of the parent is equal to the rate of decay of the daughter. A closed system will approach this equilibrium state at a predictable rate with the ratio of the parent to daughter being proportional to time until equilibrium is reached. Once equilibrium has been achieved, the ratio of parent to daughter no longer depends on the time elapsed, and we can calculate only a minimum age[5].

---

[5] In principle, a system will approach equilibrium asymptotically, and will only achieve equilibrium after an infinite amount of time. In practice, an effective equilibrium is achieved when the difference between the measured ratio of parent to daughter is less than the analytical uncertainty of the measurement. This typically will occur after a maximum of 5–10 half-lives of the nuclide with the shortest half-life.

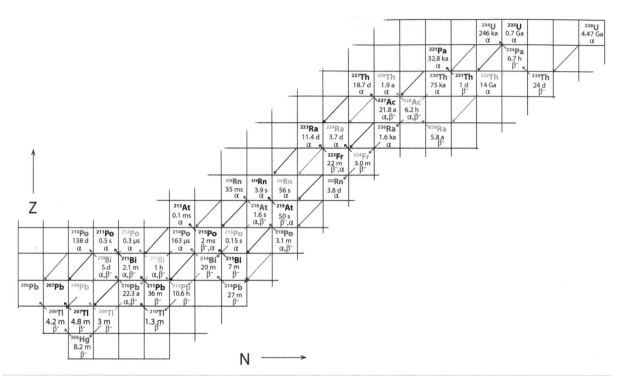

Figure 3.11 Part of the chart of the nuclides showing the series of decays that occur as ^{238}U, ^{235}U, and ^{232}Th are transformed to ^{206}Pb, ^{207}Pb, and ^{208}Pb, respectively. Red denotes the ^{238}U-decay chain, black the ^{235}U-decay chain, and green the ^{232}Th-decay chain. ^{232}Th has no intermediate daughters with sufficiently long half-lives to be geologically useful and ^{231}Pa is the only decay product of ^{235}U with geological applications.

Decay products of U and Th pass through many intermediate radioactive daughters (Figure 3.11) before becoming stable Pb isotopes. Most of these daughters have very short half-lives, ranging from milliseconds to hours, and are of little use in the study of the Earth. However, a number of the intermediate daughters have half-lives ranging from days to hundreds of thousands of years and do provide useful information about geological processes. Table 3.2 lists half-lives and decay constants of some of the most useful of these isotopes. As one might guess from the short half-lives, U-decay series isotopes are used to date relatively recent events.

The half-lives of all of these daughter nuclides are short enough so that any atoms present when the Earth formed have long since decayed (to Pb). They exist in the Earth (and in all other bodies of the Solar System) only because they are continually produced by the decay of the long-lived isotopes of U and Th.

Table 3.2 Half-lives and decay constants of long-lived U and Th daughters.

Nuclide	Half-life (yr)	Decay constant (yr^{-1})	Series
^{234}U	245 620	2.822×10^{-6}	^{238}U
^{231}Pa	32 670	2.116×10^{-5}	^{235}U
^{230}Th	75 584	9.171×10^{-6}	^{238}U
^{226}Ra	1600	4.332×10^{-4}	^{238}U
^{228}Ra	5.75	1.201×10^{-1}	^{232}Th
^{210}Pb	22.23	3.118×10^{-2}	^{238}U
^{210}Po	0.3789	1.829	^{238}U

Source: ^{234}U and ^{230}Th from Cheng et al. (2013); others are from Laboratoire National Henri Becquerel (www.nucleide.org)

The abundance of such a nuclide depends on the balance between its own radioactive decay and the rate at which it is produced by the decay of its parent:

$$\frac{dN_D}{dt} = \lambda_P N_P - \lambda_D N_D \qquad (3.12)$$

where subscripts P and D refer to parent and daughter, respectively. This equation states simply that the rate of change of the abundance of the daughter isotope is equal to the rate of its production minus the rate of its decay. This can be integrated to give:

$$N_D = \frac{\lambda_P}{\lambda_D - \lambda_P} N_P^0 \left(e^{-\lambda_P t} - e^{-\lambda_D t} \right) + N_D^0 e^{-\lambda_D t}$$

$$(3.13)$$

Scientists dealing with these nuclides generally work with *activities* rather than atomic abundances. By activity, we mean rate of decay, dN/dt, measured in disintegrations per unit time[6]. One reason for this is that the abundance of these isotopes was traditionally determined by detecting their decay. Today, abundances of the longer-lived nuclides can be measured by mass spectrometry, but the shorter-lived ones are so rare that they are still most readily detected by their decay. The other reason we work with activities is that it simplifies the math, as will become apparent shortly. We will follow the standard convention of denoting activities by enclosing the isotope or isotope ratio in parentheses. Thus, (^{230}Th) denotes the activity of ^{230}Th and $(^{230}Th/^{238}U)$ denotes the ratio of activities of ^{230}Th and ^{238}U. Activities are related to atomic (or molar) abundances by the basic equation of radioactive decay:

$$\frac{dN}{dt} = -\lambda N \qquad (1.12)$$

Hence, if we know the activity, the molar abundance can be calculated and vice versa.

The *radioactive equilibrium* state of the daughter and the parent is the condition where their *activities* are equal, i.e.:

$$\frac{dN_D}{dt} = \frac{dN_P}{dt} \qquad (3.14)$$

This is the state that will be eventually achieved by any system if it is not perturbed (remains closed).

We can demonstrate that this is so in two ways. The first is a simple mathematical demonstration. The equilibrium state is the steady state where the abundance of the daughter does not change, i.e., where the left-hand side of 3.12 is 0:

$$0 = \lambda_P N_P - \lambda_D N_D \qquad (3.15)$$

We substitute the λN for the dN/dt terms in Equation 3.14, rearrange, and obtain Equation 3.15; *QED*.

The second demonstration is a thought experiment. Imagine a hopper, a grain hopper for example, with an open top and a door in the bottom. The door is spring-loaded such that the more weight placed on the door, the wider it opens. Suppose, we start dropping marbles into the hopper at a constant rate. The weight of marbles accumulating in the hopper will force the door open slightly and marbles will start falling out at a slow rate. Because the marbles are falling out more slowly than they are falling in, the number and weight of marbles in the hopper will continue to increase. As a result, the door will continue to open. At some point, the door will be open so wide that marbles are falling out as fast as they are falling in. This is the steady, or equilibrium, state. Marbles no longer accumulate in the hopper and hence the door is not forced to open any wider. The marbles falling into the door are like the decay of the parent isotope. The marbles in the hopper represent the population of daughter isotopes. Their decay is represented by their passing through the bottom door. Just as the number of marbles

[6] The SI, and therefore official, unit of radioactivity is the Becquerel (abbreviated as Beq) and is equal to one disintegration per second. An older and still used unit is the Curie (abbreviated as Ci) equal to 3.7×10^{10} disintegrations per second – an enormous and dangerous level of radioactivity. Because concentrations of these nuclides in nature are generally low, in practice activity is often expressed in disintegrations per minute (dpm).

passing through the door depends on the number of marbles in the hopper, the activity (number of decays per unit time) of an isotope depends on the number of atoms present.

If the rate of marbles dropping into the hopper decreases for some reason, marbles will fall out of the hopper faster than they fall in. The number of marbles in the hopper will decrease; as a result, the weight on the door decreases and it starts to close. It continues to close (as the number of marbles decreases) until the rate at which marbles fall out equals the rate at which marbles fall in. At that point, there is no longer a change in the number of marbles in the hopper and the position of the door stabilizes. Again equilibrium has been achieved – this time with fewer marbles in the hopper, but nevertheless at the point where the rate of marbles going in equals the rate of marbles going out. The analogy to radioactive decay is exact.

Thus, when a system is disturbed by addition or loss of parent or daughter, it will ultimately return to equilibrium. The rate at which it returns to equilibrium is determined by the decay constants of the parent and daughter. If we know how far out of equilibrium the system was when it was disturbed, we can determine the amount of time that has passed since it was disturbed by measuring the present rate of decay of the parent and daughter.

Improvements in mass spectrometry over the past several decades have made it possible to measure the longer-lived radionuclides, including ^{234}U, ^{230}Th, ^{226}Ra, and ^{231}Pa, by mass spectrometry on small quantities of material and with better precision than by α counting, although the shorter-lived nuclides, such as ^{210}Pb, are still measured by decay counting. Over the last several decades, it has become possible to also measure these radionuclides on the scale of tens of microns using the kinds of ion microprobes used for zircon dating.

3.4.2 ^{234}U–^{238}U dating

The equilibrium, or more precisely the lack thereof, between ^{234}U and ^{238}U can be used to date carbonates precipitated from seawater. As it turns out, (^{234}U) and (^{238}U) in seawater are not in equilibrium; that is, the

(^{234}U/^{238}U) ratio is not 1. It is uniform, however, with a ratio of about 1.146 ± 0.03[7]. ^{234}U occupies damaged lattice sites in crystals of those rocks as a consequence of the α-decay of ^{238}U. This damage results from both the energy of the α-particle and the recoil of the nucleus. Since it occupies a damaged site, ^{234}U is more easily removed from the crystal by weathering than ^{238}U. The oceans collect this "leachate"; hence, they are enriched in ^{234}U. When U precipitates from seawater into, for example, the calcium carbonate in a coral skeleton, the coral will initially have the same (^{234}U/^{238}U) as seawater, but ^{234}U will decay faster than it is created by decay of ^{238}U, so (^{234}U/^{238}U) will slowly return to the equilibrium condition where (^{234}U/^{238}U) = 1. Deviations of the (^{234}U/^{238}U) ratio from the equilibrium value (1) are usually expressed in per mil units and denoted as δ^{234}U. Thus, a (^{234}U/^{238}U) value of 1.145 would be expressed as δ^{234}U = 145.

Let us see how we can take advantage of this to determine geologic time. For a system equilibrium, Equation 3.15 becomes:

$$\frac{dN_D}{dt} = \lambda_P N_p - \lambda_D N_D \qquad (3.16)$$

That is, the change is the difference in rates of production from the parent and decay of the daughter. On integrating this equation to find the number of daughters after time t, we obtain:

$$N_D = \frac{\lambda_P}{\lambda_D - \lambda_P} N_P^0 \left(e^{-\lambda_P t} - e^{-\lambda_D t} \right) + N_D^0 e^{-\lambda_D t}$$

$$(3.17)$$

In the case where the half-life of the parent is very much greater than that of the daughter, i.e., $\lambda_P \ll \lambda_D$, then Equation 3.17 becomes:

$$N_D = \frac{\lambda_P}{\lambda_D - \lambda_P} N_P^0 \left(1 - e^{-\lambda_D t} \right) + N_D^0 e^{-\lambda_D t} \quad (3.18)$$

Noting that activity is simply equal to λN, we can recast this equation in terms of activity as:

$$A_D = \frac{\lambda_D}{\lambda_D - \lambda_P} A_P^0 \left(1 - e^{-\lambda_D t} \right) + A_D^0 e^{-\lambda_D t} \quad (3.19)$$

---

[7] The ratio is uniform in space, but there is evidence to suggest it has varied slightly with time, particularly between glacial and interglacial periods.

where A denotes activity. Furthermore, the $\lambda_D/(\lambda_D - \lambda_P) \cong 1$. Thus, the equation simplifies in such case to:

$$A_D = A_P^0\left(1 - e^{-\lambda_D t}\right) + A_D^0 e^{-\lambda_D t} \qquad (3.20)$$

Just as for other isotope systems, it is generally most convenient to deal with ratios rather than absolute activities (among other things, this allows us to ignore detector efficiency, provided the detector is equally efficient at all energies of interest[8]). Thus, for example, we ratio the activity of ^{234}U to that of ^{238}U:

$$\left(\frac{^{234}U}{^{238}U}\right) = 1 + \left\{\frac{\left(^{234}U\right)^0 - \left(^{238}U\right)}{\left(^{238}U\right)}\right\}e^{-\lambda_{234}t}$$

$$(3.21)$$

or since $^{238}U = {^{238}U^0}$:

$$\left(\frac{^{234}U}{^{238}U}\right) = 1 + \left\{\left(\frac{^{234}U}{^{238}U}\right)^0 - 1\right\}e^{-\lambda_{234}t} \qquad (3.22)$$

Thus, the present activity ratio can be expressed in terms of the initial activity ratio, the decay constant of ^{234}U, and time. For material, such as a coral, in (isotopic) equilibrium with seawater at some time $t = 0$, we know the initial activity ratio was 1.145. Carbonates, for example, concentrate U. If we measure the $(^{234}U/^{238}U)$ ratio of an ancient coral and assume that the seawater in which that coral grew had $(^{234}U/^{238}U)$ the same as modern seawater, the age of the coral can be obtained by solving Equation 3.22 for t. The age determined is the time since the material last reached isotopic equilibrium with seawater.

The application of $^{234}U/^{238}U$ has been largely restricted to corals. It is not generally useful for freshwater carbonates because of uncertainty in the initial activity ratio. Mollusk shells and pelagic biogenic carbonate (e.g., foraminiferal ooze) often take up U after initial deposition of the carbonate and death of the organism, thus violating our closed-system assumption. The technique is typically useful up to about four times the half-life of ^{234}U when alpha spectrometry is the analytical method, but it can be applied to longer times with mass spectrometry because of higher precision. It is often also used in combination with ^{230}Th–^{238}U dating as a check on the "concordancy" of Th–U ages.

The ^{234}U–^{238}U technique does not have high-temperature applications because at high temperature, ^{234}U and ^{238}U do not fractionate as they do at low temperature. The reason is that radiation damage, which is the reason ^{234}U is removed in weathering more easily than ^{238}U, anneals quite rapidly at high temperature.

3.4.3 ^{230}Th–^{238}U dating

Disequilibrium between ^{230}Th and its U parents can provide useful geochronological and geochemical information in both high- and low-temperature systems. We will begin by considering the latter.

3.4.3.1   Low-temperature applications

Looking at Figure 3.11, we see that ^{230}Th is the great-great-great granddaughter of ^{238}U. Two of the intermediate decay products, ^{234}Th and ^{234}Pa, have very short half-lives and will quickly come to equilibrium with ^{238}U. However, ^{234}U has a substantial half-life and, as we discussed in the previous section, will not necessarily be in equilibrium with ^{238}U at low temperatures. In that case, we must take account of the activities of both ^{234}U and ^{238}U. In the case where no ^{230}Th is initially present, the relevant equation, which we give without derivation, is:

$$\left(\frac{^{230}Th}{^{238}U}\right) = 1 - e^{-\lambda_{230}t}$$
$$+ \left[\left(\frac{^{234}U}{^{238}U}\right) - 1\right]\frac{\lambda_{230}}{\lambda_{230} - \lambda_{234}}\left(1 - e^{-(\lambda_{234} - \lambda_{230})t}\right)$$

$$(3.23)$$

(see Ivanovich et al., 1992 for the derivation). Where some ^{230}Th is initially present, we need to account for its decay and our equation becomes:

$$\left(\frac{^{230}Th}{^{238}U}\right) = \left(\frac{^{232}Th}{^{238}U}\right)\left(\frac{^{230}Th}{^{232}Th}\right)^0\left(e^{-\lambda_{230}t}\right)$$

$$+ 1 - e^{-\lambda_{230}t}\left[\left(\frac{^{234}U}{^{238}U}\right) - 1\right]$$

$$\frac{\lambda_{230}}{\lambda_{230} - \lambda_{234}}\left(1 - e^{-(\lambda_{230} - \lambda_{234})t}\right)$$

$$(3.24)$$

---

[8] In the case of ^{238}U and ^{234}U, the α energies are quite similar (4.2 and 4.7 MeV).

One can correct for any ^{230}Th initially present by measuring the ^{232}Th and assuming an initial value of (^{230}Th/^{232}Th), e.g., a value equal to the modern value of the solution from which the calcite precipitated, such as modern seawater or modern cave water. Quite often, however, the initial Th comes from silicate grains included in the carbonates. In this case, we can assume a typical ^{232}Th/^{238}U for crustal rocks of ≈ 3.8, which implies a (^{230}Th/^{232}Th) of 0.814, assuming secular equilibrium. Both Equations 3.23 and 3.24 are yet additional examples of an equation that must be solved indirectly.

Let us first consider a simpler case where no U is initially present. We can start with an equation analogous to Equation (2.4):

$$\left(^{230}\text{Th}\right)_u = \left(^{230}\text{Th}\right)^0 e^{-\lambda_{230}t} \qquad (3.25)$$

where the subscript u denotes that it is "unsupported" by U decay. As usual practice, we normalize to another isotope, and in this case ^{232}Th is the only available choice, so our equation becomes:

$$\left(\frac{^{230}\text{Th}}{^{232}\text{Th}}\right)_u = \left(\frac{^{230}\text{Th}}{^{232}\text{Th}}\right)^0 e^{-\lambda_{230}t} \qquad (3.26)$$

^{232}Th, of course, is radioactive but, with a half-life of 14Ga, its abundance will not change on time scales comparable to the half-life of ^{230}Th.

In seawater, U is in its oxidized state and is quite soluble. Th, however, is quite insoluble: its seawater residence time is 300 years or less compared to about 500 000 years for U. (It should be noted here that solubility in seawater does not control concentrations or residence times. Nevertheless solubility is a good guide to both of these.) Once a ^{234}U atom decays to ^{230}Th, it is quickly absorbed onto particles that in turn are quickly incorporated into sediment. As a result, relatively high concentrations of unsupported ^{230}Th can be present in authigenic sediments, such as manganese nodules, and can also be removed (by leaching) from some sediments. In cases where the amount of leached ^{234}U is negligible and where the (^{230}Th/^{232}Th)^0 is known (from,

e.g., zero-age sediment at the seawater–sediment interface), Equation 3.26 can be used to determine the age of the sediment.

Manganese nodules grow by precipitation of Mn–Fe oxides and hydroxides from seawater and they incorporate Th and exclude U. They are known to grow very slowly, but how slowly? If we assume the rate of growth is constant, then depth in the nodule should be proportional to time. If z is the depth in the nodule and s is the growth (sedimentation) rate, then time is:

$$t = z/s \qquad (3.27)$$

and Equation 3.26 becomes:

$$\left(\frac{^{230}\text{Th}}{^{232}\text{Th}}\right) \cong \left(\frac{^{230}\text{Th}}{^{232}\text{Th}}\right)^0 e^{-\lambda_{230}z/s} \qquad (3.28)$$

Let us consider a specific example of a nodule from the central equatorial Pacific analyzed by Han et al. (2003). They analyzed Th and U isotopes in thirteen 0.1 mm thick layers of a nodule through a depth of 1.3 mm, recovering between 0.4 and 3 mg per sample. (^{230}Th/^{232}Th) ratios as a function of depth are shown in Figure 3.12. (^{238}U/^{232}Th) ratios were small (≈ 0.005), smaller than the error on the (^{230}Th/^{232}Th) ratios.

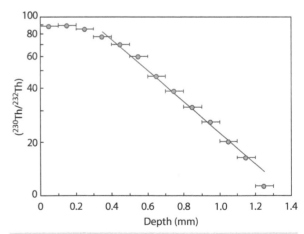

Figure 3.12 (^{230}Th/^{232}Th) as a function in depth in a manganese nodule from the central Pacific studied by Han et al. (2003). The red line denotes the regression fit using Equation 3.29. Data from Han et al. (2003).

Taking the log of Equation 3.28, we can see that the slope of a line drawn through the data is equal to $-\lambda/s$ and the intercept is the log of the initial ratio (the ratio at the very surface):

$$\ln\left(\frac{^{230}Th}{^{232}Th}\right) \cong \ln\left(\frac{^{230}Th}{^{232}Th}\right)^0 - \frac{\lambda_{230}}{s}z \quad (3.29)$$

We can determine the slope and intercept by regressing the natural log of the measured ($^{230}Th/^{232}Th$) ratios against depth. The topmost three samples contained younger and older material due to the curvature of the nodule and there was evidence of a depositional hiatus between the last and the second to last sample, so Han et al. (2003) discarded the first three samples and the last sample. Using the remaining nine samples in our regression, we calculate an average sedimentation rate of 4.58 mm/Ma and an initial ($^{230}Th/^{232}Th$) of 166. In detail, the nodule is clearly banded, which Han et al. (2003) inferred to be related to Milankovitch climate cycles. Assuming the banding was indeed Milankovitch related, the timing of which is known, Han et al. used astronomical calibration to calculate a sedimentation rate of 4.5 mm/Ma.

Another simple case occurs where no Th is initially present. This is often the case for carbonates, which concentrate U and exclude Th. Some of the most successful applications of ^{230}Th dating have been in determining the age of carbonates, such as corals, mussel shells,

and speleothems (carbonates precipitated from water moving through limestone caves, including stalactites, stalagmites, and flowstone). Studies of speleothems have provided useful paleoclimatic information as the rate and composition of water flowing through caves have varied with past climate. The limestone Sanbao cave in Central China has been studied by a variety of workers and provides a good example of what can be learned by high-precision dating of cores taken from speleothems. Figure 3.13 shows the results of Cheng et al. (2016) of Th–U dating samples taken from stalagmite at intervals of centimeters and less from this cave. In most cases, correction for initial ^{230}Th was less than the analytical error, but in some instances, significant initial ^{230}Th was present, which accounts for the unevenness of the $^{230}Th/^{238}U$ curve. After correction for this initial ^{230}Th, a smoother curve with age is obtained, going back 600 000 years. Measurement of oxygen isotope ratios in these samples provides a record of changing climate and Asian monsoon rainfall over this period. We will return to a discussion of the Sanbao cave $\delta^{18}O$ data and climate in Chapter 12.

The range of ages over which $^{230}Th–^{238}U$ dating can be used covers the development of our species and closest *Homo* relatives, and consequently this has become an important tool in human evolution and archeology. Caves are interesting in this respect because they contain some of the oldest examples of art. The ability to produce and appreciate art

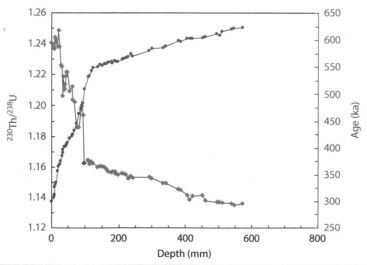

Figure 3.13 Measured ($^{230}Th/^{238}Th$) ratios and ages (corrected for initial ^{230}Th) calculated using Equation 3.24 for samples taken from stalactite SB-14 of Sanbao cave. Data of Cheng et al. (2016).

seems to be a uniquely human trait, and one that apparently developed rather late in our evolution. Where the pigment is an organic substance, such as charcoal and the art young enough, carbon-14 dating can be used. However, many cave paintings were done with mineral pigments, such as red ocher, which cannot be so dated. Pike et al. (2012) sampled thin layers of calcite flowstone that precipitated over Paleolithic cave paintings and engravings in a variety of caves in northern Spain and determined ^{230}Th–^{238}U ages ranging from 22 000 to 41 000 years (see Problem 7), which are minimum ages of the art since the flowstone was deposited over the art.

The study of Hoffmann et al. (2018) indicates that cave art in Maltravieso Cave in Spain, including the hand stencil in Figure 3.14, is even older. Hoffmann et al. carefully scraped thin layers of flowstone overlying the art and performed Th–U analysis on 53 samples. Although some of the flowstone is as young as 32 000 years, they concluded the oldest art has a minimum age of 64 800 years. Because modern *Homo sapiens* did not arrive in Europe until 40 000–45 000 years ago, Hoffmann et al. concluded that this art was made by Neanderthals, indicating they were capable of abstract and symbolic thinking. These results, unsurprisingly, have proved controversial, challenging the conventional wisdom that art and abstract thinking are unique to modern *Homo sapiens,* but they are consistent with a number of even older

Neanderthal artifacts that could only have served as "jewelry," which is also indicative of abstract thinking.

^{230}Th dating of fossil reef corals has proved extremely useful in a number of ways; for example, it has been used to calibrate the ^{14}C timescale (in the next chapter, we will discuss ^{14}C dating and why it must be calibrated) and has provided a timescale for Pleistocene climate change. Pleistocene climate swings have been accompanied by the fall and rise of sea level as ice sheets advance and retreat. We will discuss climate change, the Pleistocene ice ages, and their causes more extensively in Chapter 12; herein, will focus only on deducing its timescales. Reef-building corals contain photosynthetic symbiotic algae, which require that they grow near the ocean surface to maximize light. Thus, dating reef corals allows us to date change in sea level and put a precise timescale on changing climate. This approach was pioneered by Edwards et al. (1987) and many other groups of scientists have expanded on this work (e.g., Bard et al., 1990; Fairbanks et al., 2005; Peltier and Fairbanks, 2006). These studies have revealed that the last glacial maximum occurred around 26 000 years ago when sea level was 140 m below the present one, that the glacial episode began to end around 14 000 years ago and was briefly interrupted by the Younger Dryas event at around 12 000 years ago, and that sea level has been *comparatively* stable over the last few millennia.

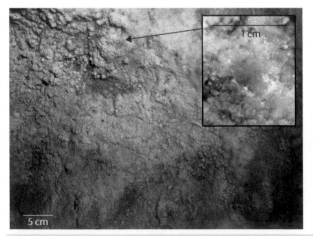 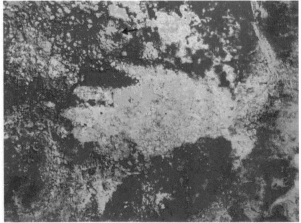

Figure 3.14 Left: A photograph of flowstone over a hand stencil in Maltravieso Cave in Spain. The inset shows one area of carbonate that was sampled for U–Th dating. Right: The same picture digitally processed to enhance color contrast to bring out the underlying image. Source: Hoffmann et al. (2018), American Association for the Advancement of Science.

3.4.3.2 High-temperature applications

In high-temperature systems, we can generally assume that $(^{234}U) = (^{238}U)$, which simplifies the mathematics somewhat. With this assumption, we can treat the production of ^{230}Th as if it were the direct decay product of ^{238}U. We write an equation analogous to Equation 3.13 and from it derive:

$$\left(\frac{^{230}Th}{^{232}Th}\right) = \left(\frac{^{230}Th}{^{232}Th}\right)^0 e^{-\lambda_{230}t} + \left(\frac{^{238}U}{^{232}Th}\right)\left(1 - e^{-\lambda_{230}t}\right)$$

(3.30)

(the tricks to this derivation are to make the approximations $\lambda_{230} - \lambda_{238} = \lambda_{230}$ and $e^{-\lambda_{238}t} = 1$; that is, assume $\lambda_{238} \approx 0$; this is the mathematical equivalent of assuming the activity of ^{238}U does not change with time). The first term on the right describes the decay of unsupported ^{230}Th while the second term describes the growth of supported ^{230}Th. Note that this equation has the form of a straight line in $(^{230}Th/^{232}Th) - (^{238}U/^{232}Th)$ space where the first term is the intercept and $(1 - e^{-\lambda_{230}t})$ is the slope. This is illustrated in Figure 3.15.

Imagine a crystallizing magma with homogeneous $(^{230}Th/^{232}Th)$ and $(^{238}U/^{232}Th)$ ratios. Th and U will partition into different minerals to different degrees. The minerals will have homogeneous $(^{230}Th/^{232}Th)$ (assuming crystallization occurs quickly compared to the half-life of Th), since these two isotopes are chemically identical, but variable $(^{238}U/^{232}Th)$ ratios. Thus, the minerals will plot on a horizontal line in Figure 3.15. After the system closes, ^{238}U and ^{230}Th will begin to come to radioactive equilibrium (either ^{230}Th will decay faster than it is produced or vice versa, depending on whether $(^{230}Th/^{238}U)$ is greater than or less than 1, the equilibrium value). Thus, the original horizontal line will rotate, as it does in a conventional isochron diagram. The rotation occurs about the point where $(^{230}Th/^{232}Th) = (^{238}U/^{232}Th)$ known as the *equipoint*. As t approaches infinity, the exponential terms approach 1 and:

$$\lim_{t \to \infty} \left(\frac{^{230}Th}{^{232}Th}\right) = \left(\frac{^{238}U}{^{232}Th}\right)$$

(3.31)

Thus, the *equilibrium* situation, the situation at $t = \infty$, is $(^{230}Th/^{232}Th) = (^{238}U/^{232}Th)$. In this case, all the minerals will fall on a line,

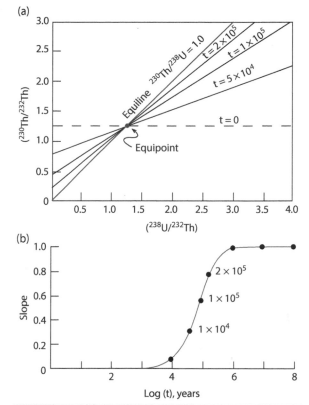

Figure 3.15 (a) $^{230}Th-^{238}U$ isochron diagram. The initial $(^{238}U/^{232}Th)$ is given by the intersection of the isochron with the equiline. (b) This shows how the slope changes as a function of time. Source: Faure (1986)/with permission of Wiley.

having a slope of 1. This line is known as the *equiline*.

The study of Jicha et al. (2005) of the Seguam Volcano in the Aleutian Island arc provides an excellent example of Th–U dating. They carried out Th–U and $^{40}Ar/^{39}Ar$ dating on 16 lava samples, including two historic flows (1977 and 1993). The Th–U isochrons for two of these are shown in Figure 3.16: the 1993 basaltic andesite flow and a rhyolite flow that yielded a $^{40}Ar/^{39}Ar$ of 76.8 ± 1.1 ka. The slope of the Th/U isochron for the 1993 flow is horizontal, consistent with its 0 age, while the rhyolite has a slope that yields an age calculated from Equation 3.30 of 75.6 ± 5.4 ka; this agrees well within error with the $^{40}Ar/^{39}Ar$ age. The two lavas also have distinctly different initial $(^{238}U/^{232}Th)$ ratios; the difference is consistent with the differing composition of the lavas and with compositional trends over time in this volcano.

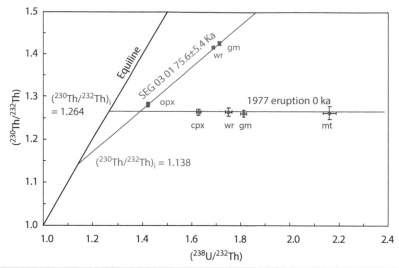

Figure 3.16 $(^{230}Th/^{232}U)$–$(^{238}U/^{232}Th)$ isochron diagram for two mineral separates from the 1993 basaltic andesite lava and rhyolite sample SEG 03 01 from Seguam Island volcano in the Aleutians. gm: groundmass; wr: whole rock; cpx: clinopyroxene; opx: orthopyroxene; mt: magnetite.

The Seguam Volcano case is a particularly simple one; Th–U ages do not always agree with ages obtained by other techniques. The disagreement between Th/U ages and those of other techniques can reflect different closure times and temperatures. For example, the K–Ar system closes only upon eruption while the Th/U system gives the age of crystallization of the minerals analyzed, which often precede eruption. While this might seem problematic, it can be quite useful understanding volcanic and magmatic evolution. Indeed, understanding magma processes beneath volcanoes has become one of the focuses of U-series dating. We will consider only a few examples here; Schmitt (2011) and Cooper (2015, 2019) provide more extensive reviews.

One of the first such studies was that of Reid et al. (1997) on the Long Valley magmatic system, a large silicic "supervolcano" on the western side of the Sierra Nevada in California. They analyzed U–Th disequilibria in zircons in two rhyolite flows, namely, Deer Mountain with an eruption age of 115 ka and Deadman Dome with an eruption age of ≈600 years. While some zircons had model U–Th ages matching the age of the Deer Mountain flow, ages of zircons from both flows clustered around 230 ka, indicating the magma erupted in those flows had been resident in the magma chamber for >100 ka.

One of the largest eruptions over the period of human existence was the eruption of Toba volcano in Sumatra, Indonesia, 73 000 years ago, which erupted 3000 km³ of magma. The resulting caldera, now filled by Lake Toba, is enormous, 100 × 30 km. How long had that magma resided in the magma chamber? Using ion probe ^{230}Th–^{238}U dating of allanite crystals from the pyroclastic products of the 73 ka before present (BP) eruption, Vazquez and Reid (2004) found that allanite cores had crystallized between 100 and 225 thousand years ago, whereas most rims had ages identical with error of the eruption age.

When combined with chemical analyses, Th–U dating of accessory minerals, such as zircon and allanite, can reveal much about the workings and history of magma chambers. The study of Barboni et al. (2016) of the Soufrière Volcanic Complex of St. Lucia in the Lesser Antilles provides an example. The eruption of the Micoud pyroclastic flow at around 640 ka marks a shift in volcanic style from primarily effusive eruption of basaltic and basaltic andesite lavas from widely dispersed vents to dacitic eruptions, including both domes and large Plinian explosive eruptions, from the Soufrière Volcanic Complex on the west coast of the island. The most recent eruption, at 13.6 ± 0.4 ka, is the dacitic Belfond Dome, and it contains numerous cognate mafic enclaves of basaltic andesite and andesitic composition. Barboni et al. (2016) determined U–Th ages, rare earth element (REE) concentrations, and crystallization temperatures of zircons using the titanium-in-zircon geothermometer (Watson et al., 2006) in both the host lava and the mafic enclaves. The lava contains zircons ranging in age from the eruption age to ≈240 ka which indicates that the dacitic magma body was largely maintained at temperatures between 700 and 800°C over this period. Europium anomalies measured by Eu/Eu* show a decrease systematically indicating continual plagioclase crystallization over this period. Zircons from the mafic enclaves tell a somewhat different story with temperature increases and Eu/Eu* increases indicative or magma recharge at around the time of the La Point pyroclastic flow at ≈60 ka and again in the last ≈20 ka corresponding to the eruption of the Belfond pyroclastic flow and the Belfond and Terre Blanche domes. This clearly implies a heterogeneous magma chamber in which recharge repeatedly reheated the more mafic, and presumably deeper, parts but did not reheat the dacitic, and presumably upper, parts of the chamber despite triggering eruptions.

3.4.4 ^{226}Ra dating

^{226}Ra, the decay product of ^{230}Th, is another relatively long-lived nuclide ($\tau_{1/2} = 1600$ yr) that has proved useful in constraining the timing of recent volcanic eruptions and magmatic processes. The fundamentals are precisely analogous to those we have discussed for ^{234}U and ^{230}Th, and the half-life of ^{226}Ra is sufficiently short that in most cases we can assume the abundance of ^{230}Th is constant and Equation 3.20 applies (cases where the initial Th/U ratio is very small, such as corals would be exceptions). Unfortunately, Ra has no stable isotope to which one can ratio ^{226}Ra, so the assumption is sometimes made that Ra behaves as Ba (which sits directly above Ra in the periodic table), and the abundance of Ba is used to form a ratio. This assumption, however, does not hold in the exact. Both Ra and Ba are alkaline earth elements with a charge of 2+, but the ionic radius of Ra (162 pm) is greater than that of Ba (149 pm), which results in it being less compatible in most minerals. As a consequence, precise isochron ages cannot usually be obtained, but useful constraints on ages can nevertheless be derived. To begin with, where ^{226}Ra and

^{230}Th are not in isotopic equilibrium, we can infer that the age is less than ≈10 000 years (depending on analytical precision), assuming the system has not been disturbed (e.g., by weathering), but even stronger constraints can generally be derived.

As with U–Th, ^{226}Ra–^{230}Th mineral ages are often older than eruption ages, recording the time of crystallization rather than eruption. Cooper et al. (2001) analyzed ^{230}Th, ^{226}Ra, and Ba in augitic pyroxene, plagioclase, and groundmass in lava from the 1955 East Rift eruption of Kilauea. They used available partition coefficients and partitioning models to calculate the melt (^{226}Ra)/Ba ratio in equilibrium with the plagioclase and augite, and assumed that the groundmass (^{226}Ra)/Ba ratio represents the present ratio of the melt. Figure 3.17 illustrates how ^{226}Ra/Ba ratios in these phases evolve over time. Because augite is so strongly depleted in ^{226}Ra with respect to ^{230}Th, the (^{226}Ra)/Ba of augite evolves rapidly such that ≈900 years of radiogenic ingrowth from a ratio that would be in equilibrium with the groundmass is required. The age of crystallization is given with the calculated plagioclase-melt, and augite-melt, and groundmass curves intersect. With uncertainties, the intersection occurs at $1000^{+300}/_{-400}$ yr (see the hashed area on Figure 3.17). This is presumably the time the magma was resident in the shallow crust.

Using a similar approach, Eppich et al. (2012) obtained Ra–Th model ages for large plagioclases in the Timberline (≈1500 years) and Old Maid (250 years) eruption of Mt. Hood in the Oregon Cascades of >4.5 and >5.5 ka, respectively.

3.4.5 ^{231}Pa–^{235}U dating

As Figure 3.11 shows, ^{231}Pa, which has a half-life of 32 000 years, is the granddaughter of ^{235}U, but the intervening nuclide ^{231}Th has a sufficiently short half-life that we can always assume it is in equilibrium with its parent and hence can ignore it. Protactinium is typically in the 5+ valance state with an ionic radius of 92 pm and partitions readily into the same igneous and metamorphic accessory minerals that concentrate U and Th, but like Th, it is excluded from carbonates. Also like Th and unlike U, Pa is generally immobile. Precisely, the same principles apply to

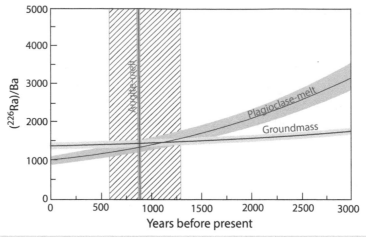

Figure 3.17 Evolution diagram for (^{226}Ra)/Ba ratios in the 1955 East Rift lava of Kilauea. Curves shown are for the groundmass (red) and calculated ratio of melt in equilibrium with augite (green) and plagioclase (blue). The intersection of the three curves gives the age of crystallization, which is between 600 and 1300 years (hashed area). Source: Adapted from Cooper et al. (2001).

^{231}Pa–^{235}U dating as to ^{238}U-decay series dating. We can derive the following relationship of the dependence of the (^{231}Pa/^{235}U) ratio on age from Equation 3.13:

$$\left(\frac{^{231}\text{Pa}}{^{235}\text{U}}\right) = \frac{\lambda_{231}}{\lambda_{231} - \lambda_{235}}\left(e^{-\lambda_{235}t} - e^{-\lambda_{231}t}\right)$$

$$+ \left(\frac{^{231}\text{Pa}}{^{235}\text{U}}\right)^0 e^{-\lambda_{231}t} \qquad (3.32)$$

In almost all instances, we can take λ_{235} to be 0, so that Equation 3.32 simplifies to:

$$\left(\frac{^{231}\text{Pa}}{^{235}\text{U}}\right) = \left(1 - e^{-\lambda_{231}t}\right) + \left(\frac{^{231}\text{Pa}}{^{235}\text{U}}\right)^0 e^{-\lambda_{231}t}$$

$$(3.33)$$

Compared to dating based on the ^{238}U–^{230}Th decay series, the ^{231}Pa–^{235}U decay has a few advantages: there is no long-lived intermediate comparable to ^{234}U, the half-life of ^{235}U is less than a sixth of that of ^{238}U and thus it decays more rapidly, and ^{231}Pa decays more rapidly than ^{230}Th. These are outweighed by a number of disadvantages: the abundance of ^{235}U is less than 1% of that of ^{238}U, ^{231}Pa is less abundant than ^{230}Th, ^{231}Pa decay constant is less well known than that of ^{234}U or ^{230}Th, and the short half-life means it is useful over a smaller range of time. Therefore, why would anyone bother?

The answer is that, just as for the ^{235}U–^{207}Pb chronometer, its power comes not from using it in isolation, but by using it in combination with the ^{238}U-decay chain (Edwards et al., 1997; Grün et al., 2010). Just as was the case for U–Pb, we expect ages based on the ^{235}U decay to be the same as those based on the ^{238}U decay if the system is undisturbed. Indeed, we can create a diagram, Figure 3.18, which has many similarities to the concordia diagram shown in Figure 3.4. The curved line on Figure 3.18 is the concordia curve, which is the locus of points where the ^{231}Pa∗/^{235}U age equals the ^{230}Th∗/^{234}U age (as is the case in the concordia diagram in Figure 3.4, the ratio is corrected for any initially present daughter). The ^{231}Pa∗/^{235}U–^{230}Th∗/^{234}U concordia diagram is similar to the ^{2071}Pb∗/^{235}U–^{206}Pb∗/^{238}U one in that U-loss will move points away from the origin and U-gain will move them toward the origin. However, Pa and Th need not be lost or gained in proportions in which they are present in the specimen, so loss or gain of daughters can result in points moving in any direction. However, in its oxidized state at the Earth's surface, U is more mobile than either Pa or Th, so U-disturbance is probably the most likely form of open-system behavior.

One other difference with the U–Pb concordia diagram is that the concordia curve depends on the (^{234}U/^{238}U) initial ratio, because the ^{230}Th∗/^{234}U age depends on it. This is illustrated in the inset in Figure 3.18: the blue curve

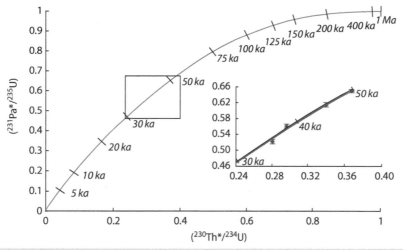

Figure 3.18 A $^{231}Pa*/^{235}U$–$^{230}Th*/^{234}U$ concordia diagram. The line is the locus of points where the $^{231}Pa*/^{235}U$ age equals the $^{230}Th*/^{234}U$ age. The inset is the rectangular region blown up; the blue concordia curve denotes $(^{234}U/^{238}U)^0 = 1.0$ and the red one $(^{234}U/^{238}U)^0 = 1.145$. Also shown are coral data from Araki Island from Chiu et al. (2006).

denotes $(^{234}U/^{238}U)^0 = 1.0$, as might be relevant to a magmatic system, while the red curve denotes $(^{234}U/^{238}U)^0 = 1.145$, relevant to carbonates precipitated from seawater, such as corals. The difference is not great; indeed, at the scale of the larger diagram, the two curves are virtually indistinguishable.

The key feature of the diagram is that Pa–Th–U ages should, if undisturbed, plot on the concordia curve. Shown in the inset are data from corals from the Araki Island, Vanuatu, New Hebrides archipelago, analyzed by Chiu et al. (2006). The data plot within error on the concordia diagram, assuring us that the corals are likely undisturbed and the ages accurate. To date, there are still few examples of combined $^{231}Pa*/^{235}U$–$^{230}Th*/^{234}U$ dating, largely because of the challenges associated with measuring ^{231}Pa accurately. As analytical techniques continue to improve, however, we can expect to see these systems used in combination more commonly.

3.4.6 ^{210}Pb dating

^{210}Pb has a much shorter half-life (22.2 years) than the nuclides we have discussed thus far, so it is useful for much shorter time intervals. It is the great, great … granddaughter (six intermediate nuclides) of ^{226}Ra, but all the intermediate nuclides are quite short-lived. ^{222}Rn, with a half-life of three days, is the longest-lived, and its existence in the decay chain is significant

because it is a noble gas and hence can diffuse out of soils, sediments, magmas, etc., and into the atmosphere. ^{210}Pb produced by decay of atmospheric radon will quickly be removed from the atmosphere by absorption onto aerosols and find its way into sediments, either through wet or dry deposition. Consequently, young sediments commonly have ^{210}Pb activities in excess of that in equilibrium with its long-lived parent, ^{226}Ra. The decay of this "excess" or "unsupported" ^{210}Pb thus provides a convenient means of dating young sediments (including snow and ice), i.e., those less than 100–200 years old. Because its abundance is low, its activity is always determined by counting rather than mass spectrometry. This can be done directly on sediment cores using gamma-ray spectrometers (Kirchner, 2011). Because its pathway into sediments can differ from that of stable Pb, it is not useful to ratio ^{210}Pb to one of the other isotopes of Pb.

We can write an equation analogous to the one we wrote for unsupported ^{230}Th (Equation [3.25]):

$$\left(^{210}Pb\right)_u = \left(^{210}Pb\right)_u^0 e^{-\lambda t} \qquad (3.34)$$

We might continue by using the same approach to determine sedimentation rates that we used in the case of ^{230}Th dating of a manganese nodule, but there are several reasons why we should not. First, detrital sediments, unlike chemical ones, such as manganese nodules, inevitably undergo compaction as they accumulate, meaning that density will increase downward in a core and hence the sedimentary mass above a given depth, say z_i, will not be a linear function of z. Thus, in attempting to calculate ages or sedimentation rates of accumulating detrital sediments, we need to correct for compaction by measuring and taking account of down-core changes in density (or, equivalently, porosity). To correct for compaction, we replace depth, z, with a function called mass depth, which has units of mass per unit area. We define Δm_i as the mass of sediment in the core in depth interval Δz_i and the mass depth, m_i, at depth z_i as the total mass (per cross-sectional area of the core) above z_i.

Next, we need to think about how ^{210}Pb accumulates in sediment. In the case of the manganese nodules discussed previously, we assumed a constant sedimentation rate and a constant flux of ^{230}Th to the surface of the nodule. These are reasonable assumptions in that case, because the process of Mn nodule growth is one of chemical precipitation and adsorption and occurs in the remote areas of the deep ocean where environmental conditions are near-constant. Accumulation of ^{210}Pb in detrital sediments in lakes and coastal areas might not meet those conditions. For example, if the ^{210}Pb in the sediment derived entirely or nearly so from dry or wet deposition of atmospheric ^{210}Pb on the surface of the lake, we might assume that the flux of ^{210}Pb to the sediment surface is constant and independent of the rate at which sedimentary mass accumulates. This is known as the *constant flux* or *constant rate of supply* model. Alternatively, if Pb absorbed onto sedimentary particles carried into the lake by streams is the primary ^{210}Pb source, we might assume that these particles always had the same concentration of ^{210}Pb. This is known as the *constant activity* or *constant initial concentration* model. Yet another possibility is that the sedimentary flux (mass accumulation rate) is constant, but the Pb flux varies, known as the *constant sedimentation* model (Sanchez-Cabeza and Ruiz-Fernandez, 2012). Yet another possibility is the *constant flux–constant sedimentation* model. This is the simplest model because it predicts a simple exponential decrease of activity with depth, once compaction has been accounted for. The advantages and disadvantages of these approaches have been discussed by a number

of authors over the years (e.g., Appleby and Oldfield, 1983; Sanchez-Cabeza and Ruiz-Fernandez, 2012). Let us briefly review them in a bit more detail.

In the *constant activity* model, the activity at the surface, $(^{210}\text{Pb})^0$, is constant so that the activity in any layer i at depth is given by Equation 3.34, and the age of that layer may be calculated as:

$$t_i = \frac{1}{\lambda} \ln \frac{(^{210}\text{Pb})^0}{(^{210}\text{Pb})_i} \qquad (3.35)$$

The initial activity can be found by regressing $\ln(^{210}\text{Pb})$ versus (compaction-corrected) depth and finding the intercept at 0 depth.

In the *constant flux* model, we assume that the flux of ^{210}Pb to the surface of the sediment, f, is constant and independent of the mass accumulation rate, r. Thus, at the time of deposition of layer i, the total accumulated ^{210}Pb in the sediment column below it, $A(0)$, will be constant, where $A(0)$ is defined as:

$$A(0) = \int_0^\infty (^{210}\text{Pb}) \, dm \qquad (3.36)$$

where m is mass depth as defined above. The total activity below layer i is then:

$$A(i) = \int_{m_i}^\infty (^{210}\text{Pb})_i \, dm_i \qquad (3.37)$$

It follows that:

$$A(i) = A(0)e^{-\lambda t_i} \qquad (3.38)$$

and the age of layer i is then:

$$t_i = \frac{1}{\lambda} \ln \frac{A(0)}{A(i)} \qquad (3.39)$$

This model requires determining the entire inventory of ^{210}Pb in the core. In practice, this would mean over the last 100–200 years (depending on the sensitivity of the analytical method used). If this entire interval is not sampled, it may be possible to extrapolate to the depth where unsupported (^{210}Pb) is 0.

In the *constant flux–constant sedimentation* model, the initial concentration is assumed to be constant and activity will decrease exponentially with mass depth. With r as the mass accumulation rate and m as mass depth, in a manner analogous to Equation 3.27, we let $t = m/r$, and we obtain:

$$\left(^{210}\text{Pb}\right) = \left(^{210}\text{Pb}\right)^0 e^{-\lambda_{210}m/r} \qquad (3.40)$$

Thus, we would expect the unsupported ^{210}Pb to decay exponentially with depth. Taking the log of both sides, we obtain:

$$\ln\left(^{210}\text{Pb}\right)_i = \ln\left(^{210}\text{Pb}\right)^0 + \frac{-\lambda_{210}}{r}m_i \qquad (3.41)$$

Equation 3.41 is the equation of a straight line on a plot of $\ln(^{210}\text{Pb})$ versus mass depth, where $\ln(^{210}\text{Pb})^0$ is the intercept and $-\lambda_{210}/r$ is the slope. Applying linear regression to the data in this form, we can determine both the slope and the intercept. From the slope, we can easily solve for r, the mass accumulation rate.

Let us consider an example from Sanchez-Cabeza and Ruiz-Fernandez (2012) based on the data of Ruiz-Fernandez et al. (2009) from the Gulf of Tehuantepec, Mexico. Figure 3.19 shows the measured unsupported or excess (^{210}Pb) as function of mass and mass depth. The data clearly do not show a simple exponential decrease in depth and suggest at least two times when sedimentation rates changed significantly. Using the constant flux model, Sanchez-Cabeza and Ruiz-Fernandez (2012) date the changes to about 1972 and 1930. Breaking the core into three sections applying the constant flux–constant sedimentation model to each, Sanchez-Cabeza and Ruiz-Fernandez (2012) calculated mass accumulation rates of 0.66, 0.22, and 0.11 g/cm^2 –yr for the upper, middle, and lower sections of the core. Sanchez-Cabeza and Ruiz-Fernandez (2012) attribute these changes in sedimentation rates to significant changes in land use, demographic changes, and channelization of the Tehuantepec River over this time.

Disequilibrium between ^{210}Pb and its longest-lived progenitor, ^{226}Ra, is frequently observed in young lavas. The mere observation of ^{210}Pb–^{226}Ra disequilibria constraints eruption ages to $\lesssim 100$ years. In most cases, those ages are known from observation; submarine eruptions, however, are rarely directly observed and these cases of ^{210}Pb–^{226}Ra disequilibria provide useful constraints on age (e.g., Waters et al., 2013).

^{210}Pb–^{226}Ra disequilibria can also provide insights into magmatic processes. Available

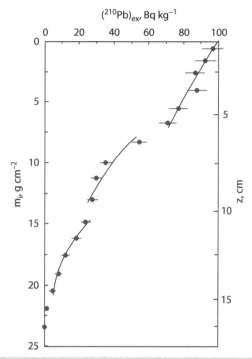

Figure 3.19 Unsupported ^{210}Pb activity measured on sediment samples from the Gulf of Tehuantepec, Mexico as a function of mass depth (depth is shown on the right). Three mass accumulation rates for three parts of the core calculated using the constant flux–constant sedimentation model are shown as solid lines. Data from Ruiz-Fernandez et al. (2009); model from Sanchez-Cabeza and Ruiz Fernandez (2012).

data indicate that ^{210}Pb–^{226}Ra disequilibria are larger in subduction zone volcanoes than in mid-ocean ridge basalts (MORBs) and ocean island basalts (OIBs) (Berlo and Turner, 2010). In all three settings, both ^{210}Pb excesses and deficits are observed, but the mean (^{210}Pb)/(^{226}Ra) is >1 for subduction-related eruptions while it is <1 for MORBs and OIBs. Recent fractional crystallization will produce ^{10}Pb–^{226}Ra disequilibrium and in most instances result in (^{210}Pb)/(^{226}Ra) < 1 as Ra is significantly more incompatible than Pb and hence will be enriched in the remaining liquid, although fractional crystallization of K-feldspar can have the opposite effect. In either case, however, the effect is somewhat limited and gas transfer is likely the more significant process. The daughter of ^{226}Ra is ^{222}Rn with a half-life of 3.8 days. As radon is a noble gas, it will readily partition into any gas phase that forms in the magma, which can then transport ^{226}Ra in gas bubbles rising through the magma column producing deficits in parts of the magma body and excesses in others. As subduction-related magmas are generally gas-rich, there is greater potential for such gas transfer. Consistent with this, Berlo and Turner (2010) also found that ^{210}Pb excesses are more common in explosive volcanic eruptions than quiescent ones.

3.4.7 ^{210}Po–^{210}Pb dating

^{210}Pb β^- decays through ^{210}Bi $(t_{1/2} = 5$ d) to ^{210}Po, which then α-decays to ^{206}Pb with a half-life of 138.4 days or 0.379 years, which is sufficiently short that it has limited geological applications. Although not a gas at room temperature, ^{210}Po is quite volatile at magmatic temperatures and is often nearly quantitatively lost through degassing during eruption (probably as halide species, such as $PoCl_3$). Girard et al. (2017) found that lavas erupted over the 40 years beginning in 1982 on Kilauea had $\approx$ 0 ^{210}Po at the time of eruption, but ash and Pele's hair (fibers of volcanic glass) had excess ^{210}Po and Pb concentrations as a result of condensation on particles from the gas phase.

^{210}Po has also been used to date unobserved submarine eruptions when those eruptions occurred shortly before sample collection. We can write the following equation:

$$\frac{(^{210}Po)}{(^{210}Pb)} = \left(1 - e^{-\lambda_{210_{Po}}t}\right) + \frac{(^{210}Po)^0}{(^{210}Pb)} e^{-\lambda_{210_{Po}}t}$$

$$(3.42)$$

To determine t_0, the eruption age, we need to know $(^{210}Po/^{210}Pb)_0$, the initial polonium–lead disequilibrium, which we do not. A simple assumption is that it is 0, which corresponds to complete degassing. An age calculated on this assumption will be a maximum. Rubin et al. (1994, 1998) made a series of measurements of (^{210}Pb) over a period of years following sample collection of submarine lava flows, and then used non-linear regression to fit the data under the assumption that $(^{210}Po/^{210}Pb)_0$ = 0. The minimum age is, of course, the date of sample collection, but Rubin et al. (1994) considered the most likely "eruption window" corresponded to 75–100% degassing of Po. Figure 3.20 shows the data for two samples collected from 9°N on the East Pacific Rise in 1991 and analyzed in this way. The results in

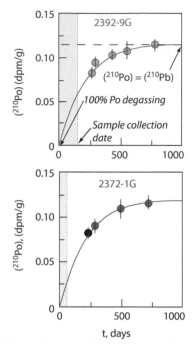

Figure 3.20   Measured ^{210}Po activities as a function of time for two samples from the East Pacific Rise. A curve through the data shows the best fit assuming $(^{210}Po) = 0$. The intersection gives the maximum age; the collection date gives the minimum age. Source: Adapted from Rubin et al. (1994). Reproduced with permission of Nature Publishing Group.

these cases indicate the eruptions occurred within three to four months prior to collection. Interesting, these samples were collected from the same area about a month apart, yet their "eruption windows" do not overlap, suggesting two separate eruptions occurred in the area. Subsequently, several other recent eruptions, most recently in 2005–2006, occurred in this area. Po dating of samples collected after that event also suggested several eruptions occurred over a period of six months or so (Tolstoy et al., 2006).

^{210}Po can also provide insights into magmatic processes. The study of the 2010 Eyjafjallajökull eruption by Sigmarsson et al. (2015) provides an example. The eruption began with alkali basalt on the flanks of the volcano on 20 March 2010, which follows seismic activity indicative of sill injections that had begun the year before. Explosive eruptions of a heterogeneous magma composition began on 14 April (the ash from which resulted in the shutdown of air traffic over northern Europe

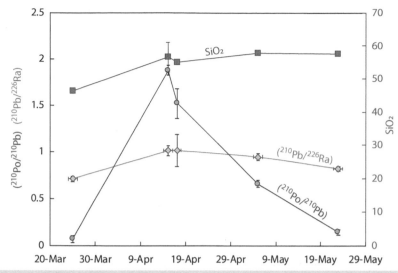

Figure 3.21   Variation in (^{210}Pb/^{226}Ra) and (^{210}Po/^{210}Pb) activity ratios and SiO$_2$ in lava and ash from the 2010 Eyjafjallajökull eruption through the spring of 2010. The composition of ash from the initial explosive eruption on 15 April was heterogeneous with SiO$_2$ ranging from ≈50 to ≈60%. Data from Sigmarsson et al. (2015).

for many days) when basalt mixed with evolved magma in a shallow (2–5 km) chamber. The explosive eruptions of trachytic magma continued until late May of that year. The basalt had low (^{210}Pb/^{226}Ra) and (^{210}Po/^{210}Pb), indicative of volatile loss while the initial explosive magma had elevated ratios that declined over time (Figure 3.21). This is best explained by transfer of gas from the

basalt injected into the overlying differentiated magma chamber followed by gas loss from the magma in the shallow chamber. Sigmarsson et al. estimated that the ratio of degassing basaltic magma to the mass of the magma in the chamber accumulating excess gas was initially ≈20 and that the duration of pre-eruptive gas accumulation in the chamber was approximately one year.

REFERENCES

Amelin, Y., Kaltenbach, A., Iizuka, T., et al. 2010. U–Pb chronology of the Solar System's oldest solids with variable ^{238}U/^{235}U. *Earth and Planetary Science Letters* 300: 343–50. doi: 10.1016/j.epsl.2010.10.015.

Appleby, P. G. and Oldfield, F. 1983. The assessment of ^{210}Pb data from sites with varying sediment accumulation rates. *Hydrobiologia* 103: 29–35. doi: 10.1007/bf00028424.

Barboni, M., Boehnke, P., Schmitt, A.K., et al. 2016. Warm storage for arc magmas. *Proceedings of the National Academy of Sciences* 113: 13959–64. doi: 10.1073/pnas.1616129113.

Bard, E., Hamelin, B., Fairbanks, R.G. et al. 1990. Calibration of the ^{14}C timescale over the past 30,000 years using mass spectrometric U-Th ages from Barbados corals. *Nature* 345: 405–10. doi: 10.1038/345405a0

Berlo, K. and Turner, S. 2010. ^{210}Pb–^{226}Ra disequilibria in volcanic rocks. *Earth and Planetary Science Letters* 296: 155–64. doi: 10.1016/j.epsl.2010.05.023.

Bourdon, B., Turner, S., Henderson, G.M. et al. 2003. *Uranium Series Geochemistry: reviews in Mineralogy and Geochemistry*, vol. 53. Washington: Mineralogical Association of America.

Bowring, S. A., Williams, I.S. and Compston, W. 1989. 3.96 Ga gneisses from the Slave province, Northwest Territories, Canada. *Geology* 17: 971–5.

Cheng, H., Edwards, R.L., Sinha, A., et al. 2016. The Asian monsoon over the past 640,000 years and ice age terminations. *Nature* 534: 640–6. doi: 10.1038/nature18591.

Cheng, H., Lawrence Edwards, R., Shen, C.-C., et al. 2013. Improvements in ^{230}Th dating, ^{230}Th and ^{234}U half-life values, and U–Th isotopic measurements by multi-collector inductively coupled plasma mass spectrometry. *Earth and Planetary Science Letters* 371–2: 82–91. doi: 10.1016/j.epsl.2013.04.006.

Chiu, T.-C., Fairbanks, R.G., Mortlock, R.A., et al. 2006. Redundant $^{230}Th/^{234}U/^{238}U$, $^{231}Pa/^{235}U$ and ^{14}C dating of fossil corals for accurate radiocarbon age calibration. *Quaternary Science Reviews* 25: 2431–40. doi: 10.1016/j. quascirev.2006.01.025.

Compston, W. and Pidgeon, R.T. 1986. Jack Hills, evidence of more very old detrital zircons in Western Australia. *Nature* 321: 766–9.

Cooper, K.M. 2015. Timescales of crustal magma reservoir processes: insights from U-series crystal ages. *Journal of the Geological Society, London, Special Publications* 422. doi: 10.1144/SP422.7.

Cooper, K.M. 2019. Time scales and temperatures of crystal storage in magma reservoirs: implications for magma reservoir dynamics. *Philosophical Transactions of the Royal Society A* 377: 20180009. doi: 10.1098/rsta.2018.0009.

Cooper, K.M., Reid, M.R., Murrell, M.T., et al. 2001. Crystal and magma residence at Kilauea Volcano, Hawaii: $^{230}Th–^{226}Ra$ dating of the 1955 East Rift eruption. *Earth and Planetary Science Letters* 184: 703–18. doi: 10.1016/S0012-821X(00)00341-1.

Edwards, R.L., Chen, J.H., Ku, T.-L., et al. 1987. Precise timing of the last interglacial period from mass spectrometric determination of thorium-230 in corals. *Science* 236: 1547–53. doi: 10.1126/science.236.4808.1547.

Edwards, R.L., Cheng, H., Murrell, M.T., et al. 1997. Protactinium-231 dating of carbonates by thermal ionization mass spectrometry: implications for quaternary climate change. *Science* 276: 782–6. doi: 10.1126/science.276.5313.782.

Eppich, G.R., Cooper, K.M., Kent, A.J.R., et al. 2012. Constraints on crystal storage timescales in mixed magmas: Uranium-series disequilibria in plagioclase from Holocene magmas at Mount Hood, Oregon. *Earth and Planetary Science Letters* 317–8: 319–30. doi: 10.1016/j.epsl.2011.11.019.

Fairbanks, R.G., Mortlock, R.A., Chiu, T.-C., et al. 2005. Radiocarbon calibration curve spanning 0 to 50,000 years BP based on paired $^{230}Th/^{234}U/^{238}U$ and ^{14}C dates on pristine corals. *Quaternary Science Reviews* 24: 1781–96. doi: 10.1016/j.quascirev.2005.04.007.

Faure, G., 1986. *Principles of Isotope Geology*, 2nd ed. New York: John Wiley & Sons, 589 p.

Froude, D.O., Ireland, T.R., Kinny, P.D., et al. 1983. Ion microprobe identification of 4100–4200 Myr-old terrestrial zircons. *Nature* 304: 616–8.

Girard, G., Reagan, M.K., Sims, K.W., et al. 2017. $^{238}U–^{230}Th–^{226}Ra–^{210}Pb–^{210}Po$ Disequilibria Constraints on Magma Generation, Ascent, and Degassing during the Ongoing Eruption of Kīlauea. *Journal of Petrology* 58: 1199–226. doi: 10.1093/petrology/egx051.

Goldmann, A., Brennecka, G.A., Noordmann, J., et al. 2013. $^{238}U/^{235}U$ of the Earth and the Solar System. *Mineralogical Magazine* 77: 1118. doi:10.1180/minmag.2013.077.5.1.

Grün, R., Aubert, M., Hellstrom, J., et al. 2010. The challenge of direct dating old human fossils. *Quaternary International* 223: 87–93, 224. doi: 10.1016/j.quaint.2009.10.005.

Han, X., Jin, X., Yang, S., et al. 2003. Rhythmic growth of Pacific ferromanganese nodules and their Milankovitch climatic origin. *Earth and Planetary Science Letters* 211: 143–57. doi: https://doi.org/10.1016/S0012-821X(03)00169-9.

Hiess, J., Condon, D.J., McLean, N., et al. 2012. $^{238}U/^{235}U$ systematics in terrestrial uranium-bearing minerals. *Science* 335: 1610–14. doi: 10.1126/science.1215507.

Hoffmann, D.L., Standish, C.D., García-Diez, M., et al. 2018. U-Th dating of carbonate crusts reveals Neandertal origin of Iberian cave art. *Science* 359: 912–5. doi: 10.1126/science.aap7778.

Ivanovich, M., Latham, A.G. and Ku, T.-L. 1992. Uranium-series disequilibrium applications on geochronology, in Ivanovich, M. and Harmon, R.S. (eds). *Uranium-Series Disequilibrium Applications to Earth, Marine and Environmental Sciences*, 2nd ed. Oxford: Oxford University Press, 62–94 pp.

Jaffey, A.H., Flynn, K.F., Glendenin, L.E., et al. 1971. Precision measurement of the half-lives and specific activities of ^{235}U and ^{238}U. *Physical Review C* 4: 1889–906.

Jicha, B. R., Singer, B. S., Beard, B. L. and Johnson, C. M. 2005. Contrasting timescales of crystallization and magma storage beneath the Aleutian Island arc. *Earth and Planetary Science Letters* 236: 195–210. doi: https://doi.org/10.1016/j.epsl.2005.05.002.

Kirchner, G. (2011). ^{210}Pb as a tool for establishing sediment chronologies: examples of potentials and limitations of conventional dating models. *Journal of Environmental Radioactivity* 102: 490-494. doi: 10.1016/j.jenvrad.2010.11.010.

Krogh, T.E. 1982. Improved accuracy of U-Pb zircon ages by the creation of more concordant systems using an air abrasion technique. *Geochimica et Cosmochimica Acta* 46: 637–49. doi: 10.1016/0016-7037(82)90165-x.

Lahtinen, R. and Nironen, M. 2010. Paleoproterozoic lateritic paleosol-ultra-mature/mature quartzite-meta-arkose successions in southern Fennoscandia--intra-orogenic stage during the Svecofennian orogeny. *Precambrian Research* 183: 770–90. doi: 10.1016/j.precamres.2010.09.006.

Ludwig, K.R. 1998. On the treatment of concordant uranium-lead ages. *Geochimica et Cosmochimica Acta* 62: 665–76. doi 10.1016/S0016-7037(98)00059-3.

Mattinson, J.M. 2005. Zircon U-Pb chemical abrasion ("CA-TIMS") method: combined annealing and multi-step partial dissolution analysis for improved precision and accuracy of zircon ages. *Chemical Geology* 220: 47–66. doi: 10.1016/j.chemgeo.2005.03.011.

Mattinson, J.M. 2010. Analysis of the relative decay constants of ^{235}U and ^{238}U by multi-step CA-TIMS measurements of closed-system natural zircon samples. *Chemical Geology* 275: 186–98. doi: 10.1016/j.chemgeo.2010.05.007.

Peltier, W.R. and Fairbanks, R.G. 2006. Global glacial ice volume and Last Glacial Maximum duration from an extended Barbados sea level record. *Quaternary Science Reviews* 25: 3322–37. doi: 10.1016/j.quascirev.2006.04.010.

Pike, A.W.G., Hoffmann, D.L., García-Diez, M., et al. 2012. U-series dating of Paleolithic art in 11 caves in Spain. *Science* 336: 1409–13. doi:10.1126/science.1219957.

Reid, M.R., Coath, C.D., Mark Harrison, T., et al. 1997. Prolonged residence times for the youngest rhyolites associated with Long Valley Caldera: ^{230}Th—^{238}U ion microprobe dating of young zircons. *Earth and Planetary Science Letters* 150: 27–39. doi: 10.1016/S0012-821X(97)00077-0.

Rubin, K.H., Macdougall, J.D. and Perfit, M.R. 1994. ^{210}Po-^{210}Pb dating of recent volcanic eruptions on the sea floor. *Nature* 368: 841–4.

Rubin, K.H., Smith, M.C., Perfit, M.R., et al. 1998. Geochronology and geochemistry of lavas from the 1996 North Gorda Ridge eruption. *Deep Sea Research Part II* 45: 2571–97. doi: 10.1016/S0967–0645(98)00084–8.

Ruiz-Fernández, A.C., Hillaire-Marcel, C., de Vernal, A., et al. 2009. Changes of coastal sedimentation in the Gulf of Tehuantepec, South Pacific Mexico, over the last 100 years from short-lived radionuclide measurements. *Estuarine, Coastal and Shelf Science* 82: 525–36. doi: 10.1016/j.ecss.2009.02.019.

Sanchez-Cabeza, J.A. and Ruiz-Fernández, A.C. 2012. ^{210}Pb sediment radiochronology: An integrated formulation and classification of dating models. *Geochimica et Cosmochimica Acta* 82: 183–200. doi: 10.1016/j.gca.2010.12.024.

Schmitt, A.K. 2011. Uranium series accessory crystal dating of magmatic processes. *Annual Review of Earth and Planetary Sciences* 39: 321–49. doi: 10.1146/annurev-earth-040610-133330.

Sigmarsson, O., Condomines, M. and Gauthier, P.-J. 2015. Excess ^{210}Po in 2010 Eyjafjallajökull tephra (Iceland): evidence for pre-eruptive gas accumulation. *Earth and Planetary Science Letters* 427: 66–73. doi: 10.1016/j.epsl.2015.06.054.

Steiger, R.H. and Jäger, E. 1977. Subcommision on geochronology: conventions on the use of decay constants in geo- and cosmochronology. *Earth and Planetary Science Letters* 36: 359–62.

Stirling, C.H., Andersen, M.B., Potter, E.-K. et al. 2007. Low-temperature isotopic fractionation of uranium. *Earth and Planetary Science Letters* 264: 208–25. doi: 10.1016/j.epsl.2007.09.019.

Tera, F. and Wasserburg, G.J. 1972. U-Th-Pb systematics in three Apollo 14 basalts and the problem of initial Pb in lunar rocks. *Earth and Planetary Science Letters* 14: 281–304. doi: 10.1016/0012–821X(72)90128–8.

Tolstoy, M., Cowen, J.P., Baker, E.T., et al. 2006. A sea-floor spreading event captured by seismometers. *Science* 314: 1920–2. doi: 10.1126/science.1133950.

Vazquez, J.A. and Reid, M.R. 2004. Probing the accumulation history of the voluminous Toba magma. *Science* 305: 991–4. doi:10.1126/science.1096994.

Vervoort, J.D., White, W.M., Thorpe, R.I., et al. 1993. Postmagmatic thermal activity in the Abitibi Greenstone Belt, Noranda and Matagami Districts: evidence from whole rock Pb isotope data. *Economic Geology* 88: 1598–614.

Waters, C.L., Sims, K.W.W., Klein, E.M., et al. 2013. Sill to surface: linking young off-axis volcanism with subsurface melt at the overlapping spreading center at 9°03′N East Pacific Rise. *Earth and Planetary Science Letters* 369–70: 59–70. doi: 10.1016/j.epsl.2013.03.006.

Watson, E., Wark, D. and Thomas, J. 2006. Crystallization thermometers for zircon and rutile. *Contributions to Mineralogy and Petrology* 151: 413–33. doi: 10.1007/s00410-006-0068-5.

Wetherill, G.W. 1956. Discordant U-Pb ages. *Transactions of the American Geophysical Union* 37: 320.

Weyer, S., Anbar, A.D., Gerdes, A., et al. 2008. Natural fractionation of ^{238}U/^{235}U. *Geochimica et Cosmochimica Acta* 72: 345–59. doi: 10.1016/j.gca.2007.11.012.

Wilde, S.A., Valley, J.W., Peck, W.H., et al. 2001. Evidence from detrital zircons for the existence of continental crust and oceans on the Earth 4.4 Gyr ago. *Nature* 409: 175–8.

Wipperfurth, S.A., Guo, M., Šrámek, O., et al. 2018. Earth's chondritic Th/U: negligible fractionation during accretion, core formation, and crust–mantle differentiation. *Earth and Planetary Science Letters* 498: 196–202. doi: https://doi.org/10.1016/j.epsl.2018.06.029.

PROBLEMS

1. Assuming present ratio of ^{238}U–^{235}U in the Earth is 137.82, what was this ratio 1.7 billion years ago when the Oklo natural reactor was operating?

2. Derive Equation 3.6.

3. The following were measured on whole-rock samples from the Seminoe Mountains of Wyoming. Plot the data on an isochron diagram and calculate the age and errors on the age. Assume that the errors on both isotope ratios are 0.1% (2 standard deviations) and that there is a 50% correlation in the errors. This and subsequent problems are best done using IsoplotR described in Chapter 2.

$^{206}Pb/^{204}Pb$	$^{207}Pb/^{204}Pb$
30.09	18.03
31.49	18.30
32.02	18.40
32.20	18.43
34.04	18.86
36.80	19.41

4. The following data were reported on zircons separated from a Paleozoic granite that gave an Rb–Sr age of 542 million years. The granite is located in Archean terrane that was reactivated in the early Paleozoic/late Precambrian. Calculate the age using a concordia diagram. You should obtain both a lower and an upper intercept with concordia. What is your interpretation of the ages?

   (*Hint: You will need to correct for the initial Pb isotopic compositions shown here.*)

Zircon	$^{206}Pb/^{204}Pb$	$^{238}U/^{204}Pb$	$^{207}Pb/^{204}Pb$	$^{235}U/^{204}Pb$
1	796	6800	88.46	49.3
2	1563	6649	257.1	48.2
3	1931	6432	340.1	46.6
4	2554	6302	481.7	45.7
5	2914	6250	560.3	45.3
Initial ratios	16.25		15.51	

Assume 0.1% 2 standard deviation errors.

5. The following data were reported for zircons from the felsic norite of the Sudbury Intrusive Complex. Using these data, calculate the crystallization age and the error on the age of the intrusion. Also determine the age of the lower intercept.

Sample	Fraction	$^{207}Pb*/^{235}U$	2σ	$^{206}Pb*/^{238}U$	2σ	Error correl.
DWD5538	Eq, 1 h HF	0.4385	0.0036	0.028	0.0001	0.61098
DWD5537	Frag, 5h Hf	3.1073	0.0096	0.1991	0.0006	0.92907
AD5	Eq, 1 h HF-HCl	2.4352	0.0205	0.1561	0.0013	0.99179
AD4	Anneal 2 h HF-HCl	2.8698	0.024	0.1841	0.0015	0.99165
AD3	Anneal 2 h HF-HCl	4.0724	0.0189	0.2614	0.0012	0.97925
AD1	Anneal 1 h HF-HCl	3.0066	0.0093	0.1932	0.0004	0.84911
AD2	Anneal 1 h HF-HCl	2.0764	0.008	0.1343	0.0004	0.83521

6. The deepest sample of the central Pacific manganese nodule analyzed by Han et al. (2003), at a depth of 1.2–1.3 mm, had a $^{230}Th/^{232}Th$ (atomic) ratio of 5×10^{-5}.

   (a) What is the $^{230}Th/^{232}Th$ activity ratio of this sample?
   (b) Using the initial ($^{230}Th/^{232}Th$) given in Section 3.4.3, calculate the age of this layer.

7. The ^{210}Pb data shown were measured on a core from coastal pond. Determine the sedimentation rate.

Depth (cm)	(^{210}Pb) (pCi)
0–2	3.9
4–6	3.7
8–10	3.3
14–16	2.6
18–20	2.2
24–26	1.8
28–30	1.8

8. Pike et al. (2012) measured the following activity ratios on flowstone covering two examples of Paleolithic art from the El Castillo cave in northern Spain. Calculate the ages of the two flowstones.

Sample	(^{230}Th/^{238}U)	(^{234}U/^{238}U)	(^{230}Th/^{232}Th)
O–87	0.7969 ± 0.0038	2.7432 ± 0.0051	61.24 ± 0.61
O–69	0.7512 ± 0.0029	2.7072 ± 0.0051	788.2 ± 5.5

9. Given the following data on a Mt. St. Helens lava flow, calculate the age. You may use either simple linear regression for obtaining your solution or the IsoplotR program.

	(^{238}U/^{232}Th)	(^{230}Th/^{232}Th)
Whole rock	1.245	1.184
Plagioclase	1.128	1.126
Magnetite	1.316	1.205
Groundmass	1.335	1.214

Chapter 4

Geochronology III: Other Dating Methods

4.1 INTRODUCTION

Unlike its sister planets, the Earth is geologically active and its surface is in a state of constant change and renewal. In this chapter, we will conclude our review of geochronology by examining a variety of techniques that let us quantify the rates of these surface processes, as well as determining the timing of other events in the recent geologic past, including human, biological, and social evolution. We first review how the buildup and, in most cases, decay of nuclides produced by interaction with cosmic rays can be used as a chronometer. Radiocarbon dating is the most common and well known of these and a key tool of archeologists in attempting to reconstruct unwritten human history. There are, however, other useful cosmogenic nuclides you may not be aware of that, among other things, are helping us unravel how ice sheets receded at the end of the last Ice Age. We then turn to so-called "thermochronometers," used to determine not when rocks formed, but rather when they cooled to surface temperatures. We have already covered one of these thermochronometers, i.e. K–Ar dating, in Chapter 2; herein, we will introduce two additional thermochronometers based on the decay of U and Th: fission track dating and (U–Th)/He dating. Because temperatures increase with depth in the Earth, rocks cool as they are uplifted and overlying rock is removed by erosion. Thermochronometers allow us to calculate the rates of these processes. These chronometers depend on the same fundamental principle as that of those we have already introduced: the rate of radioactive decay is invariant. In this chapter, we will introduce a number of new equations; however, the point of departure for most of them remains Equation (1.12).

4.2 COSMOGENIC NUCLIDES

4.2.1 Cosmic rays in the atmosphere

As the name implies, cosmogenic nuclides are produced by cosmic rays colliding with atoms in the atmosphere and the surface of the solid Earth. Nuclides so created may be stable or radioactive. Radioactive cosmogenic nuclides, like the U-decay series nuclides, have half-lives sufficiently short that they would not exist in the Earth if they were not continually produced. Assuming that the production rate is constant through time, the abundance of a cosmogenic nuclide in a reservoir isolated from cosmic ray production is simply denoted by:

$$N = N_0 e^{-\lambda t} \qquad (4.1)$$

Hence, if we know N_0 and measure N, we can calculate t. Table 4.1 lists the radioactive cosmogenic nuclides of principal interest. As we shall see, cosmic ray interactions can also produce rare stable nuclides, and their abundance can also be used to measure geologic time.

A number of different nuclear reactions create cosmogenic nuclides. "Cosmic rays" are

Isotope Geochemistry, Second Edition. William M. White.
© 2023 John Wiley & Sons Ltd. Published 2023 by John Wiley & Sons Ltd.
Companion Website: www.wiley.com/go/white/isotopegeochem2

Table 4.1 Data on cosmogenic nuclides.

Nuclide	Half-life (yr)	Decay constant (yr^{-1})
^{14}C	5730	1.209×10^{-4}
^{3}H	12.33	5.62×10^{-2}
^{10}Be	1.387×10^{6}	4.950×10^{-7}
^{26}Al	7.02×10^{5}	9.87×10^{-5}
^{36}Cl	3.08×10^{5}	2.25×10^{-6}
^{32}Si	276	2.51×10^{-2}

high-energy (≈ 1 GeV to $>10^{20}$ eV!) atomic nuclei, mainly of H and He (because these constitute most of the matter in the universe), but nuclei of all the elements have been recognized. To put these kinds of energies in perspective, the previous generation of accelerators for physics experiments, such as the Cornell Electron Storage Ring, produces energies in the tens of giga-electron volts (10^{10} eV), while the Large Hadron Collider of the European Organization for Nuclear Research (known as CERN), mankind's most powerful accelerator, located on the Franco-Swiss border near Geneva, produces energies of ≈ 10 TeV range (10^{13} eV). These energies can greatly exceed the binding energies of nuclei (Figure 1.3) and hence can readily initiate nuclear reactions. A significant fraction of cosmic rays originates in the Sun, although these are mainly of energies too low (<0.1 GeV) to generate cosmogenic nuclides. The origin of the remainder is unclear; they most likely originate in supernovae, kilonovas, or similar high-energy environments in the cosmos.

The cosmic ray flux decreases exponentially with depth in the atmosphere as these particles interact with matter in the atmosphere. This observation has an interesting history. Shortly after the discovery of radioactivity, investigators noticed the presence of radiation even when no known sources were present. They reasonably surmised that this resulted from radioactivity in the Earth. In 1910, an Austrian physicist, named Victor Hess, carried his detector (an electroscope consisting of a pair of charged gold leaves: the leaves would be discharged and caused to collapse by the passage of charged particles) aloft in a balloon. To his surprise, the background radiation increased rather than decreasing as he went up. It thus became clear that this radiation originated from the outside, rather than the inside, of the Earth.

The primary reaction that occurs when cosmic rays encounter the Earth is *spallation*, in which a nucleus struck by a high-energy particle shatters into two or more pieces, including stable and unstable nuclei, as well as protons and neutrons. Short-lived particles, such as muons, pions, and so on, are also created. The interaction of a cosmic ray with a nucleus sets off a chain reaction of sorts as the secondary particles and nuclear fragments, which themselves have very high energies and then strike other nuclei producing additional reactions of lower energy. ^{14}C is actually produced primarily by reactions with secondary particles, mainly by the ^{14}N$(n,p)^{14}$C reaction involving relatively slow neutrons.

4.2.2 ^{14}C dating

Carbon-14 is by far the most familiar and useful of the cosmogenic dating schemes. Its usefulness results from its relatively short half-life, a relatively high production rate, and the high concentration of carbon in biological material. The short half-life has the advantage of producing accurate dates of young (geologically speaking) materials and events in human history leading to easy determination by counting the decays. The traditional method of ^{14}C determination is counting of the β–rays produced in its decay. ^{14}C decays without emitting a gamma, which is unfortunate because γ-rays are more readily detected (but, of course, very fortunate for living things). Difficulties arise because the rate of decay of a reasonable sample of carbon (a few grams) is low. One of the first problems that had to be solved was eliminating "counts" that arose from cosmic rays rather than the decay of the carbon sample. This is done with a combination of shielding and "coincidence counting."[1] Counting techniques make use of either liquid scintillation counters or gas proportional counters. A gas proportional counter is simply a metal tube containing a gas, with a negatively charged wire (the anode) running through it. The

---

[1] Beta detectors are subject to electronic and other noise – in other words, pulses that originate from something other than a beta particle. Coincidence counting is a technique that employs two detectors. Only those pulses that are registered on both detectors are counted – other pulses are considered noise.

emitted beta particle causes ionization of the gas, and the electrons released in this way drift to the wire (the anode), producing a measurable pulse of current. In this technique, carbon in the sample is converted into CO_2, purified (in some cases, it is converted into methane), and then admitted into the counting chamber. Generally, several grams of carbon are required. In liquid scintillation counting, the carbon is extracted from a sample and converted into CO_2, purified, and then eventually converted into benzene (C_6H_6) and mixed with a liquid scintillator (generally an organic liquid). When a carbon atom decays, the beta particle interacts with the scintillator, causing it to give off a photon, which is then detected with a photomultiplier, and the resulting electrical pulse is sent to a counter. Liquid scintillation is the newer and more sensitive of these two techniques and it has now largely replaced gas proportional counting. Liquid scintillation analysis can be done with less than 1 g of carbon.

In the last 30 years or so, a newer method, i.e., accelerator mass spectrometry (AMS), has revolutionized ^{14}C dating. It is far more sensitive so that greater accuracy is achieved with considerably less sample (analysis requires a few milligrams to a few tens of milligrams, depending on the nature of the sample compared to a few grams to a few tens of grams). As you can imagine, the smaller sample size is a very considerable advantage for valuable archeological artifacts. Greater sensitivity also pushes back the maximum ages that can be accurately determined. Initially, this work was done with accelerators built for high-energy physics experiments, but subsequently accelerators have been designed and built exclusively for radiocarbon dating. There are now dozens of them in the world. The disadvantage is that AMS dating is significantly more expensive. Initially, it was far more expensive and rarer, but the AMS technique has become more competitive in terms of cost[2] and consequently AMS age determinations have begun to dominate.

Because of the way in which it was traditionally measured, ^{14}C is generally reported in units of *specific activity*, or disintegrations per minute per gram carbon rather than as a ratio to ^{12}C or ^{13}C (in SI units, one unit of specific activity is equal to $1/60 = 0.017$ Beq/g C). Atmospheric carbon and carbon in equilibrium with the atmosphere (carbon in living plant tissue and the surface of the oceans) have a specific activity of 13.56 dpm/g. This is, in effect, the value of N_0 in Equation (4.1). By historical convention, radiocarbon ages are reported in years before 1950 (the year the first ^{14}C age determination; consequently, in radiocarbon dating, the abbreviation BP thus means not *before present*, but before 1950) and assuming a half-life of 5568 years (instead of the currently accepted value of 5730 years). However, studies frequent restate their results in calendar years.

The actual reaction that produces ^{14}C is:

$$^{14}N + n \rightarrow {}^{14}C + p$$

This is a charge exchange reaction in which the neutron gives up unit of negative charge to a proton in the nucleus and becomes a proton. The neutron involved comes from a previous interaction between a cosmic ray and a nucleus in the atmosphere; in other words, it is a *secondary* particle. As charged particles, cosmic rays interact with the Earth's magnetic field and are deflected by it (but not those with energies above 10^{14} eV). Consequently, the production rate of ^{14}C is higher at the poles. However, because the mixing time of the atmosphere is short compared to the residence time of ^{14}C in the atmosphere, the ^{14}C concentration in the atmosphere is uniform.

However, can we really assume that the atmosphere-specific activity today is the same as it was in the past? From Equation (4.1), we can see that knowing the initial specific activity is essential to determine an age. To investigate the variation of the specific activity of ^{14}C with time in the atmosphere, the specific activity of ^{14}C in wood of old trees has been examined. The absolute age of the wood is determined by counting tree rings (dendrochronology). The result of such studies shows that the specific activity has indeed not been constant, but has varied with time (Figure 4.1). There are a number of effects involved. Beginning in 1945 and lasting through several subsequent decades, injection

[2] For example, one commercial lab charges US\$300 per sample for conventional analysis and US\$575 for AMC analysis in 2020.

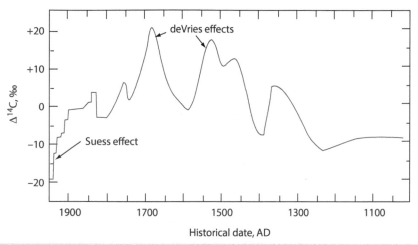

Figure 4.1   Variation of initial specific activity of ^{14}C in the past 1000 years. Source: Faure (1986)/ Reproduced with permission of John Wiley & Sons.

of nuclear-bomb-produced ^{14}C into the atmosphere raised the specific activity. Then, over the past 200 years, the specific activity has decreased because of addition of "old" (i.e., ^{14}C-free) carbon to CO_2 produced by fossil fuel burning (*the Suess effect*). Similar variations occurred further back in time. Indeed, between 33 000 and 45 000 years ago, atmosphere-specific activity appears to have been twice what it was in 1950 (Beck et al., 2001). These changes in atmosphere-specific activity almost certainly occurred as a consequence of the changing balance between CO_2 in the atmosphere and the ocean. Far more CO_2 is dissolved in the ocean, particularly in deep water, than in the atmosphere. Shifting CO_2 from the atmosphere to the ocean results in an increase in the specific activity of ^{14}C in the atmosphere. These are termed *reservoir effects*.

There are also longer-term variations in atmospheric ^{14}C, known as the deVries events that result from variation in the cosmic ray flux that in turn results from variations in solar activity. Changes in solar activity, whose most obvious manifestation is the 11 year sunspot cycle, produce changes in the intensity of the solar wind, with particles (again, mainly H and He nuclei and electrons) blown off the surface of the Sun. At the height of the sunspot cycle, the solar wind is greatly enhanced, i.e., up to a factor of 10^6. The solar wind in turn modulates the solar magnetosphere, which deflects galactic cosmic rays at the outer edge of the Solar System (the heliopause).

A stronger solar wind results in few high-energy cosmic rays reaching the Earth. While the sunspot cycle frequency is stable, its intensity, as measured by the number of sunspots, is not and it varies on a roughly century scale. Variation in the Earth's magnetic field also affects the cosmic ray flux and consequently cosmogenic nuclide production.

Cosmogenic nuclides represent an archive of these combined effects so that if we have an independent timescale, we can retrieve the production rate. Dendrochronology represents one such archive, although as we noted it is also subject to reservoir effects. Another archive is the abundance of another cosmogenic nuclide, ^{10}Be, which we will discuss in the following section, in ice cores, which is a direct measure of atmospheric ^{10}Be abundance and the cosmic ray flux. Steinhilber et al. (2012) combined the ^{10}Be record in Greenland and Antarctic ice cores with the dendrochronologically calibrated ^{14}C record to calculate the cosmic ray intensity history shown in Figure 4.2. The longer-period (≈ 8000 year) variations in intensity inversely correlate with geomagnetic intensity record of Knudsen et al. (2008), which is also shown in this figure. Shorter-period variations correlate inversely with sunspot activity: cosmic ray activity is high when the sunspot cycle is subdued.

Because of these variations in the specific activity of atmospheric carbon, a correction must be applied to ^{14}C dates due to the variation in initial specific activity with time. Calibration of ^{14}C through dendrochronology

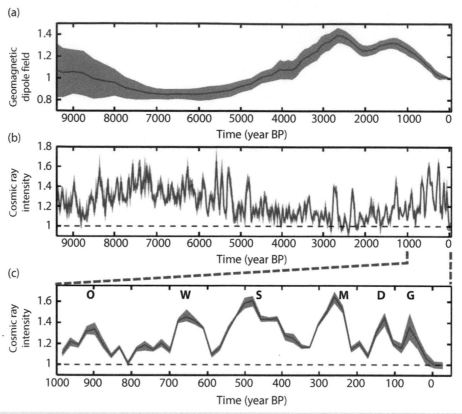

Figure 4.2   (a) Variation in geomagnetic dipole intensity from Knudsen et al. (2008). Source: Knudsen et al. (2008)/with permission of Elsevier. (b) Variation of cosmic ray intensity deduced from ^{10}Be in the Greenland and Antarctic ice cores and ^{14}C in tree rings by Steinhilber et al. (2012). (c) Expanded view of the last 1000 years of cosmic ray intensity. Capital letters mark grand solar minima in sunspot activity – O: Oort; W: Wolf; S: Spörer; M: Maunder; D: Dalton; G: Gleissberg. Source: Steinhilber et al. (2012)/National Academy of Sciences.

has now been done over the last 13 900 years using waterlogged oaks in swamps and bogs of Germany and Ireland and bristlecone pines in the United States (US) and has be extended to 45 000 years through comparison of ^{14}C and ^{238}U/^{230}Th ages, e.g., in speleothems (e.g., Fairbanks et al., 2005). The calibration is periodically updated by the IntCal international consortium, with the most recent calibration being the IntCal20 described in a series of papers published in *Radiocarbon*. The Northern Hemisphere (Reimer et al., 2020) and the Southern Hemisphere are calibrated separately, as is marine carbon (Heaton et al., 2020). Figure 4.3 illustrates the Northern Hemisphere calibration.

The variation in initial specific activity makes the conversion from radiocarbon ages to calendar years somewhat complex. However, computer programs are available for this task, e.g., CALIB 8.2 at http://calib.org (Stuiver et al., 2020). Figure 4.4 illustrates the case of a hypothetical radiocarbon age of 3000 years BP with a 1σ uncertainty of 30 years. On the left is shown the uncertainty function in red. The broad blue curve encloses the uncertainty on the calibration between radiocarbon age and calendar years. The black histogram on the bottom shows the probable correspondence of calendar dates to this radiocarbon age. The irregularity of the histogram reflects the irregularity of the calibration curve, which in turn reflects the variation in reservoir effects, cosmic ray flux, etc., mentioned previously. In this case, the 2σ uncertainty encompasses a range of calendar dates from 1390 to 1130 BP.

Geological applications of ^{14}C dating include volcanology (^{14}C dating is an important part of volcanic hazard assessment),

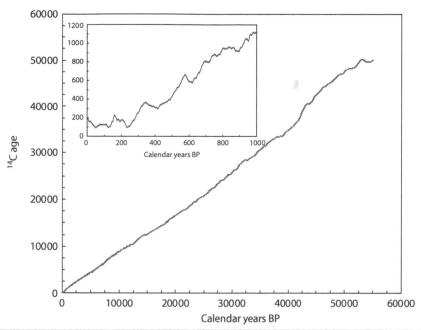

Figure 4.3 The InCal ^{14}C Northern Hemisphere calibration of Reimer et al. (2020). Data from InCal.org (http://intcal.org/curves/intcal20.14c). The inset illustrates the calibration for the last 1000 years. Present-day carbon would have a radiocarbon age of 199 years for the reasons discussed in the text. At 55 000 years, ^{14}C ages are almost 4000 years too young.

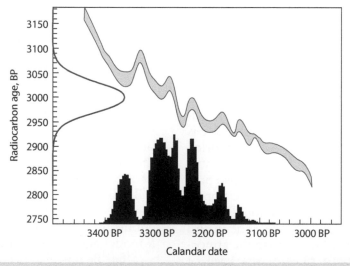

Figure 4.4 Conversion of radiocarbon dates to calendar years. The red curve on the left shows the uncertainty histogram for a radiocarbon age of 3000 ± 30 (1σ) years BP. The blue curve shows the calibration, with uncertainty between radiocarbon age and calendar date. The black histogram at the bottom shows the probably corresponding calendar date for this radiocarbon age.

Holocene stratigraphy, paleoclimatology, and oceanography. It is also used extensively in paleoseismology to determine the time of prehistoric earthquakes, and thus part of earthquake hazard assessment. Knowing the frequency of past earthquakes is useful in predicting future earthquake probabilities.

One oceanographic application, which utilizes AMS, is the determination of paleo-bottom water ages. The ice ages affected ocean circulation, but the exact nature of the effect is still uncertain. Of particular interest are the changes in the ocean circulation pattern that occurred as the last glaciation ended. Because planktonic foraminifera live in the upper part of the ocean, which is in equilibrium with the atmosphere, one can date a sedimentary stratum by measuring ^{14}C in planktonic foraminiferal tests. Benthic foraminifera live at the bottom of the ocean and build their tests from CO_2 dissolved in bottom water. Thus, a ^{14}C date of a modern benthic foram would give the age of the bottom water. By comparing the ^{14}C ages of planktonic and benthic foraminifera, one may determine the "ages" of bottom waters in the past. This in turn reveals something of paleocirculation patterns. We will return to this topic in Chapter 12.

4.2.3 Applications of meteoric ^{10}Be

4.2.3.1 Sediment geochronology

^{10}Be is created by spallation reactions in the atmosphere between cosmic rays and N and O nuclei. Since these are the most abundant nuclei in the atmosphere, the production rate of ^{10}Be is comparatively high ($\approx 10^6$ atoms/cm^2/yr). Once it is created in the atmosphere, ^{10}Be is quickly extracted from the atmosphere by rain from which it is scavenged and absorbed on particle surfaces. On land, these particles become part of the soil. In the ocean, these particles are quickly incorporated in sediment (the residence time of ^{10}Be in seawater is estimated to be 1000 years). This "meteoric" ^{10}Be is commonly designated $^{10}Be_m$. As we will discuss in a subsequent section, ^{10}Be is also produced by cosmic ray interactions in the uppermost meter or so of rocks and soil. This in situ production is designated $^{10}Be_i$. The in situ production is, however, far less than the atmospheric production, with rates at sea level of approximately five atoms/gram quartz/year. Although as for other cosmogenic nuclides, analysis is usually performed by AMS, in some cases, meteoric $^9Be/^{10}Be$ ratios are sufficiently high ($\approx 10^{-7}$) that they can also be determined by secondary ion mass spectrometry.

On continents, both $^{10}Be_m$ and $^{10}Be_i$ show the expected latitudinal dependence resulting from the dipole component of Earth's geomagnetic field. Because it is advected by ocean currents, marine ^{10}Be does not show a simple latitudinal dependence; instead, ^{10}Be concentrations vary between water masses and ocean basins mainly as a consequence of varying input of terrestrial Be from rivers. As does the ^{14}C production rate, the ^{10}Be production rate has varied with time as a result in variations in solar activity and the Earth's magnetic field. Frank et al. (1997) used ($^{230}Th/^{234}U$) ages and ^{10}Be abundances in marine sediments to calibrate the ^{10}Be production rate over the past 200 kyr.

One application of $^{10}Be_m$ has been determination of Mn nodule growth rates. As we found in the previous chapter, Th–U isotopes can be used for this purpose. Therefore, you might ask, "Why to use ^{10}Be, which is a more difficult and expensive technique?" The answer is simple: many Mn nodules have been growing for times significantly longer than the useful range (500 000 years) of ^{230}Th dating; the useful range of ^{10}Be is considerably longer.

Mn nodules generally contain substantial amounts of ^9Be, so the abundance of ^{10}Be is commonly expressed as the $^{10}Be/^9Be$ ratio. The relevant equation is:

$$\frac{^{10}Be}{^9Be} = \left(\frac{^{10}Be}{^9Be}\right)_0 e^{-\lambda t} \qquad (4.2)$$

We can determine age based on this equation if we can make an assumption about the initial $^{10}Be/^9Be$ ratio (since it involves an assumption, such an age would be considered a "model age"). For example, Nishi et al. (2017) found that the lowest analyzed layer of Mn nodule sample JA06BMSO2A from the western North Pacific had $^{10}Be/^9Be$ of $2.58 \pm 0.58 \times 10^{-10}$. Using the modern $^{10}Be/^9Be$ ratio of $\approx 1.19 \times 10^{-7}$ of deep water in the North Pacific Ocean from von Blanckenburg and Bouchez (2014), they calculated a model age of 12.3 Ma.

$^{10}Be/^9Be$ has also been used in dating lacustrine sediments. Fragments of jawbone of a hominid, nicknamed *Abel*, were found in Pliocene sediments of the Chad Basin in northern Africa in 1995 and assigned to a new species, *Australopithecus bahrelghazali*. Subsequently, a partial skull of an older hominid, nicknamed *Toumaï*, was found in the same region in 2002 and assigned to the new species, *Sahelanthropus tchadensis* (Figure 4.5). Based on shape and dimensions of the cranium, *S. tchadensis* was judged to be either the first human

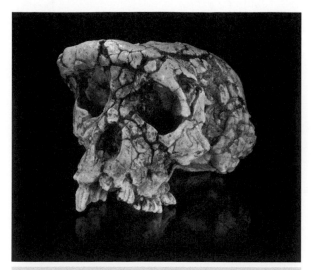

Figure 4.5 Cast of the skull of *S. tchadensis*. Source: Didier Descouens/Wikipedia Commons/ CC BY-SA 4.0.

ancestor following the split between hominins and chimpanzees or the last common ancestor of both. Either way, this fossil is remarkably important in reconstructing human evolution. Correlation based on fossils between these deposits suggested a late Miocene age of about 6–7.5 Ma, but it was not possible to better constrain the age.

Lebatard et al. (2008) analyzed $^{10}Be/^9Be$ in 32 paleosols and lacustrine sediments from this area. The model age of sediment in which *A. bahrelghazali* was found was 3.58 ± 0.27 Ma, making *Abel* a contemporary of the famous *Australopithecus afarensis* fossil "Lucy" (indeed, some have argued that *Toumaï* should be assigned to *A. afarensis*). Using $^{10}Be/^9Be$ ratios measured in 4–7 ka old Lake Chad sediment to establish the local $(^{10}Be/^9Be)_0$ as 2.54×10^{-8}, averaged $^{10}B/^9Be$ model ages of sediments above and below it constrain the age of *S. tchadensis* to between 6.8 and 7.2 Ma (see Problem 2). This age is widely taken to be the time of divergence of human ancestors from other great apes.

A somewhat more sophisticated approach follows that we used for Th–U dating (see Section 3.4.3.1) where we assumed a constant growth or sedimentation rate, *s*, in which case depth, *z*, is related to time as:

$$z = s \times t \qquad (4.3)$$

and hence

$$t = \frac{z}{s} \qquad (4.4)$$

We can substitute Equation (4.4) into Equation (4.2) to obtain:

$$\frac{^{10}Be}{^9Be} = \left(\frac{^{10}Be}{^9Be}\right)_0 e^{-\lambda z/s} \qquad (4.5)$$

Taking the log of both sides, we obtain:

$$\ln\left(\frac{^{10}Be}{^9Be}\right) \cong \ln\left(\frac{^{10}Be}{^9Be}\right)_0 - \frac{\lambda}{s}z \qquad (4.6)$$

Regressing $^{10}Be/^9Be$ against depth, the slope is λ/s, which we can readily solve for growth rate, and the initial ratio is the intercept. Figure 4.6 shows the full data of Nishi et al. (2017) for Mn nodule sample JA06BMSO2A, as well as the regression lines calculated in this way. Having determined the growth rate, we can then calculate the age of each layer form Equation (4.4). Doing so, we predict an age of 12.6 Ma for the deepest layer, which is only slightly different than the model age approach. Calculating the growth rate using this approach is left to you as Problem 3.

It is nevertheless apparent from Figure 4.6 that the constant growth rate model does not fit the data within error. There are several reasons that this could be the case. The first and simplest is that the growth rate has not been constant due to changing environmental conditions in the deep ocean at this location. Another possibility is that the $^{10}Be/^9Be$ ratio of seawater has varied with time. An obvious reason why this might be the case is that the atmospheric production rate has varied because the cosmic ray flux has varied. As Figure 4.2 shows this has clearly been the case. The residence time of Be in seawater is sufficiently short (600–1000 years) that this would be reflected in the $^{10}Be/^9Be$ of seawater. In addition, $^{10}Be/^9Be$ in seawater also varies because of varying input of Be to the oceans from continents. The $^{10}Be/^9Be$ of this riverine flux is low, typically in the range of 10^{-8} to 10^{-9} (von Blanckenburg et al., 2012), because it is diluted by eroded non-cosmogenic ^9Be. As a consequence of varying riverine fluxes, $^{10}Be/^9Be$ varies between ocean basins, ranging from 1.47×10^{-7} in Pacific Ocean surface water to 1.1×10^{-8} in Mediterranean deep water. This suggests yet another explanation: changing

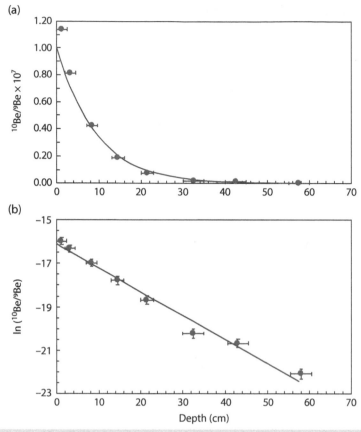

Figure 4.6 (a) Variation of $^{10}Be/^9Be$ with depth in a manganese nodule from the Pacific Ocean. (b) Natural log of $^{10}Be/^9Be$ with depth in the same nodule. In both, the red line has been calculated by regressing $\ln(^{10}Be/^9Be)$ against depth.

deep ocean currents, which would in turn reflect larger-scale environmental and climatic conditions.

4.2.3.2   Erosion and denudation rates

In addition to geochronological applications, ^{10}Be has been used to estimate erosion or denudation rates. The first such study was that of Brown et al. (1988) who measured ^{10}Be in sediments in a number of rivers from 48 drainage basins on the Atlantic US seaboard. They defined an erosion index I as:

$$I = \frac{M\left[^{10}Be\right]_s}{Aq} \qquad (4.7)$$

where M is the mass of sediment being transported out of the basin, $[^{10}Be]_s$ is the ^{10}Be concentration in that sediment, A is the area of the basin, and q is the atmospheric deposition rate (including both wet and dry deposition) of $^{10}Be_m$. The numerator is the loss rate of $^{10}Be_m$ from the basin and the denominator is the production rate, so a value of 1 represents the steady-state case where erosion equals production. Values less than 1 indicate increased sediment accumulation while values greater than 1 indicate increased erosion beyond steady state. The authors found that basins located in the Atlantic coastal plain had an average index of 0.3. Basins located in the Piedmont, an area further inland between the fall line[3] and the Appalachian Mountains, had an average value of 2.2 with individual values ranging from 0.6 to 6.7. They interpreted this as a result of two centuries of farming on Piedmont, which, because of its

---

[3] The "fall line" denotes a break in slope characterized by waterfalls and cataracts that mark the inland limit of river navigation.

higher gradient, is more susceptible to erosion. Their reconnaissance survey of rivers from Asia, North and South America, and Europe also showed evidence of increased erosion.

^{10}Be is entirely derived from cosmogenic production while ^{9}Be is entirely derived from weathering, so subsequent studies, such as that of von Blanckenburg et al. (2012), have utilized the ^{10}Be/^{9}Be ratio to constrain denudation rates. While ^{10}Be is also produced in situ when cosmic rays strike the surface of the Earth, in situ production per unit area is orders of magnitude smaller than the deposition via precipitation or dry deposition of ^{10}Be$_m$. Consequently, we can assume that ^{10}Be in soils, sediments, and streams is exclusively derived from meteoric deposition.

Once it reaches the soil, ^{10}Be$_m$ will partition between solution, ^{10}Be$_{diss}$, being adsorbed on particle surfaces, or incorporated in newly formed authigenic minerals; the latter is referred to as reactive, ^{10}Be$_{reac}$. At steady state, the flux of ^{10}Be$_m$ from a watershed, F, will equal the meteoric flux into it, hence:

$$F[^{10}Be]_m = E \times [^{10}Be]_{reac} + Q \times [^9Be]_{diss}$$

(4.8)

where E is the mass erosion rate (in kg m^{-2} yr^{-1}) and Q is the runoff rate (in the same units or equivalently, L m^{-2} yr^{-1}) in a watershed. We can define a partition coefficient describing the partitioning of Be between reactive and dissolved as:

$$K_d = \frac{[Be]_{reac}}{[Be]_{diss}}$$

(4.9)

where K_d is pH dependent and ranges from 10^4 L/kg at pH 4–5 to 10^5 at pH $\geq$ 6. As most, but not all, rivers and streams have pH $\geq$ 6, most Be will be absorbed on particle surfaces even in moderately acidic streams and essentially all Be will be absorbed in neutral to alkaline rivers, a range that encompasses most, but not all, major rivers. Substituting Equation (4.9) into Equation (4.8), the erosion rate is:

$$E = \frac{F^{10Be}_m}{[^{10}Be]_{reac}} - \frac{Q}{K_d}$$

(4.10)

This allows us to calculate the erosion rate from only the reactive ^{10}Be in stream and river sediments if we know F_m, Q, and K_d.

In contrast to ^{10}Be, weathering of bedrock is the sole source of ^{9}Be in soils and the soils solution (atmospheric deposition of ^{9}Be through precipitation or dust is trivial in almost all cases). Weathering tends to be incongruent such that some ^{9}Be in the parent rock, ^{9}Be$_{parent}$, will be released by mineral breakdown and some will be retained in residual minerals; we will designate the latter as ^{9}Be$_{min}$. As with ^{10}Be, ^{9}Be released by weathering reactions will partition between reactive, ^{9}Be$_{reac}$, and dissolved, ^{9}Be$_{diss}$, forms.

Now suppose that a system is at steady state, and with the weathering front advancing downward at the same rate, material is removed from the surface, which is to say that weathered material is being removed as fast as it is being produced. The rate at which material is being removed is the denudation rate (in units of kg km^{-2} yr^{-1}), denoted by D. In this case, we can write the following equation:

$$D[^9Be]_{parent} = E\left([^9Be]_{min} + [^9Be]_{reac}\right) + Q[^9Be]_{diss}$$

(4.11)

Substituting Equation (4.9) into Equation (4.11), we obtain:

$$[^9Be]_{reac} = \frac{D[^9Be]_{parent} - E[^9Be]_{min}}{E + Q/K_d}$$

(4.12)

Assuming K_d is the same for both ^{10}Be and ^{9}Be, which is equivalent to assuming no isotopic fractionation, we can combine Equation (4.10) and Equation (4.12) to yield:

$$\frac{[^9Be]_{reac}}{[^{10}Be]_{reac}} = \frac{F^{10Be}_m}{D[^9Be]_{parent} - E[^9Be]_{min}}$$

(4.13)

This is the ratio of atmospheric deposition of ^{10}Be to ^{9}Be being released and removed from the Earth's surface by weathering. In Equation (4.13), we subtracted the amount of ^{9}Be retained in primary minerals in the denominator. We can instead express the ^{9}Be flux as the fraction of reactive and dissolved ^{9}Be to total ^{9}Be:

$$\left(\frac{[^9Be]}{[^{10}Be]}\right)_{reac} = \frac{F^{10Be}_{met}}{D[^9Be]_{parent} \times f^{[^9Be]}_{reac+diss}}$$

(4.14)

where $f^{[^9Be]}_{reac+diss}$ is the fraction of ^{9}Be released by weathering to reactive and dissolved forms and provides a measure of the intensity of weathering. With little weathering before removal by erosion, most ^{9}Be will remain in primary minerals while intense weathering will result in release of most ^{10}Be to reactive or dissolved forms.

Furthermore, assuming no isotopic fractionation between dissolved and reactive forms, this relationship holds for both reactive and dissolved forms:

$$\left(\frac{[^9Be]}{[^{10}Be]}\right)_{diss} = \left(\frac{[^9Be]}{[^{10}Be]}\right)_{reac}$$

$$= \frac{F_{met}^{^{10}Be}}{D[^9Be]_{parent} \times f_{reac+diss}^{[^9Be]}} \quad (4.15)$$

Either of Equations (4.14) and (4.15) can be solved for the denudation rate if we know the meteoric ^{10}Be flux, the ^{9}Be concentration in the parent, and the fraction of dissolved and reactive ^{9}Be. von Blanckenburg et al. (2012) compared denudation rates determined from this approach with rates determined from river loads or in situ measurements of ^{10}Be, and found good agreement overall, taking account of uncertainties, such as the ^{10}Be meteoric flux, [^{9}Be}$_{parent}$ (they assumed an average crustal concentration of 2.5 ppm), and their assumed value of $f_{reac+diss}^{[^9Be]}$ of 0.2. Figure 4.7 compares (^{10}Be/^{9}Be) in reactive or dissolved Be in a number of South American rivers with denudation rates calculated from in situ ^{10}Be measured on quartz grains or from sediment load. The bold red line shows the case where the denudation rates predicted from (^{10}Be/^{9}Be) agree with those estimated from sediment load or in situ ^{10}Be measurements. Overall, there is qualitative agreement, but there are also clear deviations. Denudation rates predicted by (^{10}Be/^{9}Be) are lower than those determined from other methods in the Amazon and its Andean tributaries and plot above the line. Better agreement would be obtained using a $f_{reac+diss}^{[^9Be]}$ <0.2. This seems likely since sediments in the Amazon and its Andean tributaries are derived from regions of high relief and rapid erosion and with less intense weathering. On the other hand, denudation rates predicted from (^{10}Be/^{9}Be) for the Guiana Shield (drained by the Orinoco River) appear too high and a

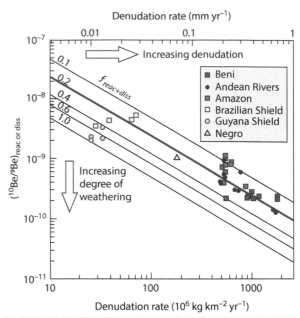

Figure 4.7 (^{10}Be/^{9}Be) in South American rivers plotted against denudation rate calculated from river sediment loads or in situ ^{10}Be exposure ages. The bold red line shows the denudation rate calculated from Equation (4.15) assuming the average crustal ^{9}Be concentration of 2.5 ppm and $f_{reac+diss}^{[^9Be]}$= 0.2. Black lines show the denudation rate calculated for other values of $f_{reac+diss}^{[^9Be]}$. Andean rivers and the Amazon plot above the red line, implying lower $f_{reac+diss}^{[^9Be]}$ while Guyana Shield rivers plot below it, implying higher $f_{reac+diss}^{[^9Be]}$ and more intense weathering. Source: Adapted from Von Blanckenburg et al. (2012).

value of $f_{reac+diss}^{[^9Be]}$ >0.4 would produce better agreement. The Guyana Shield is a region of intense tropical weathering and low relief; the latter limits transport of weathered material, which would certainly be consistent with a higher value of $f_{reac+diss}^{[^9Be]}$.

Because continents are the sole source of ^{9}Be$_{diss}$ in ocean while direct atmospheric deposition is the main source of marine ^{10}Be$_{diss}$, the marine ^{10}Be/^{9}Be ratio provides a measure of continental denudation rates. As we noted earlier, ^{10}Be/^{9}Be$_{diss}$ in modern seawater varies from 1 to 1.5 × 10^{-7} as a consequence of varying continental input (which includes both river water and dust, both of which have

^{10}Be/^{9}Be $<10^{-8}$). Brown et al. (1992) suggested that ^{10}Be/^{9}Be ratio in authigenic sediments whose Be is derived from seawater can provide an archive of continental erosion rates over time. How weathering and erosion have varied with time is an important question because silicate weathering consumes atmospheric CO_2 through a series of reactions known as the Urey reactions and some have argued that cooling during the Quaternary is due to increased weathering and erosion. Willenbring and von Blanckenburg (2010) examined ^{10}Be/^{9}Be in ferromanganese crusts and authigenic sediments in both the Pacific Ocean and the Atlantic Ocean. After correcting for ^{10}Be-decay based either in assumed constant growth rates or independent age estimates, they found that marine ^{10}Be/^{9}Be has not varied outside the present range for these oceans over the last 10 million years (Figure 4.8 shows three examples). This suggests weathering and erosion rates have not increased over this period and are consistent with isotopic proxies for atmospheric CO_2 that show atmospheric CO_2 has not declined systematically over this period (we will return to this topic in Chapter 12).

4.2.4   Cosmogenic radionuclides in hydrology

Determining the age of water in underground aquifers is an important problem because of the increasing demands placed in many parts of the world on limited groundwater resources. A prudent policy for water resource management is to withdraw water from a reservoir at a rate no greater than the recharge rate. Determination of recharge rate is thus prerequisite to wise management. Cosmogenic radionuclides, most notably ^{14}C and ^{36}Cl in this context, are swept out of the atmosphere by rain and into the groundwater system. Since their abundances decay with time, they provide a means of estimating the "age" of groundwater, i.e., the time since the water fell as precipitation and penetrated into the groundwater system.

If we know both the present and initial activity or concentration of a radionuclide in rainwater and if we can assume that it is not produced within the Earth or lost from solution (this is our closed-system requirement), then the "age" of water in an aquifer is determined simply from Equation (4.1), where we define age as the time since the water left the

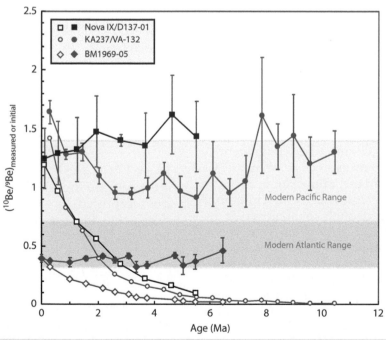

Figure 4.8   Measured (^{10}Be/^{9}Be) (open symbols) and calculated (^{10}Be/^{9}Be)$_0$ (closed symbols) in manganese crusts KA237/VA-132 (red circles) and Nova IX/D137-01 from the Pacific Ocean (black squares) and BM1969-05 (blue diamonds) from the Atlantic Ocean plotted against age assuming constant growth rate. Error bars include errors on age. Data from Willenbring and von Blanckenburg (2010).

atmosphere and entered the groundwater system. However, groundwater systems can be complex. As Bethke and Johnson (2008) emphasize, water may enter an aquifer at different points along a flow path and water within an aquifer can be a mixture of water added to it at different times. Consequently, a short-lived radionuclide, such as ^{14}C, can be present and indicate one age, while a longer-lived one, such as ^{36}Cl, can indicate an older age. It is perhaps best to think in terms of "residence time," the average time a molecule has resided in an aquifer.

^{14}C has been used successful for dating groundwater for decades. However, there are several problems with ^{14}C dating. The first is that ^{14}C is present in water principally as HCO_3^- and CO_3^{2-}. Both undergo isotopic exchange reactions with carbonates in soils and the aquifer matrix; further, precipitation and dissolution of carbonates will alter the concentration of ^{14}C in groundwater. This, of course, violates the closed-system requirement. While the half-life of ^{14}C is sufficient for most groundwater systems, which tend to be shallow and localized, larger, regional systems can contain water older than the $\approx 25\ 000$ year useful range of ^{14}C.

^{36}Cl, with a half-live of $\approx 300\ 000$ years, provides a means of dating these older groundwater systems. In the atmosphere, ^{36}Cl is primarily produced by spallation of ^{40}Ar and neutron capture by ^{35}Cl. The former process has been estimated to produce about 11 atoms $m^{-2}\ s^{-1}$, while the latter produces about half that, for a total production of about 15 atoms $m^{-2}\ s^{-1}$, which is considerably lower than production rates of ^{14}C and ^{10}Be; consequently, ^{36}Cl can only be measured by AMS. ^{36}Cl typically resides in the atmosphere about a week before it is removed by precipitation, which is not long enough to homogenize its concentration, so the deposition rate varies with latitude as shown in Figure 4.9. Chlorine is highly soluble in water (as the chloride ion), which makes it ideal for hydrological studies. A disadvantage, in addition to its low abundance, is that small amounts of ^{36}Cl are also produced within the Earth by ^{35}Cl capture of neutrons produced by ^{238}U fission, as well as secondary neutrons produced by (α,n) reactions. In contrast to ^{14}C, ^{36}Cl is essentially conservative in groundwater solutions, and

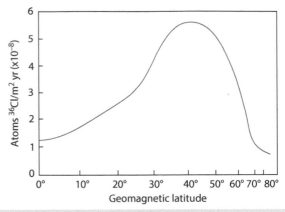

Figure 4.9 Variation of ^{36}Cl atmospheric production rate with latitude. Source: Bentley et al. (1986)/Reproduced with permission from John Wiley & Sons.

has a half-life suitable for dating water in regional aquifers as old as 1 Ma.

^{36}Cl atoms' abundances are generally reported as the $^{36}Cl/Cl$ ratio (Cl has two stable isotopes: ^{35}Cl and ^{37}Cl). Because of the high solubility of the chloride ion, chlorine tends to be highly conservative in natural waters, although some can be leached from rocks, particularly evaporites. Further complications arise from the presence of nuclear-bomb-produced ^{36}Cl in systems containing water younger than the mid-twentieth century and neutron capture by ^{35}Cl as noted above. Thus, ^{36}Cl will build up in groundwater according to:

$$[^{36}Cl] = \frac{\phi_n f\,[^{35}Cl]}{\lambda_{36}} \left(1 - e^{-\lambda_{36}t}\right) \qquad (4.16)$$

where $[^{36}Cl]$ and $[^{35}Cl]$ are concentrations, ϕ_n is the neutron flux, and f is the fraction of neutrons captured by ^{35}Cl. The secular equilibrium value, i.e., the concentration at $t = \infty$, is simply:

$$\frac{[^{36}Cl]}{[Cl]} = 0.7576 \times \frac{\phi_n f}{\lambda_{36}} \qquad (4.17)$$

where 0.7576 is the fraction of Cl that is ^{35}Cl. This in situ production must be taken into account.

Stable Cl derived from sea spray is also present in the atmosphere and in precipitation. Its concentration decreases exponentially from coasts to continental interiors. Thus, the initial $^{36}Cl/Cl$ ratio in precipitation will be variable and must be determined or estimated locally before groundwater ages can be estimated. The age of groundwater may then be determined from:

$$t = \frac{-1}{\lambda} \ln \left\{ \frac{[Cl]\left(^{36}Cl/Cl - {}^{36}Cl/Cl_{se}\right)}{[Cl]_0\left(^{36}Cl/Cl_0 - {}^{36}Cl/Cl_{se}\right)} \right\}$$

(4.18)

where [Cl] is the chloride concentration and the subscripts "0" and "se" denote initial and secular equilibrium values, respectively.

Groundwater in the Katmandu Valley of Nepal provides an example of the importance of understanding groundwater ages and recharge rates. While the rural population depends on shallow groundwater (which is often unsafe due to fecal contamination), urban areas exploit a deeper ($\gtrsim$200 m) aquifer, in which better-quality water flows slowly to the southwest through Pliocene sands confined by overlying Plio-Pleistocene clays. Based on a reconnaissance survey of ^{36}Cl in water from wells in the region, Cresswell et al. (2001) estimated groundwater ages of between 200 and 400 ka and recharge rates of 40 to 120 × 10^{11} m^3 yr^{-1}. Current withdrawals from the aquifer are estimated to be 20 times that amount, implying the resource will be exhausted within 100 years.

The more extensive study of Bentley et al. (1986) of groundwater in the Great Artesian Basin aquifer provides another example. The Great Artesian Basin aquifer is one of the largest artesian aquifers in the world and underlies about a fifth of Australia (Figure 4.10). Much of central Australia is extremely dry, so this groundwater is a critical resource for residents of the region. The primary aquifer is the Jurassic Hooray sandstone, which outcrops and is recharged in areas of higher elevation along the eastern edge of the basin. Bentley et al. (1986) sampled 28 wells from the system. They estimated an initial $^{36}Cl/Cl$ ratio of 110 × 10^{-15} and a secular equilibrium value of 9 × 10^{-15} atoms per liter. Some well samples showed evidence of Cl addition; $^{35}Cl/^{37}Cl$ isotope ratios indicate the primary source is pore waters of the confining shale diffusing into the aquifer. Other wells, particularly those in the recharge area, showed

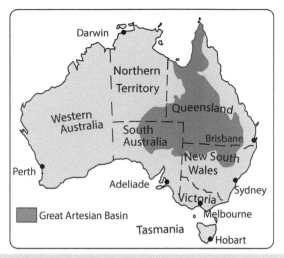

Figure 4.10   Extent of the Great Artesian Basin aquifer in Australia. Source: Adapted from Bentley et al. (1986).

evidence of evaporation, which increases Cl concentrations before the water penetrated the groundwater system, which is not surprising in an arid environment. Furthermore, some wells on the southwestern side of the basin have ^{14}C ages of ≈25–30 ka and ^{36}Cl ages of ≈75–150 ka, indicating local recharge (Abu Risha, 2016). On the whole, however, ^{36}Cl ages were comparable to calculated hydrodynamic ages, as illustrated in Figure 4.11.

^{4}He data can provide a useful compliment to ^{36}Cl studies. ^{4}He accumulates in rocks as a consequence of α-decay. Assuming that water flows uniformly through an aquifer as though it were pushed along by a piston ("piston flow"), accumulation of ^{4}He may be described by:

$$[^4He] = [^4He]_0 + \frac{R_\alpha t}{\phi}$$

(4.19)

where $[^4He]_0$ is the initial concentration, R_α is the production rate (a function of the Th and U concentration), ϕ is the aquifer porosity (volume fraction of the aquifer not occupied by rock), and t is time. In the Great Artesian aquifer, however, Torgersen and Clarke (1985) found that ^{4}He appears to accumulate non-linearly. Bethke and Johnson (2008) explain this with a model in which ^{4}He diffuses into the aquifer from underlying crystalline basement. Beds of fine-grained, low-porosity rocks within the aquifer (which consists mainly of

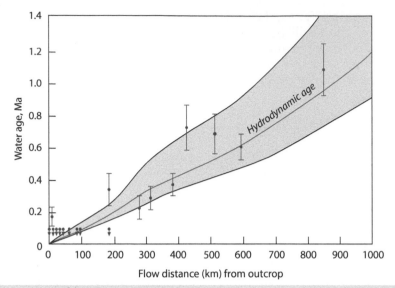

Figure 4.11 Comparison of ^{36}Cl ages with "hydrodynamic age," i.e., the age estimated from hydrologic flow parameters. Source: Bentley et al. 1986)/Reproduced with permission of John Wiley & Sons.

sandstone as noted previously) limit the upward mobility of ^{4}He, resulting in a strong vertical gradient within the aquifer. Eventually upwelling of water near the downstream end of the aquifer allows the ^{4}He to be mixed vertically.

4.2.5   In situ produced cosmogenic nuclides

Some cosmic rays and secondary particles manage to pass entirely through the atmosphere where they interact with rock at the surface of the Earth. These interactions produce a great variety of stable and unstable nuclei. Because cosmic rays penetrate only to quite shallow depth in rocks and soil, their buildup can be used to determine the length of time materials have been at the surface: so-called exposure ages. They can also be used to determine the rates at which material is being removed from the surface, i.e., erosion rates. We will examine these applications in this section.

The atmosphere is a very effective cosmic ray shield (a good thing for terrestrial life!) and consequently in situ cosmogenic nuclide production is rare. In addition to the radioactive cosmogenic nuclides listed in Table 4.1, the rare isotopes of noble gases, ^{3}He and ^{21}Ne, are also of interest. Approximate in situ production rates are listed in Table 4.2. In addition to cosmic-ray-induced spallation, these nuclides can also be produced by

reactions with lower-energy secondary particles, including neutrons, muons, and alpha particles. The most important of these reactions are also listed in Table 4.2.

4.2.5.1   Exposure ages

For a nuclide being both produced by cosmic ray interactions and lost by radioactive decay, our basic equation becomes:

$$\frac{dN}{dt} = P - \lambda N \qquad (4.20)$$

where P is the production rate. As we found for cosmogenic nuclides produced in the atmosphere, the in situ production rate is a function of time and latitude; in addition, it depends on rock composition, altitude and mean atmospheric pressure, and depth below the surface. The time dependence (Figure 4.2) results from both variations in solar activity and the strength of Earth's dipole magnetic field; the latitudinal dependence results from the cosmic ray deflection by the magnetic field (Figure 4.9). The compositional dependence reflects varying abundances of target nuclei in the sample. For ^{10}Be and ^{26}Al, analysis is typically done on quartz because of its simple and consistent composition and abundance in crustal rocks. For ^{36}Cl, the chemical composition, particularly

Table 4.2   Isotopes with appreciable in situ production rates in terrestrial rocks.

Isotope	Half-life years	Spallation		Thermal neutrons		Capture of μ^- target
		Target	Rate (atoms/g/yr)	Target	Reaction	
^{3}He	Stable	O, Mg, Al, Si, Fe	100–150	^{6}Li	(n,α)	
^{10}Be	1.6×10^6	O, Mg, Al, Si, Fe	6	—	—	^{10}B, C, N, O
^{14}C	5730	O, Si, Mg, Fe	20	^{14}N, ^{17}O	(n,p), (n,α)	N, O
^{21}Ne	Stable	Mg, Al, Si, Fe	80–160	—		Na, Mg, Al
^{26}Al	7.1×10^5	Mg, Al, Si, Fe	35	—	—	Si, S
^{36}Cl	3.0×10^5	K, Ca, Cl	10	^{35}Cl, ^{39}K	(n,γ), (n,α)	K, Ca, Sc
^{129}I	1.6×10^5		<< 1	^{128}Te	(n,γ)	^{130}Te, Ba

the concentrations of the main targets, K and Ca, must be determined.

The cosmic ray flux through both the atmosphere and the solid Earth decreases with depth as:

$$P(x) = P_0 e^{-z\rho/\Lambda} \qquad (4.21)$$

where z is depth, P_0, is the production rate at the surface, Λ is a constant that depends on the nature and energy of the particle (and weakly on the material it penetrates), and ρ is the density. For the nucleonic component of cosmic rays, Λ is approximately 160 g/cm^2. In the atmosphere, ρ is in itself a function of atmospheric depth, h, (or more conveniently, altitude) that can be approximated as:

$$\rho \approx \rho_0 e^{-h/H_0} \qquad (4.22)$$

where ρ_0 is the surface density and H_0 is a scale height, ≈ 10 km. In detail, it will depend on surface temperature and the temperature lapse rate, as well as surface pressure. This results in a decrease in the cosmic ray flux with depth in the atmosphere by a factor of ≈ 5 from an altitude of 5 km to sea level. Furthermore, just as a hill can shade a surface from sunlight, so too can it shield a surface from cosmic rays and this shielding must be accounted for in calculating the production rate.

Since density can generally be taken as constant, the depth dependence through the soil and rock is simpler than in the atmosphere;

however, due to the much higher density of solids, the flux decrease is much more rapid through them. For a material, such as a typical rock, having a density of 2.5 g/cc, the ratio ρ/Λ is about 64 cm^{-1}. Thus, at a depth of 64 cm, the cosmic ray flux would be $1/e$ or ≈ 0.37 times the flux at the surface and is referred to as the *characteristic penetration depth*. For the μ (muon[4]) component, Λ is about 1000 g/cm^2 (for neutrinos, Λ is nearly infinite because neutrinos interact so weakly with matter, but the production rate is nearly infinitesimal for the same reason). Most of the cosmic ray interactions are with the nucleonic component. Consequently, cosmogenic nuclides will be produced only on the surface (mainly the top meter and almost entirely within the top 2 meters) of a solid body. Figure 4.12 illustrates the decreased abundance in cosmogenic ^{3}He in a core through a lava flow on Haleakala volcano in Hawaii reflecting this decreasing flux.

To obtain the abundance, N, of a cosmogenic radionuclide at some time t, we integrate Equation (4.2):

$$N = \frac{P}{\lambda}\left(1 - e^{-\lambda t}\right) \qquad (4.23)$$

where P is now the time-integrated production rate. For $t >> 1/\lambda$; that is, after many half-lives, a steady state is reached where:

$$N = \frac{P}{\lambda} \qquad (4.24)$$

---

[4] The muon belongs to the family of particles known as *leptons*, the most familiar members of which are the electron and positron. Like the electron, it may be positively or negatively charged and has a spin of $^1/_2$. However, its mass is about 100 MeV, more than two orders of magnitude greater than that of the electron, and about one order of magnitude less than the proton. It is produced mainly by decay of pions, which are also leptons and are created by high-energy cosmic ray interactions. Muons are unstable, decaying to electrons and positrons and ν_u (muonic neutrino) with an average lifetime of 2×10^{-6} s. Because muons are leptons, they are not affected by the strong force, and hence interact more weakly with matter than the nucleonic component of cosmic rays.

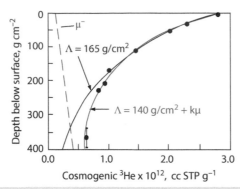

Figure 4.12 Comparison of the variation of cosmogenic ^{3}He with depth in a core from Haleakala volcano in Hawaii with the predicted decrease in the nucleonic and muon cosmic ray flux. Source: Kurz (1986)/Reproduced with permission of Elsevier.

Once this steady state is reached, an exposure age can no longer be calculated, but useful information can still be obtained, as we will see in the following section. For shorter times, where t is less than 5–10 times the half-life, we can solve Equation (4.22) for t. This is the exposure age, the time the rock has been exposed to cosmic rays at the surface of the Earth.

To solve for t, however, we must know P, which is a function of the many variables mentioned above. Very considerable effort has been made to determine P as a function of these variables, beginning with Lal (1991). Typically, the complexity of these variables results in P being expressed as polynomial equations based on a combination of theoretical physics, experiments, and calibrations from observed cosmogenic fluxes and cosmogenic nuclides in surface samples of known age around the globe. Examples of the latter include the shoreline of glacial Lake Bonneville in Utah, US, dated by radiocarbon and U-Th at 18.36 ± 0.3 ka, and the Huancané II moraines of the Quelccaya Ice Cap in southern Peru dated by radiocarbon at 12.3 ± 0.1 ka. The most recent of such "scaling model" is that of Lifton et al. (2014). An international collaboration, known as the CRONUS-Earth Project (Phillips et al., 2016), continues to improve the calibration of exposure ages and their accuracy. Online calculators and MATLAB scripts are available to perform the complex calculations at: http://web1.ittc.ku.edu:8888/1.0/ or

http://hess.ess.washington.edu (Balco et al., 2008). At present, the absolute uncertainties in these scaling models are about 6% (Balco, 2020); relative uncertainties, where one wishes to compare different exposure ages from the same area, are less.

Much of the research using in situ cosmogenic isotopes has focused on Quaternary glacial history. As we found in previous sections, sea level history has been reconstructed from ^{14}C and Th–U dating of coral reefs, and, as we will find in subsequent chapters, the temperature history of the Pleistocene glaciations has been revealed by stable isotopes in marine sediments and ice cores. This, however, does not tell us the history of continental ice sheets and mountain glaciers, which is where in situ cosmogenic isotopes are useful. Geologists have recognized for well over a century that terminal and recessional moraines across North America mark the demise of the Pleistocene Laurentide ice sheet. Cosmogenic nuclides are revealing their chronology.

The maximum extent of the Laurentide ice sheet in eastern North America is marked by a semi-continuous line of moraines that extends southward from New York into northern Pennsylvania and New Jersey and then eastward along Long Island and along the southern New England coast to the islands of Martha's Vineyard and Nantucket (Figure 4.13). In situ ^{10}Be and ^{26}Al dating of exposed glacial erratic boulders within and just behind these moraines establishes the age of this feature as ≈26 ka (e.g., Corbett et al., 2017). This corresponds well with sea level minimum recorded by U–Th and ^{14}C dating of coral reefs, minimum temperatures recorded by stable isotopes in Greenland ice cores, and maximum ice sheet extents elsewhere that indicate the last glacial maximum (LGM) began around 33 ka and ended around 26.5 ka (e.g., Clark et al, 2009). Several moraines in southern New England have exposure ages of ≈20 ka, indicating a slow retreat of the ice sheet following the LGM. To the north and west, cosmogenic exposure ages of 13–15 ka predominate, suggesting quite rapid retreat of the ice sheet around this time. This corresponds to the Bølling–Allerød interstadial, a period recognized in paleoclimate records as one of rapid warming.

To the north, complexities in cosmogenic exposure age dating have emerged. Bierman

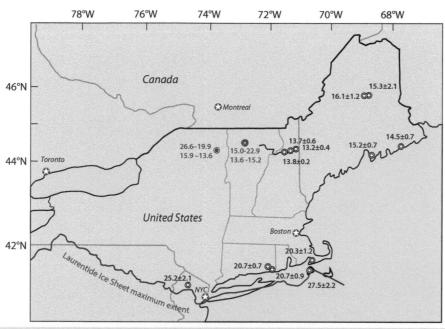

Figure 4.13   Cosmic ray exposure ages of bedrock and moraines in the Northeastern United States. The bold purple line marks the maximum extent of the Laurentide Ice Sheet at the LGM around 26 ka. Moraines just a few tens of kilometers to the north of this line in southern New England have exposure ages of ≈20 ka, indicating slow retreat of the ice sheet over 6000 years. With the exception of ages on the higher elevations of the Adirondacks (red) and Mount Mansfield (blue), ages from northern New England center around 13–16 ka, corresponding to the Bølling–Allerød warming. Ages from higher elevations in the Adirondacks and Mount Mansfield are older, reflecting thinning of the ice sheet as it slowly retreated prior to the Bølling–Allerød. Source: Adapted from Corbett et al. (2017).

et al. (2015) reported one ^{26}Al and ^{10}Be anomalously old exposure age from the summit of Mount Katahdin in Maine of 36.5 ka and three ^{26}Al–^{10}Be ages ranging from 59.6 to 152.6 ka of boulders from the summit (≈1900 m) of Mount Washington, New Hampshire. However, ^{14}C exposure ages were much younger, i.e., 11.0 ± 2.2–12.7 ± 2.8 ka. Bierman et al. argued that that glaciers at lower elevations were warm-based, resulting in significant erosion, but were cold-based at high elevations resulting in little erosion and consequently those boulders retain ^{10}Be from previous exposures, which, with its shorter half-life, ^{14}C did not. Corbett et al. (2019) also reported disagreement between in situ ^{10}Be and ^{14}C ages in bedrock and boulder surfaces at high elevations on Mount Mansfield, Vermont. Boulders in moraines at elevations of 400–1200 m all have indistinguishable ^{10}Be exposure ages of 13.9 ± 0.6 ka and are consistent with ^{14}C exposure ages. However, bedrock exposures above

1200 m range from 15.0 ± 3 to 22.9 ± 5 ka whereas ^{14}C ages were younger.

In the Adirondack Mountains in New York to the west of Mount Mansfield, Barth et al. (2019) reported that, after excluding two much older samples that likely contained inherited ^{36}Cl, most (75%) exposure ages from elevations below 1300 m were between 15.9 and 13.6 ka while six of the seven ages from higher elevations fell between 26.4 and 19.9 ka (Figure 4.14). The data from Mount Mansfield also fit this pattern. They argued that ages at the high elevations show that following the LGM, the Laurentide Ice Sheet began to slowly thin, exposing the highest elevations. The landscape then would have looked much like parts of Antarctica today, with ice-free peaks rising above ice-filled lowland. This thinning would be consistent with the slow retreat along southern New England. This scene remained much unchanged until the Bølling–Allerød period around 15 ka when the ice sheet rapidly

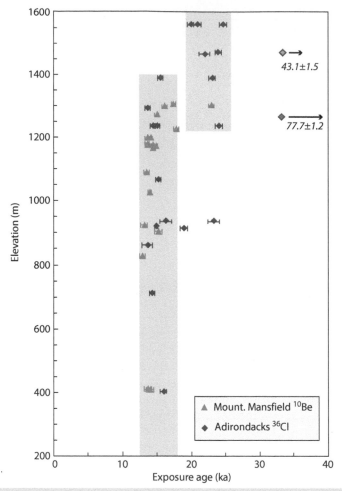

Figure 4.14 ^{36}Cl ^{10}Be exposure ages from the Adirondack Mountains (red) and Mount Mansfield (green) as a function of elevation. Most exposure ages from elevations below 1300 m range between 15.9 and 13.6 ka, corresponding to the Bølling–Allerød warm interval and rapid retreat of the Laurentide Ice Sheet while six of the seven ages from higher elevations of the Adirondacks fall between 26.4 and 19.9 ka, indicating that the ice sheet began to thin following the LGM. Mount Mansfield data also fit this pattern. Data from Corebett et al. (2019) and Barth et al. (2019).

thinned and retreated, finally exposing the lowlands.

4.2.5.2   Erosion rates

Measurement of in situ produced cosmogenic nuclides can also be used to estimate erosion rates. Let us first consider how this can be done using a rare stable isotope, such as ^{3}He, because this simplifies the math somewhat. (He is rare on Earth because it escapes from the atmosphere readily. ^{3}He is even rarer, being some six orders of magnitude less abundant than ^{4}He in the atmosphere and even

rarer in the crust.) For a stable cosmogenic nuclide, Equation (4.20) simplifies to:

$$\frac{dN}{dt} = P_0 \qquad (4.25)$$

where P_0 is the production rate at the surface. The total number of atoms after an exposure of time t is given by the integral of the production rate over time:

$$N_t = \int_0^t P(t)dt \qquad (4.26)$$

If we can assume the production rate is constant, then the right-hand side is simply the product Pt. As we found earlier, however, P is a function of the cosmic ray flux that varies with time (as well as elevation, geomagnetic latitude, and reaction cross-section).

Most cosmogenic ^{3}He is produced by spallation of abundant elements, such as O, Mg, and Si, with a minor component produced by ^{6}Li(n,α) and ^{7}Li(μ,α) reactions[5]. Figure 4.12 shows the decrease in cosmogenic ^{3}He with depth in a core from Haleakala (Maui, Hawaii) compared with the predicted decrease for $\Lambda = 165$ g cm^{-2}. The dashed line shows the depth dependence of the μ stopping rate needed to explain the discrepancy that Kurz (1986) found between the predicted and observed depth dependence. If erosion occurs, then z will be a function of t. Ignoring the small contribution from muon interactions, the concentration, C, of ^{3}He as a function of depth, z, and exposure time, t, is given by:

$$C(z, t) = \int_0^t P_0 e^{-z(t)\rho/\Lambda} dt \qquad (4.27)$$

We obtain the simplest relationship between time and depth by assuming the erosion rate is constant:

$$z = z_0 - \varepsilon t \qquad (4.28)$$

where ε is the erosion rate (although this is generally termed an erosion rate in much of the literature, it is more properly referred to as the *denudation rate* as it is measured in units of depth per time, rather than mass per time; the erosion rate is simply the denudation rate multiplied by density) and z_0 is the original depth. Substituting for z in Equation (4.27) and integrating, we obtain:

$$C(z, t) = P_0 \frac{\Lambda}{\varepsilon\rho} e^{-z_0\rho/\Lambda} \left(e^{t\varepsilon\rho/\Lambda} - 1\right) \qquad (4.29)$$

Substituting $z_0 = z + \varepsilon t$, Equation (4.29) becomes:

$$C(z, t) = P_0 \frac{\Lambda}{\varepsilon\rho} e^{-z\rho/\Lambda} \left(1 - e^{-\varepsilon t\rho/\Lambda}\right) \qquad (4.30)$$

For a sample at the surface, $z = 0$ and this equation reduces to:

$$C_0 = P_0 \frac{\Lambda}{\varepsilon\rho} \left(1 - e^{-\varepsilon t\rho/\Lambda}\right) \qquad (4.31)$$

Neither Equation (4.29) nor Equation (4.30) can be solved directly. However, if the age of the rock, t, is known, one can measure a series of values, C, as a function of depth and make a series of guesses of the erosion rate until a curve based on 4.30 fits the data. Using this procedure, Kurz estimated an erosion rate of 10 m/Ma for Haleakala (Kurz noted that for higher erosion rates, it would be necessary to take account of the muon-produced ^{3}He).

For a radioactive nuclide, such as ^{26}Al or ^{36}Cl, we need to consider its decay, as well as its production. The concentration of such nuclide as a function of time and depth is given by:

$$C(z, t) = \frac{P_0 e^{-z\rho/\Lambda}}{\lambda + \varepsilon\rho/\Lambda} \left(1 - e^{(-\lambda + \varepsilon\rho/\Lambda)t}\right) \qquad (4.32)$$

where λ is the decay constant and t is age of the rock. If the rock is much older than the half-life of the nuclide, i.e., $\lambda t \gg 1$, then the last term tends to be 1 (for example, for ^{26}Al, this would be the case for a rock $\gtrsim 4$ Ma old, but roughly twice that for ^{10}Be). Eventually, production of the nuclide, its decay, and erosion will reach steady state (assuming cosmic ray flux and erosion rate are time independent). In this case, the concentration at the surface will be given by:

$$C_0 = \frac{P_0}{\lambda + \varepsilon\rho/\Lambda} \qquad (4.33)$$

Since this equation does not contain a time term, we cannot deduce anything about time in this situation. Nevertheless, assuming P_0 is time independent (which will be approximately true over long periods as variations will average out) and knowing Λ, λ, and ρ, we can solve Equation (4.33) for the erosion rate:

$$\varepsilon = \frac{\Lambda}{\rho} \left(\frac{P_0}{C_0} - \lambda\right) \qquad (4.34)$$

Interestingly enough, and perhaps counterintuitively, Equation (4.34) applies to soil, as well as bedrock, assuming again that $\lambda t \gg 1$ and steady state has developed (see Reiners et al.

---

[5] These reactions of course produce ^{4}He as well, but the amount is trivial compared to production from α-decay. Small amounts of ^{3}He are also produced in the Earth through ^{6}Li(n,α) reactions initiated by neutrons from spontaneous fission of ^{238}U. This results in typical crustal ^{3}He/^{4}He ratios of $\lesssim 10^{-7}$.

(2017) for the derivation for soils). No matter whether it is bedrock or soil, P_0, will differ between nuclides and vary with the material, geomagnetic latitude, attitude, etc., and hence must be calculated for each site.

Portenga and Bierman (2011) calculated ^{10}Be erosion rates based on this equation and production rates calculated using the CHRONUS calculator mentioned above for a compilation of 1559 ^{10}Be measurements from 87 sites around the world to develop a global picture of erosion rates. Their erosion rates correlated well with previous estimates based on sediment load. Multivariant statistical analysis of the date revealed the following observations:

- Basins erode more rapidly (mean: 218 ± 35 m/Ma; median: 54 m/Ma) than do outcrops (mean: 12 ± 3 m/Ma; median: 5.4 m/Ma). As the difference between mean and median indicates, both outcrop and drainage basin erosion rates have highly skewed distributions, with most samples indicating relatively slow rates of erosion.
- Outcrop erosion rates correlate with mean annual precipitation and polar regions erode less rapidly than those in temperature and tropical regions. Basin erosion rates are also higher in polar regions, but do not correlate with mean annual precipitation.
- Outcrop erosion rates are higher for sedimentary rocks than igneous or metamorphic ones, but basin erosion rates are independent of rock type.
- Basin erosion rates depend most strongly on basin slope and relief. They also correlate with seismic activity (a proxy for tectonic activity), but outcrop erosion rates are independent of seismicity.

Insights into erosion rates and sediment transport can be obtained by analyzing the ratio of two cosmogenic nuclides. Beginning with Equation (4.33), we can derive the following equation:

$$\left(\frac{^{26}Al}{^{10}Be}\right)_0 = \frac{\left(P^{26}Al / P^{10}Be\right)_0}{(\lambda_{26Al} + \varepsilon\rho/\Lambda_{26Al})/(\lambda_{10Be} + \varepsilon\rho/\Lambda_{10Be})} \quad (4.35)$$

The penetration length scale, Λ, appears to be approximately the same for ^{26}Al and ^{10}Be

within uncertainties. Assuming that it is and for $\varepsilon\rho/\Lambda \gg \lambda$, which will be the case where the erosion rate is $\gtrsim 1$ m/Ma, then Equation (4.35) reduces to:

$$\left(\frac{^{26}Al}{^{10}Be}\right)_0 = \left(\frac{P^{26}Al}{P^{10}Be}\right)_0 \quad (4.36)$$

In other words, the observed ratio in an exposed surface should equal the production ratio. However, where sediment is buried and stored in a drainage system and shielded from new production for significant amounts of time relative to the ^{26}Al decay constant, the ^{26}Al/^{10}Be ratio will decrease due to the more rapid decay of ^{26}Al.

Wittmann et al. (2020) examined in situ ^{26}Al and ^{10}Be in quartz grains in sediment from >50 large rivers to estimate global denudation rates. They found that in roughly two-thirds of these rivers, ^{26}Al/^{10}Be ratios in quartz were equal or close to the production ratio of about 6.75. For these rivers, they used Equation (4.34) and ^{10}Be concentrations to calculate an average denudation rate of 54 m/Ma, which is quite similar to the median rate found by Portenga and Bierman (2011) for basins. Extrapolating to the global land surface, they calculate a global erosion rate of 15.2 Gt/yr integrated over the last 11 000 years. In the remaining third of rivers, ^{26}Al/^{10}Be ratios are significantly lower than production ratio, indicating radioactive decay over periods exceeding 0.5 Myr. They argued that this reflected slow erosion, shielding in the source area, and sediment storage and burial during long-distance transport in those river systems.

4.3   THERMOCHRONOLOGY

Thermochronology comprises a set of geochronological tools used to determine thermal histories rather than formation ages of a rocks, particularly at geologically low to moderate temperatures, i.e., below 350–500°C. The tools used in these studies are thus ones with relatively low closure temperatures. One of these is the K–Ar system, which we have already introduced in Chapter 2. Here, we will introduce two additional tools: fission track dating and U/Th–He dating. The primary focus of these studies is on determining rates of uplift and erosion to understand ongoing tectonic processes. Conductive cooling results in temperatures increasing with depth in the Earth's crust. Consequently, rocks cool as they are uplifted, and the overlying rock is removed by erosion. As they do so, these chronometers pass through their closure temperatures. Closure temperatures of these systems range from a few tens of degrees for U–He in apatite to $\gtrsim 550°C$ for K–Ar in hornblende, depending, as we found in Chapter 2, on cooling rates. Using these systems in concert and on different minerals allows reconstruction of thermal histories and hence rates of uplift and erosion. That, in turn, allows us to quantify rates of ongoing tectonic processes.

4.3.1 Fission tracks

As we have already found, a fraction of ^{238}U atoms undergo spontaneous fission rather than α-decay. The sum of the masses of the fragments is less than that of the parent U atom. The missing mass has been converted into kinetic energy of the fission fragments. Typically, this energy, which totals about 200 MeV and divided approximately equally between the two nuclei, is a considerable amount of energy on the atomic scale. Much of the energy is deposited in the crystal by stripping electrons from atoms in the crystal lattice through which the fission fragments

pass. The ionized atoms repel each other and together with thermal energy produced through collisions of the nuclei with atoms in the lattice disorder the lattice and produce a small channel, something like several nanometers (10^{-9} m) in width and ≈ 10 µm long, and a longer and wider stressed region in the crystal. The damage is visible as tracks can be seen with an electron microscope operating at magnifications of 50 000× or greater. However, the stressed region is more readily attacked and dissolved by acid so that the tracks can be enlarged by acid etching to the point where they are visible under an optical microscope; Figure 4.15 is an example. Because fission is a rare event in any case, fission track dating is performed exclusively on uranium-rich minerals. Most work has been done on apatite and zircon, but other minerals, such as titanite, are also used.

Just as the daughter nuclides of U alpha decays will build up over time, so will fission tracks, providing an analogous method of dating. An important difference is that fission tracks will *anneal,* or self-repair, over time. The rate of annealing is vanishingly small at room temperature but increases with temperature and becomes significant at geologically moderate temperatures. In the absence of such annealing, the number of tracks is a simple

Figure 4.15   Fission tracks in a polished and etched zircon. Source: Reproduced with permission from J. M. Bird.

function of time and the uranium content of the sample:

$$F_s = (\lambda_f/\lambda_\alpha)\left[^{238}U\right]\left(e^{\lambda_\alpha t} - 1\right) \qquad (4.37)$$

where F_s is the number of tracks produced by spontaneous fission, $[^{238}U]$ is the number of atoms of ^{238}U, λ_α is the α–decay constant for ^{238}U, and λ_f is the spontaneous fission decay constant, the best estimate for which is $8.46 \pm 0.06 \times 10^{-17}$ yr^{-1}. Thus, about 5×10^{-7} U atoms undergo spontaneous fission for every one that undergoes α-decay (while the exponent in Equation (4.37) should properly be $(\lambda_\alpha + \lambda_f)t$, the difference between $(\lambda_\alpha + \lambda_f)$ and λ_α is far smaller than the uncertainty on λ_α). Equation (4.37) can be solved directly for t simply by determining the number of tracks and the number of U atoms per volume of sample. In this case, t is the time elapsed since temperatures were high enough for all tracks to anneal. This is the basis for fission track dating. The temperatures required to anneal fission damage to a crystal are lower than those required to isotopically homogenize the crystal. Thus, fission track dating is typically used to "date" lower-temperature events than conventional geochronometers. It is part of a suite of chronometers, including ^{40}Ar–^{39}Ar and U–Th–He dating, which are reset over a range of fairly low temperatures that are used in t thermochronology.

As is the case for conventional radiometric dating, fission track dating measures the time elapsed since some thermal episode in which fission tracks annealed, which is often not same time as the formation of the crystal. Annealing results in the track shrinking from both ends; however, in detail, the process is somewhat irregular, with the morphology of the track becoming ragged and gaps developing before annealing is complete. Different minerals will anneal at different rates. In laboratory experiments, apatite begins to anneal at significant rates around 70°C and anneals entirely on geologically short times at 175°C. Titanite ($CaTiSiO_5$), on the other hand, only begins to anneal at 275°C and does not entirely anneal until temperatures of 420°C are reached. At higher temperatures, these minerals anneal very quickly in nature: no fission tracks are retained. The temperature at which tracks anneal completely is the *closure temperature*, but just as we found for the ^{40}K–^{40}Ar and other decay systems, closure temperatures depend on the cooling rate and are higher for rapid cooling than that for slow cooling.

4.3.1.1   Analytical procedures

Determining fission track density involves a relatively straightforward procedure of polishing and etching a thin section or grain mount, and then counting the number of tracks per unit area under a microscope. A number of etching procedures have been developed for various materials; these are listed in Table 4.3. Track densities of up to several thousands per square centimeter have been recorded. A minimum density of 10 tracks per square centimeter is required for the results to be statistically meaningful. A fission track, which is typically 10–20 μm long, must intersect the surface to be counted. Thus, Equation (4.37) becomes:

$$\rho_s = F_s q = (\lambda_f/\lambda_\alpha)\left[^{238}U\right]\left(e^{\lambda_\alpha t} - 1\right)q \qquad (4.38)$$

where ρ is the track density, q is the fraction of tracks intersecting the surface, and $[^{238}U]$ is now the concentration of ^{238}U per unit area.

The second step is to determine the U concentration of the sample. Historically, this has usually been done by neutron

Table 4.3   Etching procedures for fission track dating.

Mineral	Etching solution	Temperature (°C)	Duration
Apatite	5% HNO_3	21	10–30 s
Epidote	37.5 M NaOH	159	150 min
Muscovite	48% HF	20	20 min
Sphene	Conc. HCl	90	30–90 min
Volcanic glass	24% HF	25	1 min
Zircon	100 M NaOH	220	1.25 h

irradiation and counting of the tracks resulting from neutron-induced fission. There are variations to this procedure. In one method, spontaneous fission tracks are counted, then the sample is heated to anneal the tracks, and irradiated and recounted (this is necessary because irradiation heats the sample and results in partial annealing). Alternatively, a "detector," either a U-free muscovite sample or a plastic sheet, is placed over the polished surface that has previously been etched and counted. The sample together with the detector is irradiated, and the tracks in the detector are counted. This avoids having to heat and anneal the sample. This latter method is more commonly employed.

Whereas ^{238}U is the isotope that fissions in nature, it is actually ^{236}U, produced by neutron capture by ^{235}U that undergoes neutron-induced fission. The number of ^{235}U fission events induced by thermal neutron irradiation is:

$$F_i = \left[{}^{235}\text{U}\right]\phi\sigma \tag{4.39}$$

where ϕ is the thermal neutron dose (neutron flux times time) and σ is the reaction cross-section (about 580 barns for thermal neutrons). The induced track density is:

$$\rho_i = Fq_i = \left[{}^{235}\text{U}\right]\phi\sigma q \tag{4.40}$$

By dividing Equation (4.38) by Equation (4.40), we obtain:

$$\frac{\rho_s}{\rho_i} = \frac{\lambda_f}{\lambda_\alpha}\frac{137.82}{\phi\sigma}\left(e^{\lambda_\alpha t} - 1\right) \tag{4.41}$$

In the detector method, Equation (4.41) must be modified slightly to become:

$$\frac{\rho_s}{\rho_i} = \frac{\lambda_f}{\lambda_\alpha}\frac{137.82}{2\phi\sigma}\left(e^{\lambda_\alpha t} - 1\right) \tag{4.42}$$

The factor of 2 arises because surface-intersecting tracks produced by spontaneous fission originate both from U within the sample and from that part of the sample removed from etching. However, tracks in the detector can only originate in the remaining sample. This is illustrated in Figure 4.16.

One of the most difficult problems in this procedure is correctly measuring the neutron dose. This can be done by including a gold or aluminum foil and counting the decays of the radioisotope produced by neutron capture.

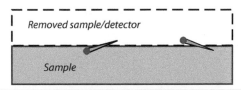

Figure 4.16   Geometry of the fission tracks in the detector method of U determination. Spontaneous fission tracks in the sample surface could have originated from either the existing sample volume or the part of the sample removed by polishing. Tracks in the detector can only originate from the existing sample volume.

Nevertheless, the neutron flux can be quite variable within a small space and it remains a significant source of error.

We can readily solve Equation (4.42) for t:

$$t = \frac{1}{\lambda_\alpha}\ln\left[1 + \frac{\rho_s}{\rho_i}\frac{\lambda_a}{\lambda_f}\frac{2\phi\sigma}{137.82}\right] \tag{4.43}$$

and thus determine the time since the tracks last annealed.

A widely used alternative method is the *zeta method*, which involves comparison of spontaneous and induced fission track density against a standard of known age. This approach is quite similar to that used in ^{40}Ar/^{39}Ar dating, where comparison to a standard eliminates some of the more poorly controlled variables. In the zeta method, the dose, cross-section, spontaneous fission decay constant, and U isotope ratio are combined into a single constant:

$$\zeta = \frac{\phi\sigma^{235}\text{U}}{\lambda_f{}^{238}\text{U}\rho_d} = \frac{\phi\sigma}{\lambda_f 137.82\rho_d} \tag{4.44}$$

where ρ_d is the density of tracks measured in a U-doped glass standard. ζ is thus very much analogous to the J parameter in ^{40}Ar/^{39}Ar dating. The value of ζ is determined by analyzing standards of known age with every sample batch and is determined from:

$$\zeta = \frac{e^{\lambda_\alpha t} - 1}{\lambda_\alpha(\rho_s/\rho_i)\rho_d} \tag{4.45}$$

The age is then calculated from:

$$t = \frac{1}{\lambda_\alpha}\ln\left(1 + \frac{\zeta\lambda_\alpha\rho_s\rho_d}{\rho_i}\right) \tag{4.46}$$

Standards used in the zeta method include zircon from the Fish Canyon Tuff (27.9 Ma), the

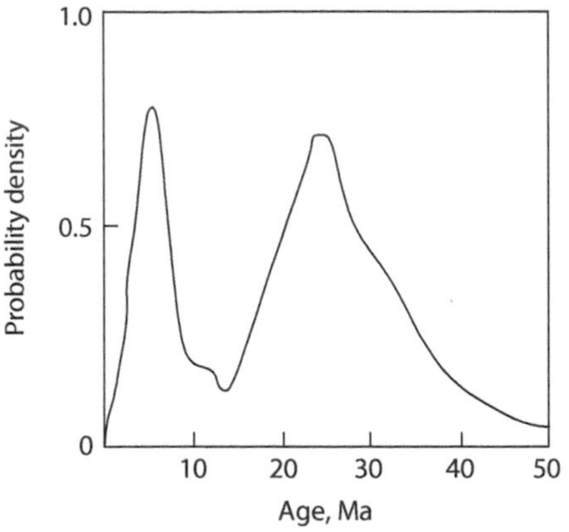

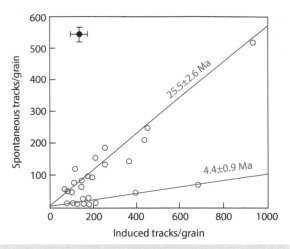

Figure 4.17 Probability density plot of fission track ages of 30 detrital zircon grains from the reworked El Ocote tephra from Mexico. The data show a bimodal distribution. Source: Adapted from Kowallis et al. (1986).

Figure 4.18 Spontaneous track density versus induced track density for the same set of zircon grains as in Figure 4.17. On this plot, the slope of the correlation is proportional to time.

Tardree Rhyolite of Ireland (58.7 Ma), and South African kimberlites (82 Ma).

Often, a population of separated mineral grains are randomly divided into aliquots. Grains in one aliquot are polished and the tracks counted as above while grains of the second aliquot are irradiated and U concentrations are measured as above. Alternatively, U concentrations in the second aliquot may be measured by other techniques, such as laser ablation inductively coupled plasma mass spectrometry.

Usually, fission track ages on a number of grains must be measured for the results to be significant. The results are often presented as histograms. Alternatively, when the errors are also considered, the results may be presented as a probability density diagram, such as Figure 4.17. Yet another approach is to plot the spontaneous track density (ρ_s) versus the induced track density (ρ_i), such as in Figure 4.18. From Equation (4.46), we see that the log of the slope on such a diagram is proportional to time. Thus, these kinds of plots are exactly analogous to conventional isochron diagrams; however, there is a difference. On a plot of ρ_s versus ρ_i, the intercept should be 0.

4.3.1.2 Fission track lengths

Except at very high temperatures, fission tracks do not disappear suddenly, but anneal over time and shorten as they do so. Additional information about the thermal history of a sample can be recovered from measurements of fission track lengths. Whereas in simple fission track analysis only tracks intersecting the surface are counted, in track length analysis only tracks *not* intersecting the surface are counted (because tracks intersecting the surface will be shortened by polishing). Such tracks will only be visible if they intersect a pathway, such as a cleavage plane or another track, for the etching solution to reach them. Furthermore, because both etching rates and annealing rates depend on crystallographic orientation, track length should be measured only on tracks having the same crystallographic orientation. Let us first consider how length will vary with time and temperature, and then discuss interpretation of lengths.

As is typically the case for chemical reaction rates, we expect that the temperature dependence of the annealing rate, da/dt, can be expressed by the *Arrhenius relation*:

$$\frac{da}{dt} \propto Ae^{-E_A/RT} \qquad (4.47)$$

where T is thermodynamic temperature (in kelvin), A is the frequency factor (usually, but not necessarily, independent of T), R is the gas constant (some equations use k, Boltzmann's constant, which is simply R divided by Avogadro's number), and E_A is the activation energy. Estimates of activation energies for fission track annealing range from 140–190 kJ/mol for apatite to 200–325 kJ/mol for zircon (Reiners and

Brandon, 2006). In detail, activation energies depend on the composition of the mineral, crystallographic direction, and radiation damage. Experiments suggest that the annealing rate also depends on the extent of annealing, such that the rate accelerates as it proceeds, and can be expressed as:

$$\frac{da}{dt} = Ae^{-E_A/RT}(1-a)^n \qquad (4.48)$$

where n has a value around -3 or -4 (Laslett et al., 1987). This latter results from a dependence of the activation energy on the extent of annealing, measured in experiments as the reduction in track length r (i.e., the ratio of observed to original track length). When r is plotted as a function of time (t) and temperature (T) (Figure 4.19), the dependence of E_A results in "fanning," that is, contours of equal values of r tend to spread out at low T and long t and converge on a point at high T and short t. Laslett et al. (1987) developed the following empirical model:

$$\frac{\left[(1-r^b)/b\right]^a - 1}{a} = c_0 + c_1 \left[\frac{\ln(t)-c_2}{1/T-c_3}\right] \quad (4.49)$$

For the experimental data of Green et al. (1985) for apatite, they found $a = 0.34$,

$b = 2.7$, $c_0 = -4.87$, $c_1 = 0.00168$, $c_2 = -28.13$, and $c_3 = 0$, in which case Equation (4.50) becomes:

$$\frac{\left[(1-r^{2.7})/2.7\right]^{0.34} - 1}{0.34}$$
$$= -4.87 + -0.000168T(\ln(t)-28.12)$$

$$(4.50)$$

Figure 4.19 compares contours of r, the reduction in track length, in experimental data (symbols) of Green et al. (1985) as a function of time and temperature with predictions by this "fanning model" (solid lines) with a simple "parallel model" (dashed lines). Subsequent studies have developed curvilinear models; see Reiners et al. (2017) for details.

Because tracks tend to have a constant initial length (controlled by the energy liberated in the fission) and tracks become progressively shorter during annealing, and each track is actually a different age and has experienced a different fraction of the thermal history of the sample, the length distribution records information about the thermal history of the sample. Figure 4.20 illustrates how track lengths are expected to vary for a variety of hypothetical time–temperature paths. Uniform track lengths suggest a simple thermal history

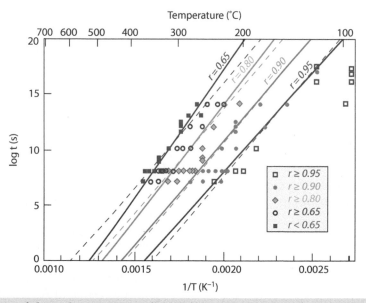

Figure 4.19 Contours of the reduction in track length, $r = l/l_0$, in experimental data (symbols) of Green et al. (1985) as a function of time and temperature with predictions by this "fanning model" (solid lines) of Laslett et al. (1987) with a simple "parallel model" (dashed lines). The solid lines converge at $T_0 = -3 \times 10^{-6}$ and $\ln t_0 = 28.12$.

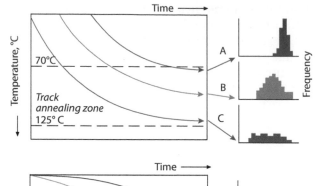

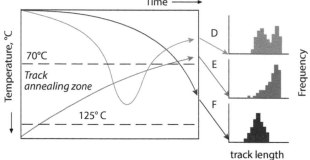

Figure 4.20 Hypothetical time–temperature paths and the distribution of track lengths that should result from these paths. Source: Ravenhurst and Donelick (1992)/with permission of Mineralogical Association of Canada.

of rapid cooling and subsequent low temperatures (such as might be expected for a volcanic rock), while a broad distribution of track lengths suggests slower cooling. A skewed distribution suggests initial slow cooling and subsequent low temperatures.

4.3.2 (U–Th)/He

In Chapter 3, we reviewed the U–Th decay systems, including both U–Th–Pb and U-series geochronology. However, we have thus far ignored the other decay product, ^{4}He. Each alpha decay in these chains produces a ^{4}He atom: eight in total from the decay of ^{238}U, seven from ^{235}U, and six from ^{232}Th. ^{147}Sm decay also produces ^{4}He. Consequently, assuming a closed system, ^{4}He will accumulate according to:

$$^4He = 8 \times {}^{238}U\left(e^{\lambda_{238}t} - 1\right)$$
$$+ 7 \times {}^{235}U\left(e^{\lambda_{235}t} - 1\right) + 7 \times {}^{232}Th\left(e^{\lambda_{235}t} - 1\right)$$
$$+ {}^{147}Sm\left(e^{\lambda_{147}t} - 1\right)$$

$$(4.51)$$

As we mentioned in Chapter 2, Rutherford produced the first U–He age by measuring U and He in a fergusonite sample[6] and a uraninite sample, concluding they were at least 500 Ma old (Rutherford, 1906). Rutherford recognized that these were minimum ages as a consequence of expected diffusional loss of He. Here, we will examine how accumulation of ^{4}He in minerals and its diffusional loss can also be used as a thermochronological tool. The most common targets for (Th–U)/He dating are apatite, zircon, titanite, xenotime, and monazite, all of which tend to concentrate U and/or Th, although some studies have utilized iron or titanium minerals, such as magnetite and goethite, which can also concentrate U or Th. ^{147}Sm production of ^{4}He is far less than that of U and Th, typically less than 0.5% of the total. However, in rare-earth-rich minerals, such as monazite and xenotime, ^{147}Sm production must be taken into account.

He exists as a monatomic gas and it is a very small atom, with a radius of 28 pm (for comparison, the radius of Ar is 106 pm). Furthermore, it is uncharged, and the result is that it readily

[6] Fergusonite has the nominal composition of $YNbO_4$, but both lanthanide and actinide REE readily substitute for Y and Rutherford's specimen contained 7% U.

diffuses through and out of crystal lattices. As a consequence of this ready diffusion, attempts to follow up on Rutherford's work to determine mineral ages largely failed. Diffusion rates, however, depend exponentially on temperature (Equation [2.35]), so at sufficiently low temperatures, He can remain trapped in a crystal lattice almost indefinitely, but at sufficiently high temperatures will be lost as quickly as it is produced. Following the work of Zeitler et al. (1987), the (Th/U)–He decay system has found a use as a thermochronometer.

The rapid diffusion of He together with its origin as an energetic α-particle presents several challenges that are not present in other decay systems. The first is that the α-particle travels some distance before it stops, meaning that ^{4}He in exterior regions will immediately be lost from the crystal lattice. Mean stopping distances vary with α-particle energy and mineral density and are typically in the range of 15–20 μm (alpha energy for the ^{147}Sm decay is lower and stopping distances are typically ≈5 μm). The fraction of He retained depends on the size and geometry of the crystal; for a sphere, the fraction retained, F_T, varies as:

$$F_T = 1 - \left(\frac{3}{4}\right)\left(\frac{s}{r}\right) + \left(\frac{1}{16}\right)\left(\frac{s^3}{r}\right) \quad (4.52)$$

where s is the mean stopping distance and r is the radius. Few crystals are spherical and other equations are available for other geometries (see Reiners et al., 2017), but this equation works in approximation for most crystals; only when the surface to volume ratio differs substantially from that of a sphere will F_T vary by more than 10% from Equation (4.52). Alpha particles that are not stopped within the crystals in which they originate can be implanted in adjacent ones, which can present a problem when other U or Th–rich minerals are present. Just as can be done in U–Pb analysis of zircon, mineral grains can be abraded to remove the outer portions, from which or into which α-particles are most likely to have been lost or implanted.

As with K–Ar ages, (Th–U)/He ages reflect the last time that the rate of diffusional loss of the daughter was equal to its rate of production. And just as for Ar, we expect diffusion to be thermally activated and to follow an Arrhenius-type temperature dependence:

$$D = D_0 e^{-E_A/RT} \quad (2.43)$$

We found in Chapter 2 that diffusional fractional loss depends on crystal size and shape; the key parameter is Dt/a^2, where a is a characteristic dimension (e.g., radius in the case of a sphere), t is time, and D is the diffusion coefficient in Equation 2.43. We also found that we can use equations, such as Equation (2.46), to predict the diffusional loss over time (e.g., Figure 2.3). However, (Th–U)/He adds several complexities to this problem.

First, whereas K–Ar dating generally involves major phases, such as feldspar and hornblende, which typically have dimensions of at least a few millimeters and often significantly more, (Th–U)/He dating generally targets accessory minerals whose dimensions are typically in the range of tens to hundreds of micrometers. Because loss depends on the square of the dimensional term, a, crystal size is a more important factor in (Th–U)/He system than the K–Ar one, crystals of different sizes but identical thermal histories can yield different ages.

Second, He diffusion coefficients have been shown to depend on thermal history. Perhaps counterintuitively, He diffusion is slower through crystals with significant radiation damage, which is a function of both the total U+Th+Sm concentration (denoted by eU) and the time since and extent to which damage was last annealed. Higher eU results in greater damage while higher temperatures tend to anneal it. This damage includes not only that done by fission tracks, which we discussed in the previous section, but also damage from alpha recoil. While each fission produces significantly more damage than each alpha, fission is far, far less likely, so it is primarily alpha decay that damages crystals. As we noted in Chapter 1, conservation of momentum requires that the daughter of an alpha decay recoil with momentum equal to that of the alpha. The distance traveled by the daughter is typically only ≈20 nm, which is far smaller than the distance traveled by the α-particle, but this is nevertheless ≥100 times a typical lattice spacing and consequently can displace thousands of atoms and produce significant local damage. In addition, U and Th atoms will go through six to eight alpha decays before finally becoming Pb, so the same area will be repeatedly damaged this way (think of striking a crystal repeatedly with a hammer!). Both fission and alpha recoil produce lattice dislocations that can serve as "traps" for He atoms. Atoms dislocated from these regions will then concentrate in surrounding areas, clogging diffusion channels. The net result is that additional activation energy is necessary for He to escape from these "traps." However, there does seem to be a limit to this. Experiments show that diffusivity increases in crystals that have suffered very high radiation damage. This effect has been observed only in minerals (such as zircon) that have very high Th and U concentrations. Empirical models and computer codes (e.g., Flowers et al., 2009) have been developed to account for the effects of radiation damage on diffusivity. Delving into these complexities will take us too far afield; interested readers are referred to Flowers et al. (2009) and Reiners et al. (2017). While radiation damage can also be an issue for U–Pb dating of U-rich minerals, such as zircon, as we found in Chapter 3, beta decay does not result in significant damage and electron capture produces none, so this has no parallel in K–Ar dating.

Finally, zircon, one of the most commonly used minerals in (U–Th)/He thermochronology, shows significant diffusional anisotropy, which manifests itself in the pre-exponential factor rather than the activation energy. Cherniak et al. (2009) found that the activation energies parallel and perpendicular to the c-axis were identical (perpendicular: 146 ± 11; parallel: 148 ± 17 kJ/mol), but D_0 perpendicular was 2.3×10^{-7} m^2/s and parallel was 1.7×10^{-5} m^2/s. Thus, diffusion parallel to the c-axis is some two orders of magnitude greater regardless of temperature and consequently a simply spherical geometry cannot be assumed. In contrast, Cherniak et al. found no perceptible anisotropy in apatite with $D_0 = 2.1 \times 10^{-6}$ m^2/s and $E_A = 117 \pm 6$ kJ/mol.

Figure 4.21 compares (U–Th)/He closure temperatures for several minerals with those of fission tracks and K–Ar as a function of cooling rate (see Section 2.3.1 for the relationship between closure temperature and cooling rate). In addition to the caveats discussed above, diffusion coefficients and consequently closure temperature will depend on mineral compositions. This is particularly true for apatite, where substitution of F$^-$ and Cl$^-$ in the OH$^-$ site appears to significantly affect both He retentivity and fission track annealing rate: high Cl slows the annealing rate and correlates with greater He retentivity; high F does the opposite. This can result in differences in closure temperatures of up to $\approx 60°$C for the same cooling rate.

^{4}He/^{3}He dating is a variation on (U–Th)/He that has parallels to ^{40}Ar/^{39}Ar dating. In most

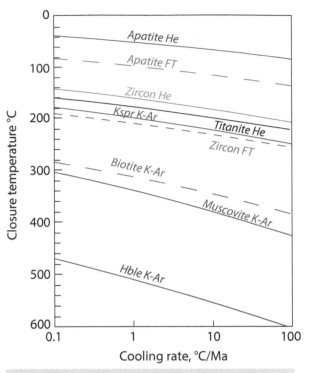

Figure 4.21 Comparison of apparent closure temperatures of fission tracks, (U–Th)/He, and K–Ar as a function of cooling rate for a variety of minerals. Source: Reiners and Brandon (2006)/ Reproduced with permission of Annual Reviews.

crustal rocks and minerals, ^{3}He is so rare that it can be considered non-existent, as is ^{39}Ar. Both can be produced artificially: ^{39}Ar by neutron bombardment and ^{3}He by proton bombardment. In the ^{4}He/^{3}He technique, a uniform concentration of ^{3}He of 10^9 to 10^{10} atoms/mg is produced by spallation of heavier elements by bombarding the sample with a 150 MeV proton beam. As in ^{40}Ar/^{39}Ar, the sample is then heated, and He released in steps and the ^{4}He/^{3}He ratio measured in each step in a gas source mass spectrometer. The ^{4}He/^{3}He ratio is a measure of the relative ^{4}He content in each step. As with ^{40}Ar, we expect the initial, low-temperature steps to release He from the rim of the crystal. An age can in principle be calculated for each step, but only by assuming values for eU concentrations since unlike ^{40}Ar/^{39}Ar dating, parent nuclide concentrations are not simultaneously measured. Nevertheless, the release pattern provides constraints on the thermal history of the sample, although with some ambiguity. In practice, the release pattern is iteratively compared with that calculated thermal histories to find a best fit. We will examine this below in more detail.

4.3.3   Uplift and erosion rates

The thermochronometers just discussed in combination with estimates of geothermal gradients can be used to estimate uplift and erosion rates, particularly when the various chronometers are used in combination. Let us see how this works. Consider a layer of crust L km thick undergoing denudation at a rate ε. Let the temperature at the base and top of the layer (the surface) be T_L and T_S, respectively, assuming that internal radiogenic heat production provides heating of H_T °C/Ma with thermal conductivity κ. Temperature as a function of depth is then given by:

$$T(z) = T_S + \left(T_L - T_S + \frac{H_T L}{\varepsilon}\right) \frac{1 - e^{-\varepsilon z/\kappa}}{-e^{-\varepsilon L/\kappa}} - \frac{H_T z}{\varepsilon}$$

$$(4.53)$$

When erosion rates are high, this results in non-linear geothermal gradients, as can be seen in Figure 4.22. When erosion rates are small, the exponential terms disappear, and the gradient is nearly linear. To determine erosion rates, what we really wish to know is the closure depth, rather than closure temperature. By iteratively simultaneously solving Equation (4.54), as well as Dodson's closure temperature equation (Equation [2.38]), Reiners and Brandon

(2006) determined the "closure depths" for several minerals as a function of cooling rates shown by the bold lines in Figure 4.22.

Let us consider a simple example. We measure an apatite fission track age of 12.5 Ma. We chose 10°C/Ma for a first-order estimate of cooling rate and determine the closure temperature from Figure 4.22 to be 120°C. Assuming an average surface temperature of 10°C, we calculate the cooling rate to be:

$$\frac{dT}{dt} = \frac{120 - 10}{12.5} = 8.8°C/Ma \qquad (4.54)$$

We could iteratively refine our result value by recalculating the closure temperature based on previously calculated cooling rate and obtain a rate of 8°C/Ma. If we assume the geothermal gradient to be 30°C/km, we can calculate the exhumation rate to be:

$$\frac{dz}{dt} = \frac{dT/dt}{dT/dz} = \frac{8°C/Ma}{0.030°C/m} = 267m/Ma$$

$$(4.55)$$

Using this approach, exhumation rates have been estimated as 500 m/Ma over the past 10 Ma for the Alps and 800 m/Ma for the Himalayas. Figure 4.23 shows an example of the results of one such study of the Himalayas from northern India (Kashmir). A plot of fission track ages versus the altitude at which the samples were collected indicates an

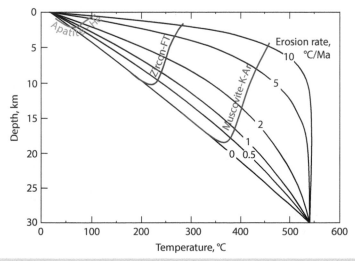

Figure 4.22   Geothermal gradients calculated from Equation (4.47) for T_S =14°C, T_L = 540°C, κ = 27 km/Ma, and H_T = 4.5°C/Ma. Bold lines show the "closure depths" for apatite UTh–He, zircon fission track, and muscovite K–Ar chronometers. Source: Reiners and Brandon (2006)/Reproduced with permission of Annual Reviews.

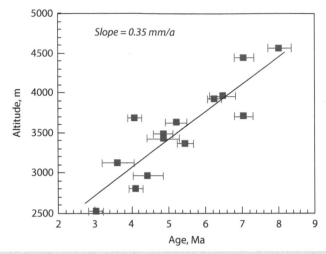

Figure 4.23 Apatite fission track ages versus altitude for metamorphic rocks of the Higher Himalayan Crystalline Belt of Kashmir. The correlation coefficient is 0.88. The slope indicates an uplift rate of 350 m/Ma. Source: Kumar et al. (1995)/Reproduced with permission of the Geological Society of India.

exhumation rate of 0.35 mm/a or 350 m/Ma over the last seven million years.

Now let us consider an example where several thermochronometers are used in concert. The age of the Grand Canyon in the Southwestern United States, which is locally as deep as 1850 m, has been the subject of speculation and controversy since John Wesley Powell (later to become director of the US Geological Survey) made expeditions through it in 1869 and 1871–1872. Powell (1875) argued for an "old" origin of the Canyon, beginning with or before what would later become known as the Laramide Orogeny at around 70 Ma when the region drained by the Colorado River was uplifted[7]. Others have pointed out that sediments derived from the upstream reaches of the Colorado River first appeared in Grand Wash Trough, where the river exits the Colorado Plateau and Grand Canyon at around 5–6 Ma, suggesting a young age for the Canyon, very much post-dating Laramide uplift.

The Canyon has eroded through a sequence of Permian to Cambrian sediments into Neoproterozoic basement and ultimately the Paleoproterozoic Vishnu Schist. $^{40}Ar/^{39}Ar$ ages of the Proterozoic rocks indicate they were at temperatures of >850°C in the Paleoproterozoic. Subsequent erosion evidence by the "Great Unconformity" between Proterozoic basement and the overlying Paleozoic sediments indicate the basement was exhumed to the surface and eroded by the Cambrian. Subsequent sedimentation continued through the Paleozoic, laying down ≈2000 m of sediment. Apatite fission track data indicate that the Permian strata then underwent burial to depths of ≈2.7–4.5 km and temperatures of ≈90–100 °C in the Late Cretaceous. Cooling subsequently began about 75 Ma (Dumitru et al., 1994) as a consequence of Laramide uplift of the Colorado Plateau and consequent erosion. This thermal history provides a baseline for interpreting its more recent thermal history.

Flowers and Farley (2012) reported $^4He/^3He$ release spectra for apatite samples taken from the Proterozoic Separation Pluton[8] exposed at bottom of the western part of the Canyon, an example of which is shown in

[7] Powell, of course, meant "old" only in a relative sense and could not have put an absolute age on the Canyon as radiometric dating would not begin until the following century.

[8] So named because it was here that three members of Powell's 1869 expedition left the party out of fear that the rapids there could not be run. The three were never heard from again. Powell did successfully run the rapids with his remaining five crew members and emerged from the Canyon two days later, 98 days after they had started.

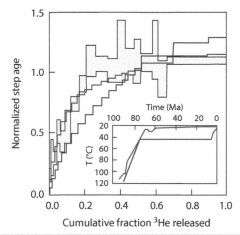

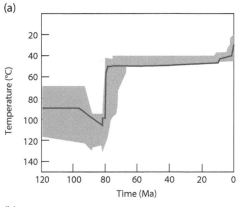

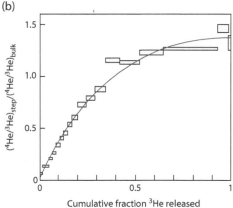

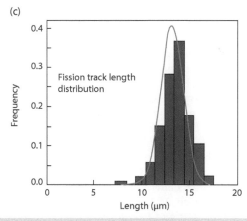

Figure 4.24 ^{4}He/^{3}He step-heating release spectrum for apatite sample from the Separation pluton in the Grand Canyon. Gray area shows the analytical uncertainty for each step. Red and blue lines show the predicted release patterns for "old canyon" (red) and "young canyon" (blue) thermal evolution models shown in the inset. Source: Adapted from Flowers and Farley (2012).

Figure 4.25   (a) Thermal evolution models for the western Grand Canyon generated by Winn et al. (2017) to fit (U+Th)/He, ^{4}He/^{3}He, and fission track thermochronology. Green field shows the range of the 25 best-fit models out of more than 3 million generated. Solid green line is the model fit to the ^{4}He/^{3}He and fission track data. (b) ^{4}He/^{3}He release spectrum for apatite from the Separation pluton in the western Grand Canyon. Gray squares show the analytical uncertainty range in ^{4}He/^{3}He relative to the bulk ^{4}He/^{3}He for each step. Green line shows the predicted pattern from the thermal model in (a). (c) Fission track length distribution for the same apatite sample. Green line shows the predicted distribution for the thermal model in (a). Source: Modified from Winn et al. (2017).

Figure 4.24. Just as is done in ^{40}Ar/^{39}Ar spectra, the age normalized to the total (U+Th)/He age or the ^{4}He/^{3}He ratio or for each step is plotted as a function of the cumulative fraction released. Flowers and Farley then compared these spectra with those predicted from a variety of thermal evolution models that assume temperatures of 110–120°C at 80–85 Ma and complete (U–Th)/He resetting and a geothermal gradient of ≈30°C/km. Two of these thermal evolution models, one corresponding to an "old" canyon (red) and one corresponding to a "young" canyon (blue), are shown in the inset. An "old" canyon thermal model in which the Canyon was excavated to within a few hundred meters of modern depths by ≈70 million years ago (Ma) provided the best fit to the release spectrum.

Subsequently, Winn et al. (2017) interpreted a new set of apatite thermochronology data as favoring a young age for western part of the Grand Canyon, but an old age for the eastern part. They resampled apatites from the Separation Pluton at the same location as Flowers and Farley (2012) and found that the ^{4}He/^{3}He release spectrum of this sample performed at high precision better fits a "young" canyon model in which cooling begins with the Laramide orogeny, but temperatures only fall to the range of 60–50°C with surface temperatures reached only in the last 3–6 Ma. They then attempted to find a best fit for the combined fission track, total (U–Th)/He and ^{4}He/^{3}He approaches. Figure 4.25a shows the range of the 25 best-fitting thermal models (among >3 000 000 models evaluated) to the ^{4}He/^{3}He release spectra. Figure 25b shows the fit to the ^{4}He/^{3}He release spectrum of the thermal model shown as the dark line in Figure 25a. Figure 25c shows the fission track distribution for this apatite and the green line shows the spectrum predicted by this thermal model. The Winn et al. thermal model differs from that of the Flowers and Farley model in part because pre-Laramide temperatures in the Winn model are not hot enough to fully anneal radiation damage as Flowers and Farley

had assumed. The fits are imperfect, probably in part reflecting the present imperfect understanding of He-retention and fission track annealing.

In contrast to samples from the western Grand Canyon, Winn et al. (2017) found that eastern canyon did fit an old Canyon model, cooling to surface temperatures by 50 Ma. This is consistent with some models that hypothesize that the eastern part of the Canyon was carved by a north-flowing Paleocene river that was later captured by the Colorado. Our understanding of the evolution of the Grand Canyon remains imperfect, but future thermochronological studies will no doubt improve it.

REFERENCES

Abu Risha, U.A. 2016. Radiocarbon dating and the ^{36}Cl/Cl evolution of three Great Artesian Basin wells at Dalhousie, South Australia. *Hydrogeology Journal* **24**: 987–1000. doi: 10.1007/s10040-016-1364-4.

Balco, G. 2020. Glacier change and paleoclimate applications of cosmogenic-nuclide exposure dating. *Annual Review of Earth and Planetary Sciences* **48**: 21–48. doi: 10.1146/annurev-earth-081619-052609.

Balco, G., Stone, J.O., Lifton, N.A., et al. 2008. A complete and easily accessible means of calculating surface exposure ages or erosion rates from ^{10}Be and ^{26}Al measurements. *Quaternary Geochronology* **3**: 174–95. doi: 10.1016/j.quageo.2007.12.001.

Barth, A.M., Marcott, S.A., Licciardi, J.M., et al. 2019. Deglacial thinning of the Laurentide Ice Sheet in the Adirondack Mountains, New York, USA, revealed by ^{36}Cl exposure dating. *Paleoceanography and Paleoclimatology* **34**: 946–53. doi: 10.1029/2018pa003477.

Beck, J.W., Richards, D.A., Edwards, R.L. et al. 2001. Extremely large variations of atmospheric ^{14}C concentration during the last glacial period. *Science* **292**: 2453–8.

Bentley, H.W., Phillips, F.M., Davis, S.N., et al. 1986. Chlorine-36 dating of very old groundwater 1. The Great Artesian Basin, Australia. *Water Resources Research.* **22**: 1991–2001. doi: 10.1029/WR022i013p01991.

Bethke, C.M. and Johnson, T.M. 2008. Groundwater age and groundwater age dating. *Annual Review of Earth and Planetary Sciences* **36**: 121–52. doi: 10.1146/annurev.earth.36.031207.124210.

Bierman, P.R., Davis, P.T., Corbett, L.B., et al. 2015. Cold-based Laurentide ice covered New England's highest summits during the Last Glacial Maximum. *Geology* **43**: 1059–62. doi: 10.1130/g37225.1.

Brown, E.T., Measures, C.I., Edmond, J.M., et al. 1992. Continental inputs of beryllium to the oceans. *Earth and Planetary Science Letters* **114**: 101–11. doi: 10.1016/0012-821X(92)90154-N.

Brown, L., Pavich, M.J., Hickman, R. E., et al. 1988. Erosion of the eastern United States observed with ^{10}Be. *Earth Surface Processes and Landforms* **13**: 441–57. doi: 10.1002/esp.3290130509.

Cherniak, D.J., Watson, E.B. and Thomas, J.B. 2009. Diffusion of helium in zircon and apatite. *Chemical Geology* **268**: 155–66. doi: 10.1016/j.chemgeo.2009.08.011.

Clark, P.U., Dyke, A.S., Shakun, J.D., et al. 2009. The Last Glacial Maximum. *Science* **325**: 710–4. doi: 10.1126/science.1172873.

Corbett, L.B., Bierman, P.R., Stone, B.D., et al. 2017. Cosmogenic nuclide age estimate for Laurentide Ice Sheet recession from the terminal moraine, New Jersey, USA, and constraints on latest Pleistocene ice sheet history. *Quaternary Research* **87**: 482–98. doi: 10.1017/qua.2017.11.

Corbett, L.B., Bierman, P.R., Wright, S.F., et al. 2019. Analysis of multiple cosmogenic nuclides constrains Laurentide Ice Sheet history and process on Mt. Mansfield, Vermont's highest peak. *Quaternary Science Reviews* **205**: 234–46. doi: 10.1016/j.quascirev.2018.12.014.

Cresswell, R.G., Bauld, J., Jacobson, G., et al. 2001. A first estimate of ground water ages for the deep aquifer of the Kathmandu Basin, Nepal, using the radioisotope chlorine-36. *Groundwater* **39**: 449–57. doi: 10.1111/j.1745-6584.2001.tb02329.x.

Dumitru, T.A., Duddy, I.R. and Green, P.F. 1994. Mesozoic-Cenozoic burial, uplift, and erosion history of the west-central Colorado Plateau. *Geology* **22**: 499–502. doi: 10.1130/0091-7613(1994)022<0499:Mcbuae>2.3.Co;2.

Fairbanks, R.G., Mortlock, R.A., Chiu, et al. 2005. Radiocarbon calibration curve spanning 0 to 50,000 years BP based on paired ^{230}Th/^{234}U/^{238}U and ^{14}C dates on pristine corals. *Quaternary Science Reviews* 24: 1781–96. doi: 10.1016/j.quascirev.2005.04.007.

Faure, G. 1986. *Principles of Isotope Geology*, 2nd ed. New York: John Wiley & Sons, 589 p.

Flowers, R.M. and Farley, K.A. 2012. Apatite ^{4}He/^{3}He and (U-Th)/He evidence for an ancient Grand Canyon. *Science* 338: 1616–9. doi: 10.1126/science.1229390.

Flowers, R.M., Ketcham, R.A., Shuster, D.L., et al. 2009. Apatite (U–Th)/He thermochronometry using a radiation damage accumulation and annealing model. *Geochimica et Cosmochimica Acta* 73: 2347–65. doi: 10.1016/j.gca.2009.01.015.

Frank, M., Schwarz, B., Baumann, S., et al. 1997. A 200 kyr record of cosmogenic radionuclide production rate and geomagnetic field intensity from ^{10}Be in globally stacked deep-sea sediments. *Earth and Planetary Science Letters* 149: 121–9. doi: 10.1016/S0012–821X(97)00070–8.

Green, P. F., Duddy, I. R., Gleadow, et al. 1985. Fission-track annealing in apatite: Track length measurements and the form of the Arrhenius plot. *Nuclear Tracks and Radiation Measurements (1982)*, 10, 323–328. doi: https://doi.org/10.1016/0735-245X(85)90121-8

Heaton, T.J., Köhler, P., Butzin, M., et al. 2020. MARINE20—the marine radiocarbon age calibration curve (0–55,000 cal bp). *Radiocarbon*. doi: 10.1017/RDC.2020.68.

Knudsen, M.F., Riisager, P., Donadini, F., et al. 2008. Variations in the geomagnetic dipole moment during the Holocene and the past 50 kyr. *Earth and Planetary Science Letters* 272: 319–29. doi: 10.1016/j.epsl.2008.04.048.

Kowallis, B.J., Heaton, J.S. and Bringhurst, K. 1986. Fission-track dating of volcanically derived sedimentary rocks. *Geology* 14: 19–22. doi:10.1130/0091-7613(1986)14<19: fdovds>2.0.co;2.

Kumar, A., Lal, N., Jain, A.K., et al. 1995. Late Cenozoic-Quaternary thermo-tectonic history of Higher Himalaya Crystalline (HHC) in Kishtwar-Padar-Zanskar region, NW Himalaya: evidence from fission track ages. *Journal of the Geological Society of India* 45: 375–91.

Kurz, M.D. 1986. *In Situ* production of terrestrial cosmogenic helium and some applications to geochronology. *Geochim. Cosmochim. Acta* 50: 2855–62.

Lal, D. 1991. Cosmic ray labeling of erosion surfaces: in situ nuclide production rates and erosion models. *Earth and Planetary Science Letters* 104: 424–39. doi: 10.1016/0012-821X(91)90220-C.

Laslett, G.M., Green, P.F., Duddy, I.R., et al. 1987. Thermal annealing of fission tracks in apatite 2. A quantitative analysis. *Chemical Geology* 65: 1–13. doi: 10.1016/0168–9622(87)90057–1.

Lebatard, A.-E., Bourlès, D.L., Duringer, P., et al. 2008. Cosmogenic nuclide dating of *Sahelanthropus tchadensis* and *Australopithecus bahrelghazali*: Mio-Pliocene hominids from Chad. *Proceedings of the National Academy of Sciences* 105: 3226–31. doi: 10.1073/pnas.0708015105.

Lifton, N., Sato, T. and Dunai, T.J. 2014. Scaling *in situ* cosmogenic nuclide production rates using analytical approximations to atmospheric cosmic-ray fluxes. *Earth and Planetary Science Letters* 386: 149–60. doi: 10.1016/j.epsl.2013.10.052.

Nishi, K., Usui, A., Nakasato, Y., et al. 2017. Formation age of the dual structure and environmental change recorded in hydrogenetic ferromanganese crusts from Northwest and Central Pacific seamounts. *Ore Geology Reviews* 87: 62–70. doi: https://doi.org/10.1016/j.oregeorev.2016.09.004.

Phillips, F.M., Argento, D.C., Balco, G., et al. 2016. The CRONUS-Earth Project: a synthesis. *Quaternary Geochronology* 31: 119–54. doi: 10.1016/j.quageo.2015.09.006.

Portenga, E.W. and Bierman, P.R. 2011. Understanding Earth's eroding surface with ^{10}Be. *GSA Today* 21: 4–10.

Powell, J.W. 1875. Exploration of the Colorado River of the West and its tributaries: explored in 1869, 1870, 1871, and 1872, *Smithsonian Institution Annual Report*, US government printing office.

Ravenhurst, C.E. and Donelick, R.A. 1992. Fission track thermochronology, in Zentilli, M. and Reynolds, P.H. (eds). *Short Course Handbook on Low Temperature Thermochronology*, Nepean, Ontario: Mineralogical Society Canada, 21–42 pp.

Reimer, P.J., Austin, W.E.., Bard, E., et al. 2020. The IntCal20 northern hemisphere radiocarbon age calibration curve (0–55 cal kBP). *Radiocarbon* 62: 725–57. doi: 10.1017/RDC.2020.41.

Reiners, P.W. and Brandon, M.T. 2006. Using thermochronology to understand orogenic erosion. *Annual Review of Earth and Planetary Sciences* 34: 419–66. doi: 10.1146/annurev.earth.34.031405.125202.

Reiners, P.W., Carlson, R.W., Renne, P. R., et al. 2017. *Geochronology and Thermochronology*. Oxford: John Wiley & Sons.

Rutherford, E. 1906. *Radioactive transformations*. New York: C. Scribner's Sons.

Steinhilber, F., Abreu, J.A., Beer, J., et al. 2012. 9,400 years of cosmic radiation and solar activity from ice cores and tree rings. *Proceedings of the National Academy of Sciences* 109: 5967–71. doi: 10.1073/pnas.1118965109.

Stuiver, M., Reimer, P.J., and Reimer, R.W., 2020. CALIB 8.2 [WWW program] at http://calib.org accessed 2020-9-3.

Torgersen, T. and Clarke, W.B. 1985. Helium accumulation in groundwater, I: An evaluation of sources and the continental flux of crustal 4He in the Great Artesian Basin, Australia. *Geochimica et Cosmochimica Acta* 49: 1211–1218. doi: 10.1016/0016-7037(85)90011-0.

von Blanckenburg, F. and Bouchez, J. 2014. River fluxes to the sea from the ocean's ^{10}Be/^{9}Be ratio. *Earth and Planetary Science Letters* 387: 34–43. doi: https://doi.org/10.1016/j.epsl.2013.11.004.

von Blanckenburg, F., Bouchez, J. and Wittmann, H. 2012. Earth surface erosion and weathering from the ^{10}Be(meteoric)/^{9}Be ratio. *Earth and Planetary Science Letters* 351–2: 295–305. doi: 10.1016/j.epsl.2012.07.022.

Willenbring, J.K. and von Blanckenburg, F. 2010. Long-term stability of global erosion rates and weathering during late-Cenozoic cooling. *Nature* 465: 211–4. doi: 10.1038/nature09044.

Winn, C., Karlstrom, K.E., Shuster, D.L., et al. 2017. 6 Ma age of carving Westernmost Grand Canyon: Reconciling geologic data with combined AFT, (U–Th)/He, and ^{4}He/^{3}He thermochronologic data. *Earth and Planetary Science Letters* 474: 257–71. doi: 10.1016/j.epsl.2017.06.051.

Wittmann, H., Oelze, M., Gaillardet, J., et al. 2020. A global rate of denudation from cosmogenic nuclides in the Earth's largest rivers. *Earth-Science Reviews* 204: 103147. doi: 10.1016/j.earscirev.2020.103147.

Zeitler, P.K., Herczeg, A.L., Mcdougall, I., et al. 1987. U-Th-He dating of apatite: a potential thermochronometer. *Geochimica et Cosmochimica Acta* 51: 2865–8. doi: 10.1016/0016-7037(87)90164-5.

PROBLEMS

1. Suppose you determine that a sample of bone from southern England has a specific carbon-14 activity of 11.85 ± 0.7 dpm/g C.

   (a) Use Equation (2.1) to determine the radiocarbon age and the associated error on the age.
   (b) Use the CALIB 8.2 program at http://calib.org to calculate the age in calendar years and produce a probability graph similar to Figure 4.3 (ignore inputs for deltas and fractionation factors).
   (c) An archeologist judges from artifacts found with the bones that the bones are pre-Norman (i.e., pre-1066 AD). Is he wrong?

2. Lebatard et al. (2008) measured a ^{10}Be/^{9}Be ratio of 7.32×10^{-10} in the sediment directly overlying the horizon in which the *S. tchadensis* fossil was found. Assuming an initial ratio of 2.54×10^{-8}, calculate the ^{10}Be/^{9}Be ratio model age of this sediment.

3. Using the adjacent data from Mn nodule JA06BMS02A of Nishi et al. (2017) in the table below, calculate the growth rate assuming constant flux and constant sedimentation. (This is the data set plotted in Figure 4.6.)

Interval (cm)	^{10}Be/^{9}Be	±
0–2.5	1.14×10^{-7}	7.0×10^{-9}
2.5–5.0	8.13×10^{-8}	4.6×10^{-9}
7.5–10.0	4.24×10^{-8}	2.2×10^{-9}
13.0–16.0	1.88×10^{-8}	1.1×10^{-9}
19.0–22.0	7.68×10^{-9}	4.5×10^{-10}
30.0–35.0	1.65×10^{-9}	1.1×10^{-10}
40.0–45.0	1.06×10^{-9}	7.0×10^{-11}
55.0–60.0	2.58×10^{-10}	5.8×10^{-11}

4. The $(^{10}$Be/^{9}Be$)_{diss}$ of the Columbia River has been measured as 1.06×10^{-8} and the meteoric flux, F_m, of ^{10}Be is estimated to be 1×10^{16} atoms/km^2/yr. Assuming a ^{9}Be$_{parent}$ concentration of 2.5 ppm and a $f^{[^9Be]}_{reac + diss}$ of 0.20, use Equation (4.14) to calculate the denudation rate (in kg/km^2/yr).

5. Using the values for ε, ρ, and Λ given in the text and a concentration of 2.76×10^{-13} cc STP^9/g for the cosmogenic ^{3}He in the surface sample from Core 33 (Figure 4.12) and an age of 650 000 years, calculate the production rate (your answer will be in cc STP/g).

---

^{9} STP is an abbreviation for standard temperature and pressure (25°C and 1 bar pressure).

6. Corbett et al. (2017) reported the following cosmogenic nuclide data on quartz samples from New Jersey. Assuming a density of 2.6 g/cc, a shielding factor of 1, and an erosion rate of 0, use the CRONUS-Earth online exposure age calculator (http://hess.ess.washington.edu) to calculate exposure ages. Enter *std* for elevation/pressure handling and 07*KNSTD* for ^{10}Be standardization and *KNSTD* for ^{26}Al standardization, 1 for shielding, 0 for erosion rate, and 1994 for collection date. Be sure to look at the link to input format before you begin.

Sample name	Latitude	Longitude	Elevation (m)	Thickness (cm)	^{10}Be (atoms/g)	±	^{26}Al (atoms/g)	±
SPA-1	40.96619	−74.53888	342	1.5	138000	4890	831000	32100
SPA-2	40.96619	−74.53888	342	5	141000	4490	864000	35500
SPA-6	40.97425	−74.52339	336	4.5	133000	40500	673000	49000

HINT: *Each sample requires an input line for sample data, the format for which is:*
Sample Name Latitude Longitude elevation elevation/pressure handling thickness density shielding erosion rate collection year;
Each isotope is then entered on a separate line, the format for which is
Sample Name isotope material atoms/gram uncertainty standard;
In both, terms can be separated by spaces, commas, and tabs and lines are terminated by semicolons. For example, your first two lines should be:
SPA-1 40.96619 -74.53888 342 std 1.5 2.6 1 0 1994;
SPA-1 Be-10 quartz 138000 4890 07KNSTD;

7. Given average upper crustal abundances of Sm, Th, and U of 4.7, 10.5, and 2.7 ppm (by weight) and decay constants listed in Table 2.1, what fraction of radiogenic ^{4}He in the upper crust is currently being produced by decay of ^{147}Sm?

8. A series of apatite (U–Th)/He ages were measured on samples from a variety of elevations in southeastern Alaska listed in the table below. Assuming that the erosion rates exactly balance uplift rates so that the surface of the present landscape would have risen through a horizontal closure isotherm, with lower elevations passed through the isotherm most recently.

Elevation (m)	Age (Ma)
300	17
600	21
900	26
1200	30
2000	38

(a) Use the data to estimate the uplift/erosion rate.
(b) Assuming a geothermal gradient of 30°C/km, calculate the cooling rate and use this to estimate the apatite He closure temperature from Figure 4.21.

9. Suppose you have data, such as the series of elevations and ages in Figure 4.23, which suggest an uplift rate of 50 m/Ma. Assume a geothermal gradient of 30°C/Ma, an activation energy, E_A, of 170 kJ/mol, a frequency factor, $D_{0,}$, of 0.46 cm^2/s, a radius of 60 μm, a geometric factor, A, of 27, and R = 8.314 J/K-mol.

(a) Calculate the zircon fission track closure temperature from Equation (2.38) using Figure 4.22 for the initial guess. Iterate this procedure as necessary. *(HINT: Convert temperatures to kelvins and convert all the values above to consistent units; iterative calculations are easiest done in a spreadsheet).*

(b) Suppose the fission track age is 3 Ma. Assuming a surface temperature of 10°C, the same geothermal gradient and the closure temperature you calculated earlier estimate a new uplift rate as described in Section 4.2.2. Iterate calculation of the closure temperature and uplift rate until your answer converges.

Chapter 5

Fractionation of Isotopes

5.1 INTRODUCTION

The chemical behavior of elements is almost entirely dictated by their electronic configurations. Ions form when atoms loose or gain electrons, which changes their behavior and can lead to chemical bonding; bonds also form when atoms share electrons. The nuclear charge, dictated by the number of protons, is responsible for why different elements behave differently. The third atomic particle, the neutron, plays no direct role in all this and consequently we expect atoms with the same proton number but differing neutron number, i.e., isotopes, to behave the same; to a good approximation, they do. However, atomic mass does depend on the neutron number, and this has an indirect effect on chemical behavior. Neutron number also affects the volume and shape of the nucleus, as well as its spin, and this too has a slight effect on the electron cloud surrounding the nucleus and consequently the behavior of heavy atoms. The resulting small differences in chemical behavior are the basis for stable isotope geochemistry. The resulting changes in isotope ratios are referred to as *fractionations*.

You might then wonder: how can we distinguish these chemically produced variations in isotope ratios from those produced by radioactive decay? The answer is all measurements of radiogenic isotope ratios require a correction for these effects (see Section 2.2.3). Indeed, fractionations that occur during mass spectrometry are generally much larger than natural ones, requiring a correction even in the absence of natural isotopic variations.

In this chapter, we restrict ourselves to the physiochemical basis for stable isotope fractionations and leave discussion of the myriad applications of stable isotope geochemistry to subsequent chapters. In the earlier edition of this book, stable isotopes were not discussed until Chapter 8. However, stable and radiogenic isotope geochemistry is not so easily pigeon-holed, with many discoveries about the Earth and the cosmos involving simultaneous application of radiogenic and stable isotope geochemistry. This necessitates some understanding of stable isotope geochemistry before we go on to consider meteorites and the early Solar System and the Earth's mantle, crust, and surface. Although the emphasis in the next several chapters will be on radiogenic isotope geochemistry, we will also make reference to stable isotope studies.

Stable isotope geochemistry began in the 1950s with a focus on light elements H, C, N, O, and S, with much of the initial work carried out by students and postdocs of Harold Urey. We will refer to these as the *traditional* elements. Subsequent studies in the 1980s and 1990s added Li and B to this list. All of these exhibit large and readily measured isotopic variations. The reasons for this are as follows:

- They have low atomic mass and consequently the relative mass difference between the isotopes is large.

Isotope Geochemistry, Second Edition. William M. White.
© 2023 John Wiley & Sons Ltd. Published 2023 by John Wiley & Sons Ltd.
Companion Website: www.wiley.com/go/white/isotopegeochem2

- They form bonds with a high degree of covalent character.
- They exist in more than one oxidation state (C, N, and S), form a wide variety of compounds (O), or are important constituents of naturally occurring solids and fluids.
- The abundance of the rare isotope is sufficiently high (generally at least tenths of a percent) to facilitate analysis.

Furthermore, isotope ratios of all these elements are most readily measured in a gas source mass spectrometer. As a consequence, isotopic variations are both large and readily measured. Lithium and boron are exceptions in that both Li and B exist in nature only in a single-oxidation state, cannot be analyzed in a gas source mass spectrometer, and Li forms purely ionic bonds and B forms ones with a significant ionic character. Boron, does, however, have two quite different bonding geometries that, as we will see, contribute substantially to B isotopic variations.

The dam broke, so to speak, around the turn of the century as new techniques offering greater sensitivity and higher precision were developed, and with techniques and instruments capable of measuring elements in nongaseous form (most notably multi-collector inductively coupled plasma mass spectrometer), stable isotope geochemistry expanded to include all multi-isotopic elements; we will refer to these as the non-traditional stable isotopes. We will mainly focus on the traditional elements to explain the principles in this chapter, but these principles remain the same for other elements.

5.2 NOTATION, DEFINITIONS, AND STANDARDS

5.2.1 The δ notation

Variations in stable isotope ratios are usually reported in δ notation, which are *permil* variations from a standard, e.g.:

$$\delta^{18}O = \left[\frac{\left(^{18}O/^{16}O\right)_{sam} - \left(^{18}O/^{16}O\right)_{std}}{\left(^{18}O/^{16}O\right)_{std}} \right] \times 10^3$$

$$= \left[\frac{\left(^{18}O/^{16}O\right)_{sam}}{\left(^{18}O/^{16}O\right)_{std}} - 1 \right] \times 10^3$$

$$(5.1)$$

where "sam" denotes sample and "std" denotes standard. This, of course, is analogous to the ε, μ, and γ notations introduced in Chapter 2 (while values for δ are generally followed by the parts per thousand sign, ‰, this is actually superfluous, since the parts per thousand is incorporated in the definition). For non-traditional stable isotopes where variations are small, the ε and μ notations are also used (denoting variations in parts in ten thousand and parts per million from a standard, respectively).

Table 5.1 lists standards and isotopic compositions of the traditional elements; those for other isotope systems are listed on Table 5.2. For the most part, however, you need not concern yourself with these values and work instead with the δ notation. The initial standards for traditional stable isotopes initially were natural ones – Standard

Table 5.1 Standards and isotope ratios of traditional stable isotopes.

Element	Notation	Ratio	Standard	Absolute ratio
Hydrogen	δD	D/H ($^2H/^1H$)	(V)SMOW	1.557×10^{-4}
Carbon	$\delta^{13}C$	$^{13}C/^{12}C$	PDB	1.122×10^{-2}
Nitrogen	$\delta^{15}N$	$^{15}N/^{14}N$	ATM	3.613×10^{-3}
Oxygen	$\delta^{18}O$	$^{18}O/^{16}O$	(V)SMOW	2.005×10^{-3}
	$\delta^{17}O$	$^{17}O/^{16}O$	(V)SMOW	3.829×10^{-4}
Sulfur	$\delta^{34}S$	$^{34}S/^{32}S$	CDT	4.390×10^{-2}
	$\delta^{33}S$	$^{33}S/^{32}S$	CDT	4.9788×10^{-3}
	$\delta^{36}S$	$^{36}S/^{32}S$	CDT	1.5362×10^{-4}

Table 5.2   Reference values of non-conventional stable isotope ratios.

Element	Notation	Ratio	Standard	Absolute ratio
Lithium	$\delta^7\text{Li}$	$^7\text{Li}/^6\text{Li}$	NIST L-SVEC	12.1735
Boron	$\delta^{11}\text{B}$	$^{11}\text{B}/^{10}\text{B}$	NIST 951	4.0436
Magnesium	$\delta^{26}\text{Mg}$	$^{26}\text{Mg}/^{24}\text{Mg}$	DSM3	0.13979
Silicon	$\delta^{30}\text{Si}$	$^{30}\text{Si}/^{28}\text{Si}$	NBS28 (NIST-RM8546)	0.033532
	$\delta^{29}\text{Si}$	$^{29}\text{Si}/^{28}\text{Si}$		0.050804
Chlorine	$\delta^{37}\text{Cl}$	$^{37}\text{Cl}/^{35}\text{Cl}$	Sea water (SMOC)	0.319627
			NIST-SRM 975a	0.319770
Calcium	$\delta^{44/42}\text{Ca}$	$^{44}\text{Ca}/^{42}\text{Ca}$	NIST SRM 915a	0.310163
	$\delta^{44/40}\text{Ca}$	$^{44}\text{Ca}/^{40}\text{Ca}$		0.021518
	$\delta^{43/42}\text{Ca}$	$^{43}\text{Ca}/^{42}\text{Ca}$		0.208655
Titanium	$\delta^{49}\text{Ti}$	$^{49}\text{Ti}/^{47}\text{Ti}$	OL-Ti	0.729437
Chromium	$\delta^{53}\text{Cr}$	$^{53}\text{Cr}/^{52}\text{Cr}$	NIST-SRM-979	0.11339
		$^{54}\text{Cr}/^{52}\text{Cr}$		0.02822
		$^{54}\text{Cr}/^{52}\text{Cr}$		0.05186
Iron	$\delta^{56}\text{Fe}$	$^{56}\text{Fe}/^{54}\text{Fe}$	IRMM-14 (IRMM-524a)	15.698
	$\delta^{57}\text{Fe}$	$^{57}\text{Fe}/^{54}\text{Fe}$		0.363255
Copper	$\delta^{65}\text{Cu}$	$^{65}\text{Cu}/^{63}\text{Cu}$	NIST SRM976	0.44562
Zinc	$\delta^{68}\text{Zn}$	$^{68}\text{Zn}/^{64}\text{Zn}$	JMC3-0749L	0.37441
	$\delta^{66}\text{Zn}$	$^{66}\text{Zn}/^{64}\text{Zn}$		0.56502
Germanium	$\delta^{74}\text{Ge}$	$^{74}\text{Ge}/^{70}\text{Ge}$	NIST-SRM3120a	1.7609
	$\delta^{73}\text{Ge}$	$^{73}\text{Ge}/^{70}\text{Ge}$		0.3734
	$\delta^{72}\text{Ge}$	$^{72}\text{Ge}/^{70}\text{Ge}$		1.3290
Selenium	$\delta^{82}\text{Se}$	$^{82}\text{Se}/^{76}\text{Se}$	NIST-SRM 3149	0.95657
	$\delta^{82/78}\text{Se}$	$^{82}\text{Se}/^{78}\text{Se}$		0.37327
Molybdenum	$\delta^{98}\text{Mo}$	$^{98}\text{Se}/^{95}\text{Se}$	NIST-SRM 3134	1.5303
Tin	$\delta^{124}\text{Sn}$	$^{124}\text{Sn}/^{116}\text{Sn}$	NIST-SRM 3161a	0.39817
	$\delta^{124}\text{Sn}$	$^{124}\text{Sn}/^{118}\text{Sn}$		0.23903
		$^{122}\text{Sn}/^{118}\text{Sn}$		0.19156
	$\delta^{122/118}\text{Sn}$	$^{122}\text{Sn}/^{118}\text{Sn}$	Sn_IPGP	–
Mercury	$\delta^{199}\text{Hg}$	$^{199}\text{Hg}/^{198}\text{Hg}$	NIST- SRM 3133	1.6872
	$\delta^{200}\text{Hg}$	$^{200}\text{Hg}/^{198}\text{Hg}$		2.3047
	$\delta^{201}\text{Hg}$	$^{201}\text{Hg}/^{198}\text{Hg}$		1.3121
	$\delta^{202}\text{Hg}$	$^{202}\text{Hg}/^{198}\text{Hg}$		2.9614
	$\delta^{204}\text{Hg}$	$^{204}\text{Hg}/^{198}\text{Hg}$		0.68012
Thallium	$\varepsilon^{205}\text{Tl}$	$^{205}\text{Tl}/^{203}\text{Tl}$	NIST-SRM 997	2.38714
Uranium	$\delta^{238}\text{U}$	$^{238}\text{U}/^{235}\text{U}$	CRM145/112a (NIST 960)	137.849
			NIST-SRM950a	137.852

Mean Ocean Water[1] or (SMOW) for H and O, Pee Dee Belemnite (PDB), a fossil cephalopod carbonate shell from the Pee Dee Formation of North Carolina for C as well as O in carbonates, atmospheric nitrogen (ATM) for N, and troilite in the Canyon Diablo (CDT) iron meteorite for sulfur (this was the meteorite responsible for the crater in Meteorite Crater in Arizona, the United States). Because the original SMOW, PDB, and CTD standards have been exhausted, as have several

---

[1] The story goes that when Harmon Craig, who had moved to Scripps Institute of Oceanography after completing his PhD with Harold Urey at the University of Chicago, needed an isotopic standard for O and H, he simply retrieved a bucket of Pacific sea water from the end of Scripps Pier, which became the first SMOW standard.

other original standards, and, in some cases, proven to be isotopically inhomogeneous, they have been replaced by artificial standards of the same composition, which are isotopically uniform and available to all. The substitute for SMOW is a standard issued by the International Atomic Energy Agency (IAEA) in Vienna and is designated VSMOW (but in principle, VSMOW is formulated to match the original composition of SMOW, so the V is somewhat superfluous). Standards for non-traditional stable isotopes are often materials issued by governmental or international organizations, e.g., the US National Institute of Standards and Technology designated as NIST (formerly the National Bureau of Standards [NBS]) or the European Union's Institute for Reference Materials and Measurements designated as IRMM. Some, however, are shared informally among laboratories around the world. We should also point out that working analytical standards are often different from reference standards. For example, while sea water remains the reference standard for Cl, the working analytical standard is NIST 975a, whose isotopic composition is slightly different, and while Li isotope ratios continue to be reported as δ^7Li relative to the NIST 8545 L-SVEC, this has been exhausted and IRMM-16 is used as the practical analytical standard. (Prior to 1996, Li isotope ratios were reported as δ^6Li, i.e., deviations from the $^6Li/^7Li$ ratio of that standard. For variations of less than about 10‰, $\delta^7Li \approx -\delta^6Li$.) For isotope ratios exhibiting wide variations, accurate analytical calibration can require analysis of more than one standard. For example, for hydrogen and oxygen in precipitation, a second standard, Standard Light Antarctic Precipitation (SLAP), has been issued by the IAEA. The current version of this standard, SLAP2 (NIST 8735a), has a $\delta^{18}O_{SMOW}$ value of −55.5‰ and a δD_{SMOW} value of −427.5‰. Nitrogen in air has proved to be isotopically homogeneous, and inexhaustible, so that it remains both a reference and an analytical standard.

For historical and analytical reasons, O isotopes in carbonates are reported relative to PDB. These values are related to SMOW normalized values as:

$$\delta^{18}O_{PDB} = 1.03091 \times \delta^{18}O_{SMOW} + 30.91$$

$$(5.2)$$

Oxygen in silicates is actually measured relative to two other standards, NBS 28 (quartz)

with a defined $\delta^{18}O = 9.57‰$ and NBS 30 (biotite) with a $\delta^{18}O = 5.12‰$.

5.2.2 The fractionation factor

An important parameter in stable isotope geochemistry is the *fractionation factor*, α. It is defined as:

$$\alpha_{A-B} \equiv \frac{R_A}{R_B}$$

$$(5.3)$$

where R_A and R_B are the isotope ratios of two phases, A and B. Fractionation of isotopes between two phases is often also reported as $\Delta_{A-B} = \delta_A - \delta_B$. The relationship between Δ and α is:

$$\Delta \approx (\alpha - 1)10^3 \text{ or } \Delta \approx 10^3 \ln \alpha \qquad (5.4)$$

This approximation holds for small values of Δ_{A-B}, up to a few tens of per mil. We derive it as follows. Rearranging Equation (5.1), we obtain:

$$R_A = (\delta_A + 10^3) \times R_{STD}/10^3 \qquad (5.5)$$

where R denotes an isotope ratio. Thus, α may be expressed as:

$$\alpha = \frac{(\delta_A + 10^3)R_{STD}/10^3}{(\delta_B + 10^3)R_{STD}/10^3} = \frac{(\delta_A + 10^3)}{(\delta_B + 10^3)} \quad (5.6)$$

Subtracting 1 from each side and rearranging, and since δ is generally $\ll 10^3$, we obtain:

$$\alpha - 1 = \frac{(\delta_A - \delta_B)}{(\delta_B + 10^3)} \cong \frac{(\delta_A - \delta_B)}{10^3} = \Delta \times 10^{-3}$$

$$(5.7)$$

The second part of Equation (5.4) results from the approximation that for $x \approx 1$, $\ln x \approx x - 1$. As we will see, α is related to the equilibrium constant of thermodynamics by:

$$\alpha_{A-B} = K^{1/n} \qquad (5.8)$$

where n is the number of atoms exchanged.

Unfortunately, there are two other ways in which the Δ symbol is used in stable isotope geochemistry, namely, to denote deviations from mass-dependent fractionation and the extent of isotopic clumping, both of which we will discuss shortly. Subscripts denoting the phases involved, as in Δ_{A-B}, should help distinguish this use from others.

As we while see shortly, an additional notation, β, is useful for theoretically calculated equilibrium fractionations. We will define β once we review that theory.

5.3   THEORY OF EQUILIBRIUM ISOTOPIC FRACTIONATIONS

Isotope fractionation can originate from either *kinetic* or *equilibrium* effects, or both. The former might be intuitively expected (since, for example, we can readily understand that a lighter isotope will diffuse faster than a heavier one), but the latter may be somewhat surprising. After all, we were taught in introductory chemistry that oxygen is oxygen, and its properties are dictated by its electronic structure. In the following sections, we will see that quantum mechanics predicts that mass affects the strength of chemical bonds and the vibrational, rotational, and translational motions of atoms. These quantum mechanical effects predict the small differences in the chemical properties of isotopes that arise as a consequence quite accurately.

The electronic structures of all isotopes of an element are identical and since the electronic structure governs chemical properties, these properties are generally identical as well. Nevertheless, small differences in chemical behavior arise when this behavior depends on the frequencies of atomic and molecular vibrations. The energy of a molecule can be described in terms of several components: electronic, nuclear volume and spin, translational, rotational, and vibrational. The first two terms are generally negligible; nuclear volume and spin come into play only for very heavy elements and we will review those in due course. The last three terms are the modes of motion available to an atom or molecule and are the primary cause of differences in chemical behavior among isotopes of the same element. Of the three, vibration motion plays the most important role in isotopic fractionations. Translational and rotational motion can be described by classical mechanics, but an adequate description of vibrational motions of atoms in a lattice or molecule requires the application of quantum theory. As we shall see, *temperature-dependent equilibrium isotope fractionations arise from quantum mechanical effects on vibrational motions*. These effects are, as one might expect, generally small. For example, the equilibrium constant for the reaction:

$$\tfrac{1}{2}C^{16}O_2 + H_2^{18}O \rightleftharpoons \tfrac{1}{2}C^{18}O^{16}O + H_2^{16}O$$

is only slightly different from 1, ≈ 1.04, at 25°C.

Figure 5.1 is a plot of the potential energy of a diatomic hydrogen molecule as a function of distance between the two atoms. This plot looks broadly similar to one we might construct for two masses connected by a spring. When the distance between masses is small, the spring is compressed, and the potential energy of the system is correspondingly high. At great distances between the masses, the spring is stretched and the energy of the system is also high. At some intermediate distance, there is no stress on the spring, and the potential energy of the system is at a minimum while the kinetic energy is at a maximum. A diatomic oscillator consisting of a Na and a Cl ion works in an analogous way. At small interatomic distances, the electron clouds repel each other (the atoms are compressed); at large distances, the atoms are attracted to each other by the net charge on atoms. The energy and the distance over which the atoms vibrate are quantized and increase in steps as temperature increases.

Quantum theory dictates that a diatomic oscillator cannot assume just any energy: only discrete energy levels may be occupied, which is true for all types of energy as well. The permissible energy levels, as we shall see, depend on mass. Quantum theory also tells us that even at absolute 0, the atoms will vibrate at a ground frequency ν_0. The system will have energy of $^1/_2 h\nu_0$, where h is Planck's constant. This energy level is called the zero point energy or *ZPE*. Its value depends on the electronic arrangements, the nuclear charges, and the positions of the atoms in the molecule or lattice, all of which will be identical for isotopes of the same element. However, the energy also depends on the masses of the atoms involved and thus will be different for isotopes. *The vibrational frequency will be lower for a bond involving a heavier isotope of an element, which in turn lowers the vibrational energy of the molecule or crystal*, as suggested in Figure 5.1. Because their energy is lower, *bonds involving heavier isotopes will be stronger*. If a system consists of two possible atomic sites with different bond energies and two isotopes of an element available to fill those sites, *the energy of the system is minimized when the heavy isotope occupies the site with the stronger bond*. This, in brief, is why equilibrium fractionations arise. Because bonds involving lighter isotopes are weaker and more readily broken, the lighter isotopes of an element participate more readily in a given chemical reaction, which is an example

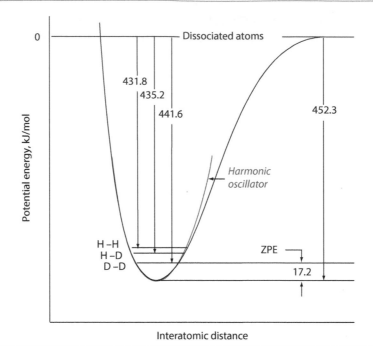

Figure 5.1 Energy-level diagram for the hydrogen atom. Fundamental vibration frequencies are 4405 cm^{-1} for H2, 3817 cm^{-1} for HD, and 3119 cm^{-1} for D$_2$. The zero-point energy (ZPE) of H$_2$ is greater than that for HD, which is greater than that for D$_2$. Source: After O'Neil (1986)/Reproduced with permission of the Mineralogical Society of America.

of a kinetic effect. If the reaction fails to go to completion, which is often the case, this tendency gives rise to kinetic fractionations of isotopes. There are other causes of kinetic fractionations as well and we will consider them in due course. We will now consider in greater detail the basis for equilibrium fractionation and see that it can be predicted from statistical mechanics.

5.3.1   Partition functions and thermodynamics

Harold Urey (1947) and his associates at the University of Chicago, Jacob Bigeleisen and Maria Mayer (1947) pointed out the possibility of calculating the equilibrium constant for isotopic exchange reactions from the *partition function, q,* of statistical mechanics. At equilibrium, the ratio of the number of molecules having internal energy E_i to the number having the ZPE E_0 is:

$$\frac{n_i}{n_0} = g_i e^{-E_i/kT} \qquad (5.9)$$

where n_0 is the number of molecules with ground-state or ZPE, n_i is the number of

molecules with energy E_i and k is Boltzmann's constant, T is the thermodynamic, or absolute, temperature, and g is a statistical weight factor used to account for possible degenerate energy levels[2] (g is equal to the number of states having energy E_i). The average energy (per molecule) in a system is given by the *Boltzmann distribution function*, which is just the sum of the energy of all possible states multiplied by the number of particles in that state divided by the number of particles in those states:

$$E = \frac{\sum\limits_i n_i E_i}{\sum\limits_i n_i} = \frac{\sum g_i E_i e^{-E_i/kT}}{\sum g_i e^{-E_i/kT}} \qquad (5.10)$$

The denominator of this equation, which is the sum of all energy states accessible to the system, is called the *partition function, q*:

$$q = \sum g_i e^{-E_i/kT} \qquad (5.11)$$

Substituting Equation (5.11) into Equation (5.10), we can rewrite Equation (5.10) in terms of the partial derivatives of q:

---

[2] Energy levels or states are said to be "degenerate" if two or more have the same energy level E_i.

$$E = kT^2 \frac{\partial \ln q}{\partial T} \qquad (5.12)$$

We will return to these equations shortly, but first let us see how all this relates to some parameters that are more familiar from thermodynamics and physical chemistry. It can also be shown (but we will not) from statistical mechanics that *entropy*[3] is related to energy and q by:

$$S = \frac{U}{T} R \ln q \qquad (5.13)$$

where R is the ideal gas constant and U is the internal energy of a system. We can rearrange this as:

$$U - TS = -R \ln q \qquad (5.14)$$

and for the entropy and energy changes of a reaction, we have:

$$\Delta U - T\Delta S = -R \ln \prod q_n^{\xi} \qquad (5.15)$$

where ξ in this case is the stoichiometric coefficient. In this notation, the stoichiometric coefficient is taken to have a negative sign for reactants (left side of reaction) and a positive sign for products (right side of reaction). The left-hand side of this equation is simply the Gibbs-free energy change of reaction under conditions of constant volume (as would be the case for an isotopic exchange reaction), so that

$$\Delta G = -R \ln \prod q_n^{\xi} \qquad (5.16)$$

The Gibbs-free energy change is related to the equilibrium constant, K, by:

$$\Delta G = -RT \ln K \qquad (5.17)$$

so the equilibrium constant for an isotope exchange reaction is related to the partition function as:

$$K = \prod_n q_n^{\xi} \qquad (5.18)$$

For example, in the reaction involving exchange of ^{18}O between H_2O and CO_2, the equilibrium constant is simply:

$$K = \frac{q_{C^{18}O^{16}O}^{1/2} q_{H_2^{16}O}}{q_{C^{16}O_2}^{1/2} q_{H_2^{18}O}} \qquad (5.19)$$

The point of all this is simply that the usefulness of the partition function is that it can be calculated from quantum mechanics, and from it we can calculate equilibrium fractionations of isotopes.

The partition function involving energies of interest can be written as approximately:

$$q_{total} = q_{trans} q_{rot} q_{vib} \qquad (5.20)$$

i.e., the product of the translational, rotational, and vibrational partition functions (we are ignoring contributions from anharmonic vibrations, rotational–vibrational interactions, and electronic energies). We should note here that since rotational and translational motions are not available to atoms in a solid, the partition function reduces to the vibrational partition function only.

It is convenient to treat these three modes of motion separately. Let us now do so.

5.3.1.1   Translational partition function

Writing a version of Equation (5.11) for translational energy, q_{trans} is expressed as:

$$q_{trans} = \sum_i g_{tr,i} e^{-E_{tr,i}/kT} \qquad (5.21)$$

Now all that remains is to find an expression for translational energy and a way to do the summation. At temperatures above about 2 K, translational energy levels are so closely spaced that they essentially form a continuum.

---

[3] Entropy is defined in the second law of thermodynamics, which states:

$$dS = \frac{dQ_{rev}}{T}$$

where Q_{rev} is heat gained by a system in a reversible process. Entropy can be thought of as a measure of the randomness of a system.

The quantum translational energy of a particle in a cubical box is given by:

$$E_{trans} = \frac{n^2 h^2}{8 M d^2} \quad (5.22)$$

where n is the quantum energy level, h is Planck's constant, d is the length of the side of the cube, and M is mass of the particle. Substituting Equation (5.22) into Equation (5.21) and integrating:

$$q_{trans} = \int_0^\infty e^{-n^2 h^2/8 M d^2 kT} = \frac{(2\pi M k T)^{1/2}}{h} d \quad (5.23)$$

gives an expression for q_{trans} for each dimension. The total three-dimensional translational partition function is then:

$$q_{trans} = \frac{(2\pi M k T)^{3/2}}{h^3} V \quad (5.24)$$

where V is volume and is equal to d^3. It may seem odd that the volume should enter into the calculation, but since it is the ratio of partition functions that are important in equations such as Equation (5.19), the volume term cancels, and since two isotopes in the same box will have the same temperature and all other terms in Equation (5.24) are constants, all terms except for mass cancel. If translation motion were the only component of energy, the equilibrium constant for exchange of isotopes would be simply the ratio of the molecular weights raised to the $^3/_2$ power.

If we define the translational contribution to the equilibrium constant as K_{tr} as:

$$K_{tr} = \prod q_{tr}^\xi \quad (5.25)$$

K_{tr} reduces to the product of the molecular masses raised to three-halves of the stoichiometric coefficient:

$$K_{tr} = \prod \left(M_{tr}^\xi\right)^{3/2} \quad (5.26)$$

For polyatomic molecules, M is the mass of the isotopic version of the molecule, the isotopologue, not the isotope alone. We see that the translational contribution to the total partition function depends on the mass of the isotopologue and is independent of temperature.

5.3.1.2   Rotational partition function

The allowed quantum *rotational* energy states are:

$$E_{rot} = \frac{j(j + 1)h^2}{8\pi^2 I} \quad (5.27)$$

where j is the rotational quantum number and I is the moment of inertia. For a diatomic molecule, $I = \mu d^2$, where d is the bond length and μ is reduced mass:

$$\mu = \frac{m_1 m_2}{m_1 + m_2} \quad (5.28)$$

A diatomic molecule will have two rotational axes, i.e., one along the bond axis and the other perpendicular to it. Hence, in a diatomic molecule, j quanta of energy may be distributed $2j + 1$ ways because there are two possibilities for every value of j except $j = 0$, for which there is only one possible way. The statistical weight factor is therefore $2j + 1$. Hence:

$$q_{rot} = \sum (2j + 1) e^{j(j + 1)h^2/8\pi^2 I kT} \quad (5.29)$$

Again the spacing between energy levels is relatively small (except for hydrogen) and Equation (5.29) may be evaluated as an integral. For a diatomic molecule, the partition function for rotation is given by:

$$q_{rot} = \frac{8\pi^2 I kT}{\sigma h^2} \quad (5.30)$$

where σ is the symmetry number and is equal to the number of equivalent ways the molecule can be oriented in space. It is 1 for a heteronuclear diatomic molecule (such as CO or $^{18}O^{16}O$), and 2 for a homonuclear diatomic molecule (such as $^{16}O_2$) or a symmetric triatomic molecule (such as $^{16}O^{12}C^{16}O$) (more complex molecules will have higher symmetry numbers, e.g., 12 for $^{12}C^1H_4$). Equation (5.30) also holds for linear polyatomic molecules with the symmetry factor equal to 2 if the molecule has a plane of symmetry (e.g., CO_2) and 1 if it does not.

For non-linear polyatomic molecules, the partition function is given by:

$$q_{\text{rot}} = \frac{8\pi^2 \left(8\pi^3 I_A I_B I_C\right)^{1/2} (kT)^{3/2}}{\sigma h^3} \qquad (5.31)$$

where I_A, I_B, and I_C are the principal moments of inertia. In calculating the rotational contribution to the equilibrium constant, all terms cancel except for moments of inertia and the symmetry factor, and hence the contribution of rotational motion to isotope fractionation is also independent of temperature. For diatomic molecules, the equilibrium constant calculated from the ratios of partition functions reduces to:

$$K_{\text{rot}} = \prod_i \left(\frac{I_i}{\sigma_i}\right)^{\xi_i} \qquad (5.32)$$

In general, bond lengths are approximately, although not entirely, independent of the isotope involved, so the moment of inertia term may be replaced by the reduced masses (μ):

$$K_{\text{rot}} \cong \prod_i \left(\frac{\mu_i}{\sigma_i}\right)^{\xi_i} \qquad (5.32a)$$

5.3.1.3 Vibrational partition function

We will simplify the calculation of the vibrational partition function by treating the diatomic molecule as a harmonic oscillator (as Figure 5.1 suggests, this is a good approximation in most, but not all, cases). In this case, the quantum energy levels are given by:

$$E_{\text{vib}} = \left(n + \frac{1}{2}\right)h\nu \qquad (5.33)$$

where n is the vibrational quantum number and ν is vibrational frequency. Unlike rotational and vibrational energies, the spacing between vibrational energy levels is large even at geologic temperatures, so the partition function cannot be integrated. Instead, it must be summed over all available energy levels. Fortunately, the sum has a simple form: for diatomic molecules, the summation is simply equal to:

$$q_{\text{vib}} = \frac{e^{-h\nu/2kT}}{1 - e^{-h\nu/kT}} \qquad (5.34)$$

For molecules consisting of more than two atoms, there are many vibrational motions possible. In this case, the vibrational partition function is the product of the partition functions for each mode of motion, with

the individual partition functions given by Equation (5.34). For a non-linear polyatomic molecule consisting of i atoms and the product is performed over all vibrational modes, ℓ, the partition function is given by:

$$q_{\text{vib}} = \prod_\ell^{3\ell - n} \frac{e^{-h\nu_\ell/2kT}}{1 - e^{-h\nu_\ell/kT}} \qquad (5.35)$$

where n is equal to 6 for *non-linear* polyatomic molecules and 5 for *linear* polyatomic molecules.

The vibrational energy contribution to the equilibrium constant for diatomic molecules is thus:

$$K_{\text{vib}} = \prod_i \left(\frac{e^{-h\nu_i/2kT}}{1 - e^{-h\nu_i/kT}}\right)^{\xi_i} \qquad (5.36)$$

We can think of the exponential term in the numerator of Equation (5.36) as describing isotopic effects on ground state energies and the $(1 - e)$ term in the denominator as describing isotopic effects of excited states. At room temperature and below, most atoms will be in their ground state and the exponential term in the denominator approximates to 0, and the denominator therefore approximates to 1, so the relation simplifies to:

$$q_{\text{vib}} \cong\sim e^{-h\nu/2kT} \qquad (5.37)$$

Thus, at low temperature, the vibrational contribution to the equilibrium constant approximates to:

$$K_{\text{vib}} \cong \prod_\ell e^{-\xi_\ell h\nu_\ell/2kT} \qquad (5.38)$$

which has an exponential temperature dependence.

5.3.2 Reduced partition functions and β factors

The full expression for the equilibrium constant calculated from partition functions for diatomic molecules is then:

$$K = \prod_i \left(M_i^{3/2} \left[\frac{I_i}{\sigma_i}\right] \frac{e^{-h\nu_i/2kT}}{1 - e^{-h\nu_i/kT}}\right)^{\xi_i} \qquad (5.39)$$

where again ξ is the stoichiometric coefficient. This equation can be simplified through use of the Teller–Redlich spectroscopic theorem[4] to:

$$K = \prod_i \left(\frac{m_i^{3r_i/2}}{\sigma_i} \frac{h\nu_i}{kT} \frac{e^{-h\nu_i/2kT}}{1 - e^{-h\nu_i/kT}} \right)^{\xi_i} \quad (5.40)$$

where r_i is the number of atoms being exchanged in the molecule.

Returning to our reaction between CO_2 and water, we can rearrange the equilibrium constant expression (Equation [(5.19)]) as:

$$K = \frac{\left(q_{18_O}/q_{16_O} \right)_{CO_2}^{1/2}}{\left(q_{16_O}/q_{18_O} \right)_{H_2O}} \quad (5.41)$$

Thus, the equilibrium constant is the ratio of partition function ratios of the two isotopic versions of the two substances. We can see that the mass terms in Equation (5.40) will cancel in the computation and consequently we can omit them. The partition function ratio with the mass terms omitted is sometimes referred to as the *reduced partition function*.

When theoretical calculations, such as these, are involved, isotope fractionation is also often expressed in terms of a β-factor defined as the ratio at equilibrium of the isotope ratio of the substance of interest to the isotope ratio of dissociated atoms:

$$\beta_{H_2O}^{18_O/16_O} \equiv \frac{\left({}^{18}O/{}^{16}O \right)_{H_2O}}{\left({}^{18}O/{}^{16}O \right)_O} \quad (5.42)$$

In this case, the denominator is the isotope ratio in a gas of monatomic oxygen and the numerator is the isotope ratio in water in equilibrium with it.

The β-factor for CO_2 in an oxygen isotope exchange reaction would also have the isotope ratio of dissociated oxygen atoms in the denominator and thus in our example of CO_2–water exchange, the denominators cancel so that:

$$\alpha = \frac{\beta_{CO_2}}{\beta_{H_2O}} \quad (5.43)$$

This is the value of the β-factor: we can readily calculate fractionation factors from them. We need to calculate a different β-factor for each different pair of isotopes of interest. We have focused on the ${}^{18}O/{}^{16}O$ ratio; β-factors for ${}^{17}O/{}^{16}O$ would be different (although the math is the same). If we were interested in a carbon isotopic exchange reaction, e.g., between CO_2 and methane, we could calculate β-factors relative to monatomic carbon atoms in a gas in a similar way.

β-factors are closely related to the reduced partition function and are equal to it where the molecule of interest contains only one atom of the element of interest (this would be the case for water, e.g., which contains only one oxygen atom). Where this is not the case, i.e., CO_2, the β-factor differs from the reduced partition function by what Richet et al. (1977) termed an "excess factor," and arises from our interest in atomic isotopic ratios rather than of isotopic molecular abundances. The reason for this will become clear in the following example.

5.3.3   Example of fractionation factor calculated from partition functions

To illustrate the use of partition functions in calculating theoretical fractionation factors, we will do the calculation for a very simple reaction: the exchange of ${}^{18}O$ and ${}^{16}O$ between O_2 and CO:

$$C^{16}O + {}^{18}O^{16}O \rightleftharpoons C^{18}O + {}^{16}O_2 \quad (5.44)$$

The choice of diatomic molecules greatly simplifies the equations. Choosing even a slightly more complex molecule, such as CO_2, would complicate the calculation because there are more vibrational modes possible. Richet et al. (1977) and Chacko et al. (2001) provide examples of the calculation for more complex molecules.

Let us first consider the relationship between the equilibrium constant and the fractionation factor for this reaction. The equilibrium constant for our reaction is:

$$K = \frac{\left[{}^{16}O_2 \right]\left[C^{18}O \right]}{\left[{}^{18}O^{16}O \right]\left[C^{16}O \right]} \quad (5.45)$$

---

[4] The Teller–Redlich Theorem relates the products of the frequencies for each symmetry type of the two isotopes to the ratios of their masses and moments of inertia:

$$\left(\frac{m_2}{m_1} \right)^{3/2} \frac{I_1}{I_2} \left(\frac{M_1}{M_2} \right)^{3/2} = \frac{h\nu_1/kT}{h\nu_2/kT}$$

where m is the isotope mass and M is the molecular mass. We need not concern ourselves with its details.

where we are using the brackets in the usual chemical sense to denote concentration. We can use concentrations rather than activities or fugacities because the activity coefficient of a phase is independent of its isotopic composition. The fractionation factor, α, is defined as:

$$\alpha = \frac{\left(^{18}O/^{16}O\right)_{CO}}{\left(^{18}O/^{16}O\right)_{O_2}} \qquad (5.46)$$

We must also consider the exchange reaction:

$$^{18}O^{18}O + {}^{16}O^{16}O \rightleftharpoons 2\,{}^{16}O^{18}O$$

for which we can write a second equilibrium constant, K_2. It turns out that when both reactions are considered, $\alpha \approx 2K$. The reason for this is as follows. The isotope ratio in molecular oxygen is related to the concentration of the two molecular oxygen species as:

$$\left(\frac{^{18}O}{^{16}O}\right)_{O_2} = \frac{[^{18}O^{16}O]}{[^{18}O^{16}O] + 2[^{16}O_2]} \qquad (5.47)$$

Brackets denote concentration and parentheses denote isotope ratios. Therefore, $[^{18}O^{16}O]$ is the concentration of the $^{18}O^{16}O$ version, or isotopologue, of the O_2 molecule and $(^{18}O/^{16}O)_{O2}$ is the isotope ratio of O_2 in the reacting gas. Note that $^{16}O_2$ has 2 ^{16}O atoms, so it must be counted twice, whereas the ratio in CO is simply:

$$\left(\frac{^{18}O}{^{16}O}\right)_{CO} = \frac{[C^{18}O]}{[C^{16}O]} \qquad (5.48)$$

Designating the isotope ratio as R, we can rearrange Equation (5.47) and solve for $[^{18}O/^{16}O]$:

$$[^{18}O/^{16}O] = 2\frac{[^{16}O_2]R_{O_2}}{1 - R_{O_2}} \qquad (5.49)$$

and substitute it into Equation (5.45):

$$K = \frac{(1 - R_{O_2})[C^{18}O]}{2R_{O_2}[C^{16}O]} = \frac{(1 - R_{O_2})R_{CO}}{2R_{O_2}} \qquad (5.50)$$

If the isotope ratio is a small number as it generally is, the term $(1 - R) \approx 1$, so that:

$$K \cong \frac{R_{CO}}{2R_{O_2}} = \frac{\alpha}{2} \qquad (5.51)$$

Relationships, such as these, are also why the β factor can differ from the reduced partition function.

Now, let us return to the problem of calculating K from the partition functions:

$$K = \frac{q_{^{16}O_2}q_{C^{18}O}}{q_{^{18}O^{16}O}q_{C^{16}O}} \qquad (5.52)$$

where each partition function is the product of the translational, rotational, and vibrational partition functions. Since the reaction involves only diatomic molecules, we could simply use Equation (5.41). However, it is informative to see how the three separate modes of motion contribute to the overall equilibrium constant, so we will proceed by calculating the equilibrium constant for each mode of motion. The total equilibrium constant will then be the product of all three partial equilibrium constants.

For translational motion, we noted the ratio of partition functions reduces to the ratio of molecular masses raised to the 3/2 power. Hence:

$$K_{tr} = \frac{q_{^{16}O_2}q_{C^{18}O}}{q_{^{18}O^{16}O}q_{C^{16}O}} = \left(\frac{M_{^{16}O_2}M_{C^{18}O}}{M_{^{18}O^{16}O}M_{C^{16}O}}\right)^{3/2}$$

$$= \left(\frac{32 \times 30}{34 \times 28}\right)^{3/2} = 1.0126 \qquad (5.53)$$

We find that CO would be 12.6‰ richer in ^{18}O if translational motions were the only modes of energy available.

In the expression for the ratio of rotational partition functions, all terms cancel except for the moment of inertia and the symmetry factors. The symmetry factor is 1 for all the molecules involved in this reaction except for $^{16}O_2$. In this case, the terms for bond length also cancel, so the expression involves only the reduced masses. Therefore, the expression for the rotational equilibrium constant becomes:

$$K_{rot} = \frac{q_{^{16}O_2}q_{C^{18}O}}{q_{^{18}O^{16}O}q_{C^{16}O}} = \left(\frac{\mu_{^{16}O_2}\mu_{C^{18}O}}{2\mu_{^{18}O^{16}O}\mu_{C^{16}O}}\right)$$

$$= \frac{1}{2}\left(\frac{\dfrac{16 \times 16}{16 + 16} \times \dfrac{12 \times 18}{12 + 18}}{\dfrac{18 \times 16}{18 + 16} \times \dfrac{12 \times 16}{12 + 16}}\right) = \frac{0.9916}{2}$$

$$(5.54)$$

(we will ignore the 1/2; it will cancel out later when we calculate the fractionation factor). If rotation were the only mode of motion,

CO would be 8‰ *poorer* in ^{18}O. Notice that both the translational and rotational equilibrium constants do not depend on temperature.

We will do the calculation for low temperature, which will allow us to use Equation (5.38) to calculate the vibrational equilibrium constant:

$$K_{vib} = \frac{q_{^{16}O_2}q_{C^{18}O}}{q_{^{18}O^{16}O}q_{C^{16}O}} = e^{\dfrac{-h\left(\nu_{^{16}O_2}+\nu_{C^{18}O}-\nu_{^{18}O^{16}O}-\nu_{C^{16}O}\right)}{2kT}}$$

(5.55)

Further, since we expect the difference in vibrational frequencies to be quite small, we may make the approximation $e^x = x + 1$. Hence:

$$K_{vib}=1+\frac{-h}{2kT}\left[\left\{\nu_{^{16}O_2}+\nu_{C^{18}O}\right\}-\left\{\nu_{^{18}O^{16}O}+\nu_{C^{16}O}\right\}\right]$$

(5.56)

Let us make the simplification that the vibration frequencies are related to reduced mass as in a simple Hooke's law harmonic oscillator:

$$\nu = \frac{1}{2\pi}\sqrt{\frac{\kappa}{\mu}}$$

(5.57)

where κ is the forcing constant and depends on the nature of the bond and hence will be independent of isotopic composition. In this case, we may write:

$$\nu_{C^{18}O} = \nu_{C^{16}O}\sqrt{\frac{\mu_{C^{16}O}}{\mu_{C^{18}O}}} = \nu_{C^{16}O}\sqrt{\frac{6.857}{7.2}}$$

(5.58)

$$= 0.976\nu_{C^{16}O}$$

Alternatively, we could measure the frequencies spectroscopically (or simply look them up in tables). A similar expression may be written relating the vibrational frequencies of the oxygen molecule:

$$\nu_{^{16}O^{18}O} = 0.9718\nu_{^{16}O_2}$$

Substituting these expressions in the equilibrium constant expression, we obtain:

$$K_{vib}=1+\frac{-h}{2kT}\left[\nu_{^{16}O_2}\{1-0.9718\}-\nu_{C^{16}O}\{1-0.976\}\right]$$

The measured vibrational frequencies of CO and O_2 are 6.50×10^{13} s^{-1} and 4.74×10^{13} s^{-1}. Substituting these values and values for the Planck and Boltzmann constants, we obtain:

$$K_{vib} = 1 + \frac{5.544}{T}$$

(5.59)

At 300 K (room temperature), this evaluates to 1.0185.

We may now write the total equilibrium constant expression as:

$$K=K_{tr}K_{rot}K_{vib} \cong \left(\frac{M_{^{16}O_2}M_{C^{18}O}}{M_{^{18}O^{16}O}M_{C^{16}O}}\right)^{3/2}$$

$$\times\left(\frac{\mu_{^{16}O_2}\mu_{C^{18}O}}{2\mu_{^{18}O^{16}O}\mu_{C^{16}O}}\right)$$

$$\times\left\{1+\frac{h}{4\pi kT}\left[\left(\sqrt{\frac{\kappa}{\mu_{C^{16}O}}}-\sqrt{\frac{\kappa}{\mu_{C^{18}O}}}\right)\right.\right.$$

$$\left.\left.-\left(\sqrt{\frac{\kappa}{\mu_{^{16}O_2}}}-\sqrt{\frac{\kappa}{\mu_{^{18}O^{16}O}}}\right)\right]\right\}$$

(5.60)

Evaluating this at 300 K, we obtain:

$$K = 1.0126 \times \frac{0.9916}{2} \times 1.0185 = \frac{1.023}{2}$$

Since $\alpha = 2$ K, the fractionation factor is 1.023 at 300 K and would decrease by about 6 per mil per 100°C temperature increase (however, we must bear in mind that our approximations hold only at low temperature; at high temperature, equilibrium constants depend on the inverse square of temperature). This temperature dependence is illustrated in Figure 5.2. Thus, CO would be 23 per mil richer in the heavy isotope, ^{18}O, than O_2.

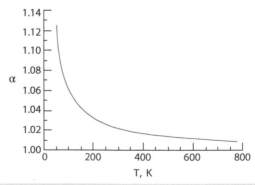

Figure 5.2 Fractionation factor, $\alpha = (^{18}O/^{16}O)_{CO}/(^{18}O/^{16}O)_{O2}$, calculated from partition functions as a function of temperature.

Oxygen binds more strongly to carbon than it does to itself and this illustrates an important rule of stable isotope fractionations: *the heavy isotope goes preferentially in the chemical compound in which the element is most strongly bound.*

Translational and rotational energy modes are, of course, not available to solids. Thus, isotopic fractionations between solids are entirely controlled by the vibrational partition function. Fractionations between coexisting solids can be calculated as we have done above. A few decades ago, the task was daunting because the varieties of vibrational modes available to atoms in a lattice make the task computationally intensive. With the computer power available today, theoretically computed fractionation factors have become more common. The lattice can be treated as a large polyatomic molecule having 3N-6 vibrational modes, where N is the number of atoms in the unit cell. For large N, this approximates to 3N. The unit cell differs from the chemical formula; in the case of quartz, one of the most compositionally simply minerals, the unit cell consists of 3 silicons and 6 oxygens, hence there are 21 vibrational modes that must be considered.

Vibrational frequency and heat capacity are closely related because thermal energy in a crystal is stored as vibrational energy of the atoms in the lattice. Einstein and Debye independently treated the problem by assuming the vibrations arise from independent harmonic oscillations. Their models can be used to predict heat capacities in solids. The vibrational motions available to a lattice may be divided into two types, the first of which is "internal" or "optical" vibrations between individual radicals or atomic groupings, such as CO_3 and Si–O, within the lattice. The vibrational frequencies of these groups can be calculated from the Einstein function and can be measured by optical spectroscopy. The second type are vibrations of the lattice as a whole, called "acoustical" vibrations, which can be measured, but may also be calculated from the Debye function.

In some cases, lattice vibrational frequencies can be measured spectroscopically through techniques, such as Raman or Mössbauer spectroscopy, but even in those cases only for the dominant isotope. The general case for solids then is that vibrational frequencies are not known and must be estimated. Lattice structures are, however, known, and hence bond lengths and the number and location of coordinating atoms are known, as are the atomic environment of dissolved substances, and these allow calculation of vibrational frequencies and hence partition functions through techniques, such as force field modeling or density function theory. These techniques are described in the review by Schauble (2004). The details of those techniques would take us too far afield, but we can state a few general rules that can help us anticipate how isotopes will be distributed between solids when we keep in mind that heavier isotopes are concentrated in sites in which it is most strongly bond.

- In general, the strongest bonds are also the stiffest and the shortest.
- Bonds of a particular atom to atoms of light elements are stronger than bonds to heavy ones.
- Bond strength is proportional to the valence divided by the number of coordinating atoms so that lower coordination numbers and higher oxidation state are associated with stronger bonds.
- Covalent bonds are stronger than ionic ones.

From either calculated or observed vibrational frequencies, partition function ratios may be calculated, which in turn are directly related to the fractionation factor. Generally, the optical modes are the primary contribution to the partition function ratios. For example, for partitioning of ^{18}O between water and quartz, the contribution of the acoustical modes is less than 10%. The ability to calculate fractionation factors is particularly important at low temperatures where reaction rates are quite slow and experimental determination of fractionation is therefore difficult. Figure 5.3 shows the calculated fractionation factor between quartz and water as a function of temperature. On the other hand, theoretically determined fractionation factors have the disadvantage that they generally assume perfect stoichiometry and crystallinity, whereas most natural crystals are chemically more complex. Thus, experimental determination of fractionation factor remains important.

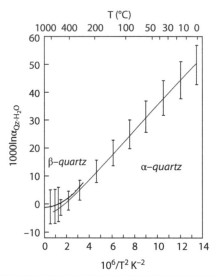

Figure 5.3 Calculated temperature dependencies of the fractionation of oxygen between water and quartz. Source: After Kawabe (1978)/Reproduced with permission of Elsevier.

5.3.4 Isotopologues and isotopic "Clumping"

In the example we just considered, we were concerned only with how ^{18}O was distributed between CO and O_2. However, the CO and O_2 gas will consist of a variety of molecules of distinct isotopic composition, or isotopologues. Indeed, there will be 12 such isotopologues: specifically, $^{12}C^{16}O$, $^{12}C^{17}O$, $^{12}C^{18}O$, $^{13}C^{16}O$, $^{13}C^{17}O$, $^{13}C^{18}O$, $^{16}O_2$, $^{16}O^{17}O$, $^{16}O^{18}O$, $^{17}O_2$, $^{17}O^{18}O$, and $^{18}O_2$. The statistical mechanical theory we just considered predicts that the distribution of isotopes within a species will not be random but rather that some of these isotopologues will be thermodynamically favored; in particular, the heavier isotopes of elements, ^{13}C and ^{18}O in this case, are more likely to be "clumped" together in a single molecule than to be randomly distributed among all molecules. The reason for this is that double heavy isotope substitution generally leads to a reduction in vibrational frequency and energy greater than twice that of single substitution. Thus, there is an energy advantage to bonding, or *clumping*, heavy isotopes together as opposed to simply distributing them randomly (Eiler, 2013).

As with the distribution of isotopes between chemical species, this *clumping* of isotopes within a species will be temperature dependent. We can use this as a geothermometer, and one that is independent of the isotopic composition of other phases. This, as we shall see in subsequent chapters, is an important advantage.

Let us begin by considering the distribution of isotopes between the isotopologues of CO. There are six isotopologues and they can be related through the following two reactions:

$$^{12}C^{16}O + {}^{13}C^{17}O \rightleftharpoons {}^{13}C^{16}O + {}^{12}C^{17}O \quad (5.61)$$

$$^{12}C^{16}O + {}^{13}C^{18}O \rightleftharpoons {}^{13}C^{16}O + {}^{12}C^{18}O \quad (5.62)$$

(Since we can relate the six isotopologues through two reactions, we only need to choose four of these isotopologues as the components of our system.) The equilibrium constant for reaction (5.62) can be calculated from:

$$K = \frac{q_{^{13}C^{16}O} \, q_{^{12}C^{18}O}}{q_{^{12}C^{16}O} \, q_{^{13}C^{18}O}} \quad (5.63)$$

A similar equation can be written for the equilibrium constant for Equation (5.61). The individual partition functions can be calculated just as described in the previous section. Doing so, we find that the two heaviest species, $^{13}C^{17}O$ and $^{13}C^{18}O$, will be more abundant than if isotopes were merely randomly distributed among the six isotopologues; that is, the heavy isotopes tend to "clump." Wang et al. (2004) introduced a delta notation to describe this effect:

$$\Delta_i = \left(\frac{R_{i\text{-}e}}{R_{i\text{-}r}} - 1 \right) \times 1000 \quad (5.64)$$

where the numerator $R_{i\text{-}e}$ is that ratio of the observed or calculated equilibrium abundance of isotopologue i to the isotopologue containing no rare isotopes and the denominator $R_{i\text{-}r}$ is that same ratio if isotopes were distributed among isotopologues randomly. Thus, for example, in the $CO\text{-}O_2$ system,

$$\Delta_{^{13}C^{18}O} = \left\{ \frac{\left([^{13}C^{18}O] \big/ [^{12}C^{16}O] \right)_e}{\left([^{13}C^{18}O] \big/ [^{12}C^{16}O] \right)_r} - 1 \right\} \times 1000$$

$$(5.65)$$

Delta used to describe clumping can be distinguished from other uses through a subscript designating either the isotopologue of interest, such as $^{13}C^{18}O$ in Equation (5.65), or its mass number, 31 in this case or 47 in the case of $^{13}C^{16}O^{18}O$.

Since $R_{i\text{-}r}$ is the random distribution, it can be calculated directly as the probability of choosing isotopes randomly to form species. In the case of $^{13}C^{18}O$, it is:

$$R_{^{13}C^{18}O\text{-}r} = \left(\frac{[^{13}C^{18}O]}{[^{12}C^{16}O]}\right)_r = \frac{[^{13}C][^{18}O]}{[^{12}C][^{16}O]}$$

(5.66)

It gets a little more complex for molecules with more than two atoms. In most cases, we are interested in combinations of isotopes rather than permutations, which is to say we do not care about order. This will not be the case for highly asymmetric molecules, such as nitrous oxide, N_2O. The structure of this molecule is N–N–O and $^{14}N^{15}N^{16}O$ will have different properties than $^{15}N^{14}N^{16}O$, so in that case, order does matter. The CO_2 molecule is, however, symmetric and we cannot distinguish $^{16}O^{12}C^{18}O$ from $^{18}O^{12}C^{16}O$. Its random abundance would be calculated as:

$$R_{^{13}C^{16}O^{18}O\text{-}r} = \left(\frac{[^{13}C^{16}O^{18}O]}{[^{12}C^{16}O_2]}\right)_r$$
$$= \frac{2[^{13}C][^{16}O][^{18}O]}{[^{12}C][^{16}O]^2} = \frac{2[^{13}C][^{18}O]}{[^{12}C][^{16}O]}$$

(5.67)

The factor of 2 is in the numerator to take account of both $^{12}C^{16}O^{18}O$ and $^{12}C^{18}O^{16}O$.

As Wang et al. (2004) showed that value of Δ as defined in Equation (5.65) is related to the equilibrium constant (Equation [(5.63)]) for the exchange reaction (Equation [(5.61)]) as:

$$\Delta_i \approx -1000 \ln \frac{K}{K_r}$$

(5.68)

Since K_r refers to the case of random distribution of isotopes, it is equal to 1 and since K will have a value close to 1, we may use the approximation $\ln x \approx 1 - x$, so that Equation (5.68) reduces to:

$$\Delta \approx (K - 1) \times 1000$$

(5.69)

Figure 5.4 shows the Δ_i values calculated for three isotopologues of CO as a function of temperature by Wang et al. (2004). Δ_i values vary with the inverse of temperature, a direct consequence of the inverse temperature dependence of the vibrational equilibrium constant expressed in Equation (5.59). The calculations predict that at room temperature, $^{13}C^{18}O$ and $^{13}C^{17}O$

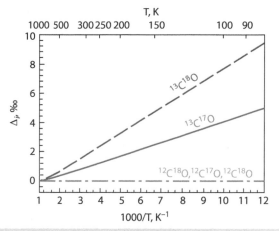

Figure 5.4 Predicted enrichment of isotopologues in CO_2 gas based on statistical mechanical calculations. Enrichments in $^{12}C^{18}O$, $^{12}C^{17}O$, and $^{13}C^{16}O$ above random are all negligible. Source: After Wang et al. (2004)/ Reproduced with permission of Elsevier.

concentrations will be 1–2 per mil above a purely random distribution of isotopes. Both are quite rare compared to the most abundant isotopologue, $^{12}C^{16}O$; the concentration of $^{13}C^{18}O$ will be less than 0.003% that of $^{13}C^{17}O$ about 0.0004%. Very high precision and sensitivity is required to analyze the abundance of these isotopes and detect the small enrichment resulting from clumping.

CO is a relatively rare species on Earth (but not in the cosmos; it is the second most common diatomic molecule in the interstellar medium, and only H_2 is more abundant) and it is of limited geochemical interest, at least compared to CO_2 and its related forms, such as carbonate and bicarbonate ions and, particularly, carbonate minerals. The latter are among the most common sedimentary minerals. Most form by precipitation from water (often biologically mediated), and the oxygen isotopic fractionation resulting from this precipitation reaction is temperature dependent. Thus, the isotopic composition of carbonates has proved to be a useful geothermometer and a very useful paleoclimatic tool. However, the isotopic composition of carbonate precipitated from water depends both on temperature and the isotopic composition of the water; calculating temperature thus requires knowing the isotopic composition of the water. In contrast, the relative abundance of carbonate isotopologues depends only on temperature and the isotopic composition of the carbonate.

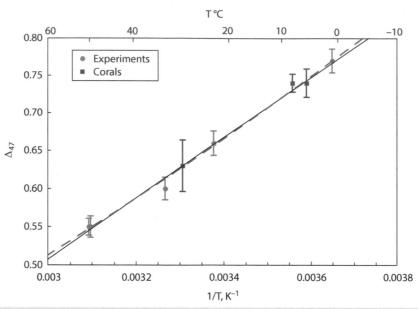

Figure 5.5 Data for Δ_{47} measured in experimental and natural carbonates by Ghosh et al. (2006) as a function of inverse temperature. Both $1/T$ (solid line) and $1/T^2$ (dashed line) are fitted to the data.

Since the latter is readily measured, temperature can be calculated from the abundance of the isotopologues.

The difficulty is twofold. First, the abundance of carbonate ion isotopologues containing two or more rare isotopes is very small; the most abundant will be $^{13}C^{18}O^{16}O$ with a relative abundance of about 67 ppm. Second, no analytical technique can directly measure the abundance of these rare isotopologues in carbonates. Carbonates are analyzed by first digesting the material in anhydrous phosphoric acid, which generates CO_2 gas. Carbon and oxygen isotopic compositions are then determined by analyzing the CO_2 as the CO_2^+ ion in a gas source mass spectrometer. Ghosh et al. (2006) showed that the abundance of $^{13}C^{18}O^{16}O$ isotopologue of analyzed CO_2 gas was proportional to the abundance of $^{13}C-^{18}O$ bonds in the carbonate. They found that the Δ_i value of the $^{13}C^{18}O^{16}O$ isotopologue in the CO_2 gas, referred to as Δ_{47}, was about 0.2‰ higher than that of the carbonate, but this fractionation appears to be nearly constant (two other CO_2 isotopologues also have mass 47, but their concentration is much, much lower than that of $^{13}C^{18}O^{16}O$). Analyzing both natural carbonate from corals known to have grown at different temperatures and carbonates precipitated from laboratory solutions, Ghosh et al. (2006) showed Δ_{47} was inversely proportional to temperature as predicted by statistical mechanics (Figure 5.5). Ghosh et al. (2006) concluded that "clumped" isotopic analysis could be used as a geothermometer with a precision of about ±2°C.

To further improve application of the clumping technique, workers from four laboratories (Dennis et al., 2011) proposed a method for standardizing and reporting clumped isotopic species to an absolute reference frame to avoid interlaboratory biases that result, among other things, from differences in fractionation that occur during acid digestion and ionization in the mass spectrometer. The approach involves standardization against prepared CO_2 gases whose Δ_{47} can be theoretically predicted. These include gases that vary in C and O isotopic composition that are equilibrated to a common temperature (typically 1000°C), and hence should have a common (and theoretically predicted) Δ_{47}, and a series of gases with the same isotopic composition equilibrated with water at known temperatures, and hence should also have theoretically predicted Δ_{47} values. The gases are then analyzed and the measured values regressed against theoretical ones to provide calibration functions. We will consider specific examples of applications of the clumping technique in Chapters 10 and 13.

equilibrium isotope ratios. Where A and B are not pure phases, the exponent on the right must be divided by the mole fraction of B (Criss, 1999). Notice that the left-hand side is a measure of the progress of the reaction (the reaction progress variable). This tells us that a reaction will reach equilibrium asymptotically.

We can come to a similar conclusion about rates by making use of transition state theory. A full discussion of this theory would take us too far afield (a brief overview can be found in White [2020] and a more detailed description in Zhang [2008]), so we will simply introduce the following equation without derivation:

$$\mathfrak{R}_{net} = \mathfrak{R}_+ \left(1 - e^{n\Delta G/RT}\right) \quad (5.86)$$

Where $\mathfrak{R}_{net}$ is the net rate of reaction, $\mathfrak{R}_+$ is the forward rate as in Equation (5.78), ΔG is the free energy difference between products and reactants, and n can be any number but is equal to 1 for elementary reactions. This equation tells us that as equilibrium is approached, the reaction rate slows because ΔG decreases. It also tells us that reactions with small ΔG of reaction will be slow from the outset. Isotopic exchange reactions typically have ΔG of reaction values of only a few kilojoules, whereas free energies of an order of magnitude or greater are typical of other types of reactions. ΔG represents the chemical energy available to drive the reaction. If the system is otherwise in equilibrium, isotope exchange reactions will simply not occur, leaving it out of equilibrium at low temperature. There are numerous examples of this, such as isotopic disequilibrium between atmospheric CO_2 and CH_4. Another is O and S isotopic disequilibrium among hydrothermal ore deposits precipitated below $\approx 200°C$, as we will find in Chapter 9.

Now consider Figure 5.1 once again. The energy required to raise the D_2 molecule to the energy where the atoms dissociate is 441.6 kJ/mole, whereas the energy required to dissociate the H_2 molecule is 431.8 kJ/mole. Therefore, it is easier to break bonds, such as H—H and C—H, than D—D and C—D. This is true generally: molecules containing the heavy isotope are more stable and have higher dissociation energies than those containing the light isotope. Breaking those bonds is a necessary first step before the hydrogen molecule can react to form other compounds. Indeed, we can consider that this energy is a barrier energy contributing to, and perhaps dominating, the E_B term in Equation (5.81).

We can examine this in a more quantitative sense. Referring to Figure 5.1, let us assume that barrier energy in Equation (5.81) is the difference between the dissociation energy, ε, and the ZPE. The constant A is independent of isotopic composition, thus the ratio of reaction rates between the HD molecule and the H_2 molecule is:

$$\frac{R_D}{R_H} = \frac{e^{-(\varepsilon - 1/2h\nu_D)kT}}{e^{-(\varepsilon - 1/2h\nu_H)/kT}} \quad (5.87)$$

or

$$\frac{R_D}{R_H} = e^{(\nu_H - \nu_D)h/2kT} \quad (5.88)$$

Substituting for the various constants, and using the wavenumbers given in the caption to Figure 5.1 (remembering that $\omega = c\nu$ where c is the speed of light), the ratio is calculated as 0.24; in other words, we expect the H_2 molecule to react four times faster than the HD molecule, a very large difference. For heavier elements, the rate differences are smaller. For example, the same ratio calculated for $^{16}O_2$ and $^{18}O^{16}O$ shows that the ^{16}O will react about 15% faster than the $^{18}O^{16}O$ molecule.

Where reactions go to completion, this difference in bonding energy plays no role: equilibrium isotopic fractionations will be governed by the considerations of equilibrium discussed in the previous section. *Where reactions do not achieve equilibrium, the lighter isotope will be preferentially concentrated in the reaction products*, because of this effect of bonds involving light isotopes in the reactants being more easily broken. Large kinetic effects are associated with biologically mediated reactions (e.g., bacterial reduction), because such reactions generally do not achieve equilibrium. Perhaps the best and most common example of this is the ^{12}C-enriched nature of organic matter, essentially all of which is produced through photosynthesis. ^{12}C is enriched relative to ^{13}C in the products of photosynthesis in plants relative to atmospheric CO_2 because the ^{12}C—O bond requires less energy to break than the ^{13}C—O bond and the photosynthesis reaction in plant cells does not consume all CO_2 within the cell; in other words, it does not go to completion. We noted above that the more rapid diffusion of $^{12}CO_2$ into plant cell interiors results in an $\approx 4‰$ enrichment in ^{12}C. Kinetic fractionation increases this by another $\approx 10-20‰$, depending on how efficiently CO_2 in the cell interior is converted to organic matter. Thus, terrestrial organic matter typically has $\delta^{13}C$ of -15 to $-25‰$. Other examples include ^{32}S enrichment in H_2S produced by bacterial reduction of sulfate that is well beyond equilibrium values. We will examine biological fractionations in much more detail in subsequent chapters.

5.4.3 Kinetic fractionation during evaporation and condensation

Let us now consider evaporation of water. δ^2H and $\delta^{18}O$ are well correlated in meteoric water (rain and snow) along a line $\delta^2H = 8\delta^{18}O +10$, which is known as the *meteoric water line* (*MWL*), or, because meteoric water can vary from this locally, the *global meteoric water line* (*GMWL*). Ocean water, which has $\delta^2H_{SMOW} \approx 0$ and $\delta^{18}O_{SMOW} \approx 0$ as consequence of the choice or normalization (although the isotopic composition of ocean surface waters does vary), supplies virtually all meteoric water through evaporation, yet the GMWL does not pass through 0, but is offset to higher δ^2H by 10‰. Furthermore, water vapor at the ocean surface is invariably isotopically lighter than the experimentally determined equilibrium fractionation. For example, Craig and Gordon (1965) found that the $\delta^{18}O$ of water vapor collected at mast height above the North Pacific Ocean varied from $-11‰$ to $-14‰$, depending on humidity, which was several per mil lighter than predicted by equilibrium evaporation. We will discuss the GMWL and the hydrologic cycle in detail in Chapter 10; here, we will only explore the contribution of kinetic fractionation to it.

Assuming that the activity of liquid water undergoing evaporation is unity, then the net rate of evaporation of $H_2^{16}O$, dN/dt, is:

$$\frac{dN}{dt} = k_+ - k_- P_v \qquad (5.89)$$

where k_+ is the rate constant for evaporation, k_- is the rate constant for condensation, and P_V is the vapor pressure. The hydrogen bonds between water molecules are weaker and more readily broken if ^{16}O or ^1H is involved rather than ^{18}O or ^2H, meaning isotopically light species are more readily available to transform into the gas phase than heavy ones. In addition, while air in a thin, vapor-saturated layer immediately above the ocean surface may be in isotopic equilibrium with the water, vapor molecules in this layer must first transit a diffusive sublayer before being incorporated in the turbulently mixed free atmosphere. The ratio of the molecular diffusivities in air of the pairs $H_2^{18}O/H_2^{16}O$ and $^1H^2HO/^1H_2O$ is 0.9723 and 0.9755, respectively (Merlivat, 1978), so the lighter isotopic species will more readily escape the boundary layer into the free atmosphere. For these reasons, the rate constants for evaporation of $H_2^{16}O$ and $H_2^{18}O$ will be different. We can write an equivalent equation for the rate of evaporation of $H_2^{18}O$:

$$\frac{dN^*}{dt} = k_+^* - k_-^* P_v^* \qquad (5.90)$$

where we are using the asterisk to denote the variables for the minor isotope, ^{18}O in this case. Following from the derivation of Equation (5.80), the equilibrium fractionation factor is:

$$\alpha_{eq} = \frac{k_+ k_-^*}{k_- k_+^*} \qquad (5.91)$$

Consider first the case where both vapor pressures are 0 (i.e., zero humidity) and hence P_v is 0 and there is no condensation. The rate of change is:

$$\frac{dN^*}{dN} = \frac{k_+^*}{k_+} R_W \qquad (5.92)$$

where R_W is the isotope ratio of the water. The ratio dN^*/dN is the instantaneous isotope ratio of evaporating water and it differs from that of the water by the ratio of the rate constants. If we denote that ratio as R_{ev}, then the ratio R_W/R_{Ev} is the fractionation factor under conditions of 0 humidity, which we will designate α^0:

$$\alpha^0 = \frac{k_+^*}{k_+} \qquad (5.93)$$

Comparing Equations (5.91) and (5.93), we can see that the kinetic fractionation factor for dry air differs from the equilibrium one:

$$\alpha^0 = \alpha_{eq} \frac{k_-}{k_-^*} \qquad (5.94)$$

Of course, while evaporation into a perfectly dry atmosphere might be an appropriate approximation for Mars, it is not for the Earth where the atmosphere always contains some finite amount of water vapor. Defining humidity, h, as the ratio of the vapor pressure to the saturation vapor pressure (a ratio such that $h \leq 1$, not the normally reported percent), the isotope ratio of evaporating water is:

$$R_{ev} = \frac{R_w - \alpha_{eq} h R_v}{\alpha^0 (1 - h)} \quad (5.95)$$

Dividing both sides by R_w and rearranging, we have:

$$\frac{R_w}{R_{ev}} = \frac{\alpha^0 (1 - h)}{1 - \alpha_{eq} h R_v / R_w} = \alpha_{ev}^* \quad (5.96)$$

where α_{ev}^* is the non-equilibrium or kinetic fractionation factor. In the model of Craig and Gordon (1965), water evaporates into a microlayer at the ocean surface that is vapor saturated and in isotopic equilibrium with the water. This vapor must then diffuse through a combination of molecular and eddy diffusion into the free atmosphere and this diffusion is the dominant kinetic effect. Their equation is:

$$\alpha_{L-V}^* = \frac{\alpha_{diff} \alpha_{eq} (1 - h)}{1 - \alpha_{eq} h (R_A / R_W)} \quad (5.97a)$$

where α_{L-V}^* is the effective fractionation factor. We see that Equations (5.96) and (5.97a) are identical when $\alpha^0 = \alpha_{diff} \alpha_{eq}$

For the ocean, where $\delta^{18}O \approx \delta^2H \approx 0$, Equation (5.96) can be rewritten in delta notation as:

$$\alpha_{ev}^* \cong \frac{\alpha^0 (1 - h)}{1 - \alpha_{eq} h (1 + \delta_v / 1000)} \quad (5.97b)$$

where δ_v is the isotopic composition of the vapor (and Equation [5.97a] can be similarly rearranged). Solving for δ_v:

$$\delta_v \cong \left\{ \frac{1 - \alpha^0 (1 - h) / \alpha_{ev}^*}{\alpha_{eq} h} - 1 \right\} 1000 \quad (5.97c)$$

Based on the estimated average meteoric precipitation of $\delta^{18}O = -4.5$ and $\delta^2H = -26$ (which by mass balance must equal the average composition of water vapor evaporated from the ocean), Criss (1999) inferred that α_{ev}^* was 1.0045 and 1.0267 for O and H, respectively. α^0 can be determined experimentally to be 1.0260 and 1.094 for O and H, respectively. At $h = 1$, Equation 5.97b reduces to $(1/\alpha_{eq} - 1) \times 1000$, the equilibrium fractionation. It goes to infinity at $h = 0$, but in this case the fractionation is given by α^0.

Horita and Wesolowski (1994) experimentally determined the temperature dependence of the equilibrium fractionation between water vapor and liquid (v-l) to be:

$$103 \ln \alpha_{l-v}^{18O} = -7.685 + \frac{6.7123 \times 10^3}{T}$$
$$- \frac{1.6664 \times 10^6}{T^2} + \frac{3.5041 \times 10^8}{T^3}$$
$$(5.98)$$

$$103 \ln \alpha_{l-v}^{2H} = 1.1588 \times 10^{-9} T^3$$
$$- 1.6201 \times 10^{-3} T^2 + 0.79484T$$
$$- 161.04 + \frac{2.9992 \times 10^9}{T^3}$$
$$(5.99)$$

where T is in kelvins (and recall that $10^3 \alpha \cong \Delta$). This equation predicts that at 20°C evaporating water should have $\delta^{18}O$ and δ^2H lighter than the liquid by 9.7‰ and 81‰ per mil, respectively, while at 30°C, it should have $\delta^{18}O$ and δ^2H 8.9‰ and 71‰ lighter. The kinetic effects discussed above result in vapor being several per mil lighter than this, depending on humidity. We can use the inverse of these, i.e., $\alpha_{l/v}$ rather than $\alpha_{v/l}$ to compute the isotopic composition of water evaporating from the ocean at various temperatures and humidities. For example, at 20°C, $\alpha_{l/v}$ is 1.00978 and 1.0843 for $^{18}O/^{16}O$ and $^2H/^1H$, respectively. Computing $\delta^{18}O$ and δ^2H for various humidities, we find that the results do not lie on the GMWL but instead define a much shallower slope of ≈ 2.8 rather than 8. The line, however, does cross the GMWL at roughly 89% humidity. Repeating the process at other temperatures, we find lines of similar slope but offset to lower $\delta^{18}O$ and δ^2H with decreasing temperature and crossing the GMWL at humidities that decrease with temperature (Figure 5.8). Figure 5.8 also shows the predicted composition of water vapor produced under equilibrium conditions, which has a slope of ≈ 12 over the range of 0–30°C. Thus, equilibrium fractionation produces water vapor that is displaced from liquid water to lighter isotopic composition along a slope of ≈ 12; kinetic fractionation then shifts this vapor to the left by amounts that depend on humidity along shallower slopes. The GMWL then results from the process occurring at variable temperature.

As water evaporates from the ocean surface, the remaining water becomes

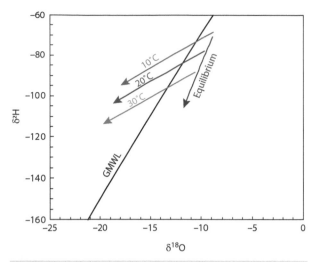

Figure 5.8 Calculated composition of water evaporated from sea water ($\delta^2H = \delta^{18}O = 0$) at varying humidities and temperatures based on Equation 5.97b. The vapor shifts further to the left with decreasing humidity. Also shown is the trajectory of equilibrium fractionation from sea water with decreasing temperature. We see then that the GMWL is not a result of equilibrium evaporation, but is a result of kinetic fractionation under varying conditions of temperature and humidity.

isotopically heavier, which in turn shifts the isotopic composition of water vapor to heavier values. This effect is relatively small for the ocean, but can be large for smaller bodies of water undergoing extensive evaporation, such as saline lakes.

Cappa et al. (2003) identified an additional kinetic effect in water evaporation, namely, that evaporation consumes thermal energy of the water, and this necessarily results in cooling of the surface. The extent of cooling will depend on evaporation rates and is limited to the top few millimeters or less of the water, but it is this thin surface layer where evaporation is occurring and will result in a shift of both kinetic and equilibrium fractionation factors toward greater fractionation.

5.4.4 Rayleigh fractionation

5.4.4.1 *Rayleigh fractionation and the meteoric water line*

Further fractionations during condensation of water vapor produce much larger variations in $\delta^{18}O$ and δD in global precipitation but

nevertheless along the MWL. Snow and rain condensing from water vapor will be isotopically heavier than the vapor and as condensation in clouds occurs at much lower temperatures than evaporation from the ocean surface, the fractionations will be larger than during evaporation. Once droplets grow large enough to form rain droplets and fall, they can no longer equilibrate with the vapor from which they formed. As ^{18}O and ^2H are preferentially removed from the vapor by condensation, the remaining vapor becomes progressively lighter according to:

$$\delta = 1000\left(f^{\alpha-1} - 1\right) \qquad (5.100)$$

where δ is the isotopic composition of remaining vapor, f is the fraction of vapor remaining, and α is the water–vapor fractionation factor. Such a process is an example of *Rayleigh fractionation*, and we will encounter other examples of it, including fractional crystallization of magma, which we will discuss in the following section. Assuming a value of α of 1.01 and an initial $\delta_0 = 0\permil$, δ will vary with f, the fraction of vapor remaining, as shown in Figure 5.9. This equation is a general one for any Rayleigh process, including the reverse process, Rayleigh evaporation, which might occur, e.g., of water evaporating from a small lake or pond (ignoring the humidity-related kinetic effects discussed in the previous section). In that case we would use the inverse of α, the vapor–water fractionation factor.

Even if the vapor and liquid remain in equilibrium throughout the condensation process, the isotopic composition of the remaining vapor will change continuously. The relevant equation is:

$$\delta = \left(1 - \frac{1}{(1-f)/\alpha + f}\right) \times 1000 \qquad (5.101)$$

The effect of equilibrium condensation is also shown in Figure 5.9.

5.4.4.2 *Rayleigh fractionation in magmatic processes*

As we found earlier, the equilibrium constant of isotope exchange reactions, K, is proportional to the inverse square temperature and consequently isotopic fractionation at

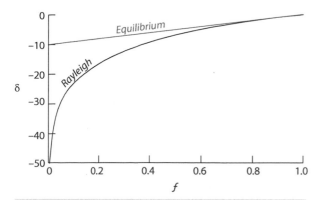

Figure 5.9 Fractionation of isotope ratios during Rayleigh and equilibrium condensation. δ is the per mil difference between the isotopic composition of original vapor and the isotopic composition as a function of f, the fraction of vapor remaining. A value of $\alpha = 1.01$ is assumed for both.

high temperature is small. In magmatic systems, another factor limiting oxygen isotope fractionation is the limited variety of bonds that O is likely to form. In both silicate minerals and in silicate liquids oxygen bonded to silicon atoms to form silica tetrahedra, consequently the fractionation of oxygen isotopes between silicate magmas and crystallizing silicate minerals from those liquids is rather limited. There is somewhat greater fractionation when non-silicates, such as magnetite (Fe_3O_4), crystallize. Zhao and Zheng (2003) calculated the temperature dependence coefficients, A, of mineral–melt fractionation factors, α, according to:

$$\Delta_{\varphi-melt} \cong 10^3 \ln \alpha = \frac{A \times 10^6}{T^2} \quad (5.102)$$

These coefficients are listed in Table 5.3. A fundamental tenant of thermodynamics

is that two phases each independently at equilibrium with a third phase are also in equilibrium with each other. As a consequence, the data in Table 5.3 can be used to calculate the fractionation between any two phases listed. For example, the fractionation between magnetite and quartz is:

$$\Delta_{Qz-Mt} \cong \frac{A_{Qz} \times 10^6}{T^2} - \frac{A_{Mt} \times 10^6}{T^2} \quad (5.102a)$$

Figure 5.10 shows the fractionation factors calculated from Table 5.3 as a function of temperature. In general, crystallization of quartz will lead to a depletion of ^{18}O in the melt, crystallization of silicates, such as olivine, pyroxene, hornblende, and biotite, will lead to slight enrichment of the melt in ^{18}O, and crystallization of oxides, such as magnetite and ilmenite, will lead to a more pronounced enrichment of the melt in ^{18}O. However, oxides, such as magnetite, are generally only present at the level of a few percent in igneous rocks, which limits their effect. Crystallization of feldspars can lead to either enrichment or depletion of ^{18}O, depending on the temperature and the composition of the feldspar and the melt.

The variation in O isotope composition produced by crystallization of magma will depend on the how crystallization proceeds. The simplest, and most unlikely, case is *equilibrium* crystallization. In this situation, the crystallizing minerals remain in isotopic equilibrium with the melt until crystallization is complete. At any stage during crystallization, the isotopic composition of a mineral and the melt will be related by the fractionation factor, α. Upon complete crystallization, the rock will have precisely the same isotopic composition as the melt initially had. At any time during the

Table 5.3 Coefficients for phenocryst–lava fractionation factors, A, where $10^3 \ln \alpha = A \times 10^6/T^2$.

Rock	Q	Or	Ab	An	Di	Hy	Ol	Mt	Il	Ap
Rhyolite	0.77	−0.32	−0.24	−1.26				−4.67	−4.99	−1
Andesite		0.37	0.46	−0.57	−1.3	−1.11		−3.97	−4.29	−0.31
Basalt		0.93	1.02	−0.01	−0.74	−0.55	−1.77	−3.41	−3.73	0.25

Q, quartz; Or, orthoclase; Ab, albite; An, anorthite; Di, diopside; Hy, hypersthene; Ol, olivine; Mt, magnetite; Il, ilmenite; Ap, apatite.
Source: Data from Zhao and Zheng (2003).

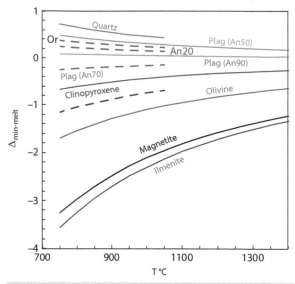

Figure 5.10 Mineral–melt fractionation factors, $\Delta = 10^3 \ln \alpha$ as a function of temperature calculated from Table 5.3.

crystallization, the isotope ratio in the remaining melt will be related to the original isotope ratio as:

$$\frac{R_\ell}{R_0} = \frac{1}{f + \alpha(1-f)}; \qquad \alpha = \frac{R_s}{R_\ell} \qquad (5.103)$$

where R_ℓ is the ratio in the liquid, R_s is the isotope ratio of the solid, R_0 is the isotope ratio of the original magma, and f is the fraction of melt remaining. This equation is readily derived from mass balance, the definition of α, and the assumption that the O concentration in the magma is equal to that in the crystals – an assumption valid to about 10%. Since we generally do not work with absolute ratios of stable isotopes, it is more convenient to express Equation (5.103), in terms of δ:

$$\Delta = \delta_{melt} - \delta_0 \cong \left[\frac{1}{f + \alpha(1-f)} - 1 \right] \times 1000$$

$$(5.104)$$

where δ_{melt} is the value of the magma after a fraction $f-1$ has crystallized and δ_0 is the value of the original magma. For silicates, α is not likely to be much less than 0.998 (i.e., $\Delta = \delta^{18}O_{melt} - \delta^{18}O_{xtals} \leq 2$). For $\alpha = 0.999$, even after 99% crystallization, the isotope ratio in the remaining melt will change by only 1 per mil.

Fractional crystallization is a process analogous to Rayleigh condensation/distillation.

Indeed, it is governed by the same equation (Equation [(5.100)]), which we can rewrite as:

$$\Delta = 1000 \left(f^{\alpha-1} - 1 \right) \qquad (5.105)$$

Typically, several minerals will be crystallizing from a magma simultaneously, in which case α in this equation is the weighed mean (by mole fraction of the element of interest, O in this case) of the individual fractionation factors.

The key to the operation of Rayleigh processes is that the product of the reaction (vapor in the case of distillation and crystals in the case of crystallization) is only instantaneously in equilibrium with the original phase. Once it is produced, it is removed from further opportunity to equilibrate with the original phase. This process is more efficient at producing isotopic variations in igneous rocks than equilibrium crystallization, but its effect remains limited because α is generally not greatly different from 1.

Figure 5.11 shows $\delta^{18}O$ as a function of f calculated using Equation (5.105) for three cases assuming an initial $\delta^{18}O = 5.6$‰. The first is a "basalt" crystallizing 10% olivine, and 45% each of plagioclase (An90–Ab10) and clinopyroxene (80Di–20Hy) at 1200°C. The second is for an "andesite" crystallizing 45% each of An20 plagioclase and clinopyroxene (Di50), 8% magnetite, and 2% alkali feldspar at 900°C. The third is for a composition that evolves from these cases and to a quartz-precipitating rhyolite at 750°C with changing mineral precipitation and temperature. Interestingly, for much of the sequence, the opposing effects of crystallizing magnetite and plagioclase result in nearly no net fractionation.

We can generalize by stating that stable isotopes of magmas will be only minimally affected by fractional crystallization, reflecting both the high temperatures and the limited difference in sites between minerals and melts occupied by atoms. This generalization leads to an axiom stated by Taylor and Sheppard(1986): "igneous rocks whose oxygen isotopic compositions show significant variations from the primordial value (6) must either have been affected by low temperature processes or must contain a component that was at one time at the surface of the Earth." While we have focused exclusively on oxygen here, this holds for the stable isotopes of other

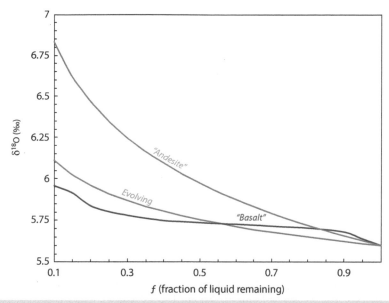

Figure 5.11 Evolution of $\delta^{18}O$ in a magma with an initial value of 5.6‰ undergoing fractional crystallization. The lines labelled andesite and basalt use constant weighted fractionation factors at 1200°C and 900°C as described in the text while the "evolving" line accounts approximately for changing temperatures and minerals crystallizing.

elements as well. When we find evidence of the latter in igneous rocks, the question we will want to ask whether that "component that was at one time at the surface of the Earth" was inherited from the mantle source, acquired as the magma transited the crust, or reflects subsequent reactions at the Earth's surface. We will consider the case where magmas assimilate surrounding country rock in Chapter 9.

5.5 MASS-INDEPENDENT VERSUS MASS-DEPENDENT FRACTIONATION

5.5.1 Mass dependence of equilibrium and kinetic fractionations

Most isotopic studies have focused on the fractionation between the two most abundant isotopes of an element, e.g., ^{16}O and ^{18}O. Some elements, however, have three or more isotopes. For example, O consists of ^{17}O, as well as ^{16}O and ^{18}O. However, ^{17}O is an order of magnitude less abundant than ^{18}O (which is two orders of magnitude less abundant than ^{16}O), so useful data about ^{17}O required higher sensitivity than was possible a few decades ago. With the advancement of analytical instrumentation and techniques, analysis of $^{17}O/^{16}O$ as

well as the minor isotopes ratios of S (and metals as well) has become common place.

Based on the theory that we have just reviewed, mass fractionation should depend on mass difference. This is referred to as *mass-dependent fractionation*. The mass difference between ^{17}O and ^{16}O is half the difference between ^{18}O and ^{16}O; hence, we expect the fractionation between ^{17}O and ^{16}O to be about half that between ^{18}O and ^{16}O. In the example of fractionation between CO and O_2 in the previous section, it can be shown from Equation (5.60) that through the range of temperatures we expect near the surface of the Earth (or Mars), the ratio of fractionation factors $\Delta^{17}O/\Delta^{18}O$ should be ≈0.53 (doing so is left to you as Problem 5.3). In the limit of infinite temperature (and using exact atomic masses rather than mass numbers):

$$\Delta^{17}O/\Delta^{18}O \approx \frac{1/m_{16_O} - 1/m_{17_O}}{1/m_{16_O} - 1/m_{18_O}} = 0.5305$$

(5.106)

The empirically observed ratio for oxygen in most terrestrial materials (and also within classes of meteorites) is close to this value with $\Delta^{17}O/\Delta^{18}O \approx 0.528$. This is the value of the

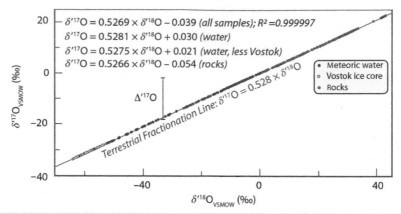

Figure 5.12 Nearly all terrestrial materials have $\delta^{17}O$ and $\delta^{18}O$ that plot along a line known as the TFL with a slope of 0.528. In detail, different processes lead to mass-dependent fractionations with slightly differing slopes such that deviations from the line of a few tens of ppm are ubiquitous. The $\Delta'^{17}O$ parameter is the amount by which $\delta^{17}O$ falls off the TFL for a given $\delta^{18}O$. Much larger deviations occur in the atmosphere and meteorites as a result of mass-independent fractionations (MIFs) induced by photochemical effects. Source: Adapted from Sharp et al. (2018).

slope of data plotted on a $\delta^{17}O$ versus $\delta^{18}O$ diagram; we refer to this as *terrestrial fractionation line* (TFL) illustrated in Figure 5.12. A similar equation can be written for other elements, such as sulfur, where three or more isotopes can be analyzed. More broadly for any two isotope ratios of element, we define the ratio of fractionation factors and hence slope on a two-isotope ratio diagram as λ *when the slope is empirically determined or* θ *when it is theoretically calculated.* In our example of the CO–O$_2$ fractionation, Equation (5.60) provides the value of θ = 0.5305.

The value of λ = 0.528 holds only in approximation as O isotope ratios of many substances show small variations from the TFL. Indeed, we expect different processes should lead to slightly different slope. For equilibrium fractionations, we can write a general form of Equation (5.106) as:

$$\theta = \frac{\ln \alpha_{A-B}^{2/1}}{\ln \alpha_{A-B}^{3/1}} = \frac{1/m_1 - 1/m_2}{1/m_1 - 1/m_3} \qquad (5.107)$$

where numerical subscripts denote isotopes (e.g., 1 is ^{16}O and 2/1 denotes the $^{17}O/^{16}O$ ratio). From this, we can readily derive:

$$\alpha_{A-B}^{1/2} = \left(\alpha_{A-B}^{3/1}\right)^{\theta} \qquad (5.108)$$

Young et al. (2002) showed that kinetic fractionations can lead to different values of θ. For example, for some kinetically controlled reactions, the expected relationship is:

$$\theta = \frac{\ln \alpha_{A-B}^{2/1}}{\ln \alpha_{A-B}^{3/1}} = \frac{\ln \left(m_1^*/m_2^*\right)}{\ln \left(m_1^*/m_3^*\right)} \qquad (5.109)$$

where the m^* values in Equation (5.109) could refer to the reduced masses of vibrating diatomic molecules containing the isotope rather than atomic mass. In the case where fractionation arises from different molecular velocities, e.g., diffusion, the masses in Equation (5.109) are simply the molecular masses. In many cases, observed fractionations are a consequence of a combination of kinetic and equilibrium effects. Quite often, the kinetic contribution result in value of a λ less than the expected equilibrium one, including evaporation of water, photosynthesis, and respiration. Fractionation during evaporation of water, Equation 5.97a, provides an example: for equilibrium $^{17}O/^{16}O$–$^{18}O/^{16}O$ fractionation, θ is 0.529 but is 0.5185 for molecular diffusion. In another example, the value of θ for CO$_2$–H$_2$O equilibrium exchange has been found experimentally to be 0.5229, but $\delta^{17}O$ and $\delta^{18}O$ in atmospheric CO$_2$ define a λ of 0.516. More complex forms of 5.107 are also possible. Thus, the 0.528 slope of the TFL is simply the average of θ values for a variety of processes. Young et al. (2002) and Dauphas and Schauble (2016) discuss this in greater detail.

The amount by which $\delta^{17}O$ deviates from an expected fractionation line for a given $\delta^{18}O$ is quantified using an upper-case delta notation to denote the deviation from the expected mass-dependent fractionation, for oxygen, e.g.:

$$\Delta^{17}O = \delta^{17}O - \lambda\delta^{18}O \qquad (5.110)$$

$\Delta^{17}O$ can have two meanings. In the first case, it is simply the deviation from a reference line of slope λ, such as the TFL with $\lambda = 0.528$. In the second case, it is the intercept of an observed or predicted fractionation line. For example, most meteoric water falls along a $\delta^{17}O_{VSMOW}$–$\delta^{18}O_{VSMOW}$ line with a slope quite close to the TFL value of 0.528, but with an intercept at $\delta^{18}O_{VSMOW} = 0$ of $\delta^{17}O_{VSMOW} = 0.033‰$. This is principally a consequence of kinetic effects in evaporation of water. Similarly, the combined effects of photosynthesis and respiration result in O_2 defining a $\delta^{17}O_{VSMOW}$–$\delta^{18}O_{VSMOW}$ line with a slope of 0.518 and an intercept of $\Delta^{17}O = 0.26‰$. *Because $\Delta^{17}O$ can be used in these different ways, it is important to always define the value of λ used to compute $\Delta^{17}O$*. Mathematically we can convert $\Delta^{17}O$ defined by one slope to another by multiplying by the ratio of slopes. Bao et al. (2016) provide a far more detailed discussion of $\Delta^{17}O$, its computation, and its significance.

Note that Equations (5.107) and (5.109) imply that when fractionated samples are plotted on isotope ratio–isotope ratio plot, e.g., $^{33}S/^{32}S$ versus $^{34}S/^{32}S$, the data will define a curve with:

$$^{2/1}R = {}^{3/1}R^\theta\left({}^{2/1}R_{ref}/{}^{3/1}R^\theta\right) \qquad (5.111)$$

In δ notation, the equation becomes:

$$\delta^{2/1} = \left(10^3 + \delta^{2/1}_{ref}\right)\left(\frac{10^3 + \delta^{3/1}}{10^3 + \delta^{3/1}_{ref}}\right)^\theta - 10^3$$
$$(5.112)$$

(Young et al., 2002). For a small range of δ, however, the curvature may be too slight to notice.

Most deviations from the O isotope TFL or comparable correlations for isotopes of other elements, such as S, are small and only become apparent with high-precision analysis. One consequence of high-precision analyses is that inaccuracies due to approximations become important, e.g., in the relationship between fractionation factors α and Δ in Equation (5.4) ($\delta_A - \delta_B = \Delta_{A-B} \approx 10^3 \ln \alpha_{A/B}$). This has led to the introduction of δ' notation:

$$\delta' = 10^3 \ln\left(R_{sam}/R_{Std}\right) \qquad (5.113)$$

and therefore

$$\delta' = 10^3 \ln\left(1 + \delta/10^3\right) \qquad (5.114)$$

The difference between δ and δ' is generally quite small, of the order of a percent (or 10 ppm or so relative to absolute ratios), and is not important in conventional studies but is important in triple-isotope studies where ppm differences matter. In addition, some fractionations have a small, constant offset from the reference line. These factors lead to a more precise definition of MIF:

$$\Delta'^{17}O = \delta'^{17}O - \lambda\delta'^{18}O - \gamma' \qquad (5.115)$$

The γ' term is an intercept and accounts for these constant offsets. This is apparent in Figure 5.12, where different sets of materials have λ varying from 0.5266 to 0.5281 and γ' varying from −0.54 to +0.30‰. $\Delta'^{17}O$ is also sometimes referred to as ^{17}O *excess*. It is worth pointing out that interlaboratory calibration issues have not been entirely resolved at the level of 10 ppm or so, leading to slight uncertainties in the values of λ and γ in many cases.

As instrumentation has improved enabling higher-precision analysis, triple oxygen isotope measurements have begun to provide new insights into geochemical processes; the book edited by Bindeman, I. and Pack, A. (2021) provides an overview of these advances and we will explore them when we return to these topics in subsequent chapters.

5.5.2 Mass-dependence of fractionations in biological processes

Farquhar et al. (2003) found that microbial dissimilatory sulfate reduction (DSR) can lead to $\delta^{33}S$–$\delta^{34}S$ relationships whose slope, λ, differs from the widely observed value of 0.515. DSR is so-called because in contrast to *assimilatory sulfate reduction* where sulfur is reduced and incorporated in biomolecules, such as the amino acid cystine ((SCH_2CH (NH_2)CO_2H)$_2$), sulfate is merely used as a metabolic electron receptor in anaerobic organisms. The DSR process is a chain reaction involving several branches in which the fractionation differs and the final product is a mixture of reduced sulfur produced along the different pathways. The combination of chain and branched gives rise to an average λ of 0.5117, as demonstrated by experimental studies of Farquhar et al. (2003) with the sulfur-reducing archaea *Archaeglobus fulgidus*.

The results of the Farquhar et al. (2003) experiments are shown in Figure 5.13 using this notation where $\Delta'^{33}S$ is defined as:

$$\Delta'^{33}S = \delta^{33}S - 0.515\delta^{34}S \qquad (5.116)$$

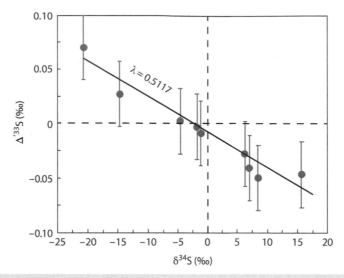

Figure 5.13 Sulfur isotope fractionation in DSR experiments using the archaea *Archaeglobus fulgidus* results in apparent MIF. $\Delta'^{33}S$ is the deviation from the canonical fractionation of $\lambda = 0.515$ as defined in Equation (7.101c). Data from Farquhar et al. (2003).

The deviation from $\lambda = 0.515$, although small, provides a fingerprint of microbial sulfate reduction. Ono et al. (2012) documented $\Delta'^{33}S$ in basalts from the Juan de Fuca Ridge ranging from –0.06 to +0.04‰ and peridotites from the Hess Deep in the Pacific near the Galapagos Spreading Center – East Pacific Rise triple junction and the Iberian margin ranging from 0.00 to 0.16‰ with $\delta^{34}S$ values ranging from 0 to –44‰, providing evidence that sulfate reduction of the oceanic crust is biologically mediated. These $\Delta'^{33}S$ values are considerably smaller than those resulting from other mechanisms, which we now turn to.

5.5.3 Mass-independent fractionations

Most deviations from the TFL, i.e., $\Delta^{17}O \neq 0$, arise because the mass dependence of the fractionation factors varies with the process involved as we have just discussed. These deviations are typically quite small. In other cases, isotope ratios show large deviations from expected mass-dependent fractionations. MIF, where isotopic fractionations of multiple isotopes are independent of mass difference, has been demonstrated to occur in both laboratory experiments and nature, and indeed may provide important clues to Earth and Solar System processes and history. First observed in meteorites as we will find in the following chapter, it has subsequently been observed in oxygen isotope ratios of atmospheric gases, most dramatically in stratospheric ozone, and a variety of elements in other materials.

5.5.3.1 Photochemical-produced fractionations

Clayton et al. (1973) found that a class of primitive meteorites, carbonaceous chondrites, and components with in them, most notably calcium–aluminum inclusions (CAIs), has O isotopes that fall on a slope of ≈ 1 on a plot of $\delta^{17}O$ versus $\delta^{18}O$. At least some CAIs condensed from high-temperature gas and they have been shown to be the earliest formed solids in the Solar System. Their isotopic compositions extend to very low $\delta^{17}O$ and $\delta^{18}O$ values. Clayton et al. (1973) interpreted as addition of pure ^{16}O derived from interstellar dust with a different nucleosynthetic history.

Ten years after Clayton's discovery, Thiemens and Heidenreich (1983) reproduced the 1 : 1 correlation between $\delta^{17}O$ and $\delta^{18}O$ in laboratory experiments in which they produced ozone from O_2 by electrical discharge. Their results are shown in Figure 5.14. This opened the possibility of oxygen isotope variations of chemical origin rather than nucleosynthetic. Mauersberger et al. (1987) found strong enrichments of stratospheric ozone in ^{17}O and ^{18}O, demonstrating that MIF was occurring in nature, as well as in the laboratory.

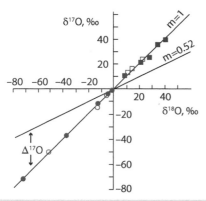

Figure 5.14 Variation in oxygen isotope ratios in experiments in which O_3 was produced from O_2 by electrical discharge relative to the starting O_2 ($\delta^{17}O = \delta^{18}O = 0$). Blue squares are ozone; red circles are O_2 at the end of the experiments. Solid symbols are measurements done at UC San Diego; open symbols are measurements done at the University of Chicago. Experimental results fall along a line with a slope of 1 compared with a slope of 0.528 expected for mass-dependent fractionation. The per mil deviation from the mass-dependent fractionation line is denoted by $\Delta^{17}O$. Source: Adapted from Thiemens and Heidenreich (1983).

Thiemens and Heidenreich interpreted this MIF as a result of self-shielding. According to this idea, O_2 absorbed ultraviolet (UV) light produced by the discharge, resulting in dissociation of O_2, producing monatomic O available for reaction to produce O_3. As we found earlier in this chapter, different isotopologues of molecules have slightly different vibrational frequencies and hence absorb slightly light at slightly different frequencies. Because $^{16}O^{16}O$ is several orders of magnitude more abundant than $^{18}O^{16}O$ and $^{17}O^{16}O$, the light of the frequency necessary to dissociate $^{16}O^{16}O$ was quickly absorbed, so that $^{18}O^{16}O$ and $^{17}O^{16}O$ molecules were preferentially dissociated to form ozone. A similar hypothesis is now the consensus explanation for the variations observed in meteorites with the forming Sun as the possible source of UV radiation and may provide an explanation for other natural examples of MIF, which we will discuss in subsequent chapters. However, subsequent experiments and theoretical analysis showed that self-shielding cannot explain MIF Thiemens and Heidenreich observed in their experiments.

Heidenreich and Thiemens (1986) then proposed a different mechanism for MIF during ozone formation. Their theory can be roughly explained as follows. Formation of ozone involves the energetic collision of monatomic and molecular oxygen, i.e.:

$$O + O_2 \rightarrow O_3^*$$

The ozone molecule thus formed is in a vibrationally excited state (designated by the asterisk) and, consequently, subject to dissociation if it cannot lose this excess energy. The excess vibrational energy can be lost either by collisions with other molecules or by partitioning to rotational energy. At high pressures, such as those that prevail in the troposphere, ozone molecules can readily lose energy through collisions, lessening the importance of the vibrational to rotational energy conversion. At lower pressures, such as those prevailing in the stratosphere or in the solar nebula (with estimated pressures of 10^{-4} atm), collisions are comparatively infrequent and repartitioning of vibrational to rotational energy becomes a more important pathway to stability. Symmetric ozone, $^{16}O_3$, molecules have few available rotational states than asymmetric, e.g., $^{18}O^{16}O^{16}O$, ones and hence the latter can more readily repartition excess vibrational energy. Thus O_{3*} containing either ^{18}O or ^{17}O is more likely to transition to the stable state before dissociating.

While Heidenreich and Thiemens' explanation did not hold up in detail, subsequent studies have refined this idea and confirmed the role of symmetry. Gao and Marcus (2001) proposed a similar solution, pointing out that there are fewer dynamical coupling terms between vibrational and rotational motions in the symmetric molecules than in the asymmetric ones and hence the density of quantum states available in the transition to stable O_3 is expected to be greater for asymmetric molecules than for symmetric ones. They were able to closely match observed experimental fractionations, but their approach was in part empirical rather than theoretical. The theoretical basis remains incompletely understood, but a consensus exists that UV photodissociation and the role of symmetry stabilizing excited molecules is an important mechanism of MIF of stratospheric ozone. As we will find in Chapter 10, this MIF signature can then be transferred to other O-bearing molecules, including NO_3 and CO_2, through subsequent reactions.

UV photolysis has also been shown to produce MIF of S isotopes in SO_2 (Farquhar et al., 2001). Exposed to UV light, SO_2 breaks down to produce oxidized and reduced sulfur phases that show MIF. The nature of these phases and the isotope fractionation depends on wavelength. Of particular interest are experiments with wavelengths less than 190 nm, where SO_2 can be dissociated through reactions:

$$SO_2 \overset{uv}{\rightarrow} SO + O \qquad (5.117)$$

$$SO \overset{uv}{\rightarrow} S + O \qquad (5.118)$$

Oxygen liberated in these reactions can then recombine with some of the sulfur to form sulfates. Farquhar et al. found that this produced elemental sulfur depleted in ^{34}S and enriched in ^{33}S, leaving residual SO_2 enriched in ^{34}S and depleted in ^{33}S. These experimental results match sulfur isotopic compositions observed in Archean sedimentary rocks, with pyrites having positive $\Delta'^{33}S$ and barites ($BaSO_4$) and hydrothermal sulfides (formed by reduction of dissolved SO_4) having negative $\Delta'^{33}S$ (Farquhar et al., 2000). This is of great significant in understanding the evolution of oxygen in the Earth's atmosphere, a topic we will return to in subsequent chapters. SO_2 exposed to a wavelength of 248 nm also experiences MIF but with a different pattern. In this case, there was positive correlation between $\delta^{33}S$ and $\delta^{34}S$ with a slope, λ, of ≈ 0.65. This in turn matched sulfur isotopic compositions of sulfuric acid aerosols lofted into the stratosphere by large volcanic eruptions, such as Mount Pinatubo (Savarino et al., 2003).

5.5.3.2 Nuclear magnetic and volume effects

As stable isotope geochemistry has expanded from a few light elements to the heaviest ones, it became apparent that fractionations were much larger in some of these elements, including Hg, Tl, and U, than predicted from the mass-related effects we have reviewed so far. As hinted at in the beginning of this chapter, the magnetic moment (spin), shape, and volume of the nucleus can also influence chemical behavior and can lead to MIFs. The first of these is referred to as the *magnetic isotope effect* and the latter two are collectively known as *nuclear field shift effects*. For the most part, these effects are much smaller, and usually negligible, in light elements, such as O. They are more important in heavy elements partly because the relative mass difference between isotopes decreases and partly because nuclear volume increases with increasing atomic number. For the heaviest elements, such as Hg, Tl, and U, these effects can be more important than the mass effect on vibrational frequencies.

Let us begin with the magnetic isotope effect, which is fundamentally a kinetic one. As we found in Chapter 1, nucleons have "spin" or magnetic moment, which can be either positive or negative. Nucleons are paired in the nucleus such that spins of pairs of protons and neutrons cancel, so even–even nuclei have 0 spin and 0 magnetic moment while odd–odd nuclei have integral spin and even–odd nuclei have half integral spin and nuclear magnetic moments. Although small, nuclear spin contributes to the total spin and magnetic moment of the atom, which is dominated by electron spin. Quantum spin must be conserved in chemical reactions such that the total spin of products is equal to that of reactants. Consequently, some reaction paths are forbidden, which requires spin conversion before reaction (which can be achieved in various ways). Specifically, in reactions involving a pair of excited radical intermediates that undergo spin conversion from the "triplet" state (total spin angular momentum of 1) to "singlet state" (total spin angular momentum of 0), the reaction will occur more rapidly for pairs with magnetic nuclei than pairs with nonmagnetic (zero spin) nuclei. The effect has been observed in the laboratory, e.g., in carbon isotopes where it has been shown that relative reaction rates of ^{12}C- and ^{13}C-containing reactants are sensitive to magnetic field intensity (Buchachenko, 2001). In nature, the magnetic isotope effect is most apparent in Hg isotopes. Mercury has an even number of protons, so that even numbered isotopes, such as ^{198}Hg, have 0 nuclear spin while odd numbered ones have half integral spin. This result is a pattern of odd–even isotopic fractionations, most particularly in photochemical redox reactions as it cycles between Hg^0 and Hg^{2+} and organic and inorganic forms. These dominate the terrestrial mercury cycle at the Earth's surface and produce a variety of Hg isotopic compositions at the Earth's surface (e.g., Blum et al., 2014). We will examine this in detail in Chapter 11.

Nuclear field shift effects are quantum effects that result from the overlap of the electronic and nuclear wave functions (i.e., the quantum probability density functions that describe the probability of finding the particle at a particular position) and produce displacement of the ground state electronic energy of an atom or molecule due to the differences in nuclear size and shape and distribution of charge within the nucleus of the isotopes of an element (King, 1984; Bigeleisen, 1996a). Information about the volume and shape of nuclei can be obtained by electron-scattering experiments and from the quite small frequency/wavelength shifts (typically 2.5 pm at optical wavelengths of 500 nm) between isotopes in optical and X-ray emission spectra (which is the origin of the term "shift").

As a consequence of the nuclear shell structure, nuclear volume is a discontinuous function of neutron number; volume decreases to a minimum at magic neutron numbers (28, 50, 82, 128) and increases sharply when additional neutrons are added. Volume also depends on whether protons and neutrons are paired (that is, there are even–odd effects). The shape of the nucleus can also vary and contributes to overlap of wave functions; it is often spherical but can take on other shapes, such as oblate spheroid in neutron-rich ones, or more exotic shapes, such as that of a pear in unstable nuclei. In samarium, for example, deformation of the nucleus into an oblate spheroid produces a greater field shift between isotopes than do volume differences.

Assuming nucleus is simply a charged point of infinitesimal volume, the coulomb potential acting on an electron varies with radial distance as $-Ze^2/r$ where Z is the proton number, e is the charge of the electron, and r is radial distance. However, the nucleus occupies a finite volume, so the potential does not go to negative infinity at $r = 0$ but instead reaches a finite minimum, which increases with nuclear radius as illustrated in Figure 5.15. The s-orbital electrons have large probability densities at the nucleus while the p-, d-, and f-orbital electrons have an insignificant probability density at the nucleus (arising from spin polarization). The primary effect is then that s-orbital electrons are bound less tightly to a large nucleus than to a small one, while the screening of p-, d-, and f-orbital electrons by s-electrons and vice versa results in the former being stabilized around a large nucleus.

Energy will be minimized where electrons are more strongly bound to the nucleus. The field shift energy difference, δE^{FS}, between phases A and B can be approximately expressed as:

$$\delta E_{A-B}^{FS} \cong \frac{2\pi Z e^2}{3}\left(|\Psi(0)_A|^2 - |\Psi(0)_B|^2\right)\Delta r^2$$

(5.119)

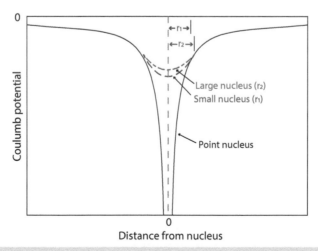

Figure 5.15   Cartoon illustrating the effect of nuclear radius on the coulomb potential on an electron. Electrons with a finite probability density at the nucleus will be more strongly bound to a small nucleus. Source: Schauble (2007)/with permission of Elsevier.

where Z is the atomic number, e is the electron charge, the $|\Psi(0)|^2$ terms are the electron densities at the center of the nucleus in phases A and B, and Δr^2 is the difference in root-mean-square radii of the nuclei in phases A and B (root-mean-square values are taken because the nuclear volume oscillates). The reduced partition function for fractionation between A and B is then:

$$\ln\left(\beta_{A-B}^{FS}\right) \cong \frac{\delta E_{A-B}^{FS}}{kT} \qquad (5.120)$$

Energy is minimized when isotopes with the larger nucleus partition into the phase where it has lower electron density at the nucleus (i.e., fewer s-electrons or more p-, d-, and f-electrons), and isotopes with the smaller nucleus partition into phases with high electron density at the nucleus (more s-electrons). The largest shifts are found when the number of valence s-electrons decreases in the transition. Unlike the magnetic isotope effect, the nuclear field shift effect can result in both equilibrium and kinetic fractionations.

The overlap of wave functions and consequent nuclear field shift effects increase with the square of nuclear radius, which in turn increases with atomic number. Hence field shift isotope fractionation is most significantly greatest for heavy elements, decreasing from ≈1‰ for Tl, to ≈0.2‰ for Ru and ≈0.02‰

for S per unit of mass difference. In contrast, fractionations arising from the mass effect on electronic vibrations decrease approximately with the square of mass and hence are least significant for heavy elements.

As mentioned above, nuclear volume depends on neutron pairing so that nuclear field shifts can produce odd–even mass-independent isotope fractionations because nuclear radius and volume are not smooth functions of mass or neutron number. As Figure 5.16 shows, isotopes with an odd number of neutrons tend to have radii less than expected if radius increased linearly with neutron number. Consequently, s-orbital electrons in different isotopes of an element will have slightly different energy levels, particularly between even and odd numbered isotopes. This effect is most dramatically seen in mercury, which has seven isotopes, of which four are with even mass numbers and three with odd ones.

For heavy elements, the nuclear field shift effect can be larger than the effect of vibrational frequency mass effect (Section 5.3). Figure 5.17 compares the ab initio calculated nuclear volume fractionation factor with the calculated mass-dependent fractionation for reduction of aqueous Tl^{3+} to Tl^+. The nuclear volume effect is roughly 2‰ at 25°C and exceeds the mass effect at all temperatures.

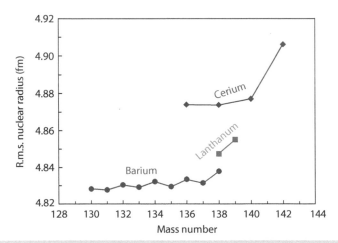

Figure 5.16 Root-mean-square nuclear radii of Ba, La, and Ce (atomic numbers 56 through 58) as a function of mass number. Odd mass numbered isotopes have smaller radii (and smaller volumes) than expected if radius increased linearly with the number of neutrons in the nucleus. This can lead to odd–even effects in isotope fractionation factors. Adding protons to the nucleus increases volume more than adding neutrons due to the repulsive interaction between them. Data from Angeli and Marinova (2013).

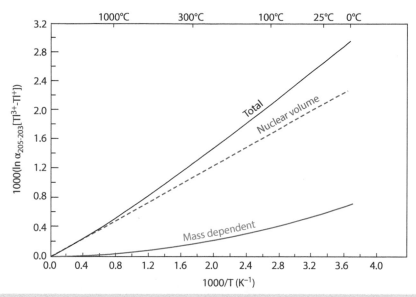

Figure 5.17 Calculated nuclear volume and mass-dependent fractionation factors as a function of temperature for reduction of aqueous Tl^{3+} to Tl^{+}. Source: Schauble (2007)/with permission of Elsevier.

The nuclear field shift effect can result in fractionations opposite those due to mass-dependent fractionation: the field shift leads to the heavy isotope partitioning into the chemical species with the smallest number of s electrons in the bonding or valence orbitals while the nuclear field shift effect favors the heavy isotope being most strongly bound. Bigeleisen (1996a) found that the electron density at the nucleus of the hydrated U^{6+} ion, UO_2^{2+} (uranyl), is larger than for the hydrated or complexed U^{4+} ion. Consequently, the field shift leads to a preference of ^{238}U for the U^{4+} species in isotope exchange reactions. For the UO_2^{2+}–U^{4+} isotope exchange reaction, Bigeleisen (1996a) calculated the $\Delta_{235-238}$ fractionation factor due to the nuclear field shift effect to be 2.06‰ compared to a fractionation factor for the mass effect of –0.76‰ at 308 K, consistent with experimental observation. The net fractionation is thus +1.3‰, with opposite sign to what we would calculate in Section 5.3. The positive nuclear field fractionation factor reflects the larger number of s-electrons and smaller number of f–electrons

in the valence shell of the uranyl ion compared with the hydrated or complexed U^{4+} ion. The negative mass-dependent fractionation factor results from the bonding of the uranium atom in the uranyl ion, which is stronger than the hydration and complexing of the U(IV) due to the triple bond to each of the two oxygen atoms.

Bigeleisen (1996b) found that nuclear field shift fractionations have an approximately $1/T$ temperature dependence, whereas mass-dependent fractionations have a $1/T^2$ dependence. Consequently, the contribution of the nuclear field shift increases relative to the mass effect as temperature increases. In the example above of UO_2^{2+}–U^{4+} exchange, the nuclear field shift effect becomes increasingly dominant as temperature increases while at sufficiently low temperature (too low to be geologically relevant in this case) the mass effect becomes dominant and the net fractionation becomes negative. Schauble (2007) and Dauphas and Schauble (2016) provide more detailed reviews of the nuclear volume effect.

REFERENCES

Angeli, I. and Marinova, K.P. 2013. Table of experimental nuclear ground state charge radii: an update. *Atomic Data and Nuclear Data Tables* **99**: 69–95. doi: https://doi.org/10.1016/j.adt.2011.12.006.

Bao, H., Cao, X. and Hayles, J. A. 2016. Triple oxygen isotopes: fundamental relationships and applications. *Annual Review of Earth and Planetary Sciences* **44**: 463–492. doi: 10.1146/annurev-earth-060115-012340.

Bigeleisen, J. 1996a. Nuclear size and shape effects in chemical reactions. Isotope chemistry of the heavy elements. *Journal of the American Chemical Society* **118**: 3676–80. doi: 10.1021/ja954076k.

Bigeleisen, J. 1996b. Temperature dependence of the isotope chemistry of the heavy elements. *Proceedings of the National Academy of Sciences* **93**: 9393–6. doi: 10.1073/pnas.93.18.9393.

Bigeleisen, J. and Mayer, M.G. 1947. Calculation of equilibrium constants for isotopic exchange reactions. *Journal of Chemical Physics* **15**: 261–7.

Bindeman, I. and Pack, A. (eds.) 2021. *Triple oxygen isotope geochemistry: reviews in Mineralogy and Geochemistry 86.* Washington, DC: Mineralogical Society of America.

Blum, J.D., Sherman, L.S. and Johnson, M.W. 2014. Mercury isotopes in earth and environmental sciences. *Annual Reviews of Earth and Planetary Sciences* **42**: 249–69. doi: 10.1146/annurev-earth-050212-124107.

Buchachenko, A. L. 2001. Magnetic isotope effect: nuclear spin control of chemical reactions. *The Journal of Physical Chemistry A* **105**: 9995–10011. doi: 10.1021/jp011261d.

Cappa, C. D., Hendricks, M. B., DePaolo, D. J., et al. 2003. Isotopic fractionation of water during evaporation. *Journal of Geophysical Research: Atmospheres* **108**. doi: https://doi.org/10.1029/2003JD003597.

Chacko, T., Cole, D. and Horita, J. 2001. Equilibrium oxygen, hydrogen and carbon isotope fractionation factors applicable to geologic systems, in Valley, J. W. and Cole, D. (eds). *Reviews in Mineralogy and Geochemistry, vol. 43. Stable Isotope Geochemistry.* Chantilly, USA: Mineralogical Society of America.

Clayton, R.N., Grossman, L. and Mayeda, T.K. 1973. A component of primitive nuclear composition in carbonaceous meteorites. *Science* **182**: 485–8. doi: 10.1126/science.182.4111.485.

Craig, H.G. and Gordon, L.I. 1965. Deuterium and oxygen 18 variations in the ocean and the marine atmosphere. *Stable Isotopes in Oceanographic Studies and Paleo-temperatures.* Piza: Consiglio Nazionale Delle Ricerche Laboratoric di Geologia Nucleare.

Crank, J. 1975. *The Mathematics of Diffusion.* Oxford: Oxford Univiersity Press.

Criss, R.E. 1999. *Principles of Stable Isotope Distribution.* Oxford; New York: Oxford University Press.

Dauphas, N. and Schauble, E.A. 2016. Mass fractionation laws, mass-independent effects, and isotopic anomalies. *Annual Review of Earth and Planetary Sciences* **44**: 709–83. doi: 10.1146/annurev-earth-060115-012157.

Dennis, K.J., Affek, H.P., Passey, B.H., et al. 2011. Defining an absolute reference frame for "clumped," isotope studies of CO_2. *Geochimica et Cosmochimica Acta* **75**: 7117–131.

Eiler, J.M. 2013. The isotopic anatomies of molecules and minerals. *Annual Review of Earth and Planetary Sciences* **41**: 411–41. doi: 10.1146/annurev-earth-042711–105348.

Gao, Y.Q. and Marcus, R.A. 2001. Strange and unconventional isotope effects in ozone formation, *Science* **293**: 259–63.

Ghosh, P., Adkins, J., Affek, H., et al. 2006. ^{13}C–^{18}O bonds in carbonate minerals: a new kind of paleothermometer. *Geochimica et Cosmochimica Acta* **70**: 1439–56. doi: 10.1016/j.gca.2005.11.014.

Farquhar, J., Bao, H. and Thiemens, M. 2000. Atmospheric influence of earth's earliest sulfur cycle. *Science* **289**: 756–758. doi: 10.1126/science.289.5480.756.

Farquhar, J., Savarino, J., Airieau, S., et al. 2001. Observation of wavelength-sensitive mass-independent sulfur isotope effects during SO2 photolysis: implications for the early atmosphere. *Journal of Geophysical Research* **106**: 32829–32839. doi:10.1029/2000JE001437.

Farquhar, J., Johnston, D.T., Wing, B.A., et al. 2003. Multiple sulphur isotopic interpretations of biosynthetic pathways: implications for biological signatures in the sulphur isotope record. *Geobiology* **1**: 27–36. doi: https://doi.org/10.1046/j.1472-4669.2003.00007.x.

Heidenreich, J.E.T. and Thiemens, M.H. 1986. A non-mass-dependent oxygen isotope effect in the production of ozone from molecular oxygen: the role of molecular symmetry in isotope chemistry. *The Journal of Chemical Physics* **84**: 2129–136. doi: 10.1063/1.450373.

Horita, J. and Wesolowski, D.J. 1994. Liquid-vapor fractionation of oxygen and hydrogen isotopes of water from the freezing to the critical temperature. *Geochimica et Cosmochimica Acta* **58**: 3425–3437.doi: https://doi.org/10.1016/0016-7037(94)90096-5.

Kawabe, I. 1978. Calculation of oxygen isotope fractionation in quartz-water system with special reference to low temperature fractionation. *Geochemica et Cosmochimica Acta* **43**: 613–621. doi: 10.1016/0016-7037(78)90006-6.

King, W.H. 1984. *Isotope Shifts in Atomic Spectra.* New York: Plenum Press.

Merlivat, L. 1978. Molecular diffusivities of $H_2{}^{16}O$, $HD^{16}O$, and $H_2{}^{18}O$ in gases. *The Journal of Chemical Physics* **69**: 2864–2871. doi: 10.1063/1.436884

O'Neil, J.R. 1986. Theoretical and experimental aspects of isotopic fractionation, in J.W. Valley, (ed). *Stable Isotopes in High Temperature Geologic Processes, Reviews in Mineralogy Vol.* **16**. Washington: Mineralogical Society of America, 1–40 pp.

Richet, P., Bottinga, Y. and Javoy, M. 1977. A review of hydrogen, carbon, nitrogen, oxygen, sulphur, and chlorine stable isotope fractionation among gaseous molecules. *Annual Review of Earth and Planetary Sciences* **5**: 65–110. doi: 10.1146/annurev.ea.05.050177.000433.

Richter, F.M., Davis, A.M., DePaolo, D.J., et al. 2003. Isotope fractionation by chemical diffusion between molten basalt and rhyolite. *Geochimica et Cosmochimica Acta* **67**: 3905–23. doi: 10.1016/s0016-7037(03)00174-1.

Savarino, J., Romero, A., Cole-Dai, J., et al. 2003. UV induced mass-independent sulfur isotope fractionation in stratospheric volcanic sulfate. *Geophysical Research Letters* **30**. doi: https://doi.org/10.1029/2003GL018134.

Schauble, E.A. 2004. Applying stable isotope fractionation theory to new systems. *Reviews in Mineralogy and Geochemistry* **55**: 65–111. doi: 10.2138/gsrmg.55.1.65.

Schauble, E.A. 2007. Role of nuclear volume in driving equilibrium stable isotope fractionation of mercury, thallium, and other very heavy elements. *Geochimica et Cosmochimica Acta* **71**: 2170–89. doi: https://doi.org/10.1016/j.gca.2007.02.004.

Sharp, Z., Wostbrock, J. and Pack, A. 2018. Mass-dependent triple oxygen isotope variations in terrestrial materials. *Geochemical Perspectives Letters* **7**: 27–31. doi: 10.7185/geochemlet.1815.

Taylor, H.P. and Sheppard, S.M.F. 1986. Igneous rocks: I. Processes of isotopic fractionation and isotope systematics, in Valley, J.W., Taylor, H.P. and O'Neil, J.R. (eds). *Stable Isotopes in High Temperature Geological Processes Reviews in Mineralogy*. Washington: Mineral. Soc. Am., 227–71 pp.

Teng, F.-Z., McDonough, W.F., Rudnick, R.L., et al. 2006. Diffusion-driven extreme lithium isotopic fractionation in country rocks of the tin mountain pegmatite. *Earth and Planetary Science Letters* **243**: 701–10. doi: 10.1016/j.epsl.2006.01.036.

Thiemens, M.H. and Heidenreich, J.E. 1983. The mass-independent fractionation of oxygen: a novel isotope effect and its possible cosmochemical implications. *Science* **219**: 1073–1075. doi: 10.1126/science.219.4588.1073.

Underwood, S.J. and Clynne, M.A. 2017. Oxygen isotope geochemistry of mafic phenocrysts in primitive mafic lavas from the southernmost Cascade Range, California. *American Mineralogist* **102**: 252–61. doi: 10.2138/am-2017-5588.

Urey, H. 1947. The thermodynamic properties of isotopic substances. *Journal of the Chemical Society (London)* **1947**: 562–81.

Wang, Z., Schauble, E.A. and Eiler, J.M. 2004. Equilibrium thermodynamics of multiply substituted isotopologues of molecular gases. *Geochimica et Cosmochimica Acta* **68**: 4779–97. doi: 10.1016/j.gca.2004.05.039.

White, W.M. 2020. *Geochemistry*. 2nd edn. Oxford: Wiley-Blackwell.

Young, E.D., Galy, A. and Nagahara, H. 2002. Kinetic and equilibrium mass-dependent isotope fractionation laws in nature and their geochemical and cosmochemical significance. *Geochimica et Cosmochimica Acta* **66**: 1095–104. doi: 10.1016/S0016-7037(01)00832-8.

Zhang, Y. 2008. *Geochemical Kinetics*. Princeton, NJ: Princeton University Press, 631 pp.

Zhao, Z.-F. and Zheng, Y.-F. 2003. Calculation of oxygen isotope fractionation in magmatic rocks. *Chemical Geology* **193**(1): 59–80. doi: 10.1016/S0009-2541(02)00226-7.

PROBLEMS

1. Using the statistical mechanical approach outlined in this chapter, calculate the fractionation factor ΔCO–O_2 for the $^{17}O/^{16}O$ ratio at 273 K. What is the expected ratio of fractionation of $^{17}O/^{16}O$ to that of $^{18}O/^{16}O$ at this temperature?

2. Show that at 300 K, equilibrium fractionation between O_2 and CO should lead to $\Delta^{17}O/\Delta^{18}O \approx 0.52$.

3. Derive a temperature-dependent expression for the equilibrium constant in Equation (5.63):

$$K = \frac{q_{^{13}C^{16}O}\, q_{^{12}C^{17}O}}{q_{^{12}C^{16}O}\, q_{^{13}C^{17}O}}$$

What is the value at 300 K? What is the value at 1000 K?

4. Calculate the ratio $[^{13}C^{16}O^{18}O]/[^{12}C^{16}O_2]$ in CO_2 gas with a bulk isotopic composition of $\delta^{13}C_{PDB} = -6‰$ and $\delta^{18}O_{SMOW} = 0$ if the distribution is random. Assume that the standards have the isotopic compositions listed in Table 5.1.

5. Use the data and Equation (5.102a) to calculate the fractionation between olivine and plagioclase with composition An50 as a function of temperatures from 1200 to 900°C. Assume that the plagioclase fractionation factor is a linear combination of the listed An and Ab fractionation factors.
   Make a plot of your results.

6. Calculate the isotopic composition of water vapor in equilibrium with sea water, $\delta^{18}O = \delta^2H = 0$, at 5°C using Equations (5.98) and (5.99).

7. Using your result from Problem 7 as the value of α_{eq}, α^0 of 1.0260 and 1.094 and α^*_{ev} of 1.0045 and 1.0267 for O and H, respectively, use Equation (5.97b) to calculate $\delta^{18}O$ and δ^2H at 5°C for humidities between 0.95 and 0.7.
   At what humidity do your calculated $\delta^{18}O$ and δ^2H fall on the MWL with $\delta^2H = 8\delta^{18}O + 10$?

Chapter 6

Isotope Cosmochemistry

6.1 INTRODUCTION

The Earth is one of a family of objects ranging from dust to gas giant planets that rotate around a rather ordinary star, the Sun. All of these objects formed more or less simultaneously, astronomically speaking, about four and a half billion years ago. To understand the Earth and its origins, we need to understand its context within this Solar System and, indeed, within the broader cosmos. We have already reviewed some of this context in Chapter 1. In this chapter, we will take a more detailed look at the evolution of our Solar System and how the Earth formed within it from the perspective of isotope geochemistry.

As we will find in this chapter, isotopic studies have played and continue to play a remarkably key role in understanding the formation and evolution of the Solar System. To begin with, they allow us to date it. Beyond that, fractionations of stable isotopes provide insights into processes that occurred within the cloud of gas and dust, the solar nebula, from which the Solar System formed, and during planet formation. Isotopic studies have also allowed us to identify so-called presolar grains within meteorites: tiny mineral grains blown off red giant stars or supernovae. They have demonstrated that the nebula was an incompletely homogenized mix of material synthesized in different nucleosynthetic environments and this is helping us understand the dynamics of how the solar nebula evolved into a solar system. Isotopic studies also demonstrate that

some of this material was synthesized only shortly before, or perhaps even while, the Solar System was forming.

While much can be learned from spectral studies of stars and Solar System objects, as well as automated chemical analyses carried out by robotic spacecraft, such as the ones now operating on Mars, high-precision isotopic analyses require samples in hand. Only the latter allow us to, for example, date the beginning of the Solar System with an accuracy of only a few hundred thousand years and to identify presolar grains and deduce their origins. The inventory of extraterrestrial samples in hand includes returned lunar samples, very small samples returned by the *Hayabusa* missions from the small asteroids *25143 Itokawa* in 2010 (<1 g) and *162173 Ryugu* in 2020 (≈5 g), a small sample (roughly 1000 very small grains) of dust from the comet *Wild 2* returned by the *Stardust* mission in 2006, an even smaller sample of solar wind particles returned by the *Genesis* mission in 2004, interplanetary dust and solar wind particles collected by spacecraft and high-flying aircraft and sieved from Antarctic ice, and finally meteorites, of which many thousands are in scientific collections.

While future missions, such as the *OSIRIS-REx* mission now on its way back from the small asteroid 101955 *Bennu* with as much as 1 kg, will no doubt return additional extraterrestrial material, meteorites are our primary source of information about Solar System evolution: the chemical, isotopic, and petrologic

Isotope Geochemistry, Second Edition. William M. White.
© 2023 John Wiley & Sons Ltd. Published 2023 by John Wiley & Sons Ltd.
Companion Website: www.wiley.com/go/white/isotopegeochem2

features of meteorites reflect events that occurred in the first few tens of millions of years of Solar System history.

Observations on meteorites, together with astronomical observations on the birth of stars and the laws of physics, are the basis for our ideas on how the Solar System, as well as the Earth, formed. To establish a context for what we observe in meteorites, we will first review what we know about star birth from astronomical observations. We will then proceed to a brief overview of the nature, composition, and classification of meteorites before we review what their isotopic compositions reveal about the birth of our own star and Solar System.

6.2   STAR BIRTH

We found in Chapter 1 that stars continually die, often spectacularly, adding newly synthesized nuclides to the cosmos. New stars are also continually being born; as geochronological studies of meteorites have shown, our own star was born some 8.1 billion years after the Big Bang. We can observe the process of star birth as it is occurring today in large molecular clouds, such as the Great Nebula in Orion, visible in the night sky in the Northern Hemisphere winter. Such clouds may have dimensions in excess of 10^6 AU[1] and masses greater than 10^6 $M_\odot$[2]. Typically, about 1% of the mass of these clouds consists of submicron-sized dust and the remainder is gas, 99% of which is H_2 and He. Such massive clouds are inherently gravitationally unstable, but thermal and rotational motion, turbulence, and magnetic fields can stabilize them indefinitely. When this careful balance of the forces is broken locally, e.g., through a shock wave or the removal of magnetic fields, part of nebula will begin to gravitationally collapse on itself. The Taurus–Auriga cloud complex is a good example of a region in which low-mass stars similar to the Sun are currently forming. The cloud is about 6×10^5 AU across, has a mass of roughly 10^4 $M_\odot$, a density of 10^2–10^3 atoms/cm^3, and a temperature around 10 K. Embedded within the cloud are clumps of gas and dust with densities two orders of

magnitude higher than the surrounding cloud. Within some of these clumps are luminous protostars. About 100 stars with mass in the range of 0.2–3 $M_\odot$ have been formed in this cloud in the past few million years.

These cloud fragments will heat up adiabatically as they collapse, resulting in thermal pressure that opposes and slows collapse. Magnetic fields inherited from the larger nebula will intensify as the system contracts and as an increasing fraction of the material ionizes as temperature increases. Even small amounts of net angular momentum inherited from the larger nebula will cause the system to spin at an increasing rate as it contracts. Indeed, for a cloud to collapse and create an isolated star, it must rid itself of over 99% of its angular momentum in the process. Otherwise, the resulting centrifugal force will break up the star before it can form. This angular momentum flattens the collapsing cloud into a central spinning disk.

Model calculations show that once a cloud fragment or clump becomes unstable, supersonic inward motion develops and proceeds rapidly as long as the cloud remains transparent and the energy released by gravitational collapse can be radiated away. While the initial stage, where a fragment of the cloud with a dimension of >8000 AU becomes gravitationally unstable and begins to collapse, might occur over a period of 10^5–10^6 years, things proceed rapidly once the fragment becomes optically dense. Angular momentum progressively flattens the envelope into a rotating disk, the stellar nebula. A protostellar core forms within 3000 years; temperatures in the inner part of the nebula that rise rapidly are sufficient to evaporate all or nearly all of the dust inside 1 AU by this time (Tscharnuter et al., 2009). Material from the surrounding envelope continues to accrete to the disk while the protostellar core continues to accrete material from the surrounding disk.

The object L1551 IRS5 in the constellation Taurus consists of two protostars (recall that most stars are binaries) with a combined mass of about 1.5 M_0, which are accreting mass at a rate of $\approx 10^{-5}$ M_0 yr^{-1}. They are embedded in circumstellar disks that have diameters of about 20 AU, which are in turn embedded in

---

[1] AU stands for astronomical unit, which is the Earth–Sun distance or 1.49×10^8 km.
[2] $\odot$ is the astronomical symbol for the Sun; M_0 is mass of the Sun.

an $\approx$200 AU envelope. Infrared observations indicate disk surface temperatures in the range 50–400 K at 1 AU from the central protostars decreasing exponentially with radial distance. Numerical models indicate that these disk surface temperatures imply interior temperatures of up to 1500 K in the central region, which are hot enough to vaporize silicates. (NASA released a spectacular James Webb Telescope image of another such forming star in Taurus, L1527, as this book was going to press in November, 2022.)

An interesting feature of L1551 and other protostars at this stage is strong "bipolar flows" or jets oriented perpendicular to the circumstellar disks in which gas is moving outward at velocities of 200–400 km/s and temperatures may reach 100 000 K. As the material in the jets collides with the interstellar medium (ISM), it creates a shock wave that in turn generates X-rays. The physics that generates these jets is incompletely understood, but magnetic fields undoubtedly play a dominant role. In the X-wind model (e.g., Shu et al., 1997), the bipolar outflows emerge from the innermost part of the circumstellar disk as it interacts with the strong magnetic field of the central protostar. The jets and associated X-wind remove both mass and angular momentum from the system. Shang et al. (2000) estimated that about a third of the mass accreted to the disk and a larger fraction of the angular momentum is carried away by the X-wind, which helps to stabilize the star.

Eventually, in the T-Tauri phase, a visible star begins to emerge from its cocoon of gas and dust, although it remains surrounded by its circumstellar disk. The star surface is relatively cool (4000 K) but several times more luminous than mature stars of similar mass (Figure 1.9). The luminosity is entirely due to continued accretion and gravitational collapse – fusion has not yet ignited in its interior. X-ray bursts, highly variable luminosity, and very strong stellar winds also characterize this phase. The surrounding disk is still warm enough to give off measurable infrared radiation. Astronomical surveys of young stellar objects indicate that the circumstellar disk typically dissipates after about three million years.

The central star continues to slowly accrete remaining material from the disk and collapse until temperatures reach $\approx 10^6$ K igniting deuterium burning in its core, which can continue for several million years (see Section 1.4.2.1).

The thermal energy produced resists further collapse of the stellar core delaying the onset of hydrogen burning (the pp process). Eventually, any material that has not accreted to the central star or formed planetary bodies is blown out of the system by these stellar winds and the disk clears. Deuterium burning ceases allowing further collapse of the central core until temperatures reach 10^7 K, upon which hydrogen burning begins and the stars settle down on the main sequence (see Figure 1.9).

In the meantime, planets are forming within the spiraling disk. As the disk is opaque, we have few direct observations of this process, so our understanding relies heavily on numerical modelling, such as that of Weidenschilling (2000). Much of the dust in the innermost part of the disk, perhaps as far out as several AU (and even further out in some models), is initially evaporated; even two or three times this distance temperature maximums are high enough for the most volatile components to be evaporated. As temperatures drop, microscopic dust particles condense and then begin to stick together through surface tension and electrostatic forces to form fluffy aggregates. Transient high temperatures result in thermal processing of these into solid grains – chondrules, which are the dominant constituents of primitive meteorites. The grains settle to the mid-plane of the disk. Because of gas drag, larger grains settle more rapidly than smaller ones, allowing the larger grains to sweep up smaller ones producing centimeter-sized aggregates. Within the dust-rich, turbulent mid-plane layer, dispersion in radial velocities results in further collisions between grains ultimately producing bodies of a meter or larger that settle into a thinner sublayer of a few hundred kilometer thickness where particle density is several hundred times the gas density and accumulation into large bodies continues. Particles can also be aerodynamically concentrated into unstable streamers that can then feed growth of larger aggregates.

Once these bodies reach kilometer size, which occurs in 2000 years or less in the inner part of the nebula, gravity begins to play a role. Gas drag damps eccentricity and these "planetesimals" move into Keplerian orbits, but gravitational interactions between them accentuate eccentricity, more so for smaller bodies than for larger ones. This can lead an increase in collisions and to runaway growth with the largest bodies accumulating mass more rapidly than smaller ones, producing "planetary embryos" with masses of 0.02 $M_\oplus$[3] within 10^5 yr at 1 AU. As the orbital feeding zones of these bodies are cleared out, growth slows and occurs only through infrequent and energetic collisions of fairly large bodies. Although this "oligarchic growth," in which large planetary embryos and planets grow by collisions of progressive larger bodies may dominate, continued accretion of particles, so-called "pebble accretion," may also be important (Johansen et al., 2015).

Within the inner Solar System, nebular gas clears before planetary embryos attain enough mass to gravitationally accrete significant amounts of gas. This, of course, is obviously not the case for Jupiter and Saturn. At sufficient distance from the protostar, temperatures will be cool enough, ≈160 K, at prevailing pressures for water and other ices to condense, a point known as the "snowline." Beyond the snowline, which migrates inward over time as the nebula cools, the abundance and "stickiness" of icy particles resulted in much more rapid growth of planetary embryos. Once they reached ≈20 Earth masses, they would have had sufficient gravity to capture gas from the solar nebula and eventually become gas giants. In an alternative theory of giant planet formation, density perturbation in the disk could cause a clump of gas to become massive enough to be self-gravitating (e.g., Boss, 2003). Once that happens, the clump could collapse into a gas giant planet on timescales of 10^3–10^4 yr.

Regardless of how the giant planets form, hydrodynamic models show that they carve out gaps in the nebular disk and migrate inward. Such inward migration explains why some extrasolar gas giant planets have been found quite close to their stars where they are very unlikely to have formed. Models show that with two giant planets, such as Jupiter and Saturn, both would migrate inward, but with Saturn migrating more rapidly. Eventually when Jupiter is orbiting at about 1.5 AU, they become locked in a 2 : 3 orbital resonance and the migration reverses. This is known as the Grand Tack model (Walsh *et al.*, 2011). At this time, the inner Solar System was populated by a large number of planetesimals. In the model, these are shepherded by Jupiter's gravity into a tight distribution around 1 AU. As Saturn and Jupiter began to migrate outward, many of these planetesimals are scattered outward. At the same time, planetesimals that had formed outward of Jupiter's orbit are scattered inward. These inner solar planetesimals eventually accrete to form the terrestrial planets. The Grand Tack model successfully predicts the formation of the four terrestrial planets and the zonation in asteroid belt. As we will find later in this chapter, it also explains some of the isotopic features of meteorites.

6.3   METEORITES

Meteorites can be divided into two broad groups: *primitive* and *differentiated*. The former, chondritic meteorites, are essentially collections of the nebular dust from which the planets formed. Although inevitably modified to varying degrees in their asteroidal parent bodies, from which meteorites are derived by collisions, their chemical, isotopic, and petrological features primarily reflect processes that occurred in the cloud of gas and dust that we refer to as the solar nebula. They are far more commonly observed to fall than differentiated meteorites.[4] The differentiated meteorites, which include the achondrites, stony irons, and irons, were processed so extensively in parent bodies by melting and metamorphism that information about nebular processes has largely been lost. On the other hand, the differentiated meteorites provide insights into the early stages of planet formation. However, we will provide only a brief overview here. More details of meteoritics (the study of meteorites) can be found in McSween and Huss (2010) and White (2020).

Chondrites are so-called because they contain "chondrules," small (typically a few millimeters in diameter) round bodies that were clearly once molten droplets (Figure 6.1). They often constitute more than 50% of chondrites and in some cases up to 80%; the cause of this melting has been a matter of debate for over 100 years, but there is still little consensus.

---

[3] $\oplus$ is the astronomical symbol for Earth.
[4] Among so-called "finds," where meteorites are found but not associated with an observed fireball, irons are more common because they are more likely to be recognized as meteorites. Meteorites that can be associated with an observed fireball are referred to as "falls."

Figure 6.1 Allende (CV3) meteorite. Round objects are chondrules. Light-colored irregularly shaped objects are CAIs. Source: Jon Taylor/ Wikimedia Commons/CC BY-SA 2.0.

What is clear is that heating must have been transient because liquids are unstable at nebular pressures and that they cool quite rapidly, i.e., within minutes to hours. At present, shock waves within the nebula produced by gravitational instabilities or planetary bow shocks are the leading hypothesis. Another possibility is that they are produced in impact plumes produced by high-velocity collisions between planetesimals. However, the best evidence for this comes from chondrules in CB chondrites, which are a couple of million years younger than most chondrules.

The other main constituents of chondrites are amoeboid olivine aggregates (AOAs) and calcium–aluminum inclusions (generally called CAIs), which are together called refractory inclusions, and fine-grained matrix. The CAIs are so-called because the principal minerals, which include melilite ($Ca_2Al_2SiO_7$–$Ca_2Mg_2Si_2O_7$), perovskite ($CaTiO_3$), hibonite ($CaAl_{12}O_{19}$), anorthite ($CaAl_2Si_2O_8$), spinel ($MgAl_2O_4$), and calcic pyroxene ($CaMg_2Si_2O_6$), are rich in Ca or Al, or both. These are the phases that are thermodynamically predicted to be the first to condense from a hot gas of solar composition. In detail, CAIs may have formed by a variety of mechanisms, and some appear to have melted; however, we can generalize and say that all these are grains or aggregates of grains that equilibrated (or partially so) with nebular gas at high temperature through condensation and/or evaporation. As we will see, the CAIs were the first formed objects in the Solar System and their ages are used to define the birth of the Solar System, i.e., time 0. AOAs also appear to have condensed directly from hot nebular gas. As their name implies, they are aggregates of fine-grained, rapidly crystallized olivine, but they often contain additional minerals, such as pyroxenes, spinel, anorthite, and Fe–Ni metal.

CAIs almost certainly formed in the hottest part of the nebula near the proto-Sun. Yet they are most common in carbonaceous chondrites that almost certainly formed much further out in the nebula. They may have been cycled through the X-wind and bipolar flows and then have fallen back on the cooler outer parts of the disk. Alternatively, although flow is dominantly inward in the disk, numerical models reveal an outward counterflow in the central part of the disk capable of transporting inner disk material to regions beyond 5 AU.

Most chondrites can be divided into carbonaceous (C), ordinary, and enstatite classes[5]. The carbonaceous chondrites are, as their name implies, rich in carbon (as carbonate, organic matter, graphite, and, rarely, microdiamonds) and other volatiles, and are further divided into classes CI, CV, CM, CO, CR, CH, and CB. Of these, meteorites from the CM, CV, and CO groups are the most common. CI chondrites are rare but are nevertheless of great significance. They lack chondrules and refractory inclusions and appear to be collections of bulk nebular dust that escaped the high-temperature processing. Their compositions for condensable elements match those of measured spectrographically in the Sun's photosphere and are considered the compositionally most primitive objects. Their composition is taken to represent that of the Solar System as a whole, excluding H, C, N, and the rare gases. The sample returned by the Hayabusa2 mission from the asteroid *162173 Ryugu* falls into CI chondrite class, confirming that the C class of asteroids is indeed carbonaceous chondrites (Yokoyama et al., 2022). CH and CB chondrites are often

---

[5] In the last two decades or so, additional classes have been added that are defined by rarer meteorites.

grouped into a single CB/CH class. They are relatively rare and compositionally distinct, consisting of up to 70% of Fe–Ni metal by volume. They are highly reduced and contain elemental carbon, and are depleted in both volatile and refractory elements when compared to other carbonaceous chondrites.

The ordinary chondrites (OCs), so-called because they are by far the most common type, are divided into classes H, L, and LL, which denote high iron, low iron, and both low total iron and low metallic iron content. Enstatite chondrites can be subdivided into EH and EL, also based on iron content (again high and low iron, respectively). Enstatite chondrites formed in a highly reducing environment in which most of the iron is present as metal; due to this lack of ferrous iron, enstatite is the primary mineral rather than olivine. Rumuruti chondrites are rare, with only one documented fall (although several hundred finds are known); they have similarities with OCs, but with nearly all metal in oxidized form. Each class of chondrites is likely ultimately derived from a single asteroidal parent body.

Chondrites are further assigned a petrographic grade based on the extent of modification they have experienced in parent bodies. Grades 4, 5, and 6 have experienced increasing degrees of high-temperature metamorphism, while grades 1 and 2 have experienced low-temperature aqueous alteration; for example, Yokoyama et al. (2022) concluded that the Ryugu samples returned by the Hayabusa mission have experienced aqueous alteration at temperatures of $37 \pm 10°C$. Grade 3 is the least altered. Carbonaceous chondrites usually have grades 1 through 3 (only CIs are grade 1), while ordinary and enstatite chondrites have grades 3 through 6. The numerical petrologic grade (to which an additional decimal, e.g., 3.1, is sometimes appended) is typically appended to the class designation, e.g., CM-2 and LL-6.

The various groups of differentiated meteorites can also be further subdivided. *Achondrites* are in most cases igneous rocks, including both extrusive and intrusive, of roughly basaltic composition. Among achondrites, the *acapulcoites, lodranites, winonaites,* and *ureilites* are referred to as *primitive achondrites* because they retrain some similarity in composition and mineralogy to chondrites

and, in some cases, relict chondrules. The *diogenites, eucrites,* and *howardites,* collectively called the HED meteorites, were long known to be related and thought to come from the asteroid 4 *Vesta*. NASA's *DAWN* mission orbited Vesta from 2011 to 2012 and confirmed this inference based on spectroscopic measurements of Vesta's surface. (DAWN subsequently traveled on to study *1 Ceres,* the largest asteroid; *Ceres,* with a diameter of ≈ 1000 km, has been reclassified as a minor planet. *Ceres* surface does not match any known meteorite class, but shows greatest similarity to carbonaceous chondrites.) The diogenites are intrusive igneous rocks and the eucrites are extrusive ones (lavas), while the howardites are highly brecciated mixtures of both. The *angrites* constitute another group of achondrites likely derived from a single parent body; they are considerably rarer than the HED group. They too include both intrusive and extrusive igneous rocks, which appear to be derived from a chondritic parent by partial melting under highly oxidizing conditions. They are highly depleted in moderately volatile elements, such as Na, which has led to speculation that they might come from Mercury. Based on their reflectance spectra, the small asteroids *239 Nenetta* and *3819 Robinson* are also parent body candidates. In contrast, *ureilites* are highly reduced and seem more closely related to enstatite chondrites.

Irons, as their name implies, consist mainly of Fe–Ni metal (typically $\approx 95\%$ Fe and 5% Ni). Based on composition, they are divided into a dozen or so groups. Most of these, referred to as the *magmatic irons,* represent cores of disrupted asteroids. A few, which includes the IAB, IIICD, and IIE irons, are referred to as *non-magmatic irons,* but the term is misleading because they too solidified from liquid iron. The IIE irons have oxygen isotope ratios and trace element ratios similar to H chondrites, suggesting they may derive from the same parent body (the asteroid 8 *Hebe* is the leading suspect). The non-magmatic irons are rich in silicate inclusions and this and other evidence suggest they formed as pools of iron liquid segregated from a silicate matrix following impact melting.

Stony-irons, also called siderites, are, as their name implies, mixtures of iron metal and silicates. They are divided between mesosiderites and pallasites. Pallasites have silicates (mostly olivine) embedded in a matrix of iron metal and are likely derived from the core–mantle boundary of disrupted asteroids. Mesosiderites are metamorphosed breccias with clasts of both metal and silicates and presumably formed in an impact.

A variety of evidence, most notably their noble gas composition that matches that of the Martian atmosphere and ages that are in most cases far younger than other achondrites, have established that the *SNC meteorites*, a group that includes the *shergottites*, *nakhlites*, and *chassignites*, are derived from Mars, having been blasted off the surface by impacts. *Shergottites* are the most common of this group and include basaltic lavas and related intrusive and cumulate lherzolitic rocks. *Nakhlites* are clinopyroxenites and *chassignites* are dunites. ALH84001 (a meteorite found in the Allan Hills of Antarctica in 1984) is a unique Martian meteorite that does not fall into any of these groups and is an orthopyroxenite that contains secondary carbonates. In addition, a few achondrites have been shown to be derived from the Moon, again blasted by impacts from the Moon and later collided with Earth.

6.4 COSMOCHRONOLOGY

6.4.1 Conventional methods

The oft-cited value for the age of the Solar System is 4.567 Ga and is based on dating of meteorites, or more precisely on their components. Before we discuss meteorite ages further,

we need to consider the question of precisely what event is being dated by radiometric chronometers. Radioactive clocks record the last time the isotope ratio of the daughter element, e.g., $^{87}Sr/^{86}Sr$, was homogenized. This is usually some thermal event. In the context of what we know of early Solar System history, the event dated might be (1) the time solid particles condensed from a homogeneous solar nebula, (2) thermal metamorphism in meteorite parent bodies, or (3) crystallization (in the case of chondrules and achondrites), or (4) impact metamorphism of meteorites on their parent bodies. In some cases, the nature of the event being dated is unclear.

The most precise ages of meteorites have been obtained using the U–Pb chronometer (Figure 6.2). Advances in analytical techniques have remarkably improved precision over the couple of decades, to the point that ages with uncertainties of only a few 100 000 years can be obtained. Some of the issues that traditionally plague geochronology come into focus, including lack of complete initial isotopic homogeneity, terrestrial contamination, and deviations from closed-system behavior. In addition, new issues arise at this level of precision, including uncertainties in half-lives of the parents and, in particular, variation of the $^{238}U/^{235}U$ ratio (Brennecka et al., 2010). Progress is being made in resolving these issues, but further research remains necessary.

As Figure 6.2 shows, the oldest objects are CAIs; CAIs are good U–Pb dating targets since they are rich in refractory elements like U and

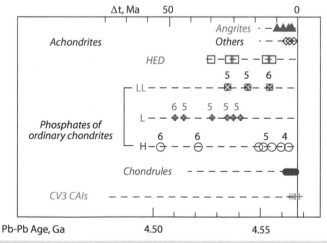

Figure 6.2 Summary of U–Pb ages of chondritic and achondritic meteorites.

depleted in volatile elements like Pb; they are largest (up to 1 cm) and most abundant in CV chondrites but have been found in all chondrites except for the CI group. The oldest high-precision date of a CAI is 4568.67 ± 0.17 Ma for a CAI from the CV3 meteorite *NWA2364*[6] calculated using the "canonical" $^{238}U/^{235}U$ ratio of 137.88 (Bouvier and Wadhwa, 2010). The next oldest age is a CAI from *Allende*, also from CV3 meteorite, whose age, calculated using the canonical $^{238}U/^{235}U$ value, is 4567.59 ± 0.11 Ma (Bouvier et al., 2007). However, it is now clear that the $^{238}U/^{235}U$ ratio is variable in CAIs: Brennecka et al.'s (2010) found $^{238}U/^{235}U$ ratios in CAIs from Allende (a CV3 carbonaceous chondrite) ranging from 137.409 ± 0.039 to 137.885 ± 0.009, while bulk *Allende* has a $^{238}U/^{235}U$ ratio of 137.818 ± 0.012. Subsequently, Connelly et al. (2012) measured $^{238}U/^{235}U$ in bulk chondrites and achondrites as well as chondrules from Allende and found all fell within error of 137.782 ± 0.13, but they confirmed that $^{238}U/^{235}U$ ratios were variable in CAIs and generally lower than those in bulk meteorites. Brennecka et al. (2010) concluded that the cause of the $^{235}U/^{238}U$ variability was decay of ^{247}Cm, which decays to ^{235}U with a half-life of 13.6 Ma. Connelly et al. (2012) pointed out that while CAIs have variable $^{238}U/^{235}U$, this ratio appears to be uniform in chondrules and achondrites – an observation inconsistent with ^{247}Cm decay. They argued that the cause was mass-dependent fractionation instead.

Because of the variability in $^{238}U/^{235}U$ in CAIs, ages calculated from the canonical $^{238}U/^{235}U$ ratio are probably too old. The oldest high-precision U–Pb ages for which $^{238}U/^{235}U$ ratios were measured include a CAI from Allende (Amelin et al., 2010) and three CAIs from *Efremovka* (CV3) (Connelly et al., 2012), which define a tight range of 4567.23 ± 0.29 to 4567.35 ± 0.28 Ma and give a pooled age of 4567.35 ± 0.28 Ma, which stands as the best estimate for the oldest Solar System objects and the age of the Solar System.

High-precision Pb–Pb ages of pooled chondrules range from 4563.66 ± 0.63 to 4564 ± 0.81 Ma, suggesting an approximately three million years gap between CAI and chondrule

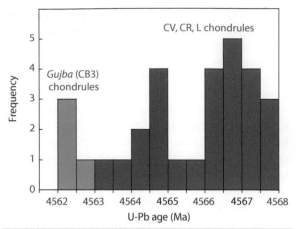

Figure 6.3 Distribution of U–Pb ages of individual chondrules. The initial chondrule-forming event coincides, or nearly so, with CAI formation, but continued for several million years thereafter. *Gujba* chondrules formed much later, much likely in an ejecta plume resulting from collision of planetesimals. Data from Bollard et al. (2017).

formation. However, dating of individual chondrules appears to reveal a more complex picture. Bollard et al. (2017) analyzed chondrules from four petrographic grade-3 chondrites, representing CV, CR, and L classes. Figure 6.3 shows the distribution of these U–Pb ages, as well as earlier reported analyses of individual chondrules from *Allende* (CV) (Connelly et al., 2012) and *Gujba* (CB) (Bollard et al., 2015). The oldest chondrule ages overlap within error of the age of CAIs but extend downward in age to 4562.5 Ma. Roughly half of the chondrules analyzed by Bollard et al. (2017) formed within the first million years following CAI formation. The oldest chondrules have initial Pb isotope ratios closest to the estimated Solar System initial while younger ones have more evolved initial ratios. Because the Solar System μ value (i.e., $^{238}U/^{204}Pb$) is low, ≈ 0.13, Pb isotope ratios in unprocessed nebular dust will evolve only slowly. Because Pb is volatile while U is refractory, thermally processed materials, such as CAIs and chondrules, typically have μ values several orders of magnitude higher and hence making Pb isotope ratios higher. Based on this, Bollard et al. (2017) argued that main

---

[6] NWA stands for Northwest Africa. This is one of many meteorites found in the Sahara desert over the last couple of decades.

chondrule-forming period was nearly simultaneous with CAI formation and subsequent events merely reprocessed this material to produce the younger chondrules.

This conclusion is, however, contentious, as abundant evidence from the short-lived radionuclide ^{27}Al, discussed below, indicates the main chondrule formation period followed CAI formation by $\approx$2 Ma. Pape et al. (2019) point out that the chondrules analyzed by Bollard et al. are on average an order of magnitude larger than average and hence may be atypical. They also suggest the possibility of loss of ^{222}Rn. ^{222}Rn is in the ^{238}U decay chain and its loss would ultimately raise the $^{207}Pb/^{206}Pb$, resulting in apparent older ages because of the progressive leaching technique used by Bollard et al. The first leaches would dissolve the glassy mesostasis, which would be the richest in U and therefore define the upper end of the isochron and its slope. The mesostasis would also be the least radon retentive. As little as 0.8% loss of ^{222}Rn would increase the apparent $^{207}Pb/^{206}Pb$ age by 1 Ma.

The youngest chondrules are from *Gujba* (CB3). Four chondrules from this meteorite have identical ages within error and average of 4562.49 ± 0.21 Ma (Bollard et al., 2015). CB meteorites are relatively rare (only six known specimens) and distinguished by a high-metal content and particularly large chondrules. They may have been produced in impact plumes following a collision of planetesimals (and are sometimes referred to as "splash chondrules") and the ages of the *Gujba* chondrules do not correspond to the principal chondrule-forming period. Excluding these, the youngest chondrule has an age of 4563.34 ± 0.63 Ma and there is little or no systematic variation in chondrule ages between meteorite classes.

Phosphates also have high U/Pb ratios and these were analyzed by Göpel et al. (1994) to obtain high-precision ages of a variety of equilibrated (i.e., petrologic classes 4–6) OCs, whose ages range from 4.563 to 4.502 Ga. The phosphates are thought to have formed during metamorphism; thus, these ages represent the age of metamorphism of these meteorites. The oldest of these meteorites is the H4 chondrite *Ste. Marguerite*. Bouvier et al. (2007) subsequently reported a Pb–Pb isochron age of 4562.7 Ma, which is in excellent agreement with the age determined by Göpel et al. (1994). As can be seen in Figure 6.2, there is no obvious relationship among chondrites between meteorite class and age. H chondrites do seem a bit older than other OCs and Göpel et al. (1994) did find an inverse correlation between petrologic type and age (the least metamorphosed chondrites are oldest).

Among achondrites, the chronology of the angrites is perhaps best documented. The oldest high-precision Pb–Pb age is 4564.58 ± 0.14 Ma for the angrite SAH99555 (Connelly et al., 2008). An age nearly as old as 4563.37 ± 0.25 was reported for the angrite *D'Orbigny* (Amelin, 2008; Brennecka and Wadhwa, 2012). *Angra dos Reis*, the type meteorite of the class, has a Pb–Pb age of 4557.65 ± 0.13 Ma, and *Lewis Cliff 86010*, a coarse grained "plutonic" angrite, has an age of 4558.55 ± 0.15 Ma. Thus, differentiation, cooling, and crystallization of the angrite parent body apparently lasted some six million years. Wadhwa et al. (2009) reported an age of 4566.5 ± 0.2 Ma for the unusual basaltic achondrite, *Asuka 881394*. Bouvier et al. (2011) determined an age of 4562.89 ± 0.59 Ma for another unusual basaltic achondrite, *NWA2976*. Amelin (2019) reported U-Pb ages of 4562.76 ± 0.26 and 4562.63 ± 0.25 Ma for the ungrouped achondrites *NWA 6704* and *NWA 6693*, respectively. The two meteorites appear to be derived from the same parent body and are most similar to primitive achondrites of the acapulcoite–lodranite clan.

Perhaps surprisingly, achondrites appear to be older than chondrites; the parent body of these objects formed, melted, and differentiated, and the outer parts crystallized within a few million years of the birth of the Solar System. Indeed, based on the initial $^{87}Sr/^{86}Sr$ ratio in plagioclase of NWA 6704, Amelin et al. (2019) concluded that its parent body accreted within 3.6 million years of CAI formation. However, not all achondrites are quite so old. A few other high-precision ages (those with quoted errors of less than 10 Ma) are available and they range from this value down to 4.529 ± 5 Ma for *Nuevo Laredo* and 4510 ± 4 Ma for the *Bouvante*, both eucrites derived from *Vesta*. Thus, the total range of the few high-precision ages in achondrites is about 50 million years.

Iron meteorites appear to be similarly old. Blichert-Toft et al. (2010) determined Pb–Pb ages in troilite (FeS) of 4565.3 ± 0.1 Ma and

4544 ± 7 Ma in *Muonionalusta* and *Gibeon*, respectively, both iron meteorites belonging to the IVA family. However, Brennecka et al. (2018) found that the $^{238}U/^{235}U$ ratio in troilite from *Muonionalusta* was variable and concluded that, based on the range of $^{238}U/^{235}U$ ratios they observed, the actual age was likely between 4558 and 4563. More recently, Connelly et al. (2019) reported an age of 4565.37 ± 0.20 Ma for silicate inclusions in the IVA iron *Steinbach*; measured $^{238}U/^{235}U$ ratios were within error of the 137.782 value observed in non-CAI material. They also recalculated Blichert-Toft et al.'s age for *Muonionalusta* using this $^{238}U/^{235}U$ ratio and obtained an age of 4564.1 ± 2.6 Ma.

Re–Os fractionation in iron meteorites is somewhat limited, which in turn limits the precision that can be achieved with the Re–Os system. Smoliar et al. (1996) reported Re–Os ages of 4558 ± 12 and 4537 ± 8 Ma for IIIA and IIA irons, respectively, while McCoy et al. (2016) reported an age of 4540 ± 17 Ma for IVA irons. Re–Os ages of other irons range from 4456 to 4569 Ma, but are of lower precision.

The present state of conventional radiometric chronology of the early Solar System may be summarized by saying that it appears the oldest meteorite parent bodies, those of achondrites and irons, formed within about 2 Ma of CAI formation, which occurred around 4567.3 Ga. Sm–Nd, Lu–Hf, Re–Os, and Rb–Sr ages are less precise than the U–Pb ages and are often a bit younger, and perhaps date peak metamorphism of chondrite parent bodies, and melting or metamorphism on achondrite parent bodies.

Most *shergottites* yield consistent Rb–Sr, Sm–Nd, Lu–Hf ages in the range of 160–180 Ma; one pair, QUE94201 and DAG476, has Rb–Sr and Sm–Nd ages of 474 Ma. However, Bouvier et al. (2009) obtained Pb–Pb ages of 4.3 and 4.1 Ga for two groups of shergottites. Because rather poor fit to isochrons of the Pb–Pb data (mean squared weighted deviations (MSWD) values of 11 and 34) together with their inconsistency with other chronometers has led to a consensus (e.g., Grott et al., 2013) that the younger ages are preferred and the old ages reflect mixing lines or, perhaps, terrestrial contamination (terrestrial common Pb is nearly colinear with the shergottite data). *Nakhlites* yield consistent Rb–Sr and Sm–Nd ages of 1.27–1.32 Ma and Bouvier et al. (2009) obtained a well-fitted Pb–Pb isochron (MSWD = 0.31) corresponding to an age of 1.33 ± 0.14 Ma. Analyses of *chassignites* have produced an age almost identical to this. The unique Martian meteorite ALH84001 has a Pb–Pb age of 4.1 Ga. These ages are clearly distinct from all other meteorite ages and strengthen the case that they are derived from a large terrestrial body, such as Mars. They also demonstrate that geologic and magmatic activity continued on Mars long after its formation (and indeed may continue today).

In contrast to the tightly clustered ages for the chronometers discussed above, K–Ar (including ^{40}Ar–^{39}Ar) ages show a much wider range. While K–Ar ages older than 4.4 Ga dominate, ages as young as a few tens of millions of years also occur. As we found in Chapter 2, closure temperatures for K–Ar are generally much lower than that for other chronometers, so these younger ages generally reflect heating events resulting from impacts. Many meteorites are brecciated, providing clear evidence that impacts and collisions have been common among meteorite parent bodies. Heavy cratering of asteroids imaged by spacecraft, as well as larger bodies including the Moon, Mercury, and Mars, also provides additional evidence of impacts. Figure 6.3 shows the probability distribution of 100 $^{40}Ar/^{39}Ar$ ages of OCs compiled by Swindle et al. (2014). There is a clear prevalence of ages just a few hundred million years younger than the CAI forming event, particularly among LL and H chondrites, but many ages are far younger. Indeed, ages <1.2 Ga are also common. Interestingly, there is a dearth of ages between about 2.2 and 3.6 Ga. Distinct peaks are also apparent, likely associated with breakup of the parent bodies and their remnants.

The peak in ages of the L chondrites around 470 Ma is particularly interesting. Abundant "fossil" meteorites have been found in an Ordovician marine limestone horizon dated at 467.3 ± 1.6 Ma. These have proved to be L chondrites. Schmitz et al. (1997) estimate from their abundance that meteorite accretion rates were one to two orders of magnitude greater at that time than now and speculate that this influx might record the breakup of the L chondrite parent body. Subsequent ^{40}Ar–^{39}Ar dating of shocked L chondrites, particularly *Ghubara*, a highly brecciated meteorite, dates this breakup event to 470 ± 6 Ma (Korochantseva et al., 2007). Today,

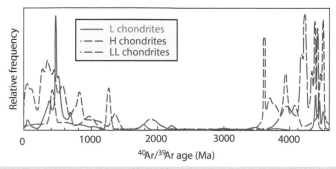

Figure 6.4   Probability density plot of ≈100 ^{40}Ar/^{39}Ar ages of OCs. Source: Adapted from Swindle et al. (2014).

fragments of the parent body continue to rain down on us; L chondrites account for more than a third of all meteorite falls.

Grains from the asteroid *Itokawa*, likely a fragment of the LL chondrite parent body, yielded a flat ^{40}Ar/^{39}Ar release spectrum indicating complete degassing at 1.32 ± 3 Ga (Park et al., 2015). The most likely explanation is impact and consequent breakup of the LL parent body (or a fragment of it) at this time. Although solar heating when *Itokawa* may have been in an orbit closer to the Sun cannot be entirely ruled out, this age corresponds to a peak in ^{40}Ar/^{39}Ar ages in LL chondrites apparent in Figure 6.4.

Other groups of meteorites also yield a variety ^{40}Ar–^{39}Ar ages. Some primitive achondrites

(acapulcoites and lodranites) have ages that tightly cluster around 4.51 Ga. The ungrouped primitive achondrite NWA 6704 has a ^{40}Ar/^{39}Ar age of 4199 ± 32 Ma and the release spectrum suggests a subsequent heating event at ≤2.12 Ga (Amelin et al., 2019). Silicate from IAB irons (which are among the so-called non-magmatic irons) gives a continual distribution of ages between 4.53 and 4.32 Ga. Eucrites (from Vesta), many of which are brecciated, yield a wide range of ages with peaks in the distribution occurring at 4.5 Ga, 4.0–3.7 Ga, and 3.45–3.55 Ga, which date heating from large impacts on the surface of Vesta, and the view of Vesta from the DAWN spacecraft (Figure 6.5) certainly shows there have been many of those.

Figure 6.5   A mosaic image of Vesta based on photographs taken by the DAWN spacecraft as it orbited Vesta from 2011 to 2012. Two very large craters in the Southern Hemisphere appear to be less than 2 Ga, and the largest and most recent of which, the Rheasilvia basin, may have been the impact that launched the HED meteorites into Earth-crossing orbits. Lindsay et al. (2105) have speculated that a ^{40}Ar/^{39}Ar plateau in feldspar grains from the howardite Kapota records the impact that formed this basin. NASA photo.

6.4.2 Extinct radionuclides

Anomalous abundance of nuclides, e.g., ^{129}Xe, known to be produced by the decay of short-lived radionuclides, such as ^{129}I, and correlations between the abundance of the radiogenic *isotope* and the parent *element* in meteorites and their phases, provides compelling evidence that those short-lived nuclides were present when the Solar System formed. Table 6.1 lists the primary short-lived nuclides discovered in this manner.

6.4.2.1 ^{53}Mn–^{53}Cr

For example, consider ^{53}Cr, which is the decay product of ^{53}Mn. The half-life of ^{53}Mn, only 3.7 million years, is so short that any ^{53}Mn produced by nucleosynthesis has long since decayed. If ^{53}Mn is no longer present, how do we know that the anomalous ^{53}Cr is due to decay of ^{53}Mn? We reason that the abundance of ^{53}Mn, when and if it was present, should have correlated with the abundance of ^{55}Mn – the only stable isotope of Mn. Therefore, we construct a plot similar to a conventional isochron diagram (isotope ratios versus parent/daughter ratio), but use the stable isotope, in this case ^{55}Mn, as a proxy for the radioactive one, ^{53}Mn; an example is shown in Figure 6.6. Starting from our basic equation of radioactive decay, we can derive the following equation:

$$D = D_0 + N_0\left(1 - e^{-\lambda t}\right) \qquad (6.1)$$

This is similar to the isochron equation we derived earlier, i.e., Equation (2.8) in Chapter 2, but not identical. In particular, notice that Equation (6.1) contains N_0, the initial abundance of the parent, whereas Equation (2.8) in Chapter 2 contains the present abundance of the parent. Notice also that the exponential term is $(1-e^{-\lambda t})$ rather than $(e^{\lambda t}-1)$. Written for the example of the decay of ^{53}Mn to ^{53}Cr, we obtain:

$$\frac{^{53}\text{Cr}}{^{52}\text{Cr}} = \left(\frac{^{53}\text{Cr}}{^{52}\text{Cr}}\right)_0 + \left(\frac{^{53}\text{Mn}}{^{52}\text{Cr}}\right)_0\left(1 - e^{-\lambda t}\right) \quad (6.2)$$

where, as usual, the subscript naught denotes the initial ratio. The problem we face is that we do not know the initial ^{53}Mn/^{52}Cr ratio. We can, however, measure the ^{55}Mn/^{53}Cr ratio. Assuming that initial isotopic composition of Mn was homogeneous in all the reservoirs of interest; i.e., ^{53}Mn/^{55}Mn$_0$ is constant, the initial ^{53}Mn/^{52}Cr ratio is just:

$$\left(\frac{^{53}\text{Mn}}{^{52}\text{Cr}}\right)_0 = \left(\frac{^{55}\text{Mn}}{^{52}\text{Cr}}\right)_0\left(\frac{^{53}\text{Mn}}{^{55}\text{Mn}}\right)_0 \qquad (6.3)$$

Table 6.1 Important short-lived radionuclides in the early Solar System.

Radionuclide	Half-life (Ma)	λ	Decay	Daughter	Initial abundance ratio
^{10}Be	1.387	4.50×10^{-7}	β	^{10}B	^{10}Be/^{9}Be $\approx 5 \times 10^{-4} - 1.3 \times 10^{-4}$
^{26}Al	0.702	9.87×10^{-7}	β	^{26}Mg	^{26}Al/^{27}Al $\approx 2.8 - 5.2 \times 10^{-5}$
^{36}Cl	3.08×10^5	2.30×10^{-6}	β	^{36}Ar/^{36}S	^{36}Cl/^{35}Cl $\approx 1.6 \times 10^{-4}$
^{41}Ca	0.102	6.80×10^{-6}	β	^{41}K	^{41}Ca/^{40}Ca $\approx 4.9 \times 10^{-9}$
^{53}Mn	3.7	1.87×10^{-7}	β	^{53}Cr	^{53}Mn/^{55}Mn $\approx 9.1 \times 10^{-6}$
^{60}Fe	2.6	2.62×10^{-7}	β	^{60}Ni	^{60}Fe/^{56}Fe $\approx 3.7 \times 10^{-7}$
^{107}Pd	6.5	6.5×10^{-7}	β	^{107}Ag	^{107}Pd/^{108}Pd $\approx 3 \times 10^{-5}$
^{129}I	15.7[†]	4.41×10^{-8}	β	^{129}Xe	^{129}I/^{127}I $\approx 1.35 \times 10^{-4}$
^{146}Sm	≈103* (≈68)	6.73×10^{-9} (1.02×10^{-8})	α	^{142}Nd	^{146}Sm/^{144}Sm ≈ 0.0085
^{182}Hf	8.9	7.79×10^{-8}	β	^{182}W	^{182}Hf/^{180}Hf $\approx 1 \times 10^{-4}$
^{244}Pu	81.2	8.54×10^{-9}	α, SF	Xe	^{244}Pu/^{238}U ≈ 0.0068

*103 Ma is the most commonly used half-life and is based on alpha-countering measurements of Meissner et al. (1987). However, Kinoshita et al. (2012) reported a half-life of 68 ± 7 Ma based on alpha-counting. The International Union of Pure and Applied Chemistry–International Union of Geological Sciences (IUPAC–IUGS) joint Task Group on Isotopes in the Geosciences concluded that no recommendation could be made on the half-life of ^{146}Sm at this time (Villa et al., 2020). More recently, however, Fang et al. (2022) obtained a half-life of 102 ± 9 Ma by calibration against other chronometers in the meteorite *Erg Chech 002*, apparently confirming the longer half-life.

†Glimour and Crowther (2017) found that a half-life of 16.1 Ma better fit a calibration against U–Pb ages.

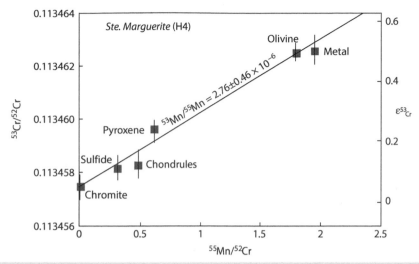

Figure 6.6 Correlation of the $^{53}Cr/^{52}Cr$ ratio with $^{55}Mn/^{52}Cr$ ratio in components of the OC *Ste. Marguerite*. Source: Trinquier et al. (2008)/with permission of Elsevier.

Of course, since ^{55}Mn and ^{52}Cr are both non-radioactive and non-radiogenic, the initial $^5Mn/^{52}Cr$ ratio is equal to the present ratio (i.e., this ratio is constant through time), so we can drop the naught. Substituting Equation (6.3) into Equation (6.2), we obtain:

$$\frac{^{53}Cr}{^{52}Cr} = \left(\frac{^{53}Cr}{^{52}Cr}\right)_0 + \left(\frac{^{55}Mn}{^{52}Cr}\right)\left(\frac{^{53}Mn}{^{55}Mn}\right)_0 \left(1 - e^{-\lambda t}\right)$$

(6.4)

Finally, for a short-lived nuclide like ^{53}Mn, the term λt becomes very large after 4.56 Ga, so the term $e^{-\lambda t}$ is 0 (this is equivalent to saying all the ^{53}Mn has decayed away). Thus, we are left with:

$$\frac{^{53}Cr}{^{52}Cr} = \left(\frac{^{53}Cr}{^{52}Cr}\right)_0 + \left(\frac{^{55}Mn}{^{52}Cr}\right)\left(\frac{^{53}Mn}{^{55}Mn}\right)_0$$

(6.5)

As does Equation (2.8) in Chapter 2, this equation has the form of $y = a + bx$, so $^{53}Mn/^{52}Cr$ defines a slope on a plot of $^{53}Cr/^{52}Cr$ versus $^{55}Mn/^{52}Cr$; however, in this case, the slope is not proportional to time as in a conventional isochron diagram, but it is proportional to the initial $^{53}Mn/^{55}Mn$ ratio. $^{53}Mn-^{53}Cr$ dating revealed an age of 5.3 Ma of hydrothermal alteration of the asteroid *Ryugu* after CAI formation (Yokoyama et al., 2022). In this way, many extinct radionuclides have been identified in meteorites from variations in the abundance of their decay products.

The most important of these are listed in Table 6.1.

These short-lived radionuclides must have been synthesized shortly before the Solar System formed. To understand why, consider the example of ^{53}Mn. Its half-life is 3.7 Ma. Hence, 3.7 Ma after it was created, only 50% of the original number of atoms would remain. After two half-lives, or 7.4 Ma, only 25% would remain, after four half-lives, or 14.8 Ma, only 6.125% of the original ^{53}Mn would remain, etc. After 10 half-lives, or 37 Ma, only $1/2^{10}$ (0.1%) of the original amount would remain. The correlation between the Mn/Cr ratio and the abundance of ^{53}Cr indicates some ^{53}Mn was present when the meteorite, or its parent body, formed. From this we can conclude that ^{53}Mn was synthesized not more than roughly 30 million years before the meteorite formed. The shorter-lived nuclides provide even tighter constraints on this timescale. We will return to this later in the chapter.

6.4.2.2 $^{129}I-^{129}Xe$ and ^{244}Pu

The first of these short-lived radionuclides discovered was ^{129}I, which decays to ^{129}Xe (Reynolds, 1960). Figure 6.7 shows the example of the analysis of the LL5 chondrite *Tuxtuac*. In this case, the analysis is done in a manner very analogous to $^{40}Ar-^{39}Ar$ dating: the sample is first irradiated with neutrons so that ^{128}Xe is produced by neutron capture by

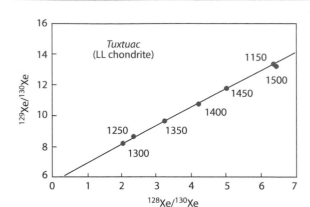

Figure 6.7 Correlation of $^{129}Xe/^{130}Xe$ with $^{128}Xe/^{130}Xe$. The ^{128}Xe is produced from ^{127}I by irradiation in a reactor so that the $^{128}Xe/^{130}Xe$ ratio is proportional to the $^{127}I/^{130}Xe$ ratio. Numbers adjacent to data points correspond to temperature (°C) of the release step. Source: Adapted from Bernatowicz et al. (1988).

^{127}I creating ^{128}I, which subsequently decays to ^{128}Xe. The amount of ^{128}Xe produced is proportional to the amount of ^{127}I present (as well as the neutron flux and reaction cross-section). Unlike ^{39}Ar, ^{128}Xe is a stable, naturally occurring nuclide, necessitating a correction for it. The sample is then heated in vacuum through a series of steps and the Xe released at each step is analyzed in a mass spectrometer. As was the case in Figure 6.6, the slope is proportional to the $^{129}I/^{127}I$ ratio at the time the meteorite formed.

In addition to ^{129}Xe produced by decay of ^{129}I, ^{130}Xe ^{132}Xe, ^{134}Xe, and ^{136}Xe are produced by fission of ^{238}U and ^{244}Pu; the latter, with a half-life of 81 Ma is another extinct radionuclide. Fission does not produce a single nuclide, rather a statistical distribution of many nuclides with masses roughly half that of the parent. Each fissionable isotope produces a different distribution. The distribution produced by U is similar to that produced by ^{244}Pu, but the difference is great enough to demonstrate the existence of ^{244}Pu in meteorites, as is shown in Figure 6.8. Fission tracks in excess of the expected number of tracks for a known uranium concentration are also indicative of the former presence of ^{244}Pu. Terrestrial Xe isotopes indicate that ^{129}I and ^{244}Pu were also present in the Earth after it formed – a topic we will return to in Chapter 14.

These extinct radionuclides provide a means of relative dating of meteorites and other bodies. Figure 6.9 shows relative ages based on this ^{129}I–^{129}Xe decay system. These ages are calculated from $^{129}I/^{127}I$ ratios, which are in turn calculated from the ratio of excess ^{129}Xe to ^{127}I. I–Xe ages are often reported relative to the age of the *Shallowater* aubrite (achondrite). Whitby et al. (2000) determined the initial Solar System $^{129}I/^{127}I$ from halite in the H chondrite *Zag* to be 1.35×10^{-4}. That, together with I–Xe dating of phosphates also dated by Pb–Pb, allows I–Xe ages to be calibrated to an absolute timescale; the absolute

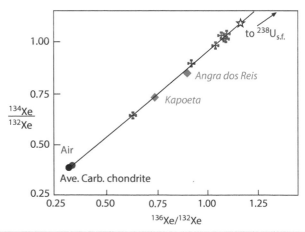

Figure 6.8 Variation of $^{134}Xe/^{132}Xe$ and $^{136}Xe/^{132}Xe$ in the *St. Séverin* (LL6) chondrite (5) and achondrites *Kapoeta* and *Angra dos Reis* (◆). The isotopic composition of fission products of man-made ^{244}Pu is shown as a star (☆). Source: Adapted from Lewis (1975).

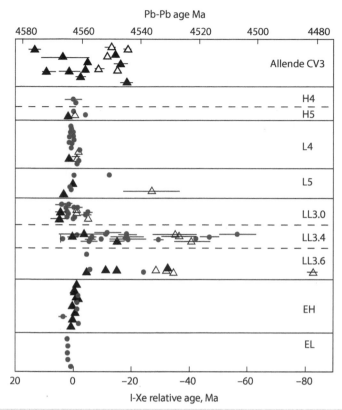

Figure 6.9 Summary of I–Xe ages of meteorites relative to the aubrite *Shallowater*. Solid red symbols are "whole-rock" ages; open triangles are chondrules with I–Xe isochrons released at low temperature; and solid triangles are chondrules or clasts with high-temperature isochrons. Source: Adapted from Hohenberg et al. (2008), Swindle et al. (1983), Whit-by et al. (2002), and Crowther et al. (2018).

age of *Shallowater* is 4562.4 ± 0.4 Ma on this timescale (Pravdivtseva et al. 2017).

Because both I and Xe are mobile elements, the I–Xe system is readily disturbed. Consequently, they more likely date processes occurring on meteorite parent bodies rather than nebular events. Most chondrules have I–Xe ages that are no more than a few million years younger than the age of CAIs, although some, particularly those with low-temperature Xe release patterns are tens of millions of years younger. Two igneous clasts in the Barwell (L5) meteorite gave ages of 4564.96 ±0.33 Ma and 4565.60 ± 0.33 Ma, which are consistent with ^{27}Al ages for the period of chondrule formation and similar in age to the angrite parent body (Crowther et al., 2018). Differentiated meteorites show a similar spread of ages, with some, such as the aubrite *Shallowater* and the acapulcoite *Acapulco*, being only a few million years younger than CAIs.

6.4.2.3 ^{107}Pd–^{107}Ag

^{107}Pd decays to ^{107}Ag with a half-life of 6.5 million years. Palladium is a refractory and highly siderophile element and hence concentrates strongly in metal phases, such as iron meteorites. Silver is also siderophile, but less so and also chalcophile and can be concentrated in sulfide phases, such as troilite. Silver is also moderately volatile. Thus, one might expect variations in the abundance of ^{107}Ag, quantified as the ^{107}Ag/^{109}Ag ratio, which correlate with Pd concentrations. This is what Chen and Wasserburg (1990) found when they analyzed metal from the IVA iron *Gibeon*; the data defined a fossil isochron indicating an initial ^{107}Pd/^{108}Pd ratio of 2.4 ± 0.5 × 10^{-5}. The IV irons are depleted in volatile elements, which results in quite high Pd/Ag ratios, allowing for relatively precise initial ^{107}Pd/^{108}Pd ratios. Subsequent studies of IAB and IIIAB irons, as well as several anomalous irons, show ^{107}Ag/^{109}Ag–^{108}Pd/^{109}Ag correlations that indicate $^{107}Pd/^{108}Pd$ ratios between 1.5 and 2.4 × 10^{-5}.

There are, however, several issues with Pd–Ag chronology, some of which have not yet been entirely resolved. The first of which is the poorly constrained nature of the Solar System initial $^{107}Pd/^{108}Pd$ and $^{107}Ag/^{109}Ag$ ratios. Schönbächler et al. (2008) used analyzed $^{108}Pd/^{109}Ag$ and $^{107}Ag/^{109}Ag$ ratios in carbonaceous chondrites to infer a Solar System initial $^{107}Pd/^{108}Pd$ ratio of 5.9 ± 2.2 × 10^{-5}. This implies an age of 8.5 Ma for *Gibeon* and ages of 13 Ma for the IIIAB iron *Grant* and 19 Ma for the IAB iron *Canyon Diablo*. Horan et al. (2012) analyzed $^{108}Pd/^{109}Ag$ and $^{107}Ag/^{109}Ag$ ratios in the IVA iron *Muonionalusta* and obtained an initial $^{107}Pd/^{108}Pd$ ratio of 2.1 ± 0.3 × 10^{-5}. Using the 4565.3 ± 0.1 Ma U–Pb age obtained by Blichert-Toft (2010) for troilite in this iron, they calculated an initial Solar System $^{107}Pd/^{108}Pd$ ratio of 2.8 ± 0.5 × 10^{-5}. However, as we have seen, variable $^{235}U/^{238}U$ ratios in this troilite have raised questions about this age. Depending on the actual $^{235}U/^{238}U$, the age could be anywhere from 4558 to 4564 Ma. The younger age would yield a Solar System initial in good agreement with Schönbächler et al.'s value (calculation of the exact value is left to the reader as Problem 6.1). Using the recalculated age for the troilite of Connelly et al. (2019) implies a Solar System initial $^{107}Pd/^{108}Pd$ of 3.0 (+1/−0.7) × 10^{-5}.

A second issue is cosmic radiation. Several secondary neutron-capture reactions can affect the isotopic composition of both Pd and Ag; in particular, ^{107}Ag can capture a neutron and decay to ^{108}Pd (or alternatively to ^{108}Cd), affecting both the isotopic composition of Pd and Ag. This is particularly a problem for iron meteorites that often have cosmic-ray exposure ages of the order of 100 Ma (exposure ages of stones, which are typically an order of magnitude lower). Matthes et al. (2018) found that these effects are insignificant for IVA irons and they calculated $(^{107}Pt/^{108}Pt)_0$ values of 2.57 ± 0.3 × 10^{-5} and 2.57 ± 0.11 × 10^{-5} for *Muonionalutra* and *Gideon*, respectively, which are in reasonable agreement with previous work of Chen and Wasserburg (1990) and Horan et al. (2012). For IIIAB irons, which are not volatile-depleted and have much lower Pd/Ag ratios than IVAB irons, cosmogenic effects are significant. Using Pt isotope ratios as neutron flux monitors to correct for neutron-capture effects, Matthes et al. (2015) were able to obtain $^{107}Pd/^{108}Pd_0$ values ranging from 1.76 to 2.76 × 10^{-5} for a number of IIAB irons, and Matthes et al. (2020) obtained a 2.05 ± 0.05 × 10^{-5} $^{107}Pt/^{108}Pt_0$ value for the IIIAB iron *Cape York*, implying an age of 5–11 Ma after CAI formation, depending on the initial $^{107}Pt/^{108}Pt$ of the Solar System. $^{53}Mn–^{53}Cr$ ages for phosphates in IIIAB irons provide somewhat tighter constraints; these ages range from 4563.6 to 4561.8 Ma, or 3.7–5.5 Ma after CAI formation. These ages and the Cape York $(^{107}Pt/^{108}Pt)_0$ imply a Solar System initial $^{107}Pt/^{108}Pt$ ratio of 3.03–3.67 × 10^{-5}.

Since both Pd and Ag are siderophile, these ages reflect the time of cooling through closure temperatures of originally liquid metal, not the time of formation of the iron magma; in Section 6.4.2.5, we will find that the $^{182}Hf–^{182}W$ decay system constrains the timing of metal-silicate fractionation to very early in Solar System history.

6.4.2.4 $^{26}Al–^{26}Mg$

The existence of ^{26}Al in the early Solar System was the first to be clearly demonstrated following the discovery of ^{129}I (Lee et al., 1976). Because of its short half-life (0.702 Ma), the $^{26}Al–^{26}Mg$ decay system potentially provides a much more detailed chronology of early Solar System events and stronger constraints on the amount of time that could have passed between nucleosynthesis and processes that occurred in the early Solar System. Furthermore, the abundance of ^{26}Al was sufficiently high that its decay would have been a significant source of heat and that it very likely played a role in heating, melting, and metamorphism of meteorite parent bodies; indeed, this energy is now widely viewed as the principal reason some asteroidal-sized bodies underwent melting and differentiation into silicate mantles and metallic cores. For these reasons, considerable effort has been devoted to measurement of Mg isotope ratios in meteorites. Much of this work has been carried out with ion microprobes, which allow the simultaneous measurement of $^{26}Mg/^{24}Mg$ and $^{27}Al/^{24}Mg$ on spatial scales as small as 10 μ. As a result, there are thousands of measurements on more than 60 meteorites and their components in the literature.

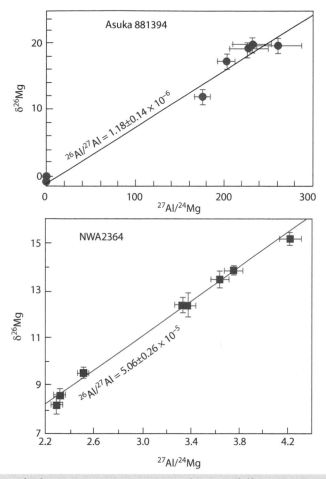

Figure 6.10   Comparison of Al–Mg isotope systematics for two different meteorites. The lower diagram shows minerals separated from a CAI in CV3 *NWA2364* – one of the oldest objects in the Solar System. Source: Adapted from Bouvier and Wadhwa (2010). The upper diagram shows plagioclase and pyroxene separate from the eucrite *Asuka 881394*. Source: Adapted from Nyquist et al. (2003). The later has an initial $^{27}Al/^{26}Al$ more than 40 times lower than *NWA2364*; assuming a uniform initial ^{26}Al in the Solar System implies that *Asuka 881394* is some four million years younger than the CAI.

Initial $^{26}Al/^{27}Al$ ratios can be deduced in a manner exactly analogous to the ^{53}Mn–^{53}Cr system: by regressing the $^{26}Mg/^{24}Mg$ ratio against $^{27}Al/^{24}Mg$. Figure 6.10 shows two examples. In this diagram, the $^{26}Mg/^{24}Mg$ ratio is expressed as $\delta^{26}Mg$, which is per mil deviations from a terrestrial standard (typically DSM3, where $^{26}Mg/^{24}Mg = 0.13979$) calculated as:

$$\delta^{26}Mg = \left(\frac{^{26}Mg/^{24}Mg_{sample} - {}^{26}Mg/^{24}Mg_{std}}{^{26}Mg/^{24}Mg_{std}} \right) \times 1000$$

$$(6.6)$$

As analytical precision has improved, quite small variations in $^{26}Mg/^{24}Mg$ are now often being reported in μ notation, in which the 1000 term in Equation (6.6) is replaced by 10^6. Because Mg isotopes can undergo fractionation, a correction for this must be made by also measuring the $^{25}Mg/^{24}Mg$ ratio.

The highest inferred $^{26}Al/^{27}Al$ ratios are found in CAIs, consistent with conventional U–Pb chronology that shows that these are the earliest formed objects. Most chondrules and other objects have lower inferred $^{26}Al/^{27}Al$ ratios or lack evidence of $^{26}Al_0$ altogether, suggesting they formed at least several million years later, after ^{26}Al had largely decayed. Advances in analytical technology have allowed higher-precision measurements that have recently revealed that there are systematic variations in inferred $^{26}Al/^{27}Al_0$ ratios even among CAIs. MacPherson et al. (2012) found that primitive, unmelted CAIs have initial $^{26}Al/^{27}Al$ ratios of $5.2 \pm 0.1 \times 10^{-5}$, melted CAIs range from 5.17×10^{-5} to 4.24×10^{-5}, and mineral grains in a single CAI have a range of $4.77–2.77 \times 10^{-5}$. Assuming a uniform initial $^{26}Al/^{27}Al$ ratio, the range in unmelted CAIs corresponds to a time of only 40 000 years while the entire range, a factor of two, corresponds to a time span of ≈ 0.7 Ma over which the inner Solar System was hot enough for CAI formation and reprocessing.

While most CAIs apparently formed with an initial $^{26}Al/^{27}Al$ ratio of 5.2×10^{-5}, some CAIs, notably those from the rare CH and CB classes, appear to have formed with initial $^{26}Al/^{27}Al$ ratios an order of magnitude or more lower. Rare CAIs with anomalous isotopic compositions of many elements, referred to as FUN CAIs (FUN is an acronym for Fractionation and Unknown Nuclear isotopic anomalies), have lower $(^{26}Al/^{27}Al)_0$. Significantly, those CAIs with lower $(^{26}Al/^{27}Al)_0$ also differ in their oxygen isotopic compositions from those with the canonical $(^{26}Al/^{27}Al)_0$ of $5.2 \pm 0.1 \times 10^{-5}$ (Krot et al., 2020).

Most AOAs appear to have initial $^{26}Al/^{27}Al$ ratios similar or identical to the "canonical" CAI value of $5.2 \pm 0.1 \times 10^{-5}$, suggesting they formed in the same region at around the same time as CAIs (Larsen et al., 2011). Although Krot et al. (2014) found that anorthite from AOAs in the CH3 chondrite *Acfer 214* had $(^{26}Al/^{27}Al)_0$ of $\approx 4 \times 10^{-5}$, close to the canonical CAI ratio, one had $(^{26}Al/^{27}Al)_0$ of $\approx 1.7 \times 10^{-5}$ and one had no resolvable ^{26}Mg excess at all. Hibonite, grossite, and melilite in CAIs within the AOAs also lacked resolvable ^{26}Mg excesses. They interpreted this as evidence of inhomogeneous distribution of ^{26}Al in the early solar nebula.

Most chondrules have $(^{26}Al/^{27}Al)_0$ lower than those of CAIs and AOAs, implying they formed later. Pape et al. (2019) reported new analyses of 31 chondrules from L and LL OCs. Their data, together with previously published data, show that chondrules in ordinary chondrite chondrules have a well-defined distribution of $^{26}A/^{27}Al$ ages. Individual meteorites contain chondrules of a range of ages, but there is no age distinction between these groups (no chondrule ages are available for H chondrites or enstatite chondrites). The vast majority have $(^{26}Al/^{27}Al)_0$ between 1.3×10^{-5} and 3×10^{-6}, implying formation ages between 1.6 and 2.9 Ma after CAI formation with a peak in the distribution around 2 Ma (Figure 6.11). CO and CV carbonaceous chondrites show a very similar distribution although slightly shifted to younger ages (there are no data on CM or CK chondrites).

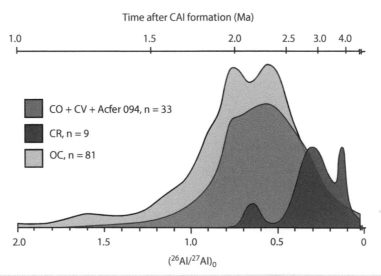

Figure 6.11 Probability density function of $(^{26}Al/^{27}Al)_0$ determined in chondrules from OCs, carbonaceous chondrites of classes CO and CV plus Acfer 094, a unique ungrouped carbonaceous chondrite, and CR chondrites. Source: Pape et al. (2019)/with permission of Elsevier.

Chondrules from CR chondrites form a distinctly younger population with most having formed >3 Ma after CAI formation.

As we found earlier, however, some chondrules have Pb–Pb ages indicating that they formed essentially simultaneously with CAIs, although these appear to be atypical. Bollard et al. (2019) found that chondrules from Allende (CV3) and NWA5697 (L3) with Pb–Pb ages close to those of CAIs ($\approx$4567 Ma) had a mean of $(^{26}Al/^{27}Al)_0 = 4.75 \times 10^{-6}$ while chondrules with Pb–Pb ages about a million years younger had a mean of $(^{26}Al/^{27}Al)_0 = 1.82 \times 10^{-5}$ (Figure 6.12).

This clearly raises the question of whether the $^{26}Al/^{27}Al$ ratio in the early Solar System was uniform. Bollard et al. (2019) interpret the variability in $(^{26}Al/^{27}Al)_0$ values as a consequence of progressive inward transport and admixing of ^{26}Al-rich dust from the outer disk during the period of chondrule formation. This model could be consistent with other evidence of isotopic heterogeneity that we will explore in a subsequent section. There is also some evidence that many meteorite parent bodies formed from material with initial $^{26}Al/^{27}Al$ ratios lower than the "canonical" $(^{26}Al/^{27}Al)_0$

ratio of 5.2×10^{-5}. Angrite achondrites, which have Pb–Pb ages as old as 4564 Ma, appear to have formed from material with an $(^{26}Al/^{27}Al)_0$ of 1.33×10^{-5} (Schiller et al., 2015) and Main Group pallasites appear to have formed from similarly ^{26}Al-poor material (Larsen et al., 2016).

This question of whether ^{26}Al was homogeneously distributed in the early Solar System has important implications for its chronology, particularly for formation of chondrite parent bodies, and it has been debated for decades. With the exception of those of high petrologic grade that have been significantly metamorphosed well after parent body accretion, the materials that form chondrites are not isotopically equilibrated and hence cannot be directly dated. However, they must be younger than the chondrules they contain. If chondrules formed in a reservoir with $(^{26}Al/^{27}Al)_0$ lower than that of CAIs, it shortens formation times of meteorite parent bodies and their components ("by how much" is left to the reader as Problem 3). One alternative approach is to anchor $^{26}Al–^{26}Mg$ ages on meteorites with both precise $^{26}Al–^{26}Mg$ isochrons and Pb–Pb absolute ages, e.g., the angrite D'Orbigny,

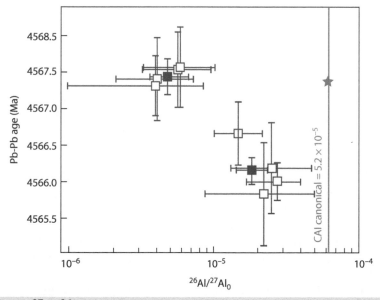

Figure 6.12 Calculated $^{27}Al/^{26}Al$ initial ratios in chondrules from Allende and NWA5697 plotted against their Pb–Pb ages. The oldest chondrules formed simultaneously with CAIs, but in a region of the nebula much poorer in ^{27}Al than the CAI-forming region. Chondrules formed about a million years later had higher $(^{26}Al/^{27}Al)_0$, but still lower than most CAIs. Solid symbols are the averages of the two groups. Star shows the age and $(^{26}Al/^{27}Al)_0$ of CAIs. Data from Bollard et al. (2019).

for which $^{26}Al/^{27}Al_0$ has been precisely determined as $3.98 \pm 0.21 \times 10^{-7}$ and has a Pb–Pb age of 4563.37 ± 0.25 Ma.

The distribution of ^{26}Al also has important implications for planetary differentiation. At levels of 10^{-5} and 10^{-6}, ^{26}Al is abundant enough to be a significant source of heat for young planetary embryos. The existence of igneous achondrites and iron meteorites, as well as differentiated asteroids, such as *Vesta*, shows that temperatures in these relatively small bodies reached the point where both silicates and metal at least partially melted, allowing segregation of metal cores and silicate mantles. Release of gravitational energy during accretion is probably insufficient to raise temperatures to this point, but ^{26}Al could have provided sufficient heat if initial concentrations were high enough and the planetary embryos formed early enough. For these reasons, there is still a need to better understand how ^{26}Al was distributed in the young Solar System and to resolve apparent contradictions.

6.4.2.5 ^{182}Hf–^{182}W

The longer half-life of ^{192}Hf, about nine million years, means that the Hf–W decay system does not have the resolution of the ^{26}Al–^{26}Mg and ^{53}Mn–^{53}Cr systems, but it has its own advantages. Hf is lithophile while W is moderately siderophile; consequently, Hf concentrates strongly in silicate phases while W concentrates strongly in metal, resulting in large variation in Hf/W ratios. As a consequence, the ^{182}Hf–^{182}W decay system has proved extremely useful in dating metal–silicate fractionation and core formation in asteroids; the relatively longer half-life means we can also use the system to constrain the time of core formation in the Earth and Mars.

However, this system has its own set of problems. First, W isotopes are subject to cosmic-ray-induced neutron capture. This is particularly a problem for iron meteorites, which can have quite long cosmic-ray exposure ages. Corrections for this can be made by monitoring the $^{196}Pt/^{195}Pt$ or $^{189}Os/^{188}Os$ ratio. Second, the W isotopic composition is not uniform, with small but measurable variations in s- and r-process contributions to several isotopes, including ^{182}W. This too can be corrected for by monitoring the ^{183}W relative abundance.

Tungsten isotope ratios are reported either in the epsilon notation, e.g., ε_{182W}, which

indicates deviations in parts per 10 000, or the mu notation, μ_{182W}, which indicates deviations in parts per million of the $^{182}W/^{184}W$ from the terrestrial standard value of $^{182}W/^{184}W = 0.866412$. Thus, on this scale, the terrestrial ε_{182W} and μ_{182W} values are 0. Kruijer et al. (2014) analyzed CAIs and determined a Solar System initial $^{182}Hf/^{180}Hf = 1.02 \pm 0.04 \times 10^{-4}$ and an initial $\varepsilon_{182W} = -3.49 \pm 0.07$ ($^{182}W/^{184}W = 0.863388$). To date at least, it appears that the initial $^{182}Hf/^{180}Hf$ ratio was uniform throughout the Solar System, implying a uniform distribution of ^{182}Hf.

With a Solar System initial $^{182}Hf/^{180}Hf = 1.02 \times 10^{-4}$ and ε_{182W}, it is possible to construct a relative timescale based on these values. Budde et al. (2016) found that, after the corrections for nucleogenic and cosmogenic effects mentioned above, chondrules from Allende (CV3) produced a regression-derived $^{182}Hf/^{180}Hf = 8.56 \pm 0.39 \times 10^{-5}$, corresponding to an age of 2.2 ± 0.8 Ma after CAI formation. Carbonaceous chondrites have an average $\varepsilon_{182W} = -1.9 \pm 0.1$; this is presumably the Solar System ratio subsequent to complete decay ("burnout") of ^{182}Hf and a $^{182}Hf/^{184}W = 1.34 \pm 0.11$ (Kleine and Walker, 2017). OCs show a larger spread of ε_{182W} from -2.5 to -1.5, as expected from their more variable metal/silicate and $^{182}Hf/^{184}W$ ratios.

Kleine et al. (2008) used the Hf–W system to date metamorphism on the H chondrite parent body. They obtained internal isochrons of various H chondrites (Figure 6.13) and found that the $^{182}Hf_0/^{180}Hf_0$ ratios decreased with increasing petrologic grades, and hence intensity of metamorphism, corresponding to post-CAI ages increasing from 0.7 Ma for *Ste. Marguerite* (H4) to 9.6 Ma for *Kernouvé* and *Estacado* (both H6). Presumably, this reflects increasing temperatures and slower cooling rates with depth in the H chondrite parent body. The age of the least metamorphosed, *Ste. Marguerite*, overlaps with that of chondrule formation and indeed probably reflects that event. These Hf–W ages are older than the Pb–Pb ages (Figure 6.2), probably reflecting the higher closure temperature of Hf–W compared to U–Pb in phosphates.

Because the Hf–W system is particularly sensitive to metal–silicate fractionation, it is particularly useful in constraining the time of core formation in planets and planetesimals. Kleine et al. (2012) analyzed Hf–W in eight angrites.

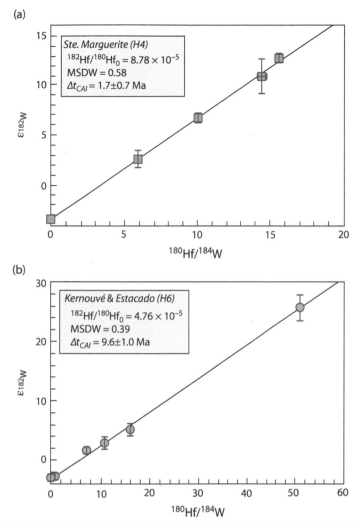

(a)

Ste. Marguerite (H4)
$^{182}Hf/^{180}Hf_0 = 8.78 \times 10^{-5}$
MSDW = 0.58
$\Delta t_{CAI} = 1.7 \pm 0.7$ Ma

(b)

Kernouvé & Estacado (H6)
$^{182}Hf/^{180}Hf_0 = 4.76 \times 10^{-5}$
MSDW = 0.39
$\Delta t_{CAI} = 9.6 \pm 1.0$ Ma

Figure 6.13 $^{182}W/^{184}W-^{182}Hf/^{184}W$ isochrons ($^{182}W/^{184}W$ within ε_{182W} units) for variously metamorphosed H chondrites. (a) *Ste. Marguerite* is modestly metamorphosed and its $(^{182}Hf/^{180}Hf)_0$ corresponds to a relative age of 1.7 Ma after CAI formation. H chondrites consist of 60–80% chondrules, so this age likely represents chondrule formation. (b) More severely metamorphosed meteorites *Kernouvé* and *Estacado* yield a combined $^{182}Hf/^{180}Hf_0$ corresponding to a relative age of 9.6 Ma after CAI formation, representing cooling through the Hf–W closure temperature following metamorphism.

Calculated initial $^{182}Hf/^{180}Hf_0$ fell into two groups. Fine-grained meteorites, such as D'Orbigny, which are effectively lava flows and shallow intrusives, had $^{182}Hf/^{180}Hf_0$ of $\approx 7 \times 10^{-5}$ corresponding to ages of ≈ 4 Ma after CAI formation, while plutonic angrites, such as Lewis Hills 86010, had $^{182}Hf/^{180}Hf_0$ of $\approx 4.4 \times 10^{-5}$ corresponding to ages of 9–11 Ma after CAI formation. They interpreted these results as evidence of two distinct metal–silicate fractionation and core-formation events.

Because their Hf concentrations are effectively 0, model ages of iron meteorites can be calculated directly from measured ε_{182W} (equivalently from $^{182}W/^{184}W$ ratios) as:

$$t = -\frac{1}{\lambda} \ln \left(\frac{\varepsilon^{182}W_{iron} - \varepsilon^{182}W_{chondrites}}{\varepsilon^{182}W_{SSI} - \varepsilon^{182}W_{chondrites}} \right) \quad (6.7)$$

where SSI is Solar System initial. The denominator is fixed at $-3.49 - (-1.91) = -1.59$, so the lower the ε_{182W} of the iron, the older the metal segregated from silicates. Since, with the exceptions noted earlier of the IAB, IIICD, and IIE Group irons, iron meteorites are thought to represent the cores of asteroids, these model ages represent the time of core formation. ε_{182W} in magmatic irons ranges from –3.40 to –3.18. These yield surprisingly early core-formation ages, ranging from 0.3 ± 0.5 to 2.8 ± 0.7 Ma. Assuming that ^{27}Al was the principal source of heat to drive melting and core segregation, these ages imply parent body accretion times of 0.3–1.4 Ma (Kruijer et al., 2017), meaning that planetesimal accretion began almost immediately after CAI formation and was simultaneous with chondrule formation. Equation (6.7) implicitly assumes that the tungsten resided in a reservoir with chondritic ^{180}Hf/^{184}W until the metal segregated; these short timescales imply this assumption is likely approximately valid.

Based on nucleosynthetic isotopic anomalies that we will discuss subsequently, iron meteorites can be associated with either carbonaceous chondrite meteorites (CC type) or non-carbonaceous meteorites (NC type), which include enstatite and OCs, as well as most achondrites. Kruijer et al. (2017) found that those in the CC group are systematically younger than those in the NC group (Figure 6.14). Furthermore, implied parent body accretion times based on the assumption that ^{26}Al decay provided the energy for melting are ≤1 Ma for the NC group while those for the CC group are ≥1 Ma. We will return to this topic in a subsequent section.

The half-life of ^{180}Hf is sufficiently long that ^{180}Hf–^{182}W decay can also be used to constrain, but not precisely date, planetary core formation. Assuming that the age of a planetary core matches the accretion age of the planet, this provides a constraint on the time the terrestrial planets formed. The SNC meteorites provide samples of the silicate portion of Mars allowing us to constrain the age of the Martian core. ε_{182W} values in the meteorites are variable, particularly among nakhlites, and some of this variation correlates with variation in ^{142}Nd, the product of the short-lived nuclide ^{146}Sm, and hence likely relates to silicate–silicate melt fractionation. Shergottites have somewhat less variable ε_{182W} with a mean value of ≈0.37 ± 0.04. In the case of planet-sized objects, the simple two-stage model of residence in a chondritic reservoir prior to metal–silicate

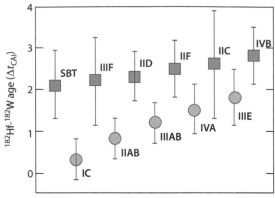

Figure 6.14 ^{182}Hf–^{182}W model ages relative to CAI formation of planetesimal differentiation of core formation of the parent bodies of the main groups of iron meteorites and the rarer South Biron Trio (SBT) group. Iron meteorites that share nucleosynthetic anomalies with carbonaceous chondrites (blue) formed systematically later than those sharing isotopic compositions with non-carbonaceous chondrites and related achondrites (red). Later planetesimal formation and differentiation of the carbonaceous group is consistent with their formation at greater heliocentric distance. Source: Adapted from Hilton et al. (2020).

segregation is not appropriate. We can still calculate a model age, however, provided we know both the ε_{182W} of the Martian mantle and its ^{180}Hf/^{184}W ratio as:

$$t = \frac{1}{\lambda} \ln \left(\frac{\left(\varepsilon^{182}W_{chondrites} - \varepsilon^{182}W_{SSI} \right) f^{Hf/W}}{\varepsilon^{182}W_{mantle} - \varepsilon^{182}W_{SSI}} \right)$$

(6.8)

where $f^{Hf/W}$ is the Hf/W fractionation (fractional difference) relative to chondrites defined as:

$$f^{Hf/W} = \frac{\left(^{180}Hf/^{184}W \right)_{mantle}}{\left(^{180}Hf/^{184}W \right)_{chondrites}} - 1 \quad (6.9)$$

(ignoring small nucleosynthetic variations in the abundance of ^{184}W, the value of f calculated from the ^{180}Hf/^{184}W ratio is identical to that calculated from the Hf/W atomic ratio). Dauphas and Pourmand (2011) estimated the ^{180}Hf/^{184}W ratio of Mars to be 3.8. Substituting these values for Mars and the chondritic and

Solar System initial values mentioned above into Equation (6.8) yields a Martian core-formation time of 4.1 Ma.

This model is, however, probably oversimplistic, as we have found planetesimals in the inner Solar System have already formed and differentiated into silicate mantles and iron cores. To the extent that Mars formed by accretion of these already differentiated bodies, this model would underestimate the age of final Martian core because the cores of the planetesimals would likely have merged as Mars accreted. Furthermore, it is reasonable that the process of accretion would slow as a planet clears out its feeding zone. A reasonable model might be expressed as:

$$F = 1 - e^{-t/\tau} \tag{6.10}$$

where F is the fraction accreted and τ is the accretion timescale such that $F = 63\%$ at $t = \tau$. Dauphas and Pourmand (2011) modeled such a scenario and found that a value $\tau = 2.4$ Ma reproduced the Martian data. In that model, Mars achieves $\approx 50\%$ of its final mass in two million years and $\approx 90\%$ in four million years, but its full mass only after approximately eight million years.

Mars has only a tenth the mass of the Earth and may well have formed more rapidly. Simulations of planetary accretion from the solar nebula have long suggested that the full assembly of large terrestrial planets, Venus and Earth, could be a protracted process that lagged formation of the Solar System by tens of millions of years or more (e.g., Wetherill, 1990). And because all evidence of the accretion process has been erased from the Earth, the well-established age of the Solar System does not establish the age of the Earth. Just as we did for Mars, we should be able to use the Hf–W system to constrain the age of the Earth's core and therefore that of the Earth itself. This requires we first estimate both the $^{182}W/^{184}W$ (ε_{182W}) and Hf/W ratios of the silicate Earth.

Despite discovery of small, typically a few parts per million, variations in $^{182}W/^{184}W$ in the Earth, particularly in Archean rocks, we can say that the ε_{182W} value of the silicate Earth is well established as 0 (variations in ε_{182W} are nonetheless of great significance, and we will examine them in Chapter 7). Establishing the Hf/W value of the silicate Earth is somewhat more problematic as this ratio varies widely due to their differing behavior (among other differences, W is distinctly more incompatible than Hf). Estimates range from 18.8 to 21.5, corresponding to a range of 22.2–25.4 in $^{180}Hf/^{184}W$. Choosing a round number for Hf/W = 20 ± 4 ($^{180}Hf/^{184}W = 23.6 ± 5$; $f^{Hf/W} = 17.5$), a simple two-stage model (Equation [6.9]) yields an age of 34 ± 3 Ma for formation of the Earth's core (Kleine and Walker, 2017). This is nearly 10 times older than what we found for Mars and is consistent with accretion models indicating an extended interval for formation of large terrestrial planets. Yet such a simple two-stage model is likely to be even more oversimplistic for the Earth than for Mars.

Continuous formation models, such as Equation (6.10), suggest an accretion time-scale, τ, of 11 Ma. However, more realistic accretion models in which the Earth and its core form through a series of large collisions produce a large range of results, anything between 30 and 200 Ma. A key unknown in these models is the extent to which silicate and melt equilibrate after each collision and core-merging event, and this is not well constrained.

Accretion of the Earth is widely believed to have culminated with a large impact that formed the Moon (known as the *giant impact hypothesis*). In this case, the Hf–W systematics and the age of the Moon should provide additional constraints on the age of the Earth. As is the case with the isotopic composition of most elements, the W isotopic composition of the Moon is nearly identical to that of the Earth, with $\varepsilon_{182W} = +0.26$ ($\mu_{182W} = +26$), and appears to be nearly uniform despite variations in the Hf/W ratio (Touboul et al., 2015; Kruijer et al., 2017), implying a late formation of the Moon. Touboul et al. (2015) argued that because the Hf/W ratios of the Earth and Moon appear to be nearly identical, the Earth–Moon difference in ε_{182W} is best explained by a greater amount of additional chondritic material accreted to the Earth ($\approx 0.5\%$) than the Moon ($\approx 0.05\%$) after the Moon-forming giant impact. However, Thiemens et al. (2019) analyzed a variety of lunar samples and, together with fractional crystallization modeling, concluded that Hf/W ratio of the Moon is between 30.2 and 48.5, which is much higher than the terrestrial value of

≈ 20. This difference in Hf/W combined with the difference in ε_{182W} implies a time for the giant impact of 40–60 million years after CAI formation.

Geochronological constraints on the age of the Moon are similarly ambiguous. Nyquist et al. (2010) obtained concordant Sm–Nd and Rb–Sr ages of 4.47 ± 0.7 and 4.49 ±0.07, respectively, on lunar anorthosite 67075. Barboni et al. (2017) calculated Lu–Hf model ages as old as 4.51 Ma on lunar zircons, whose U–Pb ages range from 4.335 to 3.969 Ga. Other constraints come from the extinct radionuclide ^{146}Sm and its decay product ^{142}Nd, which we will discuss in the following section. Planetary isochrons give ages of 4336–4385 Ma.

In summary, while we now know the age of the Solar System with exquisite precision, the age of the Earth and its Moon is not known with much better precision than the estimate of Claire Patterson (1956), who concluded based on comparing Pb isotopes in terrestrial materials with meteorites that the age of the Earth was 4550 ± 70 Ma. At best, we can refine this to 4500 ± 60 Ma.

6.4.2.6 ^{146}Sm–^{142}Nd

^{142}Nd is the product of α-decay of the extinct radionuclide ^{146}Sm. Because the behavior of rare earths is so well understood and because ^{146}Sm–^{142}Nd chronology can be combined with ^{147}Sm–^{143}Nd chronology, this is a particularly powerful system. Although the longer half-life of ^{146}Sm limits the utility of ^{146}Sm–^{142}Nd in elucidating very early Solar System processes, it has proved useful in understanding early differentiation of planets, including Mars and the Earth (we will continue with this topic in the next two chapters). As with the other short-lived decay systems we have discussed, however, there are issues to overcome.

The first of these is the half-life of ^{146}Sm. As noted in Table 6.1, Kinoshita et al. (2012) reported a half-life of 68 ± 12 Ma based on alpha-counting experiments (this was the value used in first edition of this book), which differed significantly from previously used 103 ± 5 Ma value based on two earlier alpha-counting determinations. Subsequent calibration against other chronometers by Marks et al. (2014) and Fang et al. (2022) have

yielded half-lives consistent with the 103 Ma half-live of Meissner et al. (1987) and inconsistent with the shorter one of Kinoshita et al. (2012). Marks et al. (2014) calculated a Solar System initial ^{146}Sm/^{144}Sm of 0.00828 ± 0.00044; Fang et al. (2022) determined an identical and slightly more precise value of 0.00840 ± 0.00032. As a consequence, the 103 Ma half-life is the one more widely accepted at present and used here.

A second issue is cosmogenic reactions. This is mainly an issue for lunar samples, some of which have been exposed on the lunar surface to cosmic rays for more than four billion years. Cosmic-ray exposure ages of stony meteorites are generally a few tens of millions of years at most, which are usually not enough to affect Nd isotopic composition and it is not an issue for terrestrial samples.

The third issue is that there are variations in the Nd isotopic compositions of meteorites that are clearly nucleosynthetic, including variations in the abundance of ^{142}Nd, which presents a challenge in distinguishing nucleosynthetic variations in ^{142}Nd/^{144}Nd from radiogenic ones. We consider this issue in more detail below.

Lugmair et al. (1975) reported a ^{146}Sm–^{142}Nd isochron from the eucrite Juvinas indicating an initial ^{146}Sm/^{144}Sm ratio of 0.0054. Since the uncertainty was ± 0.0072, this was not significantly different from 0, but did raise the possibility of the presence of ^{146}Sm for the early Solar System. Prinzholder et al. (1992) confirmed this with ^{146}Sm–^{142}Nd isochrons in the eucrite *Ibitira* and the mesosiderite *Morristown* corresponding to initial ^{146}Sm/^{144}Sm ratios of 0.0090 ± 0.0010 and 0.0075 ± 0.0011, respectively (within error of the current best estimate of the initial ^{146}Sm/^{144}Sm is 0.0084). The half-life of ^{146}Sm is sufficiently long that some should have persisted after the Earth formed and therefore that variations in ^{142}Nd/^{144}Nd terrestrial samples should also be expected. A study by Harper and Jacobsen (1992) reported a 33 ppm excess of ^{142}Nd compared to laboratory standards in one 3.8 Ga old metavolcanic rock from Isua. The result was controversial for almost a decade as other workers failed to confirm it until Caro et al. (2003) and Boyet et al. (2003) both documented ^{142}Nd/^{144}Nd ratios in Isua rocks higher than modern terrestrial standards. Subsequently, variations in ^{142}Nd/^{144}Nd ratio have been found in many Archean rocks around the world. These are reported in the epsilon, $\varepsilon^{142}{}_{Nd}$, or mu, $\mu^{142}{}_{Nd}$, notations, which like $\varepsilon^{182}{}_W$ and $\mu^{182}{}_W$, are deviations in parts in 10 000 or 1 000 000, respectively, from a *terrestrial standard*. It is important to distinguish this notation from ε_{Nd}, which is deviations from the $^{143}Nd/^{144}Nd$ *chondritic* value.

As it turns out the modern terrestrial $\varepsilon^{142}{}_{Nd}$ differs from the chondritic one. This unexpected result came when Boyet and Carlson (2005) analyzed the $^{142}Nd/^{144}Nd$ ratios of meteorites and found that modern terrestrial igneous rocks had $^{142}Nd/^{144}Nd$ ratios that average 20 ppm or 0.2 epsilon units higher than OCs. This was surprising as the Earth is thought to have a chondritic Sm/Nd, which should lead to a chondritic $\varepsilon^{142}{}_{Nd}$. The 50% nebular condensation temperatures of Nd and Sm are very similar, 1602 and 1590 K, respectively, and the variation of the Sm/Nd ratio in carbonaceous and OCs is quite limited ($\approx 3\%$) (enstatite chondrites show more variation). It therefore seems unlikely that nebular processes could produce significant fractionation of Sm from Nd, which raised the question of how the difference in $\varepsilon^{142}{}_{Nd}$ between the Earth and chondrites could arise.

Boyet and Carlson (2005) suggested that cooling and differentiation of an early terrestrial magma ocean produced a dense basaltic crust that was more enriched in Nd than Sm, which they referred to as an *early enriched reservoir*. They hypothesized that this crust sank into the deep mantle where it remains because of its high density (HD), perhaps in enigmatic structures known as large low shear-wave velocity provinces or LLSVPs subsequently identified through seismic tomography. Alternatively, Caro et al. (2008) and O'Neill and Palme (2008) suggested that rather than sinking into the deep mantle, this early crust was blasted away as the growing Earth collided with other bodies, a process known as "collisional erosion." The later stages of planetary accretion involve infrequent, energetic collisions between large bodies producing extensive melting of the Earth. Caro et al. (2008) hypothesized that a substantial fraction of an incompatible element-enriched crust formed during terrestrial magma ocean differentiation was blasted away in these collisions, leaving the Earth depleted in elements that were concentrated in that crust, Nd more so than Sm.

Subsequently, small variations in the relative abundance of non-radiogenic isotopes of Nd and Sm (and Ba and Sr as well) were found in chondrites (e.g., Andreasen and Sharma, 2006; Carlson et al. 2007). These have now been amply confirmed by many studies (e.g., Burkhardt et al, 2016; Bouvier et al., 2016). The carbonaceous, ordinary, and enstatite chondrite groups each appear to have different $^{142}Nd/^{144}Nd$ after correction for decay of ^{146}Sm (Figure 6.14). Mean radiogenic-corrected $\mu^{142}{}_{Nd}$ is -32 ± 13 for carbonaceous chondrites, -15 ± 8 for OCs, and -10.5 ± 14 for enstatite chondrites (Boyet et al., 2018). Furthermore, radiogenic-corrected $^{142}Nd/^{144}Nd$ ratios correlate with other Nd isotope ratios, such as $^{148}Nd/^{144}Nd$ ratios (Figure 6.14). We can interpret these correlations in terms of the nucleosynthetic processes discussed in Chapter 1. ^{142}Nd is primarily an s-process nuclide produced in red giants, with a small fraction ($<5\%$) produced by the p-process. ^{144}Nd and ^{145}Nd are produced both by the s- and r-processes, but the s-process produces relatively more ^{144}Nd while the r-process produces relatively more ^{145}Nd. ^{148}Nd and ^{150}Sm are essentially exclusively produced by the r-process. Indeed, Saji et al. (2020) argue that three nucleosynthetic components are necessary to explain the variation in abundance of ^{142}Nd: variations in the s-process component, those resulting from variations in the pure p-process component, and those resulting from coupled s-process and p-process variation. The correlation in Figure 6.15 thus likely reflects incomplete mixing of material synthesized in different stellar environments (e.g., r-process material from supernovas and kilonovas and s-process material from red giants).

Both Qin et al. (2011) and Boyet and Gannoun (2013) performed stepwise dissolution experiments that demonstrated internal isotopic heterogeneity in meteorites, indicating these distinct nuclear components were present prior to accretion of meteorite parent bodies. In both cases, the most resistance fractions were enriched in s-process nuclides (and deficient in r-process nuclides). Although neither study identified the carrier phase, a variety of presolar grains (discussed in Section 6.5) that are enriched in s-process nuclides have been identified, including SiC and graphite, both of which are resistant to dissolution.

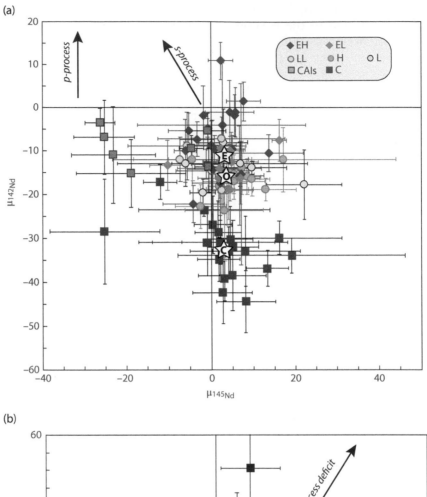

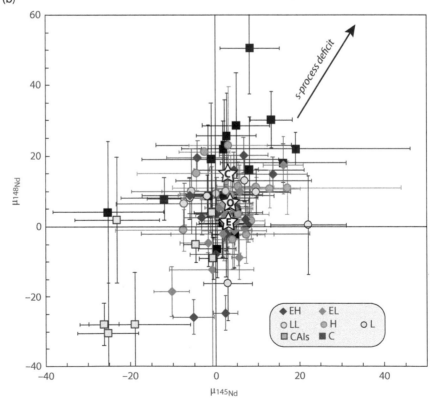

Figure 6.15 Variations in the Nd isotopic composition of chondrites and CAIs. (a) μ^{142}_{Nd} versus μ^{145}_{Nd} (b) μ^{148}_{Nd} versus μ^{145}_{Nd}, and (c) μ^{142}_{Nd} versus μ^{148}_{Nd} (μ^{148}_{Nd}, μ^{145}_{Nd}, and μ^{142}_{Nd} are the deviations in parts per million from the $^{148}Nd/^{144}Nd$, $^{145}Nd/^{144}Nd$, and $^{142}Nd/^{144}Nd$ ratios, respectively). Stars with the letters E, O, and C represent average of these ratios reported by Boyet et al. (2018) for enstatite, ordinary, and carbonaceous chondrites, respectively. Arrows indicate the effects of additions or deficits in r-, p-, and s-process nucleosynthetic products. Source: Adapted from Boyet et al. (2018).

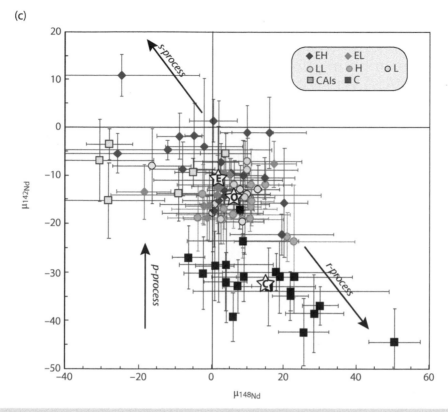

(c)

Figure 6.15 (Continued)

Nucleosynthetic anomalies have now been found in many heavy elements. We will postpone discussion of their origin and distribution to a following section, but we note here that some provide a sort of genetic isotopic fingerprint allowing us to relate different meteorite groups and their parent bodies, much as DNA can delineate evolutionary relationships between organisms. Based on this, the Earth and Moon seem most closely related to enstatite chondrites. As Figure 6.15 shows, the terrestrial μ^{142}_{Nd} falls within the enstatite chondrite range although on average $^{142}Nd/^{144}Nd$ in enstatite chondrites differs from the terrestrial value. However, while much of the difference between $^{142}Nd/^{144}Nd$ in chondrites and the Earth can be explained by Solar System isotopic heterogeneity, based on new high precision analyses, Johnson et al. (2022) concluded that Earth's Sm/Nd ratio is 2.4% higher than the average for

chondrites, which is consistent with recent mass balance calculations of Hofmann et al. (2022).

Despite these complications, the $^{146}Sm-^{142}Nd$ system can provide insights into the early history of planetary bodies, such as Mars. The Martian meteorites exhibit heterogeneity in both ε^{182}_W and ε^{142}_{Nd}. As we pointed out in the previous section, tungsten isotopes suggest a quite short timescale for Martian accretion and core formation of ≈ 2 Ma. Debaille et al. (2007) found that most shergottites fell into two groups: "enriched" shergottites with $\varepsilon^{142}_{Nd} \approx -0.2$ and depleted shergottites with $\varepsilon^{142}_{Nd} \approx +0.6$. *Chassigny* and the nakhlites also have $\varepsilon^{142}_{Nd} \approx +0.6$. Shergottites have $\varepsilon^{182}_W \approx 0.45$ whereas *Chassigny* and the nakhlites have ε^{182}_W of +2 to +3.2.

Despite their relatively young ages (≈ 200 and 500 Ma for shergottites and ≈ 1.3 Ga for chassignites and nakhlites), these Martian igneous rocks retain a memory of mantle differentiation that occurred within the first 100 Ma. The Martian meteorites crystalized long after ^{146}Sm was extinct; can we still constrain when that early differentiation occurred?

Caro et al. (2008) attempted to estimate that age by assuming a simple two-stage evolution of the Sm–Nd system for the evolution of the shergottites. In the first stage, Mars evolves with a chondritic Sm–Nd ratio (^{147}Sm/^{144}Nd = 0.1966) from the Solar System initial ratios of ^{143}Nd/^{144}Nd = 0.50668 and ^{146}Sm/^{144}Sm = 0.008. Following mantle differentiation at t_1, the mantle source of each meteorite evolved to its ^{143}Nd/^{144}Nd at the time of crystallization (i.e., ^{143}Nd/^{144}Nd$_i$), t_2, with a fixed ^{147}Sm/^{144}Nd$_{source}$, which can be calculated as:

$$\left(\frac{^{147}\text{Sm}}{^{144}\text{Nd}}\right)_{source} = \frac{\left(^{143}\text{Nd}/^{144}\text{Nd}\right)^{sample}_{t_2} - \left(^{143}\text{Nd}/^{144}\text{Nd}\right)^{Mars}_{t_1}}{e^{\lambda t_1} - e^{\lambda t_2}}$$

(6.11)

They then made a first guess of t_1 and plotted the measured ^{142}Nd/^{144}Nd against the calculated ^{147}Sm/^{144}Nd$_{source}$ (Figure 6.16). This plot is analogous to plots, such as Figure 6.6, where the slope is proportional to ^{146}Sm/^{147}Sm at t (^{146}Sm/^{144}Sm is then readily calculated from ^{146}Sm/^{147}Sm by multiplying by ^{147}Sm/^{144}Sm = 4.88899). Excluding *NWA1183*, which falls off the correlation, they then iteratively refined t_1 to produce the best fit to the ^{142}Nd/^{144}Nd–^{147}Sm/^{144}Nd$_{source}$ correlation. Figure 6.15 shows these data with a regression slope = 0.001225 calculated using IsoplotR (Versmeesch, 2018). The slope corresponds to a ^{146}Sm/^{147}Sm = 0.0060. Assuming a Solar System initial ^{146}Sm/^{147}Sm = 0.0085 and a half-life of 103 Ma, this corresponds to 52 Ma after CAI formation and t_1 = 4515.3 Ma (Caro et al. found t_1 = 40 Ma using a Solar System initial ^{146}Sm/^{147}Sm = 0.0080). Caro et al. (2008) interpreted this as the age of the crystallization of a Martian magma ocean. Chassigny and the nakhlite fall off the correlation, indicating that their sources experienced more complex evolution.

6.4.2.7 Other extinct radionuclides: ^{10}Be, ^{36}Cl, ^{41}Ca, and ^{60}Fe

The three remaining short-lived radionuclides listed in have not found application in quantitative cosmochronology, but they do place

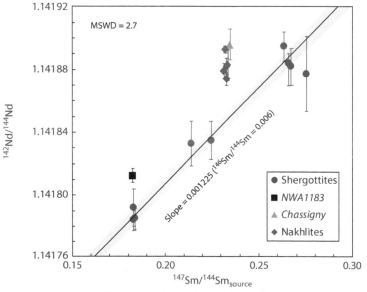

Figure 6.16 ^{142}Nd/^{144}Nd plotted against ^{147}Sm/^{144}Nd$_{source}$ of SNC meteorites where ^{147}Sm/^{144}Nd$_{source}$ is calculated as described in the text following Caro et al. (2008). The regression slope (black line, gray area shows the uncertainty) is 0.001225 ± 39 and can be converted to ^{146}Sm/^{144}Sm by multiplying by ^{147}Sm/^{144}Sm = 4.88899. Assuming a Solar System ^{146}Sm/^{147}Sm = 0.0085 at the time of CAI formation, the ^{146}Sm/^{144}Sm value corresponds to 52 Ma after CAI formation. Data from Debaille et al. (2007) and Caro et al. (2008).

constraints on the environment and timing of synthesis of short-lived radionuclides.

^{10}Be, which decays to ^{10}B with a half-life of 1.39 Ma, differs from the other radionuclides we have considered because it is not synthesized in stellar interiors; indeed, it is consumed in them. It is instead exclusively synthesized by spallation reactions, including neutrino spallation reactions. McKeegan et al. (2000) found that ^{10}B/^{11}B correlated with Be/B ratios with the slope of the correlation indicating an initial ^{10}Be/^{9}Be of 9.5 ± 1.9 × 10^{-4} in *Allende* CAIs. Subsequent studies found ubiquitous evidence of the presence of ^{10}Be at the time of their formation of CAIs in CV chondrites, but found a range of initial (^{10}Be/^{9}Be) from 5 to 9 × 10^{-4}; a similar ratio of 5 × 10^{-4} was inferred in hibonite from *Murchison* (CM2). Fukuda et al. (2021) found initial ^{10}Be/^{9}Be in individual melilites in CAIs in *Yamato-81020* (CO3) from 2.2 ± 1.0 to 4.4 ± 0.8 × 10^{-3} (Figure 6.17) with overall initial ^{10}Be/^{9}Be ratios for the two studied CAIs of 2.9 ± 0.6 and 2.2 ± 1.0 × 10^{-3}, significantly higher than in CAIs from CV chondrites. Gounelle et al. (2013) found that only 5 of 21 CAIs in *Isheyevo* (CH/CB3) showed evidence of ^{10}Be; those 5, however, lie on a common isochron corresponding to (^{10}Be/^{9}Be)$_0$ = 1.31 ± 0.43 × 10^{-3}. As we noted in Section 6.4.1, CB chondrites are unusual in that their Pb–Pb ages of their chondrules are some 2–4 Ma younger than those of other chondrites and are thought to have formed

through impact jetting. Thus, their high (^{10}Be/^{9}Be)$_0$ are particularly significant.

The wide range of inferred initial ^{10}Be/^{9}Be ratios in CAIs suggests production of ^{10}Be was ongoing as the CAIs were forming. This inference is supported by evidence for the presence of ^{7}Be as well, also a spallogenic nuclide but with a half-life of only 53 days and by large variations in Li isotope ratios in CAIs. That in turn suggests ^{10}Be was not inherited from the molecular cloud from which the Solar System formed, but rather was produced in high-energy environments within it.

^{36}Cl, with a half-life of ≈300 000 years, is another very short-lived radionuclide that was present in the early Solar System. Ninety-eight percent of decays are to ^{36}Ar and the remaining 2% to ^{36}S. As we learned in Chapter 4, it can be produced through neutron capture by ^{35}Cl; it can also be produced in a variety of stellar environments, such as supernovae and AGB and Wolf–Rayet stars. A number of studies have found correlations between ^{36}S/^{34}S and ^{35}Cl/^{34}S in Cl-rich minerals, such as sodalite (Na$_8$(Al$_6$Si$_6$O$_{24}$)Cl$_2$) and wadalite (Ca$_6$Al$_5$Si$_2$O$_{16}$Cl$_3$) in CAIs from carbonaceous chondrites, implying (^{36}Cl/^{35}Cl)$_0$ in the range of 5 × 10^{-6} to 1.8 × 10^{-5} (e.g., Lin et al., 2005). These minerals are secondary and formed as Cl-bearing hydrous fluids reacted with primary Ca–Al–Si minerals, such as hibonite in the CAIs after they formed, most likely on the meteorite parent bodies or alternatively with Cl condensing directly from cooling nebular gas. Low inferred initial ^{26}Al abundances in these minerals indicate the alteration occurred well after (>1 Ma) the CAIs originally formed. Although ^{36}Ar is the primary product of ^{36}Cl decay, attempts to find evidence of radiogenic ^{36}Ar failed with the exception of Turner et al. (2013), who calculated a much lower (^{36}Cl/^{35}Cl)$_0$ of 1.9 ± 0.5 × 10^{-8} in sodalite in an Allende CAI. The sodalite yielded an I–Xe formation age of 4559.4 Ma, confirming that it formed long after the CAI. Turner et al. speculated that the lower inferred (^{36}Cl/^{35}Cl)$_0$ could reflect diffusive loss of ^{36}Ar for ≈1 Ma following sodalite formation.

Since ^{36}Cl has been inferred only in secondary minerals formed well after CAI formation, the ^{36}Cl/^{35}Cl would have been orders of magnitude higher (10^{-3} to 10^{-2}) at the time of CAI formation. While ^{36}Cl can be produced in

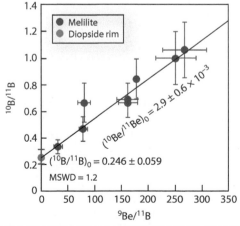

Figure 6.17 ^{10}Be/^{9}Be isochron for melilites and diopside in a CAI from Yamato-81020 (CO3) determined by ion probe. Source: Adapted from Fukada et al. (2021).

evolved or exploding stars, these ratios are well above those predicted for stellar synthesis. Thus, irradiation by energetic particles from the proto-Sun is generally favored for ^{36}Cl production. The ^{36}Cl was likely produced in nebular gas or icy dust before it was incorporated in these secondary minerals (Leya et al., 2018). Sulfur is also a late condensing element and the ^{36}S might also have been introduced well after it was produced by ^{36}Cl decay, in which case the ^{36}S/^{34}S–^{35}Cl/^{34}S correlations may be mixing lines rather than isochrons.

^{41}Ca is another identified extinct radionuclide with a quite short half-life (100 000 years) and an unrealized potential for cosmochronology. As is the case with ^{36}Cl, it can be produced both in a variety of stellar environments and through reactions with energetic particles, such as:

$$^{40}Ca\left(\alpha, {}^{3}He\right){}^{41}Ca$$

It then beta decays to ^{41}K. Early studies identified excesses of ^{41}K that correlated with Ca/K ratios in CAIs from CV and CM chondrites. These correlations indicated ^{41}Ca/^{40}Ca ratios of $\approx 1.4 \times 10^{-8}$ at the time of CAI formation. Evidence of ^{41}Ca was found only in CAIs with $(^{26}Al/^{27}Al)_0$ near the "canonical" value of $\approx 5 \times 10^{-5}$. However, with improved ion probe instrumentation, Liu et al. (2012) determined $(^{41}Ca/^{40}Ca)_0$ ratios in two CAIs from the CV3 chondrite *Efremovka* of $\approx 2 \times 10^{-9}$. When age corrected for $(^{26}Al/^{27}Al)$ ratio to the canonical value of $\approx 5 \times 10^{-5}$, this implied a Solar System initial ratio of 4×10^{-9}. Subsequently, Liu (2017) determined $(^{41}Ca/^{40}Ca)_0$ ratios in two CAIs from different CV3 chondrites, *NWA3118* and *Vigarano*. The CAI from NWA3118 yielded a $(^{41}Ca/^{40}Ca)_0$ of 4.6×10^{-9}, but found no resolvable radiogenic ^{41}K excess in the *Vigarano* CAI despite this CAI having a canonical $(^{26}Al/^{27}Al)_0$ ratio. This implies that the distribution of ^{41}Ca in the early Solar System was heterogeneous, which certainly compromises its cosmochemical value but provides further constraints on the origin of short-lived radionuclides in the early Solar System.

Unlike ^{10}Be, ^{36}Cl, and ^{41}Ca, ^{60}Fe is synthesized only in stellar interiors. It decays via β^- emission with a half-life of 5.3 years to ^{60}Co, which then decays to ^{60}Ni with a half-life of 2.6 Ma. The first evidence of ^{60}Fe in the early

Solar System was reported by Shukolyukov and Lugmair (1993) in the eucrite *Chervony Kut*. A poorly constrained pseudo-isochron indicated $(^{60}Fe/^{56}Fe)_0$ of 4×10^{-9}, which would imply a $(^{60}Fe/^{56}Fe)$ at the time of CAI formation of $\approx 10^{-8}$. Subsequent studies of the Fe–Ni system, however, have reported a range of initial ^{60}Fe/^{58}Fe and a number of them failed to find evidence of radiogenic ^{60}Ni. There are a number of factors responsible for this. The first is the analytical challenges as Ni concentrations in many materials are quite low. In particular, CAIs lack Fe because they formed at temperatures above the condensation temperature of Fe. Second, non-radiogenic nucleosynthetic variations have been demonstrated in Ni in early Solar System materials. Third, subsequent thermal and hydrothermal processing has in many cases disturbed Fe–Ni distribution, which can lead to both underestimate and overestimate of initial $(^{60}Fe/^{56}Fe)_0$ ratios.

The most recent studies using improved analytical techniques leave the question of the ^{60}Fe/^{56}Fe ratio in the early Solar System incompletely resolved, but suggest the value was low. Analysis of a chondrule from *Semarkona* (LL3) by Trappitsch et al. (2018) yielded a $(^{60}Fe/^{56}Fe)_0$ ratio of $3.8 \pm 6.9 \times 10^{-8}$. Telus et al. (2018) found that of 24 chondrules examined from unequilibrated OCs (i.e., petrologic grade 3), only a few had resolvable excesses of ^{60}Ni. They concluded that the $(^{60}Fe/^{56}Fe)_0$ in the chondrule-forming region was between 0.5 and 3×10^{-7}, implying a $(^{60}Fe/^{56}Fe)_0$ at the time of CAI formation of 0.9 and 5×10^{-7}.

6.4.2.8 Origin of short-lived nuclides

Without question, the short-lived radionuclides provide evidence of nucleosynthesis relatively shortly before, and perhaps during, Solar System formation. It is also beyond question that at least two distinct nucleosynthetic environments are required: ^{10}Be is synthesized only through reactions with high-energy particles while ^{60}Fe and heavier elements are synthesized only in the interiors of large stars or cataclysmic events, such as kilonovas and supernovae. Some nuclides, ^{36}Cl and ^{41}Ca, can be synthesized in both ways. Beyond that, however, there is considerable debate over several possibilities outlined below.

First, the presence of these nuclides may simply reflect a steady-state galactic background concentration of the nuclides supported by more or less continuous production, on a galactic scale, in dying large stars followed by injection into the ISM by stellar winds and supernovae or, in the case of ^{10}Be, production by galactic cosmic rays.

Second, they could reflect enhanced production within the giant molecular cloud in which the Solar System formed because massive, short-lived stars are concentrated within such clouds and may have seeded the cloud with enhanced levels of these short-lived nuclides in their death throws. This could include ^{10}Be as magnetic fields could serve to trap and focus cosmic rays within the cloud.

Third, their synthesis could have been directly linked to formation of the Solar System. A nearby supernova explosion or enhanced winds from a red giant or Wolf–Rayet star may have triggered collapse of part of the giant molecular clouds and initiated Solar System formation. There is evidence for such an event in a star-forming cloud L1251 in the constellation Taurus. Finally, high-energy particles from the proto-Sun itself may have been responsible for ^{10}Be (as well as ^{36}Cl and ^{41}Ca) synthesis.

These possibilities are not mutually exclusive, and several papers have indeed argued that multiple sources are required. The keys to discriminating between these possibilities are their initial abundance in the early Solar System and the homogeneity or lack thereof in their distribution. High abundance and heterogeneous distribution favors local production. For example, the galactic background ^{60}Fe/^{56}Fe from gamma-ray surveys and chemical evolution models has been estimate at $\approx 3 \times 10^{-8}$; initial Solar System production at or below this level would imply that extant ^{60}Fe in the forming Solar System was simply inherited from the galactic background with no need for local synthesis.

A useful approach to deciding among these possibilities is to consider the relationship between Solar System initial isotope ratios, such as ^{26}Al/^{27}Al, to astrophysical estimates of production rates in various environments (Wasserburg et al., 1996; Jacobsen, 2005).

Let us consider the first possibility, namely, that the short-lived radionuclides present when the Solar System formed simply reflect their abundances in the ISM as a consequence of continuous synthesis within the galaxy. For simplicity, we will assume that both stable and radionuclides are added to the ISM at a constant rate. For a stable nuclide, such as ^{27}Al, the abundance is simply:

$$dN_S = \int_0^T P_S dt = P_S T \qquad (6.12)$$

where N_S is the number of atoms of the stable nuclide, P_S is its production rate, and T is the age of the galaxy at the time of Solar System formation. Assuming the Milky Way formed along with the first galaxies about 500 Ma after the Big Bang, this would be 7.7 Ga.

For a radioactive nuclide, the math is exactly analogous to production by cosmic rays that we discussed in Chapter 4 and we merely need to integrate Equation (4.20):

$$dN_R = \int_0^T (P_R - \lambda N_R) dt \qquad (6.13)$$

where $1/\lambda \ll T$, which is for a radionuclide whose mean life is much shorter than the age of the galaxy, and this reduces to:

$$N_R = \frac{P_R}{\lambda} \qquad (6.14)$$

which is the same as Equation (4.24) in Chapter 4. Dividing Equation (6.14) by Equation (6.12), we expect the isotope ratio of a radioactive element in the ISM to be simply:

$$\frac{N_R}{N_S} = \frac{P_R}{P_S} \frac{1}{T\lambda} \qquad (6.15)$$

We can compare the observed isotope ratios from this simple model with those predicted in chondrites. The ratio of the observed to predicted is simply:

$$\frac{N_R/N_S}{P_R/P_S} = \frac{1}{T\lambda} \qquad (6.16)$$

Comparing this with the ratios of observed radiogenic ratios we found in the previous sections to production rates from Young (2014), this model overpredicts most short-lived radioactive isotope ratios in the early Solar System. However, while continuous production may be appropriate on the 10^9 yr timescale, nucleosynthesis of heavy elements occurs primarily in dying massive stars, which is very definitely a discontinuous on timescales of the mean lives of the shorter-lived radionuclides.

Let us consider a somewhat more sophisticated model that takes account of this intermittent production. We imagine that nucleosynthesis occurs at discrete intervals of δt over the age of the galaxy, T. In this case, the ratio of abundances to production rates is (Lugaro et al., 2014):

$$\frac{N_R/N_S}{P_R/P_S} = \frac{\delta t}{T}\left(1 + \frac{e^{-\lambda \delta t}}{1 - e^{-\lambda \delta t}}\right) \qquad (6.17)$$

If the interval between nucleosynthetic events is long compared to the mean life of the radioactive nuclide, i.e., if $\delta t \gg 1/\lambda$, then it will matter how much time has passed since the most recent event. We can account for this with the additional term $e^{-\lambda \Delta t}$ where Δt is the time since the last nucleosynthetic event (Wasserburg et al., 2006), and we obtain:

$$\frac{N_R/N_S}{P_R/P_S} = \frac{\delta t}{T}\left(1 + \frac{e^{-\lambda \delta t}}{1 - e^{-\lambda \delta t}}\right)\left(e^{-\lambda \Delta t}\right) \qquad (6.18)$$

Using this formulation, Young (2016) calculated abundance to production rate ratios for various combinations of δt and Δt, which are compared with observed abundance/calculated production ratios in Figure 6.18 (^{232}Th is used as the normalizing nuclide because its half-life is very long, greater than the age of the universe). Different groups of nuclides can be fit to this model using different parameters. The abundances of the longest-lived nuclides, ^{238}U, ^{235}U, ^{146}Sm, ^{244}Pu, and ^{129}I (green), are largely insensitive to the nucleosynthesis frequency ($\delta t = 10$ Ma used in the example in Figure 6.18) and can be fit to a model where the time since the last event, Δt, is 100 Ma. Significantly, these are all r- or p-process nuclides produced in supernovas and kilonovas. ^{182}Hf and ^{107}Pd, which are shorter-lived, require a model with a somewhat more recent last event, 40 Ma in Figure 6.18. ^{107}Pd is an s-process nuclide produced primarily in AGB stars. ^{182}Hf was thought to be exclusively an

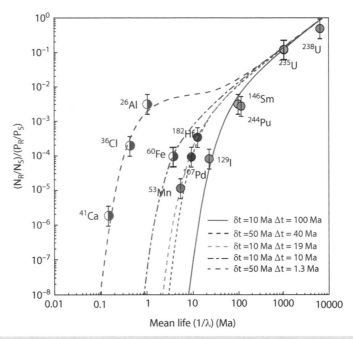

Figure 6.18 Abundance/production ratios of short-lived radionuclides found in chondrites as a function of their mean lives. Solid symbols are based on abundance ratios in Table 6.1 and production ratios from Young (2014). Green symbols are nuclides produced by the r- and p-processes in explosive nucleosynthesis. Orange and brown symbols are nuclides produced by the r-process and s-process in the giant phase of stars (^{53}Mn is produced by silicon burning). ^{60}Fe (blue) is produced by silicon burning and the s-process. Red symbols are nuclides produced by the s-process (and also by spallation). Dashed lines show the prediction from Equation (6.18) using various combinations of parameters. Half-filled symbols for ^{56}Fe, ^{26}Al, ^{36}Cl, and ^{41}Ca are production rates modified by enhanced trapping of Wolf–Rayet winds within giant molecular clouds.

r-process nuclide, but Lugaro et al. (2014) found that the intervening nuclide, ^{181}Hf, is sufficiently stable (half-life 42.1 days) that significant s-process production can also occur in AGB and Wolf–Rayet stars. ^{53}Mn also fits this case, but it is produced primarily during Si burning. ^{60}Fe, produced primarily in supernovae, requires both a relatively frequent production (δt = 10 Ma) and relatively recent last event (10 Ma). The abundances of the shortest-lived nuclides ^{41}Ca, ^{36}Cl, and ^{26}Al are relatively insensitive to synthesis frequency, but require a relatively recent event; Δt = 1.3 Ma is used in the example in Figure 6.18. They all are also primarily produced by Wolf–Rayet stars. Thus, while all the short-lived radionuclides (except for ^{10}Be) can be explained by various combinations of model parameters, no one set of model parameters can simultaneously explain all of them.

Stars do not form in empty interstellar space; they form in giant molecular clouds, which themselves cluster in distinct star-forming regions, such as the spiral arms of galaxies, so perhaps it is no surprise these models do not fit. (Indeed, spiral arms of galaxies are bright because of the concentrations of massive, highly luminous O and B class stars within them.) Let us now consider the second case of nucleosynthesis within star-forming giant molecular clouds. These are stellar nurseries and may give birth to hundreds or thousands of stars over the course of their $\approx 10^8$ yr of existence before they disperse. While most of these stars will be low mass, long-lived stars, such as the Sun, which eventually "fledge" and wander out of their nurseries, some will be massive O and B class stars with life expectancies of only a few million years and hence most will spend their entire existence within the cloud (the red supergiant Betelgeuse seems to be an exception and appears to have escaped its birth environment the Orion nebula in its short ≈ 10 Ma life), seeding it with newly formed nuclei in their death throes. Thus, nucleosynthesis will be concentrated in such clouds. The newly formed nuclei not incorporated in new stars will eventually disperse into the ISM. Nonetheless, giant molecular clouds and ISM represent two distinct reservoirs with higher concentrations of short-lived radionuclides in giant molecular clouds than the ISM. Assuming a constant flux of nuclei between giant molecular clouds and the ISM,

Jacobsen (2005) modeled this with the following equation:

$$\frac{N_R/N_S}{P_R/P_S} = \frac{1}{[(1-x_{MC})\tau_{MC}]/\lambda + 1} \times \frac{1}{\lambda T} \quad (6.19)$$

where τ_{MC} is the residence time of the nuclides in the molecular cloud, x_{MC} is the mass fraction of the ISM comprised of giant molecular clouds (about 17%), and T is again the age of the galaxy.

Young (2016) argued that s-process nuclei produced by Wolf–Rayet stars are more likely to be retained within molecular clouds than r- and p-process nuclides produced in explosive nucleosynthesis since they are less energetically injected. Young adjusted production rates in Equation (6.19) by assuming that nuclei produced in Wolf–Rayet stars are more efficiently trapped within giant molecular clouds than those produced by supernovae. Assuming a residence time, τ_{MC}, in giant molecular clouds of 200 Ma, he was able to approximately reproduce the abundance/production ratios of all the short-lived radioactive isotope ratios (excepting again ^{10}Be).

Thus, a plausible case can be made that the Solar System's inventory of at least those short-lived radionuclides synthesized in stellar environments could simply have been inherited from the molecular cloud rather than a specific nucleosynthetic event occurring shortly before or during Solar System formation. On the other hand, heterogeneous distribution of such nuclides would seem to favor, but not necessarily require, a specific nucleosynthetic event. As we will see in the following sections, stable isotope ratios of many elements exhibit systematic variations as well. Debate on whether a specific nucleosynthetic event occurred shortly before or during Solar System formation continues.

6.5 STARDUST

In addition to the isotopic anomalies that resulted from decay of short-lived radionuclides, there are other isotopic anomalies in meteorites that are not due to such in situ decay. Many of these anomalies, like those created by decay of extinct radionuclides, may reflect the injection of newly synthesized material into the cloud of dust and gas from which the Solar System ultimately formed. Alternatively, they may reflect isotopic inhomogeneity within this cloud, and the variable abundance of exotic gas and grains of material synthesized at various times and places in the galaxy. Still other isotopic anomalies may reflect chemical fractionations within this cloud. We focus on these anomalies in this section.

6.5.1 Neon alphabet soup and "presolar" noble gases in meteorites

Noble gases were the first group of elements in which isotopic variations in meteorites were identified, and these variations occur in virtually all of the carbonaceous chondrites that have not experienced extensive metamorphism. The reason for this is that noble gases can be readily extracted simply by heating the sample sufficiently. Much of the isotopically distinct noble gas is contained in the matrix that accreted at low temperature (below 100–200°C). Noble gases are present in meteorites at concentrations that are often as low as one part in 10^{10}. Although they can be isolated and analyzed at these concentrations, their isotopic compositions are nonetheless partly sensitive to change due to processes, such as radioactive decay (for He, Ar, and Xe), spallation and other cosmic-ray-induced nuclear processes, and solar wind implantation. In addition, mass fractionation can significantly affect the isotopic compositions of the lighter noble gases (He and Ne) and the heavier ones to a lesser extent. Up to the late 1960s, it was thought that all isotopic variations in meteoritic noble gases were related to these processes. For example, Ne isotopic variations could be described as mixtures of three components, "Neon A" or "planetary," "Neon B," or "solar" – now recognized to originate from solar wind implantation – and "Neon S," or spallogenic (cosmogenic) (Figure 6.19). In 1969, evidence of a ^{22}Ne-rich component, named "Neon E," was found in the high-temperature (900–1100°C) release fractions of six carbonaceous chondrites. Its release at high temperature indicated it was efficiently trapped in a phase that breaks down only at high temperature.

Simply releasing gases by heating bulk meteor samples revealed the existence of nucleosynthetic isotopic variations but did not reveal which phase or phases contained them. Many scientists participated in an intensive search over nearly two decades for the carrier

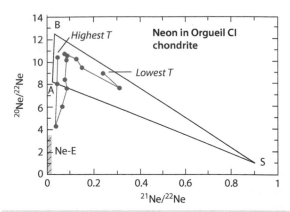

Figure 6.19 Neon isotopic compositions in a step-heating experiment on *Orgueil* (CI), which produced the first evidence of "presolar" or exotic Ne. The points connected by the line show the changing Ne isotope ratios with increasing temperature. Shaded area is the original estimate of the composition of the pure Ne–E component. Also shown are the compositions of Ne–A ("planetary"), Ne–B ("solar"), and Ne–E ("spallogenic"). After Black and Pepin (1969).

phase of these components. The search quickly focused on the matrix, particularly that of CM2 chondrites. However, the fine-grained nature of the matrix, together with the abundance of sticky and refractory organic compounds, made work with the groundmass difficult. In the late 1980s, E. Anders and his colleagues at the University of Chicago (e.g., Ming and Anders, 1988) found that Neon-E is associated with fine-grained (<6 μm) graphite and SiC (silicon carbide) of the matrix. They found that Ne-E actually consisted of two isotopically distinct components: Ne-E(L), which was released at low temperature (designated by the L) and ultimately found to reside in graphite, and Ne-E(H), which is released at high temperature (designated by the H) and resides in SiC. The ^{20}Ne/^{22}Ne ratio of Ne-E (L) is less than 0.01, while that of Ne-E(H) is less than 0.2.

The origin of Ne-E, and Ne-E(L), which is almost pure ^{22}Ne, posed something of a mystery. It was originally thought that it was a decay product of ^{22}Na, which has a half-life of 2.6 years, produced in red giants. Na could readily separate from Ne and other noble gases by condensation into grains. However, this hypothesis was subsequently rejected. For one thing, Huss et al. (1997) found that SiC grains in *Orgueil* (CI) had far too little Na to account for the observed amount of ^{22}Ne. More detailed analytical work (Lewis et al., 1990) found that the Ne isotopic abundances actually match rather well that expected for nucleosynthesis in the He-burning shells of low-mass, carbon-rich, thermally pulsing AGB stars. ^{22}Ne is synthesized from ^{14}N, which is synthesized from C and O nuclei during the previous hydrogen-burning phase, through the sequence ^{14}N$(\alpha,\gamma)^{18}$F$(\beta^{+},\nu)^{18}$O$(\alpha,\gamma)^{22}$Ne (Gallino et al., 1990).

The plot thickened with further work, however, and led to an even longer list of recognized components, each identified by yet another letter of the alphabet. Neon in SiC, named "Neon G," does indeed appear to come from AGB stars, but neon in graphite appears to come from a range of sources, including decay of ^{22}Na, named "Neon R" (Amari et al., 1995). Neon A turns out to be a mixture of components, including neon in presolar microdiamonds, named "Neon P3," as well as Neon P6, Neon A2, and Neon HL. The "Neon B," named "solar" by Black and Pepin (1969), has ^{20}Ne/^{22}Ne lower than true solar, which the Genesis mission found to be 13.35 (Heber et al., 2012), but it is consistent with the isotopic composition of neon implanted in nebular dust by solar wind in the early Solar System.

The other key noble gas in this context was xenon. Having nine isotopes rather than three and with contributions from both ^{129}I decay and fission of ^{244}Pu and ^{238}U, isotopic variation in xenon is bound to be much more complex than those of Ne. On the other hand, its high mass minimizes mass fractionation effects, so "solar" (more properly solar wind) and "planetary" Xe are less isotopically dissimilar than Ne. The first evidence of isotopic variations in Xe came in the early 1960s, but these variations were thought to be fissiogenic (at one time it was argued they were produced by fission of short-lived superheavy elements). Subsequently, several isotopically distinct Xe components were identified. One of these is associated with Ne-E(H) in SiC and is enriched in the s-process-only isotopes of Xe (^{128}Xe and ^{130}Xe) and is called, appropriately enough, Xe–S. The isotopic pattern of Xe–S is shown

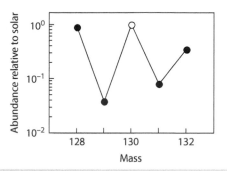

Figure 6.20 Isotopic composition of Xe–S (relative to normal solar Xe with ^{130}Xe $\equiv$ 1). Xe–S is found in silicon carbide and associated with Ne-E(H). ^{128}Xe and ^{130}Xe are synthesized only in the s-process; hence, the most likely site for its synthesis is giant stars.

in Figure 6.20. This is most likely synthesized in giant phase stars. Indeed, there is a striking similarity of the isotopic abundances to the calculated production of s-process nuclides in AGB stars. Comparison of the isotopic composition of Kr, which is also anomalous in the SiC, with theoretical calculations further narrows the site of synthesis to low-mass AGB stars, consistent with the inferences made for ^{22}Ne synthesis. SiC grains (Figure 6.21) thus apparently condensate from material ejected from giant stars, which have very strong solar winds. Thus, in a very real sense, these grains are truly stardust.

The Ne in the SiC is a little richer in ^{21}Ne than the predicted products of AGB stars. This is presumably due to cosmogenic production of ^{21}Ne. If so, some 130 Ma of cosmic-ray irradiation would be required to produce the observed ^{21}Ne, indicating the grains predate meteorite parent body formation by this amount of time (Lewis et al., 1994). If they had been degassed, however, the grains could be much older. Subsequent work has identified SiC grains in *Murchison* (CM2) with ^{21}Ne and ^{3}He cosmic-ray exposure ages ranging from as little as 4 Ma to as much as $\approx$3 Ga before being incorporated in meteorite parent bodies as the Solar System was forming (Heck et al., 2020). The majority of grains have exposure ages <300 Ma before the start of the Solar System, but a few have ages > 1 Ga before the start of the Solar System.

Another isotopically distinct component, identified in acid-dissolution residues of

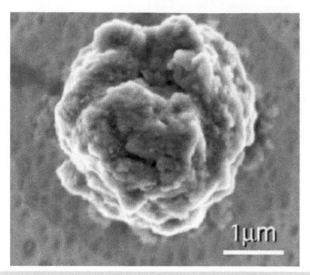

Figure 6.21   A presolar SiC grain. Source: Nittler et al. (2003)/with permission of Elsevier.

Allende (CV3) named Xe-HL because it is enriched in both heavy and light Xe isotopes, was released in the 700–1000°C temperature step. This particular enrichment pattern can be produced only by a combination of the p- and r-processes (Figure 6.22). As we found in Chapter 1, these processes operate only in supernovae and kilonovas. Unlike Ne-E, Xe-HL is accompanied by the other noble gases, of which Ne, Ar, and Kr all show enrichment in their heavier isotopes. Eventually, Anders' group identified the carrier of Xe-HL as nanodiamonds. These diamonds are extraordinarily

fine, averaging only 1 nm in diameter and containing typically only 10^3 to 10^6 or so atoms. Roughly one in every four atoms is at the surface. As a result, the properties of this material differ significantly from that of normal diamond, which considerably complicated the effort to isolate it and nearly impossible to study individual grains. Subsequent studies have shown that the nanodiamond concentrates are actually a mixture of glassy carbon and nanodiamond, and it is unclear which phase contains the Xe. Furthermore, they have carbon isotopic compositions within a few percent of the solar value, which is quite surprising given the isotopic variability observed in other kinds of presolar grains. While there is little doubt of the supernova origin of the Xe, the grains might have originated elsewhere and the Xe subsequently implanted in them. We will return to noble gas isotopic compositions in Chapter 14.

6.5.2 Isotopic composition of other elements in presolar grains

Many additional studies have followed these noble gas studies once presolar grains were identified. The sub-micrometer size of most of these grains represents an analytical challenge. In some meteorites, such as *Murchison* (CM2), the grains are abundant enough that can be separated and analyzed in bulk by conventional mass spectrometry. Progress in the development of ion probes, particularly those

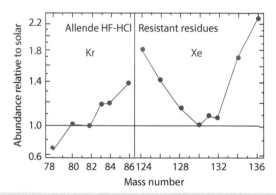

Figure 6.22   The isotopic composition of Kr and Xe of the "Xe-HL" component in the matrix of *Allende* (CV3) carbonaceous chondrite. Xe-HL is characteristically enriched in both the light and heavy isotopes while the lighter noble gases show enrichment only in the heavy isotopes. Data from Ming and Anders (1988).

with very fine spatial resolution (e.g., Nano-SIMS), and laser ablation inductively coupled plasma mass spectrometry (ICP-MS) has also enabled isotopic analysis of many elements in individual presolar grains (which have also been identified among interplanetary dust particles collected by aircraft and spacecraft). In addition to the SiC and graphite carriers of presolar noble gases, presolar silicates, oxides, silicon nitride, metal carbides, and kamacite grains (some of these are present as inclusions in SiC) have also been identified. Although silicates are the most abundant of presolar grains, they were the last to be discovered (Nguyen and Zinner, 2004). They cannot easily be extracted from the chondrite matrix in the same way as SiC, graphite, and nanodiamond, which typically involves extreme acid digestion, so that analytical data on silicon carbide and graphite predominate, with over 20 000 analyses in a presolar grain database maintained at Washington University of St. Louis (https://presolar.physics.wustl.edu/presolar-grain-database/) (Hynes and Gyngard, 2009), but there are also several thousand analyses

of graphite, silicates, and oxides. Based on their isotopic compositions, mainly C, N, O, and Si, presolar grains can be grouped into a number of types, each created in a distinct stellar environment. The challenge is to understand these isotopic variations in terms of distinct nucleosynthetic astrophysical environments. Zinner (2014) provides a detailed summary of these grains.

We begin by considering the SiC grains. Based on their carbon and nitrogen isotopic compositions, they have been grouped into mainstream grains, AB, C, X, Y, and Z classes (Figure 6.23). These are briefly described below.

Mainstream are by far the most common, comprising ≈93% of all analyzed grains. They, along with rarer Y and Z grains, have isotopic compositions consistent with an origin in the outflow of carbon-rich AGB stars. Mainstream grains tend to be enriched in ^{13}C relative to terrestrial or solar carbon, consistent with C isotopic compositions determined spectrographically in such stars. They are also enriched in the heavy Si isotopes. Calculated $(^{26}Al/^{27}Al)_0$

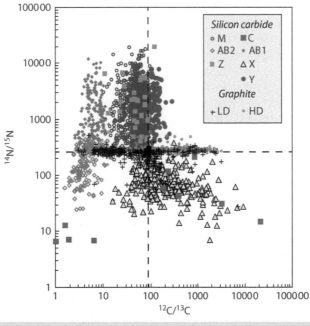

Figure 6.23 Carbon and nitrogen isotope ratios of presolar SiC (mainstream [M], C, AB, X, Y, Z) and graphite (LD and HD) grains. X, C, and LD grains are likely derived from supernova outflow, with the remaining groups are derived from AGB stars of different masses and metallicities. Nitrogen isotopic composition of graphite is close to terrestrial and likely reflects secondary equilibration or contamination. Dashed line shows the isotope ratios of terrestrial standards. Data from the presolar grain database of Hynes and Gyngard (2009).

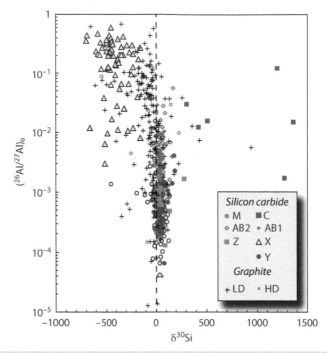

Figure 6.24 $\delta^{30}Si$ (per mil deviations of the $^{30}Si/^{28}Si$ ratio from an international standard) and calculated (from ^{26}Mg excesses) initial $^{26}Al/^{27}Al$ ratios of presolar SiC. Most AGB-derived grains have Si isotopic compositions close to or within error of the terrestrial value. Supernova-derived X and C grains show great deviations in Si isotope ratios. Data from the presolar grain database of Hynes and Gyngard (2009).

are higher than in CAIs with some values $> 10^{-2}$. Some, but not all, contain excess ^{22}Ne. Heavy elements, such as the rare earths, exhibit enrichment in s-process nuclides. These isotopic features agree in most aspects with theoretical predictions of nucleosynthesis in low mass (1.5–3 M_O) of solar or greater metallicity AGB stars. Type Y grains are in many ways similar to those of mainstream grains, but $^{12}C/^{13}C$ and $^{14}N/^{15}N$ ratios are systematically higher than the terrestrial ratios. They are thought to have originated in low-to-intermediate mass AGB stars of lower metallicity than the Sun. Z grains are thought to originate in low-mass AGB stars of even lower metallicity.

AB grains, which account for roughly 5% of all grains, are enriched in ^{13}C, have higher $^2(^{26}Al/^{27}Al)_0$ and, like mainstream grains, many but not all are enriched in the heavy Si isotopes (Figure 6.24). Most are not enriched in s-process nuclides. They are subdivided into AB1 with $^{14}N/^{15}N >$ solar (440) and AB2 with $^{14}N/^{15}N <$ solar. Spectroscopic analyses have shown that a rare subclass of carbon-rich giant stars known as J-stars strongly enriched in ^{13}C and AB grains may have originated in such

stars. Synthesis in the He-burning shell of supernovae has also been suggested, but a consensus on their origin has yet to evolve.

X grains, which account for only 1% of all grains, are enriched in ^{12}C, ^{15}N, and ^{28}Si relative to solar and have $(^{26}Al/^{27}Al)_0$ ratios as high as 0.6 (compared to 5×10^{-5} in CAIs) (Figure 6.24). A key feature in some of these grains is large ^{44}Ca excesses produced by decay of ^{44}Ti ($t_{1/2} = 60$ yr.) As ^{44}Ti can only be produced in supernovae, this indicates X grains are derived from supernovae. Interesting, these grains appear to be heterogeneous mixtures of material synthesized in different shells of the pre-supernova star (Figure 1.10). While ^{44}Ti is synthesized in the deep Si-burning core, ^{28}Si is synthesized in the Ne-burning layer and ^{26}Al is synthesized in the Ne-rich layer, as well as the H-burning layer. Analysis of particularly large (25 µ in diameter) grain, called "Bonanza," confirmed the supernova origin of X grains (Gyngard et al., 2018). It has an inferred $(^{27}Al/^{27}Al)_0$ of 0.9, which is higher than any other grain analyzed thus far. The large size of the grain allowed isotopic analysis of Li, B Mg, S, Ca, Ti, Fe, and Ni in addition to

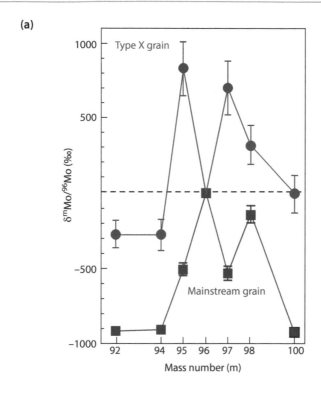

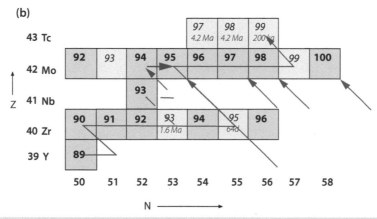

Figure 6.25   (a) Mo isotope ratios (in delta notation relative to terrestrial) in an X and a mainstream SiC grain from the *Murchison* meteorite. Source: Adapted from Pellin et al. (1999). (b) Part of the chart of the nuclides illustrating s- and r-process production. Half-lives of longer-lived radionuclides are shown in italics below the mass number. ^{95}Mo through ^{98}Mo are all produced by the s-process; the s-process branches at ^{93}Zr such that there can be some production of ^{94}Mo as well, particularly at low neutron fluences. As can ^{94}Mo to a lesser extent, depending on the branch at ^{93}Zr, ^{95}Mo, ^{97}Mo, ^{98}Mo, and ^{100}Mo can all be produced by the r-process. ^{92}Mo is produced exclusively by the p-process and ^{94}Mo is mainly produced by the s-process. Comparing this with the isotopic compositions in (a), we see that the X grain is enriched in r-process nuclides whereas the mainstream grain is depleted in them. This leads to the conclusion that the X grain represents supernova ejecta whereas the mainstream grain condensed from the ejecta of an AGB star.

C, N, and Si, and these exhibit large isotopic variations consistent with mixtures of material synthesized in different shells of the star prior to the core collapse supernova.

Figure 6.25 compares Mo isotopic compositions of a mainstream and an X grain. Looking at Figure 6.25b, we see that ^{96}Mo is exclusively an s-process nuclide, ^{92}Mo is exclusively a p-process nuclide, and ^{100}Mo is exclusively an r-process nuclide. The mainstream grain is enriched in ^{96}Mo relative to other isotopes, indicative of s-process enrichment, consist with an origin in an AGB star. Compared to terrestrial Mo, the X grain is strongly enriched in ^{95}Mo, ^{97}Mo, and, to a lesser extent, ^{98}Mo relative to ^{96}Mo, indicative of r-process enrichment. The patterns are consistent with the inference that X grains are derived from supernovae while mainstream grains are derived from AGB stars. The exception is ^{100}Mo, which we expect should be more strongly enriched as it is an r-process-only nuclide. However, Meyer et al. (2000) found that this could be modeled as the result of a very short, ≈ 1 s, neutron burst in an exploding He-rich star. This results in a buildup of ^{95}Zr and ^{97}Zr, which then decay to ^{95}Mo and ^{97}Mo, but not to the precursors of ^{100}Mo. Furthermore, original ^{100}Mo is slightly depleted by the neutron capture during the burst.

Finally, even rarer C grains show an extreme range of ^{12}C/^{13}C are ^{14}N depleted, and show extreme enrichment in ^{29}Si and ^{30}Si. They also have high inferred $(^{26}$Al/^{27}Al$)_0$. They too appear to have formed as supernova ejecta.

Graphite grains are divided into two types: HD and low density (LD). Essentially all analyses have been made on grains from just two meteorites: *Orgueil* (CI1) and *Murchison* (CM2). Most have one of two morphologies: "cauliflowers" and "onions." The former consists of concentrically packed scales of poorly crystallized carbon (Figure 6.21 is an example); the onions consist of platy well-crystallized graphite throughout, although some have a small core of graphene. Both contain inclusions of TiC, which in some cases appear to have acted as condensation nuclei. Both types have a wide range of ^{12}C/^{13}C ratios, with HD grains having predominantly ^{12}C/^{13}C lower than solar while the opposite is true for LD grains (Figure 6.23). Nitrogen in both types is close to atmospheric suggesting it is not intrinsic to the grain and results from equilibration with surrounding material in the parent body or terrestrial contamination. The majority of graphite grains have Si isotope ratios within error of solar values (errors are large because the analytical precision on Si isotope ratios in graphite is poor due to low concentrations), but many LD grains are enriched in ^{28}Si and few are ^{28}Si poor. $(^{26}$Al/^{27}Al$)_0$ ratios in LD grains are as high as those in SiC

X grains while the few $(^{26}$Al/^{27}Al$)_0$ ratios determined in HD grains tend to be similar to those in SiC AB grains (Figure 6.23). In contrast to SiC grains where ^{22}Ne excesses appear to be entirely due to the s-process, ^{22}Ne is some graphite grains apparently derived from the decay of ^{22}Na $(t_{1/2} = 2.5$ yr). Isotopic compositions of heavier elements, including Si, K, Ca, and Ti, favor a supernova origin for the LD grains while an origin in AGB stars appears to best explain the HD grains.

Silicates and oxides are the most abundant presolar grains, up to a few hundred ppm in some meteorites compared to tens of ppm for some other types of grains, but since they exist within a predominately silicate matrix, they are less easily identified and separated than SiC or graphite grains. Many have been identified only through anomalous O isotope ratios detected using automated searches with secondary ion mass spectrometry (SIMS) instruments. Oxides include corundum (Al_2O_3), spinel ($MgAl_2O_4$), hibonite ($CaAl_{12}O_{19}$), and Fe- and Ti-oxides. Silicates include olivine and pyroxene, but the mineralogy of the majority has not been determined. Oxygen isotopic compositions of these grains are shown in Figure 6.26. Relative to solar, these grains exhibit both depletions and enrichments in heavy oxygen isotopes. Nittler et al. (1997) divided them in to four groups.

Group 1 is enriched in ^{17}O and has solar or slightly lower than solar ^{18}O/^{16}O (1×10^{-3} to 2.01×10^{-3}). This is consistent both with observations of moderate-mass (>1.3 M$_\odot$) AGB stars and evolutionary models of the initiation of H burning in the shell surrounding the He core in such stars (Figure 1.10). As we discussed in Section 1.4.2.1, as H burning ignites in red giants, the convective envelope can expand downward into the radiative region. In doing so, it will dredge up material, the soc-called "first dredge up," that had previously experienced mild main-sequence H burning in which the abundance of ^{17}O was enhanced by the CNO cycles and ^{18}O was reduced slightly by the reaction ^{18}O$(p,\alpha)^{15}$N.

Group 2 grains are strongly depleted in ^{18}O, have ^{17}O/^{16}O higher than solar, and have very high ^{26}Al/^{27}Al$_0$ ratios. They are also thought to be products of the H-burning shell of AGB stars, because ^{18}O is depleted by the reaction:

$$^{18}\text{O} + {}^1\text{H} \rightarrow {}^{15}\text{N} + \alpha + \gamma$$

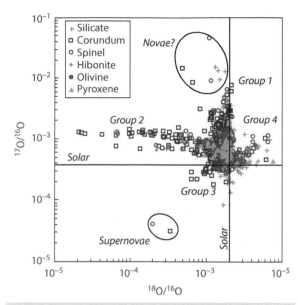

Figure 6.26 Oxygen isotope ratios of presolar silicate and oxide grains. Groups are those defined by Nittler et al. (1997). Group 1 is thought to be derived from moderate-mass AGB stars, Group 2 by hot bottom burning in more massive AGB stars, Group 3 in low-mass, low-metallicity AGB stars, and Group 4 by helium burning of ^{14}N during thermal pulses in AGB stars. Data from the presolar grain database of Hynes and Gyngard (2009).

In intermediate-mass stars ($M > 5\ M_\odot$), temperatures are hot enough in the H-burning shell to initiate the Mg–Al chain (described in Section 1.4.2) in which ^{26}Al is a significant byproduct. In the *hot bottom processing*, the convective outer envelope of the star can extend down to the top of the H-burning shell (Figure 1.10) where these reactions are occurring and dredge this material to the surface. Until recently, models that reproduced the low ^{18}O/^{16}O and high ^{26}Al/^{27}Al also predicted higher than observed ^{17}O/^{16}O ratios as ^{17}O is also in the H-burning shell through:

$$^{16}O + {}^{1}H \rightarrow {}^{17}F + \gamma \rightarrow {}^{17}O + \beta^-$$

However, recent experiments' measurements have shown that the rate of the reaction:

$$^{17}O + {}^{1}H \rightarrow {}^{15}N + \alpha$$

is twice that previously assumed and is sufficient to bring predicted ^{17}O/^{16}O ratios into the range

of Group 2 grains (Lugaro et al., 2017). An alternative mechanism is "cool bottom processing" in which material in the convective envelope circulates close to the hot regions of the underlying hydrogen-burning shell before being mixed back to the outer regions of low-mass ($M < 1.65\ M_\odot$) AGB stars. Thus, Group 2 grains also appear to be produced in AGB stars, but in larger quantity than Group 1 grains.

Group 3 grains have low ^{18}O/^{16}O and low ^{17}O/^{16}O relative to solar; most also lack evidence of ^{26}Al. They are thought to be derived from AGB stars of low mass and low metallicity. As low-mass stars are long-lived, these stars likely formed when the galaxy was still young and relatively poor in elements heavier than He ("metals").

Group 4 grains have both high ^{18}O/^{16}O and high ^{17}O/^{16}O relative to solar. One possibility is that they come from low-mass AGB stars in which ^{18}O produced by helium burning of ^{14}N during thermal pulses that was dredged up into the convective envelope during subsequent pulses. Alternatively, they could come from AGB stars with high metallicity and higher initial ^{18}O/^{16}O and ^{17}O/^{16}O. The Group 4 grains with the largest ^{18}O excesses may come from ^{18}O-rich material from the He/C zone mixed with material from oxygen-rich zones during supernova explosions.

Two grains that are exceptionally depleted in ^{17}O and ^{18}O were probably produced in supernova explosions and the few grains most enriched in ^{17}O may have been produced in novae.

In summary, presolar grains provide insights into the makeup of the dust of the giant molecular cloud, which gave birth to the Solar System. This material was isotopically heterogeneous, reflecting an array of stellar sources. Most of it seems to derive from AGB stars, but supernovae contributed material as well. Short-lived radionuclides indicate some of this material was produced relatively shortly before the Solar System formed; indeed, it is likely there was a constant production of dust as massive stars that formed within this giant molecular cloud also grow old within it, seeding it with dust to form the next generation of stars in their death throes. Some of it was inherited from the ISM. This was the raw material for formation of all other Solar System materials, with Solar System formation having reprocessed it into gas, CAIs, chondrules, asteroids, comets, planets, and the Sun, and mostly homogenizing it isotopically. What we recognize as presolar grains is the small fraction that escapes reprocessing. As we will see in the following section, this homogenization was not complete, however, as the isotopic composition of bulk meteorites reveals.

6.6 ISOTOPIC VARIATIONS IN BULK METEORITES

6.6.1 Oxygen isotope variations and nebular processes

Isotopic variations in oxygen are ubiquitous in meteorites. Until 1973, O isotope variations in meteorites were thought to be simply the result of fractionation, as they generally are on Earth. However, when Robert Clayton of the University of Chicago went to the trouble of measuring ^{17}O (0.037% of O), as well as ^{18}O and ^{16}O, he found that these variations were not consistent with simple mass-dependent fractionation (Clayton et al., 1973); this is illustrated in Figure 6.27. As we found in Chapter 5, almost all terrestrial materials (atmospheric ozone is the most notable exception) plot on a line with

a slope of ≈ 0.52 – the *terrestrial fractionation line (TFL)*. Lunar samples fall on this same line, but meteorites and meteoritic components do not. In fact, anhydrous minerals from carbonaceous chondrites scatter about a line (CCAMs: carbonaceous chondrite anhydrous minerals) with a slope of ≈ 1. Most CAIs also fall along the same line and extend to extremely ^{16}O-enriched compositions ($\delta^{18}O$ as low as $-50‰$). CI chondrites are a notable exception to the carbonaceous chondrite trend, with extreme ^{16}O depletion and falling close to the TFL. Most non-carbonaceous materials could be derived from CI materials mainly by mass-dependent fractionation.

An initial interpretation was that the CCAM line reflected mixing between a nearly pure ^{16}O component, such as that might be created by helium burning, and a component of "normal" isotopic composition close to that of CI chondrites. However, as we found in Chapter 5, the experiments of Thiemens and Heidenreich (1983) showed that "mass-independent fractionation" can result during production of ozone. This results from a kinetic fractionation mechanism, which arises because non-symmetric (e.g., $^{16}O^{17}O$ or $^{18}O^{16}O$)

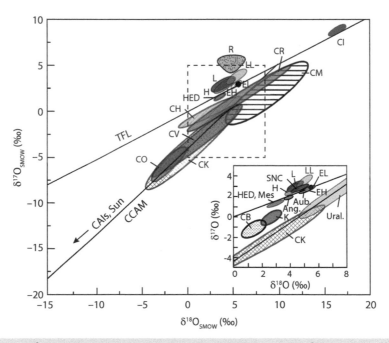

Figure 6.27   Variation of O isotope ratios in meteorites. CO, CK, etc.: carbonaceous chondrites; H, L, LL: OCs; HED: howardites, eucrites, diogenites; Ural.: ureilites; SNC: shergottites, nakhlites, chassignites; E: enstatite chondrites. The Earth, Moon, and aubrites have the essentially same isotopic composition as enstatite chondrites. Source: Adapted from Clayton (1993).

molecules have more available energy levels than symmetric (e.g., $^{16}O^{16}O$) molecules.

A different, and now generally accepted, mechanism was proposed by Clayton (2002). He suggested that the anomalies arose through radiation self-shielding in the solar nebula. In his model, ultraviolet (UV) radiation from the early proto-Sun dissociated carbon monoxide, which would have been among the most abundant gases in the solar nebula and a major reservoir of O. Because $C^{16}O$ was far more abundant than either $C^{17}O$ or $C^{18}O$, the radiation of the wavelength necessary to dissociate $C^{16}O$ would have been quickly absorbed as it traveled outward from the Sun. Consequently, radiation of this frequency would have been absent at greater distance from the Sun, while that needed to dissociate $C^{17}O$ and $C^{18}O$ would still be available. At those distances, $C^{17}O$ and $C^{18}O$ would have been dissociated by UV radiation, and equally so, making ^{17}O and ^{18}O available for reaction to form the condensable phases, most notably silicates, which accreted to produce meteorite parent bodies. CAIs, which can be extremely enriched in ^{16}O with $\delta^{18}O$ approaching $-30‰$, would have formed in the ^{16}O-rich environment close to the Sun before being expelled back out into the nebular disk. Clayton's model also predicted that the Sun itself should be poor in ^{18}O and ^{17}O compared with meteorites and the Earth – closer in composition to the CAIs. Based on the analysis of *Genesis* Mission solar wind samples, McKeegan et al. (2011) estimated the composition of the solar wind as $\delta^{18}O_{SMOW} = -103 \pm 3.3‰$ and $\delta^{17}O_{SMOW} = -80.8 \pm 5‰$, plotting on an extension of the CCAM line well to the lower left of Figure 6.27. This appears to confirm Clayton's self-shielding hypothesis, although some problems persist with the self-shielding model, as Thiemens (2006) points out.

Krot et al. (2020) found that the CAs with $(^{26}Al/^{27}Al)_0 < 5 \times 10^{-6}$, which are mainly found in metal-rich CB and CH chondrites, are internally heterogeneous in O isotopic compositions extending to extreme ^{16}O enrichment while those with canonical $(^{26}Al/^{27}Al)_0$ of $\approx 5 \times 10^{-5}$ show less internal heterogeneity and are less enriched in ^{16}O. Because the CAIs with low $(^{26}Al/^{27}Al)_0$ are thought to predate injection of newly synthesized ^{26}Al sampled by CAIs with canonical $(^{26}Al/^{27}Al)_0$, they argued that isotopic heterogeneity in the early Solar System derives from self-shielding of CO in the parent molecular cloud from which the Solar System formed rather than self-shielding from the proto-Sun.

While variations in O isotopic compositions *between* classes are mostly mass independent, variations *within* OC and achondrite classes fall along mass-dependent fractionation lines (MDFLs). This strongly suggests that, for the most part, each class of meteorites derived from parent bodies that formed in different parts of the solar nebula. There are a few exceptions: IIE irons fall on a MDFL with H-chondrites, IVA irons plot on an MDFL with L and LL chondrites, and HED achondrites plot on an MDFL with IIIAB irons and some stony-irons. This suggests a genetic relationship between these objects, perhaps derivation from a single parent body. The Moon and the Earth plot on a single MDFL, which is evidence of their close genetic relationship. Intriguingly, the enstatite chondrites and aubrites also plot essentially along the TFL. This suggests that the aubrite achondrites derive from a differentiated enstatite chondrite parent body and, further, that enstatite chondrites might be a better compositional model for the Earth than either carbonaceous or OCs.

6.6.2 Isotopic variations in other elements

Following the discovery of oxygen isotope heterogeneity, subsequent studies beginning with Gunter Lugmair of the University of California San Diego, Jerry Wasserburg[†] at California Institute of Technology and his students and colleagues, and Claude Allègre and his students and colleagues at the University of Paris began to reveal systematic nucleosynthetic anomalies in

---

[†]Gerald Wasserburg (1927–2016) completed his PhD dissertation on K–Ar dating under Harold Urey at the University of Chicago in 1954. He joined the faculty at California Institute of Technology in 1955 and remained there for the rest of his long career. In the 1960s, he established the "Lunatic Asylum" mass spectrometry laboratory in anticipation of the Apollo lunar samples and then rapidly produced high-precision lunar ages. He was responsible for a remarkable number of discoveries, including the evidence of ^{26}Al in meteorites, and many analytical advances in mass spectrometry. He won the Crafoord Prize in Geosciences jointly with Claude Allègre in 1982. Nevertheless, he once told me that one of his proudest achievements was serving as a rifleman in the US Army in Europe during World War II.

a variety of elements in bulk samples of meteorites. The earliest discoveries were in Ti and Cr isotopic compositions; advances in analytical instrumentation, most notably multi-collector ion probe and Multi-collector Inductively Coupled Plasma–Mass Spectrometry (MC-ICP-MS), have greatly accelerated this work and isotopic variations in many elements have now been documented. Some of these variations, such as in Ca and Ti, appear to be correlated over all meteorite classes. In contrast, Cr and Ti isotope ratios appear to define two correlations: one among carbonaceous chondrites and the other among all other objects (Figure 6.28).

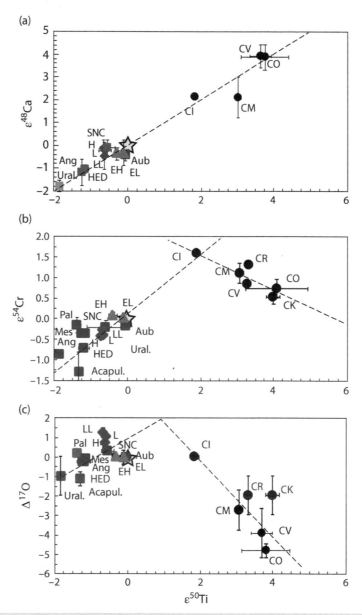

Figure 6.28 Isotopic variations in meteorites. $\varepsilon^{50}Ti$, $\varepsilon^{54}Cr$, and $\varepsilon^{48}Ca$ are the deviation in parts per 10 000 from the $^{50}Ti/^{47}Ti$, $^{54}Cr/^{52}Cr$, and $^{48}Ca/^{44}Ca$ ratios, respectively, in terrestrial standards. $\Delta^{17}O$ is the displacement of $\delta^{17}O$ in per mil from the TFL. CO, CK, etc.: carbonaceous chondrites (circles); H, L, LL: OCs (diamonds); EL and EH: enstatite chondrites (triangles); squares are achondrites and stony-irons: HED, Ural. (uralites), SNC, Aub (aubrites) Ang: angrites, Mes (mesosiderites), Pal (pallasites). Error bars are 1 standard deviation (in some cases, only a single analysis exists, and error bars represent analytical uncertainty). The star shows the isotopic composition of the Earth and Moon.

6.6.3 The Great Isotopic Solar System Divide

Warren (2011) pointed out that isotopic compositions reveal a dichotomy between carbonaceous chondrites and other Solar System material and that they correlate, at least in part, with oxygen isotope variations. Carbonaceous chondrites have ε^{50}Ti and ε^{48}Ca, and ε^{54}Cr > 0 while other Solar System solids have the opposite with a large gap between the two (Figure 6.28). CI chondrites are notably more similar isotopically to non-carbonaceous material than are other carbonaceous chondrites. The two groups define distinct correlations in some cases (e.g., ε^{50}Ti versus ε^{54}Cr). The origin of these correlations is somewhat enigmatic: both ^{54}Cr and ^{50}Ti are produced primarily by the s-process in AGB stars; in contrast, ^{48}Ca can only be produced in supernovae.

Nucleosynthetic isotopic variations have also been found in heavier elements, such as Nd, and siderophile ones, such as Mo and Ru iron meteorites as well (Figure 6.29). The inverse correlation between ^{92}Mo anomalies and ^{100}Ru suggests these variations result from incomplete mixing of nuclides synthesized in different cosmic environments: ^{92}Mo is a p-process nuclide and hence produced by explosive nucleosynthesis (kilonovas and supernovae) while ^{100}Ru is synthesized by the s-process in red giants. Thus, to a first approximation, carbonaceous chondrites appear to be enriched in debris from neutron star mergers or supernovae relative to non-carbonaceous

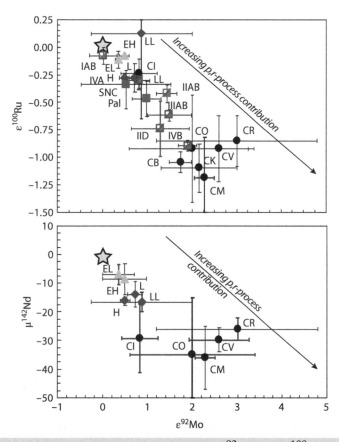

Figure 6.29 Heavy element isotopic variations in meteorites. ε^{92}Mo and ε^{100}Ru are the deviation in parts per 10 000 from ^{92}Mo/^{97}Mo and ^{100}Ru/^{101}Ru in terrestrial standards, respectively. $\mu_{142\text{Nd}}$ is the deviation in parts per million from ^{142}Nd/^{144}Nd in a terrestrial standard after correction for the radiogenic contribution from extinct ^{146}Sm. CO, CK, etc.: carbonaceous chondrites (circles); H, L, LL: OCs (diamonds); EL and EH: enstatite chondrites (triangles); square: SNC; half-filled squares: iron meteorites, IVA, IIAB, etc. Error bars are 1 standard deviation (in some cases, only a single analysis exists, and error bars represent analytical uncertainty). The star shows the isotopic composition of the Earth and Moon.

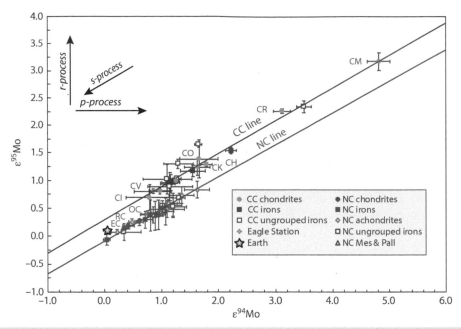

Figure 6.30 ε^{95}Mo versus ε^{94}Mo in meteorites and the bulk Earth (ε is deviations in ppm from a laboratory standard after correction for mass fractionation to a constant ^{98}Mo/^{96}Mo ratio). The data define two correlations: the first for carbonaceous chondrites and other materials having a carbonaceous chondrite like signature in other isotope compositions (e.g., oxygen), and the second for all other meteorite groups. The Earth (star) falls between the two correlations. Arrows on the upper left show the effect of additions of r-, p-, and s-nucleosynthetic processes. Data are from the compilation of Budde et al. (2019).

material, with lesser contributions of this debris in other Solar System bodies. The Earth is in contrast relatively enriched in s-process nuclides: Qin and Carlson (2016) point out that nearly all meteorites are deficient in s-process nuclides of Sr, Zr, Mo, Ba, and Nd relative to the Earth. These deficits are quite small, however, ranging from 0.02 to 0.004%. As is the case for O, Ca, Cr, Ti, Earth plots closest to enstatite chondrites, suggesting a genetic affinity.

Let us consider the Mo isotopes in more detail. There are several reasons why molybdenum is a useful element in understanding isotopic heterogeneity in the Solar System. First, it is siderophile but not strongly so and is abundant enough to be analyzed in both silicate-rich and iron-rich meteorites (in contrast, Nd is strongly lithophile and challenging to analyze in iron meteorites while Ru is strongly siderophile and challenging to analyze in achondrites). Second, it has seven stable isotopes created by different nucleosynthetic processes. We see from Figure 6.25 that ^{94}Mo is produced by a combination of the p- and s-processes, and

^{95}Mo is produced by a combination of the r- and s-process. Figure 6.30 shows a plot of the deviation in abundance in parts per million (epsilon units) from a laboratory standard of these two isotopes in meteorites. The data fall along two trends: the first for carbonaceous chondrites and other materials having a carbonaceous chondrite-like signature (CC) in other isotope compositions (e.g., ^{17}O/^{16}O and ^{54}Cr/^{52}Cr), and the second for non-carbonaceous chondrites and other meteorites having a non-carbonaceous chondrite-like signature (NC), including ordinary and enstatite chondrites. The offset can be explained either by relative enrichment of the CC group in r-process nuclides or enrichment of the NC group in p-process nuclides. As we discussed in Chapter 1, the nature of the p-process is poorly understood, but probably occurs in explosive nucleosynthesis, including supernovae and kilonovas. Thus, the difference appears to relate to addition of nuclides in these explosive nucleosynthesis events. Within each group, ε^{95}Mo and ε^{95}Mo vary in a correlated manner along two parallel trends with constant relative contributions of r- and p-process nucleosynthesis (the slopes of the regression lines are identical within error; Budde et al., 2019). Since both nuclides are produced by the s-process, this suggests the along-slope variation is due to variable abundance of s-process nuclides produced in giant stage stars.

As we have seen, mainstream presolar SiC grains were produced by AGB stars while X grains were produced by supernovae, so we might wonder whether the variation simply reflects variable amounts of these types of grains. However, the abundance of identifiable presolar grains is too small, a few hundred ppm at most and generally much less, to explain the difference between the NC and CC clans or the variation along the trends.

The isotopic distinction between these two groups means that they formed from two distinct isotopic reservoirs, each of which contained variable amounts of s-process nuclides, but with little mixing between the two reservoirs. Given their enrichment in volatile elements, carbonaceous chondrites are assumed to have been formed at great heliocentric distances within the solar nebula. This is supported to some extent by spectral observations of asteroids: those most similar to OCs are found in the inner asteroid belt while those similar to carbonaceous chondrites, the C-class asteroids, are found further out. If this is the case, then the dichotomy between NC and CC objects reflects an isotopically zoned solar nebula.

If the solar nebula was isotopically zoned, the sharp distinction between the CC and NC clan still requires explanation. Dynamic models of nebular evolution suggest that the most likely explanation is the formation of Jupiter. As described in Section 6.2, Jupiter likely formed near the "snowline" where water ice condenses and would have grown quickly after that. Jupiter's accretion of gas and dust produces an orbital zone in the nebula of low gas density, which in turn produces density maxima at roughly 80 and 120% of its orbital radius. These act as traps that inhibit movement of dust (including planetesimal-sized objects) across this zone (Desch et al., 2018) and would serve to separate the NC inner nebula from the CC outer nebula. In most models, this causes Jupiter to migrate inward, before eventually migrating outward to its present position. In the Grand Tack model (Walsh et al., 2011), this occurs when Saturn forms and also migrates inward until orbital 3:2 resonance causes them to reverse and migrate outward. That outward migration scatters some of the planetesimals in the inner zone outward and some from the outer zone inward. C-type asteroids, likely the parent bodies of carbonaceous chondrites, exist inside the present orbit of Jupiter, providing evidence of this mixing.

Since Mo isotopes show no evidence of mixing between meteorite parent bodies of the two zones, the outward migration and consequent mixing could only have happened after those parent bodies have formed. The best constraint on time of that is perhaps the ^{182}Hf–^{182}W ages of iron meteorites (Figure 6.14). While the IC irons, part of the NC clan, have ^{182}Hf–^{182}W ages of $\approx$1.5 Ma and less, the oldest irons of the CC clan, the IVB group have ages of $\approx$2.8 Ma. Mixing could have occurred only after that; Kruijer et al. (2020) suggest that this mixing only occurred after 4 Ma.

Significantly, the Earth lies between the two trends (the bulk silicate Earth Mo isotopic composition is thought to be slightly different from that of the laboratory standard, with $\varepsilon_{94Mo} = 0.04 \pm 0.06$ and $\varepsilon_{95Mo} = 0.10 \pm 0.06$, a difference hardly discernable on Figure 6.30). As for other isotopic compositions, the Earth plots closest to enstatite chondrites, but the Earth could not consist entirely of enstatite chondrite material and would seem to require a contribution from carbonaceous chondrite-like material. It is reasonable to assume this occurred once the barrier created by Jupiter was broken down and CC planetesimals were scattered into the inner Solar System.

Estimating the necessary amount of CC-like material added to the Earth is complicated, however, because Mo is siderophile and most of the Earth's inventory is in its core and its isotopic composition is not known, so we cannot assume the silicate Earth Mo isotopic composition is the same as the total Earth. In particular, if the CC material was added late and did not entirely equilibrate with the core, only a relatively small amount of CC material would be required to explain the Mo isotopic composition of the silicate Earth. Once scenario suggested by Budde et al. (2019) is that it was added by the Moon-forming impactor, Theia. This event marks the final formation of the

Earth and occurred well after Jupiter had reached its present mass and position, likely tens of millions of years after CAI formation. Depending on the extent to which the cores of the Earth and Theia equilibrated, this impact may have added between 2 and 10% of CC material to the Earth (Budde et al., 2019). Regardless of when it occurred, this mixing of volatile-rich CC material may account for much of the volatile and water inventory of the terrestrial planets, including Earth.

6.7 COSMIC-RAY EXPOSURE AGES OF METEORITES

All meteorites are derived from larger bodies (primarily asteroids) by impacts. Prior to the impact, they would have been shielded from cosmic rays by overlying rock. Once freed, they are exposed to a significantly greater cosmic-ray flux than the Earth's surface, which is substantially shielded by the atmosphere. The time between freeing of the meteorite from its parent body and colliding with Earth can be determined by analyzing the buildup of the cosmic-ray-produced nuclides.

The principles are precisely the same as those we reviewed for terrestrial materials in Chapter 4. The nuclides used include those used in terrestrial studies as well as a few others, including ^{38}Ar, ^{40}K, and ^{81}Kr. Just as is the case for terrestrial materials, the composition of the target material as well as shielding effects must be taken into consideration. Here, we will simply summarize the results and not go into details of the procedures and calculations, which are reviewed in Herzog and Caffee (2014).

Figure 6.31 summarizes cosmic-ray exposure ages of iron meteorites. There is a wide spread of ages, but comparison with Figure 6.13 shows that they are much younger than the ^{182}Hf–^{182}W iron segregation ages. Thus, the current supply of meteoroids crossing Earth's orbit results from collisions that occurred long after parent body formation, but nevertheless ones that occurred over a considerable range of Solar System history. For some groups, there are distinct clusters in ages; for example, most IVA cluster between 350 and 500 Ma, the IIIAB iron ages cluster around 600–800 Ma, which may reflect specific major breakup events. There is no systematic

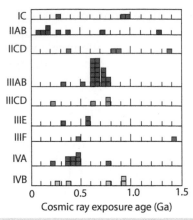

Figure 6.31 Summary of cosmic-ray exposure ages of iron meteorites based on the $^{41}K/K$ method. Source: Adapted from data summary of Voshage et al. (1983).

difference in exposure ages between the NC clan (red shades) and the CC clan (blue shades).

Figure 6.32 shows the distribution of cosmic-ray ages of OCs, which are remarkably shorter than those of iron meteorites. This is believed to reflect the great strength of iron meteorite parent bodies, resulting in a lower likelihood of shattering in collisions. All show a spectrum of ages within the last ≈50 million

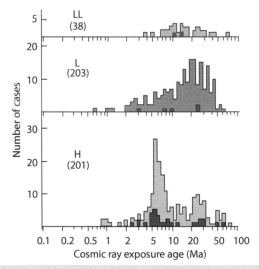

Figure 6.32 Cosmic-ray exposure ages for three classes of OCs. Filled histogram is for meteorites with regolith histories (i.e., brecciated meteorites). Source: Crabb and Schultz (1981)/ with permission of Elsevier.

years. Other class of chondrites shows a similar exposure age distribution. For example, average exposure ages for carbonaceous chondrites range from ≈2 Ma for CM chondrites to ≈25 Ma for CO chondrites with ages ranging from ≈70 Ma to ≈200 ka (Herzog and Caffee, 2014). We noted in Section 6.4 that there is a peak in K–Ar ages of L chondrites at around 50 Ma. In contrast, exposure ages of L chondrites show a distribution around a mean of ≈20 Ma, suggesting either a subsequent impact on the parent body or modern L chondrites derived from breakup of the fragments originally produced at around 500 Ma.

Achondrites show a similar age range as well. HED meteorites, which are derived from *4 Vesta* (or smaller members of the Vestoid asteroid family, themselves likely derived from *Vesta*), show peaks in distribution at around 20 and 30 Ma. Two exceptions are lunar meteorites and the SNC meteorites, which tend to have much younger exposure ages. Most lunar meteorites have exposure ages of less than 1 Ma and some less than 0.1 Ma; this of course is not surprising since they begin their short transit to Earth within the reach of Earth's gravity. SNC meteorite exposure ages are typically several million years and can be tied to events at around 2–3.5 Ma and another at around 4 Ma (Herzog and Caffee, 2014).

Except for the SNC and lunar meteorites, cosmic-ray exposure ages most likely reflect additional breakup events of fragments of the original parent bodies. H chondrites show a sharp peak at around five million years, while L chondrites show a broad peak at around 20–30 million years, reflecting such events. The near-Earth asteroid *25143 Itokawa* sampled by the Hayabusa spacecraft may be a good example of such a fragment. Returned samples have an LL5–6 chondrite composition. This object has a diameter of roughly 330 m and its low density and appearance indicate it is a little more than a "rubble pile" consisting of rock loosely held together by gravity. The ^{21}Ne exposure age of the returned grains from the surface is ≈8 Ma, suggesting it is continuously shedding material to space (Nagao et al., 2011). In contrast to these generally young ages, ^{40}Ar/^{39}Ar ages of shocked and brecciated meteorites indicate much earlier parent body breakup, perhaps 1300 Ma for the LL chondrite parent body and 470 Ma for the L chondrite parent body.

REFERENCES

Amari, S., Lewis, R.S. and Anders, E. 1995. Interstellar grains in meteorites: III. Graphite and its noble gases. *Geochimica et Cosmochimica Acta* **59**: 1411–26. doi: https://doi.org/10.1016/0016-7037(95)00053-3.

Amelin, Y. 2008. U-Pb ages of angrites. *Geochimica et Cosmochimica Acta* **72**: 221–32. doi: 10.1016/j.gca.2007.09.034.

Amelin, Y., Kaltenbach, A., Iizuka, T., et al. 2010. U–Pb chronology of the Solar System's oldest solids with variable ^{238}U/^{235}U. *Earth and Planetary Science Letters* **300**: 343–50. doi: 10.1016/j.epsl.2010.10.015.

Amelin, Y., Koefoed, P., Iizuka, T., et al. 2019. U-Pb, Rb-Sr and Ar-Ar systematics of the ungrouped achondrites Northwest Africa 6704 and Northwest Africa 6693. *Geochimica et Cosmochimica Acta* **245**: 628–642. doi: https://doi.org/10.1016/j.gca.2018.09.021.

Andreasen, R. and Sharma, M. 2006. Solar nebula heterogeneity in p-process samarium and neodymium isotopes. *Science* **314**:806–9.

Barboni, M., Boehnke, P., Keller, B., et al. 2017. Early formation of the Moon 4.51 billion years ago. *Science Advances* **3**: e1602365. doi: 10.1126/sciadv.1602365.

Black, D.C. and Pepin, R.O. 1969. Trapped neon in meteorites — II. *Earth and Planetary Science Letters* **6**: 395–405. doi: 10.1016/0012-821X(69)90190-3.

Bollard, J., Connelly, J.N. and Bizzarro, M. 2015. Pb-Pb dating of individual chondrules from the CBa chondrite Gujba: Assessment of the impact plume formation model. *Meteoritics & Planetary Science* **50**: 1197–1216. doi: 10.1111/maps.12461.

Bollard, J., Connelly, J.N., Whitehouse, M.J., et al. 2017. Early formation of planetary building blocks inferred from Pb isotopic ages of chondrules. *Science Advances* **3**. doi: 10.1126/sciadv.1700407.

Bollard, J., Kawasaki, N., Sakamoto, N., et al. 2019. Combined U-corrected Pb-Pb dating and ^{26}Al-^{26}Mg systematics of individual chondrules – Evidence for a reduced initial abundance of ^{26}Al amongst inner Solar System chondrules. *Geochimica et Cosmochimica Acta* **260**: 62–83. doi: https://doi.org/10.1016/j.gca.2019.06.025.

Boss, A.P. 2003. Rapid formation of outer giant planets by disk instability. *The Astrophysical Journal* **599**: 577–81. doi: 10.1086/379163

Bouvier, A., Blichert-Toft, J., Moynier, F., et al. 2007. Pb–Pb dating constraints on the accretion and cooling history of chondrites. *Geochimica et Cosmochimica Acta* **71**: 1583–604. doi:10.1016/j.gca.2006.12.005.

Bouvier, A., Blichert-Toft, J. and Albarède, F. 2009. Martian meteorite chronology and the evolution of the interior of Mars. *Earth and Planetary Science Letters* **280**: 285–295. doi: 10.1016/j.epsl.2009.01.042.

Bouvier, A. and Boyet, M. 2016. Primitive Solar System materials and Earth share a common initial [142]Nd abundance. *Nature* **537**: 399–402. doi: 10.1038/nature19351

Bouvier, A. and Wadhwa, M. 2010. The age of the Solar System redefined by the oldest Pb-Pb age of a meteoritic inclusion. *Nature Geoscience* **3**: 637–41. doi:10.1038/ngeo941.

Bouvier, A., Spivak-Birndorf, L.J., Brennecka, G.A., et al. 2011. New constraints on early Solar System chronology from Al–Mg and U–Pb isotope systematics in the unique basaltic achondrite Northwest Africa 2976. *Geochimica et Cosmochimica Acta* **75**: 5310–5323. doi: 10.1016/j.gca.2011.06.033.

Boyet, M. and Gannoun, A. 2013. Nucleosynthetic Nd isotope anomalies in primitive enstatite chondrites. *Geochimica et Cosmochimica Acta* **121**: 652–66. doi: 10.1016/j.gca.2013.07.036.

Boyet, M. and Carlson, R.L. 2005. [142]Nd evidence for early (>4.3 Ga) global differentiation of the silicate Earth. *Science* **309**: 576–81.

Boyet, M., Blichert-Toft, J., Rosing, M., et al. 2003. [142]Nd evidence for early Earth differentiation. *Earth and Planetary Science Letters* **214**: 427–42.

Boyet, M., Bouvier, A., Frossard, P., et al. 2018. Enstatite chondrites EL3 as building blocks for the Earth: The debate over the [146]Sm–[142]Nd systematics. *Earth and Planetary Science Letters* **488**: 68–78. doi: https://doi.org/10.1016/j.epsl.2018.02.004.

Brennecka, G.A., Amelin, Y. and Kleine, T. 2018. Uranium isotope ratios of *Muonionalusta* troilite and complications for the absolute age of the IVa iron meteorite core. *Earth and Planetary Science Letters* **490**: 1–10. doi: 10.1016/j.epsl.2018.03.010.

Brennecka, G.A. and Wadhwa, M. 2012. Uranium isotope compositions of the basaltic angrite meteorites and the chronological implications for the early Solar System. *Proceedings of the National Academy of Sciences* **109**: 9299–303. doi: 10.1073/pnas.1114043109.

Brennecka, G.A., Weyer, S., Wadhwa, M., et al. 2010. [238]U/[235]U variations in meteorites: extant [247]Cm and implications for Pb-Pb dating. *Science* **327**: 449–51. doi:10.1126/science.1180871.

Budde, G., Burkhardt, C. and Kleine, T. 2019. Molybdenum isotopic evidence for the late accretion of outer solar system material to earth. *Nature Astronomy* **3**: 736–41. doi: 10.1038/s41550-019-0779-y.

Budde, G., Kleine, T., Kruijer, T.S., et al. 2016. Tungsten isotopic constraints on the age and origin of chondrules. *Proceedings of the National Academy of Sciences* **113**: 2886–91. doi: 10.1073/pnas.1524980113.

Burkhardt, C., Borg, L.E., Brennecka, G.A., et al. 2016. A nucleosynthetic origin for the Earth's anomalous [142]Nd composition. *Nature* **537**: 394–8. doi: 10.1038/nature18956.

Carlson, R.W., Boyet, M. and Horan, M. 2007. Chondrite barium, neodymium, and samarium isotopic heterogeneity and early earth differentiation. *Science* **316**: 1175–8. doi: 10.1126/science.1140189.

Caro, G., Bourdon, B., Halliday, A.N., et al. 2008. Super-chondritic Sm/Nd ratios in Mars, the Earth, and the Moon. *Nature* **452**: 336–9. doi: 10.1038/nature06760.

Caro, G., Bourdon, B., Birck, J.-L. and Moorbath, S. 2003. [146]Sm-[142]Nd evidence from Isua metamorpohosed sediments for early differentiation of the Earth's mantle. *Nature* **423**: 428–32.

Chen, J.H. and Wasserburg, G.J. 1990. The isotopic composition of Ag in meteorites and the presence of [107]Pd in protoplanets. *Geochimica et Cosmochimica Acta* **54**:1729–43.

Clayton, R.N. 1993. Oxygen isotopes in meteorites. *Annual Review of Earth and Planetary Sciences* **21**: 115–49. doi:10.1146/annurev.ea.21.050193.000555.

Clayton, R.N., Grossman, L. and Mayeda, T.K. 1973. A component of primitive nuclear composition in carbonaceous meteorites. *Science* **182**: 485–8. doi:10.1126/science.182.4111.485.

Clayton, R.N. 2002. Self-shielding in the solar nebula. *Nature* **415**: 860–1.

Connelly, J.N., Bizzarro, M., Krot, A.N., et al. 2012. The absolute chronology and thermal processing of solids in the solar protoplanetary disk. *Science* **338**: 651–5. doi:10.1126/science.1226919.

Connelly, J.N., Bizzarro, M., Thrane, K., et al. 2008. The Pb–Pb age of angrite *SAH99555* revisited. *Geochimica et Cosmochimica Acta* **72**: 4813–24. doi: 10.1016/j.gca.2008.06.007.

Connelly, J.N., Schiller, M. and Bizzarro, M. 2019. Pb isotope evidence for rapid accretion and differentiation of planetary embryos. *Earth and Planetary Science Letters* **525**: 115722. doi: 10.1016/j.epsl.2019.115722.

Crabb, J., and Schultz, L. 1981. Cosmic-ray exposure ages of the ordinary chondrites and their significance for parent body stratigraphy, *Geochimica et Cosmochimica Acta* **45**: 2151–60.

Crowther, S.A., Filtness, M.J., Jones, R.H., et al. 2018. Old formation ages of igneous clasts on the L chondrite parent body reflect an early generation of planetesimals or chondrule formation. *Earth and Planetary Science Letters* **481**: 372–86. doi: 10.1016/j.epsl.2017.10.047.

Dauphas, N. and Chaussidon, M. 2011. A perspective from extinct radionuclides on a young stellar object: the Sun and its accretion disk. *Annual Reviews of Earth and Planetary Sciences* **39**: 351–86. doi: 10.1146/annurev-earth-040610-133428.

Debaille, V., Brandon, A.D., Yin, Q.Z., et al. 2007. Coupled ^{142}Nd–^{143}Nd evidence for a protracted magma ocean in Mars. *Nature* **450**: 525–8. doi: 10.1038/nature06317.

Desch, S.J., Kalyaan, A. and Alexander, C.M.O.D. 2018. The effect of Jupiter's formation on the distribution of refractory elements and inclusions in meteorites. *The Astrophysical Journal Supplement Series* **238**: 11. doi: 10.3847/1538-4365/aad95f.

Fang, L., Frossard, P., Boyet, M., et al. 2022. Half-life and initial Solar System abundance of ^{146}Sm determined from the oldest andesitic meteorite. *Proceedings of the National Academy of Sciences* **119**: e2120933119. doi: 10.1073/pnas.2120933119.

Fukuda, K., Hiyagon, H., Fujiya, W., et al. 2021. Irradiation origin of ^{10}Be in the solar nebula: evidence from Li-Be-B and Al-Mg isotope systematics, and REE abundances of CAIs from *Yamato-81020* CO3.05 chondrite. *Geochimica et Cosmochimica Acta* **293**: 187–204. doi: 10.1016/j.gca.2020.10.011.

Gallino, R., Busso, M., Picchio, G., et al. 1990. On the astrophysical interpretation of isotope anomalies in meteoritic SiC grains. *Nature* **348**: 298–302.

Göpel, C., Manhès, G. and Allègre, C. 1994. U-Pb systematics of phosphates from equilibrated ordinary chondrites. *Earth and Planetary Science Letters* **121**: 153–71.

Gounelle, M., Chaussidon, M. and Rollion-Bard, C. 2013. Variable and extreme irradiation conditions in the early solar system inferred from the initial abundance of ^{10}Be in *Isheyevo* CAIs. *The Astrophysical Journal* **763**: L33. doi: 10.1088/2041-8205/763/2/l33.

Gyngard, F., Jadhav, M., Nittler, L.R., et al. 2018. Bonanza: an extremely large dust grain from a supernova. *Geochimica et Cosmochimica Acta* **221**: 60–86. doi: 10.1016/j.gca.2017.09.002.

Harper, C.L. and Jacobsen, S.B. 1992. Evidence from coupled ^{147}Sm-^{143}Nd and ^{146}Sm-^{142}Nd systematics for very early (4.5-Gyr) differentiation of the Earth's mantle. *Nature* **360**: 728–32.

Heber, V.S., Baur, H., Bochsler, P., et al. 2012. Isotopic mass fractionation of solar wind: evidence from fast and slow solar wind collected by the *Genesis* Mission. *The Astrophysical Journal* **759**:121. doi: 10.1088/0004-637x/759/2/121.

Heck, P.R., Greer, J., Kööp, L., et al. 2020. Lifetimes of interstellar dust from cosmic ray exposure ages of presolar silicon carbide. *Proceedings of the National Academy of Sciences* **117**: 1884–9. doi: 10.1073/pnas.1904573117.

Herzog, G.F. and Caffee, M.W. 2014. 1.13 - cosmic-ray exposure ages of meteorites, in Holland, H.D. and Turekian, K. K. (eds). *Treatise on geochemistry*, 2nd ed. Oxford: Elsevier.

Hilton, C.D. and Walker, R.J. 2020. New implications for the origin of the IAB main group iron meteorites and the isotopic evolution of the non-carbonaceous (NC) reservoir. *Earth and Planetary Science Letters* **540**: 116248. doi: 10.1016/j.epsl.2020.116248.

Hofmann, A.W., Class, C., Goldstein, S.L., 2022. Size and Composition of the MORB+OIB Mantle Reservoir. *Geochemistry, Geophysics, Geosystems* **23**, e2022GC010339.doi: https://doi.org/10.1029/2022GC010339

Hohenberg, C.M. and Pravdivtseva, O.V. 2008. I–Xe dating: from adolescence to maturity. *Chemie der Erde - Geochemistry* **68**: 339–51. doi: 10.1016/j.chemer.2008.06.002.

Horan, M.F., Carlson, R.W. and Blichert-Toft, J. 2012. Pd–Ag chronology of volatile depletion, crystallization and shock in the *Muonionalusta* IVA iron meteorite and implications for its parent body. *Earth and Planetary Science Letters* **351–2**: 215–22. doi: 10.1016/j.epsl.2012.07.028.

Huss, G.R., Hutcheon, I.D. and Wasserburg, G.J. 1997. Isotopic systematics of presolar silicon carbide from the *Orgueil* (CI) chondrite: implications for solar system formation and stellar nucleosynthesis. *Geochimica et Cosmochimica Acta* **61**: 5117–48.

Hynes, K. and Gyngard, F. 2009. The presolar grain database: http://presolar.wustl.edu/~pgd. Lunar and Planetary Science Conference.

Jacobsen, S.B. 2005. The birth of the solar system in a molecular cloud: evidence from the isotopic pattern of short-lived nuclides in the early solar system, in Krot, A., Scott, E. and Reipurth, B., (eds). *Chondrites and the protoplanetary disk*. Astronomical Society of the Pacific, 548–57 pp.

Johansen, A., Low, M.-M.M., Lacerda, P., et al. 2015. Growth of asteroids, planetary embryos, and Kuiper belt objects by chondrule accretion. *Science Advances* **1**: e1500109. doi: 10.1126/sciadv.1500109.

Johnston, S., Brandon, A., Mcleod, C., et al. 2022. Nd isotope variation between the Earth–Moon system and enstatite chondrites. *Nature* **611**: 501–506. doi: 10.1038/s41586-022-05265-0.

Kinoshita, N., Paul, M., Kashiv, Y., et al. 2012. A shorter ^{146}Sm half-life measured and implications for ^{146}Sm-^{142}Nd chronology in the solar system. *Science* **335**: 1614–7. doi: 10.1126/science.1215510.

Kleine, T., Touboul, M., Van Orman, J.A., et al. 2008. Hf–W thermochronometry: closure temperature and constraints on the accretion and cooling history of the H chondrite parent body. *Earth and Planetary Science Letters* **270**: 106–18. doi: 10.1016/j.epsl.2008.03.013.

Kleine, T., Hans, U., Irving, A.J., et al. 2012. Chronology of the angrite parent body and implications for core formation in protoplanets. *Geochimica et Cosmochimica Acta* **84**: 186–203. doi: https://doi.org/10.1016/j.gca.2012.01.032.

Kleine, T. and Walker, R.J. 2017. Tungsten isotopes in planets. *Annual Review of Earth and Planetary Sciences* **45**: 389–417. doi: 10.1146/annurev-earth-063016-020037.

Korochantseva, E.V., Trieloff, M., Lorenz, C.A., et al. 2007. L-chondrite asteroid breakup tied to Ordovician meteorite shower by multiple isochron ^{40}Ar-^{39}Ar dating. *Meteoritics and Planetary Science* **42**: 113–30. doi: https://doi.org/10.1111/j.1945-5100.2007.tb00221.x.

Krot, A.N., Park, C. and Nagashima, K. 2014. Amoeboid olivine aggregates from CH carbonaceous chondrites. *Geochimica et Cosmochimica Acta* **139**: 131–53. doi: 10.1016/j.gca.2014.04.050.

Krot, A.N., Nagashima, K., Lyons, J.R., et al. 2020. Oxygen isotopic heterogeneity in the early Solar System inherited from the protosolar molecular cloud. *Science Advances* **6**: eaay2724. doi: 10.1126/sciadv.aay2724.

Kruijer, T.S., Kleine, T., Fischer-Gödde, M., et al. 2014. Nucleosynthetic W isotope anomalies and the Hf–W chronometry of Ca–Al-rich inclusions. *Earth and Planetary Science Letters* **403**: 317–27. doi: 10.1016/j.epsl.2014.07.003.

Kruijer, T.S., Burkhardt, C., Budde, G., et al. 2017. Age of Jupiter inferred from the distinct genetics and formation times of meteorites. *Proceedings of the National Academy of Sciences* **114**: 6712–6. doi: 10.1073/pnas.1704461114.

Kruijer, T.S., Kleine, T. and Borg, L.E. 2020. The great isotopic dichotomy of the early Solar System. *Nature Astronomy* **4**: 32–40. doi: 10.1038/s41550-019-0959-9.

Larsen, K.K., Schiller, M. and Bizzarro, M. 2016. Accretion timescales and style of asteroidal differentiation in an ^{26}Al-poor protoplanetary disk. *Geochimica et Cosmochimica Acta* **176**: 295–315. doi: https://doi.org/10.1016/j.gca.2015.10.036.

Larsen, K.K., Trinquier, A., Paton, C., et al. 2011. Evidence for magnesium isotope heterogeneity in the solar protoplanetary disk. *The Astrophysical Journal* **735**: L37. doi: 10.1088/2041-8205/735/2/l37.

Liu, M.-C., Chaussidon, M., Srinivasan, G., et al. 2012. A Lower initial abundance of short-lived ^{41}Ca in the early solar system and its implications for solar system formation. *The Astrophysical Journal* **761**: 137. doi: 10.1088/0004-637x/761/2/137.

Liu, M.-C. 2017. The initial ^{41}Ca/^{40}Ca ratios in two type-A Ca–Al-rich inclusions: Implications for the origin of short-lived ^{41}Ca. *Geochimica et Cosmochimica Acta* **201**: 123–35. doi: 10.1016/j.gca.2016.10.011.

Lee, T., Papanastassiou, D.A. and Wasserburg, G.J. 1976. Demonstration of ^{26}Mg excess in Allende and evidence for ^{26}Al *Geophysical Research Letters* **3**: 41–4.

Lewis, R.S., Amari, S. and Anders, E. 1990. Meteorite silicon carbide: pristine material from carbon stars. *Nature* **348**: 293–8.

Lewis, R.S., Amari, S. and Anders, E. 1994. Interstellar grains in meteorites: II. SiC and its noble gases. *Geochim. Cosmochim. Acta* **58**: 471–94.

Lewis, R.S., Srinivasan, B. and Anders, E. 1975. Host phase of a strange xenon component in *Allende*. *Science* **190**: 1251–62. doi: 10.2307/1741803.

Leya, I., Masarik, J. and Lin, Y. 2018. Alteration of CAIs as recorded by ^{36}S/^{34}S as a function of ^{35}Cl/^{34}S. *Meteoritics and Planetary Science* **53**: 1252–66. doi: i10.1111/maps.13070.

Lugaro, M., Heger, A., Osrin, D., et al. 2014. Stellar origin of the ^{182}Hf cosmochronometer and the presolar history of solar system matter. *Science* **345**: 650–3. doi: 10.1126/science.1253338.

Lugaro, M., Karakas, A.I., Bruno, C.G., et al. 2017. Origin of meteoritic stardust unveiled by a revised proton-capture rate of ^{17}O. *Nature Astronomy* **1**: 0027. doi: 10.1038/s41550-016-0027.

Lugmair, G.W., Scheinin, N.B. and Marti, K. 1975. Search for extinct ^{146}Sm, I. The isotopic abundance of ^{142}Nd in the *Juvinas* meteorite. *Earth and Planetary Science Letters* **27**: 79–84. doi: 10.1016/0012-821X(75)90163-6.

MacPherson, G.J., Kita, N.T., Ushikubo, T., et al. 2012. Well-resolved variations in the formation ages for Ca–Al-rich inclusions in the early Solar System. *Earth and Planetary Science Letters* **331**: 43–54. doi: 10.1016/j.epsl.2012.03.010.

Marks, N.E., Borg, L.E., Hutcheon, I.D., et al. 2014. Samarium–neodymium chronology and rubidium–strontium systematics of an Allende calcium–aluminum-rich inclusion with implications for 146Sm half-life. *Earth and Planetary Science Letters* **405**: 15–24. doi: 10.1016/j.epsl.2014.08.017.

Matthes, M., Fischer-Gödde, M., Kruijer, T.S., et al. 2018. Pd-Ag chronometry of IVA iron meteorites and the crystallization and cooling of a protoplanetary core. *Geochimica et Cosmochimica Acta* **220**: 82–95. doi: 10.1016/j.gca.2017.09.009.

Matthes, M., Fischer-Gödde, M., Kruijer, T.S., et al. 2015. Pd–Ag chronometry of iron meteorites: correction of neutron capture-effects and application to the cooling history of differentiated protoplanets. *Geochimica et Cosmochimica Acta* **169**: 45–62. doi: https://doi.org/10.1016/j.gca.2015.07.027.

Matthes, M., Van Orman, J.A. and Kleine, T. 2020. Closure temperature of the Pd-Ag system and the crystallization and cooling history of IIIAB iron meteorites. *Geochimica et Cosmochimica Acta* **285**: 193–206. doi: https://doi.org/10.1016/j.gca.2020.07.009.

McKeegan, K.D., Chaussidon, M., Robert, F., 2000. Incorporation of short-lived [10]Be in a calcium-aluminum-rich inclusion from the Allende meteorite. *Science* **289**, 1334–1337. doi: 10.1126/science.289.5483.1334

McKeegan, K.D., Kallio, A.P.A., Heber, V.S., et al. 2011. The oxygen isotopic composition of the Sun inferred from captured solar wind. *Science* **332**: 1528–32. doi: 10.1126/science.1204636.

McSween, H.Y. and Huss, G.R. 2010. *Cosmochemistry*. Cambridge: Cambridge University Press.

Meissner, F., Schmidt-Ott, W.-D., Ziegeler, L., 1987. Half-life and α-ray energy of [146]Sm. *Zeitschrift für Physik A Atomic Nuclei* **327**, 171–174.

Meyer, B.S., Clayton, D.D. and The, L.S. 2000. Molybdenum and zirconium isotopes from a supernova neutron burst. *The Astrophysical Journal* **540**: L49–52. doi: 10.1086/312865.

Ming, T. and Anders, E. 1988. Isotopic anomalies of Ne, Xe, and C in meteorites. II. Interstellar diamond and SiC: carriers of exotic noble gases. *Geochimica et Cosmochimica Acta* **52**: 1235–44. doi: 10.1016/0016-7037(88)90277-3.

Nagao, K., Okazaki, R., Nakamura, T., et al. 2011. Irradiation history of *Itokawa* regolith material deduced from noble gases in the *Hayabusa* samples. *Science* **333**: 1128–31. doi: 10.1126/science.1207785.

Nguyen, A.N. and Zinner, E. 2004. Discovery of ancient silicate stardust in a meteorite. *Science* **303**: 1496–9.

Nittler, L.R. 2003. Presolar stardust in meteorites: recent advances and scientific frontiers. *Earth and Planetary Science Letters* **209**: 259–73.

Nittler, L.R., Alexander, C.M.O.D., Gao, X., et al. 1997. Stellar sapphires: the properties and origins of presolar Al_2O_3 in meteorites. *The Astrophysical Journal* **483**: 475–95. doi: 10.1086/304234.

Nyquist, L.E., Reese, Y., Wiesmann, H., et al. 2003. Fossil [26]Al and [53]Mn in the *Asuka 881394* eucrite: evidence of the earliest crust on asteroid 4 Vesta. *Earth and Planetary Science Letters* **214**: 11–25. doi: 10.1016/s0012-821x(03) 00371-6.

Nyquist, L.E., Shih, C.-Y., Reese, D., et al. 2010. Lunar crustal history recorded in lunar anorthosites. Lunar and Planetary Science Conference.

O'Neill, H.S.C. and Palme, H. 2008. Collisional erosion and the non-chondritic composition of the terrestrial planets. *Philosophical Transactions of the Royal Society A* **366**: 4205–38. doi: 10.1098/rsta.2008.0111.

Pape, J., Mezger, K., Bouvier, A.S., et al. 2019. Time and duration of chondrule formation: constraints from [26]Al-[26]Mg ages of individual chondrules. *Geochimica et Cosmochimica Acta* **244**: 416–36. doi: 10.1016/j.gca.2018.10.017.

Park, J., Turrin, B.D., Herzog, G.F., et al. 2015. [40]Ar/[39]Ar age of material returned from asteroid *25143 Itokawa*. *Meteoritics and Planetary Science* **50**: 2087–98. doi: 10.1111/maps.12564.

Patterson, C. 1956. Age of meteorites and the Earth. *Geochimica et Cosmochimica Acta* **16**: 230–7.

Pellin, M., Davis, A., Lewis, R., et al. 1999. *Molybdenum isotopic composition of single silicon carbide grains from supernovae*. Lunar and Planetary Science Conference.

Pravdivtseva, O., Meshik, A., Hohenberg, C.M., et al. 2017. I–Xe systematics of the impact plume produced chondrules from the CB carbonaceous chondrites: implications for the half-life value of [129]I and absolute age normalization of [129]I–[129]Xe chronometer. *Geochimica et Cosmochimica Acta* **201**: 320–30. doi: 10.1016/j.gca.2016.01.012.

Qin, L. and Carlson, R.W. 2016. Nucleosynthetic isotope anomalies and their cosmochemical significance. *Geochemical Journal* **50**: 43–65. doi: 10.2343/geochemj.2.0401.

Qin, L., Carlson, R.W. and Alexander, C.M.O.D. 2011. Correlated nucleosynthetic isotopic variability in Cr, Sr, Ba, Sm, Nd and Hf in Murchison and QUE 97008. *Geochimica et Cosmochimica Acta* **75**: 7806–28. doi: 10.1016/j. gca.2011.10.009.

Reynolds, J.R. 1960. Isotopic composition of xenon from enstatite chondrites. *Zeitshrift für Naturforschung* **15a**: 1112–4.

Schiller, M., Connelly, J.N., Glad, A.C., et al. 2015. Early accretion of protoplanets inferred from a reduced inner solar system [26]Al inventory. *Earth and Planetary Science Letters* **420**: 45–54. doi: 10.1016/j.epsl.2015.03.028.

Schmitz, B., Peucker-Ehrenbrink, B., Lindström, M., et al. 1997. Accretion rates of meteorites and cosmic dust in the early Ordovician. *Science* **278**: 88–90. doi: 10.1126/science.278.5335.88.

Shukolyukov, A., and Lugmair, G.W. 1993. [60]Fe in eucrites. *Earth and Planetary Science Letters* **119**: 159–66.

Swindle, T.D., Caffee, M.W., Hohenberg, C.M., et al. 1983. I-Xe studies of individual Allende chondrules. *Geochimica et Cosmochimica Acta* **47**: 2157–77. doi:10.1016/0016-7037(83)90040-6.

Swindle, T.D., Kring, D.A. and Weirich, J.R. 2014. [40]Ar/[39]Ar ages of impacts involving ordinary chondrite meteorites. *Geological Society, London, Special Publications* **378**: 333–47. doi: 10.1144/sp378.6.

Telus, M., Huss, G.R., Nagashima, K., Ogliore, R.C., et al. 2018. *In situ* [60]Fe-[60]Ni systematics of chondrules from unequilibrated ordinary chondrites. *Geochimica et Cosmochimica Acta* **221**: 342–57. doi: https://doi.org/10.1016/j. gca.2017.06.013.

Thiemens, M.H., and Heidenreich, J.E. 1983. The mass independent fractionation of oxygen — A novel isotopic effect and its cosmochemical implications. *Science* **219**: 1073–5.

Thiemens, M.H. 2006. History and applications of mass-independent isotope effects. *Annual Review of Earth and Planetary Sciences* **34**: 217–62.

Thiemens, M.M., Sprung, P., Fonseca, R.O.C., et al. 2019. Early Moon formation inferred from hafnium–tungsten systematics. *Nature Geoscience* **12**: 696–700. doi: 10.1038/s41561-019-0398-3.

Touboul, M., Puchtel, I.S. and Walker, R.J. 2015. Tungsten isotopic evidence for disproportional late accretion to the Earth and Moon. *Nature* **520**: 530–3. doi: 10.1038/nature14355.

Trappitsch, R., Boehnke, P., Stephan, T., et al. 2018. New constraints on the abundance of ^{60}Fe in the early solar system. *The Astrophysical Journal* **857**: L15. doi: 10.3847/2041-8213/aabba9.

Trinquier, A., Birck, J.L., Allègre, C.J., et al. 2008. ^{53}Mn–Cr systematics of the early Solar System revisited. *Geochimica et Cosmochimica Acta* **72**: 5146–63. doi: 10.1016/j.gca.2008.03.023.

Tscharnuter, W.M., Schönke, J., Gail, H.-P., et al. 2009. Protostellar collapse: rotation and disk formation. *Astronomy and Astrophysics* **504**: 109–113. doi: 10.1051/0004-6361/200912120.

Turner, G., Crowther, S.A., Burgess, R., et al. 2013. Short lived ^{36}Cl and its decay products ^{36}Ar and ^{36}S in the early solar system. *Geochimica et Cosmochimica Acta* **123**: 358–67. doi: 10.1016/j.gca.2013.06.022.

Vermeesch, P., 2018. IsoplotR: A free and open toolbox for geochronology. *Geoscience Frontiers* **9**, 1479–1493.

Villa, I.M., Holden, N.E., Possolo, A., et al. 2020. IUPAC-IUGS recommendation on the half-lives of ^{147}Sm and ^{146}Sm. *Geochimica et Cosmochimica Acta* **285**: 70–7. doi: 10.1016/j.gca.2020.06.022.

Voshage, H., Feldmann, H. and Braun, O. 1983. Investigations of cosmic-ray-produced nuclides in iron meteorites: 5. More data on the nuclides of potassium and noble gases, on exposure ages and meteoroid sizes. *Zeitschrift für Naturforschung A* **38**: 273–80. doi: https://doi.org/10.1515/zna-1983-0227.

Wadhwa, M., Amelin, Y., Bogdanovski, O., et al. 2009. Ancient relative and absolute ages for a basaltic meteorite: implications for timescales of planetesimal accretion and differentiation. *Geochimica et Cosmochimica Acta* **73**: 5189–201. doi: 10.1016/j.gca.2009.04.043.

Walsh, K.J., Morbidelli, A., Raymond, S.N., et al. 2011. A low mass for Mars from Jupiter's early gas-driven migration. *Nature* **475**: 206. doi: 10.1038/nature10201

Warren, P.H. 2011. Stable-isotopic anomalies and the accretionary assemblage of the Earth and Mars: a subordinate role for carbonaceous chondrites. *Earth and Planetary Science Letters* **311**: 93–100. doi: 10.1016/j.epsl.2011.08.047.

Wasserburg, G.J., Busso, M., Gallino, R., et al. 2006. Short-lived nuclei in the early Solar System: possible AGB sources. *Nuclear Physics A* **777**: 5–69. doi: 10.1016/j.nuclphysa.2005.07.015.

Weidenschilling, S.J. 2000. Formation of planetesimals and accretion of the terrestrial planets. *Space Science Reviews* **92**: 295–310. doi: 10.1023/A:1005259615299.

Wetherill, G.W. 1990. Formation of the Earth. *Annual Review of Earth and Planetary Sciences* **18**: 205–56. doi: 10.1146/annurev.ea.18.050190.001225.

Whitby, J., Burgess, R., Turner, G., et al. 2000. Extinct ^{129}I in halite from a primitive meteorite: evidence for evaporite formation in the early Solar System. *Science* **288**: 1819–21. doi: 10.1126/science.288.5472.1819.

Whitby, J.A., Gilmour, J.D., Turner, G., et al. 2002. Iodine-Xenon dating of chondrules from the *Qingzhen* and *Kota Kota* enstatite chondrites. *Geochimica et Cosmochimica Acta* **66**: 347–59. doi: 10.1016/S0016-7037(01)00783-9.

White, W.M. 2020. *Geochemistry*, 2nd ed. Oxford: Wiley Blackwell.

Yokoyama, T., Nagashima, K., Nakai, I., et al. 2022. Samples returned from the asteroid *Ryugu* are similar to Ivuna-type carbonaceous meteorites. *Science*: eabn7850. Epub. doi:10.1126/science.abn7850.

Young, E.D. 2014. Inheritance of solar short- and long-lived radionuclides from molecular clouds and the unexceptional nature of the solar system. *Earth and Planetary Science Letters* **392**: 16–27. doi: 10.1016/j.epsl.2014.02.014.

Young, E.D. 2016. Bayes' theorem and early solar short-lived radionuclides: the case for an unexceptional origin for the solar system. *The Astrophysical Journal* **826**: 129. doi: 10.3847/0004-637x/826/2/129.

Zinner, E. 2014. 1.4 - presolar grains, in Holland, H.D. and Turekian, K.K. (eds). *Treatise on Geochemistry*, 2nd ed. Oxford: Elsevier.

PROBLEMS

1. Horan et al. (2012) determined an initial ^{107}Pd/^{108}Pd ratio for the iron meteorite *Muonionalustra* of 2.15×10^{-5}. The decay constant of ^{107}Pd is 1.06×10^{-4} yr^{-1}. Assuming an age for the meteorite of 4558 Ma and an age for the Solar System of 4567.3 Ma, calculate the initial ^{107}Pd/^{108}Pd ratio of the Solar System.

2. The following data were measured on the ungrouped achondrite NWA7325 by Koefoed et al. (2016). δ^{26}Mg* is the fractionation corrected per mil deviation from the standard DSM3, calculated as in Equation (6.6) where ^{26}Mg/^{24}Mg$_{std}$ = 0.13979. Calculate the ^{27}Al/^{26}Al$_0$ for this meteorite. Assuming an initial ^{27}Al/^{26}Al equal to that of CAIs (5.25×10^{-5}) at the start of the Solar System, how much younger is it than CAIs?

	$^{27}Al/^{24}Mg$	$\delta^{26}Mg^*$
PL1	78.08	0.267
PL2	48.84	0.194
PL3	41.45	0.174
PL4	43.56	0.178
Px	0.187	0.091
Ol	0.078	0.083
WR	1.47	0.099

3. The calculated initial $^{27}Al/^{26}Al$ of the ungrouped achondrite NWA2976 is 3.94×10^{-7}. Assume a ^{27}Al half-life of 7.07×10^5 yr.

   (a) Assuming an initial $^{27}Al/^{26}Al$ equal to that of CAI's (5.25×10^{-5}) at the start of the Solar System, how much younger is it than CAIs?

   (b) Now use the angrite D'Orbigny, for which $^{26}Al/^{27}Al_0$ is $3.98 \pm 0.21 \times 10^{-7}$ at an absolute Pb–Pb age of 4563.37 ± 0.25 Ma as an anchor. If CAIs formed at 4567.3 Ma, how much younger is NWA2976 than CAIs?

4. Suppose you measure the $^{129}Xe/^{130}Xe$ and I/Xe ratios in two chondritic meteorites in step-heating gas release experiments. From the first, you deduce a $^{129}I/^{127}I$ ratio of 0.53×10^{-4} at the time of formation. In the second, you deduce a $^{129}I/^{127}I$ ratio of 1.35×10^{-4} at the time of formation. How much time elapsed between formation of these meteorites, and which formed first?

5. Bouvier et al. measured the following data on the H4 meteorite *Ste. Marguerite*:

	$^{206}Pb/^{204}Pb$	$^{207}Pb/^{204}Pb$
Whole-rock R0	41.6593	30.4466
L0	24.8906	20.0734
Chondrule R0	35.8299	26.8506
L0	18.5585	16.0838
Px+Ol R2	47.2008	33.9302
L1	20.3812	17.2259
L2	21.9916	18.2169

Calculate the $^{207}Pb/^{204}Pb$ age of this meteorite assuming a $^{238}U/^{235}U$ ratio of 137.88. Then calculate the age assuming a $^{238}U/^{235}U$ ratio of 138.79.

6. The initial $^{206}Pb/^{204}Pb$ and $^{207}Pb/^{204}Pb$ ratios of the Solar System at 4.567 Ga are 9.307 and 10.294, respectively. If the value of $^{238}U/^{204}Pb$ in the solar nebula was 0.14, calculate the evolution of $^{206}Pb/^{204}Pb$ and $^{207}Pb/^{204}Pb$ of the solar nebula at 100 Ma intervals from 4.567 to 4.45 Ga. (*Hint: this calculation is a little trickier than you might think.*)

7. In which cosmic environments could the following short-lived radionuclides be made? Justify your answer.

   (a) ^{182}Hf
   (b) ^{146}Sm
   (c) ^{107}Pd
   (d) ^{129}I
   (e) ^{247}Cm

8. Excesses of ^{135}Ba, ^{137}Ba, and ^{138}Ba have been found in several chondrites. How would you explain these in terms of the different nucleosynthetic processes occurring in different stellar environments? *(Hint: look at a Chart of the Nuclides and refer to Chapter 1.)*

9. The following data were measured on the ungrouped basaltic achondrite NWA 2976. Calculate the initial ^{26}Al/^{27}Al using the isochron method (adapt Equation [6.5] for the ^{26}Al–^{26}Mg decay). Assuming the ^{26}Al/^{27}Al abundance ratio listed in Table 6.1 was the initial ratio of the Solar System, what is the age of this meteorite relative to that initial time?

	^{27}Al/^{24}Mg	^{26}Mg/^{24}Mg
WR5	1.87	0.13979
PX3	0.67	0.13979
PL3	513.60	0.13999
PL4	143.00	0.13984
PL5	121.60	0.13984
PL6	82.27	0.13983

Chapter 7

Isotope Geochemistry of the Mantle

7.1 INTRODUCTION

In the preceding chapters, we learned how and why unstable nuclei undergo radioactive decay and how we can use the decay products of this process to determine geologic time. We also learned how ratios of stable isotopes can change. We learned how the various chemical elements and their isotopes were produced in various cosmic environments and how the Sun, the Earth, and various other Solar System bodies formed from a cloud of gas and dust leaving them with distinctive isotopic compositions. We now have the tools that we need to use isotope ratios to understand the Earth, its evolution, and processes operating in it and on it. Beginning with this chapter, we will do just that, starting with the Earth's mantle. In the next chapter, we will turn to the geochemistry of the continental crust, whose chemical evolution is closely related to that of the mantle.

Radiogenic isotopes will be the primary focus of this chapter, but we will also find that stable isotopes are useful in understanding mantle evolution. Stable isotope fractionations generally decrease with the square of temperature so that little stable isotope fractionation is expected at mantle temperatures. Because of this, we can use stable isotopes as passive tracers of material that has experienced isotope fractionations in environments at or near the Earth's surface in the Earth's mantle. The evidence provided by stable isotopes that material from the Earth's surface can be carried deep

into the mantle is a critical insight into how the mantle and indeed the Earth as a whole work. We will explore the use of stable isotopes in understanding the solid earth more fully in Chapter 9.

The initial use of radioactive and radiogenic isotopes in geology was directed exclusively toward geochronology. The potential geochemical applications became apparent only later. One of the first to recognize the potential of radiogenic isotope studies was Paul Gast, who was a student of Al Nier, the inventor of magnetic sector mass spectrometer widely used today and an early pioneer of isotope geochemistry. Since Gast's pioneering work, understanding mantle composition and evolution has become one of the most important applications of isotope geochemistry and we will devote considerable space to this topic. And since the mantle constitutes most of the mass and volume of the Earth, perhaps this is appropriate.

Radiogenic isotopes are particularly useful in studying the mantle for several reasons. First, basalts are our most primary mantle sample, but because they are partial melts, their composition differs from that of the mantle, but their isotopic composition does not. Second, basalts lack the detailed stratigraphic or structural context that a sample of crust might have, which is also true of another important mantle sample: the xenolith basalts sometimes carry to the surface. The time-integrated nature of radiogenic isotope ratios, however, provides important context. In what was one of the first

Isotope Geochemistry, Second Edition. William M. White.
© 2023 John Wiley & Sons Ltd. Published 2023 by John Wiley & Sons Ltd.
Companion Website: www.wiley.com/go/white/isotopegeochem2

papers to apply radiogenic isotopes to understanding the nature and composition of the mantle, Gast (1960) summarized these points as follows:

In a given chemical system the isotopic abundance of ^{87}Sr is determined by four parameters: the isotopic abundance at a given initial time, the Rb/Sr ratio of the system, the decay constant of ^{87}Rb, and the time elapsed since the initial time. The isotopic composition of a particular sample of strontium, whose history may or may not be known, may be the result of time spent in a number of such systems or environments. In any case the isotopic composition is the time-integrated result of the Rb/Sr ratios in all the past environments. Local differences in the Rb/Sr will, in time, result in local differences in the abundance of ^{87}Sr. Mixing of material during processes will tend to homogenize these local variations. Once homogenization occurs, the isotopic composition is not further affected by these processes. Because of this property and because of the time-integrating effect, isotopic compositions lead to useful inferences concerning the Rb/Sr ratio of the crust and of the upper mantle. It should be noted that similar arguments can be made for the radiogenic isotopes of lead, which are related to the U/Pb ratio and time.

This statement can just as easily be applied to the U–Th–Pb system as well as the decay systems developed since Gast's time, namely, Nd, Ce, Hf, and Os.

Gast's first sentence is simply a statement of the radiogenic growth equation for the Rb–Sr system:

$$\frac{^{87}Sr}{^{86}Sr} = \left(\frac{^{87}Sr}{^{86}Sr}\right)_0 + \frac{^{87}Rb}{^{86}Sr}\left(e^{\lambda t} - 1\right) \qquad (7.1)$$

It remains a valid and succinct summary of the principles of radiogenic isotope geochemistry. A principal objective of geology is to understand how the Earth evolved from its initial state to its present one. Radiogenic isotope geochemistry is uniquely suited for this sort of study because an isotope ratio, such as $^{87}Sr/^{86}Sr$, is a function not only of the differentiation processes that fractionate Rb from Sr, but also of the time at which the fractionation occurred. On a continuously evolving Earth, ancient features tend to be destroyed by subsequent processes. Isotope ratios, however, preserve, albeit incompletely, information about both chemistry and time.

As Gast said, the $^{87}Sr/^{86}Sr$ ratio is a function of the time-integrated Rb/Sr. Ultimately, we can draw much broader inferences than merely the time-integrated Rb/Sr ratio. Rb and Sr are both trace elements, and together account for only a few ppm of the mass of the Earth. However, Rb and Sr share some of their properties with other elements of their group: Rb with the alkalis and Sr with the alkaline earths. So Rb/Sr fractionations tell us something about the relative abundances of alkali and alkaline earth elements and fractionations between them. Indeed, in that same paper, Gast inferred from isotope ratios in basalts, "the upper mantle and crust of the Earth do not contain K, Rb, Cs, U, Ba, and Sr, in the proportions found in chondrites."

In addition, since Rb is highly incompatible and Sr is only moderately so, Rb/Sr fractionations tell us something about the fractionation of incompatible elements from less incompatible ones. Similarly, Sm/Nd variations are related to variations between light and heavy rare earths, as well as incompatible/less incompatible element variations. We have some knowledge of the processes that fractionate the alkalis and alkaline earths and the light and heavy rare earths. Thus, knowledge of variations in these element ratios allows us to limit the range of possible processes occurring within the Earth; the time parameter in Equation (7.1) allows us to limit the range of possible times at which this fractionation occurred.

7.1.1 Definitions: time-integrated and time-averaged

Gast stated that the $^{87}Sr/^{86}Sr$ is a function of the *time-integrated* Rb/Sr ratio. What did he mean by "time-integrated"?

Suppose, the $^{87}Rb/^{86}Sr$ ratio evolves in some reservoir in some complex way. Let us allow the $^{87}Rb/^{86}Sr$ to be an arbitrary function of time, such as $^{87}Rb/^{86}Sr = t + \sin(5\ t/\pi) + 1$; this is shown plotted in Figure 7.1a. If we integrate $^{87}Rb/^{86}Sr$ with respect to time, we get the area under the curve. From that, we can find the average $^{87}Rb/^{86}Sr$ simply by dividing the area under the curve by $(t - t_0)$, which is 3.3. The $^{87}Sr/^{86}Sr$ would evolve as shown by the solid red line in Figure 7.1b. Just as we can calculate the average $^{87}Rb/^{86}Sr$ ratio from the area under the curve in Figure 7.1a, we can calculate the average $^{87}Rb/^{86}Sr$ in a reservoir from Equation (7.1) if we know the initial and final values of t and

(a)

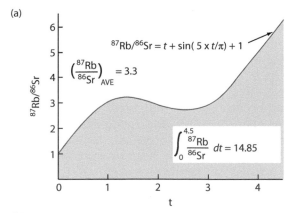

$$^{87}Rb/^{86}Sr = t + \sin(5 \times t/\pi) + 1$$

$$\left(\frac{^{87}Rb}{^{86}Sr}\right)_{AVE} = 3.3$$

$$\int_0^{4.5} \frac{^{87}Rb}{^{86}Sr}\, dt = 14.85$$

(b)

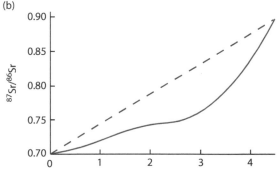

Figure 7.1 (a) $^{87}Rb/^{86}Sr$ is shown as changing in a hypothetical reservoir as some arbitrary function of time. The average $^{87}Rb/^{86}Sr$ may be calculated by integrating $^{87}Rb/^{86}Sr$ with respect to time and then by dividing by $(t - t_0)$. (b) Evolution of $^{87}Sr/^{86}Sr$ for the case where $^{87}Rb/^{86}Sr$ is a function of time as shown in (a). The dashed line shows the growth of $^{87}Sr/^{86}Sr$ if $^{87}Rb/^{86}Sr$ is constant and equal to 3.3.

$^{87}Sr/^{86}Sr$. It is in this sense that $^{87}Sr/^{86}Sr$ yields a *time-integrated* value of $^{87}Rb/^{86}Sr$.

Notice that the same final $^{87}Sr/^{86}Sr$ value in Figure 7.1b would have been reached if $^{87}Rb/^{86}Sr$ had a constant value of 3.3 ($^{87}Sr/^{86}Sr$ evolves along the dashed line). It could have been reached by an infinite number of other paths as well. Thus, while we can deduce the time-averaged $^{87}Rb/^{86}Sr$ of a reservoir over its history from the $^{87}Sr/^{86}Sr$, we cannot decipher the details of the evolution (i.e., the exact path) of the $^{87}Rb/^{86}Sr$ ratio. An interesting feature of the U–Pb system is that we can constrain, though not specify exactly, the evolutionary path. This is because the two isotopes of U decay to Pb with very different half-lives.

7.2 COMPOSITION OF THE EARTH'S MANTLE

The mantle constitutes two-thirds of the mass of the Earth and very nearly the entire mass of the silicate Earth. The relative abundances of many elements in the mantle should, therefore, be close to that of the Earth as a whole, most particularly for those elements that do not partition into either the core or the crust. We will begin with a working hypothesis that the crust, which constitutes only 0.5% of the mass of the Earth, has been created by partial melting of the mantle. This is one of only two possible alternatives: either the crust was formed at the same time as the Earth or it was grown over time from the mantle. There is no a priori reason why either of these two must be the case; we simply arbitrarily chose the second (in due course, we will see that isotopic data substantiate our choice). Doing so allows us to define a reservoir that we shall call *primitive mantle* (PM) and is *equivalent to the composition of the bulk Earth less the composition of the core*, the bulk silicate Earth (BSE). The *relative* abundances of lithophile elements in the PM are the same as in the BSE. The average modern mantle composition is then equal to the PM less present composition of the crust. The mass of the crust is sufficiently small (~0.5% of the mass of the mantle) that the difference between PM and BSE is trivial, although it is not for incompatible elements. This formulation implicitly assumes the core was formed before the crust. Again, this is a somewhat arbitrary choice, but we found in Chapter 6 from ^{182}Hf–^{182}W systematics that planetary cores, including the Earth's, formed early. Mass balance equations may also be written for isotope ratios and we will do so in due course.

The major element composition of the mantle is constrained by several observations. A combination of density, seismic velocities, cosmological considerations, and observations of xenoliths and outcrops all constrains the mantle to be composed of a rock known as peridotite[1], which consists of greater than 50% olivine, $[Mg,Fe]_2SiO_4$. Other essential minerals in the upper mantle are clinopyroxene ($[Mg,Fe]CaSi_2O_6$) and orthopyroxene ($[Mg,Fe]SiO_3$), and an aluminous phase, either plagioclase ($CaAl_2Si_2O_8$), spinel ($MgAlSiO_4$), or garnet ($[Mg,Fe,Ca]_3Al_2Si_2O_{12}$) depending

---

[1] The name peridotite derives from the gem name of olivine, peridot.

on pressure. Most other elements will be dissolved in these four phases. In the lower mantle, this assemblage is replaced by one consisting primarily of bridgmanite (Mg,Fe, Ca)SiO_3 and magnesiowüsitite (Mg,Fe)O (also called ferripericlase).

Although mantle rock is occasionally exposed at the surface, including so-called alpine peridotites in thrust zones, sections of oceanic crust known as ophiolites, and abyssal peridotites in fracture zones along mid-ocean ridges, its utility for isotopic studies is limited by its rarity as well as weathering and metamorphic and metasomatic processes associated with emplacement. Consequently, geochemists must resort to indirect samples. Much of the information we have about the geochemistry of the Earth's mantle comes from studies of mantle-derived magmas, namely, basalts[2], as well as xenoliths[3], which are carried to the surface by magmas. The information from xenoliths is limited both by their rarity and their small size; the information from basalts is limited because, as we noted earlier, all structural information is lost and much of the chemical information is "distorted" by the partial melting and subsequent fractional crystallization processes. We will make use of all three types of samples in this chapter. Together, they provide a picture of the composition and structure of the mantle complimentary to the information derived from geophysical observations, such as seismic waves, free oscillations, and so on.

Geoneutrinos, Heat Production, and Composition of the Earth

In addition to their importance in geochronology, U, Th, and K are also important sources of energy in the Earth. Indeed, except for residual heat left over from accretion of the Earth, radioactive decay is the *only* source of energy to drive processes, such as mantle convection, which in turn drives plate tectonics, volcanism, faulting, folding, and mountain building. Decay of potassium presently accounts for $\approx 19\%$ of radioactive heat production, thorium 42%, uranium 39%, samarium 0.4%, and rubidium 0.1%. To know how much heat the Earth is producing, we need to know the concentrations of these elements. Knowing the concentrations of U and Th is important for another reason that is these elements are refractory lithophile elements, a group of elements whose relative concentrations in chondritic meteorites vary little. Many models of the composition of the Earth (e.g., McDonough and Sun, 1995; Palme and O'Neill, 2003) are built from the assumption that the relative concentrations of these elements in the Earth are the same as those in chondrites. Thus, independently determining the concentrations of U and Th provides a test of these models (Bellini et al., 2013).

As we found in Chapter 1, β-decay also produces neutrinos. β^+ decay produces an electron neutrino, ν_e, as does electron capture, while β^- decay produces an electron antineutrino, $\bar{\nu}_e$. Six electron antineutrinos are produced in the decay of ^{238}U to ^{206}Pb, while four are produced in both the decay of ^{235}U to ^{207}Pb and that of ^{232}Th to ^{208}Pb. Most ^{40}K decays to ^{40}Ca with the production of an electron antineutrino while roughly 10% of it decays to ^{40}Ar and produces an electron neutrino. ^{87}Rb and ^{187}Re produce about 5% of the total neutrino production, or "luminosity." Assuming the U, Th, and K terrestrial concentrations of McDonough and Sun (1995), the Earth produces something like 5.5×10^{25} *geoneutrinos* per second. If we could detect those geoneutrinos and determine their rate of production, we could determine the concentrations of radioactive elements in the Earth.

[2] By definition, basalt is an extrusive igneous rock (i.e., lava) with less than 52% SiO_2. Basalt is the primary product of melting of the mantle. Mid-ocean ridge basalts are simply those lavas erupted at mid-ocean ridge spreading centers, i.e., divergent plate boundaries.

[3] A xenolith is any foreign rock found in an igneous rock and may be of either mantle or crustal origin.

That is easier said than done. Neutrinos, having no electric charge and with virtually no mass, interact so weakly with matter that essentially they all pass through the Earth unnoticed. Indeed, Wolfgang Pauli, having proposed the existence of the neutrino to solve the beta decay spin and energy conservation conundrums (Chapter 1) lamented, "I have done a terrible thing, I have postulated a particle that cannot be detected." To make matters worse, neutrinos are also produced by the p–p and other fusion processes in stars, including the Sun, as well as in nuclear reactors. Something like 7.7×10^{28} solar neutrinos pass through the Earth every second. Neutrinos are also produced by cosmic ray interactions at the surface of the Earth and in the atmosphere and are produced in enormous numbers in supernovae and other extreme environments in the cosmos. Making things even more complicated is the ability of neutrinos to oscillate between three flavors: electron, tau, and muon. Fortunately, the oscillations average out when distributed over a broad range of distances to uniform a survival probability of the electron neutrino of ≈ 0.55.

The very small probability of a neutrino interacting with matter through the weak nuclear force provides an opportunity for its detection. One such reaction is the so-called *neutral current reaction*, in which the neutrino dissociates a ^2H nucleus into a ^1H nucleus (a proton and a neutron) and continues on with somewhat less energy:

$$^2H + \nu \rightarrow {}^1H + n + \nu$$

The reaction can be detected by the gamma radiation given off when the neutron is captured by another nucleus. All flavors of neutrinos participate in this reaction. Another interaction is the so-called *inverse beta decay* reaction:

$$^1H + \bar{\nu}_e \rightarrow e^+ + n$$

(the bar over the neutrino indicates that it is an antiparticle) in which a proton is converted into a neutron. Only electron antineutrinos participate in this reaction.

There are a number of neutrino experiments around the world actively detecting neutrinos, but most are focused on neutrino problems related to fundamental physics or astrophysics. Only two of these, KamLAND in Japan and Borexino in Italy, have had active programs investigating geoneutrinos. The Borexino detector has shut down, but two other detectors, SNO+ in Sudbury, Ontario, Canada and the JUNO detector in Jiangmen, Guangdong province, China will also contribute data in the near future, and several others are in the planning or construction stages. These detectors, located deep underground to shield against cosmic rays, are designed to detect the flash of energy resulting from the almost instantaneous annihilation of the positron produced by the inverse beta reaction that releases energy approximately equal to the mass of the positron–electron pair (≈ 1.02 MeV) plus most of the kinetic energy of the neutrino allowing the energy of the neutrino to be determined from the energy of this radiation. A small fraction of the energy is carried off as kinetic energy by the neutron, which is subsequently captured by a proton to form deuterium. The latter reaction produces a 2.2 MeV γ-ray. Since the neutron must lose some of its kinetic energy through collisions before it can be captured, neutron capture typically follows the positron annihilation by ≈ 200 μs, during which time the neutron has moved only about a meter. This double burst of energy in a small volume provides a nearly unambiguous signal of the inverse beta reaction.

The detectors consist of large volumes of transparent hydrocarbon fluid spiked with wavelength shifting organic compounds (scintillators) and surrounded by large numbers of photomultiplier tubes. When electromagnetic radiation is produced in these reactions, they cause the liquid scintillator to produce a pulse of light that is then detected by the photomultiplier tubes (Figure 7.2).

Figure 7.2 A 3D view of the KamLAND neutrino detector. Source: U.S. Department of Energy National Laboratory Managed by the University of California, http:kamland.lbl.gov, last accessed July 11, 2022.

There are some limitations and complications:

- First, only neutrinos with energies above 1.8 MeV can produce inverse beta decay. Unfortunately, the maximum neutrino energy of ^{40}K decay is only 1.3 MeV and all neutrinos emitted in the decay chain of ^{235}U are also below this energy threshold. Thus, the method is limited to detecting neutrinos from the ^{228}Ac and ^{212}Bi β^- decays of the ^{232}Th chain and the ^{234}Pa and ^{214}Bi decays of the ^{238}U chain.
- Second, the flux of geoneutrinos of appropriate energy at the surface of the Earth is of the order of 10^6 cm^{-2} s^{-1} and the reaction cross-section is $\approx 10^{-44}$ cm^2; therefore, a large numbers of protons are necessary to provide useful detection rates. One thousand tons of detector fluid (whose composition is approximately CH$_2$) contains $\approx 10^{32}$ hydrogens and results in a geoneutrino detection rate of ≈ 10 geoneutrinos per year assuming 100% detector efficiency (Dye, 2012). Results from these experiments are reported in *terrestrial neutrino units* (TNUs), which are events per 10^{32} protons (≈ 1 kiloton) per year. The KamLAND detector employs 1000 tons of liquid scintillator surrounded by 1879 large photomultiplier tubes and the Borexino detector employs 300 tons and 1800 photomultiplier tubes. The SNO+ detector will employ 780 tons of fluid and 9500 photomultiplier tubes.

- Third, nuclear reactors produce large fluxes of electron antineutrinos, the energy spectrum of which overlaps with the geoneutrino spectrum. The reactor flux depends on the proximity of the detectors to nuclear reactors; to estimate the geoneutrino flux, the local reactor flux must be estimated and subtracted. In both the KamLAND and Borexino observatories, the reactor flux accounts for ≈30% of the total antineutrino flux.

- Finally, just as is the case for light, the neutrino flux decreases with the inverse square of distance. Consequently, most of the neutrinos originate locally (≈500 km radius), and because U and Th are concentrated in the continental crust, the detected neutrino flux is dominated by the continental crust in the region. At the KamLAND and Borexino detectors, the local continental crust accounts for roughly 75% of the total neutrino flux. To arrive at a mantle geoneutrino flux and estimates of U and Th concentrations and mantle heat production, a correction based on local geology and U and Th concentrations must be made (e.g., Šrámek et al., 2013).

As of 2020, the Borexino detector had captured 52 ± 9 geoneutrinos over 3262 days of observation; the KamLAND detector had captured 169 ± 26 events over 4397 days of observation, which translates to 47 and 32 TNUs and neutrino fluxes of $\approx 5.5 \pm 1 \times 10^6$ and $\approx 4 \pm 1.2 \times 10^6$ neutrinos/cm²/s, respectively. After correcting for contributions from local geology, and assuming a bulk crustal heat production of 7 ± 2 terawatts (TW) and a K/U ratio of the Earth of 13 400, Wipperfurth et al. (2020) estimated mantle heat production of 13 ± 4 TW and a BSE heat production of 21.5 ± 10.4 TW. Heat loss from the Earth is estimated to be about 47 ± 2 TW; thus, these measurements imply a ratio of heat production to heat loss, a value known as the *Urey ratio*, of about 0.46. The remainder of the terrestrial heat loss reflects loss of initial heat and secular cooling of the Earth. The heat production corresponds to a BSE U and Th of 21.7 ± 10.5 ppb and 81.7 ± 39.5 ppb, respectively, and assuming the above K/U ratio, a K concentration of 308 ± 140 ppm. The mean estimates are in good agreement with the canonical estimates of BSE composition by McDonough and Sun (1995), but the large uncertainties allow other estimates, such as Lyubetskaya and Korenaga (2007) and O'Neil and Palme (2008).

The large uncertainties reflect the relatively few geoneutrinos detected to date and will improve as more detectors come online and more geoneutrinos are detected. Current uncertainties on the geoneutrino flux are $\approx\pm20\%$, but the large corrections for the local neutrino contributions, which also have large ($\approx\pm25\%$) uncertainties, must be added to that in translating this into terrestrial heat production. To resolve the problem of the crustal contribution, Tohoku University and the Japan Agency for Marine-Earth Science are studying the possibility of ocean bottom neutrino detector that could be deployed by the Japanese ocean drilling ship *Chikyū*. Deployed far from continents and nuclear reactors and shielded by 4 km of ocean water, most of the geoneutrino flux would be from the mantle, which would significantly reduce uncertainties in determining the mantle neutrino flux and heat production. Furthermore, this ocean bottom detector could be deployed successively in various locations to look for geographic variation in U and Th concentrations in the mantle.

7.3 RADIOGENIC ISOTOPES IN OCEANIC BASALTS

7.3.1 Sr, Nd, Ce, and Hf isotope geochemistry of the mantle

7.3.1.1 Sr and Nd isotope ratios

Figure 7.3 shows the Sr and Nd isotopic characteristics of the Earth's major silicate reservoirs. We begin by focusing on our attention on the composition and evolution of the convecting mantle, which constitutes the vast bulk of the mantle. In doing so, we will rely primarily on the isotopic compositions of basalts derived from the suboceanic mantle. There are two reasons for this initial focus on oceanic basalts: (1) many continental basalts are contaminated by the continental crust through which they ascend, and (2) the subcontinental lithosphere, from which some of the

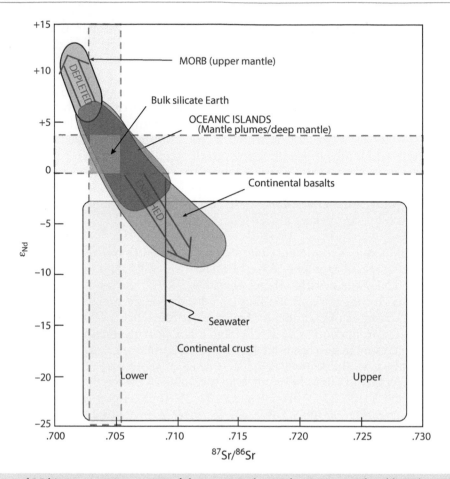

Figure 7.3   Sr and Nd isotopic systematics of the crust and mantle. Oceanic island basalts (OIBs) and mid-ocean ridge basalts (MORBs) sample major reservoirs in the mantle. Continental basalts are derived from a variety of mantle components, including mantle plumes and subcontinental lithosphere, plus, in many cases, assimilated continental crust. Horizontal and vertical hashed bands show the range of possible ε_{Nd} and $^{87}Sr/^{86}Sr$ of the *observable Earth*. Arrows indicate predicted evolution of reservoirs *depleted* and *enriched* in incompatible elements as a consequence of melting.

continental basalts are derived, does not convect and, hence, is not as well mixed and is less representative of the mantle as oceanic basalts derived from the convecting mantle. We will then turn our attention to the subcontinental lithosphere in Section 7.5, making use of isotope data from xenoliths and peridotites as well as continental basalts to examine its nature and evolution.

We see in this figure that most mantle-derived magmas (as well as other mantle samples) have higher $^{143}Nd/^{144}Nd$ (expressed here as positive ε_{Nd}) and lower $^{87}Sr/^{86}Sr$ ratios than estimated values for the BSE. In contrast, the continental crust has lower ε_{Nd} and higher $^{87}Sr/^{86}Sr$ ratios than the BSE. This is easily understood, at least qualitatively, if we think

back to Chapter 2 about the relative behavior of the parents and daughters. Nd is more incompatible than Sm, and Rb is more incompatible than Sr. When the mantle partially melts (and it only ever *partially* melts), we expect more incompatible elements to be enriched relative to less incompatible in the melt phase. When these partial melts subsequently rise to the surface as basaltic magmas, they produce a continental crust with low Sm/Nd and high Rb/Sr that over time result in low ε_{Nd} and high $^{87}Sr/^{86}Sr$ and leave behind a mantle with high Sm/Nd and low Rb/Sr, generating over time the opposite isotopic signal (Figure 2.16). This provides one piece of evidence for our working hypothesis stated above that the crust was produced from the mantle. Within the continental crust, we see large ranges and poor correlations between isotope ratios, which is indicative of less systematic solid/melt partitioning as well as of processes other than melting affecting the continental crust.

Some basalts do have lower ε_{Nd} and higher $^{87}Sr/^{86}Sr$ ratios than the BSE. This is less easily understood, but we can infer that some process or processes have enriched some regions of the mantle in Nd over Sm and Rb over Sr. What those processes might be is a very interesting question and one will be taken up later in this chapter.

Now let us focus in on oceanic basalts in more detail in Figure 7.4. MORBs are those erupted at plate boundaries along the Earth's 50 000 km mid-ocean ridge system. OIBs are those that erupted on oceanic island volcanoes and volcanic seamounts, such as the Hawaiian ones, and include oceanic islands, such as Iceland, which lie astride mid-ocean ridges. The first observation is that Nd and Sr isotope ratios are inversely correlated, but with

considerable scatter; the region occupied by the oceanic basalt data is often referred to as the "mantle array." The second observation is made clear in Figure 7.5: although there is overlap, MORBs generally have lower $^{87}Sr/^{86}Sr$ ratios and higher ε_{Nd} than OIB: the mean MORB ε_{Nd} is 8.5 while that of OIB is 5.0. Third, most OIBs (as well as nearly all MORBs) have ε_{Nd} values above the likely BSE value, indicative of net long-term incompatible element depletion. However, when Sm/Nd ratios of these basalts are also considered, these high ε_{Nd} values very likely reflect a more complex *time-integrated* history of incompatible-element enrichment event following earlier incompatible element depletion because estimated Sm/Nd ratios in their sources are too low, often lower than the BSE ratio, to accommodate a simple two-stage evolution. Finally, MORBs also have a more uniform isotopic compositions with less dispersion than OIB, implying the reservoir from which MORBs are derived is less heterogeneous. There is, however, variation in MORB

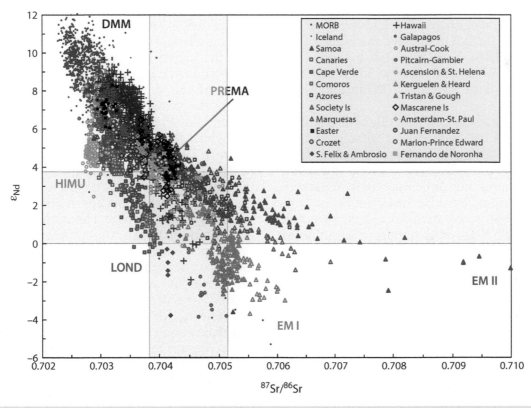

Figure 7.4 Sr and Nd isotope ratios of OIBs. Gray bands show the probable Sr and Nd isotopic compositions of the BSE. Labels refer to the isotopic genera acronyms of Zindler and Hart (1986) and Hart et al. (1986).

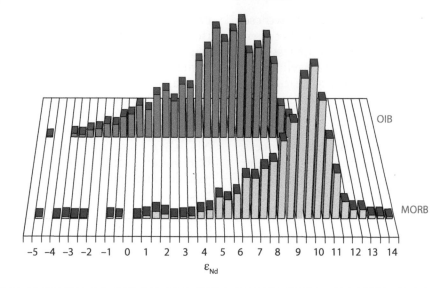

Figure 7.5   Comparison of Nd isotope ratio analyses of 1679 MORB and 2380 OIB from the EarthChem (https://www.earthchem.org) and GEOROC (https://georoc.eu/georoc/new-start.asp) databases. Although they overlap, the two groups have distinct distributions, with the differences between them being highly statistically significant.

isotopic compositions at all scales. At the scale of ocean basins, MORBs define several overlapping, parallel arrays. For a given $^{87}Sr/^{86}Sr$, North Atlantic MORBs have the highest ε_{Nd}, Indian MORBs have the next highest, then Pacific MORBs, and finally South Atlantic MORBs with the lowest ε_{Nd} (White and Klein, 2014).

We can infer that, in general, MORBs come from a source with lower time-integrated $^{87}Rb/^{86}Sr$ and higher $^{147}Sm/^{144}Nd$ than OIB. Variations in radiogenic isotope ratios in basalts require not only variations in parent–daughter ratios but also time, and a lot of it. For example, to create a 1 ε_{Nd} unit variation in $^{143}Nd/^{144}Nd$ would require a variation in $^{147}Sm/^{144}Nd$ of 0.02, or about 10% of the bulk Earth ratio, to have existed for 1.4 Ga. Thus, the isotopic distinctions observed in oceanic basalts indicate distinct chemical reservoirs have existed in the mantle for a considerable fraction of Earth's history.

While there is scatter in the OIB data, the scatter is not entirely random. We can see in Figure 7.4 that individual oceanic volcanic chains occupy distinct regions of ε_{Nd}–$^{87}Sr/^{86}Sr$ space and in many cases define correlations with distinct slopes. Thus, the mantle regions giving rise to magmas erupted on various oceanic island volcanic chains seem to have

experienced distinct histories of Sm/Nd–Rb/Sr fractionations. Zooming out just a bit from individual island chains, we again see that there is also some higher-level order. For example, the Society Islands and Samoa define parallel and overlapping correlations and that most of the data from the Marquesas plot along the same trend. The Kerguelen–Heard Islands in the Indian Ocean, Tristan and Gough in the South Atlantic Ocean, and the Pitcairn Islands in the Pacific Ocean define a distinctly different trend. Similarly, many but not all, islands of the Austral–Cook chain in the Pacific basalts occupy the same isotopic space as basalts from St. Helena and Ascension Islands in the South Atlantic Ocean.

These close geochemical relationships between geographically separated volcanic island chains were first pointed out by White (1985), who denoted them by "type" localities, St. Helena, Societies, Kerguelen, Hawaii, and MORB. Zindler and Hart (1986) gave them the acronyms that are now widely used: HIMU (hi-μ), EMI (enriched mantle I), EMII (enriched mantle II), PREMA (prevalent mantle), and DMM (depleted MORB mantle). Hart et al. (1986) subsequently identified a sixth group named LOND (low-Nd) that includes Cape Verde and the Canaries in the Atlantic, the Comoros in the Indian Ocean, and San Felix and San Ambrosio in the Pacific Ocean. These groups have been referred to as "isotopic species," but if we are to use an analogy to biological taxonomy, a better term would be *genera* because each group consists of isotopically distinctive individual volcanic chains that constitute the *species*.

The ubiquity of MORB-like isotopic compositions at divergent plate boundaries, where mantle upwelling and melting is partly or mostly a passive response to plate motion, strongly indicates that the MORB reservoir is located in the upper mantle. In contrast, it is now well established that oceanic island volcanoes are products of buoyancy-driven mantle plumes rising from the deep mantle, quite likely near the core–mantle boundary, an idea first proposed by Morgan (1971). Thus, a reasonable and widely held assumption is that MORB and OIB isotopic compositions represent the upper and the lower mantle, respectively. These OIB isotopic genera therefore provide evidence of distinct geochemical reservoirs in the deep mantle.

As we noted, there is considerable overlap in isotopic compositions between OIBs and MORBs. OIBs with MORB-like isotopic compositions come mostly from Iceland and the Galapagos, which sit on or near mid-ocean ridges, so these volcanoes may be tapping a

mixed source. Most MORBs with OIB-like isotopic signatures, but not all, come from ridge segments near oceanic island volcanoes, suggesting the upper mantle in these areas is contaminated by plume material (Schilling, 1973). Nevertheless, some of these MORBs with low ε_{Nd} do occur well away from oceanic islands.

7.3.1.2   *Ce isotope ratios*

The limited Ce isotope data paint a somewhat similar picture. While the first studies on Ce isotopes were carried out in the 1980s, only in the last 10 years or so has instrumentation advanced to the point where high-precision analysis of oceanic basalts has become more abundant, although as Figure 7.6 shows, there are still far fewer data than for other radiogenic isotope systems. Overall, there is a better correlation of ε_{Nd} with ε_{Ce} than with $^{87}Sr/^{86}Sr$, which is not surprising as we expect fractionation between pairs of rare earth elements (REE) should be more coherent than between Sm/Nd and Rb/Sr. Most of the data fall in the positive ε_{Nd} and the negative ε_{Nd} quadrant, which is indicative of net long-term depletion in La relative to Ce and Nd relative to Sm.

Although the data are still limited, MORBs define a shallower ε_{Nd}–ε_{Ce} slope (–1.55 ± 1.68)

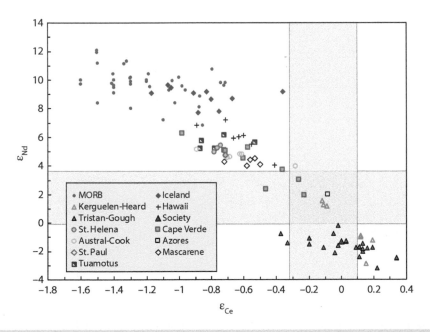

Figure 7.6   Nd and Ce isotope ratios in oceanic basalts. Data from Makishima and Masuda (1994), Israel et al. (2020), Boyet et al. (2019), Willig and Stracke (2019), Willig et al. (2020), Bellot et al. (2015), and Doucelance et al. (2014).

than OIBs (−8.48 ± 0.35); Iceland falls on the shallow MORB trend rather than the OIB one. Indeed, the overall array appears to consist of individual arrays whose slope appears to differ from that of the overall array, just as was the case for ε_{Nd}–$^{87}Sr/^{86}Sr$.

7.3.1.3 Hf isotope ratios

Now let us turn to Hf isotope ratios and see what they can tell us. Figure 7.7 shows ε_{Hf} and ε_{Nd} in oceanic basalts. We first see that ε_{Nd} is better correlated with ε_{Hf} than with $^{87}Sr/^{86}Sr$, with $\varepsilon_{Hf} \approx 1.5 \times \varepsilon_{Nd}$. The better correlation reflects the greater similarity of Lu/Hf and Sm/Nd fractionation during melting, so it is no surprise. Once again, we see that most of the data fall in the "depleted" quadrant with positive ε_{Hf} and ε_{Nd}, corresponding to long-term depletion in Hf relative to Lu and Nd relative to Sm and indicating a dominant role for melt removal in the chemical evolution of the mantle.

MORBs again anchor the more depleted end of the array. And on this array, it is MORBs rather than OIBs that show the greater dispersion whereas the opposite is true in the Sr–Nd array. In detail, however, the MORB field consists of many regional arrays within which ε_{Hf} and ε_{Nd} are well correlated (Salters et al., 2011). As is the case for Sr and Nd isotopes, there appear to be distinctions between ocean basins, with Atlantic MORBs having higher ε_{Hf} for a given ε_{Nd} than Pacific MORBs, with the latter being more uniform as well (Chauvel and Blichert-Toft, 2001; Salters et al., 2011). The dispersion in MORBs likely reflects the difference in Sm–Nd and Lu–Hf partitioning in garnet-present and garnet-absent regimes because Lu is highly compatible in garnet, which leaves melting residues with very high Lu/Hf ratios (Salters and Hart, 1989). On the other hand, the relative compatibilities of Sm and Nd between garnet and clinopyroxene, the other upper mantle mineral in which these elements primarily reside, are less dramatically different. Depending on temperature, melting of rising mantle beneath mid-ocean ridges can begin within the garnet stability field (depths ≳ 60–70 km) and continues into shallower depths where spinel replaces garnet as the aluminous phase. This produces melting residues in the upper mantle that will, depending on the extent of melting and the amount of residual garnet, leave

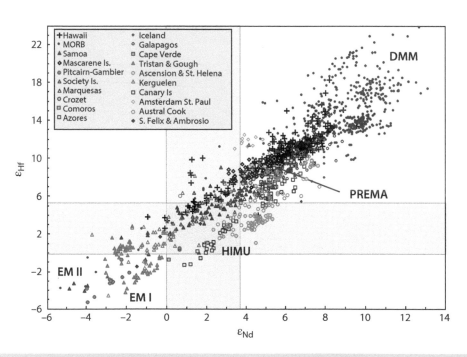

Figure 7.7 Hf and Nd isotope ratios of the suboceanic mantle as sampled by oceanic basalts. Gray bands show the inferred isotopic composition of the BSE. Data from GEOROC database.

variably high Lu/Hf and Sm/Nd that evolve to high and variable ε_{Hf} and ε_{Nd} (Salters et al., 2011).

The isotopic clans of OIBs we identified in the Sr–Nd array are less distinct in the ε_{Hf}–ε_{Nd} array. EMI and EMII clans do not appear to differ in Hf–Nd isotope systematics. We can, however, clearly identify the HIMU clan that defines an array extending to lower ε_{Hf} for a given ε_{Nd} than the main array.

7.3.2   Pb isotope ratios

Pb is by far the most powerful of the isotopic tools available to us because three parent isotopes decay to three isotopes of Pb. Let us consider the special features of the Pb isotope system before we examine the data. We noted in Chapter 3 that the slope on a plot of $^{207}Pb/^{204}Pb$–$^{206}Pb/^{204}Pb$ is proportional to time. Since Pb is a volatile element and therefore likely depleted in the Earth relative to chondrites, and also somewhat siderophile and chalcophile with a significant fraction of the Earth's Pb inventory likely sequestered in the core, the U/Pb ratio of the silicate Earth is lower than the chondritic one. Hence, the Pb isotope ratios of the BSE are not readily constrained from chondritic values.

Assuming for the moment that core formation happened simultaneously with the formation of the Earth, Pb isotope ratios of the silicate Earth are constrained by it having evolved since its formation as a closed system. If we assume that (1) the solar nebula had a uniform Pb isotopic composition when it formed, which we take to be equal to the composition of Pb in troilite (FeS) in the *Nanton* (IAB-IICD) iron meteorite (Blichert-Toft et al., 2010), and (2) the Earth formed from this nebula 4.567 Ga ago, the Earth's present composition and the *Nanton* iron must lie on an isochron in $^{207}Pb/^{204}Pb$–$^{206}Pb/^{204}Pb$ space whose slope is proportional to 4.567 Ga. This unique line is known as the *Geochron* (Figure 7.8a; Table 7.1). Indeed, all planetary bodies that formed from the solar nebula at that time (4.567 Ga ago) and have remained closed systems since then must plot on this isochron.

Many of those Solar System bodies, certainly including the Earth, have undergone

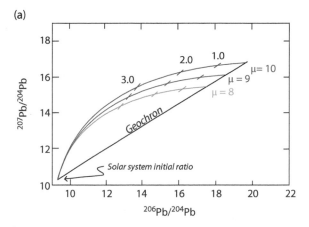
(a)

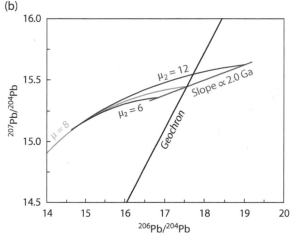
(b)

Figure 7.8   (a) Closed-system evolution of Pb isotope ratios over 4.567 Ga from the Solar System initial ratios. The curve lines represent the evolutionary paths for systems having μ values of 8, 9, and 10. The hash marks on the evolution curves mark Pb isotope compositions 1.0, 2.0, and 3.0 Ga ago. (b) Evolution of a reservoir that evolved in a closed system with $\mu = 8$ until 2.0 Ga when U/Pb fractionation produces two new reservoirs with μ of 6 and 12. Decrease in μ results in present-day Pb isotopic compositions to the left of the Geochron while an increase in μ results in isotopic compositions to the right of the Geochron.

Table 7.1   Pb isotope ratios in Nantan troilite.

$^{206}Pb/^{204}Pb$	9.306
$^{207}Pb/^{204}Pb$	10.307
$^{208}Pb/^{204}Pb$	29.532

internal differentiation since they formed, and the resulting reservoirs need not lie on the Geochron. Reservoirs that have experienced an increase in μ would now lie to the right (high $^{206}Pb/^{204}Pb$) side of the Geochron while a reservoir experiencing a decrease in μ would like to the left as illustrated in Figure 7.8b. Note that the new reservoirs with $\mu = 6$ and $\mu = 12$ along with original $\mu = 8$ reservoir lie on a secondary isochron whose slope is proportional to the time of differentiation. The average of all such reservoirs must nonetheless lie on the Geochron.

The Solar System certainly formed 4.567 Ga ago, and while, judging from the chronology we discussed in Chapter 6, planetesimal and planetary embryos accreted rapidly, the final accretion of large terrestrial planets may have required a significant amount of time. Indeed, as we found in Chapter 6, $^{182}Hf-^{182}W$ systematics suggest the Earth is younger than the Solar System itself, probably by some tens of millions of years. Pb isotope ratios would have evolved over time preceding Earth's formation, so the Earth may have both a different age and initial ratio, although if that evolution occurred in a reservoir with chondritic μ (≈ 0.19), Pb isotope ratios would have increased only slowly. The point is that the bulk Earth Pb isotope ratios need not lie on the Geochron as shown in Figure 7.8, although it certainly must lie near it.

With this in mind, we can now consider the available Pb isotopic data on oceanic basalts, which is shown in Figure 7.9. What we find is that virtually all fall on the high $^{206}Pb/^{204}Pb$ side of the 4.567 Ga Geochron. Taken together, these basalts likely represent the isotopic composition of the convecting mantle. As we shall see in the next chapter, the average isotopic composition of the bulk continental crust also plots to the high $^{206}Pb/^{204}Pb$ side of the Geochron (average lower continental crust probably plots slightly to the low $^{206}Pb/^{204}Pb$ side). Thus, the terrestrial reservoirs available to us – the *accessible* Earth – seem to have a mean isotopic composition falling off the 4.567 Ga Geochron. Halliday (2004) compiled 10 estimates of the Pb isotopic composition of the BSE. These estimates vary widely, from $^{206}Pb/^{204}Pb = 17.44$ and $^{207}Pb/^{204}Pb = 15.16$ to $^{206}Pb/^{204}Pb = 18.62$ and $^{206}Pb/^{204}Pb = 15.565$; however, all of these estimates plot significantly to the high $^{206}Pb/^{204}Pb$ side of the Geochron. The mean of these estimates, $^{206}Pb/^{204}Pb = 18.1$ and $^{207}Pb/^{204}Pb = 15.5$, is

shown as a star in Figure 7.9. If this is indeed the isotopic composition of the BSE, it implies a μ for the BSE of ≈ 8.9. This fact that both the mantle and the BSE appear to have a Pb isotopic composition falling to the high $^{206}Pb/^{204}Pb$ side of the Geochron has long been known and constitutes the first part of what is known as the *Pb paradox*.

In our review of Sr, Nd, Ce, and Hf radiogenic isotope ratio, we found that MORBs, and indeed many OIBs, have isotopic signatures of past incompatible element depletion. Since U is more incompatible than Pb, we would predict from this that MORBs should fall to the low $^{206}Pb/^{204}Pb$ side of the 4.567 Geochron. Instead, we see that most MORBs lie to the high $^{206}Pb/^{204}Pb$ side implying a past increase in U relative to Pb, which is inconsistent with the incompatible element *enrichment* implied by other isotope systems. This is the second part of the Pb paradox.

Part of the solution is certainly that the Pb age of the Earth is younger than 4.567 Ga. In this context, the "age" of the Earth is the time it closed to U and Pb loss or gain. Since some, and very possibly most, of the Earth's Pb inventory is in the core, the Pb isotope age of the Earth is the age of final core–mantle equilibration, which is often assumed to be the time of the Giant Impact. This is also the time of $^{182}Hf-^{182}W$ closure, which we found in the previous chapter occurred tens of millions of years after Solar System formation.

With this in mind, we can devise several models to explain the silicate Pb isotopic composition. In the simplest case, we can assume that prior to Earth's formation, its Pb inventory resided in the solar nebula with a carbonaceous chondrite composition, which implies a μ ($^{238}U/^{204}Pb$) value of ≈ 0.19 and increased instantaneously at the time of Earth's formation to its present value. Assuming the Earth is younger than the Solar System, we can calculate an "alternative Geochron" by calculating the Pb isotope initial ratios at the time of Earth's formation. These assumptions probably represent the most extreme shift of the Geochron to higher $^{206}Pb/^{204}Pb$ values.

The calculation of the initial ratio is a bit less straightforward than what you might think, because the μ value mentioned above (as well as the $^{207}U/^{204}Pb$ and $^{238}Th/^{204}Pb$ ratios derived from it) is the present-day one, which is lower than the ratio when the Earth formed because of its decrease through radioactive

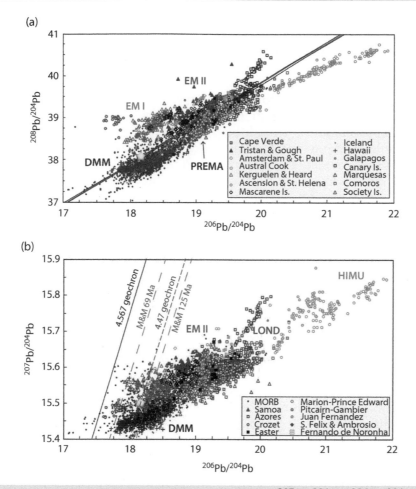

Figure 7.9   Pb isotope systematics of oceanic basalts on the (b) $^{207}Pb/^{204}Pb$–$^{206}Pb/^{204}Pb$ and (a) $^{208}Pb/^{204}Pb$–$^{206}Pb/^{204}Pb$ diagrams. Star is the mean of 10 estimates of BSE adapted from Halliday (2004). Different Geochrons are based on different assumptions about the formation of the Earth; those labeled M&M are based on models of Maltese and Metzger (2020) in which final formation of the Earth occurred 69 and 125 Ma after the start of the solar system. Line in the $^{208}Pb/^{204}Pb$–$^{206}Pb/^{204}Pb$ diagram are calculated for the ages of the Geochrons shown in the $^{207}Pb/^{204}Pb$–$^{206}Pb/^{204}Pb$ plot and the chondritic κ of 3.814.

decay. To calculate the initial values at Earth's formation, we calculate the growth from initial ratios to the present and then subtract from that the amount of growth since the Earth's formation. For $^{206}Pb/^{204}Pb$ or equation is:

$$\left(\frac{^{206}Pb}{^{204}Pb}\right)_{t_2} = \left(\frac{^{206}Pb}{^{204}Pb}\right)_{t_1} + \mu\left(e^{\lambda_{238}t_1} - e^{\lambda_{238}t_2}\right)$$

(7.2)

where t_1 is the age of the Solar System (4.567 Ga), t_2 is the age of the Earth, and μ is the value between those times. Equation (7.2) is a general solution to any two-stage growth model.

Let us assume final assembly of the Earth, the Giant Impact, occurred 100 Ma after initial Solar System formation and set t_2 as 4.47 Ma and calculate initial ratios at that time (e.g., resulting in 10.818 for $^{206}Pb/^{204}Pb$). We can then calculate our alternative Geochron simply by using our decay equation (Equation [2.14] in Chapter 2) with this initial value and different μ values (e.g., 6, 7, 8, etc.). This line is shown in Figure 7.9 and is labeled 4.47 Ga isochron. In this case, the mean of estimates of BSE Pb isotopic compositions of Halliday falls on our 4.47 Ga Geochron, but most MORBs and OIBs lie to the right of the Geochron.

Maltese and Metzger (2020) have proposed a more complex, but still highly hypothetical, model of early Earth Pb isotopic evolution in which the Earth formed in the highly volatile-depleted inner Solar System with $\mu \approx 100$. They posit that the Moon-forming impactor, *Theia*, formed further out in the Solar System with approximately chondritic abundances of volatile elements and a chondritic μ of ≈ 0.19. In their model, *Theia*, adds an only 15% of the Earth's mass, but because it is far richer in Pb, it accounts for the bulk of the terrestrial ^{204}Pb inventory. Figure 7.9 shows a Geochron based on their preferred model (labeled M&M Geochron) age for the Giant Impact of 69 Ma after CAI formation. An alternative version of the model with the Giant Impact event set at 125 Ma after the start of the Solar System shifts the Geochron further to the right. Even in this case, however, most MORBs and OIBs fall to the right of the Geochron. Indeed, while some MORBs fall to the left, which is consistent with incompatible element depletion indicated by other isotope systems, even the mean value for MORB (^{206}Pb/^{204}Pb = 18.40, ^{207}Pb/^{204}Pb = 15.51; White and Klein, 2014) falls to the right, which is indicative of net time-integrated enrichment of U over Pb. This may well reflect how U and Pb partition between mantle and crust in subduction zones where fluid transport as well as melting is involved, and the dependence of U valence state and behavior on the seawater oxidation state. We will discuss this further in the next chapter.

The various OIB clans we identified in Sr–Nd isotope space also occupy distinct fields in Pb isotope space in Figure 7.9. We can clearly see here how the HIMU group has acquired its name: these OIBs have ^{206}Pb/^{204}Pb > 20, which is indicative of high time-integrated μ. At the opposite end of the spectrum, EMI basalts are characterized by low ^{206}Pb/^{204}Pb and high ^{207}Pb/^{204}Pb and ^{208}Pb/^{204}Pb for a given ^{206}Pb/^{204}Pb. Most EMII basalts tend to plot within the main array at intermediate ^{206}Pb/^{204}Pb of $\approx 19 \pm 0.25$. The LOND clan falls within the main array, but extends to more radiogenic Pb than EMII.

Since slopes on ^{207}Pb/^{204}Pb–^{206}Pb/^{204}Pb plots are proportional to time, we can associate an age with the overall slope of the general array in Figure 7.9. The slope corresponds to an age of ≈ 1.68 Ga (White, 2010). Exactly what this age means, if indeed it is meaningful

at all, is unclear. The array in Figure 7.9 can be interpreted as a mixing line between components at each end (or more likely multiple mixing arrays), in which case, the age is only the minimum time that the two end members must have been isolated. Alternatively, the age may date a single differentiation event or represent the average age of a series of differentiation events, with the latter being the more likely. Perhaps the one thing we can conclude from it is that the isotopic heterogeneity in the mantle is ancient, of the order of 10^9 yr. On the other hand, it is not as old as the Earth itself, indicating it resulted from processes occurring within the Earth since its formation.

Turning to the ^{208}Pb/^{204}Pb–^{206}Pb/^{204}Pb relationship, for a closed system, the slope on this diagram is proportional to ^{232}Th/^{238}U, κ, as given by Eqn. (3.11) in Chapter 3:

$$\frac{\Delta\left(^{208}\text{Pb}/^{204}\text{Pb}\right)}{\Delta\left(^{206}\text{Pb}/^{204}\text{Pb}\right)} = \frac{\kappa\left(e^{\lambda_{232}t} - 1\right)}{\left(e^{\lambda_{238}t} - 1\right)}. \quad (3.11)$$

The three virtually indistinguishable lines in Figure 7.9a are calculated from the "Geochrons" in the ^{207}Pb/^{204}Pb–^{206}Pb/^{204}Pb plot (Figure 7.9b) assuming $\kappa = 3.9$, and all provide a reasonable approximate fit to the data. MORBs and many OIBs fall below this line, implying a time-integrated $\kappa < 3.9$, while some OIBs plot above it, implying a time-integrated $\kappa > 3.9$. As with other isotope ratios, there are distinctions in MORBs from different ocean basins, with those from the S. Atlantic Ocean and the Indian Ocean tending to have higher ^{207}Pb/^{204}Pb and higher ^{208}Pb/^{204}Pb for a given ^{206}Pb/^{204}Pb than MORBs from the N. Atlantic Ocean, which in turn tend to plot above the Pacific MORB.

Now let us look at how one of the other systems, Sm–Nd, relates to the U–Pb system. Figure 7.10 shows that the correlations we saw between Sr–Nd, Ce–Nd, Hf–Nd, and ^{206}Pb/^{204}Pb–^{207}Pb/^{204}Pb degenerate into a broad scatter in ε_{Nd}–^{206}Pb/^{204}Pb space. However, a closer look reveals that there is some order to the chaos. MORBs define a reasonably strong negative correlation, although the Indian Ocean MORBs are displaced to low ^{206}Pb/^{204}Pb relative to the Atlantic and Pacific MORBs. We also see that the HIMU group forms a similar negative correlation, but of somewhat lower slope. The EMII group forms a nearly vertical array with widely varying ε_{Nd} with more limited variation in ^{206}Pb/^{204}Pb. In

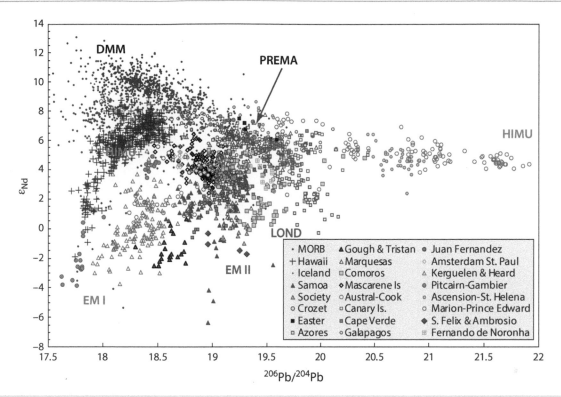

Figure 7.10 ε_{Nd} and $^{206}Pb/^{204}Pb$ in oceanic basalts. The overall correlation is poorer than for other isotope ratio pairs, but MORBs and individual island chains do define distinct correlations. Correlations of $^{87}Sr/^{86}Sr$ and ε_{Hf} with Pb isotope ratios are similar in these respects.

contrast to MORBs, the EMI group comprises several distinct, weak positive correlations. Here, in particular, we can see why we should consider these as "genera" rather than species: Hawaii, Tristan–Gough, and Kerguelen–Heard all display positive correlations in the same region, but each one is distinct. Notably, MORBs, EM I, EM II, HIMU, and LOND all form loosely defined arrays that extend from the region of highest data density, PREMA, toward more extreme compositions. This is an important observation that we will take up later.

Also of note, some of the data from the Austral–Cook chain plot in the EMI region. It turns out, however, that the Austral–Cook volcanic chain results from the near superposition of two and perhaps as many as four hot spot tracks (e.g., Chauvel, et al. 1997; Rose and Koppers, 2018), and several islands have experienced two distinct episodes of volcanism. On Rurutu, e.g., the 13–10.8 Ma basalts carry the HIMU signature whereas the young 1.8–1.1 Ma volcanics do not (Chauvel et al., 1997). Both the HIMU and EMI volcanics in the Austral–Cook form arrays pointing toward PREMA.

Sm–Nd, La–Ce, Lu–Hf, and Rb–Sr all appear to be fractionated in a generally coherent and expected manner in the mantle, but one or all of U, Th, and Pb appear to behave differently. We know that the $^{207}Pb/^{204}Pb$ and $^{206}Pb/^{204}Pb$ ratios provide information about the time-integrated U/Pb ratio, or μ, and $^{208}Pb/^{204}Pb$ provides information about time-integrated Th/Pb. The Pb isotope system can also provide information about the time-integrated Th/U ratio, or κ. This is done as follows. We can write two equations:

$$^{208}Pb^* = {}^{232}Th\left(e^{\lambda_{232}t} - 1\right) \qquad (7.3)$$

and

$$^{206}Pb^* = {}^{238}U\left(e^{\lambda_{238}t} - 1\right) \qquad (7.4)$$

where, as usual, the asterisks denote the radiogenic component. Dividing Equation (7.3) by Equation (7.4), we obtain:

$$\frac{^{208}Pb^*}{^{206}Pb^*} = \frac{^{232}Th}{^{238}U} = \kappa\frac{\left(e^{\lambda_{232}t} - 1\right)}{\left(e^{\lambda_{238}t} - 1\right)} \qquad (7.5)$$

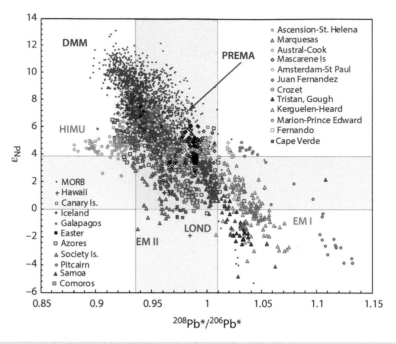

Figure 7.11 ε_{Nd} versus $^{208}Pb^*/^{206}Pb^*$ ratios of the suboceanic mantle as sampled by oceanic basalts. Gray bands show the estimated composition of the silicate Earth.

Thus, the ratio of radiogenic ^{208}Pb to radiogenic ^{206}Pb is proportional to the time-integrated value of κ. This ratio may be computed as:

$$\frac{^{208}Pb^*}{^{206}Pb^*} = \frac{^{208}Pb/^{204}Pb - \left(^{208}Pb/^{204}Pb\right)_i}{^{206}Pb/^{204}Pb - \left(^{206}Pb/^{204}Pb\right)_i} \quad (7.6)$$

where the subscript i denotes the initial ratio. By substituting a value for time in Equation (7.5) and picking appropriate initial values for Equation (7.5), we can calculate the time-integrated value of κ over that time. For example, picking $t = 4.567$ Ga and initial ratios equal to the *Nantan* meteorite, we calculate the time-averaged κ over the past 4.567 Ga.

Now let us see how $^{208}Pb^*/^{206}Pb^*$, and hence κ relates to other isotope ratios. Figure 7.11 shows ε_{Nd} plotted against $^{208}Pb^*/^{206}Pb^*$. We can see that the two are reasonably well correlated, implying the fractionations of Sm from Nd and U from Th in the mantle have been closely related. From this, we conclude that the lack of correlation of "first-order" Pb isotope ratios with Sr, Nd, and Hf isotope ratios is due to "anomalous" behavior of Pb. Once again in this plot, the various clans define distinct, although often diffuse, arrays.

7.3.3   Os isotope ratios

As we noted in Chapter 2, Os differs from other radiogenic elements in being highly siderophile and being highly compatible in silicate systems. As a consequence of the former, most of the Earth's Os inventory is in the Earth's core and its concentration in the mantle is low. As a consequence of the latter, Os concentrations in basalts and the crust are even lower than that in the mantle. Low concentrations by themselves present not only serious analytical challenges, but also serious opportunities for contamination and assimilation of crust as magma rises through it, and consequently serious challenges in using basalts as mantle samples. This is particular true of MORBs, which tend to have low Os concentrations.

Based on their incompatible element-depleted Sr, Nd, Ce, and Hf isotopic signatures, we expect MORBs to have depleted Os isotope signatures because Os is significantly more compatible than Re. Thus, we expect $^{187}Os/^{188}Os$ values less than the primitive upper mantle (PUM) estimate of 0.1296 and in the γ_{Os} notation we introduced in

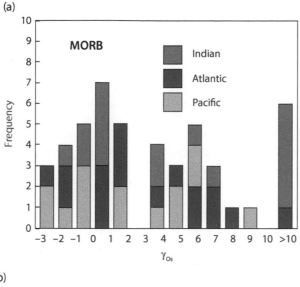

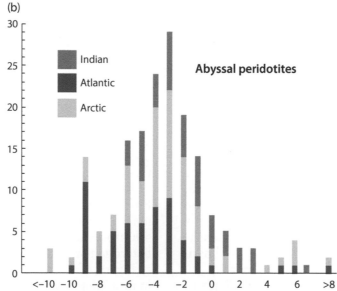

Figure 7.12 Comparison of γ_{Os} in (a) MORB and (b) abyssal peridotites. While MORBs tend to have $\gamma_{Os} > 0$, indicating enrichment in Re over Os, abyssal peridotites recovered from mid-ocean ridges and associated fracture zones have γ_{Os} predominantly < 0, which is consistent with the depleted nature of the upper mantle. The most likely explanation for this inconsistency is the very low Os concentrations in MORBs and frequent assimilation of altered oceanic crust containing seawater derived Os.

Chapter 2, $\gamma_{Os} < 0$. In contrast, most MORBs have $\gamma_{Os} > 0$ (Figure 7.12).

Abyssal peridotites, those dredged from fracture zones and fault scarps along mid-ocean ridges, provide an alternative, and indeed more direct, sample of the depleted mantle and their Os concentrations are orders of magnitude higher than basalts. As shown in Figure 7.12, abyssal peridotites have predominantly negative γ_{Os}, with a mean of −3.1 and a mode of −2.7. This is consistent with the

depleted signatures of other radiogenic elements. If both peridotites and basalts are samples of the upper mantle directly beneath mid-ocean ridges, why is there the difference in γ_{Os}?

One issue can be radiogenic ingrowth; Re/Os ratios in a few MORBs are sufficiently high that significant increase of $^{187}Os/^{188}Os$ can occur in as little as a few tens of thousands of years, but most samples have too low Re/Os ratios or are too young for this to be significant. Another explanation is that Os isotopic equilibrium is not achieved during melting. As we note below, individual sulfide grains in abyssal peridotites, which are the melting residues of MORBs, can display considerable Os isotopic heterogeneity. Melting may preferentially sample the more liable intergranular sulfides, which tend to have more radiogenic Os than sulfides included in silicate grains.

The most likely explanation, however, is that MORBs have frequently assimilated oceanic crust that has exchanged Os with seawater. Massive sulfide deposits associated with modern ridge crest hydrothermal systems have $^{187}Os/^{188}Os$ ranging from 0.645 to 1.209 ($\gamma_{Os} \approx 400$–800), with the latter values being similar to that of seawater, with an average Os concentration of 17 ppt, which is equivalent to or greater than Os concentrations in MORBs (often less than 10 ppt); they can also have very high Re/Os ratios (Zeng et al., 2014). Assimilation of only small amounts of this material would account for the elevated $^{187}Os/^{188}Os$ observed in MORBs with essentially no effect on Nd, Ce, Hf, and Pb isotope ratios and minimal effect on $^{87}Sr/^{86}Sr$. Furthermore, $^{187}Os/^{188}Os$ correlates positively with $\delta^{11}B$ (Gannoun et al., 2016). Seawater boron, which has $\delta^{11}B = +39.5$‰ compared to ≈ -10‰ for the mantle, is readily taken up in the oceanic crust, so this correlation strongly supports the notion that most MORBs with $\gamma_{Os} > 0$ have likely assimilated altered oceanic crust. Consequently, we will rely on Os isotope ratios in abyssal peridotites as a proxy for the depleted mantle.

The abyssal peridotites provide evidence of the long-term Re depletion of upper mantle. To address the question of how long, we can calculate rhenium depletion ages, T_{RD} (Equation [2.58]) and model ages, T_{Ma} (Equation [2.59]). Average T_{RD} in abyssal peridotites is 750 Ma with a mode of 490 Ma. This provides only a minimum age of Re depletion because the rhenium depletion age assumes Re concentrations of 0, which is not the case. The average model age is older, 990 Ma, with a mode of 605 Ma. This, however, assumes no disturbances to Re/Os ratios, which is unlikely, given the melt extraction beneath mid-ocean ridges experienced by these peridotites. Nonetheless, we can conclude that depletion of the mantle beneath mid-ocean ridges is ancient.

However, some caution is necessary in interpreting Os isotope ratios in abyssal peridotites.

First, they are not entirely immune to change as a result of seawater interaction (Snow and Riesberg, 1995). Second, even fresh peridotites can exhibit small-scale heterogeneity. Os in peridotites is hosted mainly by sulfides and, to a lesser degree, by tiny grains of platinum group alloys, such as osmiridium. Separated grains can display considerably greater $^{187}Os/^{188}Os$ variation than whole rocks. The difference, however, cannot generally be explained by the difference in Re/Os ratios, suggesting that Os isotopic equilibrium is not achieved during melting.

Because concentrations of other radiogenic elements in abyssal peridotites are quite low and because abyssal peridotites have invariably reacted with seawater to some extent, reliable data on these other radiogenic isotope ratios are comparatively rare. Some Sr, Nd, and Hf data on separated pyroxenes do exist from abyssal peridotites and, for the most part, are similar to those ratios in MORBs, but extend to more depleted signatures (e.g., Warren et al., 2009). This is particularly true for $^{176}Hf/^{177}Hf$ where ε_{Hf} values of +60 and +104 were found in pyroxenes from peridotites dredged from the Gakkel Ridge in the Arctic Ocean (Stracke et al., 2011).

The presence of both highly depleted Os and Hf isotopic signatures suggests that MORB magmas may not sample the mantle in a representative way. This may be a result of lithologic heterogeneity in the mantle, with lenses or stringers of recycled oceanic crust embedded in more depleted peridotite. The former would melt at lower temperatures and may not fully equilibrate with peridotite. Furthermore, because of its low solubility in silicate melts, Os may be slow to equilibrate during melting. In this respect, however, it should be noted that most abyssal peridotites are recovered from slow-spreading ridges whose greater depth suggests lower mantle temperatures and smaller melt fractions (Klein and Langmuir, 1987). The Gakkel Ridge is particularly slow spreading, ≈ 1 cm/yr. In the Eastern Magmatic Zone, where the peridotites with high ε_{Hf} values were recovered, less than 20% of the rift valley appears to be covered by basalt, suggesting very little melt production. In regions where spreading rates and magma production are higher, basalts may sample the mantle in a more representative way. Nevertheless, these data certainly provide evidence of highly

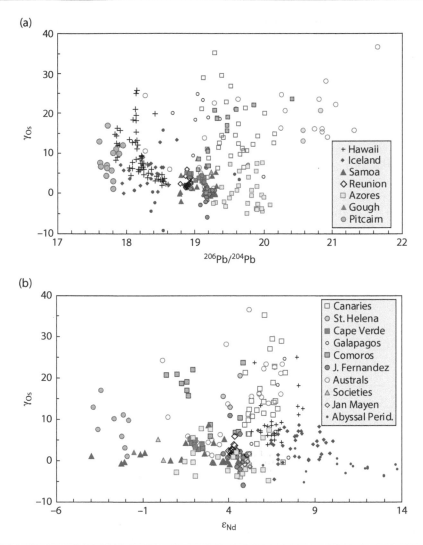

Figure 7.13 γ_{Os} versus (a) $^{206}Pb/^{204}Pb$ and (b) ε_{Nd} in oceanic basalts. Only analyses with > 50 ppt Os are plotted.

depleted enclaves within the upper oceanic mantle.

As oceanic island volcanoes are constructed on ocean crust, assimilation of altered oceanic crust may also influence Os isotope ratios in OIB magmas. Furthermore, because the oceanic upper crust has high Re/Os ratios that will evolve rapidly to high $^{187}Os/^{188}Os$, even assimilation of unaltered oceanic crust could influence Os isotope ratios in OIBs. The association in many cases of low Os concentrations with high $^{187}Os/^{188}Os$ ratios in OIBs is clear evidence that this occurs (e.g., Day, 2013). However, many OIBs have considerably higher Os concentrations than in MORBs, often in the tens to hundreds of ppt and by

considering only those samples with sufficiently high Os concentrations, typically more than 30–50 ppt (Day, 2013), the assimilation/contamination problems that plague the MORB data set can be minimized, although not necessarily eliminated entirely in OIBs.

Figure 7.13 shows γ_{Os} in OIBs with Os concentrations > 50 ppb as a function of $^{206}Pb/^{204}Pb$ and ε_{Nd}. The first thing we notice is that OIBs have predominantly positive γ_{Os}, although some negative values from some island chains have been reported. This is consistent with their less incompatible element-depleted Sr, Nd, Ce, and Hf isotopic signatures than MORBs and abyssal peridotites. We also note that there are systematic differences between island chains, with some having quite restricted ranges of γ_{Os}, such as Reunion and Juan Fernandez, and others showing a much greater range, notably the Canaries, the Australs, Hawaii, and Pitcairn. Third, while overall there is no correlation between these ratios, there are clear suggestions of correlations within several chains, particularly bearing in mind the possibility that some of the high γ_{Os} could reflect assimilation. Additional study will be required to confirm whether these correlations are real. Finally, the highest γ_{Os} tend to be associated with EMI and HIMU isotopic signatures, while γ_{Os} close to PUM values are most common among samples with intermediate, PREMA-like Pb and Nd isotope ratios, although this too requires confirmation of additional data.

7.4 INFERENCES ON MANTLE STRUCTURE AND EVOLUTION

The systematic differences in isotopic composition between MORBs and OIBs lead to the inference that there are at least two major reservoirs in the mantle, with the OIB reservoir consisting of a variety of sub-reservoirs. However, simply recognizing the existence of mantle reservoirs tells us little about where they are and how big they are. The consensus interpretation is that MORBs are derived from the upper mantle, which we can see is the most depleted of the reservoirs sampled by oceanic volcanism. This interpretation is based on the observation that seafloor spreading is primarily a response to "slab pull,"; in other words, it is the gravitational energy of the sinking of old, cold, and dense oceanic lithosphere in subduction zones that primarily drives plate motions. Thus, when plates are pulled apart, it is upper mantle that rises and melts to produce oceanic crust.

On the other hand, oceanic island volcanoes (and a number of continental ones as well) are thought to be surface manifestations of mantle plumes, columns of hot rock buoyantly rising from the deep mantle that partially melt upon reaching the uppermost mantle, an idea first proposed by Morgan (1971). Among other pieces of evidence, Morgan pointed to the geochemical distinctions between OIBs and MORBs. The idea was highly controversial at first, as the debate between Schilling

(1973b) and O'Hara (1973) illustrates. Indeed, the debate continued for decades, but a broad consensus has formed around Morgan's idea in light of an array of geophysical evidence, including geoid and topographic anomalies (e.g., Sleep, 1990), high temperatures of primary OIB magmas (e.g., Herzberg et al., 2007), and slow seismic anomalies beneath these volcanoes that extend into the lower mantle (Montelli et al., 2004; French and Romanowicz, 2015). Thus, while MORBs represent samples of the shallow mantle, OIBs primarily sample the deep mantle (some intraplate volcanism, however, is clearly not associated mantle plumes). Each of these is chemically heterogeneous and we use the terms upper and lower mantle only in the sense of relative position, rather in the seismically defined sense of the upper and lower mantle being, respectively, above and below the 670 km discontinuity. It is also possible that neither the upper nor the lower mantle is representatively sampled by volcanism. The non-convecting subcontinental lithosphere comprises a third reservoir. We will now consider these in turn.

7.4.1 The depleted MORB mantle

7.4.1.1 Mass of depleted MORB mantle

Most of the geochemistry of the MORB source can be described in terms of depletion in incompatible elements due to partial melting and removal of the melt to form the crust, both continental and oceanic. Indeed, it is the process of forming oceanic crust that produces MORBs and provides the sample of the underlying mantle. One of the enduring questions about the reservoir from which MORBs are derived is: how much of the mantle does it occupy? Following DePaolo (1980), we can address this question through mass balance. We begin by writing a series of mass balance equations. The first is mass of the reservoirs:

$$\sum_j M_j = 1 \qquad (7.7)$$

where M_j is the mass of reservoir j as a fraction of the total mass of the system, in this case the silicate Earth. We can also write a mass balance equation for any element i as:

$$\sum_j M_j C_j^i = C_0^i \qquad (7.8)$$

where C_0 is the concentration in the BSE. For an isotope ratio, R, of element i, or for an elemental ratio of which element i is the denominator, the mass balance equation is:

$$\sum_j M_j C_j^i R_j^i = C_0^i R_0^i \qquad (7.9)$$

Our problem assumes the existence of three reservoirs: the continental crust, the depleted mantle by crust formation, which we equate with DMM, and we assume that the remainder of the mantle is the PM whose composition equals that of BSE. We explicitly assume OIB reservoirs and continental lithosphere have trivial mass in this calculation. These mass balance equations can be combined to solve for the mass ratio of DMM to continental crust:

$$\frac{M_{DMM}}{M_{CC}} = \frac{C_{CC}^i}{C_0^i} \frac{(R_{CC}^i - R_{DM}^i)}{(R_0^i - R_{DM}^i)} - 1 \qquad (7.10)$$

where the subscripts DM and CC refer to depleted mantle and continental crust, respectively. A number of solutions to the mass balance equations are possible; we want an expression based on values that are relatively well constrained. Isotopic compositions of mantle reservoirs are well constrained, but this is not true of elemental concentrations. We do have good estimates of concentrations in the crust because great numbers of samples can be analyzed and an average value computed.

We can use these equations with any isotope system; DePaolo chose the Sm–Nd isotope system because these elements are both refractory and lithophile, and hence the Sm/Nd and $^{143}Nd/^{144}Nd$ ratios in the BSE could be assumed to be the same as chondritic ratios, i.e., when expressed in epsilon notation the R_0 in Equation (7.9) is 0. However, the discovery of Boyet and Carlson (2005) that the terrestrial $^{142}Nd/^{144}Nd$ is higher than chondritic challenged this assumption (recall that ^{142}Nd is the daughter of ^{146}Sm, which has a half-life of 103 Ma). If this difference was due to radioactive decay of ^{146}Sm, it implied the terrestrial Sm/Nd and $^{143}Nd/^{144}Nd$ ratios are also higher than chondritic. As we discussed in Chapter 6, it is now clear that there are nucleosynthetic variations in the $^{142}Nd/^{144}Nd$ ratio among chondrites and much, and perhaps all, difference between Earth and chondrites in $^{142}Nd/^{144}Nd$ is nucleosynthetic rather than

radiogenic. As we noted, the Earth's $^{142}Nd/^{144}Nd$ falls with the range of enstatite chondrites, although near the extreme of the range (see Figure 6.15 in Chapter 6). The Earth and enstatite chondrites share similar O isotope ratios, but differ in other aspects, such as Si, Ca, and Ti isotope compositions. Dauphas (2017) modeled the Earth's composition as 60% enstatite chondrite and 40% O and C chondrite; Maltese and Metzger (2020) modeled it as 85% enstatite chondrite and 15% carbonaceous chondrites. These combinations would produce an Earth with a lower initial $^{142}Nd/^{144}Nd$ than both enstatite chondrites and the observed $^{142}Nd/^{144}Nd$ of the modern Earth, requiring the difference to be make up by radioactive decay of ^{146}Sm, in other words, a higher Sm/Nd. As we will discuss subsequently, there are other reasons to think the BSE $^{143}Nd/^{144}Nd$ is higher than chondritic, i.e., $\varepsilon_{Nd} > 0$. We cannot exclude the value of 0 for the terrestrial ε_{Nd}, but we need to consider the possibility that it is higher.

The Nd concentration and the Sm/Nd ratio of the continental crust are also better constrained than many other elements. The Sm/Nd ratio and $^{143}Nd/^{144}Nd$ of the crust are related through isotopic evolution, specifically:

$$\frac{^{143}Nd}{^{144}Nd} = \left(\frac{^{143}Nd}{^{144}Nd}\right)_0 + \left(\frac{^{147}Sm}{^{144}Nd}\right)(e^{\lambda t} - 1) \qquad (7.11)$$

Because the half-life of ^{147}Sm is long compared with the age of the Earth, we can linearize this equation as:

$$\frac{^{143}Nd}{^{144}Nd} = \left(\frac{^{143}Nd}{^{144}Nd}\right)_0 + \left(\frac{^{147}Sm}{^{144}Nd}\right)\lambda t \qquad (7.12)$$

The continental crust was certainly not created at single time t. However, because it is linear, this equation remains valid for an average crustal age, T. Hence, we may write:

$$\left(\frac{^{143}Nd}{^{144}Nd}\right)_{CC} = \left(\frac{^{143}Nd}{^{144}Nd}\right)_{BSE}^T + \left(\frac{^{147}Sm}{^{144}Nd}\right)_{CC}\lambda T \qquad (7.13)$$

where the superscript T denotes the value at time T. The $^{143}Nd/^{144}Nd$ of the continental crust calculated in Equation (7.13) can then be used in Equation (7.10) to calculate the mass fraction of DMM.

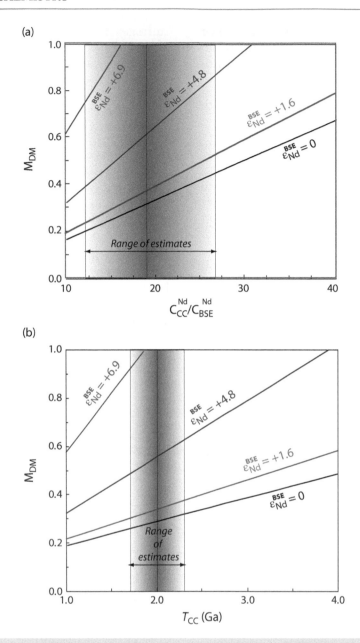

Figure 7.14 Calculated mass of the DMM as a function of the ε_{Nd} of the BSE and (a) the enrichment of Nd in the continental crust relative to the BSE (C_{CC}/C_{BSE}) and (b) the average age of continental crust (T_{CC}).

Now let us assign some values to these equations. The mass of the continental crust as a fraction of the mass of the silicate Earth, M_{CC}, is about 0.0055. For the $^{143}Nd/^{144}Nd$ value of the depleted mantle, we will choose 0.51310 (or $\varepsilon_{Nd} \approx +9$; which is equal to the median value for MORB, which is somewhat higher than the mean value of 8.5; White and Klein, 2014). Based on assessments of the composition of continental crust, a good estimate for its $^{147}Sm/^{144}Nd$ ratio is 0.123. We can

use these values to calculate the mass of depleted mantle as a fraction of the silicate Earth as a function of the ratio of Nd concentration in the crust to PM, C_{CC}/C_0, the average age of the continents, T, and the $^{143}Nd/^{144}Nd$ of the silicate Earth (expressed in epsilon units); Figure 7.14 shows the results. A good estimate of C_{CC}/C_{PM} is about 19, but there is considerable uncertainty. A good estimate of T is about 2 Ga, but there is easily 10% uncertainty in this value. If the terrestrial Sm/Nd ratio is chondritic ($\varepsilon_{Nd} = 0$), Figure 7.14 suggests that the depleted mantle constitutes 40% or less, and more likely only about 25% of the mantle. These values correspond roughly to the mass of the mantle above the 660 km seismic discontinuity and encouraged the view of a stratified mantle; as mentioned earlier, however, this seems untenable in view of the seismic evidence of convection across this boundary.

Now let us consider the case where the Sm/Nd ratio of the Earth, or the observable part of it, is different from chondritic value of 0.1960. First, in assessing the various nucleosynthetic contributions to Nd isotopic compositions, Saji et al. (2020) predict a terrestrial ε_{142Nd} of −3.6 from the correlation between $^{142}Nd/^{144}Nd$ and $^{148}Nd/^{144}Nd$ in meteorites, which corresponds to an initial $^{142}Nd/^{144}Nd$ = 1.141846. The difference between this and the observed value of $\varepsilon_{142Nd} = 0$ ($^{142}Nd/^{144}Nd$ = 1.141850) must then be due to radiogenic decay of ^{146}Sm. Next, we need to know the value of the $^{147}Sm/^{144}Nd$ and $^{143}Nd/^{144}Nd$ ratios this difference implies, assuming single-stage evolution (i.e., constant Sm/Nd) can begin by writing a version of the equation we derived for extinct radionuclides (Equation [7.5]):

$$\frac{^{142}Nd}{^{144}Nd} = \left(\frac{^{142}Nd}{^{144}Nd}\right)_0 + \left(\frac{^{147}Sm}{^{144}Nd}\right)\left(\frac{^{146}Sm}{^{147}Sm}\right)_0$$

(7.14)

From which we can derive:

$$\left(\frac{^{147}Sm}{^{144}Nd}\right) = \frac{\left[\frac{^{142}Nd}{^{144}Nd} - \left(\frac{^{142}Nd}{^{144}Nd}\right)_0\right]}{\left(\frac{^{146}Sm}{^{147}Sm}\right)_0}$$

(7.15)

The current best estimate of the Solar System initial $^{146}Sm/^{147}Sm$ at 4.567 Ga is 0.0017386 and the terrestrial $^{142}Nd/^{144}Nd$ = 1.14185, and we calculate a $^{147}Sm/^{144}Nd$ of 0.1984, which is 1.2% greater than the canonical chondritic value and corresponds to a BSE $^{143}Nd/^{144}Nd$ = 0.51271 or ε_{Nd} = +1.6.

Next let us suppose that average enstatite chondrites represent the raw material from which the Earth formed and the ≈12 ppm difference in the terrestrial and average enstatite chondrite $^{142}Nd/^{144}Nd$ is due to radiogenic decay of ^{146}Sm. The enstatite chondrite initial $^{142}Nd/^{144}Nd$ is 1.14145 and we calculate a

terrestrial $^{146}Sm/^{144}Nd$ of 0.2039, which is 4% greater than the chondritic value of 0.1960. This implies a BSE $^{143}Nd/^{144}Nd$ of 0.51272 or ε_{Nd} = +4.8. As Figure 7.12 shows, this implies that the depleted mantle constitutes ≈40–60% of the entire mantle. In both these calculations, we assume that the Sm/Nd fraction leading to the terrestrial larger value occurred at 4.567 Ga (time of CAI formation); if it occurred about one half-life of ^{146}Sm, or about 100 Ma, later, it implies a terrestrial ε_{Nd} of twice these values.

Now suppose, following Dauphas (2017) that the Earth formed from 60% enstatite chondrite and the rest a 50–50% mixture of ordinary and carbonaceous chondrite material. This implies an ε_{Nd} of +6.9 for the BSE and that the depleted mantle constitutes the entire mantle – a value that would appear to require a young crust and one minimally enriched in Nd and may lie outside the possible range.

There are a number of caveats to this sort of calculation. The first caveat is that this calculation implicitly assumes that the mantle consists of only the DMM reservoir and a primitive one and ignores other mantle reservoirs, such the source of OIBs. OIBs have an average ε_{Nd} ≈ 4.75; if the BSE has ε_{Nd} = 0 or +1.6, including OIB would reduce our estimate of DMM mass fraction; it would have no effect on the case where ε_{Nd} = 4.8 and would increase the DMM mass fraction it in the case where ε_{Nd} = 6.9. It is also possible that domains exist within the mantle that are so highly depleted that they do not melt significantly and hence are not sampled; their presence would reduce our estimate of DMM. Another caveat is that, as Metzger et al. (2020) write, "a close investigation of the currently available chemical and isotopic data on samples from the rocky planets Earth and Mars and the different meteorite classes shows that it is impossible to account for all chemical and isotopic characteristics, particularly of the Earth, by combining different proportions of known materials from the solar system." To put it plainly, given the existence of nucleosynthetic variations in $^{142}Nd/^{144}Nd$ in the early Solar System, we cannot be certain what the BSE value is and hence the mass of DMM.

Finally, we should point out that if an unsampled early formed incompatible element-enriched reservoir exists as postulated by Boyet and Carlson (2005), then the discussion above of the $^{142}Nd/^{144}Nd$ of the Earth and the mass fraction of the depleted reservoir applies only to the observable Earth, i.e., excluding this unsampled early enriched reservoir.

7.4.1.2 Evolution of DMM

Regardless of its mass, the evolution of DMM has without doubt been more complex than the simple extraction of continental crust. As we pointed out above, the primary mechanism by which incompatible elements are removed from DMM is through creation of the oceanic crust. The present rate of melt extraction to produce oceanic crust is 6×10^{14} kg/yr. Assuming the melt fraction is 10%, this extracts melt and incompatible elements from 6×10^{15} kg/yr. Over Earth's history, this would extract melt from over six times the entire mantle mass at this rate.

Clearly then, there must also be a flux of relatively incompatible element-rich material into DMM to keep it from becoming completely infertile, as well as a flux out of it. In other words, DMM is necessarily an open system. To make this point, let us consider the Th/U ratio of DMM, the time-integrated value of which is recorded by the $^{208}Pb*/^{206}Pb*$ ratio. Galer and O'Nions (1985) found that the average $^{208}Pb*/^{206}Pb*$ in MORBs corresponded to a time-integrated Th/U ratio of about 3.75. The presumed BSE (chondritic $^{232}Th/^{238}U$) ratio (κ) is about 3.9. Based on $^{230}Th/^{232}Th$ activity ratios in MORBs, they assumed that κ in DMM was presently about 2.5. In a simple two-stage model, Galer and O'Nions (1985) calculated that the decrease in κ from 3.9 to 2.5 would have occurred only 600 Ma ago. They argued that there must be fluxes into the depleted mantle as well as out of it, and the apparent depletion time of 600 Ma was in reality simply the *residence time* of Pb in the upper mantle. The steady-state residence time of some element i in a reservoir is defined as:

$$\tau_i = \frac{C_i M_i}{f_i} \qquad (7.16)$$

where τ is the residence time, C_i is the concentration of element i in the reservoir, M_i is the mass i in the reservoir, and f_i is the flux of i into or out of the reservoir. The DMM $^{232}Th/^{238}U$ ratio Galer and O'Nions assumed is almost certainly too low; a better value would be

≈ 3, but this still implies a residence time of only 800–900 Ma.

We can now ask: what is the source of this flux of incompatible elements into DMM? A large fraction of the material resupplying the depleted mantle may simply be subducted oceanic crust. Once subducted, the oceanic crust converts into eclogite, a rock composed of garnet and pyroxene with high density; the underlying lithosphere from which melt has been extracted consists of harzburgite, a rock composed of olivine and orthopyroxene with low density. This excess density of eclogite allows it to separate from the rest of the slab and makes it more likely to sink into the deep mantle and potentially pool near the core–mantle boundary (Brandenburg et al., 2008) while the lower-density harzburgite is more likely recycled into the ambient mantle. This separation is likely to be inefficient with at least some oceanic crust remixing with the upper mantle.

Another important source of incompatible elements to the DMM is likely to be mantle plumes rising through and into the upper mantle. Because rock on the periphery of the plume will be horizontally deflected upon reaching the lithosphere, only material near the center of the plume will rise to the base of the lithosphere and melt to its full potential. That fraction of the plume not extracted as melt becomes part of the DMM. The collinearity and overlap of isotope ratios we see in the figures in Section 7.3 also testifies to a close relationship between MORBs and OIBs.

Along-ocean ridge geochemical profiles also demonstrate mixing of plume material with ambient upper mantle. The first such study was conducted by Schilling (1973a) who used rare-earth element ratios to demonstrate a geochemical gradient of increasing incompatible element enrichment along the Mid-Atlantic Ridge toward Iceland. Subsequent analyses of $^{87}Sr/^{86}Sr$ demonstrated this gradient must reflect mantle source composition and was not an effect of melting (Hart et al., 1973). A similar geochemical gradient was also found along the Mid-Atlantic Ridge around the Azores with isotopic gradients extending up to 1000 km from the plume (White et al., 1976). Similar gradients were subsequently documented even on mid-ocean ridges where the plume was not centered on the ridge, including Easter Island, the Galapagos, and several of the islands in the South

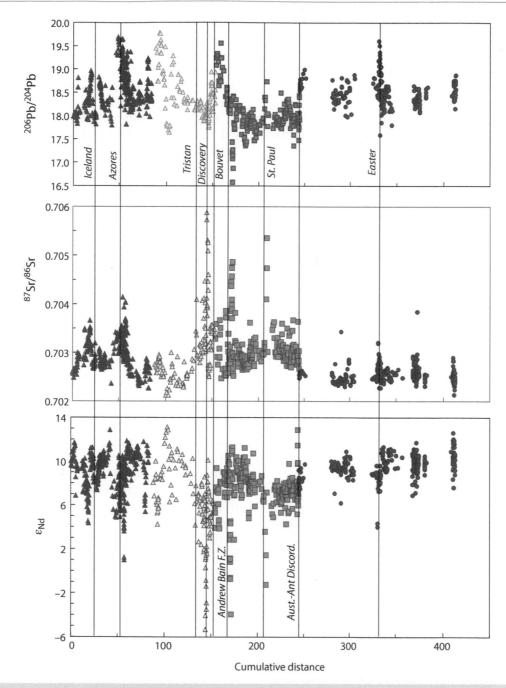

Figure 7.15 Variation of Sr, Pb, and Nd isotope ratios with angular distance along the mid-ocean ridge system. The "0" point is the location of the northernmost sample site of the Gakkel Ridge at 85.64°N, 85.05°E. Filled triangles: North Atlantic MORBs; open triangles: South Atlantic MORBs; squares: Indian Ocean MORBs; and circles: Pacific Ocean MORBs. Representing MORB data in this way was first done by Meyzen et al. (2007).

Atlantic Ocean and the Indian Ocean (Schilling, 1985). These variations can be seen in Figure 7.15.

Although proximity to oceanic islands and mantle plumes is an important source of isotopic variation in MORBs, in numerous other cases, isotopic variations do not appear to be directly related to mantle plumes and reflect intrinsic heterogeneity in the upper mantle.

Isotopic data from MORBs also provide evidence of large-scale provinces in the upper mantle. Perhaps the first such "province" to be identified was the Indian Ocean geochemical province. Data published as early as the 1970s suggested MORBs from the Indian Ocean were distinct from those of the Pacific and the Atlantic, having higher $^{87}Sr/^{86}Sr$ ratios (Hedge et al., 1973). Subsequent work has showed that isotopic distinctions exist between North Atlantic, South Atlantic, and Pacific Ocean MORBs as well. Pacific Ocean MORBs have lower $^{87}Sr/^{86}Sr$ ratios for a given ε_{Nd} than does either group of Atlantic MORBs, while high $^{208}Pb/^{204}Pb$ for a given $^{206}Pb/^{204}Pb$ and low ε_{Nd} and ε_{Hf} relative to $^{87}Sr/^{86}Sr$ distinguish South Atlantic MORBs from North Atlantic MORBs (e.g., Meyzen et al., 2007; White and Klein, 2014). In contrast, there appears to be no distinction between ocean basins in O isotope ratios (Cooper et al., 2009), although this could reflect the limited size of the data set.

Boundaries between these domains can be sharp or diffuse (Figure 7.15). The Indian–Pacific boundary is located at a small discontinuity of the Southeast Indian Ridge (SEIR) within the Australian–Antarctic Discordance and manifests a very sharp change in isotopic characteristics (e.g., Klein et al., 1988). By contrast, the boundary between the South Atlantic and Indian domains, which occurs west of the Andrew Bain Fracture Zone (the Antarctic–Nubian–Somalian triple junction) located at 30°E on the Southwest Indian Ridge, is gradual (Meyzen et al., 2007). The boundary between the North and South Atlantic provinces, located near 23°S, is also diffuse.

7.4.2 Mantle plumes and the lower mantle

Hart (1984) noticed that oceanic island volcanoes with particularly usual radiogenic isotope ratios come mainly from a belt centered at about 30°S, which he named the DUPAL anomaly based on the study of B. Dupré and C. Allègre, (1983). He pointed out that this region correlated with anomalies in the geoid (the Earth's equipotential gravitational surface). Subsequently, Castillo (1989) argued that Hart's "DUPAL anomaly" actually consisted of two separate regions: the "DUPAL" in the Indian Ocean and the "SOPITA" (South Pacific Isotope and Thermal Anomaly) in the South Pacific. He also pointed out these regions corresponded to

poorly defined regions of slow seismic velocities in the deep mantle.

Over the subsequent decades, seismologists greatly refined seismic velocity maps of the deep mantle, using tomography to bring those regions of slow seismic velocities, now referred to as *large low-shear-wave velocity provinces* (LLSVPs), into focus (Figure 7.16) and to clearly relate them to geoid anomalies. Although variations in mantle seismic velocity have generally been interpreted in terms of temperature (slow equals hot), a number of other factors suggest that the LLSVPs must also be compositionally distinct. In particular, s-wave speed and the bulk sound wave speed should covary with temperature in homogeneous media, but they do not and perhaps even inversely correlate in the lowermost mantle (McNamera, 2019). The boundaries of the LLSVPs, although poorly defined, also appear to be too sharp (tens of kilometers) based on waveform and travel-time studies to simply be a result of temperature variation. Remarkably, the LLSVPs are almost antipodal and centered near the equatorial plane, leading Dziewonski et al. (2010) to suggest they play a role in stabilizing the Earth's rotational axis and perhaps have done so for geologically long periods, perhaps the entire history of the Earth. The South Pacific LLSVP extends at least 400 km above the core–mantle boundary while the African LLSVP extends at least 1000 km above it. Together, they cover a substantial ($\approx 30\%$) fraction of the core–mantle boundary, but represent only a small fraction ($\approx 2.5\%$) of total mantle volume.

Located on the margins of LLSVPs are features known as ultralow-velocity zones (ULVZs). These are small (up to a few hundred kilometer lateral dimension) and thin (10–40 km deep) regions at the base of the mantle where p- and s-wave velocities are up to 10 and 30%, respectively, lower than in surrounding regions and lie directly above the core–mantle boundary. These seismic velocities are most easily explained as increases of up to 10% in density. The smaller reduction in p-wave velocities led Williams and Garnero (1996) to suggest these were regions of partial melting. In part, because iron would preferentially partition into them, such high-pressure melts may be denser than surrounding solid and simply pond near the base of the mantle. Alternatively, the ULVZs may be compositionally distinct layers, possibly due to

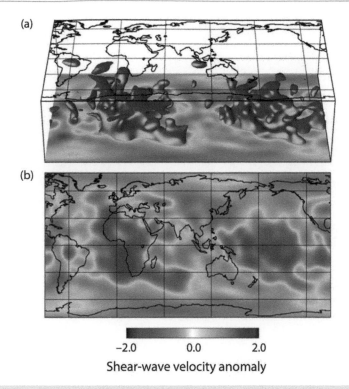

Figure 7.16 (a) Three dimensional seismic tomographic images of the LLSVPs in the lower mantle. (b) They cover a substantial (≈30%) fraction of the core–mantle boundary, but represent only a small fraction (<2.5%) of total mantle volume. The tops of these features are poorly defined and may be diffuse. Source: McNamara (2019)/with permission of Elsevier.

iron enrichment produced by reactions between the core and mantle.

Mantle plumes appear to preferentially rise from the margins of the LLSVPs (Figure 7.17) and, in at least some cases, appear to be associated with the ULVZs (e.g., French and Romanowicz, 2015). Based on the position through time of large igneous provinces (LIPs), presumably produced by mantle plumes, Burke (2011) argued that these LLSVPs are effectively permanent and stationary. It now seems clear that mantle plumes are somehow related to the LLSVPs and probably the ULVZs as well. What that relationship is and whether LLSVPs and ULVZs provide material to plumes or merely control their location remains unclear.

7.4.2.1 Radiogenic isotopes and the recycling paradigm

What processes have led to the distinct geochemical identities of mantle reservoirs? As with the MORB reservoir, at least some OIB reservoirs have been affected by ancient extraction of partial melt, but others have clearly been enriched in incompatible elements relative to any plausible BSE composition. Differences in the major and compatible element chemistry between the MORB and mantle plume reservoirs are difficult to demonstrate because these elements are strongly affected by magma genesis and differentiation (although there has been some suggestion that such differences exist; e.g., Hauri, 1996; Sobolev et al., 2005). We will find in Chapter 9, however, that there are distinctions in the stable isotope geochemistry of major elements such as Mg and Si.

Incompatible element and isotope ratio variations in oceanic basalts point to melting as a key process in creating mantle heterogeneity. For the most part, melting in the silicate Earth is restricted to the upper few hundred kilometers of the Earth, and mostly within the upper 100 km: this is where temperatures can exceed the solidus temperature. ULVZs in the deepest mantle suggest the possibility that melting may locally occur there as well. However, high-pressure melting experiments show that melting in the deep mantle produces elemental fractionations quite different from

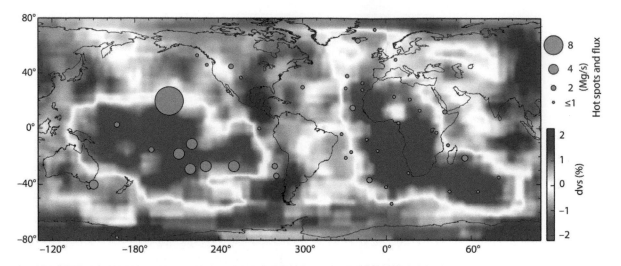

Figure 7.17 Relationship between LLSVPs, defined by reduction in shear-wave velocity (dVs) and mantle plumes, shown as orange circles whose size is proportional to the buoyancy flux in 10^6 g/s. Source: McNamara (2019)/with permission of Elsevier.

those in the upper mantle (e.g., Corgne et al., 2005). The incompatible enrichment patterns in OIBs match those created by melting processes in the upper rather than deep mantle (White, 2015). Thus, while mantle plumes sample deep mantle reservoirs, their geochemical characteristics were acquired in the upper mantle. There is wide agreement on this point.

Hofmann and White (1982) suggested mantle plumes obtain their unique geochemical signature through deep subduction and recycling of oceanic crust (Figure 7.18). Partial melting at

mid-ocean ridges creates oceanic crust that is enriched in incompatible elements relative to its mantle source. Despite the vast amounts of oceanic crust that have been created over geologic time, very little survives at the surface, demonstrating that oceanic crust is efficiently subducted and recycled back into the mantle. The question is: what becomes of it? Once oceanic crust reaches depths of about 60 km or so, it converts into eclogite, which is particularly dense and it remains so at all greater depths (except, perhaps, just at the 660 discontinuity

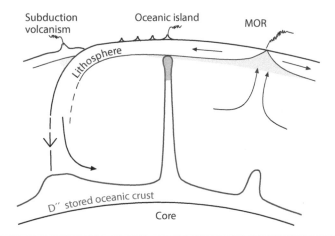

Figure 7.18 Cartoon illustrating the oceanic crustal recycling model of Hofmann and White (1982). Oceanic crust is transformed into eclogite and post-eclogite assemblages upon subduction. It separates from the less dense underlying lithosphere and sinks to the deep mantle where it accumulates. Eventually, it becomes sufficiently hot to form plumes that rise to the surface, producing oceanic island volcanism. Source: Adapted from Hofmann and White (1982).

due to the negative Clapeyron slope). Once separated from underlying low-density harzburgite, it can sink to the base of the mantle where it can be sequestered. Upon sufficient heating, Hofmann and White postulated that it can be incorporated in rising mantle plumes, melting near the surface to create intraplate volcanoes.

Two other similar hypotheses have been proposed. McKenzie and O'Nions (1983) noted the common evidence for incompatible element enrichment in the subcontinental lithosphere, a consequence of a process called *mantle metasomatism,* which we will discuss in a subsequent section, and suggested this material may, because it is cold, occasionally founder and sink to the deep mantle. As in the case of the Hofmann and White model, it would be stored in a thermal boundary layer, heated, and rise in the form of mantle plumes. However, we will find in Section 7.5 that the subcontinental lithosphere is characterized by a combination of highly negative γ_{Os} and highly negative ε_{Nd} which is absent in OIBs, suggesting that subcontinental lithosphere is not an important contributor to mantle plumes. Workman et al. (2004) noted that mantle xenoliths from oceanic regions also show evidence of metasomatic enrichment and suggested that incompatible element-enriched oceanic lithosphere, which is inevitably subducted, might later rise as mantle plumes. It is hard to rule out this process, but at the same time to imagine that oceanic lithospheric enrichment is great enough and widespread enough to account for that seen in mantle plumes.

The incompatible element enrichment of ocean crust and of lithosphere is limited and it is difficult to see how this alone could account for the extreme isotopic compositions seen in some oceanic island volcanoes. An obvious source of incompatible element-enriched material is the continental crust and as Armstrong (1968) pointed out very early on, continent-derived sediment can be transported into the mantle in subduction zones. This certainly seems to be the case in the EM II clan. The Pb/Ce ratio is a particularly useful indicator of the presence of sediment because this ratio is an order of magnitude and higher in sediment than in the mantle. The $^{207}Pb/^{204}Pb$ ratio is also a good indicator of a crustal component because, as we shall see in the next chapter, $^{207}Pb/^{204}Pb$ is consistently higher in the crust than in the mantle (whereas

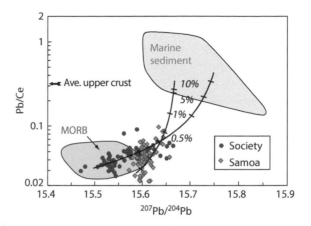

Figure 7.19   Pb/Ce and $^{207}Pb/^{204}Pb$ in basalts from the Society Islands and the Samoan Islands. Two calculated mixing lines between an MORB-like depleted mantle and sediment passes through the data.

this is not true of $^{206}Pb/^{204}Pb$). Basalts from both the Society Islands and the Samoan Islands have Pb/Ce ratios well above the mantle background level. As Figure 7.19 shows, the relationship between $^{207}Pb/^{204}Pb$ and Pb/Ce is consistent with mixing between recycled sediment and mantle. Extremely high $^{87}Sr/^{86}Sr$ (>0.72) and low ε_{Nd} ($\approx$−7) values in some volcanics from the Samoan chain provide further strong evidence that EM II compositions reflect recycling of continental material, most likely through sediment subduction (Jackson et al., 2007).

Two other mechanisms might also feed continental crust into the mantle and produce incompatible element-enriched reservoirs within it. The first is "subduction erosion," first proposed by Scholl et al. (1980), in which the subducting plate removes material from the upper plate in subduction zones through a variety of mechanisms (Straub et al., 2020). Evidence pointed out by Scholl et al. included migration of magmatic arcs away from the trench axis and truncation of geologic trend features on the overriding plate. Subsequent studies have documented additional evidence, including crystalline basement in the present in landward trench wall, tilting and subsiding erosional surfaces on the overriding plate, erosional angular unconformities in the forearc, and the presence of "ghost" detrital zircons, i.e., zircons in Japanese sandstones derived from Paleozoic and Mesozoic intrusions no longer present in the islands. There is also some geochemical evidence for this process in the Andes and Central America (e.g., Kay et al., 2005). Forearc erosion may abrade any material that comprises forearc basement, including upper and lower continental crust, oceanic crust, accreted sediments, and allochthonous terranes. Estimates of the flux of continental to the mantle as a consequence of subduction erosion exceed that due to sediment subduction (e.g., Stern and Scholl, 2010; Straub et al., 2020).

A third mechanism for transporting continental crust into the mantle is lower crustal floundering (often less accurately called "delamination"). This includes both the apparently common loss of dense young mafic arc lower crust and loss of older, mature continental crust that occurs when continental crust is tectonically thickened and the lower crust transforms to eclogite and becomes denser than the underlying mantle (Kay and Kay, 1993). In addition to seismic imaging, evidence includes eruption of strongly light REE-enriched andesitic and dacitic magmas, sometimes with lower crustal isotopic signatures and old xenocrystic zircons. Magmatism is often accompanied or followed by topographic uplift as the dense continental root is lost. Examples include the North China Craton, the Western United States, the Western Mediterranean, and the Andes. While sediment subduction primarily recycles upper continental crust into the mantle, continental foundering recycles lower crust, and subduction erosion can do both. Several studies (e.g., Paul et al., 2002; Willbold and Stracke, 2010) have emphasized the need to recycle lower continental crust (in addition to upper crust) to explain isotopic compositions of OIBs.

Os isotope ratios are also consistent with the recycling paradigm. The crust, both oceanic and continental, is characterized by positive γ_{Os} while the convecting upper mantle and the subcontinental lithosphere are characterized by low $^{187}Os/^{188}Os$. Hence, the nearly ubiquitous positive γ_{Os} in OIBs indicates that the mantle "isotopic heterogeneity observed in hotspot volcanism derives in part from subduction of sediments, crust and mantle lithosphere into a heterogeneous convecting mantle" (Day, 2013), with negative γ_{Os} in some cases, such as the Juan Fernandez and the Azores reflecting a contribution of subducted oceanic mantle lithosphere.

7.4.2.2 Stable isotope evidence

Stable isotope ratios provide an important test for the recycling hypothesis. We found in Chapter 5 that stable isotope fractionations decrease with the square of temperature; as a consequence, fractionations of stable isotope ratios at mantle temperatures ($\gtrsim 1000°C$) should be small. Furthermore, the variety of atomic environments and bonds in mantle minerals is limited, again limiting potential fractionations. Because of this, stable isotopes can be used as tracers of surficial material that has undergone large fractionations in the mantle.

Using stable isotopes in this way also poses challenges because it requires identifying and avoiding those samples that have been affected by secondary low-temperature processes or have assimilated crust or uppermost mantle that has interacted with fluids. In addition, volatile elements, such as H, C, N, and, to a lesser degree, S, undergo fractionation during magma degassing. Finally, while the mantle signal can be compromised for all isotope systems when magmas assimilate crustal material, some elements, such as B and Cl, are extremely sensitive to this process.

We begin with oxygen. Taylor and Sheppard (1986) expressed it this way: "*igneous rocks whose oxygen isotopic compositions show significant variations from the primordial value (+6) must either have been affected by low temperature processes or must contain a component that was at one time at the surface of the Earth.*" In the decades since this was written, the estimated average $\delta^{18}O$ of the mantle has been refined somewhat to about +5.2–5.5‰ and there are fractionations of a few tenths of a per mil between mantle minerals and between minerals and melt, with the latter resulting in basalts having $\delta^{18}O_{SMOW}$ around +5.6‰. In contrast, crustal rocks have $\delta^{18}O$ ranging over many tens of per mil. Silicate rocks that have reacted with hydrothermal fluids can have $\delta^{18}O \leq 0‰$ while carbonate rocks can have $\delta^{18}O$ of $\geq +30‰$.

Cooper et al. (2004) determined $\delta^{18}O$ fresh MORB glasses and although the range was small, ≈+5.2 to +5.7‰, $\delta^{18}O$ correlated with radiogenic isotope ratios, strongly suggesting the upper mantle has been "polluted" by subducted crustal materials. Even larger variations have been found in OIBs. To avoid potential

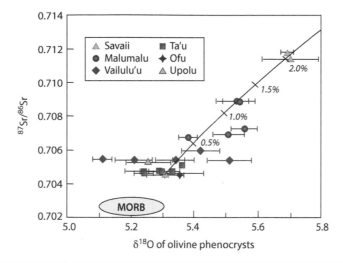

Figure 7.20   Correlation between $^{87}Sr/^{86}Sr$ in lavas and $\delta^{18}O$ of their olivine phenocrysts from the Samoan Islands and seamounts. The line is the upper continental crust–mantle mixing model of Workman et al. (2008); hash marks and numbers in italics are the percentage of upper continental crust with $\delta^{18}O = 10‰$, Sr = 52 ppm, and $^{87}Sr/^{86}Sr = 0.75$. Source: Adapted from Workman et al. (2008).

weathering effects, Eiler et al. (1997) analyzed oxygen in individual olivine phenocrysts from OIBs and reported $\delta^{18}O_{SMOW}$ values as high as 6.1‰ in basalts from the Society Islands. Assuming a recycled sedimentary component with $\delta^{18}O_{SMOW}$ of +15‰, Eiler et al. calculated that the mantle source of the Society Islands lavas contained up to 5% of a sedimentary component. Subsequently, Workman et al. (2008) reported variations of $\delta^{18}O_{SMOW}$ in olivines in Samoan lavas that correlated positively with $^{87}Sr/^{86}Sr$ and $^{207}Pb/^{204}Pb$ and incompatible element ratios, confirming the presence of recycled material in the source of Samoan lavas (Figure 7.20). Samoa represents the extreme of the EM II clan, so this discovery confirmed that their isotopic compositions reflect recycled crustal material in the sources of these lavas.

Correlations between $\delta^{18}O$ and radiogenic isotopes have been documented in other oceanic islands as well. Day et al. (2009) found $\delta^{18}O$ in olivines from the Canary Islands of La Palma defined a negative correlation with $^{187}Os/^{188}Os$ (Figure 7.21) while olivines from the neighboring island of El Hierro showed more limited $\delta^{18}O$ variation in spite of variation in $^{187}Os/^{188}Os$. They modeled these as variable presences of 1 Ga old subducted lower and upper oceanic crust, respectively. They also modeled $\delta^{18}O$ from Hawaii as reflecting mixtures of depleted mantle and 1 Ga old

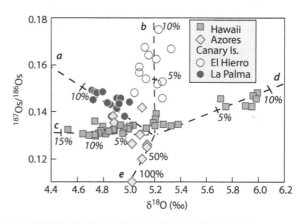

Figure 7.21 $\delta^{18}O$ and $^{187}Os/^{188}Os$ in olivines from the Canary Islands of La Palma and El Hierro, the Azores, and Hawaii. Dashed lines show modeled mixing between ambient upper mantle (DMM: $\delta^{18}O$ 5.2‰; $^{187}Os/^{188}Os$ 0.127) and various recycled materials a: 1 Ga old lower oceanic crust ($\delta^{18}O \approx 0‰$; $^{187}Os/^{188}Os$ 0.4; [Os] 2 ppb); b: 1 Ga old recycled oceanic upper crust ($\delta^{18}O \approx 5.2‰$; $^{187}Os/^{188}Os$ 0.83; [Os] 2 ppb); c: 1 Ga old oceanic altered lithospheric peridotite ($\delta^{18}O \sim 0‰$; $^{187}Os/^{188}Os$ 0.124; [Os] 3.3 ppb); d: 1 Ga old pelagic sediment ($\delta^{18}O \approx 20‰$; $^{187}Os/^{188}Os$ 2; [Os] 0.5 ppb); and e: 2.5 Ga old recycled oceanic lithosphere ($\delta^{18}O \approx 5.0‰$; $^{187}Os/^{188}Os$ 0.1100; [Os] 3.3 ppb). Values in italics are the percentages of these components in the mixture. Source: Day et al. (2009)/with permission of Geological Society of America.

sediment and ancient oceanic lithosphere, while $\delta^{18}O$ values from the Azores were modeled assuming 2.5 Ga old oceanic lithosphere contributed unradiogenic Os.

While the Canary and Hawaii data exhibit good correlations between $\delta^{18}O$ and $^{187}Os/^{188}Os$ as shown in Figure 7.21, the Azores data are not well correlated. Instead, Genske et al. (2013) found that $\delta^{18}O$ correlated with the MgO and CaO content of the olivines. This strongly suggests that the O isotopic variations in the Azores case are due to assimilation during fractional crystallization, most likely of hydrothermally altered lower oceanic crust with $\delta^{18}O < 4.5$. Other instances where O isotopic variations in OIBs correlate with indices of magmatic differentiation or assimilation, or where disequilibrium between minerals exists, have been documented and also point to assimilation.

Let us now turn to Li isotope ratios. For background, the continental crust appears to have average δ^7Li of +1.2‰, modern seawater δ^7Li = +31‰, fresh mantle peridotite has δ^7Li = +3.5±0.5‰, which is similar to the chondritic value and presumably represents the BSE, and fresh MORB has +3.7±1‰. Basalt that has reacted with seawater at low temperature has an average δ^7Li of +11.3‰. In contrast, hydrothermal metamorphism decreases δ^7Li in the oceanic crust and the net effect of low- and high-temperature seawater interaction appears to decrease δ^7Li of the oceanic crust on average. δ^7Li in clastic sediment tends to be lighter (–3 to +5‰) than that of biogenic sediment (+6 to +15). Thus, subducting lithosphere can potentially have both higher and lower δ^7Li than PM. As with $\delta^{18}O$, δ^7Li in oceanic basalts is sensitive to assimilation and weathering, and some caution is necessary in interpreting the data.

Figure 7.22 shows δ^7Li in oceanic basalts as a function of $^{206}Pb/^{204}Pb$ and ε_{Nd}. Values both below and above the PM occur and there is a trend of increasing δ^7Li with increasing $^{206}Pb/^{204}Pb$ and decreasing ε_{Nd}, which is statistically significant. A similar trend can be seen for MORBs, but the correlation between δ^7Li and ε_{Nd} (or either $^{87}Sr/^{86}Sr$ or $^{206}Pb/^{204}Pb$) in MORBs is not statistically significant. Hawaiian basalts show a particularly large range in δ^7Li, which correlates significantly with $^{206}Pb/^{204}Pb$, but not ε_{Nd}. There are considerable data from the Austral–Cook Islands and, as

Chan et al. (2009) noted, the δ^7Li–$^{206}Pb/^{204}Pb$ correlation exists in this data set alone. Chan et al. also analyzed olivine compositions in Cook–Austral samples and found some examples of disequilibrium, which they attributed to post-eruptional alteration (those samples and others where secondary effects are suspected were excluded from Figure 7.22). They nevertheless concluded that these "lavas derive from a source containing recycled dehydrated oceanic crust." In contrast to the isotopically heavy lithium observed in HIMU lavas, EM I and EM II type lavas from the Pitcairn, the Koolau series of Oahu, the Societies, the Marquesas, and the Azores tend to exhibit decreasing ^7Li with decreasing ε_{Nd}, suggesting the presence of a continental or marine sedimentary component. Krienitz et al. (2012) concluded that "The Li contents and isotope characteristics of HIMU-type lavas are consistent with recycling of altered and dehydrated oceanic crust, whereas those of the EM1-type lavas can be attributed to sediment recycling." While it is speculative at this point to associate specific Li isotopic compositions in OIBs with specific crustal components, the Li isotope data nonetheless provide evidence that the mantle has been extensively polluted by material from the surface of the Earth.

Turning to hydrogen isotope ratios, we note that, depending on its concentration, water becomes saturated in basaltic magmas and begins to degas at pressures of 10–20 MPa (1–2 km of water depth), so most subaerial erupted basalts have lost substantial amounts of water during eruption with preferential loss of ^1H driving δD to heavier values. Thus, most of the data on δD come from glassy rims of basalts erupted at depths of >1000 m. Reaction with seawater may also affect δD during or after eruption, so again caution in interpreting the data is necessary (Kyser, 1986). Nevertheless, there is greater variation in δD in oceanic basalts than can be explained by these factors, once again testifying to the presence of material from the Earth's surface being present in the mantle.

MORBs have an average δD of –66‰ and a mode of –69% and a range of –98 to –28.6‰. If only samples analyzed with the more recent Thermal Conversion-Element Analyzer technique are considered, the mean decreases to –75‰. There appears to be significant regional variation, even outside the influence of mantle

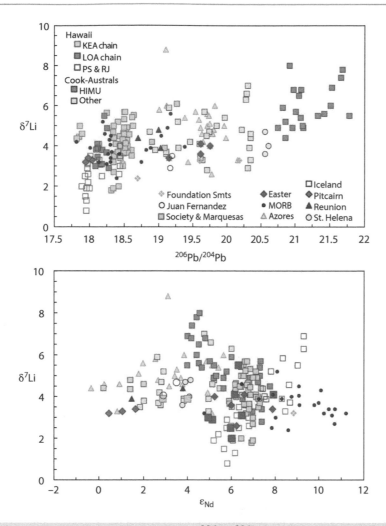

Figure 7.22 δ^7Li in oceanic basalts as a function of ^{206}Pb/^{204}Pb and ε_{Nd}. Data from Harrison et al. (2015); Chan et al. (2009) and Krienitz et al. (2012).

plumes; depleted MORBs in the North Atlantic have a δD of -90 ± 10‰ while Pacific MORBs have a δD of $\approx -60 \pm 10$‰, and the most depleted Pacific MORBs with ^{206}Pb/^{204}Pb < 18.2 have δD values of ≈ 75‰ (Dixon et al., 2017); this can be seen in Figure 7.23.

Dredged samples from ridges and seamounts associated with the Azores, Easter Island, and St. Paul–Amsterdam Islands mantle plumes exhibit significant correlations between δD and radiogenic isotope ratios, such as ^{206}Pb/^{204}Pb (Figure 7.23). In those cases, basalts with the greatest plume component have the most radiogenic Pb and the heaviest H. There is a similar positive correlation between δD and ^{206}Pb/^{204}Pb on the Arctic portions of the Mid-Atlantic Ridge, which in some

regions is influenced by the Iceland plume and Jan Mayen. In contrast, the EM I clan, which has unradiogenic Pb and δD close to mean MORB values, including Discovery and Shona seamounts in the South Atlantic and Loihi (the youngest and still submarine Hawaiian volcano), shows no correlation between ^{206}Pb/^{204}Pb and δD.

Dixon et al. (2017) found systematic variability both in δ^2H and the H$_2$O/Ce ratio in MORBs associated with mantle plumes. Compared to MORBs away from hot spots (H$_2$O/Ce $\approx$ 200; δ^2H ≈ -75‰), Pacific and North Atlantic PREMA-type basalts are somewhat wetter H$_2$O/Ce $\approx$ 225 and isotopically heavy with δ^2H $\approx$ -40‰. Basalts with EM-type signatures have regionally variable volatile compositions. Northern Atlantic EM-type basalts

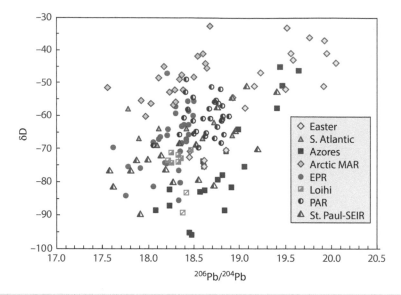

Figure 7.23 δD versus $^{206}Pb/^{204}Pb$ in submarine basalts. The Arctic Mid-Atlantic Ridge (MAR) group includes samples from 77°N southward to Iceland (Dixon et al. 2017), "Easter" includes samples from the Easter Microplate and the Salas y Gómez seamount chain (Kingsley, et al., 2002). The South Atlantic includes samples from the southern MAR in the vicinity of the Discovery and Shona seamount chains from 47 to 52°S (Dixon et al., 2017), "Azores" are samples from the MAR near and on the Azores platform, East Pacific Rise (EPR) includes samples from 8° to 15°N (Dixon et al., 2017), Pacific–Antarctic Ridge (PAR) samples includes ones from 110° W to 178°W (Clog et al., 2013), St. Paul-SEIR includes samples from the SEIR and seamounts associated with the St. Paul–Amsterdam hot spot (Loewen et al., 2019). Loihi data from Loewen et al. (2019); open symbols are highly vesiculated lavas erupted at shallower depth where degassing and H isotope fractionation are suspected.

are wet ($H_2O/Ce \approx 330$) and isotopically heavy ($\delta^2H \approx -57‰$) while Northern Pacific EM-type basalts are dry ($H_2O/Ce \approx 110$) and isotopically light ($\delta^2H \approx -94$).

Sulfur is another volatile element subject to degassing loss and fractionation from magma, so again the focus is on submarine eruptions or inclusions in minerals. Labidi et al. (2013) reported high-precision sulfur isotope data for MORBs and showed that $\delta^{34}S_{CDT}$ in MORBs from the South Atlantic correlates positively with $^{87}Sr/^{86}Sr$ and negatively with $^{143}Nd/^{144}Nd$ (Figure 7.24). Significant sulfur isotope fractionation is not expected to occur within the mantle. Labidi et al. interpreted these correlations as evidence of mixing between a DMM component with $\delta^{34}S_{CDT} \approx -1.5‰$ and an "enriched" component associated with the Discovery and Shona mantle plumes. They argue the data are most consistent with the enriched component being recycled sediment with $\delta^{34}S \approx +10‰$. Subsequently, Labidi et al. (2015) reported $\delta^{34}S$ values ranging from +0.11 to +2.79‰ in the

reduced sulfur fraction of Samoan lavas (the subordinate sulfate fraction has higher $\delta^{34}S$). Furthermore, $\delta^{34}S$ in the reduced sulfur correlated with $^{87}Sr/^{86}Sr$ in glasses (Figure 7.24). They argued the correlation "requires the EM-2 endmember to be relatively S-rich, and only sediments can account for these isotopic characteristics." More recently, Beaudry et al. (2018) reported elevated $\delta^{34}S$ in sulfides, silicate melt inclusions, and matrix glasses lavas of El Hierro volcano, Canary Islands, indicative of recycled surficial material in the Canary Islands plume.

Even more dramatic was the discovery by Cabral et al. (2013) of mass independently fractionated (MIF) sulfur in olivines from Mangaia island lavas of the Cook-Austral chain (Figure 7.25). Small mass-independent fractionations of sulfur isotopes occur in modern stratospheric aerosols as a consequence of ultraviolet mass photolysis (e.g., Savino et al., 2003) and in the upper oceanic crust as a consequence of microbial dissimilatory sulfate reduction (Ono et al., 2012). However, in both cases, deviations

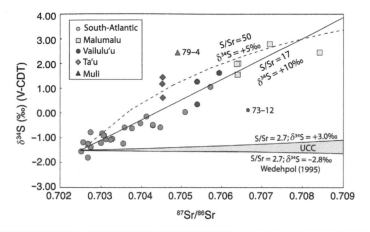

Figure 7.24   Sulfur isotopes as a function of $^{87}Sr/^{86}Sr$ in South Atlantic MORBs and lavas from Samoan seamounts. Lines are mixing models between a DMM component and sedimentary components with high $\delta'^{33}S_{CDT}$ and high elemental S/Sr ratios. Mixing with average upper crust fails to predict the correlations. Source: Adapted from Labidi et al. (2015).

from mass-dependent fractionations are small (a few tenths of per mil) and this MIF sulfur is a vanishingly small fraction of sulfur at the surface of Earth. In contrast, MIF sulfur isotope ratios with $\Delta'^{33}S$[4] as large as several per mil are common in sedimentary and hydrothermal sulfides and sulfates in rocks older than 2.3 Ga, but disappear by 2.0 Ga (Farquhar and Wing, 2003). It is now widely agreed that significant atmospheric oxygen first developed around 2.4–2.3 Ga. Prior to that time, known as the *Great Oxidation Event*, ultraviolet radiation could penetrate through the whole atmosphere and photolyze atmospheric SO_2 producing reduced species, such as S_8 with positive $\Delta^{33}S$, and oxidized species, such as H_2SO_4 with negative $\Delta^{33}S$. Once the atmosphere was oxygenated, stratospheric ozone severely limited the penetration of ultraviolet radiation into the troposphere, ending widespread mass-independent fractionation of sulfur. We will return to the Great Oxidation Event and discuss it in more detail in Chapter 12.

Farquhar et al. (2002) documented MIF sulfur in sulfide inclusions in diamonds from the Orapa kimberlite mine of South Africa (Figure 7.25). The diamonds are associated with eclogite and have two distinct sets of Re–Os model ages of 1.0 and 2.9 Ga. The MIF sulfur along with variable O, N, and C isotope ratios indicates that the eclogite was subducted oceanic crust and associated

sediment and stored in the subcontinental lithosphere until this material was sampled by kimberlite magmas and carried to the surface ≈100 Ma ago.

The discovery of MIF sulfur in Mangaia lavas (Figure 7.25) is significant in several important ways. First, it demonstrates the transport of surface mantle into the deep mantle, not merely the lithospheric mantle. Second, Mangaia lavas represent the extreme isotopic compositions of the HIMU clan, confirming that this clan also contains recycled materteral from the Earth's surface. Third, it demonstrates the ancient nature of recycled material and its long residence time in the mantle. Fourth, these data demonstrate a tectonic regime capable of subducting surficial material into the deep mantle has operated since at least the earliest Proterozoic.

The discovery of MIF sulfur in Mangaia lavas was quickly followed by the discovery of MIF sulfur in lavas from Pitcairn Island (also shown in Figure 7.25) by Delavault et al. (2016), another oceanic island volcano in the central South Pacific (famous as the refuge of the *HMS Bounty* mutineers who settled there with their Tahitian wives in 1790). The Pitcairn–Gambier chain volcanoes are part of the EM I clan; hence, these data indicate the EMI source also contains recycled material from the Earth's surface. Both the Pitcairn

---

[4] Recall from Chapter 5 that the Δ' denotes the per mil deviation from mass-dependent fractionation.

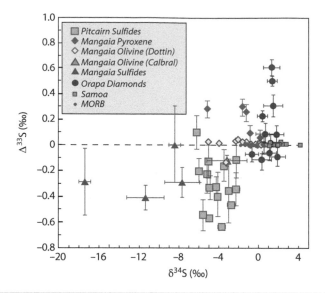

Figure 7.25 MIF sulfur isotope diamonds from the Orapa kimberlite mine in Botswana and sulfides and olivines from Mangaia and Pitcairn Islands. Dashed lines indicate estimated bulk mantle composition. Mass-independent fractionation of sulfur was restricted to the Archean/early Proterozoic atmosphere. This sulfur must have been subducted and stored in the mantle for 2.3 billion years. Data from Farquhar et al., (2002), Cabral et al. (2013), Delavault et al. (2016), and Dottin et al. (2020a).

and Mangaia sulfur has predominantly negative Δ'^{33}S, which Farquhar and Wing (2003) predicted should characterized Archean oceanic crust after hydrothermally reacting with dissolved marine sulfate. This suggests Archean subducted oceanic crust is the most likely source of the MIF sulfur. More recently, Dottin et al. (2020a) found examples of small positive Δ'^{33}S values in olivines from Mangaia.

Dottin et al. (2020b) reported positive Δ'^{33}S values in basalts from Samoa. As was the case in the more recently reported values from Mangaia, the Δ'^{33}S values were small (<0.03) and are uncorrelated with δ^{34}S. The Samoan samples have positive δ^{34}S and compositions similar to the Orapa diamond inclusions. Interestingly, Δ'^{33}S values correlate positively with ^{206}Pb/^{204}Pb and negatively with ^{3}He/^{4}He. We discuss the significance of ^{3}He/^{4}He in oceanic basalts in Section 7.4.2.4.

7.4.2.3 Heterogeneity and zoning in mantle plumes

Some mantle plumes, such as the Reunion plume, form chains of single volcanoes conforming to the classic Wilson hot spot model. In other cases, plumes produce more than one volcanic chain; Hawaii is the classic example. Kilauea is the most recent volcano of the Kea chain, so-called because it includes Mauna Kea, which has not erupted for 10 000 years. The seamount Loihi is the most recent volcano of the Loa chain, so-called because it includes Mauna Loa, which erupted multiple times in the twentieth century, and it also includes Hualalai, which is near the end of its life and has erupted only once historically. This double chain can be traced back to the northwest through the older Hawaiian Islands and seamounts as well (Figure 7.26). Tatsumoto (1978) found that the two chains differed in Pb isotopic compositions and Stille et al. (1983) showed that they could be distinguished on the basis of other radiogenic isotopic compositions as well, and many subsequent studies have confirmed this. Figure 7.27 summarizes the differences between the chains. The chains also differ in Li isotopic compositions (Harrison et al., 2015). The double chain continues for ≈9 Ma to Nihoa seamount, the last Loa volcano; beyond that, the Loa composition dwindles and eventually the Emperor seamounts show only Kea compositions. It is perhaps significant that the magma production rate of the Hawaiian plume has also increased by roughly a factor of two in the last 5–10 Ma.

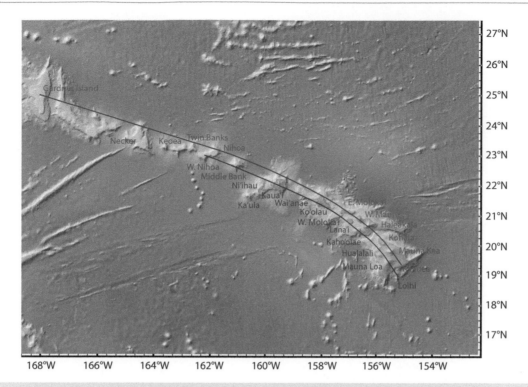

Figure 7.26   Map illustrating the double chain of volcanoes in the Hawaiian archipelago. Red is the "Kea" chain whose youngest member is Kilauea and extends back through the Hawaiian–Emperor seamounts; blue is the "Loa" chain whose youngest member is Loihi and extends back only through West Nihoa. Base map produced using GeoMapApp (www.geomapapp.org).

The Galapagos archipelago includes at least a dozen volcanoes that have erupted in the Holocene and are arranged in a rectilinear pattern rather than a chain. Prior to 5 Ma ago, the hot spot was ridge-centered and this is reflected in a westerly increase in the age of the oldest volcanism on two chains of extinct volcanoes, the Cocos and the Carnegie Ridges, extending in the direction of Cocos and Nazca plate motions, as expected of volcanism above a fixed long-lived mantle plume.

As Figure 7.28 shows, within the archipelago, there is a distinct geographic pattern in magma geochemistry (White et al., 1993). Harpp and White (2001) found that four distinct isotopic components were required to explain this pattern. These are geographically distributed and show up principally in the northern, central, southern, and eastern sections, respectively, of the archipelago. Based on analyses from seamounts of the Cocos Ridge, this spatial zonation appears to have persisted for at least 14 Ma (e.g., Hoernle et al., 2000).

Subsequent studies have recognized similar long-lived zonation is a number of other hot spot tracks, including Samoa, the Marquesas, Tristan–Gough, Easter–Salas y Gómez, Discovery seamount chain, and possibly the Society Islands. Most interpretations of these isotopically and chemically zoned hot spot tracks involve contributions from LLSVPs and surrounding mantle. The Hawaiian plume is located at the northeast margin of the Pacific LLSVP, and Weis et al. (2011) have argued that the Loa chain, which is located southwest of the Kea chain, samples this LLSVP. Harpp and Weis (2020) point out that the Galapagos also overlies the northeastern edge of the Pacific LLSVP and argue that the plume's tapping of material from both the LLSVP and surrounding lower mantle account for the zonation in the Galapagos plume as well (Figure 7.29).

7.4.2.4   A common component and a primordial signal

We note earlier how the arrays formed by the various oceanic basalt clans tend to converge in radiogenic isotope ratio plots (Figures 7.4–7.10 and Figure 7.13). Hart

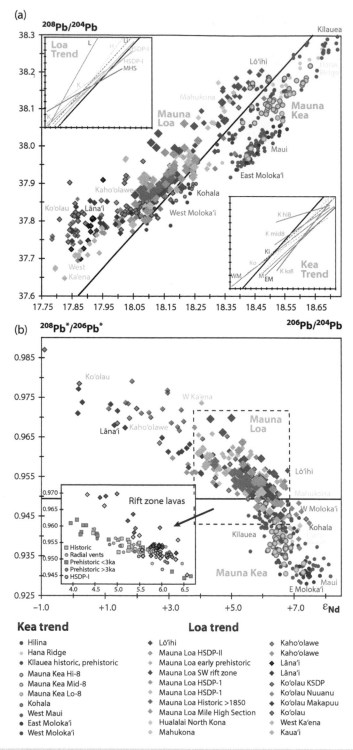

Figure 7.27 (a) ^{208}Pb/^{204}Pb versus ^{206}Pb/^{204}Pb and (b) ^{208}Pb*/^{206}Pb* versus ε_{Nd} in Hawaiian basalts. The Kea and Loa chains can be distinguished bases on Pb and Nd isotope ratios: the Loa chain has higher ^{208}Pb/^{204}Pb at a given ^{206}Pb/^{204}Pb than the Kea chain and also has lower ε_{Nd} than the Kea chain. Source: Weis et al. (2011)/with permission of Springer Nature.

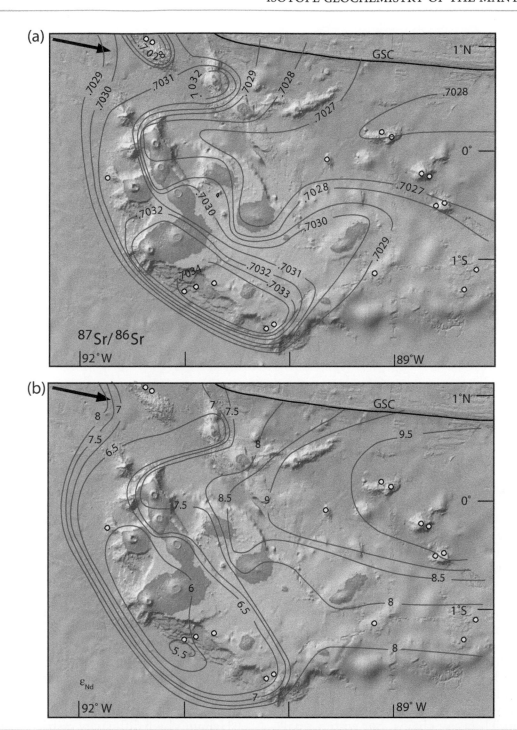

Figure 7.28   Contours of (a) $^{87}Sr/^{86}Sr$ and (b) ε_{Nd} in lavas from the Galapagos archipelago. Yellow dots are dredge locations. Source: Base map produced with GeoMapApp, and contours are adapted from Harpp and White (2001).

et al. (1992) were the first to note this, pointing out that oceanic basalt isotope data formed three-dimensional $^{87}Sr/^{86}Sr$–$^{143}Nd/^{144}Nd$–$^{206}Pb/^{204}Pb$ arrays that tended to converge toward a region they called "FOZO" (an acronym for

Focal Zone). They suggested that FOZO is the isotopic composition of the lower mantle and that plumes rising from the core mantle boundary entrain and mix with this lower mantle material.

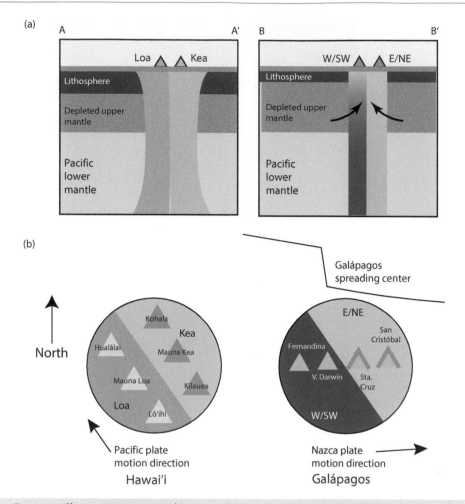

Figure 7.29   Cartoon illustrating geographic variation in isotopic compositions in the Galapagos and Hawaiian archipelagos in (a) cross section and (b) in plan view. Both lie above the edge of the Pacific LLSVPs, and Harpp and Weis suggest material in the southwest side of both plumes is drawn from the LLSVP. Source: Adapted from Harpp and Weis (2020).

Let us now bring in an important radiogenic element that we have neglected thus far: He. We will postpone a full discussion of He isotopes to Chapter 14, but He is a key to unraveling mantle isotope geochemistry, so some discussion is necessary here. ^{4}He is, of course, produced by alpha decay, almost entirely from the U and Th decay chains, but with a small contribution from ^{147}Sm. According to convention, we should report He isotope ratios as the ratio of the radiogenic to non-radiogenic isotope; i.e., ^{4}He/^{3}He, which turns out to be very big number, $\approx 10^6$ and greater, and indeed some geochemists report this ratio in this way. However, many mantle geochemists break with convention and report the ^{3}He/^{4}He ratio normalized to this ratio in air (1.38×10^{-6}; ^{4}He/^{3}He = 724 638), denoted as R/R_A. This notation was first introduced when anomalously ^{3}He-rich helium was discovered in seawater above the East Pacific Rise (e.g., Clarke et al., 1969; Craig et al., 1975) and atmospheric He provides a useful analytical standard since it is isotopically uniform and readily available to all.

Helium's high diffusivity restricts its use as a geochronometer to the low-temperature applications we discussed in Chapter 4. This high diffusivity also means there is a continual flux from the crustal rocks to pore and ground-waters, and eventually to the atmosphere. From the atmosphere, He is continually lost to space; He is unique in that respect. And unlike the other radiogenic elements we have discussed thus far, it is not significantly recycled into the mantle. It is, however,

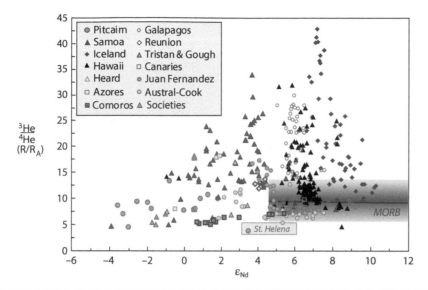

Figure 7.30 ^{3}He/^{4}He and ε_{Nd} in OIBs. The highest ^{3}He/^{4}He values tend to be associated with ε_{Nd} in the range of $\approx$7 to $\approx$3. Extremes of the EM I, EM II, and HIMU clans tend to have the lowest ^{3}He/^{4}He.

continually lost from the mantle through degassing, as the discovery of elevated ^{3}He/^{4}He in Pacific deep water by Clarke et al. (1969) demonstrated.

Like hydrogen, He is lost through degassing of subaerially erupted basalt, meaning the data set is largely restricted to submarine basalts, although He can sometimes be extracted and measured from melt inclusions in olivine phenocrysts. Unfortunately, this means that very often He isotope ratios are not measured on the same samples as other isotope ratios. Nevertheless, there are now enough cases where both He and other radiogenic isotope ratios have been measured that we can discern a pattern (Figure 7.30). High ^{3}He/^{4}He ratios are not associated with the most extremely compositions of other radiogenic isotope ratios, but rather with those with median values. This was first pointed out by Farley et al. (1992), based on work with Samoan samples, and who hypothesized a component, which they called "PHEM" (primitive helium mantle) containing the high ^{3}He/^{4}He. Since then, many other analyses of Samoan lavas have accumulated, and we can see in Figure 7.30 that Samoan basalts with the highest ^{3}He/^{4}He have $\varepsilon_{Nd} \approx 4$ while those with more extreme ε_{Nd} have ^{3}He/^{4}He equal to or lower than MORB ^{3}He/^{4}He ratios. Class and Goldstein (2005) demonstrated this is generally true: the highest He/^{4}He in oceanic basalts tends to be

associated with intermediate Sr, Nd, and Pb isotope ratios. In contrast, the extremes of the EM I (e.g., Pitcairn, Tristan, and Gough), EM II (Society and the extreme Samoan compositions), and HIMU (St. Helena, the Australs) have low ^{3}He/^{4}He, often lower than MORB values, which is what we expect of material having been degassed at the Earth's surface. In addition to Samoa, high ^{3}He/^{4}He ratios also occur in lavas from Hawaii, Galapagos, and Iceland with ε_{Nd} values ranging from 3 to 7.

Hanan and Graham (1996) found that Pb isotope ratios form three-dimensional arrays that converged on an intermediate composition, which is associated with the highest He isotope ratios. They argued for the existence of a component common to all or most plumes, which they called "C." Although FOZO, PHEM, and C were all defined differently, they all have compositions similar to PREMA of Zindler and Hart (1986) and we will use that term because it has precedence. The convergence of isotopic arrays we see in Figures 7.4–7.11 suggests a PREMA component may indeed be common to many plumes.

Tungsten isotope ratios provide additional evidence of a relatively primitive component in mantle plumes. We learned in Chapter 6 that the short-lived radionuclide ^{182}Hf decays to ^{182}W with a half-life of 8.9 Ma. Hf is lithophile and partitions into the silicate portions of planets while W is siderophile and partitions into planetary cores. This produces mantles with high Hf/W ratios and high $^{182}W/^{184}W$ and cores with low Hf/W and low $^{182}W/^{184}W$. Over the last decade or so, a number of studies have identified W isotope anomalies, in most cases positive $\varepsilon^{182}W$ or $\mu^{182}W$ values[5], in Archean continental rocks (post-Archean continental rocks lack W isotope variations), but given the short half-life of ^{182}Hf, these isotopic variations must have arisen in the first <100 Ma of Earth's history and we will discuss this in more detail in the following chapter.

More recently, W isotope anomalies have also been identified in OIBs (Mundl, et al., 2017; Rizo et al., 2019; Mundl-Petermeier et al., 2020 and Jackson et al., 2020). These are smaller, less than 25 ppm, and almost exclusively negative. They are clearly associated with high $^3He/^4He$ (Figure 7.30a). In contrast, the single high-precision analysis of a MORB sample yielded $\mu^{182}W = 0.2 \pm 4.9$, indistinguishable from the standard value, which is presumed to be the BSE value. As Figure 7.31a shows, the correlation between $\mu^{182}W$ and $^3He/^4He$ is imperfect; in particular, Icelandic basalts, which have the highest $^3He/^4He$, appear to define a different correlation than other samples. Mundl-Petermeier et al. (2020) and Jackson et al. (2020) argued this requires several reservoirs in the deep Earth with different $\mu^{182}W$ and $^3He/^4He$. As we saw with $^3He/^4He$, the largest anomalies are associated with intermediate Nd, Sr, Hf, and Pb isotope ratios (Figure 7.31b), i.e., PREMA-like isotopic compositions; those samples with extreme EM I, EM II, and HIMU have the smallest W isotopic anomalies.

Given this association of high $^3He/^4He$ with LLSVPs in the lower mantle, $\mu^{182}W$ anomalies are apparently also associated with them. ULVZs have been identified at the base of several mantle plumes, including Hawaii, Iceland, and Samoa. One interpretation of the ULVZs is that they consist of dense, Fe-rich material exsolved from the core. Mundl-Petermeier et al. (2020) suggested that the most negative $\mu^{182}W$ values, which are associated with more modestly high $^3He/^4He$ and are observed in the Galapagos and Hawaii, derive from these

ULVZs. They suggested the more modest μ_{182} that are associated with the more extreme $^3He/^4He$ and notably higher ε_{Nd} observed in Iceland are derived from an "early formed mantle reservoir" that could represent early crystallization products of early mantle differentiation and is now contained with the LLSVPs.

The association of intermediate ε_{Nd} values with high $^3He/^4He$ (Figure 7.30) and low $\mu^{182}W$ (Figure 7.31), both of which are indicative of relatively primitive, less processed mantle material, suggests that the bulk Earth ε_{Nd} is not 0, but rather falls in this intermediate range of $\approx+3$ to +7. This range of values is also the most densely populated part of the radiogenic isotope space of mantle-derived rocks, which motivated the acronym PREMA (prevalent mantle). That in turn suggests that the Earth is depleted in the most incompatible elements (e.g., O'Neill and Palme, 2008) relative to the conventionally accepted composition of McDonough and Sun (1995).

These variations must have arisen very early in Earth's history and subsequently been sequestered deep in the Earth. The early Earth was likely extensively, if not completely, melted as a consequence of the energy released by collisions with large planetary embryos in the late stages of its accretion, with the Moon-forming impactor, *Theia*, being only the last of these. One possibility to explain these W isotopic variations is that crystallization of a magma ocean resulted in Hf/W fractionation as W is more incompatible in silicate minerals than is Hf. A difficulty with any model involving early mantle differentiation is that it should have fractionated Sm/Nd as well as Hf/W, and this would have produced variation in ε_{142Nd} resulting from decay of the short-lived radionuclide ^{146}Sm. While variations in ε_{142Nd} have indeed been observed in Archean crustal rocks, measurements made at the highest precision reveal that ε_{142Nd} in the mantle is uniform within ± 1.7 ppm of the standard (Hyung and Jacobsen, 2020). It is also possible that cores of these impactors had segregated with varying efficiency leaving the Earth's early mantle with variable Hf/W ratios.

Since we expect low $^{182}W/^{184}W$ ratios to be associated with planetary cores, another explanation is that this tungsten is somehow derived

---

[5] Recall that ε_{182W} is deviations in parts in 10 000 and $\mu^{182}W$ is deviations in parts in 10^6 from a terrestrial standard.

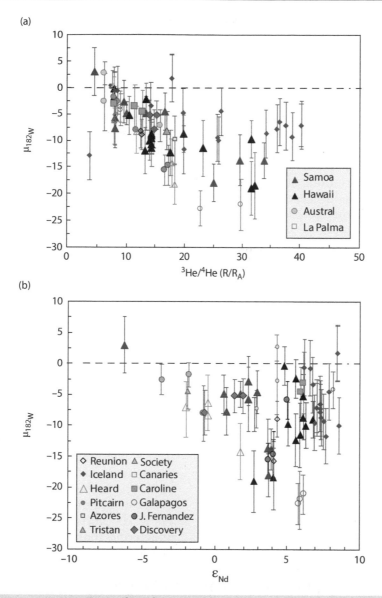

Figure 7.31 μ^{182}W as a function of (a) ^{3}He/^{4}He and (b) ε_{Nd} in oceanic basalts. μ^{182}W variations result from decay of the extinct radionuclide ^{182}Hf. Because of its short half-life, 8.9 Ma, we can conclude these variations must have arisen very early in Earth's history. The largest μ^{182}W anomalies are associated with more primordial (i.e., high) ^{3}He/^{4}He values and ε_{Nd} in the range of 3–6. Data from Mundl, et al. (2017), Rizo et al. (2019), Mundl-Petermeier et al. (2020), and Jackson et al. (2020).

from the Earth's core (Rizo, et al., 2019; Mundl-Petermeier et al., 2020), perhaps somehow related to exsolution of O-rich metal at the top of the core. Although there is no consensus on this, several studies have suggested that significant He may have dissolved in the core during its formation (e.g., Roth et al., 2019), which would be consistent with the observed ^{3}He/^{4}He–μ^{182}W correlation. This idea predicts that there should be a correlation between μ^{182}W and highly siderophile element abundances, but no such correlation has been found. However, as the core is expected to have μ^{182}W $\approx$ –220, only a very small amount of core material (<0.3%) would be needed to account for the μ^{182}W values observed. Also, Rizo et al. (2019) note there is a possible negative correlation between μ^{182}W and Fe/Mn ratios in OIBs.

We can summarize by saying the dominant process shaping the isotopic composition of the Earth mantle has been extraction of partial melts to form the crust, both continental and oceanic. However, many of the most obvious isotopic variations have resulted from recycling of material from the Earth's surface, no doubt through subduction of oceanic crust and its sedimentary veneer, but also through subduction erosion and continental floundering, over geologic time. This recycled material is present in each of the isotopic clans: HIMU, EM I, EM II, and PREMA as well as in the DMM. Different combinations of this material returned to the mantle as different times likely account for the different isotopic flavors of this clan.

In addition, there is also clear evidence that some isotopic variations must have originated within the first 100 Ma or less of Earth history and that these have been preserved in the deep mantle or perhaps in the Earth's core. We will find in Chapter 14 that xenon isotopes provide particularly strong evidence of preserved initial heterogeneity. This "primordial" material appears primarily associated with the PREMA clan, which seems to contribute to most plumes.

7.5 THE SUBCONTINENTAL LITHOSPHERE

The lithosphere is that region where heat is transported by conduction rather than convection. This includes the crust, but our focus in this chapter will be on mantle lithosphere. New oceanic lithosphere is continuously produced at mid-ocean ridges as upwelling mantle cools, slowly thickening with time eventually reaching thicknesses of 60 km and more. It is also continuously consumed in subduction zones and recycled into the deeper mantle. In contrast, the isotopic evidence shows that the lithosphere beneath continents is more permanent and can grow to thicknesses of 200 km or so, although in regions such as the Great Basin of the Western United States it can be thinned to only a few tens of kilometers.

Mantle-derived magma erupted on continents includes highly alkaline, volatile-rich magmas, such as alkali basalts, lamproites, and kimberlites erupted in small volumes in tectonically stable regions, those related to intracontinental rifts, flood basalts of LIPs, and those related to subduction. We will consider the latter in the following chapter, LIPs are primarily products of mantle plumes rather than lithosphere. While rift-related magmatism, such as Rhine Rift in Europe and the Rio Grande Rift in North America, does sample underlying subcontinental lithospheric mantle (SCLM), it often also reveals intrusion of ambient asthenosphere. Consequently, information about the SCLM is provided primarily by alkaline magmatism. These display not only a wide variety of Sr, Nd, Hf, and Pb isotopic compositions that overlap a considerable degree those observed in OIBs, but can also have more extreme compositions, revealing that the SCLM is compositionally diverse. They nonetheless differ from OIBs in one important respect: they have $^3He/^4He$ ratios less than that MORBs have (Day et al., 2005).

There are, however, several obstacles to relying on intracontinental alkaline magmatism to sample the SCLM. First, recent intracontinental magmatism is rare and while there are many older examples, they have been exposed to weathering and reaction with surface water for considerable lengths of time, compromising their value as mantle samples. Second, many, though not all, show evidence of having assimilated continental crust (this is true of other types of intracontinental magmatism as well). Fortunately, many of these eruptions have carried mantle xenoliths to the surface, a consequence of their gas-rich nature of these magmas that drives rapid ascent. These xenoliths provide direct samples of the underlying mantle and we will rely primarily on these xenoliths in assessing the isotope geochemistry of the SCLM.

Xenoliths from the SCLM are often pyroxene-poor and olivine-rich; this and their Fe-poor olivine compositions demonstrate that the SCLM is often highly refractory and depleted in basaltic components through previous melting episodes. Consistent with this, some xenoliths have extremely depleted incompatible-element isotopic signatures, with ε_{Nd} and ε_{Hf} values exceeding +500 in some cases. Peridotites from beneath Archean cratons (areas of ancient stable continental crust) tend to be particularly melt-depleted. Their compositions suggest up to 30–45% melt extraction, which substantially exceeds that occurring under modern mid-ocean ridges (Pearson and Wittig, 2014). One possible explanation is that this reflects melting in mantle plumes, which are demonstrably hotter than the mantle beneath mid-ocean ridges.

On the other hand, thermal models suggest ambient mantle temperatures would have been 100–150°C hotter in the Archean, and Herzberg and Rudnick (2012) and Pearson and Wittig (2014) argue that the extreme depletion of the Archean SCLM simply reflects melting beneath mid-ocean ridges in the hotter Archean regime.

Geochronological studies, particularly Re–Os, show that the timing of these melt extraction events is generally related to the age to the overlying crust above it and there appears to be long-term coupling of ancient continental crust with the underlying mantle lithosphere. Figure 7.32 shows Re-depletion model ages (T_{RD}) calculated from $^{187}Os/^{188}Os$ ratios in peridotite xenoliths from beneath the Kaapvaal, Slave, and Siberian cratons. Rhenium depletion ages (Equation [2.58]) assume quantitative removal of Re and because Re extraction is often not quantitative, these ages can underestimate actual ages somewhat. Although some ages are younger, the mode in both the Kaapvaal and Slave cratons is Archean. There is a sharp drop-off in ages at above 3 Ga, even though some of the overlying crustal rocks are older, suggesting that the subcontinental lithosphere began to stabilize around this time (Pearson and Wittig, 2014).

In the Siberian craton, the largest on Earth, Re–Os systematics reveal a more complex history. Crustal basement rocks cluster around two main ages: 2.0 Ga and 2.8–3.0 Ga. Peridotites from the Paleozoic Udachnaya kimberlite in the center of the craton show a mode in T_{RD} distribution at 2 Ga. Clinopyroxenes in harzburgites among the Udachnaya xenoliths have extremely depleted Hf and Nd isotopic signatures with ε_{Hf} and ε_{Nd} as high as +2084 and +123, respectively; Lu–Hf model ages range from 1.7 to 3.0 Ga (Doucet et al., 2015). Further to the north, peridotites from the Late Jurassic Obnazhennaya kimberlite exhibit a bimodal distribution with one mode around 2 Ga matching that of the Udachnaya peridotites and a second mode at around 3 Ga. Ionov et al. (2015) concluded that the Siberian lithosphere formed as early as 2.8 Ga, nearly simultaneously with lithosphere beneath Archean cratons elsewhere. A subsequent Paleoproterozoic event produced additional crust and modified the lithosphere adding new material to it, as well as encapsulating some older material.

Seismic tomographic studies have revealed that anomalously high seismic velocities, indicative of stiff, buoyant lithosphere, extend to depths below 200 km beneath Southern Africa (Fouch et al., 2004). Figure 7.33 shows that Kaapvaal kimberlite xenolith suites with Archean T_{RD} ages are concentrated in regions of fast s-wave seismic velocities at 200 km depth in Southern Africa. The fast s-wave

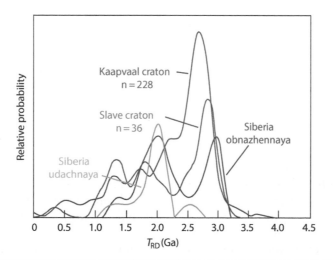

Figure 7.32 Probability density functions of Re–Os model ages for peridotite whole rocks for xenoliths in kimberlites from the Kaapvaal (Southern Africa), Slave (Canada), and Siberia (Russia) provinces. Although younger ages occur, the mode in both Kaapvaal and Slave is Archean, while the Siberian xenoliths show modes of both Archean and Paleoproterozoic. Source: Adapted from Pearson and Wittig (2014) and Ionov et al. (2015).

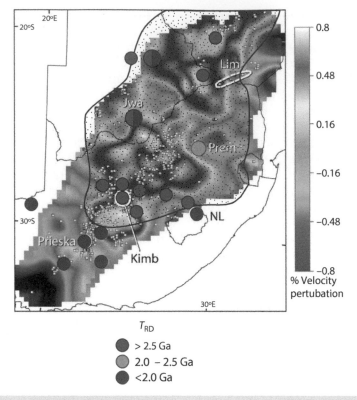

Figure 7.33 Comparison of mean rhenium depletion ages (T_{RD}) in kimberlite xenolith suites from the Kaapvaal craton and surrounding areas with s-wave velocities anomalies at 200 km depth. Small squares indicate the location of kimberlites; blue line and stippled area delineate the region of Archean crustal basement. The Orapa and Jwaneng kimberlite suites have a bimodal distribution of T_{RD} ages. Age distribution from Pearson and Wittig (2014); seismic velocity map from Griffin et al. (2009).

velocities in turn reflect the extensive melting recorded in the peridotite xenoliths. Extensive melting leaves the residue depleted in iron and dense minerals, such as garnet, and consequently significantly less dense that fertile mantle. Melting also extracted water, increasing viscosity, with the combination of these effects leaving the Archean SCLM buoyant and stiff and hence stabilized against convection.

In contrast to the strong melt depletion recorded by major element chemistry, mineral compositions, and Re–Os isotope systematics, many peridotite xenoliths are enriched in incompatible elements. In many cases, this incompatible element enrichment is not reflected in Nd, Sr, or Hf isotopic compositions, indicating that the enrichment must have occurred shortly before the eruption that carried the xenoliths to the surface. In these cases, the agent of enrichment is most likely fluids or small-degree melts that infiltrated the area in association with magmatism that carried the peridotites to the surface. In other cases,

particularly beneath Archean cratons, this enrichment is ancient, as evidenced by unradiogenic Nd isotopic compositions. This is illustrated in Figure 7.34 that shows that Archean cratons are often characterized by a combination of negative γ_{Os} and ε_{Nd} values. Often, this enrichment is cryptic in the sense that it is not recorded by mineral assemblages. In other cases, the presence of phases, such as phlogopite mica, amphibole, rutile, ilmenite and diopside (this combination of minerals carries the acronym MARID), in some cases concentrated in veins, records reaction of the peridotites with metasomatic H_2O- and CO_2-rich fluids and/or melts. In most cases, this metasomatism has apparently not affected Re–Os systematics, which retain depleted signatures. In some cases, however, notably some of the Wyoming xenoliths, Os isotopes also record enrichment. The difference is likely related to the concentration of dissolved sulfide in the fluids and/or melts that can transport and precipitate Re.

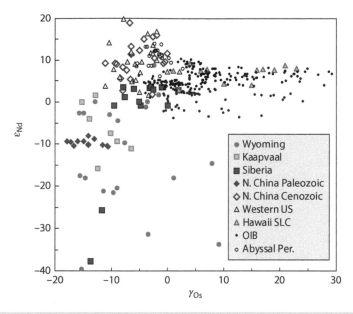

Figure 7.34 γ_{Os} and ε_{Nd} in peridotite xenoliths compared with abyssal peridotites and OIBs. Xenoliths from Salt Lake Crater on Oahu, Hawaii overlap the OIB field (although extending to higher γ_{Os} than observed in Hawaiian basalts). Peridotites from the Western United States (including the Great Basin and Rio Grande Rift) overlap the abyssal peridotite field, but extend to higher ε_{Nd}, consistent with seismic studies that reveal the lithosphere here is particularly thin. In contrast, peridotite xenoliths from the Wyoming, Kaapvaal, and Siberia Archean and Paleoproterozoic cratons have a unique combination of negative γ_{Os} and negative ε_{Nd} reflecting metasomatism of highly depleted lithosphere. That signature is also present in peridotite xenoliths from Paleozoic kimberlites from the North China craton, but not in peridotites carried to the surface in the Cenozoic eruptions, indicating the Archean cratonic lithosphere was replaced by asthenosphere in the Mesozoic.

There are several possible sources of these metasomatizing agents. One would be upward percolation of small-degree melts from the asthenosphere directly below the lithosphere where low seismic velocities allow the possibility of the presence of a small melt fraction. Another would be small-degree melting derived from mantle plumes impinging on the base of the lithosphere, but too weak to penetrate or erode it. A third would be fluids derived from dehydration of subducting oceanic lithosphere or melts of the overlying mantle wedge. The latter seems to be the case for some eclogite xenoliths with surficial O and S isotopic compositions in South African kimberlites, indicating that subduction was active in the Archean.

Younger lithosphere, including the oceanic lithosphere beneath Hawaii and the lithosphere in regions of active extension and rifting in the Western United States, lacks the combination of negative γ_{Os} and ε_{Nd} values that characterizes Archean lithosphere, although they may nevertheless record incompatible element enrichment.

The North China craton is an interesting case. Scattered surface exposures of Archean crustal rocks across the region testify to the antiquity of the crust there. The thick crust and lithosphere and low heat flow in western region of the craton are characteristics of Archean lithosphere, but to the east heat flow and crust and lithospheric are atypically higher and thinner, respectively. Peridotite xenoliths in Paleozoic kimberlites from the eastern region have old T_{RD} ages and negative γ_{Os} and ε_{Nd} characteristics of Archean SCLM. In contrast, most xenoliths in Cenozoic and Cretaceous basalts from the same region have γ_{Os} and ε_{Nd} falling within the field of modern abyssal peridotites and display a range of Re depletion ages from Paleoproterozoic to the present (e.g., Wu et al., 2006; Zhang et al., 2008 and Liu et al., 2019). There is a widespread consensus that this reflects replacement of the Archean mantle lithosphere during the Mesozoic by modern ambient asthenospheric mantle, although the mechanism is debated. Eastern Asia has been tectonically active throughout most of the Phanerozoic as a series of terranes have accreted to its southern margin through the Paleozoic and early Mesozoic. Subduction of the Pacific plate from the east began in the Early Jurassic and mechanisms have been proposed that result in various extents of replacement of the North China lithosphere by asthenosphere (e.g., Liu et al., 2019).

Peridotites, including lherzolites, harzburgites, and dunites, from the lithosphere underlying Archean cratons have remarkably uniform olivine oxygen isotope ratios of 5.22 ± 0.22‰ (Regier, et al., 2018). Factoring in an olivine–melt fractionation factor of ≈0.3‰, this indicates the SCLM is indistinguishable in $\delta^{18}O$ from the MORB source mantle (MORB mean: +5.5‰).

Eclogites, on the other hand, show quite significant variation in stable isotope ratios, and we will consider their stable isotopic compositions, as well as those of diamonds, in more detail in Chapter 9. Fitzpayne et al. (2019) also found significant O and N isotopic variability in MARID metasomatic assemblages within peridotite xenoliths from the Bultfontein kimberlite of South Africa. $\delta^{18}O$ in secondary clinopyroxene and oxides are below typical mantle values and distinctly out of equilibrium with olivine. $\delta^{15}N$ values ranged from +4 to +7‰, compared to an asthenospheric value of about –5‰. Since O isotopic equilibrium should be achieved quickly at mantle temperatures, Fitzpayne et al. concluded the observed metamorphism occurred shortly before eruption. They speculated that the metasomatic fluids were derived from eclogites that had originated as subducted oceanic crust.

Most peridotites also have $^3He/^4He$ ratios equal to or lower than the MORB average after correction for radiogenic ingrowth since eruption, consistent with the generally low $^3He/^4He$ ratios observed in continental alkaline volcanics (Day et al., 2015). Thus, high $^3He/^4He$ seems exclusively associated with volcanism tied to mantle plumes, strengthening the case for a deep mantle origin.

7.6 U-SERIES ISOTOPES AND MELT GENERATION

We introduced the concepts of activity and radioactive equilibrium in the uranium decay series in Chapter 3. We noted that a system that has remained undisturbed for a sufficiently long time will achieve radioactive equilibrium – a condition such that:

$$a_p = a_d$$

where a is the radioactivity (more often simply called activity) and the subscripts p and d refer to the parent and daughter, respectively. Measurement of U series isotopes in young volcanic rocks reveals that radioactive *disequilibrium* among U decay series nuclides is common. Figure 7.35 shows that on a $^{230}Th/^{232}Th$–$^{238}U/^{232}Th$ plot, most ratios for oceanic basalts plot to the low ^{238}U side of the equiline (defined in Section 3.4.3.2). The disequilibrium is not entirely surprising, since we expect radioactive equilibrium will be disturbed by the melting process if the parent and daughter are partitioned between solid and melt differently. The magnitude of the disequilibrium is surprising. $^{230}Th/^{238}U$ ratios can exceed 1.4, although the mean value is somewhat lower, 1.15, and is similar for both OIBs and MORBs. On the other hand, disequilibrium between ^{234}U and ^{238}U is not expected; consequently, the ($^{234}U/^{238}U$) ratio can be used as a monitor of any post-eruptional disturbance through weathering and assimilation of crust.

The data from active Hawaiian volcanoes shown in the inset reveal that the volcanoes in their active tholeiitic shield-building and pre-shield-building states, Mauna Loa, Kilauea, and Loihi, plot close to the equiline in Figure 7.35, while volcanoes in their post-shield alkalic phase, Haleakala on Maui and Hualalai and Mauna Kea on Hawaii, show greater disequilibrium.

Even greater disequilibrium is apparent in Figure 7.36 between ^{230}Th and its daughter, ^{226}Ra, with ($^{226}Ra/^{230}Th$) ratios as high as 4 and, although not shown here, between ^{231}Pa and ^{235}U, with ($^{231}Pa/^{235}U$) ratios as high as 3.5. Given the short half-life of ^{226}Ra, 1600 years, this large disequilibrium is intriguing and requires very short time intervals between production of the disequilibrium and eruption. As Figure 7.36 shows, the highest ($^{226}Ra/^{230}Th$) ratios occur in MORBs, in particular from the fast-spreading East Pacific Rise and the very slow spreading Gakkel Ridge in the Arctic. The latter also have high ($^{230}Th/^{230}U$) ratios, but the East Pacific Rise data do not; instead, there is a rough negative

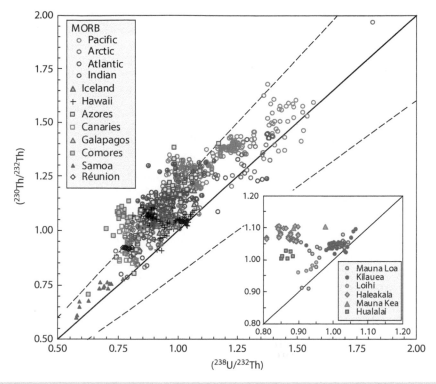

Figure 7.35 $(^{230}\text{Th}/^{232}\text{Th})$ versus $(^{238}\text{U}/^{232}\text{Th})$ ratios for oceanic basalts. Dashed lines show 20% disequilibria. Data from the compilation of Elkins et al. (2019) and the literature restricted to eruption ages less than 8000 years Before Present (BP). Plume-influenced MOR segments are shown with the $\oplus$ symbol.

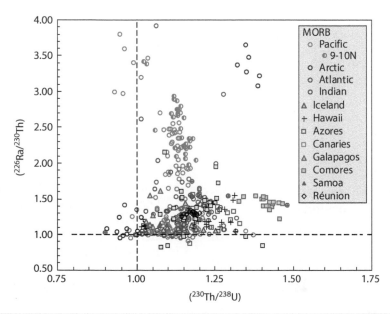

Figure 7.36 $(^{230}\text{Ra}/^{230}\text{Th})$ versus $(^{230}\text{Th}/^{238}\text{U})$ ratios for oceanic basalts. Data from the compilation of Elkins et al. (2019) and the literature restricted to eruption ages of OIBs less than 650 years BP. Plume-influenced MOR segments are shown with the $\oplus$ symbol.

correlation between $(^{226}\text{Ra}/^{230}\text{Th})$ and $(^{230}\text{Th}/^{230}\text{U})$ in the EPR data. An even clearer negative correlation is apparent, and we consider only the highest $(^{226}\text{Ra}/^{230}\text{Th})$ for a given $(^{230}\text{Th}/^{230}\text{U})$; doing so is justified because the measured $(^{226}\text{Ra}/^{230}\text{Th})$ activity in a sample is a minimum value of that produced in the mantle due to its rapid decay. Indeed, when the age-constrained data from the 9–10°N region of the EPR are considered separately (half-filled circles), there is a statistically significant negative correlation between $(^{226}\text{Ra}/^{230}\text{Th})$ and $(^{230}\text{Th}/^{230}\text{U})$.

Let us consider generation of U-series disequilibrium during melting in more detail. The simplest model of melting is *batch melting*, in which a fraction of melt, F, is created, and remains in contact and in equilibrium with the solid until melting is complete (melt fraction reaches F). Only then is the melt extracted. For this situation, we may write the following mass balance equation:

$$c_i^o = c_i^s(1 - F) + c_i^l F \qquad (7.17)$$

where c is the concentration of element i, F is the fraction of melt by weight, and the superscripts o, s, and l refer to the original solid (= system as a whole), the solid, and the melt, respectively. The partition or distribution coefficient is defined as:

$$D_i^{s/l} = c_i^s / c_i^l \qquad (7.18)$$

Substituting this into Equation (7.17) and upon rearranging, we obtain:

$$\frac{c_i^l}{c_i^o} = \frac{1}{D_i(1 - F) + F} \qquad (7.19)$$

Since the activity of i is simply $a_i = \lambda_i c_i$, we can substitute activities for concentrations.

If we assume that the system was in radioactive equilibrium before melting occurs, then:

$$a_P^o = a_D^o \qquad (7.20)$$

Then, the activity ratio of the parent and daughter in the melt is given by:

$$\frac{a_D^l}{a_P^l} = \frac{D_P(1 - F) + F}{D_D(1 - F) + F} \qquad (7.21)$$

From this, we can easily solve for F if we know the values of the partition coefficients.

Table 7.2 Partition coefficients for U-series nuclides. Note that Pa and Ra are highly incompatible and partition coefficients for them are very poorly known.

	Cpx	Gnt	Opx	Ol
Th	0.012	0.003	0.00002	0.00001
U	0.015	0.016	0.00005	0.00006
Pa	0.00001	0.00001	0.00001	0.00001
Ra	0.00002	0.00002	0.00002	0.00001

Source: Adapted from Salters et al. (2002) and Blundy and Wood (2003).

Table 7.2 lists solid/liquid partition coefficients for U and Th for major mantle minerals. In peridotites, essentially all U and Th will be in clinopyroxene and garnet and these will control fractionation between solid and melt. The U and Th partition coefficients for olivine and orthopyroxene are so low that virtually no U or Th is retained in them, regardless of the exact values of the partition coefficients. Experimentally determined partition coefficients for U and Th in clinopyroxene and garnet show a wide range (which is not surprising as we theoretically expect they should depend on a number of factors, including temperature, pressure, and composition), but there does seem to be a consistent relationship between them such that the Th and U partition coefficients for clinopyroxene are equal or nearly so while the Th partition coefficient for garnet is three to four times lower than that of U. Consequently, radioactive disequilibrium between ^{238}U and ^{230}Th will primarily occur where melting occurs in the presence of garnet; for peridotites, this will mainly be below ≈ 65 km (≈ 2 GPa). Both radium and protactinium are highly incompatible, significantly more so than U or Th, so larger disequilibria can be generated for ^{226}Ra–^{230}Th and ^{231}Pa–^{235}U.

The bulk partition coefficient, which we want to use here, is simply the average for each element weighted by the fraction of the mineral in the solid (i.e., weighted by the mineral *mode*):

$$\overline{D}_i = \sum_\chi m_\chi D_i^{\chi/l} \qquad (7.22)$$

where m_χ is the modal abundance of phase χ. Using the partition coefficients in Table 7.2, we find that producing $\approx 40\%$ disequilibrium between ^{238}U and ^{230}Th with simple batch melting requires very low extents of melting ($\approx 0.2\%$). While such low extents of melting might be reasonable under some circumstances, they certainly are not at mid-ocean ridges and vigorous volcanic systems, such a Kilauea. The oceanic crust is typically 6 km thick. To produce it by 0.2% melting would require efficient extraction of melt from 6/0.002 = 3000 km of mantle – the entire mantle! Clearly, if our partition coefficients are correct (there is certainly plenty of variability in partition coefficients – but even the most favorable ones would require <1% melting), then radioactive disequilibrium must involve other factors beyond the extent of melting. Let us now consider those.

7.6.1 Spiegelman and Elliot model of reactive melt transport

The problem of large disequilibrium between ^{230}Th and ^{238}U was first addressed theoretically by McKenzie (1985). He concluded that the large disequilibrium observed required slow melting compared to the half-life of ^{230}Th and no more than a few percent melt present in the mantle at any time. Spiegelman and Elliot (1993) developed a more complete model of melting and showed that large isotopic disequilibrium can result from differences in transport velocities of the elements and continued solid–melt equilibria and exchange as melt percolates upward through the melting column. The key to the model is that as melt rises by porous flow through rock, elements continually shuttle between melt and

solid to maintain chemical equilibrium. Let us consider the Spiegelman and Elliot model in more detail.

We begin by considering the physics of melting beneath mid-ocean ridges (the physics are similar for mantle plumes). Melting occurs at mid-ocean ridges as a consequence of decompression. Because the slope of the solidus is steeper than that of the adiabat (Figure 7.37), rising mantle eventually crosses the solidus. Melting begins at the point the mantle reaches the solidus and continues as the mantle continues to rise until it reaches the base of the lithosphere where temperature decreases due to conductive cooling. As melt is created, it will buoyantly rise. However, surface tension will force some fraction of melt to remain in the solid, resulting in a small but finite porosity of the solid. Melts streaming upward out of the mantle will eventually aggregate and mix in crustal magma chambers before erupting.

Spiegelman and Elliot (1993) began by writing a conservation equation for each parent–daughter pair as:

$$\frac{\partial[\phi \rho_m + (1-\phi)\rho_s D_i]}{\partial t} c_i^m + \nabla \cdot [\rho_m \phi v + \rho_s (1-\phi) D_i V] c_i^m =$$
$$\lambda_{i-1}[\rho_m \phi + \rho_s (1-\phi) D_{i-1}] c_{i-1}^m - \lambda_i [\rho_m \phi + \rho_s (1-\phi) D_i] c_i^m$$

$$(7.23)$$

where the subscript i denotes the element, c^m is the concentration of the element of interest in the melt, ∇ is the gradient operator, ρ_m is the density of the melt, ρ_s is the density of the solid, v is the velocity of the melt, V is the velocity of

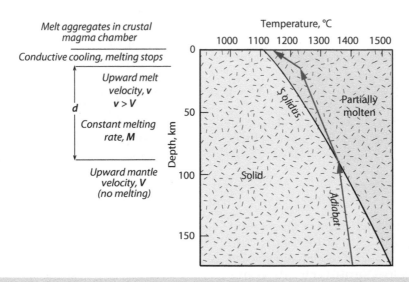

Figure 7.37 Schematic model of decompression melting beneath mid-ocean ridges and oceanic islands.

the solid, D is the partition coefficient, ϕ is the porosity, effectively the melt *volume* fraction, and, as usual, λ is the decay constant. The first nuclides in the chain, ^{238}U, have no parent, so that $\lambda_0 = 0$. In a one-dimensional steady-state system, the melt flux is simply the melt density times porosity times melt velocity:

$$\text{Melt flux} = \rho_m \phi v \qquad (7.24)$$

The mid-ocean ridge approximates such a one-dimensional (vertical) system, and for simplicity, we will restrict ourselves to one dimension. In this case, the melt flux is simply:

$$\rho_m \phi v = \dot{M}_0 z \qquad (7.25)$$

where $\dot{M}_0$ is the (constant) melting rate and z is height above the base of the melting column. Equation (7.25) is simply a statement that for a constant melting rate, the melt flux in an upwelling column of mantle must increase linearly with height. Mass balance then requires that upward flux of solid is given by:

$$\rho_s(1 - \phi)V = \rho_s V_0 - \dot{M}z \qquad (7.26)$$

where V_0 is the initial upwelling velocity of the solid (i.e., at the base of the melting column). Since melting is driven by decompression, the melting rate depends on upward velocity. Melting rate $\dot{M}_0$ can be expressed as:

$$\dot{M}_0 = \frac{\rho_s V_0 F_{max}}{h} \qquad (7.27)$$

where F_{max} is the maximum extent of melting (which occurs at the top of the column) and h is the height over which melting occurs (height of the melting column). We assume that the extent of melting increases with height so that:

$$F = F_{max} \frac{z}{h} \qquad (7.28)$$

From these relationships, we can derive the following relationship between melt velocity and porosity:

$$v = V_0 \frac{\rho_s F}{\rho_m \phi} \qquad (7.29)$$

Using Darcy's law to determine the relationship between porosity, permeability, and depth, Spiegelman and Elliot derived the following expression for porosity, effectively melt fraction, as a function of height:

$$\frac{F_{max} z}{h} = \frac{\rho_m}{\bar{\rho}} \left\{ \phi + \frac{\phi^n}{\phi_{max}^{n-1}(1 - \phi_{max})^2} \left[\frac{\bar{\rho} F_{max}}{\rho_m} - 1 \right] (1 - \phi)^2 \right\} \qquad (7.30)$$

where ϕ_{max} is the maximum porosity at the top of the column, $\bar{\rho}$ is the mean density at ϕ_{max}, and permeability is assumed to be proportional to the square of porosity. Thus, if ϕ_{max} and F_{max} are chosen, then the distribution of ϕ throughout the column is fixed.

Equation (7.27) may now be rewritten as:

$$\frac{dc_i'}{d\zeta} = c_i' \frac{(D_i - 1)F_{max}}{D_i + (1 - D_i)F_{max}\zeta}$$

$$+ \lambda_i h \left\{ \left[\frac{D_i[D_{i-1} + (1 - D_{i-1})F_{max}\zeta]}{D_{i-1}[D_i + (1 - D_i)F_{max}\zeta]} \right] \frac{c_{i-1}'}{v_{eff}^{i-1}} - \frac{c_i'}{v_{eff}^i} \right\} \qquad (7.31)$$

where $\zeta = z/h$ is the fractional height in the column, c' is the concentration of the element in the melt relative to its *initial* concentration in the melt, and v_{eff} is the *effective velocity* of the element through the melting column:

$$v_{eff}^i = \frac{\rho_m \phi v + \rho_s(1 - \phi)D_i V}{\rho_m \phi + \rho_s(1 - \phi)D_i} \approx V + \frac{v - V}{1 + D_i/\phi} \qquad (7.32)$$

The effective velocity of an element is just the average of the velocity of the melt and the solid, weighted by the fraction of the element in each. Equation (7.32) reveals that if $D_i \ll \phi$ then the element will spend most of its time in the melt and its effective velocity will approach that of the melt velocity. For an element with $D_i \gg \phi$, most of the element will be in the solid and the effective velocity will approach that of the solid. Thus, elements travel upward at different velocities depending on the partition coefficient. Very incompatible elements travel up through the melting column near the velocity of the melt; compatible elements travel upward more slowly at velocities near the solid velocity.

Assuming equilibrium between melt and solid, the initial concentration of element i in the melt, c_{i0}^m, is readily assessed from Equation (7.17) by setting $F = 0$. In that case, we see that:

$$\frac{c_{i0}^m}{c_i^0} = \frac{1}{D_i} \qquad (7.33)$$

Equation (7.35) written for each element of interest, i, forms a system or ordinary differential equations that can be solved numerically.

This model is calculation intensive, and Spiegelman (2000) produced a web-based calculator, *UserCalc*, for it. This has now been replaced by a more capable open-source, cloud-based Jupyter Notebook coded in Python, *pyUserCal* described by Elkins and Spiegelman (2021) and available in a Git repository (https://gitlab.com/ENKI338portal/pyUsercalc). The program includes several additional options over the original *UserCalc*: the ability to model varying amounts of disequilibrium between solid and melt and to include a lithosphere layer in which no melting occurs.

Let us consider an example using *pyUserCalc* in which melting begins at 40 kb (4 GPa, 123 km) and F_{max} is set to be 20% at the top of the column ($P = 0$). We assume initial radioactive equilibrium in the solid and full chemical equilibrium between melt and solid, and use *pyUserCalc*'s default bulk partition coefficients of $D_U = 0.009$, $D_{Th} = 0.005$ for the garnet peridotite facies below 20 kb, and $D_U = 0.005$, $D_{Th} = 0.004$ in the spinel peridotite facies, and $D_{Ra} = 0.0002$, $D_{Pa} = 0.0001$ throughout. We use default values for initial upwelling velocity, $V_0 = 3$ cm/yr, maximum porosity $\phi_{max} = 0.008$, $\rho_s = 3300$ kg/m^3, $\rho_m = 2800$ kg/m^3, and the permeability exponent $n = 2$. We also impose a 5 kb (18.5 km) thick lithosphere in which no melting occurs, but the melt can still interact with the rock using default partition coefficients of $D_U = 0.001$, $D_{Th} = 0.0014$, $D_{Ra} = 0.00001$, and $D_{Pa} = 0.00001$. The solid lines in Figure 7.38 show the results of this model. Note that melt fraction does not reach 20% because of the imposed lithosphere. The final results are $(^{230}Th/^{238}U) = 1.12$, $(^{231}Pa/^{235}U) = 1.74$, and $(^{226}Ra/^{230}Th) = 1.21$. Repeating this experiment with different values of F_{max}, we find that the results do not strongly depend on extent of melting. What do they depend on?

Radioactive disequilibrium does depend on the relative effective velocities (Equation [7.39]) of the parent and daughter through the melting column. In this example, the melt velocity, which increases as F and ϕ increase, reaches a maximum of 72 cm/yr, or 24 times greater than the mantle upwelling velocity. The effective velocity of Th is 44% of the melt velocity, while that of U is only 33% of the melt velocity. In other words, the upward velocity of Th is a third greater than that of U. The longer residence time of U in the column leads to an excess of the daughter nuclide.

The times required for solid and melt to traverse the melting zone are ≈4 Ma and ≈0.35 Ma, respectively, but both Th and U travel more slowly than the melt velocity. U transits the melting column (123 km) in ≈1 Ma while Th transits it in 0.77 Ma. The latter time is approximately the half-life of ^{230}Th, so half of the ^{230}Th atoms present at the bottom of the column have decayed in this time. Because it is highly incompatible, almost all the ^{231}Pa is in the melt phase, and the transit time for ^{231}Pa is much faster than that of Th, ≈0.37 Ma, approaching that of the melt. This is more than 10 times the half-life of ^{231}Pa, 32480 years, so most of the ^{231}Pa has been produced in the upper part of the melting column; indeed, about half would have been produced during transit of the lithosphere. ^{226}Ra is similarly highly incompatible with a transit time nearly as fast, 0.38 Ma. Given it short half-life, 1622 years, essentially all the ^{226}Ra would have been produced while transiting the lithosphere (and during magma chamber storage, which is not included in the model).

In the model shown in Figure 7.38, the $(^{226}Ra/^{230}Th)$ ratio decreases rapidly in the lithosphere because we have defined a Th partition coefficient in the lithosphere that is much lower than in the melting region so that melt–rock reaction in the lithosphere produced little additional disequilibrium; meanwhile, ^{226}Ra in the melt is decaying and approaching equilibrium with ^{230}Th. If we instead run the model with only a 6 km thick crust, we obtain a higher final $(^{226}Ra/^{230}Th)$ of 1.79. This value is still substantially lower than some ratios observed in Pacific MORBs. This high $(^{226}Ra/^{230}Th)$ may reflect high spreading rates, which in turn results in high eruption rates and rapid transport of magma through thin crust and lithosphere.

We can compare these results with cases where full chemical equilibrium is not achieved, which might occur due to slow diffusion within mineral grains or wide spacing between melt channels. The Damköhler number, Da, is a measure of this disequilibrium; Da is 0 at full disequilibrium and infinity at full

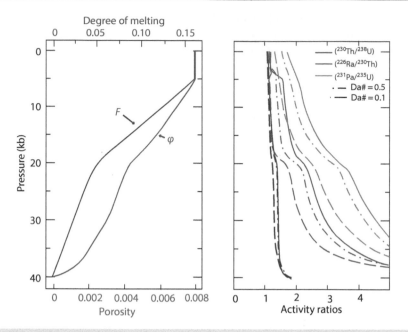

Figure 7.38 Output adapted from Elkins and Spiegleman (2021) *pyUserCalc*. The kink reflects the transition from garnet to spinel peridotite at 20 kb where D_U and D_{Th} change from 0.009 to 0.005 and from 0.005 to 0.004, respectively. Maximum melting is set at 20%, upwelling velocity at 3 cm/yr, and maximum porosity at 0.008. A 5 kb thick lithosphere is also imposed. Kinks in the curves reflect the phase changes. Solid lines illustrate the case of full equilibrium and dashed lines show partial disequilibrium as indicated by the Damköhler number, *Da*.

equilibrium. Figure 7.38 also illustrates cases for *Da* = 0.5 and *Da* = 0.1. Chemical disequilibrium leads to less radioactive disequilibrium because the melt is not equilibrating with the solid.

PyUserCalc also allows us to explore the upwelling rate and porosity parameter space in "batch operations" by running the model multiple times while varying these parameters. Figure 7.39 shows the results for this same model run. This reveals that maximum $(^{230}Th/^{238}U)$ disequilibrium occurs at low porosities and relatively high upwelling rates of 4–10 cm/yr. Large $(^{226}Ra/^{230}Th)$ disequilibrium also requires low porosities, but it is favored by very high upwelling rates. $^{231}Pa/^{235}U$ disequilibrium (not shown) is maximum at low porosities and upwelling rates of ≈2 cm/yr.

7.6.2 Variations on a theme: other models of U-series disequilibrium

The combination of very high $(^{226}Ra/^{230}Th)$ and low $(^{230}Th/^{238}U)$ observed in EPR MORBs from the 9–10°N region, however, is not readily explained by any combination of upwelling rates and porosities in this model. Jull et al. (2002) proposed a model in which melt is transported by a combination of diffuse intergranular flow at low porosity and channeled flow at high porosity to explain these observations. In their model, high porosity (≳7%) channels develop near the garnet–spinel transition at around 70 km depth when melt fraction reaches ≈3%, and transport ≈40% of the melt to the surface rapidly with little or no equilibration with surrounding rock while the remaining 60% travels via very low porosity (<1%) intergranular flow and continually equilibrates with surrounding rock. Mixing between the channelized flow with high $(^{226}Ra/^{230}Th)$ and low $(^{230}Th/^{238}U)$ and the intergranular flow with low $(^{226}Ra/^{230}Th)$ and high $(^{230}Th/^{238}U)$ results in the negative correlation. Alternatively, Stracke et al. (2006) noted that melting is not strictly one-dimensional beneath mid-ocean ridges (nor beneath other volcanoes), but rather occurs throughout a region of finite width (which conductivity measurements suggest might be as wide as 100 km beneath the EPR). Upwelling

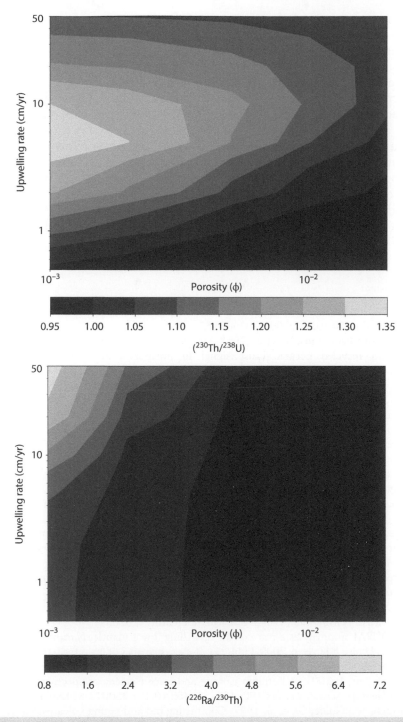

Figure 7.39 Output adapted from Elkins and Spiegleman (2021) *pyUserCalc* model showing the dependence of $(^{230}\text{Th}/^{238}\text{U})$ and $(^{226}\text{Ra}/^{230}\text{Th})$ on upwelling rate and porosity. Maximum $(^{230}\text{Th}/^{238}\text{U})$ disequilibrium occurs at low porosity and intermediate upwelling rates while maximum $(^{226}\text{Ra}/^{230}\text{Th})$ occurs at low porosity and high upwelling rates.

velocity necessarily decreases with distance from the ridge axis and melts created far from the ridge axis travel over greater distances than melts created close to the ridge axis. They found that the inverse correlation could be explained by mixing of melts produced at high upwelling rates and short transport times with high $(^{226}\text{Ra}/^{230}\text{Th})$ and low $(^{230}\text{Th}/^{238}\text{U})$

directly beneath ridge crests and those produced as off axis at lower upwelling rates and long transport times and low (^{226}Ra/^{230}Th) and high (^{230}Th/^{238} U).

Although the models discussed up to this point can explain much of the radioactive disequilibrium observed in oceanic basalts, it is difficult to explain all of it. There has been considerable speculation and debate over the last two decades in the literature over the role of lithologic heterogeneity in the mantle, specifically regions of eclogite, also called garnet pyroxenite, embedded within the dominant mantle peridotite. Given the enormous amount of oceanic crust that has been subducted, we certainly expect the presence of recycled crust, now transformed to eclogite, in the mantle. Because partitioning of U and Th is controlled primarily by clinopyroxene and garnet, garnet-rich lithologies will lead to larger U–Th fractionations than will peridotite. For example, Elkins et al. (2019) showed that reactive porous flow and two-porosity models of melting of peridotite-dominated mantle regimes with contributions from garnet pyroxenite partial melts to aggregated melts could produce (^{230}Th/^{238} U) as high as 1.4 and (^{226}Ra/^{230}Th) ratios well above those observed in oceanic basalts.

REFERENCES

Armstrong, R.L. 1968. A model for the evolution of strontium and lead isotopes in a dynamic earth. *Reviews of Geophysics* **6**: 175–99. doi: 10.1029/RG006i002p00175.

Beaudry, P., Longpré, M.-A., Economos, R., et al. 2018. Degassing-induced fractionation of multiple sulphur isotopes unveils post-Archaean recycled oceanic crust signal in hotspot lava. *Nature Communications* **9**: 5093. doi: 10.1038/s41467-018-07527-w.

Bellini, G., Ianni, A., Ludhova, L., et al. 2013. Geo-neutrinos. *Progress in Particle and Nuclear Physics* **73**: 1–34 doi: 10.1016/j.ppnp.2013.07.001.

Bellot, N., Boyet, M., Doucelance, R., et al. 2015. Ce isotope systematics of island arc lavas from the Lesser Antilles. *Geochimica et Cosmochimica Acta* **168**: 261–79. doi: http://dx.doi.org/10.1016/j.gca.2015.07.002.

Blichert-Toft, J., Zanda, B., Ebel, D.S., et al. 2010. The Solar System primordial lead. *Earth and Planetary Science Letters* 300: 152–63. doi: 10.1016/j.epsl.2010.10.001.

Blundy, J. and Wood, B. 2003. Mineral-melt partitioning of uranium, thorium and their daughters. *Reviews in Mineralogy and Geochemistry* **52**: 59–123. doi: 10.2113/0520059.

Boyet, M. and Carlson, R.L. 2005. ^{142}Nd evidence for early (>4.3 Ga) global differentiation of the silicate Earth. *Science* **309**: 576–81.

Boyet, M., Doucelance, R., Israel, C., et al. 2019. New constraints on the origin of the em-1 component revealed by the measurement of the La-Ce isotope systematics in Gough Island lavas. *Geochemistry, Geophysics, Geosystems.* **20**: 2484– 98. doi: 10.1029/2019gc008228.

Burke, K. 2011. Plate tectonics, the Wilson cycle, and mantle plumes: geodynamics from the top. *Annual Review of Earth and Planetary Sciences* **39**: 1–29. doi: 10.1146/annurev-earth-040809-152521.

Cabral, R.A., Jackson, M.G., Rose-Koga, E.F., et al. 2013. Anomalous sulphur isotopes in plume lavas reveal deep mantle storage of Archaean crust. *Nature* **496**: 490–3. doi: 10.1038/nature12020.

Castillo, P. 1989. The DUPAL anomaly as a trace of the upwelling lower mantle. *Nature* **336**: 667–70.

Chan, L.-H., Lassiter, J.C., Hauri, E.H., et al. 2009. Lithium isotope systematics of lavas from the Cook–Austral islands: constraints on the origin of HIMU mantle. *Earth and Planetary Science Letters* **277**: 433–42. doi: 10.1016/j.epsl.2008.11.009.

Chauvel, C. and Blichert-Toft, J. 2001. A hafnium isotope and trace element perspective on melting of the depleted mantle. *Earth and Planetary Science Letters* **190**: 137–51. doi: 10.1016/S0012-821X(01)00379-X

Chauvel, C., McDonough, W., Guille, G., et al. 1997. Contrasting old and young volcanism in Rurutu island, Austral chain. *Chemical Geology* **139**: 125–43. doi: 10.1016/S0009-2541(97)00029-6.

Clarke, W.B., Beg, M.A. and Craig, H. 1969. Excess ^{3}He in the sea: evidence for terrestrial primodal helium. *Earth and Planetary Science Letters* **6**: 213–20. doi: 10.1016/0012-821X(69)90093-4.

Class, C. and Goldstein, S.L. 2005. Evolution of helium isotopes in the Earth's mantle. *Nature* **436**: 1107–12. doi: 10.1038/nature03930.

Clog, M., Aubaud, C., Cartigny, P., et al. 2013. The hydrogen isotopic composition and water content of southern Pacific MORB: A reassessment of the D/H ratio of the depleted mantle reservoir. *Earth and Planetary Science Letters* **381**: 156–65. doi: 10.1016/j.epsl.2013.08.043.

Cooper, K.M., Eiler, J.M., Sims, K.W.W., et al. 2009. Distribution of recycled crust within the upper mantle: Insights from the oxygen isotope composition of MORB from the Australian-Antarctic Discordance. *Geochemistry, Geophysics, Geosystems* **10**. doi: 10.1029/2009GC002728.

Corgne, A., Liebske, C., Wood, B.J., et al. 2005. Silicate perovskite-melt partitioning of trace elements and geochemical signature of a deep perovskitic reservoir. *Geochimica et Cosmochimica Acta* **69**: 485–96.

Craig, H., Clarke, W.B. and Beg, M.A. 1975. Excess ^{3}He in deep water on the East Pacific Rise. *Earth and Planetary Science Letters* **26**: 125–132. doi: 10.1016/0012-821X(75)90079-5.

Dauphas, N. 2017. The isotopic nature of the Earth's accreting material through time. *Nature* **541**: 521–4. doi: 10.1038/nature20830

Day, J.M.D. 2013. Hotspot volcanism and highly siderophile elements. *Chemical Geology* **341**: 50–74. doi: 10.1016/j.chemgeo.2012.12.010.

Day, J.M.D., Hilton, D.R., Pearson, D.G., et al. 2005. Absence of a high time-integrated ^{3}He/(U+Th) source in the mantle beneath continents. *Geology* **33**: 733–6. doi: 10.1130/g21625.1.

Day, J.M.D., Pearson, D.G., Macpherson, C.G., et al. 2009. Pyroxenite-rich mantle formed by recycled oceanic lithosphere: oxygen-osmium isotope evidence from Canary Island lavas. *Geology* **37**: 555–8. doi: 10.1130/g25613a.1.

Delavault, H., Chauvel, C., Thomassot, E., et al. 2016. Sulfur and lead isotopic evidence of relic Archean sediments in the Pitcairn mantle plume. *Proceedings of the National Academy of Sciences* **113**: 12952–6. doi: 10.1073/pnas.1523805113.

DePaolo, D.J. 1980. Crustal growth and mantle evolution: inferences from models of element transport and Nd and Sr isotopes. *Geochimica et Cosmochimica Acta* **44**: 1185–96.

Dixon, J.E., Bindeman, I.N., Kingsley, R.H., et al. 2017. Light stable isotopic compositions of enriched mantle sources: Resolving the dehydration paradox. *Geochemistry, Geophysics, Geosystems* **18**: 3801–39. doi: 10.1002/2016GC006743.

Dottin, J.W., Labidi, J., Jackson, M.G., et al. 2020a. Isotopic evidence for multiple recycled sulfur reservoirs in the Mangaia mantle plume. *Geochemistry, Geophysics, Geosystems* **21**: e2020GC009081. doi: 10.1029/2020GC009081.

Dottin, J.W., Labidi, J., Lekic, V., et al. 2020b. Sulfur isotope characterization of primordial and recycled sources feeding the Samoan mantle plume. *Earth and Planetary Science Letters* **534**: 116073. doi: 10.1016/j.epsl.2020.116073.

Doucelance, R., Bellot, N., Boyet, M., et al. 2014. What coupled cerium and neodymium isotopes tell us about the deep source of oceanic carbonatites. *Earth and Planetary Science Letters* **407**: 175–86. doi: 10.1016/j.epsl.2014.09.042.

Doucet, L.S., Ionov, D.A. and Golovin, A.V. 2015. Paleoproterozoic formation age for the siberian cratonic mantle: Hf and Nd isotope data on refractory peridotite xenoliths from the Udachnaya kimberlite. *Chemical Geology* **391**: 42–55. doi: 10.1016/j.chemgeo.2014.10.018.

Dupré, B. and Allègre, C.J. 1983. Pb-Sr isotope variations in Indian Ocean basalts and mixing phenomena. *Nature* **303**: 142–6.

Dye, S.T. 2012. Geoneutrinos and the radioactive power of the Earth. *Reviews of Geophysics* **50**: RG3007. doi: 10.1029/2012rg000400.

Dziewonski, A.M., Lekic, V. and Romanowicz, B.A. 2010. Mantle anchor structure: an argument for bottom up tectonics. *Earth and Planetary Science Letters* **299**: 69–79. doi: 10.1016/j.epsl.2010.08.013.

Elkins, L.J. and Spiegelman, M. 2021. pyUserCalc: a revised Jupyter notebook calculator for uranium-series disequilibria in basalts. *Earth and Space Science* **8**: e2020EA001619. doi: https://doi.org/10.1029/2020EA001619.

Elkins, L.J., Bourdon, B. and Lambart, S. 2019. Testing pyroxenite versus peridotite sources for marine basalts using u-series isotopes. *Lithos* **332–3**: 226–44. doi: https://doi.org/10.1016/j.lithos.2019.02.011.

Farley, K.A., Natland, J.H. and Craig, H. 1992. Binary mixing of enriched and undegassed (primitive?) mantle components (He, Sr, Nd, Pb) in Samoan lavas. *Earth and Planetary Science Letters* **111**: 183–99. doi: 10.1016/0012-821X(92)90178-X.

Farquhar, J. and Wing, B.A. 2003. Multiple sulfur isotopes and the evolution of atmospheric oxygen. *Earth and Planetary Science Letters* **213**: 1–13. doi: 10.1126/science.1078617.

Farquhar, J., Wing, B.A., Mckeegan, K.D., et al. 2002. Mass-independent fractionation sulfur of inclusions in diamond and sulfur recycling on early earth. *Science* **298**: 2369–72. doi: 10.1126/science.1078617.

Fitzpayne, A., Giuliani, A., Harris, C., et al. 2019. Evidence for subduction-related signatures in the southern African lithosphere from the N-O isotopic composition of metasomatic mantle minerals. *Geochimica et Cosmochimica Acta* **266**: 237–57. doi: https://doi.org/10.1016/j.gca.2019.02.037.

Fouch, M.J., James, D.E., Vandecar, J.C., et al. 2004. Mantle seismic structure beneath the Kaapvaal and Zimbabwe cratons. *South African Journal of Geology* **107**: 33–44. doi: 10.2113/107.1-2.33.

French, S.W. and Romanowicz, B. 2015. Broad plumes rooted at the base of the earth's mantle beneath major hotspots. *Nature* **525**: 95–9. doi: 10.1038/nature14876.

Galer, S.J.G. and O'Nions, R.K. 1985. Residence time of thorium, uranium and lead in the mantle with implications for mantle convection. *Nature* **316**: 778–82.

Gannoun, A., Burton, K.W., Day, J.M.D., et al. 2016. Highly siderophile element and Os isotope systematics of volcanic rocks at divergent and convergent plate boundaries and in intraplate settings. *Reviews in Mineralogy and Geochemistry* **81**: 651–724. doi: 10.2138/rmg.2016.81.11.

Gast, P.W. 1960. Limitations on the composition of the upper mantle. *Journal of Geophysical Research* **65**: 1287–97.

Genske, F.S., Beier, C., Haase, K.M., et al. 2013. Oxygen isotopes in the Azores Islands: crustal assimilation recorded in olivine. *Geology* **41**: 491–4. doi: 10.1130/g33911.1.

Griffin, W.L., O'Reilly, S.Y., Afonso, J.C., et al. 2009. The composition and evolution of lithospheric mantle: a re-evaluation and its tectonic implications. *Journal of Petrology* **50**: 1185–204. doi: 10.1093/petrology/egn033.

Halliday, A.N. 2004. Mixing, volatile loss and compositional change during impact-driven accretion of the Earth. *Nature* **427**: 505–9.

Hanan, B.B. and Graham, D.W. 1996. Lead and helium isotope evidence from oceanic basalts for a common deep source of mantle plumes. *Science* **272**: 991–5.

Harpp, K.S. and Weis, D. 2020. Insights into the origins and compositions of mantle plumes: a comparison of Galápagos and Hawai'i. *Geochemistry, Geophysics, Geosystems* **21**. doi: 10.1029/2019gc008887.

Harpp, K.S. and White, W.M. 2001. Tracing a mantle plume: isotopic and trace element variations of Galápagos seamounts. *Geochemistry, Geophysics, Geosystems* **2**: 1042. doi: 10.1029/2000GC000137.

Harrison, L., Weis, D., Hanano, D., et al. 2015. The lithium isotope signature of Hawaiian basalts, in Carey, R., Cayol, V., Poland, M., et al. (eds). *Hawaiian Volcanoes from Source to Surface*. Washington: AGU.

Hart, S.R. 1984. The DUPAL anomaly: a large-scale isotopic mantle anomaly in the Southern Hemisphere. *Nature* **309**: 753–57.

Hart, S.R., Gerlach, D.C. and White, W.M. 1986. A possible new Sr-Nd-Pb mantle array and consequences for mantle mixing. *Geochimica et Cosmochimica Acta* **50**: 1551–7. doi: 10.1016/0016-7037(86)90329-7.

Hart, S.R., Hauri, E.H., Oschmann, L.A., et al. 1992. Mantle plumes and entrainment: isotopic evidence. *Science* **256**: 517–20.

Hart, S.R., Schilling, J.-G. and Powell, J.L. 1973. Basalts from Iceland and along the Reykjanes Ridge: Sr isotope geochemistry. *Nature Physical Science* **246**:104–7.

Hauri, E. 1996. Major element variability in the Hawaiian mantle plume. *Nature* **382**: 415–9.

Hedge, C.E., Watkins, N.D., Hildreth, R.A., et al. 1973. $^{87}Sr/^{86}Sr$ ratios in basalts from islands in the Indian ocean. *Earth and Planetary Science Letters* **21**: 29–34. doi: 10.1016/0012-821X(73)90222-7.

Herzberg, C. and Rudnick, R. 2012. Formation of cratonic lithosphere: an integrated thermal and petrological model. *Lithos* **149**: 4–15. doi: 10.1016/j.lithos.2012.01.010.

Herzberg, C., Asimow, P.D., Arndt, N., et al. 2007. Temperatures in ambient mantle and plumes: Constraints from basalts, picrites, and komatiites. *Geochemistry, Geophysics, Geosystems* **8**. doi: 10.1029/2006gc001390.

Hoernle, K., Werner, R., Morgan, J.P., et al. 2000. Existence of complex spatial zonation in the Galapagos plume for at least 14 M.Y. *Geology* **28**: 435–8. doi: 10.1130/0091-7613(2000)28<435:EOCSZI>2.0.CO;2

Hofmann, A.W. and White, W.M. 1982. Mantle Plumes from ancient oceanic crust. *Earth and Planetary Science Letters* **57**: 421–36.

Hyung, E. and Jacobsen, S.B. 2020. The $^{142}Nd/^{144}Nd$ variations in mantle-derived rocks provide constraints on the stirring rate of the mantle from the Hadean to the present. *Proceedings of the National Academy of Sciences* **117**: 14738–44. doi: 10.1073/pnas.2006950117.

Ionov, D.A., Carlson, R.W., Doucet, L.S., et al. 2015. The age and history of the lithospheric mantle of the Siberian craton: re–Os and PGE study of peridotite xenoliths from the Obnazhennaya kimberlite. *Earth and Planetary Science Letters* **428**: 108–19. doi: https://doi.org/10.1016/j.epsl.2015.07.007.

Israel, C., Boyet, M., Doucelance, R., et al. 2020. Formation of the Ce-Nd mantle array: crustal extraction vs. recycling by subduction. *Earth and Planetary Science Letters* **530**: 115941. doi: 10.1016/j.epsl.2019.115941.

Jackson, M.G., Blichert-Toft, J., Halldórsson, S.A., et al. 2020. Ancient helium and tungsten isotopic signatures preserved in mantle domains least modified by crustal recycling. *Proceedings of the National Academy of Sciences* **117**: 30993–31001. doi: 10.1073/pnas.2009663117.

Jackson, M.G., Hart, S.R., Koppers, A.A.P., et al. 2007. The return of subducted continental crust in Samoan lavas. *Nature* **448**: 684–7.

Jull, M., Kelemen, P.B. and Sims, K. 2002. Consequences of diffuse and channelled porous melt migration on uranium series disequilibria. *Geochimica et Cosmochimica Acta* **66**: 4133–48. doi: 10.1016/S0016-7037(02)00984-5.

Kay, R.W. and Kay, S.M. 1993. Delamination and delamination magmatism. *Tectonophysics* **219**: 177–89. doi: https://doi.org/10.1016/0040-1951(93)90295-U.

Kay, S.M., Godoy, E. and Kurtz, A. 2005. Episodic arc migration, crustal thickening, subduction erosion, and magmatism in the south-central Andes. *Geological Society of America Bulletin* **117**: 67–88. doi: 10.1130/b25431.1.

Kingsley, R.H., Schilling, J.-G., Dixon, J.E., et al. 2002. D/H ratios in basalt glasses from the Salas y Gomez mantle plume interacting with the East Pacific Rise: water from old d-rich recycled crust or primordial water from the lower mantle? *Geochemistry, Geophysics, Geosystems* **3**: 1–26. doi: 10.1029/2001GC000199.

Klein, E.M. and Langmuir, C.H. 1987. Ocean ridge basalt chemistry, axial depth, crustal thickness, and temperature variations in the mantle. *Journal of Geophysical Research* **92**: 8089–115.

Klein, E.M., Langmuir, C.H., Zindler, A., et al. 1988. Isotope evidence of a mantle convection boundary at the Australian-Antarctic Discordance. *Nature* **333**: 623–9.

Krienitz, M.-S., Garbe-Schönberg, C.-D., Romer, R.L., et al. 2012. Lithium isotope variations in ocean island basalts—implications for the development of mantle heterogeneity. *Journal of Petrology* **53**: 2333–47. doi: 10.1093/petrology/egs052.

Kyser, K.T. 1986. Stable isotope variations in the mantle, in Valley, J.W., Taylor, H.P. and O'neil, J.R. (eds). *Stable Isotopes in High Temperature Geologic Processes*. Washington: Mineralogical Association of America.

Labidi, J., Cartigny, P. and Jackson, M.G. 2015. Multiple sulfur isotope composition of oxidized Samoan melts and the implications of a sulfur isotope 'mantle array' in chemical geodynamics. *Earth and Planetary Science Letters* **417**: 28–39. doi: 10.1016/j.epsl.2015.02.004.

Labidi, J., Cartigny, P. and Moreira, M. 2013. Non-chondritic sulphur isotope composition of the terrestrial mantle. *Nature* **501**: 208–11. doi: 10.1038/nature12490.

Liu, J., Cai, R., Pearson, D.G., et al. 2019. Thinning and destruction of the lithospheric mantle root beneath the North China craton: a review. *Earth-Science Reviews* **196**: 102873. doi: https://doi.org/10.1016/j.earscirev.2019.05.017.

Loewen, M.W., Graham, D.W., Bindeman, I.N., et al. 2019. Hydrogen isotopes in high ^{3}He/^{4}He submarine basalts: primordial vs. recycled water and the veil of mantle enrichment. *Earth and Planetary Science Letters* **508**: 62–73. doi: 0.1016/j.epsl.2018.12.012.

Makishima, A. and Masuda, A. 1994. Ce isotope ratios of N-type MORB. *Chemical Geology* **118**: 1–8. doi: 10.1016/0009-2541(94)90166-X.

Maltese, A. and Mezger, K. 2020. The Pb isotope evolution of bulk silicate Earth: constraints from its accretion and early differentiation history. *Geochimica et Cosmochimica Acta* **271**: 179–93. doi: https://doi.org/10.1016/j.gca.2019.12.021.

McDonough, W.F. and Sun, S.-S. 1995. The composition of the Earth. *Chemical Geology* **120**: 223–53. doi: 10.1016/0009-2541(94)00140-4.

McKenzie, D.P. and O'Nions, R.K. 1983. Mantle reservoirs and ocean island basalts. *Nature* **301**: 229–31.

McKenzie, D.P. 1985. ^{230}Th-^{238}U disequilibrium and the melting processes beneath ridge axes *Earth and Planetary Science Letters* **72**: 149–57.

McNamara, A.K. 2019. A review of large low shear velocity provinces and ultra-low velocity zones. *Tectonophysics* **760**: 199–220. doi: 10.1016/j.tecto.2018.04.015.

Meyzen, C.M., Blichert-Toft, J., Ludden, J.N., et al. 2007. Isotopic portrayal of the Earth's upper mantle flow field. *Nature* **447**: 1069–74. doi: 10.1038/nature05920

Montelli, R., Nolet, G., Dahlen, F.A., et al. 2004. Finite-frequency tomography reveals a variety of plumes in the mantle. *Science* **303**: 338–43. doi: 10.1126/science.1092485.

Morgan, W.J. 1971. Convection plumes in the lower mantle. *Nature* **230**: 42–3.

Mundl-Petermeier, A., Walker, R.J., Fischer, R.A., et al. 2020. Anomalous ^{182}W in high ^{3}He/^{4}He ocean island basalts: fingerprints of earth's core? *Geochimica et Cosmochimica Acta* **271**: 194–211. doi: 10.1016/j.gca.2019.12.020.

Mundl, A., Touboul, M., Jackson, M.G., et al. 2017. Tungsten-182 heterogeneity in modern ocean island basalts. *Science* **356**: 66–9. doi: 10.1126/science.aal4179.

O'Hara, M.J. 1973. Non-primary magmas and dubious mantle plume beneath Iceland. *Nature* **243**: 507–8. doi: 10.1038/243507a0.

O'Neill, H.S.C. and Palme, H. 2008. Collisional erosion and the non-chondritic composition of the terrestrial planets. *Philosophical Transactions of the Royal Society A* **366**: 4205–38. doi: 10.1098/rsta.2008.0111.

Ono, S., Keller, N.S., Rouxel, O., et al. 2012. Sulfur-33 constraints on the origin of secondary pyrite in altered oceanic basement. *Geochimica et Cosmochimica Acta* **87**: 323–40. doi: https://doi.org/10.1016/j.gca.2012.04.016.

Palme, H. and O'Neill, H.S.C. 2003. Cosmochemical estimates of mantle composition, in Heinrich, D.H. and Karl, K.T. (eds). *Treatise on Geochemistry: the mantle and core*. Amsterdam: Elsevier.

Paul, D., White, W.M. and Turcotte, D.L. 2002. Modelling the Pb isotopic composition of the Earth. *Philosophical Transactions of the Royal Society of London A* **360**: 2433–74.

Pearson, D.G. and Wittig, N. 2014. The formation and evolution of cratonic mantle lithosphere – evidence from mantle xenoliths, in Holland, H.D. and Turekian, K.K. (eds). *Treatise on Geochemistry*, 2nd ed. Oxford: Elsevier.

Regier, M.E., Mišković, A., Ickert, R.B., et al. 2018. An oxygen isotope test for the origin of Archean mantle roots. *Geochemical Perspectives Letters* **9**: 6–10. doi: 10.7185/geochemlet.1830.

Rizo, H., Andrault, D., Bennett, N.R., et al. 2019. ^{182}W evidence for core-mantle interaction in the source of mantle plumes. *Geochemical Perspectives Letters* **11**: 6–11. doi: http://dx.doi.org/10.7185/geochemlet.1917.

Rose, J. and Koppers, A.a.P. 2019. Simplifying age progressions within the Cook-Austral Islands using ARGUS-VI high-resolution ^{40}Ar/^{39}Ar incremental heating ages. *Geochemistry, Geophysics, Geosystems* **20**: 4756–4778. doi: https://doi.org/10.1029/2019GC008302.

Roth, A.S.G., Liebske, C., Maden, C., et al. 2019. The primordial He budget of the earth set by percolative core formation in planetesimals. *Geochemical Perspectives Letters* **9**: 26–31. doi: 10.7185/geochemlet.1901.

Saji, N.S., Wielandt, D., Holst, J.C., et al. 2020. Solar system Nd isotope heterogeneity: insights into nucleosynthetic components and protoplanetary disk evolution. *Geochimica et Cosmochimica Acta* **281**: 135–48. doi: 10.1016/j. gca.2020.05.006.

Salters, V.J.M. and Hart, S.R. 1989. The hafnium paradox and the role of garnet in the source of mid-ocean-ridge basalts. *Nature* **342**: 420–422. doi: 10.1038/342420a0.

Salters, V.J.M., Longhi, J.E. and Bizimis, M. 2002. Near mantle solidus trace element partitioning at pressures up to 3.4 GPa. *Geochemistry, Geophysics, Geosystems* **3**: 1–23. doi: 10.1029/2001GC000148.

Salters, V.J.M., Mallick, S., Hart, S.R., et al. 2011. Domains of depleted mantle: new evidence from hafnium and neodymium isotopes. *Geochemistry, Geophysics, Geosystems* **12**. doi: 10.1029/2011GC003617.

Savarino, J., Bekki, S., Cole-Dai, J., et al. 2003. UV induced mass-independent sulfur isotope fractionation in stratospheric volcanic sulfate. *Geophysical Research Letters* **30**: 11–14. doi: 10.1029/2003GL018134

Schilling, J.-G. 1973a. Iceland mantle plume: geochemical study of the Reykjanes ridge. *Nature* **242**: 565–71.

Schilling, J.-G. 1973b. Iceland mantle plume. *Nature* **246**: 141–3.

Schilling, J.-G. 1985. Upper mantle heterogeneities and dynamics. *Nature* **314**: 62–7.

Scholl, D.W., Von Huene, R., Vallier, T.L., et al. 1980. Sedimentary masses and concepts about tectonic processes at underthrust ocean margins. *Geology* **8**: 564–8. doi: 10.1130/0091-7613(1980)8<564:smacat>2.0.co;2.

Sleep, N.H. 1990. Hotspots and mantle plumes: some phenomenology. *Journal of Geophysical Research* **95**: 6715–36. doi: 10.1029/JB095iB05p06715.

Snow, J.E. and Reisberg, L. 1995. Os isotopic systematics of the MORB mantle: results from altered abyssal peridotites. *Earth and Planetary Science Letters* **133**: 411–21. doi: 10.1016/0012-821x(95)00099-x.

Sobolev, A.V., Hofmann, A.W., Sobolev, S.V., et al. 2005. An olivine-free mantle source of Hawaiian shield basalts. *Nature* **434**: 590–7. doi: 10.1038/nature03411.

Spiegelman, M. 2000. UserCalc: a web-based uranium series calculator for magma migration problems. *Geochemistry Geophysics, Geosystems* **1**: paper number 1999GC00030.

Spiegelman, M. and Elliot, T. 1993. Consequences of melt transport for uranium series disequilibrium in young lavas. *Earth and Planetary Science Letters* **118**: 1–20.

Šrámek, O., McDonough, W.F., Kite, E.S., et al. 2013. Geophysical and geochemical constraints on geoneutrino fluxes from Earth's mantle. *Earth and Planetary Science Letters* **361**: 356–66. doi: 10.1016/j.epsl.2012.11.001.

Stern, R.J. and Scholl, D.W. 2010. Yin and yang of continental crust creation and destruction by plate tectonic processes. *International Geology Review* **52**: 1–31. doi: 10.1080/00206810903332322.

Stille, P., Unruh, P.M. and Tatsumoto, M. 1983. Pb, Sr, Nd and Hf isotopic evidence of multiple sources for Oahu basalts, *Hawaii. Nature* **304**: 25–9.

Stracke, A., Bourdon, B. and Mckenzie, D. 2006. Melt extraction in the Earth's mantle; constraints from U-Th-Pa-Ra studies in oceanic basalts. *Earth and Planetary Science Letters* **244**: 97–112. doi: 10.1016/j.epsl.2006.01.057

Stracke, A., Snow, J.E., Hellebrand, E., et al. 2011. Abyssal peridotite Hf isotopes identify extreme mantle depletion. *Earth and Planetary Science Letters* **308**: 359–68. doi: 10.1016/j.epsl.2011.06.012.

Straub, S.M., Gómez-Tuena, A. and Vannucchi, P. 2020. Subduction erosion and arc volcanism. *Nature Reviews Earth & Environment* **1**: 574–89. doi: 10.1038/s43017-020-0095-1.

Tatsumoto, M. 1978. Isotopic composition of lead in oceanic basalt and its implication to mantle evolution. *Earth and Planetary Science Letters* **38**: 63–87. doi: 10.1016/0012-821X(78)90126-7.

Taylor, H.P. and Sheppard, S.M.F. 1986. Igneous rocks: I. Processes of isotopic fractionation and isotope systematics, in Valley, J.W., Taylor, H.P. and O'Neil, J.R. (eds). *Stable Isotopes in High Temperature Geological Processes*. Washington: Mineralogical Association of America.

Warren, J.M., Shimizu, N., Sakaguchi, C., et al. 2009. An assessment of upper mantle heterogeneity based on abyssal peridotite isotopic compositions. *Journal of Geophysical Research* **114**. doi: 10.1029/2008jb006186.

Weis, D., Garcia, M.O., Rhodes, J.M., et al. 2011. Role of the deep mantle in generating the compositional asymmetry of the Hawaiian mantle plume. *Nature Geoscience* **4**: 831–8. doi: 10.1038/ngeo1328.

White, W.M. 2015. Probing the Earth's deep interior through geochemistry. *Geochemical Perspectives* **4**: 95–251. doi: 10.7185/geochempersp.4.2.

White, W.M. and Klein, E.M. 2014. The composition of the oceanic crust, in Rudnick, R. (ed). *Treatise on Geochemistry, vol. 2: The Crust*. Amsterdam: Elsevier. doi: 10.1016/B978-0-08-095975-7.00315-6.

White, W.M. 1985. Sources of oceanic basalts: radiogenic isotope evidence. *Geology* **13**: 115–8. doi: 10.1130/0091-7613 (1985)13<115:soobri>2.0.co;2.

White, W.M., 2010. Oceanic island basalts and mantle plumes: the geochemical perspective. *Annual Review of Earth and Planetary Sciences* **38**: 133–60. doi: 10.1146/annurev-earth-040809-152450.

White, W.M., McBirney, A.R. and Duncan, R.A. 1993. Petrology and geochemistry of the Galapagos: portrait of a pathological mantle plume. *Journal of Geophysical Research* **93**: 19533–63.

White, W.M., Schilling, J.-G. and Hart, S.R. 1976. Evidence for the Azores mantle plume from strontium isotope geochemistry of the Central North Atlantic. *Nature* **263**: 659–63. doi:

Willbold, M. and Stracke, A. 2010. Formation of enriched mantle components by recycling of upper and lower continental crust. *Chemical Geology* **276**: 188–97. doi: 10.1016/j.chemgeo.2010.06.005.

Williams, Q. and Garnero, E.J. 1996. Seismic evidence for partial melt at the base of earth's mantle. *Science* **273**: 1528–30. doi: 10.2307/2891048.

Willig, M. and Stracke, A. 2019. Earth's chondritic light rare earth element composition: evidence from the Ce–Nd isotope systematics of chondrites and oceanic basalts. *Earth and Planetary Science Letters* **509**: 55–65. doi: 10.1016/j.epsl.2018.12.004.

Willig, M., Stracke, A., Beier, C., et al. 2020. Constraints on mantle evolution from Ce-Nd-Hf isotope systematics. *Geochimica et Cosmochimica Acta* **272**: 36–53. doi: 10.1016/j.gca.2019.12.029.

Wipperfurth, S.A., Šrámek, O. and Mcdonough, W.F. 2020. Reference models for lithospheric geoneutrino signal. *Journal of Geophysical Research* **125**: e2019JB018433. doi: https://doi.org/10.1029/2019JB018433.

Workman, R.K., Eiler, J.M., Hart, S.R., et al. 2008. Oxygen isotopes in Samoan lavas: confirmation of continent recycling. *Geology* **36**: 551–4. doi: 10.1130/g24558a.1.

Workman, R., Hart, S.R., Jackson, M., et al. 2004. Recycled metasomatized lithosphere as the origin of the Enriched Mantle II (EM2) end member: evidence from the Samoan Volcanic Chain. *Geochemistry Geophysics, Geosystems* **5**: doi: 10.1029/2003GC000623.

Wu, F.-Y., Walker, R.J., Yang, Y.-H., et al. 2006. The chemical-temporal evolution of lithospheric mantle underlying the North China craton. *Geochimica et Cosmochimica Acta* **70**: 5013–34. doi: 10.1016/j.gca.2006.07.014.

Zhang, H.-F., Goldstein, S.L., Zhou, X.-H., et al. 2008. Evolution of subcontinental lithospheric mantle beneath eastern china: re–Os isotopic evidence from mantle xenoliths in Paleozoic kimberlites and Mesozoic basalts. *Contributions to Mineralogy and Petrology* **155**: 271–93. doi: 10.1007/s00410-007-0241-5.

Zindler, A. and Hart, S.R. 1986. Chemical Geodynamics. *Annual Review of Earth and Planetary Sciences* **14**: 493–571.

PROBLEMS

1. Assume that the depleted mantle has a ε_{Nd} = +8, that the Earth is 4.56 billion years old, and that the depleted mantle has evolved with constant $^{147}Sm/^{144}Sm$ since that time. What is the value of that $^{147}Sm/^{144}Nd$ ratio? Chondritic $^{143}Nd/^{144}Nd$ and $^{147}Sm/^{144}Nd$ are 0.512630 and 0.1960, respectively.

2. Assume that the depleted mantle has a $^{87}Sr/^{86}Sr$ = 0.7029, that the Earth is 4.56 billion years old, that the depleted mantle has evolved with constant $^{87}Rb/^{86}Sr$ since that time, and that its initial $^{87}Sr/^{86}Sr$ = 0.7000. What is the value of that $^{87}Rb/^{86}Sr$ ratio?

3. Plot the evolution of a reservoir that has μ = 8 from 4.456 to 2.7 Ga and a μ = 10 from 2.7 Ga to the present on a $^{207}Pb/^{204}Pb$–$^{206}Pb/^{204}Pb$ diagram. Assume a present $^{238}U/^{235}U$ of 137.82. Also plot the 4.456 Ga Geochron on this diagram. *(Hint: μ is defined as the present $^{238}U/^{204}Pb$ ratio. The latter changes with time due to radioactive decay – you will have to modify them, e.g., Equation 7.2, for this problem.)*

4. Assuming the following: the concentration of Nd in the BSE is 1 ppm, that of the continental crust is 22 ppm, the $^{147}Sm/^{144}Nd$ of the continents is 0.12, average age of the continents is 2.2 Ga, the present chondritic $^{143}Nd/^{144}Nd$ is 0.51263, the present chondritic $^{147}Sm/^{144}Nd$ is 0.1960, the ε_{Nd} of the depleted mantle is +8, and mass of the continents is 0.5% of the mass of the silicate Earth. Calculate mass of the depleted mantle as a fraction of the mass of the silicate Earth assuming the silicate Earth has ε_{Nd} = 0 and ε_{Nd}= +4.

5. The most extreme HIMU Pb isotope compositions are about $^{206}Pb/^{204}Pb$ = 22 and $^{208}Pb/^{204}Pb$ = 40.5. What is the corresponding $^{208}Pb*/^{206}Pb*$ ratio/what is the time-integrated value of κ?

6. Using the default values in pyUserCalc (available at https://mybinder.org/v2/gl/ENKI-portal%2FpyUsercalc/master?filepath=pyUserCalc_manuscript.ipynb or https://gitlab.com/ENKI-portal/pyUsercalc [the latter requires a free GitLab registration at https://gitlab.com/

ENKI-portal]), calculate activity ratios $(^{230}Th/^{238}U)$, $(^{226}Ra/^{230}Th)$, and $(^{231}Pa/^{235}U)$ and produce a one-dimensional column plot similar to Figure 7.38. If this basalt has a $^{232}Th/^{238}U$ atomic ratio (i.e., κ) of 1, where would it plot on Figure 7.35?

7. Assume the HIMU reservoir with present $^{206}Pb/^{204}Pb = 21$ and $^{207}Pb/^{204}Pb = 15.85$ evolved by a single differentiation event (a single change in μ) from the BSE, whose present $^{206}Pb/^{204}Pb = 18.1$ and $^{207}Pb/^{204}Pb = 15.5$. When would this hypothetical differentiation event have occurred?

Chapter 8

Isotope Geochemistry of the Continental Crust

8.1 INTRODUCTION

The continental crust is the most accessible part of the solid Earth and, therefore, certainly the part we know most about. That, however, is not to say that we know it well; there is still uncertainty about the overall composition, age, and evolution of the crust: there is still much work left for future generations of earth scientists, and geochemists in particular. This is because the continental crust is remarkably complex tapestry of terrains with different histories. In the previous chapter, we assumed that the crust has been created through time (rather than all at the beginning, as the core was). In this chapter, we will review the evidence that this was indeed the case. Because of the time-dependent nature of radiogenic isotope ratios, isotope geochemistry is a particularly powerful tool in unraveling crustal genesis, precisely because the crust has evolved and changed over time. The Earth has, of course, two kinds of crust: continental and oceanic. The continental crust is thick, low density, and persistent. The oceanic crust is thin, dense, and ephemeral, continually being created and destroyed such that its average age is only about 60 million years. It consists almost entirely of mid-ocean ridge basalts and related intrusive rocks. It is compositionally and structurally simple in comparison to the continental crust. In this chapter, we will focus on the continental crust, which, being thicker and more persistent, better qualifies and an important terrestrial reservoir.

Subduction-related volcanism appears to be the principal way in which continent crust has been created, at least during the Phanerozoic, so we will examine that process in detail in this chapter as well.

8.2 MECHANISMS OF CRUSTAL GROWTH

There is unanimous agreement the crust has formed through magmatism: partial melting of the mantle followed by buoyant rise of those melts to the surface. There is less agreement on the details. The composition of the continental crust, which is approximately that of granodiorite or andesite, is problematic in the context of this hypothesis because it does not have the composition of a mantle-derived magma, which is fundamentally basaltic. This suggests that the evolution of the crust is more complex than simple melt extraction from the mantle.

We should begin by noting that the Earth's first crust was likely quite different from modern continental crust and generated through different mechanisms. Once the Earth reached the planetary embryo stage, the size of the Moon or Mars, continued growth likely involved energetic collisions that resulted in melting followed by the crystallization of new crust that in turn was at least partially destroyed by the next collision. This process culminated in the Giant Impact, which resulted in extensive, and perhaps complete, melting of the Earth. Solidification would have been rapid

Isotope Geochemistry, Second Edition. William M. White.
© 2023 John Wiley & Sons Ltd. Published 2023 by John Wiley & Sons Ltd.
Companion Website: www.wiley.com/go/white/isotopegeochem2

($< 10^7$ years, Elkins-Tanton, 2008) and would have produced the Earth's first crust by around 4.5 ± 0.05 Ga. This crust might have been mafic or might have been similar to the anorthositic lunar crust, which resulted from plagioclase floatation in the lunar magma ocean. That crust would likely have been repeatedly punctured by impacts generating large sheets of basaltic magma, which might have undergone internal differentiation much as occurred following the Paleoproterozoic Sudbury impact.

Whatever its nature, this proto-crust was ultimately replaced by the continental crust produced by different mechanisms in a cooler Earth where impacts play little role. We can identify a number of possible mechanisms, all involving magmatism, that would result in the creation of new continental crust. Most of these mechanisms suffer from the problem that they result in a more mafic crust than that observed.

- *Plume and rift-related volcanism*: For example, flood basalts have occasionally been erupted in tremendous volumes forming *large igneous provinces* (LIPs).
- *Underplating*: Because of the low density of the continental crust, magmas have difficulty rising through it and may become trapped at the crust mantle boundary; consequently, volcanism may significantly underestimate the growth of the continents. Magma trapped at great depth forms new basaltic lower crust, which upon remelting would produce a granitic upper crust. Underplating and volcanism and may also have occurred simultaneously.
- *Accretion of oceanic crust and oceanic plateaus*: The oceanic crust is generally subducted and returned to the mantle. It might in unusual situations be thrust upon or under the continental crust. Subsequent melting of the basalt could produce granite. Thick oceanic plateaus produced by mantle plumes such as Ontong-Java, a good example of an oceanic LIP, and Iceland would be subducted less readily than normal oceanic crust. Their isotopic composition would be less depleted. We found in Chapter 7 that lithosphere underlying Archean crustal cratons formed through extensive melt loss, with melt fractions greater than beneath modern mid-ocean ridges and similar to what we might expect of hotter Archean asthenosphere. This would support an oceanic LIP origin for the Archean crust.
- *Subduction-related volcanism*: Volcanism is usually present along active continental margins. Most of the magma is of mantle derivation. The accretion of intraoceanic island arcs to continents is a closely related mechanism.

It is clear that at present, and almost certainly throughout the Phanerozoic, the last mechanism has produced the greatest additions to the continental crust. The continental crust and subduction-related magmas share key compositional features, including relative depletions in Nb and Ta and enrichment in Pb. It is thus tempting to assume that it has been the principal mechanism of continent creation throughout geologic time, but this has not been demonstrated and indeed is a matter of debate. There is also no consensus on how long plate tectonics has operated, with some arguing that it has operated throughout Earth's history and others arguing modern plate tectonic began as late as the Neoproterozoic; we will briefly review this question in Section 8.4.2. If plate tectonics have not operated throughout most of Earth's history, the continental crust must have been produced by some other (magmatic) process as in most estimates more than half of the continental crust was produced in the first half of that history.

In all these processes, there must be some additional mechanism by which the crust loses a mafic component. In the case of subduction-related magmatism, while parental magmas are predominantly basaltic, the erupted magma is predominantly andesitic. It is possible that mafic cumulates produced by fractional crystallization during the evolution from basaltic to andesitic are retained in the mantle. It is perhaps more likely that internal crustal differentiation is followed eventually by lower crustal foundering: the lower crust rich in mafic mineral cumulates of fractional crystallization or melting residues under some circumstance becomes so dense that it sinks into the mantle. There does seem to be some geophysical evidence of this in the case of the Sierra Nevada batholith in California, which was originally produced above a Jurassic subduction zone. In addition, weathering and erosion tend to remove Mg in preference to Si and Al. Rivers then carry Mg to the ocean where hydrothermal activity transfers it into the oceanic crust.

8.3 THE EARLIEST CONTINENTAL CRUST

The Hadean eon is defined as the time of Earth's formation and preceding the geologic record (the name itself implying a hellish time when no rocks could survive). Throughout much of the twentieth century, discoveries of progressively older rocks pushed the end of the Hadean further back in time, but in the last three decades or so, it has changed only by a few tens of millions of years. The Hadean–Archean boundary is now defined as the age of the oldest rocks within the Acasta basement complex, located in the extreme western part of the Slave Craton of Canada's Northwest Territory. The complex consists of a heterogeneous assemblage highly metamorphosed tonalites, granodiorites, and granites, along with amphibolites and ultramafic rocks that have been deeply buried and experienced complex history of intrusion, deformation, and high-grade metamorphism with igneous crystallization ages ranging from 4.02 to 3.40 Ga (Bowring and Williams, 1999). The oldest of these is the metamorphosed tonalites[1] of the Idiwhaa gneiss ("ancient times" in the local aboriginal language), zircons from which define a 4.02 Ga age (Reimink et al., 2014).

By definition, then, there is no geologic record of the Hadean, and with seemingly little prospect of finding older rocks, this might seem to place the Hadean beyond scientific investigation. Although no rocks survive from that time, some zircon grains do. While never abundant, zircon is a common accessory mineral in almost all felsic igneous rocks as well as some slowly crystallized mafic ones and many metamorphic rocks. Zircon is a remarkably tough mineral: it is nearly as chemically inert and hard as diamond, so it is perhaps unsurprising that the oldest surviving mineral grains are zircons. Fortunately, for science, zircon readily incorporates U (and Th) and excludes Pb, making zircon an excellent recorder of geologic time, as we discussed in Section 3.3. In addition, zircon contains oxygen and readily accommodates a variety of trace elements, most notably Hf, nearly the chemical twin of Zr, and these elements and their isotopes, along with minerals present as inclusions in zircons provide key insights into the Earth's earliest history. The time-integrating nature of radiogenic isotopes provides additional key insights into the Hadean period.

8.3.1 The Hadean zircon record

The first zircons ages older than 4 Ga were detrital grains in a metamorphosed sandstone (quartzite) from Mt. Narryer in the Yilgarn Block of Western Australia (Froude et al., 1983), and because the analysis was done using Australian National University's then novel custom-built ion microprobe known as SHRIMP, they were initially highly controversial. Subsequently, Hadean zircons were found in greater abundance in a heavy mineral-rich metasedimentary conglomerate in the nearby Jack Hills (Compston and Pidgeon, 1986), where eventually zircons as old as ~4.4 Ga were found (Valley et al., 2014). Textural characteristics, zoning, Th/U ratios, and inclusion suites in them point to an igneous origin. Most Jack Hills zircons, however, are much younger, with a mode in the distribution of around 3.4 Ga; only 3% of the >100 000 analyzed zircons are older than 4 Ga and most of these are younger than 4.1 Ga. The depositional age of the conglomerate is constrained to be between 2.65 Ga by the U-Pb age of metamorphic monazite in the quartzite, and 3.08 Ga, the

---

[1] Tonalites and trondhjemites are silicic intrusive igneous rocks that differ from the diorite-granodiorite-granite suite primarily in having lower K_2O/Na_2O ratio and, hence, are poorer in alkali feldspar and proportionally richer in plagioclase. Suites of tonalite–trondhjemites–granodiorite (TTG) intrusive rocks are common in the Archean and often occur in association with "greenstone belts" consisting of metamorphosed mafic and ultramafic (e.g. komatiites) lava flows and intrusives. Modern suites, such as the coastal batholith of Peru, are associated with subduction-related magmatism. Archean TTGs differ from modern ones in being strongly depleted in heavy rare earths and lacking Eu-anomalies, suggesting that they were generated by melting of mafic crustal rocks in the garnet rather than plagioclase stability field.

age of the youngest detrital zircons (Rasmussen et al., 2010).

Since that discovery, the *in situ* analysis of individual grains and parts of grains has led to the discovery of Hadean zircons in rocks from around the world, including as xenocrysts in the Acasta Gneisses of Canada, xenocrysts in a quartz schist in Tibet, xenocrysts in Ordovician volcanics of the North China Craton, detrital zircons from the Central Asian Orogenic Belt of northwestern China, detrital zircons in Neoproterozoic metasediments of the Cathaysia Block of Southern China, a xenocryst in Paleoproterozoic igneous rocks from Southern Guyana, detrital zircons in Archean metasediments of the São Francisco Craton in northeastern Brazil, and detrital zircons from the Barberton Greenstone Belt of in the northeast of the Kaapvaal craton in Southern Africa. The oldest of these are xenocrysts, which yielded concordant ages as old as 4.24 Ga in the 3.4 Ga Older Metamorphic Tonalitic Gneiss of the Singhbhum Craton in Eastern India (Chaudhuri, et al., 2018). As with the Australian zircons, the vast majority of zircons in these localities are of Archean age or younger, and distribution drops off sharply beyond 4 Ga.

8.3.1.1 The Zircon Hf Isotope Record

Additional insights into the formation of early continental crust can be gained by the Hf isotopic analysis of zircons. Zircons have very low Lu/Hf ratios, and hence, their initial $^{176}Hf/^{177}Hf$ ratios can be calculated with only a small correction for ^{176}Lu decay. The Hadean grains are all of igneous origin so that initial ε_{Hf} values of the zircons effectively reflect the Lu/Hf ratios of the precursors of the magmas from which they formed. Before looking at the data, let us review their interpretation on an initial ε_{Hf} versus age diagram (Figure 8.1). Zero or positive initial ε_{Hf} values imply that the zircon formed from material with recent mantle heritage (high Lu/Hf ratio); here, we use "recent" in a relative sense; it could be several hundred million years earlier. And since zircons do not generally form in mantle-derived magmas (i.e. basalts), the zircons probably formed from a felsic magma formed by the remelting of the basalt. Negative initial ε_{Hf} values imply a crustal (low Lu/Hf) heritage. Zircons forming a trend of decreasing ε_{Hf} with decreasing age are indicative of repeatedly reprocessing older crust. The slope is a function of the Lu/Hf ratio of this older crust, with steeper slopes indicative of lower Lu/Hf. On the other hand, a vertical trend that is variable ε_{Hf} at fixed age is indicative of mixing between mantle-derived magma and older crust, most likely through assimilation.

The cores of these ancient zircon grains are usually highly concordant, but they are almost always zoned (Figure 3.3) with the original igneous cores surrounded by younger overgrowths. Consequently, the calculation of

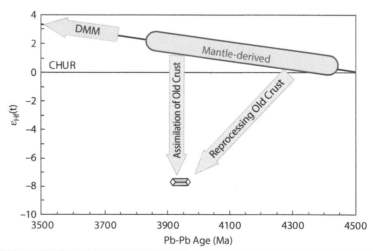

Figure 8.1 Cartoon illustrating the interpretation of ε_{Hf} versus age diagram. The systematics of this diagram are the same for whole rocks as well as for *in situ* analysis of zircons. They are also the same for ε_{Nd}, although whole rocks are more subject to secondary disturbance than zircons.

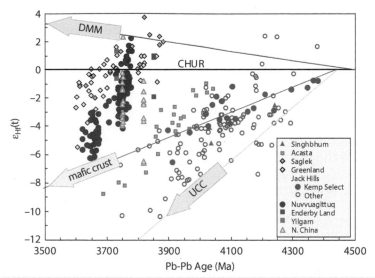

Figure 8.2   Initial ε_{Hf} determined in Hadean and Eoarchean zircons from five regions of ancient continental crust. CHUR line shows the evolution of a reservoir with chondritic $^{176}Lu/^{177}Hf$ (0.0338), DMM shows the evolution of the depleted mantle reservoir with $^{176}Lu/^{177}Hf = 0.039$, UCC shows the evolution of the upper continental crust with $^{176}Lu/^{177}Hf = 0.008$, and "mafic crust" shows the evolution of crust with $^{176}Lu/^{177}Hf = 0.022$. Most initial ε_{Hf} values are negative, indicative of preexisting crust. Filled red circles are zircons adapted from Kemp et al. (2010) to be least disturbed based on well-preserved oscillatory zoning, Th/U $\geq$ 0.3, and concordant Pb/U isotope ages.

initial ratios requires that $^{176}Hf/^{177}Hf$ be determined on the same part of the zircon as the Pb isotope ratios, which can be done with *in situ* microanalytical techniques. Particularly useful is "split stream analysis" in which the particle stream produced by the laser is split such that the U-Pb system is analyzed in one instrument and the Lu–Hf system in another.

Figure 8.2 shows initial ε_{Hf} in Hadean and Eoarchean zircons as a function of Pb-Pb age determined on the same spot. The majority of these zircons have negative ε_{Hf}, indicative of low Lu/Hf continental crustal precursors. Indeed, ε_{Hf} values in many of the Jack Hills zircons suggests that precursors with continental Lu/Hf ratios formed as early as 4.5 Ga. Data from the 3.7 Ga Meeberrie Gneiss, located only 10 km from the Jack Hills in the Yilgarn block, however, suggest that it was derived from a new addition to the crust from the mantle around 3.7 Ga. The Acasta data form a pattern of decreasing ε_{Hf} with decreasing age that is consistent with that of new continental crust that originally formed around 4.1–4.2 Ga followed by episodes of remelting of this crust at 3.6–3.8 Ga without new additions from the mantle. In contrast, zircons from the Nuvvuagittuq terrane of Labrador form two vertical

arrays suggesting that they crystallized from magmas derived from depleted mantle around ~3.8 and ~3.65 Ga that assimilated much older Hadean crust. The limited data on zircons from the Singhbhum Craton in India also suggest that crust in this region first formed very early in the Hadean. Data from the lower crustal xenoliths in Jurassic igneous rocks of the North China Craton suggest that mantle-derived magmas assimilated likely Hadean crust at around 3820 and 3750 Ma.

Not all cratons have Hadean roots. Southwest Greenland and the Saglek-Hebron Complex, exposed on the east coast Labrador, together form the North Atlantic Craton that was split by rifting in the early Tertiary. Data from these regions form an array of increasing ε_{Hf} with age that intersects the depleted mantle curve at around 3900 Ma, suggesting that crust in this region first formed around that time in the earliest Archean.

8.3.1.2   Stable isotopes in Hadean zircons

Stable isotope ratios of Hadean zircons provide additional evidence of early formed continental crust and, additionally, evidence that this crust reacted with liquid water. Zircons in xenoliths and other mantle material have an average has $\delta^{18}O_{SMOW}$ of $\approx$ +5.3 ± 0.2‰. Fractionation between zircon and magma it forms from varies from ~−0.5‰ for mafic rocks to ~−2‰ for granites. Thus, zircons with $\delta^{18}O$ significantly higher than ~6.3‰ in igneous zircons require a component of supracrustal rock that had interacted with water at relatively low temperatures. Cavosie et al. (2005) analyzed $\delta^{18}O$ in the Jack Hills Hadean zircon cores and found that the very oldest had $\delta^{18}O$ within the mantle range, those with ages of 4325–4200 Ma had $\delta^{18}O$ as high as 6.5‰, and those with ages between 4200 and 3950 Ma had $\delta^{18}O$ up to 7.3‰ (zircons with disturbed zoning patterns had $\delta^{18}O$ both above and below this range but these were not regarded as primary values). These data suggest that the magmatic protoliths had experienced increasing low temperature reaction with liquid water, providing evidence not only of the presence of water at the Earth's surface but that its interaction with rock to produce secondary ^{18}O-rich minerals.

Silicon isotopes provide additional evidence of such interactions. Mantle zircons have $\delta^{30}Si_{NBS28}$ of −0.38 ± 0.02‰, a little lower than the presumed bulk silicate Earth value of ~−0.3‰. Silicon isotopes are only slightly sensitive to magmatic processes (not surprising given the very similar atomic environment of Si in magma and most silicate minerals), so that $\delta^{30}Si$ is expected to increase by only ~0.06‰ in the evolution from basalt to granite. On the other hand, interaction between silicate minerals and water during weathering does lead to significant Si isotope fractionation such that Si in Si-poor secondary clay minerals can be more than 1‰ lighter than the parent rock and Si precipitated from solution typically has positive $\delta^{30}Si$. Trail et al. (2018) found that $\delta^{30}Si$ in Hadean Jack Hills zircons ranged from −0.8‰ to −1.5‰, a range comparable with that observed in almost the entire range for Phanerozoic samples. $\delta^{18}O$ measured on the same zircons shows a range from mantle-like up to +6.6‰. Trail et al. (2018) concluded that these zircons crystallized from magmas produced by remelting of crustal rocks including sedimentary ones as well as mafic igneous ones.

Other observations support the inference that the magmas from which the Jack Hills zircons formed were produced by the melting of preexisting crustal materials. The most common primary inclusions in the Jack Hills zircons include muscovite, Fe–Ti oxides, biotite, and quartz, a suite of minerals that are typical of granitoid magmas. Crystallization temperatures of these magmas, which can be estimated from the Ti-in-zircon geothermometer (Watson et al., 2006), are in the range of 650–750 °C, a range typical of hydrous granitoid magmas (Harrison, 2009). Finally, incompatible element abundances suggest the magmas from which the zircons crystallized were predominantly produced by the remelting of mafic igneous crust.

While no pre-4 Ga crust survives at the Earth's surface (xenocrystic zircons suggest some may survive at depth), zircons from six continents with cores having concordant Pb–Pb ages >4 Ga testify that the Hadean crust was widespread, although not necessarily voluminous, and that it consisted, in part, of some of the same rock types, including granitoid igneous rocks and sediments, as does the modern crust. While it may be unlikely that outcrops of the Hadean crust will be found, it seems quite likely that additional localities will be found containing Hadean zircons, and isotopic studies of them will no doubt tell us more about the Earth's early history.

8.3.2 Echoes of early continental crust from extinct radionuclides

In addition to the Acasta gneisses, Eoarchean (>3.6 Ga) crust has been found in West Greenland, Northeast Canada, Wyoming USA, Western Australia, Southern Africa, Antarctica, and Northern China, and detrital and xenocrystic zircons provide evidence that it exists or existed in India and South America as well. Were these rocks produced by the remelting of the preexisting Hadean crust or do they represent new additions to the continents from the mantle? Because of their time-integrating nature, radiogenic nuclides can help to answer these questions. For example, if they are new additions to crust, we would expect Hf and Nd model ages (τ_{CHUR} or τ_{DM}) to match crystallization ages but be older if they were produced by remelting the existing crust. The difficulty is that essentially all the Eoarchean rocks have experienced metamorphism, usually severe, which can disturb these radiogenic isotope systems. This, however, would not be the case for the ^{146}Sm-^{142}Nd system because ^{146}Sm, with a half-life of 103 Ma, would have been effectively extinct by the end of the Hadean, while the ^{182}Hf-^{182}W system, with a half-life of 9 Ma, would have been effectively extinct after the first ~50 Ma of solar system existence. Thus, these systems provide unique insights into the Earth's earliest history.

8.3.2.1 ^{142}Nd evidence of early crust formation

Following the discovery of anomalies as great as 5 ppm in ^{142}Nd/^{144}Nd ratios in 3.8 Ga rocks from the Isua Supracrustal Belt of Southwest Greenland, once the oldest known rocks, in 1992 by Harper and Jacobsen (1992) and subsequent confirmation by Caro et al. (2003) and Boyet et al. (2003), Bennett et al. (2007) reported μ_{142Nd} values as high as +20 in metabasalts from Isua and in amphibolites from the neighboring Itsaq complex. Just as is the case for ε_{Nd}, we expect rocks derived from an incompatible element-depleted precursor to have positive μ_{142Nd}, and those produced from incompatible element-enriched sources to have negative μ_{142Nd}. Hence, the Isua samples were derived from a mantle already depleted in incompatible elements.

Following that, O'Neil et al. (2008) reported negative μ_{142Nd} values in amphibolites from the Nuvvuagittuq Greenstone Belt located within in the Hudson Bay Block of the Superior Province in Northern Quebec; these were subsequently confirmed by additional analyses of O'Neil et al. (2012) and Roth et al. (2014). O'Neil et al. (2008) found a strong correlation between ^{142}Nd/^{144}Nd and ^{147}Sm/^{144}Nd in cummingtonite amphibolites from the area (Figure 8.3). As we found in Chapter 6, when an extinct radionuclide is involved, the slope of the line on plots such as this is proportional to the parent isotope ratio, in this case ^{146}Sm/^{144}Sm, at the time the rocks formed. Using both the data from O'Neil et al. and Roth et al., the slope in the case for the cummingtonite amphibolites corresponds to ^{146}Sm/^{144}Sm = 0.00116. Using a ^{146}Sm half-life of 103 Ma and the solar system initial ^{146}Sm/^{144}Nd of 0.0094, the apparent age is 4.26 Ga. O'Neil et al. (2012) reported ^{142}Nd/^{144}Nd–^{147}Sm/^{144}Nd pseudo-isochrons corresponding to ages ranging from 4.31 to 4.41 Ga using a half-life of 68 Ma for other

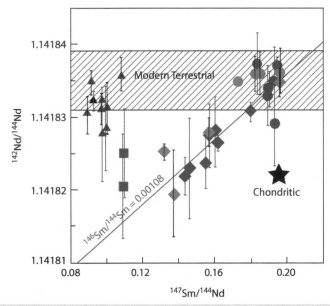

Figure 8.3 ^{142}Nd/^{144}Nd versus ^{147}Sm/^{144}Nd for rocks from the Nuvvuagittuq Belt of northwestern Labrador. The slope of the line corresponds to a ^{146}Sm/^{144}Nd ratio of 0.00116. If this line is interpreted as an isochron, it implies a formation age of 4.26 Ga assuming a half-life of ^{146}Sm of 103 Ma. Blue triangles: felsic gneisses, diamonds: cummingtonite amphibolites (red from O'Neil et al. (2008) green from Roth et al., 2014), circles: hornblende amphibolites, squares tonalites: from O'Neil et al. (2008); blue and green from O'Neil et al. (2012); and green from Roth et al. (2014).

suites of rocks from the region. However, a $^{147}Sm/^{144}Nd$–$^{143}Nd/^{144}Nd$ isochron for the cummingtonite amphibolites gives an age of only 3.89 Ga with considerable uncertainty, and other suites in the area give even younger ages. There are no zircons in the mafic rocks, but the crystallization age of zircons in felsic rocks as well as detrital zircons in metasediments give ages that range from 3.75 to 3.78 Ga, which is likely the maximum age of the Nuvvuagittuq supracrustal belt (Cates et al., 2013).

Two component mixtures can produce correlations between isotope ratios that exactly mimic isochrons (as we will discuss later in this chapter). Given the evidence for an Eoarchean age of the Nuvvuagittuq Greenstone Belt, it is more likely that $^{142}Nd/^{144}Nd$–$^{147}Sm/^{144}Nd$ correlations are mixing lines resulting from assimilation of early crust by much later mantle-derived basaltic magma. Modeling done by Cates et al. (2013) suggests this assimilated crust could have formed as earlier as 4.5 Ga and was assimilated by the amphibolite precursor, which they speculate were originally pyroclastic rocks, around 3.8–4.0 Ga. Metamorphism at around 3.8 Ga resulted in a decoupling of the $^{142}Nd/^{144}Nd$ and $^{143}Nd/^{144}Nd$ ages. The $^{142}Nd/^{144}Nd$–$^{147}Sm/^{144}Nd$ correlation and

negative μ_{142Nd}, nevertheless, provide strong evidence of an early-formed incompatible element-enriched crust in this region.

Subsequently, a large body of data on $^{142}Nd/^{144}Nd$ in Archean rocks has accumulated and is shown in Figure 8.4. Other Eoarchean terranes in the northeastern Superior Province, Tikkerutuk and Inukjuak, share the negative μ_{142Nd} of the Nuvvuagittuq Greenstone Belt, suggesting that they too formed in part from the preexisting Hadean crust. In the Inukjuak terrane, μ_{142Nd} tends to increase toward 0 with time, such that the 2700-Ma granites have μ_{142Nd} close to 0. On the other hand, the low μ_{142Nd} persists in the much younger ∼2750-Ma tonalite-trondhjemite–granodiorite (TTG) suite of the Tikkerutuk terrane surrounding the Nuvvuagittuq Greenstone Belt, indicating substantial recycling of much older crust continued through the Archean in the Hudson Block of the Superior Province. The Neoarchean komatiites from the Abitibi Belt in the southwestern part of the Superior craton have ε_{142Nd} ranging from positive to negative, suggesting that these mantle-derived magmas had in some cases assimilated older crust although there is no evidence of Eoarchean crust in the region.

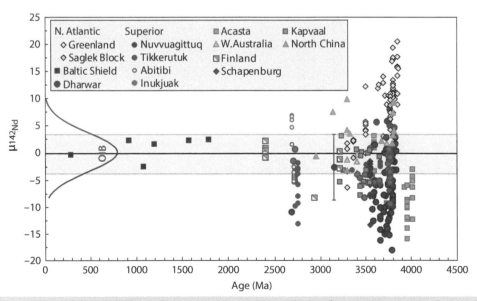

Figure 8.4 μ_{142Nd} as a function of age determined by U-Pb analysis of zircons in various continental rocks. Gray band shows the typical analytical uncertainty (but which varies between laboratories and for each individual analysis); data falling within the band can be considered indistinguishable from the standard. Because of lower analytical precision, the Dharwar data are shown only as mean and range. Red curve at left is the probability density of 85 analyses of modern oceanic island basalts (OIB).

Systematically low μ_{142Nd} found in the Eoarchean Acasta gneisses, which range as low as −16.2, suggests that they were substantially derived from the preexisting Hadean crust, consistent with the Hadean zircons with negative ε_{Hf} found in these rocks. This negative μ_{142Nd} disappears in younger Paleoarchean rocks of the Slave Province, whose μ_{142Nd} is not resolvable from 0. Rocks from the Kapvaal Craton, including the Schapenburg komatiites, show something of bimodal distribution with many falling within analytical error of μ_{142Nd} ≈ 0 and others with resolvable negative anomalies. This provides hints that the Hadean crust existed in the region. Alternatively, some of the Archean magmas may have been derived from a mantle source already enriched in incompatible elements, perhaps a consequence of fractional crystallization of a magma ocean.

Overall, the picture provided by μ_{142Nd} is one of the considerable heterogeneities in the Eoarchean, indicative of melting and crust production and consequent incompatible element depletion of the mantle, as well, perhaps, of the fractional crystallization of a primordial magma ocean, in the Earth's first 500 million years. This heterogeneity then decreases throughout the Archean and disappears almost entirely by its end. The probability distribution function for modern oceanic basalts is shown as a red curve in Figure 8.3; the mean and standard deviation of all the modern oceanic basalts are 0.30 and ±3.6 (n = 41), respectively. While some modern lavas from Reúnion Island appear to have resolvable μ_{142Nd} anomalies, these still await independent confirmation. Otherwise, all Proterozoic and younger have μ_{142Nd} indistinguishable standard. If confirmed, from Reúnion, anomalies point to preserved heterogeneity in the deep mantle as does μ_{182W}. With this possible exception of the deep mantle, all μ_{142Nd} heterogeneity in the mantle appears to have been erased through mixing (at least within present analytical resolution, but with continued improvement in analytical techniques, this may change).

8.3.2.2 Tungsten isotope anomalies: evidence of a late accretionary veneer?

As we found in Chapter 6, ^{182}Hf was present in the early solar system and the silicate Earth's ^{182}W/^{184}W is about two epsilon units higher than chondrites. The difference reflects the sequestration of W in the Earth's core (or the cores of the planetary embryos that accreted to form the Earth) before ^{182}Hf had completely decayed, leaving the silicate Earth and Moon with high Hf/W ratios, and consequently enriched in ^{182}W. The silicate Earth was assumed to have a uniform ^{182}W/^{184}W until Willbold et al. (2011) reported an average μ_{182W} of +13±4 (2σ) for seven samples of 3.8-Ga Isua Belt of southwest Greenland including gneisses, metabasalts, and metasediments (recall that μ_{182W} is the ppm deviation from a terrestrial standard). The following year, Touboul et al. (2012) reported μ_{182W} averaging +15±5 in 2.8-billion-year-old komatiites from Kostomucksha in the Karelia region the Baltic Shield of northeastern Russia, but μ_{182W} values in 3.5-billion-year-old komatiites from the Komati Formation of the Barberton Greenstone Belt were not resolvable from the bulk silicate Earth (BSE) value outside of analytical error. Subsequently, tungsten isotope anomalies have been reported in a variety of Archean rocks as well as 2.4-billion-year-old Vetreny komatiites from Finland. With the notable exception of the 3.55-billion-year-old Schapenburg komatiites of the Barberton Greenstone Belt of the Kapvaal craton in Southern Africa, all the resolvable anomalies are positive (Figure 8.5). These positive μ_{182W} in ancient crustal rocks contrast with modern OIB, whose probability distribution is skewed to negative values shown by the red curve in Figure 8.3.

To look at how widely tungsten isotopic anomalies were distributed in the upper continental crust, we can turn to sedimentary rocks, which are natural averages of compositions of wide areas. Among these are diamictites, which are essentially lithified glacial till. Diamictites occur sporadically through time in the geologic record and are important in reconstructing climate history, but they also provide snapshots of average surface composition at the time of their deposition. Figure 8.5 also shows μ_{182W} in diamictites ranging in age from ~2950 to 600 Ma reported by Mundl et al. (2018). Archean diamictites from four localities in the Kapvaal Craton show resolvable negative μ_{182W} suggesting that rocks with compositions similar to the Schapenburg komatiites were widespread in that region at around 3 Ga. Indeed, these diamictites are notably rich in Ni and other highly compatible

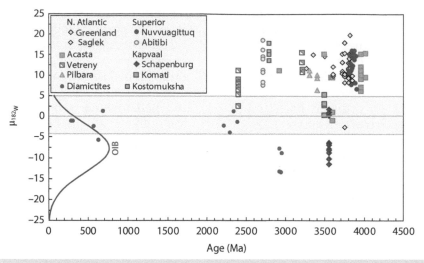

Figure 8.5 μ_{182W} as a function of age determined by the U-Pb analysis of zircons in various continental rocks. Gray band shows the typical analytical uncertainty (but which varies between laboratories and for each individual analysis); data falling within the band are indistinguishable from those of the standard. Red curve at left is the probability density of 93 analyses of modern OIB. Also shown are the glacial diamictite analyses adapted from Mundl et al. (2018).

elements, suggesting mafic rocks predominated in the area from which they were eroded. With the possible exception of the 600-Ma-old Gaub diamictites from Namibia, which also has a negative μ_{182W}, all the subsequently deposited diamictites have μ_{182W} identical within error to BSE.

The similarity of μ_{182W} in the Schapenburg komatiites to modern mantle plume-derived basalts is striking and might speculate that the former were produced by a deep mantle plume. We might further speculate that positive μ_{182W} values characterized the crust in the Earth's early history, while the lower mantle was, as it is today, characterized by negative μ_{182W}. We noted in the previous chapter that the incorporation of a very small amount of core-derived material into modern mantle plumes might explain the negative μ_{182W}; perhaps, this explains the Schapenburg data as well.

How do we explain the positive μ_{182W} anomalies? Early metal–silicate fractionation is the most obvious way to explain variation in μ_{182W}: metal-rich material should have low Hf/W and consequently lower μ_{182W} than material from which metal has been extracted; silicate material from which metal had been efficiently stripped should have high μ_{182W}. Varying efficiency of metal–silicate segregation during the formation of the Earth's core could might explain the data, which would imply the material from which ancient crust was extracted had more effectively stripped of siderophile elements that the modern upper mantle.

Another possibility is the addition of chondritic material, from which metal had not been segregated, after core segregation was complete. A key observation in this respect is that the silicate Earth is richer in highly siderophile elements and these metals are present in the silicate Earth in relative proportions closer to chondritic than one would expect from equilibrium partitioning between mantle and core. A long-standing hypothesis to explain these observations is that a "late veneer" of chondritic material was accreted to the Earth after the final segregation of the core in the Moon-forming giant impact (Chou, 1978; Wänke et al., 1984). An amount of chondritic material equaling less than 1% of the mass of Earth would explain the high siderophile abundances in the silicate Earth. If this late veneer were undifferentiated chondritic material, it would have ε_W of –2 (or +200 μ_{182W}), and adding it would decrease the ε_W of the Earth. If the Earth today has μ_{182W} of 0, it must have had a higher value prior to addition of the late accretionary veneer.

Willbold et al. (2011) and several studies have proposed that the Archean rock with positive μ_{182W} retains a memory of Earth's tungsten isotopic composition before the addition of this late accretionary veneer or at least before it was thoroughly mixed with the mantle. The decline of Earth's μ_W from ~+13 to 0 is consistent with the addition of just under 0.5% by mass of chondritic material, roughly consistent with what would be needed to raise the Earth's noble metals to observed abundances. The data do not directly constrain the timing of the addition but are most easily explained if the addition occurred after ^{182}Hf was effectively extinct, or more than ~50 million years after the start of the solar system. This model predicts a positive correlation between μ_{182W} and Os isotope ratios as well as inferred highly siderophile element abundances (such as those of the noble metals) in mantle sources; however, such correlations have not been observed.

Another possibility proposed by Rizo et al. (2016) is magmatic fractionation. Tungsten is substantially more incompatible in mantle silicates (roughly as incompatible as Th) than is Hf (roughly as incompatible as Sm). Consequently, fractionation between silicate liquids and solids could lead to considerable Hf/W fractionation. This must have occurred very early in Earth's history; Rizo et al. suggested that it has occurred during the crystallization of a magma ocean following the Moon-forming impact. This model predicts a correlation between μ_{182W} and μ_{142Nd}, but the two appear completely uncorrelated. As μ_{182W} is sensitive only to processes occurring in the first 50 Ma, while μ_{142Nd} is sensitive to processes occurring in the first 500 Ma, it is possible that later processes in the Hadean destroyed the predicted correlations, so it may be premature to rule out either hypothesis.

While there is no consensus on the cause of the Hf/W fractionation that produced the heterogeneity in tungsten isotopic composition, it must have occurred within the first 50–60 million years of solar system history. Unlike μ_{142Nd}, variability in μ_{182W} does not appear to taper off through the Archean, but like μ_{142Nd}, this variability disappears after the earliest Proterozoic, with the exception of the Gaub diamictites from Namibia, which likely partly sample Archean crust, and modern OIB. Thus, Archean silicate Earth may retain

a memory of processes because convection had not yet succeeded in mixing in a late accretionary veneer into the mantle or to fully homogenize the mantle following the extraction of early Hadean crust or the crystallization of a magma ocean. Subsequent convection apparently homogenized most of the mantle such that this memory has been lost.

For many decades now, geologists and geochemists have noted that the structure and composition of Archean crust, often characterized by extensive basaltic and komatiitic lavas forming "greenstone belts" and punctuated by silicic intrusions, particularly suites of the TTG association, differs from that of post-Archean crust. The disappearance of μ_{142Nd} and μ_{182W} anomalies at the end of the Archean strengthens the case that the Earth operated in some fundamentally different manner in the Archean; some arguing that modern plate tectonics only began around this time. Ironically, the Archean-Proterozoic boundary remains the only one arbitrarily defined (at 2500 Ma). The persistence of μ_{182W} anomalies to slightly later (2407 Ma) is one of a number of observations, including the disappearance of Δ^{33}S anomalies, that this important boundary should be moved forward to 2.3–2.4 Ma.

8.3.3 What happened to the Hadean crust?

The zircon record now reveals that the Hadean crust must have existed in a variety of locations throughout the world, yet none of it survives. Despite the intensive search for the most ancient ones, zircons with Hadean crystallization ages constitute only 0.13% of the >400 000 analyses compiled by Puetz et al. (2018), and even zircons of Eoarchean age (3600–4000 Ma) constitute only 0.7% of the zircon population. One of the many questions we might ask about the Hadean crust and the early Archean crust is whether these numbers represent the amount of early crust present and if not, what happened to it?

Part of the answer to this question is already apparent in Figure 8.2. Looking more closely at the Jack Hills data and see a trend toward decreasing ε_{Hf} with decreasing age. This becomes clearer when we focus on just those zircons of the Kemp et al. data set selected to be least disturbed. These define a slope corresponding to ^{176}Lu/^{177}Hf = 0.020±0.004, a ratio typical of mafic to intermediate igneous crust. The very limited data from the Singhbhum Craton in India fall along this same trend. In these cases, the data suggest that the original crust produced around 4.4 Ga was repeatedly reprocessed over the succeeding 500 Ma into new crust with minimal additions from the mantle.

The Eoarchean zircons from the Acasta Gneiss Complex show a similar pattern, although they define a steeper ε_{Hf} versus age pattern, one consistent with $^{176}Lu/^{177}Hf \approx 0.08$, a ratio typical of the upper continental crust and an age for the original crust of ~4.2 Ga. Data from both parts of the North Atlantic Craton also define a slope corresponding roughly to an upper crustal Lu/Hf ratio, but with an even younger age of initial crust production of 3.8–3.9 Ga. Figure 8.2 tells a different story about the Hudson Block of the Superior Craton, one in which the Hadean crust was variably assimilated by mantle-derived magmas in pulses around 3750–3800 Ma and 3675–3625 Ma. In all these cases, early crust was simply reprocessed into somewhat younger crust.

8.4 THE CONTINENTAL CRUST THROUGH TIME

8.4.1 The Zircon record

8.4.1.1 The early Archean

Let us now move forward in time. The Slave Craton experienced a longer history of magmatism, metamorphism, and crustal growth than shown in Figure 8.2. While the Eoarchean period appears to involve only reworking of existing crust that first formed at the end of the Hadean, the analysis of Hf isotope ratios in zircons by Reimink et al. (2019) revealed that subsequent episodes of crust formation in the Acasta Gneiss Complex, which began with a pulse between about 3.6 and 3.4 Ga and continued to 2.9 Ga, involved new additions of mantle-derived magma that assimilated that preexisting crust. To the East, the oldest rocks of the Central Slave Basement Complex formed around 3.3 Ga. Their zircons have positive ε_{Hf}, indicating that they were new additions to mantle. Subsequent pulses of crust formation occurred from 3.16 to 3.1 Ga, 3.06 to 3.01Ga, 2.99 to 2.93 Ga, and 2.9 to 2.85 Ga also consisted of mantle-derived

magmas, in some cases with minor assimilation of preexisting crust.

Although some mafic lithologies occur, the rocks of both the basement complexes from which the zircons are derived are metamorphosed tonalites and trondhjemites and not direct mantle melts. Rather they were produced by the remelting of mantle-derived basalts at depth. An estimate of the maximum time between basalt emplacement can be estimated from the difference between the U-Pb crystallization age and the Hf τ_{DM}. These residence times depend on the assumed Lu/Hf ratio of both the crustal precursor and the depleted mantle, but they are relatively brief, <200–300 Ma for the Paleo- and Mesoarchean TTG suites of the Central Slave Basement Complex, particularly in comparison to the >400 Ma estimates for the Eoarchean rocks of the Acasta basement complex.

Detrital zircons collected in small streams in Southwest Greenland by Kirkland et al. (2021) reveal a history similar to that of the Slave Craton, although with crust formation beginning in the Eoarchean (Figure 8.6). The oldest zircons form a trend of decreasing ε_{Hf} with decreasing age consistent with the reprocessing of crust that evolved with $^{176}Lu/^{177}Hf \approx 0.01$ and which that first formed around 3.9 Ga without subsequent new additions from the mantle. Beginning about 3.2 Ga, positive ε_{Hf} values signal that new mantle-derived magmas were being added and were assimilating and mixing with older crust. Distinct pulses occurred around 3.2, 3.0, and 2.8–2.7 Ga, with minor components at around 1.8 Ga and as old as c. 3.8 Ga (Kirkland et al., 2021). Zircons from the western part of the craton composed of the Saglek Block in Labrador analyzed by Wasilewski et al. (2021) show the same pattern, although absent the 1.8 Ga pulse, indicating that the entire North Atlantic Craton experienced a similar history.

The pulses of magmatic activity beginning in the Paleoarchean in the Slave and North Atlantic Cratons appear to be the rule, not the exception. Figure 8.7 shows the global distribution of Pb ages of >13 000 detrital and >11 000 igneous zircons older than 2800 Ma from (Condie et al., 2018). The igneous zircons show distinct peaks in abundance at around 3800 Ma, 3400-3200 Ma, and again beginning around 2850 Ma. Not surprisingly, there appears to be a time lag between the igneous and detrital zircons.

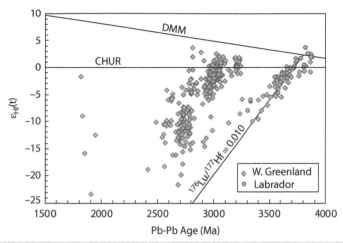

Figure 8.6   Initial ε_{Hf} in zircons from the North Atlantic Craton, including detrital zircons collected from streams in southwest Greenland by Kirkland et al. (2021) and zircons separated from rocks of the Saglek Block of coastal Labrador by Wasilewski et al. (2021). The oldest detrital zircons indicate crust in the area initially formed from depleted mantle around 3.9 Ga and was reprocessed over the following 500 Ma with little or no new additions to mantle. Beginning around 3.2 Ga, new mantle-derived magmas assimilated some of this preexisting crust. Subsequent pulses of crustal growth occurred around 3.0, 2.8–2.7, and 1.9–1.8 Ga.

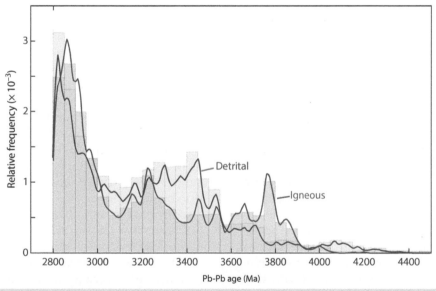

Figure 8.7   Histograms and kernel density functions for the detrital (blue) and igneous (red dashed) zircons older than 2800 Ma, respectively. Data from (Condie et al., 2018).

8.4.1.2   The zircon record through geologic time

It has been apparent for decades that the distribution of radiometric ages of continental rocks is uneven, with ages, particularly for Precambrian rocks, concentrated in relatively few

short periods. With the development of secondary ion mass spectrometry (SIMS) and laser ablation-multiple collector-inductively coupled plasma–mass spectrometry (LA-MC-ICP-MS) techniques allowing for the rapid determination of U-Pb ages in U-rich minerals in the last decade or so, the concentration of

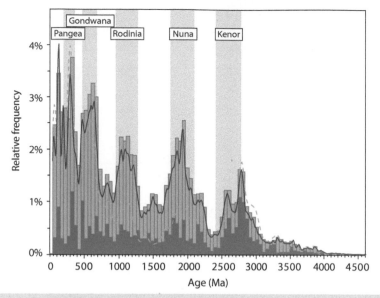

Figure 8.8 Histogram of more than 400 000 U-Pb ages obtained on zircons and other U-rich minerals. Red line shows the kernel density probability distributions for all data (red line) and igneous zircons only (dashed orange line). Darker blue shows the frequency of zircons with mantle-like $\delta^{18}O$ (5.3±0.6‰) adapted from the dataset of Spencer et al. (2014) and Puetz et al. (2018).

U-Pb ages in peaks has become even more apparent. Figure 8.8 shows the recent comprehensive compilation of Puetz et al. (2018) of >400 000 Pb ages for all of Earth's history, ~98% of which were obtained on zircons and ~1% are monazite, with most of the remainder being titanite, baddeleyite, apatite, and xenotime. It is immediately apparent that the data shown in Figure 8.7 are a very small fraction of all zircons, indicating that very little crust older than 2800 Ma survives; indeed, the data in Figure 8.7 represent <6% of all the data in Figure 8.8. The most important aspect of these data is the peaks in the distribution, which tells us that the production of surviving continental crust has not been in steady state. The peaks occur at 2700, 2500–2400, 2200, 1900–1850, 1650–1600, 1100, 800, 600, and 250 Ma and appear to coincide with the existence of supercontinents (the gray bands).

The 2700-Ma peak appears to be global, with rocks of this age present in abundance on all the continents, as is the trough from 2.5 to 2.1 Ga. Other peaks are restricted to certain areas, for example, the 2.5-Ga event is represented in China, India, and Australia, the 2.1-Ga peak is apparent rocks from South America, western Africa, and in northwestern

North America, and the 1.85-Ga peak occurs in North America, Scandinavia, and Australia.

It might be tempting to interpret Figure 8.8 as the change in crustal volume over time. However, when we consider only zircons with mantle-like $\delta^{18}O$ (5.3±0.6‰; Valley et al., 2014), the distribution of which are shown in a darker shade in Figure 8.8, we see that at most times, the zircon ages reflect the reprocessing of the existing continental crust, the exception being around 2700 Ma when a majority of zircons have mantle-like $\delta^{18}O$.

Figure 8.9 shows ε_{Hf} in ~42 000 zircons compiled by Roberts and Spencer (2015) and $\delta^{18}O$ in ~6300 zircons. There is a notable increase in average $\delta^{18}O$ in zircons after ~2700 Ma indicative of increased crustal reprocessing, although ε_{Hf} shows little change. Overall, average ε_{Hf} is negative at most times and average $\delta^{18}O$ is substantially higher than the mantle value, reflecting substantial contributions of the preexisting crust to new additions to crust. Calculating the proportions of new mantle-derived material and older crust from ε_{Hf} values would be straightforward if we knew the compositions of the two end-members, but we do not. In particular, there is no reason to believe that continents must

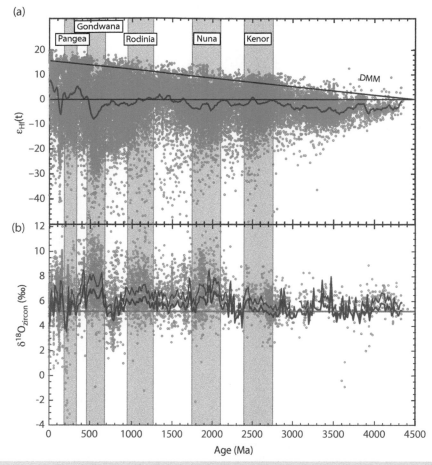

Figure 8.9   (a) Initial ε_{Hf} in ~42 000 zircons as a function of measured U-Pb ages compiled by Roberts and Spencer (2015). Horizontal line is ε_{Hf} = 0; also shown is the depleted mantle evolution line. (b) $\delta^{18}O$ in ~6300 zircons adapted from Spencer et al. (2014). Horizontal line is the mantle value of +5.3‰. Red lines in both show the average values smoothed with a 50-Ma bandwidth. Gray band represents times of supercontinents.

be derived from a MORB-like source; indeed, an OIB-like or island-arc-like source might be more appropriate, both of which tend to have positive ε_{Hf}, but not has high as MORB. We see that many zircons have $\delta^{18}O$ and initial ε_{Hf} within the mantle range. Thus, while Figure 8.9 reveals that there has been considerable reprocessing of existing crust, it also demonstrates that there have been new additions to crust throughout Earth's history.

Yet another way to look at these data is to examine the Hf τ_{DM} values through time, as shown in Figure 8.10a. Nearly all τ_{DM} ages are older than the crystallization age, again suggesting a significant reprocessing of older crust. The lag increases through time, which may only reflect the aging of the continents and increasing amounts of much older crust

available for reprocessing. Superimposed on this trend appears to be an excursion to a greater lag between crystallization age and depleted mantle model ages during the Gondwana period and a sharp spike following the breakup of Pangea.

Figure 8.10b shows the distribution of τ_{DM} ages for the Roberts and Spencer (2015) data set. Principal modes occur at 2700–2800 and 1400–1600 Ma with smaller modes at 600–500 Ma corresponding to the Gondwana supercontinent and around 4000 Ma, the period when the oldest rocks appear in the geologic record, suggesting major new additions to crust at these times.

There are a number of caveats we should bear in mind in examining Figure 8.10, particularly when we compare Figure 8.10b to

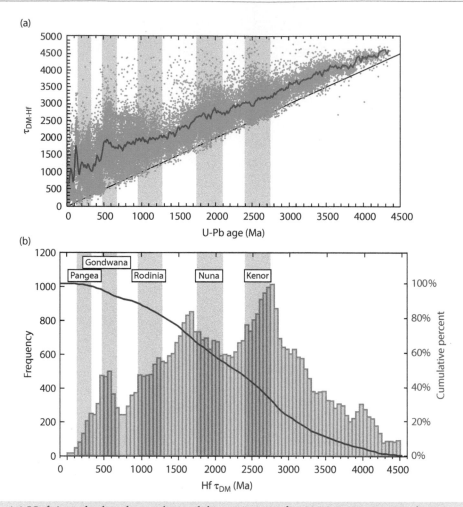

Figure 8.10   (a) Hafnium depleted mantle model ages (τ_{DM}) of ~42 000 zircons as a function of measured U-Pb ages adapted from Roberts and Spencer (2015). Black line is where τ_{DM} equals crystallization age; red line shows the mean values smoothed over a 50-Ma bandwidth. (b) Histogram of hafnium depleted mantle model ages (τ_{DM}) for the Roberts and Spencer (2015) data set. Dark blue line shows the cumulative percent of τ_{DM} ages.

Figure 8.8. First, depleted mantle model ages are just that: *models*. We are assuming that the mantle uniformly evolved along a line extending from the CHUR value at 4500 Ma to the present MORB having ε_{Hf} (~ +14) and that new additions to crust were derived from this reservoir. That need not be the case and indeed Guitreau et al. (2012) argue that the precursors to Archean TTG suites were derived from the deep mantle with near chondritic Lu/Hf and $\varepsilon_{Hf} \approx 0$ If so, we should use τ_{CHUR} ages rather than τ_{DM} ages that would be younger, more so for younger rocks than old ones (Guitreau et al., 2012; Payne et al., 2016). Dhuime et al. (2017) argue that new crust is mainly produced by subduction-related magmas that have lower ε_{Hf} than MORB.

There are also questions about the accuracy of Hf initial ratios and model ages determined on zircons. The accuracy of both depends on the accuracy not only of the $^{176}Hf/^{177}Hf$ ratio, but also of the $^{176}Lu/^{177}Hf$ ratio assumed to project back to DM or CHUR evolution; for example, in Figure 8.2, assuming a $^{176}Lu/^{177}$ Hf appropriate for mafic crust would produce an older τ_{DM} than a $^{176}Lu/^{177}Hf$ appropriate for upper crust. Model age also depends on the U-Pb age; if the latter has been disturbed, the initial $^{176}Hf/^{177}Hf$ ratio is incorrect and so is $\varepsilon_{Hf}(t)$ and τ_{DM}. The point is that we should not attach too much importance to exact values. Second, the Roberts and Spencer data set used to produce Figure 8.10 is about an order of magnitude smaller than the Puetz et al. (2018) data set used to produce Figure 8.8, raising the question of how representative the former is. That said, Figure 8.10 paints a dramatically different portrait of crustal growth than Figure 8.8: peaks in the τ_{DM} distribution do not correspond particularly well with peaks in zircons ages or with the existence of supercontinents, except Gondwana. More significantly, while less than 50% of zircons are older than 1100 Ma, 50% of their Hf depleted mantle model ages are older than 2350 Ma. Finally, we also need to bear in mind that both Figures 8.8 and 8.10 represent only surviving continental crust. We found in the previous chapter that the isotopic signal of crustal material in the mantle is pervasive, suggesting significant losses of continental crust over time through subduction and floundering. In that sense, the cumulative distribution curve in Figure 8.10b likely represents a minimum estimate of crustal volumes in the past.

8.4.2   The Debate over Continental Growth

The zircon record shown in Figures 8.7–8.10 lies at the heart of several key debates over continental evolution:

- Are zircons recording episodic continental growth or do they reflect a preservation bias?
- What controls continental growth? Which of the mechanisms of crustal growth listed in Section 8.2 has been most important and has the dominant mechanism varied over time?
- What has been the net growth in continental mass, which is the difference between additions to the continents and continental destruction through erosion, sediment subduction, subduction erosion, and continental floundering?
- When did modern plate tectonics begin and what was the mechanism of continental growth before it began?

The first two questions are intimately intertwined, so we will discuss them together.

8.4.2.1   Growth or preservation?

The most straightforward interpretation of Figure 8.8 is that the continental crust has grown in spurts, and this is the interpretation championed by Albarède (1998), Arndt and

Davaille (2013), Arndt (2013), and Condie et al. (2015), among others. In their interpretation, the distribution of ages in Figure 8.8 reflects the episodic convection of the mantle driving accelerated crustal growth. Arndt and Davaille (2013) argue that heat from the core in the early Earth periodically destabilized the lowermost mantle resulting in episodes of thermal convection and subduction. High upper mantle temperatures associated with this enhanced convection also result in an enhanced melting of the deep crust and generation of granitoid magmas. Condie et al. (2015) argue that the eruption of LIPs correlates with zircon distribution peaks in Figure 8.8 at 2700, 2500, 2100, 1900, 1750, 1100, and 600 Ma and also probably at 3450, 3000, 2000, and 300 Ma. Since LIPs are thought to be produced when new mantle plumes reach the surface, this correlation would support the argument that crustal production is essentially controlled by events in the lower mantle. They argue that the apparent correlation with supercontinents reflects the influence of these plumes on continent collisions and breakup.

The alternative view emphasizes that continental growth reflects the balance between rates of continent generation and destruction, the association of peaks in the zircon distribution with supercontinents, and the chemical similarities (such as Nb–Ta depletion) of continents to subduction-related magmas. In this model, the production of the continental crust is in approximate steady state, and the peaks in zircon age distribution record periods of enhanced preservation of new crust during the assembly of supercontinents (Hawkesworth et al., 2010, 2019; Condie and Aster, 2010; Dhuime et al., 2012). These authors note that while magma generation rates are high in subduction zones, so also are rates at which the continental crust is destroyed by erosion, subduction, and floundering, so that new crust produced in this setting is less likely to be preserved. On the other hand, while magma generation rates in the final stages of continental convergence and collision are lower, the chances of preservation are higher. Furthermore, they argue that peaks in the zircon distribution are also associated with high $\delta^{18}O$ and low ε_{Hf} in zircons, consistent with extensive continental reworking in supercontinent assembly. While the geology of the putative Nuna (alternatively named Columbia) and Kenor (alternatively Kenorland or Superia) supercontinents is poorly understood, Hawkesworth et al. (2017) also argue that the more recent peaks in the zircon distribution (Figure 8.8) are clearly associated with the Grenville, Pan-African, and Variscan-Alleghenian orogenies, and the assembly of the Rodinia, Gondwana, and Pangea supercontinents rather than times of unusual volumes of magma generation.

The debate over of the extent to which the zircon age distribution in Figure 8.8 represents growth or preservation continues without resolution. Both sides agree that a significant fraction of the Earth's early crust has been lost.

8.4.2.2 When did plate tectonics begin?

An even larger debate over the role of plate tectonics in the production of continental crust is also unresolved. As we noted in Section 8.2, subduction-related magmatism appears to be the dominant present mechanism of production of continental crust through Phanerozoic and hence directly related to plate tectonics, but there is little consensus on when modern-style plate tectonics began and on the nature of mantle convection prior to its beginning. This is an extremely broad topic, and an entire book could be written about this question (indeed, at least one has: Condie and Pease, 2008). Our discussion here will necessarily be brief; Brown et al. (2020) provide a recent in-depth review of this question. Here, we will only briefly review the question in the context of crustal evolution and the isotopic evidence that can be brought to bear on it.

The debate is animated in large part by the widespread recognition that the Archean continental crust is fundamentally different from the post-Archean crust. It is more mafic, thicker, and underlain by thick, stable subcontinental lithospheric mantle. The so-called granite-greenstone belts, which consist of metamorphosed basalts and komatiites punctuated by intrusions of TTG plutons, predominate. This geology is unusual if not entirely absence in younger crust. It is chemically distinct as well. Keller and Schoene (2012, 2018) document systematic changes in trace element basalt chemistry, such as an increase in La/Sm that occur globally between 3 and

2 Ga, with the sharpest transition around 2.5 Ga.

Before we go further, we should define what is meant by "modern-style" plate tectonics and how it might differ from forms of mantle convection that preceded it. As Stern (2008) explains, "Plate motions result mostly from the negative buoyancy of oceanic lithosphere, which slides down and away from mid-ocean ridges (ridge push) toward the trench and sinks in subduction zones (Forsyth and Uyeda, 1975)." Plates themselves (at least oceanic ones) provide the principal driving force for plate motions through slab pull and organize mantle convection that is characterized by sheet-like upwellings and downwellings.

This style of mantle convection appears unique to the Earth among the terrestrial planets. Mantle convection on Mars appears dominated by plumes. The Venusian surface is estimated to be on average less than 1 Ga old and possibly much younger, which implies continued mantle convection, volcanism, and tectonics, but it lacks the linear features characteristic of plate tectonics. Venus has inspired a variety of ideas of about a preplate tectonics Earth. One of these is "stagnant-lid tectonics" in which there are bidirectional fluxes between crust and mantle but without lateral motion of the lithosphere (e.g. Davaille et al., 2017). Another of these is sometimes referred to as "sagduction" (Bédard, 2018), in which thick Archean crust could lead to Rayleigh–Taylor instabilities and drips of dense lower crust into the mantle. Some preplate tectonic models involve episodes of mantle overturn triggered either by cooling from the top or heating from below, and many involve occasional episodes of short-lived subduction. An important point here is that the lack of modern plate tectonics does not necessarily preclude return of crustal material to the mantle.

At one extreme, Stern (2008) argued that modern-style plate tectonics only began in the Neoproterozoic (<1 Ga). At the other extreme, Kusky et al. (2018) argue that convergent-margin-related sequences can be identified in rocks of all ages; for example, rocks of the Nulliak Belt in the Saglek Block of Labrador include MORB-like basalt and metamorphosed marine sediments, which they argue are an accreted ophiolite sequence. While there is no consensus on when plate tectonics began, many scientists would place their bets on an answer between these extremes, particularly in that time between 3 and 2 Ga when the nature of the continental lithosphere appears to change (e.g. Brown et al., 2020, Hawkesworth et al., 2020). The transition from whatever preceded it to modern plate tectonics was likely not an instantaneous one and may have occupied much of this period. It is worth noting that another important transition occurred during this time as well: oxygenation of the atmosphere, which we will discuss at greater length in Chapter 12.

Isotopic evidence is unlikely to entirely resolve this debate, but it can place some constraints. Shirey and Richardson (2011) pointed out that diamonds older than 3200 Ma contain only peridotite minerals as inclusions, whereas diamonds with inclusions of eclogitic minerals become prevalent around 3000 Ma. The latter have high $^{187}Os/^{188}Os$, consistent with a crustal origin, in contrast to peridotite diamonds. Eclogitic diamonds also often have C and N isotopic compositions indicative of a crustal origin as well and, as we found in the previous chapter, can contain sulfide inclusions with mass independently fractionated (MIF) sulfur, which as we found in Section 5.4 originated in the Archean atmosphere. Shirey and Richardson argued that the eclogites originated as subducted oceanic crust, and their presence in the subcontinental lithosphere marks the beginning of modern-style plate tectonics. Further support for this theory was reported by Smit et al. (2019) who demonstrated that MIF sulfur can be found in a variety of diamonds from the Kapvaal and Zimbabwe cratons but is absent in 3500-Ga diamonds from the Slave Province.

8.4.2.3 Titanium isotopes and plate tectonics

Titanium isotopes can provide constraints on the nature and origin of ancient continental crust. Millet et al. (2016) demonstrated that $\delta^{49}Ti$, which is the deviation of the $^{49}Ti/^{47}Ti$ ratio from that of the OL–Ti standard, correlates very strongly with the SiO_2 content of igneous rocks, indicating a fractionation factor for fractional crystallization of $\Delta^{49}Ti_{oxide-melt}$ = $-0.23 \times 10^6/T^2$. On the other hand, they found that $\delta^{49}Ti$ did not change during partial melting. The likely cause of this fractionation is the difference between Ti coordination in silicate melts and in Fe–Ti oxides, which are the principal Ti-hosting minerals. In the latter, Ti is exclusively sixfold-coordinated, whereas in silicate melts, Ti is present in four-, five-, and sixfold-coordination. Our rule of thumb in Chapter 5 was that the heavy isotope will preferentially partition into the stronger and stiffer low-coordination environment (the silicate melt) leaving the Fe–Ti minerals enriched in isotopically light Ti. Once Ti–Fe oxides begin to crystallize, fractional crystallization will thus drive the residual melt to heavier $\delta^{49}Ti$. In principle then, $\delta^{49}Ti$ could be a proxy for SiO_2.

Ti is a highly insoluble element, and $\delta^{49}Ti$ can be expected to be unaffected by weathering, so that the Ti isotopic signature of sediments should reflect that of their original igneous source and their SiO_2 concentration. Deng et al. (2019) showed that tholeiitic and calc-alkaline magmas defined two distinct $\delta^{49}Ti$–SiO_2 correlations and argued that $\delta^{49}Ti$ could not be used as a simple proxy for SiO_2 of ancient crust, but that $\delta^{49}Ti$ in combination with SiO_2 could be used to distinguish between the calc-alkaline and tholeiitic series despite those rocks having experienced significant metamorphism, which is often the case for Archean rocks. Subsequently, Hoare et al. (2020) showed that alkaline magmas, such as those from Ascension and Heard Islands, define yet another distinct trend with $\delta^{49}Ti$ increasing with increasing SiO_2 or decreasing MgO even more rapidly than for calc-alkaline magmas, with the most extremely fractionated compositions having $\delta^{49}Ti > 2‰$. The reason for this is the generally higher TiO_2 content of the alkaline magmas, which results in the earlier saturation of Ti-oxides and greater removal of them.

What is the difference between the calc-alkaline and tholeiitic evolution and how does this help us understand crustal evolution and plate tectonics? Magmas generated in subduction zones are richer in water and more oxidized that other magmas, and consequently, Fe–Ti oxides precipitate early in the fractional crystallization sequence, which leads to rapid increase in SiO_2 and depletion in Fe and Ti; this is calc-alkaline trend. In contrast, in more reduced and drier magmas generated at mid-ocean ridges and above mantle plumes evolve along the tholeiitic trend in which Fe–Ti oxide precipitation is delayed, producing enrichments and Fe and Ti and more modest increase

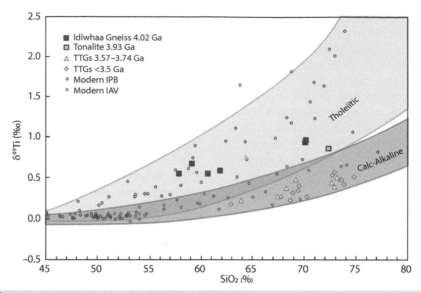

Figure 8.11 δ^{49}Ti versus SiO_2 for rocks from the Acasta Gneiss Complex showing the differing trends for the tholeiitic and calc-alkaline magma series. IAV: island arc basalts, IPB: intraplate basalts. The oldest rocks from the Acasta Gneiss Complex fall within the region of the tholeiitic trend, while rocks younger than 3.75-Ga plot within the calc-alkaline trend, suggesting that they were generated in an environment similar to modern subduction zones. Data from Aarons et al. (2020).

in SiO_2. The more rapid increase in SiO_2 in the calc-alkaline trend leads to higher SiO_2 at a given δ^{49}Ti than in the tholeiitic trend. Demonstrating that a suite of Archean rocks belong to the calc-alkaline series implies the existing of subduction zones and subduction zone magmatism at the time.

Aarons et al. (2020) measured Ti and Fe isotopes and SiO_2 in rocks from the Acasta Gneiss Complex that ranged in age from 4.02 to 2.95 Ga. As Figure 8.11 shows, the oldest rocks, the Idiwaa Gneiss unit and a tonalite dated at 3.93 Ma, fall in the tholeiitic field on a δ^{49}Ti versus SiO_2 plot, while all the rocks younger than 3.75 Ga fall on the calc-alkaline trend. This suggests that the new additions to crust in the Slave craton after 3.8 Ga apparent in Figure 8.2 were water-rich magmas produced in a plate-tectonic-like subduction regime. Merely finding examples of rocks that may have been generated by possible plate tectonic processes does not mean that it was the dominant tectonic regime at the time as other models of early Earth tectonics include intervals of subduction. Nevertheless, even the intermittent operation of subduction and subduction-related volcanism could mean that this has been an important mechanism of crust production even in the Archean and help explain the geochemical similarities of continental crust and subduction zone magmas.

8.4.3 Case studies of crustal growth

Let us now focus on regional studies of how the continental crust has grown making use of Nd and Hf isotope systematics. We begin by examining the classic study of Bennett and DePaolo (1987) of the Western USA. Figure 8.12 shows a map of Western USA showing contours of Nd depleted mantle ages or crustal residence times (τ_{DM}). The data define three distinct provinces, Mojave, Yavapi, and Mazatzal, and suggest the existence of several others. Figure 8.13 shows the initial ε_{Nd} values of the granites from the three numbered provinces plotted as a function of crystallization age. Only in province 3 do we find rocks, tholeiitic and calc-alkaline greenstones, whose crustal residence time is equal to their crystallization ages. In the other regions, the oldest rocks have initial ε_{Nd} values that plot below the depleted mantle evolution curve. This suggests that they contain significant amounts of reprocessed preexisting crust. We should emphasize at this point that the crustal residence time gives the average crustal residence time of Nd in the material. Thus, if a

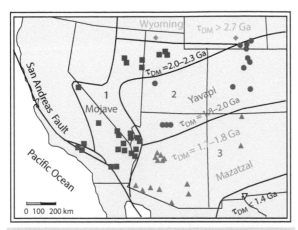

Figure 8.12 Map of the western USA showing isotopic provinces varying Nd depleted mantle model ages. Model ages define three provinces that become progressively younger to the south away from the Archean Wyoming Craton (green).

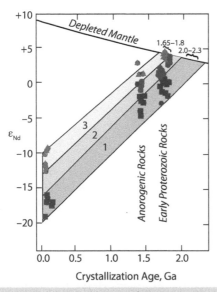

Figure 8.13 Initial ε_{Nd} as a function of crystallization age of Western USA. Groupings 1, 2, and 3 refer to provinces shown in Figure 8.12. (Source: Adapted from Bennett and DePaolo, 1987).

continental rock formed at 1.0 Ga contained Nd derived in equal proportions from the mantle and 2.0 Ga crust, its crustal residence time would be 1.5 Ga. In each province, there have been subsequent episodes of magmatism, but in those subsequent episodes, there have been

no new additions of crust (they plot along the same evolutionary array as the older material in the province). A subsequent study by Chapman et al. (2018) found that the pattern of initial ε_{Hf} in zircons was broadly similar, although Hf model ages in some cases are older than equivalent Nd model ages.

All three provinces apparently formed between 1.8 and 1.65 Ga, though rocks from province 1 may be slightly older. A scenario suggested by Bennett and DePaolo that is consistent with the observations is successive accretion or growth of island arcs to the preexisting Archean Wyoming Craton to the north. The earliest formed arcs or at least those closest to the craton received a substantial component of older crust from the craton. This could have occurred through erosion and subduction, or, if the arc was built directly on the continent, through the assimilation of crust. As new Proterozoic crust was built outward from the continent, it screened subsequent arcs from the contribution of material from the Archean crust. A similar effect has been observed in the Proterozoic provinces of Canada.

Our second case study is the pioneering work by Hawkesworth and Kemp (2006) using combined O and Hf isotopic analysis of zircons to reveal details of local crustal growth. The Lachlan Fold Belt of southeastern Australia consists of metamorphosed Ordovician to Silurian turbidites and granitoids and spatially related volcanic rocks, almost all of which were emplaced between 430 and 350 Ma. Hawkesworth and Kemp analyzed detrital zircons from the metasediments and found that the crystallization age spectra of the detrital zircons are dominated by peaks at 450–600 Ma and 0.9–1.2 Ga (Figure 8.14). Hf model ages of zircons, calculated assuming the zircons crystallized from magmas with Lu/Hf ratios similar to average bulk continental crust (0.08), are much older and show two distinct peaks in distribution. The τ_{DM} peak centered around 1.7–1.9 Ga consists primarily of zircons with elevated $\delta^{18}O$, implying the zircons crystallized from magmas with substantial amounts of reworked preexisting upper crust. The older peak at around 2.9–3.1 Ga consists mainly of zircons with more mantle-like O isotope ratios ($\delta^{18}O < 6.5$). Presumably, the latter group reflects the reprocessing or melting of deeper crustal material that had not interacted with water at low and moderate

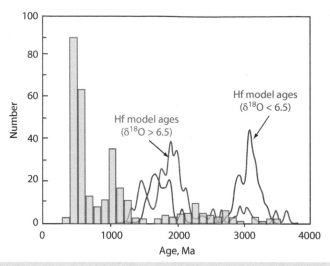

Figure 8.14 Histogram of U-Pb ages of zircons in S-type granites and detrital zircons in Ordovician metasediments of the Lachlan Fold Belts of Southeastern Australia. Superimposed on this is the probability distribution of τ_{DM} of these zircons; the red curve represents zircons with $\delta^{18}O > 6.5‰$ and the blue curve represents zircons with more mantle-like $\delta^{18}O$ ($< 6.5‰$). (Source: Hawkesworth and Kemp (2006)/ Reproduced with permission from Elsevier.)

temperatures. The important result is that the events at 450–600 Ma and 0.9–1.2 Ga recorded by zircon ages represent mainly reprocessing of the existing crust rather than net crustal growth.

Hawkesworth and Kemp (2006) also determine ages and Hf and O isotopic compositions of two suites of granites from the Lachlan Fold Belt. In a classic earlier study, Chappell and White (1974) had divided these granites into I- and S-types, introducing what has become a widely used terminology. The I (for igneous) types are metaluminous $(((Ca+Na+K)/Al)_{atomic} > 1)$ to weakly peraluminous $(((Ca+Na+K)/Al)_{atomic} < 1)$, with high CaO and Ca/Na, usually contain hornblende, and are thought to be derived by the melting of a metaigneous protolith. The S (for sedimentary) types are strongly peraluminous, typically cordierite-bearing and have generally lower abundances of elements enriched in seawater (Na, Ca, and Sr) and are thought to be derived by the melting of a metasedimentary protolith. Despite the difference in origin, the trace element compositions are similar, and they define a single overlapping array on an ε_{Nd}-$^{87}Sr/^{86}Sr$ diagram. As a result, in contrast to the Southwestern USA discussed earlier, the question remained whether the low ε_{Nd} of some of the I-type

granites reflect the contamination of mantle-derived magmas by metasediments, or melts thereof, or derivation from older metaigneous protoliths.

Hawkesworth and Kemp analyzed O and Hf in zircons from two suites of I-type granites and found that the answer to the above question was both. The two studied suites have overlapping ranges of $\delta^{18}O$ values extending down to mantle-like values ($\sim +5.5‰$), and both the suites define curved arrays suggesting mixing between two end-members. In both the cases, ε_{Hf} decreases from core to rim in individual zircons suggesting the assimilation of "crustal" or sedimentary component as crystallization proceeded. The Cobargo Granite, however, has higher ε_{Hf}, and the $\delta^{18}O$–ε_{Hf} relationship suggests that the high ε_{Hf} end-member was juvenile mantle-derived magma. In the Jindabyne Suite, located inland of the Cobargo granite, minimum $\delta^{18}O$ and maximum ε_{Hf} values are quite similar to locally outcropping gabbro (denoted by the star), suggesting that the Cobargo granite was produced by the melting of older mafic crust that has evolved to lower ε_{Hf} that mixed with or assimilated melts of a sedimentary protolith (Figure 8.15). Thus, the Cobargo suite reflects a new addition to crust, while the Jinabyne suite merely reflects the reprocessing of older crust.

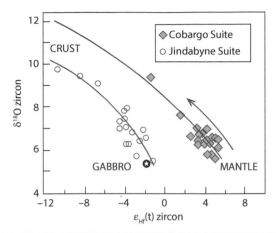

Figure 8.15 $\delta^{18}O$ versus initial ε_{Hf} for two I-type granite suites from the Lachlan Fold Belt of southeastern Australia. They define two distinct mixing curves with a "sedimentary" (high $\delta^{18}O$) material. The average Hf and O isotope (zircon) composition of a gabbro that is spatially associated with the Jindabyne Suite is shown by the star symbol. Arrows show the direction of crystallization deduced from core-to-rim variations in individual zircons. (Source: Hawkesworth and Kemp (2006)/Reproduced with permission from Elsevier.)

8.4.4 Growth of the continental crust through time

One of the most enduring debates in geology is the question of how crustal mass has changed over time and what the *net* rate of crustal growth has been. A first way to look at this is to determine the age of exposed crustal rocks, as was done by Hurley and Rand (1969). In an approach that foreshadowed the use of Nd and Hf τ_{DM} model ages, Hurley and Rand used average initial $^{87}Sr/^{86}Sr$ ratios to estimate the age of large tracts of crust from all the continents and concluded that continental growth had accelerated through time, mainly by the accretion of new crust to the margins of older cratons. They found that 30% of the crust was younger than 450 Ma and 63% younger than 900 Ma. Their work was, however, largely based on K–Ar ages, which are readily reset. In addition, few ages older than 3 Ga had been reported by that time, and while they had extensive sampling from many regions, they had no samples from many other regions, including China and western Russia. For all its faults, the results of the

Hurley and Rand study are not greatly different from what we might infer from looking at the distribution of zircon ages in Figure 8.8: 30% of zircons are Phanerozoic (<550 Ma) and 43% are younger than 900 Ma.

We have found in this chapter, however, that ages of rocks do not necessarily reflect the time that material has been part of the continental crust as old crust is frequently reprocessed into new crust (as James Hutton recognized in 1788). Furthermore, continental crust can be recycled into the mantle. Indeed, Armstrong (1968, 1981) argued that despite evidence of new additions to crust throughout geologic time, the mass or volume of continents has remained essentially unchanged since 4 Ga because the rate of destruction of continental crust has matched the rate of production. While this perhaps represents an extreme view of crustal volume through time, there is widespread recognition that crust has been destroyed as well as produced through geologic time. Continental mass can be lost through weathering, erosion, and transport, and ultimately the subduction of sediment, subduction erosion, and lower crustal foundering. There is good evidence that all three mechanisms operate, and some have argued that the rates of the first two processes are sufficient to balance rates of new crustal addition and, therefore, produce a steady-state crustal volume, just as Armstrong proposed (e.g. Stern and Scholl, 2010). However, no consensus has formed around these views, and the question of the net change in crust mass is still debated.

Recognizing this, Condie and Aster (2010) used Hf isotopes and statistical methods to produce a globally representative distribution of surviving juvenile crust shown in Figure 8.16. In contrast to what Hurley and Rand concluded, it shows that the rate of production of new crust has decreased with time. In their accounting, about a third of the present crust was produced in the Archean and less than 14% during the Phanerozoic. Even assuming that Condie and Aster succeeded in estimating the present age distribution of surviving crust, this does not tell us how continental volume (or mass) has changed with time because we do not know how much was lost.

A different approach to estimating continental growth is to examine sediments. This has several advantages. First, sediments are natural

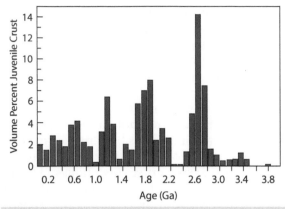

Figure 8.16 Estimated volume of juvenile crustal preserved as a function of geologic age based on igneous and detrital U/Pb zircon ages, whole-rock Nd isotope ratios and/or zircon Hf isotope ratios, and areal distributions of rocks of known ages. Source: Condie and Aster (2010)/with permission of Elsevier.

averages of the rocks being eroded within a drainage basin, so much can be learned from a minimum number of samples. Second, sediments are deposited in lowland areas and are consequently less likely to be lost to erosion. As the Jack Hills metasediments so strikingly demonstrate, sediments can survive even though its source rocks have been lost to erosion.

However, the question is: Is the isotopic composition of sediments representative of the rocks being eroded? The answer depends on how the parent and daughter elements behave during weathering and subsequent transport. Zircons are virtually immune to the effects of weathering and transport so that the detrital zircon population in sediments will reflect that of their sources. As we have already seen, detrital zircons are widely exploited to study continental evolution. For bulk sediment, however, the answer would be no for the Lu–Hf system. Lu and Hf behave quite differently during weathering, with Lu concentrating in clays, while Hf will be concentrated in accessory minerals, zircon in particular, in the course sedimentary fraction. Since the course and clay fraction will be transported at different rates and deposited in different environments, Hf isotope ratios of bulk sediments will not be representative of the source. U, Th, and Pb also behave differently during

weathering as will Rb and Sr. For the Sm–Nd system, however, the answer is yes. Both Sm and Nd are highly insoluble and concentrate in the clay fraction, and several studies have demonstrated that there is little fractionation of Sm from Nd during erosion and transport of fine-grained sediments such as shales (e.g. McCulloch and Wasserburg, 1978; Goldstein et al., 1984). Consequently, we will concentration on the Sm–Nd system and Nd model ages.

Figure 8.17a shows a plot of the crustal residence age, τ_{CR}, which is the same as τ_{DM}, versus the stratigraphic age (τ_{ST}) of sediments. Note that, in general, we expect that the crustal residence age will be somewhat older than the stratigraphic age. Only when a rock is eroded into the sedimentary mass immediately after its derivation from the mantle will its stratigraphic (τ_{ST}) and crustal residence age (τ_{CR}) be equal. We can consider how several scenarios of crustal growth would plot on this diagram, assuming that new material added to the continents is reflected in new material added to the sedimentary mass. If the continents had been created 4.0 Ga ago and if there had been no new additions to the continental crust since that time, then the crustal residence time of all the sediments should be 4.0 Ga regardless of stratigraphic age, which is illustrated by the line labeled "No new input." If, on the other hand, the rate of continent growth through time has been uniform since 4.0 Ga, then τ_{ST} and τ_{CR} of the sedimentary mass should lie along a line with slope of 1/2, which is the line labeled "Uniform Rate."

Allègre and Rousseau (1984) determined Nd crustal residence times of a series of eight Australian shales with stratigraphic ages ranging from 0.2 to 3.3 Ga (Figure 8.17b). They performed a similar analysis on shales from the circum-North Atlantic. For modeling purposes, they then averaged these into uniformly spaced time bins of 500 Ma: the resulting models are represented by lines in Figure 8.17b. The Australian data define a trend that falls below the uniform rate line but eventually flatten to a much lower slope implying a decreasing rate or perhaps even no new input. The North Atlantic data are more scattered but appear to imply more rapid initial growth but then also flatten in the Phanerozoic.

Allègre and Rousseau then noted that continents do not erode uniformly: topographic

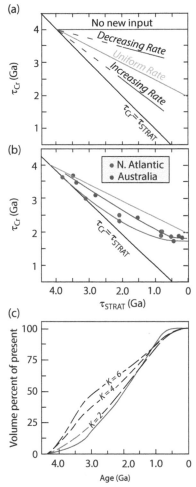

(a)

(b)

• N. Atlantic
• Australia

(c)

Figure 8.17 (a) Relationship between the stratigraphic age of sedimentary rocks (τ_{ST}) and Nd crustal residence age and the crustal residence age ($\tau_{CR} = \tau_{DM}$) of material in those sediments. A slope of ½ implies that the continental growth rate is constant, a slope < ½ implies a decreasing rate, and a slope > ½ implies an increasing rate of crustal growth. Source: Adapted from (Goldstein et al., 1984). (b) Nd crustal residence ages as a function of stratigraphic age for shales analyzed from Australia (red) and the circum-North Atlantic (blue) by Allègre and Rousseau (1984). Lines are their modeled growth rates. Source: Adapted from (Allègre and Roussseau, 1984). (c) Australian shale model of Allègre and Rousseau converted to volume percent of the total percent of the continental crust as a function of geologic age showing the effect of the young crust bias factor, K defined in 8.1. Red line is the case where $K = 0$. Higher bias factors imply more rapid early crustal growth. Source: Adapted from (Dhuime et al., 2017).

highs will contribute disproportionally to sedimentary mass. Indeed, this is no doubt why some >4 Ga detrital zircons survive, while the rocks in which they formed did not: the parent rocks occupied topographic highs, while the eroded zircons were deposited in topographic lows. New mantle-derived material is more likely to be found at topographic highs (e.g. volcanoes), while surviving older rocks are more likely to be found in topographic lows. Hence, sediments will be biased toward sampling younger rocks. They defined a parameter K, which is the ratio of the ratio of the fraction sediment derived from the rocks formed in that time bin, x, to sediments derived from all other rocks, 1–x, to the ratio of the youngest rocks, y, to all other rocks, 1– y:

$$K = \frac{x/(1-x)}{y/(1-y)} \qquad (8.1)$$

At each time step, n, at time t, the crustal residence time of the shale, $\tau_{CR}(n)$, can then be calculated as

$$\tau_{CR}(n) = xt + (1-x)\tau_{CR}(n-1). \qquad (8.2)$$

And the mean age of the continent at that time can be calculated as

$$\langle T(n)\rangle = yt_n + (1-y)\langle T(n-1)\rangle. \qquad (8.3)$$

There are few constraints on the value of K, so they computed their model for values of 2, 4, and 6 and computed an average age of the continents for each case and obtained 1.9, 2.0, and 2.3 for K equal to 2, 4, and 6, respectively. Figure 8.17c shows the Australian shale data translated into crustal volume as a function of time by Dhuime et al. (2017) and illustrates the effect of the sampling bias. We can see that if shales preferentially sample young crust in topographic highs, older crust is underrepresented resulting in the underestimation of early crustal growth rates. Subsequently, Dhuime et al. (2017) determined Nd crustal residence times in a global sample of 645 fine-grained sediments. Their data show considerably more scatter, but the running median matched the Allègre and Rousseau (1984) curve quite well.

Detrital zircons, of course, can also be extracted from sediments, and their Hf model ages can be used to constrain crustal growth. To do so, Belousova et al. (2010) calculated

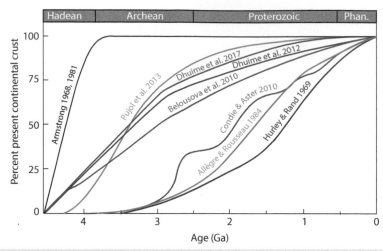

Figure 8.18 Summary of isotopic models of the growth of continental crust. Source: Adapted from Dhuime et al. (2017) and Hawkesworth et al. (2019).

a crustal growth curve based on the fraction of zircons derived from juvenile crust through time in a data set of 13 000 detrital zircons. The juvenile fraction was calculated in 100-Ma time slices as

$$X_{juv} = N_{model\,age}/\left(N_{U\,-\,Pb\,age} + N_{model\,age}\right)$$
(8.4)

where $N_{model\,age}$ is the number of zircons having an Hf model age within the time slice and $N_{U\text{-}Pb\,age}$ is the number of zircons having a U-Pb crystallization age within the time slice. They constructed the continental growth curve from the cumulative proportion of juvenile crust through time (Figure 8.18), which indicates that ~50% of the present crustal volume was present by 3 Ga.

Dhuime et al. (2012) took a similar approach but noted that the Hf isotope composition of the zircons may reflect mixtures of both juvenile and older reprocessed crust and, hence, may not represent true periods of crust formation. To address this, they also used $\delta^{18}O$ correct for the bias due to zircons with hybrid Hf model ages. They found that the proportion of true new crust to hybrid Hf model ages was ~0.73 prior to 3.2 Ga, which subsequently gradually decreased to a minimum of ~0.2 in the Mesoproterozoic before increasing again to ~1 at present. Dhuime et al. (2017) then modeled the sedimentary Nd τ_{CR} data in the same way as the zircon data assuming a $K = 2$ (also shown in Figure 8.18).

With these constraints, they produced the model growth curve shown in Figure 8.18 in which ~65% of the present crustal volume was present by 3 Ga.

It is important to point out that the approach of Belousova et al. (2010) and Dhuime et al. (2017), which estimate the proportions of juvenile crust at different times over crustal history, differ from the approach of Condie and Aster (2010) and Allègre and Rousseau (1984), which estimate the proportion of different ages in the present-day crust, because the latter does not account for potential losses of crust.

Yet another isotopic system that can constraint crustal growth is K–Ar. When new crust forms through magmatism, ^{40}Ar that has accumulated in the mantle source from ^{40}K decay is released to the atmosphere (degassed). As a consequence, roughly half of the Earth's inventory of ^{40}Ar is now in the atmosphere and the $^{40}Ar/^{36}Ar$ ratio of the atmosphere has grown through time from the primordial value of ~3×10^{-4} to the present value of ~296 (we will discuss Ar isotopic evolution in more detail in Chapter 14). Pujol et al. (2013) analyzed noble gases in fluid inclusions in hydrothermal quartz in pillow basalts in the 3.5-Ga Dresser Formation of the Pilbara Craton of Western Australia. The fluids proved to be mixtures of a hydrothermal end-member with high Cl and high ^{40}Ar and a low-salinity, low-Cl fluid poor in ^{40}Ar. They argued that the low-salinity fluid was likely derived from meteoric water.

Although the age of the basalts is well constrained by several dating methods, the fluid could be younger. Barites from the same formation have U-Xe$_{fission}$ ages of 3.5±0.5 Ga (we will discuss fissiogenic Xe in Chapter 14), which indicate that the secondary mineralization is Archean, but do not strongly constrain the actual age. Pujol et al. (2013) calculated initial $^{40}Ar/^{36}Ar$ for three assumed ages of 3.5, 3.0, and 2.7 Ga of 143±24, 189±21, and 211±21 (compared to the modern value of 296), respectively. Taking account of the relatively short half-life of ^{40}K (1.25 billion years), all of these ratios imply substantial early outgassing of ^{40}Ar from the Earth's interior. Pujol et al. (2013) calculated the fraction of crust that must have formed by the time the fluid formed for outgassing efficiencies associated with crust formation of 1% and 50% for the three different assumed fluid ages. The results imply a fraction of present crustal volume of 31–55% by 3.5 Ga, 59–78% by 3.0 Ga, and 69–87% by 2.7 Ga. The implied crustal evolution curve is also shown in Figure 8.18.

Figure 8.18 provides a summary of the estimates of crustal growth based on isotope ratios. We should emphasize that the Hurley and Rand, Allègre and Rousseau, and Condie and Aster curves are based on the amount of crust of various ages found in the continents at present, whereas the other curves are based on estimates of the amount of new crust added to the continents through geologic history. As Hawkesworth et al. (2019) point out, the difference between these approaches is a measure of the minimum volumes of continental crust that have been destroyed through recycling back into the mantle. It is interesting to note that the latter group have tended to approach Armstrong's initial proposal of steady-state crustal mass. Armstrong (1968) initially proposed his model to explain contrasting interpretations of Sr and Pb isotope ratios of the crust and mantle. He argued that these could be reconciled if one allowed not only the transfer of mantle material to the continental crust but the transfer of crust to the mantle as well. It was a revolutionary idea at the time, particularly given that plate tectonic theory was in its infancy and the mechanism of crust-to-mantle transfer, subduction, was only beginning to be understood.

Fifty years later, there is near-unanimous recognition that significant amounts of continental crust have been recycled into the mantle over Earth's history. Estimates of the amount of crust recycled into the mantle range up to several times the present crustal mass (Hawkesworth et al., 2019), but there is little consensus on this as the constraints are few. On the other hand, few accept Armstrong's idea that the present crustal mass had developed by 4 Ga; nevertheless, most models now involve a rapid growth of continental crust during the Archean, with estimates ranging from ~40% Condie and Aster (2010) to 60–65% (Dhuime et al., 2017).

8.5   ISOTOPIC COMPOSITION OF THE CONTINENTAL CRUST

As Chauvel et al. (2014) point out, knowing of the average composition of the continental crust is crucial to knowing how and when it formed: the models of crustal growth must sum to give the present composition of the crust. Knowing the composition of the crust is also key to understanding the evolution of the mantle, from which the crust is derived, and understanding the evolution of seawater, as most of the dissolved salts are derived from the crust. The upper crust is largely accessible to direct sampling, so establishing its composition is, in principle at least, a less challenging problem than establishing mantle composition. On the other hand, the upper crust is extremely compositionally diverse, so that establishing its average composition is difficult. The lower crust is far less accessible, and establishing its composition presents the same kind of difficulties as establishing mantle composition.

8.5.1   Sediments as samples of the upper crust

A commonly used approach to sampling the upper crust is to let nature do some of the work for us. Weathering constantly breaks down rock, which is then transported by rivers and streams in both the dissolved and particulate forms. Goldstein and Jacobsen (1987) determined the Nd and Sr isotopic composition in dissolved load of rivers from North America, Australia, Japan, the Philippines, India, and Venezuela. These and other data provided a weighted average runoff of $^{87}Sr/^{86}Sr$ = 0.7101 and ε_{Nd} = −10.1. Palmer and Edmond (1989) did a more thorough job of measuring the Sr isotopic compositions of the dissolved load of rivers and obtained a somewhat higher average $^{87}Sr/^{86}Sr$ of 0.7119. Goldstein and Jacobsen (1988) measured the riverine Sr and Nd isotopic fluxes by measured isotopic compositions of *suspended load* in a subset of rivers (mainly North American) and attempting to extrapolate their results to obtain a global average. They estimated the $^{87}Sr/^{86}Sr$ and ε_{Nd} of the continental crust exposed to weathering as 0.716 and –10.6, respectively. However, they had no data on a number of major rivers, notably the Brahmaputra, Ganges, and Yangtze. Goldstein and Jacobsen (1988) also calculated the bulk load (dissolved plus suspended) carried by rivers. A small but significant fraction of the Sr in rivers is in dissolved form, whereas the amount of dissolved Nd is insignificant compared to that in the suspended load so that the total load is essentially the suspended load; their estimate of the $^{87}Sr/^{86}Sr$ of the bulk load was 0.7133. The lower $^{87}Sr/^{86}Sr$ in the dissolved fraction reflects the influence of dissolving carbonates, which have lower $^{87}Sr/^{86}Sr$ than silicate rocks because their Rb/Sr is low and seawater, from which they precipitate, is influenced by hydrothermal activity at mid-ocean ridges. Bayon et al. (2021) carried out a far more extensive survey of the riverine suspended load which included 61 of the world's largest rivers. The also examined the variation between the clay- and silt-sized fractions and found that it varied with mean annual temperature as well as elevation, reflecting differing temperature-dependent breakdown of biotite and feldspar and differences in transport- versus kinetic-limited weathering and erosion conditions. Their average global weighted riverine suspended $^{87}Sr/^{86}Sr$ flux was 0.716, identical to the value of Goldstein and Jacobsen (1988).

Asmeron and Jacobsen (1993) estimated the Pb isotopic composition of the upper crust by measuring Pb isotope ratios in the suspended load of sediments and then estimating the global average from the correlation between Pb isotope ratios and ε_{Nd} in suspended loads. Their estimated composition of the upper crust exposed to weathering is $^{206}Pb/^{204}Pb$ = 19.32, $^{207}Pb/^{204}Pb$ = 15.76, and $^{208}Pb/^{204}Pb$ = 39.33. Hemming and McLennan (2001) obtained a similar estimate of the eroding upper crust ($^{206}Pb/^{204}Pb$ = 19.32, $^{207}Pb/^{204}Pb$ = 15.78, and $^{208}Pb/^{204}Pb$ = 39.58) by analyzing modern marine turbidites sediments. Millot et al. (2004) analyzed Pb isotopes in major rivers draining 45% of the continental surface area and obtained somewhat less radiogenic values. Averaging the results by surface area, their estimate of the Pb isotopic composition of the upper continental crust was somewhat less radiogenic: $^{206}Pb/^{204}Pb$ = 18.93, $^{207}Pb/^{204}Pb$ = 15.71, and $^{208}Pb/^{204}Pb$ = 39.03.

There are a number of limitations to using sediments to sample the crust. The first of these is that, as we saw the previous section, sediments disproportionately sample topographic high areas, which tend to be younger and more mafic than low areas. The second is that weathering is incongruent, with some mineral weathering faster than others and elements partition differently into solution, clays, and residual minerals, which then move downstream at different rates in the dissolved, suspended, and bed load fractions. In a study of river sediments of the Ganges River, Garçon et al. (2014) documented different isotopic compositions of the suspended and bed load fractions. ε_{Nd} differed only slightly between the fractions (–17.2 versus –17.0, respectively), but hafnium was distinctly more radiogenic in the suspended load (ε_{Hf} = –20.5) than the bed load (ε_{Hf} = –28.4) and the opposite was true of Pb ($^{206}Pb/^{204}Pb$ = 20.0 versus 21.0), reflecting the concentration of heavy minerals such as zircon and monazite in the bed load. These averages disguise much greater variability in these sediments, however. Garçon et al. (2013) found strong correlations "pseudo-isochrons" among Pb isotopes in Ganges sediments. These are effectively produced by varying proportions of that heavy minerals, with heavy mineral concentrates having $^{206}Pb/^{204}Pb$ >250. Their preferred Pb isotopic composition of the upper crust based on modeling was effectively identical to that of Millot et al. (2004).

Wind-deposited sediments provide yet another means of estimating upper crustal isotopic composition. Pleistocene glaciation extensively eroded high-latitude regions as well as mountainous areas at lower latitudes. Some of this material was ground to very fine grain size and, as glaciers and ice sheets retreated, was transported by wind to produce thick deposits of loess. Unlike river deposits, little chemical weathering is involved in the production of this sediment. Desert sands represent yet another type of sediment that is produced with relatively little chemical fractionation and deposited by wind. Analyzing samples from a large number of these deposits, Chauvel et al. (2014) estimated the average ε_{Nd} of the upper crust as -10.3 ± 1.2, a value indistinguishable from that of Goldstein and Jacobsen's (1988) value of -10.6. Chauvel et al. (2014) also estimated the average ε_{Hf} of the upper crust as -13.2 ± 2. Israel et al. (2020) analyzed a subsample of the loess samples of Chauvel et al. and obtained average values $\varepsilon_{Ce} = 1.8\pm0.3$ and $\varepsilon_{Nd} = -11.2\pm3.0$. Willig et al. (2020) found that ε_{Ce} and ε_{Nd} are related in oceanic basalts (the "Mantle Array") as $\varepsilon_{Ce} = -0.11\times \varepsilon_{Nd} + 0.08$. If this relationship holds for the crust, the ε_{Ce} value predicted for Chauvel et al.'s ε_{Nd} of -10.3 is 1.21.

Peucker-Ehrenbrink and Jahn (2001) also relied on loess samples to estimate the $^{187}Os/^{188}Os$ ratio of the upper continental crust. They found an average $^{187}Os/^{188}Os$ of 1.05 ± 0.23 and $^{186}Os/^{188}Os = 0.119871$ for six loess samples from around the world, which were surprisingly uniform. From these data, they estimated the average Os isotopic composition of the upper crust as $^{187}Os/^{188}Os$ of 1.40 and $^{186}Os/^{188}Os = 0.119885$.

Table 8.1 provides a summary of the estimated radiogenic isotopic composition of the upper continental crust. Considerable uncertainty still remains in these values, which will no doubt evolve somewhat in the future.

8.5.2   Isotopic Composition of the Lower Crust

In the preceding assessment of upper continental crust isotopic composition, we relied on the processes of erosion and sedimentation to obtain average compositions of large volumes of rocks at the surface. In assessing the isotopic composition of the mantle, we made the extensive use of basalts, each of which results from the melting of many cubic kilometers and, hence, tends to average out small-scale heterogeneities. These approaches are not possible for the lower continental crust. We can infer some aspects of lower crustal composition from remote geophysical techniques (seismic waves, gravity, heat flow, etc.), including that it is almost certainly denser, more mafic, and poorer in radioactive elements, but they are of limited use in defining its isotopic composition. As with the mantle, three kinds of samples of lower crust are available: terrains or massifs that have been tectonically emplaced in the upper crust, xenoliths in igneous rocks, and, in principle at least, magmas produced by the partial melting of the lower crust. However,

Table 8.1   Estimated average radiogenic isotopic composition of the continental crust.

		Lower crust	
	Upper crust	Xenolith mean	Xenolith median
$^{87}Sr/^{86}Sr$	0.716[1]	0.7095	0.7061
$^{143}Nd/^{144}Nd$	0.51210[2]	0.51224	0.51228
ε_{Nd}	−10.3	−7.6	−6.74
$^{176}Hf/^{177}Hf$	0.28242[2]	0.28252	0.28271
ε_{Hf}	−13.2	−4.95	−2.94
$^{138}Ce/^{136}Ce$	1.337059	–	–
$^{138}Ce/^{142}Ce$	0.0225685	–	–
ε_{Ce}	1.21[3]	–	–
$^{187}Os/^{188}Os$	1.4[4]	0.7366	0.3104
γ_{Os}	980	468	139
$^{186}Os/^{188}Os$	0.119885	–	–
$^{206}Pb/^{204}Pb$	18.93[5]	17.96	18.25
$^{207}Pb/^{204}Pb$	15.71	15.56	15.57
$^{208}Pb/^{204}Pb$	39.06	38.22	38.31

[1] Bayon et al. (2021); Goldstein and Jacobsen (1988)
[2] Chauvel et al. (2014)
[3] extrapolated from Willig et al. (2020)
[4] Peucker-Ehrenbrink and Jahn (2001)
[5] Millot et al. (2004)

the utility of the latter is compromised by several factors: first, lower crustal melting is often triggered by the intrusion of mantle-derived magmas, which mix with crustal melts, and second, lower temperatures, higher viscosities, and the presence of unreactive accessory minerals such as zircons in crustal melts raise questions about the extent to which isotopic equilibrium can be achieved. Consequently, we will focus on granulite xenoliths and terrains.

Figure 8.19 summarizes the distribution of Sr, Nd, and Hf isotope ratios of lower crustal granulite xenoliths. $^{87}Sr/^{86}Sr$ ratios, while spread over a similar range to that observed in the upper continental crust, are distinctly skewed to low values, with a mode of 0.7055 and a mean of 0.7095, which are closer to mantle values than upper crustal ones. In contrast, ε_{Nd} shows a much more even distribution ranging from mantle-like values to extremely unradiogenic ones with no clear mode and a mean of $\varepsilon_{Nd} = -7.6$. Hf isotopic compositions show a similar wide range from even more radiogenic than typical depleted mantle values to extremely unradiogenic, but they do show a weak mode at $\varepsilon_{Hf} \approx +5$ with a mean of -5. Thus, consistent with its more mafic inferred composition, the lower continental crust has less enriched isotopic signatures than the upper crust, particularly for Sr isotope ratios.

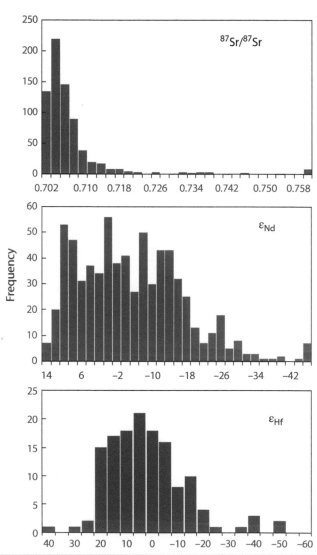

Figure 8.19 Distribution of $^{87}Sr/^{86}Sr$, ε_{Nd}, and ε_{Hf} in lower crustal granulite xenoliths. Data from the Earthchem.org (http://portal.earthchem.org/) database of 720 Sr, 697 Nd, and 140 Hf isotopic analyses.

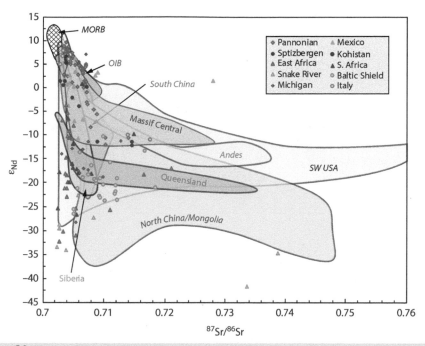

Figure 8.20 $^{87}Sr/^{86}Sr$–ε_{Nd} relationships in lower crustal granulite xenoliths. Source: Data from Earthchem.org.

Figure 8.20 shows the relationship between Sr and Nd isotopic ratios in these granulite xenoliths. While the least radiogenic $^{87}Sr/^{86}Sr$ ratios are associated with the most radiogenic ε_{Nd} values and plot near or close to the mantle array, this correlation breaks down for more enriched isotopic signature ratios with a wide divergence of values. Some granulites from the Southwestern US, North China, and the Andes have extremely radiogenic Sr with more modestly unradiogenic Nd, while some granulites from the Snake River volcanic province (Idaho, USA) and southern Africa have extremely unradiogenic Nd and also unradiogenic Sr.

Figure 8.21 shows ε_{Hf}–ε_{Nd} relationships in lower crustal xenoliths. Most of the data lie on an extension of correlation defined by oceanic basalts, the mantle array, and indeed overlap the OIB field to a considerable extent, indicating that the lower crust is *"broadly similar to the upper crust in terms of both its present-day parent/daughter values and time-integrated Lu–Hf and Sm–Nd evolution"* (Vervoort et al., 2000). Some rocks, however, diverge quite widely from this correlation, most notably metasediments from southern Africa.

The only studies of the Re–Os system in lower crustal rocks are those of Esperança et al. (1997) of xenoliths from the Southwestern US and Saal et al. (1998) of xenoliths from Queensland. There is a wide range of $^{187}Os/^{188}Os$ from 0.165 to 2.01 (γ_{Os} 27.5–1450) with a mean of 0.737. However, the data are skewed to lower, more mantle-like values and have a distinct mode at $^{187}Os/^{188}Os$ = 0.31((γ_{Os} =140).

Figure 8.22 shows the distribution of Pb isotopic compositions in granulite xenoliths, again revealing a wide range of values, but with distinct modes at $^{206}Pb/^{204}Pb$ = 18.6, $^{207}Pb/^{204}Pb$ = 15.7, and $^{208}Pb/^{204}Pb$ = 38.5. The data are skewed somewhat such that mean values are lower: $^{206}Pb/^{204}Pb$ = 18.0, $^{207}Pb/^{204}Pb$ = 15.56, and $^{208}Pb/^{204}Pb$ = 38.30. Figure 8.23 shows that much of the data plots close to the mantle arrays, although offset to higher $^{207}Pb/^{204}Pb$.

The Pb isotopic composition of the lower crust is a particularly important question because it bears on the question of the composition of the bulk silicate Earth and its age. The upper crust, the upper mantle, and mantle plumes all have Pb isotopic compositions lying to the right of the 4.57-Ga Geochron. If the

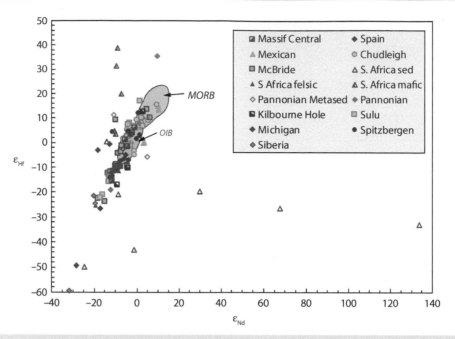

Figure 8.21 ε_{Hf}–ε_{Nd} relationships in lower crustal granulite xenoliths. Source: Data from (Vervoort et al., 2000), (Schmitz et al., 2004), and earthchem.org database (http://portal.earthchem.org/).

Earth is this old, mass balance requires a significant reservoir of unradiogenic Pb somewhere in the Earth. Many Archean granulite terrains tectonically exposed at the surface, such as the Scourian in Scotland, are indeed characterized by unradiogenic Pb, which is consistent with the low heat production inferred for the lower crust, which, in turn, implies low concentrations of U and Th.

Rudnick and Goldstein (1990), however, found that while most Archean lower crustal terrains did indeed have very unradiogenic Pb, post-Archean ones did not (Figure 8.24). They also analyzed granulite xenoliths in basalts in the Chudleigh and McBride Provinces of Queensland, Australia and the Eifel region of Germany and found that they plotted to the right of the Geochron, partly overlapping the mantle array. The additional analyses of granulite xenoliths published over the last 30 years show a much larger range, but the average value is only slightly lower than that of the much smaller data set of Rudnick and Goldstein. Rudnick and Goldstein (1990) concluded that unradiogenic Pb can only develop in regions that have remained stable for long time periods, that is, only in cratons. In areas where orogenies have occurred subsequent to crust formation, the Pb isotopic composition

of the lower crust is rejuvenated through mixing with radiogenic Pb from upper crust and mantle-derived magmas. An important conclusion of Rudnick and Goldstein (1990) was the average lower crustal Pb isotopic composition plots to the right of the 4.57-Ga Geochron and, hence, cannot be the missing reservoir of unradiogenic Pb. This remains true of the mean, mode, and median values for the much larger data set considered here. However, these values do plot slightly to the left of a 4.47-Ga Geochron.

The dispersion in the lower crustal isotopic compositions reflects both the wide variety of rock types found there, from metasediments to mafic meta-igneous ones, as well as the variety of processes that they have experienced. Unlike the mantle where trace element fractionation is largely controlled by mineral-melt partitioning between just a few ferromagnesian minerals, a number of other minerals become important in the lower crust. For example, when plagioclase is present, Sr becomes a compatible element, which leads to large Rb–Sr fractionations and ultimately to unradiogenic Sr in melting residues. Hydrous fluids also play a more important role in lower crustal evolution than in the mantle and fluids lost from the lower crust

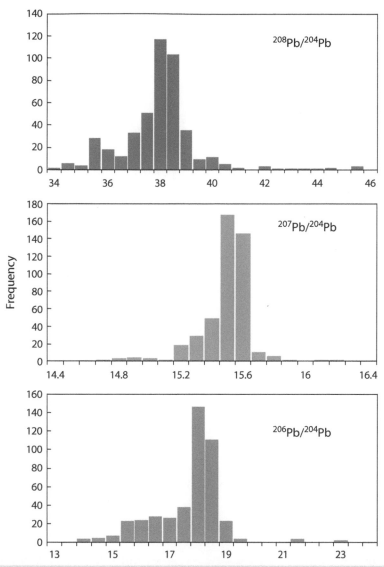

Figure 8.22   Distribution of Pb isotope ratios in 424 lower crustal granulite xenoliths. Siurce: Data from Rudnick and Goldstein (1990) the Earthchem.org database (http://portal.earthchem.org/).

can carry away fluid-mobile elements such as Rb and U, leading to unradiogenic Sr and Pb. Minor minerals including zircon and a variety of oxides can control fluid–solid and melt–solid partitioning of trace elements as well, leading to the decoupling of Hf and Nd isotope systematics. Small degree melts or fluids infiltrating from the mantle below may also modify lower crustal composition, as Kempton et al. (2001) suggested occurred in the Northern Baltic Shield.

Although many hundreds of isotopic analyses of lower crustal xenoliths are now available in the literature, the overall isotopic composition of the lower crust must still be considered poorly constrained. Lower crustal averages summarized in Table 8.1 need to be considered with some caution as they represent rather limited samples of a vast volume. In addition, Geng et al. (2020) have documented that granulite xenoliths can be infiltrated by the magmas that bring them to the surface, which can bias isotopic compositions toward mantle values. Nevertheless, the data do show that the lower crust has a substantially less incompatible element-enriched isotopic signature than the upper crust, which is consistent with its more mafic character inferred from geophysical observations and xenolith compositions.

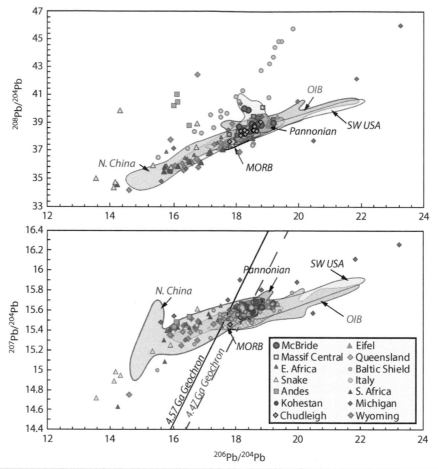

Figure 8.23   Pb isotope ratios in lower crustal xenoliths. Source: Data from (Rudnick and Goldstein, 1990) and the Earthchem.org database.

8.6   SUBDUCTION ZONES

8.6.1   Geochemistry of two-component mixtures

Subduction-related magmatism is probably the principal way in which new material is added to the continental crust at present. Such magmas are, however, often mixtures of mantle-derived and crust-derived components, including both subducted sediment, perhaps subduction-eroded crust, and crust assimilated as the magma ascended through or was stored in it. Thus, before exploring their isotope geochemistry, we need to consider the effects of mixing on isotope ratios.

When two components contribute material to magmas, we might expect that the proportion contributed by each might vary. If we plot the concentration of any two elements in different samples of this mixture against each other,

they must lie on a straight line between the two end-members. However, if we plot ratios of either elements or isotopes, they need not lie on a straight line. Indeed, in the general case, they do not; rather, they will define a curve whose equation is

$$A\left(\frac{x}{X}\right) + B\left(\frac{x}{X}\right)\left(\frac{y}{Y}\right) + C\left(\frac{y}{Y}\right) + D = 0 \quad (8.5)$$

where ratios x/X and y/Y are the elemental or isotopic ratios of the abscissa and ordinate, respectively. If end-members are designated 1 and 2 and have ratios $(x/X)_1$ and $(y/Y)_1$ and $(x/X)_2$ and $(y/Y)_2$, respectively, then

$$A = X_2 Y_1 \left(\frac{x}{X}\right)_2 - Y_1 X_2 \left(\frac{x}{X}\right)_1 \quad (8.6)$$

$$B = X_1 Y_2 - X_2 Y_1 \quad (8.7)$$

$$C = X_2 Y_1 \left(\frac{y}{Y}\right)_2 - X_1 Y_2 \left(\frac{y}{Y}\right)_1 \quad (8.8)$$

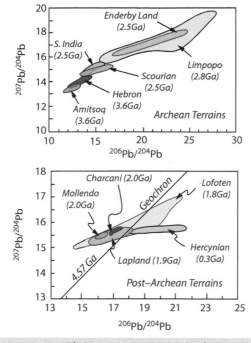

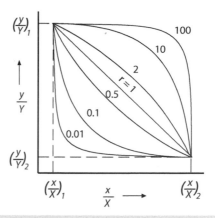

Figure 8.25   Plots of ratios of elements or isotopes x/X versus y/Y for mixing of end-members 1 and 2. The numbers along the curves are the values for r. Source: Langmuir et al. (1978)/with permission of Elsevier.

Figure 8.24   Pb isotope ratios in Archean and post-Archean granulite (i.e. lower crustal) terrains. The 4.57-Ga Geochron is shown for reference. While Archean terrains appear to be characterized by unradiogenic Pb, this is less true of post-Archean terrains. Source: Rudnick and Goldstein (1990)/Reproduced with permission from Elsevier.

$$D = X_1 Y_2 \left(\frac{y}{Y}\right)_2 \left(\frac{x}{X}\right)_2 - X_2 Y_1 \left(\frac{y}{Y}\right)_1 \left(\frac{x}{X}\right)_1 .$$

$$(8.9)$$

The curvature of the mixing line will depend on the ratio r:

$$r = (X_1 Y_2)/(X_2 Y_1). \qquad (8.10)$$

The greater the value of r, the greater the curvature. Only in the special case where $r = 1$, there is the line straight. This is illustrated in Figure 8.25. This result is completely general and applies to the mixing of river water and seawater, as well as the mixing of magmas.

Taking a concrete example, if our plot is $^{143}Nd/^{144}Nd$ versus $^{87}Sr/^{86}Sr$, then the curvature depends on the ratio of $(^{144}Nd_1{}^{86}Sr_2)/(^{144}Nd_2{}^{86}Sr_1)$. Since in most instances the amounts of ^{144}Nd and ^{86}Sr are to a very good approximation proportional to total Nd and Sr, respectively, r is approximated by $(Nd_1 Sr_2)/(Nd_2 Sr_1)$. If we re-express this ratio

as $r = (Nd/Sr)_1/(Nd/Sr)_2$, we see that the curvature depends on the ratio of the Nd/Sr ratio in the two end-members. Most mantle-derived rocks have similar Sr/Nd ratios of about 16, so mixing curves typically approximate straight lines. In crustal rocks and sediments, deviations from $r = 1$ are more likely, and curved mixing lines are, therefore, more common.

Note that r will always be 1 where $X = Y$, that is, where the two denominators are the same. Consequently, on $^{207}Pb/^{204}Pb$—$^{206}Pb/^{204}Pb$ plots, mixing curves will always be straight lines because the denominators are the same (i.e. $X = Y = {}^{204}Pb$).

Two component mixtures will also form straight lines when we plot a radiogenic isotope ratio versus a parent–daughter ratio, that is, on isochron plots, for example, $^{87}Sr/^{86}Sr$—$^{87}Rb/^{86}Sr$, because the denominators are the same. Thus, mixing lines can be mistaken for isochrons and vice versa. One way to distinguish the two is a ratio–element plot. A ratio–element plot, for example $^{87}Sr/^{86}Sr$ versus Sr, will also in general be a curved line described by 8.5 (because the denominators are ^{86}Sr and 1), but a ratio plotted against the inverse of the denominator, for example $^{87}Sr/^{86}Sr$—$1/Sr$, will be a straight line (at least to the degree that ^{86}Sr is proportional to total Sr, which will be the case where the range in $^{87}Sr/^{86}Sr$ ratios is small). Such a plot can be a useful discriminator between isochrons and mixing lines because only in the

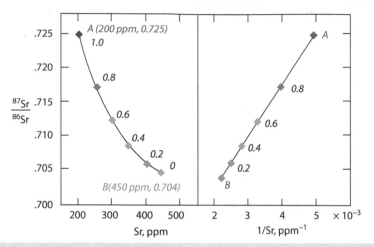

Figure 8.26   Mixing hyperbola formed by components A and B. Mixing between of two components with different concentrations will always define a hyperbola on concentration versus isotope ratio plot, but a straight line on an isotope ratio versus inverse concentration plot. Source: Faure (1986)/Reproduced with permission from John Wiley & Sons.

latter case will $^{87}Sr/^{86}Sr$—$1/Sr$ necessarily define a straight line (Figure 8.26). Again, this result is completely general, and while the general principles have been illustrated with isotope ratios, they apply equally well to elemental ratios.

When the compositions of a magma or series of magmas appear to reflect mixing, we are often faced with having to decide whether (i) two mantle-derived magmas are mixing, (ii) two distinct mantle sources are mixing or a mantle-derived magma is mixing with assimilated crust. Magaritz et al. (1978) pointed out that by combining radiogenic isotope and oxygen isotope analyses, it is possible to distinguish between the first two possibilities. They noted that all rocks have similar oxygen concentrations. Because of this, the addition of 10% sediment with $\delta^{18}O$ of ~+20 to the mantle (~+6) followed by the melting of that mantle produces a magma with the same $\delta^{18}O$ (+8.4), as adding 10% sediment directly to the magma (i.e. the magma also ends up with $\delta^{18}O$ = +8.4). However, the addition of 10% sediment to the mantle, followed by the melting of that mantle produces a magma with a very different Sr isotope ratio than does adding 10% sediment directly to the magma because the Sr concentration of the magma is an order of magnitude or more higher than that of the mantle. Indeed, most subduction-related magmas are richer in Sr (~ 400 ppm) than most

continental materials and sediments, but sediments invariably have more Sr than does the mantle (10–40 ppm). Thus, the addition of sediment to magma affects the Sr isotope ratio less than the addition of sediment to the mantle.

This point is illustrated in a more quantitative fashion in Figure 8.27. When the ratio of Sr in end-member M (magma or mantle) to Sr in end-member C (crust or sediment) is greater than 1, a concave downward mixing curve results. If that ratio is less than 1 (i.e. concentration of Sr in C is greater than

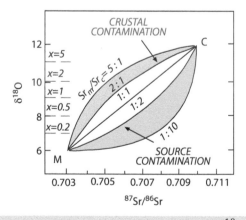

Figure 8.27   Mixing curves on a plot of $\delta^{18}O$ versus $^{87}Sr/^{86}Sr$. The labels on the lines refer to different ratios of Sr concentration in the two end-members. Source: James (1981)/Reproduced with permission from Annual Reviews.

in M), a concave upward curve results. A simple assimilation of crust by magma essentially corresponds to the concave downward case ($Sr_M > Sr_C$), and the mixing of subducted sediment and mantle corresponds to the concave upward case ($Sr_M < Sr_C$). In principle then, these two processes are readily distinguished on such a plot. However, assimilation is almost always accompanied by fractionation crystallization, and the combination produces more complicated variations, which can mimic the mixing of sediment and mantle. We will discuss assimilation in Section 8.6.3.

8.6.2 Radiogenic isotopic compositions of subduction-related magmas

Subduction zones are not only a principal mechanism by which new crust is created but also regions in which oceanic crust and its veneer of sediment are recycled into the mantle. Thus, subduction zones play a critical role in the chemical evolution of the Earth.

Island arc and continental margins volcanics (we will use the abbreviation IAV for both) are distinctive in many of their geochemical features in part because of the unique fluid-transfer processes occurring in subduction zones and in part because these processes carry crustal components, subducted oceanic crust and sediment, into the magma genesis zone.

Armstrong (1971) and Armstrong and Cooper (1971) were the first to recognize that subducted sediment contributed to IAV based on Pb isotopes, particularly the radiogenic Pb found in the Lesser Antilles Island Arc. Since that work, a considerable amount of Pb isotope and other data has accumulated to support Armstrong's inference. Figure 8.28 summarizes the Pb isotopic data from nine intraoceanic island arcs. Most arcs form arrays with higher slopes on the $^{206}Pb/^{204}Pb$—$^{206}Pb/^{204}Pb$ than MORB and OIB, suggesting that their sources are indeed mixtures of depleted mantle and ^{207}Pb-rich continent-derived sediment. Nevertheless, Pb isotope ratios vary considerably between subduction zones. This can partly be attributed to the differing amounts of sediment as well as their differing Pb isotopic composition being carried into the subduction zone (differences in mantle composition also contribute). Yellow circles show the average composition of sediment being carried into the subduction zone compiled by Plank (2014). We see that those arcs with the most radiogenic Pb, Lesser Antilles and Sunda, are also those where the subducting sediment is most radiogenic. The Tonga-Kermadec and Izu-Bonin arcs appear to be exceptions of sorts: both form $^{206}Pb/^{204}Pb$—$^{206}Pb/^{204}Pb$ arrays that are parallel to and overlap the MORB array.

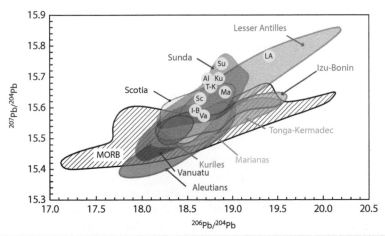

Figure 8.28 $^{207}Pb/^{204}Pb$ versus $^{206}Pb/^{204}Pb$ in lavas from nine intraoceanic arcs. The average composition of sediment on the subducting plate in front of the arcs from Plank (2014) is shown in the yellow circles: Va: Vanuatu, I-B: Izu-Bonin, Sc: Scotia/South Sandwich, T-K: Tonga-Kermedec, Al: Aleutians, Ku: Kuriles, Ma: Marianas, LA: Lesser Antilles, Su: Sunda. Data from Sunda and Aleutian arcs in this and subsequent figures in this section include only those parts of the arc that are intraoceanic (central and western Aleutians, eastern Sunda).

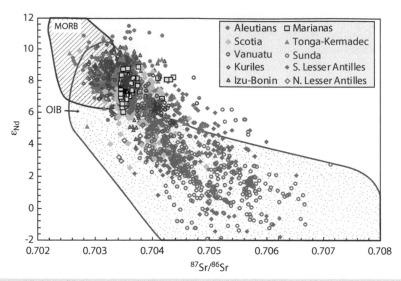

Figure 8.29 ε_{Nd} versus $^{87}Sr/^{86}Sr$ in nine intraoceanic island arcs. Source: Data from the GEOROC database (http://georoc.mpch-mainz.gwdg.de/georoc/).

Relatively radiogenic Pb found in Tonga IAV may reflect contributions from subducted seamounts of the Louisville Seamount chain (Beier, et al., 2017). In the case of the Izu-Bonin arc, Ishizuka et al. (2007) suggest that Pb isotopic compositions reflect a contribution from volcanogenic sediment derived from Pacific oceanic islands.

In terms of their Sr–Nd systematic, island arcs overlap the MORB and OIB fields considerably (Figure 8.29), although they have some-tendency to plot to the high $^{87}Sr/^{86}Sr$ side. This reflects the contribution of subducted oceanic crust to IAV sources. Hydrothermal activity at mid-ocean ridges and low-temperature basalt–seawater reaction shift the $^{87}Sr/^{86}Sr$ of the oceanic crust to more radiogenic values (the mean value of oceanic crust being subducted is probably in the range of 0.703–0.7035), while the Nd of even old (< 150 Ma) oceanic crust will be only a little less radiogenic than modern MORB. We can also see considerable variation in Sr and Nd isotopic compositions between arcs: the Aleutians and Marianas differ only modestly from MORB, while Sunda and the Lesser Antilles data extend to quite radiogenic Sr and unradiogenic Nd. This in part reflects the difference in subducting sediment: for example, the average subduction sediment in the Aleutians and Marianas have ε_{Nd} of –3.9 and –2.15, respectively, while the Southern

Lesser Antilles and Sunda subducting sediments average –11.9 and –9.17, respectively. Volcaniclastic sediment, derived from the arc volcanoes themselves and well as oceanic islands, can make up a considerable fraction of the subducting sediment in some arcs with the result that average subducting sediment $^{87}Sr/^{86}Sr$ can be closer to mantle values than crustal ones, most notably Vanuatu and the Marianas, where average $^7Sr/^{86}Sr$ is 0.70548 and 0.70617, respectively.

Figure 8.30 shows ε_{Hf}–ε_{Nd} relationships in IAV. The overlap with MORB (and OIB) is considerable, demonstrating that the primary source of IAV is depleted asthenospheric mantle similar to the DMM. There is, however, some tendency for IAV to plot to the high side of the ε_{Hf}–ε_{Nd} mantle array. This may reflect several factors. First is the so-called zircon effect, which results in some marine sediments, particularly hydrogenous ones, having relatively radiogenic Hf the reason for which we discussed earlier (and will discuss further in Chapter 11). The second is the role of fluids in transporting material from subducted oceanic crust and sediment into the magma genesis zone. As Hf is less soluble in such fluids, it is transported this way less effectively than Nd. This may account for the anomalously high ε_{Hf} in some magmas from the Tonga-Kermadec arc. The mantle beneath this arc appears to be particularly incompatible

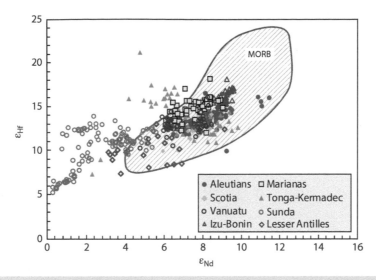

Figure 8.30 ε_{Hf} versus ε_{Nd} in eight intraoceanic island arcs. Source: Data from the GEOROC database (http://georoc.mpch-mainz.gwdg.de/georoc/).

element-depleted material and fluids derived from the subducting slab readily shift its ε_{Nd} to lower values but have less effect on ε_{Hf} (Pearce et al., 2007).

Looking in more detail at the Lesser Antilles, we can see how sediment composition influences the isotopic compositions of arc magmas. The arc is built perpendicular to the continental margin (an unusual situation). The Guiana Highland in northeastern South American consists in part of Archean crust and is the drained by the Orinoco River, which has deposited a considerable volume of sediment in front of the arc. Because it drains ancient cratonic crust, the sediment of the

Orinoco contains particularly radiogenic Pb and Sr and unradiogenic Nd and Hf. Isotopic compositions in the sediment grade northward (Figure 8.31). This northward variation is generally mirrored by in the isotopic composition of arc lavas and reflects a decreasing continental contribution with distance from the continent.

8.6.3   Assessing crustal assimilation in island arc magmas

The islands of St. Lucia and Martinique in the central part of the arc are exceptions to the regional isotopic pattern in the Lesser Antilles

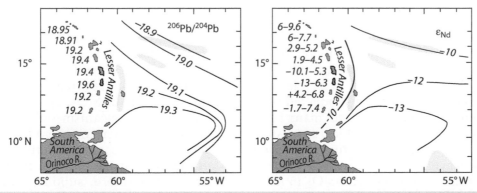

Figure 8.31 Contours of $^{206}Pb/^{204}Pb$ and ε_{Nd} in sediment in front of the Lesser Antilles Island arc. Range or mean of these parameters in Lesser Antilles arc volcanics is written adjacent to each island. The islands of St. Lucia and Martinique discussed in the text have a bolder outline. Source: White and Dupré (1986)/Reproduced with permission from John Wiley & Sons.

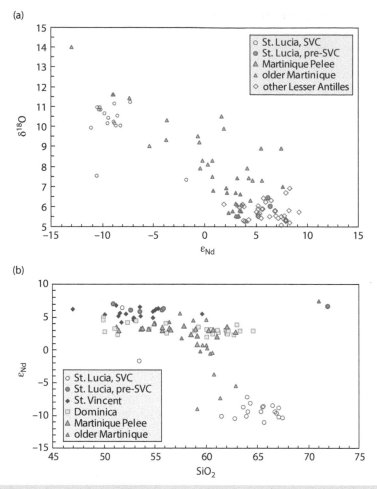

Figure 8.32 $\delta^{18}O$ versus ε_{Nd} and SiO_2 in the Lesser Antilles volcanics. Older volcanics from Martinique and those erupted from the Soufriere Volcanic Complex (SVC) on western coast of St. Lucia stand out by having $\delta^{18}O$ higher than in other Lesser Antilles volcanics. Correlations between ε_{Nd} and SiO_2 and between $\delta^{18}O$ and ε_{Nd} indicate that assimilation-fractional crystallization is occurring. Two rhyolites have ε_{Nd} identical to the basalts, indicating that they have evolved without assimilation. Source: Data from the GEOROC database (http://georoc.mpch-mainz.gwdg.de/georoc/) and W. M. White (unpublished).

with elevated $^{206}Pb/^{204}Pb$ and ε_{Nd} extending to highly negative values (Figure 8.31). Figure 8.32a shows the low ε_{Nd} values in these islands are associated with elevated $\delta^{18}O$. This could result from subducted sediment in their sources, and indeed, many IAVs have $\delta^{18}O$ above mantle values, consistent with a percent or two of sediment in their sources. However, we see in Figure 8.32b that ε_{Nd} is inversely correlated with SiO_2 in Martinique and St. Lucia magmas (and to a lesser extent in Dominica) in contrast to St. Vincent, located south of St. Lucia. This strongly suggests that the anomalous Nd, Pb, and Sr isotope ratios on these islands result from the

assimilation of sediment within the crust accompanied by fractional crystallization, rather than sediment subduction because the latter is not expected to produce significant variation in SiO_2. In contrast to mid-ocean ridge volcanism where preexisting crust is nonexistent or oceanic islands developed on thin oceanic crust, the crust beneath intraoceanic island arcs can be 15 or 20 km thick. Furthermore, subduction-related volcanism can persist in the same locality for many tens of millions of years. The produces a situation ripe for magma to assimilate the crust it transits. Let us now consider the isotope geochemistry of crustal assimilation.

8.6.3.1   Combined fractional crystallization and assimilation

Because oxygen isotope ratios of mantle-derived magmas are relatively uniform and generally different from rocks that have equilibrated with water at the surface of the Earth, oxygen isotopes are a useful tool in identifying and studying the assimilation of country rock by intruding magma. This is always at least a three-component problem, involving country rock, magma, and minerals crystallizing from the magma. The reason for this is that magmas are essentially never superheated; hence, the heat required to melt and assimilate surrounding rock can only come from the latent heat of crystallization of the magma. Approximately 1000 J/g would be required to heat rock from 150 °C to 1150 °C and another 300 J/g would be required to melt it. If the latent heat of crystallization is 400 J/g, the crystallization of 3.25 g of magma would be required to assimilate 1 g of cold country rock. Since some heat will be lost by simple conduction to the surface, we can conclude that the amount of crystallization will inevitably be greater than the amount of assimilation (the limiting case where mass crystallized equals mass assimilated could occur only at very deep layers of the crust where the rock is at its melting point to begin with).

We considered the effect of fractional crystallization in Section 5.3.5.2 (Equation 5.105). If magma simultaneously assimilates country rock, a process known as assimilation-fractional crystallization (AFC), oxygen isotope ratios will change according to

$$\delta_m - \delta_0 = \left([\delta_a - \delta_0] - \frac{\Delta}{R} \right) \left\{ 1 - f^{-R/(R-1)} \right\}$$

$$(8.11)$$

where R is the mass ratio of material assimilated to material crystallized, Δ is the fractionation factor (i.e. the difference in isotope ratio between the crystals and the magma, δ_{magma}-δ_{crystals}), f is the fraction of liquid remaining, δ_m is the $\delta^{18}O$ of the magma, δ_o is the initial $\delta^{18}O$ of the magma, and δ_a is the $\delta^{18}O$ of the material being assimilated. (This equation is from (DePaolo, 1981) but differs slightly because we changed the definition of Δ.) The assumption is made that the concentration of oxygen is the same in the crystals, magma,

and assimilant, which is a reasonable assumption. This equation breaks down at $R = 1$, but as discussed earlier, this is unlikely: R will always be less than 1. Figure 8.33 shows the variation of $\delta^{18}O$ of a magma with an initial $\delta^{18}O = 5.7$ as crystallization and assimilation proceed.

8.6.3.2   Combining radiogenic and oxygen isotopes

Combining O isotopes with a radiogenic isotope ratio, such as $^{87}Sr/^{86}Sr$, provides an even more powerful tool for the study of assimilation processes. There are several reasons for this. In the case of basaltic magmas, radiogenic elements, particularly Nd and Pb, often have lower concentrations in the magma than in the assimilant. This means a small amount of assimilant can have a large effect on the radiogenic isotope ratios. On the other hand, oxygen will be present in the magma and assimilant at nearly the same concentration, making calculation of the mass assimilated fairly straightforward. Also, it is easier to uniquely characterize the assimilant using both the radiogenic and stable isotope ratios, as suggested in Figure 8.33.

The equation governing a radiogenic isotope ratio in a magma during AFC is different from 8.11 because we cannot assume that the concentration of the element is the same in all the components. On the other hand, there is no fractionation between crystals and melt. The general equation describing the variation of the *concentration* of an element in a magma during AFC is

$$\frac{C_m}{C_m^0} = f^{-z} + \left(\frac{R}{R-1} \right) \frac{C_a}{C_m^0} (1 - f^{-z}) \quad (8.12)$$

where C_m and the concentration of the element in the magma, C^o is the original magma concentration, f is as defined for 8.11 earlier, and z is as

$$z = \frac{R + D - 1}{R - 1} \quad (8.13)$$

where D is the solid/liquid partition coefficient. The isotopic composition of the magma is then given by (DePaolo, 1981)

$$\varepsilon_m = \frac{\left(\frac{R}{R-1} \right) \frac{C_a}{z} (1 - f^{-z}) \varepsilon_a + C_m^0 f^{-z} \varepsilon^0}{\left(\frac{R}{R-1} \right) \frac{C_a}{z} (1 - f^{-z}) + C_m^0 f^{-z}} \quad (8.14)$$

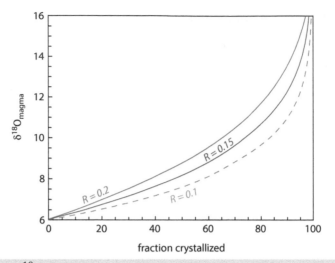

Figure 8.33 Variation in $\delta^{18}O$ of a magma undergoing AFC versus fraction crystallized computed using 8.11. Initial $\delta^{18}O$ of the magma is +6, $\delta^{18}O$ of the assimilant is +18, and Δ = +1. R is the ratio of mass assimilated to mass crystallized.

where ε is the isotope ratio with subscripts m, a, and 0 denoting the magma, the assimilant, and the original magma, respectively, and the other parameters are defined as in 8.12 and 8.13. Figure 8.34 shows calculated AFC curves on a plot of $\delta^{18}O$ versus $^{87}Sr/^{86}Sr$. Note that, except in the case where z = 1 where the

problem simplifies to one of simple mixing, the mixing lines end at f = 0.01 (99% crystallized).

Figure 8.35 shows an actual case, the Adamello Massif in Italy. Actual analyses are plotted as dots with the concentrations of Sr in the sample shown adjacent to the dot. AFC lines are computed assuming the original magma had

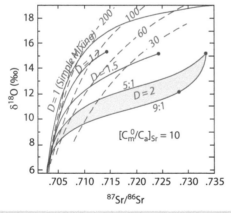

Figure 8.34 Variation of $\delta^{18}O$ with $^{87}Sr/^{86}Sr$ during AFC for a magma with an initial $\delta^{18}O$ = 5.7 and $^{87}Sr/^{86}Sr$ = 0.703 and an assimilant with $\delta^{18}O$ = 19 and $^{87}Sr/^{86}Sr$ = 0.735. All the curves are for R = 0.2 except for one with D = 2 for which R = 0.11 (labeled 9:1). Dashed red lines are calculated Sr concentrations (ppm) assuming an initial Sr concentration of 500 ppm in the magma. Source: Talyor (1980)/with permission of Mineralogical Society of America.

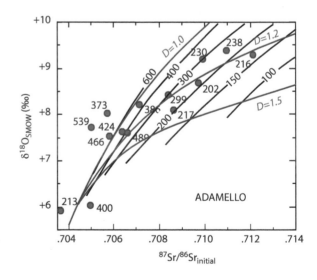

Figure 8.35 Variation of $\delta^{18}O$ with $^{87}Sr/^{86}Sr$ in the Adamello Massif in Italy compared with the model AFC process using 8.11–8.14. Source: Taylor and Sheppard (1986)/Reproduced with permission from Mineralogical Society of America.

$\delta^{18}O$ = 5.6, $^{87}Sr/^{86}Sr$ = 0.704, and 750 ppm Sr, and the country rocks have $^{87}Sr/^{86}Sr$ = 0.736, $\delta^{18}O$ = +13.6, and 150 ppm Sr. Dashed lines are contours of *calculated* Sr concentrations in the magma. There is reasonably good agreement between the calculated model and the actual data, if we assume that the bulk partition coefficient varied a bit (which would certainly be the case).

8.6.3.3 Sediment subduction versus assimilation

James and Murcia (1984) used this combined O and radiogenic isotopes to distinguish between sediment subduction and assimilation in the northern Andes. As shown in Figure 8.36, andesites from Nevado del Ruiz and Galeras volcanoes in Columbia define a steep array on plots of $\delta^{18}O$ versus $^{87}Sr/^{86}Sr$ and $^{143}Nd/^{144}Nd$. Comparing this with Figure 8.27, we see that such steep arrays imply mixing between magma and crust rather than mantle and sediment. As we observed in the previous section, assimilation will inevitably be accompanied by fractionation crystallization. James and Murcia (1984) modeled this assimilation using the equations similar to 8.11–8.14. The model fits the data reasonably well and implies assimilation of up to about 12% country rock. An R of 0.3 provided the best fit, though James and Murcia (1984) noted that this parameter was not well constrained. A low value of R suggests that the assimilation is occurring at moderate depth (<10 km), because it suggests only moderate input of heat in melting the country rock.

Not all rocks from St. Lucia and Martinique islands have unusually low ε_{Nd}; indeed, as Figure 8.32 shows, some rhyolites from both islands have ε_{Nd} indistinguishable from basalts. Thus, assimilation is not inevitable on these islands but instead appears restricted to certain volcanic centers. The case of St. Lucia has been particularly well studied. Most of the island consisting mainly of basalts and basaltic andesite lava flows erupted over the last 7 Ma or so (Lindsay et al., 2013). These have ε_{Nd} ranging from +5.8 to +6.9, similar to ε_{Nd} values from St. Vincent, the neighboring volcanic island to the south (and where a lack of correlation between isotope ratios and SiO_2 indicating assimilation is not significant). Roughly 650 000 years ago, magmatism on St. Lucia became confined to the Soufriere Volcanic Complex (SVC), a large collapsed caldera on the western side of the island, which has erupted andesites and dacites, often explosively, with low ε_{Nd} and elevated $\delta^{18}O$, and some of which have metasedimentary xenoliths. Apparently, a large magma chamber has developed beneath the SVC with a sedimentary layer intercepting rising mafic magma and assimilating the surrounding sediment (Bezard et al., 2014).

The situation on Martinique appears to be in some sense the opposite of St. Lucia in that elevated $\delta^{18}O$ and anomalously low ε_{Nd} values are associated more with older volcanic centers (Davidson and Harmon, 1989). However, samples from the current eruptive center of Mt. Pelée, infamous for its 1902 eruption that killed 29 000 people, exhibit statistically

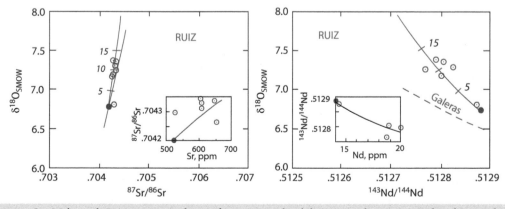

Figure 8.36 Sr, Nd, and O isotopes in lavas from Nevado del Ruiz Volcano in Columbia and an AFC model calculation. Dashed line shows AFC trajectory for another Colombian volcano, Galeras. Source: James and Murcia (1984)/with permission of Elsevier.

significant correlations between $^{87}Sr/^{86}Sr$, $\delta^{18}O$, and SiO_2, indicating assimilation continues, although not as severe as with older eruption; indeed, the dacite pumice from the 1902 eruption has normal mantle-like $\delta^{18}O$. Modest but statistically significant correlations also exist between SiO_2 and ϵ_{Nd} and $^{87}Sr/^{86}Sr$ among samples from Dominica (Figure 8.32), the next island to the north, suggesting that some assimilation is occurring there as well. In contrast, on the island to the south of St. Lucia, St. Vincent, there is no correlation between isotope ratios and major elements, indicating a lack of sediment assimilation.

The islands of the central Lesser Antilles may be the most spectacular example of crustal assimilation in island arc magmas, but they are not the only ones. Assimilation has also been documented in the intraoceanic part of the Sunda arc (e.g. Gasperon et al., 1994). Often, the assimilant is older igneous rocks of the arc crust. Thirlwall and Graham (1984) argued that crustal sediment assimilation explained the very radiogenic Sr and Pb and unradiogenic Nd in the southernmost Lesser Antilles Island of Grenada. However, White et al. (2017) showed that this was inconsistent with the chemistry of these lavas and instead that the assimilation of igneous crust explained the evolutionary trends observed. Similarly, Barker et al. (2013) found that open system mixing of magmas with gabbros and hydrothermally altered tonalites in the Kermadec arc is ubiquitous. Singer et al. (1992) found that the assimilation of igneous crust during fractional crystallization explains O isotope ratio variations in Kanaga Volcano of the Aleutians. More generally, Bezard et al. (2015) showed that $^{187}Os/^{188}Os$ correlates negatively with Os concentrations in IAV, a correlation also observed in MORB, and as we pointed out in Section 7.3.2.5 likely reflects oceanic crustal assimilation. Lopevi Volcano in the Vanuatu Arc, where $^{87}Sr/^{86}Sr$ decreases with increasing SiO_2 provides yet another of assimilation of igneous crust in an island arc setting (Handley et al., 2008).

8.6.4 ^{10}Be in subduction-related magmas

If further evidence of the presence of subducted sediment in arc magmas is needed, it is provided by yet another isotopic system: ^{10}Be. We found in Chapter 4 that ^{10}Be is produced by spallation in the atmosphere and then washed out and accumulated in soils and sediment. Because its half-life is only 1.6 Ma and cosmic rays penetrate solid matter so poorly, we do not expect ^{10}Be to be present in the interior of the Earth. Yet it is present in IAV (Figure 8.37). Note that the quantities are quite small, however, of the order of 10^6–10^7 atoms per gram; since rock will have something of the order of 10^{22} atoms per gram, these concentrations are of the order of less than 10^{-15} and were done with accelerator mass spectrometry following preparative chemistry of large samples. Making these measurements was quite an achievement by Fouad Tera, Lou Brown, Julie Morris, and others at Carnegie Institution.

A skeptic might suppose that some unknown neutron reaction can create ^{10}Be in the Earth's interior. In addition, cosmogenic

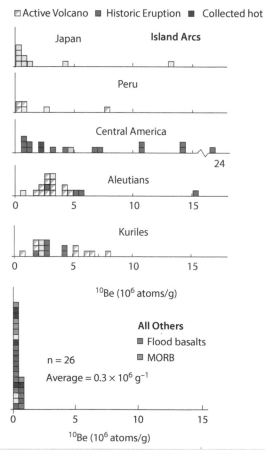

Figure 8.37 Comparison of ^{10}Be contents in arc and nonarc lavas. Source: Adapted from Tera et al. (1986) and Tsvetkov et al. (1989).

[10]Be in rain is rapidly absorbed onto clays, and that skeptic might suppose that even very young lavas might also absorb [10]Be. For these reasons, it was important to do control experiments by measuring [10]Be in nonarc lavas. As Figure 8.37 shows, [10]Be is not present in nonarc lavas. Thus, the only reasonable interpretation of [10]Be in arc magmas is that it is derived from subducted sediment.

Not all arc lavas have [10]Be. For example, no [10]Be was found in lavas from the Lesser Antilles, where Pb and other isotopes suggest a significant contribution from sediment. The same is true of the Sunda arc. In both these arcs, however, the sediment pile is so thick that most sediment is accreted in a forearc wedge rather than subducted. In the Lesser Antilles, seismic and other studies of the forearc show that only the lowermost 100 m or so of sediment is carried into the subduction zone and possibly subducted. These are pre-Miocene sediments. Using our rule of thumb that a radioactive isotope will be gone after 5–10 half-lives, we can predict that sediment older than 8–16 Ma should not have measurable [10]Be. Thus, it is no surprise that [10]Be is not present in Lesser Antilles magmas.

8.6.5 U-Decay series geochemistry of arc magmas

Another isotope system that has contributed significantly to our understanding of island arc processes is the U decay series. This system has been important in confirming the role of fluids in arc magma genesis. As we found in Chapter 3, the equilibrium situation is that the activity of parent and daughter is equal, for example, ^{230}Th activity is equal to the activity of ^{238}U, and hence, the ratio $(^{230}\text{Th}/^{232}\text{Th})$ will be equal to the $(^{238}\text{U}/^{232}\text{Th})$ ratio. Equilibrium should characterize the mantle before melting (and any other material left undisturbed for $\gtrsim$350 000 years). Because Th is more incompatible than U, the $(^{238}\text{U}/^{232}\text{Th})$ ratio in a melt should decrease, but the $(^{230}\text{Th}/^{232}\text{Th})$ ratio of a melt will be the same as that of its source. Thus, on a conventional plot of $(^{230}\text{Th}/^{232}\text{Th})$ against $(^{238}\text{U}/^{232}\text{Th})$, the melt should be driven to the left of the equiline. As we found in Chapter 7, this is indeed what we observe in most MORB and OIB (Figure 7.35).

As Figure 8.38 shows, although many arc magmas are close to equilibrium and some do plot to the left of the equiline, most arcs lavas have $(^{238}\text{U}/^{232}\text{Th})$—$(^{230}\text{Th}/^{232}\text{Th})$ values that plot to the right of the equiline. The implication is that in subduction zone magmas, U is enriched over Th, an observation first pointed out by Newman et al. (1984) and Gill and Williams (1990). Experiments have shown that under modestly oxidizing conditions, U becomes more soluble in hydrous fluids than Th and other high-field-strength elements, and

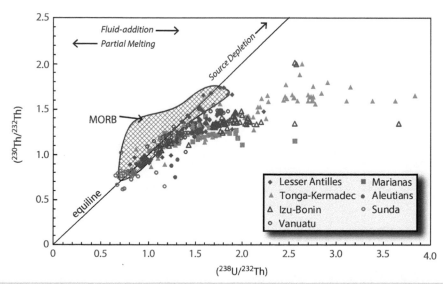

Figure 8.38 $(^{230}\text{Th}/^{232}\text{Th})$ versus $(^{238}\text{U}/^{232}\text{Th})$ in island arc magmas. Source: Data from the GEOROC database.

studies of peridotites from subduction zones show that the subarc mantle is typically more oxidizing than the mantle elsewhere. As a consequence, U can be oxidized and more readily transported by hydrous fluids from the subducting lithospheric (the sediments and basalts of the oceanic crust) into the subarc mantle than Th. These data confirm the hypothesis, proposed on other grounds, that fluids play a role transporting material from the slab to the magma genesis zone.

Figure 8.38 shows, however, that this U enrichment variable and is most apparent in the Marianas, Izu-Bonin, and Tonga-Kermadec subduction zones. Tonga-Kermadec arc magmas show a particularly wide range of $(^{238}U/^{232}Th)$ at similar $(^{230}Th/^{232}Th)$. At the other extreme, Sunda arc magmas have low $(^{230}Th/^{232}Th)$ and generally show only slight departure from (^{230}Th)-(^{238}U) equilibrium.

Elliott et al. (1997) noted that (^{230}Th)-(^{238}U) disequilibrium in the Marianas varied from island to island and correlated negatively with Th/Nb ratios and positively with Ba/Nb ratios such that those most enriched in Th and least enriched in Ba were closest to isotopic equilibrium. They also found that $(^{230}Th/^{238}U)$ correlated negatively with ε_{Nd} such that magmas with the most depleted Nd isotopic signatures showed the greatest disequilibrium and recent U enrichment (Figure 8.39). They proposed that the mantle wedge beneath the volcanoes is modified by two discrete processes: the addition of a melt derived from subducted sediments, which is relatively enriched in Th and close to radioactive equilibrium and a fluid enriched in U, Ba, and other fluid-mobile elements and out of radioactive equilibrium. Freymuth et al. (2016), however, argue that in the Izu arc, the addition of hydrous melts, rather than fluids, of the uppermost oceanic crust, which had been enriched in U during ridge-crest hydrothermal activity, produces the radioactive disequilibrium.

Figure 8.39 shows that there is a statistically significant negative correlation between $(^{230}Th/^{238}U)$ and ε_{Nd} in IAV overall and within Izu-Bonin, and Sunda arcs as well as the Marianas: those IAV with the most depleted Nd isotopic signatures tend to have the greatest (^{230}Th)-(^{238}U) disequilibrium. In addition to the effect of the subducted sediment contribution identified by Elliott et al., the composition of the mantle wedge before modification by material from the slab also likely plays a role in variable (^{230}Th)-(^{238}U) disequilibrium. In general, the highest ^{238}U excesses are occur in samples with the lowest Th concentrations, implying that where the mantle wedge is more incompatible element-depleted, slab-derived material more readily dominates the Th and U budgets, which appears to explain the $(^{230}Th/^{238}U)$ and ε_{Nd} correlation in Izu-Bonin. The extreme disequilibrium observed in the

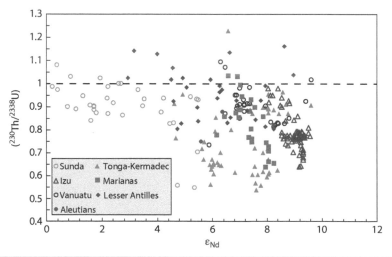

Figure 8.39 $(^{230}Th/^{238}U)$ versus ε_{Nd} in intraoceanic island arc magmas. There are statistically significant correlations among Sunda, Marianas, and Izu-Bonin samples as well as in the data set as a whole ($r = -0.355$, $n = 245$, $p = 0.037$). Source: Data from the GEOROC database.

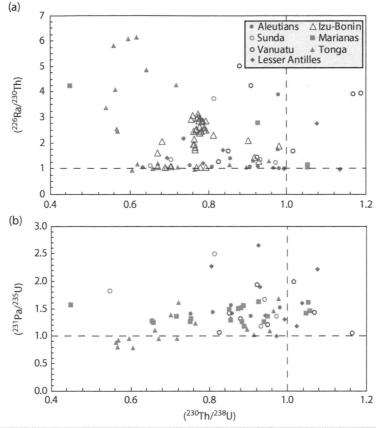

Figure 8.40 $(^{226}Ra/^{230}Th)$ and $(^{231}Pa/^{235}U)$ versus $(^{230}Th/^{238}U)$ in intraoceanic island arc magmas. Dashing lines denote secular radioactive equilibrium. Most IAVs have $(^{226}Ra/^{230}Th)$ and $(^{231}Pa/^{235}U) \geq 1$. Source: Data from the GEOROC database.

Tonga part of the Tonga-Kermadec arc appears to reflect the extreme depletion of the mantle wedge that occurred during recent back-arc spreading.

Excesses of ^{226}Ra over ^{230}Th and of ^{231}Pa over ^{235}U are also common among IAV (Figure 8.40). Excesses of ^{226}Ra over ^{230}Th are also observed in MORB, but those observed in IAV are sometimes far greater than observed in MORB. The half-lives of ^{226}Ra and ^{231}Pa are significantly shorter than that of ^{230}Th (1600 and 33 000 years, respectively), implying that disequilibrium must have been generated relatively shortly before eruption. Since (^{226}Ra) excesses decrease with increasing SiO_2 among individual arc islands and volcanoes, it is unlikely that this disequilibrium is generated during fractional crystallization. Ra is an alkali Earth and expected to be much more fluid-mobile than Th, which suggests that the ^{226}Ra enrichment, like that of U, may be due to the fluid enrichment of the melting zone

in the mantle wedge. If so, it implies very short times (thousands of years) between fluid enrichment and eruption (e.g. Turner et al., 2001).

^{231}Pa enrichment provides a different enigma: because its valance is 5+, it is expected to behave similarly to high-field-strength, fluid-immobile elements such as Nb and Zr, which are characteristically depleted in IAV compared to other incompatible elements. Yet excesses of ^{231}Pa over its parent ^{235}U are the rule rather than the exception in IAV. MORB also generally show an excess of ^{231}Pa over ^{235}U, which is expected as Pa should be more incompatible during mantle melting than U. Turner et al. (2006) suggested that melting within the mantle wedge can overprint the expected U-enrichment from slab fluids and that $(^{231}Pa/^{235}U)$ is higher in sediment-rich arc lavas where the effects of fluid addition are muted, and there is less of a ^{231}Pa deficit for melting to overprint. Avanzinelli et al. (2012)

argued that the ^{231}Pa is derived from sub-ducted sediment that melted <150 ka before the eruption of the lavas.

Some studies have questioned these short time scales of fluid addition to the mantle wedge and eruption implied by high ^{226}Ra excesses. Reubi et al. (2014) noted that some IAVs have equilibrium (^{230}Th/^{238}U) ratios yet have signif-icant ^{231}Pa and ^{226}Ra excesses. They argue that U-series activity ratios in IAV are best explained by models in which the fluid-metasomatized mantle returns to secular equilibrium before melting, implying a time lag ≥350 kyr, with

subsequent production of ^{231}Pa and ^{226}Ra excesses by in-growth during melting rather than by addition of slab fluids. They, as well as Huang et al. (2016), argue that (^{230}Th/^{238}U) disequilibrium observed in IAV is produced during the melting of the mantle wedge instead of fluid addition, and that ^{228}U excesses reflect melting under relatively oxidiz-ing conditions that result from the addition of slab-derived fluids and melts. These authors conclude that the time lag of fluid addition and melting can vary from hundreds of years to hundreds of thousands of years.

REFERENCES

Aarons, S.M., Reimink, J.R., Greber, N.D., et al. 2020. Titanium isotopes constrain a magmatic transition at the Hadean-Archean boundary in the Acasta gneiss complex. *Science Advances* 6: eabc9959. doi: 10.1126/sciadv.abc9959.

Albarède, F. 1998. The growth of continental crust. *Tectonophysics* 296: 1–14. doi: 10.1016/S0040-1951(98)00133-4.

Allègre, C.J. and Rousseau, D. 1984. The growth of the continent through geological time studied by Nd isotope analysis of shales. *Earth and Planetary Science Letters* 67: 19–34. doi: 10.1016/0012-821X(84)90035-9.

Armstrong, R.L. and Cooper, J.A. 1971. Lead isotopes in island arcs. *Bulletin Volcanologique* 35: 27–63. doi: 10.1007/BF02596807.

Armstrong, R.L. 1968. A model for the evolution of strontium and lead isotopes in a dynamic Earth. *Reviews of Geophysics* 6: 175–99.

Armstrong, R.L. 1971. Isotopic and chemical constraints on models of magma genesis in volcanic arcs. *Earth and Planetary Science Letters* 12: 137–42.

Armstrong, R.L. 1981. Radiogenic isotopes: the case for crustal recycling on a near-steady-state no-continental-growth Earth, in Moorbath, S. and Windley, B.F. (eds). *The Origin and Evolution of the Earth's Continental Crust*. London: The Royal Society, 259–87 pp.

Arndt, N. and Davaille, A. 2013. Episodic Earth evolution. *Tectonophysics* 609: 661–74. doi: 10.1016/j.tecto.2013.07.002.

Arndt, N.T. 2013. The formation and evolution of the continental crust. *Geochemical Perspectives* 2(3): 405–533. doi: 10.7185/geochempersp.2.3

Asmeron, Y. and Jacobsen, S.B. 1993. The Pb isotopic evolution of the Earth: inferences from river water suspended loads. *Earth and Planetary Science Letters* 115: 245–56. doi: 10.1016/0012-821X(93)90225-X

Avanzinelli, R., Prytulak, J., Skora, S., et al. 2012. Combined ^{238}U–^{230}Th and ^{235}U–^{231}Pa constraints on the transport of slab-derived material beneath the Mariana Islands. *Geochimica et Cosmochimica Acta* 92: 308–28. doi: 10.1016/j.gca.2012.06.020.

Barker, S. J., Wilson, C.J.N., Baker, J.A., et al. 2013. Geochemistry and petrogenesis of silicic magmas in the intra-oceanic Kermadec arc. *Journal of Petrology* 54: 351–91. doi: 10.1093/petrology/egs071.

Bayon, G., Freslon, N., Germain, Y., et al. 2021. A global survey of radiogenic strontium isotopes in river sediments. *Chemical Geology* 559: 119958. doi: 10.1016/j.chemgeo.2020.119958.

Bédard, J.H. 2018. Stagnant lids and mantle overturns: implications for Archaean tectonics, magmagenesis, crustal growth, mantle evolution, and the start of plate tectonics. *Geoscience Frontiers* 9: 19–49. doi: 10.1016/j.gsf.2017.01.005.

Beier, C., Turner, S.P., Haase, K.M., et al. 2017. Trace element and isotope geochemistry of the northern and central Tongan islands with an emphasis on the genesis of high Nb/Ta signatures at the northern volcanoes of Tafahi and Niuatoputapu. *Journal of Petrology* 58: 1073–106. doi: 10.1093/petrology/egx047.

Belousova, E.A., Kostitsyn, Y.A., Griffin, W.L., et al. 2010. The growth of the continental crust: constraints from zircon Hf-isotope data. *Lithos* 119(3–4): 457–66. doi: 10.1016/j.lithos.2010.07.024.

Bennett, V.C. and DePaolo, D.J. 1987. Proterozoic crustal history of the western United States as determined by neodymium isotope mapping. *Bulletin of the Geological Society of America* 99: 674–85.

Bennett, V.C., Brandon, A.D. and Nutman, A.P. 2007. Coupled ^{142}Nd-^{143}Nd isotopic evidence for Hadean mantle dynamics. *Science* 318: 1907–10.

Bezard, R., Davidson, J.P., Turner, S., et al. 2014. Assimilation of sediments embedded in the oceanic arc crust: myth or reality? *Earth and Planetary Science Letters* 395: 51–60. doi: 10.1016/j.epsl.2014.03.038.

Bezard, R., Schaefer, B.F., Turner, S., et al. 2015. Lower crustal assimilation in oceanic arcs: Insights from an osmium isotopic study of the Lesser Antilles. *Geochimica et Cosmochimica Acta* **150**: 330–44. doi: 10.1016/j. gca.2014.11.009.

Bowring, S.A. and Williams, I.S. 1999. Priscoan (4.00–4.03 Ga) orthogneisses from northwestern Canada. *Contributions to Mineralogy and Petrology* **134**: 3–16. doi: 10.1007/s004100050465.

Boyet, M., Blichert-Toft, J., Rosing, M., et al. 2003. $^{142}Nd/^{144}Nd$ evidence for early earth differentiation. *Earth and Planetary Science Letters* **214**: 427–42. doi: 10.1016/s0012-821x(03)00423-0.

Brown, M., Johnson, T. and Gardiner, N.J., 2020. Plate tectonics and the Archean Earth. *Annual Review of Earth and Planetary Sciences* **48**: 291–320. doi: 10.1146/annurev-earth-081619-052705.

Caro, G., Bourdon, B., Birck, J.-L., et al. 2003. ^{146}Sm-^{142}Nd evidence from Isua metamorphosed sediments for early differentiation of the Earth's mantle. *Nature* **423**: 428–32.

Cates, N.L., Ziegler, K., Schmitt, A.K., et al. 2013. Reduced, reused and recycled: detrital zircons define a maximum age for the Eoarchean (ca. 3750–3780 Ma) Nuvvuagittuq supracrustal belt, Québec (Canada). *Earth and Planetary Science Letters* **362**: 283–93. doi: 10.1016/j.epsl.2012.11.054.

Cavosie, A.J., Valley, J.W., Wilde, S.A., et al. 2005. Magmatic $\delta^{18}O$ in 4400–3900 Ma detrital zircons: a record of the alteration and recycling of crust in the early Archean. *Earth and Planetary Science Letters* **235**: 663–81. doi: 10.1016/j. epsl.2005.04.028.

Chapman, J.B., Dafov, M.N., Gehrels, G., et al. 2018. Lithospheric architecture and tectonic evolution of the southwestern U.S. Cordillera: Constraints from zircon Hf and O isotopic data. *GSA Bulletin* **130**: 2031–2046. doi: 10.1130/b31937.1.

Chappell, B.W. and White, A. 1974. Two contrasting granite types. *Pacific Geology* **8**: 173–4.

Chaudhuri, T., Wan, Y., Mazumder, R., et al. 2018. Evidence of enriched, Hadean mantle reservoir from 4.2-4.0 Ga zircon xenocrysts from Paleoarchean TTGs of the Singhbhum Craton, Eastern India. *Scientific Reports* **8**: 7069. doi: 10.1038/s41598-018-25494-6.

Chauvel, C., Garçon, M., Bureau, et al. 2014. Constraints from loess on the Hf–Nd isotopic composition of the upper continental crust. *Earth and Planetary Science Letters* **388**: 48–58. doi: 10.1016/j.epsl.2013.11.045.

Chou, C.-L. 1978. Fractionation of siderphile elements in the Earth's upper mantle and lunar samples. *Proceedings of the Lunar and Planetary Science Conference* **9**: 163–5.

Compston, W. and Pidgeon, R.T. 1986. Jack hills, evidence of more very old detrital zircons in western Australia. *Nature* **321**: 766–9. doi: 10.1038/321766a0.

Condie, K.C. and Pease, V. 2008. When Did Plate Tectonics Begin on Planet Earth? *Geological Society of America Special Paper* **440**. doi: 10.1130/spe440

Condie, K.C. and Aster, R.C. 2010. Episodic zircon age spectra of orogenic granitoids: the supercontinent connection and continental growth. *Precambrian Research* **180**: 227–36. doi: 10.1016/j.precamres.2010.03.008.

Condie, K.C., Davaille, A., Aster, R.C., et al. 2015. Upstairs-downstairs: supercontinents and large igneous provinces, are they related? *International Geology Review* **57**: 1341–8. doi: 10.1080/00206814.2014.963170.

Condie, K.C., Puetz, S.J. and Davaille, A. 2018. Episodic crustal production before 2.7 Ga. *Precambrian Research* **312**: 16–22. doi: 10.1016/j.precamres.2018.05.005.

Davaille, A., Smrekar, S.E. and Tomlinson, S. 2017. Experimental and observational evidence for plume-induced subduction on Venus. *Nature Geoscience* **10**: 349–55. doi: 10.1038/ngeo2928.

Davidson, J.P. and Harmon, R.S. 1989. Oxygen isotope constraints on the petrogenesis of volcanic arc magmas from Martinique, Lesser Antilles. *Earth and Planetary Science Letters* **95**: 255–70.

Deng, Z., Chaussidon, M., Savage, P., et al. 2019. Titanium isotopes as a tracer for the plume or island arc affinity of felsic rocks. *Proceedings of the National Academy of Sciences* **116**: 1132–5. doi: 10.1073/pnas.1809164116.

Dhuime, B., Hawkesworth, C.J., Cawood, P.A., et al. 2012. A change in the geodynamics of continental growth 3 billion years ago. *Science* **335**: 1334–6. doi: 10.1126/science.1216066.

Dhuime, B., Hawkesworth, C.J., Delavault, H., et al. 2017. Continental growth seen through the sedimentary record. *Sedimentary Geology* **357**: 16–32. doi: 10.1016/j.sedgeo.2017.06.001.

Elkins-Tanton, L.T. 2008. Linked magma ocean solidification and atmospheric growth for earth and mars. *Earth and Planetary Science Letters* **271**: 181–91. doi: 10.1016/j.epsl.2008.03.062.

Elliott, T., Plank, T., Zindler, A., et al. 1997. Element transport from subducted slab to juvenile crust at the Mariana arc. *Journal of Geophysical Research* **102**: 14991–15019. doi: 10.1029/97JB00788.

Esperança, S., Carlson, R.W., Shirey, S.B., et al. 1997. Dating crust-mantle separation: Re-Os isotopic study of mafic xenoliths from central Arizona. *Geology* **25**: 651–4. doi: 10.1130/0091-7613(1997)025<0651:DCMSRO>2.3.CO;2.

Faure, G. 1986. *Principles of Isotope Geology*. New York: John Wiley & Sons, Inc.

Forsyth, D. and Uyeda, S. 1975. On the relative importance of the driving forces of plate motion. *Geophysical Journal International* **43**: 163–200. doi: 10.1111/j.1365-246X.1975.tb00631.x.

Freymuth, H., Ivko, B., Gill, J.B., et al. 2016. Thorium isotope evidence for melting of the mafic oceanic crust beneath the Izu arc. *Geochimica et Cosmochimica Acta* **186**: 49–70. doi: 10.1016/j.gca.2016.04.034.

Froude, D.O., Ireland, T.R., Kinny, P.D., et al. 1983. Ion microprobe identification of 4100–4200 Myr-old terrestrial zircons. *Nature* 304: 616–8. doi: 10.1038/304616a0

Garçon, M., Chauvel, C., France-Lanord, C., et al. 2014. Which minerals control the Nd–Hf–Sr–Pb isotopic compositions of river sediments? *Chemical Geology* 364: 42–55. doi: 10.1016/j.chemgeo.2013.11.018.

Garçon, M., Chauvel, C., France-Lanord, C., et al. 2013. Removing the "heavy mineral effect" to obtain a new Pb isotopic value for the upper crust. *Geochemistry, Geophysics, Geosystems* 14. doi: 10.1002/ggge.20219.

Geng, X., Liu, Y., Zhang, W., et al. 2020. The effect of host magma infiltration on the Pb isotopic systematics of lower crustal xenolith: an *in-situ* study from Hannuoba, *North China*. *Lithos* 366–7: 105556. doi: 10.1016/j.lithos.2020.105556.

Gill, J.B. and Williams, R.W. 1990. Th isotope and U-series studies of subduction-related volcanic rocks. *Geochimica et Cosmochimica Acta* 54: 1427–42. doi: 10.1016/0016-7037(90)90166-I.

Goldstein, S.J. and Jacobsen, S.B. 1987. The Nd and Sr isotopic systematics of river-water dissolved material: implications for the sources of Nd and Sr in seawater. *Chemical Geology* 66: 245–72.

Goldstein, S.J. and Jacobsen, S.B. 1988. Nd and Sr isotopic systematics of river-water suspended material: implications for crustal evolution. *Earth and Planetary Science Letters* 87: 249–65.

Goldstein, S.J., O'Nions, R.K. and Hamilton, P.J. 1984. A Sm–Nd isotopic study of atmospheric dusts and particulates from major river systems. *Earth and Planetary Science Letters* 70: 221–36.

Guitreau, M., Blichert-Toft, J., Martin, H., et al. 2012. Hafnium isotope evidence from Archean granitic rocks for deep-mantle origin of continental crust. *Earth and Planetary Science Letters* 337–8: 211–23. doi: 10.1016/j.epsl.2012.05.029.

Handley, H.K., Turner, S.P., Smith, I.E.M., et al. 2008. Rapid timescales of differentiation and evidence for crustal contamination at intra-oceanic arcs: geochemical and u–th–ra–sr–nd isotopic constraints from Lopevi Volcano, Vanuatu, SW Pacific. *Earth and Planetary Science Letters* 273: 184–94. doi: 10.1016/j.epsl.2008.06.032.

Harper, C.L. and Jacobsen, S.B. 1992. Evidence from coupled ^{147}Sm-^{143}Nd and ^{146}Sm-^{142}Nd systematics for very early (4.5-Gyr) differentiation of the Earth's mantle. *Nature* 360: 728–32. doi: 10.1038/360728a0.

Harrison, T.M. 2009. The Hadean crust: evidence from >4 Ga zircons. *Annual Review of Earth and Planetary Sciences* 37: 479–505. doi: 10.1146/annurev.earth.031208.100151.

Hawkesworth, C.J. and Kemp, A.I.S. 2006. Using hafnium and oxygen isotopes in zircons to unravel the record of crustal evolution. *Chemical Geology* 226: 144–62.

Hawkesworth, C.J., Cawood, P.A., Dhuime, B., et al. 2017. Earth's continental lithosphere through time. *Annual Review of Earth and Planetary Sciences* 45: 169–98. doi: 10.1146/annurev-earth-063016-020525.

Hawkesworth, C.J., Dhuime, B., Pietranik, et al. 2010. The generation and evolution of the continental crust. *Journal of the Geological Society* 167: 229–48. doi:10.1144/0016-76492009-072.

Hawkesworth, C., Cawood, P.A. and Dhuime, B. 2019. Rates of generation and growth of the continental crust. *Geoscience Frontiers* 10: 165–73. doi: 10.1016/j.gsf.2018.02.004.

Hawkesworth, C.J., Cawood, P.A. and Dhuime, B. 2020. The evolution of the continental crust and the onset of plate tectonics. *Frontiers in Earth Science*, 8. doi: 10.3389/feart.2020.00326.

Hemming, S.R. and McLennan, S.M. 2001. Pb isotope compositions of modern deep sea turbidites. *Earth and Planetary Science Letters* 184: 489–503. doi: 10.1016/S0012-821X(00)00340-X.

Hoare, L., Klaver, M., Saji, N.S., et al. 2020. Melt chemistry and redox conditions control titanium isotope fractionation during magmatic differentiation. *Geochimica et Cosmochimica Acta* 282: 38–54. doi: 10.1016/j.gca.2020.05.015.

Huang, F., Xu, J. and Zhang, J. 2016. U-series disequilibria in subduction zone lavas: Inherited from subducted slabs or produced by mantle in-growth melting? *Chemical Geology* 440: 179–90. doi: 10.1016/j.chemgeo.2016.07.005.

Hurley, P.M. and Rand, J.R. 1969. Pre-drift continental nuclei. *Science* 164: 1229–42.

Ishizuka, O., Taylor, R.N., Yuasa, M., et al. 2007. Processes controlling along-arc isotopic variation of the southern Izu-Bonin arc. *Geochemistry, Geophysics, Geosystems* 8. doi: 10.1029/2006GC001475.

Israel, C., Boyet, M., Doucelance, et al. 2020. Formation of the Ce-Nd mantle array: crustal extraction vs. recycling by subduction. *Earth and Planetary Science Letters* 530: 115941. doi: 10.1016/j.epsl.2019.115941.

James, D.E. 1981. The combined use of oxygen and radiogenic isotopes as indicators of crustal contamination. *Annual Review of Earth and Planetary Sciences* 9: 311–344. doi: 10.1146/annurev.ea.09.050181.001523.

James, D.E. and Murcia, L.A. 1984. Crustal contamination in northern Andean volcanics. *Journal of the Geological Society* 141: 823–830. doi: 10.1144/gsjgs.141.5.0823.

Keller, C.B. and Schoene, B. 2012. Statistical geochemistry reveals disruption in secular lithospheric evolution about 2.5 Gyr ago. *Nature* 485(7399): 490–3. doi: 10.1038/nature11024.

Keller, C.B., Schoene, B. and Samperton, K.M. 2018. A stochastic sampling approach to zircon eruption age interpretation. *Geochemical Perspectives Letters* 8: 31–5. doi: 10.7185/geochemlet.1826.

Kemp, A.I.S., Wilde, S.A., Hawkesworth, C.J., et al. 2010. Hadean crustal evolution revisited: new constraints from Pb-Hf isotope systematics of the Jack Hills zircons. *Earth and Planetary Science Letters* 296: 45–56. doi: 10.1016/j.epsl.2010.04.043.

Kempton, P.D., Downes, H., Neymark, L.A., et al. 2001. Garnet granulite xenoliths from the Northern Baltic Shield—the underplated lower crust of a Palaeoproterozoic large igneous province? *Journal of Petrology* **42**(4): 731–63. doi: 10.1093/petrology/42.4.731.

Kirkland, C.L., Hartnady, M.I.H., Barham, et al. 2021. Widespread reworking of Hadean-to-Eoarchean continents during Earth's thermal peak. *Nature Communications* **12**(1): 331. doi: 10.1038/s41467-020-20514-4.

Kusky, T.M., Windley, B.F. and Polat, A. 2018. Geological evidence for the operation of plate tectonics throughout the Archean: records from Archean paleo-plate boundaries. *Journal of Earth Science* **29**: 1291–303. doi: 10.1007/s12583-018-0999-6.

Langmuir, C.H., Vocke Jr, R.D., Hanson, G.N., et al. 1978. A general mixing equation with applications to Icelandic basalts. *Earth and Planetary Science Letters* **37**: 380–92.

Lindsay, J.M., Trumbull, R.B., Schmitt, A.K., et al. 2013. Volcanic stratigraphy and geochemistry of the Soufrière Volcanic Centre, Saint Lucia with implications for volcanic hazards. *Journal of Volcanology and Geothermal Research* **258**: 126–42. doi: 10.1016/j.jvolgeores.2013.04.011.

Magaritz, M., Whitford, D.J., and James, D.E. 1978. Oxygen isotopes and the origin of high-^{87}Sr/^{86}Sr andesites. *Earth and Planetary Science Letters* **40**: 220–230. doi: 10.1016/0012-821X(78)90092-4.

McCulloch, M.T. and Wasserburg, G.J. 1978. Sm-Nd and Rb-Sr chronology of continental crust formation. *Science* **200**: 1003–11. doi: 10.1126/science.200.4345.1003

Millet, M.-A., Dauphas, N., Greber, N.D., et al. 2016. Titanium stable isotope investigation of magmatic processes on the Earth and Moon. *Earth and Planetary Science Letters* **449**: 197–205. doi: 10.1016/j.epsl.2016.05.039.

Millot, R., Allègre, C.-J., Gaillardet, J., et al. 2004. Lead isotopic systematics of major river sediments: a new estimate of the Pb isotopic composition of the upper continental crust. *Chemical Geology* **203**(1): 75–90. doi: 10.1016/j.chemgeo.2003.09.002.

Mundl, A., Walker, R.J., Reimink, J.R., et al. 2018. Tungsten-182 in the upper continental crust: evidence from glacial diamictites. *Chemical Geology* **494**: 144–52. doi: 10.1016/j.chemgeo.2018.07.036.

Newman, S., Macdougall, J.D. and Finkel, R.C. 1984. ^{230}Th-^{238}U disequilibrium in island arcs: evidence from the Aleutians and the Marianas. *Nature* **308**: 268–70. doi: 10.1038/308268a0.

O'Neil, J., Carlson, R.L., Francis, D., et al. 2008. Neodymium-142 evidence for Hadean mafic crust. *Science* **321**: 1828–31.

O'Neil, J., Carlson, R.W., Paquette, J.-L., et al. 2012. Formation age and metamorphic history of the Nuvvuagittuq Greenstone Belt. *Precambrian Research* **220–1**: 23–44. doi: 10.1016/j.precamres.2012.07.009.

Payne, J.L., Mcinerney, D.J., Barovich, K.M., et al. 2016. Strengths and limitations of zircon Lu-Hf and O isotopes in modelling crustal growth. *Lithos* **248–51**: 175–192. doi: https://doi.org/10.1016/j.lithos.2015.12.015.

Pearce, J.A., Kempton, P.D. and Gill, J.B. 2007. Hf–Nd evidence for the origin and distribution of mantle domains in the SW Pacific. *Earth and Planetary Science Letters* **260**: 98–114. doi: 10.1016/j.epsl.2007.05.023.

Peucker-Ehrenbrink, B. and Jahn, B.-M. 2001. Rhenium-osmium isotope systematics and platinum group element concentrations: Loess and the upper continental crust. *Geochemistry, Geophysics, Geosystems* **2**: 1061. doi: 10.1029/2001gc000172.

Plank, T. 2014. The chemical composition of subducting sediments, in Turekian, K.K., Holland. and Heinrich D. (ed). *Treatise on Geochemistry*, 2nd ed. Oxford: Elsevier.

Puetz, S.J., Ganade, C.E., Zimmermann, U., et al. 2018. Statistical analyses of global U-Pb database 2017. *Geoscience Frontiers* **9**: 121–45. doi: 10.1016/j.gsf.2017.06.001.

Pujol, M., Marty, B., Burgess, R., et al. 2013. Argon isotopic composition of Archaean atmosphere probes early Earth geodynamics. *Nature* **498**: 87–90. doi: 10.1038/nature12152.

Rasmussen, B., Fletcher, I.R., Muhling, J.R., et al. 2010. In situ U–Th–Pb geochronology of monazite and xenotime from the Jack Hills belt: Implications for the age of deposition and metamorphism of Hadean zircons. *Precambrian Research* **180**: 26–46. doi: 10.1016/j.precamres.2010.03.004.

Reimink, J.R., Chacko, T., Stern, R.A., et al. 2014. Earth's earliest evolved crust generated in an Iceland-like setting. *Nature Geoscience* **7**: 529–33. doi: 10.1038/ngeo2170.

Reimink, J., Pearson, D., Shirey, S., et al. 2019. Onset of new, progressive crustal growth in the central Slave Craton at 3.55 Ga. *Geochemical Perspectives Letters* **14**: 8–14. doi: 10.7185/geochemlet.1907.

Reubi, O., Sims, K.W.W. and Bourdon, B. 2014. ^{238}U–^{230}Th equilibrium in arc magmas and implications for the time scales of mantle metasomatism. *Earth and Planetary Science Letters* **391**: 146–58. doi: 10.1016/j.epsl.2014.01.054.

Rizo, H., Walker, R.J., Carlson, R.W., et al. 2016. Preservation of earth-forming events in the tungsten isotopic composition of modern flood basalts. *Science* **352**: 809–12. doi: 10.1126/science.aad8563.

Roberts, N.M.W. and Spencer, C.J. 2015. The zircon archive of continent formation through time. *Geological Society of London, Special Publications* **389**: 197–225. doi: 10.1144/sp389.14.

Roth, A.S.G., Bourdon, B., Mojzsis, S.J., et al. 2014. Combined ^{147,146}Sm-^{143,142}Nd constraints on the longevity and residence time of early terrestrial crust. *Geochemistry, Geophysics, Geosystems* **15**: 2329–45. doi: 10.1002/2014GC005313.

Rudnick, R.L., and Goldstein, S.L. 1990. The Pb isotopic compositions of lower crustal xenoliths and the evolution of lower crustal Pb, *Earth and Planetary Science Letters* **98**: 192–207. doi: 10.1016/0012-821X(90)90059-7.

Saal, A.E., Rudnick, R.L., Ravizza, G.E., et al. 1998. Re–Os isotope evidence for the composition, formation and age of the lower continental crust. *Nature* **393**(6680): 58–61. doi: 10.1038/29966.

Schmitz, M.D., Vervoort, J.D., Bowring, S.A., et al. 2004. Decoupling of the Lu-Hf and Sm-Nd isotope systems during the evolution of granulitic lower crust beneath southern Africa. *Geology* **32**: 405–408. doi: 10.1130/g20241.1.

Shirey, S.B. and Richardson, S.H. 2011. Start of the Wilson Cycle at 3 Ga Shown by Diamonds from Subcontinental Mantle. *Science* **333**: 434–436. doi: 10.1126/science.1206275.

Singer, B.S., O'Neil, J.R. and Brophy, J.G. 1992. Oxygen isotope constraints on the petrogenesis of Aleutian arc magmas. *Geology* **20**: 367–70. doi: 10.1130/0091-7613(1992)020<0367:Oicotp>2.3.Co;2.

Smit, K.V., Shirey, S.B., Hauri, E.H., et al. 2019. Sulfur isotopes in diamonds reveal differences in continent construction. *Science* **364**: 383–385. doi: 10.1126/science.aaw9548.

Spencer, C.J., Cawood, P.A., Hawkesworth, C.J., et al. 2014. Proterozoic onset of crustal reworking and collisional tectonics: reappraisal of the zircon oxygen isotope record. *Geology* **42**(5): 451–4. doi: 10.1130/g35363.1.

Stern, R.J. and Scholl, D.W. 2010. Yin and Yang of continental crust creation and destruction by plate tectonic processes. *International Geology Review* **52**(1): 1–31. doi: 10.1080/00206810903332322.

Stern, R.J. 2008. Modern-style plate tectonics began in Neoproterozoic time: an alternative interpretation of Earth's tectonic history, in Condie, K.C. and Pease, V. (ed). *When did plate tectonics begin on planet Earth. Geological Society of America Special Paper* **440**: 265–80 pp.

Taylor, H.P. 1980. The effects of assimilation of country rocks by magmas on $^{18}O/^{16}O$ and $^{87}Sr/^{86}Sr$ systematics in igneous rocks. *Earth and Planetary Science Letters* **47**: 243–254. doi: https://doi.org/10.1016/0012-821X(80)90040-0.

Tera, F., Brown, L., Morris, J.D., et al. 1986. Sediment incorporation in island-arc magmas: inferences from ^{10}Be. *Geochimica et Cosmochimica Acta* **50**: 535–50.

Thirlwall, M.F. and Graham, A.M. 1984. Evolution of high-Ca, high-Sr C-series basalts from Grenada, Lesser Antilles: the effects of intra-crustal contamination. *Journal of the Geological Society* **141**: 427–45. doi: 10.1144/gsjgs.141.3.0427.

Touboul, M., Puchtel, I.S. and Walker, R.J. 2012. ^{182}W evidence for long-term preservation of early mantle differentiation products. *Science* **335**: 1065–9. doi: 10.1126/science.1216351.

Trail, D., Boehnke, P., Savage, P.S., et al. 2018. Origin and significance of Si and O isotope heterogeneities in phanerozoic, Archean, and Hadean zircon. *Proceedings of the National Academy of Sciences* **115**: 10287–92. doi: 10.1073/pnas.1808335115.

Tsvetkov, A., Gladkov, N. and Volynets, O. 1989. Problem of sediment subduction and ^{10}Be isotope in lavas of Kuril islands and Kamchatka peninsula. *Doklady Akademii Nauk SSSR* **306**: 1220–5.

Turner, S., Evans, P. and Hawkesworth, C. 2001. Ultrafast source-to-surface movement of melt at island arcs from ^{226}Ra-^{230}Th systematics. *Science* **292**: 1363–6. doi: 10.1126/science.1059904.

Turner, S., Regelous, M., Hawkesworth, C., et al. 2006. Partial melting processes above subducting plates: constraints from ^{231}Pa–^{235}U Disequilibria. *Geochimica et Cosmochimica Acta* **70**: 480–503. doi: 10.1016/j.gca.2005.09.004.

Valley, J.W., Cavosie, A.J., Ushikubo, T., et al. 2014. Hadean age for a post-magma-ocean zircon confirmed by atom-probe tomography. *Nature Geoscience* **7**: 219–23. doi: 10.1038/ngeo2075.

Vervoort, J.D., Patchett, P.J., Albarède, F., et al. 2000. Hf–Nd isotopic evolution of the lower crust. *Earth and Planetary Science Letters* **181**: 115–29. doi: 10.1016/S0012-821X(00)00170-9.

Wänke, H., Dreibus, G. and Jagoutz, E. 1984. Mantle geochemistry and accretion history of the Earth, in Kröner, A. (ed). *Archean Geochemistry*. Berlin: Springer.

Wasilewski, B., O'Neil, J., Rizo, H., et al. 2021. Over one billion years of Archean crust evolution revealed by zircon U-Pb and Hf isotopes from the Saglek-Hebron complex. *Precambrian Research* **359**: 106092. doi: 10.1016/j.precamres.2021.106092.

Watson, E., Wark, D. and Thomas, J. 2006. Crystallization thermometers for zircon and rutile. *Contributions to Mineralogy and Petrology* **151**: 413–33. doi: 10.1007/s00410-006-0068-5.

White, W.M., and Dupré, B. 1986. Sediment subduction and magma genesis in the Lesser Antilles: isotopic and trace element constraints. *Journal Geophysical Research* **91**: 5927–41. doi: 10.1029/JB091iB06p05927

White, W.M., Copeland, P., Gravatt, D.R., et al. 2017. Geochemistry and geochronology of Grenada and Union Islands, Lesser Antilles: the case for mixing between two magma series generated from distinct sources. *Geosphere* **5**: 1359–91. doi: 10.1130/GES01414.1.

Willbold, M., Elliott, T. and Moorbath, S. 2011. The tungsten isotopic composition of the Earth/'s mantle before the terminal bombardment. *Nature* **477**: 195–8. doi: 10.1038/nature10399.

Willig, M., Stracke, A., Beier, C., et al. 2020. Constraints on mantle evolution from Ce-Nd-Hf isotope systematics. *Geochimica et Cosmochimica Acta* **272**: 36–53. doi: 10.1016/j.gca.2019.12.029.

PROBLEMS

1. If we view Indian Ocean water with ε_{Nd} of -6.6 as a simple mixture of Atlantic water with an average ε_{Nd} of -11.4 and Pacific water with an average ε_{Nd} of -3.9 (which it is not), assuming equal Nd concentrations, what are the proportions of Atlantic and Pacific water in the mixture?

2. Suppose that a dacitic magma containing 100 ppm Sr and 25 ppm Nd with $^{87}Sr/^{86}Sr = 0.7076$ and $\varepsilon_{Nd} = -2$ mixes with a basaltic magma with 500 ppm Sr and 5 ppm Nd with $^{87}Sr/^{86}Sr = 0.7035$ and $\varepsilon_{Nd} = +6$. Plot, at intervals of 10% addition of the dacitic magma, the Sr and Nd isotopic composition of the mixture. What is the value of r as defined in 8.10?

3. If we assume that Sr in seawater ($^{87}Sr/^{86}Sr = 0.70925$) is a mixture of Sr from rivers ($^{87}Sr/^{86}Sr = 0.7119$) and Sr from mid-ocean ridge hydrothermal systems ($^{87}Sr/^{86}Sr = 0.7030$), what is the proportion of mid-ocean ridge derived Sr in seawater?

4. The average continental crust has $^{147}Sm/^{144}Nd$ of 0.112. Assuming that the average age of the crust is 2.2 Ga and that new crust when it forms as an ε_{Nd} of 0 (i.e. derived from a "chondritic" mantle), what should the average ε_{Nd} of crust be?

5. Assuming that the continental crust has $^{147}Sm/^{144}Nd$ of 0.112 and an ε_{Nd} equal to that of the suspended load of rivers of -10.6, what is the τ_{DM} model age of the crust?

6. The observable modern silicate Earth has today has $^{142}Nd/^{144}Nd = 1.141837$. The initial solar system $^{142}Nd/^{144}Nd$ was 1.141437. Assuming that the Earth evolved with a constant Sm/Nd from this initial value, what would its effective initial (at the formation of the solar system) $^{146}Sm/^{144}Nd$ be?

7. The present $^{142}Nd/^{144}Nd$ of the Nuvvuagittuq amphibolites is 1.141825. Assume that bulk silicate Earth today has $^{142}Nd/^{144}Nd = 1.141837$ and had effective initial you calculated in Question 6. If the Nuvvuagittuq amphibolites formed from that bulk silicate Earth reservoir 200 million years after the start of the solar system, what would their $^{146}Sm/^{144}Nd$ have been at the time they formed?

8. The (CI) chondritic concentration of W is 0.09 ppm while that of the bulk silicate Earth is 0.03 ppm. If the late accretionary veneer consisted of CI chondritic material with $\varepsilon_W = -2$ and the silicate had $\varepsilon_W = +0.13$ before the addition of this material, what must the mass fraction of this late accretionary veneer have been if the present silicate Earth has $\varepsilon_W = 0$?

9. Make a plot using 8.11 similar to Figure 8.33 comparing AFC evolution where $\Delta = 0$ and $\Delta = -2$. Use a value of $R = 0.15$ in both the cases and an initial $\delta^{18}O = 6$ and assimilant $\delta^{18}O = +18$.

10. Plot the evolution of $\delta^{18}O$ versus $^{87}Sr/^{86}Sr$ in a magma undergoing AFC. Assume that the magma initially has $\delta^{18}O = 6‰$ and $^{87}Sr/^{86}Sr = 0.703$, and that the assimilant has $\delta^{18}O = +18$ and $^{87}Sr/-Sr = 0.720$. Use $R = 0.15$, $\Delta^{18}O_{xtals-magma} = -1$, $D_{Sr} = 1.2$, and Sr concentrations for the initial magma and the assimilant of 500 and 100 ppm, respectively.

Chapter 9

Stable Isotopes in the Solid Earth

9.1 INTRODUCTION

We now shift our principal focus over the next several chapters to stable isotopes and use the principles we learned in Chapter 5. We will begin with the high-temperature environment of the mantle and the magmas it produces because this material is the feedstock, so to speak, for the surface environment that we will consider in the following chapters. Isotopic fractionation decreases with increasing temperature, so the mantle was once considered essentially homogeneous with respect to stable isotopes and was largely ignored by stable isotope geochemists. As analytical precision has increased over time, however, small but significant variations in the isotopes of nearly all multi-isotopic elements have been found. These have provided particularly important insights into how the Earth works and how it has evolved.

The dependence of isotopic fractionations on temperature is the basis of the first application we will consider, namely geothermometry: using isotope fractionations between phases to deduce the temperatures at which rocks equilibrated. We will begin with a brief overview of fractionation factors and their temperature dependencies. As temperatures approach mantle values of 1000 °C or so, fractionations become quite small, but as analytical precision has improved, it is now sometimes possible to extend isotope geothermometry even to mantle temperatures. Nevertheless, the generally small fractionation at these temperatures means that

we can often neglect them and use stable isotopes in the mantle as tracers, much as we do radiogenic isotopes. We can determine past surface temperatures as well, and we will explore palaeothermometry in Chapter 12. We will go on to review the stable isotopic composition of the mantle. Since the mantle represents the bulk of the mass of the Earth, comparing its isotopic composition with that of the solar system, represented by chondritic meteorites, helps us understand the Earth's formation. The isotopic variability we encounter in the mantle can often be attributed to material transported from the surface into the mantle. We will then move to review how stable isotopes can be used to understand the formation of ore deposits.

9.2 EQUILIBRIUM FRACTIONATIONS AMONG MINERALS

9.2.1 Compositional and structural dependence of equilibrium fractionations

Just as they are in the simple gases we considered in Chapter 5, the nature and strength of the chemical bond is of key importance in determining isotope fractionations between solids. In general, bonds to ions with a high ionic potential and low atomic mass are associated with high vibrational frequencies and have a tendency to incorporate the heavy isotope preferentially. This point is illustrated by the site-potential method of estimating fractionation factors (Smyth, 1989). The site

Isotope Geochemistry, Second Edition. William M. White.
© 2023 John Wiley & Sons Ltd. Published 2023 by John Wiley & Sons Ltd.
Companion Website: www.wiley.com/go/white/isotopegeochem2

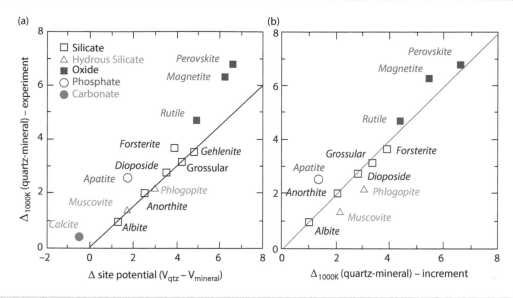

Figure 9.1 (a) Comparison of quartz–mineral fractionation factors estimated from the difference in oxygen site potential ($V_{qtz} - V_{mineral}$) and experimentally observed fractionation factors at 1000 K. (b) Comparison of fractionation factors estimated through the increment method, which also considers cation mass, and experimentally observed fractionation factors at 1000 K. Source: Adapted from (Chacko et al., 2001).

potential is simply the energy required (e.g. in electron volts) to remove an atom from its crystallographic site and is a measure of bond strength. Figure 9.1a shows that there is a strong correlation between the difference in oxygen site potential in minerals and the oxygen fractionation factor between those two minerals. The solid line shows that silicates plot along a line with the equation Δ_{1000K} (qtz-mineral) = 0.751 ($V_{qtz} - V_{mineral}$). Oxides (and, to a less extent, apatite and calcite) fall off this correlation. The deviation in the case of calcite and apatite probably reflects the more strongly covalent nature of oxygen bonds in those minerals. In silicates, oxygen is primarily bound to Si and secondarily to other cations; hence, the nature of those other cations is of secondary importance. In the case of the oxides, such as rutile, perovskite, magnetite, etc., oxygen is bound primarily to Ti, Fe, etc., and the cation mass affects bond strength.

In the "increment method" (e.g. Schütze, 1980; Zheng, 1993), the process of determining the β-factor (reduced partition function) of a mineral of interest begins with a reference mineral with similar crystal structure whose β-factor is known. Adjustments are then made by multiplying by an "increment factor,"

I–^{18}O, for different structural sites that depend on the oxidation state of the bound cation, coordination number, and ionic radius (the latter equals bond length when summed with the O ionic radius). As Figure 9.1b shows, this method produces an improved agreement of calculated and experimentally observed fractionation factors for the oxides.

As we noted, the substitution of cations in a primarily ionic site in silicates has only a minor effect on O fractionation factors. Thus, for example, the O isotope fractionation between K-feldspar and Na-feldspar is of the order of 0.1 per mil at room temperature. The substitution of Al for Si in plagioclase, which accompanies the substitution of Ca for Na, is more important, leading to a 1.1 per mil fractionation between anorthite and albite at room temperature.

Oxygen occupies a generally similar lattice site in virtually all mantle and igneous minerals, forming a tetrahedron composed of predominately covalent bonds to silicon and with predominately ionic bonds to other cations such as Mg, Fe, Ca, etc.). Consequently, the fractionation of oxygen isotopes between these phases and during melting and crystallization are relatively small, although still measurable.

Carbonates tend to be ^{18}O-rich because O is bonded to a small, highly charged atom, C^{4+}. The fractionation relative to water, $\Delta^{18}O_{cal-water}$, is about +30 for calcite at surface temperature. The cation (i.e. Ca, Mg, Sr, and Ba in carbonate) has a secondary role (because of the effects of its mass on vibrational frequency). $\Delta_{18}O$ decreases to about 25 for Ba (which has about three times the mass of Ca).

Crystal structure usually plays a secondary role. The $\Delta^{18}O$ between the two polymorphs of $CaCO_3$, aragonite and calcite, is of the order of 0.5‰. However, there is a (predicted) large fractionation (10‰) of C between graphite and diamond extrapolated to room temperature.

The effect of pressure on equilibrium constants, including fractionation factors, is

$$\frac{\partial \ln K}{\partial T} = -\frac{\Delta V_r}{RT} \tag{9.1}$$

where ΔV_r is the volume change of reaction. Because there is little difference in the atomic or ionic size of isotopes of an element, we can expect the effect of pressure to be small. However, the mass of atoms affects bond strength, and substituting a heavier isotope for a lighter one will reduce bond lengths and, hence, volume slightly. This effect tends to be small, close to analytical uncertainty over ranges of hundreds of MPa (tens of kbar) and, furthermore, tends to decrease with temperature. Since high-pressure environments are typically high-temperature ones, the effect of pressure can in most, but not all, cases be ignored at least for the upper 100 km or so of the Earth. Hydrogen isotope partitioning between water and hydrous minerals is an exception, particularly near the critical point of water (374 °C). Experiments with several minerals show that increasing pressure tends to decrease the water molecule's affinity of deuterium.

9.2.2 Theoretical and experimental determination of equilibrium fractionations

In Chapter 5, we found that the vibrational contribution varies with the inverse of absolute temperature. At higher temperature, the $e^{-h\nu/kT}$ term in the denominator of (5.36) becomes finite, this relationship breaks down, and the equilibrium constant becomes approximately proportional to the inverse square of temperature. At infinite temperature, the fractionation is unity. The actual temperature dependence of fractionation between minerals is far more complex than the ones we calculated for a simple diatomic gas in Chapter 5. The good news is that neither rotational nor translational modes of motion are available to atoms in a crystal lattice, so we do not have to worry about these terms. The bad news is that even simple crystals tend to have many vibrational modes, making such calculations considerably more complex than for gases. As the frequencies of these vibrational modes have become better known and computers and computational methods more powerful, it is now possible to calculate fractionation factors from first principles, and theoretically computed fractionation factors are increasingly available.

Nevertheless, vibrational frequencies are rarely known for minor isotopes and, therefore, must be estimated (much as we did for the $CO-O_2$ fractionation factor in Section 5.3.1.6 but account must also be taken of any anharmonicities) for each vibrational mode. In practice, this is done over at least $3N-3$ modes, where N is the number of atoms in the unit cell. Quartz is one of the simplest rock-forming minerals with are only three atoms in its chemical formula, but the unit cell consists of three linked tetrahedra with $N = 12$ so that 33 vibrational modes that must be considered. Clearly calculating fractionation factors is computationally intensive for even compositionally simple minerals. The details of first principle calculations would take us well beyond the scope of this book; the interested reader is referred to (Young et al., 2015) for an overview. The incremental method mentioned above is a slightly simpler approach.

As one might imagine, these calculations lead to quite complex temperature dependencies. As we found in Chapter 5 (Equation 5.43) the fractionation factor between two phases can be computed from the ratio of their reduced partition functions (β-factors), and theoretical fractionation factors are often based on these β-factors. For convenience, they are often fitted to a polynomial function such as

$$1000 \ln \beta = A \times \left(\frac{10^6}{T^2}\right) + B \times \left(\frac{10^6}{T^2}\right)^2$$
$$+ C \times \left(\frac{10^6}{T^2}\right)^3. \tag{9.2}$$

The actual calculated temperature functions are often more complex. In some cases, the last two terms in this equation can often be dropped, so that the fractionation factor is just a function of $1/T^2$.

Although theoretically calculated fractionation factors are increasingly available, geochemists continue to rely on empirically determined ones; furthermore, calculated fractionation factors require at least some verification against observed ones, including both those based on observations of natural samples and ones determined by the experiment. In the former, when two coexisting phases are found within a rock and there is evidence that equilibrium has been achieved and the temperature of equilibration is known from some other geothermometer, the fractionation factor can be determined by analyzing those phases. In the latter, mineral pairs are experimentally equilibrated at the temperature of interest. Solid phases, particularly silicates and oxides, tend to equilibrate slowly. Some techniques rely on the dictum of thermodynamics that two phases each in equilibrium with a third phase are in equilibrium with themselves. Consequently, some experiments determine fractionation factors with a third phase that equilibrates more rapidly such as water or calcite and then calculate the fractionation factors for phases of interest.

9.2.2.1 Silicates and oxides

Table 9.1 lists coefficients for reduced oxygen isotope partition function ratios in the form $1000\ln\beta$ for common silicate and oxide minerals based on theoretical calculations that have been fitted to 9.2 and experimental results. Although perhaps not obvious, this form is actually quite convenient since the fractionation factor between two phases is $\alpha_{A\text{-}B} = \beta_A/\beta_B$ and $\Delta_{A\text{-}B} \cong 1000\ln\beta_A - 1000\ln\beta_B$. Thus, the fractionation factor between any two minerals listed in Table 9.1 can be computed from their β-factors. Oxygen isotope fractionation factors calculated from this table are shown in Figure 9.2. The values are generally applicable above 400 K (127 °C) and are not necessarily applicable to surface processes (but most of these minerals do not form under equilibrium conditions at the Earth's surface).

Table 9.1 Coefficients for reduced partition coefficient ratios for oxygen isotope fractionation $1000\ln\beta = A \times 10^6/T^2 + B \times (10^6/T^2)^2 + C \times (10^6/T^2)^3$.

	A	B	C
Calcite	11.781	−0.42	0.0158
Quartz	12.55277	−0.41976	0.01979
Albite	11.558	−0.35325	0.01557
Anorthite	10.67645	−0.2868	0.01150
Diopside	9.88999	−0.23237	0.00862
Forsterite	9.05093	−0.16847	0.00517
Zircon	9.87022	−0.23965	0.00832
Phlogopite	9.969	−0.382	0.0194
Muscovite	10.766	−0.412	0.0209
Rutile	7.258	−0.125	0.0033
Magnetite	5.674	−0.038	0.0003

Source: From (Qin et al., 2016) and the compilation of Chacko et al. (2001).

As temperatures increase, the second and third terms in 9.2 become increasingly negligible, and the temperature dependence of empirically derived fractionation factors for oxides and silicates at higher temperature is often reported in the following form:

$$1000\ln\alpha = \frac{A \times 10^6}{T^2}. \qquad (9.3)$$

A compilation of these coefficients for common silicates and oxides is listed in Table 9.2.

9.2.2.2 Sulfides

Another isotope that has been used extensively for geothermometry is sulfur, particularly for ore deposits as many important ores are sulfides. The temperature range of ore-forming fluids is quite wide, ranging from <100 to >600 °C, but fractionation factors between major sulfide minerals can nonetheless be reasonably approximated by a $1/T^2$ relationship. Table 9.3 lists fractionation factors for common sulfide minerals calculated by Li and Liu (2006) using the incremental method. They agree reasonably well with earlier experimentally determined fractionation factors of Ohmoto and Rye (1979).

Most sulfide ores precipitate from hydrous fluids; while fractionation factors between sulfide minerals can be approximated by a simple $1/T^2$ relationship, the temperature dependence of fractionations between dissolved sulfur species is more complex.

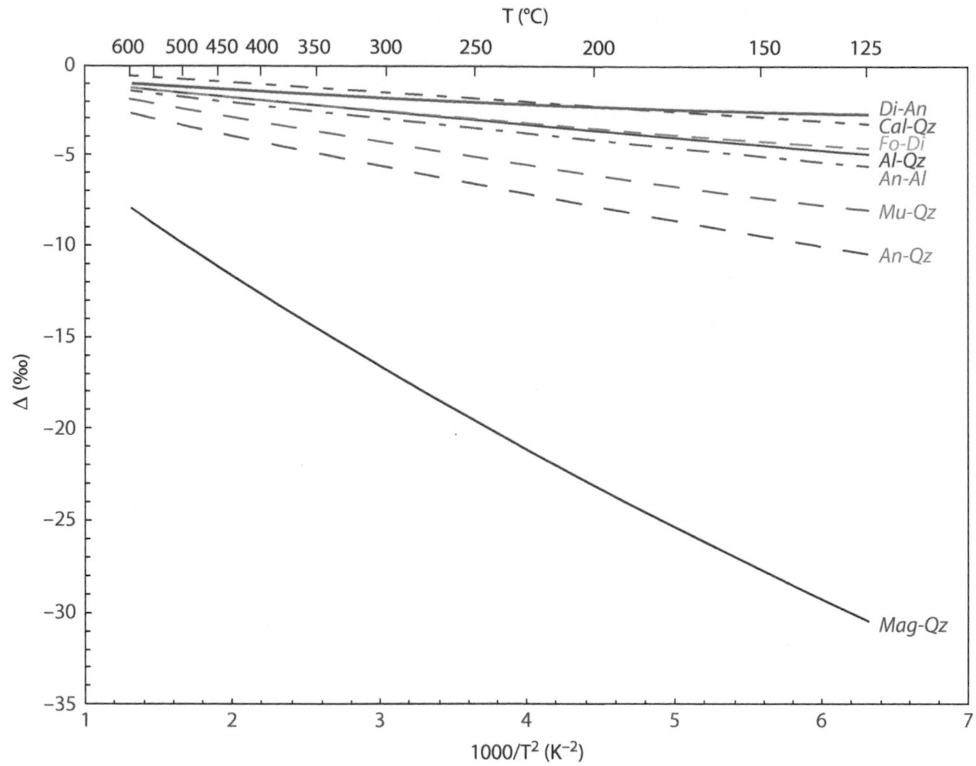

Figure 9.2 Oxygen isotope fractionation for several mineral pairs as a function of temperature based in partition coefficient ratios listed in Table 9.1.

Table 9.2 Empirically determined mineral pair coefficients, A, for oxygen isotope fractionation factors computed as $\Delta = 1000 \ln \alpha = A \times 10^6/T^2$ for temperatures >600°C. T is in kelvin.

	Calcite	Albite	Muscovite	Anorthite	Phlogopite	Diopside	Forsterite	Rutile	Magnetite	Zircon
Quartz	0.38	0.94	1.37	1.99	2.16	2.75	3.67	4.69	6.29	2.33
Calcite		0.56	0.99	1.61	1.78	2.37	3.29	4.31	5.91	1.95
Albite			0.43	1.05	1.22	1.81	2.73	3.75	5.35	1.77
Muscovite				0.62	0.79	1.38	2.30	3.32	4.92	−1.9
Anorthite					0.17	0.76	1.68	2.7	4.3	−1.71
Phlogopite						0.59	1.51	2.53	4.13	−2.19
Diopside							0.92	1.94	3.54	−1.74
Forsterite								1.02	2.62	−3.25
Rutile									1.60	−1.31
Magnetite										−0.73

Source: From the compilation of Chacko et al. (2001).

Eldridge et al. (2016) calculated theoretically reduced partition functions, β, for these species fit to the equation:

$$\beta = \frac{A \times 10^5}{T^4} + \frac{B \times 10^4}{T^3} + \frac{C \times 10^2}{T^2} + \frac{D \times 10^{-2}}{T} + E$$

(9.4)

where T is in kelvin. Table 9.4 lists a selection of these, from which fractionation factors

are readily calculated using (5.43) ($\alpha_{A\text{-}B} = \beta_A/\beta_B$). Fractionation factors between various species and $H_2S_{(aq)}$ calculated from these values are shown in Figure 9.3 as function of temperature.

Also shown in Figure 9.3 are fractionation factors between H_2S and pyrite, galena, and sphalerite calculated from the study of Zhang (2021). We can see from this figure that fractionation is a function of oxidation state,

Table 9.3 Calculated coefficients, A, for sulfide isotope fractionation factors computed as $\Delta = 1000 \ln \alpha = A \times 10^6/T^2$.

	Galena	Pyrrhotite	Chalcopyrite	Greenocite	Bornite	Cubanite	Violarite	Sulvanite
	PbS	FeS	FeCuS$_3$	CdS	CuFeS$_4$	CuFe$_2$S$_3$	FeNi$_2$S$_4$	CuVS$_4$
Sphalerite	−0.74	−0.15	0.05	−0.08	0.17	0.06	0.37	0.07
Galena		−0.89	−0.69	−0.82	−0.57	−0.68	−0.37	−0.67
Pyrrhotite			0.2	0.07	0.32	0.21	0.52	0.22
Chalcopyrite				−0.13	0.12	0.01	0.32	0.02
Greenockite					0.25	0.14	0.45	0.15
Bornite						−0.11	0.3	−0.1
Cubanite							0.31	0.01
Violarite								−0.3

Source: From (Li and Liu, 2006).

Table 9.4 Coefficients ^{34}S/^{32}S for β-factors for 9.4.

	A	B	C	D	E
SO$_4^{2-}$	590.371	−126.799	109.869	−137.516	1.000347
SO$_{2(aq)}$	692.521	−108.509	71.7931	−103.029	1.000245
SO$_{2(g)}$	754.026	−114.672	71.5395	−105.747	1.000248
SO$_2^{2-}$	11.8836	−20.417	30.3733	−24.9619	1.000066
H$_2$S$_{(aq)}$	462.536	−48.3382	20.1638	77.2695	0.999749
HS$^-$	213.257	−22.4384	12.2701	42.8805	0.999864
H$_2$S$_{(g)}$	470.075	−48.9645	19.2083	82.5595	0.999735
S^{2-}	2.46274	−2.30464	7.54757	−0.688045	1.000001

Source: From (Eldridge et al., 2016).

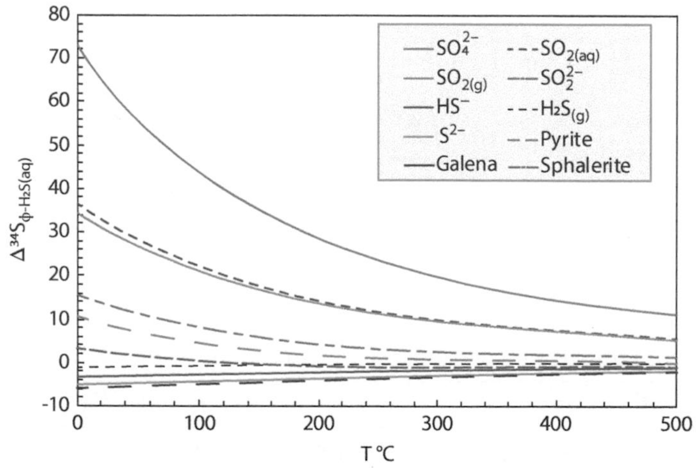

Figure 9.3 Isotope fractionation between various sulfide and sulfate species as a function of temperature calculated from the reduced partition functions in Table 9.4 and Zhang (2021).

with fractionation factors increasing with oxidation state, as we expect as S is more strongly bound to oxygen than to hydrogen. We also see that, as we expect, fractionation factors decrease with temperature. Finally, fraction between sulfide minerals and H$_2$S is much less temperature dependent than that between species of varying oxidation state. Caution is needed with respect to these mineral–H$_2$S fractionation factors, however, as equilibrium is sometimes not achieved, particularly between pyrite and dissolved H$_2$S.

Eldridge et al. also calculated reduced partition functions for $^{33}S/^{32}S$ and $^{34}S/^{32}S$ from which slopes $\theta^{33/34} = \ln\alpha^{33/32}/\ln\alpha^{34/32}$ and $\theta^{36/34} = \ln\alpha^{36/32}/\ln\alpha^{34/32}$ can be calculated. They found that these defined tight ranges over a wide range of temperature for all the aqueous compounds of $\theta^{33/34} \approx 0.5148–0.5159$ and $\theta^{36/34} \approx 1.89–1.90$ converging at infinite temperature on 0.51587 and 1.8905, respectively, compared to the canonical values of $\gamma^{33/34} = 0.515$ and $\gamma^{33/34} = 1.90$ (recall that we use θ and γ to denote theoretical observed slopes, respectively, in isotope–isotope plots such as $\delta^{33}S–\delta^{34}S$). Thus, at low and moderate temperatures, equilibrium among sulfur species is expected to produce slight *apparent* mass-independent fractionation (MIF) with both positive and negative $\Delta^{33}S$ and $\Delta^{36}S$ values.

9.2.2.3 Carbonates

Dissolved CO_2 and other carbon-bearing species are ubiquitous in meteoric and hydrothermal waters. Carbonates often precipitate from such solutions, and the fractionation between carbon species provides yet another opportunity for geothermometry. Chacko and Deines (2008) fit calculated reduced oxygen isotope partition functions (β-factors) to the equation:

$$1000\ln\beta = C_0 + C_1\left(\frac{10^6}{T^2}\right) - C_2 \times 10^{-1}\left(\frac{10^6}{T^2}\right)^2$$
$$- C_3 \times 10^{-2}\left(\frac{10^6}{T^2}\right)^3 - C_4 \times 10^{-3}\left(\frac{10^6}{T^2}\right)^4$$
$$+ C_5 \times 10^{-5}\left(\frac{10^6}{T^2}\right)^5 - C_6 \times 10^{-6}\left(\frac{10^6}{T^2}\right)^6$$

$$(9.5)$$

where T is in kelvin. Coefficients are listed Table 9.5. To repeat, fractionation factors can be computed from the ratio of β-factors. Consequently, calculating β factors from the parameters in Table 9.5 allows us to readily calculate the fractionation factor between any two of the minerals listed in Table 9.5. Figure 9.4 shows oxygen isotope fractionation factors between calcite and other carbon bearing species as a function of temperature calculated in this way.

9.3 GEOTHERMOMETRY

In principle, with temperature dependencies of fractionation factors known, a temperature can be calculated from the isotopic fractionation between any phases provided the phases achieved equilibrium. All geothermometers are based on the apparently contradictory assumptions that complete equilibrium is achieved between phases during, or perhaps after, the formation of the phases, but that the phases do not re-equilibrate as they subsequently cool to surface temperatures. Isotope geothermometers have same implicit assumptions about the achievement of equilibrium as other geothermometers. The reason why these assumptions can be made and geothermometry works at all is that reaction and diffusion rates depend exponentially on temperature (2.35). Just as we found in Section 2.3.1, diffusion rates slow to the point that re-equilibration is no longer possible.

There are several reasons why isotope geothermometers might not record valid temperatures. First, isotope geothermometers are often applied to relatively low-temperature situations where reaction rates are slow to

Table 9.5 Coefficients for temperature dependence of reduced partition coefficient ratios (β factors) in 9.5.

Mineral	C_0	C_1	C_2	C_3	C_4	C_5	C_6
Calcite $CaCO_3$	0	11.75	4.6655	3.1252	1.8869	7.4768	1.3404
Aragonite $CaCO_3$	0	12.03	5.0094	3.5385	2.2189	8.9981	1.636
Aragonite (100 kbar)	0	12.795	5.4354	3.9288	2.5053	10.259	1.8756
Magnesite [$MgCO_3$]	0	13.189	5.2359	3.5607	2.1908	8.7925	1.5889
Dolomite $CaMg(CO_3)_2$	0	12.317	4.8399	3.2448	1.9665	7.8119	1.4026
Rhodochrosite $MnCO_3$	0	12.007	4.6966	3.1079	1.8627	7.3494	1.3143
Siderite $FeCO_3$	0	12.082	4.8693	3.2974	2.0081	8.0035	1.4402
Natrite Na_2CO_3	0	11.243	4.4745	2.9631	1.7696	6.9618	1.2425
CO_2	0	16.7604	14.8378	16.9027	13.7557	63.735	12.3984
H_2O (liquid)	12.815	6.237	1.254	0.2419	0	0	0

Source: From Chacko and Deines (2008).

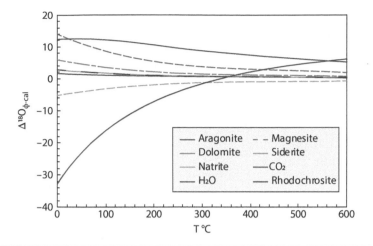

Figure 9.4 Fractionation between calcite and other carbonates species as a function of temperature calculated from 9.5 and the coefficients of Chacko and Deines (2008) listed in Table 9.5.

begin with and equilibrium is more difficult to attain. Second, kinetic fractionations may compete with equilibrium ones. If reactions do not run to completion, the isotopic differences may reflect kinetic effects as much as equilibrium. Third, a system may partially re-equilibrate at some lower temperatures during cooling. Fourth, free energies of isotope exchange reactions are low, meaning that there is little chemical energy available to drive the reaction to equilibrium. Indeed, isotopic equilibrium no doubt often depends upon other reactions occurring that mobilize the element involved in the exchange. Finally, solid-state exchange reactions are particularly slow at temperatures well below the melting point. Equilibrium between solid phases will, thus, generally depend on the presence of a fluid reacting with them (which is often the case in metamorphism and ore deposition). Of course, this is true of "conventional" chemical reactions as well.

The requirement of equilibrium means that disturbed samples should be avoided as well as minerals capable of rapid re-equilibration, such as biotite and feldspar (Sharp, 2007). In addition, several mineral pairs should be analyzed, which should all yield the same temperature. If all the minerals in a sample equilibrate at the same temperature, they should define a single line, an "isotherm," when measured fractionation, Δ, is plotted against the temperature-dependence parameter (e.g. "A" in Equation 9.3). The slope of the line, which is equal to $10^6/T^2$, can be calculated by

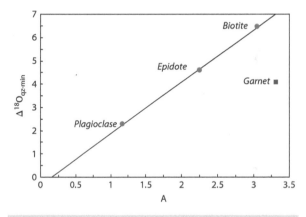

Figure 9.5 Measured quartz–mineral O isotope fractionations in a gneiss from the Dabie ultrahigh-pressure belt in east-central China by Zheng et al. (2003) plotted against the A parameter in 9.3. The slope of this "isotherm" is equal to $10^6/T^2$, which in this case is 297 °C. This temperature likely reflects retrograde metamorphism, while the garnet temperature likely records peak eclogite metamorphism.

regression and from this T. A study by Zheng et al. (2003) provides an example. They analyzed rocks from the Dabie ultrahigh-pressure belt in east-central China. Figure 9.5 shows the analyzed quartz–mineral fractionation factors for four minerals from their sample 92SH-1. Biotite, epidote, and plagioclase define a line whose slope is 2.224, from which a temperature of 670 K or 297 °C can be calculated. Garnet lies off this line and the measured quartz–garnet fractionation gives a higher temperature of 630 °C. A biotite–whole rock Rb–Sr iso-chron for the rock yielded an age of 171±3 Ma, while Sm–Nd isochron, largely defined by garnet, yielded an age of 213±5 Ma. Their interpretation was the garnet temperature and the Sm–Nd age likely records peak eclogite metamorphism, and the younger age and lower temperatures reflect re-equilibration during retrograde metamorphism with the much slower O diffusion rates in garnet compared with the other minerals inhibiting re-equilibration.

Figure 9.6 compares sphalerite–galena sul-fur isotope temperatures of Mississippi Valley-type (MVT) ore deposits in the USA and the Kuroko volcanogenic massive sulfide (VMS) deposit in Japan with fluid-inclusion homogenization temperatures. Excluding the Pine Point data, the best fit to the data is close to that expected from the sphalerite–galena frac-tionation factor $(0.74 \times 10^6/T^2)$. Many fall off the expected curve, indicating disequilibrium. Ohmoto and Rye (1979) noted a number of fac-tors that may contribute to the lack of fit, such as impure mineral separates used in the analysis; for example, 10% of the galena in sphalerite and vice versa would result in an estimated tem-perature of 215 °C if the actual equilibration

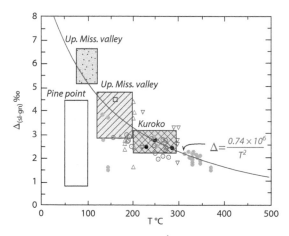

Figure 9.6 Comparison of temperatures determined from sphalerite–galena sulfur isotope fractionation with fluid-inclusion homogenization temperatures (T). Fields for MVT ore deposits in the USA and the Kuroko deposit in Japan. Solid circles: Creede, CO, inverted triangle: Sunnyside, O: circle with dot: Finlandia vein, triangle: Pasto Bueno, open square: Kuroko. Source: Ohmoto and Rye (1979)/with permission of John Wiley & Sons.

temperature was 145 °C. Different minerals may crystallize at different times and different temperatures in a hydrothermal system and, hence, would never be in equilibrium. In gen-eral, those minerals in direct contact with each other give the most reliable temperatures. Real disequilibrium may also occur if crystallization is kinetically controlled. The generally good fit at temperature >300 °C and poor fit at tempera-tures <200 °C suggests that the temperature dependence of reaction kinetics is the most important factor.

Isotope geothermometers do have several advantages over conventional chemical ones. First, as we have noted, there is no volume change associated with isotopic exchange reactions and hence little pressure dependence of the equilibrium constant (however, Rumble has suggested an indirect pressure dependence, wherein the fractionation factor depends on fluid composition, which, in turn, depends on pressure). Second, whereas conventional chemical geothermometers are generally based on solid solution, isotope geothermometers can make use of pure phases such as SiO_2, etc. Generally, any dependence on the compo-sition of phases in isotope geothermometers involved is of relatively second-order impor-tance. For example, isotopic exchange between calcite and water is independent of the concen-tration of CO_2 in the water. Compositional effects can be expected only where composi-tion affects bonds formed by the element involved in the exchange. For example, we noted that the substitution of Al for Si into plagioclase affects O isotope fractionation fac-tors because of the nature of the bond with oxygen. The composition of a CO_2 bearing solution, however, should not affect isotopic fractionation between calcite and dissolved carbonate because the oxygen is bonded with C regardless of the presence of other ions (if we define the fractionation as between water and calcite, some effect is possible if O in the carbonate radical exchanges with other radi-cals present in the solution; small effects are also possible due to complexes formed by both water and CO_2).

9.4 STABLE ISOTOPE COMPOSITION OF THE MANTLE

We have already found in Chapters 7 and 8 that stable isotope ratios can provide key insights into the evolution of the Earth's crust and mantle. For example, we found that we could use O isotopes in zircon to distinguish whether magmatic rocks represented new additions to continents or merely recycled older crust. We also found that we could use stable isotopes to trace continental material subducted into the mantle. There are, however, reasons to explore the stable isotopic composition of the mantle more fully. The mantle contains the majority of the Earth's mass and largest fraction of most elements, the notable exceptions being the siderophile elements concentrated in the Earth's core and perhaps a few other elements including some of the noble gases, nitrogen, and chlorine, and highly incompatible elements in the continental crust. We need to know the isotopic composition of the mantle to know the isotopic composition of the Earth. The Earth's isotopic composition, in turn, places constraints on its formation within the solar nebula, as we found in Chapter 6.

We will have to proceed with some caution in assessing the stable isotopic composition of the mantle. Basalts are certainly the most widely available mantle sample, and we relied on them heavily in defining the radiogenic isotope composition of the mantle. We could do so because radiogenic isotope ratios are not changed in the magma generation process. This will not be true for stable isotope ratios, which can be changed by chemical processes. Although the effects of the melting process on most stable isotope ratios of interest are small, they are not always negligible. Degassing significantly affects isotope ratios of volatile elements such as carbon and hydrogen, which compromises the value of magmas as a mantle sample. Once minerals begin to crystallize, fractional crystallization will affect isotope ratios, although the resulting changes are again generally small. Finally, weathering and hydrothermal processes can affect stable isotope ratios of basalts as well as tectonically emplaced mantle samples such as alpine and oceanic peridotites.

9.4.1 Oxygen

For the reasons stated above, xenoliths in basalts provide the best sample of mantle oxygen, but they are unfortunately considerably rarer than are basalts, so we will rely on the latter as well. Figure 9.7 shows the oxygen isotope

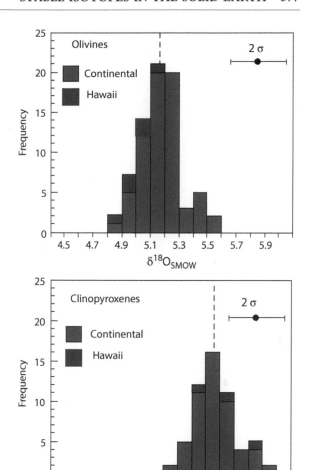

Figure 9.7 Oxygen isotope ratios in olivines and clinopyroxenes from mantle peridotite xenoliths. Source: Adapted from Mattey et al. (1994).

composition of olivine and clinopyroxene in 76 peridotite xenoliths analyzed by Mattey et al. (1994) using the laser fluorination technique. The total range of values observed is only about twice that expected from analytical error alone, suggesting that the mantle is fairly homogeneous in its isotopic composition. The difference between co-existing olivines and clinopyroxenes averages about 0.5 per mil, which is consistent with the expected fractionation between these minerals at mantle temperatures. Mattey et al. (1994) estimated the bulk composition of these samples to be about +5.5 per mil.

Figure 9.8 shows the distribution of $\delta^{18}O$ in selected basalts from four different

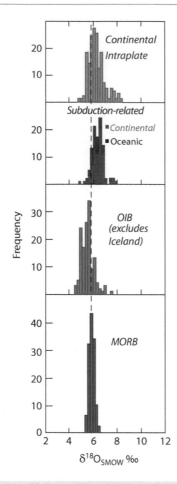

Figure 9.8 $\delta^{18}O$ in young, fresh basalts. Dashed line is at the of MORB (+5.7). Source: Adapted from Harmon and Hoefs (1995).

groupings. To avoid the weathering problems we discussed earlier, Harmon and Hoefs (1995) included only submarine basaltic glasses and basalts that had less than 0.75% water or had erupted historically in their compilation. There are several points worth noting in these data.

Mid-ocean ridge basalts (MORB) are significantly more homogeneous than are other basalts. MORB have a mean $\delta^{18}O_{SMOW}$ of +5.7‰ and a standard deviation of ±0.2‰. Thus, the depleted upper mantle appears to be a comparatively homogeneous and well-mixed reservoir for oxygen, just as it is for other elements. The difference between MORB and Mattey's estimated mantle composition, about 0.2‰, is consistent with the fractionation expected at mantle temperatures. Oceanic

island basalts (OIB), which presumably sample mantle plumes, are slightly less enriched in ^{18}O (mean $\delta^{18}O_{SMOW}$ = +5.5‰) on average and are also more variable (1σ = 0.5‰). Low values tend to be associated with high $^{3}He/^{4}He$ (Starkey et al., 2016). We found in Chapter 7 that there are clear correlations of $\delta^{18}O$ with radiogenic isotope ratios in both MORB and OIB in some cases. This variability most likely reflects the subduction of material that has interacted with water at the Earth's surface into the mantle.

The histogram shown excludes Iceland, because Icelandic basalts are quite anomalous in their low $\delta^{18}O$ (mean ~ 4.5‰). This is in part due to the assimilation of older igneous crust that has equilibrated with meteoric water, which is quite ^{18}O depleted at the latitude of Iceland. However, a number of studies such as that of Thirlwall et al. (2006) showed that while such assimilation has indeed occurred in some instances, the Icelandic mantle source does have notably low $\delta^{18}O$, as low as or lower than +4.5‰.

$\delta^{18}O$ in olivines from subduction-related basalts (i.e. island arc basalts and their continental equivalents) vary between 4.88‰ and 6.78‰ (e.g. Bindeman et al., 2005), corresponding to calculated melt values of 6.36–8.17‰, and on average have more positive $\delta^{18}O$ than MORB. This also reflects the subduction of material that has interacted with water at the Earth's surface. Finally, continental intraplate volcanics are more enriched in ^{18}O than are OIB, suggestive of crustal assimilation.

Defining the precise $\delta^{17}O$ and therefore $\Delta^{17}O$ of the mantle (and consequently the Earth) is limited by interlaboratory differences and standardization. This is illustrated by two recent high-precision data sets on mantle materials from Young et al. (2016) and Cano et al. (2020) shown in Figure 9.9. The two data sets define statistically indistinguishable $\delta^{17}O/\delta^{18}O$ slopes 0.5223±0.0018 and 0.5213 ±0.0014, respectively, but the data of Cano et al. are offset to lower $\delta^{17}O$ by 0.06‰. The results demonstrate that ^{17}O and ^{18}O vary in a consistent manner in igneous processes, but the slopes are slightly lower than the canonical terrestrial fractionation line with γ =0.528. Assuming a $\delta^{18}O$ for the mantle of 5.4‰, this gives $\delta^{17}O$ = 2.85‰ for the Young et al. data set and $\delta^{17}O$ = 2.78‰ for the Cano et al. data set.

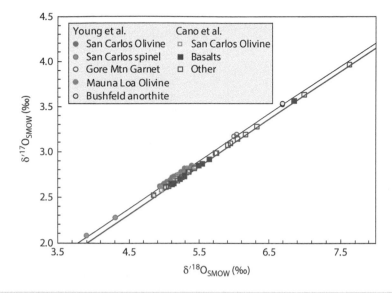

Figure 9.9 $\delta'^{18}O$ and $\delta'^{17}O$ analyses of mantle materials by Young et al. (2016) and Cano et al. (2020). Both included extensive analyses of olivines in xenoliths from the San Carlos locality in the Southwestern USA. The two data sets define statistically indistinguishable $\delta^{17}O/\delta^{18}O$ slopes of $\gamma = 0.5223\pm0.0018$ and $\gamma = 0.5213\pm0.0014$, respectively, but the data of Cano et al. are offset to lower $\delta^{17}O$ by 0.06‰.

9.4.2 Carbon

Carbon is mainly present in magmas as CO_2 and much of it exsolves and escapes even before eruption. Only basalts erupted beneath several kilometers of water provide samples of mantle carbon, but even then some degassed has occurred. As a result, the basalt data set is restricted to MORB, samples recovered from Loihi and the submarine part of Kilauea's East Rift Zone, and a few other seamounts. Thus, there are far less data on carbon isotopes in basalts, and the meaning of these data is somewhat open to interpretation.

The question of the isotopic composition of mantle carbon is further complicated by fractionation and contamination. There is a roughly 4‰ fractionation between CO_2 dissolved in basaltic melts and the gas phase, with ^{13}C enriched in the gas phase. If Rayleigh distillation occurs, that is if bubbles do not remain in equilibrium with the liquid, then the basalt that eventually erupts may have carbon that is substantially lighter than the carbon originally dissolved in the melt. Furthermore, MORB are often pervasively contaminated with a very ^{13}C-depleted carbon. This carbon is organic in origin, and recent observations of an eruption on the East Pacific Rise suggest a source. Following the 1991 eruption at 9°30' N, there was an enormous "bloom" of

bacteria stimulated by the release of H_2S. Bacterial mats covered everything. The remains of these bacteria are likely the source of this organic carbon. Fortunately, it appears possible to avoid most of this contamination by the stepwise heating procedure now used by many laboratories. Most of the contaminant carbon is released at temperatures below 600 °C, whereas most of the basaltic carbon is released above 900 °C.

Figure 9.10 shows $\delta^{13}C_{PDB}$ in various mantle and mantle-derived materials. MORB have a mean $\delta^{13}C_{PDB}$ of -5.8 ± 1.6‰; however, almost all these lavas have experienced some CO_2 degassing, which would decrease the $\delta^{13}C$ of the remaining CO_2; thus, this value is more likely an lower limit of the mean $\delta^{13}C$ of the MORB mantle. The most CO_2-rich MORB samples have $\delta^{13}C$ of about -4‰. Since they are the least degassed, they presumably best represent the isotopic composition of the depleted mantle (Javoy and Pineau, 1991). Ocean island basalts erupted under sufficient water depth to preserve some CO_2 in the vesicles have slightly heavier carbon, with a mean of -6.58 ± 1.8‰. To what extent this difference is real is not yet clear, given the effects of fractionation. Aubaud et al. (2006) concluded that the mantle source of Pitcairn Island basalts had slightly lighter carbon ($\delta^{13}C \sim -6$‰) than

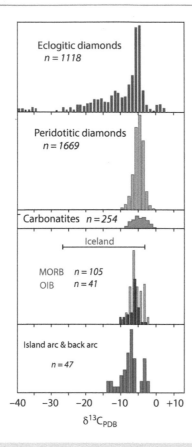

δ¹³C_PDB

Figure 9.10 Carbon isotope ratios in mantle and mantle-derived materials. Source: Adapted from Mattey (1987) and Cartigny et al. (2014).

MORB. Subglacial basalts from Iceland show a particularly large range of $\delta^{13}C$ including quite negative values, a result of both fractionation and crustal assimilation; nevertheless, Barry *et al.* (2014) estimated that the Iceland mantle had slightly heavier $\delta^{13}C$ (~ –2‰) than MORB. Subduction zone volcanics and back-arc basin basalts, which erupt behind subduction zones and are often geochemically similar to island arc basalts, show an even greater range of $\delta^{13}C$ with a standard deviation of 3, although the mean value, –7.65‰, is only modestly lighter than that of MORB. Again, whether this difference is real is unclear. Carbonatites also have a mean $\delta^{13}C$ close to -4‰, similar to that of MORB.

Diamonds, of course, represent another sample of mantle carbon. They are divided into several classes, based on factors including occurrence, shape, and inclusions present. Most gem-quality diamonds occur as xenocrysts in kimberlites or within xenoliths in

kimberlites and can be divided into peridotitic and eclogitic parageneses based on their inclusions or the minerals occurring with them. They formed by precipitation from hydrous fluids in the lithosphere or upper asthenosphere. Another group of cloudy or opaque diamonds, known as fibrous diamonds, is thought to have precipitated from the kimberlite magmas. Another, rarer group principally from the Juina mine in Brazil, and Kanka mine in Guinea, has inclusions of minerals that formed in the mantle transition zone (400–670 km), such as majorite, or the lower mantle (ferripericlase, bridgmanite, etc., or decompression products of them). Much smaller diamonds also occur in ultrahigh-pressure metamorphic belts and in impact ejecta.

While diamonds show a large range of $\delta^{13}C$, most fall within a narrow interval of –8‰ to —2‰ with a mode of –5‰±1 (Cartigny et al., 2014), similar to MORB. Thus, a likely value for mantle $\delta^{13}C$ is –5‰. While eclogitic diamonds have the same mode, –5‰, as peridotitic ones, Figure 9.10 shows that they are distinctly more heterogeneous with a distribution skewed to lower $\delta^{13}C$ values.

Three hypotheses have been put forward to explain the C isotopic heterogeneity in diamonds: primordial heterogeneity, fractionation effects, and recycling of carbonate and organic carbon from the Earth's surface into the mantle. Primordial heterogeneity seems unlikely for a number of reasons. Among these are the absence of very negative $\delta^{13}C$ in other materials, such as MORB, and the absence of any evidence for primordial heterogeneity from the stable isotopic compositions of other elements. Equilibrium fractionations during diamond formation by either oxidation or reduction of CO_2 or methane are within roughly ±2‰ of 0 and cannot explain the range in $\delta^{13}C$. Boyd and Pillinger (1994) have argued that since diamonds are kinetically sluggish (witness their stability at the surface of the Earth, where they are thermodynamically out of equilibrium), isotopic equilibrium might not be achieved during their growth. Large fractionations might, therefore, occur due to kinetic effects. However, these kinetic fractionations have not been demonstrated, and fractionations of this magnitude (20‰ or so) would be surprising at mantle temperatures. A third alternative is based on the observations that their $\delta^{13}C$ values are similar to organic carbon and that these diamonds with are almost exclusively of eclogitic paragenesis, that is, their inclusions consist of eclogitic minerals. Eclogite is the high-pressure equivalent of basalt. The subduction of oceanic crust continuously carries large amounts of basalt into the mantle. Oxygen and sulfur isotope heterogeneity observed in some eclogite xenoliths suggests these eclogites do indeed represent subducted oceanic crust. Hence, the heterogeneous $\delta^{13}C$ of eclogitic diamonds likely reflects the subduction of surficial carbon, including organic carbon, into the mantle.

9.4.3   Hydrogen

Like carbon, hydrogen can be lost from basalts during degassing, but the problem is somewhat less severe than for carbon because H is primarily present as water whose solubility in basalt is greater than that of CO_2. Basalts erupted beneath a kilometer of more of water retain most of their dissolved water. However, basalts, particularly submarine basalts, are far more readily contaminated with hydrogen (i.e. with water) than with carbon. Furthermore, the effect on hydrogen isotopic composition depends on the mode of contamination, as Figure 9.11 indicates. The direct addition of water or hydrothermal reactions will raise δD (because there is little fractionation during this process), while low-temperature weathering and hydration will lower δD, because

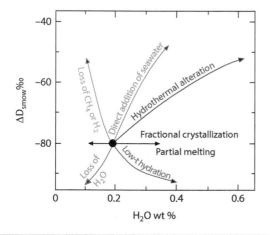

Figure 9.11   Effect of degassing and posteruptive processes on the water content and δD of basalts. Source: Kyser and O'Neil (1984)/ Reproduced with permission from Elsevier.

hydrogen, rather than deuterium, is preferentially incorporated into alteration phases. Loss of H_2 and CH_4, which may partition into a CO_2 gas phase when it forms, could also affect the hydrogen isotopic composition of basalts. However, the available evidence suggests that these species constitute only a small fraction of the hydrogen in basalts, so this effect is likely to be minor.

The first attempt to assess the hydrogen isotopic composition of the mantle materials was that of Sheppard and Epstein (1970), who analyzed hydrous minerals in xenoliths and concluded that δD varied in the mantle. Since then, many additional studies have been carried out. The mean of 75 MORB samples analyzed using the thermal conversion elemental analyzer technique is −75‰, which is close to the mode of −76‰, with a standard deviation of ±12‰ (Loewen et al., 2019). As we found in Chapter 7, δ^2H is correlated with radiogenic isotope ratios in some cases, indicative of H isotopic heterogeneity in the mantle. This appears to be the result of recycling of water within the oceanic crust and associated sediments as well as various dehydration and metasomatic processes occurring during subduction (Dixon et al., 2017). Loihi Seamount, which has notably high $^3He/^4He$, indicative of a primordial component has δ^2H of −70 to −90‰ (Loewen et al., 2019).

Hydrous minerals in xenoliths also provide a sample of mantle hydrogen. As Figure 9.12 shows, phlogopites ($K(Mg,Fe)_3AlSi_3O_{10}(OH)_2$) have δD that is generally similar to that of MORB, though some lighter values also occur. Amphiboles have much more variable δD and have heavier hydrogen on average. Part of this difference probably reflects equilibrium fractionation. The hydrogen fractionation between water and phlogopite is close to 0‰ in the temperature range 800–1000 °C, whereas the fractionation between water and amphibole is about −15‰. However, equilibrium fractionation alone cannot explain either the variability of amphiboles or the deference between the mean δD of phlogopites and amphiboles. Complex processes involving in amphibole formation that might include Rayleigh distillation may be involved in the formation of mantle amphiboles. This would be consistent with the more variable water content of amphiboles compared to phlogopites.

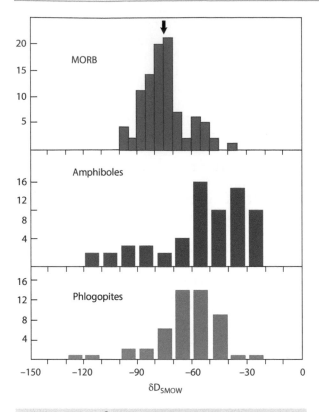

Figure 9.12 δ^2H in MORB (thermal conversion elemental analyzer data from Dixon et al., 2017 and Loewen et al., 2019) and in mantle phlogopites and amphiboles. The MORB and phlogopite data suggest that the mantle has δD_{SMOW} of about –60‰ to–80‰. Arrow indicates the mean MORB value.

9.4.4 Nitrogen

With strong triple covalent bonds holding the two atoms together, atmospheric dinitrogen is nearly as unreactive as noble gases. Although it highly volatile and concentrated in the atmosphere, roughly 10–15% of the Earth's inventory is in the crust and ~60% is thought to be in the mantle (Bebout et al., 2013). Microbes reduce N_2 to ammonium, which is incorporated into a variety of organic molecules. While most of this fixed nitrogen is recycled, some is retained in soil organic matter and sediments. The ammonium ion has an ionic radius similar to that of K so that during diagenesis, a fraction of this ammonium will be incorporated in clays, and consequently, sediments and metasediments can contain more than 1000 ppm N, although most contain only a few hundred ppm. During subduction,

isotopically light nitrogen is lost to fluids, leaving the flux into the mantle somewhat enriched in ^{15}N relative to sediments. How much is lost to a fluid phase depends on the P-T path taken. Over Earth's history, Förster et al. (2019) estimate that a mass of nitrogen equivalent to half the present atmosphere has been subducted into the mantle. Nitrogen is returned to the atmosphere through volcanic degassing so that the atmospheric nitrogen mass depends on the difference between subduction and degassing rates.

Figure 9.13 illustrates the nitrogen isotopic composition in igneous, sedimentary and metamorphic rocks and in diamonds. There are far less data for $\delta^{15}N$ than for other stable isotope ratios. The solubility of N_2 in basalts is very limited, though much of the nitrogen may be present as NH_4^+, which is somewhat more soluble. Hence of volcanic rocks, once again only submarine basalts provide useful samples of mantle N. There are both contamination and analytical problems with determining nitrogen in basalts, which, combined with its low abundance (generally less than a ppm), mean that accurate measurements are difficult to make. Measurements of $\delta^{15}N_{ATM}$ in MORB range from about –10 to +8‰, with most in the range of –2 to –4‰. OIB tend to be heavier; Cartigny and Marty (2013) estimate a mean value of +2‰.

Diamonds can contain up to 2000 ppm of N and, hence, provide an excellent sample of mantle N. Earlier studies had found that diamonds of peridotitic paragenesis (those containing peridotitic rather than eclogitic mineral inclusions) had lower $\delta^{15}N$ than eclogitic diamonds, which often have low $\delta^{13}C$. As Figure 9.13 shows, however, there is considerable overlap between both classes, with most diamonds having $\delta^{15}N$ between 0 and –10‰. Most diamonds are formed in the lithosphere or upper asthenosphere, but a rare subset of diamonds have inclusions of peridotitic minerals that must have formed in the mantle transition zone (400–670 km), such as majorite, or the lower mantle (ferripericlase, bridgmanite, etc., or decompression products of them). $\delta^{15}N$ in the deep mantle diamonds from the Juina, Brazil, and Kanka, Guinea, are similar to peridotitic diamonds formed at shallower depth, with the exception of one diamond from the Kankan mine which has $\delta^{15}N$ of –24.9‰ (Palot et al., 2012). Fibrous

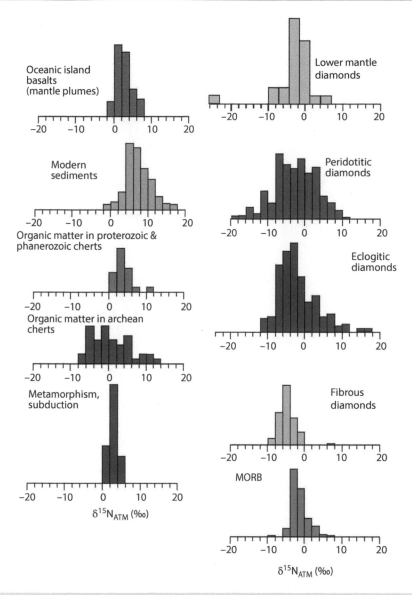

Figure 9.13   Isotopic composition of nitrogen in rocks and minerals of the crust and mantle. Source: Based on (Marty and Dauphas, 2003; Cartigny and Marty, 2013; and Palot et al., 2012).

diamonds, whose growth may be directly related to the kimberlite eruptions that carry them to the surface (Boyd and Pillinger, 1994), have more uniform $\delta^{15}N$, with a mean of about −5‰. Since there can be significant isotopic fractionations involved in the incorporation of nitrogen into diamond, the meaning of the diamond data is also uncertain, and the nitrogen isotopic composition of the mantle remains poorly constrained.

Overall, Cartigny and Marty (2013) estimate the $\delta^{15}N$ of the mantle to be −5‰ and that of the crust to be ~ +6‰.

9.4.5   Sulfur and selenium

Although sulfur is more soluble in magma than H_2O, CO_2, or N_2, H_2S and other sulfur species are also lost from magma at low pressure, consequently, the reliable data on sulfur isotopes are also somewhat limited. The available data are shown in Figure 9.14. MORB have a mean $\delta^{34}S_{CDT}$ of −0.74‰ width and a standard deviation of 0.49‰. $\delta^{33}S/\delta^{34}S$ and $\delta^{36}S/\delta^{34}S$ do not depart significantly from the canonical slopes of 0.515 and 1.91, with average $\Delta^{33}S$ = 0.002±0.16‰ and $\Delta^{36}S$ = −0.021 ±0.11‰. There is negligible fractionation of

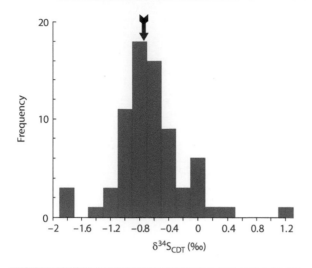

Figure 9.14 Distribution of $\delta^{34}S_{CDT}$ in MORB. Arrow indicates the average value of –0.74‰.

sulfur isotopes during partial melting, and Labibi and Cartigny (2016) concluded that the MORB value approximately represents the depleted upper mantle value. As we found in Chapter 7, $\delta^{34}S$ in South Atlantic MORB correlates with radiogenic isotope ratios (e.g. Labidi et al., 2013), clearly demonstrating that the sulfur isotopic composition of the mantle is variable.

Sulfur isotope ratios vary even more in OIB, although the data are as yet limited. $\delta^{34}S$ of sulfides in Samoan submarine lavas varied between +0.11‰ and +4.23‰ and correlated with $^{87}Sr/^{86}Sr$, $^{3}He/^{4}He$, and $^{182}W/^{184}W$ (Labidi et al., 2015; Dottin et al., 2020). Small but resolvable variations also occur in $\Delta^{33}S$. Ten to twenty percent of sulfur is these lavas is present as sulfate, and the sulfate is isotopically heavier, with $\delta^{34}S$ ranging from +4.19‰ to +9.71‰, consistent with the fractionation during oxidation. In Chapter 7, we noted the discovery of mass independently fractionated sulfur in sulfide inclusions in diamonds from the Orapa kimberlite mine (Farquhar et al., 2002) as well in sulfide inclusions in olivines from lavas of Mangaia Island and Pitcairn Island (Delavault et al., 2016) also in the central South Pacific. These testify to the incorporation of Archean surficial sulfur in the deep mantle source of mantle plumes. Data on subduction-related magmatism are also limited. $\delta^{34}S_{CDT}$ in historic eruptions of Indonesian volcanoes average 4.5±1.5‰ (de Hoog et al., 2001).

Selenium has six stable isotopes, ^{74}Se, ^{76}Se, ^{77}Se, ^{78}Se, ^{80}Se, and ^{82}Se, with isotopic abundances of 0.86%, 9.23%, 7.60%, 23.69%, 49.80%, and 8.82%, respectively. Most commonly, the $^{82}Se/^{76}Se$ ratio is reported in per-mil deviations from NIST SRM 3149 as $\delta^{82/76}Se$ (or $\delta^{82}Se$), but some studies have reported $^{82}Se/^{78}Se$ as $\delta^{82/78}Se$ (also with respect to NIST3149). Earlier $\delta^{82/76}Se$ data reported relative to the MERCK standard can be converted to as $\delta^{82/76}Se_{NIST3149} = \delta^{82/76}Se_{MERCK} + 1.54‰$ (Carignan and Wen, 2007). All variation appears to be mass dependent, including in meteorites, so that $\delta^{82/76}Se = \delta^{82/78}Se \times 2.562$. In this book, all data are presented as $\delta^{82/76}Se_{NIST3149}$.

$\delta^{82}Se$ exhibits a substantial natural range of almost 15‰. Selenium sits directly below sulfur in the periodic table and unsurprisingly shares many of its properties including being chalcophile, volatile, and redox sensitive with common valences of VI, IV, 0, and –II. As with other volatile and chalcophile elements, Se is depleted in the silicate Earth relative to chondrites, with the majority of the Earth's Se inventory likely to be in the core.

$\delta^{82}Se$ in carbonaceous chondrites varies by class from –0.09‰ in CI to –1.04‰ in CV, with an average of –0.53‰. Differences between classes of ordinary and enstatite chondrites are not statistically significant with averages of –0.35‰ and –0.62‰, respectively (e.g. Vollstaedt et al., 2016; Labidi et al., 2018). Iron meteorites analyzed to data are slightly heavier on average with $\delta^{82/76}Se = -0.2$. $\delta^{82}Se$ in peridotites define a narrow range of –0.09‰ to +0.03‰ with an average of –0.03‰ and standard deviation of 0.04‰ (Varas-Reus et al., 2019). If this is the silicate Earth value, it is heavier than average chondrites, but does fall within the chondritic range. $\delta^{82}Se$ in peridotites is independent of Se and Al_2O_3 concentration, suggesting a lack of isotopic fractionation during partial melting.

As is the case for sulfur, degassing can result in loss of Se from subaerial basalts, so the attention focuses on submarine ones. $\delta^{82}Se$ in MORB from the North Atlantic and the Pacific-Antarctic Ridge unaffected by mantle plumes (NMORB) range from –0.05 to –0.30‰ with a mean value of –0.16‰ and are uncorrelated with radiogenic isotope ratios (Yierpan et al., 2019, 2021). The offset of MORB to lighter $\delta^{82}Se$ contradicts the

(a)

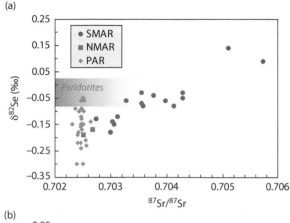

(b)

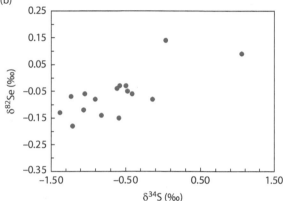

Figure 9.15 (a) $\delta^{82}Se$ versus $^{87}Sr/^{86}Sr$ in MORB from the Northern Mid-Atlantic Ridge (NMAR) and Pacific-Antarctic Ridge (PAR), and Southern Mid-Atlantic Ridge; the latter show the influence of the Shona and Discovery hot spots (data from Yierpan et al., 2018, 2019). Also shown is the range of $\delta^{82}Se$ in peridotites analyzed by Varas-Reus et al. (2019). (b) $\delta^{82}Se$ versus $\delta^{34}S$ from the Southern MAR. $\delta^{34}S$ data from Labidi et al. (2013).

inference from peridotites that Se isotopes are not fractionated during melting: an issue that future research will have to resolve. Yierpan et al. (2020) found that MORB from the South Atlantic in the vicinity of the Shona and Discovery hot spots, however, vary to $\delta^{82}Se$ as heavy as +0.14‰ and correlate with radiogenic isotope ratios (Figure 9.15) as well as $\delta^{34}S$ previously reported by Labidi et al. (2013) and discussed in Chapter 7. Following the interpretation of Labidi et al., Yierpan et al. (2020) argue that this variation results from the incorporation of subducted sediment into the source of these mantle plumes. As we will find in a subsequent chapter,

Phanerozoic marine sediments have $\delta^{82}Se$ of ~0 and, hence, are unlikely to produce the variations observed. However, in the less oxidizing conditions of the Mesoproterozoic sediment, marine sedimentary $\delta^{82}Se$ averaged ~+0.5‰. Based on this, Yierpan et al. (2020) argued that the sediment in the plume source is likely Mesoproterozoic. Kurzawa et al. (2019) found that $\delta^{82}Se$ in submarine erupted volcanics from the Marianas arc range from 0.03‰ to −0.33‰, with an average of −0.13‰. Hence, they overlap with MORB but extend to more negative compositions. Kurzawa et al. suggested that this isotopically light Se derived from hydrothermal sulfides within the subducting oceanic crust.

9.4.6 Li, B, and Cl

9.4.6.1 Lithium isotopes

Terrestrial lithium isotopic variation is dominated by the strong fractionation that occurs between minerals, particularly silicates, and water. Indeed, this was first demonstrated experimentally by Urey in the 1930s. This fractionation, in turn, reflects the chemical behavior of Li. The ionic radius of Li^+ is small (76 pm) and Li can substitute for Mg^{2+}, Fe^{2+}, and Al^{3+} in crystal lattices, mainly in octahedral sites coordinated by six oxygen atoms. In aqueous solution, it is tetrahedrally coordinated by four water molecules (the solvation shell) to which it is strongly bound, judging from the high solvation energy. These differences in atomic environment and binding energies and large relative mass differences of ^6Li and ^7Li all lead to strong fractionation of Li isotopes, one consequence of which is that ^7Li is strongly enriched in seawater, which has an approximately uniform δ^7Li = +31‰ (Figure 9.16).

The Earth appears to have a chondritic Li isotopic composition as δ^7Li in unmetasomatized mantle peridotites range from +2.5‰ to +5‰ compared to bulk chondrites which range from +3‰ to +4‰. There appears to be little isotopic fractionation, <0.5‰, during fractional crystallization, and perhaps also partial melting (Tomascak et al., 2016). Fresh MORB have δ^7Li of +3.6±0.8‰, a range not much larger than that expected from analytical error alone (Figure 9.16) and indistinguishable from peridotites. OIB have higher average δ^7Li, +4.4±1.3‰, and as we found in Chapter 7,

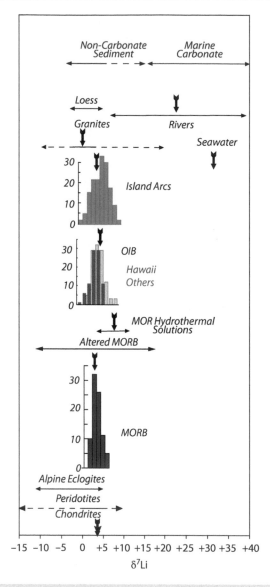

Figure 9.16 Li isotopic composition of terrestrial materials and chondrites. Arrows indicate mean values.

Lithium isotopic signatures in arc magmas can vary from the mantle value for a number of reasons: a contribution from subducted seawater-altered oceanic crust, contribution from subducted sediment, and fractionation during dehydration as oceanic crust and sediment subducts, which should produce and isotopically heavy fluid and light residue.

9.4.6.2 Boron isotopes

While there were some earlier studies, modern study of boron isotopes began with the work of Spivack and Edmond (1987), preceding the development of inductively coupled plasma-mass spectrometry (ICP-MS) instruments. The range of $^{11}B/^{10}B$ variation in terrestrial materials is ~100‰ (Figure 9.17), reflecting the large mass difference between the two isotopes (~10%). Variation is also large in chondritic meteorites, from −50‰ to +40‰, reflecting both fractionation and nucleosynthetic variations. Seawater is isotopically heavy, with $\delta^{11}B$ of +39.6±0.04‰ and is

there are clear correlations with radiogenic isotope ratios in individual island chains, demonstrating that the mantle is heterogeneous with respect to δ^7Li. The highest δ^7Li occurs on islands characterized by particularly radiogenic Pb (the so-called HIMU OIB group), such as the volcanoes of the Cook-Austral chain. Here, δ^7Li may be as high as +8‰. This may reflect a recycled crustal component in their sources (e.g. Vlastelic et al., 2009).

The average δ^7Li of island arc volcanics (IAV), ~+3.7‰, is indistinguishable from than that of MORB, but they are considerably more variable with a standard deviation of 2.4‰.

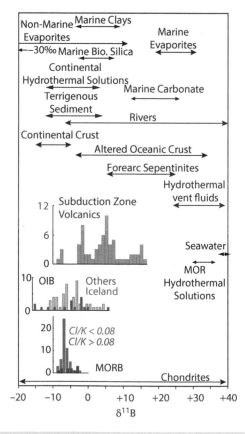

Figure 9.17 Boron isotopic composition of terrestrial materials and chondrites.

uniform within analytical error. Boron is readily incorporated into secondary phases, so that even slightly altered basalts show a dramatic increase in B concentration and $\delta^{11}B$, with altered oceanic crust having $\delta^{11}B$ in the range of −2‰ to +26‰. Smith et al. (1995) estimated that average altered oceanic crust contains 5 ppm B and $\delta^{11}B$ of +3.4‰. Marschall et al. (2017) estimate the mean $\delta^{11}B$ of the continental crust to be −9.1±2.4‰.

$\delta^{11}B$ in fresh MORB range from −9.4‰ to −2.2 ‰, but much of this range appears to reflect the assimilation of altered oceanic crust or brines stored within it. Marschall et al. (2017) found that MORB with Cl/K ratios <0.08 had nearly homogeneous $\delta^{11}B$ of −7.1 ±0.9‰ and inferred that this likely represented the mantle and bulk silicate Earth (BSE) composition. OIB have larger range of $\delta^{11}B$ with an average of −4.4‰. The heaviest values in this range have been shown to result from the assimilation of altered crust, most notably in the Azores (Genske et al., 2014) and Iceland. Excluding these samples, the OIB mean is slightly lighter than MORB, −8.2‰, but with an apparent bimodal distribution. One possible explanation is that OIB sources contain recycled sediment containing heavy boron and recycled oceanic crust containing light boron held in phengite (Palmer, 2017), a metamorphic mineral sometimes found in eclogites.

The boron isotopic composition of IAV is heavier and more variable than OIB or MORB, with a mode of ~+5‰. $\delta^{11}B$ in IAV correlates negatively with Na_2O, Nb, and Sr concentrations, and La/Sm and Nb/B ratios, and positively with slab dip, B, and Sc concentrations and B/Ce and $^{87}Sr/^{86}Sr$ ratios. As summarized by De Hoog and Savov (2018) and Palmer (2017), $\delta^{11}B$ is so high in IAV that its only plausible source is serpentinized mantle, either in the subducting slab, the hydrated mantle above the slab, or in the forearc. As Figure 9.16 shows, such forearc peridotites have quite heavy $\delta^{11}B$ and are B-rich as a result of reaction with seawater.

9.4.6.3 Chlorine isotopes

Chlorine has two isotopes ^{35}Cl and ^{37}Cl whose abundances are 75.76% and 24.24%, respectively (CIAAW, 2020). It is most commonly in the −I valance state and strongly electropositive, forming predominantly ionic bonds, but in the rare circumstance of strongly oxidizing conditions, it can have valances up to +VII. Chlorine is relatively volatile and has a strong propensity to dissolve in water; consequently, much of the Earth's inventory is in the oceans. Since the oceans are a large and isotopically uniform reservoir of chlorine isotope, ratios are reported as $\delta^{37}Cl$ relative to seawater (formally called Standard Mean Ocean Chloride or SMOC) denoted as $\delta^{37}Cl_{SMOC}$.

Because the oceans and evaporites, which both have $\delta^{37}Cl$ of ~0‰, potentially contain more than half of the Earth's chlorine inventory, the $\delta^{37}Cl$ of the Earth is probably within a few tenths of a per mil of 0‰. Chlorine isotopic analyses of meteorites bracket this value: Sharp et al. (2013) found that average $\delta^{37}Cl$ in meteorites range from −0.4‰ in ordinary chondrites through −0.2‰ in carbonaceous chondrites to +0.4‰ in enstatite chondrites.

Chlorine is substantially more soluble in magma than water, CO_2, and sulfur and, hence, is not lost as readily during magmatic degassing. Nevertheless, its concentration is low, and submarine basalts are subject to contamination from seawater. High-precision analyses of $\delta^{37}Cl$ in MORB range from −4‰ to +0.4‰, but the higher values are in question. Michael and Cornell (1998) showed that many MORB have assimilated chlorine, apparently by assimilating hydrothermally altered oceanic crust and related brines within the crust. They argued that samples with K/Cl ratios lower than 12.5 had assimilated seawater chlorine. Bonifacie et al. (2008) showed that $\delta^{37}Cl$ in MORB correlated positively with Cl concentration (and negatively with K/Cl) and concluded (i) that this trend was indeed a result of assimilation of seawater-derived Cl brines, (ii) even samples with K/Cl higher than 12.5 might be contaminated, and (iii) the $\delta^{37}Cl$ of MORB was probably less than −1.5‰. Although it is clear that some MORB have assimilated Cl, $\delta^{37}Cl$ appears to correlate with $^{206}Pb/^{204}Pb$ in low-Cl MORB (Figure 9.18), suggesting that at least some of the variation in $\delta^{37}Cl$ is due to mantle heterogeneity.

High-temperature hydrothermal vent fluids have homogeneous $\delta^{37}Cl$ identical to seawater within analytical error, although low-temperature off-axis basement fluids are lighter and more variable (−2.09 to −0.12‰). Altered basalts from Hole 504B are only

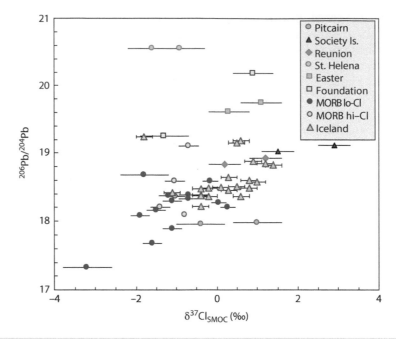

Figure 9.18 $^{206}Pb/^{204}Pb$ versus $\delta^{37}Cl$ in oceanic basalts. Source: Data from (Sharp, 2007; Bonifacie et al., 2008; John et al., 2010; Halldórsson et al., 2016).

slightly heavier than fresh MORB, with $\delta^{37}Cl$ from -1.30 to $-0.8‰$ (Bonifacie et al., 2007). Greater fractionation appears to occur in deeper parts of the oceanic crust. Altered gabbros and peridotites recovered from mid-ocean ridges and associated fractures have $\delta^{37}Cl$ ranging from -2.4 to $+1.8‰$, with talc systematically lighter than serpentine and the highest values occurring in amphibole-rich regions that have experienced higher temperature metamorphism (Barnes et al., 2009; Barnes and Cisneros, 2012), which is consistent with theoretical calculations predicting that ^{37}Cl will partition into phases with heavier 2+ cations such as Fe (Schauble et al., 2003). Altered oceanic crust is significantly enriched in Cl, containing ~100 to 200 ppm, and consequently, its subduction represents a substantial flux of Cl into the mantle.

Marine sediments overlying the oceanic crust vary only from $-1.8‰$ to $+0.7‰$. Further reaction between mantle peridotite and fluids may occur in subduction zones. Fluids expelled through accretionary prisms in subduction zones have extremely light $\delta^{37}Cl$, and serpentinite muds and clasts can have $\delta^{37}Cl$ ranging from $-0.22‰$ to $1.61‰$ (Barnes et al., 2008). Subduction-related volcanics also exhibit a large range in $\delta^{37}Cl$: $-2.6‰$ to $+2.4‰$.

OIB also show a considerable range in $\delta^{37}Cl$ and, as Figure 9.18 shows, there is a relationship between $\delta^{37}Cl$ and $^{206}Pb/^{204}Pb$ that hints at a systematic variation in $\delta^{37}Cl$ between mantle reservoirs established from radiogenic isotope ratios. Basalts from St. Helena, the "type locality" for the HIMU genus, have low $\delta^{37}Cl$; the Society Islands, the type locality of the EMII genus, appears to have the highest $\delta^{37}Cl$. Pitcairn that has low $^{206}Pb/^{204}Pb$ characteristic of EMI has intermediate $\delta^{37}Cl$. The only other island for which extensive data are available is Iceland (Halldórsson et al., 2016), where $\delta^{37}Cl$ correlates positively with $^{206}Pb/^{204}Pb$. Most MORB appear to fall on an extension of that correlation to lower $^{206}Pb/^{204}Pb$. Carbonatites have an average $\delta^{37}Cl$ of $+0.14‰$ and show only limited variation about this mean.

Since fractionation factors decrease with temperature, $\delta^{37}Cl$ variations in the mantle likely originate through the subduction of altered oceanic crust and sediment and perhaps other crustal material as well. Since, at least in the modern ocean, hydrothermal reactions in the oceanic crust result in uranium enrichment, the correlation between $\delta^{37}Cl$ and U enrichment in this process could, over time, explain the $\delta^{37}Cl$–Pb isotope ratio correlation. Further study of

Cl isotopes in mantle-derived magmas could provide important new insights into the origin of mantle reservoirs.

9.4.7 Mg, Ca, Si, and Ge Isotopes

9.4.7.1 Magnesium isotopes

Mg has three stable isotopes: ^{24}Mg, ^{25}Mg, and ^{26}Mg with relative abundances of 78.99%, 9.00%, and 11.01%, respectively (CIAAW, 2020). We have already seen that ^{26}Mg is the radiogenic product of the short-lived radionuclide, ^{26}Al, and that the ^{26}Al–^{26}Mg is the most important chronometer of events in the early Solar System. However, the Earth as well as meteorite parent bodies and planets appear to be homogeneous with respect to radiogenic ^{26}Mg, and our focus in this chapter will be on variations in Mg isotope ratios resulting from chemical fractionations. For those looking for greater detail on Mg isotopes in terrestrial and planetary materials, Teng (2017) provides an excellent review.

By convention, the isotope ratios of interest are ^{25}Mg/^{24}Mg and ^{26}Mg/^{24}Mg, and these are reported in the usual notation indicating per-mil deviations from a standard DSM3, δ^{25}Mg, and δ^{26}Mg, respectively.[1] To date, all identified Mg isotope fractionation is mass dependent with δ^{25}Mg $\approx 0.52\ \delta^{26}$Mg.

Schauble (2011) and Huang et al. (2013) fit theoretical Mg reduced partition functions for a variety of minerals as well as aqueous solution to 9.2. Fractionation factors for common upper mantle minerals calculated from them are plotted as a function of temperature in Figure 9.19. At relevant temperatures (>1000 °C), fractionation between clinopyroxene, orthopyroxene, and olivine, which are the main Mg-bearing minerals in basalts and the upper mantle minerals, are predicted to be quite small. Larger fractionations will occur between these minerals and garnet and

spinel ($MgAl_2O_4$), reflecting the differing atomic coordination and bond strength in these minerals. Mg is weakly bonded to eight coordinating oxygen atoms in garnet, so it is isotopically light and more strongly bounded to four in spinel in which Mg is isotopically heavy. Stronger fractionations occur at low temperature between water and/or silicates and carbonate, with the carbonate becoming isotopically light.

Twenty-nine fresh peridotite xenoliths analyzed by Teng et al. (2010) have average δ^{26}Mg = –0.25±0.02‰ and shows no resolvable variation with composition. This value presumably represents the average isotopic composition of the BSE. δ^{26}Mg values in chondrites range from –0.38‰ to –0.15‰, with an overall average of –0.27‰, and there are no resolvable systematic variations between carbonaceous (average δ^{26}Mg = –0.29±0.16‰ (2σ)), ordinary (average δ^{26}Mg = –0.28 ±0.13‰), and enstatite chondrites (average δ^{26}Mg = –0.26±0.08‰) (Pogge von Stradmann et al., 2011). MORB have an average δ^{26}Mg of –0.25‰ with a standard deviation of 0.03‰, which is not statistically different from peridotites, suggesting that Mg isotope fractionations at high temperature and in magmatic systems are, overall, insignificant and are consistent with our expectations based on calculated partition functions shown in Figure 9.19.

Among oceanic island and seamount basalts, small, but statistically significant differences emerge (Figure 9.20). Zhong et al. (2017) found that Hawaiian alkali basalts from Hualālai had lower δ^{26}Mg (–0.29±0.01‰) than tholeiites from Mauna Loa and Kilauea (–0.24±0.02‰) and that alkali basalts from the Louisville Seamount chain had even lower δ^{26}Mg (–0.32±0.02‰). A statistically analysis of the data reported by Teng et al. (2010) reveals further differences. While basalts from

---

[1] There has, however, been some evolution in this notation. Initially, interest in Mg isotopes focused exclusively on radiogenic ^{26}Mg and the symbol δ^{26}Mg referred to variations in ^{26}Mg due to the decay of ^{26}Al and not to mass fractionation. Mass fraction effects were expressed as Δ^{25}Mg (per mil deviations of ^{25}Mg/^{24}Mg). In order to be consistent with notation used for other elements, current notation uses δ^{26}Mg to refer to variations due to mass fractionation and δ^{26}Mg∗ to refer to radiogenic variations in ^{26}Mg. Prior to 2003, Mg isotope data were reported relative to a purified magnesium metal standard, NIST-SRM 980; however, that standard proved to be isotopically heterogeneous, and the current standard is a solution designated DSM3, derived from Mg extracted from the Dead Sea. Data can be approximately converted using δ^{26}Mg$_{DSM3}$ = δ^{26}Mg$_{SRM980}$ + 3.405 (the exact conversion is given by Young and Galy, 2004). Our discussion here will exclusively use δ^{26}Mg$_{DSM3}$.

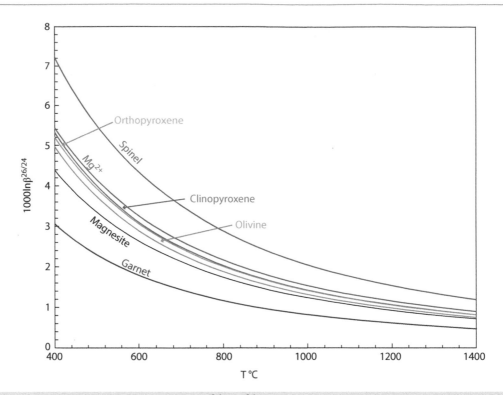

Figure 9.19 Reduced partition functions for ^{26}Mg/^{24}Mg fractionation factors for a variety of minerals as well as Mg^{2+} ion in aqueous solution from theoretical calculations of Schauble (2011) and Huang et al. (2013) fitted to 9.2. At any given temperature, the fractionation Δ^{26}Mg between can be calculated from the difference between reduced partition functions.

Kilauea are statistically indistinguishable from MORB, Koolau basalts from Oahu, Hawaii have significantly higher (at the 5% confidence level) δ^{26}Mg (−0.229±0.031‰), and Society Island basalts have significantly lighter δ^{26}Mg (−0.288±0.038‰) than MORB. The average δ^{26}Mg of Pitcairn Island basalts reported by Wang et al. (2018) is −0.31±0.019‰, significantly lower than MORB and Hawaiian basalts; the Tedside basalts of Pitcairn are even lighter with average δ^{26}Mg of −0.36‰. Zhong et al. (2017) suggested that these differences are mainly related to the extent of partial melting, alkali basalts being generally smaller degree melts than tholeiites. They noted that garnet is isotopically light compared to other mantle silicates. Garnet is preferentially consumed during melting and eventually exhausted at higher degrees of melting; hence, small degree melts where garnet is abundant in the melting residue can be expected to be isotopically lighter than large degree ones. Nonetheless, they argued that the low δ^{26}Mg in Louisville Seamounts could not be entirely explained

by low extents of melting and must reflect mantle heterogeneity. Furthermore, Wang et al. (2018) found that δ^{26}Mg correlated with ε_{Nd} and inversely with ^{87}Sr/^{86}Sr in Pitcairn basalts, which clearly indicates that at least some of the variation in OIB reflects Mg isotopic mantle heterogeneity. Furthermore, the contrast between the homogeneity in MORB and the variability of δ^{26}Mg in OIB, even between Hawaiian volcanoes, argues that the cause is mantle heterogeneity, not melting.

There is even more variability among subduction-related volcanics, which are on average heavier than MORB and peridotites (δ^{26}Mg = −0.20‰). Means for individual arcs range from considerably heavier (δ^{26}Mg = −0.08‰) for the Philippines to lighter (δ^{26}Mg = −0.30‰) for Costa Rica and Kamchatka (Figure 9.20). While some of this variation could reflect the assimilation of crustal material as arc magmas rise through and are stored in the crust, particularly continental arcs such as the Cascades, most of this variability likely reflect source heterogeneity

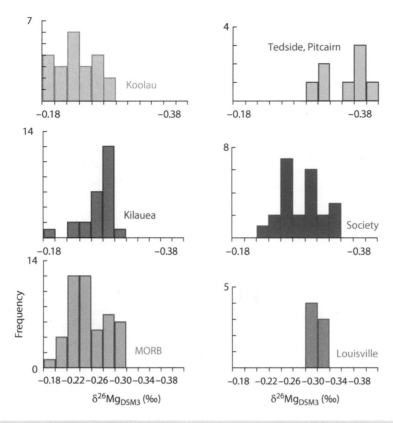

Figure 9.20   Distribution of $\delta^{26}Mg_{DSM3}$ in MORB and OIB. MORB has a mean value of –0.29‰. Tedside basalts of Pitcairn have the lowest mean $\delta^{26}Mg$ of –0.36‰ and the Koolau basalts from Oahu, Hawaii have highest $\delta^{26}Mg$ (–0.229±0.031‰) and differ significantly from MORB.

(Figure 9.21). Material being carried into the mantle in subduction zones has highly variable $\delta^{26}Mg$. Zhong et al. (2017) found that altered MORB recovered from Ocean Drilling Project (ODP) sites varies from –1.70 to +0.21‰. Heavier values are associated with saponite, a smectite mineral formed at low water–rock ratios under reducing conditions, while lighter values are associated with secondary carbonates formed under oxidizing conditions at higher water–rock ratios. Subducting sediments show an even wider range, with average $\delta^{26}Mg$ for individual subduction zones ranging from –0.61‰ (Central America) to +0.02‰ (Tonga), although the overall average of –0.34‰ is not dramatically different from the mantle value. This variety of isotopic compositions can readily explain the variation seen both in arc volcanics and in OIB.

9.4.7.2   Calcium isotopes

Calcium has six stable isotopes (a consequence of having a magic number of protons): ^{40}Ca,

^{42}Ca, and ^{43}Ca, ^{44}Ca, ^{46}Ca, and ^{48}Ca with relative abundances of 96.94%, 0.647%, 0.135%, 2.086%, 0.0004%, and 0.187%, respectively (CIAAW, 2020). ^{40}Ca is the principal decay product of ^{40}K (89.5% of ^{40}K decays). However, because ^{40}Ca is vastly more abundant than ^{40}K (^{40}Ca is doubly magic), variation in ^{40}Ca due to radioactive decay is usually less than 0.1‰ (K-rich/Ca-poor materials, such as salt deposits, are an exception). Chemical fractionations of Ca isotope ratios are, thus, the principal interest. There has been, however, inconsistency both in the isotope ratio reported and in the standard to which the ratio is normalized in the δ notation. Early studies measured the $^{44}Ca/^{40}Ca$ ratio reported as $\delta^{44}Ca$ (or $\delta^{44/40}Ca$). There are several difficulties with this; however: (i) the $^{44}Ca/^{40}Ca$ ratio is small, ~0.02, making it more difficult to measure accurately, (ii) this ratio cannot be measured by MC-ICP-MS due to the interference of ^{40}Ar (the plasma in ICP instruments is formed by ionizing Ar gas), and (iii) radiogenic ^{40}Ca

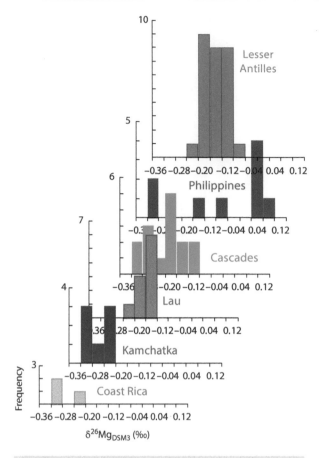

Figure 9.21 Distribution of $\delta^{26}Mg_{DSM3}$ in IAV. Means for individual arcs range from considerably heavier ($\delta^{26}Mg = -0.08‰$) for the Philippines to lighter ($\delta^{26}Mg = -0.30‰$) for Costa Rica and Kamchatka.

can in some cases be significant. The latter point is illustrated by a study by Ryu et. al. (2011), who found that $\delta^{44/40}Ca$ in various minerals of the Boulder Creek granodiorites varied by 8.8‰ (K-feldspar had the lowest $\delta^{44/40}Ca$) while $\delta^{44/42}Ca$ values varied only by 0.5‰. Consequently, many laboratories have chosen to measure the $^{44}Ca/^{42}Ca$ ratio, reported as $\delta^{44/42}Ca$. Studies to date suggest that all the observed calcium isotope fractionations are mass dependent: in other words, $\delta^{44/40}Ca$, $\delta^{43/42}Ca$, and $\delta^{44/42}Ca$ are strongly correlated, and since the observed range in these isotope ratios does not exceed a few permil, one Ca isotope ratio is readily converted to any other (excepting any *radiogenic* variations in ^{40}Ca). $\delta^{44/42}Ca$ can be converted to $\delta^{44/40}Ca$ by multiplying by 2.05.

A consensus has also not entirely emerged on the standard used for the delta notation. Early studies from the UC Berkeley (UCB) laboratory reported ratios relative to BSE. Subsequently, the NIST SRM 915a $CaCO_3$ standard, whose isotope ratios are listed in Table 5.2, was adopted. Relative to NIST SRM 915a, the UCB BSE value is $\delta^{44/40}Ca_{SRM915a} = +0.97‰$, $\delta^{43/40}Ca_{SRM915a} = +0.88‰$, and $\delta^{44/42}Ca_{SRM915a} = +0.46‰$. Other laboratories have reported values relative to seawater, whose isotopic composition is $\delta^{44/40}Ca_{SRM915a} = 1.88±0.04‰$. Although it continues to be used as the normalization standard, the supply of SRM 915a was eventually exhausted and has been replaced as an analytical standard by SRM 915b whose $\delta^{44/40}Ca_{SRM915a} = +0.72±0.04‰$ and $\delta^{42/44}Ca = +0.34±0.02‰$ relative to SMR 915a.

Simon and DePaolo (2010) found that the Earth, Moon, Mars, and differentiated asteroids (4-Vesta and the angrite and aubrite parent bodies) have $\delta^{44/40}Ca$ indistinguishable from primitive ordinary chondritic meteorites, while enstatite chondrites are 0.5‰ enriched and primitive carbonaceous chondrites 0.5‰ depleted relative to ordinary chondrites. They speculate that the variations observed in enstatite and carbonaceous chondrites reflect fractionation during high-temperature evaporation and condensation in the early solar nebula. On the other hand, $\delta^{44/40}Ca$ in ordinary chondrites was invariant within analytical error and closely matched their estimated $\delta^{44/40}Ca$ of BSE ($\delta^{44}Ca_{SRM915a} = 0.97‰$). A subsequent study by Kang et al. (2017) found that "fertile peridotites," those least affected by metasomatism or melting, had an average $\delta^{44/40}Ca_{SRM915a} = 0.94±0.05‰$, indistinguishable from earlier estimates of BSE and close to the average for oceanic basalts (0.90±0.28‰). Melt-depleted peridotites (those with low Al_2O_3) had higher average $\delta^{44/40}Ca_{SRM915a} = 1.06‰$; metasomatized peridotites have highly variable but on average lower $\delta^{44}Ca$ (0.25 to 0.96‰). Kang et al. reported that the average $\delta^{44/40}Ca_{SRM915a}$ for oceanic basalts was 0.90±0.28‰. MORB have an average $\delta^{44/40}Ca_{SRM915a}$ of 0.84±0.06‰, which is indistinguishable from the average of gabbros (0.85‰) from the lower oceanic crust (e.g. Chen et al., 2020).

To date, there are no experimental constraints on Ca isotope partitioning between

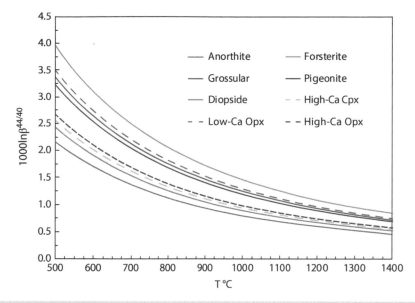

Figure 9.22   Calculated reduced partition functions for Ca-bearing silicate minerals as a function of temperature from the compilation of *ab intio* fitting factors in (adapted from Antonelli and Simon, 2020).

mantle minerals, but there have been several theoretical studies. Vibrational frequencies of Ca bonds are not known, so bond length is used as a proxy for bond strength (shorter bonds = stronger bonds) in *ab initio* studies. The calculated temperature dependence of reduced partition functions from some of these studies is shown as a function of temperature in Figure 9.22. In general, Ca-poor mafic minerals, such as olivine and Ca-poor orthopyroxene, will be enriched in ^{40}Ca, while Ca-rich minerals, such as diopside and anorthite, will be relatively depleted in ^{40}Ca, with pyroxenes having variable Ca/Mg ratios fall between these extremes. Basalts generally are about ~0.1‰ lighter than fertile peridotites, and presumed BSE, suggesting a fractionation of this order during partial melting, which is also consistent with the ^{40}Ca enrichment of melt-depleted peridotites. On the other hand, there appears to be negligible Ca isotope fractionation during fractionation crystallization from basalt through andesitic compositions, but dacites and rhyolites can be isotopically lighter (e.g. Valdes et al., 2019).

There is some evidence that $\delta^{44/40}Ca$ does vary in the mantle. Huang et al. (2011) analyzed Hawaiian lavas and found that $\delta^{44/40}Ca_{SRM915a}$ in that calcium isotope ratios correlated inversely with $^{87}Sr/^{86}Sr$ (Figure 9.23) and with Sr/Nb ratios, but not with other trace element ratios. Huang et al. argued that the calcium isotope variations and these correlations reflected the presence of marine carbonates in ancient recycled oceanic crust that makes up part of the Hawaiian mantle plume.

Carbonatites typically have mantle-like $\delta^{13}C$ and $^{87}Sr/^{86}Sr$ and, with a few exceptions, are undoubtedly mantle-derived magmas. Amsellem et al. (2020) reported $\delta^{44/40}Ca_{SRM915a}$ for 74 carbonatites ranging in age from 3 billion years to historic. Excepting the historic Oldoinyo Legnai natrocarbonatites,[2] carbonatites have systematically lower $\delta^{44/40}Ca_{SRM915a}$ than mantle-derived silicate magmas, with a mean +0.26±25‰. The Oldoinyo Legnai natrocarbonatites have higher $\delta^{44}Ca$, ranging from 0.65‰ to 0.82‰, but still lower than most mantle-derived silicate magmatic rocks. Amsellem et al. argued that these low $\delta^{44}Ca$ values reflect extensive recycling of surficial carbonates, which typically have $\delta^{44}Ca_{SRM915a}$ of 0 to +0.2‰, into the mantle throughout Earth history.

Subsequently, Sun et al. (2021) reported $\delta^{44/42}Ca_{SRM915a}$ for 47 carbonatites with

---

[2] Natrocarbonatites are unusual and differ from other carbonatites in being richer in Na_2O and poorer in CaO, MgO, and FeO.

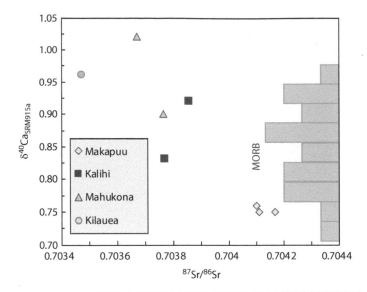

Figure 9.23 $\delta^{44/40}Ca_{SRM915a}$ versus $^{87}Sr/^{86}Sr$ in Hawaiian lavas adapted from Huang et al. (2011). Huang et al. interpreted this correlation as mixing between recycled sedimentary carbonate and mantle. Blue histogram shows the distribution of MORB values.

an average value of 0.37‰, which is equivalent to $\delta^{44/40}Ca_{SRM915a}$ = 0.76‰ (assuming $\delta^{44/40}Ca$ = 2.05×$\delta^{44/42}Ca$), still lower than typically mantle-derived silicate magmas, but less so. Sun et al. noted that carbonatite magmas have distinctly lower temperatures than silicate ones and argued that the low $\delta^{44/42}Ca$ results from greater fractionation at lower temperatures during melting.

Finally, Banerjee et al. (2021) reported $\delta^{44}Ca_{SRM915a}$ in 41 carbonatites and found that, apart from four samples from India and South Africa with anomalously high $^{87}Sr/^{86}Sr$, those older than 300 Ma display limited variation (~0.18‰, n = 12) with $\delta^{44/40}Ca_{SRM915a}$ values overlapping the BSE value. Carbonatites younger than 300 Ma displayed much wider variations in $\delta^{44/40}Ca$ values (0.63‰, n = 29) with median value of ~0.70‰. The younger carbonatites also have higher and more variable $^{87}Sr/^{86}Sr$. Banerjee et al. rejected the conclusion of Sun et al. that the difference can be explained by fractionation during melting pointing out an error in their calculations, and Amsellem et al. argue that the low $\delta^{44}Ca$ of carbonatites reflects the recycling of surficial carbonates into the mantle, but only in more recent Earth history.

The carbonatite story does not yet have an ending but highlights several unresolved issues in calcium isotope studies. These include interlaboratory biases (ratios reported by Amsellem et al. are systematically lower than those of Sun et al. or Banerjee et al. for the same samples) and the lack of experimentally partition coefficients between carbonates and silicates, between minerals and melts, and between silicate and carbonatite melts.

To date, there is only very limited Ca isotope data on subduction-related basalts. Wang et al. (2021) found that basalts and dacites from the Tonga rear arc had average $\delta^{40}Ca$ of 0.84‰ and Mariana arc lavas had average $\delta^{40/40}Ca$ of 0.79‰, values indistinguishable from MORB.

9.4.7.3 Silicon and germanium isotopes

Silicon, which is the third or fourth most abundant element on Earth (depending on how much may be in the core), has three isotopes: ^{28}Si (92.23%), ^{29}Si (4.67%), and ^{30}Si (3.10%). By convention, $^{30}Si/^{28}Si$ ratios are reported as $\delta^{30}Si$ relative to the standard NBS28 (now called NIST-RM8546) (Table 5.2); $^{29}Si/^{28}Si$ is also sometimes analyzed and reported as $\delta^{29}Si$. The study of silicon isotope geochemistry began in the 1950s, but the field has rapidly expanded only since the advent of the high-precision MC-ICP-MS in the last two decades. Silica is surprisingly important in biogeochemical cycles and is utilized by terrestrial plants as well as freshwater and marine algae and protists. Although abundant in the solid Earth, its concentration in solution is limited due in part to its low solubility but also to extensive bioutilization. Silicon isotopes, thus, have potential for understanding biogeochemical cycling, a topic we will cover in Chapter 11.

The BSE appears to have a $\delta^{30}Si$ of −0.29 ±0.07‰ (e.g. Fitoussi et al., 2009), which is heavier than that of carbonaceous chondritic meteorites (−0.36‰ to −0.56‰) and ordinary chondrites (−0.41‰ to −0.49‰) and much heavier than enstatite chondrites (−0.52‰ to −0.82‰). The Moon appears to have the same Si isotopic composition as the silicate Earth. The difference in Si isotopic composition between the silicate Earth and chondrites results from two factors. The first is nebular fractionation: volatile-poor meteorites have high Mg/Si and high $\delta^{30}Si$. The Earth's high Mg/Si is indicative of ^{28}Si depletion related to volatile loss (Dauphas et al., 2015). Second, it results from Si isotopic fractionation between silicate and iron during core formation. Experiments by Shahar et al. (2011) found that the fractionation factor varies with temperature as

$$\Delta^{30}Si_{silicate-metal} = \frac{7.45 \times 10^6}{T^2}. \quad (9.6)$$

Geophysical constraints require that the core contain several percent of one or more light elements. Based on metal–silicate partition coefficients and depending on whether the core formed in a single event or continuous during Earth accretion, the core could contain from 2 to 9 wt.% Si, which is consistent with fractionation between silicate and metal grains in meteorites. Moynier et al. (2020) concluded that even under the most extreme plausible conditions of minimum temperature and maximum amount of Si in the core, core–mantle isotope fractionation could not explain all the difference in $\delta^{30}Si$ between the Earth and chondrites and that ^{28}Si depletion due to volatility is the main factor accounting for the Earth's heavy Si.

The average $\delta^{30}Si$ of mantle-derived ultramafic xenoliths is −0.30±0.04‰. MORB appear to be isotopically uniform within analytical error with a mean of −0.28±0.03‰. OIB

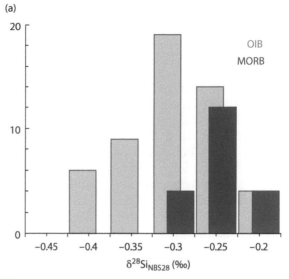

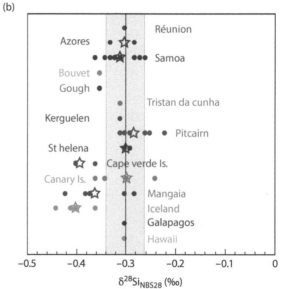

Figure 9.24   (a) Histogram of $\delta^{28}Si$ in MORB and OIB. (b) Distribution of $\delta^{28}Si$ in various oceanic island volcanoes. Stars show mean values. Source: Data from (Pringle et al., 2016).

are slightly lighter on average and more variable: −0.32±0.05‰. While these differences are small, OIB are clearly skewed to lighter values (Figure 9.24), and this is a result of particularly light compositions of basalts from a just a few island chains (Pringle et al., 2016). Tantalizingly, Pitcairn, with an extreme EMI radiogenic isotopic composition, has notably heavy Si, while Mangaia, with an extreme HIMU radiogenic isotopic composition, has notably light Si. Samoa, with extreme EMII radiogenic isotopic composition, has quite variable δ^{28}Si, but the average value falls within the MORB range. This suggests, as do other stable isotopes, that OIB and the mantle plumes from which they are produced contain a component of surficial material that has been recycled through subduction from the Earth's surface.

There is a slight fractionation associated with melting and fractional crystallization, with lighter silicon isotopes partitioning preferentially into SiO_2-poor mafic minerals such as olivine, and consequently, granites are heavier than basalts: −0.23±0.13‰. Savage et al. (2012) found that δ^{30}Si in igneous rocks depended on SiO_2 composition approximately as

$$\delta^{30}Si(‰) = 0.0056 \times [SiO_2] - 0.567. \quad (9.7)$$

Compared to fractionations occurring at the Earth's surface, however, Si isotope fractionation in igneous processes is quite small.

In contrast to silicon, germanium is trace element with a silicate Earth concentration of ~1 ppm. Sitting directly below Si on the periodic table, its geochemical behavior mimics that of Si in many respects. For example, it substitutes for Si in tetrahedral coordination and forms a hydroxy acid, $Ge(OH)_4$, in solution. It differs from Si in that it is ionic radius and is slightly larger than that of Si, making it somewhat incompatible and can also be found in octahedral coordination. It is also distinctly more volatile and siderophile as well as somewhat chalcophile. These properties account for the germanium's ~30-fold depletion on the silicate Earth relative to carbonaceous chondrites; roughly 90% of the Earth's Ge inventory is thought to be in the core.

Germanium is an important component in many semiconductors and fiber optics, and demand for it has increased over the last several decades. It has five stable isotopes: ^{70}Ge (20.52%), ^{72}Ge (27.45%), ^{73}Ge (7.76%), ^{74}Ge (36.52%), and ^{76}Ge (7.75%) (CIAAW, 2020). The usually analyzed isotope ratio is ^{74}Ge/^{70}Ge, reported as δ^{74}Ge in per-mil deviations from a standard, most commonly NIST SRM3120a whose isotopic composition is listed in Table 5.2. Two other less commonly used standards are JMC 301230 "JMC" and Aldrich Ge metal. Values reported relative to these standards can be converted to δ^{74}Ge$_{NIST3120a}$ as δ^{74}Ge$_{NIST3120a}$ = δ^{74}Ge$_{JMC}$ + 0.32‰ and δ^{74}Ge$_{NIST3120a}$ = δ^{74}Ge$_{Aldrich}$ +

1.97‰ While δ^{72}Ge and δ^{73}Ge are also often reported, isotopic fractionations are mass dependent, so all information is contained in the δ^{74}Ge value alone.

To date, there have been no analyses of Ge isotopes in carbonaceous chondrites. Ge is siderophile during high-temperature condensation and becomes lithophile at low temperatures. Nearly, all Ge in ordinary chondrites is in the metal phase such that bulk δ^{74}Ge is equal to that of the metal. Average δ^{74}Ge$_{NIST3120a}$ ranges from −0.51±0.09‰ in H chondrites through −0.31±0.06‰ in L chondrites to −0.26±0.09‰ in LL chondrites and correlates with O isotopes and silicate iron content, so this variation reflects nebular fractionation (Florin et al., 2020). The mean metal–silicate fractionation in these meteorites is Δ^{74}Ge$_{met-sil}$ = +0.22±0.35‰. This is qualitatively consistent with the heavier Ge observed in iron meteorites: magmatic irons have a mean δ^{74}Ge$_{NIST2120a}$ = 1.41‰; nonmagmatic irons range from −0.63‰ to +1.04‰ (Luais, 2012).

There are few isotopic analyses of Ge in igneous rocks, but the data that do exist indicate that terrestrial Ge is isotopically heavier than in chondrites and lighter than that in magmatic irons. Two peridotite standards have δ^{74}Ge of 0.64‰ and 0.62‰; seven analyzed basalts and andesites have an average δ^{74}Ge of 0.55‰ and a standard deviation of 0.07‰, and three analyzed granites range from 0.54‰ to 0.69‰. Most metamorphic and sedimentary rocks have δ^{74}Ge in the range of 0–1‰, although cherts and biogenic silica show a somewhat wider range.

Sulfides have isotopically light Ge, with δ^{74}Ge ranging from −5 to +2, consistent with theoretical prediction based on Ge–S bond lengths being longer than G–O or Ge–OH bonds. Ge concentrations in sphalerite are often an order of magnitude greater than in pyrite or galena and generally δ^{74}Ge$_{pyrite}$ < δ^{74}Ge$_{sphalerite}$ < δ^{74}Ge$_{galena}$, but this is not systematic, and this likely is a kinetic/Rayleigh fractionation rather than equilibrium.

Rouxel and Luias (2017) estimate the average δ^{74}Ge$_{NIST2120a}$ of the silicate Earth to be 0.58±0.21‰, which is much heavier than the chondritic range. Judging from iron meteorites, the Earth's core, Ge in the Earth's core is also likely heavy, so that the Earth as a whole likely has heavier Ge than chondrites.

9.4.8 Transition metal stable isotopes

9.4.8.1 Titanium isotopes

Titanium has five stable isotopes, ^{46}Ti, ^{47}Ti, ^{48}Ti, ^{49}Ti, and ^{50}Ti with abundances of 8.25%, 7.44%, 73.72%, 5.41%, and 5.18%, respectively (CIAAW, 2020). The $^{49}Ti/^{47}Ti$ is the ratio typically analyzed and reported as $\delta^{49}Ti$ in per-mil deviations from the OL-Ti standard (Table 5.2). In the Earth, Ti is overwhelmingly present in the 4+ valence; only in highly reducing conditions will it be present as 3+, so that redox reactions usually play no role in isotopic fractionations. Ti also tends to be highly insoluble, so water–mineral fractionations are also limited, and fractionations are largely governed by crystal lattice coordination and bond strength.

Several studies have found that chondrites have an average $\delta^{49}Ti$ of +0.004±0.010‰ with no detectable systematic variation between chondrite classes (e.g. Williams et al., 2021), despite the systematic nucleosynthetic variations in the abundance of ^{50}Ti we noted in Chapter 6. Achondrites show slightly more variable $\delta^{49}Ti$ as a consequence of fractional crystallization and, in the case of aubrites, highly reducing conditions where some Ti^{4+} is reduced to Ti^{3+}. The few analyzed peridotites have a similar range: +0.012‰ to –0.003‰; komatiites have an average $\delta^{49}Ti$ of –0.001±0.019‰. We can reasonably infer that the bulk Earth and the mantle have $\delta^{49}Ti_{OL-Ti} \cong 0‰$, indistinguishable from chondrites.

In the upper mantle, Ti is mainly hosted in pyroxenes and where it substitutes for Si in the tetrahedral site or in garnet in coupled substitutions into an octahedral site normally occupied by Al. Wang et al. (2020) used *ab initio* calculations based on density function theory to conclude that fractionation factors involving Ti in the tetrahedral site were negligible at mantle temperatures (<0.1‰), although slightly large fractionations (~0.5‰) were possible in substituting for Al in garnet. This as well as observational data indicates that Ti isotope fractionation during peridotite melting is likely to be very small (e.g. Deng et al., 2018).

MORB have an average $\delta^{49}Ti$ of 0.006‰ with a standard deviation of 0.022‰, but there is a statistically significant difference between the so-called NMORB ("normal" MORB with La/Sm$_N$ <1 and $^{87}Sr/^{86}Sr$ <0.703) with an average of –0.005±0.003‰ and EMORB ("enriched" MORB with La/Sm$_N$>1 and $^{87}Sr/^{86}Sr$ >0.703) with an average of 0.034±0.022‰. Deng et al. (2018) also found that 2.7–2.9 Ga komatiites had $\delta^{49}Ti$ similar to that of N-MORB, while komatiites older than 3 Ga had higher $\delta^{49}Ti$ and similar to EMORB. They argued that this secular change reflected the evolution of the depleted upper mantle due to the extraction of the continental crust. This hints at heterogeneity in $\delta^{49}Ti$ within the mantle, but additional research is necessary to confirm this.

Fractionation between Ti–Fe oxides and the silicate liquids can lead to significant fractionation when those minerals crystallize from evolving basaltic magma. This leads to the distinct trends in $\delta^{49}Ti$ as a function of indices of fractional crystallization such as SiO_2 and Mg# for tholeiitic, calc-alkaline, and alkaline magma series as a consequence of their differing Ti contents and oxygen fugacity, which we have already discussed in Chapter 8.

9.4.8.2 Vanadium isotopes

Vanadium has two stable isotopes, ^{50}V and ^{51}V with abundances of 0.25% and 99.75%, respectively (CIAAW, 2020). Conventionally, the $^{51}V/^{50}V$ is reported as $\delta^{51}V$ as per-mil deviation from the Alfa Aesar standard. The interest in vanadium stems from its multiple valence states of 2+ through 5+, as a measure of oxygen fugacity in the mantle and igneous rocks. In the mantle and mafic magmas with oxygen fugacity close to the fayalite–magnetite–quartz (FMQ) buffer, V is predominantly (~75%) in the 4+ valence with the remainder as V^{5+} and V^{3+} in subequal amounts depending on oxygen fugacity. Most crystallographic sites in mantle minerals do not readily accommodate V in the higher valence states, making vanadium a modestly incompatible element.

Chondrites have an average $\delta^{51}V$ of –1.2‰ with a standard deviation of ±0.105‰ (Nielsen et al., 2019). While carbonaceous and ordinary chondrites have indistinguishable means, $\delta^{51}V$ in carbonaceous chondrites correlates with $\delta^{54/52}Cr$, indicative of nucleosynthetic anomalies in them, but ordinary chondrites display no such correlation (Nielsen et al., 2019).

Qi et al. (2019) found that the range of fertile peridotites was only –0.89 to –0.95‰ (mean –0.91‰), and the mean of refractory peridotites

(–0.93‰) did not differ significantly; komatiites also had a mean of –0.91‰; from this, we can infer a mantle value of ~–0.92‰, indicating that the silicate Earth has somewhat heavier $\delta^{51}V$ than chondrites. Vanadium is siderophile and over half of the Earth's V inventory is in the core. While the difference in $\delta^{51}V$ between silicate Earth and chondrites might reflect fractionation during core formation, a sufficiently large fractionation factor seems unlikely (Nielsen et al., 2019). As we found in Chapter 6, however, the Earth often matches the isotopic composition of enstatite chondrites better than that of other chondrites. Only a single enstatite chondrite has been analyzed to date, and it is slightly heavier ($\delta^{51}V$ = –1.05‰) than the chondritic average, and we cannot yet rule out the possibility that enstatite chondrite range will overlap the terrestrial $\delta^{51}V$ value.

MORB are somewhat heavier than peridotites, with average $\delta^{51}V$ of –0.84‰, and $\delta^{51}V$ in MORBs from individual segments is correlated with the mean ridge depth and Na_2O of that segment, suggesting that $\delta^{51}V$ in basalts is sensitive to melting extent (Wu et al., 2018). Altered MORB are more variable with $\delta^{51}V$ from –0.101‰ to –0.77‰ but on average are not different from fresh MORB. The limited analyses to date of OIB from Iceland and Hawaii indicate that they have $\delta^{51}V$ indistinguishable from MORB.

Owing to the orientation of the d electron orbitals, V^{3+} has a strong preference for sixfold coordination (i.e. a high octahedral site preference energy), and hence, in crystallizing magmas, V strongly partitions into magnetite during magmatic crystallization. Magnetite also incorporates V^{4+}, and electron exchange between Fe and V results in a constant V^{3+}/V^{4+} ratio in magnetite. Sossi et al. (2018) found experimentally that partitioning of V into magnetite decreases with increasing oxygen fugacity and V in magnetite becomes increasingly isotopically light relative to melt with an oxygen fugacity dependence of the fractionation factor $\Delta^{51}V_{mag-melt}$ (‰) = (–0.045±0.021 × ΔFMQ – 0.70) ±0.05. However, these experiments were carried out with titanium-free compositions, whereas natural magnetite is solid solution of magnetite (Fe_3O_4) and ulvospinel (Fe_2TiO_4). This effects the Fe^{2+}/Fe^{3+} balance, and it is unclear whether this effects the V fractionation factor.

Magnetite is not a crystallizing phase in primitive mantle-derived basalts, and there appears to be little systematic variation in $\delta^{51}V$ among them. Once Fe–Ti oxides crystallize, $\delta^{51}V$ increases, reaching values of ~ +1‰ in a dacite with ~65% SiO_2 from Anatahan Volcano in the Mariana (Prytulak et al., 2017). For magmas crystallizing magnetite, Prytulak et al. estimated the bulk mineral–melt fractionation factor $\Delta^{51}V_{min-melt}$ to be –0.4‰ to –0.5‰, Ding et al. (2020) estimated it to be –0.15‰ for Kilauea Iki magmas, and Wu et al. (2018) estimated $\Delta^{51}V_{min-mel}$ to be –0.3‰ and vary with temperature as –0.15 × $10^6/T$ (‰) for MORB.

Vanadium isotope studies remain in an early stage, and it is not yet clear what they will be able to reveal about mantle oxidation state.

9.4.8.3 Chromium isotopes

Chromium has four stable isotopes, ^{50}Cr, ^{52}Cr, ^{53}Cr, and ^{54}Cr with abundances of 4.345%, 83.799%, 9.501%, and 2.365%, respectively (CIAAW, 2020). The $^{53}Cr/^{52}Cr$ ratio is most commonly analyzed, although $^{54}Cr/^{52}Cr$ and $^{50}Cr/^{52}Cr$ are sometimes analyzed; these are reported in the usual per mil notation as $\delta^{53}Cr$ and $\delta^{54}Cr$, and $\delta^{50}Cr$ (some earlier studies used $\delta^{53/52}Cr$ and $\delta^{54/52}Cr$, and $\delta^{50/52}Cr$) relative to the NIST-SRM-979 standard. Chromium has three oxidation states, Cr^{2+}, Cr^{3+}, and Cr^{6+}, but only Cr^{2+} and Cr^{3+} are present in the mantle, with the $Cr^{2+}/\Sigma Cr$ ratio varying from 0 to 1 depending on oxygen fugacity. Fractionations between oxidation states can be expected. Because of its abundance, the oxidation state iron controls the oxidation state of the mantle, while the Cr^{2+}/Cr^{3+} ratio will merely reflect that oxidation state. At the Earth's surface, Cr is oxidized to Cr^{6+} during weathering and the latter form is dominant in the hydrosphere.

As we found in Chapter 6, ^{53}Cr is the decay product of the short-lived radionuclide ^{53}Mn, and there are also nucleosynthetic distinctions in ^{54}Cr between carbonaceous chondrites and other meteorites. However, all the chondrite groups appear to have indistinguishable $\delta^{53/52}Cr$ with a mean of –0.115‰ and standard 0.025‰ (Bonnand et al., 2016; Schoenberg et al., 2016). Schoenberg et al. (2008, 2016) found that the mean $\delta^{53}Cr$ of mantle peridotites was −0.134±0.103‰, presumably the silicate Earth value, which is indistinguishable from chondrites. Based on the analysis of komatiites, Jerram et al. (2020) concluded the silicate Earth $\delta^{53}Cr$ was −0.12±0.04‰ in agreement with the previous estimates. Farkaš et al. (2013) and Xia et al. (2017) found large ranges of $\delta^{53}Cr$ in mantle xenoliths: −0.51‰ to +0.88‰. From a negative correlation with Al_2O_3 and CaO, they inferred that isotopically light Cr is preferentially extracted during partial melting, shifting the residua to higher $\delta^{53}Cr$. However, they concluded that the largest variations resulted from kinetic fractionation during melt–rock or melt–fluid interactions. Their estimate for $\delta^{53}Cr$ of undepleted mantle was −0.14±0.12‰, consistent with earlier estimates.

Shen et al. (2018) used a combination of ionic modeling and observed mineral–mineral fractionations to predict the following $\Delta^{53}Cr$ fractionation factors at 870–970 °C: spinel–olivine: 0.11–0.16‰, spinel–pyrope: 0.04–0.11‰, and pyroxene–olivine: 0.05–0.1‰. These fractionations in part reflect oxidation state. Cr is predominantly in the 3+ state in spinel and mixed valence in other silicates.

Shen et al. (2020) found that $\delta^{53}Cr$ in the Kilauea Iki lava lake varied from −0.18‰ to 0‰, with the heaviest compositions associated with olivine–spinel cumulates and the lightest values with the most differentiated residual liquids. They concluded that this resulted from the crystallization and accumulation of spinel, which is dominated by Cr^{3+} and, hence, enriched in heavier Cr isotopes relative to the residual melt. Shen et al. also found that Kea-trend volcanoes Mauna Kea and Kilauea had higher $\delta^{53}Cr$ than Koolau volcanics from Oahu (part of the Loa trend) and that this reflected a difference in source rather than fractional crystallization, perhaps because the Koolau source was more reducing. Bonnand et al. (2020) analyzed basalts of Fangataufa Island in the Tuamoto Archipelago and inferred a $\Delta^{53}Cr_{melt/crystals}$ of −0.10‰ during fractional crystallization. Two magma series are present, and there are differences in $\delta^{53}Cr$ and radiogenic isotope ratios between them with higher $\delta^{53}Cr$ associated with higher ε_{Nd}. They interpreted the observed variation as a result of the source consisting of a mixture of a garnet peridotite and a more fertile "recycled" component.

9.4.8.4 Iron isotopes

Iron has four stable isotopes: ^{54}Fe, ^{56}Fe, ^{57}Fe, and ^{58}Fe, whose abundances are 5.845%, 91.754%, 2.119%, and 0.282%, respectively (CIAAW, 2020). Most research has focused on the ratio of the two most abundant isotopes, $^{56}Fe/^{54}Fe$, expressed as $\delta^{56}Fe$. Because of the interference of $^{40}Ar^{16}O^+$ on $^{56}Fe^+$ in MC-ICP-MS measurements, some studies also report $^{57}Fe/^{54}Fe$ as $\delta^{57}Fe$. Less frequently $^{57}Fe/^{56}Fe$ has been reported as $\delta^{57/56}Fe$. Values are most commonly reported relative to the IRMM-14 standard, although some workers have used average igneous rocks as the standard to define $\delta^{56}Fe$. Although IRMM-14 remains the reference standard, it has been exhausted and replaced by IRMM-524a, which has the same Fe isotopic composition. As all the fractionations measured to date are mass dependent, $\delta^{57}Fe$ values can be converted as $\delta^{56}Fe = \delta^{57}Fe \times 0.668$, and fractionation factors converted as:

$$\alpha_{57/54} = \left(\alpha_{56/54}\right)^{1.457}. \tag{9.8}$$

Iron exists in the Earth in three oxidation states: Fe^{3+}, Fe^{2+}, and Fe^0, the latter being extremely rare in nature outside of the Earth's core. Not surprisingly, the largest Fe isotope fractionations are between redox states. These redox are sometimes biologically mediated; the importance of Fe as a nutrient leads to considerably interest in iron isotopes in biogeochemical cycling, which we will discuss in Chapter 11. Fractionations involving biologically mediated reduction can be as great as 3‰. In contrast, nonredox fractionations between igneous minerals are generally <0.1‰. Dauphas et al. (2017) provide a comprehensive summary of Fe isotope fractionation factors.

Carbonaceous, ordinary, and enstatite chondrites have relatively uniform $\delta^{56}Fe$ values of −0.003±0.011‰. The three other planetary bodies from which we have samples, the Moon, Vesta, and Mars, have Fe isotopic compositions that are nearly indistinguishable from those of chondrites. Whole rock peridotite samples are variable and on average lighter than chondrites with a mean $\delta^{56}Fe_{IRMM-14}$ of −0.027‰ and a standard deviation of ±0.026‰ (Figure 9.25). A key question is whether metal–silicate fractionation during

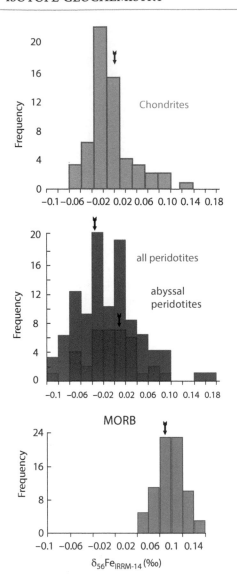

Figure 9.25 Distribution of ^{56}Fe in chondrites and peridotites (from Dauphas et al., 2017), abyssal peridotites (data from Craddock et al., 2013), and MORB (data compiled from the literature). Arrows indicate mean values.

core formation can explain the difference in δ^{56}Fe between the silicate Earth and chondrites. Iron meteorites have a mean and standard deviation δ^{56}Fe of $+0.050\pm0.101$‰ and are only slightly heavier than chondrites (Poitrasson et al., 2005) – not enough to explain the difference.

In a recent evaluation of experimental studies of metal–silicate fractionation, which have produced Δ^{57}Fe$_{\text{metal-silicate}}$ values that range from +0.4 to –0.9‰ (equivalent to Δ^{56}Fe$_{\text{metal-silicate}}$ +0.27 to –0.6‰), Shahar and Young (2020) concluded: *its complicated*. Correcting all the experimental results to the same temperature, they found that Δ^{57}Fe$_{\text{metal-silicate}}$ was a strong function of composition. For pure iron, the fractionation factor is essentially 0 but increases to +0.35‰ when the atomic fraction of other elements reaches 26%. They also found that the fractionation factor is a function of pressure (but dependence is negligible below ~5 to 10 GPa) and decreases with temperature. They conclude that for large planets such as the Earth, any core–mantle fractionation should be small.

The question then arises as to what the actual silicate Earth δ^{56}Fe is Sossi et al. (2016) noted that peridotites have more variable δ^{56}Fe than basalts. They suggested that this arises from liquid percolation and subsequent reaction, where fluid/melt transport through the lithospheric mantle is often associated with kinetic isotope fractionation. They found that when only the most primitive lherzolitic peridotites were considered, the mean was +0.05‰ with a standard deviation of ±0.04‰ and argued that this more likely represented the mantle composition. Craddock et al. (2013) argued that abyssal peridotites recovered from mid-ocean ridges were a more reliable mantle sample than continental peridotites. The mean δ^{56}Fe of these abyssal peridotites was 0.014‰ with a standard deviation of ±0.04‰ (Figure 9.24), essentially identical to Sossi's estimate and not resolvable from chondrites. Based on these more recent studies, it seems likely that the Fe composition of the mantle and the entire Earth falls within the chondritic range.

Figure 9.26 shows reduced partition functions for Fe^{2+} in a variety of silicates calculated based on density function theory by Rabin et al. (2021). In most silicates, iron is coordinated by oxygen in a similar fashion, so we expect only limited fractionation. Rabin et al. found that the most important factor following temperature and oxidation state was the second cation neighbor. Consistent with observation and experiments, the calculations show that olivine preferentially incorporates isotopically light Fe (more so for Fe-rich compositions), while magnetite incorporates isotopically heavy Fe. Clinopyroxene compositions overlap those of olivine, with Fe-rich

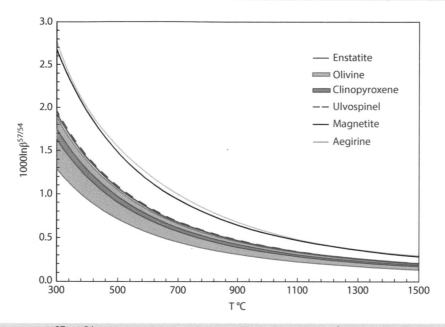

Figure 9.26   Reduced $^{57}Fe/^{54}Fe$ partition functions (β functions) for Fe^{2+} in a variety silicate and oxide minerals of Rabin et al. (2021) as a function of temperature. Fields for olivine and clinopyroxene reflect compositional variation, with Fe-rich compositions favoring light Fe isotopes more than Mg-rich compositions. Since fractionation factors are equal to the difference in the log of partition functions, the overlap of β functions of olivine and pyroxenes indicates little fractionation between them. Aegirine ($NaFeSi_2O_6$) is an uncommon sodic pyroxene restricted to highly alkaline compositions.

compositions being isotopically lightest. They did not compute reduced partition functions for garnet, but based on natural mineral pairs, garnet should be isotopically lighter than clinopyroxene.

Observations show that Fe isotopes are indeed fractionated during magmatic evolution. This was first demonstrated by Teng et al. (2008) for the Kilauea Iki lava lake, which formed in 1959 and crystallized under close system conditions and from which a 1200-m drill core was recovered. Residual melts had $\delta^{56}Fe$ 0.2‰ greater than the initial composition and olivines are up to 0.12‰ lighter. Weyer and Seitz (2012) found that olivine phenocrysts in basalts from the Canary Islands typically had $\delta^{56}Fe$ 0.1‰ lower than the matrix from which they were separated. However, some olivines from the Vogelsberg volcanics of Germany had $\delta^{56}Fe$ ~0.4‰ lower than the matrix. As olivines attempt to equilibrate with evolving melt compositions, Fe will diffuse in and Mg will diffuse out. Weyer and Seitz attributed these larger fractionations to diffusional fractionation.

As a consequence of these isotope fractionations, mantle-derived basalts are offset to heavier isotopic compositions from peridotites. Seventy-five analyses of MORB have an average $\delta^{56}Fe_{IRMM-14}$ of +0.10‰ with a standard deviation of 0.02‰ (Figure 9.25). OIB show significantly greater variation, as summarized in Figure 9.27. Correcting for the effects of fractional crystallization as Soderman et al. (2021) did reduces this variation only modestly in most cases. While corrected Hawaiian compositions are quite similar to those of MORB, $\delta^{57}Fe$ is generally lower in Galapagos and Iceland basalts, but higher in other OIB. Furthermore, there are correlations with radiogenic isotope ratios in a number of cases, demonstrating that $\delta^{57}Fe$ variation in these lavas reflects $\delta^{57}Fe$ variation in the mantle source. Soderman et al. concluded that these result from the subduction of lithospheric material into the mantle and fractionations occurring within subduction zones. It is interesting to note in the respect that IAV are lighter than oceanic basalts, with a mean $\delta^{56}Fe$ of 0.05‰ and a standard deviation of

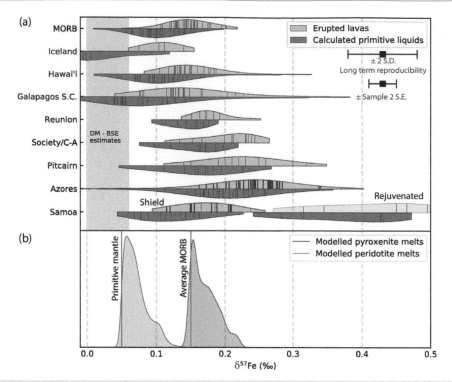

Figure 9.27 Kernel density plots of δ^{57}Fe in OIB and MORB with MgO between 5 and 16 wt.%. Each vertical line represents a sample. Correcting for the effects of fractional crystallization only modestly reduces the variability. Samoan shield lavas on average are indistinguishable from MORB, whereas the rejuvenated lavas record the heaviest δ^{57}Fe in the global data set. Source: Soderman et al. (2021)/with permission of Elsevier.

0.003‰ (Foden et al., 2018). Granitoid igneous rocks are slightly heavier and more variable δ^{56}Fe = +0.16±0.07‰.

9.4.8.5 Copper isotopes

Copper has two isotopes, ^{63}Cu and ^{65}Cu, with respective abundances of 69.15% and 30.85% (CIAAW, 2020). Isotope ratios are reported as per-mil deviations of ^{65}Cu/^{63}Cu from the NIST976 standard as δ^{65}Cu (just as in other cases, this standard has been exhausted but remains the reference value). Copper can occur in any of three valences, 0, I, and II, although the first of these, native copper, is rare. Carbonaceous chondrites display a largest range with δ^{65}Cu varying from +0.02 for CI to −1.45‰ for CV. Ordinary chondrites have a smaller range, varying by class from +0.07 to −0.44‰. Enstatite chondrites form a tighter cluster at −0.25±0.09‰. It is possible, and perhaps likely, that some of this variation is nucleosynthetic as it is for several other elements in this mass range, but with only two isotopes mass-dependent and mass-independent

variations cannot be distinguished. Cu is relatively volatile, and δ^{65}Cu correlates with volatile/refractory element ratios in meteorites indicating volatility played a role in this variability (Luck et al., 2003). Copper is, of course, a critical metal to society, and we will consider the Cu isotopic compositions of Cu ore deposits at the end of this chapter.

Figure 9.28 shows that unmetasomatized peridotites have an average δ^{65}Cu of 0.078‰ with a standard deviation of 0.1‰ (e.g. Savage et al., 2015; Liu et al., 2015; Huang et al., 2017); metasomatized peridotites have a wider range from −0.65 to +1.82‰ (Liu et al., 2015). Average δ^{65}Cu for komatiites, OIB, and MORB are, respectively, 0.06‰, 0.06‰, and 0.08‰, and all are statistically indistinguishable from unmetasomatized peridotites. Copper in the silicate Earth, thus, appears to have δ^{65}Cu ≈ 0.08‰ (Figure 9.27) and be significantly heavier than chondrites (although there is overlap with LL chondrites).

There are a several reasons why this might be the case. The Earth is depleted in moderately volatile elements such as copper, and as a

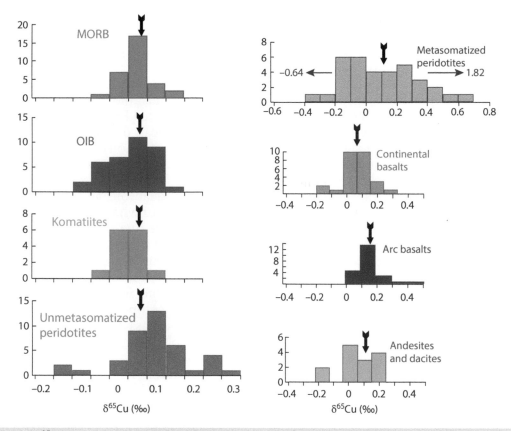

Figure 9.28 $\delta^{65}Cu_{NIST976}$ in mantle and mantle-derived rocks. Arrows show mean values. The similarity of $\delta^{65}Cu$ in unmetasomatized peridotites, komatiites, and basalts indicates a mantle value of ~0.08‰ and that little or no fractionation occurs during melting and basaltic fractional crystallization.

result, we might expect it to be depleted in the lighter isotope. However, zinc is even more volatile than Cu and is not depleted in its light isotopes. Cu is also moderately siderophile (as well as strongly chalcophile) so redox reactions and metal–silicate fractionations may also have played a role in producing this variability. Indeed, the bulk of the Earth's copper appears to be in the core. However, experiments by Savage et al. (2015) showed that liquid metal–silicate fractionation is small (<0.1‰) irrespective of temperature, and more importantly, the fractionation of the opposite sense needed to explain the difference: the metal is enriched in ^{65}Cu relative to ^{63}Cu. In contrast, they found that sulfide–silicate fractionation factor, $\Delta^{65}Cu_{sulf-sil}$, was negative, although they could not establish the precise value. These experimental results are consistent with the observational data on meteorites of Williams and Archer (2011), which show $\Delta^{65}Cu_{metal-sil} \approx 0.5$‰ and average $\Delta^{65}Cu_{metal-sulf} = 2.9 \pm 0.9$‰. Savage et al.

(2015) concluded that the Cu isotope composition of BSE seems to require that large-scale sulfide–silicate equilibration occurred sometime in Earth's history. They suggested a Cu-rich layer of Fe–O–S liquid formed at the base of the mantle (a "Hadean Matte") following the Moon-forming Giant Impact that was subsequently incorporated into the core, leaving Cu in the mantle isotopically heavy.

Huang et al. (2017) found that melt-depleted peridotites from the Iverea Zone in the Italian Alps had heavier Cu isotopes than more fertile ones and suggested that ^{63}Cu enters the melt fraction more readily than ^{65}Cu. However, the uniformity of $\delta^{65}Cu$ in komatiites, MORB, and OIB and their similarity to unmetasomatized peridotites indicates that copper isotopes are not significantly fractionated during partial melting. Given the chalcophile nature of Cu, it is likely that it is hosted in the mantle by sulfide phases. Since sulfide solubility in basaltic liquids is limited, Lee et al. (2012) argued sulfide phases may survive through the first 20% of melting. Again, the homogeneity of $\delta^{65}Cu$ over a range from small melt fractions (oceanic island alkali basalts) through high melt fractions (komatiites) suggests that this is not the case.

Island arc basalts have more variable $\delta^{65}Cu$ and are slightly heavier on average ($\delta^{65}Cu = 0.16‰$) than MORB and OIB (Liu et al., 2015). Liu et al. (2015) suggested that this resulted from the incorporation of heavier copper from the subducted slab. Wang et al. (2021) found that $\delta^{65}Cu$ was uncorrelated with Ba/Nb ratios, implying little or no addition of slab-derived Cu. They suggested instead that fractionation related to the oxidation of sulfides in the mantle wedge could produce the observed variations.

9.4.8.6 Zinc isotopes

Zinc has five stable isotopes: ^{64}Zn (49.17 %), ^{66}Zn (27.73 %), ^{67}Zn (4.04 %), ^{68}Zn (18.45 %), and ^{70}Zn (0.61 % (CIAAW, 2020). All the fractionations appear to be mass dependent, and isotopic composition is conventionally reported as $^{66}Zn/^{64}Zn$ in per-mil deviations from a Johnson-Mathey standard solution first developed at ENS Lyon: $\delta^{66}Zn_{JMC-Lyon}$. Unlike many other transition elements, redox reactions are typically not important as zinc rarely varies from the Zn^{2+} oxidation state in nature. It has a mixed lithophile–chalcophile character but lacks the siderophilic nature of many other transition metals; nevertheless, some Zn, perhaps 30% of the Earth's inventory, likely accompanied sulfur into the core (Sossi et al., 2018), but relevant experiments have found negligible isotopic fractionation between silicates and Fe metal, with or without sulfur, so core segregation is expected to have little effect on mantle isotopic composition. Zinc is substantially more volatile than other first transition series metals. Overall, zinc is only mildly incompatible, and in the mantle, it is primarily hosted in spinel, when present, and in olivine.

There is substantial variation in the Zn isotopic composition of meteorites, with $\delta^{66}Zn$ in carbonaceous chondrites varying by class from 0.48‰ for CI to 0.13‰ for CK, averaging 0.25‰. Unequilibrated (i.e. petrologic grade <4) ordinary chondrites vary by class from –0.47‰ for H to –0.11‰ for L, averaging 0.13‰ and in enstatite chondrites from 0.01‰ to 7.35‰ (Luck et al., 2005; Pringle et al., 2017). Most of the variation among enstatite chondrites is found in the highly metamorphosed EL6 meteorites and correlates with composition in manner indicative of volatile loss during metamorphism. The mean $\delta^{66}Zn$ of unequilibrated EH and EL chondrites is similar with a mean $\delta^{66}Zn$ of 0.21‰ and 0.38‰ respectively. Curiously, Luck et al. (2005) found that, with the exception of the EL6, correlations between zinc isotopes and composition of carbonaceous and ordinary chondrites was opposite of what would be expected if this variation were due to volatile loss of Zn: isotopically light zinc is associated with the most volatile-depleted compositions. They suggested that refractory material acquired Zn by subsequent reaction with a gas phase enriched in light Zn isotopes. All the variations to date appear to be mass dependent. The Moon, however, is depleted in light Zn isotopes, consistent with kinetic fractionation associated with volatile loss during its formation.

Figure 9.29 shows $\delta^{66}Zn_{JMC-Lyon}$ in mantle and mantle-derived rocks. Most peridotite have $\delta^{66}Zn$ in the narrow range of 0.1–0.26‰ with a mean of 0.16‰. A few "fertile" peridotites analyzed by Doucet et al. (2016) fall outside this range with heavier Zn, including those shifts to mean to 0.18‰. Komatiites also have a mean $\delta^{66}Zn$ of 0.18‰. Thus, $\delta^{66}Zn$ in silicate Earth appears to fall within the mean values for ordinary and enstatite chondrites.

MORB define a narrow range that is roughly what would be expected from analytical error alone, but are heavier than peridotites, with an average $\delta^{66}Zn$ of 0.28‰. This is consistent with the sense of mineral–melt fractionation factors, which should leave the melt enriched in heavy isotopes but implies a bulk $\Delta^{56}Zn_{melt-mantle}$ of ~0.1‰, which is somewhat heavier than the value estimated by Sossi et al. (2018) (~$0.15 \times 10^6/T^2$).

OIB display a somewhat larger range than MORB with a higher mean of 0.31‰; although the difference between OIB and MORB is small, it is statistically significant at the 5% level. There appear to be systematic variations between islands, ranging from 0.21‰ in Iceland to ~0.35‰ for the Crozet Islands. As Beunon et al. (2020) point out, this variation is unlikely to result from partial melting alone. They point to the correlation

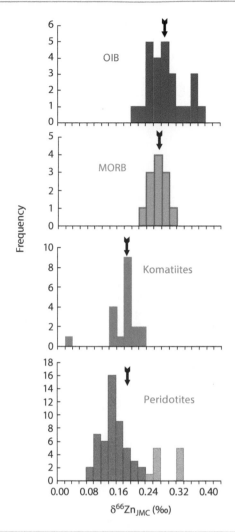

Figure 9.29 $\delta^{66}Zn_{JMC\text{-}Lyon}$ in mantle and mantle-derived rocks. Arrows show mean values.

vapor/fluid; $\delta^{66}Zn$ up to 0.88‰ in pegmatites supports this idea (Telus et al., 2012).

9.4.8.7 Molybdenum isotopes

Molybdenum has seven isotopes: ^{92}Mo (14.65%, ^{94}Mo (9.187%), ^{95}Mo (15.873%), ^{96}Mo (16.67%), ^{97}Mo (9.582%), ^{98}Mo (24.29%), and ^{100}Mo (9.74). Stable isotope ratios are now conventionally reported in $\delta^{98}Mo$ notation as per-mil deviations of the $^{98}Mo/^{95}Mo$ ratio from the NIST-SRM 3134 standard listed in Table 5.2. However, prior to 2013, data were reported relative to a variety of in-house standards, which were found to vary from NIST-SRM 3134 by up to 0.37‰ (Goldberg et al., 2013). Thus these older data need to be corrected to this common standard. Two other widely analyzed standards are useful in making this correction: USGS shale standard SDO-1 with $\delta^{98}Mo_{NIST3134} = 0.80 \pm 0.14$‰ and open ocean seawater, which has a uniform concentration of $\delta^{98}Mo_{NIST3134} = 2.09 \pm 0.1$‰. In this book, values reported relative to other standards have been corrected to $\delta^{98}Mo_{NIST3134}$ values.

Molybdenum is strongly siderophile and chalcophile and redox sensitive and has seven possible valence states ranging from 6+ to 2– but occurs primarily in the 6+ and 4+ valences in magmas and igneous rocks. As we found in Section 6.5.2, mass-independent variations in the Mo isotopic composition of meteorites occur due to varying contributions from p-, r-, and s-process nucleosynthesis (Figure 6.25), with the Earth being enriched in s-process nuclides relative to most meteorites. After correction for apparently nucleosynthetic effects, enstatite, ordinary, and carbonaceous chondrites, excepting CM and CK (which are heavier with $\delta^{98}Mo = +1$‰ and $+12$‰, respectively), form a tight distribution of $\delta^{98}Mo$ with an average of –0.15‰ and standard deviation of 0.026‰ (Burkhardt et al. 2014; Liang et al., 2017). Magmatic irons have a slightly lower $\delta^{98}Mo$, but not significantly so, mean of –0.12±0.06‰.

Because of its highly siderophile nature ($D^{met/sil} > 100$), most of the Earth's Mo is likely in the core, and consequently, the isotopic composition of bulk Earth equals that of its core. Hin et al. (2013) found that the metal–silicate fractionation factor for Mo had a temperature dependency of

between Zn concentrations and $^{87}Sr/^{86}Sr$ and ε_{Nd} in OIB and argue that the heavy Zn in OIB results from subducted carbonate-containing oceanic crust in their mantle sources.

Zn isotopes appear to be fractionated slightly during fractional crystallization; $\delta^{66}Zn$ increases from 0.26‰ to 0.36‰ during the crystallization of the Kilauea Iki lava lake (Chen et al., 2013). Continental granites overlap the range of peridotites and basalts, ranging from 0.11‰ to 0.69‰, but are on average just slightly heavier, with mean $\delta^{66}Zn$ of 0.34‰. Some of the variation among granitic rocks may be due to the partitioning of isotopically heavy Zn chloride species into an exsolved

$$\Delta^{98}\mathrm{Mo}_{\mathrm{met/sil}} = -4.70(\pm 0.59) \times 10^5/T^2.$$

$$(9.9)$$

Assuming that this is applicable to terrestrial core formation, Mo in the silicate Earth should be a few tens of ppm (hundredths of permil) heavier than the bulk Earth and core. The mean δ^{98}Mo of 15 peridotites (mainly xenoliths) analyzed by Liang et al. (2017) is −0.22‰ with a standard deviation of ±0.09‰ (Figure 9.30). If this is the silicate Earth value, it is inconsistent with a chondritic bulk Earth δ^{98}Mo and core–mantle fractionation described by 9.8. On the other hand, MORB have and mean δ^{98}Mo of −0.09‰ and a standard deviation of 0.11‰ (Bezard et al. 2016; Liang et al. 2017). OIB are lighter with a mean of −0.184‰ and more variable with a standard deviation of 0.18‰. Such δ^{98}Mo values for the silicate Earth would be consistent with a chondritic bulk Earth δ^{98}Mo. That said, data on both chondrites and peridotites are still too limited to draw firm conclusions about whether the Earth's δ^{98}Mo is chondritic.

Owing to its high charge, Mo is highly incompatible; it likely primarily occurs in tetrahedral coordination in silicate melts but likely in octahedral ones in silicates, oxides, and sulfides. Based on our rule of thumb that the heavy isotope partitions into sites of the lowest coordination, we might expect that during melting and fractional crystallization, melts will be isotopically heavier than minerals. MORB, OIB, and IAV are all slightly heavier than peridotites, which would be consistent with a very small, <0.1‰, fractionation during melting, but more data are needed. Several studies have found that the fractional crystallization of water-poor tholeiitic magmas does not produce significant changes in δ^{98}Mo (Yang et al., 2015; Bezard et al., 2016; Gaschnig et al., 2021a). On the other hand, isotopic fractionation does appear to occur in the calc-alkaline magmas found in many island arcs, which crystallize hydrous minerals such as amphibole and biotite (e.g. Wille et al., 2018). Consistent with this fractionation, Willbold and Elliott (2017) estimate the δ^{98}Mo$_{\mathrm{NIST3134}}$ of the upper continental crust to be ~0.15‰ and that of the entire crust to be between 0.1‰ and 0.3‰.

Mo^{6+} can form the soluble oxyanion MoO$_4^{2-}$ and can be mobile in oxidizing environments, and consequently, Mo isotopes can

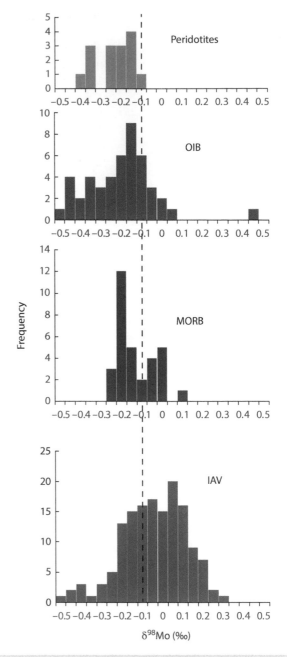

Figure 9.30 Distribution of δ^{98}Mo in peridotites, MORB, OIB, and IAV. The latter three tend to be just slightly heavier than peridotites, consistent with a quite small fractionation during melting. OIB and particularly IAV are more variable than MORB, a consequence of subduction of surficial material and fractionations in subduction zones. Dashed line is the chondritic mean.

be disturbed by weathering. Gaschnig et al. (2021b) measured δ^{98}Mo and (^{234}U)/(^{238}U) activity ratios in samples from Samoa, Pitcairn, and St. Helena and found that samples with the lowest δ^{98}Mo typically show considerable $(^{234}U)/(^{238}U)$ disequilibrium, suggesting that Mo was leached during weathering with preferential loss of heavy isotopes (recall from Chapter 3 that we expect mantle and mantle-derived rocks to be in uranium isotopic equilibrium, i.e. $(^{234}U)/(^{238}U) = 1$). It is consequently unclear how much of the variability in OIB reflects weathering effects, and these will have to be assessed carefully in future studies. Nevertheless, some analyzed samples with $(^{234}U)/(^{238}U)$ at or close to equilibrium have δ^{98}Mo significantly lighter than MORB, suggesting some of the variation does reflect source variability in δ^{98}Mo.

IAV have considerably more variable δ^{98}Mo, ranging from considerably heavier to considerably lighter compositions than MORB, and on average are slightly heavier δ^{98}Mo than MORB, with a mean of −0.026‰ and a standard deviation of 0.18‰. In some cases, such as Martinique in the Lesser Antilles Arc and Serua in the Banda Arc, correlations between δ^{18}O and δ^{98}Mo suggest the assimilation of underlying crust (see Section 8.6.2), but the majority of the variation appears to be due to contributions from the subducting slab including both sediments and fluids derived from the igneous and mantle parts of the slab.

Radiogenic isotope ratios indicate that the sediment contribution is minimal in the Mariana (δ^{98}Mo = −0.04‰) and Izu (δ^{98}Mo = −0.03‰) arcs, both of which have average δ^{98}Mo that is isotopically heavier than MORB. A variety of lines of evidence suggest that fluids derived from the subducting oceanic crust and underlying serpentinized mantle contribute this heavy Mo. Seawater is isotopically heavy with a globally uniform δ^{98}Mo of 2.34 ±0.1‰, and extensive hydration and serpentinization of Mariana forearc mantle seawater results in serpentinite mud volcanoes with δ^{98}Mo up to ~0.5‰. Experimental evidence indicates that Mo becomes highly fluid-mobile under oxidization conditions and will partition into the fluid during slab dehydration. Villalobos et al. (2020) and Li et al. (2021) argue that the fluids released from serpentinized mantle both above and below the subducting oceanic crust carry this isotopically heavy Mo into the magma genesis zone. Exhumed subducted oceanic crust in the form of blueschists and eclogites are strongly depleted in heavy Mo isotopes, consistent with a loss of an isotopically light fluid (Chen et al., 2019).

On the other hand, the Lesser Antilles demonstrates how sediments can contribute isotopically heavy Mo. The northern islands form a tight cluster of δ^{98}Mo ranging from −0.05 to −0.09‰, while the southern islands (St. Vincent, the Grenadines, and Grenada) have heavier Mo (0.1–0.34‰) (Freymuth et al., 2016) as well as distinctly more radiogenic Pb (e.g. White et al., 2017). Both the radiogenic Pb and heavy Mo can be traced to U- and Mo-rich black shales deposited during a Cretaceous oceanic anoxic event (OAE II) and now being subducted beneath the arc with δ^{98}Mo as heavy as 0.67‰ and $^{206}Pb/^{204}Pb$ as high as 27 (Freymuth et al., 2016; Carpentier et al., 2008).

Molybdenite is the primary Mo ore mineral and occurs in a variety of hydrothermal deposits, including pegmatites, and although several other minerals are known in which Mo is a stoichiometric component, economic interest is essentially exclusively focused on molybdenite. Most production comes from Mo porphyry deposits, which form in a very similar manner to Cu porphyry deposits; indeed, there is a continuum between Cu and Mo porphyries such that Mo is often present in economic concentrations in Cu porphyries and vice versa. There is a fairly wide range in δ^{98}Mo among molybdenite in ores and considerable overlap between types, with porphyries tending to be isotopically lighter than others, with an average δ^{98}Mo$_{NIST3134}$ of −0.26‰. Substantial variation occurs even in individual deposits. In the Questa porphyry deposit of northern New Mexico, Greber et al. (2014) found that δ^{98}Mo$_{NIST3134}$ varied from −0.73‰ to +0.15‰ increasing from early to late formed ores. They recognized three processes occurring between about 700 and 350 °C, affecting the Mo isotope composition of molybdenites:

- Minerals preferentially incorporate light Mo isotopes during progressive fractional crystallization in subvolcanic magma reservoirs, with the melt increasingly enriched in heavy Mo isotopes as it evolves.
- Heavy Mo isotope preferentially partitions into fluids exsolved from magma.
- Light Mo isotopes are preferentially incorporated in molybdenite during crystallization from an aqueous fluid producing

Rayleigh-type fractionation such that the hydrothermal fluid that gets heavier with progressive molybdenite crystallization in a manner similar to Cu isotopes.

9.4.8.8   Tin isotopes

Tin has 10 stable isotopes, the most of any element: ^{112}Sn (0.97%), ^{114}Sn (0.66%), ^{115}Sn (0.34%), ^{116}Sn (14.54%), ^{117}Sn (7.671%) ^{118}Sn (24.22%) ^{119}Sn (8.588%) ^{120}Sn (32.60%), ^{122}Sn (4.632%), and ^{124}Sn (5.788%) (Friebel et al., 2020). (Why does tin have some many isotopes? The answer is left to you as Problem 9.8.) Tin is moderately volatile, and like most transition metals, it is siderophile and redox sensitive with principal valances of 2+ and 4+ with both valance states likely present in the mantle. Their difference in ionic size results in difference in coordination numbers of stannic (Sn^{4+}) and stannous (Sn^{2+}) tin and hence bond stiffness and strength, which in turn leads to differences in isotopic fractionation between the two oxidation states.

High-precision analyses of meteorites and igneous rocks have only recently been reported, and there is as yet not uniformly agreed up reference standard; however, this will likely become NIST SRM 3161a, whose isotopic composition is listed in Table 5.2. Most data have come from two laboratories, the Institut de Physique du Globe at the Université Paris, which reports the $^{122}Sn/^{118}Sn$ ratio as $\delta^{122/118}Sn$ in per-mil deviations from an in-house standard (Sn_IPGP) and Laboratoire de Géologie de Lyon at ENS Lyon, which reports the $^{124}Sn/^{116}Sn$ as $\delta^{124}Sn$ in per-mil deviations from NIST 3161a. The lack of a common standard makes comparison of the data between these laboratories difficult.

Wang et al. (2021) analyzed four carbonaceous chondrites and found they had identical $\delta^{124}Sn_{NIST3161a}$ within analytical error and an average of 0.04‰. Ordinary chondrites are lighter and more variable with an average $\delta^{124}Sn_{NIST3161a}$ of −0.71‰ and standard deviation of 0.65‰ and are uncorrelated with group, petrologic grade, or Sn concentration; Creech and Moynier (2019) observed an analogous difference between carbonaceous and ordinary chondrites in $\delta^{122/118}Sn$. They found that enstatite chondrites had higher $\delta^{122/118}Sn$ than ordinary ones, but still lower on average than carbonaceous chondrites by

0.35‰. Fractionation appears to be entirely mass dependent as all the meteorites define a trend of $\delta^{124/118}Sn = \delta^{124/116}Sn \times 0.743 \pm 0.04$, consistent with theoretical expectation.

Four peridotites analyzed by Wang et al. (2018) range from $\delta^{124}Sn_{NIST3161a} = -1.04$ to 0.07‰, the most primitive of which has a value of −0.08±0.11‰. Two analyzed MORB are identical within analytical error with $\delta^{124}Sn_{NIST3161a}$ of −0.01‰ and 0.06‰. Basalts from Hawaii have average $\delta^{124}Sn_{NIST3161a}$ of 0.21‰; Azores basalts are lighter with average $\delta^{124}Sn_{NIST3161a}$ of 0.08‰. This difference is statistically significant, suggesting that tin in the mantle is isotopically heterogeneous. Basalts from Pitcairn, Society Islands, and the Galapagos fall within this range. Tin behaves as in incompatible element in the mantle and mantle-derived rocks with an approximately constant Sn/Sm ratio of 0.32 (Jochum et al., 1993). Consistent with this, the basalts have Sn concentrations about an order of magnitude higher than peridotites. Based on experimentally measured force constants, Roskosz et al. (2020) predicted a fractionation factor $\Delta^{124/116}Sn$ between melt and solid of ~0.1‰ for moderate amounts of melting (~0.1%) with the melt being enriched in heavy tin. Basalts indeed have heavier Sn than peridotites, consistent with this. Sn^{4+} readily substitutes for Ti^{4+} in ilmenite where it is in sixfold coordination. Badullovich et al. (2017) found that $\delta^{122/118}Sn$ remained constant in the Kilauea Iki crystallization sequence until ilmenite began to decrease, at which point $\delta^{122/118}Sn$ decreased with the removal of heavy Sn by ilmenite.

Considering both the data on basalts and peridotites, the Sn isotopic composition of the silicate Earth appears to better match that of carbonaceous chondrites than ordinary or enstatite ones. Kubik et al. (2021) found in metal–silicate melting experiments at 2 GPa that heavy tin isotopes preferentially partition into the metal phase. While the fractionation factor decreases with increasing temperature, $\Delta^{122/118}Sn_{met-sil}$ remains significant (0.22‰) even at 2300 °C. This suggests that core formation would result in the silicate Earth being lighter than the bulk Earth, which further eliminates ordinary and enstatite chondrites as the source of terrestrial tin. They argue that much of the Earth's tin and other volatile elements were added to the Earth by carbonaceous chondrite-like material after core formation.

9.4.8.9 Mercury isotopes

Mercury has seven stable isotopes: ^{196}Hg (0.15%), ^{198}Hg (9.97%), ^{199}Hg (16.87%), ^{200}Hg (23.1%), ^{201}Hg (13.18%), ^{202}Hg (29.86%), and ^{204}Hg (6.87%). Mercury is redox sensitive metal and is present in the silicate Earth in the Hg(II) and less commonly Hg(I) valance state but can be present in the atmosphere, aqueous solution, ore deposits, and the core as Hg^0. Mercury is also volatile and strongly chalcophile as well as siderophile. As we found in Chapter 5, the magnetic isotope effect, a consequence of the nuclear magnetic moment of nuclei with odd numbers of nucleons, and the nuclear field shift effect, a consequence of the wave functions of inner electrons overlapping with that of the nucleus, result in MIFs that can produce a pattern of odd–even isotopic fractionations. The former is most apparent in reactions involving spin conversion, most notably photochemical redox reactions, which are responsible for much of the Hg isotopic variation observed in the exogene (Chapter 11). Mercury also experiences mass-dependent fractionations (MDFs).

Because of the former, Hg isotopic variations cannot be fully characterized by a single isotope ratio. By convention, ^{198}Hg is taken as the normalizing isotope and variations are expressed as:

$$\delta^{xxx}Hg = \left[\frac{\left(^{xxx}Hg/^{198}Hg\right)_{sample} - \left(^{xxx}Hg/^{198}Hg\right)_{std}}{\left(^{xxx}Hg/^{198}Hg\right)_{std}} \right] \times 1000.$$

(9.10)

The conventional standard is NIST-SRM 3133 whose isotopic composition is listed in Table 5.2. MDFs are generally reported as δ^{202}Hg. The ^{199}Hg/^{198}Hg ratio is often used to characterize MIFs, which are reported using the same Δ notation as for oxygen and sulfur, which is the difference between the observed fractionation and the predicted mass-dependent one, e.g.:

$$\Delta^{199}Hg = \delta^{199}Hg - 0.2520 \times \delta^{202}Hg. \quad (9.11)$$

Similar equations for other isotopes can be written with other constants: ^{200}Hg: 0.5024, ^{201}Hg: 0.7520, and ^{204}Hg: 1.493. At the Earth's surface, mercury undergoes extensive MDF and MIF of >8‰ for δ^{202}Hg and >10‰ for Δ^{199}Hg (Blum et al., 2014).

Chondrites have surprisingly variable Hg concentrations, ranging from 8 ppb in some ordinary chondrites and achondrites to 23 000 ppb in CI1 Orgueil (Meier et al., 2016; Moynier et al., 2020). Much of this range, 12 to 23 000 ppb, occurs in carbonaceous chondrites alone. Concentrations are lower on average in ordinary chondrites and lower still in enstatite chondrites but nevertheless variable. Given the ubiquity of mercury in the environment, including the atmosphere, much of it anthropogenic, one might suspect that some of this variation might be due to terrestrial contamination. Meier et al. (2016) investigated this possibility and found no evidence of it, at least for recent falls and well-preserved Saharan finds; indeed, samples deliberately exposed to the atmosphere over months lost Hg rather than absorbing it from the atmosphere. They also found no correlation between concentration and isotopic composition.

δ^{202}Hg in carbonaceous chondrites shows a large range of −7.13 to −1.01‰, with an average of −3.49‰ (Meier et al., 2016; Moynier et al., 2020). There are no resolvable MIF in even numbered isotopes but Δ^{199}Hg and Δ^{201}Hg vary from −0.07 to 0.31‰ and −0.09 to 0.16‰, respectively and are strongly correlated (Figure 9.31). ^{199}Hg through ^{202}Hf are produced by both the s- and r-processes, while ^{196}Hg is a p-process only nuclide and ^{204}Hg is primarily an r-process nuclide. The absence of resolvable variation in Δ^{196}Hg and Δ^{204}Hg led Meier et al. (2016) to conclude that none of the observed isotopic variations are nucleosynthetic. Ordinary chondrites have similarly variable δ^{202}Hg, ranging from −6 to −1.19‰ with a mean of −2.27‰. Enstatite chondrites have less variable δ^{202}Hg, ranging from −2.6 to −4.4‰ with a mean of −3.58‰. Enstatite chondrites exhibit MIF variations of odd isotopes similar to those of carbonaceous chondrites. MIF variations among ordinary chondrites are more subdued, and only 2 of the 10 analyzed specimens have resolvable anomalies.

The positive anomalies in odd-numbered are consistent with the nuclear effects mentioned above, and the relationship between Δ^{199}Hg and Δ^{201}Hg can constrain the nature of the

(a)

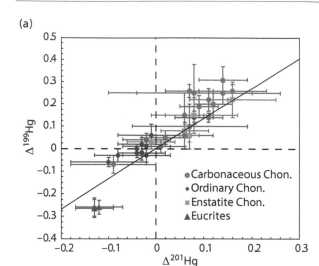

(b)

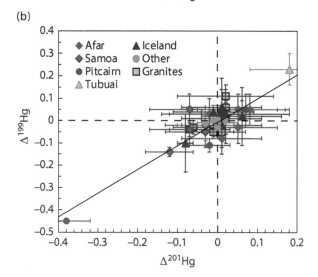

Figure 9.31 (a) $\Delta^{199}Hg$ versus $\Delta^{201}Hg$ in chondrites and eucrites. Most, but not all, ordinary chondrites have $\Delta^{199}Hg$ and $\Delta^{201}Hg$ within error of 0‰. Carbonaceous chondrites define a slope of 1.50±0.15, consistent with liquid–vapor equilibrium. (b) $\Delta^{199}Hg$ versus $\Delta^{201}Hg$ in igneous rocks. Most have $\Delta^{199}Hg$ and $\Delta^{201}Hg$ within error of 0, but the data nevertheless define a slope of 1.06±0.1. Source: Data from (Moynier et al., 2020, 2021) and (Meier et al., 2016).

process inducing them. The $\Delta^{199}Hg$–$\Delta^{201}Hg$ slope defined by carbonaceous chondrites in Figure 9.31 is 1.50±0.15. This is close to the slope observed in vapor–liquid equilibrium experiments by Ghosh et al. (2013) of 1.59 ±0.05 and the theoretically expected one for the nuclear volume effect. Kinetic magnetic

isotope effect fractionations should result in lower slopes. A somewhat lower slope is found when all the meteorite data are included, suggesting that kinetic effects are involved as well (Moynier et al., 2020). Three analyzed eucrites (from Vesta) have distinctly higher $\delta^{202}Hg$ and strong odd isotope depletions. These features suggest evaporative loss of Hg, perhaps during a magma ocean phase.

Owing to very low Hg concentrations in them, there are as yet no Hg isotope data on peridotites or MORB, but some data now exist on other igneous rocks including OIB. Hg isotopic variations in igneous rocks are considerably smaller than those produced by surficial processes but are nevertheless significant. Basalts have an average $\delta^{202}Hg$ of –1.66‰ with a standard deviation of 0.81‰ and, although within the range of carbonaceous and ordinary chondrites, the basalt mean is statistically significantly heavier than that of carbonaceous and enstatite chondrites although not for ordinary chondrites. Most basalts have $\Delta^{199}Hg$ and $\Delta^{201}Hg$ within analytical error of 0, but there are several exceptions most notably basalts from Pitcairn, which have negative $\Delta^{199}Hg$ and $\Delta^{201}Hg$, and Tubuai, which has positive $\Delta^{199}Hg$ and $\Delta^{201}Hg$. Tubuai is one of the Austral Islands with a strong HIMU radiogenic isotope signature while Pitcairn has a strong EM I signature. The Tubuai basalt also has the lightest $\delta^{202}Hg$ (–3.23‰), while the Pitcairn basalts are the heaviest. Furthermore, $\Delta^{199}Hg$ correlates positively with $^{87}Sr/^{86}Sr$ in Pitcairn lavas, confirming that this reflects mantle heterogeneity (Moynier et al., 2021). Despite overall averages of $\Delta^{199}Hg$ and $\Delta^{201}Hg$ being within error of 0, the data define a statistically significant correlation with a slope of 1.06±0.1, notably shallower than in chondrites, and intercept of 0. A slope of 1 is consistent with fractionations associated with photochemical reduction of Hg^2 to Hg^0 and most surficial material plots along this tend (Blum et al., 2014). While the correlation is clearly controlled by the extreme points, it is nevertheless apparent in that the remaining basalt data scatter about this trend. This hints that the MIF fractionations seen in these OIB originated at the Earth's surface or in the atmosphere.

Granitic rocks, including granites and rhyolites, show greater variability in $\delta^{202}Hg$ (–1.2 to –4.95‰) and are lighter on average

(−2.9‰) than basalts but are more closely grouped about $\Delta^{199}Hg = \Delta^{201}Hg = 0$, with slightly more scatter in $\Delta^{199}Hg$ Smith et al. (2008, Moynier et al., 2020, 2021).

While to date there are no data on subduction-related volcanics, Deng et al. (2020) analyzed Hg isotope ratios in epithermal gold deposits, many of which have formed in association with subduction zone magmatism. Those from subduction zone settings have $\delta^{202}Hg$ ranging from −2.2‰ to ~0‰ with generally positive $\Delta^{199}Hg$ ranging up to 0.27‰. In contrast, deposits unrelated to subduction zones have $\delta^{202}Hg$ close to 0 or positive and $\Delta^{199}Hg$ close to 0 or negative. They also analyzed the volcanic wall rock in several cases and found that the isotopic composition was similar to that of the ore. This is consistent with the finding of Smith et al. (2008) that while there is isotopic fractionation of mercury in transport and deposition in ore deposits, there is little fractionation in release of Hf from source rocks. As seawater and most marine sediments show MIF enrichment in odd isotopes, Deng et al. argued that the isotopic composition of Hg in subduction-related ore deposits was inherited from the subduction of marine sediments.

While the data are as yet quite limited, it appears that the mantle is isotopically lighter than Hg at the Earth's surface (average $\delta^{202}Hg \approx -0.62$‰; Blum et al., 2014) but heavier than chondrites. Moynier et al. (2021) point out that those basalts with highest $^3He/^4He$ tend to have $\Delta^{199}Hg$ and $\Delta^{201}Hg$ close to 0, suggesting that the Earth has a whole lacks an MIF signature unlike carbonaceous and enstatite chondrites but like ordinary ones.

9.4.8.10 Thallium isotopes

Thallium has two isotopes: ^{203}Tl and ^{205}Tl. The latter is the decay product of the extinct radionuclide, ^{205}Pb, which has a half-life of 15 million years, the presence of which in early solar system is demonstrated by correlations between $^{205}Tl/^{203}Tl$ and $^{204}Pb/^{203}Tl$ in meteorites that indicate an initial $^{205}Pb/^{204}Pb$ of ~1×10^{-3} (Baker et al., 2010). ^{205}Pb has long since decayed away of course so that interest in terrestrial variations in $^{205}Tl/^{203}Tl$ is focused on mass fractionation. Thallium, like its neighbor mercury, is both volatile and chalcophile. It has two common valance states, 1+ and 3+,

with the reduced form dominant in the mantle. The large ionic radius of Tl^{1+} makes it a highly incompatible element in the mantle. Also like mercury, the nuclear volume effect plays a role in Tl isotopic fractionation, although with only two isotopes, it is not possible to disentangle MDF and MIF. The surprisingly large range of $\varepsilon^{205}Tl$ in terrestrial materials, about 30, for a mass difference of only 1% indicates that the nuclear volume effect plays an important role, but since both the isotopes of Tl are odd, nuclear and bond-strength fractionations are not readily distinguishable. Isotope ratios are normally reported as $\varepsilon^{205}Tl$ as parts in 10 000 deviations from the NIST-SRM 997 standard $^{205}Tl/^{203}Tl$ ratio, which, like so many standards, is now exhausted. NIST-SRM 997 has a $^{205}Tl/^{203}Tl$ ratio of 2.38714±0.001.

$\varepsilon^{205}Tl$ in carbonaceous chondrites ranges from −0.4 to −4 with an average of −1.74, and much of this range is due to the decay of ^{205}Pb (Baker et al., 2010). The correlation of $\varepsilon^{205}Tl$ with $^{204}Pb/^{203}Tl$ indicates an initial solar system $\varepsilon^{205}Tl$ of −7.6. Ordinary and enstatite chondrites show a much larger range of −21.3 to 22.2. However, the range is defined by just a few outliers; these extreme values cannot be explained by ^{205}Pb decay and instead reflects fractionation during thermal processing and shock (Palk et al., 2018). Average $\varepsilon^{205}Tl$ in enstatite chondrites is 1.4, and those of L and LL chondrites are −4.9 and −1.4, respectively; excluding the outliers, the averages are −2.8 and −0.8, values similar to that of carbonaceous chondrites. While it is not possible to entirely separate variation due to ^{205}Pb decay from fractionation, the similarity of the mean $\varepsilon^{205}Tl$ in all three classes of chondrites once outliers are excluded indicates that the present solar system $\varepsilon^{205}Tl$ is close to −1.7.

Tl isotope data on peridotites are quite limited due to the low concentrations and range from −0.2±0.8 for a harzburgite from the Eifel region of Germany to −9.9 to −1.9 in metasomatized xenoliths in South African kimberlites, but the vast majority of the latter fall in the narrow range of −2.3 to −3.4.

Tl isotope data on MORB are similarly limited: five analyses fall in the range of −0.9 to −2.5 with an average of −2. On the other hand, there is an extensive data set of $\varepsilon^{205}Tl$ in OIB, which show a surprisingly large range of −10

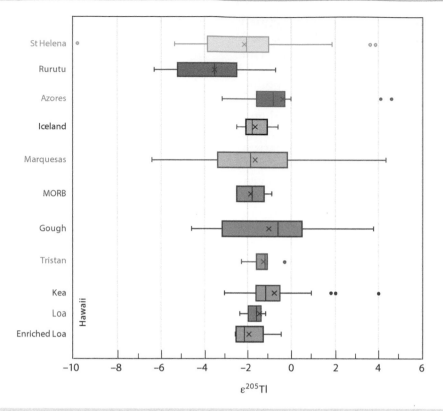

Figure 9.32 Box and whisker plot of ε^{205}Tl in OIB and MORB. Box represents the data falling in the second and third quartiles. Extreme outliers are shown as individual points. Cross shows the value of the mean; line shows the mode.

to +5 (Brett et al., 2021 and references therein), which is nearly half the total terrestrial range although most of the data fall between −0.7 and −1.7, with a mean of −1.5‰, just slightly heavier than MORB. There are some systematic differences (Figure 9.32); for example, Williamson et al. (2021) found that the Hawaiian Kea chain is statistically significantly heavier than the Loa chain, and the Cook-Austral Island of Rurutu has statistically significantly lighter Tl than the Azores, although there are nevertheless broad overlaps. Furthermore, there are no correlations with radiogenic isotope ratios that might help establish relationships with distinct mantle sources (Brett et al., 2021). Several studies have found that the effects of partial melting and fractional crystallization on Tl isotopes are likely negligible. Thallium forms sufficiently volatile chloride complexes that there can be some preferential loss of ^{203}Tl during magmatic degassing, but this effect is likely small in most

cases (Nielsen et al., 2021). Other possible explanations include fractionation during weathering, contamination by or assimilation of Mn–Fe oxides, which are enriched in Tl and can have ε^{205}Tl> 10, and assimilation of altered oceanic crust, which can have ε^{205}Tl < −10, but mantle Tl isotope heterogeneity is likely the main cause.

IAV exhibit an even greater range of ε^{205}Tl, from −11.5 to +12.3, with an overall average of −0.4‰, slightly heavier than MORB (Figure 9.33). Just as for OIB, degassing, assimilation, etc., may play some role in producing this isotopic diversity, but most of it likely reflects contributions from subducted altered oceanic crust and sediment. Pelagic clays have a typical ε^{205}Tl of +3 to +5 and concentrations in the range of 1–5 ppm, compared to a concentration in the mantle of ~1 ppb and in MORB of ~10 ppb. (Nielsen et al., 2017). Biogenic sediments, including carbonates, have variable ε^{205}Tl with far lower concentrations

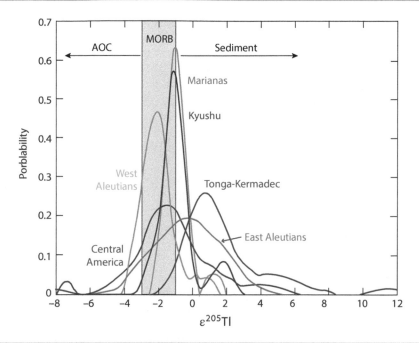

Figure 9.33   Probability distribution of $\varepsilon^{205}Tl$ in IAV. Shaded area represents the estimated composition of the mantle Arrows indicate the expected shifts due to contributions from subducted sediment and altered oceanic crust (AOC).

and terrigenous and volcanogenic sediment typically have lower $\varepsilon^{205}Tl$ and lower Tl concentrations. Nielsen et al. (2006) found that the upper ~600 m of the oceanic crust at ODP Hole 504B, which has undergone low-temperature interaction with seawater, has very light Tl, typically with $\varepsilon^{205}Tl$ of ~–7 but ranging as low as $\varepsilon^{205}Tl$ = –16 and is variably, in some cases strongly, enriched in Tl with concentrations ranging into the hundreds of ppb. In contrast, high-temperature hydrothermal reactions in the lower oceanic crust appears to simply extract Tl with little isotopic fractionation. Thus, the subduction of oceanic crust and sediment potentially supplies the subarc mantle with isotopically diverse Tl, although given that the Tl concentration in pelagic clays is an order of magnitude greater than in altered oceanic crust, the former is likely to dominate where it is present.

In the Kyushu and Marianas arcs, isotopic compositions are relatively homogeneous (excluding a single outlier) and slightly heavier on average than MORB, suggesting a modest sedimentary influence. The Tonga-Kermadec Arc is more diverse and heavier on average and rises to some extreme values, perhaps related to contributions from Fe–Mn crusts form on the Louisville Seamounts (Nielsen

et al., 2017). Central America is even more diverse, and this reflects along-arc variations related to variation in amount and kind of sediment being subducted (Nielsen et al., 2017). The Aleutians also exhibit an along-arc variation. In the west, where radiogenic isotope ratios indicate essentially no sediment contribution, the modal $\varepsilon^{205}Tl$ is close to MORB values, while in the east where radiogenic Sr and Pb and unradiogenic Nd indicate a sediment contribution of as much as 1%, $\varepsilon^{205}Tl$ rises to values >2 (Nielsen et al., 2016). Subduction into the mantle of this assemblage of material with diverse $\varepsilon^{205}Tl$ values explains much of the variation observed in OIB.

The charge and ionic radius (170 pm) of Tl^{1+} allows it to readily substitute for K and Rb in minerals, where K is a stoichiometric component such as micas and alkali feldspars. As does K, it also readily partitions into hydrothermal fluids. Rader et al. (2018) found that micas were typically enriched in Tl relative to co-existing feldspars and had systematically lower $\varepsilon^{205}Tl$ than those feldspars. They found that sulfide minerals had widely varying Tl concentrations but were typically Tl-poor and had systematically positive $\varepsilon^{205}Tl$.

Baker et al. (2010) found that in the Collahuasi porphyry copper deposit (PCD) of northern Chile ε^{205}Tl varied between −5.1 and +0.1, while Tl concentrations varied from 0.1 to 3.2 ppm. They attributed these variations to hydrothermal transport of Tl. Tl concentrations correlated with both K and Rb concentrations but not with Cu, demonstrating the limited chalcophile character of Tl. Fitzpayne et al. (2018) found that ε^{205}Tl ranged from −16.4 to +7.2 in the Bingham Canyon PCD of Utah, USA, one of the world's largest. Unbrecciated samples proximal to the Cu–Mo mineralization have ε^{205}Tl of −4.2 to +0.9, suggesting that high-temperature (>300 °C) hydrothermal alteration produces only limited Tl isotope fractionation. Larger Tl isotopic fractionations occur in hydrothermal breccias and increase with increasing distance from the main ore body. The greatest fractionation, ε^{205}Tl, from −16.4 to +6.0, occurs in samples collected ~7 km away from the main ore body in disseminated gold deposits, which contain elevated concentrations of Tl as well as Sb and As. This implies that large ε^{205}Tl fractionations can be associated with the migration of low-temperature hydrothermal fluids potentially related to sediment-hosted gold mineralization.

The Lengenbach (Switzerland) Pb–As–Tl–Zn deposit is a rare example of a sulfide deposit where Tl reaches high concentrations; it contains a variety of unusual Tl minerals, many of which are unique to the locality. It is hosted in Triassic dolomite and formed during the Cenozoic Alpine Orogeny when metamorphic temperatures exceeding 500 °C resulted in melting of a preexisting sulfide deposit (likely a sediment-hosted stratiform mineralization). The subsequent fractional crystallization of sulfosalts such as jordanite ($Pb_{14}(As,Sb)_6S_{23}$) from the sulfide melt resulted in the enrichment of As and Tl and the depletion of Pb in the remaining melt fractions. ε^{205}Tl ranges from −4.1± 0.5 to +1.9±0.5 and increases with increasing Tl/As ratios, resulting from the incorporation of the lighter Tl isotope into crystallizing solids and enrichment of the heavier isotope in the remaining sulfide melt. Isotopically heavy Tl also occurs in associated low-temperature hydrothermal mineralization.

9.4.8.11 Uranium isotopes

Uranium has no long-lived stable isotopes of course, but ^{238}U and ^{235}U have sufficiently long half-lives that over geologically short periods they can be considered stable. On longer time periods, we can account for changes in the ^{235}U/^{238}U using (2.1) knowing their half-lives. We have already explored the use of the ^{234}U/^{238}U ratio in geochronology. As we found in Chapters 3 and 5, ^{235}U/^{238}U can vary due to chemical fractionation and, in chondrites, nucleosynthetic contributions including those due to decay of ^{247}Cf, in meteorites. Here, we focus on variation in ^{235}U/^{238}U due to chemical equilibrium and kinetic fractionation. The ^{235}U/^{238}U ratio is reported in δ^{238}U notation as per mil variations from a standard, either CRM 112a (also known as CRM 145, both of which are derived from NIST SRM 960 and should have identical isotopic composition to it) or NIST SRM 950, although in some cases it is reported in epsilon notation in parts in 10 000 deviations. Uranium is lithophile and incompatible in the mantle. It has two common oxidation states, with U(VI) dominant in oxidizing environments at the Earth's surface and U(IV) dominant in reduced environments and in the mantle. Hexavalent U readily forms the uranyl UO_2^{2+} oxy-cation and soluble hydroxy- and carbonate complexes, whereas tetravalent U is insoluble. As we might expect, the largest fractionations occur in association with redox reactions. As with other heavy elements, nuclear field shift effects are important, particularly since ^{238}U is an even nuclide, whereas ^{235}U is an odd one. The nuclear field shift effect can overwhelm mass-dependent effects; for example, it results in the heavy isotope being favored in the lower oxidation state, in contrast to the expectation for mass-dependent fractionation. Nuclear field shift effects become relatively more important with increasing temperatures as they scale with $1/T$ rather than $1/T^2$.

Despite variations in ^{238}U/^{235}U in components within meteorites such as CAIs, Goldman et al. (2015) found that 20 of 27 chondrites have bulk ^{238}U/^{235}U falling within the narrow range of 137.778–137.803, corresponding to $\delta^{238}U_{CRM112a}$ = −0.33 to −0.52‰ ($\delta^{238}U_{NIST950a}$ = −0.35 to −0.54‰), as do chondrites analyzed by Andersen et al. (2015).

The mean of all analyzed chondrites excluding outliers is $\delta^{238}U_{CRM112a}$ = −0.36‰ (±0.02‰ 2 standard errors). Goldman et al. also reported the average of 14 terrestrial basalts (mostly standards), as $\delta^{238}U_{CRM112a}$ = −0.39‰, indistinguishable from the chondritic average. Based on observed ($^{234}U/^{238}U$) disequilibrium in some meteorites indicative of recent disturbance, Andersen et al. argued for a lower chondritic $\delta^{238}U_{CRM112a}$ value of −0.306±0.026‰.

MORB are significantly heavier than chondrites with average $\delta^{238}U_{CRM112a}$ of −0.27‰ and are homogeneous with a standard deviation of only 0.004‰ (Figure 9.34). Nineteen OIB analyses reported by Andersen et al. (2015) and the Kilauea Iki value reported by Gaschnig et al. (2021a) have an average $\delta^{238}U_{CRM112a}$ of −0.307‰ with a standard deviation of 0.008‰. IAV from the Marianas and Izu arcs are distinctly lighter, with an average $\delta^{238}U_{CRM112a}$ of −0.39‰ and −0.41, respectively, and more variable ranging from −0.32 to −0.48‰. Excluding subduction zones,

uranium in the mantle is exclusively in the tetravalent state and temperatures exceed 1000 °C, so that little fractionation during magmatic evolution is expected. This is consistent with the observation Gaschnig et al. (2021a) who found no fractionation of U isotopes during the fractional crystallization of the Kilauea Iki lava lake. It is also consistent with the similarity of $^{238}U_{CRM112a}$ in silicic igneous rocks, with an average $\delta^{238}U_{CRM112a}$ of −0.27‰, close oceanic basalts, although the former are more variable (Tissot and Dauphas, 2015). Consequently, the difference between OIB and MORB must reflect fractionation at low temperature near or at the surface of the Earth.

Strong fractionation can occur during seawater–basalt interaction leading to extremely variable $\delta^{238}U$ (−0.63‰ to +0.27‰) in altered oceanic crust. Andersen et al. (2015) found that samples from the uppermost section (0–110 m) of oceanic crust at ODP Site 801 display slightly lower $\delta^{238}U$ (~ −0.44 ‰) than seawater (which has $\delta^{238}U$ = 0.39 ± 0.01‰) and

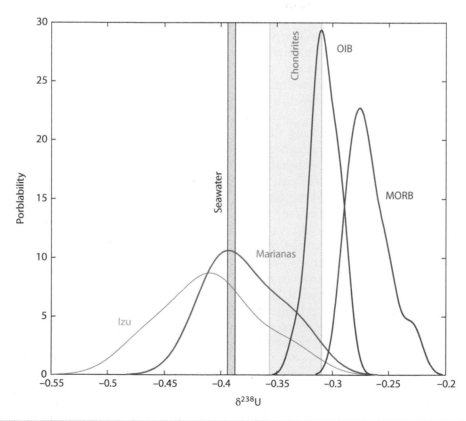

Figure 9.34 Probability distribution of $\delta^{238}U$ in MORB, OIB, and IAV from the Marianas and Izu arcs. MORB have distinctly homogeneous $\delta^{238}U$, while IAV show considerably greater variability. Source: Data from (Andersen et al., 2015) and (Freymuth et al., 2019).

that samples from two deeper sections (between 110–420 m) show significantly higher $\delta^{238}U$ with average values of +0.16‰ and –0.14‰ with U being enriched by a factor of 5–10 over fresh MORB. These data imply that in the shallowest oceanic crust, U is extracted from seawater and incorporated into oceanic crust and occurs without a significant redox change, resulting in little U isotope fractionation. At deeper levels, reducing conditions prevail and favor incorporation of ^{238}U as U^{4+} in the crust. Andersen et al. (2015) estimated the integrated $\delta^{238}U$ over the full upper 500 m of oceanic crust at ODP Site 801 to be = –0.17‰. Previous studies of high-temperature hydrothermal fluids have shown that uranium is nearly quantitatively removed from seawater in these hydrothermal systems. Noordmann et al. (2016) found that two low-temperature fluids had $\delta^{238}U$ of –0.55 and –0.42‰. Four sampled high-temperature fluids had variable $\delta^{238}U$ (–0.59 to –0.28‰) but with an average indistinguishable from seawater. Studies of deep drill sections show that the addition of seawater U to the oceanic crust is restricted to the upper 500–1000 m. Since the oceanic crust is almost invariably subducted, seawater–basalt reactions ultimately produce a flux of U from the oceans, and ultimately the continents, to the mantle, shifting the mantle U/Th ratio as well as $\delta^{238}U$ (Andersen et al., 2015).

Pelagic sediments have $\delta^{238}U$ clustering about the seawater value, although ferromagnesian crusts and nodules are lighter and terrigenous sediments somewhat heavier. In the case of the Marianas, Andersen et al. (2015) estimated the $\delta^{238}U$ of subducting sediment as –0.35‰. Altered oceanic crust, although variable, appears to have U somewhat heavier than MORB. Consequently, the extremely light U found in some lavas from the Marianas, and particularly the Izu Arc, cannot be explained simply by the incorporation of some mixture of altered oceanic crust and sediment, implying that U must be isotopically fractioned in subduction zones. We found in Chapter 8 that $(^{238}U/^{230}Th)$ systematics in IAV provides evidence that U is enriched relative to Th in the mantle wedge where these magmas are generated, and the likely mechanism is the preferential extraction of U from the subducting oceanic crust by hydrous fluids. Freymuth et al. (2019) have suggested that oxidized fluids originate in the serpentinized mantle of the subducting plate and preferentially oxidize and leach ^{235}U as they rise through the oceanic crust. This isotopically light U is then delivered to the magma genesis zone in the mantle wedge. This leaves the residual subducting oceanic lithosphere to carry isotopically heavy U into the mantle.

The U isotope fractionations occurring in the oceanic crust and subduction zones are driven by the contrasting redox state of the mantle and the Earth's surface. In particular, seawater is comparatively U-rich because it is oxidized. As we will discuss in Chapter 12, free oxygen appeared in the Earth's atmosphere and surface ocean only in the earliest Proterozoic, roughly 2500–2300 million years ago, in the Great Oxidation Event. However, redox-sensitive metal stable isotope ratios, including those of Cr, Mo, and U, indicate that atmospheric oxygen levels through most of the Proterozoic, although perhaps variable, were a small fraction of present atmospheric levels (e.g. Kendall et al., 2015). Through this period, most of the deep ocean remained anoxic and, consequently, U-depleted. In the absence of redox-driven fractionation, Andersn et al. (2015) argue that oceanic lithosphere subducted prior to the latest Proterozoic retained an approximately chondritic $\delta^{238}U$. In their model, the OIB reservoirs contain only the more anciently recycled material, consistent with ancient Pb–Pb pseudo-isochrons and MIF sulfur, while Phanerozoic subduction has polluted the MORB source, driving it to heavier $\delta^{238}U$.

9.5   OXYGEN ISOTOPES IN HYDROTHERMAL SYSTEMS

9.5.1   Ridge crest hydrothermal activity and metamorphism of the oceanic crust

Early studies of "greenstones" dredged from mid-ocean ridges and fracture zones revealed that they were depleted in ^{18}O relative to fresh basalts. Partitioning of oxygen isotopes between various minerals, such as carbonates, epidote, quartz and chlorite, in these greenstones suggested that they had equilibrated at about 300 °C (Muehlenbachs and Clayton, 1972). This was the first, but certainly not the only, evidence that the oceanic crust underwent hydrothermal metamorphism at depth. Other clues included highly variable heat flow at ridges and an imbalance in the Mg fluxes in the ocean. Nevertheless, the importance of hydrothermal processes was not generally recognized until the discovery of low-temperature (~20 °C) vents on the Galapagos Spreading Center in 1977 and high-temperature (350 °C) "black smokers" on the East Pacific Rise in 1979. Various pieces of the puzzle then began to fall rapidly into place, and it was soon clear that hydrothermal activity was a very widespread and important phenomenon. Most of the oceanic crust is affected to some degree by this process, which also plays an important role in controlling the composition of seawater.

Hydrothermal metamorphism occurs because seawater readily penetrates the highly fractured and, therefore, permeable oceanic crust. A series of chemical reactions occurs as the seawater is heated, transforming it into a reduced, acidic, and metal-rich fluid. Eventually, the fluid rises and escapes, forming the dramatic black smokers. Fluid in many "black smokers" vents at temperatures of about 350 °C.[3] This results from the density and viscosity minimum that occurs close to this temperature at pressures of 200–400 bars combined with a rapidly decreasing rock permeability above these temperatures.

Seawater entering the oceanic crust has a $\delta^{18}O_{SMOW}$ of 0; fresh igneous rock has a $\delta^{18}O$ of ~ +5.7. As seawater is heated, it will exchange O with the surrounding rock until equilibrium is reached. At temperatures in the range of 300–400 °C and for the mineral assemblage typical of greenschist facies basalt, the net water–rock fractionation is small, 1‰ or 2‰. Thus, isotopic exchange results in a decrease in the $\delta^{18}O$ of the rock and an increase in the $\delta^{18}O$ of the water. Surprisingly, there have only been a few oxygen isotope measurements of vent fluids; these indicate $\delta^{18}O$ of about +2.

At the same time as hydrothermal metamorphism occurs deep in the crust, low-temperature weathering proceeds at the surface. This also involves isotopic exchange. However, for the temperatures (~2 °C)

and minerals produced by these reactions (smectites, zeolites, etc.), fractionations can be quite large, of the order of 20‰. The result of these reactions is to increase the $\delta^{18}O$ of the shallow oceanic crust and decrease the $\delta^{18}O$ of seawater. Thus, the effects of low- and high-temperature reactions are opposing.

Muehlenbachs (1976) and Muehlenbachs and Furnas (2003) argued that these opposing reactions actually buffered the isotopic composition of seawater at a $\delta^{18}O$ of ~0. According to them, the net of low- and high-temperature fractionations was about +6, just the observed difference between the oceanic crust and the oceans. Thus, the oceanic crust ends up with an average $\delta^{18}O$ value about the same as it started with, and the net effect on seawater must also be close to zero. Could this be coincidental? One should always be suspicious of apparent coincidences in science and they were.

Let us think about this a little. Assume that the net fractionation is 6, but suppose that the $\delta^{18}O$ of the ocean was −10 rather than 0. What would happen? Assuming a sufficient amount of oceanic crust available and a simple batch reaction with a finite amount of water, the net of high- and low-temperature basalt–seawater reactions would leave the water with $\delta^{18}O$ of −10 + 6 = −4. Each time a piece of oceanic crust is allowed to equilibrate with seawater, the $\delta^{18}O$ of the ocean will increase a bit. If the process is repeated enough, the $\delta^{18}O$ of the ocean will eventually reach a value of 6 − 6 = 0. Actually, what is required of seawater–oceanic crust interaction to maintain the $\delta^{18}O$ of the ocean at 0‰ is a net increase in the isotopic composition of seawater by perhaps 1–2‰. This is because low-temperature continental weathering has the net effect of decreasing the $\delta^{18}O$ of the hydrosphere. This is what Muehlenbachs and Clayton proposed.

The "half-time," defined as the time required for the disequilibrium to decrease by half, for this process, has been estimated to be about 46 Ma. For example, if the equilibrium value of the ocean is 0‰ and the actual value is −2‰, the $\delta^{18}O$ of the ocean would

---

[3] While this is typical, temperatures of 400 °C or so have also been found. Most low-temperature vents waters, such as those on the GSC, appear to be mixtures of 350 °C hydrothermal fluid and ambient seawater, with mixing occurring at shallow depth beneath the seafloor. Although hydrothermal fluids with temperatures substantially above 400 °C have not been found, there is abundant evidence from metamorphosed rocks that water–rock reactions occur at temperatures up to 700 °C.

increase to −1‰ in 46 Ma. It would then require another 46 Ma to bring the oceans to a $\delta^{18}O$ of −0.5‰, and so on. Over long time-scales, this should keep the isotopic composition of oceanic crust constant. We will see in Chapter 12 that the oxygen isotopic composition of the ocean varies on times scales of thousands to tens of thousands of years due to changes in ice volume. The isotopic exchange with oceanic crust is much too slow to dampen these short-term variations.

This idea, that water–rock interaction produces no net change in the isotopic composition of the oceanic crust or seawater, was apparently confirmed by the first thorough oxygen isotope study of an ophiolite by Gregory and Taylor (1981). Their results for the Samail Ophiolite, an area of uplifted and exposed continental crust in Oman, are shown in Figure 9.35. As expected, they found that the upper part of the crust had higher $\delta^{18}O$ than fresh MORB, while the lower part of the section had $\delta^{18}O$ lower than MORB. Their estimate for the $\delta^{18}O$ of the entire section was +5.8, which is essentially identical to fresh MORB. At ODP Site 1256 in the equatorial eastern, the oceanic crust was cored to a depth of 1500 m, just penetrating into the gabbroic layer. Oxygen isotope ratios there show much the same pattern as the Samail ophiolite (Figure 9.36), with heavy oxygen in the volcanic zone and light oxygen in the dikes and gabbros. The shift in isotopic composition is coincident with a shift in the style of alteration from low- to high-temperature apparent from petrographic examination.

If the Muehlenbachs and Clayton hypothesis is correct and assuming a steady-state tectonic environment, the $\delta^{18}O$ of the oceans should remain constant over geologic time. Whether it has or not is controversial. Based on the analyses of marine carbonates, Jan Veizer and his colleagues have argued that it is not (e.g. Veizer et al., 1999). The basis of their argument was that $\delta^{18}O$ in marine carbonates increases over Phanerozoic time (Figure 9.37). More recently, however, Came et al. (2007) have used isotopic clumping analysis to calculate the isotopic composition of seawater from these Silurian and Carboniferous marine carbonates and found that seawater at those times indeed had values close to modern ones, implying that the Muehlenbachs and Clayton hypothesis is correct. We should clarify that

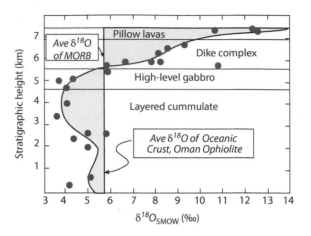

Figure 9.35 $\delta^{18}O$ of the Samail Ophiolite in Oman as a function of stratigraphic height. Source: Gregory and Taylor (1981)/Reproduced with permission from John Wiley & Sons.

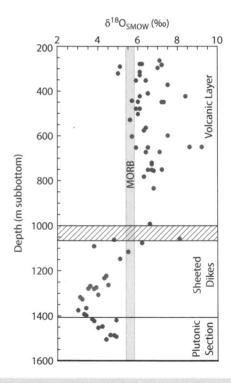

Figure 9.36 Oxygen isotope ratios measured in cores from ODP Site 1256 in the equatorial eastern Pacific. Cross-hatched area is the lava-dike transition zone. The volcanic section has mostly elevated $\delta^{18}O$ as a consequence of low-temperature interaction with seawater, while the dikes and gabbros have mostly lowered $\delta^{18}O$ as a consequence of hydrothermal interaction. Source: Data from Gao et al. (2012).

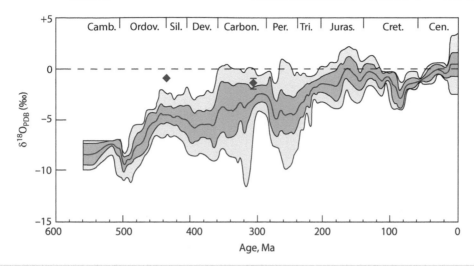

Figure 9.37 $\delta^{18}O_{PDB}$ in marine carbonates (brachiopods, belemites, oysters, and foraminifera shells) over Phanerozoic time. Note that $\delta^{18}O$ is relative to the PDB standard in this figure, whereas it was relative to SMOW in the previous ones. The difference between PDB and SMOW is approximately that of the fractionation factor between seawater and carbonates, so carbonates with $\delta^{18}O_{PDB} = 0$ would have precipitated from seawater with $\delta^{18}O_{SMOW} \approx 0$. Red diamonds are the calculated $\delta^{18}O$ of seawater using clumped isotopic analysis of carbonate shells from (Came et al., 2007). Source: Veizer et al. (1999)/Reproduced with permission from Elsevier.

there is no doubt $\delta^{18}O$ of seawater varies with time; we will find in Chapter 12 that changes in ice volume led to significant oscillations in $\delta^{18}O$ in the Pleistocene. The question is whether there has been a systematic long term increase with time superimposed on climate-driven variations.

9.5.2 Meteoric geothermal systems

Hydrothermal systems occur not only in the ocean, but just about everywhere that magma is intruded into the crust. In the 1950s, a debate raged about the rate at which the ocean and atmosphere were created by the degassing of the Earth's interior. W. W. Rubey assumed that water in hydrothermal systems such as Yellowstone was magmatic and argued that the ocean and atmosphere were created quite gradually through magmatic degassing. Rubey turned out to be wrong about hydrothermal systems. One of the first of many important contributions of stable isotope geochemistry to understanding hydrothermal systems was the demonstration by Craig (1963) that water in these systems was meteoric, not magmatic. The argument is based upon the data shown in Figure 9.38. For each geothermal system, the δD of the "chloride" type geothermal waters is nearly the same as the local

precipitation and groundwater, but the $\delta^{18}O$ is shifted to higher values. The shift in $\delta^{18}O$ results from the "high"-temperature (~300 °C) reaction of the local meteoric water with hot rock. However, because the concentration of hydrogen in rocks is nearly 0 (more precisely because ratio of the mass of hydrogen in the water to mass of hydrogen in the reacting rocks is extremely high), there is essentially no change in the hydrogen isotopic composition of the water. If the water involved in these systems was magmatic, it would not have the same δ^2H as local meteoric water (though it is possible that these systems contain a few percent magmatic water).

Acidic, sulfide-rich water from these systems does have δD that is different from local meteoric water. This shift occurs when hydrogen isotopes are fractionated during the boiling of geothermal waters. The steam produced is enriched in sulfide, accounting for their acidic character. The water then condenses from this steam and mixes with meteoric water to produce the mixing lines observed.

9.5.3 Water–rock reaction: theory

Very often in geology, it is difficult to observe the details of processes occurring today, and our understanding of many of Earth processes

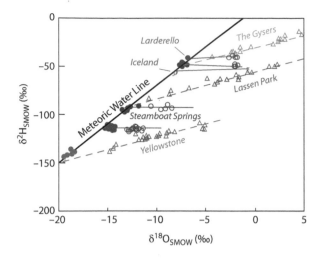

Figure 9.38 δ^2H and $\delta^{18}O$ in meteoric hydrothermal systems. Closed circles show the composition of meteoric water in the vicinity of Yellowstone, Steamboat Springs (Colorado), Mt Lassen (California), Iceland, Larderello (Italy), and The Geysers (California), and open circles show the isotopic composition of chloride-type geothermal waters at those locations. Open triangles show the composition of acidic, sulfide-rich geothermal waters at those locations. Solid lines connect the meteoric and chloride waters; dashed lines connect the meteoric and acidic waters. The "Meteoric Water Line" is correlation between δD and $\delta^{18}O$ observed in global precipitation, which we will consider in the following chapter. Source: (Craig, 1963)/ Reproduced with permission from Consiglio Nazionale delle Ricerche.

comes from observing the effects these processes have had in the past, that is, the record they have left in the rocks. So, it is with hydrothermal systems. In present systems, we can observe only the water venting; we cannot observe the reactions with rocks or the pattern of circulation. However, these, as well as temperatures involved and water–rock ratios, can be inferred from the imprint left by ancient hydrothermal systems.

Estimating temperatures at which ancient hydrothermal systems operated is a fairly straightforward application of isotope geothermometry, which we have already discussed. If we can measure the stable isotopic composition of any two phases that were in equilibrium, and if we know the fractionation factor as a function of temperature for those phases, we

can estimate the temperature of equilibration. We will focus now on water–rock ratios, which may also be estimated using oxygen isotope ratios.

For a closed hydrothermal system, we can write two fundamental equations. The first simply describes equilibrium between water and rock

$$\Delta = \delta_w^f - \delta_r^f \qquad (9.12)$$

where we use the subscript w to indicate water and r to indicate rock. The superscript f indicates the final value. Equation 9.11 just says that the difference between the final isotopic composition of water and rock is equal to the fractionation factor. The second equation is just a statement of mass balance for a closed system: the amount of the isotope present before reaction must be the same as after reaction:

$$c_w W \delta_w^i + c_r R \delta_r^i = c_w W \delta_w^f + c_r R \delta_r^f \qquad (9.13)$$

where c indicates concentration (we assume concentrations do not change, which is valid for oxygen, but not generally valid for other elements), W indicates the mass of water involved, R the mass of rock involved, and the superscript i denotes the initial isotope ratio.

Substituting 9.12 into 9.13 and rearranging, we derive the following equation:

$$\frac{W}{R} = \frac{\delta_r^f - \delta_r^i}{\delta_w^i - \delta_r^f - \Delta} \frac{c_r}{c_w}. \qquad (9.14)$$

The term on the left-hand side, W/R, is the ratio of water to rock in the reaction. Notice that the right-hand side does not include the final isotopic composition of the water, information that we would generally not have. The initial oxygen isotope composition of the water can be estimated in various ways. For example, we can determine the hydrogen isotopic composition (of rocks) and, from that, determine the oxygen isotope composition using the δD–$\delta^{18}O$ Meteoric Water Line. The initial $\delta^{18}O$ of rock can generally be inferred from unaltered samples, and the final isotopic composition of the rock can be measured. The fractionation factor can be estimated if we know the temperature and the phases in the rock. For oxygen, the ratio of concentration in the rock to water will be close to 0.5 in all the cases.

Equation 9.14 is for a closed system, that is, a batch reaction where we equilibrate W grams of water with R grams of rock. That is not very geologically realistic. In fact, a completely open system, where water makes one pass through hot rock, would be more realistic. In this case, we might suppose that a small parcel of water, dW, passes through the system and induces an incremental change in the isotopic composition of the rock, $d\delta_r$. In this case, we can write

$$Rc_r d\delta_r = \left(\delta_w^i - [\Delta + \delta_r]\right)c_w dW. \qquad (9.15)$$

This equation states that the mass of isotope exchanged by the rock is equal to the mass of isotope exchanged by the water (we have substituted $\Delta + \delta_r$ for δ). Rearranging and integrating, we have

$$\frac{W}{R} = \ln\left(\frac{\delta_r^f - \delta_r^i}{\delta_w^i - \delta_r^f - \Delta} + 1\right)\frac{c_r}{c_w}. \qquad (9.16)$$

Thus, it is possible to deduce the water rock ratio for an open system as well as a closed one.

Using this kind of approach, Gregory and Taylor (1981) estimated water–rock ratios of ≤ 0.3 for the gabbros of the Oman ophiolite. This can be done with other isotope systems as well. For example, McCulloch et al. (1981) used Sr isotope ratios to estimate water–rock ratios varying from 0.5 to 40 for different parts of the Oman ophiolite.

9.5.4 The Skaergaard intrusion

A classic example of a meteoric hydrothermal system is the Early Tertiary Skaergaard intrusion in East Greenland. The Skaergaard has been studied for over 80 years as a classic mafic layered intrusion. Perhaps ironically, the initial motivation for oxygen isotopic study of the Skaergaard was the determination of primary oxygen and hydrogen isotopic compositions of igneous rocks. The results, however, showed that the oxygen isotope composition of the Skaergaard has been pervasively altered by hydrothermal fluid flow. This was the first step in another important contribution of stable isotope geochemistry, namely the demonstration that most igneous intrusions have reacted extensively with water subsequent to crystallization.

Figure 9.39 shows a restored cross section of the intrusion with contours of $\delta^{18}O$. There are several interesting features. First, it is clear

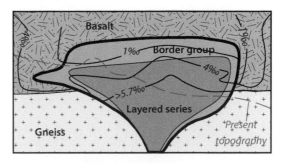

Figure 9.39 Restored cross section of the Skaergaard intrusion with contours of $\delta^{18}O$. Source: Criss and Taylor (1986)/Reproduced with permission from the Mineralogical Society of America.

that the circulation of water was strongly controlled by permeability. The impermeable basement gneiss experienced little exchange, as did the part of the intrusion beneath the contact of the gneiss with the overlying basalt. The basalt, which is typically highly fractured, is quite permeable and allowed water to flow freely through it and into the intrusion. Figures 9.39 reveals zones of low $\delta^{18}O$, which are the regions of hydrothermal upwelling. Water was apparently drawn into the sides of the intrusion and then rose through the hot center. This is just the sort of pattern observed with finite-element models of fluid flow through the intrusion.

Calculated water–rock ratios for the Skaergaard were 0.88 in the basalt, 0.52 in the upper part of the intrusion and 0.003 for the gneiss, demonstrating the importance of the basalt in conduction the water into the intrusion and the inhibiting effect of the gneiss. Models of the cooling history of the intrusion suggest that each cm^3 of rock was exposed to between 10^5 and 5×10^6 cm^3 of water over the 500 000-year cooling history of the intrusion. This would seem to conflict with the water–rock ratios estimated from oxygen isotopes. The difference is a consequence of each cm^3 of water flowing through many cm^3 of rock, but not necessarily reacting with it. Once water had flowed through enough grams of rock to come to isotopic equilibrium, it would not react further with the rock through which it subsequently flowed (assuming constant temperature and mineralogy). Thus, it is important to distinguish between W/R ratios calculated from isotopes, which reveal only the mass

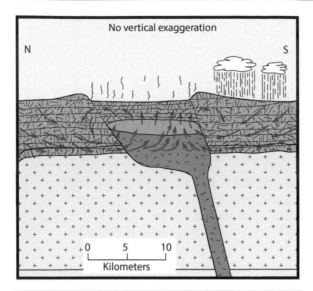

Figure 9.40 Cartoon illustrating the hydrothermal system in the Skaergaard intrusion. Source: Taylor (1974)/with permission of Society of Economic Geologists.

(or molar) ratio of water and rock in the net reaction, to flow models. Nevertheless, the flow models demonstrate that each gram of rock in such a system is exposed to an enormous amount of water. Figure 9.40 shows a cartoon illustrating the hydrothermal system deduced from the oxygen isotope study.

9.5.5 Oxygen isotopes and mineral exploration

A great variety of ores are deposited from aqueous solutions, and oxygen isotope studies can be a valuable tool in mineral exploration. Mineralization is very often (though not exclusively) associated with the region of greatest water flux, such as areas of upward moving hot water above intrusions. Such areas are likely to have the lowest values of $\delta^{18}O$. To understand this, let us solve 9.16, the final value of $\delta^{18}O$:

$$\delta_r^f = \left(\delta_r^i - \delta_w^i - \Delta\right)e^{-(W/R)(c_w/c_r)} + \delta_w^i + \Delta.$$

$$(9.17)$$

If we assume a uniform initial isotopic composition of the rocks and the water, then all the terms on the right-hand side are constants except W/R and Δ, which is a function of temperature. Thus, the final values of $\delta^{18}O$, that is, the values we measure in an area such as the Skaergaard, are functions of the temperature of equilibration and an exponential function of the W/R ratio. Figure 9.41 shows $\delta^{18}O_r^f$ plotted as a function of W/R and Δ, where

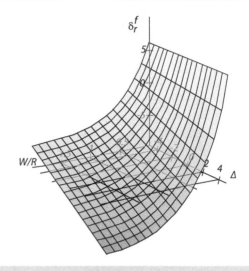

Figure 9.41 $\delta^{18}O_r^f$ as a function of W/R and Δ computed from 9.16.

$\delta^{18}O_r^i$ is assumed to be +6 and $\delta^{18}O_w^i$ is assumed to be −13.

Figure 9.42 shows another example of the $\delta^{18}O$ imprint of an ancient hydrothermal system: the Bohemia mining district in Lane County, Oregon, where Tertiary volcanic rocks of the Western Cascades have been intruded by a series of dioritic plutons. Approximately $1 000 000 worth of gold was removed from the region between 1870 and 1940. $\delta^{18}O$ contours form a bull's eye pattern, and the region of low $\delta^{18}O$ corresponds roughly with the area of prophylitic (i.e. greenstone) alteration. Notice that this region is broader than the contact metamorphic aereole (dashed red line). The primary area of mineralization occurs within the $\delta^{18}O < 0$ contour. In this area, there are relatively large volumes of gold-bearing hydrothermal solution, cooled, perhaps mixing with groundwater, and precipitated gold. This is an excellent example of the value of oxygen isotope studies to mineral exploration. Similar bull's eye patterns are found around many other hydrothermal ore deposits.

Porphyry copper deposits (PCDs) account for more than 70% of world copper ores and often also contain economic concentrations of Mo, Au, Ag, as well as other metals. They tend to be relatively young, typically of Cenozoic or Mesozoic age. They form above the shallow crustal parts of mature calc-alkaline subduction zone volcanic systems that range from dioritic to granitic. Ore is precipitated

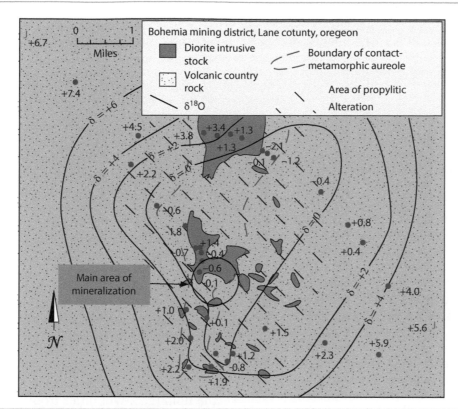

Figure 9.42 $\delta^{18}O$ variations in the Bohemia mining district, Oregon. Note the bull's eye pattern of the $\delta^{18}O$, where gold precipitated from upwelling hydrothermal fluids. Source: Taylor (1974)/with permission of Society of Economic Geologists.

at depths of 1–6 km from pulses of metal-rich hydrothermal fluids ascending from deep magmatic plutons focused into vertical pipe-like structures precipitating sulfides. Radiogenic isotope ratios demonstrate that the magmas are primarily of mantle derivation; for example, initial $^{87}Sr/^{86}Sr$ ratios of the Chilean batholiths, which include some of the world's largest PCDs, are in the range of 0.703–0.706, In other cases, however, such as the Arizona deposits, magmas have a significant crustal component.

Copper mineralization in these deposits occurs as veins formed through precipitation from magmatic water released from an underlying magma chamber. Primary chalcopyrite mineralization is associated with potassic (quartz-K-feldspar, biotite, and muscovite ± albite) hydrothermal alteration occurring at temperatures up to and in some cases exceeding 600 °C. Hydrothermal alteration typically grades concentrically outward to zones of lower temperature phyllic–propylitic zones of illite and pyrite ± kaolinite ± quartz with lower Cu concentrations and ultimately

to low-temperature argillitic zones. Veins formed at lower temperature can also cut the high-grade mineralization.

The Prebble PCD in southeast Alaska is estimated to contain 36.6 million tons of copper, 2.5 and million tons of molybdenum, and with 3341 tons of gold is one the largest deposits on Earth. In potassic altered regions of the pluton where copper concentrations average 0.55% and gold 1.28 ppm, fluid inclusion homogenization temperatures are up to 506 °C in these zones, and have $\delta^{18}O$ of +6 to +10‰ and δ^2H of –70 to –41‰, implicating a magmatic fluid source. Temperatures were lower, ~280 °C, in zones of quartz–sericite–pyrite alteration as were average copper concentrations (0.27%) but gold concentrations were higher at ~2 ppm. Fluid inclusions consist of low-salinity aqueous fluids with a significant magmatic component with $\delta^{18}O$ = +2.1 to +4.1‰ and δ^2H = –76 to –67‰. Illite alteration that overprints both potassic alteration and quartz–sericite–pyrite alteration formed from fluids with $\delta^{18}O$ –9.1 to –4.8‰ and δ^2H = –101 to –90‰ produced by the mixing of magmatic fluid with a meteoric-derived one with $\delta^{18}O$ of ~ −15‰. The highly depleted δ^2H signal is attributed to degassing as the magma cooled and crystallized.

9.6   SULFUR ISOTOPES IN MAGMATIC AND HYDROTHERMAL SYSTEMS

9.6.1   Introduction

Sulfide ores are important sources of a wide variety of metals, particularly the base metals, Cu, Pb, and Zn, but also Co, Ni, Ag, Au, and platinum group metals. These have formed in a great variety of environments and under a great variety of conditions. Sulfur isotope studies have been very valuable in sorting out the genesis of these deposits. Of the various stable isotope systems we will consider in this book, sulfur isotopes are among the most complex because of sulfur's five valence states and isotopic fractionations between them. Sulfur in each of these valence states forms a variety of compounds, and fractionations can occur between these as well.

There are two major reservoirs of sulfur on the Earth that have approximately uniform sulfur isotopic compositions: the mantle, which has $\delta^{34}S$ of ~−1‰ and in which sulfur is primarily present in reduced form, and seawater, which has $\delta^{34}S$ of +21‰ and in which sulfur is present as SO_4^{2-} (as we will find, $\delta^{34}S$ of seawater has varied through time). Sulfur in sedimentary, metamorphic, and igneous rocks of the continental crust may have $\delta^{34}S$ that is both greater and smaller than these values. All of these can be sources of sulfide in ores, and further fractionation may occur during transport and deposition of sulfides.

9.6.2   Sulfur isotope fractionations in magmatic and hydrothermal processes

Sulfur is present in peridotites as trace sulfides, and that is presumably its primary form in the mantle. At temperatures above about 400 °C, H_2S, or its dissociated form of HS^-, and SO_2 are the stable forms of sulfur in fluids and melts. As we noted earlier, there is little fractionation during melting of the mantle to produce basaltic magma (Labidi and Cartigny, 2016). The solubility of H_2S in silicate melts decreases with decreasing Fe. The solubility of H_2S in basalt appears to be only slightly less

than that of water, so that under moderate pressure, essentially all sulfur will remain dissolved in basaltic liquids.

9.6.2.1   Sulfur isotope fractionation in magmas

As basalts rise into the crust, cool, and crystallize, several processes can affect the oxidation state and solubility of sulfur and the produce isotopic fractionations. First, the decreasing pressure results in a gas or fluid phase forming into which some of the sulfide partitions. In addition, H_2 can be lost from the melt through diffusion. This increases the f_{O2} of the melt causing some of the sulfide to be oxidized to SO_2, which is very much less soluble in silicate melts than H_2S. Decreasing Fe content as a consequence of fractional crystallization will also decrease the solubility of S in the melt, increasing its concentration in a coexisting fluid or gas phase. Isotope fractionation will occur between the three species (dissolved HS^-, H_2S, SO_2). The isotopic composition of the fluid (gas) will differ from that of the melt and can be computed as

$$\delta^{34}S_{fluid} = \delta^{34}S_{melt} - \Delta_{HS^-} + \Delta_{SO_2}\left(\frac{R}{R+1}\right) \tag{9.18}$$

where Δ_{HS} is the fractionation factor between HS^- and H_2S, Δ_{SO2} is the fractionation factor between H_2S and SO_2, and R is the mole fraction ratio SO_2/H_2S and is given by

$$R = \frac{X_{SO_2}}{X_{H_2S}} = \frac{K\nu_{H_2S}f_{O_2}^{3/2}}{P_f\nu_{H_2O}X_{X_2O}\nu_{SO_2}} \tag{9.19}$$

where ν are the fugacity or activity coefficients, P_f is the fluid pressure (generally equal to total pressure), f_{O2} is oxygen fugacity, and K is the equilibrium constant for the reaction:

$$H_2S_{(g)} + \frac{3}{2}O_2 \rightleftharpoons H_2O_{(g)} + SO_{2(g)} \tag{9.20}$$

The redox state of magmas is generally referred to relative to "redox buffers" such as fayalite–quartz–magnetite, nickel–nickel oxide, and magnetite–hematite:

$$3Fe_2SiO_4 + 2O_2 \rightleftharpoons 3SiO_2 + 2Fe_3O_4 \tag{9.21}$$

$$Ni + \frac{1}{2}O \rightleftharpoons NiO \tag{9.22}$$

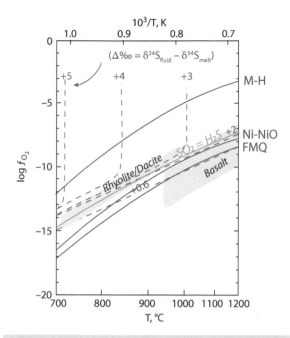

Figure 9.43 Fractionation of sulfur isotopes between fluid and melt (shown by dashed red curves) as a function of oxygen fugacity and temperature for $P_{H_2O} = 10$ MPa (1kb). Solid blue and green lines show equal concentration boundaries for fayalite $\rightleftharpoons$ quartz + magnetite (QM-F), $H_2S \rightleftharpoons SO_2$, and magnetite $\rightleftharpoons$ hematite (M-H). Source: Adapted from Ohmoto and Rye (1979).

$$2Fe_3O_4 + \frac{1}{2}O_2 \rightleftharpoons 3Fe_2O_3. \qquad (9.23)$$

Figure 9.43 shows the sulfur isotope fractionation between fluid and melt relative to these buffers calculated from 9.18 and 9.19 as a function of temperature and f_{O2} for $P_{H_2O} = 1$ kB. At the temperatures and f_{O2} of most basalts, sulfur will be present primarily as H_2S in the fluid (gas phase) and HS^- in the melt. As there is no change in oxidation state, fractionation between these species is small (~0.6‰), so the isotopic composition of fluid phase will not be very different that of the melt. For rhyolites and dacites, a significant fraction of the sulfur can be present as SO_2. Because reaction 9.20 involves a change in oxidation state, greater fractionation between melt and fluid is likely.

An interesting feature of these equations is that the fractionation between fluid and melt depends on the water pressure.

Figure 9.43 is valid only for $P_{H2O} = 10$ MPa (1kb). A decrease in P_f or X_{H2O} (the mole fraction of water in the fluid) will shift the SO_2/H_2S equal concentration boundary and the $\delta^{34}S$ contours to lower f_{O2}. Conversely, an increase in the water content will shift to boundary toward higher f_{O2}.

Both the eruptions of El Chichón in 1983 and Pinatubo in 1991 lofted substantial amounts of SO_2 into the stratosphere, which resulted in global climate cooling. The SO_2-rich nature of these eruptions is thought to result from the mixing of a mafic, S-bearing magma with a more oxidized dacitic magma, which resulted in the oxidation of the sulfur in the former, and consequent increase of SO_2 in the fluid phase. There are a number of other processes that affect the solubility and oxidation state of sulfur in the melt and, hence, isotopic fractionation. Wall rock reactions could lead to either oxidation or reduction of sulfur, and the crystallization of sulfides or sulfates could cause relatively small fractionations and additionally affect the SO_2/H_2S ratio of the fluid. Depending on the exact evolutionary path taken by the magma and fluid, $\delta^{34}S$ of H_2S may be up to 13‰ lower than that of the original magma. Thus, variations in the isotopic composition of sulfur are possible even in a mantle-derived magma whose initial $\delta^{34}S$ was that of the mantle (~ 0‰). The variability of sulfur isotopic compositions do, however, give some indication of the f_{O2} history of a magma. Constant $\delta^{34}S$ of magmatic sulfides suggests that f_{O2} remained below the SO_2/H_2S boundary; variability in isotopic composition suggests a higher f_{O2}.

9.6.2.2 Sulfur isotope fractionation in hydrothermal systems

Although some sulfide ore deposits are magmatic, most were deposited by precipitation from aqueous solution at low to moderate temperature. At temperatures below about 400 °C, sulfide species (H_2S and HS^-) are joined by sulfate ones (SO_4^{2-}, HSO_4^{1-}, KSO_4^{1-}, $NaSO_4^{1-}$, $CaSO_4$, and $MgSO_4$) as the dominant forms of aqueous sulfur. The ratio of sulfide to sulfate will depend on the oxidation state of the fluid. Small fractionations occur among the various sulfide and sulfate species, but there is a major fractionation between sulfide and sulfate. Neglecting the small fractionation between

H₂S and HS⁻ and among sulfate species, $\delta^{34}S_{H2S}$ of the fluid can be expressed as

$$\delta^{34}S_{H2S} = \delta^{34}S_{fluid} - \Delta_{SO_4^{2-}} \times \left(\frac{R'}{R'+1}\right) \quad (9.24)$$

where $\Delta_{SO_4^{2-}}$ is the fractionation between H₂S and SO_4^{2-} and R' is the molar ratio of sulfate to sulfide:

$$R' = \frac{\Sigma SO_4^{2-}}{\Sigma H_2 S}. \quad (9.25)$$

In general, R' will be a function of f_{O2}, fluid composition, and temperature. pH is important as well because there are a number of potential acid–base reactions:

$$H_2S_{aq} \rightleftharpoons H^+ + HS^- \quad (9.26)$$

$$2H^+ + SO_4^{2+} \rightleftharpoons H_2S_{aq} + 2O_2 \quad (9.27)$$

$$HSO_4^- \rightleftharpoons H^+ + SO_4^{2-}. \quad (9.28)$$

Figure 9.44 shows the difference between $\delta^{34}S$ in sulfide and $\delta^{34}S$ in the total fluid as a function of pH and f_{O2}. Only under conditions of low pH and low f_{O2}, will the $\delta^{34}S$ of pyrite (FeS₂) be the same as the total $\delta^{34}S$ of the fluid from which it precipitated. For conditions of relatively high f_{O2} or high pH, substantial differences between the $\delta^{34}S$ of pyrite and the $\delta^{34}S$

of the fluid from which it precipitated are possible. Figure 9.45 shows the difference between $\delta^{34}S$ in sulfide and $\delta^{34}S$ in the total fluid as a function of the sulfate/sulfide ratio (R') and temperature. When the fluid is sulfide-dominated, the $\delta^{34}S$ of the sulfide and that of the bulk fluid will necessarily be nearly identical. For conditions where the concentrations of sulfate and sulfide are similar, large fractionations between sulfide minerals and fluids from which they precipitate are possible.

At magmatic temperatures, reactions generally occur rapidly, and most systems appear to be close to equilibrium. This will not necessarily be the case at lower temperatures because of the strong dependence of reaction rates on temperature. While isotopic equilibration between various sulfide species and between various sulfate species seems to be readily achieved at moderate and low temperatures, isotopic equilibration between sulfate and sulfide appears to be more difficult to achieve. Sulfate–sulfide reaction rates have been shown to depend on pH (reaction is more rapid at low pH) and, in particular, on the presence

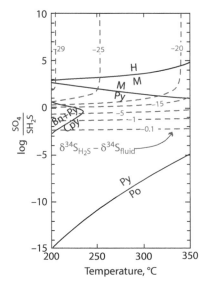

Figure 9.45 Difference in $\delta^{34}S$ between H₂S and bulk fluid as a function of temperature and sulfate/sulfide ratio in a pH neutral fluid. Equal concentration boundaries between coexisting solids: M: magnetite, H: hematite, Py: pyrite, Po: pyrrhotite, Bn: bornite (Cu₅FeS₄), Cp: chalcopyrite. Source: Ohmoto and Rye (1979)/ Reproduced with permission from John Wiley & Sons.

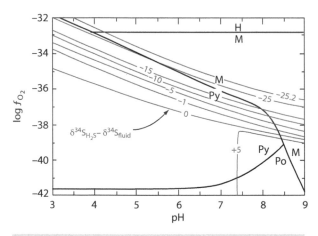

Figure 9.44 Difference in $\delta^{34}S$ between H₂S and bulk fluid as a function of pH and f_{O2} at 250 °C. Equal concentration boundaries are shown for magnetite–hematite (M–H), magnetite–pyrite (M–Py), magnetite–pyrrhotite (M–Po), and pyrite–pyrrhotite (Py–Po). Source: Ohmoto and Rye, (1979)/Reproduced with permission from John Wiley & Sons.

of sulfur species of intermediate valences. Reaction rates between widely differing valence states (e.g. sulfate and sulfide) are much slower than between species of adjacent valance states (e.g. sulfate and sulfite, SO_3^{2-}).

Hutchison et al. (2020) constructed a thermodynamic model of the evolution of sulfide and sulfate isotopic composition of cooling hydrothermal fluids at various pH values and oxygen fugacity. As a general rule, acidic and reducing conditions favor sulfide, while more alkaline and oxidizing conditions favor sulfate. Low temperature also favors sulfate so that depending on oxygen fugacity, a sulfide-dominated fluid will become increasingly sulfate-dominated as it cools. At pH of 4 and f_{O2} at the magnetite–hematite buffer or higher, nearly all sulfur is present as sulfate, and hence, sulfide $\delta^{34}S$ will change continually, while sulfate will have a constant $\delta^{34}S$ equal to that of the system (Figure 9.46a). At f_{O2} of the FMQ buffer or lower, nearly all sulfur is present as sulfide and, hence, will have a constant $\delta^{34}S$ equal to that of the system. At intermediate f_{O2} values near the Ni–NiO buffer, the sulfide–sulfate ratio will continually change as the system cools as will the $\delta^{34}S$ of both. In contrast, in a mildly alkaline system (pH 8, Figure 9.46b), nearly all sulfur will be sulfate at f_{O2} values at the Ni–NiO buffer and higher, and hence, sulfide $\delta^{34}S$ will change continually, while sulfate will have a constant $\delta^{34}S$ equal to that of the system. Only at two log units below the FMQ buffer will the system be entirely composed of sulfide. Figure 9.46c illustrates the case of a system that evolves from pH 1 to pH 8 as it cools. These systems begin as sulfide-dominated at f_{O2} below the MH buffer but eventually become sulfate-dominated above the FMQ buffer with consequent evolution of $\delta^{34}S$ of both sulfate and sulfide under all but the most initial oxidizing conditions.

The point is that in an evolving closed hydrothermal system, the $\delta^{34}S$ of precipitating sulfides will depend not only on the $\delta^{34}S$ of the system, but also on the sulfide–sulfate ratio, which in turn depends on temperature, f_{O2}, pH, and, although we have not shown it here, pressure and ionic strength of the fluid. As these parameters change, so will the sulfide/sulfate ratio and $\delta^{34}S$ of precipitating sulfides.

Low temperatures also lead to kinetic, rather than equilibrium, fractionations. As we saw, kinetic fractionation factors result from different isotopic reaction rates.

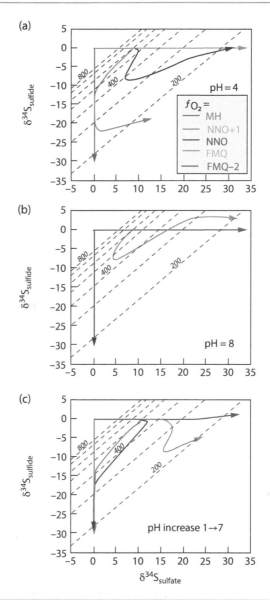

Figure 9.46 Evolution of $\delta^{34}S_{sulfide}$ and $\delta^{34}S_{sulfate}$ in closed hydrothermal systems with total $\delta^{34}S = 0‰$ cooling from an initial temperature of 900 °C to <200 °C. Dashed lines are temperature in °C. Colored arrows show the evolution of $\delta^{34}S_{sulfide}$ and $\delta^{34}S_{sulfate}$ for specific f_{O2} buffers. In panel a, FMQ-2 follows the same trajectory as FMQ; in panel b, NNO+1 and MH follow the same trajectory as NNO. Panels a and b show the evolution in, respectively, mildly acidic and mildly alkaline systems. Panel c shows a system that evolves from acidic to alkaline as it cools. In reduced systems where nearly all sulfur is present as sulfide, the sulfide will have $\delta^{34}S_{sulfide} = 0‰$ and variable $\delta^{34}S_{sulfate}$, and the opposite is true for sulfate-dominated systems. At intermediate f_{O2} where neither dominates, $\delta^{34}S$ of both sulfate and sulfide will vary during cooling. Source: Hutchison et al. (2020)/with permission of Elsevier.

Interestingly enough, the rates for oxidation of $H_2{}^{32}S$ and $H_2{}^{34}S$ appear to be nearly identical. This leads to the kinetic fractionation factor, α_k, of 1.000 ± 0.003, whereas the equilibrium fractionation between H_2S and SO_4 will be 1.025 at 250 °C and 1.075 at 25 °C. Thus, sulfate produced by the oxidation of sulfide can have $\delta^{34}S$ identical to that of the original sulfide. Kinetic fractionations for the reverse reaction, namely reduction of sulfate, are generally larger. The fractionation observed is generally less than the equilibrium fractionation and depends on the overall rate of reduction: the fractionation approaches the equilibrium value when reaction rate is slow. Disequilibrium effects have also been observed in the decomposition of sulfide minerals.

If there is disequilibrium between sulfate and sulfide in solution, it is likely that equilibrium between mineral pairs involving pyrite and chalcopyrite $(CuFeS_2)$ will not be achieved even when isotopic equilibrium is attained between other sulfides such as galena (PbS), sphalerite (ZnS), and pyrrhotite (FeS). This is because the precipitation of the former involves reactions such as

$$4Fe^{2+} + 7H_2S + SO_4^2 \rightleftharpoons 4FeS_2 + 4H_2O + 6H^+ \tag{9.29}$$

whereas the latter involve only simple combinations, for example:

$$Zn^{2+} + H_2S \rightleftharpoons ZnS + 2H^+. \tag{9.30}$$

9.6.3 Isotopic composition of sulfide ores

Sulfide ores form by precipitating metal sulfides from a fluid, most often water, but magma in some cases. Attempting to classify natural processes or products can often be a fool's errand and that is true in ore deposits since in reality there is a continuum. Nevertheless, sulfide ore deposits are generally classified as follows:

- Layered intrusions
- Sediment-hosted, including
  - Stratiform
  - Mississippi Valley type (MVT) and Irish-type
  - Sedimentary exhalative
- Volcanoclastic massive sulfide (VMS) deposits
- PCDs.

Sulfide minerals can precipitate directly from magmas in layered intrusions such as the Bushveld of South Africa or Sudbury of Canada. In these deposits, the principal sulfides of interest are generally those of nickel and platinum group metals, but they are also often exploited for a variety of metals including Ti and Cr that are present as oxides. In the remainder of these ore deposits, sulfides have precipitated from aqueous solution. The water is magmatic water in the case of PCDs (which can be source of many metals in addition to Cu); in other types of deposits, the solution was ultimately derived from seawater or meteoric water in sediment-hosted and VMS deposits. Figure 9.47 compares their $\delta^{34}S_{CDT}$ with that of various possible sources of sulfur. Beyond the source, the isotopic composition of these sulfides depends on the factors discussed above as well as whether the system was open or closed to sulfate and sulfide and the reduction mechanisms involved. Here, we will discuss in detail the sulfur isotope geochemistry of just a few

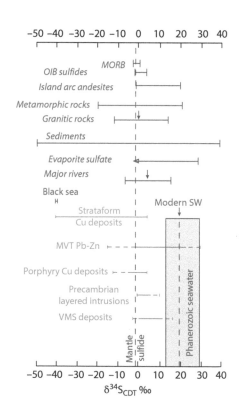

Figure 9.47 Sulfur isotopic composition of sulfide ore deposits (green) and potential sulfur sources (blue).

examples. We will consider their copper isotope geochemistry in Section 9.7.

9.6.3.1 Stratiform Cu deposits

Stratiform, or sediment-hosted deposits, are important sources of copper, but can also contain economic grades of Co, Ag, Pb, Zn, Au, and other metals. As their name implies, sulfide ores are hosted in sediments, with sulfur generally derived from evaporites whose source was in turn usually seawater. The classic example is the Permian Kupferschiefer deposit, which extends from Poland into Germany and has been mined since the Middle Ages for Cu and Ag. The Katanaga deposit, part of the Central African Copper Belt straddling the Democratic Republic of Congo and Zambia, is another example. It has been mined since prehistory and is one of the largest copper reserves in the world. It is a particularly valuable world resource in the modern economy because it presently supplies two-thirds of the world's production of cobalt, a key ingredient in lithium-ion batteries as well as jet engines, superalloys, magnets, catalysts, etc., not to mention is long history of use as the striking blue pigment "cobalt blue" ($CoAl_2O_4$).

Katanga ores formed through a complex history closely linked to the tectonic evolution of the area, which El Desouky et al. (2010) and Saintilan et al. (2018) have reconstructed through Re–Os dating. The ores are hosted by an ~880–727 Ma sedimentary sequence that includes siliciclastic and carbonate rocks as well as four thick evaporite layers and is overlain by Cryogenian (720-635 Ma) glacial diamictites. The first episode of ore formation was the precipitation of fine-grained carrollite ($CuCo_2S_4$), chalcopyrite ($CuFeS_2$), and chalcolite (Cu_2S) as well as dolomite replacing evaporite minerals such as gypsum and anhydrite and occurred at 609±5 Ma based on the Re–Os ages of the carrollites. $\delta^{34}S_{CDT}$ in these sulfides range from –10.3‰ to +3.1‰, values consistent with bacterial sulfide reduction from sulfate originating as Proterozoic marine sulfate, with a mean $\delta^{34}S_{CDT}$ of +17.5‰ (El Desouky et al., 2010). Initial $^{187}Os/^{188}Os$ of 3.2 and initial $^{87}Sr/^{86}Sr$ of 0.710–0.736, which is well above Proterozoic seawater, indicate a crustal source for the metals, inferred to be Mesoproterozoic basement. These first phase mineralizing fluids had moderate temperatures

of 115–220 °C and salinity, with 11–21% dissolved NaCl. Saintilan et al. speculate that melt water produced at the termination of the Marinoan glaciation penetrated the underlying sedimentary sequences to depths of ca. 1000 m, dissolving the evaporites. As the solubility of both Cu and Co increases exponentially with salinity, these fluids could then readily leach these metals when the collision between the Congo and Kalahari craton at the initiation the Luftian orogeny drove fluid circulation into the underlying basement. Ores then precipitated when these fluids rose, and sulfate was reduced by bacteria metabolizing organic matter in host sediments.

The second and main episode of Co–Cu mineralization occurred between 540 and 490 Ma when compressional tectonics of the Luftian orogeny drove a new episode of fluid flow. Temperatures were higher (270–385 °C), as were salinities (35–45% NaCl), than the first phase. Initial $^{187}Os/^{188}Os$ of 3.7 are similar to those of first phase mineralization, although $^{87}Sr/^{86}Sr$ was lower, 0.709–0.712. $\delta^{34}S$ of the second-stage ore is similar to that of the first stage but more variable, with $\delta^{34}S_{CDT}$ of +5.1 to –13.1‰, but with several veins with $\delta^{34}S_{CDT}$ of ~+19‰, close to Proterozoic seawater. Most likely, this sulfur originated from a combination of thermal sulfate reduction and remobilization of ore sulfides and pyrite from the first stage.

9.6.3.2 Layered intrusions

Mafic layered intrusions hosting platinum group element (PGE) ores are predominantly of late Archean and early Proterozoic age. They tended to form in periods of supercontinent amalgamation and crust formation, occur in stable cratons, and are suspected of having been produced by mantle plumes. The type example and the largest of these is the enormous 2.05-Ga Bushveld Complex of South Africa, which outcrops over an area the size of Ireland, is >6 km thick, and has an estimated volume of 0.7 to 1.0×10^6 km^3 (Maier et al., 2013). By comparison, the early Tertiary Skaergaard intrusion of Greenland that we discussed above is one of the largest comparable Phanerozoic layered intrusions and PGE prospects and has an estimated volume of only 280 km^3. The Bushveld hosts the world's largest reserves of PGEs, Cr, and V and is a significant source of other metals such as Cu, Ni, and Au.

The Bushveld intrudes the 2.67–2.07-Ga Transvaal Supergroup sedimentary sequence and, in the far north, Archean granites. PGE ores occur mainly as sulfides and less commonly as PGE alloys in localized "reefs" of which there are several types in the Bushveld, including contact reefs along the sidewalls and base of the intrusion, silicate-hosted reefs in ultramafic sequences containing 1–3% sulfides and up to 10 ppm PGEs, of which the Merensky Reef is the type example, and transgressive iron-rich ultramafic "pipes" precipitated from upward or downward percolating melts. The primary magmas are undersaturated with respect to sulfide, but 15% fractional crystallization would raise the sulfide concentration to ~600 ppm, enough to exsolve a separate sulfide liquid.

Initial $^{87}Sr/^{86}Sr$ ratios for the parental magmas (0.7034–0.7077) are well above those expected of 2-Ga mantle. This, as well as mean $\delta^{18}O$ of ~7.1‰, and initial ε_{Nd} values in the range of −5 to −8 throughout the complex are testified to extensive crustal assimilation. Initial Os isotope ratios, on the other hand, are ~0.07 γ_{Os}, indicating that the PGEs are nearly entirely mantle derived. Initial studies inferred that the sulfur was primarily magmatic; however, Penniston-Dorland et al. (2012) reported MIF of sulfur isotopes with $\Delta^{33}S$ up to 0.13‰ and $\Delta^{36}S$ as low as −0.26‰. The pattern of positive $\Delta^{33}S$ and $\Delta^{36}S$ is typical of many Archean sedimentary rocks and is consistent with fractionations produced by the atmospheric photodissociation of SO_2. Subsequently, Sharman et al. (2013) reported $\Delta^{33}S$ values >0.5‰ and $\delta^{34}S_{CDT}$ > 12‰ in the Platreef PGE-rich horizon in the northern limb of the intrusion. Contact-metamorphosed country rock and xenoliths of country rock show much larger variation with $\delta^{34}S$ ranging from −20.9 to +28.7‰ and $\Delta^{33}S$ ranging as high as 5‰. The most extreme values in the Platreef are limited to within 5 m of the contact, demonstrating local assimilation, which Sharman et al. attribute to the S-rich Duitschland Formation.

Within the Rustenburg Layered Suite of the main intrusion, Magalhães et al. (2018) showed that $\Delta^{33}S$ exhibits a systematic stratigraphic variation with occasional sharp discontinuities such as at the boundary between

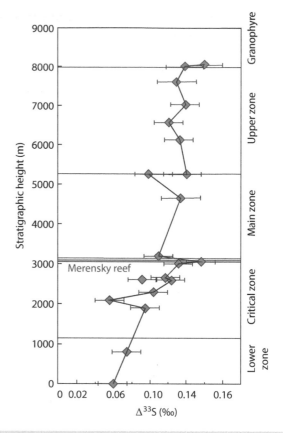

Figure 9.48 $\Delta^{33}S$ as a function of stratigraphic height in the Bushveld intrusion indicating increasing crustal assimilation with time. Discontinuities in $\Delta^{33}S$ correspond as well to discontinuities in radiogenic isotope ratios, indicating that they were produced by new pulses of magma. Source: Magalhães et al. (2018)/with permission of Elsevier.

the Main Zone and the Critical Zone below it and the Upper Zone above it (Figure 9.48). These boundaries are also marked by discontinuities in $^{87}Sr/^{86}Sr$; a lack of correlation between $\Delta^{33}S$ and $^{87}Sr/^{86}Sr$ led Magalhães et al. to conclude that different layers assimilated sulfur from different sources. While the direct source of the sulfur is unclear, the MIF signature demonstrates an original Archean atmospheric source. Importantly, increases of $\Delta^{33}S$ values coincide with the UG2 Chromitite and Merensky Reef and raise the intriguing question of whether this surface-derived sulfur played in triggering the precipitation of the PGE-rich layers. MIF sulfur has been identified in several other Precambrian ore deposits as well. On the other hand, the assimilation of MIF sulfur appears to be limited to the immediate vicinity of the wallrock in the Stillwater Intrusion of Montana, USA, which is another, although much smaller, PGE ore deposit.

9.6.3.3 Mississippi Valley and Irish-type deposits

MVT deposits are carbonate-hosted lead and zinc sulfides formed under relatively low-temperature conditions. They are epigenetic, meaning that they formed long after, tens to hundreds of million years, diagenesis of host rocks. While the type locality is the central USA, examples occur worldwide. In the Mississippi Valley, these deposits occur in a discontinuous belt from Wisconsin and Illinois to Tennessee, Southeast Missouri, and Arkansas. Mineralization occurs primarily as sphalerite and galena in platform carbonates. The few available radiometric ages range from 380 Ma, corresponding to the Arcadian orogeny, to 250 Ma, corresponding to the Alleghenian/Ouachita orogeny.

MVT deposits can be subdivided into Zn-rich and Pb-rich classes. The Pb-rich and most of the Zn-rich deposits were formed between 70 and 120 °C, while some of the Zn-rich deposits, such as those of the Upper Mississippi Valley, were formed at temperatures up to 200 °C. Co-existing sulfides of the Pb-poor Upper Mississippi Valley deposits are in isotopic equilibrium, whereas sulfur isotope equilibrium was most often not achieved in Pb-rich deposits. Isotopic disequilibrium reflects in part relatively low temperatures where isotopic exchange is slow, but likely also reflects $\delta^{34}S$ fluid variation with time. Zhou et al. (2018) found that while bulk galena and sphalerite separates from the Nayongzki MVT deposit of Guizhou Province in China were not in isotopic equilibrium, the isotopic compositions of mineral rims were, suggesting that diffusion was the limiting factor in achieving equilibrium.

Pb isotope ratios and $\delta^{34}S$ in deposits of the Mississippi Valley are heterogeneous, even within individual deposits, and these variations are often uncorrelated, suggesting that the source of metals and sulfur differed, although $\delta^{34}S$ is correlated with Pb isotope ratios in the Southeastern Missouri District. In the Viburnum Trend within this region, two mineralized zones are present. The upper, near-surface zone is a typical Pb-rich mineralization hosted by Cambrian dolomite. Textures indicate that galena replaced early chalcopyrite mineralization. Fluid inclusion temperatures of sphalerite range from 85 to 135 °C but with most falling in the narrower range of 102–118 °C. $\delta^{34}S_{CDT}$ ranges from 8.6‰ to 18.4‰, but with most of the data falling within the narrow range of 12.7–13.6‰ and with a mean of 13.99‰. Judging from mineralization textures, $\delta^{34}S$ varied through time with an overall trend toward heavier S (Shelton et al., 2020). A lower mineralized zone extends upward from the contact between the dolomite and underlying Cambrian sandstone located at a depth of 30 m. This zone is more zinc- and copper-rich than the upper zone, containing up to 29% Zn, 8% Cu, and 15% Pb, with notable enrichments in Ni, Co, and Ag. $\delta^{34}S$ in the lower zone is lighter on average but far more heterogeneous, ranging from −14.1‰ to 24‰ with $\delta^{34}S$ increasing with height above the contact. Shelton et al. (2020) concluded that as the deep ore fluid system worked its way upward, it breached less permeable units in the lower dolomite, mixing with high-$\delta^{34}S$ brines present higher in the stratigraphic section. An *in situ* secondary ion mass spectrometry (SIMS) profile through a large (several mm) banded sphalerite grain reveals a repetitive pattern of $\delta^{34}S$ values varying from ~12‰ to 5‰, punctuated by several abrupt decreases to values as low as −8‰, suggesting that mineralization occurred in pulses of distinct composition. The final thick overgrowth has a comparatively uniform $\delta^{34}S$ of ~10‰ (Shelton et al., 2020).

There is a broad consensus among several studies that multiple, isotopically distinct sulfur sources were available at different times throughout the ore mineralizing event and that precipitation resulting from the mixing of metal-rich brines with H_2S-rich ones. Low $\delta^{34}S$ in the early Cu- and Ni-rich ores of the lower ore zone suggests that sulfur was derived from underlying igneous basement, as were the metals. The higher $\delta^{34}S$ values that predominate in the upper ore zone are consistent with the thermal sulfate reduction of Paleozoic seawater and are likely derived from evaporites within the Paleozoic sedimentary sequence.

"Irish"-type deposits share many features with MVT ones including being primarily Pb–Zn sulfides deposited at low to moderate temperature and hosted in carbonates but differ in several significant respects. In Irish-type deposits, mineralization is simultaneously with diagenesis, typically involves hotter fluid temperatures (70–280 °C), is often associated with volcanism, and has uniform Pb isotopic compositions indicative of a single metal source.

The Navan Zn–Pb deposit is one of several in Midlands Basin of Ireland, which formed in the Lower Carboniferous around ~350–340 Ma and collectively constitute one of the richest Zn–Pb mineralized terrains on Earth. Mineralization is spatially and temporally associated with extensional faulting, which allowed convective flow of mineralizing brines. Fluid-inclusion homogenization temperatures range from 75 to 180 °C. Regionally, lead isotope ratios in galena of the Midlands Basin vary systematically in a manner that matches that of the underlying Lower Paleozoic rocks that in turn reflects derived from weathering of the Grenvillian and Avalonian continental blocks on either side of the Iapetus Ocean (Wilkinson, 2014). In the Navan deposit, Pb isotope ratios are uniform with $^{206}Pb/^{204}Pb$ of 18.19±0.03, demonstrating a single source of metals in the underlying Paleozoic strata.

Barite ($BaSO_4$) related to these deposits have $\delta^{34}S$ ranging from 14.1‰ to 22.6‰, indicating that sulfur was ultimately derived from Lower Carboniferous seawater. Sulfides, however, are far more heterogeneous, with $\delta^{34}S$ ranging from −13‰ to +13‰ (e.g. Gagnevin et al., 2012). A consensus has emerged that mineralization resulted from the mixing of a deep, relatively hot hydrothermal fluid with $\delta^{34}S$ of +10 ±10‰ carrying metals leached from underlying rocks and a shallow, cooler sulfide-rich brine with $\delta^{34}S$ of −15±10‰. The hydrothermal fluid provided all the metal but < 10% of the sulfur. Sulfur in the hydrothermal fluid was derived from the dissolution of diagenetic pyrite, while sulfur in the shallower field originated through the bacterial reduction of Lower Carboniferous seawater sulfate.

9.6.3.4 Volcanogenic massive sulfide deposits

VMS deposits are important sources of "base metals," primarily Cu, Pb, Zn, and some cases contain economic concentrations of Ag, Au, and chalcophile metals such as Co In, Cd, Tl, etc. They form in submarine environments from circulating hydrothermal fluids heated by volcanic activity. Ore deposition occurs when upwelling fluids cool and mix with cold seawater at or below the seafloor. Most are hosted within volcanics and shallow intrusives, but they can be hosted by overlying sedimentary successions as well. They occur throughout the geologic record with the oldest known deposits in the Pilbara of Western Australia and the Barberton of South Africa dating from 3.45 Ga. More recent examples include the Cretaceous Troodos deposit in Cyprus, which was an important source of copper for early Eastern Mediterranean civilizations. As the name implies, mineralization is primarily sulfides, including pyrite ± pyrrhotite with associated chalcopyrite ($CuFeS_2$), sphalerite (ZnS), and galena (PbS), which occur in veins, breccias, and as replacement of original minerals, often in association with chloritic, sericitic, or silicic alteration zones.

Mid-ocean ridge hydrothermal systems exemplify the way in which VMS ores are deposited, but they can form as in association with any submarine volcanism, including island arcs, back-arc basins, and intracontinental rifts such as the Red Sea. While most preserved VMS deposits likely formed in environments other than mid-ocean ridges, the extensive study of modern mid-ocean ridge hydrothermal systems have nevertheless been enormously informative with respect to VMS formation.

Sulfide in VMS deposit can be derived from both leaching igneous sulfides and reduction of seawater sulfate through reactions such as

$$6Fe_2SiO_4 + SO_4^{2-} + 2H^+ \rightleftharpoons$$
$$H_2S + 4Fe_3O_4 + 6SiO_2. \tag{9.31}$$

Where sediments are present, they represent a third potential source of sulfur. Seawater sulfate $\delta^{34}S$ has varied through time and the $\delta^{34}S$ of VMS deposits tracks this temporal variation with $\delta^{34}S$ that is ~17.5‰ lighter (Figure 9.49), consistent with the fractionation during the reduction of seawater sulfate to sulfide at temperature in the range of 250–400 °C (Huston, 1999). However, this value is also close to the mantle $\delta^{34}S$ of ~0‰, making it difficult to resolve the volcanic versus seawater contribution. $\delta^{33}S$ nevertheless provides a clue because different reactions generate different $\delta^{33}S$–$\delta^{34}S$ slopes, θ. Ono et al. (2007) used this in a $\delta^{33}S$–$\delta^{34}S$ study to estimate that 73–89%

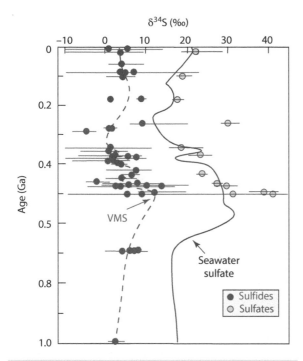

Figure 9.49 $\delta^{34}S$ in sulfide and co-existing sulfate minerals of VMS deposits through time. Solid blue line shows the isotopic composition of contemporaneous seawater sulfate. Source: Huston (1997)/with permission of Society of Economic Geologists.

of vent sulfides are derived from the leaching of basalt sulfide at two sites on the East Pacific Rise and one on the MAR. In a broader survey of modern hydrothermal vents, Zeng et al. (2017) found that hydrothermal sulfides have $\delta^{34}S$ from 0‰ to 9.6‰ with an average of 4.7‰ and concluded that sulfur is derived mainly from the associated igneous rocks, with a lesser proportion (~36%) from seawater sulfate ($\delta^{34}S = 21$‰).

McDermott et al. (2015) studied hydrothermal vent fluids and sulfides from normal spreading centers as well as back-arc basins and found $\delta^{34}S$ as light as −5.7‰ in the latter. Subduction-related magmas are more oxidizing and have a greater SO_2/H_2S ratio. Upon cooling to <400 °C, SO_2 undergoes disproportionation:

$$4SO_2 + 4H_2O \rightleftharpoons 3HSO_4^- + H_2S + 3H^+ \quad (9.32)$$

producing isotopically heavy sulfate and light sulfide. McDermott et al. concluded that these isotopically light signatures resulted from a contribution from disproportionated magmatic SO_2 gas. Similarly, Martin et al. (2020)

concluded that $\delta^{34}S$ as light as −5.5‰ in sulfides from the Troodos VMS deposit reflected also a contribution from disproportionated magmatic SO_2.

9.6.3.5 Porphyry copper deposits

Ore metals in these deposits do not appear to be derived from subducting oceanic crust and lithosphere; for example, Richards (2015) points out that Cu concentrations in IAV are no higher than in mid-ocean ridge basalt. On the other hand, we will find in following section, copper in granites hosting copper porphyries appears to be isotopically heavier than barren ones, and point to a source of copper in PCDs in the metasomatized mantle wedge. Regardless, it is clear that the subducting slab does supply two things that give rise PCDs. The first is the oxidized sulfur. Subduction-related magmas are notably richer in sulfur than mantle-derived ones in other tectonic environments and elevated $\delta^{34}S$ in some of these suggest some of this sulfur is ultimately derived from seawater sulfate (Richards, 2015). Sulfate also carries oxygen from the slab to the mantle wedge and helps to explain why IAV are more oxidized than magmas from other environments. The second thing the subducting slab provides is water. IAV are also water-rich compared to magmas in other environments and porphyry copper ore formation is triggered when the magma becomes saturated in H_2O and exsolves as a separate fluid phase that invades the overlying rock.

Sulfides in most PCDs have $\delta^{34}S$ close to 0‰, which would appear to suggest that the source of sulfur is magmatic (Figure 9.50). As we can see in Figure 9.46, except under conditions more reducing than the FMQ buffer, oxidized sulfur species will be present along with sulfides and isotopic exchange between them will result in $\delta^{34}S$ changing as the fluid cools. Figure 9.51 shows sulfate–sulfide mineral pairs from a number of well-studied PCDs and the 1982 eruption of El Chichón plotted on the same $\delta^{34}S_{sufate}$–$\delta^{34}S_{sulfide}$ plot as Figure 9.46. Hutchison et al. (2020) concluded that most, but not all, of these deposits have sulfate–sulfide compositions consistent with total sulfur $\delta^{34}S$ somewhat heavier than the mantle value (~5‰). Heavier sulfur would be consistent with a contribution of seawater sulfate from the subduction slab beneath these systems.

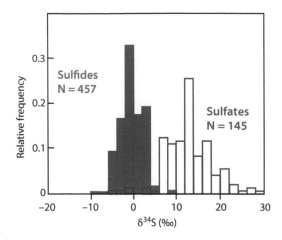

Figure 9.50 Global distribution of δ^{34}S in sulfides and co-existing sulfates in PCD PCDs. Source: Hutchison et al. (2020)/with permission of Elsevier.

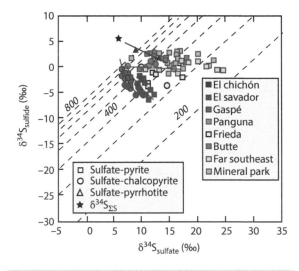

Figure 9.51 δ^{34}S of sulfate–sulfide mineral pairs from PCDs and the 1982 eruption of El Chichón on the same plot as Figure 9.46. The star represents a system with total sulfur δ^{34}S = +5‰. Colors correspond to the various deposits, while shapes indicate the mineral pairs. Trend lines for several deposits are also shown. Dashed lines are temperatures in °C.

We can also see in Figure 9.51 that these systems appear to evolve under different f_{O2} (and perhaps pH) conditions. For example, the relatively variable δ^{34}S$_{sulfate}$ at nearly constant δ^{34}S$_{sulfide}$ in the Panguna and Far Southeast deposits suggests that sulfur was predominately present as sulfide and, therefore, that conditions were relatively reducing. On the

other hand, covariation of δ^{34}S$_{sulfate}$ and δ^{34}S$_{sulfide}$ in the El Salvador and Gaspé deposits suggests that substantial fractions of both sulfide and sulfate were present. Finally, this analysis indicates that most PCDs formed between 800 and 300 °C.

9.7 COPPER ISOTOPES IN ORE DEPOSITS

One of the greatest transformational events in human history occurred when humanity emerged from the Stone Age and entered the Copper Age as people learned to smelt copper and work it into tools and weapons. In a certain sense, the age of copper continues; because of its high conductivity and stability as a metal, we make very extensive use of it for wiring and piping. Copper remains the third most produced metal (over 15 000 tons per year), following iron and aluminum. No metal has played as large a role for as long a time in human society as copper. Copper deposits quite often host economic concentrations of a number of other metals, including Zn, Mo, Ag, and A.

As we found in the previous section, most copper ores are sulfides precipitated from solution, whose origin is generally seawater in the case of VMS deposits, ultimately meteoric water in various sediment-hosted deposits, and magmatic water in the case of porphyry coppers. The majority of *primary* copper sulfide minerals in VMS deposits, such as chalcopyrite ($CuFeS_2$) and cubanite ($CuFe_2S_3$), have δ^{65}Cu in a relatively narrow range 0±0.5‰, not greatly different from igneous rocks and the BSE (Rouxel et al., 2004; Larson et al., 2003; Markl et al., 2006; Mathur et al., 2009; Mathur and Wang, 2019). Most native copper has isotopic compositions within this same range (Larson et al., 2003). Nevertheless, the total range in δ^{65}Cu is quite large (Figure 9.52).

Rouxel et al. (2004) found that primary chalcopyrite and cubanite in mid-ocean ridge active hydrothermal chimneys defined a surprisingly large range of –0.43‰ to +3.62‰, likely reflecting complex processes of leaching, sulfide precipitation, dissolution, and reprecipitation within hydrothermal systems (Rouxel et al., 2004; Li et al., 2018; Zeng et al., 2021). Secondary minerals found at inactive vents and in massive sulfide piles, which included atacamite ($Cu_2Cl(OH)_3$)

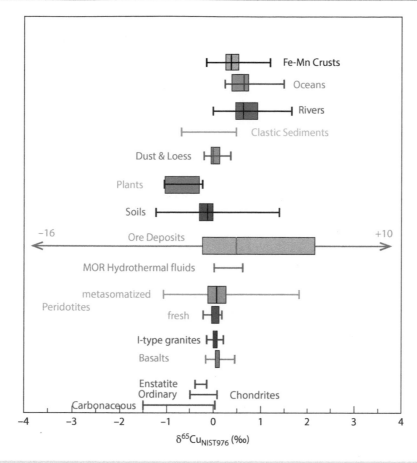

Figure 9.52 Box and whisker plot showing $\delta^{65}Cu$ (per-mil deviations of $^{65}Cu/^{63}Cu$ from the NIST976 standard) in terrestrial materials and meteorites. Source: Adapted from Moynier et al. (2017) and Savage (2018).

and bornite (Cu_5FeS_4) showed more even variable compositions (–2‰ to +4.7‰). As these systems evolve, secondary minerals can be redissolved by hydrothermal fluids and the copper reprecipitated as sulfides. Reprocessing, including the oxidation of primary sulfides and precipitation of secondary ones, appears to be a major cause of copper isotope fractionation in these systems.

The exposure of primary sulfides through uplift and erosion leads to the oxidative leaching of primary chalcopyrite in the vadose zone. Cu^{2+} is then transported downward and precipitates as chalcocite (CuS) in the reducing environment below the water table. A much wider variation in $\delta^{65}Cu$, –16 to +10‰, is observed in secondary minerals that typically develop

during the weathering of sulfide ore deposits in the near surface. Mathur et al. (2009) found that Cu in the enriched supergene[4] zones is typically 0.4–5‰ heavier than primary ore deposit (the hypogene). Markl et al. (2006) found that relict primary chalcopyrite was typically heavy isotope-depleted with $\delta^{65}Cu$ down to –3‰; secondary cuprous minerals such as cuprite (Cu_2O) were isotopically heavy ($\delta^{65}Cu$: +0.4‰ to 1.65‰), and secondary cupric mineral such as malachites ($CuCO_3(OH)_2$) were also usually isotopically heavy but showed a wider range than cuprous minerals) ($\delta^{65}Cu$: –1.5‰ to +2.4‰). Typically, a "leach cap" develops at the surface consisting of residual Fe oxides produced by the weathering of sulfides. In porphyry deposits in the Southwest

---

[4] This zone of weathering that typically occurs above an ore deposit is referred to as the *supergene*; the unweathered ore deposit is referred to as the *hypogene*. Because of oxidation, dissolution, and reprecipitation, parts of the supergene can be enriched in metals such as copper relative to the hypogene.

USA, Mathur et al. (2010) found a pattern of $\delta^{65}Cu$ in minerals with leach cap < hypogene < supergene.

In sediment-hosted stratiform copper deposits such as the Kupperschiefer, $\delta^{65}Cu$ in the principal copper minerals, including bornite, chalcolite, digenite (C_9S_5), and covellite, range from about −2.7‰ to 0‰, but most fall within the narrow range of −0.4±0.36‰. In a similar stratiform copper deposit developed in Cambrian sediments of the Timna Valley, Israel, copper sulfides have much lighter isotopic compositions: −2.0±0.4‰. Asael et al. (2009) explained the difference in terms of temperature, pH, and oxidation state of the fluids. In the Timna valley, the deposit formed under mildly oxidizing conditions (E_H values of 0.5–0.6) and low-temperature and mildly acidic (pH<6) conditions. In that case, copper is in the cupric form, and the principal species of dissolved copper is $CuCl^+$. Thus, sulfide precipitation involves the reduction of copper to the cuprous form. In contrast, the Kuipferschiefer fluids were hotter (100 °C), more reduced (E_H 0.4–0.5), and more neutral (pH 6.3), and the dissolved copper was in the cuprous form, principally as the $(CuCl_3)^{2-}$ ion, and little fractionation occurred upon precipitation.

PCDs are the world's major source of copper. Comparing Tibetan Cenozoic granites and granodiorites hosting copper porphyries with barren ones, Zheng et al. (2018) found that mineralized ones had statistically significant higher $\delta^{65}Cu$ than barren ones, with average $\delta^{65}Cu$ of 0.52‰ and 0.14‰, respectively, as well as higher Cu concentrations. Pointing to the large range and generally higher $\delta^{65}Cu$ of metasomatized peridotites compared to unmetasomatized ones, they argued that the Cu source for magmas hosting PCDs was the metasomatized mantle wedge. In this model, oxidizing fluids would have leached copper from sulfides in the subducting slab and transported heavy copper into the overlying mantle wedge.

Within the hypogene zone of PCD's, $\delta^{65}Cu$ patterns can be complex. In the Northparkes deposit of Australia, Li et al. (2010) found that $\delta^{65}Cu$ is 0.19±0.14‰ in chalcopyrite ($CuFeS_2$) of the high-grade ore in the core of the system and decreases rapidly outward to −0.25 ±0.36‰ at the margins of the orebody before increasing to 0.4–0.8‰ in the periphery. They

also found that $\delta^{65}Cu$ was uncorrelated with $\delta^{34}S$. At the Bingham Canyon PCD, the pattern is opposite.

In the Alaskan Pebble porphyry Cu–Au–Mo deposit discussed in Section 9.6, $\delta^{65}Cu$ in the hypogene copper sulfides ranges from − 2.09‰ to 1.11‰ (Gregory and Mathur, 2017). Samples from the sodic–potassic, potassic, and some illite alteration assemblages define a $\delta^{18}O$–$\delta^{65}Cu$ negative correlation with lighter $\delta^{65}Cu$ values associated with heavier $\delta^{18}O$ characteristic of magmatic fluids. The array represents a mixing line between the magmatic fluids with $\delta^{18}O$ of ~6‰ and meteoric water with $\delta^{18}O$ of −15‰. Some sample from areas where the host rock underwent significant illite alteration by meteoric water-rich fluids fall off this correlation because of the precipitation of secondary chalcopyrite. The earliest chalcopyrite precipitated in areas of sodic–potassic alteration at ~375 °C from a fluid with a magmatic heavy $\delta^{18}O$ signature and $\delta^{65}Cu$ of −2.09‰. Samples from areas of potassic alteration precipitated from a cooler (~330 °C) solution that had mixed with meteoric water have somewhat heavier $\delta^{65}Cu$. Gregory and Mathur interpreted that this increase in $\delta^{65}Cu$ is due to Rayleigh-type fractionation with $\Delta^{65}Cu_{fluid-chalcopyrite}$ of ~0.4 during cooling and chalcopyrite precipitation drove the fluid to heavier $\delta^{65}Cu$. Subsequent meteoric-rich fluids would have had isotopically heavier Cu but would have been too Cu depleted to precipitate primary chalcopyrite. Instead, they appear to have dissolved and reprecipitated preexisting chalcopyrite resulting in even heavier $\delta^{65}Cu$ signatures.

The different patterns of $\delta^{65}Cu$ in PCDs reflect the complexity of Cu transport and fractionation in these systems. Initial fluids released from the magma typically undergo phase separation as they cool and rise into a brine and a vapor phase. Copper preferentially partitions into the vapor phase, but the brine can also transport significant amounts of Cu. Fractionation between vapor and brine can occur. Seo et al. (2007) calculated vapor–brine fractionation factors using density function theory and concluded that the vapor should be isotopically heavier than the brine by about 0.6‰ at 500 °C. However, Maher et al. (2011) observed that the vapor was isotopically lighter experimentally. Maher et al. also found that the fluid–chalcopyrite fractionation factor was pH dependent, with values as low as −1‰ at low salinity and pH of ~4–7 but was in the range of +0.2 to +0.3‰ at higher pH and salinity with generally small but positive under lower pH conditions. These results suggest a dependency on the speciation of Cu in vapor and brine. Depending on pH, Cu in the brine is likely present primarily as $CuCl_2^-$ at relevant temperatures and low pH and $Cu(HS)_2^-$ at higher pH. $CuCl(H_2O)$, Cu_3Cl_3, and $CuHS^0$ are possible species in the vapor. In addition, fractionation factors will depend on how these species are coordinated by water molecules, which is not well known. Thus, we can expect that Cu isotopic fractionation will differ between different deposits and furthermore might vary within an individual deposit as it evolves.

REFERENCES

Amsellem, E., Moynier, F., Bertrand, H., et al. 2020. Calcium isotopic evidence for the mantle sources of carbonatites. *Science Advances* **6**: eaba3269. doi: 10.1126/sciadv.aba3269.

Andersen, M.B., Elliott, T., Freymuth, H., et al. 2015. The terrestrial uranium isotope cycle. *Nature* **517**(7534): 356–9. doi: 10.1038/Nature14062

Antonelli, M.A. and Simon, J.I. 2020. Calcium isotopes in high-temperature terrestrial processes. *Chemical Geology* **548**: 119651. doi: https://doi.org/10.1016/j.chemgeo.2020.119651.

Asael, D., Matthews, A., Oszczepalski, S., et al. 2009. Fluid speciation controls of low temperature copper isotope fractionation applied to the Kupferschiefer and Timna ore deposits. *Chemical Geology* **262**: 147–58. doi: 10.1016/j.chemgeo.2009.01.015.

Aubaud, C., Pineau, F., Hékinian, R., et al. 2006. Carbon and hydrogen isotope constraints on degassing of CO_2 and H_2O in submarine lavas from the Pitcairn hotspot (South Pacific). *Geophysical Research Letters* **33**: L02308. doi: 10.1029/2005GL024907.

Badullovich, N., Moynier, F., Creech, J., et al. 2017. Tin isotopic fractionation during igneous differentiation and Earth's mantle composition. *Geochemical Perspectives Letters* **5**: 24–8. doi: 10.7185/geochemlet.1741.

Baker, R.G.A., Rehkämper, M., Ihlenfeld, C., et al. 2010. Thallium isotope variations in an ore-bearing continental igneous setting: collahuasi formation, northern Chile. *Geochimica et Cosmochimica Acta* **74**: 4405–16. doi: 10.1016/j.gca.2010.04.068.

Baker, R.G.A., Schönbächler, M., Rehkämper, M., et al. 2010. The thallium isotope composition of carbonaceous chondrites — New evidence for live [205]Pb in the early solar system. *Earth and Planetary Science Letters* **291**: 39–47. doi: 10.1016/j.epsl.2009.12.044.

Banerjee, A., Chakrabarti, R. and Simonetti, A. 2021. Temporal evolution of $\delta^{44/40}Ca$ and $^{87}Sr/^{86}Sr$ of carbonatites: implications for crustal recycling through time. *Geochimica et Cosmochimica Acta* **307**: 168–91. doi: 10.1016/j.gca.2021.05.046.

Barnes, J.D. and Cisneros, M. 2012. Mineralogical control on the chlorine isotope composition of altered oceanic crust. *Chemical Geology* **326–327**: 51–60. doi: https://doi.org/10.1016/j.chemgeo.2012.07.022.

Barnes, J.D., Paulick, H., Sharp, Z.D., et al. 2009. Stable isotope ($\delta^{18}O$, δD, $\delta^{37}Cl$) evidence for multiple fluid histories in Mid-Atlantic abyssal peridotites (ODP leg 209). *Lithos* **110**: 83–94. doi: 10.1016/j.lithos.2008.12.004.

Barnes, J.D., Sharp, Z.D. and Fischer, T.P. 2008. Chlorine isotope variations across the Izu-Bonin-Mariana arc. *Geology* **36**: 883–6. doi: 10.1130/g25182a.1.

Bebout, G.E., Fogel, M.L. and Cartigny, P. 2013. Nitrogen: highly volatile yet surprisingly compatible. *Elements* **9**: 333–8. doi: 10.2113/gselements.9.5.333.

Beunon, H., Mattielli, N., Doucet, L.S., et al. 2020. Mantle heterogeneity through Zn systematics in oceanic basalts: evidence for a deep carbon cycling. *Earth-Science Reviews* **205**: 103174. doi: 10.1016/j.earscirev.2020.103174.

Bezard, R., Fischer-Gödde, M., Hamelin, C., et al. 2016. The effects of magmatic processes and crustal recycling on the molybdenum stable isotopic composition of mid-ocean ridge basalts. *Earth and Planetary Science Letters* **453**: 171–81. doi: 10.1016/j.epsl.2016.07.056.

Bindeman, I.N., Eiler, J.M., Yogodzinski, G.M., et al. 2005. Oxygen isotope evidence for slab melting in modern and ancient subduction zones. *Earth and Planetary Science Letters* **235**: 480–96. doi: 10.1016/j.epsl.2005.04.014.

Blum, J.D., Sherman, L.S. and Johnson, M.W. 2014. Mercury Isotopes in Earth and Environmental Sciences. *Annual Reviews of Earth and Planetary Sciences* **42**: 249–69. doi: 10.1146/annurev-earth-050212-124107.

Boniface, M., Busigny, V., Mével, C., et al. 2008. Chlorine isotopic composition in seafloor serpentinites and high-pressure metaperidotites. Insights into oceanic serpentinization and subduction processes. *Geochimica et Cosmochimica Acta* **72**: 126–39. doi: 10.1016/j.gca.2007.10.010.

Bonifacie, M., Jendrzejewski, N., Agrinier, P., et al. 2007. Pyrohydrolysis-IRMS determination of silicate chlorine stable isotope compositions. Application to oceanic crust and meteorite samples. *Chemical Geology* **242**: 187–201. doi: 10.1016/j.chemgeo.2007.03.012.

Bonnand, P., Doucelance, R., Boyet, M., et al. 2020. The influence of igneous processes on the chromium isotopic compositions of ocean island basalts. *Earth and Planetary Science Letters* **532**: 116028. doi: 10.1016/j.epsl.2019.116028.

Bonnand, P., Williams, H.M., Parkinson, I.J., et al. 2016. Stable chromium isotopic composition of meteorites and metal–silicate experiments: implications for fractionation during core formation. *Earth and Planetary Science Letters* **435**: 14–21. doi: 10.1016/j.epsl.2015.11.026.

Boyd, S.R. and Pillinger, C.T. 1994. A preliminary study of ^{15}N/^{14}N in octahedral growth form diamonds. *Chemical Geology* **116**: 43–59. doi: 10.1016/0009-2541(94)90157-0.

Brett, E.K.A., Prytulak, J., Rehkämper, M., et al. 2021. Thallium elemental and isotopic systematics in ocean island lavas. *Geochimica et Cosmochimica Acta* **301**: 187–210. doi: 10.1016/j.gca.2021.02.035.

Burkhardt, C., Hin, R.C., Kleine, T., et al. 2014. Evidence for Mo isotope fractionation in the solar nebula and during planetary differentiation. *Earth and Planetary Science Letters* **391**: 201–11. doi: 10.1016/j.epsl.2014.01.037.

Came, R.E., Eiler, J.M., Veizer, J., et al. 2007. Coupling of surface temperatures and atmospheric CO_2 concentrations during the Palaeozoic era. *Nature* **449**(7159): 198–201. doi: 10.1038/Nature06085.

Cano, E.J., Sharp, Z.D. and Shearer, C.K. 2020. Distinct oxygen isotope compositions of the earth and moon. *Nature GeoScience* **13**: 270–4. doi: 10.1038/s41561-020-0550-0.

Carignan, J. and Wen, H. 2007. Scaling NIST SRM 3149 for Se isotope analysis and isotopic variations of natural samples. *Chemical Geology* **242**: 347–50. doi: 10.1016/j.chemgeo.2007.03.020.

Carpentier, M., Chauvel, C. and Mattielli, N. 2008. Pb–Nd isotopic constraints on sedimentary input into the Lesser Antilles arc system. *Earth and Planetary Science Letters* **272**: 199–211. doi: 10.1016/j.epsl.2008.04.036.

Cartigny, P. and Marty, B. 2013. Nitrogen isotopes and mantle geodynamics: the emergence of life and the atmosphere-crust–mantle connection. *Elements* **9**: 359–66. doi: 10.2113/gselements.9.5.359.

Cartigny, P., Palot, M., Thomassot, E., et al. 2014. Diamond formation: a stable isotope perspective. *Annual Reviews of Earth and Planetary Science* **42**: 699–732. doi: 10.1146/annurev-earth-042711-105259.

Chacko, T. and Deines, P. 2008. Theoretical calculation of oxygen isotope fractionation factors in carbonate systems. *Geochimica et Cosmochimica Acta* **72**: 3642–60. doi: 10.1016/j.gca.2008.06.001.

Chacko, T., Cole, D. and Horita, J. 2001. Equilibrium oxygen, hydrogen and carbon isotope fractionation factors applicable to geologic systems, in Valley, J.W. (ed). *Stable Isotope Geochemistry*. Washington: Mineralogical Society of America, 1–61 pp.

Chen, C., Ciazela, J., Li, W., et al. 2020. Calcium isotopic compositions of oceanic crust at various spreading rates. *Geochimica et Cosmochimica Acta* **278**: 272–88. doi: 10.1016/j.gca.2019.07.008.

Chen, H., Savage, P.S., Teng, F.-Z., et al. 2013. Zinc isotope fractionation during magmatic differentiation and the isotopic composition of the bulk earth. *Earth and Planetary Science Letters* **369–70**: 34–42. doi: 10.1016/j.epsl.2013.02.037.

Chen, S., Hin, R.C., John, T., et al. 2019. Molybdenum systematics of subducted crust record reactive fluid flow from underlying slab serpentine dehydration. *Nature Communications* **10**: 4773. doi: 10.1038/s41467-019-12696-3.

CIAAW (Committee on Isotopic Abundances and Atomic Weights). 2020. *Atomic Weights of the Elements* [Online]. International Union of Pure and Applied Chemistry. [Accessed 2021].

Craddock, P.R., Warren, J.M. and Dauphas, N. 2013. Abyssal peridotites reveal the near-chondritic Fe isotopic composition of the earth. *Earth and Planetary Science Letters* **365**: 63–76. doi: 10.1016/j.epsl.2013.01.011.

Craig, H. 1963. The isotopic composition of water and carbon in geothermal areas, in Tongiorgi, E. (ed). *Nuclear Geology on Geothermal Areas*. Pisa: CNR Lab. Geol. Nucl, 17–53 pp.

Creech, J.B. and Moynier, F. 2019. Tin and zinc stable isotope characterisation of chondrites and implications for early solar system evolution. *Chemical Geology* **511**: 81–90. doi: 10.1016/j.chemgeo.2019.02.028.

Criss, R.E. and Taylor, H.P. 1986. Meteoric hydrothermal systems, in Valley, J.W., Taylor H.P. and O'Neil, J.R. (eds). *Stable Isotopes in High Temperature Geological Processes, Reviews in Mineralogy 16*. Washington: Mineralogical Society of America, 373–424 pp.

Dauphas, N., John, S.G. and Rouxel, O. 2017. Iron isotope systematics, in Teng, F.-Z., Dauphas, N. and Watkins, J.M. (eds). *Non-traditional Stable Isotopes*. Washington DC: Mineralogical Society of America.

Dauphas, N., Poitrasson, F., Burkhardt, C., et al. 2015. Planetary and meteoritic Mg/Si and δ^{30}Si variations inherited from solar nebula chemistry. *Earth and Planetary Science Letters* **427**: 236–48. doi: 10.1016/j.epsl.2015.07.008.

De Hoog, J.C.M. and Savov, I.P. 2018. Boron isotopes as a tracer of subduction zone processes, in Marschall, H. and Foster, G. (eds). *Boron Isotopes: The Fifth Element*. Cham: Springer International Publishing.

de Hoog, J.C.M., Taylor, B.E. and Van Bergen, M.J. 2001. Sulfur isotope systematics of basaltic lavas from Indonesia: implications for the sulfur cycle in subduction zones. *Earth and Planetary Science Letters* **189**: 237–52. doi: 10.1016/S0012-821X(01)00355-7.

Delavault, H., Chauvel, C., Thomassot, E., et al. 2016. Sulfur and lead isotopic evidence of relic Archean sediments in the Pitcairn mantle plume. *Proceedings of the National Academy of Sciences* **113**: 12952–6. doi: 10.1073/pnas.1523805113.

Deng, C., Sun, G., Rong, Y., et al. 2020. Recycling of mercury from the atmosphere-ocean system into volcanic-arc–associated epithermal gold systems. *Geology* **49**(3): 309–13. doi: 10.1130/g48132.1.

Deng, Z., Moynier, F., Sossi, P.A., et al. 2018. Bridging the depleted MORB mantle and the continental crust using titanium isotopes. *Geochemical Perspectives Letters* **9**: 11–15. doi: 10.7185/geochemlet.1831.

Ding, X., Helz, R.T., Qi, Y., et al. 2020. Vanadium isotope fractionation during differentiation of Kilauea Iki lava lake, *Hawaii*. *Geochimica et Cosmochimica Acta* **289**: 114–29. doi: 10.1016/j.gca.2020.08.023.

Dixon, J.E., Bindeman, I.N., Kingsley, R.H., et al. 2017. Light stable isotopic compositions of enriched mantle sources: resolving the dehydration paradox. *Geochemistry, Geophysics, Geosystems* **18**: 3801–39. doi: 10.1002/2016GC006743.

Doucet, L.S., Mattielli, N., Ionov, D.A., et al. 2016. Zn isotopic heterogeneity in the mantle: A melting control? *Earth and Planetary Science Letters* **451**: 232–240. doi: https://doi.org/10.1016/j.epsl.2016.06.040.

Dottin III, J.W., Labidi, J., Jackson, M.G., et al. 2020. Isotopic evidence for multiple recycled sulfur reservoirs in the Mangaia mantle plume. *Geochemistry, Geophysics, Geosystems* **21**: e2020GC009081. doi: https://doi.org/10.1029/2020GC009081.

El Desouky, H.A., Muchez, P., Boyce, A.J., et al. 2010. Genesis of sediment-hosted stratiform copper–cobalt mineralization at Luiswishi and Kamoto, Katanga Copperbelt (Democratic Republic of Congo). *Mineralium Deposita* **45**(8): 735–63. doi: 10.1007/s00126-010-0298-3.

Eldridge, D.L., Guo, W. and Farquhar, J. 2016. Theoretical estimates of equilibrium sulfur isotope effects in aqueous sulfur systems: Highlighting the role of isomers in the sulfite and sulfoxylate systems. *Geochimica et Cosmochimica Acta* **195**: 171–200. doi: 10.1016/j.gca.2016.09.021.

Farkaš, J., Chrastný, V., Novák, M., et al. 2013. Chromium isotope variations ($\delta^{53/52}$Cr) in mantle-derived sources and their weathering products: implications for environmental studies and the evolution of $\delta^{53/52}$Cr in the Earth's mantle over geologic time. *Geochimica et Cosmochimica Acta* **123**: 74–92. doi: 10.1016/j.gca.2013.08.016.

Farquhar, J., Wing, B.A., Mckeegan, K.D., et al. 2002. Mass-independent fractionation sulfur of inclusions in diamond and sulfur recycling on early earth. *Science* **298**: 2369–72. doi: 10.1126/science.1078617.

Fitoussi, C., Bourdon, B., Kleine, T., et al. 2009. Si isotope systematics of meteorites and terrestrial peridotites: implications for Mg/Si fractionation in the solar nebula and for Si in the Earth's core. *Earth and Planetary Science Letters* **287**: 77–85. doi: 10.1016/j.epsl.2009.07.038.

Fitzpayne, A., Prytulak, J., Wilkinson, J.J., et al. 2018. Assessing thallium elemental systematics and isotope ratio variations in porphyry ore systems: a case study of the Bingham Canyon district. *Minerals* **8**: 548. doi: 10.3390/min8120548

Florin, G., Luais, B., Rushmer, T., et al. 2020. Influence of redox processes on the germanium isotopic composition of ordinary chondrites. *Geochimica et Cosmochimica Acta* **269**: 270–91. doi: 10.1016/j.gca.2019.10.038.

Foden, J., Sossi, P.A. and Nebel, O. 2018. Controls on the iron isotopic composition of global arc magmas. *Earth and Planetary Science Letters* **494**: 190–201. doi: 10.1016/j.epsl.2018.04.039.

Förster, M.W., Foley, S.F., Alard, O., et al. 2019. Partitioning of nitrogen during melting and recycling in subduction zones and the evolution of atmospheric nitrogen. *Chemical Geology* **525**: 334–42. doi: 10.1016/j.chemgeo.2019.07.042.

Friebel, M., Toth, E.R., Fehr, M.A., et al. 2020. Efficient separation and high-precision analyses of tin and cadmium isotopes in geological materials. *Journal of Analytical Atomic Spectrometry* **35**: 273–292. doi:

Freymuth, H., Andersen, M.B. and Elliott, T. 2019. Uranium isotope fractionation during slab dehydration beneath the Izu arc. *Earth and Planetary Science Letters* **522**: 244–54. doi: 10.1016/j.epsl.2019.07.006.

Freymuth, H., Elliott, T., Van Soest, M., et al. 2016. Tracing subducted black shales in the lesser antilles arc using molybdenum isotope ratios. *Geology* **44**: 987–90. doi: 10.1130/G38344.1.

Gagnevin, D., Boyce, A.J., Barrie, C.D., et al. 2012. Zn, Fe and S isotope fractionation in a large hydrothermal system. *Geochimica et Cosmochimica Acta* **88**: 183–198. doi: https://doi.org/10.1016/j.gca.2012.04.031.

Gao, Y., Vils, F., Cooper, K.M., et al. 2012. Downhole variation of lithium and oxygen isotopic compositions of oceanic crust at East Pacific Rise, ODP Site 1256. *Geochemistry, Geophysics, Geosystems* **13**: Q10001. doi:10.1029/2012gc004207.

Gaschnig, R.M., Rader, S.T., Reinhard, C.T., et al. 2021a. Behavior of the Mo, Tl, and U isotope systems during differentiation in the Kilauea Iki lava lake. *Chemical Geology* **574**: 120239. doi: 10.1016/j.chemgeo.2021.120239.

Gaschnig, R.M., Reinhard, C.T., Planavsky, N.J., et al. 2021b. The impact of primary processes and secondary alteration on the stable isotope composition of ocean island basalts. *Chemical Geology* **581**: 120416. doi: 10.1016/j.chemgeo.2021.120416.

Genske, F.S., Turner, S.P., Beier, C., et al. 2014. Lithium and boron isotope systematics in lavas from the Azores Islands reveal crustal assimilation. *Chemical Geology* **373**: 27–36. doi: 10.1016/j.chemgeo.2014.02.024.

Ghosh, S., Schauble, E.A., Lacrampe Couloume, G., et al. 2013. Estimation of nuclear volume dependent fractionation of mercury isotopes in equilibrium liquid–vapor evaporation experiments. *Chemical Geology* **336**: 5–12. doi: 10.1016/j. chemgeo.2012.01.008.

Goldberg, T., Gordon, G., Izon, G., et al. 2013. Resolution of inter-laboratory discrepancies in Mo isotope data: an inter-calibration. *Journal of Analytical Atomic Spectrometry* **28**: 724–735. doi: 10.1039/C3JA30375F.

Goldman, A., Brennecka, G., Noordmann, J., et al. 2015. The uranium isotopic composition of the Earth and the Solar System. *Geochimica et Cosmochimica Acta* **148**: 145–58. doi: 10.1016/j.gca.2014.09.008.

Greber, N.D., Pettke, T. and Nägler, T.F. 2014. Magmatic–hydrothermal molybdenum isotope fractionation and its relevance to the igneous crustal signature. *Lithos* **190–1**: 104–10. doi: 10.1016/j.lithos.2013.11.006.

Gregory, M.J. and Mathur, R. 2017. Understanding copper isotope behavior in the high temperature magmatic-hydrothermal porphyry environment. *Geochemistry, Geophysics, Geosystems* **18**: 4000–15. doi: 10.1002/2017GC007026.

Gregory, R.T. and Taylor, H.P. 1981. An oxygen isotope profile in a section of Cretaceous oceanic crust, Samail Opiolite, Oman: evidence for $\delta^{18}O$ buffering of the oceans by deep (>5 km) seawater-hydrothermal circulation at mid-ocean ridges. *Journal of Geophysical Research* **86**: 2737–55.

Halldórsson, S.A., Hilton, D.R., Barry, P.H., et al. 2016. Recycling of crustal material by the Iceland mantle plume: new evidence from nitrogen elemental and isotope systematics of subglacial basalts. *Geochimica et Cosmochimica Acta* **176**: 206–26. doi: 10.1016/j.gca.2015.12.021.

Harmon, R.S. and Hoefs, J. 1995. Oxygen isotope heterogeneity of the mantle deduced from global $\delta^{18}O$ systematics of basalts from different geotectonic settings. *Contributions to Mineralogy and Petrology* **120**: 95–114. doi: 10.1007/bf00311010.

Hin, R.C., Burkhardt, C., Schmidt, M.W., et al. 2013. Experimental evidence for Mo isotope fractionation between metal and silicate liquids. *Earth and Planetary Science Letters* **379**: 38–48. doi: 10.1016/j.epsl.2013.08.003.

Huang, F., Chen, L., Wu, Z., et al. 2013. First-principles calculations of equilibrium mg isotope fractionations between garnet, clinopyroxene, orthopyroxene, and olivine: implications for Mg isotope thermometry. *Earth and Planetary Science Letters* **367**: 61–70. doi: 10.1016/j.epsl.2013.02.025.

Huang, J., Huang, F., Wang, Z., et al. 2017. Copper isotope fractionation during partial melting and melt percolation in the upper mantle: evidence from massif peridotites in Ivrea-Verbano zone, Italian alps. *Geochimica et Cosmochimica Acta* **211**: 48–63. doi: 10.1016/j.gca.2017.05.007.

Huang, S., Farkaš, J. and Jacobsen, S.B. 2011. Stable calcium isotopic compositions of Hawaiian shield lavas: evidence for recycling of ancient marine carbonates into the mantle. *Geochimica et Cosmochimica Acta* **75**: 4987–97.

Huston, D.L. 1997. Stable isotopes and their significance for understanding the genesis of volcanic-hosted massive sulfide deposits: a review, in Barrie, C.T. and Hannington, M.T. (eds). *Volcanic Associated Massive Sulfide Deposits: Processes and Examples in Modern and Ancient Settings*. Littleton, CO: Society of Economic Geologists, 157–180 pp.

Hutchison, W., Finch, A.A. and Boyce, A.J. 2020. The sulfur isotope evolution of magmatic-hydrothermal fluids: insights into ore-forming processes. *Geochimica et Cosmochimica Acta* **288**: 176–198. doi: https://doi.org/10.1016/j.gca.2020.07.042.

Javoy, M. and Pineau, F. 1991. The volatile record of a "popping" rock from the Mid-Atlantic Ridge at 14° N: chemical and isotopic composition of the gas trapped in the vesicles. *Earth and Planetary Science Letters* **107**: 598–611.

Jerram, M., Bonnand, P., Kerr, A.C., et al. 2020. The $\delta^{53}Cr$ isotope composition of komatiite flows and implications for the composition of the bulk silicate Earth. *Chemical Geology* **551**: 119761. doi: 10.1016/j.chemgeo.2020.119761.

Jochum, K.P., Hofmann, A.W. and Seufert, H.M. 1993. Tin in mantle-derived rocks: constraints on earth evolution. *Geochimica Et Cosmochimica Acta* **57**: 3585–95. doi: 10.1016/0016-7037(93)90141-I.

John, T., Layne, G.D., Haase, K.M., et al. 2010. Chlorine isotope evidence for crustal recycling into the earth's mantle. *Earth and Planetary Science Letters* **298**: 175–82. doi: 10.1016/j.epsl.2010.07.039.

Kang, J.-T., Ionov, D.A., Liu, F., et al. 2017. Calcium isotopic fractionation in mantle peridotites by melting and metasomatism and Ca isotope composition of the bulk silicate Earth. *Earth and Planetary Science Letters* **474**: 128–37. doi: 10.1016/j.epsl.2017.05.035.

Kendall, B., Komiya, T., Lyons, T.W., et al. 2015. Uranium and molybdenum isotope evidence for an episode of widespread ocean oxygenation during the late Ediacaran Period. *Geochimica et Cosmochimica Acta* **156**: 173–93. doi: 10.1016/j.gca.2015.02.025.

Kubik, E., Siebert, J., Mahan, B., et al. 2021. Tracing earth's volatile delivery with tin. *Journal of Geophysical Research: Solid Earth* **126**: e2021JB022026. doi: 10.1029/2021JB022026.

Kurzawa, T., König, S., Alt, J.C., et al. 2019. The role of subduction recycling on the selenium isotope signature of the mantle: constraints from Mariana arc lavas. *Chemical Geology* **513**: 239–49. doi: 10.1016/j.chemgeo.2019.03.011.

Kyser, K.T. and O'Neil, J.R. 1984. Hydrogen isotopes systematics of submarine basalts. *Geochimica Cosmochimica Acta* **48**: 2123–33.

Labidi, J., Cartigny, P. and Jackson, M.G. 2015. Multiple sulfur isotope composition of oxidized Samoan melts and the implications of a sulfur isotope 'mantle array' in chemical geodynamics. *Earth and Planetary Science Letters* **417**: 28–39. doi: 10.1016/j.epsl.2015.02.004.

Labidi, J., Cartigny, P. and Moreira, M. 2013. Non-chondritic sulphur isotope composition of the terrestrial mantle. *Nature* **501**: 208–11. doi: 10.1038/Nature12490

Labidi, J. and Cartigny, P. 2016. Negligible sulfur isotope fractionation during partial melting: Evidence from Garrett transform fault basalts, implications for the late-veneer and the hadean matte. *Earth and Planetary Science Letters* **451**: 196–207. doi: https://doi.org/10.1016/j.epsl.2016.07.012.

Labidi, J., König, S., Kurzawa, T., et al. 2018. The selenium isotopic variations in chondrites are mass-dependent; Implications for sulfide formation in the early solar system. *Earth and Planetary Science Letters* **481**: 212–222. doi: https://doi.org/10.1016/j.epsl.2017.10.032.

Larson, P.B., Maher, K., Ramos, F.C., et al. 2003. Copper isotope ratios in magmatic and hydrothermal ore-forming environments. *Chemical Geology* **201**: 337–50. doi: 10.1016/j.chemgeo.2003.08.006.

Lee, C.-T.A., Luffi, P., Chin, E.J., et al. 2012. Copper systematics in arc magmas and implications for crust-mantle differentiation. *Science* **336**: 64–68. doi: 10.1126/science.1217313.

Li, H.-Y., Zhao, R.-P., Li, J., et al. 2021. Molybdenum isotopes unmask slab dehydration and melting beneath the Mariana arc. *Nature Communications* **12**: 6015. doi: 10.1038/s41467-021-26322-8.

Li, W., Jackson, S.E., Pearson, N.J., et al. 2010. Copper isotopic zonation in the Northparkes porphyry Cu–Au deposit, SE Australia. *Geochimica et Cosmochimica Acta* **74**: 4078–96. doi: 10.1016/j.gca.2010.04.003.

Li, Y. and Liu, J. 2006. Calculation of sulfur isotope fractionation in sulfides. *Geochimica et Cosmochimica Acta* **70**: 1789–95. doi: https://doi.org/10.1016/j.gca.2005.12.015.

Li, X.-H., Wang, J.-Q., Chu, F.-Y., et al. 2018. Copper isotopic compositions of hydrothermal sulfides from the mid-ocean ridge and implications for later oxidation processes. *Geochemical Journal* **52**: 29–36. doi: 10.2343/geochemj.2.0494.

Liang, Y.-H., Halliday, A.N., Siebert, C., et al. 2017. Molybdenum isotope fractionation in the mantle. *Geochimica et Cosmochimica Acta* **199**: 91–111. doi: 10.1016/j.gca.2016.11.023.

Liu, S.-A., Huang, J., Liu, J., et al. 2015. Copper isotopic composition of the silicate earth. *Earth and Planetary Science Letters* **427**: 95–103. doi: 10.1016/j.epsl.2015.06.061.

Loewen, M.W., Graham, D.W., Bindeman, I.N., et al. 2019. Hydrogen isotopes in high ^{3}He/^{4}He submarine basalts: primordial vs. recycled water and the veil of mantle enrichment. *Earth and Planetary Science Letters* **508**: 62–73. doi: 10.1016/j.epsl.2018.12.012.

Luais, B. 2012. Germanium chemistry and MC-ICPMS isotopic measurements of Fe-Ni, Zn alloys and silicate matrices: insights into deep earth processes. *Chemical Geology* **334**: 295–311. doi: 10.1016/j.chemgeo.2012.10.017

Luck, J.M., Othman, D.B., Barrat, J.A., et al. 2003. Coupled ^{63}Cu and ^{16}O excesses in chondrites. *Geochimica et Cosmochimica Acta* **67**: 143–51. doi: 10.1016/s0016-7037(02)01038-4.

Luck, J.-M., Othman, D.B. and Albarède, F. 2005. Zn and Cu isotopic variations in chondrites and iron meteorites: early solar nebula reservoirs and parent-body processes. *Geochimica et Cosmochimica Acta* **69**: 5351–63. doi: 10.1016/j.gca.2005.06.018.

Magalhães, N., Penniston-Dorland, S., Farquhar, J., et al. 2018. Variable sulfur isotope composition of sulfides provide evidence for multiple sources of contamination in the Rustenburg Layered Suite, Bushveld Complex. *Earth and Planetary Science Letters* **492**: 163–73. doi: 0.1016/j.epsl.2018.04.010.

Maher, K.C., Jackson, S. and Mountain, B. 2011. Experimental evaluation of the fluid–mineral fractionation of Cu isotopes at 250°C and 300°C. *Chemical Geology* **286**: 229–39. doi: 10.1016/j.chemgeo.2011.05.008.

Maier, W.D., Barnes, S.-J. and Groves, D.I. 2013. The Bushveld Complex, South Africa: formation of platinum-palladium, chrome- and vanadium-rich layers via hydrodynamic sorting of a mobilized cumulate slurry in a large, relatively slowly cooling, subsiding magma chamber. *Mineralium Deposita* **48**(1): 1–56. doi: 10.1007/s00126-012-0436-1.

Markl, G., Lahaye, Y. and Schwinn, G. 2006. Copper isotopes as monitors of redox processes in hydrothermal mineralization. *Geochimica et Cosmochimica Acta* **70**: 4215–28. doi: 10.1016/j.gca.2006.06.1369.

Marschall, H.R., Wanless, V.D., Shimizu, N., et al. 2017. The boron and lithium isotopic composition of mid-ocean ridge basalts and the mantle. *Geochimica et Cosmochimica Acta* **207**: 102–38. doi: 10.1016/j.gca.2017.03.028.

Martin, A.J., Keith, M., Parvaz, D.B., et al. 2020. Effects of magmatic volatile influx in mafic VMS hydrothermal systems: evidence from the Troodos ophiolite, *Cyprus*. *Chemical Geology* **531**: 119325. doi: 10.1016/j.chemgeo.2019.119325.

Marty, B. and Dauphas, N. 2003. The nitrogen record of crust-mantle interaction and mantle convection from Archean to present. *Earth and Planetary Science Letters* **206**: 397–410.

Mathur, R. and Wang, D. 2019. Transition metal isotopes applied to exploration geochemistry: insights from Fe, Cu, and Zn, in Decrée, S. and Robb, L. (eds). *Ore deposits: Origin, Exploration, and Exploitation*. Washington DC: American Geophysical Union. doi: 10.1002/9781119290544.ch7.

Mathur, R., Dendas, M., Titley, S., et al. 2010. Patterns in the copper isotope composition of minerals in porphyry copper deposits in southwestern United States. *Economic Geology* **105**: 1457–67. doi: 10.2113/econgeo.105.8.1457.

Mathur, R., Titley, S., Barra, F., et al. 2009. Exploration potential of Cu isotope fractionation in porphyry copper deposits. *Journal of Geochemical Exploration* **102**: 1–6. doi: 10.1016/j.gexplo.2008.09.004.

Mattey, D.P. 1987. Carbon isotopes in the mantle. *Terra Cognita* **7**: 31–8.

Mattey, D., Lowry, D. and Macpherson, C. 1994. Oxygen isotope composition of mantle peridotite. *Earth and Planetary Science Letters* **128**: 231–41. doi: 10.1016/0012-821X(94)90147-3.

McCulloch, M.T., Gregory, R.T., Wasserburg, G.J., et al. 1981. Sm-Nd, Rb-Sr and $^{18}O/^{16}O$ isotopic systematics in an oceanic crustal section: evidence for the Samail ophiolite. *Journal of Geophysical Research* **86**: 2721.

McDermott, J.M., Ono, S., Tivey, M.K., et al. 2015. Identification of sulfur sources and isotopic equilibria in submarine hot-springs using multiple sulfur isotopes. *Geochimica et Cosmochimica Acta* **160**: 169–87. doi: 10.1016/j.gca.2015.02.016.

Meier, M.M.M., Cloquet, C. and Marty, B. 2016. Mercury (Hg) in meteorites: variations in abundance, thermal release profile, mass-dependent and mass-independent isotopic fractionation. *Geochimica et Cosmochimica Acta* **182**: 55–72. doi: 10.1016/j.gca.2016.03.007.

Michael, P.J. and Cornell, W.C. 1998. Influence of spreading rate and magma supply on crystallization and assimilation beneath mid-ocean ridges: evidence from chlorine and major element chemistry of mid-ocean ridge basalts. *Journal of Geophysical. Research* **103**: 18325–56. doi: 10.1029/98jb00791.

Moynier, F., Chen, J., Zhang, K., et al. 2020. Chondritic mercury isotopic composition of Earth and evidence for evaporative equilibrium degassing during the formation of eucrites. *Earth and Planetary Science Letters* **551**: 116544. doi: 10.1016/j.epsl.2020.116544.

Moynier, F., Deng, Z., Lanteri, A., et al. 2020. Metal-silicate silicon isotopic fractionation and the composition of the bulk Earth. *Earth and Planetary Science Letters* **549**: 116468. doi: 10.1016/j.epsl.2020.116468.

Moynier, F., Jackson, M.G., Zhang, K., et al. 2021. The mercury isotopic composition of Earth's mantle and the use of mass independently fractionated Hg to test for recycled crust. *Geophysical Research Letters* **48**(17): e2021GL094301. doi: 10.1029/2021GL094301.

Moynier, F., Vance, D., Fujii, T., et al. (2017). The isotope geochemistry of zinc and copper, in Teng, F.-Z., Dauphas, N. & Watkins, J. M. (eds.) *Non-Traditional Stable Isotopes*. Washington, DC: Mineralogical Society of America, 543–600 pp. doi: 10.2138/rmg.2017.82.13.

Muehlenbachs, C. and Furnas, H. 2003. Ophiolites as faithful records of the oxygen isotope ratio of ancient seawater: the Solund-Stavfjord Ophiolite Complex as a Late Ordovician example, in: Dilek, Y. and Robinson, P. (eds). *Ophiolites in Earth History*. London: Geological Society of London, 401–414 pp.

Muehlenbachs, K. 1976. Oxygen isotope composition of the oceanic crust and its bearing on seawater. *Journal of Geophysical Research* **81**: 4365–9.

Muehlenbachs, K. and Clayton, R. 1972. Oxygen isotope geochemistry of submarine greenstones. *Canadian Journal of Earth Sciences* **9**: 471–8.

Nielsen, S.G., Auro, M., Righter, K., et al. 2019. Nucleosynthetic vanadium isotope heterogeneity of the early solar system recorded in chondritic meteorites. *Earth and Planetary Science Letters* **505**: 131–40. doi: 10.1016/j.epsl.2018.10.029.

Nielsen, S.G., Prytulak, J., Blusztajn, J., et al. 2017. Thallium isotopes as tracers of recycled materials in subduction zones: review and new data for lavas from Tonga-Kermadec and Central America. *Journal of Volcanology and Geothermal Research* **339**(SupplC): 23–40. doi: 10.1016/j.jvolgeores.2017.04.024.

Nielsen, S.G., Rehkämper, M., Norman., et al. 2006. Thallium isotopic evidence for ferromanganese sediments in the mantle source of Hawaiian basalts. *Nature* **439**(7074): 314–7. doi: 10.1038/Nature04450.

Nielsen, S.G., Shu, Y., Wood, B.J., et al 2021. Thallium isotope fractionation during magma degassing: evidence from experiments and Kamchatka Arc lavas. *Geochemistry, Geophysics, Geosystems* **22**(5): e2020GC009608. doi: 10.1029/2020GC009608.

Nielsen, S.G., Yogodzinski, G., Prytulak, J., et al. 2016. Tracking along-arc sediment inputs to the Aleutian arc using thallium isotopes. *Geochimica et Cosmochimica Acta* **181**: 217–37. doi: 10.1016/j.gca.2016.03.010.

Noordmann, J., Weyer, S., Georg, R.B., et al. 2016. $^{238}U/^{235}U$ isotope ratios of crustal material, rivers and products of hydrothermal alteration: new insights on the oceanic U isotope mass balance. *Isotopes in Environmental and Health Studies* **52**(1–2): 141–63. doi: 10.1080/10256016.2015.1047449.

Ohmoto, H. and Rye, R.O. 1979. Isotopes of sulfur and carbon, in Barnes, H. (ed). *Geochemistry of Hydrothermal Ore Deposits*. New York: John Wiley and Sons.

Ono, S., Shanks, W.C., Rouxel, O.J., et al. 2007. S-33 constraints on the seawater sulfate contribution in modern seafloor hydrothermal vent sulfides. *Geochimica et Cosmochimica Acta* **71**(5): 1170–82. doi: 10.1016/j.gca.2006.11.017.

Palk, E., Andreasen, R., Rehkämper, M., et al. 2018. Variable Tl, Pb, and Cd concentrations and isotope compositions of enstatite and ordinary chondrites—Evidence for volatile element mobilization and decay of extinct 205Pb. *Meteoritics & Planetary Science* **53**(2): 167–86. doi: 10.1111/maps.12989.

Palmer, M.R. 2017. Boron cycling in subduction zones. *Elements* **13**: 237–42. doi: 10.2138/gselements.13.4.237.

Palot, M., Cartigny, P., Harris, J.W., et al. 2012. Evidence for deep mantle convection and primordial heterogeneity from nitrogen and carbon stable isotopes in diamond. *Earth and Planetary Science Letters* **357–8**: 179–93. doi: 10.1016/j.epsl.2012.09.015.

Penniston-Dorland, S.C., Mathez, E.A., Wing, B.A., et al. 2012. Multiple sulfur isotope evidence for surface-derived sulfur in the Bushveld Complex. *Earth and Planetary Science Letters* **337–338**: 236–242. doi: https://doi.org/10.1016/j.epsl.2012.05.013.

Pogge Von Strandmann, P.a.E., Elliott, T., Marschall, H.R., et al. 2011. Variations of LI and Mg isotope ratios in bulk chondrites and mantle xenoliths. *Geochimica et Cosmochimica Acta* **75**: 5247–68. doi: 10.1016/j.gca.2011.06.026.

Poitrasson, F., Levasseur, S. and Teutsch, N. 2005. Significance of iron isotope mineral fractionation in pallasites and iron meteorites for the core–mantle differentiation of terrestrial planets. *Earth and Planetary Science Letters* **234**: 151–64. doi: 10.1016/j.epsl.2005.02.010.

Pringle, E.A., Moynier, F., Savage, P.S., et al. 2016. Silicon isotopes reveal recycled altered oceanic crust in the mantle sources of ocean island basalts. *Geochimica et Cosmochimica Acta* **189**: 282–95. doi: 10.1016/j.gca.2016.06.008.

Prytulak, J., Sossi, P.A., Halliday, A.N., et al. 2017. Stable vanadium isotopes as a redox proxy in magmatic systems? *Geochemical Perspectives Letters* **3**: 75–84. doi: 10.7185/geochemlet.1708.

Qi, Y.-H., Wu, F., Ionov, D.A., et al. 2019. Vanadium isotope composition of the bulk silicate Earth: constraints from peridotites and komatiites. *Geochimica et Cosmochimica Acta* **259**: 288–301. doi: 10.1016/j.gca.2019.06.008.

Qin, T., Wu, F., Wu, Z., et al. 2016. First-principles calculations of equilibrium fractionation of O and Si isotopes in quartz, albite, anorthite, and zircon. *Contributions to Mineralogy and Petrology* **171**(11): 91. doi: 10.1007/s00410-016-1303-3.

Rabin, S., Blanchard, M., Pinilla, C., et al. 2021. First-principles calculation of iron and silicon isotope fractionation between Fe-bearing minerals at magmatic temperatures: the importance of second atomic neighbors. *Geochimica et Cosmochimica Acta* **304**: 101–18. doi: 10.1016/j.gca.2021.03.028.

Rader, S.T., Mazdab, F.K. and Barton, M.D. 2018. Mineralogical thallium geochemistry and isotope variations from igneous, metamorphic, and metasomatic systems. *Geochimica et Cosmochimica Acta* **243**: 42–65. doi: 10.1016/j.gca.2018.09.019.

Richards, J.P. 2015. The oxidation state, and sulfur and Cu contents of arc magmas: implications for metallogeny. *Lithos* **233**: 27–45. doi: 10.1016/j.lithos.2014.12.011.

Roskosz, M., Amet, Q., Fitoussi, C., et al. 2020. Redox and structural controls on tin isotopic fractionations among magmas. *Geochimica et Cosmochimica Acta* **268**: 42–55. doi: https://doi.org/10.1016/j.gca.2019.09.036.

Rouxel, O.J. and Luais, B. 2017. Germanium isotope geochemistry. *Reviews in Mineralogy and Geochemistry* **82**: 601–56. doi: 10.2138/rmg.2017.82.14.

Rouxel, O., Fouquet, Y. and Ludden, J.N. 2004. Copper isotope systematics of the Lucky Strike, Rainbow, and Logatchev sea-floor hydrothermal fields on the Mid-Atlantic Ridge. *Economic Geology* **99**: 585–600. doi: 10.2113/gsecongeo.99.3.585.

Ryu, J.-S., Jacobson, A.D., Holmden, C., et al. 2011. The major ion, $\delta^{44/40}$Ca, $\delta^{44/42}$Ca, and $\delta^{26/24}$Mg geochemistry of granite weathering at pH=1 and T=25°C: power-law processes and the relative reactivity of minerals. *Geochimica et Cosmochimica Acta* **75**: 6004–26. doi: 10.1016/j.gca.2011.07.025.

Saintilan, N.J., Selby, D., Creaser, R.A., et al. 2018. Sulphide Re-Os geochronology links orogenesis, salt and Cu-Co ores in the Central African Copperbelt. *Scientific Reports* **8**(1): 14946. doi: 10.1038/s41598-018-33399-7.

Savage, P.S., Georg, R.B., Williams, H.M., et al. 2012. The silicon isotope composition of granites. *Geochimica et Cosmochimica Acta* **92**: 184–202. doi: 10.1016/j.gca.2012.06.017.

Savage, P.S., Georg, R.B., Williams, H.M., et al. 2012. The silicon isotope composition of granites. *Geochimica et Cosmochimica Acta* **92**: 184–202. doi: 10.1016/j.gca.2012.06.017.

Savage, P.S., Moynier, F., Chen, H., et al. 2015. Copper isotope evidence for large-scale sulphide fractionation during earth's differentiation. *Geochemical Perspectives Letters* **1**: 53–64. doi: 10.7185/geochemlet.1506.

Savage, P.(2018). Copper isotopes in: White, W. M. (ed.) *Encyclopedia of Geochemistry*. Cham: Springer International Publishing, 305–309 pp. doi: 10.1007/978-3-319-39312-4_282.

Schauble, E.A. 2011. First-principles estimates of equilibrium magnesium isotope fractionation in silicate, oxide, carbonate and hexaaquamagnesium (2+) crystals. *Geochimica et Cosmochimica Acta* **75**: 844–69. doi: 10.1016/j.gca.2010.09.044.

Schauble, E.A., Rossman, G.R. and Taylor Jr, H.P. 2003. Theoretical estimates of equilibrium chlorine-isotope fractionations. *Geochimica et Cosmochimica Acta* **67**: 3267–81. doi: 10.1016/s0016-7037(02)01375-3.

Schoenberg, R., Merdian, A., Holmden, C., et al. 2016. The stable Cr isotopic compositions of chondrites and silicate planetary reservoirs. *Geochimica et Cosmochimica Acta* **183**: 14–30. doi: 10.1016/j.gca.2016.03.013.

Schoenberg, R., Zink, S., Staubwasser, M., et al. 2008. The stable Cr isotope inventory of solid earth reservoirs determined by double spike MC-ICP-MS. *Chemical Geology* **249**: 294–306. doi: 10.1016/j.chemgeo.2008.01.009.

Schütze, H., 1980. Der Isotopenindex-Eine Inkrementenmethode zur näherungsweisen Berechnung von Isotopenaustauschgleichgewichten zwischen kristallinen Substanzen. *Chemie der Erde* **39**: 321–4.

Seo, J.H., Lee, S.K. and Lee, I. 2007. Quantum chemical calculations of equilibrium copper(I) isotope fractionations in ore-forming fluids. *Chemical Geology* **243**: 225–37. doi: 0.1016/j.chemgeo.2007.05.025.

Shahar, A., Hillgren, V.J., Young, E.D., et al. 2011. High-temperature Si isotope fractionation between iron metal and silicate. *Geochimica et Cosmochimica Acta* **75**: 7688–97. doi: 10.1016/j.gca.2011.09.038.

Shahar, A. and Young, E.D. 2020. An assessment of iron isotope fractionation during core formation. *Chemical Geology* **554**: 119800. https://doi.org/10.1016/j.chemgeo.2020.119800.

Sharman, E.R., Penniston-Dorland, S.C., Kinnaird, J.A., et al. 2013. Primary origin of marginal Ni-Cu-(PGE) mineralization in layered intrusions: $\Delta33S$ evidence from The Platreef, Bushveld, South Africa. *Economic Geology* **108**: 365–377. doi: 10.2113/econgeo.108.2.365.

Sharp, Z.D. 2007. *Principles of Stable Isotope Geochemistry*. Upper Saddle River, NJ: Pearson Prentice Hall.

Sharp, Z.D., Mercer, J.A., Jones, R.H., et al. 2013. The chlorine isotope composition of chondrites and earth. *Geochimica et Cosmochimica Acta* **107**: 189–204. doi: 10.1016/j.gca.2013.01.003.

Shelton, K.L., Cavender, B.D., Perry, L.E., et al. 2020. Stable isotope and fluid inclusion studies of early Zn-Cu-(Ni-Co)-rich ores, lower ore zone of Brushy Creek mine, Viburnum Trend MVT district, Missouri, U.S.A.: products of multiple sulfur sources and metal-specific fluids. *Ore Geology Reviews* **118**: 103358. doi: 10.1016/j.oregeorev.2020.103358.

Shen, J., Qin, L., Fang, Z., et al. 2018. High-temperature inter-mineral Cr isotope fractionation: a comparison of ionic model predictions and experimental investigations of mantle xenoliths from the North China craton. *Earth and Planetary Science Letters* **499**: 278–90. doi: 10.1016/j.epsl.2018.07.041.

Shen, J., Xia, J., Qin, L., et al. 2020. Stable chromium isotope fractionation during magmatic differentiation: insights from Hawaiian basalts and implications for planetary redox conditions. *Geochimica et Cosmochimica Acta* **278**: 289–304. doi: 10.1016/j.gca.2019.10.003.

Sheppard, S.M.F. and Epstein, S. 1970. D/H and $^{18}O/^{16}O$ ratios of minerals of possible mantle or lower crustal origin. *Earth and Planetary Science Letters* **9**: 232–239.

Simon, J.I. and DePaolo, D.J. 2010. Stable calcium isotopic composition of meteorites and rocky planets. *Earth and Planetary Science Letters* **289**: 457–66. doi: 10.1016/j.epsl.2009.11.035.

Smith, C.N., Kesler, S.E., Blum, J.D., et al. 2008. Isotope geochemistry of mercury in source rocks, mineral deposits and spring deposits of the California Coast Ranges, USA. *Earth and Planetary Science Letters* **269**(3): 399–407. doi: 10.1016/j.epsl.2008.02.029.

Smith, H.J., Spivack, A.J., Staudigel, H., et al. 1995. The boron isotopic composition of altered oceanic crust. *Chemical Geology* **126**: 119–35.

Smyth, J.R. 1989. Electrostatic characterization of oxygen site potentials in minerals. *Geochimica et Cosmochimica Acta* **53**: 1101–10. doi: 10.1016/0016-7037(89)90215-9

Soderman, C.R., Matthews, S., Shorttle, O., et al. 2021. Heavy $\delta^{57}Fe$ in ocean island basalts: a non-unique signature of processes and source lithologies in the mantle. *Geochimica et Cosmochimica Acta* **292**: 309–32. doi: 10.1016/j.gca.2020.09.033.

Sossi, P.A., Nebel, O. and Foden, J. 2016. Iron isotope systematics in planetary reservoirs. *Earth and Planetary Science Letters* **452**: 295–308. doi: 10.1016/j.epsl.2016.07.032.

Sossi, P.A., Nebel, O., O'Neill, H.S.C., et al. 2018. Zinc isotope composition of the earth and its behaviour during planetary accretion. *Chemical Geology* **477**: 73–84. doi: 10.1016/j.chemgeo.2017.12.006.

Sossi, P.A., Prytulak, J. and O'Neill, H.S.C. 2018. Experimental calibration of vanadium partitioning and stable isotope fractionation between hydrous granitic melt and magnetite at 800 °C and 0.5 GPa. *Contributions to Mineralogy and Petrology* **173**: 27. doi: 10.1007/s00410-018-1451-8.

Starkey, N.A., Jackson, C.R.M., Greenwood, R.C., et al. 2016. Triple oxygen isotopic composition of the high-$^3He/^4He$. *Geochimica et Cosmochimica Acta* **176**: 227–38. doi: https://doi.org/10.1016/j.gca.2015.12.027.

Sun, J., Zhu, X., Belshaw, N., et al. 2021. Ca isotope systematics of carbonatites: insights into carbonatite source and evolution. *Geochemical Perspective Letters* **17**: 11–15. doi: 10.7185/geochemlet.2107.

Taylor, H.P. Jr. 1974. The application of oxygen and hydrogen studies to problems of hydrothermal alteration and ore deposition. *Economic Geology* **69**: 843–83.

Telus, M., Dauphas, N., Moynier, F., et al. 2012. Iron, zinc, magnesium and uranium isotopic fractionation during continental crust differentiation: the tale from migmatites, granitoids, and pegmatites. *Geochimica et Cosmochimica Acta* **97**: 247–65. doi: 10.1016/j.gca.2012.08.024.

Teng, F.-Z. 2017. Magnesium isotope geochemistry. *Reviews in Mineralogy and Geochemistry* **82**: 219–87. doi: 10.2138/rmg.2017.82.7.

Teng, F.-Z., Dauphas, N. and Helz, R.T. 2008. Iron isotope fractionation during magmatic differentiation in Kilauea Iki lava lake. *Science* **320**: 1620–2. doi: 10.1126/science.1157166.

Teng, F.-Z., Li, W.-Y., Ke, S., et al. 2010. Magnesium isotopic composition of the earth and chondrites. *Geochimica et Cosmochimica Acta* **74**: 4150–66. doi: 10.1016/j.gca.2010.04.019.

Thirlwall, M.F., Gee, M.a.M., Lowry, D., et al. 2006. Low $\delta^{18}O$ in the Icelandic mantle and its origins: evidence from Reykjanes ridge and Icelandic lavas. *Geochimica et Cosmochimica Acta* **70**: 993–1019. doi: 10.1016/j.gca.2005.09.008.

Tissot, F.L.H. and Dauphas, N. 2015. Uranium isotopic compositions of the crust and ocean: age corrections, U budget and global extent of modern anoxia. *Geochimica et Cosmochimica Acta* **167**: 113–43. doi: 10.1016/j.gca.2015.06.034.

Tomascak, P.B., Magna, T. and Dohmen, R. 2016. Lithium in the deep Earth: mantle and crustal systems, in Tomascak, P. B., Magna, T. and Dohmen, R. (eds). *Advances in Lithium Isotope Geochemistry*. Cham: Springer International Publishing.

Valdes, M.C., Debaille, V., Berger, J., et al. 2019. The effects of high-temperature fractional crystallization on calcium isotopic composition. *Chemical Geology* **509**: 77–91. doi: https://doi.org/10.1016/j.chemgeo.2019.01.012.

Varas-Reus, M.I., König, S., Yierpan, A., et al. 2019. Selenium isotopes as tracers of a late volatile contribution to Earth from the outer solar system. *Nature GeoScience* **12**: 779–82. doi: 10.1038/s41561-019-0414-7.

Veizer, J., Ala, D., Azmy, K., et al. 1999. $^{87}Sr/^{86}Sr$, $\delta^{13}C$, and $\delta^{18}O$ evolution of Phanerozoic seawater. *Chemical Geology* **161**: 59–88.

Villalobos-Orchard, J., Freymuth, H., O'Driscoll, B., et al. 2020. Molybdenum isotope ratios in izu arc basalts: the control of subduction zone fluids on compositional variations in arc volcanic systems. *Geochimica et Cosmochimica Acta* **288**: 68–82. doi: https://doi.org/10.1016/j.gca.2020.07.043.

Vlastelic, I., Koga, K., Chauvel, C., et al. 2009. Survival of lithium isotopic heterogeneities in the mantle supported by HIMU-lavas from Rurutu island, Austral chain. *Earth and Planetary Science Letters* **286**: 456–66. doi: 10.1016/j.epsl.2009.07.013.

Vollstaedt, H., Mezger, K. and Leya, I. 2016. The isotope composition of selenium in chondrites constrains the depletion mechanism of volatile elements in solar system materials. *Earth and Planetary Science Letters* **450**: 372–80. doi: 10.1016/j.epsl.2016.06.052.

Wang, W., Huang, S., Huang, F., et al. 2020. Equilibrium inter-mineral titanium isotope fractionation: implication for high-temperature titanium isotope geochemistry. *Geochimica et Cosmochimica Acta* **269**: 540–53. doi: 10.1016/j.gca.2019.11.008.

Wang, X.-J., Chen, L.-H., Hofmann, A.W., et al. 2018. Recycled ancient ghost carbonate in the Pitcairn mantle plume. *Proceedings of the National Academy of Sciences* **115**: 8682–7. doi: 10.1073/pnas.1719570115.

Wang, X., Amet, Q., Fitoussi, C., et al. 2018. Tin isotope fractionation during magmatic processes and the isotope composition of the bulk silicate earth. *Geochimica et Cosmochimica Acta* **228**: 320–35. doi: 10.1016/j.gca.2018.02.014.

Wang, X., Fitoussi, C., Bourdon, B., et al. 2021. The Sn isotope composition of chondrites: implications for volatile element depletion in the solar system. *Geochimica et Cosmochimica Acta* **312**: 139–57. doi: 10.1016/j.gca.2021.08.011.

Wang, X., Wang, Z., Liu, Y., et al. 2021. Calcium stable isotopes of Tonga and Mariana arc lavas: implications for slab fluid-mediated carbonate transfer in cold subduction zones. *Journal of Geophysical Research: Solid Earth* **126**: e2020JB020207. doi: 10.1029/2020JB020207.

Wang, Z., Zhang, P., Li, Y., et al. 2021. Copper recycling and redox evolution through progressive stages of oceanic subduction: insights from the Izu-Bonin-Mariana forearc. *Earth and Planetary Science Letters* **574**: 117178. doi: 10.1016/j.epsl.2021.117178.

Weyer, S. and Seitz, H.M. 2012. Coupled lithium- and iron isotope fractionation during magmatic differentiation. *Chemical Geology* **294–5**: 42–50. doi: 10.1016/j.chemgeo.2011.11.020.

White, W.M., Copeland, P., Gravatt, D.R., et al. 2017. Geochemistry and geochronology of Grenada and Union islands, Lesser Antilles: the case for mixing between two magma series generated from distinct sources. *Geosphere* **5**: 1359–91. doi: 10.1130/GES01414.1.

Wilkinson, J.J. (2014). 13.9 – Sediment-hosted zinc–lead mineralization: processes and perspectives, in Holland, H.D. and Turekian, K.K. (eds). *Treatise on Geochemistry*, Second Edition. Oxford: Elsevier, 219–249 pp.

Willbold, M. and Elliott, T. 2017. Molybdenum isotope variations in magmatic rocks. *Chemical Geology* **449**: 253–268. doi: https://doi.org/10.1016/j.chemgeo.2016.12.011.

Wille, M., Nebel, O., Pettke, T., et al. 2018. Molybdenum isotope variations in calc-alkaline lavas from the Banda arc, Indonesia: assessing the effect of crystal fractionation in creating isotopically heavy continental crust. *Chemical Geology* **485**: 1–13. doi: 10.1016/j.chemgeo.2018.02.037.

Williams, H.M. and Archer, C. 2011. Copper stable isotopes as tracers of metal–sulphide segregation and fractional crystallisation processes on iron meteorite parent bodies. *Geochimica et Cosmochimica Acta* **75**: 3166–78. doi: 10.1016/j.gca.2011.03.010.

Williams, N.H., Fehr, M.A., Parkinson, I.J., et al. 2021. Titanium isotope fractionation in solar system materials. *Chemical Geology* **568**: 120009. doi: 10.1016/j.chemgeo.2020.120009.

Williamson, N.M.B., Weis, D. and Prytulak, J. 2021. Thallium Isotopic Compositions in Hawaiian Lavas: evidence for Recycled Materials on the Kea Side of the Hawaiian Mantle Plume. *Geochemistry, Geophysics, Geosystems* **22**(9): e2021GC009765. doi: 10.1029/2021GC009765.

Wu, F., Qi, Y., Perfit, M.R., et al. 2018. Vanadium isotope compositions of mid-ocean ridge lavas and altered oceanic crust. *Earth and Planetary Science Letters* **493**: 128–39. doi: 10.1016/j.epsl.2018.04.009.

Xia, J., Qin, L., Shen, J., et al. 2017. Chromium isotope heterogeneity in the mantle. *Earth and Planetary Science Letters* **464**: 103–15. doi: 10.1016/j.epsl.2017.01.045.

Yang, J., Siebert, C., Barling, J., et al. 2015. Absence of molybdenum isotope fractionation during magmatic differentiation at Hekla volcano, Iceland. *Geochimica et Cosmochimica Acta* **162**: 126–36. doi: 10.1016/j.gca.2015.04.011.

Yierpan, A., König, S., Labidi, J., et al. 2019. Selenium isotope and S-Se-Te elemental systematics along the Pacific-Antarctic ridge: role of mantle processes. *Geochimica et Cosmochimica Acta* **249**: 199–224. doi: 10.1016/j.gca.2019.01.028.

Yierpan, A., König, S., Labidi, J., et al. 2020. Recycled selenium in hot spot-influenced lavas records ocean-atmosphere oxygenation. *Science Advances* **6**: eabb6179. doi: 10.1126/sciadv.abb6179.

Yierpan, A., König, S., Labidi, J., et al. 2018. Chemical sample processing for combined selenium isotope and selenium-tellurium elemental investigation of the earth's igneous reservoirs. *Geochemistry, Geophysics, Geosystems* **19**: 516–33. doi: 10.1002/2017GC007299.

Yierpan, A., Redlinger, J. and König, S. 2021. Selenium and tellurium in Reykjanes ridge and Icelandic basalts: evidence for degassing-induced se isotope fractionation. *Geochimica et Cosmochimica Acta* **313**: 155–72. doi: 10.1016/j.gca.2021.07.029.

Young, E.D. and Galy, A. (2004). The isotope geochemistry and cosmochemistry of magnesium. in Johnson, C.M., Beard, B.L. and Albarède, F. (eds). *Geochemistry of Non-traditional Stable Isotopes*. Washington, DC: Minerological Society of Ameria, 197230 p.

Young, E.D., Kohl, I.E., Warren, P.H., et al. 2016. Oxygen isotopic evidence for vigorous mixing during the moon-forming giant impact. *Science* **351**: 493–6. doi: 10.1126/science.aad0525.

Young, E.D., Manning, C.E., Schauble, E.A., et al. 2015. High-temperature equilibrium isotope fractionation of non-traditional stable isotopes: experiments, theory, and applications. *Chemical Geology* **395**(0): 176–95. doi: /10.1016/j.chemgeo.2014.12.013.

Zeng, Z., Li, X., Chen, S., et al. 2021. Iron, copper, and zinc isotopic fractionation in seafloor basalts and hydrothermal sulfides. *Marine Geology* **436**: 106491. doi: 10.1016/j.margeo.2021.106491.

Zeng, Z., Ma, Y., Chen, S., et al. 2017. Sulfur and lead isotopic compositions of massive sulfides from deep-sea hydrothermal systems: implications for ore genesis and fluid circulation. *Ore Geology Reviews* **87**: 155–71. doi: 10.1016/j.oregeorev.2016.10.014.

Zhang, J. 2021. Equilibrium sulfur isotope fractionations of several important sulfides. *Geochemical Journal* **55**(3): 135–47. doi:

Zheng, Y.-C., Liu, S.-A., Wu, C.-D., et al. 2018. Cu isotopes reveal initial Cu enrichment in sources of giant porphyry deposits in a collisional setting. *Geology* **47**: 135–8. doi: 10.1130/g45362.1.

Zheng, Y.-F. 1993. Calculation of oxygen isotope fractionation in anhydrous silicate minerals. *Geochimica et Cosmochimica Acta* **57**(5): 1079–91. doi: 10.1016/0016-7037(93)90042-U.

Zheng, Y.-F., Zhao, Z.-F., Li, S.-G., et al. 2003. Oxygen isotope equilibrium between ultrahigh-pressure metamorphic minerals and its constraints on Sm-Nd and Rb-Sr chronometers. *Geological Society of London Special Publications* **220**(1): 93–117. doi: 10.1144/gsl.Sp.2003.220.01.06.

Zhong, Y., Chen, L.-H., Wang, X.-J., et al. 2017. Magnesium isotopic variation of oceanic island basalts generated by partial melting and crustal recycling. *Earth and Planetary Science Letters* **463**: 127–35. doi: 10.1016/j.epsl.2017.01.040.

Zhou, J.-X., Wang, X.-C., Wilde, S.A., et al. 2018. New insights into the metallogeny of MVT Zn-Pb deposits: a case study from the Nayongzhi in South China, using field data, fluid compositions, and in situ S-Pb isotopes. *American Mineralogist* **103**(1): 91–108. doi: 10.2138/am-2018-6238.

PROBLEMS

1. $\delta^{18}O$ values in the table below were measured on minerals in a metamorphic rock. Calculate the quartz–mineral fractionations (Δ_{qz-min}) and temperature for each using (10.3) and the A coefficient listed in Table 10.2. Assume that the value of A for plagioclase of An40 is a linear function of the fractions of anorthite (An) and albite (1-An). Are these likely to be equilibrium temperatures? Construct an "isotherm" by plotting Δ_{qz-min} against A. From the slope of the line, calculate the temperature for the entire assemblage.

mineral	$\delta^{18}O$
Quartz	8
Plag (An40)	6.25
Magnetite	0.2
Diopisde	4.7

2. Šoštarić et al. (2011) measured the following $\delta^{34}S$ data on sphalerite (ZnS)–galena (PbS) pairs in a hydrothermal silver–base metal ore deposit in the Rogozna Mountains of Kosovo. Using the coefficients in Table 9.3, calculate the equilibration temperature for each pair.

	Sphalerite $\delta^{34}S$	Galena $\delta^{34}S$
3/2/V	2.9	0.4
3/6/V	4.2	1.7
12/1/V	4.2	1.2
6/1/IV	4.9	2.3
6/4/IV	4.3	2.9

3. Figure 9.36 shows that $\delta^{18}O$ reaches a minimum at Site 1256 of ~3‰ just above the base of the sheeted dike complex. Assuming that the bulk fractionation factor, Δ between water and rock is +1.5‰ and that the initial $\delta^{18}O$ of the water and rock is 0 and +5.7‰, respectively, estimate the effective water–rock ratio that these rocks experienced.

4. The fractionation between pyrrhotite and H_2S has a temperature dependence of $\Delta = 0.1 \times 10^6/T^2$; that between SO_2 and H_2S is $\Delta = -0.5 + 4.7 \times 10^6/T^2$. Estimate the $\delta^{34}S$ of pyrrhotite (FeS) crystallizing in equilibrium with an H_2S fluid that in turn is equilibrium with magmatic sulfur with $\delta^{34}S = 0$‰ at 850 °C whose mole fraction ratio SO_2/H_2S (R') is 0.1.

5. What would the R' (mole fraction ratio SO_2/H_2S) have to be for pyrite to precipitate with $\delta^{34}S = -0.49$‰ from a magmatic fluid at 700°C whose bulk $\delta^{34}S$ is 0‰? Assume that the pyrite is in equilibrium with an H_2S in the fluid and that $\Delta_{py\text{-}H2S} = 0.40 \times 10^6/T^2$ Comparing your result with Figure 9.43, roughly estimate the log f_{O2} of the magma.

6. Why does Sn have so many isotopes?

Chapter 10

Light Stable Isotopes in the Exogene

10.1 INTRODUCTION

In this and the following chapter, we will focus on the stable isotope geochemistry of the Earth's surface environment for which we will adopt the French word "exogène" and translate it as *exogene*. This encompasses atmosphere, hydrosphere, biosphere, and soil through which many elements cycle rapidly undergoing isotopic fractionations as they do. This chapter will focus on "traditional" elements, H, C, N, O, and S, the isotope geochemistry of which has been extensively studied for well over half a century. These biogeochemical cycles are closely tied to and in part driven by the hydrologic cycle. So, we begin this chapter by examining the H and O isotope geochemistry of water and the hydrologic system. We will then examine isotope systematics of the biosphere and, finally, the isotope systematics of the atmosphere. This will provide us with the necessary background understanding of isotopic fractionations associated with the biogeochemical cycles of the exogene in Chapter 11 and then to apply them to peer into past and discover how the Earth's surface and life has evolved over time in Chapters 12 and 13.

10.2 THE HYDROLOGIC SYSTEM

The oceans are the main reservoir of water on the surface of the Earth and the starting and end points of the hydrologic cycle. They are also the largest reservoir of carbon in the exogene (in the form of bicarbonate ion). They are central to many other geochemical cycles as well and exert a very strong control on climate that we will discuss in this and following two chapters. Furthermore, they are the author of the history of the planet and the history of life in the sense that sediments are that history book, and the vast majority of sediments are marine. An understanding of the oceans is essential to many of topics we will discuss in this and the subsequent two chapters, so we will digress at this point with a brief overview of the oceans.

10.2.1 Ocean chemistry, structure, and circulation: a brief overview

The oceans are an excellent example of an *open system*: ions and molecules are continually added to the oceans from a number of sources: rivers, submarine hydrothermal fluids, by diffusion out of sediments, and from the atmosphere, including dust particles that settle on the ocean surface and gas exchange. Elements are also continually lost from the oceans through a number of sinks: sedimentation, reaction with the oceanic crust, evaporation, and in droplets lofted into the atmosphere by breaking waves. To a first approximation, however, these processes balance such that the composition of the oceans does not change. We need to emphasize "to a first approximation" because changes stable and radiogenic isotope ratios in marine sediments have proved particularly useful in unraveling how ocean composition has changed over time.

Isotope Geochemistry, Second Edition. William M. White.
© 2023 John Wiley & Sons Ltd. Published 2023 by John Wiley & Sons Ltd.
Companion Website: www.wiley.com/go/white/isotopegeochem2

Despite the diversity of sources, both the concentration and the isotopic composition of many elements in seawater are uniform or nearly so throughout the open ocean. This is particularly true of the major elements, including O, H, Cl, Mg, and S but is also true of some minor and trace ones, including Li, B, Sr, Mo, and U. These elements are said to be *conservative*; they are always present in constant ratios to one and other and to salinity, and their concentrations only vary by processes, such as evaporation or rain, that concentrates or dilutes them.

Whether or not an element has uniform concentration and isotopic composition in the ocean depends on its residence time relative to the mixing time of the oceans and the extent of biological processing within the ocean. The time required to erase or reduce compositional heterogeneity in the ocean is known as the *mixing time* and is of the order of a few 10^3 years. However, this term is neither precisely defined nor precisely known; a reasonable, but not universally used, definition is the time required to reduce compositional variance by a factor of 1/e. On the other hand, *residence time*, τ, can be precisely defined for an open system such as the oceans provided it is at steady state as the ratio of the mass of an element in the ocean, A, to the flux of that element into (or out of) the ocean, dA/dt:

$$\tau = \frac{A}{dA/dt}. \qquad (10.1)$$

Conservative elements have long residence times and have approximately uniform isotopic compositions throughout the open ocean. Isotope ratios of these elements nevertheless tend to vary over long time periods in response to changes in fluxes, and they are therefore useful in understanding ocean history and related processes such as climate changes. We have already seen that systematic variation in $^{87}Sr/^{86}Sr$, which reflects the changing ratio of crustal to mantle flux though time, provides a chronological tool. We will see how variation in $\delta^{34}S$ and other stable isotope ratios in marine sediments reflect the oxygenation state of the ocean.

Elements with short residence times tend to have variable concentrations and isotopic compositions. This can reflect both geographically variable inputs, such as the case for ε_{Nd}

and processes and fractionations that occur within the oceans, which is the case for biologically utilized elements such as Si. The distribution of a surprising number of elements is also biologically controlled despite not being significantly utilized. A third category of elements with very short residence times has distributions controlled by adsorption and desorption on particle surfaces and includes Al, Pb, and Th.

Most of the ocean is stably density stratified. Density is controlled by a combination of temperature and salinity. In the modern ocean, temperature dominates, although during warmer periods in the past, salinity appears to have been dominant. The upper hundred meters or so has a relatively uniform temperature and salinity structure due to mixing by waves (the actual depth of the mixed layer varied both seasonally and geographically). Below the upper mixed layer is a region called the *thermocline*, where temperature decreases rapidly. Salinity may also change rapidly in this region; a region where salinity changes rapidly is called a *halocline*. The temperature changes cause a rapid increase in density with depth, and this region of the water column is called the *pycnocline*. Below the pycnocline, temperature and salinity vary less with depth.

The pycnocline represents a strong boundary to vertical mixing, isolating surface from deep water. Because of this, it is sometimes useful to think of the oceans in terms of two reservoirs: the surface water above the pycnocline, and the deep water below it. Fluxes in either direction can occur because of the advection of water: *upwelling* of deeper water occurs where winds or currents create a *divergence* of surface water and d*ownwelling* occurs where winds or currents produce a *convergence* of surface water. A second flux, eddy diffusion, tends to move dissolved components from the reservoir with high concentrations to that with lower ones much as chemical diffusion would. Falling particles, both organic and inorganic, constitute a third flux, but unlike the other two, this is unidirectional, moving matter only from the surface to the deep layer. The upper layer exchanges with the atmosphere and receives all the riverine input. All photosynthetic activity occurs in the upper box because light penetration is limited (only 0.5% of the incident sunlight penetrates to a depth of 100 m, even in the clearest water).

On the other hand, the flux out of the ocean (to sediment) of both particles and dissolved solids occurs mainly through the lower box. The uptake of dissolved inorganic nutrients by phytoplankton in the surface water and export of these to the deep water by falling organic remains results in strong contrasts in the concentrations of these elements and often their isotopic compositions between the upper and lower boxes.

Ocean circulation is ultimately driven by the pole-to-equator gradient in solar radiative energy or *insolation*. In response to this gradient, the atmosphere and oceans carry heat from low to high latitudes. The surface circulation of the ocean is driven by winds, and this, combined with the Coriolis effect, results in paired clockwise and counterclockwise gyres in the Northern and Southern Hemispheres, respectively, in the Pacific and Atlantic Oceans. Indian Ocean circulation has some of these aspects but is more complex and varies seasonally driven by the Indian monsoons.

The deep circulation of the ocean is driven by density differences that depend on temperature and salinity. In high-latitude regions, the density stratification is weak. Surface water cooled in winter will be more dense than underlying water and sink, in this way "forming" deep water masses. This so-called Great Conveyor Belt begins where water cooled in winter in the Norwegian, Greenland, and Labrador Seas downwells to become North Atlantic Deep Water (NADW). As it flows south, it entrains Antarctic Bottom Water (AABW) from below and Antarctic Intermediate Water from above and joins the Antarctic Circumpolar Current, making it the largest current in the ocean in terms of volume transport, as part of the Circumpolar Deep Water. AABW, the coldest and densest water in the ocean, forms when this deep water upwells along Antarctica, mainly in the Weddell Sea, during winter when extreme cooling and ice formation increase salinity and decrease temperature. AABW then flows northward into all three oceans. The flow is balanced by return flow at shallow depth. Traditionally, these water masses are identified by their temperature and salinity characteristics, which are "conservative" properties of the water mass in that once fixed at the surface, they can change only through mixing with other water masses. These water masses also have unique chemical properties as well, such as dissolved oxygen content, nutrient concentrations, and carbon isotope ratios, but these change over time, due to biologic activity. The water masses also acquire unique radiogenic isotope signatures because the isotopic composition of the sources of these elements varies geographically.

10.2.2 Hydrogen and oxygen isotope fractionation in the hydrologic system

The hydrologic system begins when water evaporates from the ocean surface, with water consisting of light H and O isotopes evaporating preferentially as a result of both equilibrium and kinetic effects (Chapter 5). As water vapor condenses to form cloud droplets and ice particles, additional fractionation occurs such that the precipitation will be isotopically heavier than the vapor and the extent of fractionation will temperature dependent. Once droplets or snowflakes grow large enough to fall, they will no longer equilibrate with vapor and the process becomes one of Rayleigh condensation described in Chapter 5.

Dansgaard (1964) summed up the controls on the isotopic composition of precipitation as *temperature, latitude, altitude, distance from source*, and *amount*. The first of these is clear from (5.98) and (5.99): fractionation during evaporation or condensation will be greater at colder temperatures. Figure 10.1 shows the global variation in mean annual $\delta^{18}O$ in precipitation as a function of mean annual air temperature. These observations show a linear dependence on temperature that is approximately $\delta^{18}O = 0.69T - 13.6$, where T is in degree Celsius; the comparable temperature dependence for hydrogen is $\delta D = 5.6T - 100$ (Dansgaard, 1964). These correlations are weaker at higher annual temperature and in summer precipitation. The latitude and altitude effects are both related to temperature as well. Mountains force air up, so that condensation will occur higher in the atmosphere and consequently at lower temperatures and with greater fractionation. The range in $\delta^{18}O$ is significantly greater than predicted by simple equilibrium fractionation alone; for example, (5.98) predicts the range of roughly 3 per mil in $\delta^{18}O$ from 0 to 30 °C, while the actual range is about 14 per mil, demonstrating the importance of other factors related to Rayleigh fractionation.

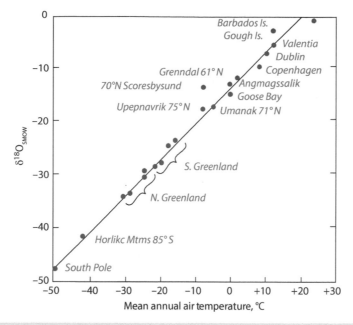

Figure 10.1   Variation of $\delta^{18}O$ in precipitation as a function of mean annual temperature. This relationship is approximately described as $\delta^{18}O = 0.69T$ °C $- 13.6$. Source: (Adapted from Dansgaard, 1964).

The further air moves from the site of evaporation (the ocean), the more water is likely to have condensed and fallen as rain, and therefore, the smaller the value of f in (5.100). Furthermore, because precipitation decreases the moisture content of a parcel of air, progressively colder temperatures will be required to achieve saturation and subsequent precipitation. Consequently, the fractionation factor, α, will covary with f, leading to greater fractionation than for simple Rayleigh fractionation. Thus, the isotopic composition of precipitation can be expected to become lighter with increasing distance from the moisture source. Mountains also tend to wring moisture out of the air, again decreasing f on the downwind side of mountain ranges. Consequently, the water vapor in air that has passed over a mountain range will be isotopically lighter than water vapor on the ocean side of a mountain range. The amount effect is also a consequence of decreasing values of f: the more rain that falls, the isotopically lighter the remaining moisture will become. These factors are illustrated in the cartoon in Figure 10.2. At the extreme, the combination of low temperature and Rayleigh fractionation leads to $\delta^{18}O$ as low as $-81.9‰$ at the Dome Fuji ice station in East Antarctica at an elevation of 3900 m where the ice is 3 km thick.

Figure 10.3 shows the variation in oxygen isotopic composition of meteoric surface waters in the continental USA. The lightest values are found in the Rocky Mountains and Sierra Nevada, and the complex pattern of $\delta^{18}O$ in the West illustrates the influence of topography. In the eastern half of the country, a much simpler pattern is apparent with precipitation becoming lighter with distance from moisture sources, particularly the Gulf of Mexico.

10.2.3   The $\delta^{18}O$–δD meteoric water line

As was pointed out in Chapter 5, $\delta^{18}O$ and δD in meteoric water are well correlated (Figure 10.4), constituting what is known as the *Meteoric Water Line* (MWL) or *Global Meteoritic Water Line* (GMWL) whose equation is $\delta D = 8\delta^{18}O + 10$. The intercept is known as the deuterium excess, d, and results from the combination of kinetic and equilibrium effects during evaporation related to humidity illustrated in Figure 5.8. In essence, these effects result in greater fractionation of O isotopes than H isotopes, which shifts water vapor to the left of the MWL as humidity decreases. Indeed, the GMWL is simply the average of individual local meteoritic water lines whose slopes and intercepts vary, with slopes for the most part lower than the global one.

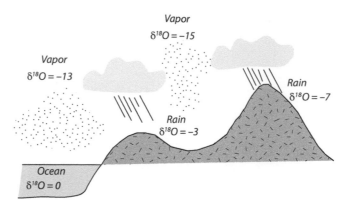

Figure 10.2 Cartoon illustrating the process of Rayleigh fractionation and the increasing fractionation of oxygen isotopes in rain as it moves inland.

Figure 10.3 Variation of $\delta^{18}O$ in precipitation in the United States (computed from analysis of tap water). $\delta^{18}O$ depends on orographic effects, mean annual temperature, and distance from the sources of water vapor. Source: Bowen et al. (2007)/Reproduced with permission from John Wiley & Sons.

Kinetic fractionation becomes dominant when evaporation occurs in low-humidity (small h) arid environments as we can see from the following equation:

$$R_{ev} = \frac{R_w - \alpha_{L-V}^{eq} h R_v}{\alpha_{L-V}^0 (1-h)} \qquad (5.95)$$

where R_{ev} is the isotope ratio of vapor produced, h is the humidity, α_{L-V}^{eq} is the equilibrium fractionation factor, and α_{L-V}^0 is the fractionation factor at zero humidity. Lake water subject to evaporation from arid environments can

sometimes deviate quite significantly from the GMWL, falling along lines of lower slope (the *evaporative trend* shown in Figure 10.4).

The evolution of the isotopic composition of such lakes will depend on humidity, the isotopic composition of atmospheric water, and whether or not it is recharged by streams or groundwater. Criss (1999) showed that for an unrecharged, closed basin lake, the isotopic composition will ultimately reach steady state at

$$R_{ss} = \frac{\alpha_{L-V}^{eq} h R_A}{1 - \alpha_{L-V}^0 (1-h)} \qquad (10.2)$$

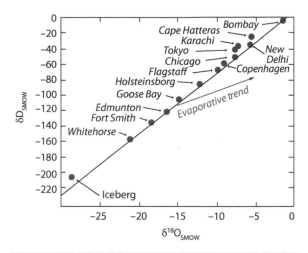

Figure 10.4 Variation in mean annual δ^2H and $\delta^{18}O$ in precipitation and meteoric waters. This relationship is known as the MWL also as the GMWL. The relationship between δ^2H and $\delta^{18}O$ is approximately $\delta^2H = 8\delta^{18}O + 10$. Humidity-controlled kinetic effects (Figure 5.8) result in waters in arid regions being shifted along the *evaporative trend* shown. Source: Adapted from Dansgaard (1964).

where R_{ss} is the steady-state isotope ratio, α^{eq}_{L-V} is the equilibrium fractionation factor, α^0_{L-V} is the effective fractionation factor at 0 humidity (Equation 5.94), h is the humidity, and R_A is the isotopic composition of atmospheric water. For a closed basin evaporative lake undergoing recharge, the equivalent steady state is

$$R_{ss} = \alpha^0_{L-V}(1-h)R_I + \alpha^{eq}_{L-V}hR_A \qquad (10.3)$$

where R_I is the isotopic composition of recharge water. For nonclosed basin lakes (those with an outlet), the relevant equation is

$$R_{ss} = \frac{\alpha^0_{L-V}(1-h)R_I + X_E h\alpha^{eq}_{L-V}R_A}{\alpha^0_{L-V}(1-h)(1-X_E) + X_E} \qquad (10.4)$$

where X_E is the fraction of water lost by evaporation. Because the steady-state conditions are different for recharged and unrecharged lakes and water evolves toward them along distinct paths, $\delta^{17}O–\delta^{18}O$ systematics can be used to understand regional hydrology (Surma et al., 2021). In particular, where isotope ratios in these equations can be measured or otherwise constrained, it is possible to solve for h. Furthermore, the O isotopic composition of

hydrated minerals such as gypsum ($Ca(SO_4)$ •2(H_2O)) precipitation from saline lakes will reflect the composition of the water, making it possible to reconstruct paleo-humidities. For example, Gázquez et al. (2018) analyzed gypsum from Lake Estanya (NE Spain). After correcting for gypsum–water fractionation, they found that humidity during the Younger Dryas (~12 000–13 000 years BP) following the last glacial termination was ~30–35% lower than today.

10.2.4   Triple oxygen isotopes in the hydrologic system

As analytical techniques and instrumentation have improved over the last several decades, data on ^{17}O have become increasingly available. Luz and Barkan (2010) found that $\delta'^{17}O$ in global meteoric water was related to $\delta'^{18}O$ as

$$\delta'^{17}O = 0.528(\pm 0.001)\delta'^{18}O + 0.033(\pm 0.004)‰.$$

$$(10.5)$$

(Recall from Chapter 5 that we defined $\delta' = 10^3 \ln(1 + \delta/10^3)$ in Equation 5.114). This slope, $\lambda = 0.528$, matches that of the *terrestrial fractionation line* (TFL) and is close to the theoretically expected one for liquid–vapor equilibrium if $\theta = 0.529$ (also recall that λ is used for observed slopes and θ for theoretical ones). Thus, meteoritic water plots along $\delta'^{17}O–\delta'^{18}O$ a line as defined by terrestrial materials generally but offset by a ^{17}O-excess or $\Delta'^{17}O$ of 0.033‰; we will refer to as the $\delta'^{17}O–\delta'^{17}O$ GMWL. A subsequent study by Sharp et al. (2018) found a slightly higher ^{17}O-excess for meteoric waters of $\Delta'^{17}O = 0.041\pm0.006‰$. Luz and Barkan also found that seawater with a range of $\delta^{18}O$ of –0.26‰ to +2.4‰ defined the same slope but with $\Delta'^{17}O = -0.05\pm0.01‰$. Thus, meteoric water has an ^{17}O-excess of ~38 ppm relative to seawater. Just as was the case for the $\delta^2H–\delta^{18}O$ MWL, the ^{17}O-excess originates from the combination of equilibrium and humidity-dependent kinetic fractionations illustrated in Figure 5.8. Indeed, Luz and Barkan showed that the ^{17}O-excess in marine water vapor depends on humidity, ranging from near 0 at a humidity of 1 to ~+50 ppm at a humidity of 0.5.

If all meteoric waters fit neatly on a single correlation line, there would be little point to further study. However, the similar

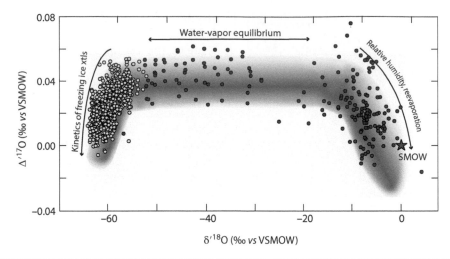

Figure 10.5 ^{17}O excess ($\Delta'^{17}O$) in meteoric water as a function of $\delta^{18}O$. Samples with $\delta'^{18}O$ values between $-55‰$ and $-15‰$ have an average $\Delta'^{17}O$ value of $0.041\pm0.006‰$ and fall on a line with slope $\lambda = 0.528$. Samples from the Vostok ice core (pale blue points) fit to a higher λ value of 0.5309 ± 0.0001. Samples with $\delta'^{18}O > -20 ‰$, show the reverse trend, with $\Delta'^{17}O$ increasing $\delta'^{18}O$ as a consequence of re-evaporation. Source: Sharp et al. (2018)/with permission of European Association of Geochemistry.

uncertainties of the slopes and intercepts of seawater and meteoric water lines reported by Luz and Barkan (2010) are deceptive; they result from the much larger range in $\delta^{18}O$ in meteoric water, >60‰, than in seawater, < 3‰. In detail, the Luz and Barkan data show considerable variation, with $\Delta'^{17}O$ ranging from −16 ppm for a sample from India to +76 ppm for a sample from Canada. Subsequent study has shown even greater variation, as illustrated in Figure 10.5. At both extremes of the $\delta^{18}O$ range, $\Delta'^{17}O$ becomes negative. The negative values in polar snow are due to kinetic effects associated with atmospheric vapor supersaturation over ice during snow formation in clear air (so-called diamond dust) at very low temperatures during the Antarctic winter (Landais et al., 2012). Sharp et al. (2018) found a higher λ for these data of 0.5309 ± 0.0001. At low and mid-latitudes waters with higher $\delta^{18}O$, the re-evaporation of moisture from local continental sources and the subcloud re-evaporation of rain result in the reverse trend with $\Delta'^{17}O$ values of meteoric waters decreasing with increasing $\delta'^{18}O$.

Raleigh fractionation during the evaporation of lake water results in the water becoming progressively isotopically heavier. In arid regions, evaporation occurs under conditions of low humidity that increases the effect of kinetic fractionations. The kinetic fractionation has a lower slope, $\theta = \sim0.518$, than the equilibrium fractionation, $\theta = \sim0.529$; this tends to push residual water to the high $\delta^{18}O$ side of the $\delta'^{17}O$-$\delta'^{18}O$ GMWL along a shallow slope in a manner exactly analogous to the δ^2H-$\delta^{18}O$ GMWL. Thus, many evaporative lakes have positive $\delta'^{18}O$ and negative $\Delta'^{17}O$, as apparent in Figure 10.6.

10.3   ISOTOPE RATIOS IN THE BIOSPHERE

10.3.1   Carbon isotope fractionation during photosynthesis

Biological processes often involve large isotopic fractionations, as we noted in Chapter 5. Indeed, biological processes are the most important cause of variations in the isotope composition of carbon, nitrogen, and sulfur. The largest carbon isotope fractionations occur during the initial production of organic matter by the so-called primary producers, or *autotrophs*. These include all plants and many kinds of bacteria and some archaea. The most important means of production of organic matter is photosynthesis, but organic matter may also be produced by chemosynthesis, for example, at mid-ocean ridge hydrothermal vents and results in similar fractionations. Large fractions of both carbon and nitrogen

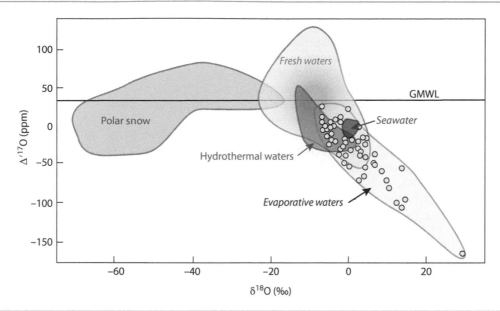

Figure 10.6 ^{17}O excess ($\Delta'^{17}O$) in various waters as a function of $\delta^{18}O$. Although there is considerable scatter, most fresh waters plot close to the GMWL with $\Delta'^{17}O$ +33 ppm. Source: Adapted from Surma et al. (2021).

occur during primary production. Additional fractionations also occur in subsequent reactions and up through the food chain as *hetrotrophs* consume primary producers, but these are generally smaller.

The photosynthetic fractionation of carbon isotopes is primarily kinetic. The early work of Park and Epstein (1960) suggested fractionation occurred in several steps. Subsequent work has elucidated the fractionations involved in these steps, which we will consider in more detail here. For terrestrial plants (those utilizing atmospheric CO_2), the first step is the diffusion of CO_2 into the boundary layer surrounding the leaf, through the stomata, and finally within the leaf. The average $\delta^{13}C$ of various species of plants has been correlated with the stomatal conductance (DeLucia et al., 1988), indicating that diffusion into the plant is indeed important in fractionating carbon isotopes. As we found in Chapter 5, the ratio of their diffusion coefficients is equal to the square root of the ratios of their reduced masses to air ($^{12}CO_2$ will diffuse more rapidly; Equation 5.71); we expect CO_2 entering leaves of terrestrial plants to be 4.4‰ lighter than air.

At this point, there is a divergence in the chemical pathways. Most plants use an enzyme called *ribulose bisphosphate carboxylase oxygenase* (RUBISCO) to catalyze a reaction in which

ribulose bisphosphate (RuBP) reacts with one molecule of CO_2 to produce two molecules of 3-phosphoglyceric acid, a compound containing three carbon atoms, in a process called *carboxylation* (Figure 10.7). Energy to drive this reaction is provided by another reaction, called *photophosphorylation,* in which electromagnetic energy is used to dissociate water, producing oxygen. The carbon is subsequently reduced, carbohydrate formed, and the RuBP

Figure 10.7 RuBP carboxylation, the reaction by which C_3 plants fix carbon during photosynthesis. Bold solid bonds are in front of the plane of the paper, dashed ones behind.

regenerated. Such plants are called C_3 plants, and this process is called the *Benson-Calvin*, or *Calvin*, cycle. C_3 plants constitute about 90% of all plants and comprise the majority of cultivated plants, including wheat, rice, and nuts. Algae and autotrophic bacteria also produce a three-carbon chain as the initial photosynthetic product. There is a kinetic fractionation associated with the carboxylation of RuBP that has been determined by several methods to be −29.4‰ in higher terrestrial plants. Bacterial carboxylation has different reaction mechanisms and a smaller fractionation of about −20‰. Thus, for terrestrial plants, a fractionation of about −34‰ is expected from the sum of the fractions. The actual observed total fractionation is in the range of −20 to −30‰.

The disparity between the observed total fractionation and that expected from the sum of the steps presented something of a conundrum. The solution is that the amount of carbon isotope fractionation expressed in the tissues of plants depends on the ratio of the concentration of CO_2 inside plants to that in the external environment (e.g. Farquhar et al., 1982). In its simplest version, this may be described by the equation:

$$\Delta_{net} = \Delta_{diff} + (c_{in}/c_{ex})(\Delta_{carbox} - \Delta_{diff}) \quad (10.6)$$

where Δ_{diff} and Δ_{carbox} are the fractionation factors for diffusion and carboxylation, respectively, and c_{in} and c_{ex} are the interior and exterior CO_2 concentrations, respectively. According to this model, where an unlimited amount of CO_2 is available (i.e. when $c_{in}/c_{ex} \approx 1$), carboxylation alone causes fractionation. At the other extreme, if the concentration of CO_2 in the cell is limiting (i.e. when $c_{in}/c_{ex} \approx 0$), essentially all carbon in the cell will be fixed, and therefore, there will be little fractionation during this step and the total fractionation is essentially just that due to diffusion alone. Both laboratory experiments and field observations provide strong support for this model.

More recent studies have shown that RUBISCO enzyme exists in at least two different forms and that these two different forms fractionate carbon isotopes to differing degrees. Form I, which is by far the most common, typically produces the fractionation mentioned earlier; fractionation produced by Form

$$
\begin{array}{ccc}
\text{CO}_2\text{H} & & \text{CO}_2\text{H} \\
| & & | \\
\text{C}-\text{OPO}_3\text{H}_2 + \text{HCO}_3^{1-} \longrightarrow & \text{C}=\text{O} + \text{H}_2\text{PO}_4^{1-} \\
| & & | \\
\text{CH}_2 & & \text{CH}_2 \\
& & | \\
& & \text{CO}_2\text{H}
\end{array}
$$

Phosphoenolpyruvate   Oxaloacetate

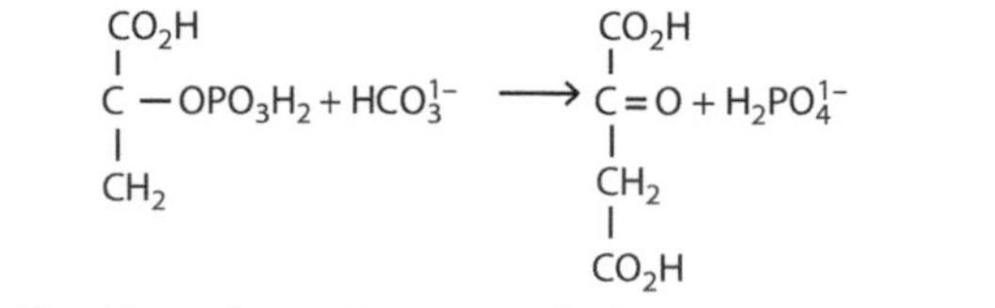

Figure 10.8 Phosphoenolpyruvate carboxylation, the reaction by which C_4 plants fix CO_2 during photosynthesis.

II, which appears to be restricted to a few autotrophic bacteria and some dinoflagellates, can be as small as 17.8‰.

The other photosynthetic pathway is the Hatch–Slack cycle, used by the C_4 plants, which include hot-region grasses and related crops such as maize, sorghum, and sugarcane. These plants use *phosphoenolpyruvate carboxylase* to initially fix the carbon and form oxaloacetate, a compound that contains four carbons (Figure 10.8). A much smaller fractionation, about −2.0‰ to −2.5‰ occurs during this step. In phosphoenolpyruvate carboxylation, the CO_2 is fixed in outer mesophyll cells as oxaloacetate and carried as part of a C_4 acid, either malate or aspartate, to inner bundle sheath cells where it is decarboxylated and refixed by RuBP (Figure 10.9). The environment in the bundle sheath cells is almost a closed system, so that virtually all the carbon carried there is refixed by RuBP. Consequently, there is little fractionation during this step. C_4 plants have average $\delta^{13}C$ of −13‰. As in the case of RuBP photosynthesis, the

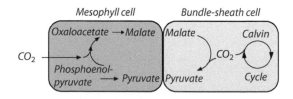

Figure 10.9 Chemical pathways in C_4 photosynthesis. Carbon is initially fixed as oxaloacetate in outer mesophyll cells then converted to malate or aspartate into inner sheath cells, decarboxylated, and refixed by RuBP. C_4 photosynthesis results in more complete fixation of carbon in cells and, consequently, produces a smaller carbon isotope fractionation.

fractionation appears to depend on the ambient concentration of CO_2. This dependence was modeled by Farquhar (1983) as

$$\Delta_{net} = \Delta_{diff} + (\Delta_{trans} + \Delta_{carbox}\phi - \Delta_{diff})(c_i/c_a)$$
$$(10.7)$$

where Δ_{trans} is the fractionation during transport into bundle-sheath cells φ is the fraction CO_2 leaked from the plant and other terms are the same as in 10.6.

The C_4 pathway is more energy expensive than the C_3 pathway but has the advantage of more efficient use of interior CO_2. Plants must open their stomata to allow CO_2 in, which allows H_2O out. Using interior CO_2 more efficiently lets plants close their stomata more, limiting H_2O loss, giving C_4 plants an advantage when CO_2 or H_2O is limiting. C_4 plants, which encompass a broad range of taxa, are found in environments where water use efficiency is important, notably regions of hot summer temperatures, inconsistent rainfall, and low nitrogen availability such as tropical and temperate grasslands.

A third group of plants, the CAM plants, have a unique metabolism called the "Crassulacean acid metabolism." These plants generally use the C_4 pathway but can use the C_3 pathway under certain conditions. They are succulents and generally adapted to arid environments and include pineapple and many cacti; as one might expect, they have $\delta^{13}C$ intermediate between C_3 and C_4 plants.

Terrestrial plants, which utilize CO_2 from the atmosphere, generally produce greater fractionations than marine and aquatic autotrophs, which utilize dissolved CO_2 and HCO_3^-, together referred to as *dissolved inorganic carbon* (DIC). Once dissolved in water, most CO_2 dissociates to form the bicarbonate ion:

$$CO_{2(g)} \rightleftharpoons CO_{2(aq)} + H_2O \rightarrow H_2CO_3$$
$$\rightleftharpoons H^+ + HCO_3^-. \qquad (10.8)$$

An equilibrium fractionation of +0.9 per mil is associated with dissolution ($^{13}CO_2$ will dissolve more readily), and an equilibrium +7 to +12‰ fractionation (depending on temperature) occurs during the hydration and dissociation of CO_2. Thus, we expect dissolved HCO_3^- to be about 8–12 per mil heavier than atmospheric CO_2. Marine algae and aquatic plants can utilize either dissolved CO_2 or HCO_3^-. Since HCO_3^- is about two orders of magnitude more abundant in seawater than dissolved CO_2, many marine algae utilize this species and, hence, tend to show a lower net fractionation during photosynthesis. Diffusion is slower in water than in air, so diffusion is often the rate-limiting step.

Most aquatic plants have some membrane-bound mechanism to pump DIC, which can be turned on when DIC is low (Sharkey and Berry, 1985). When DIC concentrations are high, fractionation in aquatic and marine plants is generally similar to that in terrestrial plants. When it is low and the plants are actively pumping DIC, the fractionation is less because most of the carbon pumped into cells is fixed. Thus, carbon isotope fractionations can be as low as 5‰ in algae. The model describing this fractionation is

$$\Delta = d + b_3(F_3/F_1) \qquad (10.9)$$

where d is the equilibrium effect between CO_2 and HCO_3^-, b_3 is the fractionation associated with carboxylation, and (F_3/F_1) is the fraction of CO_2 leaked out of the cell. Though the net fractionation varies between species and depends on factors such as light intensity and moisture stress, C_3 plants have average bulk $\delta^{13}C$ values of –27‰, C_4 plants average –13‰, and algae and lichens are typically –12 to –23‰.

In aquatic systems where the pH is lower than in seawater, dissolved CO_2 becomes more abundant and some algae can utilize this rather than HCO_3^-. In those cases, the total fractionation will be greater. An interesting illustration of this and the effect of the CO_2 concentration on net fractionation is shown in Figure 10.10, which shows data on the isotopic composition of algae and bacteria in Yellowstone hot springs. There is more fractionation at low pH simply because the concentration of CO_2 relative to HCO_3^- increases with decreasing pH.

Not surprisingly, the carbon isotope fractionation in C fixation is also temperature dependent. Higher fractionations are observed in cold-water phytoplankton than in warm water species. However, this observation also reflects a kinetic effect: there is generally less dissolved CO_2 available in warm waters because of the decreasing solubility at higher temperature (indeed, as we will see in Chapter 12, the effect of CO_2 concentration

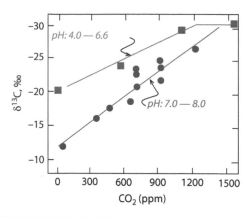

Figure 10.10 Dependence of $\delta^{13}C$ of algae and bacterial on CO_2 concentration from hydrothermal springs in Yellowstone National Park. Carbon isotope fractionation also depends on the pH of the water because this determines the species of carbon used in photosynthesis. Source: Data from (Estep, 1984), after (Fogel and Cifuentes, 1993). Reproduced with permission from Springer Science + Business Media.

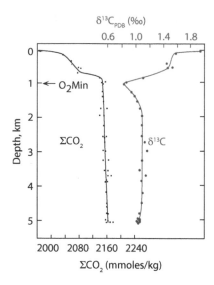

Figure 10.11 Depth profile of total DIC and $\delta^{13}C$ in the North Atlantic. Source: Adapted from (Kroopnick et al., 1972).

on fractionation in marine algae can be used to estimate ancient CO_2 levels). As a result, a larger fraction of the CO_2 is utilized, and there is consequently less fractionation. Surface waters of the ocean are generally enriched in ^{13}C because of the uptake of ^{12}C during photosynthesis (Figure 10.11). The degree of enrichment depends on the productivity: biologically productive areas show greater enrichment. Deep water, on the other hand, is depleted in ^{13}C (perhaps, it would be more accurate to say it is enriched in ^{12}C). Organic matter falls through the water column and is decomposed and "remineralized," that is, converted to inorganic carbon, by the action of bacteria, enriching deep water in ^{12}C. Thus, biological activity acts to "pump" carbon and, particularly, ^{12}C from surface to deep waters.

Nearly all organic matter originates through photosynthesis. Subsequent reactions convert the photosynthetically produced carbohydrates to the variety of other organic compounds utilized by organisms. Further fractionations occur in these reactions, as discussed in Hayes (2001). Lipids (fats, waxes, etc.) in plants tend to be isotopically lighter than other components (such as cellulose). The effect is usually small, a few per mil, compared to the fractionation during photosynthesis. These fractionations are thought to be kinetic in origin and may partly arise from

organic C–H bonds being enriched in ^{12}C, while organic C–O bonds are enriched in ^{13}C. ^{12}C is preferentially consumed in respiration (again, because bonds are weaker and it reacts faster), which would tend to enrich residual organic matter in ^{13}C. Thus, the carbon isotopic composition of organisms becomes somewhat more positive moving up the food chain.

Interestingly, although the energy source for chemosynthesis is dramatically different than for photosynthesis, the carbon-fixation process is similar and still involves the Calvin cycle. Not surprisingly then, carbon fractionation during chemosynthesis is similar to that during photosynthesis. Some chemosynthetic bacteria, notably some of the symbionts of hydrothermal vent organisms, have RUBISCO Form II that show smaller fractionations.

10.3.2 Nitrogen isotope fractionation in biological processes

Following carbon, oxygen, and hydrogen, nitrogen is the most important and most abundant element in living organisms: it is essential component of all amino acids and proteins assembled from them, as well as other key molecules such as adenosine triphosphate, nucleotides that encode genetic information, and a host of other molecules. It is often the limiting nutrient in both the terrestrial and marine ecosystems. As in the case of carbon, most of terrestrial nitrogen isotopic variation results from biological processes. These processes, however, are considerably more complex because nitrogen exists in oxidation states ranging from N(–III) in ammonia (NH_3) and ammonium (NH_4^+) to N(V) in nitrate (NO_3^-). In between, there are four other important forms of inorganic nitrogen: molecular nitrogen (N_2), nitrous oxide (N_2O), nitric oxide (NO), nitrite (NO_2^-), and nitrogen dioxide (NO_2). The latter three as well as NO_3^- are collectively referred to as NO_x.[1] In organic molecules, it is primarily present as the amine group $(NH_2)^-$ with a valence of N(–III); for example, amino acids have the general formula $NH_2CH(R)COOH$ (where R is some other organic functional group). Except for the ammonium–ammonia reaction, the reactions between these forms are all redox reactions, and they are predominantly biologically mediated. Both equilibrium and kinetic isotope fractionations occur, but the latter tend to dominate and can be quite large (Figure 10.10).

To be taken up by autotrophs, including algae, plants, and bacteria, N_2 must first be "fixed," that is be in reduced or oxidized form. Nitrogen *fixation* represents the start of the biological *nitrogen cycle* illustrated in Figure 10.12. Fixation is energy intensive and carried out by a limited cast of characters, including cyanobacteria, green sulfur bacteria, free-living azotobacters in soil and fresh waters, and rhizobium bacteria, which form an endosymbiotic relationship with the roots of legumes. Nitrogen fixation is catalyzed by *nitrogenase* enzymes, which contain either Mo-, Fe-, or V-bearing proteins that use electrons provided by other parts of the enzyme to reduce N_2, providing interesting examples of the role of trace transition metals in life. Nitrogen fixation by organisms using the Mo-bearing enzyme involves only a small fractionation of –1 to –2‰, but fractionations can be up to –7‰ by organisms utilizing V- or Fe-bearing enzymes. Lightening also produces nitrate and nitrite in the atmosphere with little fractionation, but this constitutes <3% of all naturally fixed nitrogen.

Fixed nitrogen is assimilated by organisms principally as NH_4^+ and incorporated into N-bearing molecules as NH_2 *amine* groups,

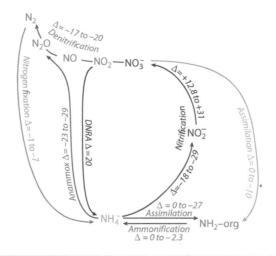

Figure 10.12   The biological nitrogen cycle. DNRA is dissimilatory nitrate reduction to ammonia, and Anammox is anaerobic ammonia oxidation. Δ values shown are combined equilibrium and kinetic fractionations.

although phytoplankton can assimilate nitrite and nitrate as well. Indeed, nitrate is the primary form of fixed nitrogen in the marine photic zone. When nitrate is taken up rather than ammonium, it must first be reduced by the action of nitrogen reductase enzymes. [14]N is preferentially assimilated and fractionation associated with assimilation by terrestrial plants is typically small, ~–0.25‰; larger fractionations, up to –27‰, can occur during uptake by algae and other microorganisms (Fogel and Cifuentes, 1993) with the fractionation being largely independent of the form assimilated. As with CO_2, the extent of fractionation depends on availability: when fixed nitrogen is abundant, fractionation tends to be large; when it is not, most available fixed nitrogen is taken up and there is little fractionation.

There is a small fractionation when ammonium is subsequently incorporated into organic molecules. There are two principal reactions by which ammonia is incorporated into organic matter: formation of the glutamic acid ($HOOC-CH(NH_2)-(CH_2)_2-COOH$) from α-ketoglutarate via the glutamate dehydrogenase reaction and formation of glutamine from glutamate via the enzyme glutamine synthetase. A positive fractionation (i.e. the product

---

[1] N_2O is sometimes included in NO_x, but we will consider it separately here.

is enriched in ^{15}N) of +2 to +4 has been measured for the glutamate dehydrogenase reaction, and the fractionation for the glutamine synthetase reaction is also expected to be positive, because N is bound more strongly in the product than in ammonia.

The net result of these various fractionations is that organic nitrogen is usually heavier than atmospheric nitrogen. The isotopic compositions of marine particulate nitrogen and non-nitrogen-fixing plankton are typically –3‰ to +12‰ $\delta^{15}N$. Terrestrial plants unaffected by artificial fertilizers generally have a narrower range of +6‰ to +13 per mil. Legumes (and a few other kinds of plants) are a special case. While they cannot fix nitrogen, they have symbiotic bacteria in their root nodules that can. As a result, legumes have distinctly lower $\delta^{15}N$ than other terrestrial plants, in the range of –2 to +4‰. Marine cyanobacteria (blue–green algae) have $\delta^{15}N$ ranging from –4 to +2, with most in the range of –4 to –2‰.

Organically bound nitrogen that is released back into soils and solution is converted to the ammonium ion, a process called *ammonification* or *mineralization*, and there is little isotopic fractionation in this process. Since fixed nitrogen is often limiting, the ammonium released is often quickly assimilated by other autotrophs. In oxidizing environments such as the oceans, the remaining ammonium will be oxidized to NO_2^- and eventually to NO_3^-. This process known as *nitrification* is carried out by consortia of nitrifying bacteria that derive energy from the process. The first step, oxidation to nitrite, has a large negative fractionation, but the process generally goes to completion resulting in little net fractionation. The second step, oxidation of nitrite to nitrate, is associated with a positive fractionation, but again the process will run to completion in oxidizing environments, and in such cases, the net production of nitrate from ammonia will result in little net fractionation. Much of this nitrate is then assimilate by organisms.

Any nitrate not assimilated by organisms will be stepwise reduced to N_2 in a process termed *denitrification* with preferential denitrification of ^{14}N. These are also exothermic reactions that supply energy to denitrifying organisms, which include both bacteria and fungi and occur principally in low-oxygen environments such as poorly drained or poorly oxidized soils, the oxygen minimum zones in the ocean, and in sediments. Denitrification has a net negative fractionation, and since the product of these reactions are gases (N_2, N_2O), denitrification tends to enrich residual soil nitrate in ^{15}N.

Figure 10.13 shows nitrate and O_2 concentrations and isotopic composition of dissolved and particulate nitrate at the ALOHA site in the central North Pacific. Uptake by phytoplankton drives nitrate to near-zero concentrations in the photic zone. Rapid and near-complete utilization and rapid recycling fixed nitrogen within this zone maintains $\delta^{15}N$ of both particulate matter (mainly organisms and their remains) nitrogen and dissolved nitrate at values only slightly above atmospheric.

Particulate matter (mainly organic remains) sinking out of the photic zone then supplies fixed nitrogen to deeper water. At 150 m, sinking organic matter (collected in sediment traps) has a $\delta^{15}N$ of +2.5‰, a value close to that of

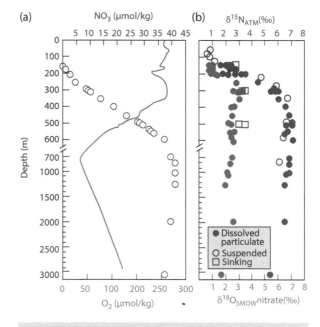

Figure 10.13 (a) Dissolved nitrate (circles) and dissolved oxygen (line) at the ALOHA site (22°45'N, 158°W) in the central north Pacific. The maximum NO_3 corresponds with the minimum O_2. (b) $\delta^{15}N$ (red) and $\delta^{18}O$ (blue) of dissolved, particulate, and sinking nitrate at the ALOHA site. "Sinking" refers to large particles, captured on sediment traps. Both particulate and sinking particles are biological. Note the change in vertical scale at ~700 m. Source: Adapted from Casciotti et al. (2008).

both suspended and dissolved nitrate. $\delta^{15}N$ of this sinking particular matter increases to +3.5‰ at 500-m depth. In contrast, dissolved and suspended matter $\delta^{15}N$ increases to +6.6‰ to 7‰ by 500 m and decreases only slightly at greater depth. The sinking organic matter represents the flux out of the photic zone with $\delta^{15}N$ of ~2.5‰, which, assuming steady state, is balanced by a combination of new fixation with $\delta^{15}N \approx 0$ to +1‰ and upward mixing diffusion of dissolved nitrate with $\delta^{15}N$ of ~3.5‰. Based on this mass balance, Casciotti et al. (2008) estimated that about 25% of this budget was supplied by new fixation. At 300–500-m depth, $\delta^{15}N$ decreases upward as a consequence of remineralization of sinking organic matter and new fixation by bacteria. Below 500 m, Casciotti et al. attributed the gradual decrease in both $\delta^{15}N$ and $\delta^{18}O$ of dissolved nitrate to a combination of remineralization and lateral advection of waters from the eastern tropical Pacific where low dissolved O_2 results in denitrification, which drives an increase in residual nitrate $\delta^{15}N$.

There are two "short-circuits" to the nitrogen cycle, the first of which is anaerobic ammonia oxidation or "*anammox*" in which certain types of bacteria can catalyze the reaction:

$$NH_4^+ + NO_2^- \rightarrow N_2 + 2H_2O \qquad (10.10)$$

in anaerobic environments. Although only documented in the 1990s, this may be a major sink for fixed nitrogen. This process appears to be associated with a negative fractionation similar to the nitrification–denitrification pathway. A second reaction, also limited to low-oxygen environments such as sediments and oxygen minimum zones, is *dissimilatory nitrate reduction to ammonia*, in which nitrate rather than O_2 is used as the electron receptor in respiration. Interestingly, it can be carried out by some types of eukaryotes such as diatoms as well as prokaryotes. Little information is available on the isotopic fractionation; Chang (2014) estimates a fractionation of ~20‰.

A caveat to all this is that most fixed nitrogen in modern terrestrial ecosystems is now derived, directly or indirectly, from artificial fertilizers. These fertilizers contain ammonia and nitrate derived from atmospheric N_2 through the Haber process, in which there is little isotopic fractionation. Consequently, modern plants, particularly those raised on artificial fertilizers, have lower $\delta^{15}N$.

10.3.3 Biological fractionation of oxygen and hydrogen isotopes

Oxygen is incorporated into biological material from CO_2, H_2O, and O_2. However, both CO_2 and O_2 are in oxygen isotopic equilibrium with water during photosynthesis, and water is the dominant form and determines the isotopic composition of oxygen taken up by the plant. Although soil water isotopic composition can be somewhat enriched in heavy isotopes relative to precipitation due to evaporation from the soil, $\delta^{18}O$ of water available to plants will in most instances be isotopically light relative to SMOW. The O and H isotopic composition of water in xylem (the water transporting tissue) generally matches that of soil water, indicating water is taken up and initially transported without fractionation. Leaf water, however, can have $\delta^{18}O$ that is ~15–35‰ heavier than soil water as a consequence of preferential loss of isotopically light water during transpiration. Transpiration is in some ways analogous to evaporation from the ocean surface with both an equilibrium fractionation associated with the liquid–vapor transition and kinetic fractionations associated with diffusional transport to, through, and from stomata. The extent of this fractionation depends on humidity and temperature and varies between species and over the course of the day and can be approximately expressed as

$$\delta_{LW} = \delta_s(1 - h) + h\delta_{AV} + \Delta_{EQ} + \Delta_K(1 - h) \qquad (10.11)$$

where δ_{LW}, δ_S, and δ_{AV} denote the isotopic composition of leaf water, stem (=soil), and atmospheric vapor, respectively, and Δ_{EQ} and Δ_K are the equilibrium and kinetic fractionation factors, and h is humidity (Burk and Stuvier, 1981). More sophisticated models have been developed for this process and are described by Barbour (2007). Since photosynthesis takes place in leaves, it is the composition of this water, which is invariably heavier than soil water, that will govern the O and H isotopic composition of the organic matter produced by terrestrial plants.

The oxygen and hydrogen of leaf water is then incorporated into organic matter, initially as sucrose through photosynthesis. Ignoring details, this reaction can be approximately described as

$$6CO_2 + 6H_2O \rightarrow 6CH_2O + 6O_2.$$

Guy et al. (1993) found experimentally that there is little (<0.5‰) O isotopic fractionation during photosynthesis. However, significant fractionation occurs during photorespiration, which is a reaction competing the carbon fixation reaction (Section 10.2.1) in which the RUBISCO enzyme oxygenates RuBP. About 25% of RuBP is consumed in this way. In addition, in the Mehler reaction, some of the oxygen produced during photosynthesis is reduced to H_2O_2, which is subsequently converted to ascorbate and water (both these reactions may have beneficial effects: nitrogen uptake in the case of photorespiration and a sink for excess excitation energy resulting from sunlight exposure in the Mehler Reaction). Guy et al. found that these reactions do result in a significant fractionation resulting in O_2 in leaves being heavier than the source water by ~21‰ and ~15‰ for photorespiration and the Mehler Reaction, respectively.

Luz et al. (1999) found using occasionally purged closed terrariums that photosynthetically produced O_2 has a $\delta^{17}O$–$\delta^{18}O$ relationship given by $\delta^{17}O = 0.51211 \times \delta^{18}O + 0.155‰$ (unconventionally, $\delta^{18}O$ and $\delta^{17}O$ were defined relative to local (Jerusalem) atmospheric O_2). Subsequent studies have found broadly similar results (although cyanobacteria fractionate oxygen along a distinctly lower slope of ~0.497). This slope is distinctly shallower than the TFL ($\lambda = 0.528$). Respiration preferentially consumes ^{16}O and shows a considerable range of fractionation of $\Delta^{18}O$ from –12‰ to –20‰, depending on the organism with an average of ~–18‰, corresponding to an α of 0.982 (Mader et al., 2017). Only a small fractionation occurs when O_2 dissolves in water, enriching the water by 0.7‰ in ^{18}O. However, Rayleigh-like fractionation as O_2 is consumed by respiration can result in a significant enrichment of water out of contact with the atmosphere and below the photic zone in ^{18}O. This is illustrated in Figure 10.14, which shows the variation of $\delta^{18}O$ and O_2 with depth in the eastern equatorial Pacific.

Young et al. (2014) concluded the slope for global photosynthesis–respiration was

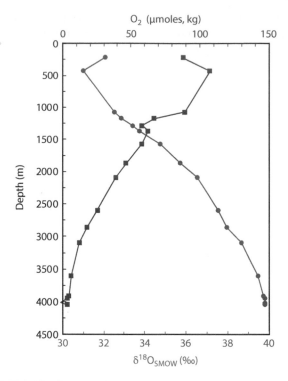

Figure 10.14   Variation of dissolved O_2 (blue squares) and $\delta^{18}O$ (red circles) at 8°07'N, 113°55'W in the eastern equatorial Pacific. Data from (Kroopnick and Craig, 1976). Note that the Kroopnick and Craig (1976) reported the original data relative to atmospheric O_2; values plotted here have been converted to values relative to SMOW.

$\lambda = \delta'^{17}O/\delta'^{18}O = 0.5149$, where $\delta'^{18}O$ is defined the natural log of the ratio of $^{18}O/^{16}O$ in the sample to that in a standard (Equation 5.113). Luz et al. also found that the photosynthesis–respiration within their experimental terrariums eventually removed the negative $\Delta^{17}O$ of air inherited from the stratosphere relative to trophospheric O_2. However, as the slope of this fractionation is lower than that of the TFL and GMWL ($\lambda = 0.528$), the fractionation during photosynthesis–respiration shifts the isotopic composition of air down relative to the TFL producing an even more negative $\Delta'^{17}O$ relative to it.

Sucrose produced in leaves is then transported elsewhere through phloem tissue to stems to synthesize to various other compounds, notably cellulose, which constitutes a large fraction of plant mass and is also more likely to survive in soil than other plant tissues.

The isotopic composition of cellulose will then additionally depend on the relative contributions of leaf and stem (=soil) water in its synthesis. Relative to soil water, $\delta^{18}O$ of cellulose varies from ~+50‰ in cold regions to +35‰ in warm regions, which reflects the variation of equilibrium and kinetic fractionation factors with temperature and humidity (Sternberg and Ellsworth, 2011).

Water is the sole source of hydrogen in plants. No fractionation occurs during uptake, but hydrogen undergoes fractionations analogous to those of oxygen in leaves due to transpiration such that leaf water can be many tens of per mil heavier than soil water. Sternberg and DiNiro (1983) found that the δ^2H of cellulose nitrate in plants collected within a small area of the Deep Canyon Desert Research Center in Riverside County, California varied by 220‰, while $\delta^{18}O$ varied by only 20‰. A good deal of this variation was due to the presence of CAM plants, which had distinctly higher δ^2H (+30 to + 80‰) than C_3 and C_4 plants (–140 to –20‰). In contrast, $\delta^{18}O$ of these groups overlapped. CAM plants are adapted to dry environments and tend to store water from wet periods in vacuoles. Subsequent transpiration in dry periods during which photosynthesis and metabolism continue progressively enriches this water in ^2H.

Significant H isotopic fractionation occurs during photosynthesis with fractionation factors as much as –200‰ (^1H is preferentially incorporated in the organic matter). Although both O and H in organic molecules can undergo postphotosynthetic isotopic exchange with cellular water, water-derived oxygen is incorporated in organic matter essentially only in the sugars produced by photosynthesis. In contrast, water-derived hydrogen is incorporated into many of the myriad other organic molecules, many of which are synthesized with attendant isotopic fractionations in other parts of the plant from sucrose exported from leaves. These reactions are species-specific, but the general result is that a significant fraction of the original hydrogen is replaced, leading to ^2H enrichment during plant metabolism (Figure 10.15). As a consequence of these factors, δ^2H in bulk plant material and cellulose is poorly correlated with

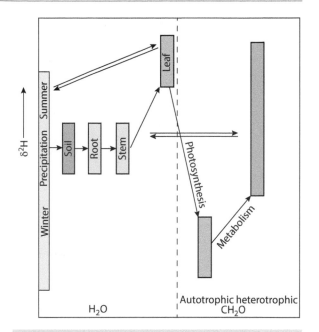

Figure 10.15 Isotopic fractionations of hydrogen during primary production in terrestrial plants. Source: Adapted from Yakir (1992).

δ^2H in precipitation. Overall, most plants are typically deuterium-depleted relative to soil water, with plants typically –80 to –160% lighter than the water in which they grown.

While the δ^2H in bulk plant material or cellulose correlates poorly with the water in which it grew, Sternberg (1988) showed that the δ^2H of lipid fraction of submerged aquatic plants correlated well. With the development of compound-specific isotopic analysis a decade later, improved correlations between specific lipids and δ^2H have been found. Englebrecht and Sachs (2005) demonstrated strong correlations between δ^2H in C37 and C38 alkenones[2] derived from *Emiliania huxleyi*, a common marine coccolithophorid algae, and δ^2H in deuterated water in which the algae were cultured (Figure 10.16 inset). A number of other studies demonstrated correlations between δ^2H in various lipids recovered from lake sediment and δ^2H in water, three examples of which are shown in Figure 10.16. Huang et al. (2004) demonstrated a correlation between δ^2H various lipids extracted from surface sediment, including the C17 n-alkanes

---

[2] Alkenones are hydrocarbon chains, in this case containing 37 or 38 carbons that are unsaturated (i.e., some of the carbons are double-bonded), and one or more carbons doubly bonded to oxygen ("ketones"). They appear to serve as energy reserves.

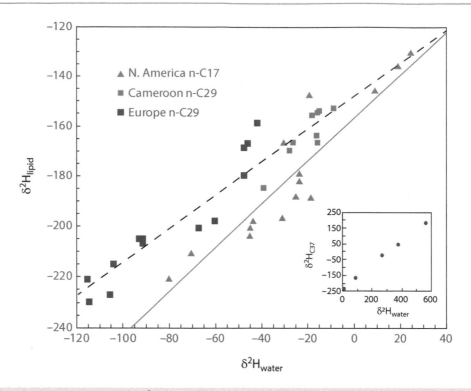

Figure 10.16 Correlation between δ^2H in various C-17 and C-29 lipids from algae and cyanobacteria and lake water in which they grew. Data from (Huang et al., 2004; Sachse et al., 2004; Garcin et al., 2012). Inset shows the correlation between C37 lipids in *Emiliania huxleyi* and the water in which they grew (Englebrecht and Sachs, 2005).

(a straight chain hydrocarbon 17 carbons long, produced primarily by algae and cyanobacteria) and δ^2H in lake water in eastern North America. Sachse et al. (2004) demonstrated correlations between δ^2H in n-C29 lipids (among others), which are derived from leaf waxes, extracted from lake sediment and δ^2H in precipitation. Garcin et al. (2012) demonstrated a similar correlation between n-C29 lipids in lacustrine sediments from Cameroon and δ^2H in lake water, which appear to be co-linear with the European data. Subsequent studies show other effects on δ^2H in lipids, including growth rates and salinity. Nevertheless, δ^2H in lacustrine sedimentary lipids has become a useful paleoclimate proxy.

10.3.4 Biological fractionation of sulfur isotopes

Sulfur is a critical but minor component of living tissue (C:S atomic ratio is about 200). Autotrophs take up sulfur as sulfate and subsequently reduce it to sulfide and incorporate into cysteine (an amino acid with the composition HOOC-CH-(NH$_2$)-CH$_2$-SH). There is apparently little fractionation of sulfur isotopes in transport across cell membranes and incorporation, but there is a fractionation of –1 to –3‰ in the reduction process, referred to as *assimilatory sulfate reduction*. This is substantially less than the expected fractionation of about –20‰, suggesting that nearly all the sulfur taken up by primary producers is reduced and incorporated into tissue.

Sulfur, however, plays two other important roles in biological processes. First, sulfur in the form of sulfate can act as an electron acceptor or oxidant and is utilized as such by sulfur-reducing bacteria. This process, in which H$_2$S is liberated, is called *dissimilatory sulfate reduction* and plays an important role in biogeochemical cycles, both as a sink for sulfur and source for atmospheric oxygen (because the sulfide is usually then buried in sediments). A large fractionation of –20 to –75‰ is associated with this process. As we found in Chapter 5, this is a multistep process, the net of which is that the ratio of $\delta^{33}S/\delta^{34}S$ fractionation factors differs from the widely observed value of 0.515, producing an apparent mass-independent fractionation (MIF).

Bacterial sulfate reduction produces by far the most significant fractionation of sulfur isotopes and, thus, governs the isotopic composition of sulfur in the exogene. Sedimentary sulfate typically has $\delta^{34}S$ of about +17‰, which is similar to the $\delta^{34}S$ of sulfate in the modern oceans (+21.24‰), while sedimentary sulfide typically has $\delta^{34}S$ of −18‰ (as we will find, however, these ratios have varied over geologic time). The living biomass has a $\delta^{34}S$ of ∼ 0‰.

The final important role of sulfur is a reductant. Sulfide is an electron donor used by variety of microbes including bacteria of the genus *Thiploca* living in euxinic or anoxic sediments and that oxidize sulfide with nitrate and green sulfur bacteria which use sulfur as an electron donor in photosynthetic reduction of CO_2 to organic carbon. This group also includes chemosynthetic bacteria of submarine hydrothermal vents, which oxidize H_2S emanating from the vents and form the base of the food chain in these unique ecosystems.

As in sulfate reduction, sulfide oxidation to sulfate is typically a multistep process with multiple pathways and carried out by a consortium of bacteria as the valence of sulfur increases from −2 to +6. The fractionation of +2 to −18‰ is associated with this process. Sulfide oxidation to sulfur typically results in fractionations of <2‰, but further oxidation can result in large fractionations, depending on the organism and pathway (Canfield, 2001). Particularly large fractionations in which sulfate is enriched in ^{34}S by ∼18‰ are associated with biological disproportionation reactions such as

$$4SO_3^{2-} + 2H^+ \rightarrow H_2S + 3SO_4^{2-} \quad (10.12)$$

$$4S° + 4H_2O \rightarrow 3H_2S + SO_4^{2-} + 2H^+. \quad (10.13)$$

The $\delta^{34}S$ of the exogene is almost entirely controlled by these microbial reduction and oxidation reactions; abiotic redox reactions are essentially restricted to temperatures environments too hot for life, such as high-temperature hydrothermal systems and to photochemical reactions. The former produces a quite large range of $\delta^{34}S$ with sulfides having $\delta^{34}S$ typically of −20‰, but occasionally even lower and sulfates typically having $\delta^{34}S$ close to the seawater value but occasionally even higher.

10.4 ISOTOPE RATIOS IN THE ATMOSPHERE

The lower 30 km of the atmosphere is divided into the troposphere, which is heated from below by infrared radiation from the Earth's surface and in which temperatures consequently decrease with height, and the stratosphere which is heated from above by solar ultraviolet (UV) radiation and in which temperatures consequently increase with height. The tropopause is the boundary between the two and is located at roughly 17 km at the equator and 9-km height at the poles, and there is only limited transport across it. Life, weather, and direct interaction with the solid Earth and hydrosphere and associated isotopic fractionations are restricted to the troposphere; on the other hand, energetic photochemical reactions in the stratosphere lead to interesting isotopic effects. Air exchange across the tropopause occurs when momentum causes strong tropical thunderstorms to overshoot and inject tropospheric air in the stratosphere. Stratospheric air then slowly mixes downward into the troposphere mainly at high latitudes such that the residence time of air in the stratosphere is ∼4 years. This downward mixing carries the isotopic imprint of photochemical reactions into the troposphere.

10.4.1 The stratosphere

10.4.1.1 Oxygen

The process of stratospheric ozone formation begins with the photodissociation of O_2 to produce monatomic oxygen that then combines with O_2 to form O_3 that is initially in an excited state. As we found in Section 5.4, for the newly formed ozone molecule to survive, this vibrational energy must be converted to kinetic energy through collisions or partitioned to rotational energy. Because the density of quantum states available in the transition to stable O_3 is greater for asymmetric molecules than such as $^{17}O^{16}O^{16}O$ or $^{18}O^{16}O^{16}O$ than for symmetric ones such as $^{16}O^{16}O^{16}O$ or $^{16}O^{18}O^{16}O$, ozone molecules containing either heavy isotope are more likely to survive the transition to become

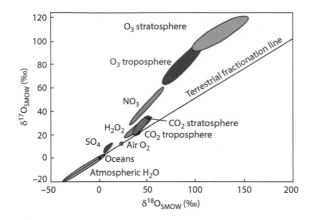

Figure 10.17   O isotopic composition of atmospheric gases. Source: Thiemens and Shaheen (2014)/with permission of Elsevier.

stable O_3. Spectroscopic measurements by Johnson et al. (2000) confirmed enrichments in $^{50}O_3$ and $^{49}O_3$ and enrichments in the asymmetric (e.g. $^{18}O^{16}O^{16}O$) over symmetric (e.g. $^{16}O^{18}O^{16}O$) versions of these molecules in stratospheric ozone (Figure 10.17).

Once produced, ozone is photolyzed by UV radiation to produce O_2 and an excited monatomic $O^{\cdot}$:

$$O_3 + h\nu \rightarrow O_2 + O^{\cdot}. \qquad (10.14)$$

This monatomic O then reacts with other O-bearing molecules, in some cases O_2 to form O_3 again, but in other cases, it reacts with CO_2:

$$CO_2 + O^{\cdot} \rightarrow CO_3. \qquad (10.15)$$

Since the monatomic O is derived from ozone enriched in ^{18}O and ^{17}O, this should impart the MIF isotopic enriched signature to CO_2. As Figure 10.18 shows, stratospheric CO_2 shows a MIF signature that increases with altitude (and O_3 concentration). In fact, Boering et al. (2004) measured $\delta^{17}O/\delta^{18}O$ enrichments in CO_2 ranging from of 1.2 to 2.1, well above the ratio of 1 expected from laboratory experiments, indicating that ^{17}O is more readily transferred to CO_2 than ^{18}O. CO_3 is unstable and dissociates:

$$CO_3 \rightarrow CO_2 + O^{\cdot}. \qquad (10.16)$$

It randomly loses one oxygen atom, so CO_3 has a 67% chance to retain its O atom derived from O_3 and release an O atom that was present in the predecessor CO_2 molecule, the

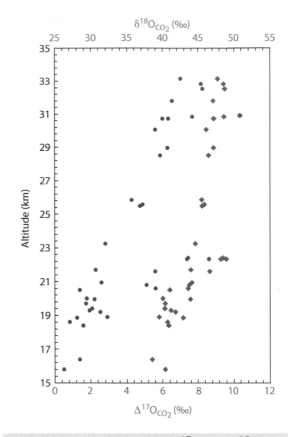

Figure 10.18   Variation of $\Delta'^{17}O$ and $\delta^{18}O$ in CO_2 with altitude in the stratosphere. Source: Data from (Boering et al., 2004) and (Lämmerzahl et al., 2002).

most likely fate of which is to react with another monatomic O to produce O_2 or with O_2 to produce O_3 again.

These photochemical reactions effectively transfer ^{17}O and ^{18}O from O_2 to CO_2 (and to a lesser extent other O-bearing species) leaving residual O_2 with a negative $\Delta^{17}O$. Pack et al. (2017) determined the $\Delta^{17}O$ of the upper atmosphere from oxidized I-type cosmic spherules (which are essentially microscopic bits of iron meteorites) to be −0.51‰ and −0.42‰, similar to tropospheric air. (Note that this anomaly is too small to be apparent on the scale of Figure 10.17.)

The magnitude of the $\Delta^{17}O$ anomaly in O_2 depends on the concentration of CO_2, or more precisely the CO_2/O_2. In the absence of CO_2 (or other reactive oxides such as SO_2 and NO_2), the monatomic O would inevitably recombine to form O_2, leaving no anomaly. This negative $\Delta^{17}O$ signature in O_2 and positive one in CO_2 is then carried into the troposphere by sinking stratospheric air.

10.4.1.2   Sulfur

As we found in Chapter 5, the UV photolysis of SO_2 can produce MIF in sulfur isotopes, but, unlike the Archean, there is limited opportunity for this in the modern Earth. In the modern atmosphere, very little UV light makes it into the troposphere due to absorption in the stratosphere. On the other hand, very little sulfur finds its way into the stratosphere. Sulfur emitted to the atmosphere is quickly converted to sulfate, which forms H_2SO_4 droplets and are mostly wrung out as precipitation before air reaches the stratosphere. The exception is large Plinian volcanic eruptions, which can feed SO_2 into the stratosphere, where it is oxidized to H_2SO_4. These sulfate aerosols can remain in the stratosphere for years, reflecting sunlight and reducing global temperatures (in contrast, sulfate is removed from the troposphere by precipitation on time scales of days).

The last such large eruption, Mt. Pinatubo in 1991, injected massive amounts (17 ± 3 Tg) of SO_2 into the stratosphere and resulted in a global average temperature decrease of 0.4 °C in the following year. Savarino et al. (2003) found that Antarctic snow from 1991 had mass-independently fractionated sulfur isotopes with $\Delta^{33}S$ of +0.67‰ and $\Delta^{36}S$ of −3.58‰. They observed the same pattern of enrichment in ^{33}S and depletion in ^{36}S in laboratory photolysis experiments with 248-nm UV lasers, essentially confirming that the observed MIF resulted from UV photolysis in the stratosphere.

The isotopic fractionation pattern observed in the modern snow and in the 248-nm photolysis experiments is different from that observed in Archean samples where the effect arises mainly from shorter wavelength (<220 nm) radiation. In the modern stratosphere, oxygen is sufficiently abundant that SO produced by the photolysis of SO_2 essentially instantly recombines with O to produce SO_2, so any MIF effect is nil. On the other hand, 248-nm radiation produces excited SO_2, which then self-reacts to form SO_3 and SO:

$$2SO_2^* \rightarrow SO + SO_3. \tag{10.17}$$

SO_3 then quickly forms H_2SO_4 and SO quickly oxidizes to SO_2. At the high SO_2 concentrations in volcanic plumes entering the stratosphere, self-shielding may also contribute to MIF (Ono, 2017). Subsequent studies of S isotopes in polar ice cores revealed 49 eruptions that injected SO_2 into the stratosphere with $\Delta^{33}S$ values >2‰ in some (e.g. Gautier et al., 2019). The largest of these (which was also identified in an initial study of Savarino et al. (2003) but whose source was not known at the time) was the eruption of Samalas Volcano on the Indonesian island of Lombok in 1257 AD. This eruption appears to have injected 158±12 Tg of SO_2 into the stratosphere, nearly 10 times that of Pinatubo. The eruption was followed by cold and rainy weather and widespread crop failures in Europe.

10.4.2   The troposphere

Table 10.1 lists the isotopic composition of the major atmospheric gases in the troposphere, excluding the rare gases which we will cover in Chapter 14. Also excluded is the composition of water vapor, which constitutes up to 3% of atmospheric gas (depending on humidity) as this was covered in Section 10.2. The isotopic composition of molecular nitrogen is uniform and that of oxygen nearly so because only a small fraction of these gases undergoes chemical reactions. In contrast, there is considerable isotopic variation in trace species, including O- and N-bearing trace species. In a number of cases, anthropogenic activity is changing their isotopic composition while changing climate. We explore these in this section.

Table 10.1   Isotopic composition of the troposphere.

N_2	78.084%	$\delta^{15}N_{ATM} = 0$
O_2	20.946%	$\delta^{18}O_{SMOW} = 23.88$‰
		$\delta^{17}O_{SMOW} = 12.03$‰
		$\Delta'^{17}O_{SMOW} = -0.51$‰
CO_2	415.00 ppm*	$\delta^{13}C_{PDB} = \sim{-8.7}$‰*
		$\delta^{18}O_{SMOW} = 40.9$‰
		$\delta^{17}O_{SMOW} = 21.26$‰
CH_4	1.89 ppm*	$\delta^{13}C_{PDB} = \sim{-47}$‰*

*Values as of late 2021. Photosynthesis and respiration impose local and seasonal variations on long-term trends due to fossil fuel and biomass burning.

10.4.2.1 Oxygen Isotopes in O_2

The existence of atmospheric molecular oxygen is entirely due to photosynthetic production combined with burial of organic matter over last 2.5 billion years or so. Since the source of photosynthetic O_2 is water and nearly all of which in the exogene resides in the ocean with $\delta^{18}O_{SMOW} \approx 0$, the positive $\delta^{18}O$ of atmospheric O_2 represents the net fractionation between biologic production consumption and burial. This difference is known as the Dole effect (Dole et al., 1954). A combination of factors accounts for the Dole effect.

- First, as we found in Section 10.2, while there is very little O isotope fractionation during actual carbon fixation, there is a significant fractionation during competing reactions, most notably photorespiration, leading to a net ^{18}O enrichment in O_2 during photosynthesis of ~6.5‰ (Luz and Barkan, 2010).
- Second, O_2 produced by photosynthesis is derived from water and in terrestrial plants, which account for ~65% of photosynthetic O_2 production; this is meteoric water that is typically ^{18}O depleted relative to SMOW.
- Third, in terrestrial systems, leaf water, and therefore photosynthetic O_2 produced from it, is enriched in ^{18}O relative to SMOW by 4–8% due to transpiration.
- Fourth, ^{16}O is preferentially consumed in respiration. Fractionation factors for the respiration vary depending on the organism and enzyme and are typically −12 ±6‰ but can be as high as −29‰.

As we found in the preceding section, O_2 exported from the stratosphere to the troposphere is depleted in ^{17}O as a consequence of photolytic reactions. This O_2 is then consumed by respiration, while new O_2 is produced by photosynthesis in the biosphere. Molecular oxygen cycles through the biosphere every 1200 years and through the stratosphere every 50 years. The net $\delta^{17}O$–$\delta^{18}O$ steady-state relationship for photosynthesis and respiration is $\delta^{17}O_{SMOW} = 0.5149 \times \delta^{18}O_{SMOW}$ ‰. This is identical to the observed variation in atmospheric O_2, demonstrating the importance of biospheric

influence on the isotopic composition of atmospheric O_2. Photosynthesis and respiration tend to erase the ^{17}O anomaly imparted by the stratosphere when air is used as the standard (Luz et al., 1999), but further decreases $\Delta^{17}O$ relative to the TFL ($\lambda = 0.528$) when SMOW is the standard.

Regardless of the reference fractionation line, biospheric and stratospheric processing controls the $\Delta^{17}O$ of tropospheric O_2. Assuming that the UV flux to the stratosphere and the rate of exchange between stratosphere and troposphere are all constant, then the $\Delta^{18}O$ of tropospheric O_2 depends on the atmospheric CO_2/O_2 concentration ratio, which controls the extent of the stratospheric ^{17}O depletion and gross primary biospheric productivity, which controls the rate at which O_2 cycles through the biosphere (Luz et al., 1999). Generally, $\Delta^{17}O$ will increase with increasing gross primary productivity at constant CO_2 and decrease with increasing CO_2 at constant gross primary productivity. This opens the possibility of determining biosphere productivity from $\Delta^{17}O$ in ice cores (e.g. Blunier et al., 2002) as well as other sedimentary archives, but the results can be model dependent, as Young et al. (2014) point out.

10.4.2.2 Oxygen isotopes in CO_2

Figure 10.19a shows the O isotopic composition of tropospheric CO_2 in various localities. We see that CO_2 is variably enriched in ^{17}O and ^{18}O relative to tropospheric O_2. In Figure 10.19b, the value of $\Delta^{17}O$ is computed relative to a slope $\lambda = 0.516$. The mean $\Delta^{17}O$ relative to this slope is 0.326‰ (calculating $\Delta^{17}O$ relative to the TFL with $\lambda = 0.528$ is left to the reader as Problem 10.5). Figure 10.20 demonstrates the origin of this $\Delta^{17}O$ value. CO_2 equilibrates with seawater water (blue oval A) with fractionation factors of $^{17}\alpha_{CO2-H2O}$ and $^{18}\alpha_{CO2-H2O}$ at 25 °C of 1.02125 and 1.04104, respectively. This results in CO_2 having a composition lying along a slope of 0.5229 in the $\delta^{18}O$–$\delta^{18}O$ plot at the position of the blue **x**. Mass-independently fractionated stratospheric CO_2 (gray oval D) contributes ^{17}O-rich CO_2 to this composition. Evapotranspiration shifts the O isotopic composition of leaf water in terrestrial plants away from meteoritic water (blue oval A)

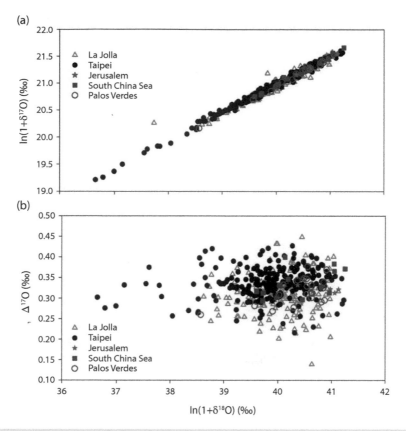

Figure 10.19   Oxygen isotopic composition of tropospheric CO_2 measured in various localities. $\Delta^{17}O$ in B is computed relative to a slope of 0.516. Source: Liang et al. (2017)/Springer Nature/CC BY.

along a lower slope of ∼0.516 to the green oval B. The equilibration of CO_2 with this leaf water then shifts the O isotopic composition to the yellow oval, C. Anthropogenic fossil fuel and biomass burning contributes CO_2 represented by the red starburst. The black diamond is the average composition of tropospheric CO_2 whose isotopic composition is a consequence of the sum of all these processes.

The O isotopic composition of atmospheric CO_2, particularly the $\Delta^{17}O$, thus depends on contributions from equilibration with seawater, the stratosphere, and terrestrial photosynthesis: increasing terrestrial photosynthesis decreases $\Delta^{17}O$. Liang et al. (2017) used $\Delta^{17}O$ to estimate the flux of CO_2 between the atmosphere and the terrestrial biosphere of $345\pm70 \times 10^{15}$ g C/year. This translates to residence time of CO_2 with respect to terrestrial biosphere cycling of 1.9 years. Others have used a similar approach to infer somewhat higher fluxes.

10.4.2.3   Nitrogen compounds

The isotopic composition of atmospheric N_2 is uniform, but the isotopic composition of minor N-bearing species is not. These species originate naturally from the biological N cycle (Figure 10.12), although physical processes in the atmosphere account for some of them; for example, lightening produces about 3% of nitrate. Those species produced by the biological N cycle tend to be isotopically light. For example, N_2O emitted from unfertilized rainforest soils has $\delta^{15}N$ ranging from ∼0‰ to −25‰. Marine emissions of ammonium and nitrate (the latter primarily present as aerosol salts) typically have $\delta^{15}N$ in the range of −3‰ to −10‰, with ammonium slightly lighter than nitrate. However, anthropogenic activity is now the major source of N-bearing molecules to the atmosphere, accounting for roughly 80% of NOx and NH_3/NH_4^- emissions and ∼40% of N_2O emissions. These

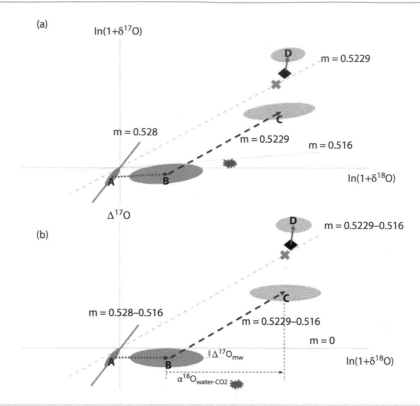

Figure 10.20 Schematic diagram (not to scale) of the evolution of the isotopic composition of atmospheric CO_2. **A** $\ln(1 + \delta^{17}O)$ versus $\ln(1 + \delta^{18}O)$ plot for meteoric water (blue line), transpiration water (green), plant equilibrated CO_2 (yellow), and stratospheric modified CO_2 (gray). Ocean water equilibrated CO_2 is shown by the blue "X" and averaged tropospheric CO_2 value by the diamond. Anthropogenic CO_2 is denoted by the red starburst symbol. Arrows indicate transport. **B** Similar to **A** but for $\Delta^{17}O$ versus $\ln(1 + \delta^{18}O)$. The corresponding slopes have been decreased by 0.516. $\Delta^{17}O_{mw}$ is the meteoric $\Delta^{17}O$, the y-intercept of the line AB. $\alpha^{18}O_{water\text{-}CO2}$ represents the fractionation in $\delta^{18}O$ of water and CO_2. Source: Liang et al. (2017)/Springer Nature/CC BY.

anthropogenic emissions are of concern because nitrate emission is a major cause of acid rain as well as eutrophication in marine and freshwater environments, while N_2O is a greenhouse gas.

Figure 10.21 illustrates the isotopic composition of NO_x and NH_3/NH_4^- emissions. In most cases, NO is the principal molecule emitted, although most studies have not discriminated between the various NO_x forms because of rapid interconversion in the atmosphere. It is apparent from Figure 10.21 that NO_x produced by combustion, particularly in coal-fired power plants, is isotopically heavier than NO_x emitted from soils (natural gas-fired power plants produce far less NO_x than coal-fired ones). The presence or absence of scrubbers on power plants and catalytic converters on vehicles tends to decrease emittance and increase $\delta^{15}N$ of emitted gases (Walter

et al., 2015). The global flux from soils is predominantly from cultivated/fertilized soils, but there appears to be little distinction between $\delta^{15}N$ in NO_x emitted by cultivated and natural ones. Instead, the dominant factor affecting $\delta^{15}N$ of emitted NO_x appears to be the soil oxidation state. Under aerobic conditions, NO_x is produced primarily by nitrification and is isotopically light, while under anaerobic conditions, NO_x is produced by denitrification and is isotopically heavier (Su et al., 2020).

Once in the troposphere, nitrite reacts with ozone to produce nitrogen dioxide:

$$NO + O_3 \rightarrow NO_2 + O_2. \qquad (10.18)$$

During the day, nitrogen dioxide can then be photolyzed by UV and visible radiation:

$$NO_2 + h\nu \rightarrow NO + O^{\cdot} \qquad (10.19)$$

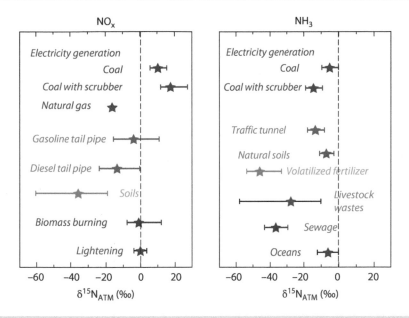

Figure 10.21 Nitrogen isotopic composition of NO_x and ammonia/ammonium emissions to the atmosphere. Source: Based on (Walters et al., 2015) and (Elliott et al., 2019).

and the free oxygen radical combines with O_2 to produce ozone:

$$O_2 + O^{\cdot} + M \rightarrow O_3 + M \qquad (10.20)$$

where M is some other molecule (usually N2 or O2). This reaction is the main source of tropospheric ozone. These reactions, known as the *Leighton Cycle,* run rapidly and quickly establish a steady-state balance of NO, NO2, and O3. After sunset, the balance shifts toward NO2 and O3 levels decline. NO and NO2 are ultimately removed from the atmosphere by irreversible conversion to nitrate, of which several pathways are possible. During daylight, reaction with the hydroxyl radical dominates:

$$NO_2 + OH^{\cdot} + M \rightarrow HNO_3 + M. \qquad (10.21)$$

At night, the production and hydrolysis of N_2O_5 dominates:

$$NO_2 + O_3 + M \rightarrow NO_3 + O_2 + M \qquad (10.22a)$$
$$NO_3 + NO_2 + M \rightarrow N_2O_5 + M \qquad (10.22b)$$
$$N_2O_5 + H_2O \rightarrow 2HNO_3 \qquad (10.22c)$$

or

$$NO_3 + RH \rightarrow HNO_3 \qquad (10.22d)$$

where RH is a hydrogen-bearing organic molecule. HNO_3 is subsequently removed by rain or on aerosol surfaces.

As an aside, it is interesting to note that when peroxy radicals such as HO_2 are present, NO_2 can be produced through alternate pathways such as

$$NO + RO_2 \rightarrow NO_2 + RO^{\cdot} \qquad (10.23)$$

where $RO^{\cdot}$ is the peroxy radical. Since O_3 is not consumed in production of NO_2 in reaction 10.23 but is still produced in reactions 10.18–10.20, the presence of peroxy radicals leads to a buildup of tropospheric O_3. Peroxy radicals are produced by the oxidation of CO and volatile organic carbons in fossil fuel burning, as is NO. This explains the association tropospheric ozone with smog.

Returning to isotope geochemistry, equilibrium, kinetic, and photochemical fractionations of both N and O are associated with these reactions. Li et al. (2020) found experimentally that the equilibrium, α_{NO-NO_2} N fractionation factor was 1.029 at room temperature, and the total fractionation factor for the Leighton cycle, combining equilibrium, kinetic, and photochemical factors, was 0.990 when O_3 alone controls the oxidation (Equation 10.18). Albertin et al. (2021) found only small diurnal variations in $\delta^{15}N$ in NO_2 in air samples from Grenoble, France associated with diurnal NO/NO_2 ratio variations. One reason, perhaps, is that most NO_x is present

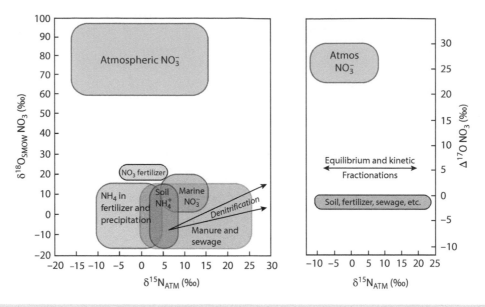

Figure 10.22 Oxygen and nitrogen isotopic composition of atmospheric nitrate and other N-bearing molecules in soils, the oceans, etc. Atmospheric nitrate is distinctive in its enrichment in ^{18}O and mass-independent enrichment in ^{17}O. Source: Based on (Michalski et al., 2003) and (Hastings et al., 2013).

as NO_2 even during daylight. In addition, in an urban environment such as Grenoble, the peroxy radical pathway (Equation 10.22) might lead to different, and as yet undetermined, fractionation. Theoretical calculations by Walters and Mikalski (2016) predict that the $\delta^{15}N$ of HNO_3 will vary depending on the reaction from +5±5‰ for reaction 10.21 to +20±5‰ for reaction 10.22c and −25±5‰ for reaction 10.22d. Consistent with these predictions, atmospheric nitrate has $\delta^{15}N$ between −15‰ and +15‰ and varies seasonally and latitudinally.

Larger variations occur in the O isotopic composition of tropospheric NO_x, which can have $\delta^{18}O$ as high as 80‰ and MIFs with $\Delta^{17}O$ of 20–31‰. This results from transfer of O primarily from tropospheric O_3 in reaction 10.18, whose $\Delta^{17}O$ is $\sim$ 39‰, as found experimentally by Savarino et al. (2008). Albertin et al. (2021) found that $\Delta^{17}O$ in NO_2 reached a maximum of ~39‰ during daytime declining at night to ~20‰ in Grenoble air, consistent with the transfer of the isotopic anomaly through reaction (10.22). Nitrate produced by bacterial nitrification in soils and the ocean typically have $\delta^{18}O_{SMOW}$ < 10‰ and lacks a significant mass-independent anomaly. Consequently, the distinctive O isotopic composition of atmospheric

nitrate allows for discrimination between atmospheric nitrate deposition and *in situ* production in the biological N cycle in soils and water (e.g. Fang et al., 2011; Hastings et al., 2013), as apparent in Figure 10.22.

N_2O emissions are of particular concern because N_2O is the fourth most important greenhouse gas (following H_2O, CO_2, and CH_4) and, with the reduction in chlorofluorocarbon emissions following the Montreal Protocol elimination of chlorofluorocarbons, is the dominant ozone-depleting gas presently being emitted. While it is produced naturally in soils, freshwaters, and the ocean in both nitrification and denitrification, emissions from soils and freshwater have increased due to the widespread use of artificial fertilizer over the twentieth century. Wastes, including sewage, landfills, and manure also contribute to increased emission. It is also produced by fossil fuel combustion and biomass burning; the former accounts for only a minor fraction, ~1%, of total emissions, while biomass burning may account for up to 20% of anthropogenic emissions. The average isotopic composition of automobile sources is estimated at $\delta^{15}N$ = −4.9 ± 8.2‰ and $\delta^{18}O$ = 43.5 ± 13.9‰ (Toyoda et al., 2008); the isotopic composition of biomass burning varies with the material being burned. N_2O is

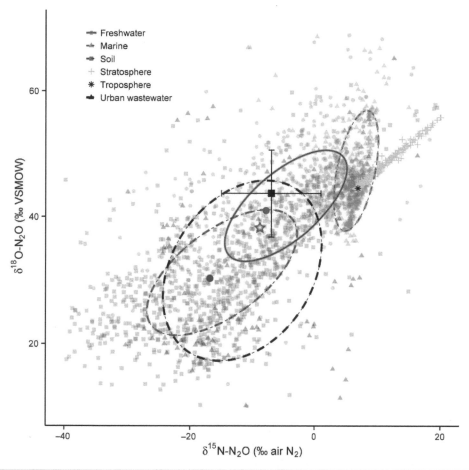

Figure 10.23 $\delta^{15}N_{ATM}$ versus $\delta^{18}O_{SMOW}$ in global atmospheric N_2O sources from various environments measured or compiled by Snider et al. (2015). Colored ellipses encompass 40% of the data for each environment and colored symbols show the mean for each environment. Black square with error bars is the estimate of vehicle exhaust of Toyoda et al. (2008). Blue star is the estimated total anthropogenic contribution. Source: Adapted from Snider et al. (2015).

slowly destroyed by UV radiation in the stratosphere, with an average lifetime of 123 years. Photolysis selectively destroys ^{14}N- and ^{16}O-isotopologues of NO_2, so that the return flow of N_2O from the stratosphere to the troposphere is enriched in ^{15}N, ^{17}O, and ^{18}O. Both concentration and isotopic composition vary regionally and seasonally, but average concentration has increased from preindustrial concentrations of 270 ppb to 334 ppb in 2021 and is currently increasing at a rate of nearly 1 ppb per year; concomitantly, its $\delta^{15}N$ is decreasing at a rate of −0.052 ±0.012‰/year to an estimated present average $\delta^{15}N$ of +6.5‰ and $\delta^{18}O$ of +44.2‰. The isotopic composition of atmospheric samples of key sources is shown in Figure 10.23.

N_2O also produced industrially and has a number of uses. It is known as "laughing gas" and used as an anesthetic in dentistry and to a lesser extent in medicine. It is also used as an oxidant in rocket fuels as well as an aerosol propellant. Industrial production, however, does not contribute significantly to increasing atmospheric concentrations.

N_2O is a linear molecule (N–N–O) with a distinct site preference, designated $\delta^{15}N^{SP}$, in which ^{15}N in enriched in the central, or α, site over the end, or β, site to varying degrees in various sources (Figure 10.24). Unfortunately, however, there is poor interlaboratory reproducibility and, at least until recently, no agreed upon reference standard for this site preference, leading to different grouping of values

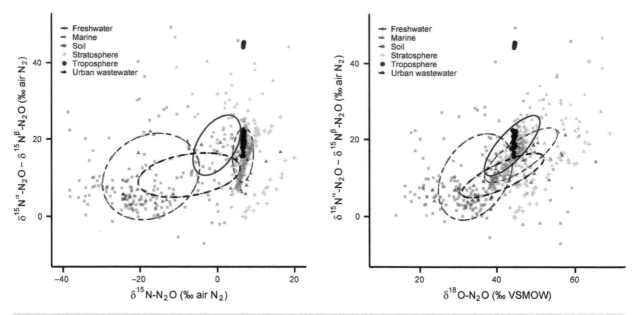

Figure 10.24 Nitrogen site preference ($\delta^{15}N^{SP} = \delta^{15}N^{\alpha}-\delta^{15}N^{\beta}$) versus $\delta^{15}N^{Total}$ and $\delta^{18}O_{SMOW}$ in global atmospheric N_2O samples from various environments measured or compiled by Snider et al. (2015). Lack of interlaboratory standardization leads to much of the scatter; nevertheless, $\delta^{15}N^{SP}$ is clearly lower in soils than in freshwater and marine environments and highest in stratospheric N_2O and correlates with $\delta^{18}O$. Source: Snider et al. (2015)/Public domain/CC BY 4.0.

for the troposphere and atmosphere. Despite the variation within sources and these analytical issues, a number of studies have used N and O isotope ratios combined with measurements of the site preference to attempt to quantify the fluxes from various sources and the overall anthropogenic contribution (e.g. Snider et al., 2015; Prokopiou et al., 2017; Yu et al., 2020). Snider et al. (2015) estimated the marine contribution to be 25% with the remaining 75% from natural and anthropogenic terrestrial sources. Yu et al. (2020) estimated anthropogenic contribution to be 90% of the total with an average isotopic signature of $-8:6\pm0:6‰$ for $^{15}N^{bulk}$, $34.8\pm3‰$ for $\delta^{18}O$, and $10.7\pm4‰$ for $\delta^{15}N^{sp}$. Improvement in analytical data, particularly interlaboratory standardization, is expected to improve estimates in the future.

10.4.2.4 C isotopes in CO_2

Carbon cycles rapidly between five reservoirs in the Earth's exogene (Figure 10.25). Of the total carbon in these five reservoirs, atmospheric CO_2 is the smallest. Roughly equal amounts of carbon are present in the terrestrial biosphere, the atmosphere, and the surface ocean, with substantially more being present in soil carbon, forest litter, permafrost, etc. The bulk of the surficial carbon, about 50 times as much as in the atmosphere, is dissolved in the deep ocean (mainly as HCO_3^-). The fluxes of carbon to and from the atmosphere are large relative to the amount of CO_2 in the atmosphere; indeed, 25% or so of the atmospheric CO_2 "turns over" in a year. The balance of these fluxes controls the concentration of atmospheric CO_2. The isotopic composition varies between these reservoirs, primarily due to the fractionation during photosynthesis.

For thousands of years, humanity has affected the carbon cycle through the clearing of forests and replacing them with cultivated fields and over the last several hundred years through burning of fossil fuels. Both these activities can be viewed as fluxes of carbon to the atmosphere: the former from the terrestrial biosphere and the latter from sedimentary organic carbon. The carbon flux from fossil fuel burning increased significantly over the twentieth century and is presently around 9 Pg/yr, a reasonably well-known value, and is growing; the deforestation flux less certain, with the current estimate being about 1.6 Pg/yr (IPCC, 2021). This has resulted in a roughly 0.5% per year annual increase in the concentration

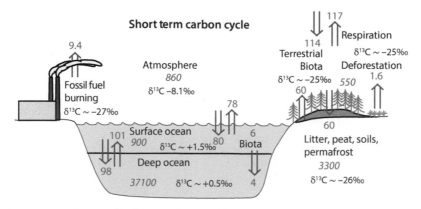

Figure 10.25   The carbon cycle. Numbers in blue italic show the amount of carbon (in pentagrams = gigatons, Gt) in the atmosphere, oceans, terrestrial biosphere, and soil (including litter, debris, etc.). Fluxes (red) between these reservoirs (arrows) are in Pg C/yr[81]. Also shown is the approximate isotopic composition of each reservoir. Magnitudes of reservoirs and fluxes are principally from (Friedlingstein et al., 2020), isotopic compositions are from (Heimann and Maier-Reimer, 1996).

of CO_2 in the atmosphere (Figure 10.26), as determined by a global system of monitoring stations (the first of which were installed by Charles Keeling in the late 1950's at Mauna Loa and the South Pole). Friedlingstein et al. (2020) estimate that at present, about 45% of the anthropogenic flux or 5.1±0.02 PgC/yr accumulates in the atmosphere, 26% (2.5±0.6 PgC/yr) is being taken up by the oceans, and 31% (3.4±0.9 PgC/yr) by the terrestrial biota. These values are used in the 2021 Intergovernmental Panel on Climate Change (IPCC) report, but the precise partitioning of anthropogenic carbon between terrestrial biosphere and oceans nevertheless remains subject to some debate.

Rising atmospheric CO_2 concentrations are a matter of concern because CO_2 is the second most important greenhouse gas (following H_2O). The importance of CO_2 and H_2O as greenhouse gases was first recognized by Eunice Foote[3] (1856) who, based on experiments she conducted by exposing cylinders of gases to sunlight, wrote, "An atmosphere of that gas [CO_2] would give to our earth a high temperature; and if as some suppose, at one period of its history the air had mixed with it a larger proportion than at present, an increased temperature… must have necessarily resulted." Recognizing this role of CO_2 in regulating climate, Arrhenius (1896) predicted that burning fossil fuels would cause the climate to warm. As the IPCC concluded in its 2021 report, "*It is unequivocal that human influence has warmed the atmosphere, ocean and land. Widespread and rapid changes in the atmosphere, ocean, cryosphere and biosphere have occurred*" (Masson-Delmotte et al., 2021).

Both the sources of the anthropogenic carbon flux, biospheric carbon released by deforestation and sedimentary organic carbon released by burning fossil fuels, have highly negative $\delta^{13}C$ (the isotopic composition of fossil fuel burned has varied over time from $\delta^{13}C \approx -24‰$ in 1850 to $\delta^{13}C \approx -28‰$ at present as coal has been partly replaced by oil and gas). Consequently, $\delta^{13}C$ of atmospheric CO_2 has been decreasing. The changing isotopic composition of CO_2 and the atmosphere is often

---

[3] While Irish physicist John Tyndall is generally given credit for discovering the greenhouse properties of CO_2, Foote's publication precedes Tyndall's by 3 years. Eunice Foote (1819-1888) was an amateur American scientist at a time when women were largely excluded from science. Her results were presented at the 1856 Annual Meeting of the American Association for the Advancement of Science on her behalf by Joseph Henry, the most famous American scientist of the time as well as the first Secretary of the Smithsonian Institution. Foote was also a woman's rights activist and a signatory of the *Declaration of Sentiments* at the 1848 First Women's Rights Convention in Seneca Falls, NY, where she resided. Her husband, Elisha Foote, a lawyer who would eventually become the U.S. Commissioner of Patents, also signed. In recognition of her pioneering work, in 2022 the American Geophysical Union established the Eunice Newton Foote Medal to be given annually to a senior scientist for research achievements at the intersection of Earth and life sciences.

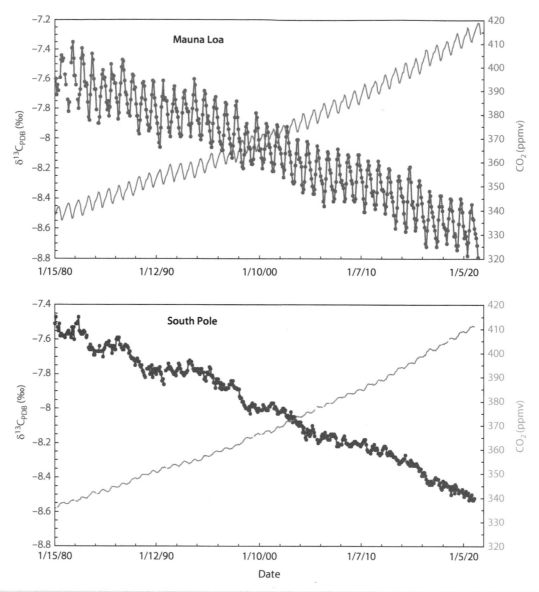

Figure 10.26   Monthly CO_2 concentrations and $\delta^{13}C_{PDB}$ of atmospheric CO_2 measured at Mauna Loa, Hawaii, and the South Pole from the Scripps CO_2 program begun by Charles Keeling in 1958 (Keeling et al., 2001). Annual cycles reflect the effects of seasonal changes in photosynthesis. Data from the Scripps CO_2 program (http://scrippsco2.ucsd.edu).

referred to as the "Suess effect." Roger Revelle and Hans Suess[4] (1957) were the first to note it and to conclude that it was due to burning fossil fuels. Aware of Svante Arrhenius's (Arrhenius, 1896) prediction that increasing atmospheric CO_2 would warm climate, they wrote, "*The increase of atmospheric CO_2 from this cause is at present small but may become significant during future decades if industrial fuel combustion continues to rise exponentially.*" They also concluded that most of the anthropogenic CO_2 was being taken up by

[4] Hans Suess (1906–1995) was an Austrian-born nuclear physicist and was part of the German nuclear program during WWII. He emigrated to the USA in 1950 and worked with Harold Urey at the University of Chicago on cosmochemical abundances. He became interested in ^{14}C dating and noted the apparent decline in the atmospheric ^{14}C over the previous 50 years and inferred this was due to fossil fuel burning. The term "Suess effect" originally referred to the decreasing ^{14}C abundance in the atmosphere but is now used to refer to changing $\delta^{13}C$ as well.

the oceans rather than remaining in the atmosphere (they overlooked the biosphere). Revelle recruited Charles Keeling to Scripps Institution of Oceanography to begin a program of measuring atmospheric CO_2 concentrations and eventually $\delta^{13}C$ as well. This step essentially initiated quantitative climate change research.

Keeling's measurements, carried on today by his son Ralph Keeling (Figure 10.26), show a clear increase in the concentration of atmospheric CO_2 and a decrease in $\delta^{13}C$ over time. Superimposed on the temporal decrease in $\delta^{13}C$ and increase in CO_2 are seasonal variations that reflect the uptake of light carbon (making atmospheric carbon heavier) in northern hemisphere spring as photosynthesis increases and the release of light carbon in fall as respiration becomes dominant over photosynthesis. Seasonal variations at the South Pole are more subdued since the terrestrial biosphere is smaller in the southern hemisphere and mixing between hemispheres is slow.

On longer time scales, measurements of $\delta^{13}C$ in ice cores complement the direct measurements (Figure 10.27) and show that $\delta^{13}C$ of atmospheric CO_2 was approximately constant over the first half of the last millennium at around −6.5‰, increased slightly around 1600, then began to decline around the time of the industrial revolution, which began with

James Watt's invention of the coal-powered steam engine in 1776. The decline is greater (up to a factor of 2 greater) than that expected from burning of fossil fuel alone, which is one line of evidence that there has been a significant destruction of the terrestrial biosphere over the last 200 years. Over the same period, atmospheric CO_2 concentrations rose from about 280 to 350 ppm in 1993.

A key constraint on how much of the anthropogenic CO_2 is being taken up by the oceans is the isotopic composition of dissolved CO_2. Figure 10.28 shows the change in $\delta^{13}C$ with depth in the North Atlantic between 1993 and 2003. Here, the anthropogenic signal has penetrated to the base of the water column in this region due to the sinking and southward flow of NADW, but this is not the case in most regions of the ocean. Eide et al. (2017) analyzed global $\delta^{13}C$ data and modeled its distribution using dissolved chlorofluorocarbons in seawater, another anthropogenic signal, and found a decrease in ^{13}C in the upper 1000 m of all basins, with strongest decrease in the subtropical gyres of the Northern Hemisphere, where $\delta^{13}C$ of DIC has decreased by more than 0.8‰ since the industrial revolution (Figure 10.29). They also found that the relationship between the Suess effect and the concentration of anthropogenic

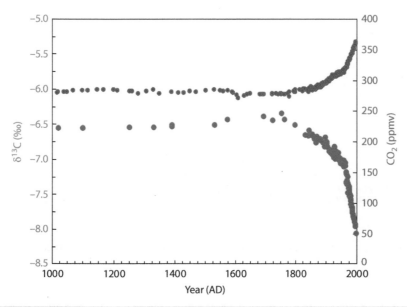

Figure 10.27 Variation in $\delta^{13}C$ and CO_2 concentrations in ice bubbles from the Law Ice Dome and South Pole, Antarctica. Source: Data from (Rubino et al., 2013).

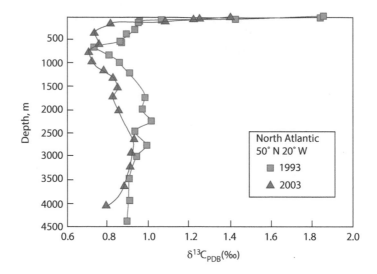

Figure 10.28 $\delta^{13}C$ as a function of water depth in the North Atlantic as sampled in 1993 and again in 2003. Source: Adapted from (Quay et al., 2007).

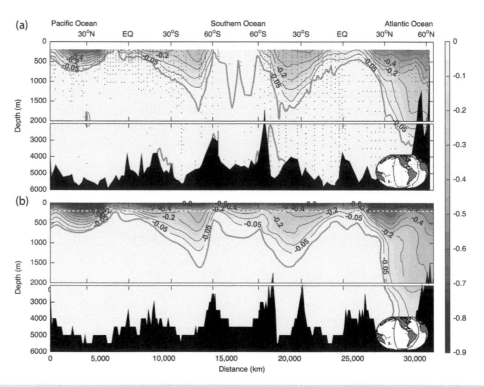

Figure 10.29   (a) Vertical ocean section of the per mil decrease in $\delta^{13}C$ (Suess effect) along the path depicted in the map inset beginning in the North Pacific reconstructed from measurements in the upper panel and predicted based dissolved CFC-12 data. Dots display sample points in the upper panel and the solid gray line marks pCFC-12 penetration depth. (b) Same as (a), but data above the stippled white line are estimated by interpolating from surface ocean ^{13}C decrease down to 200 m in each grid point. Source: Adapted from (Eide et al., 2017).

CO_2 varies strongly between water masses, because of variable equilibration with the atmospheric CO_2 before sinking. They estimated that the oceans have taken up 92 ± 46 PgC of anthropogenic CO_2.

The expansion of the northern hemisphere terrestrial biosphere at least balances, and likely exceeds, deforestation elsewhere, which now occurs mainly in the tropics. There are several possible explanations for this. These are as follows.

- Anthropogenic nitrate emission is fertilizing and enhancing the growth of the biosphere.
- Experiments show that plants photosynthesize more efficiently at higher CO_2 concentrations, so increasing atmospheric CO_2 concentrations can, in principle, stimulate plant growth. Opening stomata represents a tradeoff for plants: it lets CO_2 in but also moisture out. Higher concentrations of CO_2 allow plants to limit moisture loss by closing their stomata. However, to the extent that plant growth is limited by other factors such as the availability of nutrients rather than CO_2, this CO_2 growth effect will be limited.
- As agriculture became more efficient in the twentieth century, some lands cleared for agriculture in Europe and North America in previous centuries have been abandoned and are returning to forest, which take up more CO_2 than agricultural lands.
- Average global temperature has increased by 1 °C over the last century because of rising atmospheric CO_2 concentrations (Masson-Delmotte, et al., 2021). This temperature increase is allowing a northward expansion of boreal forests and longer growing seasons in temperate zones.

10.4.2.5   Methane

Methane is the next most important greenhouse gas following CO_2, accounting for $\sim 20\%$ of climate warming. Like CO_2, it occurs naturally in the atmosphere, and also like CO_2, its concentration has increased by anthropogenic activity from ~ 1700 ppb to >1900 ppm today. Methane is produced naturally by methanogenic microbes, all known examples of which are archaea, through two principal mechanisms:

Fermentation:

$$CH_3COOH \rightarrow CH_4 + CO_2 \qquad (10.24)$$

hydrogenotrophic methanogenesis:

$$CO_2 + 4H_2 \rightarrow CH_4 + 2H_2O. \qquad (10.25)$$

The former is also referred to as acetoclastic or methylotrophic methanogenesis and the latter as CO_2 reduction.

Both the reactions occur in a variety of in reducing environments including the gut of ruminants (cattle, sheep) and termites as well as wetlands, soils, sediments, etc.; methane from these sources is referred to as *biogenic*. Methane is also produced by breakdown, or *catagenesis*, of sedimentary kerogen during burial when it is subjected to heat and pressure, producing petroleum and ultimately "natural gas," of which methane is the primary component; this methane is known as *thermogenic*. Methane is also produced in biomass burning, both natural (wildfires) and anthropogenic (land clearing); this methane is referred to as *pyrogenic*. Finally, *abiotic methane* is also produced through a variety of inorganic reactions. Methane is not stable at magmatic temperatures, but as magmatic gas cools through 600 °C, it is produced by reaction between CO_2 and water:

$$CO_2 + 2H_2O \rightarrow CH_4 + O_2 \qquad (10.26)$$

and by oxidation of ferrous iron:

$$8FeO + CO_2 + 2H_2O \rightarrow 4Fe_2O_3 + CH_4 \qquad (10.27)$$

and metamorphic reactions such as carbonate reduction:

$$8FeO + CaCO_3 + 2H_2O \\ \rightarrow 4Fe_2O_3 + CH_4 + CaO. \qquad (10.28)$$

Methane produced by these mechanisms has different, but overlapping, C and H isotopic compositions, as illustrated in Figure 10.30 and summarized in Table 10.2. In general, abiotic methane has heavy carbon, thermogenic methane relatively high $\delta^{13}C$ and high δ^2H, microbial CO_2 reduction produces methane with the lowest $\delta^{13}C$ and low δ^2H, fermentation produces methane with the lowest δ^2H, and biomass burning produces methane with the lowest $\delta^{13}C$, but considerable overlap

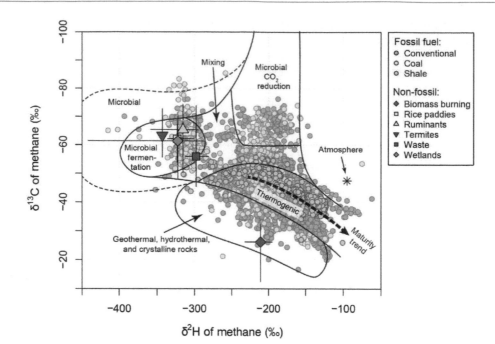

Figure 10.30 $\delta^{13}C$ of atmospheric methane sources. Source: Sherwood et al. (2017)/Earth System Science Data/CC BY 3.0.

Table 10.2 Mean carbon and hydrogen composition of methane emissions.

	$\delta^{13}C$	∂^2H
Conventional	−44.0±10.7	−194±47
Coal	−49.5±11.2	−232±52
Shale	−42.5±7.6	−167±44
All	−44.8±10.7	−197±51
Rice	−62.2±3.9	−323±16
Ruminants	−65.4±6.7	−316±29
Termites	−63.4±6.4	−343
Waste	−56±7.6	−298±11
Wetlands	−61.5±5.4	−322±42
all	−61.7±5.4	−317±33
Biomass	−26.2±4.8	−211±15

"Conventional" oil and gas recovered from in porous reservoir rock while "shale" refers to oil and gas recovered from source rocks using hydrofracturing. Source: From (Sherwood et al., 2017).

exists between sources. Part of this overlap is due to microbial association with petroleum and natural gas evolution. The first stage, referred to as *diagenesis*, involves the microbial processing of kerogen so that very immature thermogenic can have isotopic compositions extending into the biogenic field. In addition, petroleum can undergo biodegradation by microbes, which also extends the thermogenic

field into the biogenic one. Microbes also generate methane from coal, so that "coal gas" can also have a microbial isotopic signature.

Notably, the isotopic composition of atmospheric methane falls outside the fields of all sources, which is a consequence of atmospheric isotopic fractionation. Atmospheric methane has a residence time of ~10 years; most is removed by a series of oxidation reactions beginning with a reaction with the hydroxyl radical:

$$CH_4 + OH^{\cdot} \rightarrow CH_3 + H_2O \qquad (10.29)$$

followed by the oxidation of CH_3 to CO_2 + H_2O through several possible pathways (Ravishankara, 1988). Kinetic carbon and hydrogen isotope effects with $^{12/13}KIE$ = 1.0061 and $^{1/2}KIE$ = 1.3344 (corresponding to $\alpha^{13/12}$ = 0.9939 and $\alpha^{2/1}$ = 0.7447) are associated with this reaction ($^{12}CH_4$ being consumed more readily than $^{13}CH_4$). Reaction with Cl rather than OH$^{\cdot}$:

$$CH_4 + Cl^{\cdot} \rightarrow CH_3 + HCl \qquad (10.30)$$

is a subordinate pathway that results in a greater fractionation of $^{12/13}KIE$ = 1.028 and $^{1/2}KIE$ = 1.41 $\alpha^{13/12}$ = 0.9728 and $\alpha^{2/1}$ ≈ 0.7092) and in the removal of methane with isotopically light H much more readily (Gierczak et al., 1997). The result is that atmospheric methane is isotopically heavier than its sources, and any isotopic assessment of the atmospheric methane budget must take account of these fractionations (Strode et al., 2020).

Figure 10.31 shows methane concentrations, $\delta^{13}C$, and δ^2H measured over time at the Scripps Mauna Loa and South Pole observatories and the Japanese National Institute of Polar Research Svalbard station (NYA, 78°55′N, 11°56′E) and Syowa station (SYO: 69°00′S, 39°35′E) on East Ongle Island just off the Antarctic coast. Unlike CO_2, methane atmospheric concentration growth has not been steady: it grew rapidly in the 1980s and 1990s, then stabilized somewhat in the following decade, and began to increase rapidly around 2007. Seasonal swings similar to those seen in CO_2 are apparent. Routine measurements of atmospheric methane $\delta^{13}C$ are only available since 1987. These data suggest that seasonally averaged $\delta^{13}C$ was increasing until 2007 and has been decreasing since then. On the other hand, seasonally averaged δ^2H in methane appears to continually increase over this period at least in coastal Antarctica.

The increase in atmospheric CH_4 since 2007 is obviously concerning. The declining $\delta^{13}C$ associated with this increase suggests a change in the nature of source or removal fluxes. Nisbet et al. (2019) suggested three possibilities: an increase in biogenic emissions, an increase in fossil fuel emissions or a shift to fuels with more negative $\delta^{13}C$, or a change in relative removal fluxes and pathways (Equation 10.29 versus 10.30). Considering both $\delta^{13}C$ and δ^2H, Fujita et al. (2020) concluded that the stabilization of CH concentrations in the first half of the 2000s was due to a decrease in biogenic and biomass burning and that an increase in biogenic CH_4 emissions, mainly in the tropics/subtropics has contributed to the renewed growth of atmospheric CH_4 after 2006/2007. Howarth (2019) suggested that the increased extraction of shale gas accounts for the increase in CH_4 and decrease in $\delta^{13}C$, assuming that shale gas had lower $\delta^{13}C$ than other fossil fuel, but the compilation of Sherwood et al. (2017) shows that shale gas has heavier carbon on average than other fossil fuels. At this point, no consensus has formed as to the cause of this renewed growth of atmospheric CH_4 and decrease in $\delta^{13}C$.

When we look further back in time by examining the isotopic record of atmospheric methane trapped in polar ice, we find that human influence on the methane cycle has a long history. Figure 10.32 summarizes these measurements in ice cores from Greenland and Antarctica. Two millennia ago, CH_4 concentrations were steady at around 650 ppbv; this was followed by a slow and uneven rise until around 1750, when the Industrial Revolution produced a much more rapid rise. $\delta^{13}C$ appears to tell a more complex story. Sapart et al. (2012) identified three events, labeled (1), (2), and (3) in Figure 10.32. They associate the first (1) beginning 200 AD with declining biomass burning due to declining human population following the fall of the Han Dynasty in China and the Roman Empire in Europe. They associate a peak in $\delta^{13}C$ (2) with the higher temperatures of the Medieval Climate Anomaly resulting in drought and increased wildfires. The decline in $\delta^{13}C$ following (3) likely reflects increased agriculture and associated biogenic emissions. This decline is eventually reversed by fossil fuel burning associated with the Industrial Revolution. Mischler et al. (2009) reported δ^2H as well as $\delta^{13}C$ for the last 1000 years in an ice core from the West Antarctic Divide, although with a larger sampling interval and hence less resolution than the Sapart et al.'s study. δ^2H remains nearly constant up until about 1400 before declining, reaching a minimum around 1800. δ^2H is particularly sensitive to agricultural emissions, which have particularly negative δ^2H but relatively insensitive to the relative amounts of fossil fuel versus biomass burning emissions. This decline in δ^2H implicates an increase in agriculture as the cause of rising emissions and decreasing $\delta^{13}C$ prior to the industrial revolution.

10.4.2.6 Methane clumped isotope analysis

Clumped isotopic analysis provides a new and different approach to understanding methane generation, and, potentially, identifying sources of atmospheric methane. As we found in Section 5.3.2, the energy of a system is minimized when the heavy isotopes of elements are "clumped" into the same molecule. Consequently, we should expect to find the equilibrium abundance of molecular isotopologues

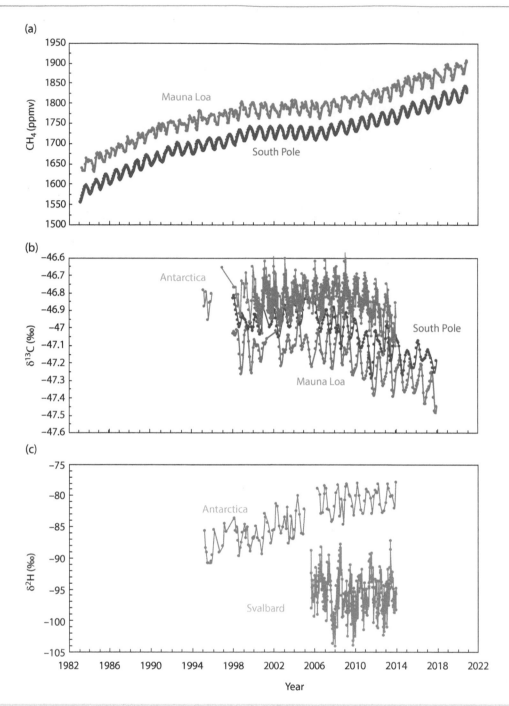

Figure 10.31   (a) Atmospheric methane concentration concentrations measured at Mauna Loa, Hawaii, and the South Pole by the Scripps CO_2 program. (b) $\delta^{13}C$ measured at these stations and the Japanese National Institute of Polar Research Syowa Antarctic (69°00′S, 39°35′E) stations. Concentrations are nearly identical at the Syowa and South Pole stations. (c) δ^2H of methane measured at Japanese Syowa Antarctic and Svalbard stations (78°55′N, 11°56′E). Source: Keeling, Walker, R.F., Piper, S.J., and Bollenbacher, A.F. Scripps CO_2 Program; http://scrippsco2.ucsd.edu and Fujita et al. (2020); World Data Centre for Greenhouse Gases; https://gaw.kishou.go.jp/.

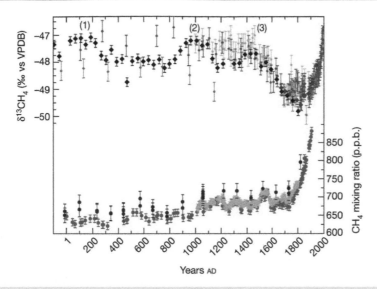

Figure 10.32 $\delta^{13}C$ of methane trapped in Greenland ice cores from NEEM (black diamonds), EUROCORE (blue diamonds), GISPII23 (green diamonds), GISPII23 (green diamonds) Greenland ice cores and Antarctic ice cores from Law Dome (red diamonds), and the WAIS divide2 (orange diamonds). (1), (2), and (3) correspond to the three excursions in the Northern Hemisphere δ13C record (see main text). CH_4 concentrations records from Greenland (GRIP5,6; black circles) and Antarctica (Law Dome4, red circles; WAIS7, orange circles). Source: Sapart et al. (2012)/with permission of Springer Nature.

such as $^{13}CH_3D$ and $^{13}CH_2D_2$, etc. to be more abundant than a mere random distribution. We introduced a new variable that measured this:

$$\Delta_i = \left(\frac{R_{i-e}}{R_{i-r}} - 1\right) \times 1000 \qquad (5.64)$$

where R_{i-e} is that ratio of the observed or calculated equilibrium abundance of isotopologue i, e.g. $^{13}CH_3D$, to the isotopologue containing no rare isotopes, e.g. $^{12}CH_4$, and R_{i-r} is that same ratio if isotopes were distributed among isotopologues randomly. In the case of methane, two version of Δ_i are possible:

$$\Delta_{^{13}CH_3D} = \left[\frac{\left([^{13}CH_3D]/[^{12}CH_4]\right)_e}{\left([^{13}CH_3D]/[^{12}CH_4]\right)_r} - 1\right] \times 1000 \qquad (10.31)$$

and

$$\Delta_{^{12}CH_2D_2} = \left[\frac{\left([^{12}CH_2D_2]/[^{12}CH_4]\right)_e}{\left([^{12}CH_2D_2]/[^{12}CH_4]\right)_r} - 1\right] \times 1000 \qquad (10.32)$$

where brackets denote molar concentrations. For $\Delta_{^{13}CH_3D}$, the random distribution in the denominator may be calculated from $\delta^{13}C$ and δ^2H as

$$\left([^{13}CH_3D]/[^{12}CH_4]\right)_r = 4[H]^3[^{13}C][D]/4[^{12}C][H]^4$$

$$= \left(\frac{^{13}C}{^{12}C}\right)\left(\frac{D}{H}\right). \qquad (10.33)$$

(The 4 arises because the deuterium can substitute for any one of four hydrogens.) From the definition of δ, we have

$$\left(\frac{^{13}C}{^{12}C}\right) = \frac{\left(\delta^{13}C \times 1000 + 1\right)}{\left(\frac{^{13}C}{^{12}C}\right)_{PDB}}. \qquad (10.34)$$

We can derive a similar expression for (D/H). Combining these, we have

$$\left([^{13}CH_3D]/[^{12}CH_4]\right)_r = \frac{\left(\delta^{13}C \times 1000 + 1\right)}{\left(\frac{^{13}C}{^{12}C}\right)_{PDB}}$$

$$\times \frac{\left(\delta^2D \times 1000 + 1\right)}{\left(\frac{D}{H}\right)_{SMOW}}. \qquad (10.35)$$

The derivation of the analogous ratio for $\Delta_{^{12}CH_2D_2}$ is left to the reader as Problem 10.4.

The measurement of the $^{13}CH_3D$ from $^{12}CH_2D_2$ isotopologues is difficult: the abundances of $^{13}CH_3D$ and $^{12}CH_2D_2$ are respectively about 5×10^{-6} and 7×10^{-7} lower than of $^{12}CH_4$. These isotopologues can now be distinguished and measured with high-resolution mass spectrometry or infrared laser absorption spectrometry, but $^{13}CH_3D$ cannot be distinguished from $^{12}CH_2D_2$, both of which have mass 18, by conventional mass spectrometry. Consequently, a parameter Δ_{18} was defined such that the isotope ratios R in 5.64 are

$$^{18}R = \frac{[^{13}CH_3D] + [^{12}CH_2D_2]}{[^{12}CH_4]}. \quad (10.36)$$

Stolper et al. (2014) calculated the equilibrium temperature dependence of Δ_{18} from molecular vibrations as

$$\Delta_{18} = -0.0117\left(\frac{10^6}{T^2}\right)^2 + 0.708\left(\frac{10^6}{T^2}\right) - 0.317 \quad (10.37)$$

where T is in kelvin. Webb and Miller (2014) calculated the Δ_{13CH_3D} temperature dependence as

$$\Delta_{13CH_3D} = -0.0141\left(\frac{10^6}{T^2}\right)^2 + 0.399\left(\frac{10^6}{T^2}\right) - 0.311. \quad (10.38)$$

Because $^{12}CH_2D_2$ is much rarer than $^{13}CH_3D$, the two curves shown in Figure 10.33, are quite similar, diverging only at low temperature. More recently, theoretical calculations by E. Schauble reported in (Young et al., 2017) predict the temperature dependence for $\Delta_{12CH_2D_2}$ as

$$\Delta_{12CH_2D_2} = 1000\ln\left(1 + \frac{0.183798}{T} - \frac{785.483}{T^2}\right. \\ + \frac{1056280}{T^3} + \frac{9.37307 \times 10^7}{T^4} \\ - \frac{8.919480 \times 10^{10}}{T^5} \\ \left. + \frac{9.901730 \times 10^{12}}{T^6}\right) \quad (10.39)$$

(Unfortunately, this equation cannot be approximated by truncating higher order terms). Experimental data fit these temperature dependencies, as shown in Figure 10.33.

Given uncertainties in both measured Δ values and temperatures, both Δ_{13CH_3D} and Δ_{18} are plotted together in Figure 10.33, which shows that observational data on methane

from sedimentary environments also fit the expected distribution, with some exceptions. For example, the Marcellus Shale, a Devonian age shale that underlies much of the northeastern USA and which is extensively exploited for natural gas plots to the left of the curves. However, the present borehole temperature of the Marcellus shale, $\sim 60\ °C$, is undoubtedly not the temperature as which the methane formed. The calculated Δ_{18} temperature of $\sim 200\ °C$ is consistent with thermal methanogenesis. That this and other examples of thermogenic methane that retain equilibrium isotopic compositions reflecting formation temperatures suggest that reaction rates drop rapidly below 150–200 °C, locking in higher temperature equilibrium compositions. In contrast, current borehole temperatures of 170–200 °C of the Jurassic Haynesville Shale, which underlies East Texas, are thought to be close to the maximum experienced since deposition. Biogenic methane from shallow water sediments in the Beaufort Sea, the Santa Monica and Santa Barbara Basins adjacent Southern California, and the Gulf of Mexico as well as methane in pore water from the coal-bearing Powder River Basin in Wyoming, USA also have Δ_{18} or Δ_{13CH_3D} close to values expected at measured temperatures. Most, but not all, temperatures derived from clumped isotope analysis of natural gas and petroleum deposits fall within the range expected (~ 75–225 °C) for catagenesis and metagenesis for the production of methane from sedimentary kerogen (Douglas et al., 2017). There appears to be little systematic difference between "shale gas" extracted through hydrofracturing directly from source rocks and "conventional gas" extracted from porous reservoir rocks in which the gas has accumulated, and no difference whether gas is associated or not with oil.

Wang et al. (2018) reported Δ_{13CH_3D} for methane in hydrothermal vent fluids from the Mid-Altantic Ridge and the Midc-Cayman Rise, which vary from 0.97‰ to 1.85‰, corresponding to equilibrium temperatures range from 450 to 269 °C. Those for the Rainbow and Lucky Strike vents are close to measured fluid temperatures (270–370 °C). Measured vent temperatures of the Lost City and Von Damm fields are cooler (96–226 °C) yet Δ_{13CH_3D} correspond to much higher equililibrium temperatures (270–350 °C). Other aspects of vent fluid chemistry indicate fluids equilibrate with the oceanic crust at

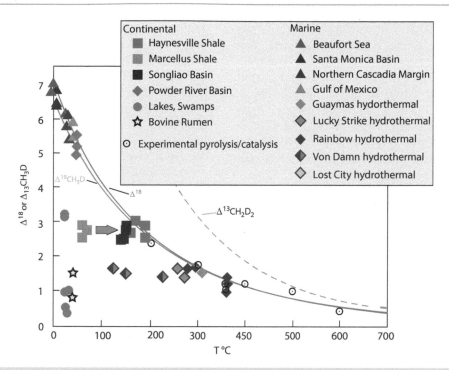

Figure 10.33 Δ_{18} or Δ_{13CH_3D} as a function of measured temperatures of methane from various sources. Thermogenic and abiotic methane generally fit the theoretical equilibrium temperature curve or yield clumped temperatures close to those expected for methane formation. Δ_{18} and Δ_{13CH_3D} biogenic methane, however, often yield temperatures well in excess of formation temperatures. Theoretical curve for $\Delta_{12CH_2D_2}$ is also shown as the dashed line.

temperatures of 300–400 °C and that lower measured exit temperatures generally reflect cooling or mixing with colder water. The reduction of CO_2 to CH_4 in the presence of magnetite and Fe suliphides is also thermodynamically favored in this temperature range. Wang et al. suggested that reaction rates drop rapidly below ~270 °C, locking clumped isotopic equilibrium at higher temperatures even in cases where fluids have cooled.

While the carbon source for methane is magmatic CO_2 in typical hydrothermal systems, in the Guaymas Basin, a sedimented portion of the East Pacific Rise in the Gulf of California, organic carbon in the sediments is the source with heat from rising hydrothermal fluids driving reactions. In this case, Δ_{13CH_3D} corresponds to temperatures of ~300 °C, close to measured vent fluid temperatures.

We noted in Chapter 5 that biological reactions are often associated with large kinetic fractionations. We should not be surprised then to find that the major exceptions to the curves shown in Figure 10.33 are biogenic methane, including lakes and swamps, bovine rumen,[5] and laboratory cultures, which often have Δ_{13CH_3D} or Δ_{18} significantly lower than expected, with the lower values for fermentation than hydrogenotrophic methanogenesis.

One goal of the study of methane clumped isotopes is to constrain contributions from the various sources to atmospheric methane, but the field remains in a very early stage. While significant progress has been made in understanding basic methane forming reactions and data now exist on the isotopic compositions of some thermogenic, biogenic, and abiotic methane sources, it can only be considered to be reconnaissance level, and direct measurements of some sources, including DMS, termites, landfills, and biomass burning, have yet to be made. More importantly, the clumped

[5] The rumen is the first compartment of the stomach of ruminants where the fermentation of ingested food and methanogenesis takes place.

isotopic composition of atmospheric CH_4 has not yet been measured. The atmospheric concentration of CH_4 is about 2 ppm and the fraction of Δ_{13CH_3D} is only a few ppm of and that of $\Delta_{12CH_2D_2}$ is nearly an order of magnitude lower, requiring extraction CH_4 from several hundred liters of air. Chung and Arnold (2021) concluded that $\Delta_{12CH_2D_2}$ is likely to place the strongest contraints on sources and argue for emphasis on developing techniques to measure $\Delta_{12CH_2D_2}$ in air at clean northern and southern hemisphere sites.

10.4.2.7 Sulfur isotopes

Atmospheric sulfur is important from an environmental perspective for several reasons. First, a significant fraction of sulfur emitted to the atmosphere is eventually oxidized to sulfuric acid, which is a primary cause of acid rain, which has had a devastating effect on some forests in Europe and northeastern North America. Second, droplets of sulfuric acid reflect sunlight and cool global climate; thus, anthropogenic sulfur emissions tend to reduce the climatic effects of anthropogenic carbon emissions. This is particularly apparent when large Plinian volcanic eruptions loft sulfur into the stratosphere where it can remain for years. For example, global climate cooled by 0.4 °C in the year following the eruption of Mt. Pinatubo in 1991. Finally, SO_2 has a negative impact on health, particularly with respect to cardiovascular diseases and respiratory disorders.

Sulfur is naturally emitted to the atmosphere in a variety of forms with varying isotopic compositions (Figure 10.34). Globally, the most natural important source is sulfate salts in sea spray. There is little fractionation in their formation, so they have $\delta^{34}S_{CDT}$ identical to that of seawater, $\delta^{34}S_{CDT} = 21.24\pm0.88\permil$ and $\Delta^{33}S = 0.050\pm0.014\permil$ (Tostevin et al., 2014). In arid continental interiors or regions downwind from them, mineral dust, primarily gypsum, can be an important source. These salts play an important role in the atmosphere as nuclei for condensation and cloud formation but are otherwise largely unreactive and quickly removed by rain.

Most other natural and anthropogenic sources of atmospheric sulfur are released as SO_2 or oxidized in the atmosphere to SO_2. Natural sources include dimethyl sulfide (DMS: $(CH_3)_2S$) from the ocean, H_2S produced by sulfate-reducing bacteria in wetlands and low-oxygen regions of the ocean, and volcanoes. Dimethyl sulfide forms from dimethylsulfoniopropionate (DMSP; $(CH_3)_2S^+CH_2CH_2COO^-$), a compound found in a wide variety of marine organisms, particularly algae and phytoplankton, and contributes to the distinct smell of marine air. Sulfur in DMS is derived from seawater sulfate, and there are small fractionations in production of DMSP and its conversion to DMS such that typical $\delta^{34}S$ values are ~+16 to +17 (Oduro et al., 2012). In reducing coastal wetlands, H_2S is formed produced by sulfate-reducing microbes with larger fractionation and is quickly converted to SO_2 in the atmosphere. Volcanoes are also a source of sulfur. As we found in the previous chapter, mantle sulfur is predominantly, although not exclusively, in reduced form with a $\delta^{34}S_{CDT}$ close to 0, but volcanoes emit sulfur primarily in the form of SO_2, which is somewhat heavier. Burning of fossil fuels, particularly coal, is the primary anthropogenic source of atmospheric sulfur, with smelting of sulfide ores a secondary source. There is some S isotope fractionation during coal combustion, with emitted SO_2 being somewhat lighter than fly ash, with the fractionation apparently greater in high-sulfur coals (Yaofa et al., 2008). Lee et al. (2002) found that SO_2 and SO_4 produced by combustion are variably enriched in ^{16}O by 10–20‰.

The tropospheric residence times of all sulfur species are relatively short: about a day for DMS and SO_2 before being oxidized or removed and ~4 days for SO_4^{2-} before being removed by rain or dry deposition. Fossil fuel combustion and industrial activities represent 68% of global non-sea-salt sulfur emissions. About 50% of SO_2 globally is converted to SO_4^{2-} aerosol (principally by in-cloud oxidation, which accounts for 85% of global SO_4^{2-}) with the remainder is removed by dry and wet deposition (Chin et al., 1996).

Atmospheric SO_2 is oxidized to H_2SO_4 through several possible pathways (Harris et al., 2012, 2013):

- gas phase by reaction with $OH^.$ radicals: $\alpha_{34} \approx 1.010$
- in cloud droplets by reaction with O_3: $\alpha_{34} \approx 1.0199$
- in cloud droplets by reaction with H_2O_2: $\alpha_{34} \approx 1.015$

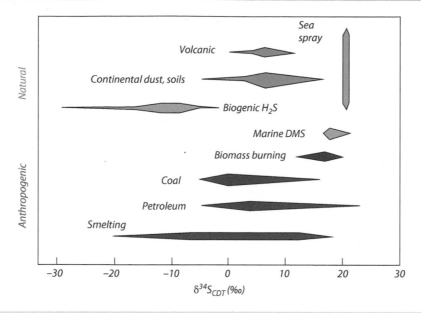

Figure 10.34   Isotopic compositions of sources of atmospheric sulfur compounds. Sulfur in sea spray is in the form of sulfate salts and is relatively inert. Sulfur from other sources is primarily emitted as SO_2 or quickly converted to it in the atmosphere. Source: Adapted from (Nielsen, 1974).

- in cloud droplets by reaction with O_2 catalyzed by transition metal ions (TMIs): $\alpha_{34} \approx 0.989$.

Both the $OH^{\cdot}$ and H_2O_2 pathways result in enrichment of H_2SO_4 in ^{34}S and ^{33}S with $\delta^{34}S$ and $\delta^{33}S$ close to the TFL with slope $\lambda = \ln(\alpha^{33}/\alpha^{34}) = 0.515$. Oxidation by O_3 has a much higher $\lambda = 0.719$ in pure water (but much lower fractionation on seasalt particles), and TMI catalysis has a lower λ of 0.498, depending on temperature. The relative contributions of these pathways vary, with the OH and H_2O_2 pathways dominant in summer and fall and the TMI pathway dominant in winter. Although these varying pathways could help explain seasonal variation observed in the isotopic composition of sulfates, most studies have concluded that variation in sources must also be involved.

Figure 10.35 shows S isotope ratios of sulfate aerosols measured in Southern California (Romero and Thiemens, 2003), Xianghe County, China (Guo et al., 2010), Mt. Everest (Lin et al., 2018), Beijing (Han et al., 2016), the South Pole (Shaheen, et al., 2014), and Montreal (Au Yang et al., 2019). Most aerosols have $\delta^{34}S_{CDT}$ within the range of +2‰ to +12‰, although the South Pole values are higher reflecting the predominance of marine aerosols and low anthropogenic inputs. The aerosols show a varying deviation from the mass-dependent $^{33}S/^{32}S$-$^{34}S/^{32}S$ fractionation ratio of 0.515 (as measured by $\Delta^{33}S$) and a diffuse trend of decreasing $\Delta^{34}S$ with increasing $\delta^{34}S_{CDT}$. Figure 10.35 shows the variation in $\Delta^{33}S$ with $\Delta^{36}S$ and compares it with the known mechanisms of MIF. None of the known mechanisms of SO_2 oxidation reproduce the observed fractionation trend. While photolysis or self-shielding in the stratosphere might explain the trend, given the low flux of air from the stratosphere, the sampling locations, and the short residence time of sulfates in the troposphere, Lin et al. (2018) and Au Yang et al. (2019) concluded that stratospheric air could account for only a small fraction of the isotopic variation observed; Au Yang et al. (2019) argued they much reflect some as yet unidentified process.

Sulfate aerosols also exhibit mass-independent O isotope fractionation with $\Delta^{17}O$ values up to +7, with the highest values found in marine air. MIF oxygen is introduced into sulfate during oxidation of SO_2 primarily through the O_3 oxidative pathway and to a less extent through reaction with H_2O_2, which also carries a small MIF signature. Neither the OH nor O_2-TMI pathways result in sulfate with a MIF oxygen signature. Variation in $\Delta^{17}O$ in

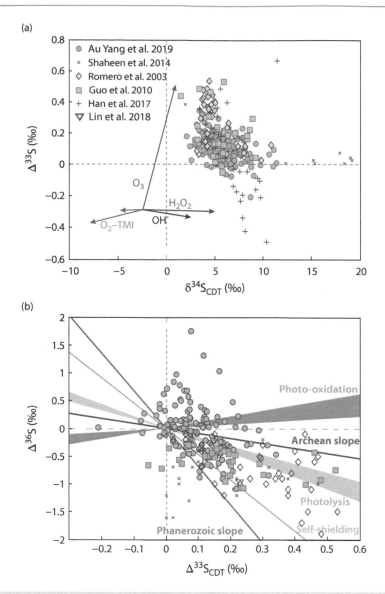

Figure 10.35 (a) $\Delta^{34}S$ relative to $\lambda = 0.515$ versus $\delta^{34}S_{CDT}$ in sulfate aerosols. Also shown are the fractionation trajectories determined experimentally by Harris et al. (2012, 2013) for gas phase oxidation by OH· radicals, and in cloud droplets by H_2O_2, O_3, and transition metal ion catalyzed oxidation by O_2 (O_2–TMI). (b) $\Delta^{34}S$ versus $\Delta^{36}S$ (relative to $\lambda = 1.889$) in sulfate aerosols. Also shown are the experimental or theoretical fractionation pathways for self-shielding, photolysis, and photo-oxidation as well as the observed Archean and Phanerozoic slopes. Source: Adapted from Au Yang et al. (2019).

sulfate, thus, depends on the proportion of sulfate produced by these oxidative pathways. Given that much of the production tropospheric O_3 is anthropogenic, one might expect to see an increase in $\Delta^{17}O$ through time, but this is not observed in sulfate from polar ice cores. Sofen et al. (2011) argued that this is a consequence of anthropogenic emission of transition metals, resulting in a high fraction of sulfur oxidation through the O_2-TMI pathway.

Environmental regulations introduced in North America and Europe in the 1970s and 1980s, which resulted in the installation of "scrubbers" on power plants and smelters to remove sulfur by reacting the combustion gas with limestone, as well as shifting electrical generation away from coal, have resulted in very large reductions in anthropogenic sulfur emissions in those regions and a global decrease as well until 2000, after which the global emissions increased due to a sharp rise in the Chinese emissions up to around 2006. Since then, Chinese emissions have decreased, particularly following the introduction of new regulations in 2013. However, increased emissions from India have more than compensated for the Chinese reductions so that emissions from Asia continue to increase.

REFERENCES

Albertin, S., Savarino, J., Bekki, S., et al. 2021. Measurement report: nitrogen isotopes (δ^{15}N) and first quantification of oxygen isotope anomalies (Δ^{17}O, δ^{18}O) in atmospheric nitrogen dioxide. *Atmospheric. Chemistry and Physics* **21**: 10477–97. doi: 10.5194/acp-21-10477-2021.

Arrhenius, S. 1896. Über den Einfluss des Atmosphärischen Kohlensäurengehalts auf die Temperatur der Erdoberfläche. *Proceedings of the Royal Swedish Academy of Science* **22**: 1–101.

Au Yang, D., Cartigny, P., Desboeufs, K., et al. 2019. Seasonality in the δ^{34}S$_{CDT}$ measured in urban aerosols highlights an additional oxidation pathway for atmospheric SO$_2$. *Atmospheric Chemistry and Physics* **19**: 3779–96. doi: 10.5194/acp-19-3779-2019.

Barbour, M.M. 2007. Stable oxygen isotope composition of plant tissue: a review. *Functional Plant Biology* **34**: 83–94. doi: 10.1071/FP06228.

Blunier, T., Barnett, B., Bender, M.L., et al. 2002. Biological oxygen productivity during the last 60,000 years from triple oxygen isotope measurements. *Global Biogeochemical Cycles* **16**: 3-1–3-13. doi: https://doi.org/10.1029/2001GB001460.

Boering, K.A., Jackson, T., Hoag, K.J., et al. 2004. Observations of the anomalous oxygen isotopic composition of carbon dioxide in the lower stratosphere and the flux of the anomaly to the troposphere. *Geophysical Research Letters* **31**. doi: 10.1029/2003GL018451.

Bowen, G.J., Ehleringer, J.R., Chesson, L.A., et al. 2007. Stable isotope ratios of tap water in the contiguous United States. *Water Resources Research* **43**: W03419. doi: 10.1029/2006wr005186.

Burk, R.L. and Stuiver, M. 1981. Oxygen isotope ratios in trees reflect mean annual temperature and humidity. *Science* **211**: 1417–9. doi: 10.1126/science.211.4489.1417.

Canfield, D. 2001. Biogeochemistry of sulfur isotopes. *Reviews in Mineralogy and Geochemistry* **43**: 607–36. doi: 10.2138/gsrmg.43.1.607

Casciotti, K.L., Trull, T.W., Glover, D.M., et al. 2008. Constraints on nitrogen cycling at the subtropical North Pacific station ALOHA from isotopic measurements of nitrate and particulate nitrogen. *Deep Sea Research Part II: Topical Studies in Oceanography* **55**: 1661–72. doi: 10.1016/j.dsr2.2008.04.017.

Chang, N. 2014. *Nitrogen Isotope Dynamics of Anammox and Denitrification in Coastal Groundwater*. Masters Thesis: University of Connecticut.

Chin, M., Jacob, D.J., Gardner, G.M., et al. 1996. A global three-dimensional model of tropospheric sulfate. *Journal of Geophysical Research: Atmospheres* **101**: 18667–90. doi: 10.1029/96JD01221.

Chung, E. and Arnold, T. 2021. Potential of clumped isotopes in constraining the global atmospheric methane budget. *Global Biogeochemical Cycles* **35**: e2020GB006883. doi: https://doi.org/10.1029/2020GB006883.

Criss, R.E. 1999. *Principles of Stable Isotope Distribution*. New York: Oxford University Press.

Dansgaard, W. 1964. Stable isotopes in precipitation. *Tellus* **16**: 436–63.

Delucia, E.H., Schlesinger, W.H. and Billings, W.D. 1988. Water relations and maintenance of Sierran conifers on hydrothermally altered rock. *Ecology* **72**: 48–51. doi: 10.2307/1940428

Dole, M., Lane, G.A., Rudd, D.P., et al. 1954. Isotopic composition of atmospheric oxygen and nitrogen. *Geochimica et Cosmochimica Acta* **6**: 65–78. doi: 10.1016/0016-7037(54)90016-2.

Douglas, P.M.J., Stolper, D.A., Eiler, J.M., et al. 2017. Methane clumped isotopes: progress and potential for a new isotopic tracer. *Organic Geochemistry* **113**: 262–82. doi: 10.1016/j.orggeochem.2017.07.016.

Eide, M., Olsen, A., Ninnemann, U.S., et al. 2017. A global estimate of the full oceanic ^{13}C Suess effect since the preindustrial. *Global Biogeochemical Cycles* **31**: 492–514. doi: 10.1002/2016GB005472.

Elliott, E.M., Yu, Z., Cole, A.S., et al. 2019. Isotopic advances in understanding reactive nitrogen deposition and atmospheric processing. *Science of The Total Environment* **662**: 393–403. doi: 10.1016/j.scitotenv.2018.12.177.

Englebrecht, A.C. and Sachs, J.P. 2005. Determination of sediment provenance at drift sites using hydrogen isotopes and unsaturation ratios in alkenones. *Geochimica et Cosmochimica Acta* **69**: 4253–65. doi: 10.1016/j.gca.2005.04.011.

Estep, M.L.F. 1984. Carbon and hydrogen isotopic compositions of algae and bacteria from hydrothermal environments, Yellowstone National Park. *Geochimica et Cosmochimica Acta* **48**: 591–9. doi: 10.1016/0016-7037(84)90287-4.

Fang, Y.T., Koba, K., Wang, X.M., et al. 2011. Anthropogenic imprints on nitrogen and oxygen isotopic composition of precipitation nitrate in a nitrogen-polluted city in southern China. *Atmospheric Chemistry and Physics* **11**: 1313–1325. doi: 10.5194/acp-11-1313-2011.

Farquhar, G.D. 1983. On the nature of carbon isotope discrimination in C_4 species. *Functional Plant Biology* 10: 205–26. doi: 0.1071/PP9830205.

Farquhar, G.D., O'Leary, M.H. and Berry, J.A. 1982. On the relationship between carbon isotope discrimination and the intercellular carbon dioxide concentration in leaves. *Functional Plant Biology* 9: 121–37. doi: 10.1071/PP9820121.

Fogel, M.L. and Cifuentes, M.L. 1993. Isotope fractionation during primary production, in Engel, M.H. and Macko, S.A. (eds). *Organic Geochemistry: principles and Applications*. New York: Plenum, 73–98 pp.

Foote, E.N. 1856. Circumstances affecting the heat of the Sun's rays. *American Journal of Science and Arts* 23: 382–3.

Friedlingstein, P., O'Sullivan, M., Jones, M.W., et al., 2020. Global carbon budget 2020. *Earth System Science Data* 12: 3269–340. doi: 10.5194/essd-12-3269-2020.

Fujita, R., Morimoto, S., Maksyutov, S., et al. 2020. Global and regional CH_4 emissions for 1995–2013 derived from atmospheric CH_4, $\delta^{13}C$- CH_4, and δD-CH_4 observations and a chemical transport model. *Journal of Geophysical Research: Atmospheres* 125: e2020JD032903. doi: https://doi.org/10.1029/2020JD032903.

Garcin, Y., Schwab, V.F., Gleixner, G., et al. 2012. Hydrogen isotope ratios of lacustrine sedimentary n-alkanes as proxies of tropical African hydrology: insights from a calibration transect across Cameroon. *Geochimica et Cosmochimica Acta* 79: 106–26. doi: https://doi.org/10.1016/j.gca.2011.11.039.

Gautier, E., Savarino, J., Hoek, J., et al. 2019. 2600-years of stratospheric volcanism through sulfate isotopes. *Nature Communications* 10: 466. doi: 10.1038/s41467-019-08357-0.

Gázquez, F., Morellón, M., Bauska, T., et al. 2018. Triple oxygen and hydrogen isotopes of gypsum hydration water for quantitative paleo-humidity reconstruction. *Earth and Planetary Science Letters* 481: 177–88. doi: 10.1016/j.epsl.2017.10.020.

Gierczak, T., Talukdar, R.K., Herndon, S.C., et al. 1997. Rate coefficients for the reactions of hydroxyl radicals with methane and deuterated methanes. *The Journal of Physical Chemistry A* 101: 3125–34. doi: 10.1021/jp963892r.

Guo, Z., Li, Z., Farquhar, J., et al. 2010. Identification of sources and formation processes of atmospheric sulfate by sulfur isotope and scanning electron microscope measurements. *Journal of Geophysical Research: Atmospheres* 115. doi: 10.1029/2009JD012893.

Guy, R.D., Fogel, M.L. and Berry, J.A. 1993. Photosynthetic fractionation of the stable isotopes of oxygen and carbon. *Plant Physiology* 101: 37–47. doi: 10.1104/pp.101.1.37.

Han, X., Guo, Q., Liu, C., et al. 2016. Using stable isotopes to trace sources and formation processes of sulfate aerosols from Beijing, *China. Scientific Reports* 6: 29958. doi: 10.1038/srep29958.

Harris, E., Sinha, B., Hoppe, P., et al. 2013. High-precision measurements of ^{33}S and ^{34}S fractionation during SO_2 oxidation reveal causes of seasonality in SO_2 and sulfate isotopic composition. *Environmental Science & Technology* 47: 12174–83. doi: 10.1021/es402824c.

Harris, E., Sinha, B., Hoppe, P., et al. 2012. Sulfur isotope fractionation during oxidation of sulfur dioxide: gas-phase oxidation by oh radicals and aqueous oxidation by H_2O_2, O_3 and iron catalysis. *Atmospheric Chemistry and Physics* 12: 407–23. doi: 10.5194/acp-12-407-2012.

Hastings, M.G., Casciotti, K.L. and Elliott, E.M. 2013. Stable isotopes as tracers of anthropogenic nitrogen sources, deposition, and impacts. *Elements* 9(5): 339–44. doi: 10.2113/gselements.9.5.339

Hayes, J.M. 2001. Fractionation of carbon and hydrogen isotopes in biosynthetic processes, in Valley, J.W. and Cole, D. (eds). *Reviews in Mineralogy and Geochemistry v. 43: Stable Isotope Geochemistry*. Washington: Mineralogical Society of America.

Heimann, M. and Maier-Reimer, E., 1996. On the relations between the oceanic uptake of CO_2 and its carbon isotopes. *Global Biogeochemical Cycles* 10: 89–110.

Howarth, R.W. 2019. Ideas and perspectives: is shale gas a major driver of recent increase in global atmospheric methane? *BiogeoSciences* 16(15): 3033–46. doi: 10.5194/bg-16-3033-2019.

Huang, Y., Shuman, B., Wang, Y., et al. 2004. Hydrogen isotope ratios of individual lipids in lake sediments as novel tracers of climatic and environmental change: a surface sediment test. *Journal of Paleolimnology* 31: 363–75. doi: 10.1023/B:JOPL.0000021855.80535.13.

Johnson, D.G., Jucks, K.W., Traub, W.A., et al. 2000. Isotopic composition of stratospheric ozone. *Journal of Geophysical Research* 105: 9025–31.

Keeling, C.D., Piper, S.C., Bacastow, R.B., et al. 2001. *Exchanges of Atmospheric CO_2 and $^{13}CO_2$ with the Terrestrial Biosphere and Oceans from 1978 to 2000. I. Global Aspects. SIO Reference Series*. San Diego, CA: Scripps Institution of Oceanography.

Kroopnick, P. and Craig, H. 1976. Oxygen isotope fractionation in dissolved oxygen in the deep sea. *Earth and Planetary Science Letters* 32: 375–88. doi: 10.1016/0012-821X(76)90078-9.

Kroopnick, P., Weiss, R.F. and Craig, H. 1972. Total CO_2, ^{13}C and dissolved oxygen-^{18}O at GEOSEC II in the North Atlantic. *Earth and Planetary Science Letters* 16: 103–10.

Lämmerzahl, P., Röckmann, T., Brenninkmeijer, C.a.M., et al. 2002. Oxygen isotope composition of stratospheric carbon dioxide. *Geophysical Research Letters* 29: 231–4. doi: https://doi.org/10.1029/2001GL014343.

Landais, A., Ekaykin, A., Barkan, E., et al. 2012. Seasonal variations of ^{17}O-excess and D-excess in snow precipitation at Vostok Station, East Antarctica. *Journal of Glaciology* **58**: 725–33. doi: 10.3189/2012JoG11J237.

Lee, C.C.W., Savarino, J., Cachier, H., et al. 2002. Sulfur (^{32}S, ^{33}S, ^{34}S, ^{36}S) and oxygen (^{16}O,^{17}O,^{18}O) isotopic ratios of primary sulfate produced from combustion processes. *Tellus B: Chemical and Physical Meteorology* **54**: 193–200. doi: 10.3402/tellusb.v54i3.16660.

Li, J., Zhang, X., Orlando, J., et al. 2020. Quantifying the nitrogen isotope effects during photochemical equilibrium between NO and NO$_2$: implications for δ^{15}N in tropospheric reactive nitrogen. *Atmospheric Chemistry and Physics* **20**: 9805–19. doi: 10.5194/acp-20-9805-2020.

Liang, M.-C., Mahata, S., Laskar, A.H., et al. 2017. Oxygen isotope anomaly in tropospheric CO$_2$ and implications for CO$_2$ residence time in the atmosphere and gross primary productivity. *Scientific Reports* **7**: 13180. doi: 10.1038/s41598-017-12774-w.

Lin, M., Kang, S., Shaheen, R., et al. 2018. Atmospheric sulfur isotopic anomalies recorded at Mt. Everest across the Anthropocene. *Proceedings of the National Academy of Sciences* **115**: 6964–9. doi: 10.1073/pnas.1801935115.

Luz, B. and Barkan, E. 2010. Variations of ^{17}O/^{16}O and ^{18}O/^{16}O in meteoric waters. *Geochimica et Cosmochimica Acta* **74**: 6276–86. doi: 10.1016/j.gca.2010.08.016.

Luz, B., Barkan, E., Bender, M.L., et al. 1999. Triple-isotope composition of atmospheric oxygen as a tracer of biosphere productivity. *Nature* **400**: 547–50. doi: 10.1038/22987.

Mader, M., Schmidt, C., Van Geldern, R., et al. 2017. Dissolved oxygen in water and its stable isotope effects: a review. *Chemical Geology* **473**: 10–21. doi: 10.1016/j.chemgeo.2017.10.003.

Masson-Delmotte, V., Zhai, P., Pirani, A., et al. (eds). 2021. *IPCC, 2021: Climate Change 2021: The Physical Science Basis. Contribution of Working Group I to the Sixth Assessment Report of the Intergovernmental Panel on Climate Change.* Cambridge UK: Cambridge University Press doi: 10.1017/9781009157896.

Michalski, G., Scott, Z., Kabiling, M., et al. 2003. First measurements and modeling of Δ^{17}O in atmospheric nitrate. *Geophysical Research Letters* **30**(16). doi: 10.1029/2003GL017015.

Mischler, J.A., Sowers, T.A., Alley, R.B., et al. 2009. Carbon and hydrogen isotopic composition of methane over the last 1000 years. *Global Biogeochemical Cycles* **23**. doi: 10.1029/2009GB003460.

Nielsen, H. 1974. Isotopic composition of the major contributors to atmospheric sulfur. *Tellus* **26**: 213–21. doi: https://doi.org/10.1111/j.2153-3490.1974.tb01969.x.

Nisbet, E.G., Manning, M.R., Dlugokencky, E.J., et al. 2019. Very strong atmospheric methane growth in the 4 years 2014–2017: implications for the Paris Agreement. *Global Biogeochemical Cycles* **33**: 318–42. doi: https://doi.org/10.1029/2018GB006009.

Oduro, H., Van Alstyne, K.L. and Farquhar, J. 2012. Sulfur isotope variability of oceanic DMSP generation and its contributions to marine biogenic sulfur emissions. *Proceedings of the National Academy of Sciences* **109**: 9012–6. doi: 10.1073/pnas.1117691109.

Ono, S. 2017. Photochemistry of sulfur dioxide and the origin of mass-independent isotope fractionation in earth's atmosphere. *Annual Review of Earth and Planetary Sciences* **45**: 301–29. doi: 10.1146/annurev-earth-060115-012324.

Pack, A., Höweling, A., Hezel, D.C., et al. 2017. Tracing the oxygen isotope composition of the upper Earth's atmosphere using cosmic spherules. *Nature Communications* **8**: 15702. doi: 10.1038/ncomms15702.

Park, R. and Epstein, S. 1960. Carbon isotope fractionation during photosynthesis. *Geochimica Cosmochimica Acta* **21**: 110–26. doi: 10.1016/S0016-7037(60)80006-3

Policy, H.W., Johnson, H.B., Marinot, B.D., et al. 1993. Increase in C$_3$ plant water-use efficiency and biomass over glacial to present CO$_2$ concentrations. *Nature* **361**: 61–4. doi: 10.1038/361061a0.

Prokopiou, M., Martinerie, P., Sapart, C.J., et al. 2017. Constraining N$_2$O emissions since 1940 using firn air isotope measurements in both hemispheres. *Atmospheric Chemistry and Physics* **17**: 4539–64. doi: 10.5194/acp-17-4539-2017.

Quay, P., Sonnerup, R., Stutsman, J., et al. 2007. Anthropogenic CO$_2$ accumulation rates in the North Atlantic Ocean from changes in the ^{13}C/^{12}C of dissolved inorganic carbon. *Global Biogeochemical Cycles* **21**: GB1009. doi: 10.1029/2006gb002761.

Rubino, M., Etheridge, D.M., Trudinger, C.M., et al. 2013. A revised 1000 year atmospheric δ^{13}C-CO$_2$ record from Law Dome and South Pole, Antarctica. *Journal of Geophysical Research: Atmospheres* **118**: 8482–99. doi: https://doi.org/10.1002/jgrd.50668.

Ravishankara, A.R. 1988. Kinetics of radical reactions in the atmospheric oxidation of CH$_4$. *Annual Review of Physical Chemistry* **39**: 367–94. doi: 10.1146/annurev.pc.39.100188.002055.

Revelle, R. and Suess, H.E. 1957. Carbon dioxide exchange between atmosphere and ocean and the question of an increase of atmospheric CO$_2$ during the past decades. *Tellus* **9**: 18–27. doi: 10.3402/tellusa.v9i1.9075.

Romero, A.B. and Thiemens, M.H. 2003. Mass-independent sulfur isotopic compositions in present-day sulfate aerosols. *Journal of Geophysical Research: Atmospheres* **108**. doi: 10.1029/2003JD003660.

Sachse, D., Radke, J. and Gleixner, G. 2004. Hydrogen isotope ratios of recent lacustrine sedimentary n-alkanes record modern climate variability. *Geochimica et Cosmochimica Acta* **68**: 4877–89. doi: 10.1016/j.gca.2004.06.004.

Sapart, C.J., Monteil, G., Prokopiou, M., et al. 2012. Natural and anthropogenic variations in methane sources during the past two millennia. *Nature* **490**: 85–8. doi: 10.1038/nature11461.

Savarino, J., Romero, A., Cole-Dai, J., et al. 2003. UV induced mass-independent sulfur isotope fractionation in stratospheric volcanic sulfate. *Geophysical Research Letters* **30**: 11–14. doi: 10.1029/2003GL018134

Savarino, J., Bhattacharya, S.K., Morin, S., et al. 2008. The $NO+O_3$ reaction: a triple oxygen isotope perspective on the reaction dynamics and atmospheric implications for the transfer of the ozone isotope anomaly. *The Journal of Chemical Physics* **128**: 194303. doi: 10.1063/1.2917581.

Shaheen, R., Abaunza, M. M., Jackson, T. L., et al. 2014. Large sulfur-isotope anomaly in nonvolcanic sulfate aerosol and its implications for the Archean atmosphere. *Proceedings of the National Academy of Sciences* **111**: 11979–83. doi: 10.1073/pnas.1406315111.

Sharkey, T.D. and Berry, J.A. 1985. Carbon isotope fractionation of algae as influenced by an inducible CO_2 concentrating mechanism, in Lucas, W.J. and Berry, J.A. (eds). *Inorganic Carbon Uptake by Aquatic Photosynthetic Organisms.* Rockville, MD: American Society of Plant Physiology, 389–401 pp.

Sharp, Z., Wostbrock, J. and Pack, A. 2018. Mass-dependent triple oxygen isotope variations in terrestrial materials. *Geochemical Perspectives Letters* **7**: 27–31. doi: 10.7185/geochemlet.1815.

Sherwood, O.A., Schwietzke, S., Arling, V.A., et al. 2017. Global inventory of gas geochemistry data from fossil fuel, microbial and burning sources, version 2017. *Earth System Science Data* **9**: 639–56. doi: 10.5194/essd-9-639-2017.

Snider, D.M., Venkiteswaran, J.J., Schiff, S.L., et al. 2015. From the ground up: global nitrous oxide sources are constrained by stable isotope values. *PLOS ONE.* **10**: e0118954. doi: 10: e0118954; doi: 10.1371/journal.pone.0118954.

Sofen, E.D., Alexander, B. and Kunasek, S.A. 2011. The impact of anthropogenic emissions on atmospheric sulfate production pathways, oxidants, and ice core $\Delta^{17}(SO_4^{2-})$. *Atmospheric Chemistry and Physics* **11**: 3565–78. doi: 10.5194/acp-11-3565-2011.

Sternberg, L. and DeNiro, M.J. 1983. Isotopic composition of cellulose from C_3, C_4, and CAM plants growing near one another. *Science* **220**: 947–9. doi: 10.1126/science.220.4600.947.

Sternberg, L. and Ellsworth, P.F.V. 2011. Divergent biochemical fractionation, not convergent temperature, explains cellulose oxygen isotope enrichment across latitudes. *PLOS ONE* **6**: e28040. doi: 10.1371/journal.pone.0028040.

Sternberg, L.D.S.L. 1988. D/H ratios of environmental water recorded by D/H ratios of plant lipids. *Nature* **333**: 59–61. doi: 10.1038/333059a0.

Stolper, D.A., Lawson, M., Davis, C.L., et al. 2014. Formation temperatures of thermogenic and biogenic methane. *Science* **344**: 1500–3. doi: 10.1126/science.1254509.

Strode, S.A., Wang, J.S., Manyin, M., et al. 2020. Strong sensitivity of the isotopic composition of methane to the plausible range of tropospheric chlorine. *Atmospheric Chemistry and Physics* **20**: 8405–19. doi: 10.5194/acp-20-8405-2020.

Su, C., Kang, R., Zhu, W., et al. 2020. $\delta^{15}N$ of nitric oxide produced under aerobic or anaerobic conditions from seven soils and their associated N isotope fractionations. *Journal of Geophysical Research: BiogeoSciences* **125**: e2020JG005705. doi: 10.1029/2020JG005705.

Surma, J., Assonov, S. and Staubwasser, M. 2021. Triple oxygen isotope systematics in the hydrologic cycle. *Reviews in Mineralogy and Geochemistry* **86**: 401–28. doi: 10.2138/rmg.2021.86.12.

Thiemens, M.H. and Shaheen, R. 2014. 5.6 - mass-independent isotopic composition of terrestrial and extraterrestrial materials, in Holland, H.D. and Turekian, K.K. (eds). *Treatise on geochemistry*, 2nd ed. Oxford: Elsevier.

Tostevin, R., Turchyn, A.V., Farquhar, J., et al. 2014. Multiple sulfur isotope constraints on the modern sulfur cycle. *Earth and Planetary Science Letters* **396**: 14–21. doi: 10.1016/j.epsl.2014.03.057.

Toyoda, S., Yamamoto, S.-i., Arai, S., et al. 2008. Isotopomeric characterization of N_2O produced, consumed, and emitted by automobiles. *Rapid Communications in Mass Spectrometry* **22**(5): 603–12. doi: 10.1002/rcm.3400.

Walters, W.W. and Michalski, G. 2016. Theoretical calculation of oxygen equilibrium isotope fractionation factors involving various NO_y molecules, ˙OH, and H_2O and its implications for isotope variations in atmospheric nitrate. *Geochimica et Cosmochimica Acta* **191**: 89–101. doi: https://doi.org/10.1016/j.gca.2016.06.039.

Walters, W.W., Tharp, B.D., Fang, H., et al. 2015. Nitrogen Isotope composition of thermally produced NO_x from various fossil-fuel combustion sources. *Environmental Science & Technology* **49**:11363–71. doi: 10.1021/acs.est.5b02769.

Wang, D.T., Reeves, E.P., Mcdermott, J.M., et al. 2018. Clumped isotopologue constraints on the origin of methane at seafloor hot springs. *Geochimica et Cosmochimica Acta* **223**: 141–58. doi: 10.1016/j.gca.2017.11.030.

Webb, M.A. and Miller, T.F. 2014. Position-specific and clumped stable isotope studies: comparison of the Urey and path-integral approaches for carbon dioxide, nitrous oxide, methane, and propane. *The Journal of Physical Chemistry A* **118**: 467–74. doi: 10.1021/jp411134v.

Yakir, D. 1992. Variations in the natural abundance of oxygen-18 and deuterium in plant carbohydrates. *Plant, Cell & Environment* **15**: 1005–20. doi: 10.1111/j.1365-3040.1992.tb01652.x.

Yaofa, J., Elswick, E.R. and Mastalerz, M. 2008. Progression in sulfur isotopic compositions from coal to fly ash: examples from single-source combustion in Indiana. *International Journal of Coal Geology* 73: 273–84. doi: 10.1016/j.coal.2007.06.004.

Young, E.D., Kohl, I.E., Lollar, B.S., et al. 2017. The relative abundances of resolved $^{12}CH_2D_2$ and $^{13}CH_3D$ and mechanisms controlling isotopic bond ordering in abiotic and biotic methane gases. *Geochimica et Cosmochimica Acta* 203: 235–64. doi: 10.1016/j.gca.2016.12.041.

Young, E.D., Yeung, L.Y. and Kohl, I.E. 2014. On the $\Delta^{17}O$ budget of atmospheric O_2. *Geochimica et Cosmochimica Acta* 135:102–25. doi: https://doi.org/10.1016/j.gca.2014.03.026.

Yu, L., Harris, E., Henne, S., et al. 2020. The isotopic composition of atmospheric nitrous oxide observed at the high-altitude research station Jungfraujoch, Switzerland. *Atmospheric Chemistry and Physics* 20: 6495–519. doi: 10.5194/acp-20-6495-2020.

PROBLEMS

1. Calculate the $\delta^{18}O$ of raindrops forming in a cloud after 80% of the original vapor has already condensed assuming the water initial evaporated from the ocean with $\delta^{18}O = 0$, and a liquid–vapor fractionation factor, $\alpha = 1.0092$.

2. For a temperature of 18 °C, calculate the following

   (a) Using Dansgaard's temperature–$\delta^{18}O$ relationship (Figure 10.1), estimate the mean annual composition of precipitation falling in this locality.
   (b) What is the equilibrium $^{18}O/^{16}O$ liquid–vapor fractionation factor at this temperature according to (5.98)?
   (c) From the composition of precipitation you estimated in a and the fractionation factor in b, calculate the $\delta^{18}O$ of water vapor from which the precipitation is derived.
   (d) Assuming the precipitation falls on the MWL ($\delta^2H = 8\delta^{18}O + 10$) and using (5.99), calculate the δ^2H of precipitation and water vapor in this locality.

3. Imagine a lake in a closed basin where the riverine input to the lake (R_I in Equation 10.3) has the isotopic compositions of the mean annual precipitation you calculated in Problem 2. Assume that α^0 is 1.0260 and 1.094 for O and H, respectively, and use the equilibrium fractionation factors you calculated in Problem 2 as well.

   (a) If the steady-state $\delta^{18}O$ of the lake is +4‰, what is the mean annual humidity?
   (b) Using the humidity you calculate in a, what is the steady-state δ^2H of the lake?

4. Using a fractionation factor of $\Delta = -18$‰ for respiration (^{16}O is preferentially consumed), make a plot of $\delta^{18}O$ in ocean water beneath the photic zone as the fractionation of dissolved oxygen remaining decreases from 1 to 0.95. Assume an initial $\delta^{18}O = 24$‰. Compare your results with Figure 9.15.

5. Using the atmospheric O isotopic composition of CO_2 given in Table 10.1.

   (a) Calculate $\Delta^{17}O$ relative to a $\delta^{18}O$–$\delta^{18}O$ slope $\lambda = 0.528$ from the composition using (5.110).
   (b) Repeat the calculation using a slope $\lambda = 0.516$.
   (c) Now calculate $\Delta'^{17}O$ using (5.115) (setting γ' to 0) after converting δ values to δ' values using (5.114) for $\lambda = 0.528$ and $\lambda = 0.516$.

6. Derive an expression analogous to 10.33 for the denominator (random distribution) in $\Delta_{12CH_2D_2}$.

7. Some natural gas samples from the Paleozoic sediments are thought to be mixtures of thermogenic and biogenic methane. Assume that the biogenic end-member has δ^2H_{SMOW} of –270‰, $\delta^{13}C_{PDB}$ of –55‰, and $\Delta_{13CH_3D} = 1.75$‰ and the thermogenic end-member has δ^2H_{SMOW} of –190‰, $\delta^{13}C_{PDB}$ of –40‰, and $\Delta_{13CH_3D} = 2.5$‰. Make plots of δ^2H versus $\delta^{13}C$ and of Δ_{13CH_3D} versus δ^2H showing the isotopic compositions of mixtures of these the two end-members at intervals of 10% of the biogenic gas. The D/H ratio of SMOW is 1.55×10^{-4} and the $^{13}C/^{12}C$ ratio of PDB is 1.122×10^{-2}.

8. Policy et al. (1993) used the carbon isotope fractionation factor between CO_2 in air and plant carbon in experimentally grown wheat (a C_3 plant) to calculate the ratio of interior leaf CO_2 concentration to that of air (c_{in}/c_{ex} in Equation 10.6) under ambient CO_2 concentrations ranging from 220 to 350 ppmv. At 220 ppm, the fractionation factor, Δ_{net}, was −19.9‰ and at 350 ppm it was −20.8‰. Using 10.6 and assuming Δ_{diff} and Δ_{carbox} are, respectively, −4.4‰ and −29‰, what was c_{in}/c_{ex} at the two concentrations.

Chapter 11

Non-Traditional Stable and Radiogenic Isotopes in the Exogene

11.1 INTRODUCTION

In this chapter, we extend our exploration of the isotope geochemistry of the exogene to radiogenic and "non-traditional" stable isotopes, those for which data have become abundant only in the last two decades or so, which in turn reflects development of new techniques and instruments. Among these instruments, multi-collector inductively coupled plasma mass spectrometry has been most significant, allowing analysis of elements that are typically solid at surface temperatures and not readily analyzed in the gas-source mass spectrometers used to analyze the elements we covered in the previous chapter. With the exception of mercury, the atmosphere plays little role in exogenic cycling of these elements, our interest will focus on the fractionations that occur as these elements are released from rock to soil by weathering and then carried by rivers to the oceans. We will give particular attention to the oceans because rock removed from the continents is deposited layer by layer as sediments and these sediments compose the layer-by-layer history of the planet that we will use to unravel Earth's history in the following chapter. One of our objectives in this chapter will be to develop the tools we will need to decode that history.

11.2 RADIOGENIC ISOTOPES IN THE MODERN OCEAN

11.2.1 Nd, Hf, Pb, and Os

Nd, Hf, and Pb have short residence times because of their high particle reactivity and low solubility: they are readily absorbed onto particles (both organic and inorganic) and removed from solution in this way. The residence time of Nd is in the range of 400–1000 years, that of Pb is in the range of 50–400 years, and that of Hf is uncertain with estimates ranging from a few hundred to several thousand years. Because these elements have residence times similar to or shorter than the mixing time, their isotopic composition varies in the ocean. Strontium, on the other hand, has a residence time in the ocean of 2.4 million years and $^{87}Sr/^{86}Sr$ of seawater in the open ocean is uniform at 0.70925, although variations occur in coastal waters. The residence time of Os in seawater is not entirely resolved; it may be as shorter as a few thousand years or less or as long as a few tens of thousands of years or more. The Os isotopic composition of deep ocean water appears to be constant within analytical error, but some variation is observed in surface waters. With the exception of Sr, all these elements are present in seawater at extremely low concentrations (parts per

Isotope Geochemistry, Second Edition. William M. White.
© 2023 John Wiley & Sons Ltd. Published 2023 by John Wiley & Sons Ltd.
Companion Website: www.wiley.com/go/white/isotopegeochem2

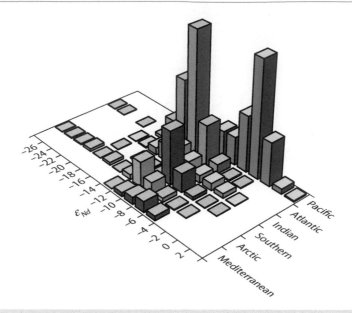

Figure 11.1 Histogram of ε_{Nd} dissolved in seawater; the Pacific and Atlantic waters have the extremes of radiogenic and unradiogenic Nd. This figure is based on over 2100 analyses adapted from van de Flierdt et al. (2016).

trillion and lower), so these isotopic analyses are challenging, although Nd and Pb analyses are now becoming routine.

The Nd isotopic composition of seawater is now well characterized. Indeed, a database published by van der Flierdt et al. (2016) contains 2100 analyses. These are illustrated in Figure 11.1, which shows the frequency distribution of ε_{Nd} in the five oceans. The most radiogenic Nd occurs in the Atlantic Ocean and the Arctic Ocean, with an average ε_{Nd} of −11.5 and −11.7 and standard deviations of 4.1 and 2.1, respectively. The least radiogenic Nd occurs in the Pacific, with an average ε_{Nd} of −4.0 and a standard deviation of 1.9. The Indian Ocean and the Southern Ocean are intermediate, with an average ε_{Nd} of −10.1 and −8.3 and standard deviations of 3.4 and 1.6, respectively. Peucker-Ehrenbrink et al. (2010) estimate the mean ε_{Nd} of seawater as −7.2 ± 0.5.

The radiogenic nature of the Pacific Ocean Nd originally suggested hydrothermal inputs might be important (Piepgras and Wasserburg, 1980) because of the greater mid-ocean ridge magmatism and hydrothermal activity there, but subsequent work shows the Nd dissolved in hydrothermal fluids is very quickly scavenged by particles, reducing the hydrothermal input to negligible levels. The difference

instead relates to the nature and age, and therefore isotopic composition, of geologic provinces supplying Nd to the ocean basins. The Pacific is surrounded by young volcanic arcs with relatively high ε_{Nd} while the Atlantic is surrounded by older terrains with lower ε_{Nd}. The most extremely unradiogenic Nd occurs in Baffin Bay and the Labrador Sea, which are surrounded by extensive outcropping of Archean crust. Seawater in this region is a major source of North Atlantic Deep Water (NADW), which accounts for the particularly unradiogenic character of this water mass ($\varepsilon_{Nd} \approx -14$). The unradiogenic nature of NADW is somewhat moderated through mixing as it flows southward, but it remains distinctive in the extreme South Atlantic where it mixes with Antarctic waters to become part of the Circumpolar Deep Water (CDW) with ε_{Nd} of −8 to −9 (Stichel et al., 2012). Waters in the North Pacific tend to have the most radiogenic Nd. Surface waters show greater variability than deep water. Attempts to reproduce the global pattern of ε_{Nd} in the oceans suggest that isotopic exchange between dissolved Nd and Nd in sedimentary particles on continental margins exerts an important control on isotopic composition (e.g., Albarède et al., 1997; Jeandel et al., 2011).

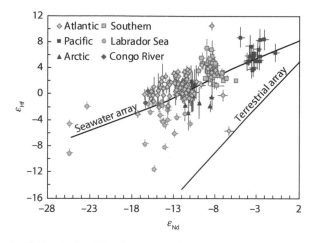

Figure 11.2 ε_{Hf} versus ε_{Nd} in seawater. The "terrestrial array" is the correlation in oceanic basalts as shown in Figure 2.21. The seawater array mirrors that observed in authigenic marine sediments, known as the "seawater array." Data from the literature.

There are far fewer data on Hf isotopic composition as it has only become possible to directly measure Hf isotope ratios in seawater in the last dozen years or so (e.g., Godfrey et al., 2009; Zimmermann et al., 2009a, b). Enough data have now been reported to show that the same isotopic distinctions between ocean basins apparent in Nd isotopes are also present in the Hf isotope data, although they are more subdued. Measured values in seawater range from ε_{Hf} –12 to 8.6 with the most radiogenic values in the Pacific and the least radiogenic value in the Labrador Sea of the Atlantic Ocean.

Figure 11.2 shows the relationship between ε_{Hf} and ε_{Nd} in seawater. The data fall along a distinctively lower slope than the mantle, or "terrestrial," array, $\varepsilon_{Hf} \cong 1.21 + 1.55 \times \varepsilon_{Nd}$ (Figure 2.21), and mirror the slope observed in marine sediments and manganese nodules (White et al., 1986; Vervoort et al., 2011), which Albarède et al., (1998) termed the "seawater array" with $\varepsilon_{Hf} \cong 7.4 + 0.62 \times \varepsilon_{Nd}$. As recognized early on, the discordance between the terrestrial and seawater arrays results from the difference in behavior of Nd and Hf in weathering, which is referred to as the "zircon effect" (Patchett et al., 1984; White et al., 1986). Whereas Nd concentrates largely in clays and other fine-grained materials, much of the budget of Hf in sedimentary rocks is in zircon. Hf in zircon is very unradiogenic due to very low Lu/Hf. Zircon is a heavy mineral that resists mechanical and chemical

weathering and transport, thus much of it remains in coarse-grained sediments of the continents and continental shelves. Clays and fine-grained accessory minerals rich in Nd and radiogenic Hf are more readily transported by a combination of rivers, currents, and winds to the deep ocean. In addition, various Lu-rich minerals, including garnet and apatite, are among those most readily weathered. This incongruent release of Hf appears to extend to Saharan dust carried to the Atlantic, as while Nd isotopic composition of surface waters is similar to that of dust, the Hf in surface waters is more radiogenic than the dust (Rickli et al., 2010). Chen et al. (2013) reported further evidence of this incongruent weathering in leaching experiments on Chinese loess and desert dust. While the grains themselves tended to plot close to the mantle array, particularly the coarser fraction, leachates of the dust tended to plot along the seawater array. Rickli et al. (2017) analyzed Hf and Nd isotopes in glacier- and non-glacier-sourced river water in Greenland and found that while ε_{Nd} was similar in both rivers, and to the underlying Proterozoic gneisses, ε_{Hf} was distinctly less radiogenic in non-glacial streams, implying that zircon was more efficiently weathered in the subglacial environment, which was a result of either glacial grinding and stress or longer times in contact with water.

The Labrador Sea, which is a key area in formation of NADW and global deepwater circulation, has particularly unradiogenic Hf and Nd, as well as considerable heterogeneity in ε_{Nd} and ε_{Hf} and in the relationship between them (Figure 11.2), with some water in this region deviating significantly from the seawater array (Filippova et al., 2017). This results from the variety of water masses present in the region and the considerable heterogeneity in the age and nature of the rock surrounding this region, ranging from young basalts in Iceland and East Greenland that weather readily and congruently to the early Archean terrains of Labrador and West Greenland, and possibly the shorter residence time of Hf compared to that of Nd in seawater. This study illustrates the potential of Hf isotopes for identifying water masses and mixing between them on relatively small scales.

Pb isotope ratios present perhaps the greatest analytical challenge because of the extremely low concentrations combined with high blank levels that result from anthropogenic lead. Consequently, data are often reported as $^{206}Pb/^{207}Pb$ and $^{208}Pb/^{207}Pb$ ratios as ^{204}Pb levels are too low to measure accurately. Much of this anthropogenic Pb came from tetraethyl lead in gasoline, the use of which has almost entirely discontinued throughout the world, beginning with the United States (US) in the 1970s followed by most other countries over the succeeding decades. However, smelting of Pb ore and coal burning continues to release Pb to the atmosphere. Pb concentrations in the North Atlantic increased by an order of magnitude between 1880 and 1970, but since have decreased drastically, to the point where natural Pb sources are beginning to again dominate (Bridgestock et al., 2016). Anthropogenic Pb in the Pacific has shown a smaller decrease of about a factor of two, in part because while China eliminated tetraethyl Pb in gasoline in 2000, Chinese Pb emissions from burning of coal have increased.

Because Pb from various sources often has distinct isotopic compositions, it is possible to identify the sources of anthropogenic Pb in the ocean and to distinguish them from natural ones. Chinese coal, e.g., has $^{206}Pb/^{207}Pb \approx 1.18$ and $^{208}Pb/^{207}Pb \approx 2.47$, whereas Chinese ores have $^{206}Pb/^{207}Pb \lesssim 1.09$ and $^{208}Pb/^{207}Pb \lesssim 2.40$. Mississippi Valley Pb ores, which were an important source for American tetraethyl Pb, have $^{206}Pb/^{207}Pb \approx 1.33$ and $^{208}Pb/^{207}Pb \approx 2.51$ while Pb from the Proterozoic Broken Hill mine in Australia was the major source

of Pb for European tetraethyl Pb, along with many other regions around the world, with $^{206}Pb/^{207}Pb \approx 1.04$ and $^{208}Pb/^{207}Pb \approx 2.22$. Natural sources of dust in the ocean include atmospheric dust, e.g., from the Sahara with $^{206}Pb/^{207}Pb \approx 1.19–1.21$ and $^{208}Pb/^{207}Pb \gtrsim 2.48$ and mid-ocean ridge hydrothermal plumes with $^{206}Pb/^{207}Pb \approx 1.19$ and $^{208}Pb/^{207}Pb \approx 2.46$, although the latter is quite quickly removed by absorption on hydrothermal particulates such that only $\approx 5\%$ of the hydrothermal flux contributes to seawater-dissolved Pb (Boyle et al., 2020).

Figure 11.3 shows how Pb concentrations and isotopic compositions have changed in the North Atlantic near Bermuda, with surface

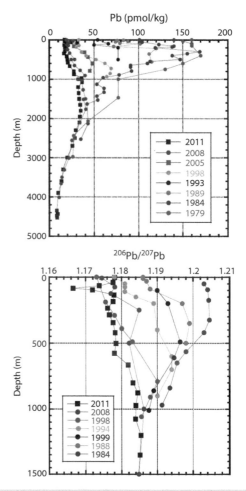

Figure 11.3 Pb depth and isotopic profiles in the North Atlantic showing decreasing concentrations with the phaseout of tetraethyl lead in gasoline. Shifts in isotopic composition reflect the earlier phaseout in the US than in Europe and an increasing relative importance of natural Pb sources. Source: Adapted from Boyle et al. (2014).

water concentrations dropping from ≈ 150 pmol/kg to ≈ 20 pmol/kg between 1979 and 2011. The isotopic profile shows that North American sources still dominated Pb in the Atlantic in the 1980s as the US was still in the process of phasing out tetraethyl Pb in gasoline. By 1988, the phaseout was essentially complete in the US but had just begun in Europe, which is reflected in a decrease in $^{206}Pb/^{207}Pb$ in surface waters. Anthropogenic emissions continued to dominate dissolved Pb in the Atlantic, mainly from coal burning; however, by 2016, the proportion of anthropogenic Pb in the Atlantic had decreased to 50–70%, with the remainder derived from natural mineral dust and the Amazon River (Bridgestock et al., 2016).

Data on the natural Pb distribution in the ocean come largely from manganese nodules on the seafloor, into which seawater Pb is incorporated at high concentration. These data show $^{206}Pb/^{204}Pb$ ratios varying from about 18.5 to 19.3 (von Blanckenburg et al., 1996). As with Nd and Hf, the more "mantle-like" isotopic signatures are found in the Pacific Ocean while more radiogenic, crustal-like signatures occur in the Atlantic Ocean, particularly the North Atlantic. Pb isotopes are well mixed in the Pacific and there is no evidence of the import of North Atlantic deepwater-derived lead into either the Pacific Ocean or the North Indian Ocean, which is a consequence of the short residence time of Pb in deep water (80–100 years). Rivers appear to be the major natural source of dissolved Pb in seawater, although dissolution of aeolian particulates may be about 12% of the total flux, but can be locally dominant in regions distant from continents (Henderson and Maier-Reimer, 2002).

Osmium is, of course, an extremely rare element and this is certainly true in seawater, where the concentration is $\approx 10^{-8}$ ppm (10^{-14} g/g or 5×10^{-14} mol/kg) and as a consequence analysis is difficult and data on the isotopic composition of seawater remain sparse. Early studies found that osmium isotopic composition of deep water was homogeneous within analytical error at $^{187}Os/^{188}Os = 1.067 \pm 0.011$ (Sharma et al., 1997; Levasseur et al., 1998; Woodhouse et al., 1999), which reflects a balance of a variety of sources.

Interestingly, interplanetary dust particles and meteorites are a small but significant (5%) source of seawater. That source and hydrothermal systems developed on abyssal peridotite provide unradiogenic Os ($^{187}Os/^{188}Os \approx 0.13$) to balance radiogenic Os from the continents (e.g., Burton et al., 2010), which has an average $^{187}Os/^{188}Os$ of 1.4. Most of this continental Os is provided by rivers, but the aeolian flux is difficult to evaluate since loess has the same $^{187}Os/^{188}Os$ as seawater (Peucker-Ehrenbrink and Jahn, 2001). Gannoun and Burton (2014) found that $^{187}Os/^{188}Os$ in a Northeast Atlantic hydrocast was essentially constant with depth at 1.024 ± 0.031. In contrast to some earlier studies, they also found that Os concentrations were essentially constant with depth at 9.7 ± 0.5 pg/kg (9.8×10^{-7} ppm). Furthermore, Os at that station is identical within analytical error to water from the Eastern Pacific with $^{187}Os/^{188}Os = 1.028 \pm 0.040$. These results suggest a longer residence time for Os in seawater than the ocean mixing time despite its extremely low concentration, implying it is present in seawater in a soluble form, such as the organometallic complex, or as an oxyanion.

11.2.2 U and Th decay series isotopes in oceanography

The decay products of U and Th also have found extensive use in oceanography. In addition to ^{230}Th, ^{231}Pa, ^{226}Ra, and ^{210}Pb that we have considered in previous chapters, interest includes even shorter-lived ones, such as ^{234}Th, ^{228}Th, ^{228}Ra, and ^{210}Po, whose half-lives have been listed in Table 11.1; Figure 11.4 shows the activities of ^{230}Th and ^{231}Pa as a function of depth in the Atlantic. The interest in these isotopes stems from their varying solubility or, more precisely, particle reactivity. Uranium is in its oxidized form everywhere in the ocean and is consequently soluble and its concentration is uniform[1] throughout the open ocean, as is $^{234}U/^{238}U$ activity ratio (which is 1.1468 rather than 1 for reasons we discussed in Section 3.4). Radium is similarly soluble, but Th, Pa, Po, and Pb are variably particle reactive. The distribution of these particle-reactive elements

---

[1] Like other "conservative" elements in the ocean, the U concentration does vary due to evaporation and precipitation, but the ratio of the U concentration to salinity and to the concentration of other conservative elements does not.

Table 11.1   U and Th decay chain nuclides used in oceanography.

	Half-life (yr)	λ (yr^{-1})	Parent	Daughter
^{234}Th	6.60×10^{-02}	1.05×10^1	^{238}U	^{234}U
^{230}Th	7.57×10^4	9.16×10^{-6}	^{234}U	^{226}Ra
^{228}Th	1.91	3.63×10^{-1}	^{228}Ra	–
^{231}Pa	3.28×10^4	2.11×10^{-5}	^{235}U	–
^{228}Ra	5.75	0.1205	^{232}Th	^{228}Th
^{226}Ra	1600	4.33×10^{-4}	^{230}Th	^{210}Pb
^{210}Po	0.3691	1.878	^{210}Pb	^{206}Pb
^{210}Pb	22.2	3.12×10^{-2}	^{226}Ra	^{210}Po
^{234}U	245,620	2.822×10^{-6}	^{234}Th	^{230}Th
^{235}U	7.04×10^8	9.857×10^{-10}	–	^{231}Pa
^{238}U	4.47×10^9	1.5513×10^{-11}	–	^{234}Th

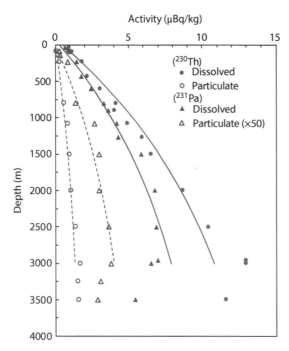

Figure 11.4   Activities of ^{230}Th and ^{231}Pa in dissolved and particulate form from GEOTRACES station GA03-24 located at 24.5°W, 17.4°N in the Central North Atlantic. All increase with depth reflecting downward transport by particles. The dissolved form dominates both elements, but much more so Pa, which is less particle reactive than Th (note that particulate ^{231}Pa activities are multiplied by 50). Solid lines show an approximate model based on Equations (11.1) through (11.6) with a settling velocity of 220 m/yr. k_1/k_{-1} ratios are 0.13 and 0.01 for ^{230}Th and ^{231}Pa, respectively. Data from GEOTRACES Intermediate Data Product 2017 (Schlitzer et al., 2018) (https://www.bodc.ac.uk/geotraces/).

provides insights into the distribution of particles in the ocean and their sinking rates. Of key interest in this respect are the workings of the so-called *biological pump* in which CO_2 is converted into organic carbon by photosynthesis in surface waters, transported into deep water as organic remains sink, and remineralized there into dissolved CO_2. Because the residence time of water in the deep ocean is long, this results in substantial buildup and sequestering of CO_2 in the deep ocean. Changes in the efficacy of the biological pump or in ocean circulation result in shifts in CO_2 to and from the deep ocean and it is these shifts (triggered by Milankovitch orbital variations) that drove the large climate swings of Pleistocene glacial cycles and we will return to this topic in Chapter 12.

Once created, these particle-reactive nuclides are absorbed onto particles on timescales comparable to or shorter than their half-lives and then begin to sink. Adsorption is a reversible process, so, ignoring advection and diffusion, the dissolved activities of a particle-reactive nuclide, such as Th, in a parcel of water will vary with time as:

$$\frac{\partial (\text{Th})_d}{\partial t} = \lambda_{\text{Th}}(\text{U}) + k_{-1}(\text{Th})_p - (\lambda_{\text{Th}} + k_1)(\text{Th})_d$$

(11.1)

and the particulate activity will be:

$$\frac{\partial (\text{Th})_p}{\partial t} = -S\frac{\partial (\text{Th})_p}{\partial z} + k_1(\text{Th})_d$$
$$- (\lambda_{\text{Th}} + k_{-1} + k_2)(\text{Th})_p$$

(11.2)

where as usual parentheses denote activity, the subscripts d and p denote dissolved and particulate, respectively, λ_{Th} is the decay constant, S

is the particle sinking rate, k_1 and k_{-1} are the absorption and desorption rate constants, respectively, and k_2 is the rate constant for loss of particulate thorium from the water column. These equations apply equally well to ^{234}Th and ^{230}Th, whose parents are ^{238}U and ^{234}U, respectively. They also apply to other parent–daughter pairs in the decay chain (Bacon and Anderson, 1982; Anderson, 2014). We will consider the steady state case, the left-hand side of Equations (11.1) and (11.2) are zero.

The ratio of dissolved to particulate activity will be:

$$\frac{(\text{Th})_p}{(\text{Th})_d} = \frac{k_1}{\lambda_{\text{Th}} + k_{-1} + k_2} \quad (11.3)$$

For the long-lived nuclides ^{230}Th and ^{231}Pa, the decay constant is small enough that it can be neglected in the $\lambda_{\text{Th}} + k_1$ and $\lambda_{\text{Th}} + k_{-1} + k_2$ terms in Equations (11.1) through (11.3). The activities in the dissolved and particular phases will be:

$$(\text{Th})_d = \frac{\lambda_{\text{Th}}(\text{U}) + k_{-1}(\text{Th})_p}{k_1} \quad (11.4)$$

$$(\text{Th})_p = \frac{\lambda_{\text{Th}}(\text{U})}{(\lambda_{\text{Th}} + k_2)(\lambda_{\text{Th}} + k_1)}\left[1 - e^{-z(\lambda + k_2)/S}\right]$$

$$(11.5)$$

For ^{231}Pa and ^{230}Th, the decay constant in the denominator and exponent may be taken as 0 so that the denominator reduces to $k_2 \times k_1$. The inverse of k_2 is the residence time of particulate Th:

$$\tau_{\text{Th}}^p = \frac{1}{k_2} \quad (11.6)$$

Figure 11.4 shows depth profiles of (^{231}Pa) and (^{230}Th) from GEOTRACES station GA03-24 in the Central North Atlantic. Both increase with depth and in both cases the dissolved form dominates, much more so Pa, indicating the reversible nature of adsorption. A simple model based on Equations (11.4) and (11.5) suggests a sinking velocity – shown as the lines in Figure 11.4 – of ≈ 220 m/yr and k_1/k_{-1} ratio of ≈ 0.13 for ^{230}Th and ≈ 0.01 for ^{231}Pa. At this sinking rate, the complete trip to the bottom would require >15 years, during which these elements would cycle through dissolved and particulate forms multiple times, with most of the time in the dissolved form

but with sufficient residence in the particulate form to produce an overall downward flux. There are clear deviations from the model, primarily as a result of lateral advection, which the model ignores. The deviation is most pronounced near the bottom because of southward flowing NADW. NADW, having relatively recently left the surface, has had less time to accumulated radionuclides. Near the bottom, currents can stir up sediment and these sedimentary particles can scavenge radionuclides. This is apparent in the poor fit of the model to the data in the lower few hundred meters in Figure 11.4.

Radionuclides can be used to estimate the total flux of particulate matter through the water column; of primary interest is the flux out of the photic zone, roughly the top 100 m, which is the region where essentially all conversion of CO_2 to organic matter occurs (the input to the biological pump). ^{234}Th is the direct daughter of ^{238}U and its half-life is sufficiently short – i.e., 24 days – that effectively all ^{234}Th has been produced in situ. Consequently, the flux of Th out of the photic zone can be calculated by integrating the deficit in the activity of ^{234}Th with respect to ^{238}U through the photic zone:

$$F\left(^{234}\text{Th}\right)_z = \int_0^z \lambda_{234}\left(^{238}\text{U} - {}^{234}\text{Th}\right)dz \quad (11.7)$$

Since flux is defined as per unit area, we need to convert concentrations to volume from mass by multiplying times seawater density (which averages 1.024 kg/l) and then, most conveniently, converting liters to cubic meters since depth is measured in meters by multiplying by 10^{-3} m^3/l. In cases where (^{230}Th) does not show systematic variation with depth in the photic zone the integral is simply the average multiplied by the depth and the ^{234}Th decay constant. Similar flux calculations can be done with other parent–daughter pairs as well, such as ^{228}Ra:^{228}Th and ^{210}Pb:^{210}Po. If the ratio of a component of interest to the daughter nuclide is measured, e.g., particulate organic carbon, the flux of that component can be calculated by multiplying by that ratio. At GEOTRACES station GA03-24, e.g., Hayes et al. (2018) calculated a ^{234}Th flux of 5.5×10^3 Bq/m^2/yr and a particulate organic carbon flux of 2.42 mmol/m^2/day out of the photic zone. Across the entire GEOTRACES GA3 transect, their calculated particulate organic carbon fluxes varied from 0.07 to 3.96 mmol/m^2/day, with the highest fluxes near the ocean margins (particularly the eastern margin) as biological productivity is highest there.

11.3   STABLE ISOTOPE RATIOS OF CONSERVATIVE ELEMENTS

11.3.1   Lithium isotopes

In aqueous solution, lithium is tetrahedrally coordinated by four water molecules (the solvation shell) to which it is strongly bound, judging from the high solvation energy. Coordination numbers of Li in minerals are generally higher and bond lengths are longer and consequently ^{7}Li preferentially partitions into the solution in mineral–water reactions. These water–rock fractionations dominate terrestrial Li isotopic variations.

Modern study of Li isotope ratios began with the work of Chan and Edmond (1988). They found that the isotopic composition of seawater was uniform within analytical error with a δ^7Li value of +33‰, which is a value subsequently revised to +31‰ based on more accurate techniques. Given that the Li concentration is uniform in the open ocean and its residence time is long, $\approx$1 Ma, a uniform isotopic composition is expected. As we found in Chapter 9, the mantle and mantle-derived rocks typically have δ^7Li in the range of 2.5–5‰. Continental crust has δ^7Li values ranging from –10 to +10‰ (Tomascak et al., 2016), and with a mean value of $\approx$0.6‰. Seawater thus represents the heavy extreme of Li isotopic composition in major terrestrial reservoirs (see Figure 9.16 in Chapter 9). Let us understand why.

River water has a wide range, from 2 to 43‰, but is distinctly heavy on average with a mean value of $\approx$23‰. Interestingly, experiments reveal that there is little or no Li isotope fractionation during mineral dissolution (e.g., Wimpenny et al., 2010). Instead, the heavier Li in rivers and streams results from formation of secondary minerals, in which Li is typically strongly enriched, with ^{6}Li partitioning preferentially into the secondary minerals. Experiments and observations show that the Δ^7Li for exchangeable Li during secondary mineral formation is 0 to –12‰ and –14 to –24‰ for structurally bound Li (Pogge von Strandmann

et al., 2019). The isotopic composition of rivers consequently depends on the extent to which weathering is congruent, in which primary minerals break down completely without secondary mineral formation, or incongruent, in which secondary minerals form and concentrate ^{6}Li. There is apparently no dependence of Δ–Li$_{solid–solution}$ on the identity of the reacting mineral or solution composition, but fractionation does decrease with temperature as we might expect.

In regions with high denudation rates where erosion quickly follows chemical weathering, typically mountainous terrains, there is little opportunity for secondary minerals to form and weather typically congruent, producing a high dissolved Li flux with low δ^7Li$_{water}$. Secondary mineral formation increases as weathering intensity increases, which leads to decreasing Li-dissolved flux and to higher δ^7Li$_{water}$. At very high weathering intensity, such as deeply weathered tropical soils, weathering becomes supply limited and secondary minerals break down, increasing Li concentrations in the solution and decreasing δ^7Li, but the overall Li flux from such terrains tends to be low and contribute minimally to seawater.

Rivers comprise just over 50% of the Li flux to seawater, with the other major flux being high-temperature hydrothermal fluids. Ridge-crest hydrothermal systems extract Li nearly quantitatively from rocks through which they pass, which results in these fluids having an average δ^7Li of $\approx$8‰ (Pogge von Strandmann et al., 2020). Thus, seawater is about 15‰ heavier than the total inputs so the fluxes out of seawater must involve a net negative fractionation of $\approx$15‰ in order to maintain this steady-state composition. This occurs through preferential adsorption of ^{6}Li on sedimentary particles, incorporation in authigenic clays, and uptake during low-temperature weathering of oceanic crust. Misra and Froelich (2012) estimate that authigenic clay formation accounts for about 70% of Li removal, with alteration of the oceanic crust accounting for most of the remainder. The net fractionation of –15‰ for these processes is consistent with experimental observation. Marine carbonate sediments, which tend to be Li-poor, typically have higher δ^7Li than non-carbonate sediments.

The isotopic composition of seawater depends on the ratio of the hydrothermal to riverine to flux, and the δ^7Li in the latter reflects the extent and nature of continental weathering. The Li isotopic composition of marine sediments will reflect variations in these factors, providing a weathering proxy. Unlike other potential tracers of weathering, such as Sr, Ca, Mg, and Si isotopes, Li is concentrated in silicate rather than carbonate minerals and is not biologically utilized or fractionated. As a consequence, Li isotope ratios in seawater and in marine sediments recorded are sensitive primarily to silicate weathering. The weathering record is important in its only right, but also because of the control it exerts on atmospheric CO_2 through the so-called "Urey reactions" in which ions released by weathering of silicate minerals ultimately sequester atmospheric CO_2 by precipitating as Ca and Mg carbonates. We will explore this more deeply in the next chapter.

Misra and Froehlich (2012) compiled lithium isotopic analyses of Cenozoic foraminifera and found that the δ^7Li of seawater has increased by 9‰ over this period. Although not identical, it tracks the rise in $^{87}Sr/^{86}Sr$ and $^{187}Os/^{188}Os$ remarkably well (Figure 11.5). Misra and Froehlich attribute the change in all three ratios to changes in weathering of the continents due primarily to changes in tectonism over the Cenozoic, which has seen the rise of the Rocky Mountains, the Andes, the Himalayas, and the Alps. As they point out, low-lying terrains where removal of weathering products is transport-limited, especially those in the tropics, undergo congruent weathering and Li isotope ratios in rivers that drain such terrains reflect those of the bedrocks. Rivers draining mountainous terrains, which undergo high weathering and denudation rates with incongruent weathering, are ^7Li-enriched.

Misra and Froehlich (2012) note the overall similarity of these curves to that of ocean bottom water $\delta^{18}O$, which records decreasing temperature and increasing ice volume. They suggest "δ^7Li_{SW} might provide alternative estimates of atmospheric CO_2 consumption by the silicate weathering."

11.3.2 Boron isotopes

Boron has a valence of +3 and is almost always bound to oxygen or hydroxyl groups in either trigonal (e.g., BO_3) or tetrahedral (e.g., $B(OH)_4^-$) coordination. Since the bond strengths and vibrational frequencies of trigonal and tetrahedral forms differ, we can expect that isotopic fractionation will occur between these two forms and this is confirmed by experiments that show a roughly 20‰ fractionation between $B(OH)_3$ and $B(OH)_4^-$, with ^{11}B preferentially found in the form with shorter bond lengths and low coordination number, $B(OH)_3$. Fractionation between these forms partially accounts for the very large range in $^{11}B/^{10}B$ in terrestrial materials of ≈100‰ (Figure 9.17).

One of the more remarkable aspects of B isotope geochemistry is this very large fractionation of B isotopes between the oceans and the silicate Earth (Figure 9.17). The isotopic composition of boron in crystalline rocks of the continental crust ranges from –20 to +10‰ with an estimated mean crustal value of –9.4 ± 2.4‰ (Marschall et al., 2017), which is slightly lighter than mid-ocean ridge basalts (MORB) and presumably mantle value of –7.1‰. Seawater lies near the heavy extreme of the spectrum, with $\delta^{11}B$ of 39.6 ± 0.04‰ with a concentration of 4.5 ppm, both of which are uniform in the open ocean. This constancy reflects its long residence time of >10 Ma in seawater.

In a manner analogous to lithium, the isotopic difference between continental crust and seawater partly reflects boron being readily absorbed on clay with an estimated $\Delta^{11}B$ of absorption of –30‰ (e.g., Schwarcz et al., 1969). This produces river water with low boron concentrations (1–40 ng/g) that is isotopically heavy ($\delta^{11}B$ = –6 to +43‰) (Foster et al., 2018). Lemarchand et al. (2002) estimated the flux-weighted mean $\delta^{11}B$ of riverine supply to the oceans as +10‰. The isotopic composition of high-temperature hydrothermal fluids is typically +24 to +37‰ suggesting that the B in these fluids is a simple mixture of seawater boron and boron derived from the oceanic crust with little or no isotopic fractionation involved (Spivack and Edmond, 1987). Thus, the net effect of hydrothermal exchange between the oceanic crust and seawater is to decrease the $\delta^{11}B$ of seawater. An additional source of B in seawater is pore fluids from sediments, particularly at convergent margins where compressive tectonic forces expel pore water. These fluids can have up to a tenfold increase in B concentrations over seawater and can have $\delta^{11}B$ both higher and lower than seawater.

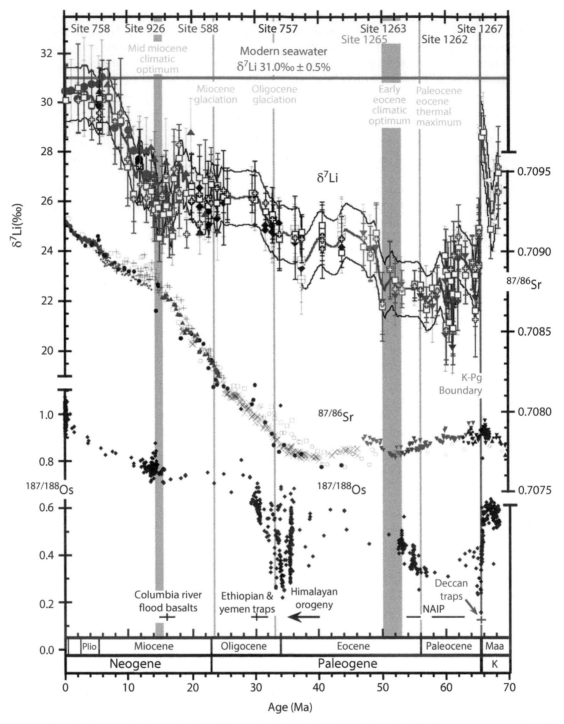

Figure 11.5 δ^7Li, $^{87}Sr/^{86}Sr$, and $^{187}Os/^{188}Os$ in seawater over the Cenozoic. Source: From Misra and Froehlich (2012).

As in the case with lithium, the net source of B in seawater is isotopically light and hence maintaining a constant elevated $\delta^{11}B$ requires that the net sinks of B seawater have a light isotopic composition identical to the net source. The primary sinks appear to be low-temperature alteration of oceanic crust, including serpentinites, which can be strongly enrich in B over fresh MORB and peridotites, and absorption on clay particles with an attendant negative fractionation (e.g., Simon et al., 2006).

Boron is readily incorporated into carbonates, with modern marine carbonates having B concentrations in the range of 15–60 ppm. The incorporation of B in carbonate is preceded by surface adsorption of $B(OH)_4^-$ (Vengosh et al., 1991; Hemming and Hanson, 1992), and consequently it is primarily $B(OH)_4^-$ rather than $B(OH)_3$ that is incorporated. We noted above that boron is present in seawater both as $B(OH)_3$ and $B(OH)_4^-$. The reaction between the two may be written as:

$$B(OH)_3 + H_2O \rightleftharpoons B(OH)_4^- + H^+ \quad (11.8)$$

and the apparent equilibrium constant for this reaction is pH dependent:

$$pK^{app} = \log \frac{B(OH)_4^-}{B(OH)_3} - pH \cong 8.6 \quad (11.9)$$

The relative abundance of these two species is thus pH-dependent. Furthermore, we can easily show that the isotopic composition of these two species must vary with pH if the isotopic composition of seawater is constant. From mass balance, we obtain:

$$\delta^{11}B_{sw} \cong \delta^{11}B_4 f + \delta^{11}B_3 (1 - f) \quad (11.10)$$

where f is the fraction of $B(OH)_3$ (and $1 - f$ is therefore the fraction of $B(OH)_4^-$), $\delta^{11}B_3$ is the isotopic composition of $B(OH)_3$, and $\delta^{11}B_4$ is the isotopic composition of $B(OH)_4^-$. If the isotopic compositions of the two species are related by a constant fractionation factor, Δ_{3-4}, we can write Equation (11.10) as:

$$\delta^{11}B_{sw} \cong \delta^{11}B_4 f + \delta^{11}B_3 - \delta^{11}B_3 f$$
$$\cong \delta^{11}B_3 + \Delta_{3-4} f \quad (11.11)$$

Solving for $\delta^{11}B_4$, we obtain:

$$\delta^{11}B_4 \cong \delta^{11}B_{sw} - \Delta_{3-4} f \quad (11.12)$$

Thus, assuming a constant fractionation factor and isotopic composition of seawater, the $\delta^{11}B$ of the two B species will depend only on f, which, as we can see in Equation (11.10), will depend on pH (the approximation here results from the use of δ and Δ values rather than ratios and α; an exact solution is given by Rae [2018]). The proof is left to the reader (Problem 11.2).

Klochko et al. (2006) experimentally determined α_{3-4} to be 1.0272 ± 0.0006

($\Delta B_{3-4}^{11} = 28.2$). There are some additional factors that must be considered: (1) there are species-specific fractionations, perhaps because they alter the pH of their microenvironment, or perhaps because $B(OH)_3$ is also incorporated to varying degrees, and (2) the equilibrium constant is temperature, pressure, and composition dependent (see review by Rae [2018]). When these factors are considered, analysis of modern planktonic and benthic foraminiferal carbonates demonstrates that they do accurately record the pH of the water in which the organism grew. The pH of seawater, in turn, is largely controlled by the carbonate equilibrium and therefore by the partial pressure of CO_2 in the atmosphere. Thus, pH of ancient seawater can be determined from $\delta^{11}B$ of carbonate sediment, and from this, it is possible to estimate p_{CO_2} of the ancient atmosphere. An additional factor is that the B isotopic composition of seawater varies with time because of variations in the isotopic composition and fluxes from the various sources. A variety of approaches have been used to constrain this.

Anagnostou et al. (2016) used an approach that takes advantage of the non-linear relationship between $\delta^{11}B$ of borate and pH, which means that for a given change in pH, a measured $\delta^{11}B$ difference will be a function of $\delta^{11}B_{sw}$. They measured $\delta^{18}O$, $\delta^{13}C$, and $\delta^{11}B$ in Eocene planktonic foraminiferal tests collected from a drill hole on the Tanzanian margin. Using $\delta^{18}O$ and $\delta^{13}C$, they could assign a depth to each test. They could then determine the relative pH gradient in the upper 300 m of water as assuming it was equal to or larger than the present one (about 0.17 pH units, which results from increased CO_2 introduced by respiration), calculate $\delta^{11}B_{sw}$ at each age. From this, they could then reconstruct Eocene seawater pH and atmospheric CO_2.

Given the concern about the relation of future climate and ocean acidification to future p_{CO_2}, it is obviously interesting to know how these factors related in the past. We will return to boron isotopes in the next chapter and see how they have been used to reconstruct atmospheric CO_2 in the Cenozoic.

11.3.3   Magnesium isotopes

$\delta^{26}Mg$ (the per mil deviation from the $^{26}Mg/^{24}Mg$ ratio in standard DSM3) in

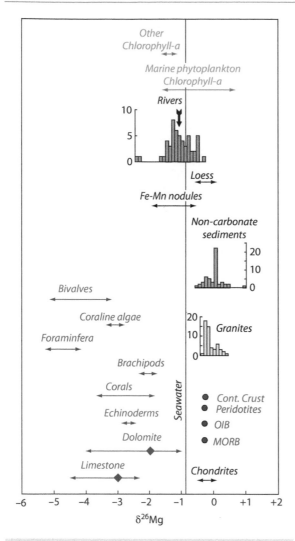

Figure 11.6 Magnesium isotopes in terrestrial materials. Flux-weighted average riverine $\delta^{26}Mg$ is shown by the arrow and is from Tipper et al. (2006).

surficial materials ranges from ≈ –5‰ in some carbonates to ≈1‰ in some clays (Figure 11.6). Magnesium is a major ion in seawater with a concentration of 52.7 mmol/kg and a residence time of ≈13 million years; unsurprisingly, it has uniform isotopic composition of $\delta^{26}Mg = -0.82 \pm 0.01$‰ (Foster et al., 2010). Rivers are the principal source of Mg to the oceans and show wide isotopic variation that reflects the $\delta^{26}Mg$ of their drainage basins: most rivers have $\delta^{26}Mg$ in the range of –1.5 to +0.5‰, but those draining carbonate terrains can have $\delta^{26}Mg$ <–2‰. However, major rivers show much less isotopic variation. Tipper et al. (2006) calculated a flux-weighted mean composition of –1.09‰ and is hence lighter than

continental crust, which has an average $\delta^{26}Mg$ of ≈–0.24‰, only slightly lighter than the mantle value. This suggests that significant Mg isotopic fractionation occurs during weathering. Teng et al. (2010) observed $\delta^{26}Mg$ values of up to +0.65‰ in saprolite developed on a diabase with $\delta^{26}Mg$ of –0.22‰. The data were consistent with Rayleigh fractionation occurring during weathering as light Mg^{2+} is progressively released to solution with heavier Mg remaining in the solid weathering products. Consistent with this, clays and other non-carbonate sediments have relatively heavy Mg, which range from about –0.5 to +0.9‰ and have an average $\delta^{26}Mg$ of 0‰.

There is considerable variation among carbonates, which range from $\delta^{26}Mg$ of –1 to –5‰. Some or much of this variation appears to be mineralogically controlled with $\delta^{26}Mg$ decreasing in the order: dolomite > aragonite > high-magnesium calcite > low-magnesium calcite (Hippler et al., 2009). Skeletal carbonates appear to show additional variation related to phylum, although within foraminifera interspecific variation appears limited (Pogge von Strandmann, 2008). For the most part, Mg isotopic variations in skeletal carbonate appear to be independent of temperature and salinity.

The principal Mg sink in the oceans is high-temperature hydrothermal activity, which is thought to remove a majority of Mg from the oceans. These systems remove Mg quantitatively from seawater – the vent fluids contain no Mg. Consequently, there can be no isotopic fractionation in this process. Low-temperature systems are less efficient at removing Mg, and hence there is potential for isotopic fractionation. Altered oceanic crust recovered from Ocean Drilling Program (ODP) Site 801 has highly variable $\delta^{26}Mg$ ranging from –2.76 to +0.21‰ (Huang et al., 2018). The range reflects the formation of both carbonate and Fe-rich silicate secondary minerals with carbonates being isotopically light, $\delta^{26}Mg$ < –2‰, and silicates being isotopically heavy, $\delta^{26}Mg$ > 0. The net effect on seawater $\delta^{26}Mg$ is likely small.

If the Mg isotopic composition of the oceans is maintained at steady state, then the isotopic composition of sinks must equal that of sources. Since hydrothermal activity removes Mg without fractionation, there must be an additional sink or sinks that preferentially remove light Mg to balance the riverine flux with δ^{26}Mg of –1.09‰. An obvious candidate is biogenic carbonate (mainly calcite) precipitation. The average δ^{26}Mg of deepwater calcareous oozes is –1.03‰ (Rose-Koga and Albarède, 2010), which is quite close to the riverine value. Since hydrothermal activity is the primary sink and occurs without fractionation, this additional sink must have δ^{26}Mg < –1.09‰. Furthermore, the Mg concentration of calcareous ooze ($\approx$0.06%) is too low for biogenic carbonate precipitation to provide that additional balance to the riverine flux. Two other potentially important sinks are dolomite precipitation and exchange reactions with clays (in which Mg^{2+} in solution replaces Ca^{2+} and other ions in clays), although the latter is likely to be minor. Unfortunately, there are as yet no empirical or theoretical fractionation factors for ion exchange reactions with clays, but both exist for dolomite precipitation. Dolomites show a fairly wide range of Mg isotopic compositions but using the mean δ^{26}Mg value of –2‰, Tipper et al. (2006) estimate that 87% of seawater Mg is removed by hydrothermal systems and 13% by dolomite precipitation.

Over the long term, Mg isotopic composition of seawater depends on (1) the riverine flux and its Mg isotopic composition, which in turn depends on the weathering rates, (2) the hydrothermal flux, which in turn depends on the rate of seafloor spreading, and (3) the rate of dolomite formation. All these three factors are likely to vary over geologic time. The question of rates of dolomite formation is interesting because judging from the amount of dolomite in the sedimentary mass, the present rate of formation appears to be slow compared to the geologic past. While the rates of seafloor spreading over the last 100 Ma can be constrained from magnetic anomaly patterns, the degree to which these rates have varied are nonetheless debated, and independent constraints would be useful. Finally, weathering rates depend on both climate (temperature and precipitation) and atmospheric CO_2 concentrations. There is intense interest in how these factors have varied. If the Mg isotopic composition of seawater through geologic time could be established, it could be potentially useful in addressing all these issues. The δ^{26}Mg of foraminiferal shells appears to be independent of temperature and water

chemistry and thus might provide a record of δ^{26}Mg through time. However, it must first be established that Mg isotopic compositions of foraminifera (or any other potential recorder of seawater δ^{26}Mg) do not change through diagenesis. The question of interspecies differences would also have to be addressed. There are a host of other issues as well. For example, what controls Mg isotopic fractionation during weathering, what controls isotopic fractionation during dolomite formation, etc.?

11.3.4   Calcium isotopes

Figure 11.7 illustrates the distribution of calcium isotope ratios in terrestrial materials (recall that δ^{44}Ca is per mil deviation from the ^{44}Ca/^{40}Ca ratio in standard NIST-915a). As was the case with B and Li, seawater represents the heavy extreme of terrestrial range with δ^{44}Ca of +1.89 ± 0.02‰, which is uniform within analytical error as expected given its residence time in seawater of one million

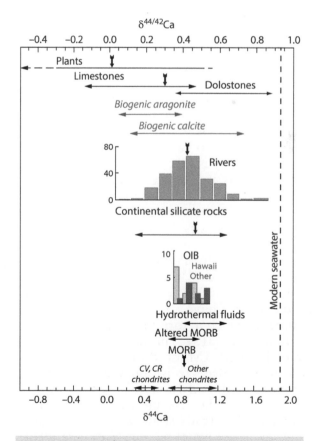

Figure 11.7   Calcium isotope ratios in terrestrial materials. Arrows indicate mean values.

year. Rivers are the primary source of Ca in the oceans, and they have a mean δ^{44}Ca of 0.88‰. As carbonate weathering is thought to contribute 75–90% of the riverine Ca flux, this presents somewhat of a conundrum because it is closer to average silicate rock ($\approx$+0.94‰) than average carbonate rock (0.60‰).

Several studies have shown that there is apparently little fractionation during mineral dissolution and weathering. In laboratory dissolution experiments, Ryu et al. (2011) found that $\delta^{42/44}$Ca of solutions produced in the experiments varied due to incongruent dissolution, but that little mass fractionation occurred. Abiotic carbonate precipitation can involve a fractionation as large as $\approx$–2‰; consequently, precipitation as secondary carbonate travertines and soil carbonates could increase the δ^{44}Ca of dissolved calcium in rivers and explain their relatively high mean δ^{40}Ca. Indeed, rivers draining carbonate terrains have higher δ^{44}Ca than those draining silicate terrains, suggesting precipitation of isotopically light carbonate may be occurring. This is potentially the case regarding some Himalayan catchments where the formation of secondary carbonates (i.e., travertines and calcretes) sequesters more than twice the mass of the dissolved Ca export flux on an annual basis (Tipper et al., 2016).

The apparent imbalance between weathering and the riverine flux suggests the possibility that the system is not at steady state and that the δ^{44}C of the oceans is changing with time. One possibility is that the biosphere is responsible for this imbalance (Fantle and Tipper, 2014). In contrast to the minimal fractionation in weathering, there is significant fractionation in uptake by plants. Plants themselves are isotopically variable, with roots and woody tissue being isotopically light compared to the nutrient sources and δ^{44}Ca in leaves. On average, the mean Δ^{44}Ca$_{\text{plant-source}}$ is about –0.7‰ and plants have an average δ^{44}Ca of $\approx$0.01‰, which in turn increases δ^{44}Ca in soil pore water to $\approx$1‰. The global biosphere contains 1.5×10^{13} M Ca. If the residence time of Ca in the biosphere is 100 years, this corresponds to a flux of 1.5×10^{13} M Ca/yr, which is similar to the riverine flux (1.38×10^{13} M Ca/yr). At steady state, net biosphere gains and losses are equal so that biosphere would have no effect, but at a non-steady state,

biosphere would affect riverine δ^{44}Ca, making it heavier or lighter depending on the magnitude of the recycling flux relative to the uptake flux. Ultimately, this will be reflected in seawater δ^{44}Ca, but only if the imbalance persists and only slowly given the long residence time of Ca in the oceans. Other non-steady-state factors that could account for the riverine imbalance include change in the size of the pedogenic carbonate reservoir. As Ca can exchange with Na and K in clays with light Ca isotopes being held in clays, a change in the size of the exchangeable soil reservoir is another possibility (Tipper et al., 2016).

High-temperature hydrothermal fluids on mid-ocean ridges are another source of Ca to the oceans and they have δ^{44}Ca ranging only from +0.95 to +0.80‰, which is up to $\approx$0.15‰ heavier than fresh MORB. However, anhydrite precipitates as these fluids exit the seafloor, which has Δ^{44}Ca$_{\text{An-fluid}}$ of –0.55‰, increasing δ^{44}Ca of the fluid by about 0.14‰. From this, Amini et al. (2008) concluded that Ca is leached from the oceanic crust without isotopic fractionation. Anhydrite precipitated in vents later dissolves at low temperature. Thus, hydrothermal activity produces a net flux of relatively light Ca (δ^{44}Ca $\approx$ +0.80‰) to the oceans and does not directly affect the Ca isotopic composition of the oceanic crust. However, significant amounts of calcite and aragonite precipitate in the oceanic crust at low temperature with an apparent Δ^{44}Ca$_{\text{min-Ca}^{2+}}$ of –1.54‰, which tends to increase in $\delta^{44/40}$Ca of seawater.

Since fluxes into the oceans are isotopically light, they must be balanced by an isotopically light flux from the oceans. This flux is biogenic carbonate precipitation, which is the principal way in which Ca is removed from the oceans. Pelagic calcite-secreting organisms, including foraminifera and autotrophs, such as the coccolithophorid *Emiliania huxleyi*, account for about 55% of the calcium flux from the ocean. Table 11.2 summarized fractionation factors for various carbonate-secreting organisms. The exact fractionation factors vary between organisms and temperature, CO_3^{2-} concentration, precipitation rate, and Mg^{2+} concentration, but there is a collective well-defined mean fractionation, Δ^{44}Ca$_{\text{min-Ca}^{2+}}$, of open ocean removal of –1.30‰ (Blättler et al., 2012). The remaining 45% is removed by

Table 11.2 Calcium isotope fractionation in biogenic carbonate.

Organism	Fractionation factors	
	$\Delta^{44/40}Ca$	$\Delta^{44/42}Ca$
Bivalves (calcite)	–1.50	–0.71
Foraminifera (calcite)	–0.94	–0.45
Coccolithophores (calcite)	–1.30	–0.62
Brachiopods (calcite)	–0.85	–0.40
Sponges (aragonite)	–1.50	–0.71
Pteropod snails (aragonite)	–1.40	–0.67
Scleractinian corals (aragonite)	–1.10	–0.52

shallow water organisms inhabiting coasts. While some of these, such as coralline algae, secrete calcite with a relatively small fractionation from seawater ($\approx$–0.9‰), most, such as reef-building corals, secrete aragonite with a larger average fractionation ($\approx$–1.6‰). The isotopic composition of the net flux out of seawater thus depends on the relative amounts removed by open ocean versus shallow water calcifiers. Planktonic calcifiers became important only in the Jurassic, so this balance could have been different in the past.

11.4   STABLE ISOTOPE RATIOS OF NUTRIENT ELEMENTS

11.4.1   Nitrogen isotopes

The oceans contain about 600 μmol/L of dissolved nitrogen, whose concentration and isotopic composition is controlled by equilibration with the atmosphere. Nitrogen is, of course, a key nutrient, but only a small fraction of total nitrogen is biologically utilized, so that N_2 is effectively conservative in the ocean with a nearly constant isotopic composition ($\delta^{15}N_{ATM} = \approx 0.5$‰). In contrast, the distribution and isotopic composition of reactive forms of nitrogen, N_r, including nitrogen oxides, ammonium, and organic nitrogen, vary widely. Of these, NO_3^- is the most abundant and averages about 30 μmol/L, followed by organic nitrogen with several micromoles per liter, and ammonium, which is typically <1 μmol/L.

The primary source of marine N_r is nitrogen fixation by cyanobacteria, principally of the genus *Trichodesmium* in the tropical and subtropical nutrient-poor surface waters. The estimated $\delta^{15}N$ of this N_r is estimated to be about –1‰. Other inputs of fixed N to the marine environment include rivers and atmospheric deposition. The riverine flux from pristine river systems has an estimated $\delta^{15}N$ of 4‰ and the pre-industrial atmospheric flux is estimated to be 0‰, but anthropogenic inputs probably raised $\delta^{15}N$ of the former while clearly decreasing the $\delta^{15}N$ of the latter (Sigman and Freipat, 2019). Jickells et al. (2017) estimate that the net terrestrial N_r flux to the oceans has quadrupled (to 39 TgN/yr from 10 TgN/yr) since 1850 due to anthropogenic activity, but the terrestrial remains substantially small than marine N fixation rate of $\approx$164 TgN/yr. While this has increased productivity and carbon export and sequestration, it also leads locally to oxygen-poor zones, such as the Gulf of Mexico and the Arabian Sea. The primary sinks are denitrification in the water column, which occurs primarily in oxygen minimum zones, and loss to the bottom followed by denitrification within the sediments. ^{14}N is preferentially denitrified, which increases $\delta^{15}N$ in the remaining dissolved NO_3^-.

The distribution of reactive nitrogen within the ocean is controlled by internal cycling. Figure 11.8 shows dissolved nitrate concentration and isotopic composition from the GEOTRACES GA03 station 24 hydrocast in the tropical North Atlantic. N_r is taken up by phytoplankton in the surface photic zone driving nitrate to near zero concentrations. Most N_r is recycled within the photic zone through ammonification and nitrification, but a fraction is carried out of the photic zone by sinking organic matter. Because isotopically light N_r is taken up, the remaining dissolved NO_3^- in the surface water is driven to high $\delta^{15}N$. Organic remains falling out of the photic zone are

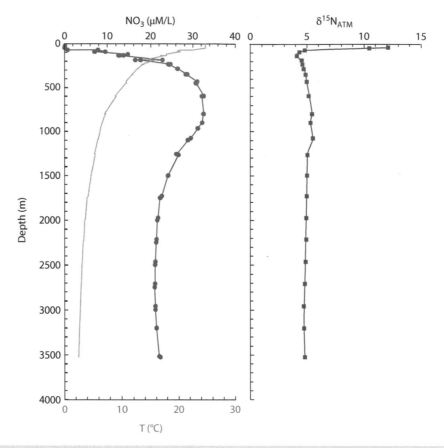

Figure 11.8 Dissolved nitrate concentration and $\delta^{15}N$ GEOTRACES GA03 station 24 hydrocast in the Central North Atlantic (21.9°W, 31.0°N). Temperature is also shown. Cruise RVKnorr K199-4, W. Jenkins and E. Boyle, chief scientists. Data from GEOTRACES Intermediate Data Project, 2017, Schlitzer et al. (2018).

remineralized within the thermocline with NO_3^- reaching a maximum concentration of ≈ 32 μM/L at depths of 600–800 m. Remineralization of organic matter produces a minimum in $\delta^{15}N$ of ≈ 4‰ just below the photic zone at ≈ 130 m depth. Below the base of the thermocline at ≈ 1000 m, both NO_3^- and $\delta^{15}N$ maintain relatively constant values of 25–30 μM/L and 5‰, respectively. Water at this depth includes North Atlantic Deep Water advected from the north, and Circumpolar Deep Water and Antarctic Bottom Water advected from the south.

11.4.2 Silicon and germanium isotopes

Figure 11.9 provides an overview of the Si isotopic composition of terrestrial and extra-terrestrial materials. The Si isotope fractionation in igneous processes discussed in Chapter 9 is quite small compared to processes

at the Earth's surface. Igneous minerals at the surface of the Earth undergo weathering reactions in which Si partitions between residual clays, quartz, and dissolved silicic acid (H_4SiO_4). Silicic acid might be more properly written as $Si\,(OH)_4$ as the Si atom strongly binds to OH groups and consequently it preferentially incorporates heavy isotopes. As a consequence, Si in seawater, as well as rivers, streams, soil water, etc., is isotopically heavier than in the solid Earth, with an estimated seawater average $\delta^{30}Si$ of 1.1 ± 3‰ (De la Rocha et al., 2000). Despite its relatively long estimated residence time of $\approx 10\,000$ years, $\delta^{30}Si$ of dissolved silica in the ocean varies from 0.5 to 4.4‰ correlating inversely with Si concentration range from <0.1 to 180 μM/Kg as a consequence of intense biological utilization and cycling within the ocean.

Significant fractionation of Si isotopes begins in the weathering regime. Méheut

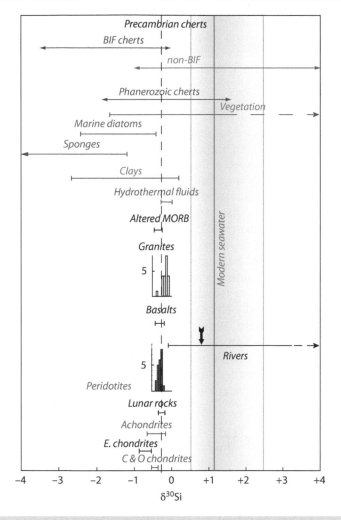

Figure 11.9 Silicon isotopic composition of terrestrial materials. The dashed line is the estimated δ^{30}Si of the bulk silicate Earth. The arrow shows estimated average composition of rivers. The blue solid line shows the estimated mean composition of seawater and the blue shaded region shows the range. "BIFs" are banded iron formations. δ^{30}Si is the per mil deviation of the ^{30}Si/^{28}Si ratio from NBS28/NIST-RM8546 standard.

et al. (2007) used a theoretical approach to calculate a Δ^{30}Si kaolinite–quartz fractionation factor of $-1.6\permil$ at 25°C. Combined with other studies, the fractionation between kaolinite and dissolved silica is inferred to be $\approx -2.5\permil$. In addition, some dissolved silica released can be absorbed on the surface of oxides and hydroxides in soils, such as ferrihydrite and goethite; experiments indicate ^{30}Si fractionation factors of Δ^{30}Si$_{\text{ferri-solu.}} \approx -1.59\permil$ and Δ^{30}Si$_{\text{geoth.-solu.}} \approx -1.06\permil$, making dissolved silica even isotopically heavier. A strong negative correlation between δ^{30}Si and the chemical index of alteration (the ratio of $Al_2O_3/(Al_2O_3 + CaO + Na_2O + K_2O)$) confirms that weathering of silicate rocks produces isotopically light

secondary minerals and an isotopically heavy solution.

Roots of terrestrial plants take up silicic acid from the soil solution with a fractionation of Δ^{30}Si$_{\text{roots-soil}} \approx -1.2\permil$ and incorporate it in various tissues, where it increases rigidity and photosynthetic efficiency, limits loss of water by evapotranspiration, and increases the resistance to pathogens and grazing by herbivores, including insects. The "phytoliths" formed in this process consist of hydrated opaline silica and vary widely in silicon isotopic composition due to Rayleigh fractionation during precipitation of the phytoliths with an apparent fractionation factor of Δ^{30}Si$_{\text{phyto-solu.}} \approx +2\permil$. Within a watershed, silica may cycle between dissolved form, incorporation in or adsorption on secondary minerals, and the living biota and dead organic matter before finally being exported, with isotopic fractionations associated with all these transitions.

Uptake by diatoms in streams, rivers, and lakes results in additional fractionations. In view of the complexity of these fractionations, it is not surprising to find that rivers have highly variable $\delta^{30}Si$ ranging from –0.17 to +4.6‰. Isotopic compositions vary not only between different rivers, but also seasonally in individual rivers depending on discharge and biological productivity. Rivers nonetheless tend to be isotopically heavier than the rock in the watershed they drain, but given this variability, the mean value is difficult to define with estimates ranging from +0.8 (De la Rocha et al., 2000) to +1.26‰ (Sutton et al., 2018a).

Rivers and submarine groundwater supply about 65% of the dissolved silica flux to the oceans, with ≈25% coming from dissolution of aeolian dust and the riverine suspended load and ≈10% from alteration of the ocean crust.

Once dissolved silica reaches the oceans, it is extensively bioutilized and can be biolimiting. Settling of biogenic silica and conversion of it to chert or authigenic clays is the main way in which silica is removed from the oceans. Diatoms, which build tests (or shells) of opaline silica and account for some >40% of primary productivity in the oceans, dominate the modern marine silica cycle. From field observations in the Southern Ocean iron fertilization experiment, Cavagna et al. (2011) estimated the fractionation between dissolved H_4SiO_4 and biogenic opaline silica indicates a near-constant $\Delta^{30}Si_{opal-diss.\ Si} \approx -1.36‰$, which is slightly greater than earlier experimental determination of –1.1 to –1.2‰. Consequently, ocean surface waters are typically depleted in Si and ^{28}Si (Figure 11.10). As they sink, the tests tend to redissolve, enriching deep waters in dissolved silica and ^{29}Si. In addition to diatoms, sponges and planktonic protists, such as silicoflagellates and radiolarians, also utilize silica. Of these, sponges appear to fractionate silica the most. De la Rocha (2003)

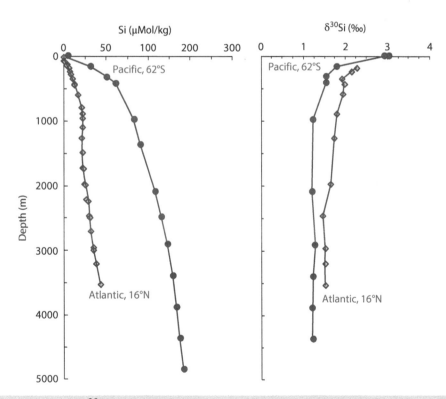

Figure 11.10 Variation of $\delta^{30}Si$ and dissolved silica concentration with depth in the Sub-Antarctic Pacific (data from de Souza et al., 2012), the Sargasso Sea, and the Central North Atlantic (data from Sutton et al., 2018b). Silica, like other nutrients, is more depleted in Atlantic deep water and has lower $\delta^{30}Si$ than other oceans as a consequence of the vertical circulation of the oceans: Atlantic deep water is younger.

found that $\delta^{30}Si$ in modern sponge spicules ranged from –1.2 to –3.7‰ and fractionation factor $\Delta^{30}Si_{sponge-sw}$, which appears to depend on growth rate and availability of dissolved Si, averaged –3.8 ± 0.8‰.

Because $\delta^{30}Si$ in the oceans is tightly coupled with biological productivity and hence the carbon cycle, there is much interest in using silicon isotopes in understanding using $\delta^{30}Si$ in siliceous biogenic sediments to reconstruct productivity and carbon export variations in the oceans.

Germanium is the chemical cousin of silicon, having a 4+ valence and being present in similar form, germanic acid H_4GeO_4, in solution and shows many similarities to Si in its behavior, including heavy Ge isotopes partitioning into solution during weathering. It nevertheless differs from Si in several respects, including being more readily complexed by organic ligands, and organically complexed Ge may be the predominant form in many natural waters. Rather than being an essential nutrient,

it is toxic to plants at high levels and autotrophs discriminate against it such that Ge/Si ratios in plants and plankton are lower than in solution. Germanium is also more readily taken up by clays during incongruent weathering of primary silicates such that the molar Ge/Si ratios in clays, typically 4–6 × 10^{-6}, are higher than in solution, typically 0.1–3 × 10^{-6}, depending on the parent rock and intensity of weathering. Ge/Si ratios in soils developed on granite are lower than those developed on basalt; in the latter case, soils have molar Ge/Si ratios as much as 10 times greater than the parent rock (Kurtz et al., 2002). Upon more intense weathering, Ge stored in the clays is released to solution at the same rate as Si. Consequently, G/Si ratios in stream water depend on the nature of the parent rock and the weathering intensity.

Figure 11.11 shows the variation of $\delta^{74}Ge$ (the per mil deviation from the $^{74}Ge/^{70}Ge$ ratio in the NIST3120a standard) in terrestrial materials. In rivers, it ranges from 0.9 to 5.5‰ and

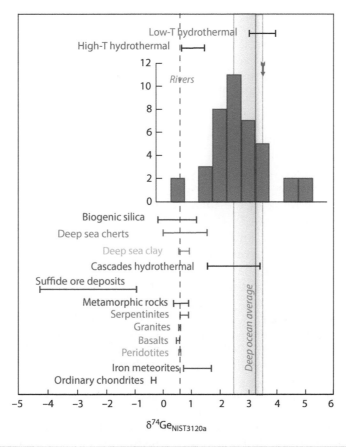

Figure 11.11   Ge isotope ratios in terrestrial materials. The dashed line is the estimated $\delta^{74}Ge$ of the bulk silicate Earth. The arrow shows estimated flux-weighted average composition of rivers.

is inversely correlated with Ge/Si ratios. Tropical soils are isotopically light, with $\delta^{74}Ge$ between –0.13 and –0.63‰ compared to parent rocks, which is consistent with uptake of light Ge isotopes in secondary minerals and partitioning of heavy Ge into solution (Baronas et al., 2018; Qi et al., 2019). The light end of the river range is a glacial river in Greenland where weathering intensity is low and is only slightly heavier than the parent rock. $\delta^{74}Ge$ of the suspended and bed loads of these rivers ranges from 0.46 to 0.7‰ and is not significantly different than igneous rocks on average, which is consistent with nearly all Ge being in particulates rather than solution. The discharge-weighted $\delta^{74}Ge$ average is 2.6 ± 0.5‰. Because a much smaller fraction of Ge is taken up by plants than Si, $\delta^{74}Ge$ may be a useful proxy for global weathering intensity.

$\delta^{74}Ge$ in seawater ranges from 2.21 to 3.48‰, the lightest being surface water from the Gulf of Mexico; water below 2000 m ranges from 3.07 to 3.48‰ and globally averages 3.24‰ (Baronas et al., 2017). High-temperature hydrothermal fluids are enriched in Ge relative to seawater by a factor of 1000–2000, reaching concentrations as high as 500 nmol/kg. Hydrothermal fluids from Loihi (161–194°C) average 0.9‰; those from Middle Valley on the Juan de Fuca Ridge (276°C) are somewhat heavier at 1.5‰ (Baronas et al., 2017). Molar Ge/Si ratios in these fluids are in the range of $30–50 \times 10^{-6}$, compared to typical basalt values of 2.5×10^{-6}; thus, Ge is being extracted by these fluids much more efficiently than Si. Ge in low-temperature fluids from the flank of the Juan de Fuca is heavier, with an average $\delta^{74}Ge$ of 3.5‰, within the range of seawater, and somewhat less elevated Ge/Si ratios but nevertheless an order of magnitude greater than basalt.

Baronas et al. (2017) estimate that the high- and low-temperature hydrothermal fluids carry 23 and 48% of the total Ge flux, respectively, to the oceans while the riverine dissolved and amorphous flux is only 20%; in comparison, the rivers carry 72% of the silica flux to the ocean and hydrothermal fluids only 10%. And while biogenic precipitation dominates silica removal, it accounts for only 30% of the removal of Ga with an average $\delta^{74}Ge$ of 3.2‰, with the remainder from authigenic mineral formation with an average $\delta^{74}Ge$ of 2.7‰.

11.4.3 Selenium isotopes

As we found in Chapter 9, selenium shares many chemical characteristics of sulfur, including chalcophile, volatile, and redox sensitive with common valences of VI, IV, 0, and –II. In the Earth's interior and in living cells, it is in the fully reduced –II selenide state and in the oxidized IV or VI states as selenite, SeO_3^{2-}, and selenate, SeO_4^{2-}, ions in systems under the oxidizing conditions of the Earth's surface. As is the case for sulfur, the oxidized forms are soluble and mobile compared to the more reduced forms of Se. Unlike the sulfur, the oxidation from selenide to selenite generally involves a transition through Se^0, which can be present in suboxic environments, and oxidation to the VI state requires a higher oxidation potential than for S(IV). It is an essential trace element for many animals while at the same time being toxic at high doses and can bioaccumulate. Both SeO_3^{2-} and SeO_4^{2-} are present in most natural waters, including seawater.

Experiments show that fractionation in adsorption of oxidized Se on Fe–Mn oxides appears to be small (≈0.2‰), although adsorption on sulfides produces greater fractionation, ≈2‰ (Mitchell et al., 2013). The largest fractionations occur during abiotic (–4.6 to –11.8‰) and biotic (–1.1 to –8.6‰) reduction of Se (VI) to Se (IV) and Se (IV) to Se (0).

At present, there is a paucity of data on Se isotopes in surface materials and Se isotopic fractionations in the surficial Se cycle remain very incompletely understood. The study by Clark and Johnson (2010) of Se isotope behavior in Sweitzer Reservoir of Colorado, US, located in a watershed draining Se-rich shale, provides some constraints on $\delta^{82/76}Se$ in the weathering environment. $\delta^{82/76}Se_{NIST3149}$ of total Se in reservoir water samples over two years averaged 3.6‰ with a standard deviation of 0.38‰ and was heavier than canal feeding it (average 2.88), which in turn was heavier than the shale (average 0.82‰), although gypsum within the shale, which likely preferentially weathers, was heavier (2.16‰). Se (IV) in the water from the canal feeding the reservoir had $\delta^{82/76}Se$ of 1.5, closer to that of the shale, while Se (VI) was heavier ($\delta^{82/76}Se$ = 3.39‰). Sediments within the reservoir were lighter on average (2.7‰) but quite variable (standard deviation 0.92‰), overlapping the water composition. Plankton sampled within the reservoir were relatively uniform and lighter on average (3.0‰) than the water.

In seawater, selenium shows a nutrient-type vertical distribution with relatively low and uniform concentrations in the surface water, increasing concentrations through the thermocline, and relatively uniform concentrations in deep water. A depth profile from the Northwest Pacific analyzed by Chang et al. (2017) had $\delta^{82/76}Se_{NIST3149}$ of 1.49 ± 0.06‰ in surface water decreasing to 0.40 ± 0.05‰ in deep water. This isotopic gradient is most likely explained by uptake of isotopically light Se by phytoplankton in the photic zone and remineralization of sinking organic matter at depth and is consistent with the average $\delta^{82/76}Se_{NIST314}$ 0.42 ± 0.22‰ for modern marine plankton found by Mitchell et al. (2012). Their data also show that Se (IV) concentrations are near 0 and Se in surface water is almost exclusively comprised of Se (VI). The Se (VI) proportion decreases to ≈60% of total dissolved Se in deep water. This suggests the biota are taking up Se (IV) in surface water and that organic Se is remineralized at depth primarily to Se (IV) rather than Se (VI). The deep water $\delta^{82/76}Se_{NIST3149}$ is similar to that of USGS standard ferromanganese nodule NOD-1, which has $\delta^{82/76}Se_{NIST3149}$ = 0.37‰, suggesting minimal fractionation during adsorption of seawater Se on ferromanganese nodules and crusts. Deep water comprises the vast bulk of seawater, so the seawater $\delta^{82/76}Se_{NIST3149}$ is probably close to 0.4‰, significantly heavier than the silicate Earth with $\delta^{82/76}Se_{NIST3149}$ ≈ −0.1 ± 0.1‰.

Stüeken et al. (2015) found that Quaternary sediments deposited from the oxic open ocean have a slightly but statistically significantly lighter isotopic composition ($\delta^{82/78}Se_{avg}$ = −0.25 ± 0.18‰) than sediments formed under anoxic conditions in restricted basins (+0.18 ± 0.61‰). Anoxic or euxinic sediments may thus better record seawater isotopic composition, perhaps because reduction of Se is nearly complete and hence there is little isotopic fractionation. Partial reduction of Se (IV) and Se (VI) within sediment may result in retention of isotopically light reduced Se with oxidized forms diffusing back into seawater. The average $\delta^{82/78}Se$ of all Phanerozoic sediments is −0.49 ± 1.5‰. $\delta^{82/78}Se$ in Precambrian sediments is higher, suggesting that Se isotope

ratios are a useful paleoredox proxy and we will consider that further in Chapter 12.

11.5 STABLE ISOTOPE RATIOS OF TRANSITION METALS

11.5.1 Chromium isotopes

As we found in Chapter 9, chromium is a redox-sensitive metal; it is most commonly in the Cr (III) state in the mantle and igneous rocks. As is common among metals in the 3+ valence state, it is relatively insoluble and readily removed by absorption in this form, but it can be oxidized to Cr (VI) at the Earth's surface, forming the highly soluble oxyanion CrO_4^{2-}, which is the predominant form in solution. While Cr may play an essential role in glucose metabolism, hexavalent chromium is carcinogenic and, because the latter has a wide range of industrial uses, such as electroplating and leather tanning, its distribution has raised considerable environmental concerns. Remediation can be accomplished through reduction to C(III). As we might expect, redox reactions generate isotopic fractionations with light isotopes preferentially reduced and redox-driven fractionation leads to a large range of $\delta^{53}Cr$ in surficial materials compared to the mantle. Consequently, there is interest in using Cr isotopes in identifying pollution sources and assessing progress in reductive remediation (e.g., Novak et al., 2018), as well as paleoenvironmental studies (e.g., Wang et al., 2016).

Figure 11.12 summarized the variation in $\delta^{53}Cr$ (the per mil deviation of the $^{53}Cr/^{52}Cr$ ratio from the NIST979 standard) in terrestrial materials. An ab initio calculation by Schauble et al. (2004) yielded a Cr (III)–Cr (VI) equilibrium fractionation factor of −6 to −7‰. Experimentally determined $\Delta^{53}Cr_{reduction}$ values for reduction of Cr (VI) to Cr (III) vary and are smaller; abiotic fractionation factors fall in the range of −1.5 to −7‰ while most bacterial reduction experiments have produced $\Delta^{53}Cr_{reduction}$ varying from −0.4 to −7.6‰ (Wei et al., 2020). This suggests the importance of kinetics. Reduction of dissolved Cr^{6+} involves breaking Cr—OH bonds, which occurs faster for light isotopes and because of its low solubility and high particle reactivity, Cr (III) will be removed from solution, which inhibits

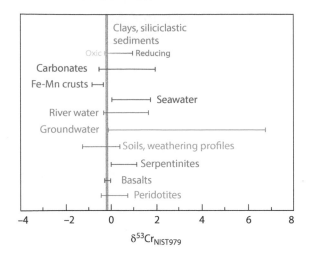

Figure 11.12 δ^{53}Cr in terrestrial materials. Gray band is the estimated composition of the silicate Earth. Compiled from the literature, principally adapted from Qin and Wang (2017) and Wei et al. (2020).

Cr (III)-Cr (VI) isotopic equilibration. This in turn leads to Rayleigh fractionation in which in which groundwater δ^{53}Cr increases as dissolved Cr^{6+} is reduced. In contrast, oxidation experiments have produced Δ^{53}Cr values generally <1‰.

Chromium is also a highly compatible element whose concentrations decrease from ≈2500 ppm in the peridotites through ≈300 ppm in basalts to ≈10 ppm in granitic rocks. Consequently, Cr in the hydrosphere derives mainly from mafic and ultramafic terranes and δ^{53}Cr studies have focused on these areas. In one such region, the Parana Basin of Argentina, Frei et al. (2014) found that δ^{53}Cr in soil profiles developed on basalts varied from –0.4 to +0.2‰ with isotopically heavy Cr found in the organic-rich topsoil and, excepting one sample, δ^{53}Cr values are lower than the unweathered basalt (δ^{53}Cr$_{NIST979}$ = –0.13‰, essentially equal to the silicate earth value) in the lower horizons. δ^{53}Cr correlated positively with Cr/Al "enrichment ratios." Such ratios are often also termed τ, with the fractional loss of an element defined as:

$$\tau_{i,m} = \left[\frac{(C_m/C_i)_h}{(C_m/C_i)_p} \right] - 1 \qquad (11.13)$$

where C_m denotes the concentration of a *mobile* element, Cr in this case, and C_i denotes the concentration of an *immobile* element, Al in this case, but some authors prefer to use Ti or Th as the immobile element. Subscripts h and p denote the measured ratios in the weathering *horizon* and *parent* rock, respectively. τ values less than 1 thus reflect loss of the mobile element in the weathering horizon; values greater than 1 indicate accumulation. Oxidative weathering produces soluble and isotopically heavy Cr^{6+}, which is removed in solution, leaving an isotopically depleted weathering residue. Isotopically heavy Cr in the topsoil likely results from reduction of mobilized Cr (VI) to Cr (III) with attendant isotopic fractionation or alternatively from bioabsorption, which produces a similar fractionation.

The extent of Cr loss during weathering depends on how Cr is hosted in bedrock. In soil developed on Parana basalts in Uruguay where Cr is hosted primarily in readily weathered pyroxene, soils are typically depleted in Cr. In soils developed on basalts of the Antrim Plateau of Northern Ireland where Cr is hosted primarily in weathering-resistant chromite ($MgCr_2O_4$), soils are typically enriched in Cr compared to bedrock (D'Arcy et al., 2016). In both areas, soil δ^{53}Cr is systematically lower than or equal to that of the bedrock.

Farkaš et al. (2013) found δ^{53}Cr values as high as 1.29‰ in altered peridotites and serpentinites from the Czech Republic and Germany, which correlated with structurally bound water (measured as loss on ignition) and other indicators of weathering intensity. Farkaš et al. suggested the heavy δ^{53}Cr values result from isotopically heavy dissolved C(VI) in oxidized meteoric/serpentinizing fluids being reduced by Fe^{2+}-bearing minerals and incorporated into serpentine.

Groundwaters can have particularly heavy Cr, with δ^{53}Cr ranging to greater than 6‰. Some of these heavy values reflect anthropogenic pollution, but heavy values also occur in uncontaminated groundwater. This reflects Rayleigh fractionation associated with reduction of dissolved Cr (VI) and loss of the isotopically light Cr (III) from solution, driving the remaining Cr (VI) to ever higher δ^{53}Cr.

River water, not surprisingly, is also generally isotopically heavy, also reflecting redox-driving fractionations. Dissolved organic carbon (DOC) can be an important influence on chromium isotopic composition. Water in the Glenariff River of Northern Ireland has

δ^{53}Cr = –0.17–1.68, a range that encompasses that of all other river water analyzed to date, which reflects the variable redox conditions of the catchment. Values with the lowest δ^{53}Cr have higher Cr and higher DOC concentrations derived from stream waters' draining bogs (D'Arcy et al., 2016). DOC likely plays the dual role of reducing oxygen fugacity and forming soluble complexes with Cr (III).

Rivers are the main source of Cr in the oceans; the average δ^{53}Cr of all sampled rivers is –0.43‰. However, δ^{53}Cr data on river water remain limited; the only major rivers for which data exist are the Parana and Brahmani, with δ^{53}Cr of 0.38 and 1.02‰, respectively. Thus, caution should be used in adopting this average as the isotopic composition of the total riverine flux.

In the oceans, dissolved Cr concentrations range from 2 to 4 nmol/kg while δ^{53}Cr ranges from ≈0 to 1.7‰ with an apparent average of 1.16 ± 0.27‰. Ninety percent or more of dissolved Cr is in the hexavalent form, with the fraction of Cr (III) highest in surface waters. Chromium shows a nutrient-type distribution pattern (Figure 11.13), albeit a weak one, in areas of high biological productivity with lowest concentration in surface waters, as is true for many other non-biologically utilized trace metals (e.g., Ge). Concentration gradients are nearly absent in low-productivity regions. Cr (III) appears to correlate with net primary production indicating biological control of Cr redox state (Janssen et al., 2020).

Laboratory studies indicate Cr depletion in surface waters primarily reflects absorption of Cr (III) onto phytoplankton surfaces and secondarily biological uptake; removal of this isotopically light Cr (III) increases δ^{53}Cr in the surface waters. In addition, abiotic redox cycling may also play a role, including photo-oxidation in surface waters and reduction through reaction with dissolved organic matter and/or ferrous iron and oxidation through reaction with Mn oxides. Phytoplankton remains then sink and are remineralized in deep water producing a decrease of δ^{53}Cr. δ^{53}Cr correlates globally with the inverse log of Cr concentration (Figure 11.13c), indicative of Rayleigh fractionation (Scheiderich et al., 2015). An effective fractionation factor for removal of –0.82 ± 0.05 can be calculated from the global trend (Janssen et al., 2020). δ^{53}Cr also correlates at least locally with the log

of Cr (III) concentration, demonstrating the importance of Cr (III) in the marine cycle (Janssen et al., 2020). Janssen et al. found that the export of Cr from the oceans by the biologic pump is 0.2–1 × 10^8 moles/yr, which roughly balances the riverine flux estimated to be 0.5–5.6 × 10^8 moles/yr. Their estimated Δ^{53}Cr for biological removal is –1.08 ± 0.25‰. Consistent with this, Wei et al. (2020) estimated Δ^{53}Cr for removal in oxic and hypoxic sediments as –0.83‰, with only a small fraction, ≈4% removed in anoxic sediments, apparently with little fractionation. Janssen et al. point out that δ^{53}Cr and Cr concentrations are uncorrelated with dissolved oxygen and fractionation and removal in oxygen minimum zones appears negligible. They conclude that δ^{53}Cr is likely to be a better proxy for biological productivity than for ocean oxygenation.

11.5.2 Iron isotopes

Iron is, of course, a redox-sensitive element; most, ≈85–90% of Fe in the mantle and mafic igneous rocks is ferrous while ferric iron is the stable state in systems in contact with the Earth's atmosphere. The surface of the Earth is thus a redox boundary. In addition, biological activity creates a variety of opportunities for redox reactions at the Earth's surface. As we have found in previous chapters, redox reactions typically involve relatively large isotopic fractionations with heavy isotopes partitioning preferentially into the phase with the higher valence. Whereas Fe isotopic fractionations in the mantle are small, typically <0.2‰, even non-redox reactions can produce Fe isotopic fractionations larger than 1‰ at the temperatures of the Earth's surface, and redox-related fractionations can be larger.

Figure 11.14 illustrates examples of fractionation factors calculated from β-factors. Temperature coefficients for calculating these β-factors are listed in Table 11.3. The equilibrium fractionation between aqueous Fe^{2+} and Fe^{3+} is approximately 3‰ at 20°C. However, ions in solution are invariably coordinated by water molecules forming the solvation shells, including some from which a hydrogen has been removed to produce OH^-. As Figure 11.14 shows, when oxidation of Fe (II) to Fe (III) involves a change in coordination, from $Fe^{2+}(H_2O)_6^{2+}$ to $Fe^{3+}(H_2O)_5OH)^{2+}$ in

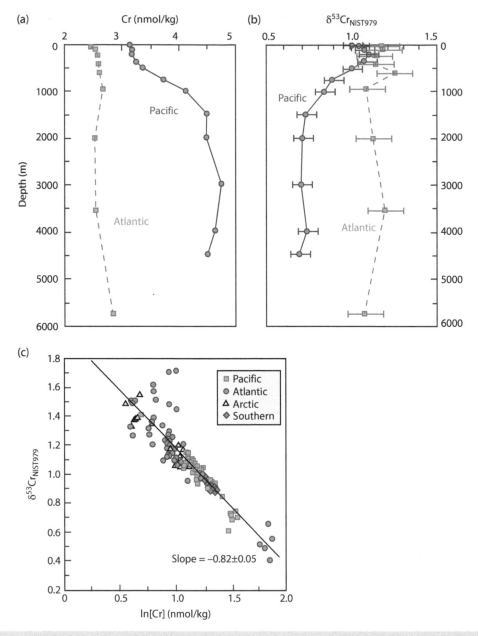

Figure 11.13 (a) Depth profiles of dissolved Cr in the eastern sub-tropical North Atlantic (data of Goring-Harford et al., 2018) and the Northeast Pacific (data of Moos and Boyle, 2019). (b) δ^{53}Cr for those same depth profiles. (c) Relationship between the log of dissolved Cr in seawater and δ^{53}Cr. Source: Adapted from Horner et al. (2021) and Janssen et al. (2020).

this example, the fractionation factor rises to ≈4.5. Significant fractionations also occur even in the absence of redox reactions. For example, there is an ≈1.3‰ fractionation between citrate complexed Fe^{3+} and $Fe^{2+}(H_2O)_6)^{2+}$ to $Fe^{3+}(H_2O)_5OH)^{2+}$, and similar fractionations of similar magnitude occur between olivine and pyrite and olivine and siderite. We should note, however, that some of the phase shown

in the figure would not coexist at equilibrium and weathering reactions between them will proceed through dissolved intermediates.

Figure 11.15 summarized δ^{56}Fe (the per mil deviation of ^{56}Fe/^{54}Fe from the IRMM-14 standard) variations in terrestrial materials. Weathering can lead to a variety of Fe isotope fractionations depending on the oxidation state and on the composition of the weathering

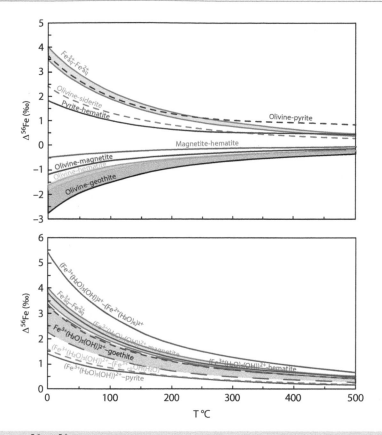

Figure 11.14 Example $^{56}Fe/^{54}Fe$ fractionation factors, $\Delta_{A-B} = 1000(\ln \beta_B - \ln \beta_A)$, as a function of temperature between minerals and dissolved Fe components common at the surface of the Earth calculated from the published β-factor temperature coefficients listed in Dauphas et al. (2017). Where two or more values were listed, a field is shown encompassing the range. Non-redox fractionations are shown as dashed lines. Fractionations between Fe-bearing silicates are generally small (Figure 9.26), so that, e.g., the pyroxene–hematite fractionation will be similar to the olivine–pyroxene fractionation.

Table 11.3 Temperature coefficients for β-factors from $1000 \ln \beta = \frac{A_1}{T^2} + \frac{A_2}{T^4} + \frac{A_3}{T^6}$.

	A_1 (×10⁶)	A_2 (×10⁹)	A_3 (×10¹²)
Olivine	0.56	−2.59	30.4
Magnetite	0.65	−3.2094	2.762
Hematite	0.70	−3.2728	7.6
Goethite	0.77	−5.301	93.18
Goethite	0.68	−3.9026	45.41467
Siderite	0.38	−1.5765	24.56
Hematite	0.68	−3.4926	51.0913
$[FeIII(H_2O)_5(OH)]^{2+}$	0.93	−11.6355	180.31175
$[FeII(H_2O)_6]^{2+}$	0.53	−2.1466	24.61976
$FeIII_{cit}OH(H_2O)^{2-}$	0.82	−5.056	89.02412
$FeIIICl(H_2O)_4]^{+}$	0.73	−4.0814	64.54233
$[FeII(H2O)_6]^{2+}$	0.49	−2.6344	23.86742
Pyrite	0.83	−3.2161	17.88867
Chalcopyrite	0.49	−1.7631	0.9529
Fe^{3+}_{aq}	0.77	−4.51	75.26
Fe^{2+}_{aq}	0.47	−1.69	17.32

From the summary of Dauphas et al. (2017). Values are either based on ab initio calculation or nuclear resonant inelastic X-ray scattering.

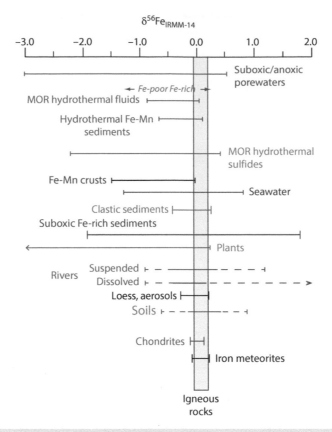

Figure 11.15   Iron isotope ratios in terrestrial materials.

solution, including both pH and the presence of other ion-dissolved organic compounds. Under oxidizing conditions, weathering of igneous and high-grade metamorphic rocks oxidizes Fe(II) to Fe(III) with attendant fractionation, but as Fe(III) produced is generally insoluble, it is likely to remain bound in solid phases and consequently there will be no change in the net isotopic composition of the rock. In the presence of organic ligands, such as oxalic acid, however, Fe (III) will form soluble complexes with them and can be lost from solution, leaving weathering horizons isotopically light. Under less oxidizing conditions, the solution can be enriched in light isotopes reflecting preferential dissolution of Fe (II)-bearing silicates, leaving the weathering horizons isotopically heavy. This contrast is illustrated by a soil profile from South Carolina studied by Liu et al. (2014) where conditions are reducing at depth. Fe^{2+} was preferentially removed relative to Fe^{3+}. $\tau_{Th,Fe}$ (Equation [11.13], the fractional loss of Fe) correlates negatively with $\delta^{56}Fe$, indicating

preferential loss of light Fe isotopes with heavy Fe isotopes preferentially left in the weathered residues.

Kinetic fractionation, which favors enrichment of light isotopes in reaction products, can oppose equilibrium ones. For example, in weathering of goethite (FeOOH), Wiederhold et al. (2006) observed an enrichment of light isotopes in the early dissolved fractions with $\Delta^{57}Fe = -2.6‰$ while later dissolved fractions in experiments in the presence of oxalic acid produced an enrichment of heavy isotopes of $\approx 0.5‰$ in solution, presumably reflecting equilibrium fractionation between Fe (III)–oxalate complexes and the goethite surface.

The role of organic acids to complex, transport, and in some cases reduction of Fe is particularly important. In a podzol[2] developed on Pleistocene sands from Northwest Germany, Wiederhold et al. (2007) found that $\delta^{56}Fe$ in different horizons ranged from −0.5 to +0.7‰ as a consequence of fractionations during weathering and translocation within

---

[2] Podzols are soils rich in organic acids and stratified by translocations within them.

the soil. In contrast, an inceptisol[3] developed on Miocene basalt from Southwest Germany was isotopically homogeneous within analytical error with $\delta^{56}Fe \approx 0.18‰$, which is only slightly heavier than the presumed $\delta^{56}Fe$ of the parent basalt ($\approx 0.1‰$). This soil is typical of ones Johnson et al. (2020) refer to as "low mobility soils," which on average exhibit slight and nearly uniform heavy isotope enrichment independent of the extent of loss or gain. The podzol is representative of "high mobility soils," which have more variable and often higher $\delta^{56}Fe$ (up to 0.9‰), which tends to increase with increasing loss of Fe.

Iron is essential for all life; it plays a vital role in oxygen and carbon dioxide transport, photosynthesis, and many enzymatic reactions. The largest fractionations evident are often microbial mediated redox reactions (Johnson et al., 2008). Two biological processes important in reducing ferric iron in anoxic environments are dissimilatory iron reduction (DIR) and bacterial sulfate reduction (BSR). In DIR, iron serves as an electron acceptor in the metabolic oxidation of organic matter to CO_2:

$$4Fe(OH)_3 + CH_2O + 8H^+ \\ \rightarrow 4Fe^{2+} + CO_2 + 11H_2O \quad (11.14)$$

In BSR, sulfur is the electron receptor in organic carbon oxidation:

$$SO_4^{2-} + 2CH_2O + 8H^+ \rightarrow 2HCO_3^- + H_2S \\ (11.15)$$

Sulfur then reacts with ferrous iron and is precipitated as iron sulfide (e.g., pyrite). Both of these reactions occur only in the absence of free oxygen. DIR produces a large decrease in $\delta^{56}Fe$ with fractionations of up to $\approx -2.6‰$. Cyanobacteria can oxidize ferrous iron and fractionations of up to +1.5‰ have been observed in this process.

Plants typically have lower $\delta^{56}Fe$ than the soil solution in which they grow, but the extent of fractionation varies depending on Fe speciation and concentration in the soil solution and the plant uptake strategy. Oats, wheat, and alpine plants analyzed to date have roots isotopically lighter than the soil solution, indicative of Fe reduction before or during uptake, which is known as Strategy I. Rice roots have slightly heavier Fe than the soil solution, indicative of Strategy II in which uptake predominantly occurred via Fe(III)-phytosiderophore[4] chelation. In both Strategy I and II plants, fractionation during transport results in stems and leaves being isotopically light relative to roots. The overall range in $\delta^{56}Fe$ in plants is $\approx 4.5‰$.

$\delta^{56}Fe$ in river-water-suspended matter varies from −0.8 to +1.2‰, with extreme values largely confined to high-latitude rivers; most rivers fall in a more restricted range of −0.4 to +0.4‰. Where coexisting data are available, dissolved Fe is usually, but not always, heavier than Fe in suspended matter (Johnson et al., 2020). Most dissolved Fe is organically complexed ferric iron or present as colloids and nanoparticles and $\delta^{56}Fe$ values depend somewhat how "dissolved" is defined. Using filtration at 0.2 or 0.4 μm, the analyzed dissolved component varies between −0.41 and 0.09‰ (Johnson et al., 2020). Fe in rivers with higher DOC contents is isotopically heavier than that in rivers with lower DOC contents. Organically complexation will increase Fe concentration and there appears to be a positive fractionation associated with organic complexation (e.g., citric acid complexation in Figure 11.14). Flocculation during mixing between river and seawater in estuaries appears to preferentially remove isotopically light Fe from the suspended pool, shifting the residual particulate flux to the oceans to heavier Fe, but appears to have minimal isotopic effect on the dissolved pool.

At present, it appears that the total dissolved riverine Fe flux to the ocean is slightly lighter than the continental crust ($\delta^{56}Fe = \approx 0.1‰$) while riverine suspended matter appears to have $\delta^{56}Fe$ close to that of average continental crust, but there are insufficient data to quantify the average fluxes precisely. The only major river analyzed to date is the Amazon, with dissolved $\delta^{56}Fe = −0.3‰$ and suspended load with $\delta^{56}Fe = −0.1‰$ (Bergquist and Boyle, 2006).

Some of the greatest interest in iron isotopes is in tracking Fe in the marine realm because Fe can be a biolimiting nutrient in as much as a quarter of the world's surface oceans. As such, it plays a role in the carbon cycle and,

---

[3] Inceptisols tend to be young soils developed on volcanic rock and are typically unstratified and organic acid-poor.
[4] Siderophores (Greek: "iron carrier") are high-affinity iron-chelating organic compounds secreted by plants, bacteria and fungi to facilitate Fe(III) uptake.

ultimately, climate. While rivers are the main source of most dissolved salts in the ocean, this is not the case for iron, in part due to the very low solubility of ferric iron and in part due to its removal by particle adsorption and flocculation of colloids in estuaries. Instead, the main sources of dissolved iron are atmospheric dust, hydrothermal fluids, and diffusion from sediments. Within the oceans, iron has a complex chemistry and is cycled between a variety of forms: the living biota; dissolved (mainly complexed by organic ligands); colloidal, organic, and inorganic particles; and iron is adsorbed on these surfaces.

Fe isotopes can be used to help distinguish between sources of Fe to the oceans and its cycling within the oceans. Mineral dust has $\delta^{56}Fe$ value close to the crustal one of $\approx +0.1‰$, but in regions of the ocean where the dust flux is the greatest, such as from the Sahara in the Central Atlantic, dissolved iron in the surface water is isotopically heavy. This appears to reflect dissolved Fe forming stronger bonds (Fe—O—C bonds) in low molecular weight organic ligands than in particulate ferri-oxy-hydroxides (Fe—O—Fe bonds) (Ilina, et al., 2013; Conway and John, 2014). In addition, the fractionation associated with phytoplankton uptake, $\Delta^{56}Fe_{phyto-diss}$, in the range of -0.13 to $-0.25‰$ would also drive surface water dissolved Fe to more positive $\delta^{56}Fe$. Primary hydrothermal fluids have $\delta^{56}Fe$ of -0.5–$0‰$, but precipitation of oxyhydroxides can drive the remaining dissolved iron in hydrothermal plumes to lower values; in contrast, sulfide precipitation can drive it to higher $\delta^{56}Fe$. These plumes can drift thousands of kilometers from their source. Reduced Fe released to the water column from low-oxygen pore waters in benthic sediments is characterized by very low $\delta^{56}Fe$ of -1.8 to $-3.5‰$. On the other hand, non-reductive dissolution of sediment releases of Fe from sediment has slightly positive $\delta^{56}Fe$ of 0 to $+0.2‰$.

The impact of these various sources is illustrated in Figure 11.16. A depth profile in the Northeast Pacific shows enhanced Fe concentrations and low $\delta^{56}Fe$ corresponding to the oxygen minimum zone. Conway and John (2015) interpreted this as Fe advected from the California margin where previous work had demonstrated a flux of isotopically light iron from reducing sediments. The depth profile from the Central Atlantic reveals the Fe-rich, isotopically light nature of the hydrothermal plume from the TAG hydrothermal field on the Mid-Atlantic Ridge. The inset shows the enrichment of the surface water in isotopically light Fe derived from Saharan dust.

Because of the large fractionation of Fe isotopes associated with redox reactions, they can help constrain the redox history of the Earth's surface. Interestingly, $\delta^{56}Fe$ in sedimentary rocks, including banded iron formations, was particularly variable around the time of the Great Oxidation Event in the early Proterozoic. In addition to their importance in understanding the oxygenation of the atmosphere, banded iron formations are the most important source of iron ore. We will discuss them and use Fe isotopes in sediments to help reveal the Earth's redox history in the next chapter.

11.5.3 Copper isotopes

Copper is another redox-sensitive element and fractionation is associated with redox reactions with the heavy isotope, ^{65}Cu, preferentially incorporated in the phase with the higher valence state. Like iron, copper is an essential nutrient for life, but can be toxic at elevated levels, particularly for phytoplankton (indeed, Cu is a commonly used algaecide), and biological utilization leads to additional fractionations. Figure 11.17 shows calculated fractionation factors between cuprous oxide, the reduced oxide, and various other forms, including tenorite (CuO), chalcopyrite ($CuFeS_2$), and various dissolved species of Cu. As we expect, the largest fractionations are associated with redox reactions, with the $\Delta^{65}Cu$ for oxidation of Cu_2O to CuO being $\approx 4‰$ at 25°C. The fractionation between dissolved Cu^{2+} and Cu^+ is approximately half that. However, significant fractionations of $>1‰$ occur in non-redox reactions, including between dissolved forms. Organically complexed dissolved Cu, e.g., citrate and oxalate in Figure 11.17, is typically somewhat heavier than Cu aquo-complex[5] ($Cu(H_2O)_5^{2+}$). Experiments have demonstrated significant

---

[5] An *aquo*-complex is one in which the ion is bound to water molecules, typically four to six of them depending on the ion's size. These water molecules are the inner *solvation shell*. While all ions in solution have solvation shells and form aquo-complexes, their existence is ignored in many chemical and thermodynamic treatments. Thus, dissolved Cu^{2+} is effectively synonymous with $Cu(H_2O)_5^{2+}$.

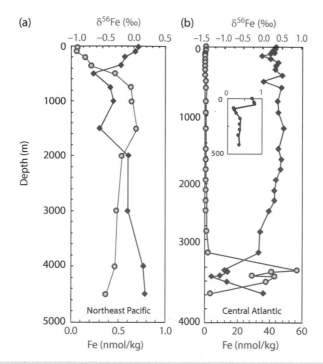

Figure 11.16 δ^{56}Fe (red) and Fe concentration profiles in the ocean. (a) Northeast Pacific Ocean (30°N, 140°W) shows a mid-depth increase in Fe and a decrease in δ^{56}Fe reflecting advection of Fe from reducing sediments on the California margin. (b) Central Atlantic (26°N, 45°W) shows a large increase below 3000 m in Fe and decrease in δ^{56}Fe resulting from hydrothermal input from the Mid-Atlantic Ridge. The inset shows the enrichment in iron in the surface water reflecting Saharan dust. Data from Conway and John (2014, 2015).

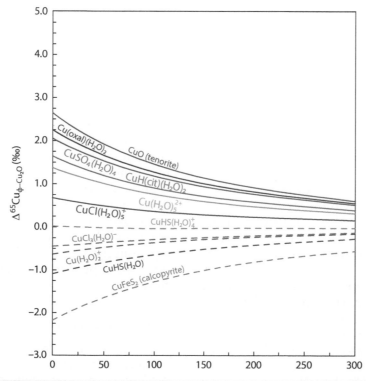

Figure 11.17 Δ^{65}Cu fractionation factors between cuprous oxide (Cu$_2$O) and other phases and aqueous species. The fractionation factor between any two species shown is the difference between their fractionation factors with Cu$_2$O. Calculated from reduced partition coefficient ratio temperature dependences adapted from Moynier et al. (2017).

fractionation during absorption of Cu onto mineral surfaces with $\Delta_{\text{sorbed-solution}}$ ranging from ≈+0.7‰ for ferrihydrite ($Fe_3O_2 \cdot 0.5H_2O$) to ≈1‰ for gibbsite (Al(OH_3) (Moynier et al., 2017). These fractionations and the lower temperatures lead to significantly greater variation in $\delta^{65}Cu$ at the Earth's surface than in its interior (Figure 9.52). Among solids, calculated $^{65}Cu/^{63}Cu$ reduced partition function ratios decrease in the order of silicates > carbonates > sulfates > oxide > native Cu > sulfides (Liu et al., 2021).

The cuprous form, Cu(I), predominates in the Earth's interior, as well as in sulfides and high-temperature hydrothermal fluids, and it can also occur organically complexed in lower-temperature solutions. The cupric, Cu (II), form predominantly at the Earth's surface. Weathering of primary minerals commonly involves oxidation from Cu(I) to Cu (II) and results in the release of isotopically heavier Cu. Soils formed on the organic-rich Marcellus Formation in Pennsylvania are depleted in ^{65}Cu and the soil solution is correspondingly isotopically heavy (Mathur et al., 2012). Cu mobility and isotopic fractionation depend on redox conditions and the presence of complexing organic ligands. In an extremely weathered soil profile developed on basalt on Hainan Island, Liu et al. (2014) found that $\delta^{65}Cu$ correlates negatively with $\tau_{\text{Th,Cu}}$, indicating preferential loss of isotopically heavy Cu, and a negative correlation between $\delta^{65}Cu$ and total organic carbon implicates organic ligands in transporting Cu.

Plants preferentially take up isotopically light Cu, with the fractionation for Strategy I plants (those that reduce Fe and presumably Cu before or during uptake) ranging from −1 to −0.5‰ and that for Strategy II plants (those that use complexing agents to accumulate Fe and presumably Cu without reduction) ranging from −0.5 to −0.1‰. Strategy II plants have a relatively constant $\delta^{65}Cu$ in all tissues, while in Strategy I plants, Cu becomes heavier upward in the plant.

Rivers are isotopically heavy with a discharged weighted average $\delta^{65}Cu$ of +0.68‰ relative to crustal rocks (which range from ≈0 to +0.2‰), reflecting preferential release of Cu (II) during oxidative weathering (Vance et al., 2008). Individual rivers have $\delta^{65}Cu$ ranging from +1.65‰ for the upper Chang Jiang (Yangtze) River to ≈0‰ for the Missouri River. The complimentary light Cu is found in riverine suspended matter with $\delta^{65}Cu$ varying from −0.24 to −1.02‰. Virtually all dissolved Cu in natural waters is complexed with organic ligands; the positive fractionation associated with organic complexation and the negative fractionation associated with Cu adsorption on particulates enhance the isotopic difference between dissolved and particulate Cu. Unlike Fe, these complexes appear stable against coagulation and adsorption so that there appears to be little loss of Cu in estuaries.

$\delta^{65}Cu$ in seawater ranges from about +0.2 to +1.5‰; however, there may be some issues with interlaboratory calibration and, in particular, uniformity of protocols with respect to sample storage and preparation and some early analyses of long-stored seawater samples appear systematically too heavy. Considering mainly the more recent data, Little et al. (2018) suggest that deep water in the open ocean is fairly homogeneous at around $\delta^{65}Cu \approx 0.64‰$. Surface water is typically lighter than deep water (Figure 11.18). This distribution is attributed to a combination of biological uptake and remineralization, benthic flux from sediments, and reversible particulate scavenging. Experiments have shown that bacteria preferentially uptake isotopically light Cu, but it is unclear whether this is the case with cyanobacteria and there are no data on fractionation associated with uptake by eukaryotic phytoplankton. Isotopically light Cu is associated with areas of high dust deposition, while a benthic flux from sediments may supply isotopically heavy Cu to deep and coastal waters (Baconnais et al., 2019). The limited data on hydrothermal vent fluids indicate they are slightly heavier than MORB with $\delta^{65}Cu \approx 0.3‰$. While these fluids are strongly enriched in Cu over seawater, Cu appears to be rapidly scavenged by Fe–Mn particulates precipitating from the fluids and hydrothermal activity does not appear to represent a significant flux to seawater.

11.5.4 Zinc isotopes

Although technically not a transition element because its d-electronic subshell is fully occupied, it is often treated together with them.

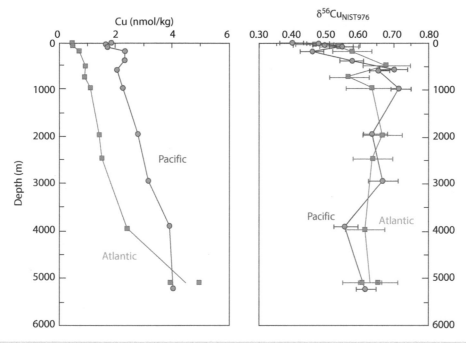

Figure 11.18 Depth profiles dissolved Cu concentrations and $\delta^{65}Cu$ in the Pacific Ocean and the Southwest Atlantic Ocean. Source: Adapted from Horner et al. (2021).

Unlike most transition elements, it is not redox sensitive, occurring in the silicate Earth strictly in the Zn (II) valence state. Like its neighbor copper, it is an essential nutrient and most Zn dissolved in natural waters is present in the form of complexes, often mainly organic, but in seawater about 50% of the Zn is complexed with inorganic ligands, such as carbonate and chloride. Zinc is an essential trace metal for all life and is an essential component of a wide variety of enzymes. It is the second most abundant transition metal, following iron in organisms ranging from phytoplankton to humans. Figure 11.19 shows fractionation factors between zinc oxide (zincite) and various dissolved species, as well as sphalerite and smithsonite. All dissolved species should be isotopically heavier than these minerals with carbonate and organic (e.g., oxalate) complexes the most enriched in heavy isotopes. Which species dominates in an environment will depend on both availability of ligands and pH. Fractionation also occurs during adsorption on mineral surfaces, with heavy isotopes preferentially adsorbed. Balistrieri et al. (2008) found that the fractionation factor

for adsorption of Zn on ferric oxyhydroxide, $\Delta^{66}Zn_{ad-soln}$, was +0.52 ± 0.04‰.

These two processes, organic complexation and adsorption along with kinetic fractionations, control Zn isotopic composition in low-temperature environments. Vance et al. (2016) found that $\delta^{66}Zn$ varied little through soil profiles developed on young (≤13 ka) alluvial terraces in Glen Feshie, Scotland and all values were generally within ≈0.1‰ of granitic parental material with $\delta^{66}Zn$ of ≈0.21‰, independent of the extent of Zn loss (which was generally small). Dissolved Zn in streams draining the area was heavier ($\delta^{66}Zn$ of ≈ 0.5‰), consistent with positive fractionation into dissolved complexes. In young soils developed on basalts in Hawaii, Vance et al. (2016) found that $\delta^{66}Zn$ was generally similar to that of the parent basalt, particularly in the drier soils, reflecting a competition between organic complexing and adsorption of Fe–Mn oxides. In soil sequences on Maui, the driest soils are slightly depleted in heavy Zn, but become enriched in them in extremely metal-depleted soils with the heaviest rainfall. Vance et al. suggest that in these reducing soils, Fe-oxides are

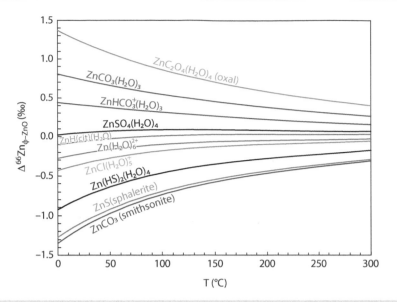

Figure 11.19 Fractionation factors between zinc oxide and variety of Zn aqueous complexes, as well as sphalerite and Smithsonite. Adapted from Ducher et al. (2018) and Moynier et al. (2017).

reduced and dissolved, releasing previously adsorbed heavy Zn. They point out that an additional complication in interpreting the Hawaiian data is the addition Zn (as well as other metals) from Asian dust deposition, which is well documented from Sr and Nd radiogenic isotope ratios (Kurtz et al., 2001). Overall, fractionation of Zn isotopes during weathering appears too small (Figure 11.20).

Zinc plays an essential role in a number of key biological processes, including carbon fixation and RNA synthesis. Fractionation during uptake by plants is species specific and concentration dependent. Roots are generally, but not always, isotopically heavier than the soil solution, more so for Strategy II plants than Strategy I ones (Moynier et al., 2017). Leaves and stems tend to be lighter than roots, more so for Strategy I plants than Strategy II ones.

$\delta^{66}Zn$ in the dissolved fraction of rivers varies from −0.51 to +0.83‰, with a discharge weighted mean $\delta^{66}Zn$ of 0.33‰ (Little et al., 2014), a value close to average continental crust. Bedrock, weathering intensity, and anthropogenic Zn no doubt all contribute variability and a study by Szynkiewicz and Borrok (2016) of the Rio Grande River in the southwestern US suggests that fractionation associated with adsorption on particulate surfaces plays an important role. At pH below 7,

$\delta^{66}Zn$ is independent of bedrock and nearly invariant at ≈0.3‰, but drops off to as low as −0.51‰ at higher pH; $\delta^{66}Zn$ also correlated inversely with dissolved Zn concentration. Absorption intensity is well known to be pH dependent because metal ions compete with H^+ for surface adsorption sites, so the most likely explanation for the drop in $\delta^{66}Zn$ is increased absorption of heavy Zn at higher pH. Petit et al. (2015) also observed ≈80% loss of dissolved Zn due to adsorption on suspended particulate matter in the Gironde Estuary of France. However, in this case, the remaining dissolved Zn becomes heavier, which they attribute to kinetically driven adsorption of light isotopes associated with strongly increasing adsorption site availability within the maximum turbidity zone. Estuaries may thus play a role in not only how much of the riverine Zn flux enters the ocean but also in its isotopic composition.

In the oceans, Zn unsurprisingly shows a typical nutrient-type distribution with uptake by phytoplankton producing strong depletions in the surface water. Zn concentrations are typically less than 0.1 nmol/kg in surface water and rise to ≈5 nmol/kg in Atlantic and ≈10 nmol/kg in Pacific deep water. Since experiments have shown that phytoplankton preferentially take up light Zn isotopes, one

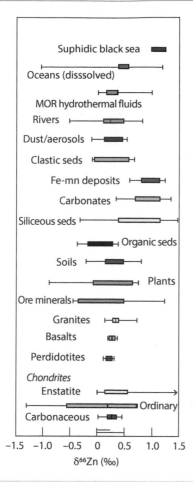

Figure 11.20 δ^{66}Zn (the per mil deviation of ^{66}Zn/^{64}Zn from the ENS-Lyon standard) in meteorites and terrestrial materials.

would expect the surface waters to be isotopically heavy compared to deep water. However, this is generally not the case; indeed, as Figure 11.21 shows, the opposite is often the case. A consensus is emerging that light isotope depletion in the shallow ocean reflects preferential adsorption of heavy isotopes on particle surfaces, particularly organic ones, and release in the deep water when the particles are remineralized (Horner et al., 2020). Input of isotopically light Zn to the surface from diffusion out of continental shelf sediments and anthropogenic aerosols might also place a role, although the latter appear to be only slightly lighter on average than seawater (Little et al., 2014).

Isotopically light dissolved Zn in hydrothermal plumes can be seen over the East Pacific Rise (EPR) and Mid-Atlantic Ridge (MAR) in the cross-sections in Figure 11.21. John et al.

(2018) estimated the δ^{66}Zn of the hydrothermal plume over the EPR as 0.24‰. In the hydrothermal fluids themselves, John et al. found that δ^{66}Zn in varies from as high 1.3‰ to as low as ≈0‰ and correlates inversely with temperature, which they suggest results from precipitation of light Zn during cooling. Most fluids fall in a much narrower range of 0.18–0.38‰ with a mean of 0.37‰. On average, fluids are thus slightly heavier than MORB and lighter than average seawater. Sulfides from vent chimneys span a range similar to fluids. Fluids are enriched in Zn by several orders of magnitude (≈50 mol/kg) and may be an important source of Zn to the oceans depending how much is removed by adsorption on particulates in hydrothermal plumes.

Most of the deep ocean has a relatively uniform δ^{66}Zn of ≈0.45‰. The principal Zn inputs to the oceans thus appear to be lighter than average seawater dissolved Zn, which implies a major sink that is isotopically heavier than seawater. Little et al. (2014) found that δ^{66}Zn in ferromanganese crusts had δ^{66}Zn varying from 0.62 to 1.42‰ with a mean of 1.12‰, which they suggested were the principal sink of isotopically heavy zinc.

δ^{66}Zn in sulfide ores varies from ≈–0.5‰ to ≈1.3‰, but most values fall in a narrow range of –0.3–0.5‰ with a mode perhaps only slightly less than the bulk silicate Earth value. There is little distinction in δ^{66}Zn between the various types of deposits discussed in Chapter 10, although relatively few deposits have been studied in detail. In those that have been studied, including the Alexandrinka VMS deposit in Russia and the Navan Irish-type deposit in Ireland, δ^{66}Zn increases from early to late deposited sulfides and from core to rim in the Alexandrinka deposit. This resulted from kinetic fractionation in which isotopically light Zn was preferentially incorporated into precipitating sulfides, leading to Rayleigh-type fractionation in which the remaining Zn in fluids became increasingly heavy over time (Mathur and Wang, 2019).

11.5.5 Molybdenum isotopes

Molybdenum is strongly chalcophile with principal oxidation states of Mo (IV) and Mo (VI). An interesting aspect of Mo chemistry and redox-sensitive metal is its low reduction potential, meaning little energy is

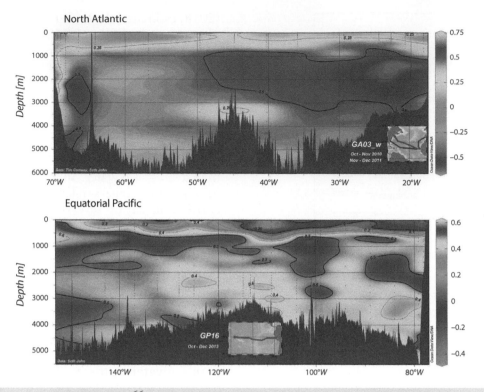

Figure 11.21   Cross-sections of δ^{66}Zn distribution in the North Atlantic and the Equatorial Eastern Pacific from GEOTRACES Expeditions GA03 and GP16 (the insets show locations). Both cross-sections illustrate the isotopically light nature of Zn in surface waters. They also reveal the presence of isotopically light Zn emanating from hydrothermal vents on the MAR and the EPR. Atlantic data are from Conway and John (2014) and Pacific data are from John et al. (2018). Cross-sections are from Schlitzer (2021) produced using Ocean Data Viewer software.

involved in transitions between oxidation states. The Mo (VI) state dominates in modern near-surface environments and it forms the highly soluble molybdate ion, MoO_4^{2+}, and consequently is a conservative ion in open, oxic seawater. Indeed, its concentration in seawater of $\approx$107 nmol/L and residence time of $\approx$500 000 years are the highest of all transition metals despite its quite low concentration in the crust. As the molybdate ion, MoO_4^{2+}, has a lower coordination number than Mo in most minerals, we expect dissolved Mo to be isotopically heavy. Figure 11.22 illustrates the range of $\delta^{98}Mo_{NIST3134}$ in terrestrial materials and shows this is indeed the case.

In euxinic environments (i.e., where HS⁻ is present), such as the deep waters of the Black Sea, molybdate can be reduced to sulfide (e.g., molybdenite (MoS_2)). Reduction is preceded by stepwise conversion of molybdate to tetrathiomolybdate, MoS_4^{2+}, through a

series of reactions in which S is progressively substituted for O:

$$MoO_4^{2+} \rightarrow MoO_3S^{2+} \rightarrow MoO_2S_2^{2+}$$
$$\rightarrow MoOS_3^{2+} \rightarrow MoS_4^{2+}$$

$$(11.16)$$

Predicted net fractionation for the net reactions is large, $\approx$−5‰, depending on temperature. However, this fractionation will only be manifest where these reactions do not go to completion. The equilibrium constants are very large so that where HS⁻ is present above about 11 mmol/L, conversion of molybdate to thiomolybdate is quantitative and no fractionation occurs. In anoxic environments lacking both free O_2 and HS⁻, these reactions stop short of producing MoS_4^{2+}, producing intermediate thiosulfides instead with intermediate fractionation. Poulson et al. (2006) found that in such anoxic environments, e.g., the Santa Monica

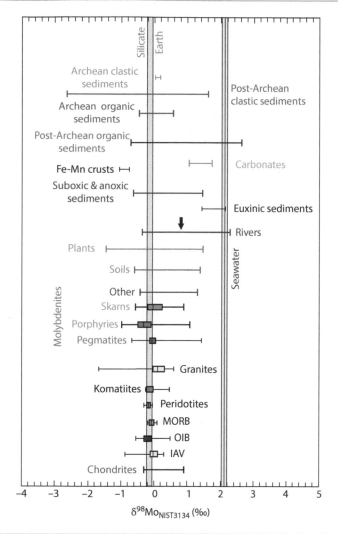

Figure 11.22 δ^{98}Mo (the per mil deviation of the ^{98}Mo/^{95}Mo ratio from the NIST3134 standard) in terrestrial materials. The arrow shows the flux-weighted riverine average. Source: Adapted from Barling et al. (2018) and Breillat et al. (2016). Data reported relative to other standards have been adjusted to δ^{98}Mo$_{NIST3134}$ = 0.

and Cariaco Basin, fractionation was consistently close to $-0.7‰$. Subsequent reduction and sequestration of Mo is incompletely understood and may occur through incorporation into iron sulfides, formation of Mo-sulfide, complexation with organic matter, reactions with oxyhydroxide and sulfide mineral surfaces, or cellular assimilation (Phillips and Xu, 2021).

Large fractionations are associated with adsorption and appear to be greater for Mn oxyhydroxides than for Fe oxyhydroxides. Thiomolybdates in Equation (11.16) are particularly prone to removal from solution by adsorption. Indeed, in the modern world, fractionation

between MoO_4^{2+} and adsorption on mineral surfaces dominates Mo isotopic variations at the Earth's surface.

Siebert et al. (2015) studied soils developed on igneous parents and found that in soils δ^{98}Mo decreased with the extent of Mo loss, implying fractionation occurs during weathering. More generally, soils can be either heavier or lighter than the parent material, depending on redox conditions, organic matter, and atmospheric inputs. In general, adsorption of Mo to Fe and Mn oxyhydroxides retains Mo in soils and preferentially scavenges light isotopes. King et al. (2018) found that the average Mo isotope fractionation between the solution and insoluble humic acid ($\Delta^{98}Mo_{solution-humic}$) was 1.39 ± 0.16‰ and that Mo absorbed on organic matter in the O-horizon of Oregon Coast Range soils had $\delta^{98}Mo$ 2‰ lighter than precipitation. Horan et al. (2020) compared $\delta^{98}Mo$ in the dissolved and solid phases of rivers and found that the dissolved component was 0.3–1.0‰ heavier, reflecting the strong bonds in MoO_4^{2+}.

Molybdenum is a key nutrient, particularly in bacteria, and is present in a number of enzymes. Most significant of these are nitrogenases, used by bacteria to "fix" nitrogen, that is, to reduce N_2 to NH_3. Soil N-fixing bacteria preferentially take up light Mo isotopes with a $\Delta^{95}Mo$ of ≈−0.45‰; cyanobacteria take up Mo with a fractionation factor in the range of 0.3 to −0.9‰. Malinovsky and Kashulin (2018) found that in three of the four plant species examined, $\delta^{98}Mo$ became progressively lighter from roots to stems to leaves by about 0.5–1‰, presumably due to fractionation during transport. In the fourth species (Rosebay willowherb), roots, stems, and leaves were identical within analytical error. However, they did not measure fractionation relative to the soil solution, so the effect of plants on Mo isotopic composition of soils remains unclear.

$\delta^{98}Mo_{NIST3134}$ in rivers ranges from −0.43 to 2.13‰ and Archer and Vance et al. (2008) calculated a flux-weighted average $\delta^{98}Mo_{NIST3134}$ discharge to the oceans of about 0.63‰. Horan et al. (2020) found that riverine dissolved Mo was 0.3–1‰ heavier than the solid load and that this correlated with the extent of Mo removal as a consequence of scavenging of light Mo isotopes on Fe–Mn (oxyhydr)oxides and/or organic matter within rivers. This scavenging, they concluded, accounts for about half the variability in rivers with the remaining half due to lithology and weathering patterns.

Open, oxic ocean water has a uniform $\delta^{98}Mo_{NIST3134} = 2.09 \pm 0.10$‰ and is thus substantially heavier than the average upper crust, which has $\delta^{98}Mo$ 0–0.15‰, and the riverine flux. McManus et al. (2002) found a $\delta^{98}Mo$ of ≈0.6‰ in low-temperature hydrothermal fluids emanating from a sedimented portion of the Juan de Fuca Ridge and estimated this represented ≈13% of the flux to the oceans. High-temperature ridge crest hydrothermal fluids are not a significant source of Mo to the oceans, as concentrations are near 0. Thus, the net flux to seawater appears to have

$\delta^{98}Mo_{NIST313}$ of ≈0.6‰, which requires an isotopically light sink to maintain the heavy isotopic composition of seawater. Siebert et al. (2003) found that Fe–Mn crusts were consistently ≈3‰ lighter than seawater in oxic environments globally, a fractionation experimentally confirmed by Barling and Anbar (2004). The reason for this large fractionation appears to be the difference in coordination environment between dissolved Mo and adsorbed Mo. Wasylenki et al. (2011) found from extended X-ray absorption fine structure analysis and density functional theory that while Mo dissolves predominantly in tetrahedral coordination as MoO_4^{2+}, it forms a polymolybdate complex on the surfaces of experimental and natural samples in distorted octahedral coordination. Since heavy isotopes partition in the phase with lowest coordination number and strongest bonds, light Mo is preferentially absorbed. The fractionation factor appears to depend on Mn and Fe content: fractionation observed for birnessite ($MnO_2 \cdot xH_2O$) is as high as −3.2‰, and the fractionation factor for Fe oxyhydroxides is −1.4 to −1.1‰.

The current Mo isotopic composition of the oceans is now maintained by a balance of removal by adsorption on ferromanganese crusts and nodules under oxic conditions with a fractionation of −3‰, removal in euxinic environments with no fractionation and removal in anoxic/suboxic environments with an average fractionation of −0.7‰. Cheng et al. (2015) estimated that 30–50% of Mo is removed by adsorption on Fe–Mn crusts and nodules in the oxic environments that cover ≈90% of the seafloor, with a fractionation of −3‰, 5–15% is removed in strongly euxinic environments, which cover 0.05–0.1% of the seafloor, with $\Delta^{98}Mo$ of 0‰, 45–60% is removed in suboxic environments, which cover the remaining 10% or so of the seafloor with $\Delta^{98}Mo$ of −0.7‰; Kendall et al. (2017) derived similar estimates.

We can see that although euxinic and suboxic regions of the ocean are presently quite small, they have an outsized influence on the Mo isotopic composition of the ocean so that small changes in their areal extent would lead to large changes in $\delta^{98}Mo$. For this reason, considerable research has focused on using the marine Mo isotope record to unravel this oxygenation history of the Earth's surface, a topic we will explore in the Chapter 12.

11.5.6 Mercury isotopes

11.5.6.1 The mercury cycle

Mercury stands out for both its volatility and its toxicity: it is the only metal that is liquid at room temperature and it along with its periodic table neighbor thallium have the lowest maximum allowable contaminant levels in drinking water allowed by the US Environmental Protection Agency of any metals. It is a highly chalcophile and redox-sensitive element and cycles continuous at the Earth's surface between (0), (I), and (II) valence states, and readily binds to organic ligands, particularly sulfur-bearing ones (thiols). With a vapor pressure of 0.26 Pa at 25°C, Hg^0 naturally evaporates into the atmosphere in gaseous form where it has an estimated atmospheric residence time of 6–12 months. Mercury also forms highly volatile organic compounds through both microbially mediated and abiotic reactions that can evaporate into the atmosphere, most notably monomethyl mercury (CH_3Hg) and dimethyl mercury (($CH_3)_2Hg$), both of which are considerably more toxic than inorganic Hg. Although concentrations in water and soil are generally low, Hg, particularly methylated forms, bioaccumulates and as a result can reach dangerous concentrations in organisms, particularly fish, at the top of the food chain.

Natural sources of atmospheric mercury include volcanoes, geothermal areas, wildfires, erosion, evasion from saline and fresh waterbodies, and soil degasification. Ninety percent or more of this mercury emitted to the atmosphere is Hg^0; much of the remainder is emitted as methyl mercury produced by sulfur- and iron-reducing bacteria in reducing environments. Hg^0 emitted to the atmosphere can then be converted to gaseous Hg (II) compounds through reactions with atmospheric oxidants, such as halogens, ozone, and OH radicals. Hg (II) is highly reactive and water soluble and is quickly and effectively removed by precipitation. Hg^0 has limited water solubility and is less effectively removed by precipitation. Both Hg^0 and Hg (II) can become associated with aerosols and particles and be removed from the atmosphere through dry deposition. Plants also uptake atmospheric Hg^0 directly through their stomata.

The global mercury cycle is now dominated by anthropogenic emissions, estimated at $\approx$2300 tons, the vast majority of which is emitted directly to the atmosphere as either Hg^0, volatile organic Hg compounds, or particulate Hg^{2+} compounds, and now pervade every part of the surficial environment. This is five times greater than primary natural emissions of $\approx$600 tons. Globally, coal burned, both in power plants and for residential heating, is the largest anthropogenic source (35%), followed by industrial emissions (21%) and artisanal gold refining (17%). While significant progress has been made in reducing emissions and the UN Minamata Convention[6] places strict limits and monitoring requirements on 126 signatory nations, the amount of anthropogenic Hg in the atmosphere has not dropped significantly. Indeed, over the last 20 years, the Hg concentration in air over Europe has remained nearly constant at around 1.5 ng/m^3 (UN Environment, 2019). This persistence reflects the way Hg repeatedly cycles through the environment, including soils plants, rivers, oceans, and atmosphere before ultimately being removed in marine sediments.

The mercury flux to the ocean includes rivers and the atmosphere, but unlike other metals, the atmospheric flux is the largest and includes precipitation, dry deposition, and gas invasion. The riverine and precipitation fluxes are primarily oxidized Hg, and dry deposition includes both oxidized and elemental, while the invasive flux consists of elemental Hg. Within the ocean, some fraction of oxidized Hg is reduced to Hg^0 so that surface waters are supersaturated in Hg^0 and hence there is an evasive flux of Hg^0 from seawater to the atmosphere that exceeds Hg^0 invasive flux. The atmospheric Hg (II) and Hg (0) fluxes to the ocean are functions of precipitation and wind speed (gas exchange increases with increasing wind speed) and hence are functions of latitude as well as proximity to sources. As a result, the Hg (II) flux is largest at low northern

---

[6] The name of this convention reflects the mid-twentieth century recognition of the health hazards of Hg following more than 1700 deaths and many more severe chronic illnesses that resulted from consumption of locally harvested seafood in Minamata, Japan where industrial Hg wastes were discharged to the Minamata Bay.

latitudes while the Hg (0) flux is greatest at high northern latitudes.

11.5.6.2   Mercury isotopic fractionations

As we found in earlier chapters, the large size of the Hg nucleus results in mass-independent fractionations (MIFs) due to the nuclear volume effect (NVE) and the magnetic isotope effect (MIE). This, along with having seven isotopes that also exhibit substantial mass-dependent fractionations, makes mercury isotope geochemistry particularly complex and fascinating. To review, the $^{202}Hg/^{198}Hg$ ratio reported as $\delta^{202}Hg$ relative to the NIST-SRM 3134 standard is used to express mass-dependent fractionation while MIF is reported in the Δ notation:

$$\Delta^{xxx}Hg = \delta^{xxx}Hg - \lambda \times \delta^{202}Hg \qquad (11.17)$$

where $\delta^{xxx}Hg$ is the per mil deviation of the $^{xxx}Hg/^{198}Hg$ ratio from the standard and λ is 0.252 for ^{199}Hg, 0.5024 for ^{200}Hg, 0.752 for ^{201}Hg, and 1.4930 for ^{204}Hg.

The largest MIF results from different reaction rates of odd- and even-numbered isotopes as a consequence of the kinetics of the MIE (Figure 11.23). Bergquist and Blum (2007) found that kinetic fractionations during photochemical reduction of Hg^{2+} to Hg^0 in aqueous solution in the presence of organic molecules result in enrichments of odd-numbered isotopes with a $\Delta^{199}Hg$–$\delta^{202}Hg$ slope of 1.15. Photochemical reduction of methyl mercury in aqueous solution results in kinetic fractionations with a $\Delta^{199}Hg/\delta^{202}Hg$ slope of 2.4 and $\Delta^{199}Hg/\Delta^{201}Hg$ of 1.36; bioaccumulated Hg in fish appears to track this fractionation. The NVE on bond strength results in both kinetic and equilibrium fractionations during evaporation with enrichment of odd-numbered isotopes, but these tend to be smaller and have $\Delta^{199}Hg/\delta^{202}Hg$ of ≈ 0.1 and $\Delta^{199}Hg/\Delta^{201}Hg \approx 2$. Arctic snow cover shows large depletions in odd-numbered Hg isotopes with $\Delta^{199}Hg/\delta^{202}Hg$ of -3.3 and $\Delta^{199}Hg/\Delta^{201}Hg \approx 1$ that result from oxidation of atmospheric Hg^0 to Hg (II)-bearing compounds followed by deposition on the snow surface and subsequent photoreduction and evaporation of Hg^0 back to the atmosphere, leaving the snow depleted in odd-numbered isotopes (Sherman et al., 2010). Finally, MIFs of even-numbered isotopes have been observed in rain and snow falling from air masses originating in the Artic with $\Delta^{200}Hg$ as high as $\approx 1.2‰$ and a $\Delta^{200}Hg/^{204}Hg$ slope of -0.5 and accompanied by odd-numbered isotope MIF as well. Because tropopause height decreases with latitude, these are thought to be generated by photochemical redox reactions near the top of the troposphere, but the cause is not yet understood

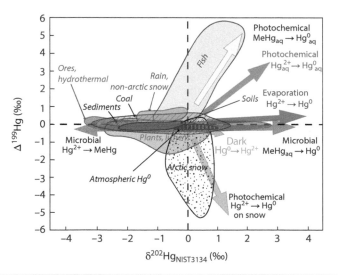

Figure 11.23   Mercury isotope fractionations and isotope compositions in terrestrial materials. $\delta^{202}Hg$ tracks mass-dependent fractionations while $\Delta^{199}Hg$ tracks mass-independent ones. Based on data in Blum and Johnson (2017) and Blum et al. (2014).

(e.g., Chen et al., 2012). As mercury cycles through atmosphere, marine and these MIF signatures propagate into terrestrial reservoirs (Bergquist, 2018).

These differing fractionations allow discrimination between differing sources and processes. The Δ^{199}Hg value can be used to estimate the proportion of a particular mercury species that has undergone photochemical reactions, and the slope of Δ^{199}Hg/Δ^{201}Hg in the reactants and products can be used to distinguish between these reactions, such as inorganic Hg and methyl mercury photoreduction in natural samples. Biological samples mainly containing methyl mercury having Δ^{199}Hg/Δ^{201}Hg slopes of 1.2–1.3 while samples containing mainly inorganic Hg such as rocks, minerals, soil, sediments, plants, and atmospheric mercury have Δ^{199}Hg/Δ^{201}Hg slopes of ≈ 1 (Kwon et al., 2020).

Figure 11.24 summarizes mercury isotopic variations in various terrestrial rocks and chondritic meteorites, which have more subdued MIFs. Igneous rocks show a substantial range in δ^{202}Hg of about 0 to –5‰, but Δ^{199}Hg, with a few exceptions, is generally less than ±0.1‰. Pre-anthropogenic sedimentary rocks have a more limited range in δ^{202}Hg of $\approx$–2.68 to 0‰, but a wider Δ^{199}Hg range of

about –0.22 to 0.3‰; coal shows a wider δ^{202}Hg range of about –4.00 to 1‰, and Δ^{199}Hg of –0.7 to 0.3‰. Ore deposits also have a wide δ^{202}Hg ranging from –4 to 1.8 ‰, and Δ^{199}Hg from –0.24 to 0.41‰.

11.5.6.3   Mercury isotopic variations in the exogene

Figure 11.25 shows mercury isotope ratios in rain, the atmosphere, seawater, and soils. Total gaseous mercury, which is >90% Hg^0, in air from regions not directly impacted by anthropogenic sources tends to have higher δ^{202}Hg and negative Δ^{199}Hg and Δ^{200}Hg (means of 0.33‰, –0.18‰, and –0.04‰, respectively) than Chinese and US urban areas heavily impacted by coal burning (e.g., Kwon et al., 2020). Mercury in coal burning undergoes a complex series of redox reactions, as well as adsorption on particulates with attendant fractionation, and is released in elemental, oxidized, and particulate forms. The isotopic composition of air in Chinese urban areas is similar to that of Chinese coal, but as Figure 11.24 shows, coals vary widely in Hg isotopic composition, so the distinction between polluted and unpolluted air is not always clear. Furthermore, because of its long

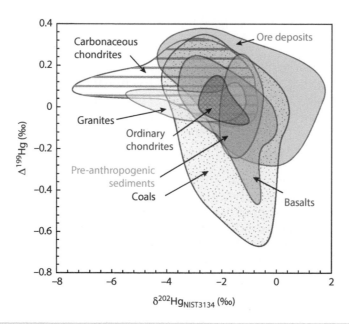

Figure 11.24   Mercury isotope compositions in rocks and chondrites. Note the change is scale compared to Figure 11.23; MIFs in these materials are generally smaller than in mercury that has cycled extensively through the atmosphere.

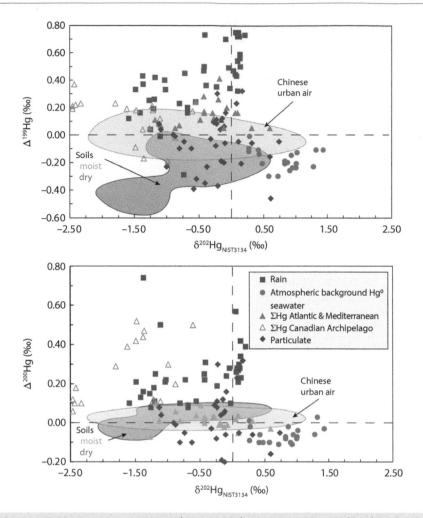

Figure 11.25   Mercury isotope ratios in rain, the atmosphere, seawater, and soils.

atmospheric residence time, mercury from coal burning influences the atmospheric composition globally. Sun et al. (2014) estimated δ^{202}Hg signatures from global coal emissions for elemental, oxidized, and particulate forms as -0.1 ± 0.5‰, -1.1 ± 0.5‰, and -1.7 ± 0.6‰, respectively.

Mercury in rain, which is predominantly Hg (II) species, displays strong MIF enrichments in both odd- and even-numbered isotopes, as well as mass-dependent depletion in heavy isotopes, with respective mean δ^{202}Hg, Δ^{199}Hg, and Δ^{200}Hg of -0.43‰, 0.4‰, and 0.16‰ (Figure 11.25). Plants take up atmospheric Hg⁰ directly through these stomata; Demers et al. (2013) found that in a forested region of Wisconsin, US, uptake by foliage resulted in large mass-dependent fractionations in δ^{202}Hg of -2.89‰ and small MIF Δ^{199}Hg fractionations of ≈-0.1‰ relative to total gaseous

atmospheric mercury, which implicates kinetic fractionation during binding of atmospheric Hg within the foliage. Forest floor δ^{202}Hg and Δ^{199}Hg were less negative than foliage due in part to natural physical mixing with underlying mineral soil. The δ^{202}Hg of soil strongly correlated with organic content, consistent with mixing between pure mineral contribution with δ^{202}Hg of -0.8 to -1‰ and organic matter with isotopic composition of forest floor litter. The isotopic composition of Hg evaded back to the atmosphere from the forest floor had more positive δ^{202}Hg than did the ambient atmospheric Hg but has similar positive odd-isotope MIF values but near-zero even isotope MIF values.

Overall, Demers et al. (2013) concluded that Hg⁰ in litterfall and dry deposition precipitation contributed most of the Hg flux from the atmosphere to the forest floor with only

16% contributed by precipitation. Zheng et al. (2016) studied Hg isotopes in 10 forested sites across the US and reached similar conclusions that Hg^0 was a significantly greater contributor to forest ecosystems than Hg (II) in rain. Observations from both these studies define a positive $\Delta^{199}Hg-\Delta^{201}Hg$ correlation with a slope of ≈ 1, implicating MIF during photochemical reduction of mercury from aqueous solutions in the presence of DOC, but do not reveal where or when this fractionation occurred. The same observation set defines a negative $\Delta^{200}Hg-\Delta^{204}Hg$ correlation with a slope of ≈ 0.5, consistent with fractionation thought to occur in the upper atmosphere.

Zheng et al. (2016) found that soil Hg concentrations and $\delta^{202}Hg$ increased with depth within the O horizon as carbon content and C/N ratios decrease. Maximum concentrations and $\delta^{202}Hg$ occur at or just below the base of the O horizon and showed little further change below it. This suggests that Hg and particularly heavy isotopes are preferentially retained as vegetation decays or that additional Hg is added directly to soil, such as direct deposition of atmospheric Hg. $\Delta^{199}Hg$ and $\Delta^{202}Hg$ generally varied little with depth, indicating a lack of MIF associated with soil processes. Much higher Hg concentrations found in O horizons compared with rock as well as isotopic composition indicate organic matter and the atmosphere are the dominant sources of soil Hg.

Interestingly, Zheng et al. (2016) found that soils in drier sites with less rainfall had lower $\delta^{202}Hg$, $\Delta^{199}Hg$, and $\Delta^{200}Hg$ than sites with greater precipitation. This cannot be explained by a greater contribution from precipitation as the isotopic differences are opposite of those if that were the case (Figure 11.25). Instead, they attributed this difference to the higher litterfall Hg input at northern wetter sites due to increased plant productivity in regions of higher rainfall.

Because of its low concentration, only few direct measurements of Hg isotope ratios in seawater have been published to date, with the available data restricted to the coastal waters of the Canadian Archipelago (Štrok et al., 2015) and open ocean waters of the Mediterranean and the Atlantic (Jiskra et al., 2021), although there is a broader array of data on marine particulates and organisms. These data reveal that Hg isotopes in seawater show considerable heterogeneity (Figure 11.26). Total $\delta^{202}Hg$ ranges from -2.85 to $0.60‰$ with a mean and mode of $-1.03‰$ and $0.1‰$, respectively. $\Delta^{199}Hg$ and $\Delta^{200}Hg$ have averages and standard deviations of $0.09 \pm 0.14‰$ and $0.11 \pm 0.16‰$, respectively. Waters from the Canadian Archipelago are shifted distinctly more negative $\delta^{202}Hg$, with a mean of $-1.78‰$ and higher $\Delta^{200}Hg$ (mean $0.22‰$) relative to open ocean waters of the Atlantic and Mediterranean (means of $\delta^{202}Hg = -0.36‰$, $\Delta^{199}Hg = 0.04‰$, and $\Delta^{200}Hg = 0.12‰$, respectively). The strongly positive $\Delta^{200}Hg$ in some of the Canadian Archipelago samples resembles the anomalies observed in high-latitude precipitation thought to be caused by photoreduction reactions high in the atmosphere and thus is likely inherited from precipitation in the region. Large riverine and erosive fluxes in these coastal waters may also contribute to their distinctive Hg isotopic composition. Štrok et al. (2015) estimated this flux at 50–80% of the input to seawater in this region.

In the open ocean, the riverine flux is less significant; Amos et al. (2014) estimate it as 27 ± 13 Mmol/yr with 27% of this reaching the open ocean, or about 30% of the atmospheric flux of ≈ 90 Mmol/yr. Data on the isotopic composition of rivers are lacking, the modern flux is likely dominated by industrial inputs from liquid mercury with $\delta^{202}Hg$ of $\approx -0.5‰$ and $\Delta^{199}Hg$ and $\Delta^{200}Hg$ of $\approx 0‰$, while the pre-anthropogenic riverine flux should match that of mineral soil with $\delta^{202}Hg$ of about $-1.4‰$. Figure 11.26 shows a depth profile of Hg concentrations and isotopic composition for GEOVIDE Site 38 in the North Atlantic reported by Jiskra et al. (2021). Total mercury concentrations are slightly depleted in surface water and increase with depth; in contrast, particulate concentrations are elevated in surface water and are low and constant in deep water. $\delta^{202}Hg$ decreases sharply from 5 to 20 m depth, but does not otherwise show significant variation with depth. $\Delta^{199}Hg$ and $\Delta^{200}Hg$ do not vary with depth outside of analytical error, which are large owing to extremely low concentrations. Depth profiles for Mediterranean sites are similar. These $\delta^{202}Hg$ values fall between those typical of rain and atmospheric gaseous Hg although they lack the strongly positive $\Delta^{199}Hg$ and $\Delta^{200}Hg$ that characterize precipitation. Both

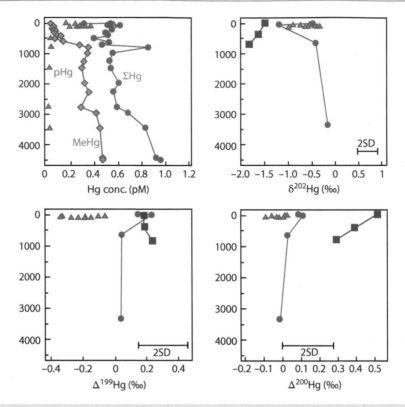

Figure 11.26 Depth profiles of mercury isotopes in North Atlantic seawater and the Canadian Arctic. Orange triangles are particulate Hg, green diamonds are methyl Hg, and blue circles are total Hg from the North Atlantic. Magenta squares are total Hg from waters of the Canadian Archipelago. North Atlantic data from Jiskra et al. (2021); Canadian Arctic data from Štrok et al. (2015).

total Hg and particulate Hg display Δ^{199}Hg/Δ^{201}Hg slopes of 0.8–1.0 indicative of photochemical MIF. While previous studies had concluded that Hg (II) dominated the atmospheric flux to the oceans, Jiskra et al. concluded from an assessment of Δ^{200}Hg mass balance that Hg (0) invasion was the major atmospheric source of both dissolved and particulate Hg in the oceans.

11.5.7 Thallium isotopes

Thallium shares some of the characteristics of mercury: it is volatile, redox sensitive, chalcophile (in low-temperature environments), highly toxic and can bioaccumulate, and has a sufficiently large nucleus that the nuclear field effect plays a role in isotopic fractionation. These characteristics are more subdued in Tl than in Hg: its vapor pressure is low enough that it volatilizes only at elevated temperatures (such as volcanic gases, coal and biomass burning, and smelters), the Tl^{1+} valence state predominates over the

oxidized Tl^{3+} state both at the Earth's surface and in its interior, and it behaves more as an incompatible lithophile element than a chalcophile one in most igneous systems, being concentrated in sulfides only at low temperatures. While highly toxic, it is not as widely dispersed in the environment as Hg. Because thallium has only two isotopes, ^{203}Tl and ^{205}Tl, and both are odd-numbered, mass-dependent and MIFs cannot be distinguished, but because the nuclear field shift effect and bond strength effects act to produce parallel ^{205}Tl/^{203}Tl fractionations, these can be quite large, up to 35 ε units (3.5‰), given the small relative mass difference, although most terrestrial materials fall in a narrower range. Figure 11.27 summarizes Tl isotope ratios in the Earth.

Nielsen et al. (2005) found that ε^{205}Tl in six Pleistocene loess samples was uniform within analytical error (±0.5 ε) with an average of –2.1 and that riverine particulate matter had an essentially identical average of –2.3. These values likely represent average upper

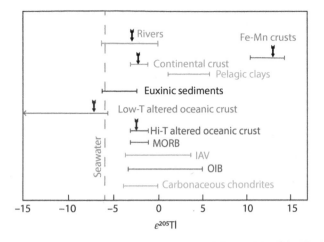

Figure 11.27   Thallium isotope ratios in terrestrial materials and the Earth. Arrows show mean values.

continental crust and are nearly identical to the MORB average of −2. $\varepsilon^{205}Tl$ in river water varied from 0 for the Nile to −6.5 for the Danube, which Nielsen et al. attribute to variable lithology: rivers containing the lightest Tl drain carbonate terranes, which contain isotopically light seawater-derived Tl. The mean $\varepsilon^{205}Tl$ for "clean" (minimally polluted) rivers was −3.0 ± 1.9 and not significantly different from the mean of −3.1 ± 1.9 for all rivers. From their study of the Amazon and Kalix estuaries, Nielsen et al. (2005) concluded that there was minimal isotopic fractionation during mixing between river and seawater and that the $\varepsilon^{205}Tl$ of the riverine flux to the ocean was −2.5 ± 1.0.

The relatively small difference between riverine particulates and dissolved Tl suggests that fractionation during weathering is small, ≈ 1 ε. There are some exceptions, however. Howarth et al. (2018) observed much larger range of −2.4–6.2 in $\varepsilon^{205}Tl$ in lateritic soils in India indicating greater fractionation in extreme weathering. Vejvodová et al. (2020) found $\varepsilon^{205}Tl$ values as high as +14 associated with pedogenic Mn-oxide enrichments in Czech soils.

Kersten et al. (2014) found that roots of green cabbage (*Brassica*[7] *oleracea*, a genus of plants is known to bioaccumulate Tl) were ≈3 ε units lighter than the soil in which they grew and leaves were up to an additional 3 ε units lighter than the roots. Rader et al.

(2019) found that Indian mustard (*Brassica juncea*) took up Tl from soil without fractionation, but this was followed by increasing discrimination against heavy Tl as the plants grew, with leaves, flowers, and seeds several ε units lighter than stems.

The isotopic composition of anthropogenic Tl can vary widely, depending on the source and process. Vaněk et al. (2016) demonstrated that Tl is fractionated in coal-burning power plants, with the volatilized fraction being as much as ≈6 ε lighter than the bottom ash. Release of this isotopically light Tl was reflected in surface soil horizons in an area of lignite mining and coal-fired power plants near the Czech–German–Poland border where topsoil was enriched in Tl with $\varepsilon^{205}Tl$ about 2 higher than deeper soil horizons. Similarly, Kersten et al. (2014) documented low $\varepsilon^{205}Tl$ in topsoils contaminated by cement kiln dust emitted from a cement plant utilized pyrite-roasting waste as a sulfur additive. Soils in the Lanmuchang area, a Tl and Hg sulfide mining district in Guizhou Province of China, are enriched in isotopically heavy Tl.

Rehkämper et al. (2002) found that marine Mn–Fe crusts have $\varepsilon^{205}Tl$ ranging from +10.4 to +14.3, implying a large isotopic fractionation ($\alpha_{crust-seawater} = 1.0021$) relative to seawater. This would be a surprising large fractionation if the mechanism were simply absorption. A clue to the fractionation mechanism is the similarity of this fractionation to

---

[7] This genus includes many widely cultivated crops, such as cauliflower, broccoli, collard greens, kale, Brussels sprouts, turnip, and rapeseed, as well as cabbage and mustard seed.

that predicted by Schauble (2007) for oxidation of Tl(I) to Tl (III) reflecting a large contribution from the NVE. Peacock et al. (2012) used X-ray adsorption spectroscopy to reveal that uptake of Tl(I) by hexagonal birnessite occurs via oxidative sorption, forming Tl (III) inner-sphere surface sorption complexes, which are then incorporated into vacant sites in the hexagonal birnessite ($MnO_2 \cdot nH_2O$) sheets. In contrast, sorption to other Mn–Fe oxyhydroxides, such as todorokite, triclinic birnessite, and ferrihydrite, does not involve oxidation and results in much smaller fractionations. Nielsen et al. (2013) found that the fractionation during sorption to hexagonal birnessite was inversely related to dissolved Tl concentrations and proposed that this reflected the presence of an additional adsorption site where Tl is adsorbed with little or no fractionation once the vacant sites associated with large fractionations are fully occupied. This is consistent with the smaller fractionation observed for Mn–Fe nodules compared to crusts as metal-rich sedimentary pore fluids contribute to the growth of the former but not the latter.

High-temperature (>300°C) ridge crest hydrothermal fluids are strongly enriched in Tl with concentrations ranging from 7 to 60 nmol/kg, corresponding to a 100-to-1000-fold enrichment over seawater with $\varepsilon^{205}Tl$ ranging from –1.2 to –3.3, averaging –1.9 (Nielsen et al., 2006), essentially identical to MORB. This implies Tl is extracted with little fractionation, which in turn reflects the high temperatures and nearly quantitative removal (≈ 90) of Tl from the affected parts of the oceanic crust.

Open ocean seawater has nearly uniform Tl concentration and isotopic composition of 64 ± 5 pM/l and $\varepsilon^{205}Tl$ of –6.0 ± 0.3 with an estimated residence time of roughly 20 000 years. Owens et al. (2017) estimate rivers constitute 23% of the flux to the oceans, subaerial volcanism with $\varepsilon^{205}Tl$ of –2 constitutes 27%, high-temperature hydrothermal fluids with $\varepsilon^{205}Tl$ of –2 constitute 17%, benthic pore water fluxes from continental margins with $\varepsilon^{205}Tl$ of 0 constitute 17%, and aerosols with $\varepsilon^{205}Tl$ of –2 are the remaining 5%. The latter two fluxes are merely inferred as there are no direct measurements of them. Because the hydrothermal flux is a major source of Tl in seawater and is extracted from the oceanic crust with essentially no fractionation, this provides a strong constraint on the total hydrothermal water flux, which Nielsen et al. (2017) estimated to be 2.5 ± 0.9 × 10^{17} kg/yr, which equates to 50–80% of the heat flux from cooling and crystallization of basalt at mid-ocean ridges.

In contrast to hydrothermal systems, seawater Tl is taken up during low-temperature alteration of oceanic crust with significant fractionation; Nielsen et al. (2006) measured $\varepsilon^{205}Tl$ as low as –15.5. This appears to be largely restricted to the volcanic layer occupying the upper few hundred meters of the oceanic. It is similar to the behavior of K and Rb, which are also lost from the oceanic crust in hydrothermal reaction but added to the oceanic crust in low-temperature reactions. Owens et al. (2017) estimate low-temperature alteration of the oceanic crust to be the principal sink (64%) of Tl in the oceans, with $\varepsilon^{205}Tl$ to be –7.2. We noted above that Tl is taken up from seawater by Mn–Fe crusts and nodules with a strong positive fractionation. Slowly accumulating pelagic clays that cover much of the deep ocean floor also absorb Tl and other transition metals from seawater (and are consequently sometimes known as "red clays"). These have $\varepsilon^{205}Tl$ up to 5, a value reflecting both intrinsic Tl (presumably with $\varepsilon^{205}Tl \approx -2$) and absorbed seawater Tl. Owens et al. estimate the absorption on pelagic clays and Mn–Fe deposits to be responsible for 32% of Tl removal from the oceans with $\varepsilon^{205}Tl \approx +16$.

In restricted basins where suboxic and euxinic conditions develop in deeper water, such as the Cariaco Basin and the Black Sea, seawater has variable $\varepsilon^{205}Tl$ ranging from between –5.6 and –2.2 in shallow oxic waters to as high as 0.4 beneath the chemocline of the Black Sea. The latter are also strongly depleted in Tl, indicating Tl is being removed by sulfides in euxinic waters. Authigenic sediments of the Black Sea have $\varepsilon^{205}Tl$ of –2.4 ± 0.4 and those of the Cariaco Basin range from –3.9 to –5.9. Owens et al. estimate the present flux from the ocean to euxinic sediments occurs without fractionation (i.e., $\varepsilon^{205}Tl = -6$) and is 4% of the total with an additional 1% extracted to suboxic sediments with $\varepsilon^{205}Tl$ of 0.

The Tl isotopic composition of the oceans is thus a function of sea floor spreading rates (which control the hydrothermal and altered oceanic crust fluxes), volcanism and weathering rates, and the relative fraction of oxic versus euxinic sedimentation. Thus, Tl isotope ratios provide an additional paleo-oceanographic tool, which we will explore in the following chapter.

11.5.8 Uranium isotopes

As we found in Chapter 10, uranium is redox-sensitive: U(VI) is the dominant form at the Earth's surface and readily forms the uranyl UO_2^{+2} oxy-cation and soluble hydroxy- and carbonate complexes whereas U(IV), which predominates in the mantle and reducing environments, is insoluble. As we expect, the largest fractionations are associated with redox reactions, but smaller fractionations occur associated with coordination changes. Uranium's large nucleus means that the nuclear field shift plays a large role in isotopic fractionation. Unlike Tl, however, the combination of an even isotope and an odd one can result in the nuclear field shift fractionation being opposite to and overwhelming that produced by electronic vibrational energies. For example, Wang et al. (2015) found that dissolved U(IV) in equilibrium with dissolved U(VI) was enriched in ^{228}U with a fractionation factor of $\Delta^{238}U_{U(IV)-U(VI)} = 1.64 \pm 0.16‰$ at 25°C, whereas mass-dependent fractionation based on vibrational energies predicts the opposite fractionation. Both laboratory experiments and observations in natural reducing marine environments indicate that abiotic U(VI) reduction by sulfide and organic matter produces only small δ^{238}U fractionations whereas microbial U(VI) reduction produces consistently heavy U(IV). In contrast, reductive absorption on Fe oxides results in ^{235}U being enriched in the product U(IV), i.e., negative fractionation factor. Small, but measurable, fractionations also occur in association with non-redox reactions, including incorporation of U(VI) in carbonates with $\Delta^{238}U_{carb-diss} \approx 0.1‰$ (Chen et al., 2017) and adsorption on Mn-oxyhydroxides with $\Delta^{238}U_{ad} \approx -0.2‰$ (Brennecka et al., 2011).

As Figure 11.28 shows, the variation in $\delta^{238}U_{CRM112a}$ in surficial materials is substantially larger than in igneous rocks. Tissot and Dauphas (2015) estimated the δ^{238}U of the continental crust to be −0.29‰, only slightly lighter than the MORB average (−0.27‰). In

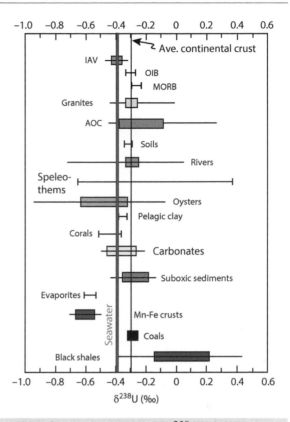

Figure 11.28 Variations in δ^{238}U (the per mil deviation of the $^{238/235}$U ratio from the CRM112a standard) in terrestrial materials. AOC is altered oceanic crust.

contrast, the range in waters, sediments, and other materials exceeds 1‰. Rivers provide >80% of the U flux to seawater and δ^{238}U in 30 rivers ranges from −0.72‰ in the Yangtze to +0.06‰ in the Watson River of Greenland, with a mean of −0.27 ± 0.16‰ and a slightly lower weighted mean flux of −0.34‰ (Andersen et al., 2016). The low value for the Yangtze reflects an abundance of evaporites and dolomites in its drainage; excluding the Yangtze reduces the mean flux to −0.26‰ – both values lie within uncertainty of the estimate of δ^{238}U in the upper continental crust. The very limited data on soils fall well within the riverine range. This implies fractionation occurring during weathering is small.

The U isotopic composition of open ocean seawater appears to be uniform as we would expect from its very long residence time (≈400 000 years), with δ^{238}U = −0.39 ± 0.01‰ (Tissot and Dauphas, 2015). Significant deviation from this value occurs in euxinic

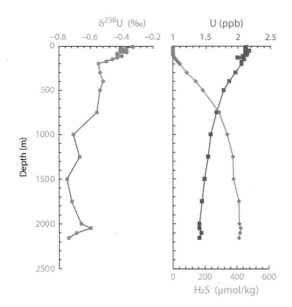

Figure 11.29 Variations in $\delta^{238}U_{CRM112a}$, U, and H_2S concentrations with depth in Black Sea waters. These variations are consistent with experimentally observed fractionations during bacterial reduction of U(VI) to U(IV). Data from Rolison et al. (2017).

waters where U(VI) is reduced to U(IV). Figure 11.29 shows the variation in $\delta^{238}U$ with depth in the Black Sea, where waters become euxinic below the top of chemocline at around 100 m depth and H_2S concentrations begin to rise and U concentrations decrease. Rolison et al. (2017) found the data fit a Rayleigh fractionation model associated with U removal by bacterial reduction of U(VI) to U(IV) with an apparent fractionation factor of 0.63‰. Interestingly, however, U(IV) is not present in the water column in either dissolved or particulate form. They concluded that U is removed at the sediment–water interface and decrease in U concentration and $\delta^{238}U$ with depth reflects upward mixing of this U-depleted bottom water. Surface sediments in the Black Sea have $\delta^{238}U = -0.01 \pm 0.1$‰, consistent with ≈43% removal of dissolved U.

Assuming that rivers are the sole source of U to the oceans with $\delta^{238}U \approx -0.3$‰ and that the oceans are in steady state with respect to U requires that $\delta^{238}U$ of the various sinks sum to –0.3‰. Andersen et al. (2017) estimated that reductive removal in anoxic basins, such as the Black Sea sediments with $\delta^{238}U \approx 0$‰, represents ≈13% of U removal from seawater; other reducing environments, such as continental shelves, represent ≈45% with $\delta^{238}U \approx 0.25$‰. ^{235}U is preferentially absorbed on Fe–Mn crusts $\Delta^{238}U_{seawater–Fe–Mn-oxide} = -0.23$‰, but this appears to be a minor sink and with little effect on the marine mass balance.

U concentrations in both low- and high-temperature hydrothermal fluids emitted at mid-ocean ridges are below the seawater value, indicating that as much as 98% of the seawater U in these fluids is removed by reaction with the oceanic crust in these systems. As we noted in Chapter 9, altered oceanic crust has highly variable $\delta^{238}U$ (Figure 11.28), with U being enriched by a factor of 5 to 10 over fresh MORB. U is also removed from seawater by precipitation in carbonate veins in the crust, which also have variable $\delta^{238}U$. This process appears to be only a minor sink for U in the oceans: Andersen et al. estimate the altered oceanic crust represents 12% of the seawater U sink with $\delta^{238}U \approx -0.2$‰.

U can partition into carbonate minerals, particularly aragonite, and this appears to be a significant sink of seawater U. Despite variable concentrations, carbonates precipitated directly from modern seawater have relatively uniform $\delta^{238}U = -0.4 \pm 0.15$, which is consistent with experimentally observed fractionation factors. Reaction with pore water, however, can lead to U enrichment and higher $\delta^{238}U$ and subsequent dolomitization can result in a decrease in $\delta^{238}U$. Andersen et al. estimate that carbonate sedimentation represents 28% of the U flux from seawater with $\delta^{238}U \approx -0.4$‰. Incorporation of Mn–Fe nodules and crusts removes the remaining 2% with $\delta^{238}U \approx -0.6$‰.

We should emphasize that other studies have produced U budget that differs somewhat from those of Andersen et al. (2017) and that future studies will no doubt produce further refinements. All, however, conclude that reductive removal is an important sink. Tissot and Dauphas (2015) estimate that anoxic/euxinic sediments presently cover 21% of the seafloor. This fraction is a function of both oxygenation of the atmosphere and ocean circulation. Longer ventilation time will lead to anoxic conditions on a greater fraction of the ocean floor, as would a decrease in atmospheric O_2 concentration. Thus, U isotopes provide yet another paleoredox proxy that we will be explored in the next chapter.

REFERENCES

Anagnostou, E., John, E.H., Edgar, K.M., et al. 2016. Changing atmospheric CO_2 concentration was the primary driver of early Cenozoic climate. *Nature* **533**: 380–384. doi: 10.1038/nature17423.

Albarède, F., Goldstein, S.L. and Dautel, D. 1997. The neodymium isotopic composition of manganese nodules from the Southern and Indian oceans, the global oceanic neodymium budget, and their bearing on deep ocean circulation. *Geochimica et Cosmochimica Acta* **61**: 1277–91. doi: 10.1016/S0016-7037(96)00404-8.

Albarède, F., Simonetti, A., Vervoort, J.D., et al. 1998. A Hf-Nd isotopic correlation in ferromanganese nodules. *Geophysical Research Letters* **25**(20): 3895–8. doi: 10.1029/1998gl900008.

Amini, M., Eisenhauer, A., Böhm, F., et al. 2008. Calcium isotope ($\delta^{44/40}Ca$) fractionation along hydrothermal pathways, Logatchev field (Mid-Atlantic Ridge, 14°45'N). *Geochimica et Cosmochimica Acta* **72**: 4107–4122. doi: 10.1016/j.gca.2008.05.055.

Amos, H.M., Jacob, D.J., Kocman, D., et al. 2014. Global biogeochemical implications of mercury discharges from rivers and sediment burial. *Environmental Science & Technology* **48**: 9514–22. doi: 10.1021/es502134t.

Andersen, M.B., Stirling, C.H. and Weyer, S. 2017. Uranium isotope fractionation. *Non-Traditional Stable Isotopes* **799–850**. doi: 10.2138/rmg.2017.82.19.

Andersen, M.B., Vance, D., Morford, J.L., et al. 2016. Closing in on the marine $^{238}U/^{235}U$ budget. *Chemical Geology* **420**: 11–22. doi: 10.1016/j.chemgeo.2015.10.041.

Anderson, R. F. 2014. Chemical tracers of particle transport, in Holland, H.D. and Turekian, K.K. (eds). *Treatise on Geochemistry*, 2nd ed. Oxford: Elsevier, 259–280 pp.

Bacon, M.P. and Anderson, R.F. 1982. Distribution of thorium isotopes between dissolved and particulate forms in the deep sea. *Journal of Geophysical Research: Oceans* **87**: 2045–56. doi: 10.1029/JC087iC03p02045.

Baconnais, I., Rouxel, O., Dulaquais, G., et al. 2019. Determination of the copper isotope composition of seawater revisited: a case study from the Mediterranean Sea. *Chemical Geology* **511**: 465–80. doi: 10.1016/j.chemgeo.2018.09.009.

Balistrieri, L.S., Borrok, D.M., Wanty, R.B., et al. 2008. Fractionation of Cu and Zn isotopes during adsorption onto amorphous Fe(iii) oxyhydroxide: Experimental mixing of acid rock drainage and ambient river water. *Geochimica et Cosmochimica Acta* **72**: 311–28. doi: 10.1016/j.gca.2007.11.013.

Barling, J. and Anbar, A.D. 2004. Molybdenum isotope fractionation during adsorption by manganese oxides. *Earth and Planetary Science Letters* **217**: 315–29. doi: 10.1016/S0012-821X(03)00608-3.

Barling, J., Yang, J. and Crystal Liang, Y.-H. 2018. Molybdenum isotopes, in White, W.M. (ed). *Encyclopedia of Geochemistry*. Cham: Springer International Publishing, 950–955 pp. doi: 10.1007/978-3-319-39312-4_219

Baronas, J. J., Hammond, D. E., Mcmanus, J., et al. 2017. A global Ge isotope budget. *Geochimica et Cosmochimica Acta* **203**: 265–283. doi: https://doi.org/10.1016/j.gca.2017.01.008.

Baronas, J.J., Torres, M. A., West, A.J., et al. 2018. Ge and Si isotope signatures in rivers: A quantitative multi-proxy approach. *Earth and Planetary Science Letters* **503**: 194–215. doi: https://doi.org/10.1016/j.epsl.2018.09.022.

Bergquist, B.A. 2018. Mercury isotopes, in White, W.M. (ed). *Encyclopedia of Geochemistry*. Cham: Springer International Publishing, pp. 900–906. doi: 10.1007/978-3-319-39312-4_122

Bergquist, B.A. and Blum, J.D. 2007. Mass-dependent and -independent fractionation of Hg isotopes by photoreduction in aquatic systems. *Science* **318**: 417–20. doi: 10.1126/science.1148050.

Bergquist, B.A. and Boyle, E.A. 2006. Iron isotopes in the Amazon River system: weathering and transport signatures. *Earth and Planetary Science Letters* **248**: 54–68. doi: 10.1016/j.epsl.2006.05.004.

Blättler, C.L., Henderson, G.M. and Jenkyns, H.C. 2012. Explaining the Phanerozoic Ca isotope history of seawater. *Geology* **40**: 843–846. doi: 10.1130/g33191.1.

Blum, J.D. and Johnson, M.W. 2017. Recent Developments in Mercury Stable Isotope Analysis, in Teng, F.-Z., Dauphas, N. and Watkins, J.M. (eds). *Non-Traditional Stable Isotopes*. Washington: Mineralogical Society of America. doi: 10.2138/rmg.2017.82.17

Blum, J.D., Sherman, L.S. and Johnson, M.W. 2014. Mercury isotopes in earth and environmental sciences. *Annual Reviews of Earth and Planetary Sciences* **42**: 249–69. doi: 10.1146/annurev-earth-050212-124107.

Burton, K.W., Gannoun, A. and Parkinson, I. J. 2010. Climate driven glacial–interglacial variations in the osmium isotope composition of seawater recorded by planktic foraminifera. *Earth and Planetary Science Letters* **295**: 58–68. doi: 10.1016/j.epsl.2010.03.026.

Boyle, E.A., Lee, J.-M., Echegoyen, Y.w., et al. 2014. Anthropogenic lead emissions in the Ocean: the Evolving Global Experiment. *Oceanography* **27**: 69–75.

Boyle, E.A., Zurbrick, C., Lee, J.-M., et al. 2020. Lead and lead isotopes in the U.S. GEOTRACES East Pacific zonal transect (GEOTRACES GP16). *Marine Chemistry* **227**: 103892. doi: 10.1016/j.marchem.2020.103892.

Breillat, N., Guerrot, C., Marcoux, E., and Négrel, P., et al. (2016). A new global database of $\delta^{98}Mo$ in molybdenites: A literature review and new data. *Journal of Geochemical Exploration* **161**: 1–15. doi:10.1016/j.gexplo.2015.07.019.

Brennecka, G.A., Wasylenki, L.E., Bargar, J.R., et al. 2011. Uranium isotope fractionation during adsorption to Mn-oxyhydroxides. *Environmental Science & Technology* **45**: 1370–5. doi: 10.1021/es103061v.

Bridgestock, L., van de Flierdt, T., Rehkämper, M., et al. 2016. Return of naturally sourced Pb to Atlantic surface waters. *Nature Communications* 7: 12921. doi: 10.1038/ncomms12921

Cavagna, A.-J., Fripiat, F., Dehairs, F., et al. 2011. Silicon uptake and supply during a Southern Ocean iron fertilization experiment (EIFEX) tracked by Si isotopes. *Limnology and Oceanography* 56: 147–160. doi: https://doi.org/10.4319/lo.2011.56.1.0147.

Chan, L.-H. and Edmond, J.M. 1988. Variation of lithium isotope composition in the marine environment: a preliminary report. *Geochimica et Cosmochimica Acta* 52: 1711–1717.

Chang, Y., Zhang, J., Qu, J.-Q., et al. 2017. Precise selenium isotope measurement in seawater by carbon-containing hydride generation-desolvation-MC-ICP-MS after thiol resin preconcentration. *Chemical Geology* 471: 65–73. doi: https://doi.org/10.1016/j.chemgeo.2017.09.011.

Chen, J., Hintelmann, H., Feng, X., et al. 2012. Unusual fractionation of both odd and even mercury isotopes in precipitation from Peterborough, ON, Canada. *Geochimica et Cosmochimica Acta* 90: 33–46. doi: 10.1016/j.gca.2012.05.005.

Chen, T.-Y., Li, G., Frank, M., et al. 2013. Hafnium isotope fractionation during continental weathering: implications for the generation of the seawater Nd-Hf isotope relationships. *Geophysical Research Letters* 40: 916–20. doi: 10.1002/grl.50217.

Chen, X., Romaniello, S.J. and Anbar, A.D. 2017. Uranium isotope fractionation induced by aqueous speciation: implications for U isotopes in marine $CaCO_3$ as a paleoredox proxy. *Geochimica et Cosmochimica Acta* 215: 162–72. doi: 10.1016/j.gca.2017.08.006.

Cheng, M., Li, C., Zhou, L., et al. 2015. Mo marine geochemistry and reconstruction of ancient ocean redox states. *Science China Earth Sciences* 58: 2123–33. doi: 10.1007/s11430-015-5177-4.

Clark, S.K. and Johnson, T.M. 2010. Selenium stable isotope investigation into selenium biogeochemical cycling in a lacustrine environment: Sweitzer lake, Colorado. *Journal of Environmental Quality* 39: 2200–10. doi: https://doi.org/10.2134/jeq2009.0380.

Conway, T.M. and John, S.G. 2014. Quantification of dissolved iron sources to the north Atlantic Ocean. *Nature* 511: 212–5. doi: 10.1038/nature13482.

Conway, T.M. and John, S.G. 2014. The biogeochemical cycling of zinc and zinc isotopes in the North Atlantic Ocean. *Global Biogeochemical Cycles* 28: 1111–28. doi: 10.1002/2014GB004862.

Conway, T.M. and John, S.G. 2015. The cycling of iron, zinc and cadmium in the north east Pacific Ocean – insights from stable isotopes. *Geochimica et Cosmochimica Acta* 164: 262–83. doi: 10.1016/j.gca.2015.05.023.

D'Arcy, J., Babechuk, M.G., Døssing, L.N., et al. 2016. Processes controlling the chromium isotopic composition of river water: constraints from basaltic river catchments. *Geochimica et Cosmochimica Acta* 186: 296–315. doi: 10.1016/j.gca.2016.04.027.

Dauphas, N., John, S.G. and Rouxel, O. 2017. Iron isotope systematics, in Teng, F.-Z., Dauphas, N. and Watkins, J.M. (eds). *Non-Traditional Stable Isotopes*. Washington DC: Mineralogical Society of America.

De La Rocha, C.L. 2003. Silicon isotope fractionation by marine sponges and the reconstruction of the silicon isotope composition of ancient deep water. *Geology* 31: 423–426. doi: 10.1130/0091-7613(2003)031<0423:sifbms>2.0.co;2.

De La Rocha, C.L., Brzezinski, M.A. and DeNiro, M.J. 2000. A first look at the distribution of the stable isotopes of silicon in natural waters. *Geochimica et Cosmochimica Acta* 64: 2467–2477. doi: 10.1016/s0016-7037(00)00373-2.

Demers, J.D., Blum, J.D. and Zak, D.R. 2013 Mercury isotopes in a forested ecosystem: implications for air-surface exchange dynamics and the global mercury cycle. *Global Biogeochemical Cycles* 27: 222–38. doi: https://doi.org/10.1002/gbc.20021.

de Souza, G.F., Reynolds, B.C., Johnson, G.C., et al. 2012. Silicon stable isotope distribution traces Southern Ocean export of Si to the eastern South Pacific thermocline. *Biogeosciences* 9: 4199–4213. doi: 10.5194/bg-9-4199-2012.

Ducher, M., Blanchard, M. and Balan, E. 2018. Equilibrium isotopic fractionation between aqueous Zn and minerals from first-principles calculations. *Chemical Geology* 483: 342–50. doi: 10.1016/j.chemgeo.2018.02.040.

Fantle, M.S. and Tipper, E.T. 2014. Calcium isotopes in the global biogeochemical Ca cycle: Implications for development of a Ca isotope proxy. *Earth-Science Reviews* 129: 148–177. doi: https://doi.org/10.1016/j.earscirev.2013.10.004.

Farkaš, J., Chrastný, V., Novák, M., et al. 2013. Chromium isotope variations ($\delta^{53/52}Cr$) in mantle-derived sources and their weathering products: Implications for environmental studies and the evolution of $\delta^{53/52}Cr$ in the Earth's mantle over geologic time. *Geochimica et Cosmochimica Acta* 123: 74–92. doi: 10.1016/j.gca.2013.08.016.

Filippova, A., Frank, M., Kienast, M., et al. 2017. Water mass circulation and weathering inputs in the Labrador Sea based on coupled Hf–Nd isotope compositions and rare earth element distributions. *Geochimica et Cosmochimica Acta* 199: 164–84. doi: 10.1016/j.gca.2016.11.024.

Foster, G.L., Lécuyer, C. and Marschall, H.R. 2018. Boron stable isotopes, in: White, W.M. (ed). *Encyclopedia of Geochemistry*. Cham: Springer International Publishing, pp. 162–166.

Foster, G.L., Pogge Von Strandmann, P.a.E. and Rae, J.W.B. 2010. Boron and magnesium isotopic composition of seawater. *Geochemistry Geophysics Geosystems* **11**. doi: 10.1029/2010GC003201.

Frei, R., Poiré, D. and Frei, K.M. 2014. Weathering on land and transport of chromium to the ocean in a subtropical region (Misiones, NW Argentina): a chromium stable isotope perspective. *Chemical Geology* **381**: 110–24. doi: 10.1016/j.chemgeo.2014.05.015.

Gannoun, A. and Burton, K.W. 2014. High precision osmium elemental and isotope measurements of North Atlantic seawater. *Journal of Analytical Atomic Spectrometry* **29**: 2330–42. doi: 10.1039/C4JA00265B.

Godfrey, L.V., Zimmermann, B., Lee, D.C., et al. 2009. Hafnium and neodymium isotope variations in NE Atlantic seawater. *Geochemistry, Geophysics, Geosystems* **10**: Q08015. doi: 10.1029/2009gc002508.

Goring-Harford, H.J., Klar, J.K., Pearce, C.R., et al. 2018. Behaviour of chromium isotopes in the eastern sub-tropical Atlantic oxygen minimum zone. *Geochimica et Cosmochimica Acta* **236**: 41–59. doi: 10.1016/j.gca.2018.03.004.

Hayes, C.T., Black, E.E., Anderson, R.F., et al. 2018. Flux of particulate elements in the north Atlantic Ocean constrained by multiple radionuclides. *Global Biogeochemical Cycles* **32**: 1738–58. doi: 10.1029/2018GB005994.

Hemming, N.G. and Hanson, G.N. 1992. Boron isotopic composition and concentration in modern marine carbonates. *Geochimica et Cosmochimica Acta* **56**: 537–543.

Henderson, G.M. and Maier-Reimer, E. 2002. Advection and removal of ^{210}Pb and stable Pb isotopes in the oceans: a general circulation model study. *Geochimica et Cosmochimica Acta* **66**: 257–72. doi: 10.1016/S0016–7037(01)00779–7.

Hippler, D., Buhl, D., Witbaard, R., et al. 2009. Towards a better understanding of magnesium-isotope ratios from marine skeletal carbonates. *Geochimica et Cosmochimica Acta* **73**: 6134–6146. doi: 10.1016/j.gca.2009.07.031.

Horan, K., Hilton, R.G., Mccoy-West, A.J., et al. 2020. Unravelling the controls on the molybdenum isotope ratios of river waters. *Geochemical Perspectives Letters* **13**: 1–6. doi: 10.7185/geochemlet.2005.

Horner, T.J., Little, S.H., Conway, T.M., et al. 2021. Bioactive trace metals and their isotopes as paleoproductivity proxies: an assessment using GEOTRACES-era data. *Global Biogeochemical Cycles* **35**: e2020GB006814. doi: 10.1029/2020GB006814.

Howarth, S., Prytulak, J., Little, S.H., et al. 2018. Thallium concentration and thallium isotope composition of lateritic terrains. *Geochimica et Cosmochimica Acta* **239**: 446–62. doi: 10.1016/j.gca.2018.04.017.

Huang, K.-J., Teng, F.-Z., Plank, T., et al. 2018. Magnesium isotopic composition of altered oceanic crust and the global Mg cycle. *Geochimica et Cosmochimica Acta* **238**: 357–373. doi: https://doi.org/10.1016/j.gca.2018.07.011.

Ilina, S.M., Poitrasson, F., Lapitskiy, S.A., et al. 2013. Extreme iron isotope fractionation between colloids and particles of boreal and temperate organic-rich waters. *Geochimica et Cosmochimica Acta* **101**: 96–111. doi: 10.1016/j.gca.2012.10.023.

Janssen, D.J., Rickli, J., Quay, P.D., et al. 2020. Biological control of chromium redox and stable isotope composition in the surface ocean. *Global Biogeochemical Cycles* **34**: e2019GB006397. doi: 10.1029/2019GB006397.

Jeandel, C., Peucker-Ehrenbrink, B., Jones, M.T. et al. 2011. Ocean margins: the missing term in oceanic element budgets? *EOS, Transactions of AGU*, **92**(26): 217–8. doi:10.1029/2011EO260001.

Jickells, T.D., Buitenhuis, E., Altieri, K., et al. 2017. A reevaluation of the magnitude and impacts of anthropogenic atmospheric nitrogen inputs on the ocean. *Global Biogeochemical Cycles* **31**: 289–305. doi: https://doi.org/10.1002/2016GB005586.

Jiskra, M., Heimbürger-Boavida, L.-E., Desgranges, M.-M., et al. 2021. Mercury stable isotopes constrain atmospheric sources to the ocean. *Nature* **597**: 678–82. doi: 10.1038/s41586-021-03859-8.

John, S.G., Helgoe, J. and Townsend, E. 2018. Biogeochemical cycling of Zn and Cd and their stable isotopes in the eastern tropical South Pacific. *Marine Chemistry* **201**: 256–62. doi: 10.1016/j.marchem.2017.06.001.

Johnson, C.M., Beard, B.L. and Roden, E.E. 2008. The iron isotope fingerprints of redox and biogeochemical cycling in modern and ancient Earth. *Annual Review of Earth and Planetary Sciences* **36**: 457–93. doi: 10.1146/annurev.earth.36.031207.124139.

Johnson, C., Beard, B. and Weyer, S. 2020. *Iron Geochemistry: an Isotopic Perspective*. Cham: Springer International Publishing.

Kendall, B., Dahl, T.W. and Anbar, A.D. 2017. The stable isotope geochemistry of molybdenum. *Non-Traditional Stable Isotopes* 683–732. doi: 10.2138/rmg.2017.82.16.

Kersten, M., Xiao, T., Kreissig, K., et al. 2014. Tracing anthropogenic thallium in soil using stable isotope compositions. *Environmental Science & Technology* **48**: 9030–6. doi: 10.1021/es501968d.

King, E.K., Perakis, S.S. and Pett-Ridge, J.C. 2018. Molybdenum isotope fractionation during adsorption to organic matter. *Geochimica et Cosmochimica Acta* **222**: 584–98. doi: 10.1016/j.gca.2017.11.014.

Klochko, K., Kaufman, A.J., Yao, W., et al. 2006. Experimental measurement of boron isotope fractionation in seawater. *Earth and Planetary Science Letters* **248**: 276–285. doi: 10.1016/j.epsl.2006.05.034.

Kurtz, A.C., Derry, L.A. and Chadwick, O.A. 2001. Accretion of Asian dust to Hawaiian soils: isotopic, elemental, and mineral mass balances. *Geochimica et Cosmochimica Acta* **65**: 1971–83. doi: https://doi.org/10.1016/S0016–7037(01)00575-0.

Kurtz, A.C., Derry, L.A. and Chadwick, O. A. 2002. Germanium-silicon fractionation in the weathering environment. *Geochimica et Cosmochimica Acta* **66**: 1525–1537. doi: https://doi.org/10.1016/S0016-7037(01)00869-9.

Kwon, S.Y., Blum, J.D., Yin, R., et al. 2020. Mercury stable isotopes for monitoring the effectiveness of the Minamata Convention on Mercury. *Earth-Science Reviews* **203**: 103111. doi: 10.1016/j.earscirev.2020.103111.

Lemarchand, D., Gaillardet, J., Lewin, É., et al. 2002. Boron isotope systematics in large rivers: implications for the marine boron budget and paleo-pH reconstruction over the Cenozoic. *Chemical Geology* **190**: 123–140. doi: https://doi.org/10.1016/S0009-2541(02)00114-6.

Levasseur, S., Birck, J.-L. and Allègre, C.J. 1998. Direct measurement of femtomoles of osmium and the $^{187}Os/^{186}Os$ ratio in seawater. *Science* **282**: 272–4. doi: 10.1126/science.282.5387.272.

Little, S.H., Archer, C., Milne, A., et al. 2018. Paired dissolved and particulate phase Cu isotope distributions in the South Atlantic. *Chemical Geology* **502**: 29–43. doi: 10.1016/j.chemgeo.2018.07.022.

Little, S.H., Vance, D., Walker-Brown, C., et al. 2014. The oceanic mass balance of copper and zinc isotopes, investigated by analysis of their inputs, and outputs to ferromanganese oxide sediments. *Geochimica et Cosmochimica Acta* **125**: 673–93. doi: 10.1016/j.gca.2013.07.046.

Liu, S.-A., Teng, F.-Z., Li, S., et al. 2014. Copper and iron isotope fractionation during weathering and pedogenesis: Insights from saprolite profiles. *Geochimica et Cosmochimica Acta* **146**: 59–75. doi: 10.1016/j.gca.2014.09.040.

Liu, S., Li, Y., Liu, J., et al. 2021. Equilibrium Cu isotope fractionation in copper minerals: a first-principles study. *Chemical Geology* **564**: 120060. doi: 10.1016/j.chemgeo.2021.120060.

Malinovsky, D. and Kashulin, N.A. 2018. Molybdenum isotope fractionation in plants measured by MC-ICPMS. *Analytical Methods* **10**: 131–7.

Marschall, H.R., Wanless, V.D., Shimizu, N., et al. 2017. The boron and lithium isotopic composition of mid-ocean ridge basalts and the mantle. *Geochimica et Cosmochimica Acta* **207**: 102–138. doi: https://doi.org/10.1016/j.gca.2017.03.028.

Mathur, R. and Wang, D. 2019. Transition metal isotopes applied to exploration geochemistry: insights from Fe, Cu, and Zn, in Decrée, S. and Robb, L. (eds). *Ore deposits: Origin, Exploration, and Exploitation*. Washington DC: American Geophysical Union. doi: 10.1002/9781119290544.ch7.

Mathur, R., Jin, L., Prush, V., et al. 2012. Cu isotopes and concentrations during weathering of black shale of the Marcellus formation, Huntingdon County, Pennsylvania (USA). *Chemical Geology* **304–305**: 175–84. doi: 10.1016/j.chemgeo.2012.02.015.

McManus, J., Nägler, T.F., Siebert, C., et al. 2002. Oceanic molybdenum isotope fractionation: Diagenesis and hydrothermal ridge-flank alteration. *Geochemistry, Geophysics, Geosystems* **3**: 1–9. doi: 10.1029/2002GC000356.

Méheut, M., Lazzeri, M., Balan, E., et al. 2007. Equilibrium isotopic fractionation in the kaolinite, quartz, water system: Prediction from first-principles density-functional theory. *Geochimica et Cosmochimica Acta* **71**: 3170–3181. doi: 10.1016/j.gca.2007.04.012.

Misra, S. and Froelich, P.N. 2012. Lithium isotope history of Cenozoic seawater: changes in silicate weathering and reverse weathering. *Science* **335**: 818–823. doi: 10.1126/science.1214697.

Mitchell, K., Couture, R.-M., Johnson, T.M., et al. 2013. Selenium sorption and isotope fractionation: Iron(III) oxides versus iron(II) sulfides. *Chemical Geology* **342**: 21–28. doi: https://doi.org/10.1016/j.chemgeo.2013.01.017.

Mitchell, K., Mason, P.R.D., Van Cappellen, P., et al. 2012. Selenium as paleo-oceanographic proxy: A first assessment. *Geochimica et Cosmochimica Acta* **89**: 302–17. doi: 10.1016/j.gca.2012.03.038.

Moos, S.B. and Boyle, E.A. 2019. Determination of accurate and precise chromium isotope ratios in seawater samples by MC-ICP-MS illustrated by analysis of safe station in the north Pacific Ocean. *Chemical Geology* **511**: 481–93. doi: 10.1016/j.chemgeo.2018.07.027.

Moynier, F., Vance, D., Fujii, T., et al. 2017. The Isotope geochemistry of zinc and copper, in Teng, F.-Z., Dauphas, N. and Watkins, J.M. (eds). *Non-Traditional Stable Isotopes*. Washington, DC: Mineralogical Society of America.

Nielsen, S.G., Rehkämper, M. and Prytulak, J. 2017. Investigation and application of thallium isotope fractionation. *Reviews in Mineralogy and Geochemistry* **82**: 759–98. doi: 10.2138/rmg.2017.82.18.

Nielsen, S.G., Rehkämper, M., Porcelli, D., et al. 2005. Thallium isotope composition of the upper continental crust and rivers—an investigation of the continental sources of dissolved marine thallium. *Geochimica et Cosmochimica Acta* **69**: 2007–19. https://doi.org/10.1016/j.gca.2004.10.025.

Nielsen, S.G., Rehkämper, M., Teagle, D.a.H., et al. 2006. Hydrothermal fluid fluxes calculated from the isotopic mass balance of thallium in the ocean crust. *Earth and Planetary Science Letters* **251**: 120–33. doi: 10.1016/j.epsl.2006.09.002.

Nielsen, S.G., Wasylenki, L.E., Rehkämper, M., et al. 2013. Towards an understanding of thallium isotope fractionation during adsorption to manganese oxides. *Geochimica et Cosmochimica Acta* **117**: 252–65. doi: 10.1016/j.gca.2013.05.004.

Novak, M., Sebek, O., Chrastny, V., et al. 2018. Comparison of $\delta^{53}Cr_{Cr(VI)}$ values of contaminated groundwater at two industrial sites in the eastern U.S. with contrasting availability of reducing agents. *Chemical Geology* **481**: 74–84. doi: 10.1016/j.chemgeo.2018.01.033.

Owens, J.D., Nielsen, S.G., Horner, T.J., et al. 2017. Thallium-isotopic compositions of euxinic sediments as a proxy for global manganese-oxide burial. *Geochimica et Cosmochimica Acta* 213: 291–307. doi: 10.1016/j.gca.2017.06.041.

Patchett, P.J., White, W.M., Feldmann, et al. 1984. Hafnium/rare earth element fractionation in the sedimentary system and crustal recycling into the Earth's mantle. *Earth and Planetary Science Letters* 69: 365–78. doi: 10.1016/0012-821X(84)90195-X.

Peacock, C.L. and Moon, E.M. 2012. Oxidative scavenging of thallium by birnessite: explanation for thallium enrichment and stable isotope fractionation in marine ferromanganese precipitates. *Geochimica et Cosmochimica Acta* 84: 297–313. doi: 10.1016/j.gca.2012.01.036.

Petit, J.C.J., Schäfer, J., Coynel, A., et al. 2015. The estuarine geochemical reactivity of Zn isotopes and its relevance for the biomonitoring of anthropogenic Zn and Cd contaminations from metallurgical activities: Example of the Gironde fluvial-estuarine system, France. *Geochimica et Cosmochimica Acta* 170: 108–25. doi: 10.1016/j.gca.2015.08.004.

Peucker-Ehrenbrink, B. and Jahn, B.-M. 2001. Rhenium-osmium isotope systematics and platinum group element concentrations: Loess and the upper continental crust. *Geochemistry, Geophysics, Geosystems* 2: 1061. doi: 10.1029/2001gc000172.

Peucker-Ehrenbrink, B., Miller, M.W., Arsouze, T., et al. 2010. Continental bedrock and riverine fluxes of strontium and neodymium isotopes to the oceans. *Geochemistry, Geophysics, Geosystems* 11(3): Q03016. doi: 10.1029/2009gc002869.

Phillips, R. and Xu, J. 2021. A critical review of molybdenum sequestration mechanisms under euxinic conditions: implications for the precision of molybdenum paleoredox proxies. *Earth-Science Reviews* 221: 103799. doi: 10.1016/j.earscirev.2021.103799.

Piepgras, D.J. and Wasserburg, G.J., 1980. Neodymium isotopic variations in seawater. *Earth and Planetary Science Letters* 50: 128–38. doi: 10.1016/0012-821X(80)90124-7.

Pogge Von Strandmann, P.a.E. 2008. Precise magnesium isotope measurements in core top planktic and benthic foraminifera. *Geochemistry Geophysics Geosystems* 9: Q12015. doi: 10.1029/2008gc002209.

Pogge Von Strandmann, P.a.E., Fraser, W.T., Hammond, S.J., et al. 2019. Experimental determination of Li isotope behaviour during basalt weathering. *Chemical Geology* 517: 34–43. doi: https://doi.org/10.1016/j.chemgeo.2019.04.020.

Pogge Von Strandmann, P.a.E., Kasemann, S.A. and Wimpenny, J.B. 2020. Lithium and lithium isotopes in earth's surface cycles. *Elements* 16: 253–258. doi: 10.2138/gselements.16.4.253.

Poulson, R.L., Siebert, C., McManus, J., et al. 2006. Authigenic molybdenum isotope signatures in marine sediments. *Geology* 34: 617–20. doi: 10.1130/g22485.1.

Qi, H.-W., Hu, R.-Z., Jiang, K., et al. 2019. Germanium isotopes and Ge/Si fractionation under extreme tropical weathering of basalts from the Hainan Island, South China. *Geochimica et Cosmochimica Acta* 253: 249–266. doi: https://doi.org/10.1016/j.gca.2019.03.022.

Qin, L. and Wang, X. 2017. Chromium isotope geochemistry. *Reviews in Mineralogy and Geochemistry* 82: 379–414. doi: 10.2138/rmg.2017.82.10.

Rader, S.T., Maier, R.M., Barton, M.D., et al. 2019. Uptake and fractionation of thallium by *Brassica juncea* in a geogenic thallium-amended substrate. *Environmental Science & Technology* 53: 2441–9. doi: 10.1021/acs.est.8b06222.

Rae, J.W.B. (2018). Boron Isotopes in foraminifera: systematics, biomineralisation, and CO$_2$ reconstruction, in Marschall, H. and Foster, G. (eds), *Boron Isotopes: The Fifth Element*. Cham: Springer International Publishing, pp. 107–143.

Rehkämper, M., Frank, M., Hein, J.R., et al. 2002. Thallium isotope variations in seawater and hydrogenetic, diagenetic, and hydrothermal ferromanganese deposits. *Earth and Planetary Science Letters* 197: 65–81. doi: 10.1016/s0012-821x(02)00462-4.

Rickli, J., Frank, M., Baker, A.R. et al. 2010. Hafnium and neodymium isotopes in surface waters of the eastern Atlantic Ocean: implications for sources and inputs of trace metals to the ocean. *Geochimica et Cosmochimica Acta* 74: 540–57. doi: 10.1016/j.gca.2009.10.006.

Rickli, J., Hindshaw, R.S., Leuthold, J., et al. 2017. Impact of glacial activity on the weathering of Hf isotopes – Observations from Southwest Greenland. *Geochimica et Cosmochimica Acta* 215: 295–316. doi: 10.1016/j.gca.2017.08.005.

Rolison, J.M., Stirling, C.H., Middag, R., et al. 2017. Uranium stable isotope fractionation in the black sea: modern calibration of the ^{238}U/^{235}U paleo-redox proxy. *Geochimica et Cosmochimica Acta* 203: 69–88. doi: 10.1016/j.gca.2016.12.014.

Rose-Koga, E.F. and Albarede, F. 2010. A data brief on magnesium isotope compositions of marine calcareous sediments and ferromanganese nodules. *Geochemistry Geophysics Geosystems* 11: Q03006. doi: 10.1029/2009GC002899.

Ryu, J.-S., Jacobson, A.D., Holmden, C., et al. 2011. The major ion, $\delta^{44/40}$Ca, $\delta^{44/42}$Ca, and $\delta^{26/24}$Mg geochemistry of granite weathering at pH=1 and T=25°C: power-law processes and the relative reactivity of minerals. *Geochimica et Cosmochimica Acta* 75: 6004–6026. doi: 10.1016/j.gca.2011.07.025.

Schauble, E.A. 2007. Role of nuclear volume in driving equilibrium stable isotope fractionation of mercury, thallium, and other very heavy elements. *Geochimica et Cosmochimica Acta* 71: 2170–89. doi: 10.1016/j.gca.2007.02.004.

Schauble, E.A. 2011. First-principles estimates of equilibrium magnesium isotope fractionation in silicate, oxide, carbonate and hexaaquamagnesium (2+) crystals. *Geochimica et Cosmochimica Acta* **75**: 844–869. doi: 10.1016/j.gca.2010.09.044.

Schauble, E., Rossman, G.R. and Taylor, H.P. 2004. Theoretical estimates of equilibrium chromium-isotope fractionations. *Chemical Geology* **205**: 99–114. doi: 10.1016/j.chemgeo.2003.12.015.

Scheiderich, K., Amini, M., Holmden, C., et al. 2015. Global variability of chromium isotopes in seawater demonstrated by Pacific, Atlantic, and Arctic Ocean samples. *Earth and Planetary Science Letters* **423**: 87–97. doi: 10.1016/j.epsl.2015.04.030.

Schlitzer, R., Anderson, R.F., Dodas, E.M., et al. 2018. The GEOTRACES intermediate data product 2017. *Chemical Geology* **493**: 210–23. doi: j.chemgeo.2018.05.040.

Schlitzer, R. 2021. GEOTRACES - Electronic Atlas of GEOTRACES Sections and Animated 3D Scenes. Available at: http://www.egeotraces.org.

Schwarcz, H.P., Agyei, E.K. and McMullen, C.C. 1969. Boron isotopic fractionation during clay adsorption from sea-water. *Earth and Planetary Science Letters* **6**: 1–5.

Sharma, M., Papanastassiou, D.A. and Wasserburg, G.J. 1997. The concentration and isotopic composition of osmium in the oceans. *Geochimica et Cosmochimica Acta* **61**: 3287–99. doi: 10.1016/S0016-7037(97)00210-X.

Sherman, L.S., Blum, J.D., Johnson, K.P., et al. 2010. Mass-independent fractionation of mercury isotopes in Arctic snow driven by sunlight. *Nature Geoscience* **3**: 173. doi: 10.1038/ngeo758.

Siebert, C., Nägler, T.F., Von Blanckenburg, F. and Kramers, J.D. 2003. Molybdenum isotope records as a potential new proxy for paleoceanography. *Earth and Planetary Science Letters* **211**: 159–71. doi: 10.1016/s0012-821x(03)00189-4.

Siebert, C., Pett-Ridge, J.C., Opfergelt, S., et al. 2015. Molybdenum isotope fractionation in soils: Influence of redox conditions, organic matter, and atmospheric inputs. *Geochimica et Cosmochimica Acta* **162**: 1–24. doi: 10.1016/j.gca.2015.04.007.

Sigman, D.M. and Fripiat, F. 2019. Nitrogen isotopes in the ocean, in Cochran, J.K., Bokuniewicz, H.J. and Yager, P.L. (eds). *Encyclopedia of Ocean Sciences* (Third Edition). Oxford: Academic Press, pp. 263–278.

Simon, L., Lécuyer, C., Maréchal, C., et al. 2006. Modelling the geochemical cycle of boron: Implications for the long-term $\delta^{11}B$ evolution of seawater and oceanic crust. *Chemical Geology* **225**: 61–76. doi: https://doi.org/10.1016/j.chemgeo.2005.08.011.

Spivack, A. and Edmond, J.M. 1987. Boron isotope exchange between seawater and the oceanic crust. *Geochimica Et Cosmochimica Acta* **51**: 1033–1044.

Stichel, T., Frank, M., Rickli, J., et al. 2012. The hafnium and neodymium isotope composition of seawater in the Atlantic sector of the Southern Ocean. *Earth and Planetary Science Letters* **317–8:**, 282–94. doi: 10.1016/j.epsl.2011.11.025.

Štrok, M., Baya, P.A. and Hintelmann, H. 2015. The mercury isotope composition of Arctic coastal seawater. *Comptes Rendus GeoScience* **347**: 368–76. doi: 10.1016/j.crte.2015.04.001.

Stüeken, E.E., Buick, R., Bekker, A., et al. 2015. The evolution of the global selenium cycle: secular trends in se isotopes and abundances. *Geochimica et Cosmochimica Acta* **162**: 109–25. doi: https://doi.org/10.1016/j.gca.2015.04.033.

Sun, R., Sonke, J.E., Heimbürger, L.-E., et al. 2014. Mercury stable isotope signatures of world coal deposits and historical coal combustion emissions. *Environmental Science & Technology* **48**: 7660–7668. doi: 10.1021/es501208a.

Sutton, J.N., André, L., Cardinal, D., et al. 2018a. A review of the stable isotope bio-geochemistry of the global silicon cycle and its associated trace elements. *Frontiers in Earth Science* **5**. doi: 10.3389/feart.2017.00112.

Sutton, J.N., De Souza, G.F., García-Ibáñez, M.I., et al. 2018b. The silicon stable isotope distribution along the GEOVIDE section (GEOTRACES GA-01) of the North Atlantic Ocean. *Biogeosciences* **15**: 5663–5676. doi: 10.5194/bg-15-5663-2018.

Szynkiewicz, A. and Borrok, D.M. 2016. Isotope variations of dissolved Zn in the Rio Grande watershed, USA: the role of adsorption on Zn isotope composition. *Earth and Planetary Science Letters* **433**: 293–302. doi: 10.1016/j.epsl.2015.10.050.

Teng, F.-Z., Li, W.-Y., Ke, S., et al. 2010. Magnesium isotopic composition of the Earth and chondrites. *Geochimica et Cosmochimica Acta* **74**: 4150–66. doi: 10.1016/j.gca.2010.04.019.

Tomascak, P.B., Magna, T. and Dohmen, R. 2016. Lithium in the deep Earth: mantle and crustal systems, in Tomascak, P. B., Magna, T. and Dohmen, R. (eds). *Advances in Lithium Isotope Geochemistry*. Cham: Springer International Publishing, pp. 119–156.

Tipper, E.T., Galy, A. and Bickle, M.J. 2006. Riverine evidence for a fractionated reservoir of Ca and Mg on the continents: Implications for the oceanic Ca cycle. *Earth and Planetary Science Letters* **247**: 267–279. doi: 10.1016/j.epsl.2006.04.033.

Tipper, E.T., Schmitt, A.-D. and Gussone, N. (2016). Global Ca cycles: Coupling of continental and oceanic processes, in Gussone, N., Schmidt, A., Hauser, A., et al. (eds). *Calcium Stable Isotope Geochemistry*. Berlin, Heidelberg: Springer Berlin Heidelberg, pp. 173–222.

Tissot, F.L.H. and Dauphas, N. 2015. Uranium isotopic compositions of the crust and ocean: age corrections, U budget and global extent of modern anoxia. *Geochimica et Cosmochimica Acta* **167**: 113–43. doi: 10.1016/j.gca.2015.06.034.

van de Flierdt, T., Griffiths, A.M., Lambelet, M., et al. 2016. Neodymium in the oceans: a global database, a regional comparison and implications for palaeoceanographic research. *Philosophical Transactions of the Royal Society A*: **374**: 20150293. doi: 10.1098/rsta.2015.0293.

Vance, D., Archer, C., Bermin, J., et al. 2008. The copper isotope geochemistry of rivers and the oceans. *Earth and Planetary Science Letters* **274**: 204–13. doi: 10.1016/j.epsl.2008.07.026.

Vance, D., Matthews, A., Keech, A., et al. 2016. The behaviour of Cu and Zn isotopes during soil development: controls on the dissolved load of rivers. *Chemical Geology* **445**: 36–53. doi: 10.1016/j.chemgeo.2016.06.002.

Vaněk, A., Grösslová, Z., Mihaljevič, M., et al. 2016. Isotopic tracing of thallium contamination in soils affected by emissions from coal-fired power plants. *Environmental Science & Technology* **50**: 9864–71. doi: 10.1021/acs.est.6b01751.

Vejvodová, K., Vaněk, A., Mihaljevič, M., et al. 2020. Thallium isotopic fractionation in soil: the key controls. *Environmental Pollution* **265**: 114822. doi: 10.1016/j.envpol.2020.114822.

Vengosh, A., Kolodny, Y., Starinsky, A., et al. 1991. Coprecipitation and isotopic fractionation of boron in modern biogenic carbonates. *Geochimica et Cosmochimica Acta* **55**: 2901–2910.

Vervoort, J.D., Plank, T. and Prytulak, J., 2011. The Hf–Nd isotopic composition of marine sediments. *Geochimica et Cosmochimica Acta*: **75**: 5903–26. doi: 10.1016/j.gca.2011.07.046.

von Blanckenburg, F., O'Nions, R.K. and Heinz, J.R. 1996. Distribution and sources of pre-anthropogenic lead isotopes in deep ocean water from Fe–Mn crusts. *Geochimica et Cosmochimica Acta* **60**: 4957–63. doi: 10.1016/S0016-7037(96)00310-9.

Wang, X., Johnson, T.M. and Lundstrom, C.C. 2015. Low temperature equilibrium isotope fractionation and isotope exchange kinetics between U(IV) and U(VI). *Geochimica et Cosmochimica Acta* **158**: 262–275. doi: https://doi.org/10.1016/j.gca.2015.03.006.

Wang, X., Reinhard, C.T., Planavsky, N.J., et al. 2016. Sedimentary chromium isotopic compositions across the cretaceous OAE2 at Demerara Rise Site 1258. *Chemical Geology* **429**: 85–92. doi: 10.1016/j.chemgeo.2016.03.006.

Wasylenki, L.E., Weeks, C.L., Bargar, J.R., et al. 2011. The molecular mechanism of Mo isotope fractionation during adsorption to birnessite. *Geochimica et Cosmochimica Acta* **75**: 5019–31. doi: 10.1016/j.gca.2011.06.020.

Wei, W., Klaebe, R., Ling, H.-F., et al. 2020. Biogeochemical cycle of chromium isotopes at the modern Earth's surface and its applications as a paleo-environment proxy. *Chemical Geology* **541**: 119570. doi: https://doi.org/10.1016/j.chemgeo.2020.119570.

White, W.M., Patchett, J. and BenOthman, D. 1986. Hf isotope ratios of marine sediments and Mn nodules: evidence for a mantle source of Hf in seawater. *Earth and Planetary Science Letters* **79**: 46–54. doi: 10.1016/0012-821X(86)90039-7.

Wiederhold, J.G., Kraemer, S.M., Teutsch, N., et al. 2006. Iron isotope fractionation during proton-promoted, ligand-controlled, and reductive dissolution of goethite. *Environmental Science & Technology* **40**: 3787–93. doi: 10.1021/es052228y.

Wiederhold, J.G., Teutsch, N., Kraemer, S.M., et al. 2007. Iron isotope fractionation in oxic soils by mineral weathering and podzolization. *Geochimica et Cosmochimica Acta* **71**: 5821–33. doi: 10.1016/j.gca.2007.07.023.

Wimpenny, J., Gíslason, S.U. R., James, R. H., et al. 2010. The behaviour of Li and Mg isotopes during primary phase dissolution and secondary mineral formation in basalt. *Geochimica et Cosmochimica Acta* **74**: 5259–5279. doi: 10.1016/j.gca.2010.06.028.

Woodhouse, O.B., Ravizza, G., Kenison Falkner, K., et al. 1999. Osmium in seawater: vertical profiles of concentration and isotopic composition in the eastern Pacific Ocean. *Earth and Planetary Science Letters* **173**: 223–33. doi: 10.1016/S0012-821X(99)00233-2.

Zheng, W., Obrist, D., Weis, D., et al. 2016. Mercury isotope compositions across North American forests. *Global Biogeochemical Cycles* **30**: 1475–92. doi: 10.1002/2015GB005323.

Zimmermann, B., Porcelli, D., Frank, M., et al. 2009a. Hafnium isotopes in Arctic Ocean water. *Geochimica et Cosmochimica Acta* **73**: 3218–33. doi: 10.1016/j.gca.2009.02.028.

Zimmermann, B., Porcelli, D., Frank, M., et al. 2009b. The hafnium isotope composition of Pacific Ocean water. *Geochimica et Cosmochimica Acta* **73**: 91–101. doi: 10.1016/j.gca.2008.09.033.

PROBLEMS

1. Using Equations (11.5) and (11.4) and assuming $k_1 = 1.5$, $k_2 = 0.1$, $k_{-1} = 6.3$ and a ^{234}U seawater activity of 0.04657 mBq/m^3, create plots of the activity of dissolved and particulate ^{230}Th using values for the sinking rate, S, of 200 m/yr and 300 m/yr.

2. Using Equations (11.9) through (11.12), derive the relationship between $\delta^{11}B_4$ and pH assuming that $\delta^{11}B_{sw}$ is constant.

3. The present ocean has a $\delta^{98}Mo$ of +2.09‰. Assume that the input to the ocean from rivers and low temperature hydrothermal vents is $\delta^{98}Mo$ = +0.6‰. Simplifying things somewhat, assume that removal of Mo from the modern oceans occurs only through oxic adsorption onto Mn-Fe crusts with $\Delta^{98}Mo_{Oxic}$ of –3.0‰ and removal in euxinic environments such as the Black Sea with $\Delta^{98}Mo_{Oxic}$ of 0‰.

   (a) What fraction of the removal occurs in euxinic environments?
   (b) During the Toacian Ocean Anoxic Event around 183 Ma, $\delta^{98}Mo$ of seawater appears to have been lower. about +1.6‰. Assume $\delta^{98}Mo$ of inputs to the oceans and the fractionation factors for the removal were the same as the modern ocean, what fraction of removal was occurring in euxinic environments in the Toacian?

4. The following are constants for the temperature dependence of reduced partition functions (β) for $^{26}Mg/^{24}Mg$ fractionation factors fit to the polynomial $1000 \ln \beta = A/T^6 + B/T^4 + C/T^2$ (T is temperature in kelvins) calculated by Schuable (2011):

	A	B	C
Dolomite	1.667×10^{14}	-1.2954×10^{10}	2.1154×10^6
Mg^{2+}_{aq}	2.3453×10^{14}	-1.9796×10^{10}	2.5007×10^6

   (a) Calculate the fractionation factor $\Delta^{26}Mg_{dol-Mg}2+$ at 15° and 25°C.
   (b) Assuming that rivers are the only source of Mg to seawater and that they have $\delta^{26}Mg$ of –1.1‰ and that hydrothermal activity and dolomite formation are the only sinks. Assuming there is no fractionation in hydrothermal removal, use the fractionation factor you calculated above for 15°C to determine the fractionation of seawater Mg removed by dolomite formation to maintain a steady state seawater of $\delta^{26}Mg$ = –0.82‰. What would the isotopic composition of this dolomite be?
   (c) Using the fraction of Mg removal by dolomite formation you calculated in (b.), how would the ^{26}Mg of seawater change if the temperature increased from 15° to 25°C?

Chapter 12

Paleoclimate, Paleoceanography, and Atmospheric History

12.1 INTRODUCTION

Over the last three quarters of a century, isotope geochemistry has revealed an enormous amount about the evolution of exogene. Much of this has come from the study of stable isotopes, which will be front and center in this chapter. Radiogenic isotopes not only have, of course, been absolutely essential in placing time stamps on this history, but also have contributed in other ways, so we will consider their contributions as well. Perhaps most strikingly, isotope geochemistry has revealed history and cause of the dramatic swings in climate that occurred during the Pleistocene, which saw ice sheets repeatedly extending as far south as St. Louis (38°N) in central North America only to retreat again to the Arctic. While slight changes in the Earth's orbit about the Sun and in its rotational axis have been the pacemaker of these climate swings, the immediate cause was changes in atmospheric CO_2 concentration. Ocean currents move vast amounts of heat around the globe and the oceans hold vastly greater amounts of CO_2 than the atmosphere, so it should not come as a surprise that these climate swings are intimately tied to changes in ocean circulation and shifts in the balance of CO_2 between oceans and the atmosphere. These changes have been documented by radiogenic and stable isotope ratios preserved in sediment.

The Pleistocene glaciations are only the last of several that occurred through Earth's history. Those not only resulted from changes in atmospheric greenhouse gas concentrations, but they are also linked, at least in some cases, to changes in the carbon cycle. The young Earth's atmosphere would have been similar to that of its neighbors, consisting predominantly of CO_2 with lesser amounts of N_2 and trace gases, but entirely lacking O_2. High atmospheric CO_2 concentrations maintained a strong greenhouse, keeping the young Earth's surface mostly above freezing despite an $\approx 30\%$ weaker Sun. The presence of O_2 in the Earth's atmosphere is entirely due to photosynthesis, which converts atmospheric CO_2 into organic matter producing O_2 as a waste product. At times, drawdown in atmospheric greenhouse gases associated with rising O_2 triggered climate crises more severe than the Pleistocene glaciations. We will find that the stable isotopic composition of redox-sensitive metals preserved in sediments, together with conventional light stable isotopes, has allowed us to begin to unravel the history of the atmosphere, and with it of life.

12.2 THE PLEISTOCENE CLIMATE RECORD IN DEEP SEA SEDIMENTS

The very first application of stable isotope geochemistry was determining how Earth's temperature had varied in the past. Beginning with work of Louis Agassiz in 1840 it became clear that the Earth's climate had been much

Isotope Geochemistry, Second Edition. William M. White.
© 2023 John Wiley & Sons Ltd. Published 2023 by John Wiley & Sons Ltd.
Companion Website: www.wiley.com/go/white/isotopegeochem2

colder in the past, with episodes in which ice sheets and mountain glaciers were much more extensive than they are at present. But until 1947, scientists had no means of quantifying paleotemperature changes. In that year, two papers were published, one by Jacob Bigeleisen and Maria Meyer (1947), entitled *Calculation of equilibrium constants for isotopic exchange reactions* and one by Harold Urey (1947) entitled *The thermodynamic properties of isotopic substances*, which effectively initiated the field of stable isotope geochemistry. Urey calculated the temperature dependence of oxygen isotope fractionation between calcium carbonate and water and proposed that the isotopic composition of carbonates could be used as a paleothermometer (Urey, 1947). Urey's postdoctoral associate Samuel Epstein and several students tested Urey's idea by growing mollusks in water of various temperatures (Epstein et al., 1953). They found the following empirical relationship:

$$\Delta = \delta^{18}O_{cal} - \delta^{18}O_{water}$$
$$= 15.36 - 2.673(16.52 + T)^{0.5} \quad (12.1)$$

This equation was in good, though not exact, agreement with the theoretical prediction of Urey (Figure 12.1). The field of paleothermometry began with a paper based on these principles by McCrea (1950).

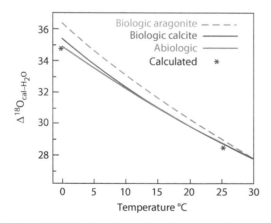

Figure 12.1 Fractionation of oxygen isotopes between calcium carbonate and water as a function of temperature for biologically precipitated calcite (mollusks), biologically precipitated aragonite, and abiologically precipitated calcite in experiments. Also shown are the calculated fractionation factors of Urey (1947) for 0 and 25°C. Source: Adapted from Epstein et al. (1953).

Over the subsequent decades, old-fashioned geological "boots on the ground" detective work combined with isotope paleothermometry and the geochronological tools we introduced in earlier chapters have not only revealed the details of the seasaw climate swings of the Pleistocene Ice Ages, but also that Earth's climate has always been in a constant state of change, including earlier episodes of severe climate change as continents have shifted and atmospheric greenhouse gas concentrations have varied. As we mentioned before, marine sediments are the history book of the planet, encoding these changes in the isotopic composition of sediments. We will begin with that record.

It is perhaps ironic that while glaciers are a continental phenomenon, our best record of them is from the oceans. In part, this is because each period of continental glaciation largely destroys the record of the previous one. In contrast, deep-sea sediments are generally not disturbed by glaciation. Thus, while much was learned by studying the effects of Pleistocene glaciation in Europe and North America (recent examples of which were described in Chapter 4), much was left unresolved, including questions such as the precise chronology, cause, temperatures, and ice volumes (ice area could of course be determined, but this is only part of the problem). The questions of temperature chronology were largely resolved through isotopic studies of deep-sea biogenic sediments. Dating of coral reefs provided the best estimates of how ice volume changed because they record ocean volume changes. These, as we shall see, provided the essential clue as to cause. While that question too has been largely resolved, the details are still being worked out.

The principles involved in paleoclimatology are simple. As Urey formulated it, the isotopic composition of calcite secreted by organisms should provide a record of paleo-ocean temperatures because the fractionation of oxygen isotopes between carbonate and water is temperature dependent. In actual practice, the problem is somewhat more complex because the isotopic composition of the shell, or test, of an organism will depend not only on temperature, but also on the isotopic composition of water in which the organism grew, vital effects (i.e., different species may fractionate oxygen isotopes somewhat differently), and post-burial isotopic exchange with sediment pore water. As it turns out, the latter two are usually not very important for carbonates, at least for Quaternary sediments, but the former is.

12.2.1 The Quaternary $\delta^{18}O$ record in sediments

The first isotopic work on deep-sea sediment cores with the goal of reconstructing the temperature history of Pleistocene glaciations was by Emiliani (1955), who was a student of Urey at the University of Chicago. Emiliani analyzed $\delta^{18}O$ in foraminifera from sediment cores from the world ocean. Remarkably, many of Emiliani's findings are still valid today, albeit in modified form. He concluded that the last glacial cycled had ended about 16 000 years ago and found that temperature increased steadily between that time until about 6000 years ago. He also recognized 14 other glacial–interglacial cycles over the last 600 000 years (we now recognize only about six major glacial intervals over this time) and found that these were global events, with notable cooling even in low latitudes. He concluded that bottom water in the Atlantic was 2°C cooler, but that bottom water in the Pacific was only 0.8°C cooler during glacial periods. He also concluded that the fundamental driving force for Quaternary climate cycles was variations in the Earth's orbital parameters.

Emiliani had the field of oxygen isotope paleoclimatology virtually to himself until about 1970. In retrospect, it is remarkable how much Emiliani got right. Early modifications to Emiliani's work included an improved timescale, initially based on extrapolated ^{14}C dating, using magnetostratigraphy and revision of the magnitude of temperature variations. Emiliani had realized that the isotopic composition of the ocean would vary between glacial and interglacial times as isotopically light water was stored in glaciers, thus enriching the oceans in ^{18}O (Figure 12.2). Assuming a $\delta^{18}O$ value of about −15‰ for glacial ice, Emiliani estimated that this factor accounted for about 20% of the observed variations. The remainder he attributed to the effect of temperature on isotope fractionation. Subsequently, Shackleton and Opdyke (1973) argued that storage of isotopically light water in glacial ice was actually the main effect. Their argument was based on the observation that

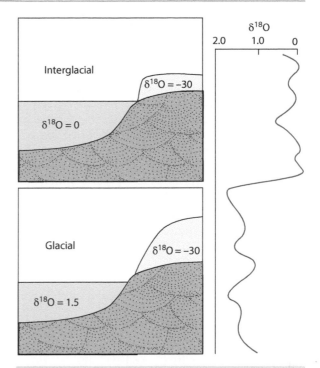

Figure 12.2 Cartoon illustrating how $\delta^{18}O$ of the ocean changes between glacial and interglacial periods as a consequence of build-up of isotopically light ice.

nearly the same isotopic variations occurred in both planktonic (surface-dwelling) and benthic (bottom-dwelling) foraminifera. Because of the way in which the deep water of the ocean is formed and circulates, they argued that deep-water temperature should not vary as much between glacial and interglacial cycles as surface water. The isotopic composition of tests of benthic organisms, i.e., those growing in deep water, was used to determine the change in seawater isotopic composition. This allowed a more precise calculation of surface water temperature change from the isotopic composition of planktonic tests. In addition, it is now clear that the average $\delta^{18}O$ of glacial ice is lower than −15‰, as Emiliani had assumed. Typical values for Greenland ice are −30 to −40‰ (relative to SMOW) and as much as −50‰ for Antarctic ice.

If the exact isotopic composition of ice and the ice volume were known, it would be a straightforward exercise to calculate the effect of continental ice build-up on ocean isotopic composition. For example, the present volume of continental ice is 27.5×10^6 km^3, while the volume of the oceans is 1350×10^6 km^3.

Assuming glacial ice has a mean $\delta^{18}O$ of $-30‰$ relative to SMOW, we can calculate the $\delta^{18}O$ of the hydrosphere as $-0.6‰$ (neglecting freshwater reservoirs, which are small). At the height of the Wisconsin Glaciation, the volume of glacial ice is thought to have increased by 42×10^6 km^3, corresponding to a lowering of sea level by 125 m. If the $\delta^{18}O$ of ice was the same as that now ($\approx -30‰$), we can readily calculate that the $\delta^{18}O$ of the ocean would have increased by $1.59‰$. This is illustrated in Figure 12.2.

To see how much this affects estimated temperature changes, we can use Craig's[1] (1965) revision of the Epstein calcite–water geothermometer:

$$T\ (°C) = 16.9 - 4.2\Delta_{cal-water} + 0.13(\Delta_{cal-water})^2 \qquad (12.2)$$

According to this equation, the fractionation should be 33‰ at 20°C. At 14°C, the fractionation is 31.5‰. If a glacial foram shell were 2‰ lighter, Emiliani would have made a correction of 0.5‰ for the change in oxygen isotopic composition of seawater and attributed the remainder of the difference, 1.5‰, to temperature. He would have concluded that the ocean was 6°C cooler. However, if the change in the isotopic composition of seawater is actually 1.5‰, leaving only a 0.5‰ difference due to temperature, the calculated temperature difference is only about 2°C. Thus, the question of the volume of glacial ice, and its isotopic composition needed to be resolved before $\delta^{18}O$ in deep-sea carbonates, could be used to calculate paleotemperatures. It is now generally assumed that the $\delta^{18}O$ of the ocean changed by 1.5–2‰ between glacial and interglacial periods, but second-order local variations also occur (due to evaporation and precipitation), leaving some uncertainty in exact temperatures and volumes. Ice volumes are better estimated from sea level curves derived from ^{14}C and U–Th dating of terraces and coral reefs, which we described in Chapters 3 and 4. These indicate that each 0.011‰ variation in $\delta^{18}O$ represents a 1 m change in sea level.

By now, thousands of deep-sea cores have been analyzed for oxygen isotope ratios.

Though most reveal the same general picture, the $\delta^{18}O$ curve varies from core to core. In addition to the changing isotopic composition of the ocean, the $\delta^{18}O$ record in a given core will depend on other factors: (1) the temperature in which the organisms grew. (2) the faunal assemblage, as the exact fractionation will vary from organism to organism. For this reason, $\delta^{18}O$ analyses are often performed on a single species. However, these "vital effects" are usually small, at least for planktonic foraminifera. (3) Local variations in water isotopic composition. This is important in the Gulf of Mexico, for example. Meltwater released at the end (termination) of glacial stages flooded the surface of the Gulf of Mexico with enough isotopically light meltwater to significantly change its isotopic composition relative to the ocean as a whole. (4) Sedimentation rate varies from core to core, so $\delta^{18}O$ as a function of depth in the core will differ between cores. Changes in sedimentation rate at a given locality will distort the appearance of the $\delta^{18}O$ curve. (5) Bioturbation, i.e., burrowing activity of seafloor animals, which may smear the record.

12.2.2 Milankovitch cycles

Figure 12.3 shows the global benthic foraminifera $\delta^{18}O$ record constructed by averaging analyses from 57 cores over the last 800 000 years (Lisiecki and Raymo, 2005). Because this curve is based on benthic foraminifera, it principally reflects ice volume rather than temperature. A cursory examination of the curve shows a periodicity of approximately 100 000 years. The same periodicity was apparent in Emiliani's initial work and led him to conclude that the glacial–interglacial cycles were due to variations in the Earth's orbital parameters. These are often referred to as the Milankovitch cycles, after Milutin Milanković, a Serbian mathematician who argued they caused the ice ages in the early part of the twentieth century (Milanković, 1920)[2].

The sawtooth record of isotopic variations is divided into marine isotope stages (MIS), which are given numbers going back in time,

---

[1] Harmon Craig was also a student of Harold Urey.

[2] While Milanković was a strong and early proponent of the idea that variations in the Earth's orbit caused ice ages, he was not the first to suggest it. J. Croll of Britain first suggested it in 1864, and published several subsequent papers on the subject.

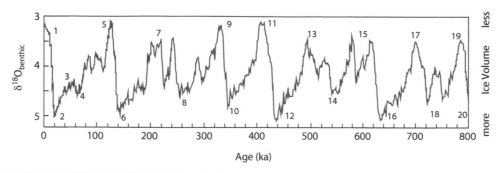

Figure 12.3 Late Pleistocene average $\delta^{18}O_{PDB}$ of benthic foraminifera in 57 globally distributed deep-sea piston and ODP drilling cores. This variation primarily reflects ice volume. Numbers 1 through 20 label marine isotope stages (MIS). Source: Adapted from Lisiecki and Raymo (2005).

a system that was first introduced by Emiliani (1955). The stages can be correlated to variable extents with terrestrial geomorphological and palaeoflora evidence of glacial retreat and advance. They are numbered from the present-day (MIS 1) backward in time, such that warm, or *interstadial*, events including interglacials are given odd numbers and cold-climate or glacial advance events, called *stadials* (also stades), are assigned even numbers. MIS 1 is the present interglacial that began with the rise in $\delta^{18}O$ around 14–16 ka, MIS 2 is the stadial corresponding to the last glacial maximum beginning around 29 ka, MIS 3 is the interstadial, a period of comparative warmth and glacial retreat that began around 57 ka, MIS 4 the stadial preceding its beginning around 71 ka, MIS 5 corresponds to the last interglacial, etc. The stages are also sometimes subdivided, particularly MIS 5 from 5a, corresponding to the interstadial around 82 ka, to 5e, corresponding to the peak of the Eemian interglacial at around 123 ka. A chart (Cohen and Gibbard, 2011) documenting the full list is available online from the International Commission on Stratigraphy (https://quaternary.stratigraphy.org/charts/). The end of major glacial periods is known as *terminations* and is labelled going backward as T_I, T_{II}, T_{III}, etc.

Three parameters describe the Milankovitch cycles – e: eccentricity; ε: obliquity (tilt), and precession: $e \sin \omega$, where ω is the longitude of perihelion (perihelion is the Earth's closest approach to the Sun). The *eccentricity* (i.e., the degree to which the orbit differs from circular) of the Earth's orbit about the Sun, and the

degree of tilt, or *obliquity*, of the Earth's rotational axis vary slightly. Precession refers to the change in the direction in which the Earth's rotational axis tilts when it is closest to the Sun (perihelion). These variations, which are illustrated in Figure 12.4, affect the pattern of solar radiation – or *insolation* – that the Earth receives. Changes in these parameters have negligible effect on the total annual insolation, but they do affect the distribution of insolation. For example, tilt of the rotational

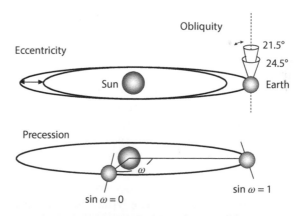

Figure 12.4 Cartoon illustrating the *Milankovitch parameters*. The eccentricity is the degree the Earth's orbit departs from circular. Obliquity is the tilt of the Earth's rotation axis with respect to the plane of the ecliptic. Obliquity varies between 21.5 and 24.5°. Precision is the variation in the direction of tilt at the Earth's closest approach to the Sun (perihelion). The parameter ω is the angle between the Earth's position on June 21 (summer solstice) and perihelion.

axis determines seasonality and the latitudinal gradient of insolation. It is this gradient that drives atmospheric and oceanic circulation. If the tilt is small, seasonality will be reduced (cooler summers and warmer winters). Precession relative to the eccentricity of the Earth's orbit also affects seasonality. For example, the Earth presently is closest to the Sun in January. As a result, Northern Hemisphere winters (and Southern Hemisphere summers) are somewhat milder than they would be otherwise. For a given latitude and season, precession will result in a ±5% difference in insolation. While the Earth's orbit is only slightly elliptical and variations in eccentricity are small, these variations are magnified because insolation varies with the inverse square of the Earth–Sun distance. These variations can change the insolation and the average annual equator-to-pole gradient.

Variation in obliquity approximates a simple sinusoidal function with a period of 41 000 years. Variations in eccentricity can be approximately described with characteristic period of 100 000 years. In actuality, variation in eccentricity is more complex, and is more accurately described with periods of 400 000, 123 000, 85 000, and 58 000 years. Similarly, variation in precession has characteristic periods of 23 000 and 19 000 years.

While Emiliani suspected $\delta^{18}O$ variations were related to variations of these "Milankovitch" parameters, the first quantitative approach to the problem was that of Hays et al. (1976). They applied Fourier analysis to the $\delta^{18}O$ curve, a mathematical tool that transforms a complex variation such as that in Figure 12.3 to the sum of a series of simple sine functions. Hays et al. then used spectral analysis to show that much of the spectral power of the $\delta^{18}O$ curve occurred at frequencies similar to those of the Milankovitch parameters. The most elegant and convincing treatment, however, was that of Imbrie (1985). Imbrie's treatment involved several refinements and extension of the earlier work of Hays et al. (1976). First, he used improved values for Milankovitch frequencies. Second, he noted these Milankovitch parameters might vary with time, as might the climate system's response to them, because the Earth's orbit and tilt are affected by the gravitational field of the Moon and other planets. In addition, other astronomical events, such as bolide

impacts, can affect them. Thus, Imbrie treated the first and second 400 000 years of Figure 12.3 separately.

Imbrie observed that climate does not respond instantaneously to forcing. For example, maximum temperatures are not reached in Ithaca, New York, until mid or late July, three to four weeks after the maximum insolation, which occurs on June 21. Thus, there is a *phase* difference between the forcing function (insolation) and climatic response (temperature). Imbrie also pointed out that the climate might respond differently to different forcing functions. As an example, he pointed to temperature variations in the Indian Ocean, which respond both to annual changes in insolation and to semiannual changes in ocean upwelling. The response to these two forcing functions differs in different localities. The extent to which climate responds to a particular forcing function is the *gain*. The phase lag may also differ from locality to locality. Mathematically, the climatic response can be expressed as:

$$y = g_1(x_1 - \phi_1) + g_2(x_2 - \phi_2) \qquad (12.3)$$

where y is the climatic response (temperature), x_1 and x_2 are the two forcing functions (insolation and upwelling), g_1 and g_2 are the gains associated with them, and ϕ_1 and ϕ_2 are the phase lags.

Imbrie (1985) constructed a model for response of global climate (as measured by the $\delta^{18}O$ curve) in which each of the six Milankovitch forcing functions was associated with a different gain and phase. The values of gain and phase for each parameter were found statistically by minimizing the residuals of the power spectrum. The resulting model is shown in comparison with the data for the past 400 000 years and the next 25 000 years in Figure 12.5. The model has a correlation coefficient, r, of 0.88 with the data. Thus, about r^2, or 77%, of the variation in $\delta^{18}O$, and therefore presumably in ice volume, can be explained by Imbrie's Milankovitch model. The correlation for the period 400 000–782 000 years is somewhat poorer, around 0.80, but nevertheless impressive. Imbrie's work has, of course, not been the last word on this and models of this type have become more sophisticated in succeeding decades (e.g., Berger, 2013). Subsequent research has identified Milankovitch signals in climate beyond waxing and waning

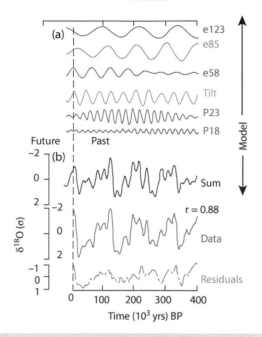

Figure 12.5 Gain and phase model of Imbrie relating to variations in eccentricity, tilt, and precession to the oxygen isotope curve.
(a) Variation in these parameters over the past 400 000 and next 25 000 years. (b) Sum of these functions with appropriated gains and phases applied and compares them with the observed data. Source: Imbrie (1985)/Geological Society of London.

of ice sheets, such as variation in the strength of the monsoon in Asia and Africa, which we will discuss below.

Since variations in the Earth's orbital parameters do not affect the average annual insolation the Earth receives, but only its pattern in space and time, one might ask how this could cause glaciation. The key factor seems to be the insolation received during summer by high northern latitudes. This is, of course, the area where large continental ice sheets develop. The Southern Hemisphere, except for Antarctica, is largely ocean and is therefore not subject to continental ice sheets. Glaciers apparently develop when summers are not warm enough to melt the winter's accumulation of snow.

Nevertheless, the total variation in insolation is small, and not enough by itself to cause the climatic variations observed. Apparently, there are feedback mechanisms at work that serve to amplify the fundamental Milankovitch forcing function. One of these feedback mechanisms was identified by Agassiz, and

that is ice albedo, or reflectance. Snow and ice reflect much of the incoming sunlight back into space. Thus as glaciers advance, they cause further cooling. Any additional accumulation of ice in Antarctica, however, does not result in increased albedo, because the continent is fully ice covered even in non-glacial periods, hence the dominant role of Northern Hemisphere insolation in driving climate cycles. Isotope geochemistry provides some insights into two additional feedback mechanisms, ocean circulation and carbon dioxide, and we discuss them in the subsequent sections.

12.2.3   Quaternary continental isotopic records

Climate change has left an isotopic record on the continents as well as in the deep sea. The record may be left directly in ice, in carbonate precipitated from water, or in clays equilibrated with water. We will consider examples of all of these in this section.

Continental δD and $\delta^{18}O$ isotopic compositions depend on the temperature in the area where the precipitation falls. However, as we found in Chapter 10, they also depend on additional factors, including the ice volume effect on ocean $\delta^{18}O$, the isotopic composition of water in the vapor source area, isotopic fractionation during condensation and evaporation of precipitation, atmospheric and oceanic circulation patterns, and seasonal temperature and precipitation patterns. Thus, climatic variations beyond temperature, e.g., precipitation, can be recovered from these records, but this requires taking all these factors into account.

12.2.3.1   Antarctic and Greenland ice cores

Climatologists recognized early on that continental ice preserves a stratigraphic record of climate change. Some of the first ice cores recovered for this purpose and analyzed for stable isotope ratios were taken from Greenland in the 1960s (e.g., Camp Century Ice Core; Dansgaard et al., 1969). Subsequent cores have been taken from Greenland, Antarctica, and various alpine glaciers. Very long ice cores that covered 150 000 years were first recovered by the Russians from the Vostok station in Antarctica in the 1980s eventually reaching back 400 000 years. In 2005, the EPICA (European Project for Ice Coring in Antarctica) project completed drilling

3270 m of ice core from ice Dome C, extending back through eight glacial cycles and more than 800 000 years. Other Antarctic cores have also been taken in regions of more rapid ice accumulation that, while not extending back as far in time, provide higher resolution.

Hydrogen isotopes show a much larger range and much greater temperature-dependent fractionation that oxygen, so in ice, interest centers of δD. Figure 12.6 compares the EPICA δD record with the marine $\delta^{18}O$ record; there is good agreement between the marine $\delta^{18}O$ record and the EPICA δD record back to ≈ 800 ky before present (BP). The lowermost 60 m of ice, however, appears to have been deformed and does not provide a reliable record. The core also provides a record of atmospheric CO_2, O_2, and N_2 from gas trapped in bubbles and we will return to this in a subsequent section.

Jouzel et al. (2007) used the deuterium excess parameter, d, (see Section 10.2.3 in Chapter 10) in combination with global circulation models that incorporate Rayleigh fractionation of water isotopes to convert δD to temperature in Antarctic snow. Their results are shown in Figure 12.7. Spectral analysis of the Vostok and EPICA isotope records shows strong peaks in variance at 41 kyr (the obliquity frequency) and at 23 kyr (the precessional frequency). Thus, the ice core data appear to confirm the importance of Milankovitch climatic forcing. It is interesting and significant

that even in this core, taken at 78°S, it is primarily insolation at 65°N that is the controlling influence. There are, however, some differences between the ice record and the marine and Greenland records, which we discuss below.

To compliment the remarkable record of the Antarctic ice cores, drilling was begun on deep ice cores at the summit of the Greenland ice cap in the late 1980s. The longest of these, NGRIP, over 3000 m deep, was completed by the European consortium in 2003 and successfully recovered a record of the entire last glacial cycle. Because snowfall rates are higher in Greenland than they are in central Antarctica, Greenland records cover less time. On the other hand, they provide more detailed climate records of the Holocene and the last glacial cycle. They also provide a record of climate in the Northern Hemisphere and the North Atlantic region, which has been ground zero Quaternary glacial cycles. Over roughly to the past 120 000 years, there is a good correlation between the NGIRP record and marine $\delta^{18}O$ and Antarctic ice records. There are differences, however. For example, interstadials in Greenland are characterized by abrupt warming whereas they are characterized by gradual warming Antarctica.

Greenland ice cores reveal that climate in the last glacial interval, spanning the period from roughly 110 000 years ago to 14 000 years ago, was much more variable, at least in the

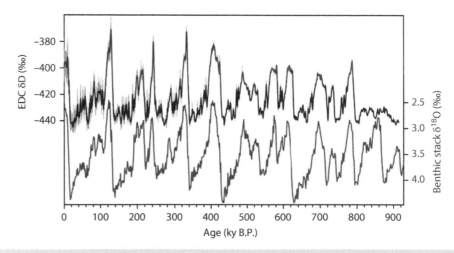

Figure 12.6 Comparison of the δD record of the EPICA ice core (top, black) with the marine benthic carbonate $\delta^{18}O$ curve of Lisiecki and Raymo (2005). Source: Jouzel et al. (2007)/American Association for the Advancement of Science.

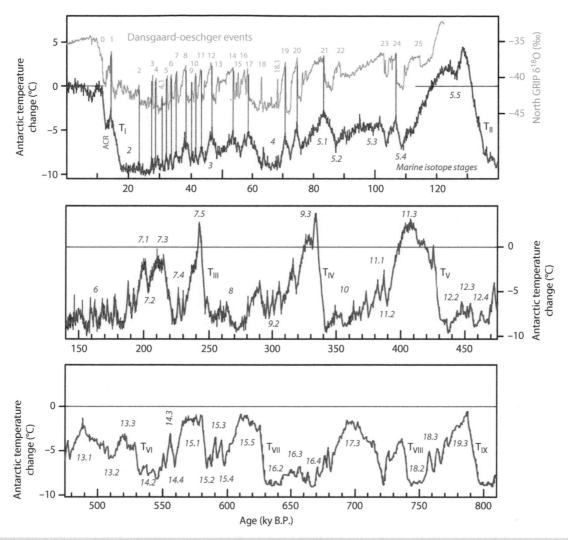

Figure 12.7   Antarctic temperature variation calculated from the EPICA ice core δD compared to the δ¹⁸O record from the NGRIP ice core from Greenland. Source: Jouzel et al. (2007)/American Association for the Advancement of Science.

Northern Hemisphere, in addition to being colder on average. There are cold periods that ended with rapid warming of 5°C or more on timescales of a few decades. The transition back to cold episodes was much slower. These rapid fluctuations in climate are known as *Dansgaard–Oeschger events*. In some cases, *Heinrich events* occur in the cold spells that precede Dansgaard–Oeschger warming. Heinrich events are characterized by layers of unusually coarse-grained sediment found in deep-sea cores from the North Atlantic. This material is interpreted as ice-rafted debris carried by icebergs originating from the Laurentide Ice Sheet by way of the Labrador Sea (and possibly Europe in some cases). They signal instability in the ice sheet and more rapid flow of glaciers draining it, but whether that

results from their reaching a critical mass initiating instability or externally driven warming is debated. Regardless, melting of the icebergs would have produced a surface layer of cold, fresh water that would have inhibited the density-driven thermohaline circulation patterns in the Atlantic, the *Atlantic meridional overturning circulation* (AMOC), which in turn would drive global-scale climate fluctuations. Ten Heinrich events have been recognized over the last glacial period. Like MIS, Dansgaard–Oeschger and Heinrich events are labelled with numbers going backward in time, as are *Terminations*, as the ends of ice ages are known, with Roman numerals used for terminations.

Figure 12.7 shows that cool periods preceding Dansgaard–Oeschger events correlate with periods of warming in Antarctica, suggesting coupling of climate between the two hemispheres. Thus, while Northern Hemisphere and Southern Hemisphere climate change is approximately in phase on glacial–interglacial timescales, it can be out of phase on millennial timescales (a phenomenon known as the *bipolar seesaw*). The causes of these events are still being debated, but the prime suspect is reduction in the AMOC that produces rapid cooling in the Northern Hemisphere, but slow warming in the Southern Ocean.

A number of other chemical and physical parameters are being or have been measured in the Greenland ice cores. One of the more important findings to date is that cold periods were also dusty periods (again, this had previously been suspected from marine records). Ice formed in glacial intervals has higher concentrations of ions, such as Ca^{2+} and Na^+, derived from sea salt and from calcite and other minerals in soils in arid regions, indicating higher atmospheric dust transport during glacial periods, reflecting conditions that were both drier and windier.

12.2.3.2 Speleothem records

Calcite precipitating from groundwater such as in veins or stalactites and stalagmites (collectively called speleothems) and flowstone in caves provide yet another climate record, as the isotopic composition of the calcite reflects that of the precipitation. A distinct advantage of these records is that, unlike ice and marine sediment records, they can be very accurately dated by U–Th decay series disequilibrium (Chapter 3). There are disadvantages as well. First, growth rates can be slow (1 mm/yr to less than 1 μm/yr) and discontinuous, limiting resolution. Second, as mentioned above, the relationship between the isotopic composition of this calcite and climate is complex, depending not only on temperature, but also on the amount of precipitation and its isotopic composition and the amount of evaporation. Third, the latter depend on local and regional climate variations that can differ significantly from global trends. This, of course, is an advantage if regional climate variations are the principal interest, and indeed much of the research of speleothems focuses on local and regional climate variations. These too, however, are influenced by Milankovitch cycles of insolation.

The study of Cheng et al. (2016) provides an example of how speleothems can record regional climate variations. Sanbao Cave is located in the Qinling Mountains of Central China, where 80% of the nearly 2 m annual precipitation occurs during the summer Asian monsoon. Cheng et al. collected four stalagmites from deep within the cave where humidity is a constant 100%. They measured $\delta^{18}O$ on 196 samples from the stalagmites and dated them using the $(^{230}Th/^{238}Th)$ chronometer (we discussed this dating in Chapter 4). Rather than temperature, $\delta^{18}O$ in the limestone mainly reflects a combination of the extent of isotopic fractionation of water vapor sourced from the Indian Ocean and the Pacific Ocean, and total annual proportion from summer monsoon rainfall – both are a measure of the strength of the Asian monsoon, with high $\delta^{18}O$ indicative of a strong monsoon.

Their results, combined with previously published data for the interval from 384 ka to present, are compared with Milankovitch variations, insolation variation, and sea level in Figure 12.8. The record reveals millennial-scale variations superimposed on longer wavelength variations that track Northern Hemisphere summer insolation.

There are several interesting aspects to these results. First, $\delta^{18}O$ is strongly correlated with Northern Hemisphere summer insolation on the longer timescales ($>10^4$ yr), clearly demonstrating the dependence of Asian monsoon strength on summer insolation on these timescales (this is not surprising as the monsoons are driving by summer warming of the land surface). Second, terminations, grey bands on Figure 12.8, are associated with weak monsoon intervals and each occurred during the rising limbs of Northern Hemisphere summer insolation. Third, glancing at both glacial and continental climate records, we immediately see an approximately 100 000 year cyclicity, and it is tempting to relate this to eccentricity. However, as Cheng et al. point out, durations between terminations were about 93, 105, 92, 92, 113, and 115 kyr, and thus the "100 kyr cycle" is merely an approximate average. Furthermore, eccentricity is high for terminations T-II, T-IIIa, T-III, T-IV, T-VIIa, and T-VII and low for terminations T-I, T-V, and T-VI. Termination timing also does not exhibit an obvious relationship to obliquity cycles. On the other hand, as Figure 12.8 shows, the terminations are separated by an integral number, 4 or 5, of

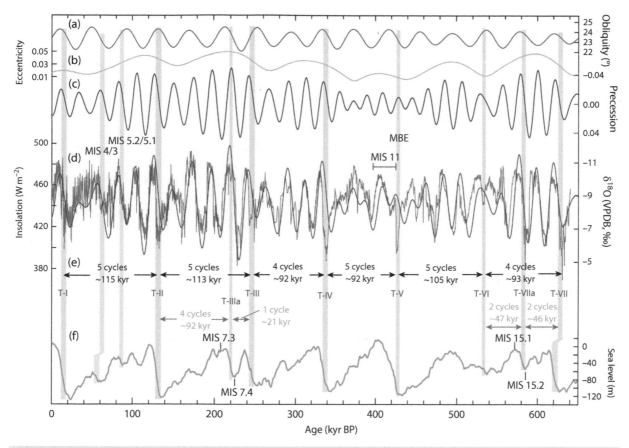

Figure 12.8   Changes in (a) obliquity, (b) eccentricity and (c) precession, (d) $\delta^{18}O$ record (green) from Sanbao Cave and (e) 21 July insolation at 65°N and termination pacing and duration, and (f) composite sea level curve. Source: Cheng et al. (2016)/with permission of Springer Nature.

precession cycles and terminations tend to occur at or shortly following precession maxima. Finally, weak monsoon intervals also coincide with Heinrich events suggesting that disruption of the AMOC provides the link between ice sheet collapse and Asian monsoon strength. Superimposed on the precession cycle, timescale variations in $\delta^{18}O$ are higher-frequency millennial-scale wiggles. When Chen et al. filtered out the longer-frequency variations to examine these high-frequency events, they found that they anti-correlated with insolation.

12.3   ISOTOPES IN PALEOCEANOGRAPHY

If the distribution of solar energy received by the Earth were the only factor in climate change, we would expect that the glaciation in the Southern Hemisphere and the Northern Hemisphere would be exactly out of phase. This, however, is not the case, at least on long timescales. Broecker (1984 and subsequent papers) argued that changes in deep circulation of the ocean play a key role in producing globally synchronous climate variation. Given the importance of ocean circulation in climate variations, we will interrupt our discussion of climate to examine how isotopes can be used to understand how ocean circulation has varied.

The role of surface ocean circulation in our present climate is clear; for example, the south-flowing California Current keeps the West Coast of the United States relatively dry and maintains more moderate temperatures in coastal regions than they would otherwise be, while the Gulf Stream maintains more moderate temperatures in Western Europe. The role of the deep, or *thermohaline*, circulation of the oceans is less obvious, but no less important. Whereas the surface ocean circulation is wind-driven, the deep circulation is driven by density, which is in turn controlled by temperature and salinity.

As we found in Chapter 10, in the present ocean, most deep ocean water masses "form" in high latitudes. Once these deep-water masses form, they do not return to the surface for nearly a thousand years. The principal site of deep-water formation today is the Southern Ocean where the Antarctic Intermediate Water (AAIW) is formed in the Antarctic Convergence and Antarctic Bottom Water (AABW), the densest of ocean water masses, is formed near the coast of Antarctica, particularly in the Weddell Sea. A lesser amount of deep water is also formed in the Labrador, Greenland, and Norwegian Seas of the far northern Atlantic when warm, salty water from the Gulf of Mexico and the Mediterranean is strongly cooled during winter; this water mass is called North Atlantic Deep Water (NADW). After formation, this water sinks to the bottom of the ocean and flows southward. The northward flux of warm surface water and southward flux of cold deep water constitute the AMOC. Today, NADW is the deepest and densest water mass in the North Atlantic. In the South Atlantic, the somewhat cold and denser AABW flows northward beneath the NADW, which is in turn overlain by AAIW.

Formation of deep water thus involves loss of thermal energy by the ocean to the atmosphere and the present thermohaline circulation of the oceans keeps high-latitude climates milder than they would otherwise be. In particular, energy extracted from the Atlantic Ocean water in the formation of NADW keeps the European climate relatively mild. As we will see, changes in the AMOC have played an important role in the Pleistocene climate seasaw.

12.3.1 Carbon isotopes in paleoceanography

We saw in Chapter 10 that $\delta^{13}C$ is lower in deep water than in surface water (Figure 12.11): photosynthesis in the surface waters discriminates against ^{13}C, leaving the dissolved inorganic carbon of surface waters with high $\delta^{13}C$ while oxidation of falling organic particles rich in ^{12}C lowers $\delta^{13}C$ of dissolved inorganic carbon in deep water: in effect, ^{12}C is "pumped" from surface to deep

water more efficiently than ^{13}C. $\delta^{13}C$ values in the deep water are not uniform, varying with the "age" and origin of deep water: the longer the time since the water was at the surface, the more enriched it becomes in ^{12}C and the lower the $\delta^{13}C$. Since this is also true of total inorganic carbon and nutrients, such as PO_4 and NO_3, $\delta^{13}C$ correlates negatively with nutrient and ΣCO_2 concentrations. NADW has relatively high $\delta^{13}C$ because it contains water that was recently at the surface (and hence depleted in ^{12}C by photosynthesis). Deep water is formed neither in the Pacific Ocean nor the Indian Ocean; all deep waters in those oceans flow in from the Southern Ocean. Hence deep water in the Pacific, being rather "old," has low $\delta^{13}C$. AABW is a mixture of young NADW and recirculated Pacific deep water and hence has lower $\delta^{13}C$ than NADW. Thus, these water masses can be distinguished on the basis of $\delta^{13}C$.

Examining $\delta^{13}C$ in benthic foraminifera in cores from a variety of locations, Oppo and Fairbanks (1987) concluded that production of NADW was lower during the last glacial maximum and increased to present levels in the interval between 15 000 and 5000 years ago. Figure 12.9 shows an example of data from core RC13–229, located in the South Atlantic. $\delta^{13}C$ values decrease as $\delta^{18}O$ increases. As we saw in the previous sections, $\delta^{18}O$ in marine carbonates is a measure of glacial ice volume and climate. As the climate warmed at the end of the last glacial interval, $\delta^{13}C$ values in bottom water in the South Atlantic increased, reflecting an increase in the proportion of NADW relative to AABW in this region. Thus, the mode of ocean circulation apparently changes between glacial and interglacial times; this change may well amplify the Milankovitch signal.

Subsequent carbon isotope studies of benthic foraminifera have established that deep circulation in the Atlantic was much different than it is today (Figure 12.10). Rather than formation of cold, dense NADW (characterized by heavy carbon) in the far northern Atlantic, a less dense water mass called "Glacial North Atlantic Intermediate Water" formed further south in the Atlantic. Rather than sinking to the bottom of the North Atlantic, it sunk only to depths of 1500 m or so and flowed southward above the north flowing glacial AABW, characterized by light carbon, which

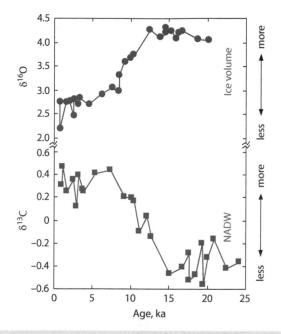

Figure 12.9 Variation in $\delta^{18}O$ and $\delta^{13}C$ in benthic foraminifera from core RC13–229 from the eastern South Atlantic. $\delta^{13}C$ data suggest the proportion of NADW in this region increased as the climate warmed. Data from Oppo and Fairbanks (1987).

penetrated all the way into the North Atlantic. Today, production of NADW releases heat to the atmosphere, warming it. In the absence of NADW production in glacial times, the North Atlantic region, and Europe in particular, would have been much colder.

12.3.2 Radiogenic isotopes in paleoceanography

Chemical sediments, including both biologically precipitated ones such as foraminiferal shells and Fe-hydroxide coatings on them can record the isotopic composition of seawater at the time of precipitation. This allows the use of Nd isotopes to assess changes in ocean circulation during the Pleistocene Ice Ages.

Labrador Sea Water contributes to NADW formation and gives it a uniquely unradiogenic Nd flavor making NADW relatively easy to track. Figure 12.11 compares Nd isotope ratios measured on sediment cores recovered from the Bermuda Rise at $\approx 33°N$ and the Cape Basin at $\approx 41°S$ with O isotopes in the NGRIP ice core from Greenland. ε_{Nd} at the Bermuda

Rise site reaches a maximum, approaching ε_{Nd} of AABW values (–7 to –9) during the last and preceding glacial maxima at $\approx 20\,000$ and $135\,000$ years BP, indicative of a near shutdown of southward flowing NADW. Also of note are the extreme negative ε_{Nd} at around $10\,000$ and $80\,000$ to $90\,000$ BP. This suggests that at these times water from the Labrador Sea was a much larger component of NADW than at present. Variations in ε_{Nd} at Cape Basin site in the South Atlantic vary in a similar but subdued way reflecting the varying contribution of NADW to AABW. Böhm et al. (2015) concluded that NADW formation persisted for much of the last glacial cycle, but that NADW formation shut off entirely in the last two glacial maxima and southern-sourced waters AABW filled the deep North Atlantic instead.

Another example of how Nd isotopes in seawater have helped unravel ocean and climate history involves the Ocean Anoxic Event 2 (OAE 2), which as the name implies was a time during which bottom water in much of the world ocean became anoxic in the late Cretaceous (≈ 94 Ma), leading to an extinction that defines the boundary between the Cenomanian and Turonian Stages. This is recorded in sediments deposited at the time as high concentrations of organic carbon (locally, as high as 25–30%). The resulting loss of isotopically light carbon (i.e., low $\delta^{13}C$) from the ocean is recorded as a shift to isotopically heavier carbon in marine carbonates deposited at this time. The question is: why did this happen?

The event was an interval of extreme greenhouse conditions, with large temperature swings and high atmospheric pCO_2 lasting approximately 850 ka, and clearly involved disruption of the ocean–atmosphere system, including both climate and ocean circulation (e.g., O'Connor, 2020). Nd isotope ratios at sites as dispersed as ODP Site 1258 in the Central Pacific and the English Chalk formation of Southeast England (which form the famous "White Cliffs of Dover") reveal a distinct shift to more positive values coincident with the carbon isotope excursion (e.g., Zheng et al., 2013). This provided the first clear evidence that the trigger for this event was massive volcanic eruptions associated with a large igneous province (LIP).

Osmium isotopes ratios show an even more dramatic shift to low values; Du Vivier et al. (2015) found that Os isotope ratios in

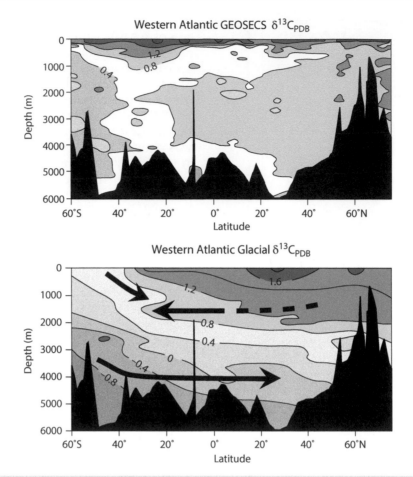

Figure 12.10 Cross-section of $\delta^{13}C$ in the North Atlantic today and during glacial times. As discussed in the text, different water masses have different $\delta^{13}C$ signatures. Source: Curry and Oppo (2005)/with permission of John Wiley & Sons.

sediments at six distinct sites, including one from Colorado, two European sites, and two ODP Atlantic sites, all record a dramatic shift in $^{187}Os/^{188}Os$ from typical Cretaceous seawater values of ≈ 0.8 to values <0.2 and in some cases approaching the mantle value of ≈ 0.13. The timing corresponds to the eruption of the Caribbean LIP, which constitutes much of the present Caribbean Plate and which can be traced to the Galapagos mantle plume, although an LIP in the high Arctic is possible alternative. The connection is complex and indirect, likely involving release of CO_2 and SO_2, which in turn produces climatic changes and increased weathering on continents and consequent enhanced nutrient delivery to the oceans (e.g., O'Connor, 2020).

As we pointed out in Chapter 2, the Os isotopic composition of seawater has varied through time as a consequence of variation in the proportion of mantle and crustal fluxes to seawater. As Figure 12.12 shows, there has been a particularly rapid increase in $^{187}Os/^{188}Os$, similar to that observed for $^{87}Sr/^{86}Sr$ (Figure 2.12). The likely cause of both is an increase of continental weathering flux resulting from Cenozoic mountain-building, most notably the rise of Himalayas, but also the Alps, Rockies, and Andes (Peucker-Ehrenbrink et al., 1995). It may also reflect a decreasing hydrothermal flux resulting from decreasing sea floor spreading rates. The geochemical behavior of both at the surface of the Earth is related to carbon, but while Sr is concentrated in carbonates, Os is concentrated in organic-rich sediments.

Very low $^{87}Os/^{188}Os$ occurred exactly at the Cretaceous–Paleogene boundary at 65.5 Ma

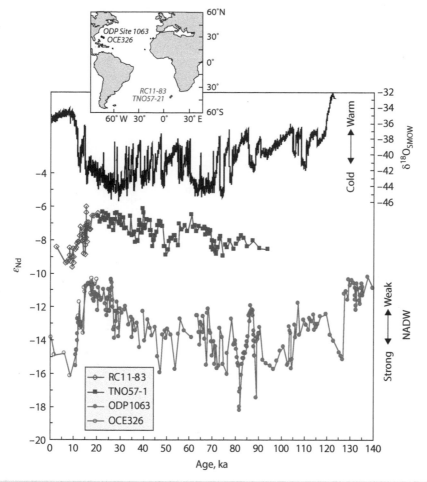

Figure 12.11 ε_{Nd} in Bermuda Rise sediment cores OCE326–GGC6 (red open symbols; Roberts et al., 2010) and ODP 1063 (red solid; Böhm et al. 2015) and cores RC11–83 and TNO57–21 from the Cape Basin of the Southeastern Atlantic (Piotrowski et al., 2005) through the last glacial cycle compared with $\delta^{18}O$ in the North Greenland Ice Core Project (NGRIP) ice core from Greenland (Andersen et al., 2004), which reflects Northern Hemisphere temperature variations over the last glacial cycle.

(previously known as the Cretaceous–Tertiary or K–T boundary). The inset of Figure 12.12 shows a high-resolution study of the Gubbio Formation in Italy by Robinson et al. (2009). The lowest ratios occur right at the Cretaceous–Paleogene boundary and are associated with elevated Ir, Os, and Pt concentrations, and are thus almost certainly due to chondritic Os ($^{187}Os/^{188}Os$ = 0.128) delivered by the Chicxulub impactor. The study shows, however, that $^{187}Os/^{188}Os$ began to decline more than 500 000 years before the K–Pg boundary and similar declines are observed in several ODP cores. One possibility is that diagenetic remobilization has smeared out the impactor signal, but based on platinum group element concentrations, Robinson et al. concluded that

in the Gubbio this affects $^{187}Os/^{188}Os$ no more than a meter (corresponding to roughly 70 000 years) from the boundary. The authors conclude the decline in $^{187}Os/^{188}Os$ was a consequence of Deccan LIP, the eruption of which marks the surfacing of the Réunion mantle plume and whose timing coincides well with this decline.

On long timescales, both Nd and Pb isotope ratios in manganese crusts suggest the present deep circulation of the Atlantic, and the characteristic radiogenic Pb and unradiogenic Nd of deep water, has only been established within the last 8 Ma (O'Nions et al., 1998). The cause of the isotopic shift is unclear. Smaller shifts in ε_{Nd} are observed in the Pacific around 3–5 Ma, which corresponds to the closure of the

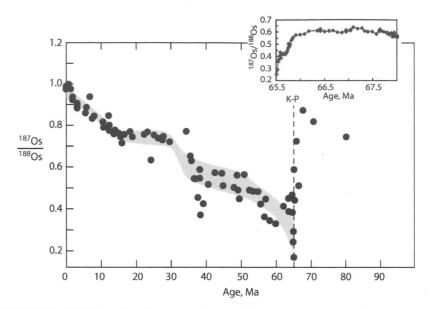

Figure 12.12 Os isotope composition of seawater over the last 80 Ma based on data in Peucker-Ehrenbrink et al. (1995) and Peucker-Ehrenbrink and Ravizza (2000). Light blue field represents Peucker-Ehrenbrink et al. (1995) best estimates of seawater Os isotopic composition. Inset shows the data from the late Cretaceous Gubbio Formation of Robinson et al. (2009). K–Pg is the Cretaceous–Paleogene boundary.

Isthmus of Panama. O'Nions et al. (1998) speculate this may reflect the flow of NADW into the Pacific. In contrast, the relatively unradiogenic nature of Pb in the North Pacific seems to have been maintained throughout the Cenozoic (Chen et al., 2013), reflecting the very short residence time of Pb.

As Figure 2.12 shows, the $^{87}Sr/^{86}Sr$ of seawater has increased dramatically through the Cenozoic from about 0.7078–0.70925 at present. This changing isotopic composition results from changes in either the hydrothermal Sr flux, the riverine Sr flux, or the $^{87}Sr/^{86}Sr$ of the riverine flux, or some combination of these. Richter et al. (1992) argued that the changes are too great to be explained by the hydrothermal flux and that the most likely explanation relates to the rise of the Himalayas. In support of that hypothesis, they noted that the most rapid change in seawater $^{87}Sr/^{86}Sr$ occurred between 15 and 20 Ma, a time of exceptionally high erosion rates in the Himalayas. Sr in rivers draining the Himalayas is also exceptionally radiogenic. The Ganges, e.g., has $^{87}Sr/^{86}Sr$ of 0.725 while that of the Brahmaputra is 0.720 compared to a global average riverine $^{87}Sr/^{86}Sr$ of 0.7111 (Peucker-Ehrenbrink et al., 2010).

Derry and France-Lanord (1996) suggested that in this case the connection between seawater $^{87}Sr/^{86}Sr$ might be the opposite of expected. They argued that reduced erosion rates but increased weathering intensity during the Pliocene released proportionally more Sr minerals with high Rb/Sr like biotite and less from low Rb/Sr minerals, such as calcite and plagioclase. The result was a decrease in the Himalayan riverine flux but an increase in its $^{87}Sr/^{86}Sr$ in the Pliocene.

12.3.3 Calcium isotopes in paleoceanography

Farkaš et al. (2007) attempted to deduce the Ca isotopic history of seawater over Phanerozoic time from the isotopic composition of marine carbonates and phosphates. The results suggest that $\delta^{44}Ca$ has increased by $\approx 0.7‰$ over the last 500 Ma, although the path has been bumpy (Figure 12.13). Additional data have supported these results while supplying somewhat clearer picture (Blättler et al., 2012). Seawater $\delta^{44/40}Ca$ was about 0.65‰ lower than present in the early Paleozoic and rose dramatically in the Carboniferous to near-modern values before declining through the Permian

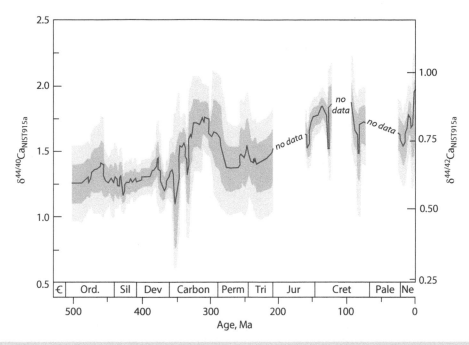

Figure 12.13 Ca isotopic composition of seawater over the last 500 Ma calculated from the isotopic composition of marine carbonates and phosphates. Red line shows a 10 point moving average; dark and light grey regions show the 1σ and 2σ bounds, respectively. Source: Farkaš et al. (2007)/with permission of Elsevier.

to values ≈ 0.4‰ lower than modern. δ^{44}Ca then began to rise again in the Jurassic reaching near-modern values in the Cretaceous. Farkaš et al. concluded these variations reflected a change in nature of the dominant Ca output fluxes from dominantly aragonitic $CaCO_3$ produced by corals and other shallow water organisms to dominantly calcic $CaCO_3$ produced by planktonic foraminifera, which first arose in the Jurassic (although benthic forams evolved much earlier). Higgins et al. (2018) suggested that instead, diagenesis and dolomitization of shallow water carbonate might be the controlling factor.

Because of the close coupling of the Ca and carbon cycles, Ca isotopes have the potential to elucidate changes in the biogeochemical carbon cycle recorded by δ^{13}C in marine carbonates, such as the decline in δ^{13}C of roughly 3‰ over several hundred thousand years at the Permian–Triassic boundary, defined by the largest mass extinction, and in which $\delta^{44/40}$Ca declines by 0.3‰ (Payne et al., 2010). This δ^{13}C excursion is indicative of a major disruption of the global biogeochemical carbon cycle, including the collapse of ocean primary productivity and extinction of many calcifiers, but Komar and Zeebe (2016)

concluded that other factors must have been involved as well, including an increase in the carbonate weathering flux, and a change in $\Delta^{44/}Ca_{min-Ca^{2+}}$ due to changing $[HCO_3^-]$, all of which could have been secondary effects of the Siberian Traps eruption. A more extreme δ^{13}C excursion, -12‰, occurs in ≈ 560 Ma old Ediacaran strata, when multi-cellular animals (metazoans) were first appearing in the fossil record. There has been debate as to whether this excursion was merely diagenetic, but the δ^{13}C excursion is also observed in the most pristine samples which preserve distinctly aragonitic values of δ^{44}Ca, indicating δ^{13}C does indeed record major disruption of the global biogeochemical carbon cycle.

12.3.4   Silicon isotopes in paleoceanography

Deep-sea cores record changes in δ^{30}Si of biogenic silica of radiolarians, diatoms, and sponges in the $\approx 25\,000$ years since the last glacial maximum that are interpreted to reflect changes in the relative abundances of biolimiting nutrients, notably Fe, N, and Si (Hendry and Brzezinski, 2014). Silica-utilizing organisms preferentially take up isotopically light Si, so assuming Rayleigh fractionation, high levels of photosynthesis and silica utilization in the surface waters will increase both $\delta^{30}Si$ surface water and in the diatom tests formed from it as Si utilization increases, such that eventually at complete utilization, $\delta^{30}Si_{diatom} = \delta^{30}Si_{sw}$. This will then be recorded in siliceous radiolarian and diatomaceous sediment on the ocean floor. Sponges live on the seafloor, so their spicules record bottom water $\delta^{30}Si$. The fractionation factor between seawater and sponge spicules depends on the silica concentration so that those growing silica-poor waters have the lowest $\delta^{30}Si$. These changes are in turn thought to reflect changes in delivery of nutrients to surface waters both from the atmosphere (wind-derived dust during glacial periods enhanced Fe concentrations) and ocean circulation (upwelling returning N and Si to the surface).

Hendry et al. (2014) analyzed $\delta^{30}Si$ in sponge spicules from core KNR140-2-56GGC taken at 1700 m depth in the western North Atlantic and found that $\delta^{30}Si$ was higher prior to the end of the last ice age 12 000–14 000 years ago (Figure 12.14), indicating deep-water Si concentrations were higher during the last glacial period. This location is presently bathed in NADW but intermediate depths of the North Atlantic were occupied by Glacial North Atlantic intermediate water during the last glacial period (Figure 12.10). Particularly elevated $\delta^{30}Si$ is found in spicules from a particularly cold period around 16 000 years ago associated with Henrich event 1. Hendry et al. suggest the implied higher Si concentrations at this site then may have been a result of northward flow of AAIW.

Alternatively, Frings et al. (2016) argue that the cooler, drier glacial climate, with larger continental shelves, large ice sheets, and altered vegetation resulted in substantially lower mean $\delta^{30}Si$ of dissolved Si delivered to the ocean from the continents. Improved understanding of the silica cycle is needed before $\delta^{30}Si$ can be fully utilized as a paleo-oceanographic proxy.

On longer timescales, Fontorbe et al. (2016) used $\delta^{30}Si$ in sponge spicules and radiolarians to document that the low dissolved silica concentrations in the Atlantic have persisted for at least 60 million years. Tatzel et al. (2017) found that $\delta^{30}Si$ in cherts decreased across the Ediacaran–Cambrian (Proterozoic–Phanerozoic) boundary, which they attribute to the expansion of the siliceous sponges and

speculate that associated nutrient balances may have led to the rise in atmospheric oxygen that occurred then. Prior to the evolution of sponges, and radiolarians, which first appeared in the Early Paleozoic, most cherts appear to be abiotic and have a large range of $\delta^{30}Si$, reaching a peak in the Mid-Proterozoic of +2.2–3.9‰ followed by a decrease to eventual modern levels. Ding et al. (2017) speculated that this reflected a decrease in ocean Si concentrations due to increased biological utilization.

12.4   CLIMATE IN THE CENOZOIC

12.4.1   The Cenozoic marine $\delta^{18}O$ record

Imbrie's (1985) analysis suggests that the climate system's response to Milankovitch forcing has changed even over the last 800 000 years. The present glacial–interglacial cycles began only two million years ago, yet the orbital variations responsible for Milankovitch forcing should be more or less stable over tens of millions of years. They should provide a steady and, hence, predictable pacing of climate change. Indeed, this pattern can be seen in high-resolution marine isotopic records, such as Figure 12.15. The amplitude of the isotopic and climatic variations, however, has not been constant, but rather has increased with time, particularly since the beginning of the mid-Pliocene (about three million years ago). The relative importance of eccentricity, precession, and obliquity in forcing climate also appears to have changed over time. During the Pliocene and early Pleistocene, the 41 000 yr, obliquity component appears to be dominant (Zachos et al., 2001). In the last of the half of the Pleistocene periodicity was roughly, but not exactly, 100 000 years, reflecting a greater influence of precession. These differences do not reflect differences in the strength of the Milankovitch signals, but rather global climate's sensitivity to it. In addition to the Milankovitch-related wiggles, one also sees a clear trend toward higher $\delta^{18}O$ with time, suggesting general global cooling.

Longer-term changes in Cenozoic climate involved additional factors. Some of these changes were driven by plate tectonics and include widening of the North Atlantic, opening and widening of the Tasmanian and Drake

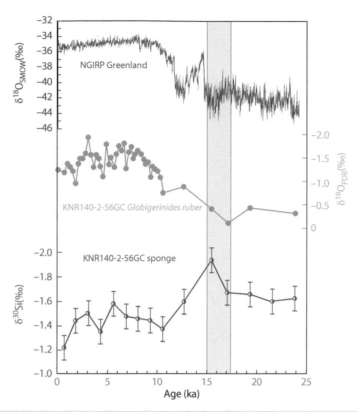

Figure 12.14 δ^{30}Si of sponge spicules in core KNR140-2-56GGC from the western North Atlantic (data of Hendry et al., 2014) compared with δ^{18}O$_{PDB}$ of the planktonic foraminifer *Globergerindes ruber* (data of Keigwin, 2001) and δ^{18}O$_{SMOW}$ in the NGRIP ice core from Greenland (data from Rasmussen et al., 2006). Shaded area is Heinrich event 1.

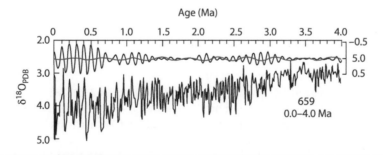

Figure 12.15 Variation in δ^{18}O in benthic foraminifera over the last four million years. Data are from Ocean Drilling Program Sites 659 in the eastern equatorial Atlantic. The upper curves represent Gaussian band-pass filters designed to isolate variance associated with the 400 (red) and 100 ky eccentricity cycles. Source: Zachos et al. (2001)/American Association for the Advancement of Science.

Passages, allowing for winds and currents in the Southern Ocean to circle the globe unrestricted, collision of India with Asia, and subsequent uplift of the Himalayas and Tibetan Plateau and uplift of Panama and closure of the Central American Seaway. Two other key factors are the growth of polar ice sheets and a decline in atmospheric CO_2.

Figure 12.16a shows the benthic foraminiferal δ^{18}O and δ^{13}C records for the entire Cenozoic. Westerhold et al. (2020) identify four distinctive climate modes: Hothouse and Warmhouse states of the Paleogene and Coolhouse and Icehouse states of the Neogene and Quaternary. During the Warmhouse and Hothouse states, global temperatures were more

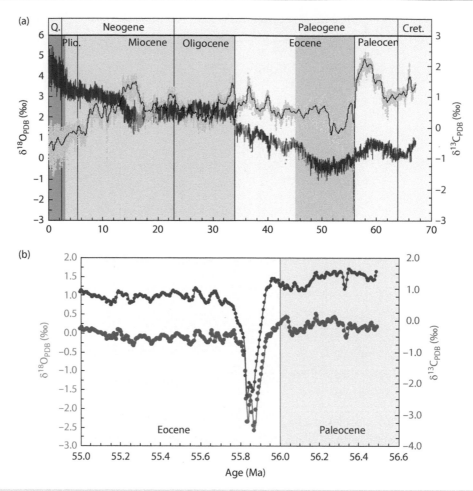

Figure 12.16 (a) Variation in $\delta^{18}O$ and $\delta^{13}C$ in benthic foraminifera during the Cenozoic from the CENGRID database of Westerhold et al. (2020). Lines show the smoothed curves. (b) Detailed view of the same data through the Paleocene–Eocene Thermal Maximum. Red shades show hothouse/warmhouse climates; blue shades show coldhouse/coolhouse climates.

than 5°C warmer than today and benthic $\delta^{13}C$ and $\delta^{18}O$ are positively correlated. The Hothouse conditions existed between the Paleocene–Eocene boundary at 56 Ma and the end of the Early Eocene Climate Optimum at 47 Ma, when temperatures were more than 10°C warmer today. This period is also characterized by transient warming events (hyperthermals), most notably the Paleocene–Eocene Thermal Maximum (PETM), a period of extreme temperature associated with mass extinction of benthic foraminifera that defines the geologic boundary between the Paleocene and Eocene epochs. Figure 12.16b shows this event in detail. This period began with an abrupt rise in global temperature by more than 5°C in less than 10 000 years. It is also characterized by a precipitous drop in marine $\delta^{13}C$,

indicative of a massive addition of isotopically light CO_2. We will return to question of how much CO_2 and its source in a subsequent section.

The Warmhouse transitioned into the Coolhouse state at the Eocene–Oligocene boundary, where $\delta^{18}O$ increases abruptly as a result of cooling temperatures and the development of large ice sheets in East Antarctic. Westerhold et al. (2020) divide the Coolhouse state into two phases based on the $\delta^{18}O$ increase at 13.9 Ma, which is related to the expansion of Antarctic ice sheets to West Antarctica. The first phase is characterized by warmer conditions culminating in the Miocene Climatic Optimum between ≈17 and 14 Ma. The second phase was characterized by cooling and increasing $\delta^{18}O$. An additional transition occurs around 7 Ma when $\delta^{13}C$ shows a large decrease. The current Icehouse state, characterized by waxing and waning of Northern Hemisphere ice sheets, begins around 3.3 Ma and was fully established by the Pliocene–Pleistocene transition.

Buried beneath these long-term trends, Milankovitch signals persist. Westerhold et al. found that in the Hothouse and Warmhouse, as well as the first Coolhouse phase, 100 000 year eccentricity-related cycles dominate. In the Coolhouse phase, the obliquity-band frequency response increases and comes to dominate climate dynamics by the late Miocene–early Pliocene. In the Icehouse state, the decrease in atmospheric CO_2 and major growth of Northern Hemisphere polar ice sheets, which enhanced variability in $\delta^{18}O$, steadily amplified the influence of complex high-latitude feedbacks with the precession signal rising in importance.

The Eocene–Oligocene shift is thought to represent the beginning of present system where temperature variations dominate thermohaline circulation in the oceans and is coincident with initiation of extensive East Antarctic glaciation. As we found previously, deep ocean water masses are formed at high latitudes and are dense mainly because they are cold. Deep water today has a temperature between 2 and –2°C. Before the Eocene, deep water appears to have been much warmer, and salinity differences played a larger role in thermohaline circulation. It was probably not until late Miocene that the present thermohaline circulation was completely established.

12.4.2 Soils and paleosols

Soil carbonates precipitate in semiarid and arid soils from infiltrating water and calcium release by weathering. The concentration of CO_2 dissolved in soil solutions is sourced from respiration within the soil with a variable contribution from atmospheric CO_2 and can reach partial pressures over 1% due to respiration within the soil. Cerling and Quade (1993) demonstrated a correlation in $\delta^{18}O$ between local average meteoric water and soil carbonate with an offset to higher $\delta^{18}O$ of the latter than one would expect from the water–calcite fractionation alone (Figure 12.17). This is because carbonates form from soil water, not precipitation, and they tend to form seasonally

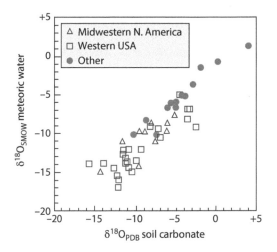

Figure 12.17 Relationship between $\delta^{18}O_{SMOW}$ in local average meteoric water and $\delta^{18}O_{PDB}$ of soil carbonate. The difference between $\delta^{18}O_{SMOW}$ and $\delta^{18}O_{PDB}$ is approximately equal to $\Delta^{18}O_{water\text{-}carbonate}$. Source: Cerling and Quade (1993)/with permission of John Wiley & Sons.

and consequently do not sample annual average precipitation.

A recent detailed study by Huth et al. (2019) of soil carbonate formation in a semi-arid region of the Colorado Plateau in southern Utah, US demonstrated that soil carbonate likely forms primarily in summer. Soil temperatures at the study site vary from $\approx 0°C$ in winter to $\approx 25°C$ in summer. Using clumped isotopic analysis of young (<10 ka) soil carbonate, they calculated temperatures of $24 \pm 4°C$. The calculated $\delta^{18}O_{SMOW}$ of soil water from which the soil carbonate formed was –8.6 to –5.1‰, which overlaps the observed $\delta^{18}O_{SMOW}$ of spring and summer soil water. By comparison, $\delta^{18}O_{SMOW}$ of precipitation at the site varied from $\approx$ –5‰ in summer to –18‰ in winter.

Soil carbonates can also record $\delta^{13}C$ of soil CO_2 but with a bias. Since most CO_2 in the soil is derived from plant matter being consumed by the soil fauna, $\delta^{13}C$ in soil carbonates primarily reflects the isotopic composition of this plant matter rather than atmospheric CO_2. Huth et al. found $\delta^{13}C$ in soil carbonates varied from –1 to –3.5‰ compared with $\delta^{13}C_{PDB}$ of soil gas that varied seasonally from $\approx$ –18‰ in summer to –24‰ in winter. In part, this reflected dominantly C_3 photosynthetic activity during winter and $C_3 + C_4 + CAM$

photosynthetic activity during the summer. In general, $\delta^{18}O$ and $\delta^{13}C$ in soil carbonates are best used to document changes over time rather than reconstructing absolute values in precipitation or soil.

Figure 12.18 shows one example of $\delta^{18}O$ in paleosol carbonates used in this way. Quade et al. (1989) documented an increase in $\delta^{18}O$ in paleosol carbonate in soils from the Siwaliks region of Pakistan beginning around 8 Ma and interpreted it as an intensification of the Monsoon system at that time, an interpretation consistent with marine paleontological evidence.

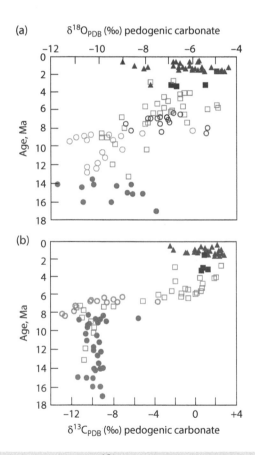

Figure 12.18 (a) $\delta^{18}O_{PDB}$ in paleosol carbonate nodules from the Potwar Plateau in northern Pakistan. The increase around 8 Ma is likely due to intensification of the Asian Summer Monsoon sourced from warm Indian Ocean water.
(b) $\delta^{13}C_{PDB}$ in the same carbonate nodules. The shift at around 6 Ma likely reflects the increasing abundance of C_4 plants. Different symbols correspond to different overlapping sections that were sampled. Source: Quade et al. (1989)/with permission of Springer Nature.

Quade et al. (1989) also measured $\delta^{13}C$ in these paleosol carbonates and notice a significant shift to heavier carbon between ≈ 7.5 and 6 Ma. They attributed this to a response to the uplift of the Tibetan Plateau and the development of the Monsoon. However, the timing differs from the shift in oxygen isotope occur about a million years earlier (at about 8 Ma). Subsequently, Cerling et al. (1993) noted that the timing coincides with C_4 grasses becoming dominant in the grasslands in Asia and North America and concluded the shift in $\delta^{13}C$ was due to a shift in vegetation from dominantly C_3 to C_4. We will explore this emergence of C_4 grasslands, and the animals that evolved with it, the *grassland biome*, further in Chapter 13.

Most soil carbonate studies analyze nodules, which typically integrate environmental conditions on timescales of 1000–1 000 000 years and consequently focus on changes over timescales longer than this. Huth et al. (2020) utilized micro-analytical techniques, microdrilling, laser ablation, and SIMS to determine ^{14}C ages, $\delta^{18}O_{PDB}$ and $\delta^{13}C_{PDB}$ on two transects through a 1–2 cm thick carbonate rind developed on the bottom of a boulder partially embedded in soil in the same Southern Utah locality as the Huth et al. (2019) study described above. From these data, they calculated variations in soil $\delta^{18}O_{SMOW}$ and soil temperature based on Δ_{47} clumped isotope analysis from 5 to 35 ka. Their results are shown in Figure 12.19. There was imperfect agreement of $\delta^{18}O$ in the two transects (red and blue), mainly between 12 and 8 thousand years, which could be due to issues with the age model, difficulty obtaining suitable analysis spots, data density (especially >15 ka), and/or small-scale variability in $\delta^{18}O$ of soil–water at the time of mineral formation, but the two transects are over all similar. They reveal significantly lower $\delta^{18}O$ during the last glacial maximum between 18 and 28 thousand years ago with a general increase over the last 10 000 years. At least in this region, soil carbonates form in summer and this region receives 45% of its precipitation in summer and early fall from the North American monsoon, which can be sourced either from the Gulf of California and the Gulf of Mexico; thus, the precipitation source could contribute to variation in $\delta^{18}O$. The pattern shows similarities to the speleothem pattern from Pink Panther Cave, New Mexico located 900 km to the southeast. Soil

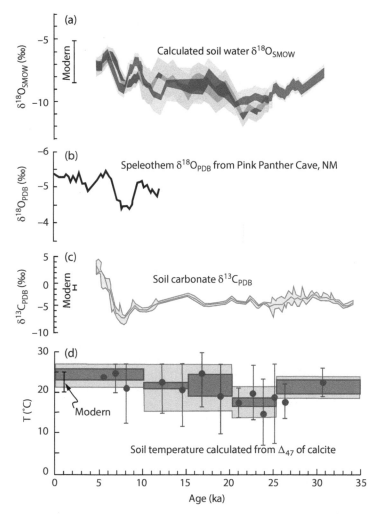

Figure 12.19   (a) Modelled soil water $\delta^{18}O_{smow}$ based on $\delta^{18}O_{PDB}$ measured along two transects (Transect 1: red; Transect 2: blue) though a carbonate rind from the Colorado Plateau. Darker fill based on the temperature estimate standard error and the lighter fill is the 95% confidence interval. (b) Moving average of $\delta^{18}O_{PDB}$ in speleothems from Pink Panther Cave, New Mexico. (c) $\delta^{13}C_{PDB}$ along Transit 1 of the carbonate rind, line is a moving average, and light color encompasses all data. Transit 2 was similar. (d) Carbonate formation temperature estimates through time for this study from Δ_{47} carbonate measurements (circles), and sample 95% confidence intervals. Dark red horizontal bars show 1 standard error and the light red is the 95% confidence interval within an interval. Ages are based on radiocarbon analysis; see Huth et al. (2020) for details of the calculation of soil $\delta^{18}O_{smow}$ and temperature. Source: Adapted from Huth et al. (2020).

carbonate appears to have been stable through early part of the record before decreasing between 7 and 10 ka, after which $\delta^{13}C$ rose rapidly. Huth et al. interpret this as changes in relative abundance of C_3 and C_4 plants. C_3 plants dominated the landscape between 10 and 7 ka (45–100% C_3 plants). After that, C_3 plants declined until they were nearly absent on the landscape by 5 ka (<10% C_3 plants).

Clays, such as kaolinites, are another important soil constituent. Savin and Epstein (1970) showed that during soil formation, kaolinite and montmorillonite form in approximate equilibrium with meteoric water so that their $\delta^{18}O$ values are systematically shifted by +27‰ relative to the local meteoric water, while δD is shifted by about 30‰. Thus, kaolinites and montmorillonites define a line parallel to the meteoric water line

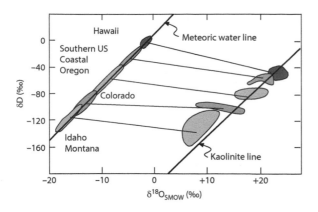

Figure 12.20 Relationship between δD and $\delta^{18}O$ in modern meteoric water and kaolinites. Kaolinites are enriched in ^{18}O by about 27‰ and ^2H by about 30‰. Source: Adapted from Lawrence and Taylor (1972).

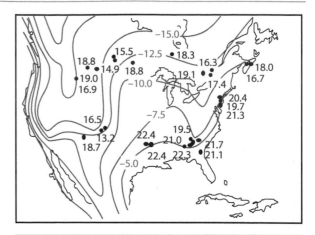

Figure 12.21 $\delta^{18}O$ in Cretaceous kaolinites from North American compared with contours of $\delta^{18}O$ (red) of present-day meteoric water. Source: Adapted from Lawrence and Meaux (1993).

(Figure 12.20), the so-called kaolinite line. From this observation, Lawrence and Taylor (1972) reasoned that one should be able to deduce the isotopic composition of rain at the time ancient kaolinites formed from their δD values. Since the isotopic composition of precipitation is climate dependent, as we have seen, ancient kaolinites provide another continental paleoclimatic record.

Lawrence and Meaux (1993) conclude, however, that most ancient kaolinites have exchanged hydrogen subsequent to their formation, and therefore a not a good paleoclimatic indicator (this conclusion is, however, controversial). On the other hand, they conclude that oxygen in kaolinite does preserve the original $\delta^{18}O$, and that can, with some caution, be used as a paleoclimatic indicator. Figure 12.21 compares the $\delta^{18}O$ of ancient Cretaceous North American kaolinites with the isotopic composition of modern precipitation. If the Cretaceous climate were the same as the present one, the kaolinites should be systematically 27‰ heavier than modern precipitation. For the southeastern US, this is approximately true, but the difference is generally less than 27‰ for other kaolinites, and the difference decreases northward. This indicates these kaolinites formed in a warmer environment than the present one. Overall, the picture provided by Cretaceous kaolinites confirms what has otherwise be deduced about Cretaceous climate: the Cretaceous climate was generally warmer, and the equator to pole temperature gradient was lower.

Many additional studies have followed up on this original work. Sheppard and Gilg (1996) determined the temperature dependence of the water–kaolinite fractionation of δD as:

$$1000 \ln \alpha_{kaol-H_2O} = -\frac{2.2 \times 10^6}{T^2} - 7.7 \quad (12.4)$$

and that for $\delta^{18}O$ as:

$$1000 \ln \alpha_{kaol-H_2O} = \frac{2.76 \times 10^6}{T^2} - 6.75$$

$$(12.5)$$

which is only slightly different from the earlier work of Lawrence and Taylor (1972). The temperature at which kaolinite recrystallizes can be determined from:

$$T = \left(\frac{3.06 \times 10^6}{\delta^{18}O_{kaol} - 0.125 \times \delta D_{kaol} + 7.04} \right)^{1/2}$$

$$(12.6)$$

Thus, kaolinites can be used as paleothermometer. When temperatures are calculated in this way from modern kaolinites, they tend to be about 3°C warmer than local mean annual temperature. This is thought to be due to crystallization primarily in the warm summer months. Consistent with other paleothermometers, kaolinites indicate that the Neogene has had temperatures several degrees cooler than that in the Cretaceous and Eocene (Sheldon and Tabor, 2009).

12.5 CARBON ISOTOPES, ATMOSPHERIC CARBON DIOXIDE, AND CLIMATE

12.5.1 CO₂ and carbon isotopes in ice cores

While Milankovitch variations are the pacemaker of climate change, they do not directly cause it. Rather, the changes in insolation trigger changes ocean circulation and partitioning of CO_2 between the ocean and atmosphere. The association between CO_2 and global temperature is based not only on an understanding of the interaction between radiation and atmospheric gases, but also on the strong correlation observed between temperatures derived from δD in Antarctic ice and the concentration of CO_2 in air trapped in bubbles within the ice over the last eight glacial cycles (Figure 12.22). Atmospheric CO_2 concentrations range from 170–180 ppm in glacial periods to ≈280–300 ppm in interglacial periods over the last 800 000 years, corresponding to a 55–60% increase in atmospheric CO_2, and consequent climate forcing between glacial and interglacial periods.

The question then becomes: where did the extra CO_2 present in interglacial atmospheres come from and why? As Figure 10.25 shows, CO_2 cycles rapidly through several reservoirs in the exogene with the deep ocean being

largest of these followed by the soil–peat reservoir. Because these reservoirs have different C isotopic compositions, knowing how $\delta^{13}C$ of atmospheric CO_2 has changed can help us answer these questions. Figure 12.23 compares a composite record of $\delta^{13}C$ in Antarctic ice from the EPICA and Talos Dome ice cores (Eggleston et al., 2016) with CO_2 concentrations in Antarctic ice and the δD-derived Antarctic temperature record over the last glacial cycle. Because of the larger sample size necessary to determine $\delta^{13}C$ from ice bubbles, the resolution of $\delta^{13}C$ over this period is less than for CO_2 or δD measurements so short-term variations are obscured. Nevertheless, we can see that while the CO_2 and temperature records are strongly correlated (although not shown, this is true of atmospheric methane as well), this is not the case for $\delta^{13}C$. Notably, while atmospheric CO_2 was slightly higher during the Eemian (i.e., the previous) interglacial than the present one, $\delta^{13}C$ was not; indeed $\delta^{13}C^{atm}$ appears to have been higher during the last glacial maximum around 18 000–28 000 years ago than the previous interglacial. This suggests multiple pathways in which carbon is shuttled between the atmosphere and other exogenic carbon reservoirs.

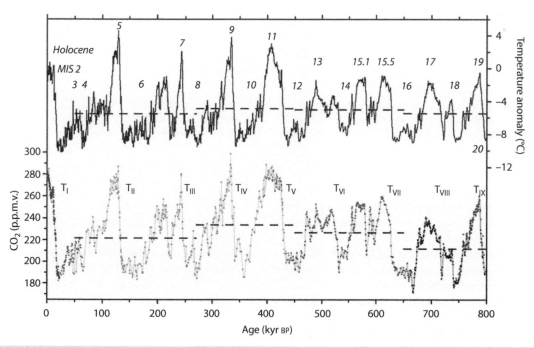

Figure 12.22   Comparison of CO_2 in bubbles in the EPICA ice core with temperatures calculated from δD. Source: Luthi et al. (2008)/with permission of Springer Nature.

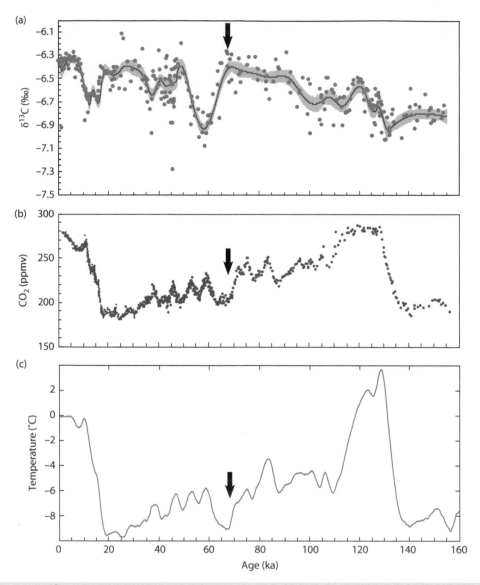

Figure 12.23 (a) $\delta^{13}C$ analyses of Eggleston et al. (2016) from the Talos and EPICA Dronning Maud Land Antarctic ice cores. Red line is Eggleston et al.'s Monte Carlo cubic spline average with associated uncertainty in pink. (b) Compiled CO_2 in Antarctic ice cores. (c) Smoothed temperature derived from δD in the EPICA Dome C ice core based on the data of Jouzel et al. (2007). Arrow shows that the decrease in $\delta^{13}C$ around 60 ka corresponds to a minimum in atmospheric CO_2 and Antarctic temperatures.

Several phenomena can lead to changes in atmospheric CO_2 and $\delta^{13}C^{atm}$ (Figure 12.24).

- First, since the solubility of CO_2 in water decreases with temperature, warming of sea surface temperatures should lead to a reduction in dissolved marine CO_2, increasing atmospheric CO_2 and increasing $\delta^{13}C^{atm}$. On the other hand, warming associated with ice sheets melting increases ocean volume and will have the opposite effect. Warming will also decrease sea ice

cover increasing atmosphere–ocean gas exchange, decreasing atmospheric CO_2 and $\delta^{13}C^{atm}$.

- Second, increases in marine productivity will increase the efficiency of the biological pump, increasing the amount of CO_2 stored in the deep ocean, decreasing atmospheric CO_2 but increasing $\delta^{13}C^{atm}$. Marine productivity is controlled by the availability of nutrients in surface waters, which in turn depends on upwelling of nutrient-rich deep water and the aeolian flux of

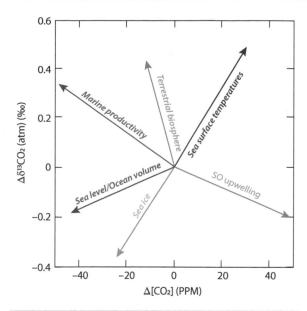

Figure 12.24 Effect of various processes on atmospheric CO_2 concentration and $\delta^{13}C^{atm}$ based on modeling of Köhler et al. (2010). Arrows indicate the changes as a consequence of an increase in each phenomenon, which can change CO_2 concentration and $\delta^{13}C^{atm}$ in opposite directions. Source: Adapted from Eggleston et al. (2016).

micronutrients, most notably Fe, from continents. The latter will increase when conditions are drier and windier, which they typically are in glacial periods.

- Third, changes in ocean circulation will change the ocean's ventilation time, affecting storage of CO_2 and the Southern Ocean is the key region of atmosphere–ocean gas exchange. Decreasing the amount of CO_2 stored in the deep ocean will in turn increase atmospheric CO_2 but decrease in $\delta^{13}C^{atm}$. As noted earlier, reduction in NADW production and the AMOC appears to lead to warming and increased stratification in the Southern Ocean, hence limiting upwelling and release of CO_2 stored in the deep ocean.
- Fourth, increasing CO_2 concentrations in the deep ocean decreases pH which in turn increases dissolution rates of carbonate shells falling through the water column. A direct result of this is a shallowing of the carbonate compensation depth (CCD) – the depth at which these shells dissolve completely and are not preserved in

sediment. When the CCD is shallow, carbonate contributes to CO_2 build-up in the deep ocean. However, removal of carbonate into sediment tends to be more important on long timescales than short ones.

- Fifth, warming and decreasing ice sheets will increase amount of CO_2 stored in the terrestrial biosphere decreasing atmospheric CO_2 but increase $\delta^{13}C^{atm}$. On the other hand, warming can also decrease stored in soils, peat, and permafrost, increasing atmospheric CO_2 but decreasing $\delta^{13}C^{atm}$.

The variation in $\delta^{13}C^{atm}$ over the last glacial interval seen in Figure 12.23 thus likely resulted from different processes or combinations of them operating at different times. In the early part of the last glacial period, there is a general increase in atmospheric $\delta^{13}C$ and decrease in atmospheric CO_2, interrupted by several excursions in both. The resolution of the ice core data is insufficient to ascribe exact causes but increase in marine productivity or decrease in upwelling could be the cause. One of the most apparent features in $\delta^{13}C$ in Figure 12.23 is the decrease beginning around 68 ka (arrows), corresponding to a decrease in both atmospheric CO_2 and Antarctic temperatures. Subsequently, Antarctic temperatures increase as does atmospheric CO_2 after a lag. Eggleston et al. (2016) attribute these changes to increased upwelling of deep water, releasing stored isotopically light carbon. $\delta^{13}C$ reaches a minimum around 59 ka and then increases until around 49 ka when atmospheric CO_2 and Antarctic temperatures both reach minima. Eggleston et al. concluded that both Southern Ocean upwelling and expansion of northern hemisphere vegetation may have influenced $\delta^{13}C$ and atmospheric CO_2 in this period.

High-resolution/high-precision $\delta^{13}C$ data for the end of the last glaciation (Termination I) from Taylor Glacier in Antarctica allow better discrimination between the various processes responsible for changing atmospheric CO_2 and climate. Figure 12.25 compares $\delta^{13}C$ with CO_2, CH_4, and δD-based temperature in Antarctic ice cores and $\delta^{18}O$ from the NGRIP Greenland ice cores. There was an initial rise in atmospheric CO_2 beginning just after the end of the last glacial

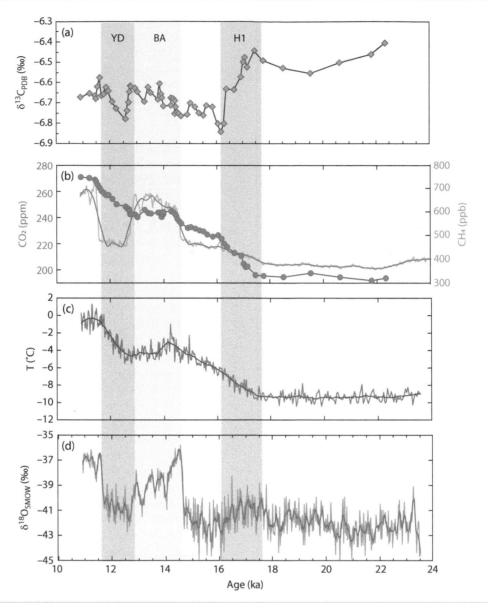

Figure 12.25   (a) $\delta^{13}C$ data of Bauska et al. (2016) from the Taylor Glacier, Antarctica ice core; (b) CO_2 in the Taylor Glacier ice core and CH_4 from the West Antarctic Ice Divide ice core (data of Rhodes et al., 2015); (c) δD-derived temperatures from the EPICA ice core (data from Jouzel et al. 2007); (d) $\delta^{18}O$ from the NGRIP ice core of Greenland (data from Rasmussen et al., 2006). Shaded areas correspond to the Heinrich Event 1 (H1) and Younger Dryas (YD) stadials and the Bølling–Allerod interstadial.

maximum at around 17.6 ka during which $\delta^{13}C$ decreases. This period also corresponds to Heinrich Event 1 and an initial temperature rise in Antarctica. Based on the slope of relationship between $\delta^{13}C$ and CO_2 (Figure 12.24), Bauska et al. (2016) interpreted this as due to a weakening of the biological pump and/or Southern Ocean upwelling. They attribute the overall subsequent rise in CO_2 and $\delta^{13}C$ from 15.5 to 11 ka to a roughly equal mix of sources from rising ocean temperature

and a weakened biological pump with relatively small contributions from $CaCO_3$ cycling and volcanic emissions. Superimposed on this are rapid decreases in $\delta^{13}C$ at the onset of the Bøllng–Allerod interstadial at 14.6 ka and the termination of the Younger Dryas stadial at 11.5 ka that correspond to an increase in CH_4 in Antarctic cores and a rapid increase in $\delta^{18}O$ of Greenland precipitation. Bauska et al. attributed these to rapid release of terrestrial carbon, most likely from tropical sources.

12.5.2   Isotopic proxies for atmospheric CO_2

Direct measurement of atmospheric CO_2 in ice cores is only available for the last 800 000 years. Beyond that, we must rely on proxies and reconstructions. Rae et al. (2021) have reviewed these in detail; we briefly summarize them here.

12.5.2.1   Boron isotopes as a CO_2 proxy

We introduce first of these proxies, based on boron isotopes ratios in foraminiferal calcite, in Section 11.3.2. To briefly review, boron is present in seawater as both $B(OH)_3$ and $B(OH)_4^-$, the relative abundance of which depends on pH. Because bonding differs in these two species, there is an isotopic fractionation between them. At constant $\delta^{11}B$ for total B in seawater, the isotopic composition $B(OH)_4^-$, the form incorporated in calcite, thus depends on pH (Equation [11.8–11.12]). In principle, then, it is possible to reconstruct the pH of seawater from $\delta^{11}B$ in calcite if the total $\delta^{11}B$ is known. Since the pH of seawater is largely controlled by the partial pressure of CO_2 in the atmosphere, it is possible to estimate p_{CO_2} of the ancient atmosphere.

There are complications. As we pointed out in Chapter 11, fractionation factors vary slightly between foraminifera species and must be separately calibrated. In addition, pH of seawater is also influenced by Ca and Mg concentrations, which must be accounted for. The fractionation factor is temperature dependent (which should come as no surprise), but the dependency is small (0.012 pH units per degree Celsius) and can be corrected for based on the known temperature dependence of Mg/Ca ratios in biogenic calcite. The variation of $\delta^{11}B$ of seawater through time, however, can only be estimated and is a major source of uncertainty. Since boron has a long residence time (11 Ma) in seawater, significant changes in $\delta^{11}B_{SW}$ over the Quaternary are unlikely, but that is not to say it cannot change over longer time periods. A variety of studies have addressed this issue; for example, Greenop et al. (2017) estimate that $\delta^{11}B_{SW}$ increased from a value of $\approx37.5‰$ in the early and middle Miocene to $40.1‰$ in the early Pliocene before declining again to the present value of $39.6‰$, a pattern similar to that of the seawater

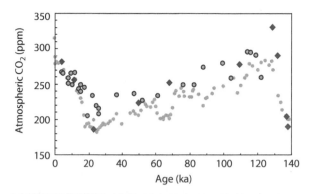

Figure 12.26   Comparison of calculated atmospheric CO_2 derived from $\delta^{11}B$ in shells of planktonic foraminifera with CO_2 measured in the EPICA Antarctic ice core (small gray circles). Adapted from Hönish and Hemming (2005) and Foster et al. (2016).

isotopic composition of $\delta^{24}Mg$, δ^7Li, and $\delta^{44}Ca$.

This approach to estimating paleo-CO_2 was pioneered by Pearson and Palmer (2000) who found dramatically lower seawater pH and higher p_{CO_2} in the Paleogene. This has been followed by a number of subsequent studies, including Hönisch and Hemming (2005) and Foster (2008), who investigated $\delta^{11}B$ during the Pleistocene. For the last two glacial cycles, their calculated pH values ranged from 8.11 to 8.32, corresponding to a p_{CO_2} range of ≈180 to ≈325 ppm. As Figure 12.26 shows, these results track the CO_2 measured in Antarctic ice, although the $\delta^{11}B$-based values tend to be offset by a few tens of ppm to higher CO_2.

12.5.2.2 $\delta^{13}C$ in C_{37} alkadienones

A second proxy is based on $\delta^{13}C$ of an organic molecule known as C_{37}-alkadienone, a 37-carbon chain including a ketone bond and two unsaturated carbons that is found in marine sediments. This molecule is a component of cell membranes of a group of haptophyte algae, coccolithophorids, such as *Emiliania huxleyi*, and is particularly resistant to diagenetic change (indeed, it survives in petroleum). As we found in Chapter 10, isotopic fractionation during photosynthesis depends on the extent to which intracellular CO_2 is fixed into organic matter. When CO_2

is abundant, photosynthesis ^{12}C is more selectively fixed, and the fractionation is large. When CO_2 is less abundant, photosynthesis is less selective, proportionally more ^{13}C is fixed into organic matter, and the fractionation is smaller. Hence, lower atmospheric CO_2 levels should result in greater isotopic fractionation between atmospheric CO_2 and organic matter produced by photosynthesis. Pagani et al. (1999) pioneered this approach by using compound-specific analysis to determine $\delta^{13}C$ in C_{37}-alkadienone. They also analyzed carbonate shells of planktonic foraminifera in Tertiary marine sediments to determine $\delta^{13}C$ of dissolved inorganic CO_2. Combining paleotemperature estimates based on $\delta^{18}O$, they then calculated $[CO_2]_{aq}$ from the difference in $\delta^{13}C$ between C_{37}-alkadienones and carbonates in the same Miocene and late Oligocene sediments. The results showed that CO_2 has been near its pre-industrial modern level throughout most of the Miocene but p_{CO_2} declined sharply at the Oligocene–Miocene boundary, coinciding with cooling documented by $\delta^{18}O$ (Figure 12.12) and extension of glaciation to West Antarctica.

Figure 12.27 compares estimates of atmospheric CO_2 concentrations based on $\delta^{11}B$ in planktonic foraminifera shells and $\delta^{13}C$ in C_{37} alkadienones with $\delta^{13}C$ and $\delta^{18}O$ in benthic foram shells. During the Paleocene, atmospheric CO_2 was generally around 1000 ppm or above. Both data sets show a decline at the Paleocene–Neogene (Eocene–Oligocene) boundary. It appears to have been somewhat variable in the first half of the Miocene but subsequently declined with shorter-term orbital variations superimposed on it.

12.5.3 The PETM and K-Pg crises

As we saw in Figure 12.12, strong negative excursions in $\delta^{13}C$ and $\delta^{18}O$ occur at the Paleocene–Eocene boundary marking the PETM. Figure 12.27 reveals a negative excursion in boron isotope ratios at this time as well. Let us focus on that era more closely. Gutjahr et al. (2017) analyzed $\delta^{18}O$, $\delta^{13}C$, and $\delta^{11}B$ in tests of the extinct planktonic foraminifer *Morozovella subbotinae* recovered from DSDP Site 401 in the northeast Atlantic (Figure 12.28). $\delta^{13}C$ drops suddenly by $\approx 3.4‰$ (other studies have suggested drops as large a 4‰), accompanied by a drop in

$\delta^{11}B$ of 1.7‰ corresponding to a decrease in pH of 0.27–0.36, depending on the calibration used, and to an at least doubling of atmospheric CO_2 from 800–900 to >1500 ppmv. The drop in $\delta^{18}O$ observed here as well as in other studies signals a sudden warming of perhaps 5°C in less than 10 000 years (Zachos et al., 2008).

The decrease in carbon isotope ratios indicates a massive addition of isotopically light carbon to the exogene, the source of which is debated. Previous studies had suggested that initial warming might have triggered release of organic carbon, such as the decomposition of methane hydrates on continental margins or thawing of polar permafrost; this carbon would have quite negative $\delta^{13}C$. Gutjahr et al. (2017) argue that the apparent drop in ocean pH implies a release of $\approx 10^{15}$ g carbon. If this were all organic carbon with $\delta^{13}C$ of −26‰ for permafrost or −60‰ for methane hydrates, it should have produced an even bigger drop in $\delta^{13}C$ than observed. They suggest volcanism as the primary source of CO_2. Previous studies had pointed out that North Atlantic Igneous Province is coincident in time with the PETM and initial release of magmatic CO_2 with $\delta^{13}C$ of $\approx −6‰$ may have then triggered warming and release of lighter organic carbon. Their mean-weighted estimate for the $\delta^{13}C$ of the released carbon was $\approx −11‰$, which would imply $\approx 75\%$ mantle carbon and 25% permafrost carbon or 90% mantle carbon and 10% methane hydrate. Regardless of the initial trigger, the PETM demonstrates the climate sensitivity to changes in atmospheric CO_2. Extinctions associated with the PETM define the Paloecene–Eocene boundary and it took 70 000–100 000 years for the oceans and ecosystems to recover and, judging from $\delta^{18}O$, temperatures remained high for much longer.

Now let us move back further in time to the mass extinction at Cretaceous–Paleogene boundary at 66 Ma. This event is famous for driving dinosaurs to extinction, but $\approx 75\%$ all terrestrial and marine species also became extinct. Calcareous nanoplankton, including foraminifera, were particularly hard hit. Curiously, benthic forams survived almost unscathed. Figure 12.27 shows that this event is associated with a positive excursion in benthic foram $\delta^{13}C$. Looking in more detail at data from ODP Site 465 in the western equatorial

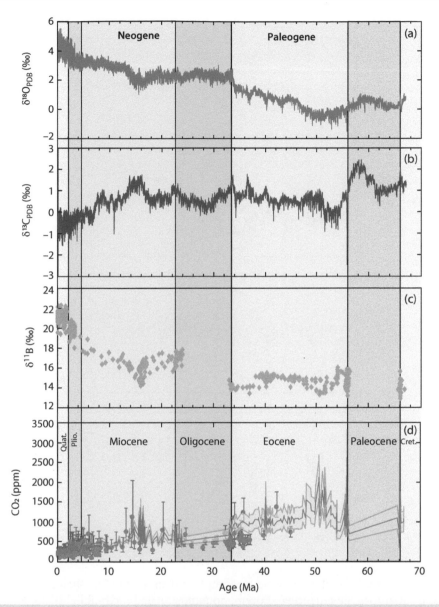

Figure 12.27 (a) $\delta^{18}O$ in benthic foraminifera and (b) $\delta^{13}C$ in benthic foraminifera from the compilation of Westerhold et al. (2020), c $\delta^{11}B$ in planktonic foraminifera compiled by Rae et al. (2021), **d** green line curve fitted to atmospheric CO_2 calculated from $\delta^{11}B$ by Rae et al. (2021) with light green field showing the uncertainty in the calculations; orange symbols are atmospheric CO_2 calculated from $\delta^{13}C$ in C_{37}-alkadienones from the compilation of Rae et al. (2021).

Pacific shown in Figure 12.29, we see that this excursion is associated with decrease in the difference between benthic foram and bulk carbonate $\delta^{13}C$. This is a global phenomenon first identified by Hsü et al. (1982) in sediment cores from the South Atlantic and indicates a decrease in the vertical ocean $\delta^{13}C$ gradient. As we found in Chapter 10 (Figure 10.11), this gradient is driven by sinking and remineralization of organic matter produced in the surface

photic zone. The decrease in the gradient therefore signals a decrease in export of organic matter from the surface layer. Figure 12.29 shows that this is also associated with a decrease in surface water $\delta^{11}B$ and pH calculated from it.

The likely explanation is sudden acidification of ocean surface waters, perhaps from a combination nitrate generated by atmospheric heating and oxidation of N_2-oxidation by the

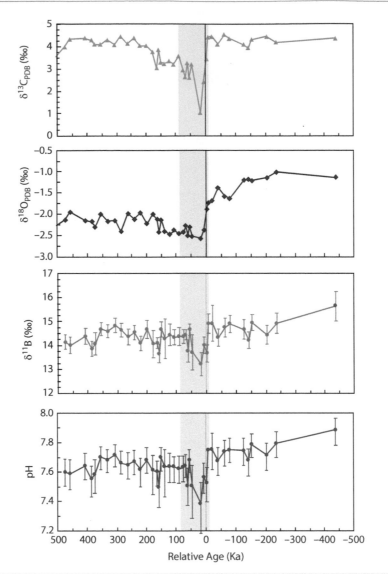

Figure 12.28 $\delta^{13}C$, $\delta^{18}O$, $\delta^{11}B$ in planktonic foraminifera from DSDP Site 401 and surface ocean pH calculated from $\delta^{11}B$ across the Paleocene–Eocene boundary. Pale blue highlights the Paleocene–Eocene Thermal Maximum (PETM). Age is relative to the initial drop in $\delta^{13}C$. Data from Gutjahr et al. (2017).

bolide and sulfates released by impact on gypsum-rich sediment at the Yucatan impact site. Fires triggered by the impact may have produced a spike in atmospheric CO_2, but H_2CO_3 is a weak acid compared to HNO_3 and H_2SO_4.

The acidification would have made it harder for calcareous nanoplankton to produce shells and explains their extinction. The consequent reduction in the surface water productivity and export of organic matter likely explains the decline in the vertical $\delta^{13}C$ gradient documented by the decrease in $\Delta^{13}C_{\text{bulk-benthic}}$. Alegret et al. (2012) argue, however, that

surface water productivity is unlikely to have shut down entirely as benthic forams and other benthic fauna, which also ultimately depend on organic matter produced in surface water, show no decline. Indeed, modeling by Henehan et al. (2019) found that a 50% reduction in the amount of organic matter exported from surface waters is sufficient to explain the decline in the vertical $\delta^{13}C$ gradient. As Henehan et al. point out, surface water pH recovered very quickly, but the vertical $\delta^{13}C$ gradient did not.

Figure 12.29 also shows there is a decrease in $\delta^{13}C$ of bulk carbonate at ODP Site 465. This

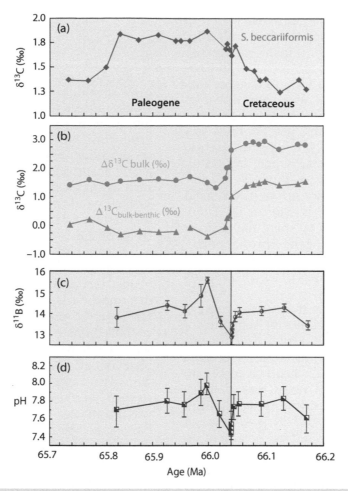

Figure 12.29   (a) and (b) δ^{13}C in tests of the benthic foraminifer *S. beccariformis*, δ^{13}C in bulk carbonate, and the difference in δ^{13}C between bulk carbonate and *S. beccariformis* at ODP Site 465 in the western Pacific. Timescale has been adjusted to that of Henehan et al. (2019). Data from Alegret et al. (2012). Panels (c) and (d) show δ^{11}B in planktonic foraminifera tests and calculated surface water pH from a variety of sites across the K–Pg boundary. Data from Henehan et al. (2019).

decrease, however, is not universal across all studied K–Pg boundary sites and probably reflects several factors. The first of these is simply a decrease in the contribution of relatively heavy planktonic carbonate relative to benthic foram carbonate, which would make bulk carbonate lighter. A second factor is injection of isotopically light carbon from fires triggered by the asteroid impact, as well, perhaps, as CO_2 from the simultaneous Deccan Traps volcanism. A third factor may have been decreases in terrestrial and marine photosynthesis. Photosynthesis removes isotopically light carbon from the atmosphere, hence driving atmospheric δ^{13}C to higher levels. Decreased productivity would result in a decline in atmospheric δ^{13}C.

12.6   TRACING THE EVOLUTION OF ATMOSPHERIC OXYGEN

Neither of our sister planets, Venus and Mars, have significant amounts of molecular oxygen in their atmospheres. Indeed, no planetary body in our solar system has significant atmospheric oxygen. Furthermore, the atmospheres of Venus and Mars are dominated by CO_2 while CO_2 is a trace gas in our atmosphere. The obvious question is why the difference? The answer is undoubtedly life. The amount of organic carbon buried in sedimentary rocks, which bears the unmistakable isotopically light carbon signature of photosynthesis, easily exceeds the amount of oxygen in the atmosphere. From this we can readily conclude that molecular oxygen is present in our atmosphere because it has been produced as a byproduct of photosynthetic reduction of CO_2 to organic carbon (e.g., Urey, 1952). The question then becomes when did this happen? Redox-sensitive elements, perhaps most notably sulfur and transition metals, behave differently in different valence states, which depend on oxygen availability. For example, Fe is soluble in its reduced state but not in its oxidized one while the opposite is true of U. These redox proxies allow us to trace the relative oxygenation of the exogene through time, but they do not allow determination of absolute oxygen concentrations, which remains considerably uncertain.

Early studies of the distribution of redox sensitive elements in sediments suggested a time around the Archean–Proterozoic boundary might be when the atmosphere first became oxidizing. For example, Archean paleosols are iron-poor whereas post-Archean ones are iron-rich, detrital uraninite and pyrite is found in Archean sediments but not in post-Archean ones, and sedimentary sulfate is absent in Archean sediments, which suggests a change in the valence state of these elements (e.g., Holland, 2006). This transition is known as the *Great Oxidation Event* (GOE).

In addition to behavioral changes, there are typically large isotopic fractionations between valence states. For this reason, the isotopic composition of redox-sensitive elements are excellent oxybarometers. We have explored the isotopic composition and behavior of these elements in previous chapters. Because of their varying electrochemical potentials and geochemical behavior, these elements will undergo oxidation at varying levels of oxygen concentrations. For example, oxidation of Cr^{3+} to Cr^{6+} requires higher O_2 levels than U does for oxidation from U^{4+} to U^{6+}, so using a variety of redox-sensitive elements in principle allows us to trace rising atmospheric and marine O_2 levels. Here we will explore the insights that the isotopic composition of these elements in sedimentary rocks have provided into the oxygenation of the Earth's surface. Several recent reviews provide more details about this critical episode in Earth's history (Kendall, 2021; Ostrander et al., 2021; Reinhard and Planasvsky, 2022).

12.6.1 Wiffs of oxygen in the Archean

Sulfur MIF data, which we discuss in Section 12.6.2, seem to suggest that atmospheric O_2 appeared quite suddenly (geologically speaking – within ≈ 100 Ma). If so, it would require not only sudden and rapid evolution of oxygenic photosynthesis, but also production of a vast amount of oxygen in a short amount of time. Before atmospheric oxygen concentrations could rise, reduced species such as sulfides and ferrous iron at the Earth's surface would first have to be oxidized. Furthermore, most O_2 produced by photosynthesis is consumed in respiration; atmospheric O_2 levels can rise only to the extent to which organic matter is buried. Development of an ozone layer would require significant amounts of atmospheric oxygen, levels at least above 10^{-5} to 10^{-6} present atmospheric levels (PAL ≈ 0.02 MPa or 20% molar). Many redox-sensitive elements will undergo partial or complete oxidation at oxygen levels lower than this with associated isotopic fractionations. Free oxygen would have first appeared where it was produced – in the surface waters of the oceans. So we now turn to stable isotopes in Archean sediments to examine the early history of oxygenation of the exogene.

12.6.1.1 Fe isotopes

Iron is the most abundant redox-sensitive element at the Earth's surface and, in the absence of biological redox cycles in a young sterile planet, it would have strongly influenced the redox state of the Earth's surface. In the reducing conditions of the early Earth, ferrous iron from hydrothermal vent fluids would have made seawater iron rich. A remarkable feature of Archean and early Paleoproterozoic sedimentary geology is the presence of vast ferric iron-rich sedimentary deposits, the most spectacular of which are layered and known as *banded iron formations* (BIFs), but more generally termed simply *iron formations*, which precipitated from this Fe-rich ocean (MacGregor, 1927). Today, these BIFs are the world's principal source of iron ore.

BIFs typically consist of layers of quartz alternating with layers of iron oxide, usually hematite or magnetite, although in some cases the iron is concentrated in clays or carbonates. Originally these would have been layers of chert alternating with goethite and related

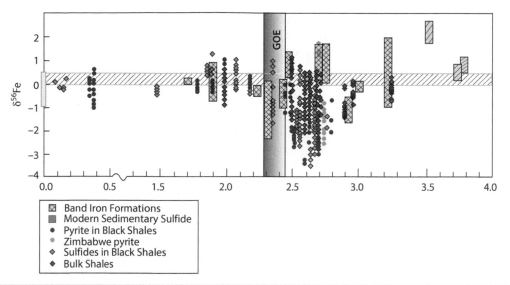

Figure 12.30 Variation of ^{56}Fe in sediments through geologic time. Source: Adapted from Anbar and Rouxel (2007); Johnson et al., (2008); Planavsky et al. (2012) and Czaja et al. (2012, 2013). Grey bar shows the time of the GOE. Diagonal lined bar shows the composition of igneous rocks. Yellow shows the range of modern black shales.

Fe-rich clays that were replaced by subsequent diagenesis and metamorphism. They are most abundant around the time of the GOE but are found throughout the Paleoproterozoic and the Archean (there are also a few examples from the Neoproterozoic). This oldest of these is found in the Isua Supracrustal Belt of Greenland, which formed around 3.7–3.8 Ga. BIFs are thought to have formed when ferrous iron-rich deep ocean water upwelled to the surface and the iron oxidized and precipitated. Layering is thought to reflect seasonal, annual, or longer-term cycles of upwelling. They typically have δ^{56}Fe in the range of 1–3‰, consistent with fractionation expected for precipitation of Fe(III) from dissolved Fe(II) (Figure 12.30). Their existence is evidence of an oxidation process in ocean surface water.

Because of its age, the Isua BIF is of particular interest. Individual bands have homogeneous δ^{56}Fe but different bands ranged from 0.4 to 1.1‰. Modelling by Czaja et al. (2013) led them to conclude that while dissolved O_2 in the surface water might be responsible for the oxidation, concentrations were probably quite low, $<10^{-3}$ PAL, and they suggested a more likely explanation is anoxygenic photosynthesis:

$$4Fe^{2+} + 7H_2O + CO_2 \rightarrow 4FeOOH$$
$$+ CH_2O + 8H^+$$

Anoxygenic photosynthesis is performed today by green and purple bacteria that use reduced sulfur as an electron donor to reduce carbon. In the reducing conditions of the Archean, sulfur would have been rare in the oceans and the abundant ferrous iron would have been a more readily available electron donor.

The Isua and other early Archean BIFs tend to be small. Much more massive ones occur around the time of the GOE. For example, the Hamersley deposit of Australia and the Transvaal deposit of South Africa, both of which formed around 2.5 Ga, are estimated to contain $>10^{18}$ moles of Fe and are of considerable higher grade than the Isua BIF, which is estimated to contain 10^{14} moles Fe (Planavsky et al., 2012). Precipitation of isotopically heavy Fe(III) in BIFs would have driven δ^{56}Fe of seawater lower and this is reflected in the negative δ^{56}Fe of shales and sulfides. Late Archean shale and sulfides are particularly light (Figure 12.30) suggesting increasing removal of Fe(III) from a large pool of dissolved Fe(II) in the oceans (e.g., Johnson et al., 2008). δ^{56}Fe becomes less variable in the Neoproterozoic, perhaps reflecting a decreasing availability of dissolved Fe^{2+}. While anoxygenic photosynthesis may be responsible for the early Archean BIFs, the presence of dissolve

O_2 produced by oxygenic photosynthesis in surface waters is almost certainly necessary to explain the massive late Archean and early Proterozoic BIFs.

12.6.1.2 U isotopes

Uranium is sensitive to oxidation at low oxygen levels and can undergo oxidation at levels as low as 10^{-7} PAL although the kinetics are quite slow and limiting and depend on pH, temperature, and p_{CO_2} as well as p_{O_2} (Wang et al. (2020). While 10^{-7} PAL might sound low, it is nevertheless orders of magnitude above levels predicted for Earth's early atmosphere in the absence of photosynthesis.

We found in Chapter 11 that while there is little U isotopic fractionation during weathering, substantial fractionation does occur in the modern exogene during reduction and absorption of U such that $\delta^{238}U$ varies by more than 1‰. In the modern ocean, U is a conservative element with a uniform $\delta^{238}U$ of -0.39 ± 0.01‰ compared to an average crustal value of -0.29‰. This low seawater $\delta^{238}U$ is maintained by removal of isotopically heavy U in black shales and altered oceanic crust. Several studies have demonstrated variability in $\delta^{238}U$ in Archean sediments and paleosols (Figure 12.31), recently summarized by Wang et al. (2020). As in the modern world, reducing sediments (shales) are heavy (enriched in ^{238}U) while Fe-rich ones are light. In a uniformly reducing environment, little U isotopic variation is expected. A statistical analysis by Wang et al. identified two times at which $\delta^{238}U$

variability significantly increased: at 2.95 Ga and again at 2.43 Ga, the latter more or less coincident with the disappearance of MIF sulfur. They argue that free O_2 was first present by ≈ 3 Ga, perhaps only fleetingly and locally in surface ocean oases where locally high nutrient levels in surface water supported relatively high levels of photosynthesis. Isotopic variability increased again in the latest Proterozoic at around 0.65 Ga. As we will see, other evidence also supports an increase in atmospheric O_2 around that time.

12.6.1.3 Mo, Tl, and Se isotopes

As we found in previous chapters, molybdenum is a siderophile element and present in the solid Earth primarily in sulfides. In surficial oxidizing environments it forms the soluble molybdate ion and is a conservative element in modern seawater with a uniform $\delta^{98}Mo$ of 2.09‰ in the open ocean compared to an average crustal value of $\approx +0.15$‰ and a mantle value of ≈ -0.1‰. The heavy seawater composition is maintained by a combination of removal by adsorption on Mn-bearing oxyhydroxides such as Fe–Mn crusts and nodules in oxic environments with a fractionation as great as -3‰ and with adsorption on and removal as sulfides in euxinic environments. When quantitative removal in euxinic environments occurs, $\delta^{98}Mo$ is equal to that of seawater; when it is less than quantitative, $\delta^{98}Mo$ will be up to 1‰ lower than seawater. Rivers are the main source of Mo to the oceans as the hydrothermal flux to the oceans is negligible.

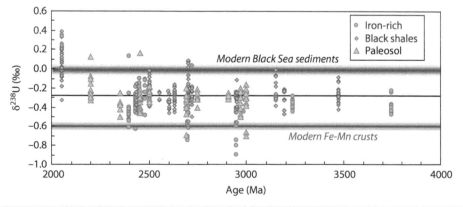

Figure 12.31 $\delta^{238}U$ in Archean and early Proterozoic sediments and paleosols. Solid line is the average $\delta^{238}U$ of the continental crust. Wang et al. identified two change points at which variation in $\delta^{238}U$ increased: 2.95 and 2.43 Ga. Source: Adapted from Wang et al. (2020).

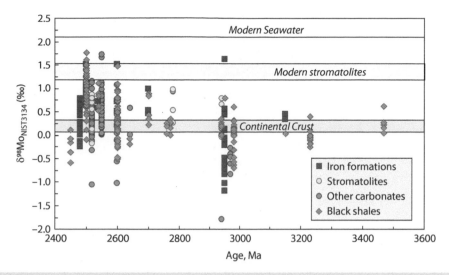

Figure 12.32 $\delta^{98}Mo_{NIST3134}$ in Archean sedimentary rocks. Shown for comparison are the composition of the upper continental crust, modern stromatolites, and modern seawater. $\delta^{98}Mo$ begins to increase around 2.9 Ga, suggesting removal of Mo by adsorption on Mn oxides precipitating from oxygenated surface waters. Data from the compilation of Thoby et al. (2019).

Figure 12.32 shows a summary of $\delta^{98}Mo$ in Archean sedimentary rocks. Up until 3.0–2.9 Ga, values are close to crustal ones indicating a lack of fractionating mechanisms in the exogene. Mo concentrations are also quite low, indicating very little Mo being delivered to the oceans through oxidative weathering. Between 2.9 and 2.5 Ga, maximum $\delta^{98}Mo$ values increase, as does the dispersion, indicating Mo was being fractionated in the exogene. Of key significance is the large fractionation that occurs during absorption on Mn-bearing oxyhydroxides. Oxidation of soluble Mn(II) to insoluble Mn(IV) requires a higher redox potential than oxidation of Fe(II) to Fe(III); essentially it requires free O_2. The large Mo isotopic fractionations indicate precipitation of Mn oxides and that in turn implies the presence of significant O_2 produced by oxygenic photosynthesis in surface waters.

Ossa Ossa et al. (2018) documented highly variable $\delta^{98}Mo$ (–0.7 to +0.16‰) in 2.95 Ga Mn-bearing sediments of the Sinqeni formation of the Pongola Super Group of South Africa deposited in a shallow water in an epicontinental sea. In contrast, they found that sediments deposited in deep water had nearly uniform $\delta^{98}Mo$ (0.34–0.56‰). This suggests the oceans were redox stratified with an oxic surface layer overlying an anoxic deep ocean.

Moving forward in time to ≈ 2.5 Ga, Ostrander et al. (2019) paired $\delta^{98}Mo$ with measurements of $\varepsilon^{205}Tl$ on the Mount McRae shales of Western Australia, which were deposited in the shallow waters of a continental shelf. Recall that Tl behavior in the modern marine environment is similar to that Mo; its isotopic composition is uniform ($\varepsilon^{205}Tl = -6.0 \pm 0.3$) and controlled by a combination of removal by absorption on Fe–Mn crusts in oxic waters and sulfides in anoxic ones. However, the fractionations are opposite to that of Mo, with isotopically heavy Tl partitioning into crusts whereas Tl partitioning into sulfides with little fractionation. Tl also differs from Mo in that the fractionation particularly sensitive to the presence of Mn oxyhydroxides as adsorption on Fe oxyhydroxides occurs without significant fractionation.

In euxinic (pyrite-rich) sedimentary sequences, Ostrander et al. found $\delta^{98}Mo$ up to 1.77‰ (average 1.31‰) and $\varepsilon^{205}Tl$ and down to –4.28 (average –2.56) (Figure 12.33); these extremes approach values for modern seawater (2.09‰ and –6, respectively). Because $\delta^{98}Mo$ and $\varepsilon^{205}Tl$ in euxinic sediments are essentially equal to that of seawater, these values are indicative of extensive removal of Mo and Tl from seawater by adsorption on Mn-bearing oxides and oxyhydroxides; a negative correlation between $\delta^{98}Mo$ and of $\varepsilon^{205}Tl$

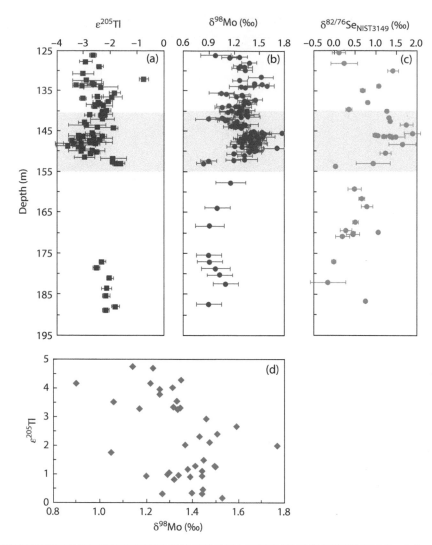

Figure 12.33 (a) ε^{205}Tl and (b) δ^{98}Mo in sediments as a function of depth in drill core samples from the Mt. McCrae formation of Western Australia. The grey area indicates the zone of pyrite-rich euxinic sediments. (c) $\delta^{82/76}$Se$_{NIST3149}$ data of Stüeken et al. (2015) in drill core samples. (d) ε^{205}Tl versus δ^{98}Mo in the euxinic sediments; Ostrander et al. found a statistically significant negative correlation. ε^{205}Tl versus δ^{98}Mo. Data from Ostrander et al. (2019).

also indicates adsorption on Mn oxyhydroxides. This requires not only that surface waters be oxygenated, but also sufficient O_2 that it penetrates well beyond centimeters below the sediment–water interface. Otherwise, even if Mn oxides precipitated in oxygenated surface waters, once buried and exposed to anoxic conditions in sediment, Mn quickly undergoes reductive dissolution and Mn (II) diffuses back into the overlying water column. Using simple mass balance calculations, Ostrander et al. concluded that oxygen penetrated sufficiently

deeply into the ocean that the water column over continental shelves was fully oxygenated.

The inference of free oxygen is supported by $\delta^{82/76}$Se$_{NIST3149}$ shown in panel (c) of Figure 12.33 as well as elevated Se concentrations in these horizons. These $\delta^{82/76}$Se$_{NIST3149}$ values overlap with those observed in modern ocean surface water; and as we noted in Chapter 11, euxinic sediments can record seawater Se isotope ratios as they do for Mo and Tl. As is the case for Mn, the high electrochemical potential involved in oxidation of reduced selenium to mobile Se(IV) essentially requires the presence of O_2.

We can summarize by saying that stable isotopes indicate O_2 was likely present in ocean surface waters by around 3 Ga and concentrations rose through the succeeding half billion years culminating in atmospheric oxygen concentrations reaching the level where a UV-blocking ozone layer developed following the Archean–Proterozoic boundary. The record is, for the most part, too sparse to tell us how steady and global this increase was. It is possible – particularly in the earlier part of the Archean – that oxygenated surface waters occurred merely "oases" or that oxygen appeared only in "wiffs" only to disappear again. Not until the latest Archean does it appear that oxygenated surface waters were widespread.

Beneath the surface mixed layer, Archean seas remained anoxic, much as the Black Sea is today. Anoxic iron-rich deep water may help explain the slow rise of oxygen in the Archean, as iron–phosphorus compounds have low solubility and phosphorus, often a productivity-limiting nutrient even in the modern world, is readily scavenged by Fe oxides. The limited land surface area through much of the Archean may also have restricted the supply of phosphorus from continental weathering. Indeed, as we will find in the following sections, nutrient availability, particularly phosphorus, likely played a key role in setting the pace of oxygenation of the exogene.

12.6.2 MIF sulfur and the GOE

Sulfur provided some of the first and quite dramatic isotopic evidence for a change in the oxidation state of the Earth's atmosphere around the Archean–Proterozoic boundary. Farquhar et al. (2000) found that sulfides (primarily pyrite) in sediments and metasediments formed prior to 2400 Ma have positive $\Delta^{33}S$ and negative $\Delta^{36}S$, while hydrothermal sulfide ores and sedimentary sulfates (mainly barite) have negative $\Delta^{33}S$ and positive $\Delta^{36}S$ (Figure 12.34). During the Archean, $\Delta^{33}S$ (i.e., deviations from mass-dependent fractionation) exceeded 3‰ in some cases. After this time, $\Delta^{33}S$ does not exceed 0.5‰. As we noted in Chapter 5, laboratory experiments show that UV photolysis of SO_2 produces mass-independent fractionation of S isotopes, producing elemental sulfur with positive $\Delta^{33}S$ and sulfate with negative $\Delta^{33}S$.

In the present atmosphere, nearly all UV radiation is absorbed in the stratosphere by

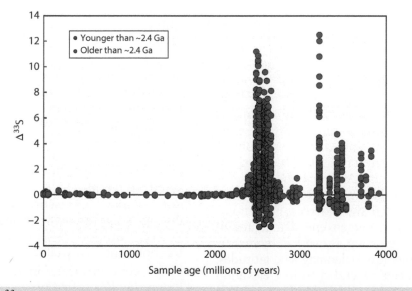

Figure 12.34 $\Delta^{33}S$ in sulfur through time. Sediments older than 3.4 Ga contain sedimentary pyrite with large positive $\Delta^{33}S$ and sulfates and hydrothermal sulfides with negative $\Delta^{33}S$. Sediments younger than 2.3–2.4 Ga lack $\Delta^{33}S$ larger than 0.5‰, indicating the development of significant atmospheric O_2 and O_3 then. Source: From Farquhar et al. (2014)/with permission of Elsevier.

ozone-forming and ozone-destroying reactions, so photolysis of SO_2 is limited to cases, such as the eruption of Pinatubo, where volcanoes loft SO_2 into the stratosphere. The common presence of MIF sulfur in Archean sediment indicates that the atmosphere lacked sufficient O_2 and O_3 to prevent UV radiation from penetrating deeply into it and photodissociating SO_2. Sulfate produced by photolysis would have dissolved in rain and ultimately found its way into the oceans. Some of this would precipitate as barite, $BaSO_4$, and some would be reduced in hydrothermal systems and precipitated as metal sulfides. The elemental sulfur would have formed particulate S_8 and also be swept out of the atmosphere by rain and ultimately incorporated into sediments, where it would react to form sedimentary sulfides. The disappearance of this MIF signature indicates a significant rise in atmospheric oxygen occurred around 2.3–2.5 Ga.

12.6.3 The Ups and downs of atmospheric oxygen in the Proterozoic

12.6.3.1 Siderean glaciations

There is widespread agreement that oxygenic photosynthesis carried out by the ancestors of cyanobacteria was widespread by the time of the GEO in the Paleoproterozoic and by around 2.4 Ga these critters had produced enough O_2 to form a UV-blocking ozone layer. Interestingly, the GEO is associated with widespread, perhaps globally severe, glaciation. In detail, at least three glaciations can be identified. These were first recognized in the Huronian Supergroup of the Superior Province in southern Ontario from the presence of diamictites (glacial sediments characterized by extreme range in grain size) and subsequently recognized in the Wyoming Craton, Kaapvaal craton in South Africa, the Pilbara craton of Western Australia and the Kola-Karelia craton of Finland and Russia. These are collectively called the Siderean glaciations. One of these, the Makgenyene diamictites of the Transvaal Supergroup, is overlain by marine sediments and has low paleomagnetic latitudes, indicating ice sheets extended to low elevation even at low latitude, which Hoffman (2013) argues was severe enough to cover the planet in ice – a "snowball Earth" glaciation. Radiometric ages constrain this event only to between 2.43 and 2.32 Ga.

Methane is the likely link between climate and oxygenation. Stars grow steadily brighter as they age, in the Sun's case by about 30% over the last 4.5 Ga, meaning that a much stronger greenhouse effect was needed to maintain liquid water on the Earth's surface in the Archean. In the reducing conditions prior to the GOE, this was likely provided by relatively high methane concentrations ($\approx$5000 ppm), as well as much higher CO_2 than present. Increasing amounts of O_2 in the atmosphere would have oxidized methane, eventually leading to a climate crisis (e.g., Kasting and Ono, 2006; Catling and Zahnle, 2020). Laakso and Schrag (2017) offered an alternative view that the glaciations caused the GEO, but Warke et al. (2020) found that in the Fennoscandian Shield in Russia, the transition from MIF to non-MIF sulfur occurs 60 m below the glacial sequences at roughly 2.43–2.50 Ga, clearly showing the GOE preceded the glaciations.

12.6.3.2 High O_2 levels in the Lomagundi–Jatuli event

Several proxies, including $\Delta^{33}S$, suggest that oxygenation of the exogene wavered for perhaps 200 million years during the GEO. By around 2.2 Ga, however, a wide range of observations indicate that the transition to a relatively oxygen-rich exogene was complete, including the appearance of thick gypsum- and anhydrite-bearing evaporites, clearly indicating a substantial marine sulfate inventory. One of these is the $\approx$2.0 Ga Tulomozero Formation of the Onega Basin in Russian Karelia, with a thickness of 800 m and an area of 18,000 km^2. $\delta^{34}S$ of sulfates are relatively uniform throughout the section, ranging from 5– 7‰, implying a seawater value about 1‰ lower than that or $\approx$4–6‰, compared to a modern seawater value of 20‰. Because of the small fractionation between seawater and sulfate minerals, $\delta^{34}S$ is relatively, although not completely, insensitive to fractionation during evaporative fractional crystallization, but Ca isotopes are sensitive to the extent of fractionation as well as the initial Ca and SO_4^{2-} concentrations in the initial seawater. Using the 1‰ difference in $\delta^{44/40}Ca$ between the least and most evaporated sections, Blättler et al. (2018) estimated a minimum seawater SO_4^{2-} concentration of $\approx$10 mmol/kg compared to a modern one of 28 mmol/kg. Assuming a modern-sized ocean, oxidation of this amount of sulfide to sulfate would have

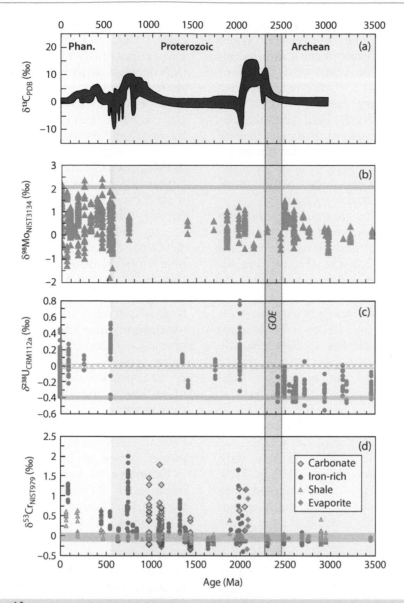

Figure 12.35 (a) δ^{13}C in marine carbonates through time (Source: Adapted from Lyons et al., 2021); (b) δ^{98}Mo in euxinic sediments through time; horizontal gray bar indicates the isotopic composition of modern seawater. Data mainly from the compilation of Thoby et al. (2019). (c) δ^{238}U in euxinic sediments through time; horizontal gray bar indicates the isotopic composition of modern seawater; hashed bar is the δ^{238}U of modern Black Sea euxinic sediments. Data from the compilation of Mänd et al., 2020). (d) δ^{53}Cr in marine sediments through time; horizontal gray bar indicates the isotopic composition of continental crust. Data from the compilations of Wei et al. (2020) and Mänd et al. (2022).

consumed an equivalent of 62% of the present atmospheric O_2 inventory.

Carbon isotope ratios provide an indication of how this O_2 was produced. Schidlowski et al. (1976) found δ^{13}C values as high as +13‰, the highest known in the geologic record, in a 300 m thick sequence of dolomites of ultimate marine origin from the ≈2.2 Ga Lomagundi Group of Zimbabwe, compared to modern marine carbonates of ≈0‰ (Figure 12.35). Similar high δ^{13}C values have been found in other carbonates of similar age around the globe, including the Jatuli Group of Fennoscandia. By ≈2.0 Ga, δ^{13}C in carbonates returned to near modern levels of ≈0‰. This period of 100–200 million years around 2.2 Ga is known as the *Lomagundi–Jatuli Excursion* or Event. Molecular oxygen in the exogene is the result of the combination of photosynthetic production and burial of organic carbon, so that $\delta^{13}C$ is effectively controlled by the relative rates of burial and weathering of isotopically light organic carbon: when burial rates exceed weathering rates, $\delta^{13}C$ increases. Thus, the Lomagundi–Jatuli Event signals the burial of large amounts of organic carbon and hence the production of large amounts of molecular oxygen.

Stable isotope ratios in marine sediments provide additional evidence of a well-oxidized exogene in the Lomagundi–Jatuli Event and following it until ≈ 2.0 Ga (Figure 12.35). $\delta^{98}Mo$ in deep-water euxinic sediments of the 2.1 Ga Francevillian Group of Gabon reaches $\approx 1‰$, suggesting oxygen was penetrating below the photic zone. In the 1.98 Ma Zaonega Formation, which lies stratigraphically above the Tulomozero Formation of the Onega Basin discussed above, euxinic sediments are enriched in soluble but redox-sensitive metals such as Mo, Re, and U and $\delta^{98}Mo_{NIST3134}$ values reach 1.24‰, representing a minimum value for seawater at the time and $\delta^{238}U$ ranges from –0.03 to 0.79‰, with an average of 0.47‰ (Mänd et al., 2020). $\delta^{53}Cr$, which prior to this time remained close to crustal values of $\approx -12 \pm 10‰$ rise to as high as 1.63‰, indicating O_2 levels had risen to the point where Cr(III) was being oxidized to Cr(VI) with attendant isotopic fractionation during weathering. Chromium is oxidized by reaction with Mn (IV) oxides, which requires oxygen levels greater than 0.1–1% of PAL. These $\delta^{238}U$ and $\delta^{53}Cr$ values are higher than at any earlier time and similarly high values would not occur again until the late Proterozoic.

$\Delta^{17}O$ values preserved in sulfates provide further constraints on atmospheric O_2 (Figure 12.36). Oxidation of reduced sulfur incorporates O both from seawater H_2O and from O_2. The latter, estimated to be 8–30% of sulfate O, will reflect atmospheric O_2 isotopic compositions. As we found in Chapter 10, negative $\Delta^{17}O$ anomalies are produced in the stratosphere by preferential capture by CO_2 of monatomic ^{17}O produced by UV photolysis of O_3 resulting in O_2 returning to the troposphere with negative $\Delta^{17}O$. This O_2 the mixes with O_2 produced by

photosynthesis, which, derived from water, carries ^{16}O, ^{17}O, and ^{18}O in isotopically normal proportions. The magnitude of the $\Delta^{17}O$ thus depends on atmospheric O_2 and CO_2 concentrations and biological productivity. All three of these factors are unknown for the Proterozoic, but modeling and varying $\Delta^{17}O$ through time provides some insights.

The earliest known sulfates are barites ($BaSO_4$), which are highly insoluble and precipitate at very low SO_4 concentrations. Barites from the ≈ 3.2 Ga Fig Tree Formation of South Africa, which contain some of the oldest known microscopic fossils, have average $\Delta^{17}O$ of –0.02‰, consistent with an atmosphere too O_2-poor to sustain an ozone layer (Figure 12.36). Between 2.4 and 2.0 Ga, the average is –0.23‰. Since not all oxygen in sulfate is derived from O_2, this is likely a maximum value and suggests $\Delta^{17}O$ of atmospheric O_2 equal to or lower than the modern value ($\approx -0.5‰$). Around 2.0 Ga, however, $\Delta^{17}O$ appears to decrease rapidly to an average of $\approx -0.54‰$, paralleling the decline in $\delta^{13}C$. Crockford et al. (2018) conclude that that the low $\Delta^{17}O$ and $\delta^{13}C$ indicate a decline of primary biological productivity to 6–41% of modern levels in the Mid-Proterozoic.

Estimates of O_2 and CO_2 concentrations immediately following the GOE vary widely, from 0.1 to 2 times PAL for O_2 and ≈ 10 to ≈ 100 times PAL for CO_2 (high CO_2 is required to maintain temperatures above 0°C when solar radiation was only $\approx 85\%$ of the present). Within these loose constraints, modelling by Crockford et al. (2019) and Hodgskiss et al. (2019) suggests a substantial increase in gross primary productivity in the early Proterozoic followed by a decrease well below present levels regardless of the assumed CO_2 and O_2. Hodgskiss et al. suggest a likely scenario is a tenfold decrease in p_{O2} levels from perhaps 1–10% in the immediate aftermath of the GEO to 0.1–1% PAL at around 2.0 Ga resulting from an approximately tenfold reduction in gross primary productivity from perhaps $\approx 60-6\%$ of modern.

There is something of a consensus, but not unanimity, that a critical aspect of the oxygenation history of this period, and indeed the entire Precambrian, was the availability of phosphorus. Phosphorus compounds play a key role in photosynthesis (Chapter 10) as well as many other cellular processes and it is a key

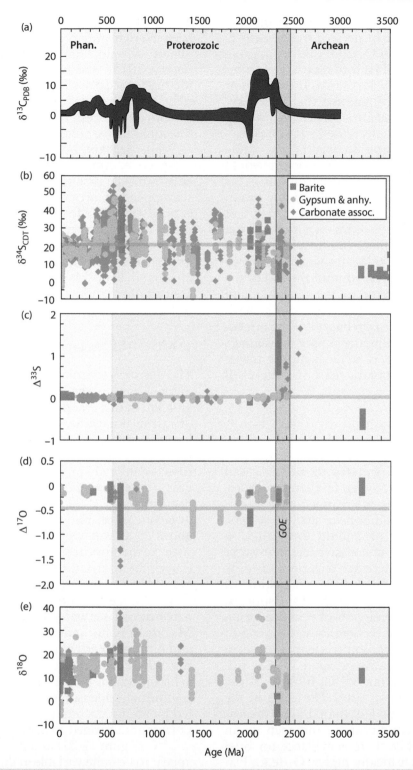

Figure 12.36 (a) $\delta^{13}C$ in marine carbonates through time as in Figure 12.35. (b) $\delta^{34}S$ in sulfates through time; gray bar is $\delta^{34}S$ of modern seawater. CAS is carbonate-associated sulfates: sulfate ions present as minor components of carbonates, (c) $\Delta^{33}S$ in sediments in sulfates through time; gray bar is $\Delta^{33}S$ of modern seawater. (d) $\Delta^{17}O$ in sulfates through time; gray bar is $\Delta^{17}O$ of modern atmospheric O_2. (e) $\delta^{18}O$ in sulfates through time; gray bar is modern atmospheric $\delta^{18}O$. Data mainly from the compilation of Crockford et al. (2019).

limiting nutrient in the modern biosphere. Phosphorus is primarily derived from weathering of apatite $(Ca_{10}(PO_4)_6(OH,F,Cl)_2)$. As we noted earlier, several factors may have further limited P availability in the Archean. Bekker and Hollard (2012) suggested that once oxygen was present in the atmosphere, weathering of pyrite would have decreased pH and consequently increased apatite weathering while lower Fe concentrations in the ocean would have allowed higher P concentrations. This combination could have led to an expansion of biological productivity. Indeed, the first appearance of phosphorites at around 2 Ga suggests high availability of phosphorus; with rare exceptions, phosphorites would not reappear in the geologic record until the end of the Proterozoic. Plate tectonics, the assembly, and breakup of supercontinent may also have played a role in the oxygenation of the atmosphere through affecting the supply of weatherable apatite as well as influencing the burial and weathering of organic carbon (Campbell and Allen, 2008).

12.6.3.3 The boring billion

By around 2 Ga $\delta^{13}C$ values had declined to ≈0‰ and $\Delta^{17}O$ in sulfates to ≈−0.8‰ and there are signs that oxygen levels were declining as well, including the reappearance of evidence for anoxic conditions in shallow waters, the disappearance of sulfate evaporites, a renewed deposition of massive shallow-water iron formations. There are a variety of ideas why this occurred, including exhaustion of readily weathered phosphorus sources, increased volcanic emission of reduced gases, and consumption of oxygen by weathering of the sedimentary organic carbon produced in the previous 200 Ma. By around 1.8 Ga, isotopic proxies including $\delta^{238}U$, $\delta^{98}Mo$, $\delta^{53}Cr$, and $\Delta^{17}O$ all suggest lower O_2 levels in the atmosphere, perhaps less than 1% of present levels and that conditions in the deep ocean were anoxic. While there is evidence for brief periods of at least locally higher O_2 (e.g., Planavsky et al., 2018; Stüeken and Kipp, 2020), these conditions continued for roughly the following billion years, hence the term *boring billion* (although one must also acknowledge a paucity of data for this period). The most likely explanation is again phosphorus limitation of biological productivity. The

anoxic deep ocean water would again have been Fe-rich and precipitation of iron phosphates would have limited productivity (Derry, 2015). The boring billion also appears for the most part to be a time of relative tectonic quiescence, which would have limited the amount of fresh rock from which phosphorus could be extracted by weathering.

While perhaps boring from a geochemical perspective, important evolutionary events occurred. The first microscopic eukaryotic fossils appear at around 1650 Ma (e.g., Miao et al., 2019). The fossils are already taxonomically diverse, suggesting eukaryotes evolved even earlier. Unambiguous multicellular algae appear in the fossil record shortly thereafter, at around 1560 Ma (Zhu et al., 2016). These are certainly landmark events in the evolution of life.

12.6.3.4 The Neoproterozoic revolution

The Neoproterozoic (1000–539 Ma) was a pivotal time in the evolution of the exogene and life. In this era, particularly the latter parts of it, the Ediacaran and Cryogenian, oxygen levels began a rise above Mesoproterozoic levels of <1% PAL. This is recorded in the isotopic paleoredox proxies shown in Figures 12.36 and 12.37. $\delta^{13}C$ begins a slow rise at around 1000 Ma, indicating increasing organic carbon burial, as do O isotope ratios and $\delta^{34}S$ as well as other paleoredox proxies such as the appearance of sulfates in massive evaporites at around 850 Ma. $\delta^{53}Cr$ in carbonates also rises around this time, although there is some question as to how well carbonates record seawater $\delta^{53}Cr$. By around 750 Ma, there is an undisputed increase in $\delta^{53}Cr$ and $\delta^{13}C$ reaches values that had not occurred since the Paleoproterozoic and would not be achieved again. A clear increase in $\delta^{34}S$ signals significant bacterial sulfate reduction, likely fed by abundant organic carbon in sediments. O isotopes changed as well.

$\delta^{13}C$ (Figure 12.37) and $\Delta^{17}O$ (Figure 12.36) appear to be quite variable in the late Neoproterozoic, suggesting variable rates of biological productivity that in turn drove variable rates of carbon burial and atmospheric oxygen levels. $\delta^{34}S$ is also quite variable, probably reflecting variable microbial sulfate reduction. Since microbial sulfate reduction is most likely to occur within sediments, this suggests at least periods

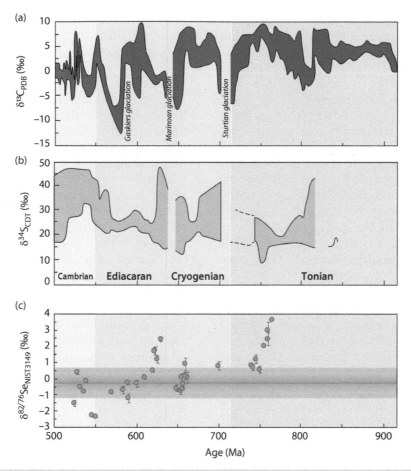

Figure 12.37 (a) $\delta^{13}C$ in marine carbonates (b) $\delta^{34}S$ in sulfates in the Neoproterozoic through time, and (c) $\delta^{82/76}Se$ in Neoprotererozoic Shales. The $\delta^{13}C$ curve is based on the compilation of Saltzman and Thomas (2012); the $\delta^{34}S$ curve is based on the compilation of Halverson and Shields-Zhou (2011), and the $\delta^{82/76}Se$ data are from Pogge von Strandman et al. (2015). Poor radiometric dating control means that the timescales are approximate.

when sulfate was present at the bottom of the water column and could penetrate into the underlying sediments. A $\delta^{13}C$ peak is quickly followed by a decline leading into the Cryogenian Period at ≈717 Ma, which opens with the Sturtian glaciation, a climate crisis if anything more severe than the Siderean glaciations during the GOE. Low paleomagnetic latitudes and marine sediments directly overlying diamictites again indicate ice sheets extended to low elevation even in the tropics. $\delta^{13}C$ then rises again before falling prior to the second Cryogenian glaciation, the Marinoan, at around 640 Ma, which is again followed by rising $\delta^{13}C$ and a maximum in $\delta^{34}S$. A third glaciation, the Gaskiers, follows at around 590 Ma; although it appears to have been less severe and less global in extent. These are often called the "Snowball Earth" glaciations as much, if not all, the planet was covered in ice,

although lacustrine sediments interbedded with diamictites in China suggest ice sheets may have waxed and waned, much as they did in the Pleistocene glaciations, driven by Milankovitch changes in insolation.

Figure 12.37 shows $\delta^{82/76}Se$ in Neoprotererozoic shales determined by Pogge von Strandman et al. (2015) from a variety of sites in North America, China, and Australia. $\delta^{82/76}Se$ decreases to modern levels, perhaps in a series of steps, through the Neoproterozoic from typical Mesoproterozoic value of +0.68 ± 0.83‰ toward Phanerozoic values of −0.30 ± 0.89‰. As we found in Chapter 10, selenium in modern seawater is dominated by oxyanions of Se(IV) and Se(VI), the formation of which from reduced forms requires high electrochemical potential, implying the presence of higher O_2, than for oxidation of its chemical cousin, sulfur. This trend toward lighter isotope ratios in sediments implies deposition from an increasingly large reservoir of oxidized Se.

Coincident with the beginning of the Cryogenian, Zn isotopes in both sulfides and organic matter become more variable (Isson et al., 2018). While not redox sensitive, Zn is key micronutrient for eukaryotes (prokaryotes also require Zn, but far less so than eukaryotes). The variability in Zn signals greater biochemical cycling and a greater relative role for eukaryotes, consistent with an increasing abundance of eukaryotic microfossils at the time. By the end of the Cryogenian, δ^{98}Mo, δ^{238}U, and δ^{53}Cr all signal a exogene that was more oxygenated than at any time since at least the Paleoproterozoic.

It is unclear how high atmospheric O_2 levels rose, with estimates ranging between 10 and 50% PAL – fully modern p_{O2} levels would have to await the development of a rich terrestrial biosphere in the mid-Paleozoic. Cause and effect relationships between glaciation and the rise of oxygen are also debated. The break-up of the Rodinia supercontinent and extensive mantle–plume related volcanism between 825 and 750 Ma may have increased delivery of phosphorus to the oceans triggering higher rates of photosynthesis, to which the increasing abundance of eukaryote autotrophs may have contributed. This might have further reduced methane levels again producing a climate crisis. On the other hand, glacial erosion would have produced an abundance of freshly ground rock from which phosphorus was readily weathered, although already high δ^{13}C prior to the Sturtian glaciations indicates this was not the primary cause.

12.6.3.5 Darwiin's dilemma

In the closing chapter of Origin of Species, Darwin wrote, "There is another…difficulty, which is much more serious. I allude to the manner in which species belonging to several of the main divisions of the animal kingdom suddenly appear in the lowest known fossiliferous rocks. If the theory be true, it is indisputable that before the lowest Cambrian stratum was deposited, long periods elapsed…and that during these vast periods, the world swarmed with living creatures. But to the question why we do not find rich fossiliferous deposits belonging to these assumed earliest periods

before the Cambrian system, I can give no satisfactory answer. The case at present must remain inexplicable; and may be truly urged as a valid argument against the views here entertained."

Since Darwin's time, many examples of microbial fossils dating as far back as 3.5 Ga have been discovered and over the latter half of the twentieth century examples of metazoan (animal) fossils of Ediacaran age ($\approx$640–540 Ma) have come to light. The oldest of undisputed metazoan fossils discovered thus far are from Newfoundland and date to 574 Ma (Matthews et al., 2020), a mere five million years after the Gaskiers glaciation. Nevertheless, Darwin's question of why it took nearly four billion years for the complex life represented by metazoans to appear remains. Molecular oxygen is essential for metazoans such as you and me, and while the exact amount necessary for the simplest metazoans is debated as are the exact O_2 levels in the Neoproterozoic, the paleoredox proxies we have discussed here seem to provide an answer to Darwin's dilemma: *it was not until the very end of the Proterozoic that oxygen levels rose to high enough to support animal life*. This, however, is not to say that oxygen in the exogene has reached modern levels. Oxygen levels would continue to grow in the Phanerozoic.

12.6.4 The Phanerozoic

12.6.4.1 Variations in δ^{13}C and δ^{34}S in a still evolving exogene

Figure 12.38 shows the evolution of δ^{13}C in carbonates and δ^{34}S in sulfates in the Phanerozoic. Overall, the δ^{13}C variation is more subdued than the Precambrian, with almost all the data falling between –3 and +6‰, compared to a total range of nearly 25‰ in the Precambrian. Nevertheless, there is considerable variability in the δ^{13}C record, particularly during the Paleozoic and this is also true of δ^{34}S, although the record is less complete. These excursions reflect a changing balance between burial and erosion of organic carbon: positive excursions resulting from dominance of burial over erosion and negative excursions the opposite. These in turn link to changes in the biosphere and signal changes to climate and redox state, indicated by varying δ^{34}S, of the

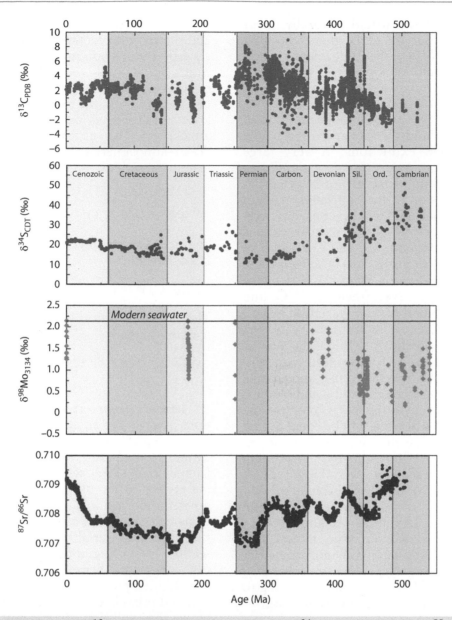

Figure 12.38   Variation of δ^{13}C in tropical marine carbonates, δ^{34}S in marine sulfate, δ^{98}Mo in marine euxinic sediments, and ^{87}Sr/^{86}Sr in marine carbonates through the Phanerozoic. δ^{13}C, δ^{34}S, and ^{87}Sr/^{86}Sr mainly from the compilation of Prokoph et al. (2008); δ^{98}Mo compiled from the literature.

exogene through relative rates of consumption and production of CO_2 and O_2 by photosynthesis and respiration.

Not surprisingly, many of the excursions are associated with boundaries between geologic periods, which are defined by appearance of new species and/or extinction of old ones. These can have any one or combination of myriad causes. For example, the spike in δ^{13}C at around 500 Ma is known as the *Steptoean positive carbon isotope excursion* or

SPICE and is associated with an extinction of trilobites living in relatively deep water, a strong positive δ^{34}S excursion, and a negative excursion in δ^{23}U of ≈ 0.19‰ (Dahl et al., 2014). The cause is thought to be anoxic and sulfidic water masses expanded over the shallow ocean, which could reflect changes in nutrient delivery in response changes in weathering or ocean circulation driven by plate tectonics. Bottom water anoxia, particularly in the shallow ocean where biologic productivity is typically highest, would naturally favor organic carbon burial and the greater variability in $\delta^{13}C$ and $\delta^{34}S$ in the Paleozoic is an indication that such anoxic areas were more common then and that oxygen had not yet reached modern levels.

The late Ordovician excursion is associated with glaciation and the second largest of the five great mass extinctions. A very large positive $\delta^{13}C$ excursion signals a significant global ocean–atmosphere–biota perturbation. An excursion at the Silurian–Devonian boundary is associated with a series of powerful extinction events, global reorganizations of biological communities, and expansion of new clades. $\delta^{13}C$ excursions also mark the Permian–Triassic and Cretaceous–Paleogene mass extinctions. Nevertheless, $\delta^{13}C$ and $\delta^{34}S$ become notably less variable overall through the Mesozoic and into the Cenozoic.

Recall that (1) the $\delta^{98}Mo$ of the oceans depends on the ratio of removal in oxic sediments (Fe–Mn nodules and crust) and euxinic sediments and (2) when reduction and removal of Mo in euxinic environments is complete, the $\delta^{98}Mo$ of those sediments will be equal to that of seawater, but less than seawater for incomplete removal. Consequently, maximum $\delta^{98}Mo$ values track seawater. Figure 12.38 suggests that prior to the Devonian $\delta^{98}Mo$ remained well below modern values and increased in the Devonian reaching modern values in the Mesozoic. True forests first appear in the Devonian. In the modern world, forest biomes are the biggest consumers of CO_2 and producers of O_2, so it is not surprising to see evidence of rising O_2 in the Devonian.

12.6.4.2 Modeling the Phanerozoic evolution of CO_2 and O_2

We saw that the evolution of atmospheric CO_2 could be reconstructed through the Cenozoic using $\delta^{11}B$ and $\delta^{13}C$ in alkenones, but they are difficult to apply to earlier times. Modelling other isotopic proxies has the potential to constrain the evolution of O_2 and CO_2 through the Phanerozoic. We will now explore one of those models.

Production and consumption of CO_2 and O_2 are linked to the production and consumption of organic matter by the reaction:

$$CO_2 + H_2O \leftrightharpoons CH_2O + O_2 \qquad (12.7)$$

where the forward reaction is photosynthesis and the reverse reaction is respiration. Imbalances in photosynthesis and respiration result from burial and oxidative weathering of organic matter leads to changes in the amount of CO_2 and O_2 in the exogene. In addition, silicate weathering is an important control on atmospheric CO_2 as it removes CO_2 from the ocean–atmosphere system through weathering reactions, such as:

$$CaMgSi_2O_6 + 2H_2CO_3 + Ca^{2+} \\ + Mg^{2+} + 2CO_3^{2-} + SiO_2 + 2H_2O \qquad (12.8)$$

and subsequent precipitation of carbonate from seawater:

$$(Mg, Ca)^{2+} + CO_3^{2-} \leftrightharpoons (CaMg)CO_3 \qquad (12.9)$$

The reverse reaction is decarbonation during metamorphism:

$$(CaMg)CO_3 + SiO_2 \leftrightharpoons (CaMg)SiO_2 + CO_2 \qquad (12.10)$$

These are often referred to as the Urey reactions (Urey, 1952), but their importance in controlling atmospheric composition had been earlier pointed out by Ebelmen (1845) and Chamberlin (1899), who, citing what was then the recent paper by Arrhenius (1896), also pointed out their role in controlling long-term climate. Using constraints from the geologic record such as burial rates of organic carbon, seafloor spreading rates, land area, volcanic degassing, etc., Berner, Lasaga and Garrels (1983) attempted to combine silicate weathering, and organic carbon burial and erosion to model the evolution of CO_2 and temperature in what is known as the BLAG model. Subsequently, Berner (1990) incorporated the Phanerozoic $\delta^{13}C$ record as a monitor of the shuffling of carbon between reservoirs into his GEOCARB model (e.g., Berner and Kothavala, 2001). We will consider only the gross aspects of this complex model here. The nature of the model is that it will not capture short-term events such as the Cambrian SPICE excursion mentioned above; it attempts only to recover long–term variations.

Berner considered the carbon fluxes between the ocean–atmosphere–biosphere, carbonate, and organic carbon sediments (Figure 12.39). He assumed that the ocean–atmosphere–biosphere was in steady state at

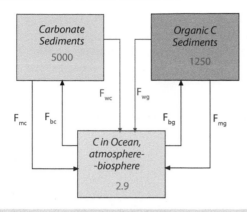

Figure 12.39   Model of carbon flow considered by Berner (1990). Masses of carbon are given in units of 10^{18} moles. Fluxes are described in the text. Source: Berner (1990)/American Association for the Advancement of Science - AAAS.

any given time, an assumption justified by the small size of the atmosphere–ocean–biosphere reservoir and the short residence time of carbon in them compared to the sedimentary ones. At steady state, one can write the following equation:

$$F_{wc} + F_{mc} + F_{wg} + F_{mg} = F_{bc} + F_{bg} \quad (12.11)$$

where F is a flux, subscript w denotes weathering, subscript m metamorphic or magmatic release of carbon, subscript b burial, subscript

c the carbonate reservoir, and subscript g organic sediments. Equation 12.11 simply states the steady-state assumption that the rate of release of carbon from organic or carbonate sediment through metamorphism, magmatism, and weathering equals the burial rate of organic carbon and carbonate sediment. The isotopic composition of the oceans and atmosphere depends on these fluxes:

$$\delta_o F_{bc} + (\delta_o - \alpha_c)F_{bg} = \delta_o(F_{wc} + F_{mc})$$
$$+ \delta_g(F_{wg} + F_{mg}) \quad (12.12)$$

where the subscript o denotes the ocean and α_c is the fractionation during photosynthesis. Because the isotopic composition of the oceans ($\delta^{13}C_o$) through time can be estimated from $\delta^{13}C$ in carbonate (e.g., Figure 12.39), Equation 12.12 provides a constraint on these fluxes: fluxes must balance to produce the observed $\delta^{13}C$ at a given time. While the carbon flux model in Figure 12.39 is simple, it depends on many other factors in complex ways as is shown in Figure 12.40.

The rate of uptake of CO_2 via the weathering of Ca and Mg silicates over time (F_{wsi}) is:

$$F_{wsi} = F_{bc} - F_{wc}$$
$$= f_B(T, CO_2)f_R(t)f_E(t)f_{AD}(t)^{0.65}F_{wsi}(0) \quad (12.13)$$

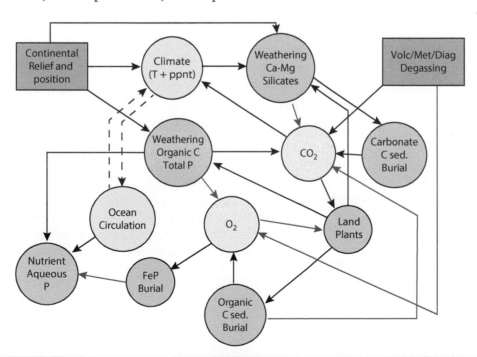

Figure 12.40   Cause–effect feedback diagram for the long-term carbon cycle. Arrows originate at causes and end at effects. Black arrows represent direct responses; red arrows represent inverse responses, e.g., as Ca–Mg silicate weathering increases, CO_2 decreases. Source: Adapted from Berner (1999).

where f_B is the feedback factor for silicates expressing the dependence of weathering on temperature and on CO_2, f_R is the mountain uplift factor (ratio of land relief at time t to present relief), f_E is a factor expressing the dependence of weathering on soil biological activity due to land plants (=1 at present), f_{AD} is the change in the ratio river discharge at time t to present river discharge due to change in paleogeography (a function of change in both land area and river runoff; the power of 0.65 reflects dilution of dissolved load at high runoff), and $F_{wsi}(0)$ is the present weathering uptake of CO_2.

The weathering feedback works in two ways. First of all, global surface temperatures should correlate with atmospheric CO_2 concentrations. Since weathering reaction rates are, in principle, temperature dependent, Berner reasoned that weathering would be more rapid when temperatures, and hence atmospheric CO_2 concentrations, are higher (these same assumptions are present in the BLAG model). Second, Berner assumes that higher atmospheric CO_2 leads to greater rates of photosynthesis and biological activity. This enhances weathering through greater production of biological acids and nutrient uptake. The weathering feedback function can be formulated as the product of these two factors:

$$f_B(T, CO_2) = f(T)fCO_2) \qquad (12.14)$$

The temperature dependence contains the usual Arrhenius exponential relationship:

$$f(T) = e^{E/RT(t) - E/RT(0)} \times [1 - \rho(T(t) - T(0))]^{0.65}$$
$$(12.15)$$

where R is the gas constant, E is the activation energy for the weathering reaction, and ρ is the coefficient expressing the effect of temperature on global river runoff. The temperature difference is in turn dependent of atmospheric CO_2 levels, solar irradiation $(W_s)^3$, and the effect of changes in paleogeography on temperature (λ):

$$T(t) - T(0) = \gamma \ln \left(\frac{CO_2(t)}{CO_2(0)} \right) - W_s(t) + \lambda(t))$$
$$(12.16)$$

where γ is the "greenhouse" coefficient and is computed from atmospheric global circulation models. Before vascular plants became important in the Devonian, f_{CO2} is expressed as:

$$f(CO_2) = \left[\frac{CO_2(t)}{CO_2(0)} \right]^{0.5} \qquad (12.17)$$

and after that as:

$$f(CO_2) = \left[\frac{CO_2(t)/CO_2(0)}{1 + CO_2(t)/CO_2(0)} \right]^{\phi} \qquad (12.18)$$

where ϕ is a fertilization factor that expresses the dependency of photosynthesis rates on CO_2 concentration and has a value near 0.4.

Other fluxes in Equations (12.11) through 12.13 are complex function of geological and biological history (e.g., volcanism, continental land area, biological productivity, and evolution) and much of the work in constructing this model comes from estimating how these have varied from the geologic record. Berner then calculated the magmatic and weathering fluxes and substituting these into Equations 12.11 and 12.12 calculated the burial fluxes in one-million-year steps. From values of F_{wc} and F_{bc}, he solved for $f_{CO_2}(t)$ in Equation 12.13 and then for $CO_2(t)$. This new value of $f_{CO_2}(t)$ was then used to iterate the calculation until a constant $f_{CO_2}(t)$ was obtained.

From this, new values for the mass of the reservoirs and their isotopic composition were calculated using mass balance equations, such as:

$$dC/dt = F_{bg} - (F_{wc} + F_{mc}) \qquad (12.19)$$

and

$$d(\delta_c C)/dt = \delta_0 F_{bc} - \delta_c(F_{wm} + F_{mc}) \qquad (12.20)$$

In a subsequent version of the model, Berner (2006) distinguished weathering of volcanic and non-volcanic rocks because young volcanic rocks weather more readily than other ones and furthermore the former are richer in Ca and Mg and hence more effective at removing atmospheric CO_2. He noted that Sr isotopic composition of seawater through time (Figure 12.38) depends on the hydrothermal flux from the oceanic crust, which mainly

---

[3] As was pointed out in Chapter 1, the Sun has grown about 30% brighter over geologic time.

depends on seafloor spreading rate and the weather flux from the continents. From the perspective of $^{87}Sr/^{86}Sr$, the weathering flux has four components: young, mantle-derived volcanic rocks have low $^{87}Sr/^{86}Sr$ ($\approx$0.703), old ones have higher ratios, and young and old marine carbonates have high and low $^{87}Sr/^{86}Sr$, respectively. $^{87}Sr/^{86}Sr$ in marine carbonates (Figure 12.38) monitor this weathering flux. A mass balance equation similar to Equation 12.11 is written for $^{87}Sr/^{86}Sr$ from this an equation for the fraction of young volcanic rock weathering $X_{volc}(t) = f(^{87}Sr/^{86}Sr(t))$ derived. The silicate weather flux (Equation [12.13]) becomes:

$$F_{wsi} = F_{bc} - F_{wc}$$
$$= f_{volc}(t)f_B(T, CO_2)f_R(t)f_E(t) \quad (12.21)$$
$$f_{AD}(t)^{0.65}F_{wsi}(0)$$

where f_{volc} is the volcanic weathering effect relative to the present. Berner assumes that volcanics weather about twice as fast as non-volcanics and the present fraction of volcanics exposed to weathering is 0.3, so that

$$f_{volc} = (X_{volc} + 1)/1.3 \quad (12.22)$$

f_{volc} reduced to 1 when the fraction of volcanics is equal to the present value of 0.3. Figure 12.41 compares the results of the older GEOCARB model (dashed green line), which does not treat

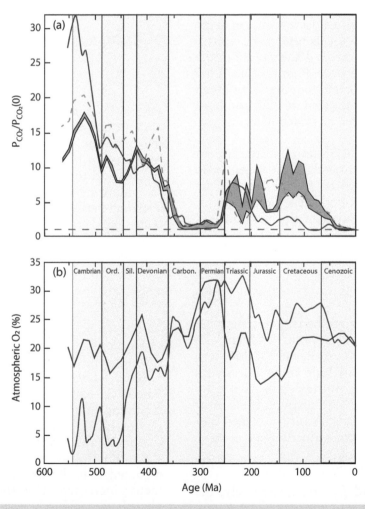

Figure 12.41   (a) Modeled partial pressure of CO_2 relative to present based on marine $\delta^{13}C$. Dashed green line: GEOCARB model of Berner (2006) without preferential volcanic weathering; pink field: GEOCARB: when preferential volcanic weathering is included with the range depending on how the non-volcanic silicate contribution to marine $^{87}Sr/^{86}Sr$ is treated in the model; blue line: GEOCARBSULFOR model of Krause et al. (2018) with updated $\delta^{13}C$ database. Dashed line is the present partial pressure of CO_2. Modified from Berner (2006). (b) Modeled atmospheric O_2 through the Phanerozoic. Purple line: GEOCARBSULF model of Berner (2006); blue line: GEOCARBSULFOR model of Krause et al. (2018).

volcanic weathering separately and the new model which does (pink fields). The range for the latter reflects two methods Berner (2006) used in treating the non-volcanic silicate weathering contribution to marine $^{87}Sr/^{86}Sr$. Also shown are the results from an updated version of the model (solid blue line), GEO-CARBSULFOR, by Krause et al. (2018), which shows much higher P_{CO_2} in the early Paleozoic.

Although the models differ in detail, all share several characteristics: high atmospheric CO_2 in the Paleozoic, sharply decreasing CO_2 beginning as trees evolve and come to dominate the terrestrial landscape in the Devonian (the modern terrestrial biosphere accounts for more than half of biological production), a minimum with levels of approaching modern one in the Carboniferous and Permian, recovery in the Mesozoic, and a decrease again in the Cenozoic. These CO_2 levels need to be considered in light of the nearly 4% increase in the Sun's brightness over the Phanerozoic. Low CO_2 in the late Paleozoic, together with much of Gondwana being located at high southern latitudes, led to the Permo–Carboniferous glaciation of Gondwana.

In addition to burial and weathering of organic carbon, atmospheric oxygen also depends on oxidation of reduce species, the most significant of which are iron and sulfur, which can be represented by the reaction:

$$15O_2 + 4FeS_2 + 8H_2O \rightleftharpoons 2Fe_2O_3 + 8SO_4^{2-}$$
$$+ 16H^+$$
$$(12.23)$$

Burial of ferric iron in red beds represents a net consumption of atmospheric oxygen. To model atmospheric O_2, Berner (2004) extended his model to include sulfur burial and oxidative weathering in the GEOCARBSULF model. To start with, he distinguished three primary burial fluxes relevant to oxygen balance: marine, coal, and redbeds:

$$F_{bg} = (f_{mar}C_{mar} + f_{cb}C_{cb} + f_{rb}C_{rb}) F_t \quad (12.24)$$

where f_{mar}, f_{cb}, and f_{rb} are, respectively, the fraction of total terrigenous sedimentation deposited as marine sediments, coal basin sediments, and redbeds, and C_{mar}, C_{cb}, C_{rb}, are the mean organic carbon contents of these and F_t is the total burial flux. Atmospheric oxygen then is related to the reduced sulfur (pyrite plus organic sulfur) and oxidized iron burial flux, F_{bp}, as:

$$F_{bp} = [O_2(0)/O_2] (f_{mar}S_{mar} + f_{cb}S_{cb} + f_{rb}S_{rb}) F_t$$
$$(12.25)$$

where $(O_2(0)/O_2)$ is the ratio of molecular oxygen mass at present to that at a prior time and S_{mar} and S_{cb} are the pyrite (+ organic) sulfur content of marine sediments and coal basin sediments, and S_{rb} is the ferric iron content of redbeds and F_t is the total terrigenous sedimentation flux. The change in the mass of sulfur in the oceans, M_S, (the mass of sulfur in the atmosphere, biosphere, and soils is negligible) is then:

$$dM_S/dt = F_{ws} + F_{wp} + F_{ms} + F_{mp} - F_{bs} - F_{bp}$$
$$(12.26)$$

and that of $\delta^{34}S$ is:

$$d\delta_S M_S/dt = \delta_{ws}F_{ws} + \delta_{wp}F_{wp} + \delta_{ms}F_{ms}$$
$$+ \delta_{mp}F_{mp} - \delta_{bs}F_{bs} - \delta_{pb} F_{bp}$$
$$(12.27)$$

where F again designates the flux, δ_S is $\delta^{34}S$ and subscripts denote: wp: oxidative weathering of pyrite, ws: weathering of Ca sulfates, mp: flux for pyrite from volcanism, metamorphism, and diagenesis, ms: degassing flux for Ca sulfates from volcanism, metamorphism, and diagenesis, bs burial of Ca sulfates in sediments, and bp: burial of pyrite in sediments. The mass of atmospheric oxygen is the difference between burial and weather of organic carbon and pyrite:

$$d[O_2]/dt = F_{bg} - F_{wg} + (15/8)(F_{bp} - F_{wp})$$
$$(12.28)$$

Figure 12.41b shows the evolution of O_2 through the Phanerozoic calculated by the GEOCARBSULF. The high levels in the early Paleozoic conflict with paleoredox proxies including among several others $\delta^{98}Mo$ in Figure 12.38. Krause et al. (2018) argue that the problem lies with the parameterization of the sulfur isotope fractionation factor. Berner assumed that the pyrite–sulfate fractionation factor increased with the molecular oxygen abundance, which he parameterized as:

$$\Delta^{34}S = 35 (O_2/O_2(0))^{1.5} \quad (12.29)$$

Krause et al. (2018) argue that this formulation results in a very strong O_2 feedback from the sulfur cycle in the model resulting in an increase in calculated pyrite burial from isotopic mass balance and an over-estimation of O_2 production when atmospheric oxygen levels are <10% PAL. Furthermore, recent studies show large fractionations even at low O_2. Given the uncertainties in the actual $\Delta^{34}S$ and its oxygen dependence, Krause et al. (2018) replaced Berner's equations that calculate the weathering and burial rates of sulfur species with ones in which pyrite burial depends on the concentration of oceanic sulfate, the supply of organic matter to sediments, and the average oceanic dissolved oxygen concentration. Their model, GEOCARBSUFOR, shows atmospheric O_2 was variable but well below modern levels in the early Proterozoic, then increased further in the Devonian as forests expanded and reached maximum values in the Triassic, better matching proxies such as $\delta^{98}Mo$.

REFERENCES

Alegret, L., Thomas, E. and Lohmann, K.C. 2012. End-Cretaceous marine mass extinction not caused by productivity collapse. *Proceedings of the National Academy of Sciences* 109: 728–32. doi: 10.1073/pnas.1110601109.

Anbar, A.D. and Rouxel, O. 2007. Metal stable isotopes in paleoceanography. *Annual Review of Earth and Planetary Sciences* 35: 717–46. doi:10.1146/annurev.earth.34.031405.125029.

Andersen, K.K., Azuma, N., Barnola., et al. 2004. High-resolution record of Northern Hemisphere climate extending into the last interglacial period. *Nature* 431: 147–51. doi: 10.1038/nature02805.

Arrhenius, S. 1896. Ueber den Einfluss des Atmosphärischen Kohlensäurengehalts auf die Temperatur der Erdoberfläche. *Proceedings of the Royal Swedish Academy of Science* 22: 1–101.

Bauska, T.K., Baggenstos, D., Brook, E.J., et al. 2016. Carbon isotopes characterize rapid changes in atmospheric carbon dioxide during the last deglaciation. *Proceedings of the National Academy of Sciences* 113: 3465–70. doi: 10.1073/pnas.1513868113.

Bekker, A. and Holland, H.D. 2012. Oxygen overshoot and recovery during the early Paleoproterozoic. *Earth and Planetary Science Letters* 317–8: 295–304. doi: https://doi.org/10.1016/j.epsl.2011.12.012.

Berger, W.H. 2013. On the Milankovitch sensitivity of the Quaternary deep-sea record. *Climate of the Past* 9: 2003–11. doi: 10.5194/cp-9-2003-2013.

Berner, R.A. 1990. Atmospheric carbon dioxide levels over Phanerozoic time. *Science* 249: 1382–6. doi: 10.1126/science.249.4975.1382

Berner, R.A. 1999. A new look at the long-term carbon cycle. *GSA Today* 9: 1–6.

Berner, R.A., 2004. *Atmospheric O_2 over Phanerozoic Time*. Oxford: University Press.

Berner, R.A. 2006. Inclusion of the weathering of volcanic rocks in the GEOCARBSULF model. *American Journal of Science* 306: 295–302. doi: 10.2475/05.2006.01.

Berner, R.A. and Kothavala, Z. 2001. Geocarb III: a revised model of atmospheric CO_2 over Phanerozoic Time. *American Journal of Science* 301: 182–204. doi: 10.2475/ajs.301.2.182.

Berner, R.A., Lasaga, A.C. and Garrells, R.M. 1983. The carbonate-silicate geochemical cycle and its effect on atmospheric carbon dioxide over the past 100 million years. *American Journal of Science* 283: 641–83. doi: 10.2475/ajs.283.7.641.

Bigeleisen, J. and Mayer, M.G. 1947. Calculation of equilibrium constants for isotopic exchange reactions. *The Journal of Chemical Physics* 15: 261. doi: 10.1063/1.1746492.

Blättler, C.L., Claire, M.W., Prave, A.R., et al. 2018. Two-billion-year-old evaporites capture Earth's great oxidation. *Science* 360: 320–3. doi:10.1126/science.aar2687.

Blättler, C.L., Henderson, G.M. and Jenkyns, H.C. 2012. Explaining the Phanerozoic Ca isotope history of seawater. *Geology* 40: 843–6. doi: 10.1130/g33191.1.

Böhm, E., Lippold, J., Gutjahr, M., et al. 2015. Strong and deep Atlantic meridional overturning circulation during the last glacial cycle. *Nature* 517: 73–6. doi: 10.1038/nature14059.

Broecker, W.S. 1984. Terminations, in Berger, A., Imbrie, J., Hays, J., et al. (eds). *Milankovitch and Climate*. Dordrecht: D. Reidel Publishing Co.

Campbell, I.H. and Allen, C.M. 2008. Formation of supercontinents linked to increases in atmospheric oxygen. *Nature Geoscience* 1: 554. doi: 10.1038/ngeo259

Catling, D.C. and Zahnle, K.J. 2020. The Archean atmosphere. *Science Advances* 6: eaax1420. doi: 10.1126/sciadv.aax1420.

Cerling, T.E. and Quade, J. 1993. Stable carbon and oxygen isotopes in soil carbonates, in Swart, P.K., Lohmann, K.C., McKenzie, J., et al. (eds). *Climate Change in Continental Isotopic Records Geophysical Monographs*. Washington, D.C: AGU, 217–31 pp.

Cerling, T.E., Wang, Y. and Quade, J. 1993. Expansion of C_4 ecosystems as an indicator of global ecological change in the late Miocene. *Nature* **361**: 344–5. doi: 10.1038/361344a0.

Chamberlin, T.C. 1899. An Attempt to Frame a Working Hypothesis of the Cause of Glacial Periods on an Atmospheric Basis. *The Journal of Geology* **7**: 545–84. doi: 10.1086/608449.

Chen, T.-Y., Ling, H.-F., Hu, R., et al. 2013. Lead isotope provinciality of central North Pacific Deep Water over the Cenozoic. *Geochemistry, Geophysics, Geosystems* **14**: 1523–37. doi:10.1002/ggge.20114.

Cheng, H., Edwards, R.L., Sinha, A., et al. 2016. The Asian monsoon over the past 640,000 years and ice age terminations. *Nature* **534**: 640–6. doi: 10.1038/nature18591.

Craig, H. 1965. Measurement of oxygen isotope paleotemperatures, in Tongiorgi, E. (ed). *Stable Isotopes in Oceanographic Studies and Paleotemperatures*. Pisa: CNR Lab. Geol. Nucl.

Cohen, K.M. and Gibbard, P. 2011. *Global Chronostratigraphical Correlation Table for the Last 2.7 Million Years*. Cambridge: Subcommission on Quaternary Stratigraphy (International Commission on Stratigraphy).

Crockford, P.W., Hayles, J.A., Bao, H., et al. 2018. Triple oxygen isotope evidence for limited mid-Proterozoic primary productivity. *Nature* **559**: 613–16. doi: 10.1038/s41586-018-0349-y.

Crockford, P.W., Kunzmann, M., Bekker, A., et al. 2019. Claypool continued: extending the isotopic record of sedimentary sulfate. *Chemical Geology* **513**: 200–25. doi: https://doi.org/10.1016/j.chemgeo.2019.02.030.

Curry, W.B. and Oppo, D.W. 2005. Glacial water mass geometry and the distribution of $\delta^{13}C$ of ΣCO_2 in the western Atlantic Ocean. *Paleoceanography* **20**: PA1017. doi: 10.1029/2004PA001021.

Czaja, A.D., Johnson, C.M., Beard, B.L., et al. 2013. Biological Fe oxidation controlled deposition of banded iron formation in the ca. 3770 Ma Isua Supracrustal Belt (West Greenland). *Earth and Planetary Science Letters* **363**: 192–203. doi: http://dx.doi.org/10.1016/j.epsl.2012.12.025.

Czaja, A.D., Johnson, C.M., Roden, E.E., et al. 2012. Evidence for free oxygen in the Neoarchean ocean based on coupled iron-molybdenum isotope fractionation. *Geochimica et Cosmochimica Acta* **86**: 118–37. doi: 10.1016/j.gca.2012.03.007.

Dahl, T.W., Boyle, R.A., Canfield, D.E., et al. 2014. Uranium isotopes distinguish two geochemically distinct stages during the later Cambrian SPICE event. *Earth and Planetary Science Letters* **401**: 313–26. doi: https://doi.org/10.1016/j.epsl.2014.05.043.

Dansgaard, W., Johnsen, S.J., Møller, J., et al. 1969. One thousand centuries of climatic record from Camp Century on the Greenland Ice Sheet. *Science* **166**: 377–80. doi: 10.1126/science.166.3903.377.

Derry, L.A. 2015. Causes and consequences of mid-Proterozoic anoxia. *Geophysical Research Letters* **42**: 8538–46. doi: https://doi.org/10.1002/2015GL065333.

Derry, L.A. and France-Lanord, C. 1996. Neogene Himalayan weathering history and river $^{87}Sr/^{86}Sr$: impact on the marine Sr record. *Earth and Planetary Science Letters* **142**: 59–74.

Ding, T. P., Gao, J. F., Tian, S. H., et al. 2017. The $\delta^{30}Si$ peak value discovered in middle Proterozoic chert and its implication for environmental variations in the ancient ocean. *Scientific Reports* **7**: 44000. doi: 10.1038/srep44000.

Du Vivier, A.D.C., Selby, D., Condon, D.J., et al. 2015. Pacific $^{187}Os/^{188}Os$ isotope chemistry and U–Pb geochronology: synchroneity of global Os isotope change across OAE 2. *Earth and Planetary Science Letters* **428**: 204–16. doi: https://doi.org/10.1016/j.epsl.2015.07.020.

Ebelmen, J.J. 1845. Sur les produits de la décomposition des espèces minèrales de la famille des silicates. *Annales des Mines* **12**: 627–54.

Eggleston, S., Schmitt, J., Bereiter, B., et al. 2016. Evolution of the stable carbon isotope composition of atmospheric CO_2 over the last glacial cycle. *Paleoceanography* **31**: 434–52. doi: https://doi.org/10.1002/2015PA002874.

Emiliani, C. 1955. Pleistocene temperatures. *Journal of Geology* **63**: 538–78.

Epstein, S., Buchbaum, H.A. and Lowenstam, H.A. 1953. Revised carbonate-water isotopic temperature scale. *Bulletin of the Geological Society of America* **64**: 1315–26.

Farkaš, J., Böhm, F., Wallmann, K., et al. 2007. Calcium isotope record of Phanerozoic oceans: implications for chemical evolution of seawater and its causative mechanisms. *Geochimica et Cosmochimica Acta* **71**: 5117–34. doi: 10.1016/j.gca.2007.09.004.

Farquhar, J., Bao, H. and Thiemens, M. 2000. Atmospheric influence of Earth's earliest sulfur cycle. *Science* **289**:756–58. doi: 10.1126/science.289.5480.756.

Farquhar, J., Zerkle, A.L. and Bekker, A. 2014. 6.4 - Geologic and geochemical constraints on Earth's early atmosphere, in Holland, H.D. and Turekian, K.K. (eds). *Treatise on Geochemistry*, 2nd ed. Oxford: Elsevier, 91–138 pp.

Fontorbe, G., Frings, P.J., De La Rocha, C.L., et al. 2016. A silicon depleted North Atlantic since the Palaeogene: evidence from sponge and radiolarian silicon isotopes. *Earth and Planetary Science Letters* **453**: 67–77. doi: https://doi.org/10.1016/j.epsl.2016.08.006.

Foster, G.L. 2008. Seawater pH, pCO_2 and $[CO_3^{2-}]$ variations in the Caribbean Sea over the last 130kyr: a boron isotope and B/Ca study of planktic foraminifera. *Earth and Planetary Science Letters* **271**: 254–66. doi: 10.1016/j.epsl.2008.04.015.

Foster, G.L. and Rae, J.W.B. 2016. Reconstructing ocean pH with boron isotopes in foraminifera. *Annual Review of Earth and Planetary Sciences* **44**: 207–37. doi: 10.1146/annurev-earth-060115-012226.

Frings, P.J., Clymans, W., Fontorbe, G., et al. 2016. The continental Si cycle and its impact on the ocean Si isotope budget. *Chemical Geology* **425**: 12–36. doi: https://doi.org/10.1016/j.chemgeo.2016.01.020.

Greenop, R., Hain, M.P., Sosdian, S.M., et al. 2017. A record of Neogene seawater $\delta^{11}B$ reconstructed from paired $\delta^{11}B$ analyses on benthic and planktic foraminifera. *Climate of the Past* **13**: 149–70. doi: 10.5194/cp-13-149-2017.

Gutjahr, M., Ridgwell, A., Sexton, P.F., et al. 2017. Very large release of mostly volcanic carbon during the Palaeocene–Eocene Thermal Maximum. *Nature* **548**: 573–7. doi: 10.1038/nature23646.

Halverson, G.P. and Shields-Zhou, G. 2011. Chapter 4 Chemostratigraphy and the Neoproterozoic glaciations. *Geological Society, London, Memoirs* **36**: 51–66. doi: 10.1144/m36.4.

Hays, J.D., Imbrie, J. and Shackleton, N.J. 1976. Variations in the Earth's orbit: Pacemaker of the ice ages. *Science* **194**: 1121–32.

Hendry, K.R. and Brzezinski, M.A. 2014. Using silicon isotopes to understand the role of the Southern Ocean in modern and ancient biogeochemistry and climate. *Quaternary Science Reviews* **89**: 13–26. doi: https://doi.org/10.1016/j.quascirev.2014.01.019.

Hendry, K.R., Robinson, L.F., McManus, J.F., et al. 2014. Silicon isotopes indicate enhanced carbon export efficiency in the North Atlantic during deglaciation. *Nature Communications* **5**: 3107. doi: 10.1038/ncomms4107

Henehan, M.J., Ridgwell, A., Thomas, E., et al. 2019. Rapid ocean acidification and protracted Earth system recovery followed the end-Cretaceous Chicxulub impact. *Proceedings of the National Academy of Sciences* **116**: 22500–4. doi: 10.1073/pnas.1905989116.

Higgins, J.A., Blättler, C.L., Lundstrom, E.A., et al. 2018. Mineralogy, early marine diagenesis, and the chemistry of shallow-water carbonate sediments. *Geochimica et Cosmochimica Acta* **220**: 512–34. doi: https://doi.org/10.1016/j.gca.2017.09.046.

Hodgskiss, M.S.W., Crockford, P.W., Peng, Y., et al. 2019. A productivity collapse to end Earth's Great Oxidation. *Proceedings of the National Academy of Sciences* **116**: 17207–12. doi: 10.1073/pnas.1900325116.

Hoffman, P.F. 2013. The Great Oxidation and a Siderian snowball Earth: MIF-S based correlation of Paleoproterozoic glacial epochs. *Chemical Geology* **362**: 143–56. doi: https://doi.org/10.1016/j.chemgeo.2013.04.018.

Holland, H.D. 2006. The oxygenation of the atmosphere and oceans. *Philosophical Transactions of the Royal Society B: Biological Sciences* **361**: 903–15. doi: 10.1098/rstb.2006.1838.

Hönisch, B. and Hemming, N.G. 2005. Surface ocean pH response to variations in pCO_2 through two full glacial cycles. *Earth and Planetary Science Letters* **236**: 305–14. doi: 10.1016/j.epsl.2005.04.027.

Hsü, K.J., He, Q., McKenzie, J.A., et al. 1982. Mass mortality and its environmental and evolutionary consequences. *Science* **216**: 249–56. doi: 10.1126/science.216.4543.249.

Huth, T.E., Cerling, T.E., Marchetti, D.W., et al. 2019. Seasonal bias in soil carbonate formation and its implications for interpreting high-resolution paleoarchives: evidence from Southern Utah. *Journal of Geophysical Research: Biogeosciences* **124**: 616–32. doi: https://doi.org/10.1029/2018JG004496.

Huth, T.E., Cerling, T.E., Marchetti, D.W., et al. 2020. Laminated soil carbonate rinds as a paleoclimate archive of the Colorado Plateau. *Geochimica et Cosmochimica Acta* **282**: 227–44. doi: https://doi.org/10.1016/j.gca.2020.05.022.

Imbrie, J. 1985. A theoretical framework for the Pleistocene ice ages: William Smith Lecture. *Journal of the Geological Society* **142**: 417–32. doi: 10.1144/gsjgs.142.3.0417.

Isson, T.T., Love, G.D., Dupont, C.L., et al. 2018. Tracking the rise of eukaryotes to ecological dominance with zinc isotopes. *Geobiology* **16**: 341–52. doi: https://doi.org/10.1111/gbi.12289.

Johnson, C.M., Beard, B.L. and Roden, E.E. 2008. The iron isotope fingerprints of redox and biogeochemical cycling in modern and ancient Earth. *Annual Review of Earth and Planetary Sciences* **36**: 457–93. doi: 10.1146/annurev.earth.36.031207.124139.

Jouzel, J., Masson-Delmotte, V., Cattani, O., et al. 2007. Orbital and millennial Antarctic climate variability over the past 800,000 years. *Science* **317**: 793–6. doi: 10.1126/science.1141038.

Kasting, J.F. and Ono, S. 2006. Palaeoclimates: the first two billion years. *Philosophical Transactions of the Royal Society B: Biological Sciences* **361**: 917–29. doi: 10.1098/rstb.2006.1839.

Keigwin, L., 2001. Data report: late Pleistocene stable isotope studies of ODP Sites 1054, 1055, and 1063, in Keigwin, L.D., Rio, D., Acton, G.D., et al. (eds). *Proceeding of the Ocean Drilling Project Scientific Results*. Texas: A and M University, College Station, TX, 9: 1–14 pp.

Kendall, B. 2021. Recent advances in geochemical paleo-oxybarometers. *Annual Review of Earth and Planetary Sciences* **49**: 399–433. doi: 10.1146/annurev-earth-071520-051637.

Köhler, P., Fischer, H. and Schmitt, J. 2010. Atmospheric $\delta^{13}C_{CO2}$ and its relation to pCO_2 and deep ocean $\delta^{13}C$ during the late Pleistocene. *Paleoceanography* **25**. doi: https://doi.org/10.1029/2008PA001703.

Komar, N. and Zeebe, R.E. 2016. Calcium and calcium isotope changes during carbon cycle perturbations at the end-Permian. *Paleoceanography* **31**: 115–30. doi: 10.1002/2015PA002834.

Krause, A.J., Mills, B.J.W., Zhang., et al. 2018. Stepwise oxygenation of the Paleozoic atmosphere. *Nature Communications* **9**: 4081. doi: 10.1038/s41467-018-06383-y.

Laakso, T.A. and Schrag, D.P. 2017. A theory of atmospheric oxygen. *Geobiology* **15**: 366–84. doi: https://doi.org/10.1111/gbi.12230.

Lawrence, J.R. and Meaux, J.R. 1993. The stable isotopic composition of ancient kaolinites of North America, in Swart, P.K., Lohmann, K.C., McKenzie, J., et al. (eds). *Climate Change in Continental Isotopic Records. Geophysical Monograph*. Washington: AGU, 249–61 pp.

Lawrence, J.R. and Taylor, H.P. 1972. Hydrogen and oxygen isotope systematics in weathering profiles. *Geochimica et Cosmochimica Acta* **36**: 1377–93. doi:

Lisiecki, L.E. and Raymo, M.E. 2005. A Pliocene-Pleistocene stack of 57 globally distributed benthic $\delta^{18}O$ records. *Paleoceanography* **20**. doi: 10.1029/2004PA001071.

Luthi, D., Le Floch, M., Bereiter, B., et al. 2008. High-resolution carbon dioxide concentration record 650,000–800,000 years before present. *Nature* **453**: 379–82. doi:

Lyons, T.W., Diamond, C.W., Planavsky, N.J., et al. 2021. oxygenation, life, and the planetary system during Earth's middle history: an overview. *Astrobiology* **21**: 906–23. doi: 10.1089/ast.2020.2418.

MacGregor, A.M. 1927. The problem of the Precambrian atmosphere. *South African Journal of Science* **24**: 155–72.

Mänd, K., Lalonde, S.V., Robbins, L.J., et al. 2020. Palaeoproterozoic oxygenated oceans following the Lomagundi–Jatuli Event. *Nature Geoscience* **13**: 302–6. doi: 10.1038/s41561-020-0558-5.

Mänd, K., Planavsky, N.J., Porter, S.M., et al. 2022. Chromium evidence for protracted oxygenation during the Paleoproterozoic. *Earth and Planetary Science Letters* **584**: 117501. doi: https://doi.org/10.1016/j.epsl.2022.117501.

Matthews, J.J., Liu, A.G., Yang, C., et al. 2020. A chronostratigraphic framework for the rise of the Ediacaran macrobiota: new constraints from Mistaken Point Ecological Reserve, Newfoundland. *Bulletin of the Geological Society of America* **133**: 612–24. doi: 10.1130/b35646.1.

McCrea, J.M. 1950. On the isotopic chemistry of carbonates and a paleotemperature scale. *The Journal of Chemical Physics* **18**: 849–57. doi: http://dx.doi.org/10.1063/1.1747785.

Miao, L., Moczydłowska, M., Zhu, S., et al. 2019. New record of organic-walled, morphologically distinct microfossils from the late Paleoproterozoic Changcheng Group in the Yanshan Range, North China. *Precambrian Research* **321**: 172–98. doi: https://doi.org/10.1016/j.precamres.2018.11.019.

Milanković, M. 1920. *Théorie Mathématique des Phénomènes Thermiques Produits par la Radiation Solaire*, Gauthier-Villars.

O'Connor, L.K., Jenkyns, H.C., Robinson., et al. 2020. A re-evaluation of the Plenus Cold Event, and the links between CO_2, temperature, and seawater chemistry during OAE 2. *Paleoceanography and Paleoclimatology* **35**: e2019PA003631. doi: https://doi.org/10.1029/2019PA003631.

O'Nions, R.K., Frank, M., von Blanckenburg, F., et al. 1998. Secular variation of Nd and Pb isotopes in ferromanganese crusts from the Atlantic, Indian and Pacific Oceans. *Earth and Planetary Science Letters* **155**: 15–28. doi: 10.1016/S0012–821X(97)00207–0.

Oppo, D.W. and Fairbanks, R.G. 1987. Variability in the deep and intermediate water circulation of the Atlantic Ocean during the past 25,000 years: Northern Hemisphere modulation of the Southern Ocean. *Earth and Planetary Science Letters* **86**: 1–15. doi: http://dx.doi.org/10.1016/0012-821X(87)90183-X.

Ossa Ossa, F., Hofmann, A., Wille, M., et al. 2018. Aerobic iron and manganese cycling in a redox-stratified Mesoarchean epicontinental sea. *Earth and Planetary Science Letters* **500**: 28–40. doi: https://doi.org/10.1016/j.epsl.2018.07.044.

Ostrander, C.M., Johnson, A.C. and Anbar, A.D. 2021. Earth's first redox revolution. *Annual Review of Earth and Planetary Sciences* **49**: 337–66. doi: 10.1146/annurev-earth-072020-055249.

Ostrander, C.M., Nielsen, S.G., Owens, J.D., et al. 2019. Fully oxygenated water columns over continental shelves before the Great Oxidation Event. *Nature Geoscience* **12**: 186–91. doi: 10.1038/s41561-019-0309-7.

Pagani, M., Arthur, M.A. and Freeman, K.H. 1999. Miocene evolution of atmospheric carbon dioxide. *Paleoceanography* **14**: 273–92. doi: 10.1029/1999pa900006.

Payne, J.L., Turchyn, A.V., Paytan, A., et al. 2010. Calcium isotope constraints on the end-Permian mass extinction. *Proceedings of the National Academy of Sciences* **107**: 8543–8. doi: 10.1073/pnas.0914065107.

Pearson, P.N. and Palmer, M.R. 2000. Atmospheric carbon dioxide concentrations over the past 60 milllion years. *Nature* **406**: 695–9.

Peucker-Ehrenbrink, B. and Ravizza, G. 2000. The marine osmium isotope record. *Terra Nova* **12**: 205–19. doi: 10.1046/j.1365-3121.2000.00295.x.

Peucker-Ehrenbrink, B., Miller, M.W., Arsouze, T., et al. 2010. Continental bedrock and riverine fluxes of strontium and neodymium isotopes to the oceans. *Geochemistry, Geophysics, Geosystems* **11**: Q03016. doi: 10.1029/2009gc002869.

Peucker-Ehrenbrink, B., Ravizza, G., Hofmann, A.W., et al. 1995. The marine $^{187}Os/^{186}Os$ record of the past 80 million years: anthropogenic osmium in coastal deposits. *Earth and Planetary Science Letters* **130**: 155–67. doi: 10.1016/0012-821X(95)00003-U.

Piotrowski, A.M., Goldstein, S.L., Hemming, S.R., et al. 2005. Temporal relationships of carbon cycling and ocean circulation at glacial boundaries. *Science* **307**: 1933–8. doi: 10.1126/science.1104883.

Planavsky, N.J., Slack, J.F., Cannon, W.F., et al. 2018. Evidence for episodic oxygenation in a weakly redox-buffered deep mid-Proterozoic ocean. *Chemical Geology* **483**: 581–94. doi: https://doi.org/10.1016/j.chemgeo.2018.03.028.

Planavsky, N., Rouxel, O.J., Bekker, A., et al. 2012. Iron isotope composition of some Archean and Proterozoic iron formations. *Geochimica et Cosmochimica Acta* **80**: 158–69. doi: 10.1016/j.gca.2011.12.001.

Pogge von Strandmann, P.A.E., Stüeken, E.E., Elliott, T., et al. 2015. Selenium isotope evidence for progressive oxidation of the Neoproterozoic biosphere. *Nature Communications* **6**: 10157. doi: 10.1038/ncomms10157.

Prokoph, A., Shields, G.A. and Veizer, J. 2008. Compilation and time-series analysis of a marine carbonate $\delta^{18}O$, $\delta^{13}C$, $^{87}Sr/^{86}Sr$ and $\delta^{34}S$ database through Earth history. *Earth-Science Reviews* **87**: 113–33. doi: 10.1016/j.earscirev.2007.12.003.

Quade, J., Cerling, T.E. and Bowman, J.R. 1989. Development of Asian monsoon revealed by marked ecological shift during the latest Miocene in northern Pakistan. *Nature* **342**: 163–6. doi:

Rae, J.W.B., Zhang, Y.G., Liu, X., et al. 2021. Atmospheric CO_2 over the past 66 million years from marine archives. *Annual Review of Earth and Planetary Sciences* **49**: 609–41. doi: 10.1146/annurev-earth-082420-063026.

Rasmussen, S.O., Andersen, K.K., Svensson, A.M., et al. 2006. A new Greenland ice core chronology for the last glacial termination. *Journal of Geophysical Research: Atmospheres* **111**. doi: https://doi.org/10.1029/2005JD006079.

Reinhard, C.T. and Planavsky, N.J. 2022. The history of ocean oxygenation. *Annual Review of Marine Science* **14**: 331–53. doi: 10.1146/annurev-marine-031721-104005.

Richter, F.M., Rowley, D.B. and DePaolo, D.J. 1992. Sr isotope evolution of seawater: the role of tectonics. *Earth and Planetary Science Letters* **109**: 11–23.

Robinson, N., Ravizza, G., Coccioni, R., et al. 2009. A high-resolution marine $^{187}Os/^{188}Os$ record for the late Maastrichtian: distinguishing the chemical fingerprints of Deccan volcanism and the KP impact event. *Earth and Planetary Science Letters* **281**: 159–68. doi: 10.1016/j.epsl.2009.02.019.

Rhodes, R.H., Brook, E.J., Chiang, J.C.H., et al. 2015. Enhanced tropical methane production in response to iceberg discharge in the North Atlantic. *Science* **348**: 1016–19. doi: 10.1126/science.1262005.

Saltzman, M.R. and Thomas, E. 2012. Chapter 11 - Carbon Isotope Stratigraphy, in Gradstein, F.M., Ogg, J.G., Schmitz, M.D., et al. (eds). *The Geologic Time Scale*. Boston: Elsevier, 207–32 pp.

Savin, S.M. and Epstein, S. 1970. The oxygen and hydrogen isotope geochemistry of clay minerals. *Geochimica et Cosmochimica Acta* **34**: 25–42. doi: https://doi.org/10.1016/0016-7037(70)90149-3

Schidlowski, M., Eichmann, R. and Junge, C.E. 1976. Carbon isotope geochemistry of the Precambrian Lomagundi carbonate province, Rhodesia. *Geochimica et Cosmochimica Acta* **40**: 449–55. doi: https://doi.org/10.1016/0016-7037(76)90010-7.

Shackleton, N.J. and Opdyke, N.D. 1973. Oxygen isotope and paleomagnetic stratigraphy of an equatorial Pacific core V28–238: oxygen isotope temperatures and ice volumes on a 10^5 and 10^6 year time scale. *Quaternary Research* **3**: 39–55.

Sheldon, N.D. and Tabor, N.J. 2009. Quantitative paleoenvironmental and paleoclimatic reconstruction using paleosols. *Earth-Science Reviews* **95**: 1–52. doi: https://doi.org/10.1016/j.earscirev.2009.03.004.

Sheppard, S.M.F. and Gilg, H.A. 1996. Stable isotope geochemistry of clay minerals. *Clay Minerals* **31**: 1–24.

Stüeken, E.E., Buick, R. and Anbar, A.D. 2015. Selenium isotopes support free O_2 in the latest Archean. *Geology* **43**: 259–62. doi: 10.1130/g36218.1.

Stüeken, E.E. and Kipp, M.A. 2020. *Selenium Isotope Paleobiogeochemistry*. Cambridge: Cambridge University Press.

Tatzel, M., von Blanckenburg, F., Oelze, M., et al. 2017. Late Neoproterozoic seawater oxygenation by siliceous sponges. *Nature Communications* **8**: 621. doi: 10.1038/s41467-017-00586-5.

Thoby, M., Konhauser, K.O., Fralick, P.W., et al. 2019. Global importance of oxic molybdenum sinks prior to 2.6 Ga revealed by the Mo isotope composition of Precambrian carbonates. *Geology* **47**: 559–62. doi: 10.1130/g45706.1.

Urey, H.C. 1947. The thermodynamics of isotopic substances. *Journal of the Chemical Society* **1947**: 562–81.

Urey, H.C. 1952. On the early chemical history of the Earth and the origin of life. *Proceedings of the National Academy of Sciences* **38**: 351–63. doi: https://doi.org/10.1073/pnas.38.4.351

Wang, X., Ossa Ossa, F., Hofmann, A., et al. 2020. Uranium isotope evidence for Mesoarchean biological oxygen production in shallow marine and continental settings. *Earth and Planetary Science Letters* **551**: 116583. doi: https://doi.org/10.1016/j.epsl.2020.116583.

Warke, M.R., Rocco, T.D., Zerkle, A.L., et al. 2020. The Great Oxidation Event preceded a Paleoproterozoic snowball Earth. *Proceedings of the National Academy of Sciences* **117**: 13314–20. doi: 10.1073/pnas.2003090117.

Wei, W., Klaebe, R., Ling, H.-F., et al. 2020. Biogeochemical cycle of chromium isotopes at the modern Earth's surface and its applications as a paleo-environment proxy. *Chemical Geology* **541**: 119570. doi: https://doi.org/10.1016/j.chemgeo.2020.119570.

Westerhold, T., Marwan, N., Drury, A.J., et al. 2020. An astronomically dated record of Earth's climate and its predictability over the last 66 million years. *Science* **369**: 1383–7. doi: 10.1126/science.aba6853.

Zachos, J.C., Dickens, G.R. and Zeebe, R.E. 2008. An early Cenozoic perspective on greenhouse warming and carbon-cycle dynamics. *Nature* **451**: 279–83. doi: 10.1038/nature06588.

Zachos, J., Pagani, M., Sloan, L., et al. 2001. Trends, rhythms, and aberrations in global climate 65 ma to present. *Science* **292**: 686–93. doi: 10.1126/science.1059412.

Zheng, X.-Y., Jenkyns, H.C., Gale, A.S., et al. 2013. Changing ocean circulation and hydrothermal inputs during Ocean Anoxic Event 2 (Cenomanian–Turonian): evidence from Nd-isotopes in the European shelf sea. *Earth and Planetary Science Letters* **375**: 338–48. doi: https://doi.org/10.1016/j.epsl.2013.05.053.

Zhu, S., Zhu, M., Knoll, A.H., et al. 2016. Decimetre-scale multicellular eukaryotes from the 1.56-billion-year-old Gaoyuzhuang Formation in North China. *Nature Communications* **7**: 11500. doi: 10.1038/ncomms11500.

Chapter 13

Life, Paleoecology, and Human History

13.1 INTRODUCTION

The study of stable isotopes, once the exclusive province of physicists, chemists, and earth scientists, has found application over the last several decades in many other fields of science, including biology, medicine, forensics, anthropology, and archeology. Exploring their use in biology and medicine would take us much too far afield. In this chapter, we will explore how stable isotopes have been put to use in studying the history of life and our own history. This too is an extremely broad field; therefore, rather than an exhaustive treatment, we will consider a few examples of how isotopes, both stable and radiogenic, have provided new insights into these topics. We will focus in particular on human evolution, tracing how our ancestors left the forests and began harvesting food from the more open landscapes of grasslands and savannahs, and later learned to grow their own food in such landscapes rather than hunting and gathering it. Without this agricultural revolution, civilization and science would not be possible.

13.2 ISOTOPES IN EVOLUTION

13.2.1 Isotopic fossils of the earliest life

We found in Chapter 10 that large carbon isotope fractionations occur during carbon fixation in both photosynthesis and chemosynthesis. Consequently, $\delta^{13}C$ values of $-20‰$ or less are generally interpreted as evidence

of biologic origin of those compounds. Schidlowski (1988) first reported $\delta^{13}C$ as low as $-26‰$ in carbonate rocks from the Isua Supracrustal Belt of West Greenland that are ostensibly older than 3.5 Ga and the oldest surviving examples of "supracrustals": rocks formed at the Earth's surface. Mojzsis et al. (1996) reported $\delta^{13}C$ between -20 and $-50‰$ for graphite inclusions in apatite grains in 3.85 Ga banded-iron formations there. In 1999, Rosing (1999) reported $\delta^{13}C$ of $-19‰$ from graphite in turbiditic and pelagic metasedimentary rocks from the Isua Supracrustals of Akilia Island (Figure 13.1 and also shown on the book cover). These rocks are thought to be older than 3.7 Ga. In each case, these negative $\delta^{13}C$ values were interpreted as evidence of a biogenic origin of the carbon and therefore that life existed on Earth at this time.

This interpretation has been controversial. There are several reasons for the controversy, but all ultimately relate to the extremely complex geological history of the area. The geology of the region includes not only the early Archean rocks, but also rocks of middle and late Archean age. Most rocks are multiply and highly deformed and metamorphosed, and the exact nature, relationships, and structure of the precursor rocks are difficult to decipher. Indeed, Rosing (1999) argued that at least some of the carbonates sampled by Schidlowski (1988) are veins deposited by metamorphic fluid flow rather than metasediments. Others have argued that the graphite in these rocks formed by thermal

Isotope Geochemistry, Second Edition. William M. White.
© 2023 John Wiley & Sons Ltd. Published 2023 by John Wiley & Sons Ltd.
Companion Website: www.wiley.com/go/white/isotopegeochem2

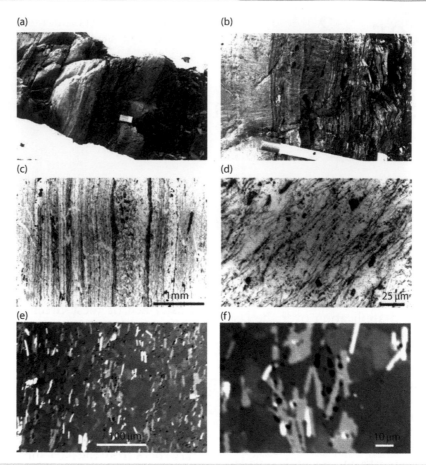

Figure 13.1 (a) Turbidite sedimentary rocks from the Isua supracrustal belt, West Greenland. For scale, the notebook is 17 cm wide. (b) A close-up of finely laminated slate representing pelagic mud. The hammer is 70 cm long. (c) Photomicrograph showing finely laminated pelagic mud. The variation in color is mainly due to variations in C abundance. (d) Photomicrograph of C grains arranged along a buckled stringer. (e) Backscattered electron image of a polished surface, showing the distribution of C grains as black areas. (f) Backscattered electron image of a polished surface (sample 810213), showing the rounded shape of C grains (black). Source: Adapted from Rosing et al. (1999), American Association for the Advancement of Science.

decomposition of siderite ($FeCO_3$) and subsequent reduction of some of the carbon. Subsequent microanalysis, however, suggests that metamorphism has increased the $\delta^{13}C$ of the graphite rather than decreasing it and the large ^{13}C-depletion of graphite was probably premetamorphic in origin (Ueno et al., 2002).

Dodd et al. (2017) reported $\delta^{13}C$ of –19.7 to –25.7‰ in carbon from hydrothermal jasper deposits in apparently Eoarchean rocks (3.95 Ga) of the Saglek Block in Labrador, Canada. Subsequent work found $\delta^{13}C$ values as low as –30.8‰ in graphite (Tashiro et al, 2017) and again that, if anything, metamorphism increased $\delta^{13}C$. Subsequent work has revised the age of

these rocks downward somewhat to ≈3.87 Ga (Wasilewski et al., 2021), making them essentially contemporaneous with the Isua rocks on the opposite side of the Labrador Straight.

Although some controversy remains, no one has been able to convincingly refute the conclusion that this isotopically light carbon is evidence of early life on Earth. Thus, the earliest evidence of life on our planet is isotopic.

Moving forward in time to 3.43 Ga, there is little, if any, controversy that the stromatolites of the Strelley Pool Formation of the Pilbara Supergroup in Western Australia provide physical evidence of microbial life. Stromatolites are structures composed mainly of carbonates distinguished by wavy layering deposited by consortia of bacteria, including photosynthetic cyanobacteria, in shallow water. They are found throughout the geologic record. Although rare today, there are examples of living ones in the Bahamas and Western Australia. Organic laminae preserved within the Strelley Pool stromatolitic layers consist of highly aromatic molecules with $\delta^{13}C$ ranging from −28.3 to −35.8‰. Flannery et al. (2018) found that the associated carbonate has $\delta^{13}C$ range of +1.8 to +3.1‰ with an average of +2.8‰, implying fractionations in the range of −29 to −45‰. These, as well as the quite negative values observed in the Saglek Block and Isua graphites, are greater than that observed in most modern autotrophs or expected from Rubisco fixation. Flannery et al. argue that these Archean autotrophs used a metabolism in addition to the Calvin–Benson cycle, such as CO_2 fixation via the Wood–Ljungdahl pathway, which is employed by modern acetogenic bacteria in a process that has been observed to produce $\delta^{13}C$ fractionations as great as −69‰ in laboratory systems.

13.2.2 Using clumped isotopes to determine dinosaur body temperatures

Whether or not dinosaurs were warm blooded is a question that has lingered for over 150 years. Reptiles are, of course, cold blooded (ectotherms) and often spend the early hours of the day warming themselves in the Sun before becoming active. Birds, however, are warm blooded and there is now almost universally agreement that birds descended from dinosaurs; indeed, many would say that birds are dinosaurs. We also know that many dinosaurs of the clade Theropoda had feathers, which make excellent insolation.

For the largest dinosaurs, the sauropods, the problem was less keeping warm than keeping cool. Because of their enormous mass, which exceeded nine tons in some cases, their internal temperatures might have exceeded 40°C if they could not efficiently shed metabolically generated internal heat. Eagle et al. (2010) analyzed the carbonate component from tooth apatite, $Ca_5(PO_4,CO_3)_3(OH,CO_3,F,Cl)$, in five modern animal species: elephant, rhino, crocodile, alligator, and sand shark, whose estimated body temperatures ranged from 37 to 23.6°C. They found that Δ_{47} (see Chapter 5)

of the carbonate component released by phosphoric acid digestion showed the same relationship to temperature as for inorganic calcite. Carbonate from the teeth of two fossil mammoths also analyzed yielded temperatures of 38.4 ± 1.8°C, within error of the body temperatures of modern elephants. Tooth dentin of the mammoths, however, yielded lower temperatures, but both $\delta^{18}O$ and rare-earth patterns of the phosphate component suggested the dentin has suffered diagenetic alteration.

Eagle et al. (2011) then analyzed tooth enamel from Jurassic sauropods. Teeth from the Tendaguru Beds of Tanzania of three *Brachiosaurus* fossils yielded temperatures of 38.2 ± 1°C and two fossils of *Diplodocinae* yielded temperatures of 33.6 ± 4°C. Three *Camarasaurus* teeth from the Morrison Formation in Oklahoma yielded temperatures of 36.9 ± 1°C, while one from Howe Quarry in Wyoming yielded a lower temperature of 32.4 ± 2.4°C. These temperatures are 5–12°C higher than modern crocodilians and, with the exception of the Howe Quarry tooth, within error of modern mammals. They are also 4– 7°C lower than predicted for animals of this size if they did not somehow thermoregulate. The authors note that this does not prove that dinosaurs were endotherms, but it does indicate that a "combination of physiological and behavioral adaptations and/or a slowing of metabolic rate prevented problems with overheating and avoided excessively high body temperatures." In modern mammals, tooth enamel is produced only in youth. If this was also true of sauropods, it indicates that they maintained high body temperatures during youth, which is consistent with observations of bone structure that suggests high growth rates.

The eggshells of oviparous animals, which include birds, reptiles, and dinosaurs, among others, mineralize within the lower oviduct of the mother. Hence, if the mother is an ectotherm, this should be reflected in the crystallization temperature of the carbonate in the eggshell. Thus, eggshells are another target to reconstruct internal temperatures of extinct animals. Eagle et al. (2015) analyzed the carbonate of fossil eggshells of two Late Cretaceous dinosaur groups: titanosaurids and oviraptorids. Titanosaurid eggshells yielded temperatures similar to large modern endotherms, but oviraptorid eggshells yielded temperatures lower than most modern endotherms, but ≈6°C higher than co-occurring abiogenic carbonates. Thus, the oviraptorids were apparently not capable of maintaining a constant body temperature like modern endotherms, but were nonetheless able to raise their body temperature above environmental temperature.

13.3 ISOTOPES AND DIET: YOU ARE WHAT YOU EAT

We now fast-forward to the last few millions of years to examine life in the modern world with a particular focus on our ancestors and what they ate. Recall from Chapter 10 that the two main photosynthetic pathways, C_3 and C_4, lead to organic carbon with different carbon isotopic compositions. Terrestrial C_3 plants have $\delta^{13}C$ values that average about –27‰ while those of C_4 plants average about –13‰. Many succulents use crassulacean acid metabolism and can effectively switch between C_4 and C_3 pathways constituting a third type, but they typically have $\delta^{13}C$ similar to C_4 plants. Marine plants (which are all C_3) utilize dissolved bicarbonate rather than atmospheric CO_2. Because seawater bicarbonate is about 8.5‰ heavier than atmospheric CO_2, marine plants average about 7.5‰ heavier than terrestrial C_3 plants. In addition, because the source of the carbon they fix is isotopically more variable, the isotopic composition of marine plants is also more variable. Finally, marine cyanobacteria (blue–green algae) tend to fractionate carbon isotopes less during photosynthesis compared to true marine plants, so they tend to average 2–3‰ higher in $\delta^{13}C$.

Plants may also be divided into two types based on their source of nitrogen: those that can utilize N_2 and those that rely on fixed nitrogen (ammonia and nitrate) present in the environment. The former include marine cyanobacteria and legumes (e.g., beans, peas, and alfalfa); the latter do not themselves fix nitrogen, but rather utilize nitrogen fixed by symbiotic bacteria (rhizobia) in their root nodules. $\delta^{15}N$ of marine cyanobacteria is mostly in the range of –4 to –2‰. Legumes have an average $\delta^{15}N$ of +1‰, whereas modern non-leguminous plants average about +3‰, but these values reflect the prevalence of chemical fertilizers; prehistoric non-leguminous plants were more positive, averaging perhaps +9‰. For both groups, there

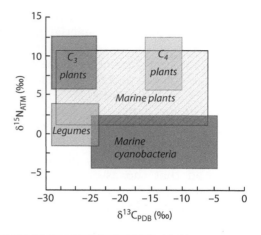

Figure 13.2   Relationship between $\delta^{13}C$ and $\delta^{15}N$ among the principal classes of autotrophs. Source: Adapted from DeNiro (1987).

was probably a range in $\delta^{15}N$ of ±4 or 5‰, because the isotopic composition of soil nitrogen varies and there is some fractionation involved in uptake. Marine plants have $\delta^{15}N$ of +7 ± 5‰. Thus, based on their $\delta^{13}C$ and $\delta^{15}N$ values, autotrophs can be divided into several groups, which are summarized in Figure 13.2.

Both $\delta^{13}C$ and $\delta^{15}N$ in autotrophs vary regionally with climate. Light intensity, temperature, salinity, CO_2 partial pressure, and humidity can influence $\delta^{13}C$. Among C_3 plants, $\delta^{13}C$ increases from dense, closed canopy rainforests through open canopy forests to dry savannah and grassland environments. This reflects a need to reduce stomatal conductance to limit water loss, and consequently fixation of a greater fraction of interior CO_2. Hence, assessment of $\delta^{13}C$ and $\delta^{15}N$ in autotrophs needs to be evaluated in a local and temporal context. Temperatures, soil moisture, and the soil biota, including both bacteria and fungi, affect N cycling in the soil and hence $\delta^{15}N$ of autotrophs.

DeNiro and Epstein (1978) studied the relationship between the carbon isotopic composition of animals and their diet. They found that there is only slight further fractionation of carbon by animals and that the carbon isotopic composition of animal tissue closely reflects that of the animal's diet. Typically, carbon in animal tissue is about 1‰ heavier than their diet. The small fractionation between animal tissue and diet is a result of the slightly weaker

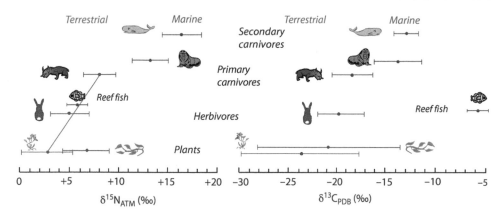

Figure 13.3   Values of $\delta^{13}C$ and $\delta^{15}N$ in various marine and terrestrial organisms. Source: Adapted from Schoeninger and DeNiro (1984).

bond formed by ^{12}C compared with ^{13}C. The weaker bonds are more readily broken during respiration, and, not surprisingly, the CO_2 respired by most animals investigated was slightly lighter than their diet. Thus, only a small fractionation in carbon isotopes occurs as organic carbon passes up the food web. Terrestrial food chains usually do not have more than three trophic levels, implying a maximum further fractionation of +3‰; marine food chains, which begin with plankton, can have up to seven trophic levels, implying a maximum carbon isotope difference of 7‰ between primary producers and top predators. These differences are smaller than the range observed in primary producers.

In another study, DeNiro and Epstein (1981) found that $\delta^{15}N$ of animal tissue reflects the $\delta^{15}N$ of the animal's diet, but in contrast to carbon, significant fractionation of nitrogen isotopes occurs as nitrogen passes up the food chain. Typically, nitrogen in herbivores is enriched by 3–7‰ relative to the plants eaten, while carnivores are enriched by 3–5‰ compared to the animals eaten. As marine food chains are typically longer, marine mammals and fish are generally enriched by 5–10‰ relative to the terrestrial mammals. Reef fish are a clear exception to the overall trends with high $\delta^{13}C$ and low $\delta^{15}N$ relative to aquatic and open ocean marine organisms. This apparently reflects a ^{13}C enrichment observed in corals and a relatively large contribution of nitrogen fixation by cyanobacteria and rapid rates of nitrogen fixation in reef environments, which leads to ^{15}N depletion of the water (Schoeninger and DeNiro,

1984). These relationships are summarized in Figure 13.3. The significance of these results is that it is possible to infer the diet of an animal from its carbon and nitrogen isotopic composition. In particular, they found that buried bone collagen and tooth enamel appear to retain their original isotopic compositions, and hence could be used to infer ancient diets.

Because of these relationships, it is possible to reconstruct ancient ecosystems and the diets of ancient humans. We will explore the discoveries based on this in the next few sections.

13.4   PALEOECOLOGY OF GRASSLANDS

13.4.1   Evolution of C_4 plants and the grassland biome

In the modern world, C_4 and mixed C_3/C_4 biomes cover much of the tropical and temperate low-elevation landscape. The rising importance of the grassland biome is intimately to the evolution of a number of mammalian species, including the modern horse and humans. C_4 crops, including maize, sorghum, millet, and sugar cane, support global societies providing food, animal feed, and sugar. However, C_4 photosynthesis is a relatively recent evolutionary invention.

As we found in the previous chapter, an increase in $\delta^{13}C$ in Pakistani paleosol carbonates occurred between 7.5 and 6 Ma recording an increasing abundance of C_4 plants there. C_4 plants occur globally in regions of low-elevation, hot growing seasons, high light

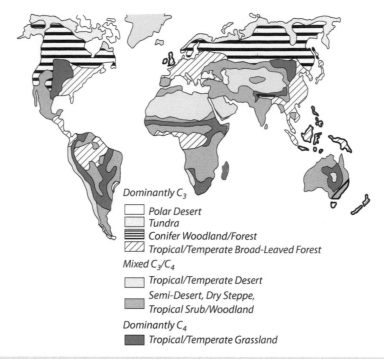

Dominantly C₃

- ☐ *Polar Desert*
- ☐ *Tundra*
- ☰ *Conifer Woodland/Forest*
- ▨ *Tropical/Temperate Broad-Leaved Forest*

Mixed C₃/C₄

- ☐ *Tropical/Temperate Desert*
- ▦ *Semi-Desert, Dry Steppe, Tropical Srub/Woodland*

Dominantly C₄

- ▨ *Tropical/Temperate Grassland*

Figure 13.4 Present global distribution of C₃ and C₄ vegetation. Source: Adapted from Cerling and Quade (1993).

levels, and inconsistent moisture, most notably in grasslands (Figure 13.4). Cerling et al. (1997) used $\delta^{13}C$ in tooth enamel of large mammals to document this global increase in C₄ vegetation in the late Miocene. Unlike other tissues, there is a significant carbon isotope fractionation between tooth enamel and diet $\delta^{13}C$ in large mammals of $\Delta^{13}C_{enamel\text{-}diet} \approx$ +14‰, but Cerling et al. found that, with this offset, mammal tooth enamel nonetheless reliably tracks diet. Figure 13.5 compares $\delta^{13}C$ of fossils of large herbivores from North America south of 37°N, South America, and East Africa over the last 20 Ma with $\delta^{13}C$ in Pakistani paleosols. There is a clear shift toward higher $\delta^{13}C$ among equids beginning roughly around 8 Ma on all three continents, matching the shift in Pakistani paleosols. Notoungulates, an extinct order of South American mammalian herbivores, also show a shift, as do East African Elephantids. Deinotheres, which are now-extinct herbivores similar to elephants but distinguish from them by inward curving tusks, which they used to strip vegetation, appear to have stuck to a diet of forest C₃ vegetation. Cerling et al. found that fossil herbivores from North America north of 37°N show a subdued increase of C₄ biomass in their diets at around 5–4 Ma and those from Europe exhibit no increase, consistent with the limited C₄ vegetation in these regions.

An obvious question is: why did C₄ plants suddenly rise to such importance? That the C₄ pathway evolved separately in over 60 different plant families (Sage et al., 2018), a striking example of convergent evolution, makes the question even more intriguing. In the C₃ photosynthetic pathway, Rubisco can catalyze not only the fixation of carbon in phosphoglycerate, but also the reverse reaction, photorespiration, where CO₂ is released. When concentrations of CO₂ are high, the forward reaction is favored and the C₃ pathway is more efficient overall than the C₄ pathway. At low CO₂ concentrations, however, the C₄ pathway, in which CO₂ is first transported into bundle-sheath cells, is more efficient, as the concentration in bundle-sheath cells is maintained at around 1000 ppm (Figure 13.6). Thus, under conditions of low CO₂, C₄ plants have a competitive advantage while at higher CO₂ conditions, C₃ plants are more efficient. The C₄ pathway is a tradeoff of greater energy expense for more efficient use of CO₂ and its widespread adoption in the late Miocene suggests a global environmental driver.

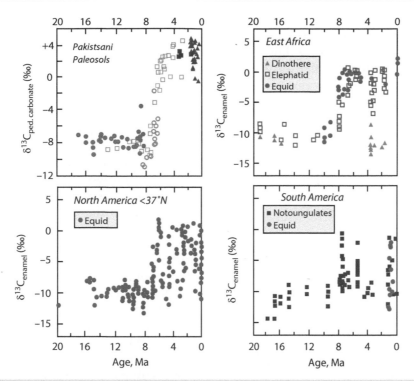

Figure 13.5 Comparison of $\delta^{13}C$ in Pakistani Siwaliks paleosols with $\delta^{13}C$ in tooth enamel from large herbivores from North and South America and East Africa. The paleosol data for Pakistan are the same as shown in Figure 12.18; tooth enamel data are from Cerling et al. (1997).

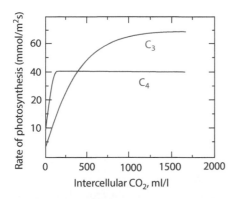

Figure 13.6 Rate of photosynthesis as a function of intercellular CO_2 concentrations in C_4 and C_3 plants. At concentrations of atmospheric CO_2 that prevailed before the Industrial Revolution, C_4 plants would have had a competitive advantage. At concentrations above the present level, C_3 plants are more efficient. Source: Ehleringer et al. (1991)/with permission of Elsevier.

Cerling et al. (1993) argued that the increasing importance of the C_4 pathway reflected a response to decreasing atmospheric CO_2 in the late Miocene. However, subsequent analyses of $\delta^{11}B$ in foraminifera and $\delta^{13}C$ in C_{37} alkadienones of marine sediments (Chapter 12) have given us a clearer picture of how atmospheric CO_2 has changed over the Cenozoic. A significant decrease occurs around the Paleocene–Neogene boundary at around 34 Ma with at most a minor decrease in the late Miocene (Figure 12.26d). While the oldest direct fossil evidence for C_4 plants (plants with enlarged bundle-sheath cells) is late Miocene, a consensus based on molecular clocks has emerged among biologists that the C_4 pathway likely first evolved in the late Oligocene some 25–30 million years ago in grasses (Sage, 2004) and subsequently in other taxa. It is thought to have evolved in Andropogoneae grasses (a tribe that includes maize, sugarcane, and sorghum) some 17 million years ago (Sage et al., 2018). Part of the explanation for why C_4 photosynthesis evolved independently in diverse families at diverse times and diverse places over the past $\approx$25 Ma is that it involves only relatively minor modification of plant enzymes and structures.

Nevertheless, while C_4 photosynthesis may have first evolved in the Oligocene following the decrease in CO_2, it clearly only became widespread in the late Miocene. The more efficient CO_2 utilization of C_4 plants allows them to limit the opening of their stomata and hence moisture loss. This and the distribution of C_4 plants in areas of high temperature and inconsistent moisture suggest that humidity and moisture may have been a factor in their rise. Sage et al. (2018) suggested that while lower atmospheric CO_2 drove evolution of the C_4 pathway, only with increased aridity in the Miocene did C_4 plants to become widespread.

Based on the $U_{37}^{k'}$ sea surface temperature[1] proxy, Herbert et al. (2016) documented global sustained late Miocene cooling, culminating with ocean temperatures dipping to near-modern values between about 7 and 5.4 million years ago. They argue that this cooling resulted in an increase in humidity and in the expansion of C_4 vegetation.

13.4.2 Evolution of horses

Horses (family *Equidae*) have been around for 58 million years. Beginning in the early Miocene, a major radiation took place and the number of genera in North America increased from 3 at 25 Ma to 12 at 10 Ma. It subsequently fell at the end of the Miocene, and the last North American species became extinct in the early Holocene (possibly due to human hunting pressure). A major change in dental morphology, from low-crowned to high-crowned (hypsodont), accompanied the Miocene radiation. For nearly 100 years, the standard textbook explanation of this dental change was associated with a change in feeding from leaf browsing to grass grazing. Grasses contain silica making them quite abrasive and soil grit taken in during grazing provides additional grit, which wears down teeth over time. A high-crowned tooth would last longer in a grazing animal and would therefore be favored in horse's evolution as it switched food sources. This change in horse diet was thought to reflect the evolution of C_4 grassland biomes. Since the carbon isotopic composition of animals reflects that of their diet, and since the $\delta^{13}C$ of dental enamel records the $\delta^{13}C$ of the animal, the change in horse dentition should parallel the change in the carbon isotopic composition of those teeth if the change in dentition were related to a change in diet associated with grazing. Carbon isotope ratios thus provide a test of the evolution of horse hypsodonty.

Wang et al. (1994) analyzed the carbon isotopic composition of dental enamel from fossil horse teeth of Eocene through Pleistocene age. They found a sharp shift in the isotopic composition of the teeth consistent with a change in diet from C_3 to C_4 vegetation, but it occurred later than the change in dental morphology (Figure 13.7). The change in dental morphology begins in the mid-Miocene (about 18 Ma), while shift in $\delta^{13}C$ occurs at around 7 Ma. This leads to an interesting dilemma. Which change, morphology, or carbon isotopic composition actually reflects the appearance of the grassland biome?

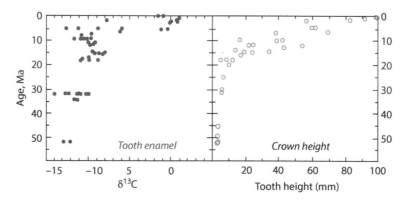

Figure 13.7 $\delta^{13}C$ and crown height in North American fossil horses as a function of age. Source: Adapted from Wang et al. (1994).

---

[1] This is based on the temperature dependence of the number of double C bonds occurring in a 37-long carbon atom alkenone produced by marine coccolithophorids such as *Emiliania huxleyi* and preserved in marine sediments.

Phytolith evidence suggests that open grass-lands first appeared in the Oligocene and only became ecologically important in the mid-Miocene (Strömberg, 2006). Nguy and Second (2022) report $\delta^{13}C$ from at least one mid-Miocene horse tooth indicative of a C_4 diet, although contemporary ungulate fossils have $\delta^{13}C$ indicative of a C_3 diet. It appears that grassland biome appeared at least by the early Miocene and was initially dominated by C_3 veg-etation but became C_4 dominated and rose to modern importance only in the Late Miocene.

13.4.3   Human evolution

13.4.3.1   The climate of hominin evolution

There is overwhelming evidence that humans – i.e., the genus *Homo* – evolved in Africa and our species, *sapiens*, evolved there as well. The evolutionary line that would lead to humans diverged from that of our nearest relatives, the chimpanzees and bonobos, some six or seven million years ago. Fossils found in East Africa, particularly areas in and around the East African

Rift, have provided a rich record of this evolu-tion. What was the climate like during this time?

The short answer is that East African envi-ronment over the last several million years was one of increasing aridity with superimpose higher-frequency changes in climate. Figure 13.8 compares $\delta^{18}O$ and $\delta^{13}C$ of soil car-bonates from two hominin fossil-rich localities, the Olduvai Gorge of Tanzania ($\approx3°S$) and the Turkana Basin of Kenya ($\approx3°N$), with benthic marine $\delta^{18}O$ from ODP Site 659 (the same as in Figure 12.11). The marine $\delta^{18}O$ record indi-cates a cooling global climate and increasing amounts of polar ice over the last four million years. The increasing $\delta^{18}O$ in paleosols at these sites suggests a trend of increasing aridity, par-ticularly over the last ≈2 Ma and the Turkana Basin as global climate changed. Pliocene rain-fall was depleted in ^{18}O relative to today, sug-gesting greater precipitation than today (Levin et al, 2011). The increase in $\delta^{13}C$ indicates an increasing proportion of C_4 plants consistent with drying conditions.

The Turkana Basin soil carbonate data show considerable high-frequency variation.

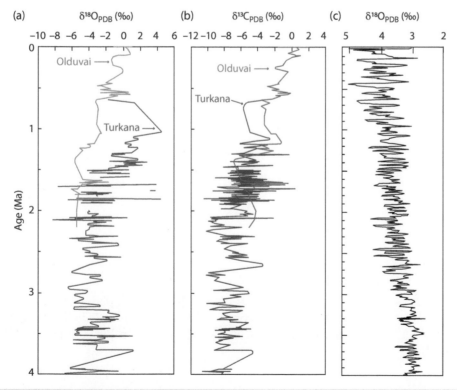

Figure 13.8   (a) $\delta^{18}O$ and (b) $\delta^{13}C$ in soil carbonates from the Turkana Basin (red) and Olduvai Gorge (blue) areas of the East Africa Rift Valley. Data from the compilation of Levin (2013). (c) $\delta^{18}O$ in benthic foraminifera from ODP site 659 in the South Atlantic (Figure 12.11). Data from Zachos et al. (2001).

Further insights into the shorter-term variations come as $\delta^{13}C$ of leaf waxes preserved in lacustrine. Magill et al. (2013) compared $\delta^{13}C$ in soil organic matter, leaf tissues, and leaf lipids for 64 plants species in ≈300 tropical and subtropical localities and found that in C_3 ecosystems $\delta^{13}C$ in leaf-derived 31-long carbon alkanes (designated nC_{31}) is ^{13}C-depleted by about 7‰ with respect to leaf tissue, while soil organic matter is ^{13}C-enriched by about 2‰. In C_4 plant soil systems, nC_{31} is ^{13}C-depleted with respect to leaf tissue by about 10‰ and soil organic matter is depleted by about 1‰. They analyzed $\delta^{13}C$ in nC_{31} and organic matter in lake sediments deposited in the Olduvai Gorge between ≈2.0 and 1.8 Ma

– a time coincident with some of the earliest *Homo* fossils found there. They found repeated shifts in $\delta^{13}C_{31}$ between about –36 and –20‰ suggesting rapid local ecosystem shifts between closed C_3 woodlands and open C_4 grasslands. These signatures correlated with precession and tropical sea-surface temperatures deduced from marine cores.

Johnson et al. (2016) recovered a 1.8 million year record of $\delta^{13}C_{31}$ from a sediment core from Lake Malawi in the southern portion (11°S) of the East African Rift. They also determined surface water temperature using the TEX_{86} paleotemperature proxy[2]. Figure 13.9 compares these with the global benthic foraminifera $\delta^{18}O$ record. They found a strong 100 000 year

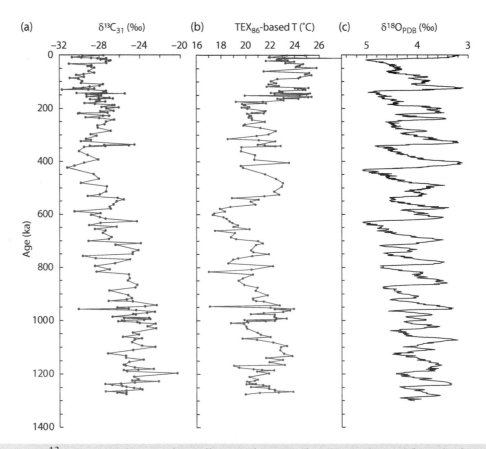

Figure 13.9 (a) $\delta^{13}C_{PDB}$ in 31-long carbon alkanes (designated as nC_{31}) derived from leaf waxes and (b) temperatures calculated from the TEX_{86} proxy from a sediment core from Lake Malawi in the southern East African Rift. Data from Johnson et al. (2016). (c) Global stack of benthic foraminifera $\delta^{18}O$ of Lisiecki and Raymo (2004).

---

[2] The TEX_{86} temperature proxy is based on the relative abundance of cylopentanes in glycerol dialkyl glycerol tetraethers (CGDTs), which are components of cell walls in planktonic archaea and which are common components of sedimentary organic matter. Schouten et al. (2002) showed that the weighted average number of cyclopentane rings in GDGTs substantially increases with growth temperature and developed the TEX_{86} based on this correlation.

(eccentricity) cyclicity of temperature and precipitation following the increased amplitude of glacial–interglacial cycles recorded in benthic forams at around 900 000 years ago. Interglacial periods in the Malawi area were relatively warm and moist, while glacial periods were cool and dry. They also found that the temperature history over the last 500 000 years correlates with the Antarctic ice core CO_2 record. In contrast to the Turkana Basin and Olduvai Gorge areas (Figure 13.8), $\delta^{13}C_{31}$ decreases, indicating an increasingly C_3-dominated flora, consistent with greater rainfall over time.

In the modern world, East and Southern Africa have differing responses to decadal oscillations in the distribution of Indian Ocean sea-surface temperatures (the Indian Ocean Dipole), so it is perhaps not surprising that they would also differ on longer timescales. In addition to orbital insolation variations, these sites all lie within the tectonically and volcanically active Earth Africa Rift system, which has resulted in considerable local climatic variations over this period. The conclusion to be drawn is that climatic and ecosystem changes are varied in the area where humans evolved.

The Turkana Basin is presently among the hottest 1% of land on the planet with a mean annual temperature of almost 30°C. Much of the basin is occupied by saline Lake Turkana, and the surround land is quite arid. But what was the climate there like as our genus evolved from Australopithecine ancestors? Some paleoclimatic proxies derived from fossil pollen assemblages suggest that late Pliocene temperatures were cooler than present, while other proxies and models suggest an even warmer climate. To resolve this, Passey et al. (2010) applied clumped isotope geothermometry to paleosol carbonates of Pliocene and Pleistocene age from the Turkana Basin. Passey et al. (2010) also measured modern soil temperatures at 50 cm depth in the Turkana Basin and found that they typically were 4°C higher than mean average air temperature (as we noted in the previous chapter, soil carbonates typically form in the warm season and reflect those, rather than annual average, temperatures). Calculated clumped isotope paleosol temperatures range from 28 to 41°C and average 33°C, quite similar to the average modern soil temperature of 35°C. Lüdecke et al. (2018) calculated soil temperatures from soil carbonate Δ_{47} from *Homo rudolfensis* and *Paranthropus boisei* fossil sites in the Malawi Lake region in the southern portion of the East Africa Rift, and although cooler than the Turkana Basin, 28 and 26°C, they too were within the range of modern soil temperatures.

A study by Cerling et al. (2015) provides insight into the evolution of the fauna of the Turkana Basin over four million years period of hominin evolution in the area. They analyzed $\delta^{13}C$ of bone apatite in >900 large herbivore fossils and compared them with data from modern herbivores. Between 4.5 and 2.4 Ma, the ecosystem was dominated by C_4–C_3 mixed feeders, but around 2.4 Ma, the region became increasingly dominated by C_4 grazers but by around 1.4 Ma, C_4 grazers had virtually disappeared from the landscape and there were very few mixed feeders as well. Today, however, C_4 grazers roughly as common as C_3 browsers with mixed feeders – the least common. Thus, the present fauna is different from past ones and that the region has not evolved in a linear unidirectional manner. Environmental variability favors adaptability, a trait conferred by, among other things, intelligence.

13.4.3.2   Walking out of the forests and into the grasslands

It is against this background of changing East and South African climates and ecosystems that the humans, the genus *Homo*, emerged. We have already seen that the late Miocene global development of C_4 grasslands and savannahs also included East Africa (Figure 13.5). Our ancestors eventually moved out of the forests and into this new grassland/savannah landscape where legs and feet adapted to upright walking rather than grasping tree limbs was an advantage. This transition left their hands free to make tools and, eventually, art. Isotopes help to trace this record.

Malone et al. (2021) calculated $\delta^{13}C$ tooth enamel-diet fractionation factor for hominids of 11.8 ± 0.3‰ based on modern chimpanzees, *Pan troglodytes* (compared to, e.g., a $\Delta^{13}C_{bone-diet}$ of 14.1‰ used by Cerling et al. [2015] for East African herbivores). They then used this value to calculate the $\delta^{13}C$ of diet for the hominin fossils shown in Figure 13.10. The earliest of these fossils, *Ardipithecus ramidus* from

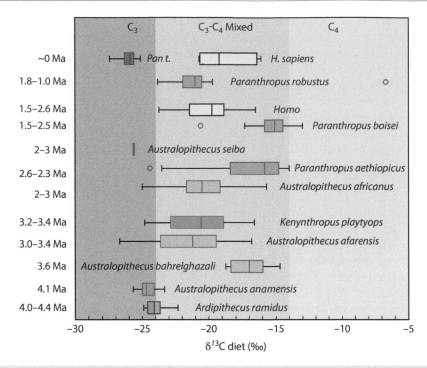

Figure 13.10 Box plot of $\delta^{13}C$ of diet calculated from $\delta^{13}C$ of tooth enamel of hominins assuming $\Delta^{13}C_{bone\text{-}diet} = 11.8‰$ (vertical line represents the mean, box encompasses 50% of the data, horizontal lines encompass the range, and extreme outliers are shown as circles). Dates listed are those of the analyzed fossils, not necessarily the age range of the species. The control group is *Pan troglodytes* (abbreviated as *Pan t.*), modern chimps. Data from the compilation of Malone et al. (2021).

Ethiopia and *Australopithecus anamensis*, found both in Ethiopia and the Turkana Basin, appear to have been capable of walking upright, but retained many adaptations for life in the trees. $\delta^{13}C$ data indicate that they had diets similar to modern chimps, who feed almost exclusively on C_3 resources, including fruit, leaves, leaf buds, seeds, blossoms, stems, pith, bark, and resin, and less frequently honey, insects, birds and their eggs, and small-to-medium-sized mammals. A shift to heavier $\delta^{13}C$ is apparent after 4 Ma beginning with *A. bahrelghazali* from Chad and *A. afarensis* from Ethiopia (some argue these fossils should be placed in the same species), indicating a change to inclusion of C_4 plant-derived foods in their diets. The change in $\delta^{13}C$ postdates major dental morphological changes that distinguish *Australopithecus* from *Ardipithecus* but is coeval with other adaptations to diets of tougher, harder foods and to committed terrestrial bipedality (Levin et al., 2015). *A. afarensis* includes the famous well-preserved fossil nick-named *Lucy*. Her anatomy indicates she

was an obligate bipedal although she retained some adaptation for tree climbing. *A. afarensis* is the likely maker of the tracks of three bipedal individuals made in fresh volcanic ash around ≈3.7 Ma and preserved as the famous Laetoli trackway in northern Tanzania. $\delta^{13}C$ indicates Lucy and her contemporaries were moving out of the forests and exploiting the food resources of the savannah. The range in $\delta^{13}C$ for *A. afarensis* is notably large, particularly in comparison to chimps, indicating that they were adaptable and exploiting a large range of C_3 and C_4 foods.

Sometime around 2.5 Ma, australopithecines split into two lines, the "robust" and "gracile." The former are generally placed in a distinct genus, *Paranthropus*, characterized by robust skulls, with a prominent gorilla-like sagittal crest to support strong chewing muscles and broad, herbivorous teeth used for grinding. These species, *P. aethiopicus* and *P. boisei* (which some argue are the same species), appear to have relied even more heavily on C_4 resources of the savannah. *P. boisei*, of which there are many specimens from the limestone caves of Gauteng Province in South Africa, appears to have obtained 75–80% of its food from grasslands and the savannah, although they also included forest foods in their diet with a good deal of variation between individuals. The last of this genus, *P. robustus*, shared these same skull features. It inhabited mixed open and wooded environments and its diet reflects mixed C_3 and C_4 resources. In contrast, *A. sediba*, known from fossils found only in Malapa Cave of South Africa, depended primarily on C_3 forest foods, although it is possible it would exhibit a wider range of $\delta^{13}C$ if more specimens were found.

Around 2.5–2.3 Ma fossils of a hominin with larger brain volume appear, but the earliest of these are otherwise sufficiently similar to australopithecines that their assignment to a new genus is somewhat controversial. Examples older than 2 Ma have been found up and down the Rift Valley from the Lake Malawi region in the south to the Afar region of Ethiopia in the north and are variously assigned to *Homo rudolfensis*, *H. ergaster*, or *H. habilis*, but whether they are several or a single *Homo* species is again controversial. The diet of these early humans appears to have been much like ours, involving a mixture of C_4 and C_3 resources. It is also around this time that the first manufactured stone tools are found in various sites along the Rift Valley, demonstrating that *Homo* was now processing foods (more recently, primitive stone tools dating to 3.3 Ma have been found in the Turkana Basin, suggesting australopithecines were using tools to process foods). Since carbon isotope ratios are insensitive to tropic level and N isotope ratios are not preserved in these ancient fossils, we cannot distinguish diets based on plants or on herbivores that fed upon them. But these early tools are sometimes found in association with butchered remains, so we know *Homo* was eating meat, but the extent to which it was scavenged or hunted is unclear.

13.5 PALEOECOLOGY OF THE PLEISTOCENE TUNDRA STEPPE

Let us now fast-forward to the last $\approx 100\ 000$ years. Early humans, such as *H. erectus*, had migrated out of Africa as early as 1.8 million years ago. There were subsequent waves of human migration out of Africa as well, the last of which was by *H. sapiens* 60 000 or more years ago. When they did, they encountered the descendants of earlier waves, most notably *H. neanderthalensis*, with whom they would share the Eurasian tundra-steppe for thousands and perhaps tens of thousands of years.

The Pleistocene Eurasian and North American Pleistocene tundra-steppe, although cool and dry, was a surprisingly productive one, supporting a number of large herbivores and carnivores, many of which have since become extinct. C and N isotope studies of plants (all of which are C_3) and bone collagen of herbivores provide insight into the ecology of this biome (Figure 13.11). Mammoths, woolly rhinoceroses, and bison fed primarily on grasses and sedges, while horses and muskox consumed substantial amounts of forbs (wildflowers) and leaves of trees and shrubs in addition to grasses and sedges (Drucker, 2022). The high $\delta^{13}C$ of reindeer reveals their heavy dependence on lichen in addition to leaves and forbs. Mammoths nevertheless stand out in their high $\delta^{15}N$ (mammoths from Eastern Europe are an exception). Why is unclear; selective feeding on dung-fertilized mature plants is the leading hypothesis. Interestingly, the isotopic distinctions among herbivores are similar among remains recovered both in the Meuse Valley of Belgium and in Siberia, showing that the herbivores occupied distinct ecological niches with different feeding strategies over a wide geographic range, with muskox being the notable exception.

13.5.1 A wide-ranging mammoth

Because of their size and spectacular tusks, mammoths are certainly the iconic mammal of the Pleistocene. Mammoths replaced an infant set of tusks at an age of about 18 months, after which the adult tusks would grow throughout life. Carbon and nitrogen isotopic analysis of these tusks can thus provide a record of what they ate, while oxygen and strontium isotopic analysis provides a record of where they ate and the climate there. Strontium is taken up by plants from the soil and hence the $^{87}Sr/^{86}Sr$ reflects that of the local bed rock while $\delta^{18}O$ reflects that of the local precipitation, which varies seasonally and hence bone and tusks records that seasonality. Wooller

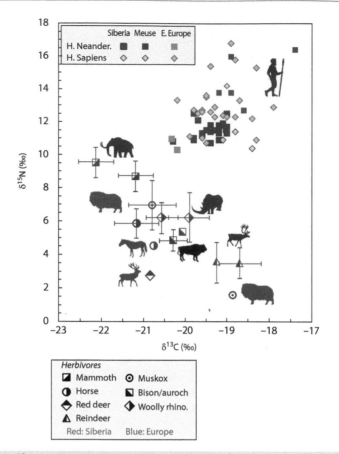

Figure 13.11 $\delta^{15}N$ and $\delta^{13}C$ in fossil bone of the inhabitants of the Eurasian tundra-steppe. Data from Wißing et al. (2019), Kuzmin et al. (2021), and Drucker (2022).

et al. (2021) analyzed the isotopic composition of the tusk of a male woolly mammoth that died above the Arctic Circle in Alaska ≈17 100 years ago. Combining micro analysis of $^{87}Sr/^{86}Sr$ and $\delta^{18}O$ along the 2.5 m length of the tusk of this individual with a database geographic variation in Alaska allowed them to reconstruct a record of his wanderings (Figure 13.12). He spent his early years in the Yukon River Valley of Central Alaska. As a juvenile, he used a larger range spanning the lowlands of interior Alaska south of the Brooks Range, undertaking regular north–south movements within this area presumably with his herd. At around age 16, the variance in $^{87}Sr/^{86}Sr$ increases and he spent much of his time north of the Brooks Range wandering as far as the Nome Peninsula. His final years were spent north of the Brooks Range, where he died at age 27. High $\delta^{15}N$ and low $\delta^{13}C$ suggest he died of starvation while $\delta^{18}O$ suggests death occurred in late winter.

13.5.2   Neanderthal diets

13.5.2.1 $\delta^{13}C$ and $\delta^{15}N$

The top predators in the Eurasian tundra-steppe, which extended as far south as the Alps in Europe, were humans (indicated by, among other things, the high $\delta^{15}N$ of human bone apparent in Figure 13.11). One of the more fascinating aspects of human evolution is that through most of our 300 000 year or so history, *Homo sapiens* shared the planet with other human species, yet we alone have survived to tell the tale. Our close cousins, the Neanderthals, have been a particular focus of interest, particularly since we learned that most of those of us who are not of strictly recent African ancestry carry a few percent of Neanderthal genes. Who were these people and how different were they from us?

Homo sapiens and *Homo neanderthalensis*, together with a third and even more enigmatic species, the *Denisovans*, shared the Eurasian

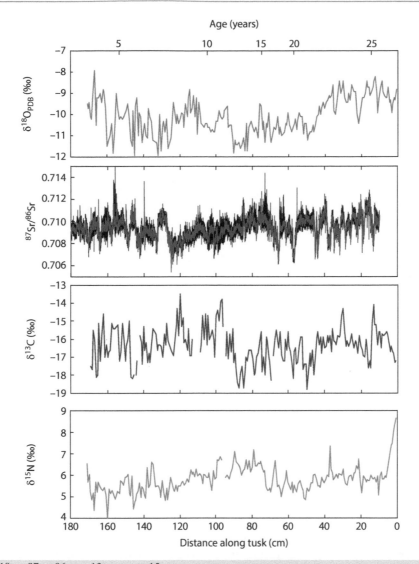

Figure 13.12 $\delta^{18}O$, $^{87}Sr/^{86}Sr$, $\delta^{13}C$, and $\delta^{15}N$ along the length of tusk of a woolly mammoth while lived in Alaska 17 100 years ago. $\delta^{13}C$ and $\delta^{15}N$ reflect diet, $^{87}Sr/^{86}Sr$ reflects that of the local geology, and $\delta^{18}O$ reflects precipitation and its seasonal variations. Data from Wooller et al. (2021).

tundra-steppe for perhaps 10 000 years prior to the last glacial maximum. How different were their diets? A study by Richards and Trinkaus (2009) of hominin bone collagen found that *H. sapiens* tended to have higher $\delta^{13}C$ and $\delta^{15}N$ than *H. neanderthalensis*, which they attributed to more extensive reliance of aquatic foods by *H. sapiens*, which is particularly true of bones found at Peçstra cu Oase, Romania. Other studies comparing bone collagen from pre-last glacial maximum remains of the two species from the Meuse Valley region in Northeast Europe and in Siberia show little distinction (Figure 13.11). Both species have similarly high $\delta^{13}C$ and $\delta^{15}N$ indicative of diets based primarily on large carnivores with a preference for mammoth and reindeer (Wißing et al., 2019; Kuzmin et al., 2021).

A significant advance in studies of diet has been the ability to determine isotope ratios specific organic compounds or *compound-specific isotopic analysis*. This is done by using chromatography to separate the compounds before mass spectrometric analysis. Neanderthals have been thought of as strict carnivores, but Naito et al. (2016) analyzed $\delta^{15}N$ analysis of two amino acids, glutamic acid and phenylalanine, in bone collagen from Neanderthals from Spy Cave in Belgium and concluded that Neanderthals may have relied on plants for up to

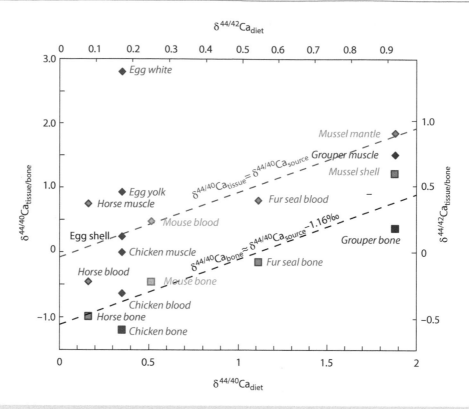

Figure 13.13   Ca isotope ratios in soft tissues and bone as a function of diet. Skulan and DePaolo found that $\delta^{44/40}$Ca in soft tissue is approximately equal to that in diet but bone is $\approx$1.2‰ lighter than diet. Chicken eggs show a particularly large range of $\delta^{44/40}$Ca with egg white > egg yolk > egg shell > chicken blood. Source: Adapted from Skulan and DePaolo (1999).

20% of their protein. A similar compound-specific analysis of N isotopes by Drucker et al. (2017) on pre-last glacial maximum *H. sapiens* bone collagen from Crimea suggests their reliance on plants was higher, but that may merely reflect the higher availability of edible plants in more southern latitudes. At least in terms of diet, we and our Neanderthal cousins appear to be not very different after all.

13.5.2.2   Ca isotopes

Calcium is an essential element of all life and performs myriad biological functions. In vertebrates, the largest store of Ca is in bone apatite, which serves not only a structural purpose but also as a store of Ca for soft tissue Ca demand as well. In tetrapods, Ca is primarily derived from diet while in fish it is also exchanged with ambient water through the gills. Skulan and DePaolo (1999) and subsequent studies have found that Ca isotope ratios of mammals and birds are approximately equal to those

of diet and that bone has $\delta^{44/40}$Ca is $\approx$1.17‰ ($\delta^{44/42}$Ca $\approx$ 0.57‰) lighter than diet (Figure 13.13). Notable exceptions are eggs and milk. Skulan and DePaolo (1999) found that $\delta^{44/40}$Ca varied by more than 3‰ in order egg white > egg yolk > egg shell > chicken blood (Figure 13.13). Reynard et al. (2010) found that in sheep, ewes had $\delta^{44/42}$Ca about 0.14‰ heavier than rams. Noting the very light isotopic composition of sheep's milk relative to their diet (Chu et al., 2006), they attributed this sexual difference to lactation in ewes. Human milk appears to be particularly isotopically light. Chu et al. (2006) found that human milk ($\delta^{44/42}$Ca $\approx$ −1.2‰) was approximately 0.6‰ lighter than commercial cow's milk ($\delta^{44/42}$Ca $\approx$ −0.6‰).

Calcium isotopes in bone apatite appear robust to diagenetic change because of the very high concentration of Ca in apatite relative to soil solutions. Dodat et al. (2021) analyzed $\delta^{42/44}$Ca in Neanderthal bone apatite as well as other mammalian bones found in the

Regourdou cave of the Dordogne region of France that date to Marine Isotope Stage V (130 000–80 000 years before present (BP)). $\delta^{42/44}Ca$ was similar to that of lions and lower than herbivores, indicating a primarily carnivorous diet.

13.6 THE AGRICULTURAL REVOLUTION

We will fast-forward once more to the last 10 000 years or so, the period within which humans began growing food. Without doubt, the most important event in human history was the development of agriculture. By domesticating and crops and animals, people no longer needed to move seasonally, allowing them to establish permanent settlements, and eventually villages and cities. Farmers could produce more food than they needed for their own sustenance, which allowed others to become specialized artisans, including, eventually, scientists.

Humans learned to process food more than two million years ago and to cook it a million years ago, long before they learned to grow it. In addition to bones and teeth, mashed grain

and vegetable charred onto potsherds during cooking provides an additional record of the diets of ancient peoples. DeNiro and Hasdorf (1985) found that vegetable matter subjected to conditions similar to burial in soil underwent large shifts in $\delta^{15}N$ and $\delta^{13}C$, but that vegetable matter that was burned or charred did not. The carbonization (charring and burning) process itself produced only small (2 or 3‰) fractionations relative to the range of isotopic compositions in various plant groups.

Since potsherds are among the most common artifacts recovered in archeological sites, this provides another valuable means of reconstructing the diets of ancient peoples. Figure 13.14 summarizes the results obtained in a number of studies of bone collagen and potsherds (DeNiro, 1987). Studies of several modern populations, including Eskimos and the Tlingit people of the Northwestern United States (US), were made as a control. Judging from the isotope data, the diet of Neolithic Europeans consisted entirely of C_3 plants and herbivores feeding on C_3 plants, in contrast to the Tehuacan people of Mexico, who depended mainly on maize, a C_4 plant. Prehistoric

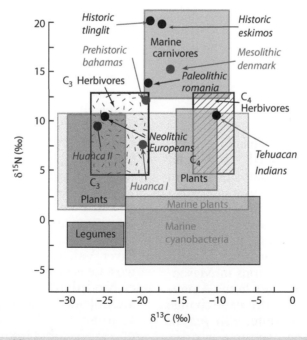

Figure 13.14 $\delta^{13}C$ and $\delta^{15}N$ of various foodstuffs and of diets reconstructed from bone collagen and vegetable matter charred onto pots by DeNiro and colleagues. The Huanca people were from the Upper Mantaro Valley of Peru. Data from potsherds of the Huanca I period (1000–800 BP) suggest both C_3 and C_4 plants were cooked in pots, but only C_3 plants during the Huanca II period (800–530 BP). Source: Adapted from DeNiro (1987).

Figure 13.15   Teosinte is on top, maize is on bottom, and the middle plant is a hybrid of the two. Source: Reproduced from John Doebley, https://en.wikipedia.org/wiki/Zea_(plant)#/media/File:Maize-teosinte.jpg.

peoples of the Bahamas and Denmark depended both on fish and agriculture. In the case of Mesolithic Denmark, other evidence indicates the crops were C_3, and the isotope data bear this out. Although there is no corroborating evidence, the isotope data suggest the Bahamians also depended on C_3 rather than C_4 plants. The Bahamians had lower $\delta^{15}N$ because as discussed earlier, $\delta^{15}N$ is anomalously low at a given tropic level in organisms of the coral reef, suggesting Bahamian relied extensively on the reefs for food.

13.6.1   Domestication of maize

In the Western Hemisphere, maize was the most significant domesticated crop and became a stable crop in many Native American cultures. Because it is a C_4 plant and readily distinguished from more common C_3 food stuffs, staple isotopes have proven particularly useful in tracing its domestication. That is of interest first because that domestication led to the development of civilizations in Mesoamerica and eventually elsewhere in the Americas, and second because of the dramatic change from its wild ancestor, teosinte (*Zea mays*), through selective breeding. Teosinte is grass with numerous branches each ultimately producing a two-ranked "ear" roughly 25 mm long with 5–10 kernels; by comparison, a modern maize ear is typically ≈250 mm long and contains as many as 600 kernels

(Figure 13.15). Based on $\delta^{13}C$ analysis of potsherds (because of the lack of human remains), DeNiro and Epstein (1981) concluded that people of the Tehuacan Valley, located in southern Mexico, had domesticated and depended heavily on maize as early as 6000 BP, whereas archeological investigations had concluded maize did not become important in their diet until perhaps 3500 BP. From $\delta^{15}N$ analyses, they also found a steady increase in the dependence on legumes (beans) from 8000 to 1000 BP and a more marked increase in legumes in the diet after 1000 B.P. Beans and maize, together with chili peppers and tomatoes, also domesticated by Mesoamericans, remain the hallmarks of Mexican cuisine.

Since these pioneering studies, subsequent work has filled in more details of the history of maize and development of civilizations in the Americas. The earliest domestication of maize appears to have occurred in the Balsas River Valley of southwestern Mexico where teosinte is native. Starch grains and phytoliths embedded in stone knives provide the earliest archeological evidence of maize domestication at around 8700 BP, although genetic analysis of maize cobs from the region as late as ≈5300 BP shows this plant was still only partially domesticated, with an assortment of wild and domesticated genes. However, maize was already being grown in the northwestern Colombia and Bolivia by ≈7000 BP and isotopic evidence from teeth indicates that it was a dietary staple on the north coast of Peru by 6000–5000 BP. The earliest examples of maize from this area have longer cobs and more kernels than the contemporary Balsas River examples, consistent with the theory that more fully domesticated versions (although still far from the modern plant) of the plant were developed in South America.

Kellner and Scheoninger (2007) showed that $\delta^{13}C$ in bone collagen primarily reflects $\delta^{13}C$ of dietary protein because the majority of carbon in collagen derives from protein. In contrast, bone apatite or carbonate forms in equilibrium with blood carbonate that is itself a product of energy metabolism. Consequently, $\delta^{13}C$ of bone carbonate and apatite reflects total diet. Kellner and Scheoninger (2007) showed that animals fed C_4 protein defined a $\delta^{13}C_{bone\ apatite}$–$\delta^{13}C_{bone\ collagen}$ correlation that was offset to higher $\delta^{13}C$ than those fed C_3 protein; both correlations, however, are imperfect with considerable scatter. A third correlation is defined by consumers of marine protein that complicates the picture somewhat as it is similar to the C_4 protein line.

Figure 13.15 illustrates the use of this combined analysis of $\delta^{13}C$ in bone collagen and apatite/carbonate. Kennett et al. (2020) analyzed well-preserved human remains spanning the last 10 000 years found in two rock shelters, Mayahak Cab Pek and Saki Tzul. The shelters are located in the Maya Mountains of Belize in the southern part of region of Mayan Classic Period and ≈1000 km to the southeast of the Balsas River Valley. Kennett et al. found that $\delta^{13}C$ in the oldest individuals had low $\delta^{13}C$ in bone collagen and apatite that fell on the C_3 protein line, indicative of minimal C_4 plant consumption. $\delta^{13}C$ in some individuals who lived between 4700 and 4000 BP is shifted toward the base of the C_4 protein line, indicating diets are still C_3-rich but with a significant C_4 component. After 4000 BP, $\delta^{13}C$ in collagen and apatite indicates a staple maize diet, which continued into the Classic Maya Period (1750–1100 BP). Subsequent genomic studies (Kennett et al. 2022) of these remains revealed that oldest individuals (9600–7300 BP) descended from an Early Holocene Native American lineage only distantly related to present-day Mayan-speaking populations. There was a gap in burials between 7300 and 5600 BP, and the genomes of all subsequent individuals indicate that more than 50% of their ancestry can be related to present-day Chibchan speakers living from Costa Rica to Colombia. The arrival of this population predates the isotopic evidence for widespread maize cultivation but does correspond to the first clear evidence for forest clearing and maize horticulture in what later became the Maya region. Kennett et al. argue that these Chibchan-related horticulturalists moved northward into the southeastern Yucatan carrying improved varieties of maize from South America that ultimately led to more intensive forms of maize agriculture and to the Mayan civilization. Most modern Mayan speakers are genetically related to these ancient Chibchan speakers.

In the meantime, maize cultivation had spread through much of South America and into the Colorado Plateau by 4000 BP and eventually into eastern North America and the Caribbean islands. A study by Harrison and Katzenberg (2003) of the remains of Native Americans shows that maize cultivation reached southern Ontario around 1500 BP. $\delta^{13}C$ analysis of the oldest remains shows that these people plot at the base of the C_3 protein line indicative of a 100% C_3-based diet based on hunting and gathering of local food stuffs (Figure 13.16). Remains younger than 1500 BP are shifted toward the C_4 protein line reflecting a maize-based diet supplemented by hunting and gathering. In the wake of the Columbian exchange, maize cultivation propagated around the world and total production now surpasses that of wheat and rice.

13.6.2   Domestication of millet

Millet is the other domesticated C_4 cereal; like maize, its distinct heavy $\delta^{13}C$ signature allows its domestication history to be reconstructed. Although not widely cultivated in Europe or the Americas, it is widely grown in semi-arid regions of Asia and Africa with a global harvest of 30 million tons and is an important part of diet in those regions. Unlike maize, millets do not comprise a single species or even genus, although most are members of the tribe *Paniceae*. The two most widely grown varieties are *Panicum miliaceum*, known as common or broomtail millet, and *Setaria italica*, known as foxtail millet. Like maize ancestors, these are mainly warm climate grasses. Millets have the

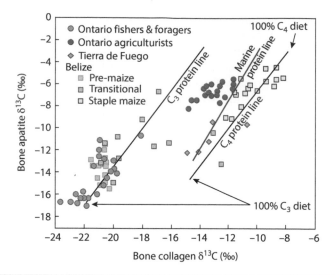

Figure 13.16　C_3 and C_4 "protein lines" defined by animal experiments (Kellner and Scheoninger, 2007). The C_3 protein line has the approximate equation of $\delta^{13}C_{bone\ apatite} \approx 1.76 \times \delta^{13}C_{bone\ collagen} +21.6‰$ while C_4 protein line is nearly parallel offset to higher $\delta^{13}C_{bone\ collagen}$ by $\approx 10.4‰$. Data from burials in Ontario of Harrison and Katzenberg (2003) and Belize caves of Kennett et al. (2020) illustrate the transition of Native American populations to maize agriculture. Tierra del Fuego data of Yesner et al. (2003) illustrate the marine-based diet of that population.

advantage over other cereals of a short growing season and drought resistance.

The earliest archeological evidence of millet cultivation is 10 300–8700 year old phytoliths and husks of common millet found in storage pits in Hebei Province of northeastern China, with evidence of foxtail millet first appearing several thousand years later. How widely it was consumed and whether it was directly consumed or used for animal feed was unclear. Liu et al. (2012) analyzed $\delta^{13}C$ in human and domesticated and wild animal bones from the Chifeng region of Inner Mongolia of northeastern China. They found that human $\delta^{13}C$ was –12 to –6% and $\delta^{15}N$ between 8 and 13‰ consistent with a C_4 terrestrial diet spanning the Early Neolithic (5800 BP) to the Bronze Age (1500 BP). In contrast, animal bones were isotopically lighter, indicating direct consumption of millet rather than use as an animal feed. The mean $\delta^{13}C$ of Bronze Age humans was higher by 2.8‰ than that of the Early Neolithic humans indicating an increasing proportion of millet in the human diet over time.

$\delta^{13}C$ of human remains from a variety of sites in China and Central Asia indicates that millet cultivation spread gradually westward from northeastern China through the Yellow Valley and North China Plain reaching the Kyzyl Bulak in eastern Kazakhstan by around 4000 BP and possibly Skala Sotiros in Greece around the same time (Wang et al., 2017), somewhat following what would become the Silk Road and which Wang et al. referred to as the *Isotopic Millet Road*. There is firm $\delta^{13}C$ of millet consumption in northern Italy in the Early and Middle Bronze Age around 3600–3200 BP.

13.6.3　Agriculture, diet, and culture in Europe

13.6.3.1　Tracing immigration and agriculture with Sr isotopes

Agriculture developed independently in many regions of the world following the end of the last Ice Age, but the first of these regions appears to be the northern part of the Levant, notably the cultivation of wheat and rye in sites, such as Abu Hureyra on the Euphrates River in Northern Syria some 11 000–12 000 years ago. Hillman et al. (2001) speculated that this occurred as a consequence of a decline in wild plants that had previously served as staple foods during the dry, cold, climatic reversal of the "Younger Dryas." There is evidence of a transition from hunting to herding of sheep in areas of southern Anatolia, such as Nevali Çori, at around 10 500 BP. Agriculture then spread eastward into Egypt and northward into the Balkans and the Aegean region and from there across Europe. Anthropologists have long debated whether this spread merely through cultural and technological diffusion or instead involved migration of peoples.

Since these domesticated plants are C_3 as are the indigenous plants of Europe, carbon isotopes are not useful in tracing this transition. On the other hand, the Sr isotopic composition of tooth enamel has helped provide answers to the questions of whether and when Neolithic migrations occurred. Tooth is particularly robust against diagenetic changes. The isotopic composition of human tooth enamel is fixed in childhood. The development of the incisors and canines as well as the crown of the first permanent adult molar tooth begins in utero and is complete by an age of approximately 4.5 years, the second molar crown forms between about 2.5 and 8.5 years of age, and third permanent molar formation begins around 8.5 is complete by an age of about 14.5 years. In contrast, bone material is renewed continually. Dense cortical bone renews over a period of decades, while trabecular bone, such as ribs, renews with turnover times as short as a few years for the ribs and hence its isotopic composition reflects the environment over the last several years of life. By comparing $^{87}Sr/^{86}Sr$ of tooth enamel and bone with $^{87}Sr/^{86}Sr$ of the environment where people were buried can tell us about their movements.

The Danube River has been an historic migration and invasion pathway from Anatolia into Europe and one might suspect may have been used by migrating Neolithic farmers. There is a rich and well-dated burial record spanning the Mesolithic–Neolithic transition in the Danube Gorges region along the border of Romania and Serbia. Borić and Price (2013) analyzed $^{87}Sr/^{86}Sr$ in teeth ranging in age from around 10 800 years BP to around 7800 years BP. Nearly all teeth from burials earlier than about 8300 years have $^{87}Sr/^{86}Sr$ in the relatively narrow range of 0.7089–0.70975, matching the range of local soils (Figure 13.17). An exception is the highest observed $^{87}Sr/^{86}Sr$ (0.7119) from a disarticulated skull that Borić and Price suggest may be an enemy trophy head. Starting around

7300 BP that range expands to 0.7067–0.7108. In addition, teeth from Mesolithic burials had high $\delta^{15}N$, consistent with archeological evidence for reliance on aquatic food sources. Lower $\delta^{15}N$ in the Neolithic teeth is consistent with a switch to greater reliance on cultivated crops. Thus, these data paint a picture of farmers moving into the Danube Valley around 7300 BP and bringing the Neolithic farming culture with them.

Further north in the Rhine Valley and its tributaries, the arrival of Neolithic farmers is signaled by many "non-local" individuals with high tooth $^{87}Sr/^{86}Sr$ values around 7500 BP as well as by the presence of pottery of the Linearbandkeramik (LBK) style and other artifact of the LBK culture. By 7000 BP, there were fewer non-locals. The establishment of Neolithic farming communities, however, appears to have brought with it the seeds of social stratification (Bentley, 2013). In the Rhine region, the uplands tend to be underlain by gneisses and granites that have significantly higher $^{87}Sr/^{86}Sr$ (>0.715) than the Jurassic and younger sedimentary rocks (<0.710) of the more desirable farmland of the regional lowlands. Bentley found that numerous Neolithic villages, high-status burials, e.g., males buried with ground-stone adzes, tend to have low $^{87}Sr/^{86}Sr$ values in teeth indicative of local lowland diets. Low-status burials, often buried in separate parts of cemeteries with no or few grave goods and in some cases simply having been thrown in ditches, had high $^{87}Sr/^{86}Sr$, indicative of having been raised in the uplands. In addition, females tend to have more variable $^{87}Sr/^{86}Sr$ than males, indicative of a patrilocal society in which brides raised elsewhere came to live with their husbands' families.

Agriculture and Neolithic culture reached Britain around 6000–5800 years ago. Cultural artifacts suggest Neolithic culture in Britain was linked to both the Picardie/Pas de Calais area of northern France and to Brittany and Lower Normandy. Neil et al. (2020) analyzed $^{87}Sr/^{86}Sr$ and $\delta^{18}O$ in teeth from a burial mound in Whitwell, located in central England in the county of Derbyshire dated to around 5780–5660 BP. The majority teeth of the majority of individuals have $^{87}Sr/^{86}Sr$ in the range of 0.715–0.721, well above the local biosphere range of ≈0.709–0.712. Particularly striking was the teeth of one young woman. $^{87}Sr/^{86}Sr$ increases from 0.716 to 0.719 from

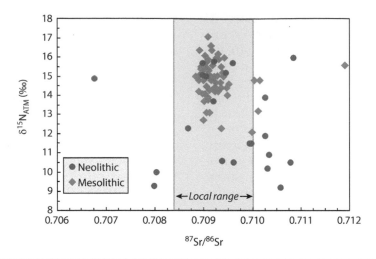

Figure 13.17 $\delta^{15}N$ and $^{87}Sr/^{86}Sr$ of teeth from Mesolithic and Neolithic burials in the Danube Gorges region of Serbia. Most Mesolithic teeth plot within a narrow range of $\delta^{15}N$ and $^{87}Sr/^{86}Sr$ consistent with individuals have been born locally. High $\delta^{15}N$ in these individuals is indicative of extensive reliance on aquatic food sources. Neolithic teeth show a wider range of $^{87}Sr/^{86}Sr$ indicating they were born elsewhere and their lower $\delta^{15}N$ indicates less reliance on aquatic food sources and more on cultivated crops. The highest $^{87}Sr/^{86}Sr$ is from a disarticulated skull that may be an enemy trophy head. Data from Borić and Price (2013).

the first to second molar and then falls to 0.7105 in the third molar, indicating she spent her first childhood years outside the local region and then moved to the local area by age 8 or 9. The elevated $^{87}Sr/^{86}Sr$ at Whitwell as well as at Penywyrlod in Wales contrasts with contemporary graves found in Southern England where most burials fall within the local range. Only a few localities in Britain, Scotland, and southwestern Britain, have such elevated $^{87}Sr/^{86}Sr$. Similar $^{87}Sr/^{86}Sr$ values do occur in the Armorican Massif in Lower Normandy and Brittany. Based on this and the similarity of cultural artifacts, Neil et al. concluded that many of the Whitwell individuals had migrated from northwestern France, perhaps as a group bringing their young children with them, to set up farms in England.

13.6.3.2 Neolithic dairying in England

Neolithic peoples of Europe raised domesticated ruminants (cattle, sheep, and goats) but it was unclear whether they were raised for meat or for milk. C_{18} fatty acids in ruminant milk are derived directly from diet and match the $\delta^{13}C$ of that diet ($-27‰$ for ruminants grazing C_3 plants), while adipose (muscle tissue) C_{18} fatty acids are synthesized by the animals

and are $\approx 2‰$ heavier than milk fat. Although shorter fatty acids degrade, longer fatty acids, such as C_{18} ones, can be preserved on pot sherds. Copley et al. (2003) used compound-specific isotopic analysis to determine $\delta^{13}C$ of fatty acids from early Neolithic pot sherds from Britain and concluded that all Neolithic, Bronze Age, and Iron Age settlements in Britain were indeed exploiting domesticated ruminants for dairy products.

13.6.3.3 Finding Ötzi the ice man's home

Let us now move forward again to the Chalcolithic period of Europe. We have already seen how oxygen and radiogenic isotope ratios have been used to determine the wanderings of an Alaskan mammoth. The case of the "Ötzi," the mummy that had been frozen and mummified in an Alpine glacier for 5200 years before his discovery in 1991, is another interesting example. His copper axe places him in the Chalcolithic period and an arrow embedded in his shoulder as well as evidence of cerebral trauma indicates he died violently. The body was discovered by Austrian hikers near the crest of the Alps and Austrian authorities took custody of it, but it actually had been found just across the border in Italy. Consequently, there were questions as to whether the Iceman was Austrian or Italian, as well as a jurisdictional dispute between the Austrian and the Italians. Müller et al. (2003) analyzed oxygen, carbon and strontium, and lead isotope ratios in the Iceman's bones and teeth. $^{87}Sr/^{86}Sr$ of his teeth is >0.720 while that of the clavicle and cortical bones are <0.7185, suggesting he migrated between childhood and adulthood. Rain falling on the Italian side of the Alps is derived from the relatively warm Mediterranean and is isotopically heavier than the rain in Austria, derived from the relatively cold North Atlantic. Iceman's tooth enamel and bone enamel $\delta^{18}O$ fall into the Italian range but bones are slightly lighter than teeth, suggesting he moved to a higher altitude. The $\delta^{18}O$ data imply he grew up and lived most of his adulthood in the Eisack/Isarco, Rienz, Hohlen, or Non valleys of the Italian side. Comparison of Sr and Pb isotopic compositions of the Iceman's bones with rocks outcropping in the region confirms his Italian origin and zeroes in on the Feldthurns archeological site in the Eisack valley as his most likely his home.

13.7   THE METALLURGICAL REVOLUTION

"The stone age did not end because the world ran out of stones"[3]; instead, it ended when people learned to mine, smelt, and work metals. The earliest known metal artifacts were fashioned from iron meteorites, but these are, of course, rare, and it did not trigger a cultural revolution. The first such metal to be widely used was copper, and copper metallurgy appears to have been independently invented in several areas, beginning roughly 9000 years ago in both the Near East and Europe, and perhaps in South Asia, and somewhat later in sub-Saharan Africa and China. Native Americans, particularly in the western Great Lakes region, were exploiting native copper around this time, but metallurgy never seems to have ignited a cultural and technological revolution in the Americas as it did elsewhere. There can be no doubt that invention of metallurgy is the second most important event in human history following the invention of agriculture.

As we found in Chapter 10, copper deposits occur in a variety of environments and are widespread. Copper is relatively easily smelted and worked but has the disadvantage of being soft. Copper tools and weapons would have needed to be frequently resharpened, reworked, or replaced. Early metalsmiths found that alloying copper with tin produces bronze, a stronger, more durable metal. Tin and copper ores, however, are formed in different environments and rarely in close association. Tin mineralization is associated with highly evolved granites that are the products of crustal anatexis or have assimilated large quantities of crustal material, most notably intensely weathered crustal precursors. Ores themselves are often hydrothermal quartz veins and pegmatites associated with these granites or placers derived from them. Production of bronze therefore required trade, often over long distances. Copper deposits often contain significant amounts of lead and the provenance of bronze artifacts can be traced with Pb isotope ratios, although not always unambiguously. As most of the lead in bronze derives from the copper ore, it is less useful in provenance studies of the tin.

One of the most remarkable examples of Bronze age metal artistry is the *Himmelsscheibe von Nebra* (Nebra Sky Disk), apparently illustrating the Sun, Moon, the seven Pleiades, and other stars (Figure 13.18). Found in 1999 near Nebra in Saxony-Anhalt, Germany, it dates to 3600 BP. The disk consists of bronze with gold inlay and very likely has religious significance, but it is not known what that might have been. It is believed to be the oldest concrete depiction of the cosmos known from anywhere in the world.

Chemical analysis revealed that the copper has been mined in Bischofshofen, Austria, but the source of the tin was initially unknown. The two most likely sources were Bronze Age mines in the Erzgebirge on the German–Czech border and Cornwall in England. Both deposits formed in association with crustal extension during the Variscan Orogeny at around 275–300 Ma. Haustein et al. (2010) analyzed Sn isotopes in tin from both Cornwall and Erzgebirge mines and found that each encompassed a distinct but slightly overlapping range

---

[3] This quote often attributed to former Saudi oil minister Yamani, but is apparently due originally to a Don Huberts, a Royal Dutch Shell engineer.

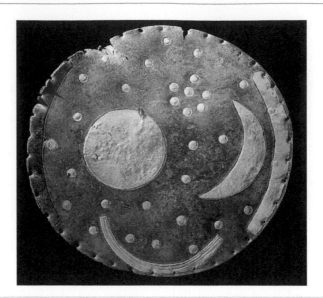

Figure 13.18   The *Himmielsshheibe von Nebra*, Bronze Age art found along with other Bronze Age objects in Nebra, Germany. It consists of bronze with gold inlay and has a diameter of 30 cm and weighs 2.2 kg. Found illegally in 1999 by individuals using metal detectors, it was sold several times for up to a million DM before being seized in 2002 by German police (archeological artifacts are by law the property of the German state). Source: Frank Vincentz/Wikimedia Commons/CC BY-SA 4.0.

of ^{122}Sn/^{116}Sn: 0.3186–0.3188 for Erzgebirge ore and 0.31875–0.31895 for Cornwall ore. Relative to the NIST-SRM3161a standard, these values are $\delta^{122}/^{116}$Sn = –1.55 to –0.93‰ and –1.08 to –0.46‰ for Erzgebirge and Cornwall, respectively. Their analysis of the Himmielssheibe revealed a ^{122}Sn/^{116}Sn of 0.31888 ± 0.00003 or $\delta^{122}/^{116}$Sn = –0.69 ± 0.01‰, placing the tin firmly in the range of ore from Cornwall.

Berger et al. (2019) combined Pb and Sn isotope analysis to determine the provenance of Late Bronze Age (≈3200 BP) tin ingots from the Eastern Mediterranean, including ones from Minoan ruins on Crete and a shipwreck off Uluburun, Turkey and several shipwrecks off the coast of Israel. The Israeli ingots had Pb model ages of 291 Ma, indicating that the ingots were derived from Variscan Age ore and therefore of European, rather than Egyptian, Turkish, or Central Asian provenance. Sn isotope analysis indicated the most likely source of the Minoan and Israeli ingots was again Cornwall (the source of the Ulubrunun ingots could not be unambiguously determined). Isotopic analysis thus confirms that tin was traded over long distances in the Bronze Age, and, in particular, that Britain was integrated into European trade networks long before the European Union.

REFERENCES

Bentley, R.A. 2013. Mobility and the diversity of early Neolithic lives: Isotopic evidence from skeletons. *Journal of Anthropological Archaeology* 32: 303–12. doi: https://doi.org/10.1016/j.jaa.2012.01.009.

Berger, D., Soles, J.S., Giumlia-Mair, A.R., et al. 2019. Isotope systematics and chemical composition of tin ingots from Mochlos (Crete) and other Late Bronze Age sites in the eastern Mediterranean Sea: an ultimate key to tin provenance? *PLOS ONE* 14: e0218326. doi: 10.1371/journal.pone.0218326.

Borić, D. and Price, T.D. 2013. Strontium isotopes document greater human mobility at the start of the Balkan Neolithic. *Proceedings of the National Academy of Sciences* 110: 3298–303. doi: 10.1073/pnas.1211474110.

Cerling, T.E., Andanje, S.A., Blumenthal, S.A., et al. 2015. Dietary changes of large herbivores in the Turkana Basin, Kenya from 4 to 1 Ma. *Proceedings of the National Academy of Sciences* 112: 11467–72. doi: 10.1073/pnas.1513075112.

Cerling, T.E., Harris, J.M., Macfadden, B.J., et al. 1997. Global vegetation change through the Miocene/Pliocene boundary. *Nature* 389: 153–8. doi: 10.1038/38229.

Cerling, T.E., Wang, Y. and Quade, J. 1993. Expansion of C_4 ecosystems as an indicator of global ecological change in the late Miocene. *Nature* **361**: 344–5. doi: 10.1038/361344a0.

Chu, N.-C., Henderson, G.M., Belshaw, N.S., et al. 2006. Establishing the potential of Ca isotopes as proxy for consumption of dairy products. *Applied Geochemistry* **21**: 1656–67. doi: https://doi.org/10.1016/j.apgeochem.2006.07.003.

Copley, M.S., Berstan, R., Dudd, S.N., et al. 2003. Direct chemical evidence for widespread dairying in prehistoric Britain. *Proceedings of the National Academy of Sciences* **100**: 1524–9. doi: 10.1073/pnas.0335955100.

DeNiro, M.J. 1987. Stable Isotopy and Archaeology. *American Scientist* **75**: 182–91. doi: 10.2307/27854539.

DeNiro, M.J. and Epstein, S. 1978. Influence of diet on the distribution of carbon isotopes in animals. *Geochimica et Cosmochimica Acta* **42**. doi: 10.1016/0016-7037(78)90199-0.

DeNiro, M.J. and Epstein, S. 1981. Influence of diet on the distribution of nitrogen isotopes in animals. *Geochimica et Cosmochimica Acta* **45**. doi: 10.1016/0016-7037(81)90244-1.

DeNiro, M.J. and Hasdorf, C.A. 1985. Alteration of $^{15}N/^{14}N$ and $^{13}C/^{12}C$ ratios of plant matter during the initial stages of diagenesis: studies utilizing archeological specimens from Peru. *Geochimica et Cosmochimica Acta* **49**: 97–115. doi: https://doi.org/10.1016/0016-7037(85)90194-2

Dodat, P.-J., Tacail, T., Albalat, E., et al. 2021. Isotopic calcium biogeochemistry of MIS 5 fossil vertebrate bones: application to the study of the dietary reconstruction of Regourdou 1 Neandertal fossil. *Journal of Human Evolution* **151**: 102925. doi: https://doi.org/10.1016/j.jhevol.2020.102925.

Dodd, M.S., Papineau, D., Grenne, T., et al. 2017. Evidence for early life in Earth's oldest hydrothermal vent precipitates. *Nature* **543**: 60–4. doi: 10.1038/nature21377

Drucker, D.G. 2022. The isotopic ecology of the mammoth steppe. *Annual Review of Earth and Planetary Sciences* **50**: 395–418. doi: 10.1146/annurev-earth-100821-081832.

Drucker, D.G., Naito, Y.I., Péan, S., et al. 2017. Isotopic analyses suggest mammoth and plant in the diet of the oldest anatomically modern humans from far southeast Europe. *Scientific Reports* **7**: 6833. doi: 10.1038/s41598-017-07065-3.

Eagle, R.A., Enriquez, M., Grellet-Tinner, G., et al. 2015. Isotopic ordering in eggshells reflects body temperatures and suggests differing thermophysiology in two Cretaceous dinosaurs. *Nature Communications* **6**: 8296. doi: 10.1038/ncomms9296.

Eagle, R.A., Schauble, E.A., Tripati, A.K., et al. 2010. Body temperatures of modern and extinct vertebrates from ^{13}C-^{18}O bond abundances in bioapatite. *Proceedings of the National Academy of Sciences* **107**: 10377–82. doi: 10.1073/pnas.0911115107.

Eagle, R.A., Tütken, T., Martin, T.S., et al. 2011. Dinosaur body temperatures determined from isotopic (^{13}C-^{18}O) ordering in fossil biominerals. *Science* **333**: 443–5. doi: 10.1126/science.1206196.

Ehleringer, J.R., Sage, R.F., Flanagan, L.B., et al. 1991. Climate change and the evolution of C_4 photosynthesis. *Trends in Ecology and Evolution* **6**: 95–9. doi: https://doi.org/10.1016/0169-5347(91)90183-X

Flannery, D.T., Allwood, A.C., Summons, R.E., et al. 2018. Spatially-resolved isotopic study of carbon trapped in ~3.43 Ga Strelley Pool Formation stromatolites. *Geochimica et Cosmochimica Acta* **223**: 21–35. doi: https://doi.org/10.1016/j.gca.2017.11.028.

Harrison, R.G. and Katzenberg, M.A. 2003. Paleodiet studies using stable carbon isotopes from bone apatite and collagen: examples from Southern Ontario and San Nicolas Island, California. *Journal of Anthropological Archaeology* **22**: 227–44. doi: https://doi.org/10.1016/S0278-4165(03)00037-0.

Haustein, M., Gillis, C. and Pernicka, E. 2010. Tin isotopy—a new method for solving old questions. *Archaeometry* **52**: 816–32. doi: 10.1111/j.1475-4754.2010.00515.x.

Herbert, T.D., Lawrence, K.T., Tzanova, A., et al. 2016. Late Miocene global cooling and the rise of modern ecosystems. *Nature Geoscience* **9**: 843–7. doi: 10.1038/ngeo2813.

Hillman, G., Hedges, R., Moore, A., et al. 2001. New evidence of Late glacial cereal cultivation at Abu Hureyra on the Euphrates. *The Holocene* **11**: 383–93. doi: 10.1191/095968301678302823.

Johnson, T.C., Werne, J.P., Brown, E.T., et al. 2016. A progressively wetter climate in southern East Africa over the past 1.3 million years. *Nature* **537**: 220–4. doi: 10.1038/nature19065.

Kellner, C.M. and Schoeninger, M.J. 2007. A simple carbon isotope model for reconstructing prehistoric human diet. *American Journal of Physical Anthropology* **133**: 1112–27. doi: https://doi.org/10.1002/ajpa.20618.

Kennett, D.J., Lipson, M., Prufer, K.M., et al. 2022. South-to-north migration preceded the advent of intensive farming in the Maya region. *Nature Communications* **13**: 1530. doi: 10.1038/s41467-022-29158-y.

Kennett, D.J., Prufer, K.M., Culleton, B.J., et al. 2020. Early isotopic evidence for maize as a staple grain in the Americas. *Science Advances* **6**: eaba3245. doi:10.1126/sciadv.aba3245.

Kuzmin, Y.V., Bondarev, A.A., Kosintsev, P.A., et al. 2021. The Paleolithic diet of Siberia and Eastern Europe: evidence based on stable isotopes ($\delta^{13}C$ and $\delta^{15}N$) in hominin and animal bone collagen. *Archaeological and Anthropological Sciences* **13**: 179. doi: 10.1007/s12520-021-01439-5.

Levin, E. 2013. Compilation of East Africa soil carbonate stable isotope data, *Interdisciplinary Earth Data Alliance (IEDA)*. doi: 10.1594/IEDA/100231

Levin, N.E., Brown, F.H., Behrensmeyer, A.K., et al. 2011. Paleosol carbonates from the Omo Group: Isotopic records of local and regional environmental change in East Africa. *Palaeogeography, Palaeoclimatology, Palaeoecology* **307**: 75–89. doi: https://doi.org/10.1016/j.palaeo.2011.04.026.

Levin, N.E., Haile-Selassie, Y., Frost, S.R., et al. 2015. Dietary change among hominins and cercopithecids in Ethiopia during the early Pliocene. *Proceedings of the National Academy of Sciences* **112**: 12304–9. doi: 10.1073/pnas.1424982112.

Lisiecki, L.E. and Raymo, M.E. 2005. A Pliocene-Pleistocene stack of 57 globally distributed benthic $\delta^{18}O$ records. *Paleoceanography* **20**. doi: 10.1029/2004PA001071.

Liu, X., Jones, M.K., Zhao, Z., et al. 2012. The earliest evidence of millet as a staple crop: New light on Neolithic foodways in North China. *American Journal of Physical Anthropology* **149**: 283–90. doi: https://doi.org/10.1002/ajpa.22127.

Lüdecke, T., Kullmer, O., Wacker, U., et al. 2018. Dietary versatility of Early Pleistocene hominins. *Proceedings of the National Academy of Sciences* **115**: 13330–5. doi: 10.1073/pnas.1809439115.

Magill, C.R., Ashley, G.M. and Freeman, K.H. 2013. Ecosystem variability and early human habitats in eastern Africa. *Proceedings of the National Academy of Sciences* **110**: 1167–74. doi: 10.1073/pnas.1206276110.

Malone, M.A., Maclatchy, L.M., Mitani, J.C., et al. 2021. A chimpanzee enamel-diet $\delta^{13}C$ enrichment factor and a refined enamel sampling strategy: implications for dietary reconstructions. *Journal of Human Evolution* **159**: 103062. doi: https://doi.org/10.1016/j.jhevol.2021.103062.

Mojzsis, S.J., Arrhenius, G., Mckeegan, K.D., et al. 1996. Evidence for life on earth before 3800 million years ago. *Nature* **384**: 55–9. doi: 10.1038/384055a0

Müller, W., Fricke, H., Halliday, A.N., et al. 2003. Origin and migration of the alpine iceman. *Science* **302**: 862–6. doi: 10.1126/science.1089837.

Naito, Y.I., Chikaraishi, Y., Drucker, D.G., et al. 2016. Ecological niche of Neanderthals from Spy Cave revealed by nitrogen isotopes of individual amino acids in collagen. *Journal of Human Evolution* **93**: 82–90. doi: https://doi.org/10.1016/j.jhevol.2016.01.009.

Neil, S., Evans, J., Montgomery, J., et al. 2020. Isotopic evidence for human movement into Central England during the Early Neolithic. *European Journal of Archaeology* **23**: 512–29. doi: 10.1017/eaa.2020.22.

Nguy, W.H. and Secord, R. 2022. Middle Miocene paleoenvironmental reconstruction in the central Great Plains, USA, from stable carbon isotopes in ungulates. *Palaeogeography, Palaeoclimatology, Palaeoecology* **594**: 110929. doi: https://doi.org/10.1016/j.palaeo.2022.110929.

Passey, B.H., Levin, N.E., Cerling, T.E., et al. 2010. High-temperature environments of human evolution in East Africa based on bond ordering in paleosol carbonates. *Proceedings of the National Academy of Sciences* **107**: 11245–9. doi: 10.1073/pnas.1001824107.

Reynard, L.M., Henderson, G.M. and Hedges, R.E.M. 2010. Calcium isotope ratios in animal and human bone. *Geochimica et Cosmochimica Acta* **74**: 3735–50. doi: 10.1016/j.gca.2010.04.002.

Richards, M.P. and Trinkaus, E. 2009. Isotopic evidence for the diets of European Neanderthals and early modern humans. *Proceedings of the National Academy of Sciences* **106**: 16034–9. doi: 10.1073/pnas.0903821106.

Rosing, M. 1999. ^{13}C-depleted carbon microparticles in >3700 ma sea-floor sedimentary rocks from West Greenland. *Science* **283**: 674–6. doi: https://doi.org/10.1126/science.283.5402.674

Sage, R.F. 2004. The evolution of C_4 photosynthesis. *New Phytologist* **161**: 341–70. doi: https://doi.org/10.1111/j.1469-8137.2004.00974.x.

Sage, R.F., Monson, R.K., Ehleringer, J.R., et al. 2018. Some like it hot: the physiological ecology of C_4 plant evolution. *Oecologia* **187**: 941–66. doi: 10.1007/s00442-018-4191-6.

Schidlowski, M. 1988. A 3800-million year isotopic record of life from carbon in sedimentary rocks. *Nature* **333**: 313–8.

Schoeninger, M.J. and Deniro, M.J. 1984. Nitrogen and carbon isotopic composition of bone collagen from marine and terrestrial animals. *Geochimica et Cosmochimica Acta* **48**: 625–39. doi: https://doi.org/10.1016/0016-7037(84)90091-7.

Schouten, S., Hopmans, E. C., Schefuß, E., et al. 2002. Distributional variations in marine crenarchaeotal membrane lipids: a new tool for reconstructing ancient sea water temperatures? *Earth and Planetary Science Letters*, **204**, 265–274. doi: https://doi.org/10.1016/S0012-821X(02)00979-2.

Skulan, J. and Depaolo, D.J. 1999. Calcium isotope fractionation between soft and mineralized tissues as a monitor of calcium use in vertebrates. *Proceedings of the National Academy of Sciences* **96**: 13709–13. doi: 10.1073/pnas.96.24.13709.

Strömberg, C.a.E. 2006. Evolution of hypsodonty in equids: testing a hypothesis of adaptation. *Paleobiology* **32**: 23, 236–58.

Tashiro, T., Ishida, A., Hori, M., et al. 2017. Early trace of life from 3.95 Ga sedimentary rocks in Labrador, Canada. *Nature* **549**: 516–8. doi: 10.1038/nature24019.

Ueno, Y., Yurimoto, H., Yoshioka, H., et al. 2002. Ion microprobe analysis of graphite from ca. 3.8 Ga metasediments, Isua Supracrustal Belt, West Greenland: relationship between metamorphism and carbon isotopic composition. *Geochimica et Cosmochimica Acta* **66**: 1257–68. doi: https://doi.org/10.1016/S0016-7037(01)00840-7.

Wang, T., Wei, D., Chang, X., et al. 2017. Tianshanbeilu and the Isotopic Millet Road: reviewing the late Neolithic/ Bronze Age radiation of human millet consumption from North China to Europe. *National Science Review* 6: 1024–39. doi: 10.1093/nsr/nwx015.

Wang, Y., Cerling, T.E. and Macfadden, B.J. 1994. Fossil horses and carbon isotopes: new evidence for Cenozoic dietary, habitat, and ecosystem changes in North America. *Palaeogeography, Palaeoclimatology, Palaeoecology* 107: 269–79. doi: 10.1016/0031-0182(94)90099-x.

Wasilewski, B., O'Neil, J., Rizo, H., et al. 2021. Over one billion years of Archean crust evolution revealed by zircon U-Pb and Hf isotopes from the Saglek-Hebron complex. *Precambrian Research* 359: 106092. doi: https://doi.org/10.1016/j. precamres.2021.106092.

Wißing, C., Rougier, H., Baumann, C., et al. 2019. Stable isotopes reveal patterns of diet and mobility in the last Neandertals and first modern humans in Europe. *Scientific Reports* 9: 4433. doi: 10.1038/s41598-019-41033-3.

Wooller, M.J., Bataille, C., Druckenmiller, P., et al. 2021. Lifetime mobility of an Arctic woolly mammoth. *Science* 373: 806–8. doi: 10.1126/science.abg1134.

Yesner, D.R., Torres, M.J.F., Guichon, R.A., et al. 2003. Stable isotope analysis of human bone and ethnohistoric subsistence patterns in Tierra del Fuego. *Journal of Anthropological Archaeology* 22: 279–91. doi: https://doi.org/ 10.1016/S0278-4165(03)00040-0.

Zachos, J., Pagani, M., Sloan, L., et al. 2001. Trends, rhythms, and aberrations in global climate 65 Ma to present. *Science* 292: 686–93. doi: 10.1126/science.1059412.

Chapter 14

Noble Gas Isotope Geochemistry

14.1 INTRODUCTION

The noble gases are the elements that are on the extreme right of the periodic table. The name is derived from their unreactive character, a consequence of having outer electron orbitals filled. Isotopes of all six noble gases are produced to some degree by nuclear processes. ^{4}He is, of course, produced by alpha decay while ^{40}Ar is produced by electron-capture decay of K. ^{129}Xe is the decay product of the extinct radionuclide ^{129}I (half-life: 15.8 Ma) and the heavy Xe isotopes were produced by fission of the extinct nuclides ^{244}Pu (half-life: 82 Ma) and ^{238}U. Fission of both ^{244}Pu and ^{238}U also produces minor amounts of the heavy isotopes of Kr. Radon consists of only short-lived isotopes of the U and Th decay series. While neon isotopes are not produced directly by radioactive decay, significant variations occur in Ne isotopes as a consequence of reaction between "fissiogenic" neutrons and magnesium, as well as between alpha particles and ^{18}O. Finally, cosmic-ray interactions also produce isotopic variations in the noble gases, as we have already seen. Cosmic-ray interactions produce a great many nuclides (see Chapter 4), but because noble gases are so rare in the solid Earth, the effects are more significant than for other elements. In addition to nuclear processes, noble gas isotope ratios also vary as a consequence of mass-dependent fractionations. The processes that produced these fractionations mainly occurred in the young Solar System and Earth, and provide important insights into that very early episode of history, as was recognized more than half a century ago (Damon and Kulp, 1958).

In the following sections, we will consider the data on He, Ne, Ar, Kr, and Xe separately (radon nuclides are too short-lived to be of interest here) and examine how noble gases and their isotope ratios vary in Solar System materials as this provides an essential background for understanding noble gas isotope variations on this planet. In later sections, we will see how all the noble gases can be used together to inform broad theories of Earth's formation, structure, and evolution. Much more details on noble gas geochemistry can be found in books by Ozima and Podosek (2002) and Porcelli et al. (2002), as well as recent reviews by Moreira (2013), Mukhopadhyay and Parai (2019), and Marty (2020, 2022).

14.1.1 Noble gas chemistry

The noble gases rarely form chemical bonds in nature although they can be induced to do so in laboratory experiments. The discovery of ArH in the Crab Nebula supernova remnant is one such exception (Barlow et al., 2013), and experiment suggests Xe *may* be reactive in nature at high pressure. They differ in their solubility in liquids (including water, oil, magma, and ices), ionization potential, polarizability, diffusivity, and the extent to which they absorb onto solid surfaces. Noble gas solubility is a complex function of a variety of factors,

Isotope Geochemistry, Second Edition. William M. White.
© 2023 John Wiley & Sons Ltd. Published 2023 by John Wiley & Sons Ltd.
Companion Website: www.wiley.com/go/white/isotopegeochem2

including temperature, pressure, and salinity, but under typical surface conditions, water solubility increases with increasing atomic number. Varying temperature dependence of noble gas solubility in water enables dissolved noble gas concentrations in groundwater to be used as a paleothermometer (e.g., Kipfer et al., 2002). In magma, solubility decreases with atomic weight (i.e., He is the most soluble, while Xe is the least soluble). In equilibration between water and petroleum, the heavier noble gases partition into the petroleum phase more extensively than the lighter noble gases. As one might expect, diffusivity increases with decreasing atomic weight. As a particularly small and electrically neutral atom, He diffuses particularly rapidly. Even at moderately elevated temperatures, radiogenic ^{4}He can diffuse rapidly out of minerals, providing a useful thermochronological tool as we found in Chapter 4. Nevertheless, on the scale of the whole crust, significant transport of even He occurs only through advection. The heavier noble gases, Xe in particular, can be strongly absorbed onto mineral surfaces. For example, Yang and Anders (1982) found mineral/gas distribution coefficients for chromite and carbon decreased from 0.1 for Xe to 0.6×10^{-3} for Ne. Helium has the highest first ionization potential of all elements (2373 kJ/mol) and ionization potential varies with atomic number such that the ionization potential of krypton (1351 kJ/mol) is only a little higher than that of hydrogen (1312 kJ/mol) and the ionization potential of xenon (1170 kJ/mol) is well below that of hydrogen.

The atmosphere is the largest terrestrial reservoir of most noble gases (He is the exception) and is well mixed and consequently isotopically homogeneous (concentrations of the noble gases are listed in Table 14.1).

Table 14.1 Abundances of noble gases in the atmosphere and sun.

Element	Atmosphere	Sun
He	5.24×10^{-6}	2.288×10^9
Ne	1.818×10^{-5}	2.146×10^6
^{4}Ar	0.0934	1.025×10^5
^{8}Kr	1.14×10^{-6}	55.15
Xe	8.7×10^{-8}	5.391
Rn	$\approx 6 \times 10^{-20}$	—

Atmospheric concentrations are mixing ratio, i.e., molar concentrations; solar concentrations are molar concentrations relative to Si $\equiv 1 \times 10^6$.

Concentrations of noble gases in rocks and minerals are extremely low and recovering them in sufficient quantities to determine their isotopic composition is a trick rivaling Moses's producing water from rock. The first part of the trick is to choose the right rock. Because they readily escape to the atmosphere when lavas erupt subaerially, data on isotopic composition of noble gases in the mantle are restricted to submarine-erupted basalts, fluid inclusions in minerals that crystallized at depth (including xenoliths and some phenocrysts), and, in the case of Iceland, subglacially erupted basalts and, more rarely, gas from deep wells, hot springs, and fumaroles. When basalts are erupted under several kilometers of water or ice, the solubility of noble gases is such that at least some of the gases remain in the melt and are trapped in the quenched glassy rims of pillow basalts and within vesicles. Several approaches can then be used to extract the noble gases from rocks and minerals. First, the rock or mineral can be fused in vacuum, allowing noble gases to be collected, cryogenically separated, and analyzed in a mass spectrometer. Second, the sample can be stepheated, terminating with complete fusion, with each temperature-release fraction analyzed separately, much as is done in ^{40}Ar/^{39}Ar dating (see Chapter 2). Third, the sample may be crushed in vacuum, which releases to only those gases trapped in vesicles and fluid inclusions; this may also be done in a stepwise fashion. The remaining material can then be step-heated to release gases trapped in the crystalline structure or glass. Finally, samples may be attacked or etched with acids, an approach largely restricted to meteorites. In some cases, distinct compositions result from different release strategies for the same sample, which in turn reflects multiple origins of the gas.

14.2 NOBLE GASES IN THE SOLAR SYSTEM

14.2.1 Noble gas abundance patterns

Figure 14.1 shows the relative abundances of the noble gases in various Solar System materials. These have been traditionally divided into two basic patterns: *solar* and *planetary*. The former includes the spectroscopically measured "solar" abundances in the Sun itself and the solar wind (as determined from spacecraft

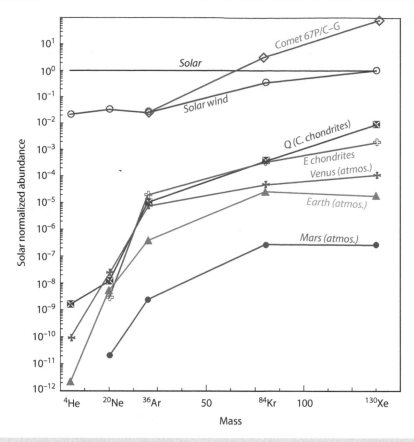

Figure 14.1   Abundance patterns of noble gases in the Solar System. Planetary patterns, which include the atmospheres of the terrestrial planets, as well as gases in chondrites, show depletions in the light noble gases relative to the Solar patterns of the Sun and solar wind. Comet 67P/Churyumov–Gerasimenko is enriched in heavy noble gases relative to the Sun. Based on data in Hunten et al. (1988), Busemann et al. (2000), Pepin and Porcelli (2002), Wieler (2002), Ozima and Podesek (2002), and Bekaert et al. (2020).

measurements), even though the latter shows small depletions in the light noble gases. This reflects elemental fractionation in the acceleration of elements into the solar wind; further fractionation can occur when the solar wind is implanted into material. *Solar* abundance patterns also occur as a component of the noble gas inventory of meteorites and in interplanetary dust particles (IDPs). The "*planetary*" patterns show much stronger depletions in the light relative to the heavy noble gases, particularly He and Ne. Planetary patterns are observed in both terrestrial planet atmospheres and as the major component of the noble gas inventory of meteorites. Step heating and successive leaching reveals carbonaceous chondrites contain a variety of noble gas components, including solar gases (implanted by the solar wind), and cosmogenic nuclides (Figure 6.19), but the most abundant components have planetary-type patterns.

One of the most ubiquitous of these planetary components is released from carbonaceous chondrites by leaching with oxidizing acids, such as HNO_3 and $HClO_4$. The carrier of these noble gases, called "Q," appears to be carbon-rich and sulfide-rich components present primarily as coatings on chondrules, but it is also found in the matrix; its exact nature is unclear (the so-called "P1" component in meteorites appears to be essentially the same as "Q"). Q hosts a "planetary" component that is depleted in light noble gases relative to heavy ones, hosting most of the inventory of Ar, Kr, and Xe, but only a small fraction of the Ne.

In contrast to ordinary chondrites, which are gas-poor, some unequilibrated (petrologic grade 3) enstatite chondrites contain a significant inventory of noble gases. Their heavy noble gas abundances are comparable to the Q component of carbonaceous chondrites, but they are even more depleted in Ne, with $^{36}Ar/^{22}Ne$ ratios of ≈4000 compared to a terrestrial atmospheric value of 18.7 and a solar value of ≈0.4 (Marty, 2022). The combination of an approximately solar heavy noble gas pattern and highly depleted light noble gases found in enstatite chondrites leads to the name "subsolar" for this noble gas component.

Comet 67P/Churyumov–Gerasimenko (hereinafter referred to as 67P/C-G) also appears to have a "planetary"-type noble gas pattern in being progressively enriched in the heavier noble gases based on measurement made by the ESA Rosetta spacecraft of its coma (Marty et al., 2017; Rubin et al., 2018). Although no data are available for He and Ne, relative solar abundances increase from ^{36}Ar through ^{84}Kr to ^{132}Xe (Figure 14.1). The mean density of 67P/C-G is ≈0.5 g/cc, indicating that it, like other comets, is composed mostly of various ices. 67P/C-G is unique in being enriched in Kr and Xe relative to solar, reflecting its ice- and volatile-rich composition. At the low temperatures of interplanetary space, noble gases are readily trapped in ices, with trapping efficiently generally increasing with atomic weight. The noble gas pattern of 67P/C-G reflects the temperature at which the gases were initially trapped and the relative abundances of CO_2, CO, CH_4, and H_2O in the ice, as well as its crystallinity (as opposed to amorphous ice). It also reflects its evaporative history, with every close approach to the Sun resulting in losses of tens of meters of surface with light noble gases being lost in preference to heavy ones. Comet 67P/C-G belongs to the group of Jupiter family of comets that originated in the outer Solar System but have been gravitationally scattered inward and are now gravitationally shepherded by Jupiter. Their orbits are unstable with lifetimes of a few tens of millions of years before they are captured by the Sun or Jupiter or scattered outward. While in its present orbit in the inner Solar System, Comet 67P/C-G may have lost hundreds of meters of ice with preferential loss of the lighter noble gases (Rubin et al., 2018).

14.2.2   Noble gas isotope ratios

Noble gases show considerably greater isotopic variability in the Solar System than most other elements we have considered thus far. Table 14.2 lists the isotopic composition of noble gases in a variety of Solar System

materials. It is not possible to measure most isotope ratios in the Sun directly so solar wind isotope ratios are used to represent the Sun. There is a difference in the extent of isotopic fractionation between "fast" solar wind (>525 km/s), which originates in coronal holes, and the "slow" solar wind (<435 km/s), which originates at the boundaries of Sun's near-equatorial streamer belts. Both are depleted in heavy isotopes of He, Ne, and Ar, with fractionation decreasing from He to Ar and with the slow solar wind being more fractionated. Based on NASA's Genesis mission samples, Heber et al. (2012) found the bulk solar wind to have $^4He/^3He = 2.15 \times 10^3$ $^{20}Ne/^{22}Ne = 13.78 \pm 0.01$ and $^{38}Ar/^{36}Ar = 0.1828 \pm 0.0001$. The varying fractionation allowed them to estimate solar photosphere isotopic compositions of $^4He/^3He = 2.76 \times 10^3$ to 3.21×10^3, $^{20}Ne/^{22}Ne = 13.36 \pm 0.10$, and $^{38}Ar/^{36}Ar = 0.1872 \pm 0.10$.

Significant isotopic variations occur even among and within chondritic meteorites and bulk isotopic compositions in meteorites differ from those of the Sun. The complexity of noble gas isotopic variations is illustrated in Figure 6.19. "Neon B" has high $^{20}Ne/^{22}Ne$ while "Neon A" has substantially lower $^{20}Ne/^{22}Ne$ and "Neon S" is a spallogenic or cosmogenic component produced by cosmic rays (and to a lesser degree by nuclear reactions initiated by radioactive decay). Neon B, although heavier than solar wind, is likely to have originated by solar wind implantation in nebular dust grains. Fractionation favoring heavy isotopes occurs during implantation because heavier isotopes have greater momentum and are implanted deeper. Because the solar wind also progressively sputters away surfaces, more deeply implanted atoms will be more likely to survive. Indeed, the mean $^{20}Ne/^{22}Ne$ of solar wind-implanted lunar soils is ≈12.64, which is quite close to meteoric Neon B. Neon A is actually a mixture of components that we discussed in Chapter 6, notably including neon "E" found in refractory pre-solar grains, such as silicon carbide and nano-diamond, phases which also contain isotopically exotic heavy noble gases. Neon A, with $^{20}Ne/^{22}Ne$ of ≈8.2 is present both in the ground mass of carbonaceous chondrites and in micro-dust particles accreted to the rims of chondrules. After correcting for the cosmogenic component, bulk

Table 14.2 Solar system noble gas isotopic compositions.

	Solar wind	Q (chondrites)	B	Earth (atm)	Mars (atm)	Comet P67/C-G
^{4}He/^{3}He	$2.15 \times 10^{3\dagger}$	6.28×10^3		7.223×10^5	—	
^{3}He/^{4}He R/R_A	336	115		1		
^{21}Ne/^{22}Ne	0.0329§	0.0294	0.032	0.0290	—	
^{20}Ne/^{22}Ne	13.78§	10.67	12.6	9.8	10.1	
^{38}Ar/^{36}Ar	0.1818	0.1873	0.1862	0.1876	0.244	0.188 ± 0.41
^{40}Ar/^{36}Ar	0.0003§	0.78$^\pounds$		296.16*	1714	
^{78}Kr/^{84}Kr	0.00640	0.0603		0.0610	0.0653	0.204 ± 0.002
^{80}Kr/^{84}Kr	0.0409	0.03937		0.0396	0.0417	0.0387 ± 0.004
^{82}Kr/^{84}Kr	0.2048	0.2018		0.2022	0.2063	0.2030 ± 0.004
^{83}Kr/^{84}Kr	0.2029	0.2018		0.2014	0.2042	0.1873 ± 0.002
^{86}Kr/^{84}Kr	0.3024	0.3095		0.3052	0.2994	0.2887 ± 0.004
^{124}Xe/^{130}Xe	0.0298	0.0251		0.0234	0.0246	–
^{126}Xe/^{130}Xe	0.0252	0.0243		0.0218	0.0214	–
^{128}Xe/^{130}Xe	0.501	0.508		0.4715	0.476	0.380 ± 0.034
^{129}Xe/^{130}Xe	6.306	6.44		6.496	15.55	7.27 ± 0.17
^{131}Xe/^{130}Xe	43998	5.06		5.210	5.14	4.68 ± 0.35
^{132}Xe/^{130}Xe	6.061	6.18		6.607	6.48	5.11 ± 0.05
^{134}Xe/^{130}Xe	2.237	2.21		2.563	2.6	1.19 ± 0.91
^{136}Xe/^{130}Xe	1.825	1.95		2.176	2.28	0.622 ± 0.064

Source: Adapted from Heber et al. (2012), Meshik et al. (2014), Rubin et al. (2018), Marty et al. (2017), and Ott (2014).

†Based on Genesis mission measurements; Heber et al. (2012) calculated ^{4}He/^{3}He of the outer convective zone of the Sun to be 2.76×10^3 (R/R_A = 262) and the primordial solar nebula ratio is inferred from Galileo mission measurements of the Jovian atmosphere to be 6.021×10^3 (R/R_A = 120.2).

*Mark et al. (2011).

§Neon in the solar wind is fractionated; solar values are ^{21}Ne/^{22}Ne = 0.0324 and ^{20}Ne/^{22}Ne = 13.34.

$^\pounds$Lowest value observed by Busemann et al. (2000).

carbonaceous chondrites have ^{20}Ne/^{22}Ne falling between Neon A and Neon B. Neon A and B have also been identified in unequilibrated enstatite chondrites. As Figures 6.20 and 6.22 show, nucleosynthetic variations are also present in Kr and Xe.

Mass spectrometric measurements of the coma of Comet 67P/C-G by the Rosetta spacecraft reveal it has approximately solar ^{38}Ar/^{36}Ar and Kr isotopic compositions within measurement errors (Rubin et al., 2018), but its xenon isotopic composition is unique: ^{128}Xe/^{132}Xe, ^{130}Xe/^{132}Xe, and ^{131}Xe/^{132}Xe are solar-like within error, but 67P/C-G Xe is strongly depleted in ^{134}Xe and ^{136}Xe, respectively, by ≈40 and ≈60% (Marty et al. 2017). ^{134}Xe and ^{136}Xe are produced only by the s nucleosynthetic process, so this difference appears to reflect primordial isotopic heterogeneity in the solar nebula. Recall from Chapter 6 that systematic differences in isotopic compositions between carbonaceous and other chondrites also point to radial isotopic heterogeneity in the solar nebula.

In addition to primordial nucleosynthetic variations, noble gas isotopic compositions also vary due to radioactive decay. This is most apparent in the ^{4}He/^{3}He and ^{40}Ar/^{36}Ar ratios, with the atmospheres of Earth and Mars having ratios orders of magnitude greater than solar due to alpha decay producing ^{4}He and decay of ^{40}K producing ^{40}Ar. ^{129}Xe is also elevated in those atmospheres, as well as P67/C-G due to decay of the short-lived radionuclide ^{129}I. Fission of ^{238}U and the extinct nuclide ^{244}Pu produces ^{131}Xe through ^{136}Xe, and to a lesser extent heavy Kr isotopes as well. As we noted above, Ne isotopes are affected by spallation, as well as nuclear reactions initiated by U decay produce ^{21}Ne and ^{22}Ne.

Helium isotope ratios vary between solar and planetary noble gases for an additional reason. The primordial ^{4}He/^{3}He ratio of the Solar System is taken to be the isotopic composition of the Jovian atmosphere as measured by the Galileo spacecraft, which is 6.021×10^3 (R/R_A = 120.2). Ratios approaching this value have also been measured in carbonaceous chondrites (Busemann et al., 2000). The ^{4}He/^{3}He ratio in the solar wind is much lower: 2.15×10^3 (R/R_A = 336.6), but He is strongly fractionated in the solar wind. Heber et al. (2012) calculated the ^{4}He/^{3}He of the outer convective zone of the Sun to be 2.76×10^3 (R/R_A = 261.9). This difference between the present solar and primordial values is a consequence of deuterium burning in the young Sun. Deuterium burning, which produces ^{3}He, occurs during the T-Tauri phase of stars when stellar temperatures reach 10^6 K while the star is still accreting mass, and before the p–p chain and other "main-sequence" fusion reactions have initiated. Importantly, this process is thought to occur while the nebula is still present and before terrestrial planets have fully accreted (although perhaps after the giant planets had formed) and largely consumes the deuterium in the star, so that subsequent ^{3}He production is limited. Furthermore, while heat is transported by radiation in the interior of main-sequence stars, in the deuterium-burning phase, heat is transported throughout the star by convection. Consequently, ^{3}He produced by deuterium burning deep within a T-Tauri star would be convectively mixed to the surface and incorporated in the solar wind, but not in a main-sequence star. T-Tauri stars have particularly strong stellar winds (orders of magnitude greater than the present solar wind). Thus, ^{3}He produced by deuterium burning would have been blown back into the solar nebula by these strong T-Tauri stellar winds where it could be implanted in dust and acquired by growing planets. If the Earth's He inventory consists of He implanted in accreting solids by these early stellar winds, its ^{3}He/^{4}He ratio would be significantly higher than if it were primordial.

14.3   HELIUM

14.3.1   The R/R_A notation

The uniform atmospheric isotope ratio provides a useful standard and reference value for He isotope ratios. He isotope ratios are often reported relative to the atmospheric value as:

$$\left(\frac{^3\text{He}}{^4\text{He}}\right)_{R/R_A} = \frac{\left(^3\text{He}/^4\text{He}\right)_{\text{sample}}}{\left(^3\text{He}/^4\text{He}\right)_{\text{atmosphere}}} \quad (14.1)$$

with the atmospheric ^{3}He/^{4}He ratio of $1.382 \pm 0.005 \times 10^{-6}$ (^{4}He/^{3}He = 7.236×10^{-4}) (Mabry et al., 2013) and the ratio on the left-hand side usually denoted simply as the R/R_A or sometimes simply R_A. This notation was introduced by Craig et al. (1975)[1] in reporting excess ^{3}He in ocean water above the East Pacific Rise, which provided the first clear evidence of primordial He being released from the Earth's mantle at spreading centers.

In these units, crustal rocks generally have He isotope ratios in the range of 0.01–0.1, mantle-derived rocks have values in the range of 5–50, the primordial value is 120, and the ratio in the present solar wind is 334 (Figure 14.2). Low values in the crust reflect degassing of He that occurred during its formation and the subsequent production of ^{4}He by

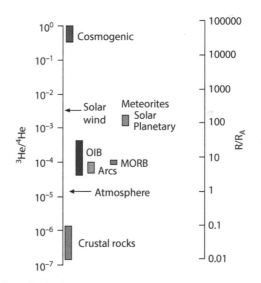

Figure 14.2   Summary of He isotope variations in the Earth and Solar System. The R/R_A scale gives the ^{3}He/^{4}He ratio relative to the atmospheric value.

---

[1] The first discovery of anomalously high ^{3}He in Pacific Ocean deep water from the southern part of the Kermadec Trench was reported by Clarke et al. (1969). They used a δ notation to report the percent deviation of the ^{3}He/^{4}He ratio from atmospheric. Craig et al. (1975) found higher ^{3}He/^{4}He ratios in water above the East Pacific Rise and replaced the δ notation with R/R_A.

radioactive decay. Higher ^{3}He/^{4}He in mantle-derived rocks indicates that the mantle has not been entirely degassed and retains some fraction of its initial, or primordial, inventory of helium. This notation has been widely used, but recently there has been a movement toward presenting He isotope ratios in the conventional manner with the radiogenic isotope in the denominator, or ^{4}He/^{3}He. Here we will use both.

14.3.2   He in the atmosphere, crust, and oceans

Helium is, of course, produced by alpha decay. Essentially all of this is from the U and Th decay series; the amount produced by other alpha emitters, such as ^{147}Sm, is relatively small. Decay of ^{238}U eventually produces 8 ^{4}He, ^{235}U produces 7, and ^{232}Th produces 6. These decays also account for $\approx$80% of heat production in the Earth, with the decay of ^{40}K producing most of the remaining $\approx$20%. The specific heat production of U and Th is 4.12×10^{-4} µW/mol and 1.14×10^{-4} µW/mol. Combining these we can estimate that the ratio of ^{4}He production to heat production is about 10^{12} ^{4}He atoms/W. In units commonly used in noble gas studies, this translates to 3.7×10^{-8} cm^3 STP/J (STP denotes standard temperature and pressure, 25°C and 1 atm; under these conditions, 1 mole of gas occupies 2.24×10^4 cm^3). Both ^{3}He and ^{4}He are also produced by nuclear reactions on Li, such as ^{6}Li(n,α)^{3}He with neutrons produced by α reactions with O, Mg, and Si. Small amounts of He are also generated by cosmic-ray interactions (including ^{3}He produced by decay of cosmogenic ^{3}H), as we found in Section 4.2.4, but this "cosmogenic" He is a trivial fraction of total He production in the Earth. As a consequence

of both alpha decay and these reactions, the ^{4}He/^{3}He production ratio in the crust is in the range of $0.25–1 \times 10^8$ (Ballentine and Burnard, 2002) and ratios in this range are typical of continental crust. The production ratio is lower in the mantle (due to lower Li and U concentrations) and any "nucleogenic" ^{3}He in that reservoir is trivial compared to primordial ^{3}He.

The atmospheric ^{4}He/^{3}He of 7.236×10^5 is well below the crustal production ratio and reflects continued release of primordial ^{3}He from the mantle. Helium is unique in that it is the only element for which the Earth is not a closed system. Its mass is sufficiently low that it can escape the upper atmosphere[2]. As a consequence, the Earth continually "bleeds" helium to space, and since ^{3}He has only ¾ the mass of ^{4}He, ^{3}He is lost even more readily than ^{4}He (the Earth also gains small amounts of He by capture of solar wind: "auroral precipitation"). Consequently, the atmospheric concentration is low, 5.2 ppmv[3], despite its being the second most abundant element in the Sun and the cosmos. The residence time of He in the atmosphere is estimated to be 10^6 to 10^7 years (see Problem 14.1). Given its short residence time, even this low concentration implies that He in the atmosphere must be continually replenished by He from the solid Earth, the first hint of which was the discovery of ^{4}He/^{3}He ratios lower than atmospheric in well gas (Aldrich and Nier, 1946).

Within the crust, He diffuses out of minerals and accumulates in pore fluids (principally water). Helium concentrations in water in equilibrium with the atmosphere (so-called air-saturated water) are around 2×10^{-12} mol/g. In old groundwaters, concentrations range as high as 10^{-7} mol/g, illustrating this accumulation. In some cases, radiogenic production

---

[2] Creationists have claimed that He cannot escape from the Earth's atmosphere because thermal velocities of He in the upper atmosphere are below escape velocity. They argue that since ^{4}He is steadily produced by decay, ^{4}He should steadily accumulate in the atmosphere if it does not escape. The atmospheric abundance, they argue, therefore fixes the age of the Earth to be young (<40 000 years). This argument is flawed because thermal escape, called *Jean's escape*, in which He is accelerated to escape velocity through thermal collisions, is the least important of three principal He escape mechanisms. Most important appears to be the "polar wind" in which He is first ionized and then accelerated along magnetic field lines, which allows flow outward at the poles. The third mechanism is acceleration by interaction with the solar wind. Satellite observations confirm the loss of He in this manner, but because of the role played by the solar wind, which varies dramatically in time, the exact flux out of the atmosphere remains somewhat uncertain, with consequent uncertainty in atmospheric residence time. H and O are also lost through this "polar wind," but those losses are trivial in proportion to the terrestrial inventories of H and O.

[3] Parts per million by volume. This is effectively a molar concentration.

within the aquifer can explain the observed concentrations, but in other cases, such as the Great Artesian Aquifer of Australia, groundwater appears to accumulate He at rates comparable to the entire production within the continental crust beneath it, about 1.5×10^{-6} mol m^{-2} yr^{-1} (Torgersen and Clark, 1985). In regions of continental extension and magmatism (the two are often associated), such as the Rhine Graben and Pannonian Basin in Europe, the Rio Grande Rift in the Western United States (US), and the Subei Basin in China, much higher He concentrations are observed in crustal fluids, including water, petroleum, and natural gas. These are always associated with ^{4}He/^{3}He ratios well below the crustal value (e.g., O'Nions and Oxburgh, 1988; Ballentine et al., 2002), and usually below the atmospheric value. The low ^{4}He/^{3}He are indicative of a flux of He from the mantle. That in turn likely reflects partial melting of the mantle and transport of He and other gases through the crust in fluids.

Tritium, ^{3}H, is produced by cosmic-ray interactions in the atmosphere and decays to ^{3}He with a half-life of 12.4 years. Tritium produced in this way reacts to form H_2O and enters the hydrologic cycle, ultimately entering the oceans. In the ocean, the ^{3}H/^{3}He ratio has been used to determine rates of mixing of water masses in and above the thermocline (e.g., Jenkins, 1987; Schlosser and Winckler, 2002). Deepwater masses are typically old enough so any ^{3}H has decayed and hence this approach is no longer useful. However, deepwater masses typically show elevated ^{3}He/^{4}He ratios as a consequence of mantle degassing. Indeed, the first evidence that the Earth continues to degas primordial He came from the observation of elevated ^{3}He/^{4}He in deep water of the Pacific (Clarke et al., 1969). Subsequent work has shown that most of this He enters the ocean via hydrothermal systems at mid-ocean ridges, with smaller contribution from submarine volcanism associated with island arcs and oceanic island chains. Based on the distribution of ^{3}He in the deep ocean and an ocean circulation model, Bianchi et al. (2010) estimated the ^{3}He flux from the oceans as 527 ± 102 mol/yr, about half the flux originally estimated by Craig et al. (1975). Assuming a ^{3}He/^{4}He ratio of this helium equal to the average value in mid-ocean ridge basalts (MORB) (see below), this corresponds to a total He flux of 4.11×10^7 mol/yr through the oceans.

Very low ^{4}He/^{3}He ratios have been found in slowly accumulating deep ocean sediments. These low ratios result from the presence of IDPs. The particles themselves have ^{4}He/^{3}He ratios 4.25×10^3 ($\approx$170 R/R_A). There is a constant rain of these particles over the entire Earth surface, but they are a significant component of sediment only at extremely slow accumulation rates, such as those occur in the Central Pacific. In those cases, bulk sediment ^{4}He/^{3}He can be $<1.5 \times 10^4$ (^{3}He/^{4}He >50 R/R_A). If one assumes that accumulation rates of IDPs are constant, ^{4}He/^{3}He ratios can be used to determine sedimentation rates. On the other hand, if accumulation rates are not constant, ^{4}He/^{3}He ratios provide a record of cosmic events, such as asteroid or comet breakup (Schlosser and Winkler, 2002).

14.3.3 He in the mantle

In Chapter 7, we briefly discussed He isotope ratios in the mantle based on data from oceanic basalts; let us now look at this in more detail. Before we do so, we must consider a couple of caveats. First, because some loss of He is almost inevitable even during submarine eruptions and ^{4}He is continually generated by alpha decay, the measured ^{4}He/^{3}He ratio measured from lavas may be higher than that of the mantle source, especially in old samples. Second, subaerial samples exposed to cosmic rays can contain cosmogenic He, in which case the ^{4}He/^{3}He could be lower than that of the mantle source. The importance of both these effects increases with the age of the sample.

Clarke et al.'s discovery of anomalous ^{3}He in deep ocean water was somewhat surprising because it showed that the mantle retained some primordial He, despite He being an extremely incompatible element and readily extracted from the mantle during melting (e.g., Graham et al., 2016). Decades of subsequent work have revealed an interesting picture. Figure 14.3 summarizes these data. The MORB data set is relatively rich because MORBs are erupted under water and pressure is sufficient that He is retained. ^{4}He/^{3}He ratio is relatively uniform in MORB and presumably therefore in the depleted upper mantle, with a mean ^{4}He/^{3}He of 90.7 ± 13.2 $\times 10^3$ (R/R_A value of 8.2 ± 1.5 and a median value of 7.3). Most of the samples with ^{4}He/^{3}He $< 72 \times 10^3$ ($R/R_A > 10$) come from parts of

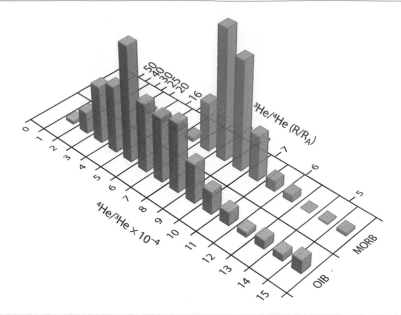

Figure 14.3   Comparison of ^{4}He/^{3}He in mid-ocean ridge basalts (MORB) and oceanic island basalts (OIB). Data from the PetDB (https://search.earthchem.org) and GEOROC (https://georoc.eu/georoc/new-start. asp) databases.

the ridge close to oceanic islands and are thus likely influenced by mantle plumes. Furthermore, there is so little He in the atmosphere and dissolved in seawater that any atmospheric contamination is usually negligible (in contrast to the other noble gases).

The OIB data set is largely restricted to basalts erupted on seamounts and from submarine fissure eruptions of oceanic islands, but complimented by data from xenoliths and phenocrysts found in the lavas. ^{4}He/^{3}He values in OIB are more variable and may be either higher or lower than in MORB, although most are higher, with a mean of ^{4}He/^{3}He 74.7 ± 8.7 × 10^3 (12.5 ± 5.6 R/R_A) and a mode of ^{4}He/^{3}He 36.1 × 10^3 (20 R/R_A). Low values likely reflect recycled and degassed surficial material, for which there is abundant radiogenic and stable isotope evidence that we reviewed in previous chapters. On the other hand, the higher ^{3}He/^{4}He ratios in mantle plumes provide evidence that they tap a part of the mantle that has been less degassed than the MORB source. The latter is generally assumed to be the upper mantle or asthenosphere. Simple logic suggests (but does not prove) that the deep mantle should have experienced less melting and degassing than the upper mantle (because melting and degassing can occur only near the surface). Hence, high He isotope

ratios in plume-derived basalts are often cited as evidence that plumes come from the deep mantle.

Figure 14.4 shows the maximum ^{3}He/^{4}He in basalts from a variety of hot spots plotted against regional seismic shear-wave velocity anomalies at 200 km depth beneath the hot spot from Boschi et al. (2007) and the calculated buoyancy flux calculated from topographic displacement (the "hot spot swell") by King and Adam (2014). For hot spots where a plume is inferred based on geophysical observations, Jackson et al. (2017) found strong correlations between the highest measured ^{3}He/^{4}He in volcanic hot spots and these geophysical parameters, which they assumed were indicative of plume temperatures. They interpreted the relationships to mean that hot plumes are more buoyant and can entrain high ^{3}He/^{4}He as well as low ^{3}He/^{4}He material, whereas cooler, less buoyant plumes do not entrain this high ^{3}He/^{4}He material. They inferred from this that the high ^{3}He/^{4}He domain is dense and can be entrained from the deep mantle only by the hottest, most buoyant plumes. Williams and Mukhopadhyay (2019) compared the highest ^{3}He/^{4}He ratios in oceanic island volcanoes with s-wave seismic velocities, at the root of the plume in the lower mantle, and found a strong

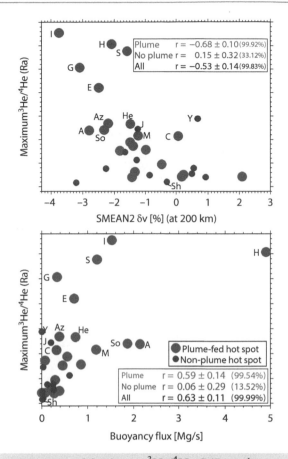

Figure 14.4 Maximum $^3He/^4He$ R/R_A values at plume-fed hot spots compared with seismic shear-wave velocity anomalies at 200 km and with hot spot buoyancy flux. The inferred presence (red circles) or absence (blue circles) of a plume is based on the presence of low-seismic velocities below the hot spot. δv is the deviation in s-wave seismic velocity at 200 km from the average in the SMEAN2 seismic model. Pearson correlation coefficients, r, are and P values (calculated with Student's t-test assuming normally distributed data). The hot spot buoyancy flux is based on regional topographic displacement. H = Hawaii; I = Iceland; S = Samoa; G = Galápagos; E = Easter; Y = Yellowstone; A = Afar; So = Societies; C = Cape Verde; Az = Azores; M = Macdonald; He = Heard; J = Juan Fernandez; and Sh = St Helena. Source: From Jackson et al. (2017).

relationship between minimum $^4He/^3He$ (maximum $^3He/^4He$) ratios in oceanic basalts and seismically slow regions in the lowermost mantle, which are generally located within the two large low s-wave velocity provinces (LLSVPs) in the lower mantle. Low s-wave velocities indicate high temperatures and/or

high densities. Low $^3He/^4He$ ratios in some plumes, such as Tristan and St. Helena, could reflect the presence or predominance of material recycled from the Earth's surface, such as oceanic crust, in these plumes.

Judging from He trapped in fluid inclusions in phenocrysts and xenoliths, the subcontinental lithospheric mantle typically has R/R_A values that range from 5 to 8, i.e., well above the atmospheric and crustal values but lower than MORB. A few xenoliths have ratios slightly higher than MORB. $^3He/^4He$ R/R_A values as high as 13 have been reported from the 250 million year old Siberian flood basalts (Basu et al., 1995), as well as the Deccan flood basalts (Basu et al., 1993). The Siberian Traps represent one of the greatest known outpourings of lava in the planet's history and are thought to have formed as the head of a large deep mantle plume reached near the surface. The highest reported values in any lavas, $R/R_A \approx 50$, were reported from early Tertiary picrites from Baffin Island and West Greenland (Starkey et al., 2009). The Baffin Island province is associated with the beginnings of the Icelandic mantle plume and the Deccan Province is associated with the beginning of the Reunion mantle plume.

Extreme $^3He/^4He$ ratios, up to 1000 R/R_A, have been reported from diamonds. Since these ratios are well above the Solar System initial value (≈ 120 R/R_A; $^4He/^3He = 6.024 \times 10^3$), they must reflect some sort of nucleogenic production of ^3He (e.g., Ozima and Zashu, 1983; Zadnik et al., 1987). In some instances where the diamonds may be from alluvial placer deposits, cosmogenic production could be the explanation. In other cases, however, cosmogenic production can be ruled out and hence production through $\alpha–n$ reactions is suspected, but a fully satisfactory explanation has not been given.

He isotope ratios in island arc volcanics are generally somewhat lower than those in MORB. If only the maximum R/R_A values in each arc are considered (because low values could reflect crustal or atmospheric contamination or radiogenic in-growth), the range is 4.5–8.9 R/R_A; the overall mean is 5.37 (Hilton et al., 2002). Values lower than MORB could reflect radiogenic He from subducting oceanic crust and sediment. If this radiogenic component has R/R_A of ≈ 0.1, then it is nonetheless clear that the primary source of He in island arcs is depleted mantle, which is an observation consistent with the idea that island arc volcanics are primarily derived from the mantle wedge above the subducting slab. Back arc basin basalts range from MORB-like values in the Mariana Trough to values higher than MORB (up to 28.1 R/R_A in the Rochambeau Bank of Lau Basin; Pető et al., 2013) in the Lau, North Fiji, and Manus Basins (Lupton et al. 2009). These higher ratios suggest a mantle plume component. There is independent evidence of a plume contribution in the North Lau Basin from shear-wave splitting measurements, which suggest southward flow from the Samoan mantle plume into the basin (Smith et al., 2001). Hilton et al. (2002) estimate the global ^{4}He flux from island arc volcanism as 1.2×10^7 mol/yr, or about 3.7% of the total flux from the solid Earth.

14.4   NEON

14.4.1   Neon isotope systematics in the Earth

Figure 14.5 provides a summary of Ne isotopic variations. The first observation is that the atmospheric ^{20}Ne/^{22}Ne ratio is lower than that of the solar wind (Table 14.2), as well as mantle Ne. Variation of neon isotope ratios in the Earth was originally thought to reflect two processes: (1) mass-dependent fractionation, which accounted for the difference between atmospheric and solar neon, and (2) production through nuclear reactions. The latter results mainly from α–n reactions, such as ^{17}O$(\alpha,n)^{20}$Ne, ^{18}O(α,n) ^{21}Ne, ^{24}Mg$(n,\alpha)^{21}$Ne, and ^{19}F$(\alpha,n)^{22}$Ne with the alpha particles derived from decay of U and Th and neutrons derived either from other α–n reactions or, more rarely, ^{238}U fission. While these reactions produce all three Ne isotopes, the largest effect is on ^{21}Ne reflecting both low abundance and production (^{21}Ne is far less abundant, 0.26% of Ne, than other Ne isotopes). The abundance of ^{20}Ne, which constitutes ≈90% of Ne is not significantly affected by these reactions in the mantle. Production of ^{22}Ne is generally small, but higher in the crust than in the mantle because of higher fluorine concentrations in the crust.

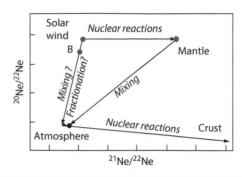

Figure 14.5   Schematic illustration of the isotopic variation of terrestrial Ne. Atmospheric ^{21}Ne/^{22}Ne is lower than solar wind Ne by about half as much as ^{20}Ne/^{22}Ne. The meteoritic B component (see also Figure 6.19) lies between the solar and the terrestrial atmosphere. This variation is roughly consistent with mass-dependent fractionation, but Marty (2022) has questioned this interpretation. ^{21}Ne production in the solid Earth by (α, n) and (n,α) reactions increases the ^{21}Ne/^{22}Ne by varying amounts, depending on the U/Ne ratio.

After accounting for a small amount of nucleogenic ^{21}Ne in the atmosphere derived from the solid Earth, atmospheric ^{21}Ne/^{22}Ne and ^{20}Ne/^{22}Ne lie close to a mass-dependent fractionation line from solar or the solar wind ratios. Hydrodynamic escape[4] of Ne and other light gases from the Earth's early atmosphere has long been the standard explanation for this (e.g., Pepin, 2006). However, Marty (2022) has questioned this interpretation based on ^{36}Ar/^{22}Ne ratios and has proposed instead that it primarily reflects mixing between two primordial components: one with solar-like Ne ratios and another enriched in ^{20}Ne similar to the Neon A associated with pre-solar grains (Figure 6.19). We will consider these theories in more detail below in a later section.

14.4.2   Neon in the solid earth

Given that the continental crust has been created by magmatism, it is not surprising to find that it has been extensively degassed and hence highly depleted in noble gases. Because of this, nucleogenic Ne is a greater fraction of Ne in the

---

[4] In hydrodynamic escape, light atoms or molecules, such as H_2 or He, at high temperature can reach velocities allowing them to escape from the atmosphere (Jean's escape) and in doing so "drag" heavier atoms or molecules, Ar, for example, which have not attained escape velocity along with them.

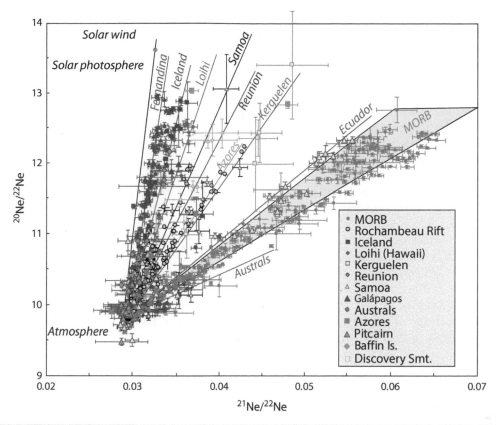

Figure 14.6 Ne isotope variations in oceanic basalts and xenoliths. Two neighboring Galápagos volcanoes, i.e., Volcán Ecuador and Volcán Fernandina, define distinctly different correlations. Data from Poreda and Farley (1992), Valbracht et al. (1996), Valbracht et al. (1997), Moreira et al. (1998), Trieloff et al. (2000), Trieloff et al. (2002), Honda and Woodhead (2005), Madureira et al. (2005), Jackson et al. (2009), Kurz et al. (2009), Hanyu et al. (2011), Tucker et al. (2012), Mukhopadhyay (2012), Parai et al. (2012), Pető et al. (2013), Péron et al. (2016), Williams and Mukhopadhyay (2019), and Horton et al. (2021).

crust than in the atmosphere or mantle. Thus, $^{21}Ne/^{22}Ne$ in crustal fluids (water, natural gas) can be an order of magnitude greater than the atmospheric value. As is the case with He, Ne isotope ratios can be used to trace the origin and age of groundwater.

Figure 14.6 shows Ne isotopic variations in oceanic basalts (including xenoliths in oceanic island basalts). The data form a variety of arrays that extend from the atmospheric value toward solar $^{20}Ne/^{22}Ne$ and more nucleogenic $^{21}Ne/^{22}Ne$. These arrays reflect ubiquitous mixing with air: the Ne/He ratio is higher in the atmosphere than in most rocks, making the neon more sensitive to contamination. Consequently, the maximum $^{20}Ne/^{22}Ne$ in each array represents a minimum value of the mantle source. The $^{21}Ne/^{22}Ne$ ratio is a function of the U–Th/Ne ratio in the mantle

source region: high U–Th/Ne leads to higher $^{21}Ne/^{22}Ne$ through nucleogenic production. We see that MORBs generally have higher $^{21}Ne/^{22}Ne$ than OIB, implying higher U–Th/ Ne, consistent with the idea suggested by the He data that the MORB reservoir is more degassed than the OIB one. The MORB data scatter somewhat suggesting some heterogeneity in the U–Th/Ne ratio. The Rochambeau Rift and Galápagos data form two distinct arrays, one MORB-like and one more OIB-like. In the Rochambeau Rift case, this is consistent with the idea of Samoan plume material invading the Lau Basin. In the Galápagos case, the arrays are consistent with differences in Sr, Nd, and Pb isotope ratios of the volcanoes sampled, with Volcán Ecuador and Volcán Wolf having more MORB-like radiogenic isotopic compositions than Volcán Fernandina.

The Austral Islands are exceptional in that their neon appears to be more rather than less nucleogenic than MORB. In this respect, He and Ne isotopes of the Australs are consistent, as $^3He/^4He$ ratios in the Australs are lower than in MORB in contrast to most OIB (Figure 7.30). The low $^3He/^4He$ and high $^{21}Ne/^{22}Ne$ ratios in Australs basalts suggest the Australs mantle plume consists of highly degassed material, such as recycled surficial material. This is also consistent with radiogenic isotope ratios and mass independently fractionated (MIF) sulfur isotope ratios indicative of recycled surficial material carried by the Australs mantle plume. In contrast, Pitcairn Island, which also has an MIF sulfur signature, fall along a steep trajectory similar to Iceland but with a lower $^{20}Ne/^{22}Ne$ maximum (11.7) and may well contain a mixture of primordial and recycled material.

Perhaps the most significant implication of the data, however, is that Ne in the atmosphere is isotopically distinct from that in the Earth's interior. Ne in some mantle plumes has a solar-like isotopic composition. The maximum reported high-precision $^{20}Ne/^{22}Ne$ in oceanic basalts is 13.03 ± 0.04. Yokochi and Marty (2004) and Williams and Mukhopadhyay (2019) reported values as high as 13.05 ± 0.20 in inclusions from Devonian alkaline magmatic rocks from the Kola Peninsula, which also had a $^3He/^4He$ of 13.5 R/R_A (higher ratios have been reported from Kerguelen xenoliths, but these have high analytical uncertainty). These values approach the estimated solar photosphere value of ≈13.35. Since there is potential atmospheric contamination in all these samples, it is possible that the $^{20}Ne/^{22}Ne$ ratio mantle source of these arrays is even higher, possibly solar. The maximum $^{20}Ne/^{22}Ne$ observed in MORB (12.75) is a single analysis with rather large error of the "popping rock" 2πD43 from the central Mid-Atlantic Ridge (Moreira et al., 1998). Excluding this analysis, the maximum $^{20}Ne/^{22}Ne$ of all other MORB values (12.5), including other analyses of 2πD43, has $^{20}Ne/^{22}Ne$ similar to the Ne–B component in chondrites, which is believed to be implanted solar wind. This hints at primordial heterogeneity in which the upper mantle has "planetary" or chondritic-like Ne while the deep mantle has solar-like Ne. We will return to this once we review the evidence from other noble gases.

Since alpha particles are, of course, ^4He nuclei, and radiogenic production of ^4He and nucleogenic production of ^{21}Ne are closely linked, with an estimated production $^{21}Ne/^4He$ ratio of 4.5 × 10^{-8} (Honda and McDougall, 1998). $^{21}Ne*/^4He*$ ratios (where, as usual, the asterisk denotes the radio/nucleogenic component) observed in MORB and OIB vary considerably, probably due to elemental fractionation during and after eruption: He is more soluble in magma than is neon and hence is less likely to be lost by degassing. On the other hand, it is much more subject to diffusive loss. The two samples with best-preserved initial gas contents, the DICE 10 sample from Iceland (a subglacially erupted glassy basalt) and the 2πD43 sample (a gas-rich "popping rock") from the Mid-Atlantic Ridge, have $^{21}Ne*/^4He*$ ratios of 4.5 × 10^{-8} and 7.2 × 10^{-8}, respectively, close to the expected ratio. We will see in a subsequent section that we can use these ratios to estimate Ne/He ratios in the mantle.

14.5   ARGON

Argon has three isotopes: ^{36}Ar, ^{38}Ar, and ^{40}Ar. The latter constitutes 99.6% of atmospheric Ar. Argon geochronology, certainly the most significant aspect of Ar isotope ratios, was discussed in Chapter 2. Here we review the geochemical aspects and implications of Ar isotope variations in the Earth. Comparing the atmospheric $^{40}Ar/^{36}Ar$ ratio of 296 to the primordial one of ≈3 × 10^{-4}, we see that atmospheric Ar is almost exclusively radiogenic. Furthermore, we infer that atmospheric Ar must owe its origin to degassing of the Earth's interior (since there is no potassium in the atmosphere). Since the half-life of ^{40}K is ≈1.4 Ga, most of this Ar must have been degassed after the Earth formed. Depending on the assumed K concentration in the Earth, something like 50–67% of the radiogenic Ar produced since the Earth formed is now in the atmosphere. As we shall see, higher than atmospheric $^{40}Ar/^{36}Ar$ ratios are observed in well gases, fluid inclusions, and oceanic basalts.

The atmospheric $^{38}Ar/^{36}Ar$ ratio of 0.1876 is higher than the solar wind value of 0.1818, but not significantly different from the inferred solar value of ≈0.187. Since nuclear processes produce neither isotope to a significant degree, the difference could be due to mass fractionation. One might then suspect that the $^{38}Ar/^{36}Ar$ ratio in the Earth's interior would differ from the atmospheric value, as is the case for neon. However, Raquin and Moreira (2009) found that both OIB and MORB had $^{38}Ar/^{36}Ar$ ratios identical to the atmospheric value within analytical error. It may be that this reflects subduction recycling of atmospheric Ar into the mantle as it appears that, unlike He and Ne, the heavier noble gases can pass through the subduction barrier (Sarda, 2004). Alternatively, future improvements in analytical precision may reveal a difference between atmospheric and mantle $^{38}Ar/^{36}Ar$ ratios.

Figure 14.7 shows $^{40}Ar/^{36}Ar$ plotted against $^{20}Ne/^{22}Ne$. The variation within individual sample suites largely reflects atmospheric contamination. Consequently, it is generally the maximum value observed in a suite of analyses of sample that is of greatest interest, but even the maximum ratio might reflect the effects of some contamination. One approach, used by Moreira et al. (1998) and Mukhopadhyay (2012), is to assume that the magmatic endmember has a solar $^{20}Ne/^{22}Ne$ ratio (assumed by them to be 13.6). A mixing hyperbola can then be fitted to the data to determine the $^{40}Ar/^{36}Ar$ ratio at $^{20}Ne/^{22}Ne$ ratio = 13.6. As Figure 14.7 shows, in the Iceland case, this value is 10 750 while in the MORB case it is 44 000. In the MORB case, the value estimated in this way is close to the maximum value observed in MORB (40 000; Burnard et al., 1997). Ballentine et al. (2005) inferred a similar value for the mantle endmember of CO_2 well gas[5] from the Bravo Dome of northeast New Mexico, US. In the Iceland case, the estimated $^{40}Ar/^{36}Ar$ is somewhat higher than any value observed in Iceland or any other oceanic island (8340 from Loihi, Hawaii).

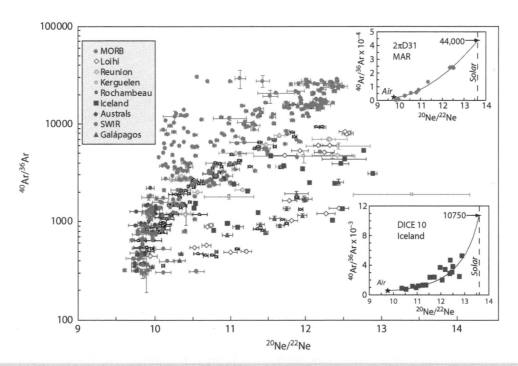

Figure 14.7 $^{40}Ar/^{36}Ar$ as plotted against $^{20}Ne/^{22}Ne$ in MORB, OIB, and related xenoliths. Variation within suites generally reflects atmospheric contamination. It is apparent that MORB have higher $^{40}Ar/^{36}Ar$ than OIB. Examples of mixing hyperbolae are shown for samples from the Mid-Atlantic Ridge (2πD31) and Iceland (DICE 10). One approach to determining the $^{40}Ar/^{44}Ar$ of mantle endmember is to assume that it has a solar $^{20}Ne/^{22}Ne$ ratio. Under this assumption, the MORB sample has $^{40}Ar/^{36}Ar$ of 44 000 and the Iceland sample has $^{40}Ar/^{36}Ar$ of 10 750.

[5] This CO_2, which is a major source of industrial CO_2, is thought to have originated from decarbonation of limestone sediments driven by intrusion of basaltic magma. The noble gases appear to be derived from the intruding magma. Helium isotope ratios in the gas range to greater than 4 R/R_A.

Nevertheless, it is clear then that there is a quite significant difference in $^{40}Ar/^{36}Ar$ between the MORB and OIB reservoirs (we should note that neither is likely to have a homogeneous $^{40}Ar/^{36}Ar$ ratio). High $^{40}Ar/^{36}Ar$ ratios imply high K/Ar ratios and a more extensively degassed source. The difference is consistent with the inference made already from He and Ne isotopes that the MORB source reservoir appears to be more degassed than the OIB one and Ar in the MORB source reservoir is more dominated by the radiogenic component.

Ratios of K, U, and Th concentrations vary to only a limited extent in most terrestrial reservoirs, with K/U ≈13 000 and Th/U ≈4. Consequently, we would expect the production rates of $^4He/^{40}Ar$ to be similar in such reservoirs. The production ratio has, however, increased over time due to the short half-life of ^{40}K relative to ^{238}U and ^{232}Th, so that for a given K/U and Th/U ratio in a closed system, the $^4He*/^{40}Ar*$ ratio is a function of age, increasing from ≈1.7 for a reservoir the age of the Solar System to a present production rate of 4.65. In principle then, the $^4He*/^{40}Ar*$ ratio could be used to deduce either the length of time the system has been closed if a fixed K/U ratio is assumed or the K/U ratio if a reservoir age is assumed. Unfortunately, however, as is true of He/Ne ratios, fractionation during and after eruption and degassing can change the He/Ar ratio, and MORB and OIB show a large range of values. The two samples with best-preserved initial gas contents, i.e., the DICE 10 sample from Iceland and the 2πD43 sample from the Mid-Atlantic Ridge, have $^4He*/^{40}Ar*$ ratios of ≈2.5 and ≈1.5, respectively, in release fractions with the highest $^{20}Ne/^{22}Ne$ ratios. Assuming, questionably, that these values represent those of the mantle, Iceland values would be consistent with a reservoir age of ≈2.5 Ga, while the MORB values suggest a reservoir age close to that of the Solar System. Alternatively, the lower MORB values could reflect a higher K/U ratio. Using the value of 19 000 that Arevalo et al. (2009) calculated for average MORB, the reservoir age decreases to 3 Ga. However, Gale et al. (2013) and White and Klein (2014) calculated lower average MORB K/U ratios of 16 500 and 12 300, respectively. Hanyu et al. (2011) found that basalts from the Australs have $^4He*/^{40}Ar*$ ratios in the least fractionated release fractions of 10–15. Assuming a reservoir age of ≈2 Ga,

this implies a much lower K/U ratio of ≈3000. These samples also have low $^{40}Ar/^{36}Ar$ (<2000) compared to other OIBs and MORBs. Assuming coherent behavior between the alkalis Rb and K, the low K/U ratio would be consistent with the low $^{87}Sr/^{86}Sr$ and high $^{206}Pb/^{204}Pb$ observed in these basalts. We should be cautious, however, not to conclude too much from $^4He*/^{40}Ar*$ ratios – firm inferences can be made only for closed systems, which mantle reservoirs certainly are not. Perhaps the only useful inferences are that these reservoirs are old and the HIMU reservoir – the source of Australs basalts – has a lower K/U than other reservoirs.

14.6 KRYPTON

Krypton has six stable isotopes: ^{78}Kr, ^{80}Kr, ^{82}Kr, ^{83}Kr, ^{84}Kr, and ^{86}Kr. The two lightest together constitute less than 3% of Kr; ^{82}Kr and ^{83}Kr both have abundances of about 11.5%; ^{84}Kr is the most abundant (57%); and ^{86}Kr is the second most with an abundance of 17%. There is some production of ^{86}Kr by fission of ^{238}U and ^{244}Pu (1 and 0.1% yields, respectively, compared to respective ^{136}Xe yields of 6.3 and 5.6% for fission of these nuclides). ^{84}Kr is also produced by fission, but the yield is seven times lower than that of ^{86}Kr. Krypton isotopic variations resulting from such fission have so far only been resolved in U-rich minerals.

Mantle krypton isotopic data are extremely limited and until quite recently were available only for well gas from the Bravo Dome, New Mexico to the east of the volcanically active Rio Grande Rift in the Southwestern US (Holland et al., 2009), and only for $^{84}Kr/^{82}Kr$ and $^{86}Kr/^{82}Kr$. Recently, Péron and Moreira (2018) reported an $^{86}Kr/^{84}Kr$ ratio of 0.3059 ± 0.0007 for an MORB sample. Broadley et al. (2020) reported Kr isotope data for two samples of fumarole gas from Brimstone Basin in Yellowstone National Park. Péron et al. (2021) reported the first full suite of Kr isotope data for basalts from Galápagos and Iceland. These data are shown in Figure 14.8. We see that Kr in both basalts and well and fumarole gas forms an array shifted away from the atmospheric and solar values to heavier isotopic compositions. Bearing in mind again that atmospheric contamination is essentially ubiquitous, the array likely at least partly results from mixing

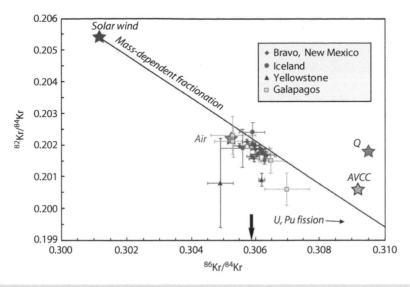

Figure 14.8 Isotopic composition of mantle krypton from CO_2 well gas from the Bravo Dome, New Mexico (Holland et al., 2009), Yellowstone fumaroles (Broadley et al., 2020), and Galápagos and Iceland basalts (Péron et al., 2021) compared to the solar wind, atmospheric Ar, average carbonaceous chondrites (AVCCs), and chondritic component Q. The black arrow shows the $^{86}Kr/^{84}Kr$ for "popping rock" $2\pi D31$ from the Mid-Atlantic Ridge (Péron and Moreira, 2018).

between atmospheric Kr and a heavier pure mantle component. This mixing could reflect the presence of subducted atmospheric Kr in the mantle as well as atmospheric contamination during and after eruption. Péron et al. estimate that the Galápagos and Iceland samples could contain up to 48 and 64%, respectively, recycled atmospheric Kr.

We also see in Figure 14.8 that there is no apparent systematic difference in $^{86}Kr/^{84}Kr$ between upper mantle, represented by the Bravo Dome data and the $2\pi D31$ analysis, and plume-derived krypton from the Galápagos, Iceland, and Yellowstone, although additional data in the future may change this. Assuming some atmospheric contamination and extrapolating away from atmospheric ratios, we see that for a given $^{82}Kr/^{84}Kr$, mantle Kr has lower $^{86}Kr/^{84}Kr$ than the chondritic values represented by both AVCCs and the Q component (Péron et al., 2021).

We can also see the low $^{86}Kr/^{84}Kr$ both in terrestrial atmospheric and mantle krypton relative to chondritic krypton in Figure 14.9. Both terrestrial and chondritic Kr are depleted in light isotopes relative to heavy ones compared to the solar wind due to apparent mass-dependent fractionation, but terrestrial ^{86}Kr is too low to be explained by mass-dependent

fractionation alone. This can also not be explained by fission of ^{238}U or ^{244}Pu as both of which increase, not decrease, $^{86}Kr/^{84}Kr$ (the effect, however, is quite small, $\lesssim 0.1\%$). And, as Péron et al. (2021) point out, the misfit would be larger for ordinary and enstatite chondrites, which have higher $^{86}Kr/^{84}Kr$ than AVCCs. There is also a hint of a deficit in ^{83}Kr relative to a fractionated solar pattern. ^{86}Kr is an r-process nuclide, but the half-life of ^{85}Kr is ≈ 11 years, so ^{86}Kr will also be produced by the s-process depending on neutron fluxes in red giants. Indeed, large variations in the relative abundance of ^{86}Kr are observed in pre-solar SiC grains in chondrites, suggesting such variable production. Figure 14.9 shows, despite large analytical uncertainties, Comet 67P/C-G has large deficits in ^{86}Kr and ^{83}Kr. This, along with the Xe data we discuss in the following section, suggests comets may have delivered a fraction ($\approx 20\%$) of the Earth's heavy noble gasses (Bekaert et al., 2020; Péron et al. 2021), although hydrogen isotopes restrict the cometary contribution to the Earth's hydrogen and carbon to a small fraction. In comparison, the Kr isotopic composition of the Martian atmosphere, based both on analysis of SNC meteorites and direct measurements by the Mars Science Laboratory on board the

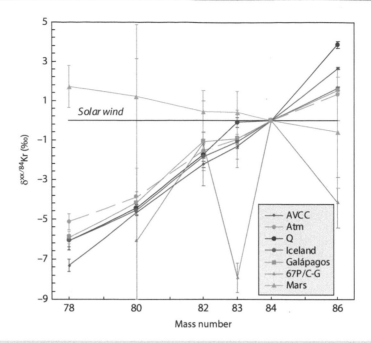

Figure 14.9   Solar wind- and ^{84}Kr-normalized krypton isotopic compositions of terrestrial atmospheric and mantle krypton compared to AVCCs, chondritic component Q, Comet P67/GC (Rubin et al., 2018), and the Martian atmosphere. Other data sources are the same as in Figure 14.8.

Curiosity Rover, is enriched in light isotopes (e.g., Avice et al., 2018). There is some debate about the extent to which this reflects cosmogenic effects, but it appears Mars Kr is more solar-like than either the Earth or chondrites.

14.7   XENON

Xenon has nine stable isotopes: ^{124}Xe, ^{126}Xe, ^{128}Xe, ^{129}Xe, ^{130}Xe, ^{131}Xe, ^{132}Xe, ^{134}Xe, and ^{136}Xe, of which several are produced by decay of extant and extinct radionuclides and as such its isotopic composition is particularly rich in information about Earth's history. Unfortunately, xenon presents an even more difficult analytical challenge than the other noble gases for several reasons. First, its concentration is far lower. For example, ^{130}Xe is two to three orders of magnitude less abundant than ^{3}He and ^{22}Ne and three to four orders of magnitude less abundant than ^{36}Ar. Furthermore, while He and Ar isotope ratios vary by an order of magnitude and more, Xe isotopic variations are in the percent to 10's of percent range, requiring higher-precision analysis. Finally, Xe turns out to be a very "stubborn" gas, readily absorbing onto walls of sampling vessels and analytical equipment,

greatly complicating analysis. Additionally, as is the case for Ne and Ar, atmospheric contamination is virtually ubiquitous. One consequence of this is that there are far fewer data on Xe isotopes than on the other noble gases, although with analytical advances, these data are becoming more common.

The two lightest Xe isotopes, ^{124}Xe and ^{126}Xe, each constitute only about 0.09% of Xe; ^{128}Xe constitutes only 1.9%. These three together with ^{130}Xe (4.1%) are the only Xe isotopes not significantly produced by nuclear processes. The atmosphere is depleted in the light non-radiogenic Xe isotopes relative to the solar wind (Figure 14.10). The extent of depletion is correlated with mass: ^{124}Xe is about three times more depleted relative to ^{130}Xe than is ^{128}Xe; consequently, mass-dependent fractionation is the suspected cause, as it was with the other noble gases. Unfortunately, because of their very low abundance, the lightest Xe isotopes are rarely included in analyses. The ^{124}Xe and ^{128}Xe data from the Earth's interior that do exist, CO$_2$ well gases, fumarole gases, and a few particularly gas-rich MORB and OIB samples, indicate that Xe in the Earth's interior is isotopically heavier than the atmosphere (Figure 14.10). Because both

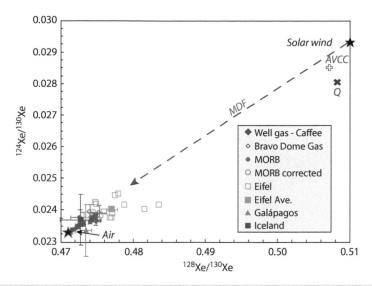

Figure 14.10 $^{124}Xe/^{130}Xe$ and $^{128}Xe/^{130}Xe$ in CO_2 well gases from the Western US (Sheep Mountain and McElmo Dome, Colorado and Bravo Dome, New Mexico) and Caroline, Australia (Caffee et al., 1999; Holland et al., 2009), fumarole gases from the Eifel region of Germany (Caracausi et al., 2016), the average of 32 analyses of gas-rich MORB from the Mid-Atlantic Ridge (Kunz et al., 1998), and average values of basalts from the Galápagos and Iceland (Péron et al, 2021). The open blue circle is Pepin and Porcelli's (2006) recomputed average of the Kunz et al. MORB data that excludes the least reliable analyses. Air, solar wind, AVCCs, and the chondritic Q component are also shown. The red dashed line shows mass-dependent fractionation from air.

the well and MORB gases are probably somewhat contaminated by atmospheric Xe, the true mantle isotope ratios are likely heavier and more "solar" or "meteoritic' in composition than the analyzed values.

The well gases sample a combination of crustal and upper mantle gas and hence are more closely related to the MORB reservoir. Eifel fumaroles, however, likely sample a mantle plume component. The data suggest that plume xenon is shifted more to meteoritic compositions than MORB xenon, but given the limited data and comparably large uncertainties, this inference is tentative. Overall, the data, particularly the Eifel data, suggest that primordial mantle Xe is derived from chondrites rather than the solar wind (Caracausi et al., 2016), but this inference is also tentative.

As we found in Chapter 5, ^{129}Xe excesses were discovered in meteorites in 1960; a few years later, Butler et al. (1963) reported excess ^{129}Xe relative to atmospheric Xe in Bravo Dome CO_2 well gas, which they attributed to decay of ^{129}I in the early Earth. They stated what at the time was a surprising conclusion: "This would mean that outgassing of the earth's interior is incomplete and that ... ^{129}Xe and possibly other daughter products are presently being added to the Earth's atmosphere and upper crust." Subsequently, Staudacher and Allègre (1982) reported elevated $^{129}Xe/^{130}Xe$ in MORB relative to the atmosphere, demonstrating that the convecting mantle has a different Xe isotopic composition from the atmosphere. Since the half-life of ^{129}Xe is only ≈16 Ma, the difference between the atmospheric and mantle $^{129}Xe/^{130}Xe$ ratio must have been established very early in Earth's history, within at most the first 100 Ma. Many subsequent studies of OIB and related xenoliths clearly demonstrate excess ^{129}Xe in OIB sources as well (e.g., Poreda and Farley, 1992; Trieloff et al, 2000). However, maximum $^{129}Xe/^{130}Xe$ ratios in OIB, ≈7, are lower than those found in MORB, ≈7.9 (Figure 14.11); that in turn implies OIBs are derived from a source that had a lower $^{129}I/^{130}Xe$ ratio than MORB. This is consistent with the conclusion drawn from the other noble gas that the OIB source is less degassed than the MORB one. This difference must have been established very early in Earth's history, given the 15.7 Ma half-life of ^{129}I.

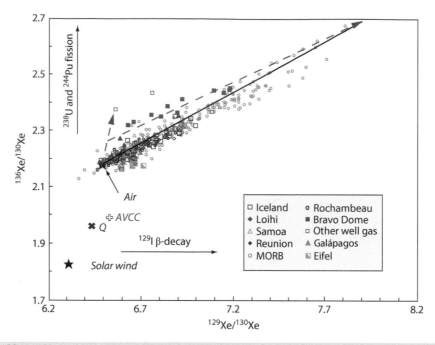

Figure 14.11 $^{129}Xe/^{130}Xe$ and $^{136}Xe/^{130}Xe$ in MORB, OIB, and in CO_2 well gases from the Western US (Sheep Mountain and McElmo Dome, Colorado and Bravo Dome, New Mexico) and Caroline, Australia from Caffee et al. (1999) and Holland and Ballentine (2008) compared to air and meteoritic components. MORB and OIB form an array extending away from air. Excesses in MORB are typically greater than observed in OIB. Some of the well gases (Caroline, Australia and Bravo Dome, Minnesota, US) fall off the trend. Holland and Ballentine argued that these well gases could be modeled as a three-component mixture of air, crustal uranogenic ^{136}Xe, and a mantle component with $^{129}Xe/^{130}Xe$ of ≈7.9 (dashed red lines).

Excesses in the heavy Xe isotopes relative to atmospheric, attributed to ^{238}U fission, were recognized in U-bearing minerals in the late 1940s and in well gas by Butler et al. (1963). Subsequent studies have confirmed these excesses in both MORB and OIB, as shown in Figure 14.11, confirming fissogenic Xe in the Earth's mantle. MORB and OIB form a single $^{136}Xe/^{130}Xe$–$^{129}Xe/^{130}Xe$ with ^{129}Xe array, but with maximum $^{136}Xe/^{130}Xe$, ≈2.5, greater in MORB than in OIB, ≈2.3. Many of the well gases, those from the Eifel region of Germany (Caracausi et al., 2016), are a notable exception, plot to the high ^{136}Xe side of this array, indicative of a high U/Xe, degassed crustal component.

Staudacher and Allègre (1982) argued excesses of $^{134}Xe/^{130}Xe$, and $^{136}Xe/^{130}Xe$ in MORB were due to decay of extinct ^{244}Pu since these ratios correlated with $^{129}Xe/^{130}Xe$. Fission of both ^{238}U and ^{244}Pu produces ^{131}Xe through ^{136}Xe. Ratios of fission yields for ^{238}U are $^{131}Xe/^{132}Xe$ = 0.155 ± 0.006,

$^{134}Xe/^{132}Xe$ = 1.458 ± 0.033, and $^{136}Xe/^{132}Xe$ = 1.76 ± 0.033, with the absolute yield for ^{132}Xe being 3.6 ± 0.4%. Fission yields for ^{244}Pu are $^{131}Xe/^{132}Xe$ = 0.278 ± 0.017, $^{134}Xe/^{132}Xe$ = 1.041 ± 0.016, and $^{136}Xe/^{132}Xe$ = 1.12 ± 0.016. The branching ratio of fission relative to total decay of ^{238}U is $\lambda_{sf}/\lambda_{238}$ = 5.45 × 10^{-7} (in other words, almost all decays occur through α-decay; less than one in a million decays occurs through fission). ^{244}Pu has an absolute yield of ^{132}Xe of 5 ± 0.4% and a branching ratio $\lambda_{sf}/\lambda_{244}$ = 1.25 × 10^{-3}. Although the chondritic initial $^{244}Pu/^{238}U$ ratio was low (≈0.0068; Table 6.1), the higher branching ratio and yield means that in a chondritic reservoir, ^{244}Pu would produce about 27 times as much Xe as ^{238}U over the history of the Solar System. Given that the ^{244}Pu half-life is significantly longer than that of ^{129}Xe, it is certainly reasonable to expect it was present in the early Earth. Nevertheless, discriminating the ^{244}Pu fission contribution from that of ^{238}U is difficult and the existence of plutogenic cannot be established from ^{136}Xe and ^{134}Xe excesses alone.

Several studies have attempted to deconvolve the fissiogenic Xe. The biggest difference in fissiogenic production is between ^{136}Xe and ^{131}Xe: ^{244}Pu produces relatively less ^{136}Xe and more than twice as much ^{131}Xe as ^{238}U; by comparing ^{136}Xe/^{130}Xe and ^{131}Xe/^{130}Xe ratios, we can more clearly distinguish ^{244}Pu production from that of ^{238}U. Comparing the "popping rock" from the Mid-Atlantic Ridge with subglacial Iceland sample DICE 10, Mukhopadhyay (2012) showed that analyses from the Icelandic sample fell along a higher ^{131}Xe/^{130}Xe and ^{136}Xe/^{130}Xe slope than did analyses from the MORB sample; these are shown along with other data in Figure 14.12. The scatter in the data is considerable, but Mukhopadhyay found the difference was significant at the 99% confidence level. He calculated that from 5 to 43% of the ^{136}Xe in MORB was produced by ^{244}Pu fission, depending on the initial composition assumed. Regardless of the initial composition assumed,

he calculated the Icelandic sample had significantly greater amounts, from 47 to 99%, of ^{136}Xe derived from ^{244}Pu fission. This result is consistent with the idea drawn from other noble gases that the Iceland source, as well as other OIB sources, is less degassed than the MORB source because while ^{238}U has continued to produce ^{136}Xe throughout Earth's history, ^{136}Xe derived from ^{244}Pu would have been produced only in the first few hundred million years of Earth's history, again indicating an ancient distinction between MORB and OIB reservoirs. Well gases from various localities define a much shallower slope on this diagram, indicating that fissogenic Xe derives primarily from ^{238}U, consistent with a degassed crustal component in these gases.

Data on other MORB samples, particularly MORB from the equatorial Mid-Atlantic Ridge analyzed by Tucker et al. (2012), show considerably more scatter, plotting both well above and below the 2πD43 regression line, implying variable contributions from ^{244}Pu and ^{238}U fission and perhaps less distinction

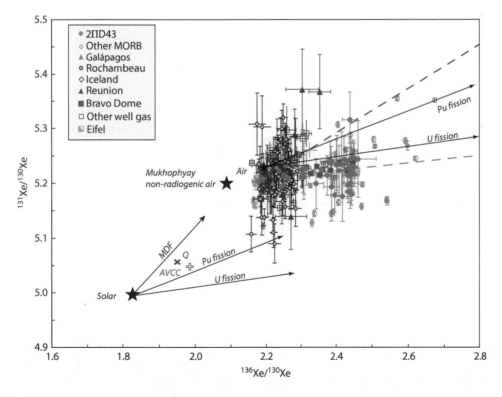

Figure 14.12 ^{131}Xe/^{130}Xe and ^{136}Xe/^{130}Xe in mantle-derived rocks and well gases. Solid arrows show the isotopic evolution of solar and atmospheric Xe resulting from ^{244}Pu and ^{238}U fission. Dashed black and red arrows show the regression slopes calculated by Mukhopadhyay (2012) for MORB sample 2πD43 and Icelandic sample DICE10, respectively.

between the MORB and OIB sources. Tucker et al. (2012) also found that those MORBs that contain an HIMU component (Chapter 6) based on radiogenic Pb isotopic compositions appear to have less radiogenic ^{129}Xe and ^{244}Pu-derived ^{136}Xe than other MORBs. They argued that this reflects a larger component recycled from the surface (through subduction) in the HIMU reservoir than in other ones, consistent with ^{3}He/^{4}He ratios lower than MORB observed in HIMU OIB.

14.8   IMPLICATIONS OF NOBLE GAS ISOTOPE RATIOS FOR THE ORIGIN AND EVOLUTION OF THE EARTH

Noble gases provide particularly useful constraints on two aspects of the Earth: the nature and relationship of chemical reservoirs in the mantle and Earth's formation and earliest history. In this and the following sections, we will summarize those constraints and the theories that have been developed to explain them. It goes without saying that these theories must also account for other chemical and physical information about the Earth and about noble gases in the Solar System.

14.8.1   Reservoirs of noble gases in the solid Earth

Because it has formed through magmatism, the continental crust has been degassed during its formation and its limited noble gas inventory is dominated by a combination of atmospheric and radiogenic gas enriched in ^{4}He, ^{21}Ne, ^{40}Ar, and heavy Xe isotopes produced by ^{238}U fission. While noble gas isotopes can be useful in geochronological studies of the crust as we found in Chapters 2 and 4, our main focus in this chapter is on what noble gases can tell us about Earth's formation and evolution. Our focus is therefore on the mantle, but we must also consider the core as a possible reservoir of primordial noble gases.

14.8.1.1   The core as a reservoir of noble gases

Porcelli and Halliday (2001), among others, have suggested the core may be a major reservoir of noble gases, helium in particular, which then leak into the lower mantle. This idea depends on how noble gases partition between silicate and metal liquids. Relatively low-pressure experiments by Matsuda et al. (1993) suggest the partition coefficients are low, but preliminary diamond anvil experiments at up to 16 GPa and 3000 K by Bouhifd et al. (2013) found He liquid metal–liquid silicate partition coefficients of 4.7×10^{-3} to 1.7×10^{-2}. Bouhifd et al. estimate that this translates into 3–12 $\times 10^{10}$ ^{3}He atoms/g for a total or roughly 1–4 $\times 10^{14}$ moles, implying that roughly 1% of the Earth's initial inventory of ^{3}He could be in core. Roth et al. (2019) found He silicate solid–FeS liquid partition coefficients of 11.8 at 1 Gpa and 1200–1450°C. The same experiments show a much smaller partition coefficient, ≈ 0.01, for Ne. Wang et al. (2022) used first-principles molecular dynamic simulations to model liquid Fe–liquid silicate partitioning at conditions up to core–mantle boundary conditions (5000 K and 135 Gpa) and found that partitioning into metal increases with temperature, rising to values of $D_{He} \approx 10^{-1}$, $D_{Ne} \approx 10^{-3}$, $D_{Ar} \approx 10^{-2}$, $D_{Kr} \approx 10^{-1}$, and $D_{Xe} \approx 10^{0}$. Thus, it seems clearly plausible that the core could contain significant amounts of noble gases. The association of high ^{3}He/^{4}He with low μ ^{182}W among OIBs (Figure 7.31) is also suggestive of a core source for primordial He (Jackson et al., 2020). Recall that ^{182}W is the product of extinct ^{182}Hf (half-life 8.9×10^{6} a) and low μ^{182}W (ppm deviations of ^{182}W/^{184}W from a terrestrial standard) imply low Hf/W ratios, as we expect for the core. If this unradiogenic W originates in the core, the unradiogenic He may also originate in the core, but other explanations are possible.

Olson and Sharp (2022) used these data to model ingassing in the presence of a solar nebula for the first 10 Ma of Earth's formation and calculated that the core could have acquired 3×10^{14} moles of ^{3}He. Although this would have been substantially less than the ^{3}He in the mantle, mantle convection and associated melting and degassing would rapidly draw down the mantle inventory such that after the first ≈ 600 Ma the mantle inventory would fall below that of the core and at present would be so by a factor of 12. This difference in concentration would drive diffusion from the core into the base of the mantle and supply He to mantle plumes generated there. Since the core should have low concentrations of radioactive parents, U, Th, and K, noble gases in the core can be expected to have near-primordial isotopic compositions.

An important possible exception is iodine, which experiments suggest becomes siderophile at high pressure, with metal/silicate partition coefficient as high as 14 (Armytage et al., 2013; Jackson et al., 2018). This of course would lead to high $^{129}Xe/^{130}Xe$ ratios. As $^{129}Xe/^{130}Xe$ ratios are higher in the MORB reservoir than the plume one, this argues against the core being a source of noble gases, or at least Xe, in the latter. Because xenon is a large atom and can be polarized, it is possible that Xe diffusion is sufficiently slower than that of the light noble gases that little would leak from the core. On the other hand, Jackson et al. (2018) point out that variable extraction of iodine during core formation could account for some of the variability in mantle $^{129}Xe/^{130}Xe$ ratios. In particular, loss of I to the core might explain the lower $^{129}Xe/^{130}Xe$ ratios in the plume reservoir than the MORB one. At this point, there is a well-established possibility of the core being a reservoir of noble gases, but more research is required to establish its probability.

14.8.1.2 Mantle reservoirs

It is clear from the previous sections that the mantle plumes, the source of oceanic island basalts, and the upper mantle source of MORB have distinct isotopic compositions. It is nevertheless difficult to establish the compositions of both, mainly due to ubiquitous atmospheric contamination. For He, where atmospheric contamination is not a problem (although radiogenic ingrowth and diffusional loss can be), there are enough data on MORB to suggest about mean $^4He/^3He$ of $90.7 \pm 13.2 \times 10^3$ ($^3He/^4He$ of value 8.2 ± 1.5 R/R_A) and the data form a nearly Gaussian distribution about this value. On the other hand, considering just the minimum $^4He/^3He$ in each island chain, mantle plumes range from 123.1×10^3 in St. Helena to 14.5×10^3 in Baffin Island (R/R_A from 5.87 to 50); hence, the plume reservoir is clearly heterogeneous reflecting variable time-integrated (U +Th)/He ratios and hence degassing extent. We want to focus on the primordial components in plumes, and we will adopt this $^4He/^3He$ value of 14.5×10^3 or 50 R/R_A as the primordial endmember component in plumes.

There is a well-defined maximum $^{20}Ne/^{22}Ne$ of 12.5 in high-precision MORB data that corresponds to a range in $^{21}Ne/^{22}Ne$ of

0.059–0.066, and we will adopt these values for the upper mantle. The maximum $^{20}Ne/^{22}Ne$ observed in OIB is 13.03 ± 0.04 in the Discovery Seamount basalts; projecting this value to the steepest $^{21}Ne/^{22}Ne-^{20}Ne/^{22}Ne$ array in Figure 14.6, we obtain a $^{21}Ne/^{22}Ne$ of 0.034. However, we need to keep in mind that various island chains define distinct $^{21}Ne/^{22}Ne-^{20}Ne/^{22}Ne$ arrays with maximum $^{20}Ne/^{22}Ne$ decreasing with decreasing $^{21}Ne/^{22}Ne$ ratios, indicative of heterogeneity in the plume reservoir.

Although Burnard et al. (1997) measured a $^{40}Ar/^{36}Ar$ ratio of 64000 ± 15000 in the popping rock, 2Dπ43, we will adopt the higher-precision value of Raquin et al. (2008) in the same rock of 36 500 as the upper mantle value, but we should also recognize there is likely considerable heterogeneity in the upper mantle $^{40}Ar/^{36}Ar$ ratio reflecting variable degassing and K/Ar ratios. The highest $^{40}Ar/^{36}Ar$ ratio measured in OIB is 8340 from Loihi, although Mukhopadhyay (2012) inferred a higher value of 10 700 in Iceland based on Ar–Ne relationships and assuming a $^{20}Ne/^{22}Ne$ of 13.8, a value likely to be too high, for the plume endmember. Available data are not precise or sufficient enough to establish a difference in $^{38}Ar/^{36}Ar$ in these reservoirs, although the mantle generally does appear to have slightly heavier Ar than the atmosphere.

At present, there are insufficient Kr data to distinguish upper mantle and plume reservoir isotopic compositions, although such a distinction may emerge in the future. The mean of all mantle data is $^{86}Kr/^{84}Kr \approx 0.306$ and $^{82}Kr/^{84}Kr \approx 0.2015$, which is clearly resolved from atmospheric Kr. In contrast, there are sufficient data on the heavier Xe isotopes to clearly establish a distinction between upper mantle and plume reservoirs. Maximum $^{129}Xe/^{130}Xe$ and $^{136}Xe/^{130}Xe$ in MORB are, respectively, 7.8 and 2.67 and Holland and Ballentine inferred slightly higher ratios of 7.9 and 2.7 from well gas data. In contrast, maximum $^{129}Xe/^{130}Xe$ and $^{136}Xe/^{130}Xe$ in OIB (Iceland) are, respectively, 7 and 2.32. Furthermore, OIB fall along a greater $^{131}Xe/^{130}Xe-^{136}Xe/^{130}Xe$ slope than MORB, implying a greater dominance of Pu fission products over U fission products.

Table 14.3 summarizes the noble gas isotopic compositions of the atmosphere and the *inferred* compositions of the upper mantle

Table 14.3   Noble gas isotopic composition of terrestrial and other reservoirs.

	Air	Upper mantle	"Primordial" OIB reservoir	Solar	Carbonaceous chondrites (Q)
^{4}He/^{3}He	7.236×10^5	90.72×10^3	$14.5\text{–}123.1 \times 10^3$	2.62×10^3	6.28×10^3
^{3}He/^{4}He (R/R_A)	$\equiv 1$	8.2 ± 2.5	$5\text{–}50$	334	120
^{21}Ne/^{22}Ne	0.0290	≈ 0.06	≈ 0.042	0.0329	0.0294
^{20}Ne/^{22}Ne	9.8	≈ 12.50	≈ 13	13.78	10.7
^{40}Ar/^{36}Ar	298.56	$\approx 36\,000$	$<10\,700$	3×10^{-4}	≈ 0.78
^{38}Ar/^{36}Ar	0.1876	0.1896 ± 0.0031	01882 ± 0.0007	0.1828	0.1880
^{82}Kr/^{84}Kr	0.2022	≈ 0.2015	≈ 0.2015	0.2048	0.2018
^{86}Kr/^{84}Kr	0.3024	≈ 0.3063	≈ 0.3063	0.3024	0.3095
^{124}Xe/^{130}Xe	0.0238	0.0239	0.024	0.0293	0.0281
^{128}Xe/^{130}Xe	0.4715	0.4752	0.477	0.5083	0.5077
^{129}Xe/^{130}Xe	6.50	7.9	7.0	6.286	6.436
^{131}Xe/^{130}Xe	5.21	5.36	5.30	4.996	5.056
^{132}Xe/^{130}Xe	6.61	7.0	6.76	6.047	6.177
^{136}Xe/^{130}Xe	2.176	2.7	2.32	1.797	1.954

source of MORB and more isolated "primordial" lower mantle reservoir sampled by the mantle plumes that produce OIB. While this reservoir may not be truly primordial, it is less processed than other ones and appears to have been isolated from the remainder of the mantle for most of Earth's history. Circumstantial evidence discussed in Chapter 7 and Section 14.3.3 suggests this reservoir is associated with the LLSVPs at the base of the mantle; alternatively, this reservoir may be the core. For comparison, Table 14.3 also lists solar wind, chondritic, and solar where it can be distinguished from solar wind, noble gas isotope ratios from Table 14.2. As we discussed in Section 14.1.2, noble gases are isotopically heterogeneous in chondritic meteorites, and furthermore, meteorites differ in noble gas isotopic composition from the solar wind. Consequently, unlike the elements we considered in previous chapters, we cannot infer the isotopic composition of the primitive Earth from a specific meteoritic value. Indeed, as we shall see, noble gases in the Earth are likely to have been derived from a mixture of several distinct isotopic components, and some of the isotopic variation we observe today may still reflect that initial isotopic heterogeneity.

Comparing noble gas isotope ratios of the mantle reservoirs in Table 14.3, we see that the upper mantle is generally enriched in isotopes produced by radioactive decay or other nuclear processes: ^{4}He, ^{21}Ne, ^{40}Ar, ^{129}Xe, and the heavy fissogenic Xe isotopes relative to the plume reservoir. The simplest way to explain all these differences is that the upper mantle has lost more of its initial inventory of noble gases than the plume reservoir, i.e., it has experienced a greater extent of degassing so that the radiogenic isotopes are more dominant. An alternative interpretation is that He and other noble gases are less incompatible, i.e., partition less readily into the melt, than U and Th and therefore that melt extraction decreases the (U+Th)/He, K/Ar, and U/Xe ratios and consequently melt-depleted mantle should have higher ^{3}He/^{4}He than more primitive mantle. Recent studies show that this is unlikely (e.g., Graham et al., 2016). In addition, the correlation of melt-depleted Sr, Nd, and Hf isotopic signatures with degassed He, Ne, Ar, and Xe signatures in MORB confirms melt removal is associated with degassing and an important control on noble gas isotope ratios. As we shall see, however, degassing alone cannot explain all the observed isotopic variation in noble gases; primordial heterogeneity also exists.

14.8.2   Subducted atmospheric noble gases in the mantle

There are some exceptions to this pattern of difference between MORB and OIB: most notably, "HIMU"-type OIB (Chapter 7) tend to have lower ^{3}He/^{4}He ratios than MORB (Figure 7.30). The Australs, the only HIMU islands for which Ne and Ar data are available, also have more nucleogenic Ne than MORB (Figure 14.6), as well as relatively low

^{20}Ne/^{22}Ne and low ^{40}Ar/^{36}Ar ratios. Low ^{3}He/^{4}He is also observed in Tristan da Cunha and appears to be present, along with lower ^{4}He/^{3}He values, in basalts from other islands, such as the Azores and Pitcairn. As we found in Chapter 7, there is clear evidence of a surficial component in many OIB, and this could well explain the examples of high ^{4}He/^{3}He, ^{40}Ar/^{36}Ar, and ^{21}Ne/^{22}Ne, because we expect material recycled from the surface to be thoroughly degassed.

As we have seen, atmospheric contamination is a pervasive problem in analysis of noble gases heavier than helium. But is all the atmospheric gas found in basalts and xenoliths simply contamination? Beginning with Sarda (2004), a consensus has emerged that while He cannot be subducted into the mantle because of its high diffusivity, heavier noble gases, Ar, Kr, and Xe can be and Ne might be. Holland and Ballentine (2006) noted the rare gas elemental patterns in the Bravo Dome gases were similar to that of seawater and argued that subduction controls the isotopic composition of subducted non-radiogenic heavy (Ar, Kr, Xe) noble gases. They suggested that over time subduction could supply the entire mantle inventory of Ar and Kr and 80% of the Xe. Serpentine–brucite inclusions within olivine from the Sangabawa metamorphic belt of Japan provide evidence to support this view. The peridotite body has been exhumed from the depths of $\approx$120 km. The olivines in the peridotite contain inclusions of serpentine and brucite, which they concluded originally formed by hydration of mantle wedge peridotite with water released from the subducting oceanic lithosphere below. Sumino et al. (2010) found that the relative abundances of noble gases and halogens in these inclusions were strikingly similar to those found in marine pore waters. Heat and pressure subsequently transformed most of the serpentinite to peridotite again. Parai and Mukhopadhyay (2015) estimate that as much as 80–90% of the Xe in the MORB and mantle plume sources is recycled atmospheric Xe. Péron et al. (2021) estimate that as much as $\approx$65% of the Kr and 95% of the Xe in the Galápagos plume source are atmospheric derived.

Finally, it is interesting to note that among MORBs, Parai et al. (2012) found that after correcting for atmospheric contamination,

^{40}Ar/^{36}Ar and ^{129}Xe/^{130}Xe ratios show extensive variation in contrast to a very limited range of ^{4}He/^{3}He. These variations of 70 and 80% of the entire mantle range of ^{40}Ar/^{36}Ar and ^{129}Xe/^{130}Xe are not consistent with variable degassing, but are readily explained by variable incorporation of recycled atmospheric heavy noble gases into the MORB source.

14.8.3 Mantle noble gas budgets

Noble gas budgets present a far more difficult challenge than those for lithophile elements for several reasons. First, as we earlier noted, chondrites provide no strong constraints on either isotopic compositions or concentrations in the bulk Earth. Second, degassing and atmospheric contamination complicates matters. However, since the concentrations of K, U and Th are at least somewhat constrained in the Earth and mantle, we can estimate the production rates and inventories of the radiogenic and nucleogenic daughters, ^{40}Ar, ^{4}He, and ^{21}Ne. In the case of argon, we also know how much ^{40}Ar is in the atmosphere and, as we previously noted, most of this is radiogenic. We can use this in a mass balance calculation to estimate how much Ar has been lost from the mantle. Because the He is lost from the atmosphere, we cannot do a similar mass balance, but the rate at which He is lost from the mantle to the atmosphere provides some useful constraints. Because the production of ^{4}He and ^{21}Ne is closely linked, we can also place constraints on the Ne budget. We will begin by considering the Ar mass balance, and then examine the He and Ne budgets.

^{40}Ar mass balance constrains the fraction of terrestrial argon present in the mantle (e.g., Allègre et al., 1996). Exact concentrations are difficult to know, except in the atmosphere, so these are "back-of-the-envelope" calculations; the results are nonetheless instructive. For example, assuming a bulk silicate Earth K concentration of 240 ppm (McDonough and Sun, 1995), we can calculate that 3.34×10^{18} moles of radiogenic ^{40}Ar have been produced in the Earth over the last 4.5 Ga. The atmosphere contains 1.69×10^{18} moles of ^{40}Ar, which is essentially entirely radiogenic. This implies that $\approx$51% of the radiogenic ^{40}Ar is now in the atmosphere. Using the Rudnick and Gao (2014) estimate of 1.8% K_2O in the continental crust and assuming a mean K/Ar age of the crust of 1 Ga, we can calculate that there should be 7.56×10^{16} moles ^{40}Ar in the crust, or only a little over 2% of the Earth's budget. Using other estimates of the crustal K concentration and perhaps a somewhat older K–Ar age of the crust, we might be able to increase the amount of ^{40}Ar in the crust to 4 or 5%, but no more. Thus, under these assumptions, more than 45% of the radiogenic Ar must be in the mantle.

We can change these numbers somewhat by changing assumptions. For example, Lyubetskaya and Korenaga (2007) estimate the bulk silicate Earth K concentration at only 190 ppm. Such a lower concentration could be consistent with the Earth's non-chondritic $^{142}Nd/^{144}Nd$. This implies a ^{40}Ar inventory in the Earth of 2.64×10^{18} moles, of which 64% would be in the atmosphere and 2.8% in the crust, leaving $\approx 33\%$ in the mantle.

Now let us calculate how much of this ^{40}Ar could be in the "upper mantle" (i.e., MORB source) reservoir. One way to approach this is to use the concentrations in the popping rock measured by Moreira et al. (1998), which was about 1.4×10^{-6} moles/kg ^{40}Ar. Assuming MORB is produced by 10% mantle melting, this implies a concentration of 1.4×10^{-7} moles $^{40}Ar/kg$; if this concentration represents the entire mantle, it corresponds to 16.8–21.2% of the Earth's ^{40}Ar inventory, depending on the assumed K concentration, but only 4.5–5.7% of the Earth's inventory if the MORB source is restricted to the upper mantle. Alternatively, we can use the estimated ^{4}He flux into the oceans from mid-ocean ridge volcanism (4.33×10^{7} moles/yr; Bianchi et al., 2010), a $^{4}He^*/^{40}Ar^*$ production rate of 2 (corresponding to a reservoir age of ≈ 2 Ga), then the ^{40}Ar flux from oceanic crust creation is 2.17×10^{7} mole/yr. The oceanic crust production rate is $\approx 5.6 \times 10^{19}$ kg/yr. From this we calculate a ^{40}Ar concentration of 3.86×10^{-7} mole/kg in the magma and again assuming 10% melting this implies a mantle concentration of about 3.86×10^{-8} mole/kg. If the MORB source occupies the entire mantle, this amounts to 1.55×10^{17} moles or about 4.6–5.9% of the Earth's Ar inventory and 1.3–1.6% of the Earth's total if the MORB source occupies only mantle above the 660 km discontinuity, this would amount to 4.17×10^{16} moles or only about 1.2% of $^{40}Ar^*$. Using this concentration of $\approx 4 \times 10^{-8}$

moles $^{40}Ar/kg$ for the depleted upper mantle and a $^{40}Ar/^{36}Ar$ ratio of 36 000, we obtain a depleted mantle ^{36}Ar concentration of 1.1×10^{-12} moles $^{36}Ar/kg$. In contrast, assuming a K concentration of 240 ppm, a primordial reservoir with a present $^{40}Ar/^{36}Ar$ of 9000 would contain 9.2×10^{-11} moles $^{36}Ar/kg$, implying a factor of ≈ 80 depletion in the upper mantle.

We should emphasize that the calculation of what fraction of the Earth's ^{40}Ar inventory is in the mantle is not affected by the question of how much has been recycled; the mass balance question is insensitive to history. However, it does affect the question of how much has been degassed. Much of the ^{40}Ar now in the upper mantle, and indeed the deep mantle as well, may be subducted atmospheric Ar. Consequently, much more of the mantle may have been degassed than the 50% estimated by Allègre et al. (1996).

It is also instructive to consider the budget for a non-radiogenic isotope, such as ^{36}Ar. In this case, there is no easy way to estimate the ^{36}Ar content of the bulk silicate Earth, but we do know how much ^{36}Ar is in the atmosphere and we can estimate how much is in the MORB source. There are 5.73×10^{15} moles of ^{36}Ar in the atmosphere. Moreira et al. (1998) measured 1.79×10^{-11} moles/kg in the popping rock, which translates to a mantle concentration of 1.79×10^{-12} moles/kg assuming again 10% melting. At these concentrations, the mantle above the 660 km discontinuity would contain 1.93×10^{13} moles, or just 0.3% of the amount in the atmosphere. If this concentration characterizes the entire mantle, it still amounts to only 1.2% of the amount in the atmosphere. Since we have already concluded that at least part of the lower mantle must be less degassed than the MORB reservoir, its concentration is likely to be higher, but nonetheless it would appear that the entire mantle has lost much of its inventory of non-radiogenic noble gas isotopes. Halliday (2013) estimates that 97% or more the Earth's ^{36}Ar inventory is in the atmosphere.

Turning to helium, it is interesting to compare the rate at which ^{4}He is being lost from the upper mantle to the rate at which it is being created. Bianchi et al. (2010) estimated the degassing flux of ^{3}He at mid-ocean ridges at 527 moles/yr. Using the average $^{3}He/^{4}He$ of 8.8 R/R_A ($^{4}He/^{3}He = 8.21 \times 10^{4}$), we calculate a ^{4}He flux of 4.33×10^{7} moles/yr. Provided we know its mass and uranium and thorium concentration, the rate at which ^{4}He is being produced is calculated as:

$$\frac{d^4He}{dt} = {}^{238}U\left\{8\lambda_{238} + \frac{{}^{235}U}{{}^{238}U}\lambda_{235} + 6\kappa\lambda_{232}\right\}M_R \tag{14.2}$$

where ^{238}U is the (molar) concentration, κ is the $^{232}Th/^{238}U$ ratio, and M_R is the reservoir mass, in this case the depleted mantle. Salters and Stracke (2004) estimated the U concentration of the depleted mantle as 4.7 ppb and κ as 2.91, which corresponds to a production rate of 1.8×10^{-19} moles/g/yr. Taking the mass of the depleted mantle as the seismically defined upper mantle (1.08×10^{27} g), the production rate is 1.94×10^8 moles/yr. In this case, the "degassing efficiency," the ratio of loss to production is 22.3%. Workman and Hart (2005) estimated a lower U concentration for the depleted mantle (3.2 ppb), which would imply a degassing efficiency of 33%. Since the geochemical evidence is that subduction-related magmas are derived largely from depleted mantle, we should add to this number the flux from subduction zone volcanoes, which Porcelli and Ballentine (2002) estimate at 0.6 to 2.3×10^7 moles/yr, which would imply a degassing efficiency as high as 34%.

We can also use Equation (14.2) to calculate the total mantle radiogenic ^{4}He production. Using the McDonough and Sun (1995) U concentration of 20 ppb less the amount in the crust (1.3 ppm; Rudnick and Gao, 2003) and a mantle mass of 4×10^{27}g, this is 2×10^9 moles/yr. If we consider that degassing occurs only at mid-ocean ridges, then the degassing efficiency is 2.2%. This rises only to 2.8% if we use the bulk silicate Earth concentrations of Lyubetskaya and Korenaga (2007). Of course, volcanism elsewhere also degasses the mantle. Porcelli and Ballentine estimate that rate of degassing through oceanic island volcanism at 0.4 to 7.3×10^7 moles/yr. Considering all volcanic degassing and using the Lyubetskaya and Korenaga (2007) U and Th concentrations, the mantle ^{4}He degassing efficiency would reach a maximum of 8.9%, so we can conclude that ^{4}He must be accumulating in the mantle. In contrast, the rate at which heat is being lost from the mantle ($\approx$30 TW) is substantially greater than it is

being produced ($\approx$12 TW using McDonough and Sun concentrations and 8.7 TW using Lyubetskaya and Korenaga concentrations). O'Nions and Oxburgh (1983) first pointed out this dichotomy between the loss of He and heat from the Earth.

We can also place some constraints on the evolution of ^{3}He/^{4}He in the less degassed reservoir sampled by mantle plumes. Let us first suppose that such a reservoir is undegassed and primordial and has the U and Th concentrations of the bulk silicate Earth. Using a primordial ^{4}He/^{3}He = 6021 (^{3}He/^{4}He = 120 R/R_A) and the highest value measured in plume material of ^{4}He/^{3}He = $\approx$14500 (50 $R/R)_A$ in 60 million year old primitive basalts produced by melting of the Iceland mantle plume and now found on Baffin Island (Starkey et al., 2009), the ratio of ^{4}He*/^{3}He is then:

$$\frac{^4He^*}{^3He} = \left(\frac{^4He}{^3He}\right)_{today} - \left(\frac{^4He}{^3He}\right)_0 = 8430 \tag{14.3}$$

Assuming 20 ppb and 80 ppb U and Th concentrations, respectively (McDonough and Sun, 1995), this reservoir would produce 1.54×10^{-9} moles/g ^{4}He over 4.5 Ga. From these values we can calculate a ^{3}He concentration of 1.82×10^{-13} moles/g. This concentration assumes a closed system, i.e., no loss of He over 4.5 Ga and is therefore a *minimum* initial concentration. Porcelli and Ballentine (2002) using a similar approach but different values calculate a concentration of 1.26×10^{-13} moles/g for the Icelandic source. For comparison, the CI chondrite He concentration is 56 nL/g or 2.5×10^{-9} moles/g (Palme and O'Neill, 2003). Assuming an initial ^{3}He/^{4}He of 120 R/R_A, we can calculate a concentration of ^{3}He 3.02×10^{-13} moles/g, implying a terrestrial concentration of 0.39 times CI chondrites. Halliday (2013) estimated the primordial terrestrial ^{3}He concentration at 1.6 to 1.9×10^{-13} moles/g, which compares to his estimated "chondritic" concentration of 3.8×10^{-13} moles/g, corresponding to a factor of $\approx$3 depletion. Given the uncertainties, the similarity in these values is more surprising than the difference.

How much of this He could be in the depleted mantle, assuming it occupies only the region above 660 km? We noted the present estimated U concentrations of 3.2–4.7 ppb U, but of course these have decreased through time from the presumed initial primordial values as U is extracted through melting. This depletion is likely a complex function of time, but building a simple model is nonetheless instructive. We could suppose that U concentration decreases exponentially from a primordial concentration of 20 ppb to a present one of 4.7 ppb. In this case, the depleted mantle would have generated 1.42×10^{-9} moles/g ^{4}He*. Using Equation (14.3), we can calculate a ^{3}He concentration of 1.87×10^{-14} moles/g, or about an order of magnitude less than what we estimated for "primordial mantle." If we assume the upper mantle attained its present low U concentration 4.5 billion years ago, in this case we estimate a concentration of 5×10^{-15} moles/g ^{3}He. In both cases, however, we assume closed-system evolution of ^{4}He and ignore losses through degassing, so these are overestimates. Taking account of the "degassing efficiency" of ^{4}He calculated above, the actual concentration would be a factor of three lower, assuming steady-state degassing. If the depleted mantle occupies only the mantle above 660 km, only 0.25–0.75% of the Earth's ^{3}He is in this reservoir. Porcelli and Ballentine (2002), again using a different approach, estimate the depleted mantle ^{3}He concentration at $2–7.6 \times 10^{-15}$ moles/g, which overlaps with the above estimates, while Porcelli and Elliott (2008) estimate a lower concentration of 1.44×10^{-15} moles/g.

It seems unlikely that either degassing of He or depletion of U and Th in the mantle has been steady state, particularly since these processes should be linked to heat production in the Earth and creation of the continental crust, both of which have declined over time. Porcelli and Elliott (2008) considered a variety of models where the degassing flux is a function the melting rate beneath mid-ocean ridges and proportional to the mantle He concentration. They assume the melting rate declines exponentially over time, so that the flux can be expressed as:

$$\dot{F}(t) = -\alpha e^{-\beta t} \left[^3\text{He}(t) \right]_M \qquad (14.4)$$

where $\dot{F}(t)$ is the ^{3}He degassing flux at time t, α is flux of mantle passing through the MOR melting regime, β is constant describing the

rate of decrease of melting through time, and $[^3\text{He}(t)]_M$ is the mantle helium concentration at time t. They assume that fractions of He and U removed from the mantle by melting are the same. An important conclusion they derive from these models is that any present reservoir with ^{3}He/^{4}He of 50 R/R_A must have been isolated from the depleted mantle and have remained closed to He loss since at least 3 Ga, and likely longer. While the "isolation time" depends on the assumed initial helium concentration, it is largely insensitive to the initial ^{3}He/^{4}He ratio.

Honda and McDougall (1998) pointed out that we can use the coupled production of radiogenic ^{4}He and nucleogenic ^{21}Ne to estimate time-integrated ^{3}He/^{22}Ne ratios. This approach depends only on Ne and He isotopic compositions, and therefore avoids the problem of differential degassing of noble gases during eruption. As we noted earlier the ^{21}Ne*/^{4}He* production ratio (we again use the asterisk to designate radio/nucleogenic) is constant at about 4.5×10^{-8}. We can calculate the ^{4}He*/^{3}He ratio from Equation (14.4). To determine the ^{21}Ne*/^{22}Ne ratio we assume that the initial ^{21}Ne/^{22}Ne falls on the mass-dependent fractionation line passing through the solar value. The ^{21}Ne*/^{22}Ne ratio is then the difference between the measured ^{21}Ne/^{22}Ne and the value of ^{21}Ne/^{22}Ne on the fractionation line corresponding to the measured ^{20}Ne/^{22}Ne. This is illustrated in Figure 14.13.

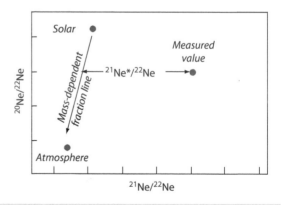

Figure 14.13 Cartoon illustrating the computation of the ^{21}Ne*/^{22}Ne ratio. The initial ratio is assumed to lie on a mass-dependent fractionation line passing through the solar isotopic composition. The ^{21}Ne*/^{22}Ne ratio is the offset from that line.

We can calculate the ^{3}He/^{22}Ne ratio as follows:

$$^3\text{He}/^{22}\text{Ne} = \frac{(^{21}\text{Ne}^*/^{22}\text{Ne})_m}{(^{21}\text{Ne}^*/^4\text{He}^*)} \times \frac{1}{(^4\text{He}^*/^3\text{He})}$$

$$(14.5)$$

where (^{4}He*/^{3}He) is calculated from Equation (14.3), (^{20}Ne/^{22}Ne)$_m$ is the measured ratio, and ^{21}Ne*/^{4}He* is the above production rate. The value we assume for the initial ^{4}He/^{3}He depends on whether we assume that the mantle's initial inventory of He was primordial planetary ^{4}He/^{3}He =6021 (R/R_A=120) or solar ^{4}He/^{3}He =2760 (i.e., post-deuterium burning, R/R_A= 262). Our result will depend on our assumption, but as we shall see, not strongly.

We also need to consider that most, if not all, samples have experienced some degree of atmospheric contamination, so as usual it is the maximum values that are of interest. Only a few data sets are complete and accurate enough to yield useful calculated ^{3}He/^{22}Ne ratios; these are shown plotted against ^{22}Ne/^{20}Ne in Figure 14.14. Each data set forms a distinct array that terminates with ^{3}He/^{22}Ne values ranging from 1.56 for the Galápagos dredges related to Volcán Fernandina and 2.61 for Iceland to 8.61 for "Popping Rock" (2ΠD3). Loihi and Baffin Island form more

poorly defined arrays terminating at ^{3}He/^{22}Ne values of 4.65 and 4.56, respectively. Galápagos dredges related to Volcán Ecuador form an array more similar to MORB, just as it does in ^{20}Ne/^{22}Ne–^{21}Ne/^{22}Ne space (Figure 14.6). If we extrapolate the trends to the solar ^{20}Ne/^{22}Ne value of 13.34, we obtain ^{3}He/^{22}Ne values of 1.69, 2.68, 5.8, and 6.8 for Fernandina, Iceland, Loihi, and Baffin, respectively. Popping Rock data extrapolated to ^{20}Ne/^{22}Ne = 12.5 have ^{3}He/^{22}Ne ≈ 8.

Now let us compare these values with actual measured ones. Dygert et al. (2018) found that the mean ^{3}He/^{22}Ne of 57 MORB samples was 7.5 ± 1.5 while the mean of 38 OIB was 4.4 ± 3.1 with a strongly skewed distribution and a mode of ≈2 (Figure 14.15). Our estimates of time-integrated ^{3}He/^{22}Ne thus fall within the range of observed present ratios.

The OIB mode is similar to the solar value (1.46) and the chondritic value (0.9), suggesting that the endmember OIB component is indeed primordial, but the MORB value is well above this. Dygert et al. proposed that during melting of mantle rising beneath mid-ocean ridges He may diffuse hundreds of meters into wall rock on timescales relevant to ocean crust generation due to its high diffusivity of He, while Ne is effectively immobile, producing an oceanic mantle lithosphere regassed with respect to He but not Ne and hence with a

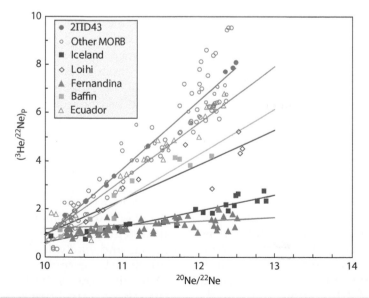

Figure 14.14 Calculated ^{3}He/^{22}Ne ratios as a function of observed ^{20}Ne/^{22}Ne assuming "primordial" isotopic compositions. Lines emphasize individual trends that reflect atmospheric contamination. Mantle plumes appear to have variable He/Ne ratios that are lower than the He/Ne ratio of the MORB reservoir.

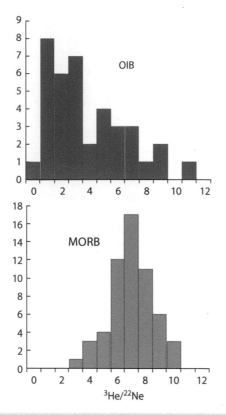

Figure 14.15 Histograms of measured ^{3}He/^{22}Ne ratios in MORB and OIB for trends extrapolated to ^{22}Ne/^{20}Ne to 12.5. Data compilation from Dygert et al. (2018).

net elevated ^{3}He/^{22}Ne. If subduction of this oceanic lithosphere feeds primarily into the depleted mantle reservoir, it would raise the ^{3}He/^{22}Ne. Mixing of this recycled lithosphere with primordial material in the deep mantle reservoir would produce the skewed distribution seen in Figure 14.15. Alternatively, because He is more soluble in melts than Ne, the high ^{3}He/^{22}Ne may be signature of prior degassing. The similarity of time-integrated ^{3}He/^{22}Ne ratios to present ones suggests this an ancient event and may reflect preferential solubility of He during degassing of a primordial magma ocean (Moreira, 2013).

Based on the above estimates of ^{3}He in the mantle, we can calculate Xe concentrations as well. Figure 14.16 shows that the data of Péron et al. (2018) for popping rocks of the Mid-Atlantic Ridge and the DICE data from Iceland of Mukopadhyay (2012) show particularly strong correlations between ^{129}Xe/^{130}Xe and ^{3}He/^{130}Xe, consistent with mixing between atmosphere and a single mantle component.

If we extrapolate regression lines for these two data sets to our inferred ^{129}Xe/^{130}Xe of 7.9 and 7 for the upper mantle and plume reservoirs, respectively, we obtain ^{3}He/^{130}Xe ratios of 872 and 924 for the upper mantle and Iceland sources. These ratios are essentially identical given uncertainties and suggest similar degassing efficiencies for the two gases. Given the variation of He/Ne ratios apparent in Figure 14.15, this is surprising. Using 2×10^{15} ^{3}He atoms/g for the MORB source we obtain a ^{130}Xe concentration of 2.3×10^{-18} moles ^{130}Xe/g for the depleted mantle and 2.3×10^{-15} moles ^{130}Xe/g for the Iceland source. Thus, despite the similarity in He/Xe ratios, the upper mantle is far more depleted in Xe than the plume reservoir.

14.8.4 Preserved noble gas heterogeneity in the Earth

Much of the variation in mantle noble gas isotope ratios can be explained by variable degassing, radioactive decay, and subduction of atmospheric and crustal gases, but not all the variability can be explained this way. Furthermore, while many previous studies had proposed that noble gases could be supplied to the degassed upper mantle from the relatively undegassed deep mantle reservoir (e.g., Kellogg and Wasserburg, 1990; Porcelli and Wasserburg, 1995), it is now clear that differences in noble gas isotope compositions between the atmosphere, the upper mantle, and plume reservoirs were established early in Earth's history and exchange between mantle reservoirs must be quite limited. Although the plume reservoir clearly contains recycled components and may indeed be dominated by these, it also contains a relatively primitive component or components that have remained isolated from the upper mantle and atmosphere for much of Earth's history. Let us briefly review this evidence.

Beginning with He, we note that any present reservoir with ^{3}He/^{4}He of 50 R/R_A must have been isolated from the depleted mantle and have remained closed to He loss since at least 3 Ga, and likely longer (Porcelli and Elliot, 2008). The association of high ^{3}He/^{4}He in OIB with low μ^{182}W (Section 7.4.2.4) and the absence of variation in μ^{182}W among MORB suggest isolation of the high ^{3}He/^{4}He reservoir from depleted mantle, essentially since the Earth's earliest history.

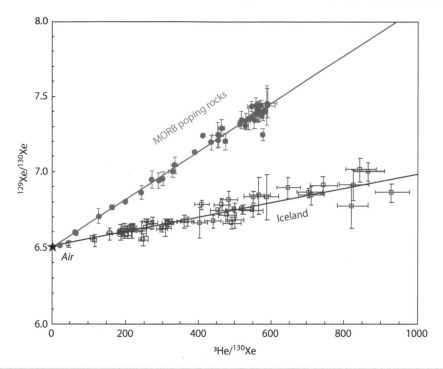

Figure 14.16 ^{129}Xe/^{130}Xe ratios plotted against ^{3}He/^{130}Xe for DICE samples from Iceland (Muckhopadhyay, 2012) and "popping rocks" from 14°N on the Mid-Atlantic Ridge (Péron et al., 2019).

Turning to Ne, we found that several OIB arrays extend to ^{20}Ne/^{22}Ne ratios of ≈13, which is well above the Neon B value found in meteorites and approaches the 13.34 solar value. In contrast, the MORB array extends only to ^{0}Ne/^{22}Ne ≈ 12.5, matching the meteoritic Neon B value. Both of these compositions are dramatically heavier than atmospheric Ne. Additionally, as Honda and McDougall (1998) pointed out, the differences in ^{3}He/^{22}Ne we calculated above are the *time-integrated* one and the differences could not survive if there were large fluxes of noble gas between these reservoirs.

The difference in ^{40}Ar/^{36}Ar between is also dramatic, with the MORB source having ^{40}Ar/^{36}Ar of at least 36 000 and the plume source having a ratio of <10 700, both of which are dramatically greater than the atmospheric value (Figure 14.7). The MORB source also appears to be a factor of ≈80 depleted in ^{36}Ar compared to the plume source. These distinctions also require long isolation times between these reservoirs. Terrestrial ^{38}Ar/^{36}Ar are heavier than solar values and mantle values appear to be slightly heavier than atmospheric.

Xenon isotopes provide some of the strongest evidence for long-term heterogeneity in the mantle. Maximum ^{129}Xe/^{130}Xe ratios in MORB are higher than those in OIB. Given the ≈16 Ma half-life of ^{129}I, this difference must have been established in the first ≈100 Ma of Earth's history through greater degassing of the MORB reservoir. The strong correlations between ^{129}Xe/^{130}Xe and ^{3}He/^{130}Xe in gas-rich popping rocks of the MAR and subglacial Iceland DICE samples (Figure 14.16) indicate that while there is mixing between both these reservoirs and the atmosphere, there has been little or no mixing between these mantle reservoirs (Mukhopadhyay, 2012). Other data sets, including Loihi and Rochambeau, show poorer correlations, which could reflect from mixing between these reservoirs, but those samples are less gas-rich and the poor correlations could reflect He–Xe fractionation during degassing. Xenon from OIB also defines a higher ^{131}Xe/^{130}Xe–^{136}Xe/^{130}Xe slope than MORB, indicative of a greater fissiogenic contribution from ^{244}Pu relative to ^{238}U (Figure 14.12). These differences in Xe isotope ratios between MORB and OIB can be explained by greater degassing of the MORB reservoir, but given the 80 Ma half-life of ^{244}Pu, this differential degassing must have begun very early in Earth's history.

While early degassing is required to explain the difference between the MORB and plume reservoirs, because of the long half-life of ^{40}K (1.25 Ga), this cannot explain the differences in $^{40}Ar/^{36}Ar$. That requires more extensive degassing of the MORB reservoir over Earth's history.

We can conclude that reservoirs with distinct noble gas concentrations and isotopic compositions have been sustained over all of Earth's history.

14.9   NOBLE GAS CONSTRAINTS ON FORMATION AND EVOLUTION OF THE EARTH

The Earth and its sister terrestrial planets are highly depleted in volatile elements, most notably the noble gases as well as H, C, and N, compared to their abundances in outer planets as well as the Sun. This no doubt reflects the accretion of planetesimals that ultimately formed the terrestrial planets in the inner Solar System while temperatures were still too hot for these volatile elements to condense. Further out in the Solar System beyond the "snow line," Jupiter was able to accrete nearly its full share of these gases. Furthermore, the final stages of accretion of the terrestrial planets likely involved violent collisions and extensive melting. Evidence of a magma ocean on the Moon indicates one existed on the Earth as well. If this occurred after the solar nebula had dissipated, it would have allowed some or much of the Earth's initial inventory of volatiles to boil off. In that case, the question becomes: how did the Earth acquire its inventory of these elements? Noble gas isotope ratios can help answer this question.

Let us begin by reviewing key observations.

Most of the Earth's non-radiogenic noble gas inventory (except for He) is in the atmosphere. Estimates of the fraction of noble gases in the mantle range from ≈15% of the Earth's Xe (Marty, 2012) to 0.6% of the Earth's Ne (Halliday, 2013). In contrast, somewhere between 33 and 47% of the Earth's radiogenic ^{40}Ar is in the mantle. This difference provides an important constraint on the timing and mechanisms of degassing of the Earth's interior.

The overall abundance of noble gases in the Earth is similar to the planetary pattern observed in chondrites with depletion relative to solar increasing from heavy to light noble gases (Figure 14.1); however, Xe is an exception, being more depleted than Kr.

- Terrestrial noble gases show variable mass-dependent fractionations from both solar and chondritic compositions. Atmospheric Ne is enriched in heavy isotopes by more than ≈130‰/u relative to solar, Ar by 13‰/u, Kr by 6‰/u, while Xe is enriched by more than 30‰/u.

Mantle noble gases differ from atmospheric ones in being richer in radiogenic and nucleogenic isotopes ^{21}Ne, ^{40}Ar, ^{129}Xe, and $^{131}Xe-^{136}Xe$, although the opposite is true of He, which reflects continual loss of He to space.

Mantle noble gases are isotopically heavier than atmospheric, suggesting they have experienced less mass-dependent fractionation.

Mantle Kr and Xe non-radiogenic isotope ratios are both displaced away from atmospheric toward "planetary" Q and AVCC (Figures 14.8 and 14.10). However, while mantle Xe isotope ratios are closer to solar than the atmosphere, the opposite is true of Kr: mantle Kr isotope ratios are displaced from atmospheric away from solar. This suggests a "planetary" rather than "solar" origin for mantle Kr and Xe.

Differences in noble gas radiogenic/nucleogenic isotope ratios between the upper and deep mantle (plume) reservoir indicate the two reservoirs have different degassing histories, which must have begun with the earliest episode of Earth's history. It furthermore requires that exchange between the upper mantle and plume reservoirs has been limited since then, despite atmospheric gases have been subducted and recycled into both.

A final observation is that while noble gases are highly depleted in the Earth relative to the Sun and primitive chondrites, they are not as depleted as one might expect from their extremely low-condensation temperatures. This is illustrated in Figure 14.17, which plots the concentrations of the elements in the silicate Earth relative to CI chondrites as a function of their 50% condensation temperature. Most elements with condensation temperatures less than 1500 K plot along a trajectory of decreasing concentration with decreasing condensation temperature, with the exception of siderophile elements, such as the platinum group elements, Fe, Ni, Ag, etc.

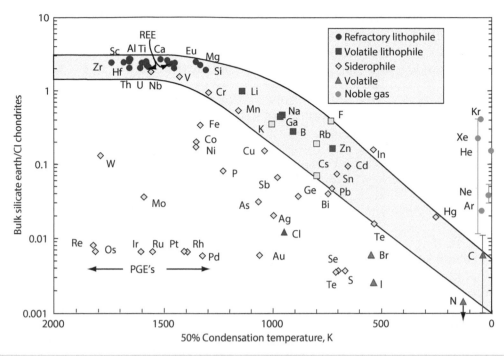

Figure 14.17 Chondrite-normalized abundances of the elements in the silicate Earth as a function of 50% nebular condensation temperature. Gray region shows the general trend of decreasing abundance with decreasing condensation temperature. Most elements plotting below the trend are siderophile and likely concentrated in the Earth core. The noble gases are an exception to the pattern. Adapted from White (2020), Halliday (2013), and Marty (2012). Error bars on Ne, Xe, and C show the range of their estimates.

Depletion of these elements is readily explained by their concentration in the Earth's core. The trend of the non-siderophile elements would predict that the noble gases should be depleted by factors of over 100, but actual depletion factors are smaller, ranging from about 2 for Kr to about 40 for Ar and showing no relationship to condensation temperature. The noble gases are also significantly less depleted than either carbon or nitrogen. The origin of the noble gas inventory in chondrites is itself a problem since these too appear to have formed at temperatures well above the condensation temperatures of the noble gases. As we noted earlier, some of this inventory resulted from solar wind implantation, but the "planetary" gases, such as Q, were acquired in some other way, perhaps absorption on to carbonaceous material.

14.9.1 Sources of terrestrial volatiles

The potential sources of terrestrial noble gases and other volatile elements include nebular gas trapped in or on grains as they formed in the inner Solar System, solar wind implanted in grains (as we noted in Section 14.2, implanted solar wind is isotopically distinct from solar at least for light noble gases), volatiles in carbonaceous chondrite planetesimals that formed in cooler regions of the Solar System, and comets. Each of these has unique isotopic compositions to varying degrees.

Neon is particularly revealing in this respect. Solar neon has $^{22}Ne/^{20}Ne = 13.34$ while implanted solar wind (Neon B) has $^{20}Ne/^{22}Ne \approx 12.6$. Carbonaceous chondrites have average $^{20}Ne/^{22}Ne$ that varies considerably with an average of ≈ 10.7, distinctly lighter than either implanted solar wind or nebular gas. The highest $^{20}Ne/^{22}Ne$ is plume-derived magmas that exceed 13, which strongly suggests that the plume neon is of solar nebular origin.

A commonly held view has been that the low $^{20}Ne/^{22}Ne$ of atmospheric Ne (9.8) compared to solar (13.34) resulted from mass fractionation during hydrodynamic escape under the influence of gravity (e.g., Pepin, 2006). The idea is that Ne and to a lesser extent even heavy noble gases could be entrained by the escape of a hot, light gas, such as hydrogen or methane. This is resisted by gravity, so that the extent to which a species is entrained and escapes depends on its mass. Ozima et al. (1998) proposed it occurred through Rayleigh fractionation during dissipation of the solar nebula, driven perhaps by intense UV radiation from the young Sun or a neighboring star. Such a general model would explain the compositional and isotopic similarity of planetary atmospheres and meteorites. Pepin (1991, 2003) suggested instead that the fractionation occurred during hydrodynamic escape of methane from transient atmospheres of planetesimals. In that case, the energy source would have been heating by short-lived radionuclides. On larger planets and planetary embryos, further fractionations could occur during hydrodynamic escape of primitive atmospheres driven either by UV radiation from the Sun or energy released by impacts. In the case of Mars, further fractionation of noble gas isotope ratios might have occurred as the atmosphere was lost by solar wind sputtering once the magnetic field ceased (Pepin, 2006).

This view has lost favor over the last decade or two. Albarède (2009) notes the lack of isotopic fractionation of moderately volatile elements, such as Zn and K, and argues against a scenario in which the Earth lost its volatile component by hydrodynamic escape following the Moon-forming collision. He argues that the terrestrial planets are likely to have initially accreted from very volatile-poor material, which is consistent with the relationship between elemental abundance and condensation temperature in Figure 12.17 and nebular models, which suggests that radiation from the young Sun would have cleared the inner nebula of gas before elements more volatile than the alkalis condensed. In this model, more volatile elements, including atmospheric noble gases, were added 100 million years or so later by accretion of a veneer of carbonaceous chondrite-like asteroids that initially formed near the orbits of the giant planets. If the atmosphere is largely derived from this late accretionary veneer, it requires that the noble gas isotopic composition of late accreting material was isotopically distinct.

As Marty (2022) points out, once corrected for cosmogenic contributions, the $^{36}Ar/^{22}Ne$ ratio variations observed in terrestrial and meteoritic noble gases cannot be explained by mass fractionation during hydrodynamic escape. And as we have found that neon in the Earth's interior has $^{20}Ne/^{22}Ne$ approaching the solar value, the implied fractionations of $^{20}Ne/^{22}Ne$, $\approx 330‰$, become implausible. Marty (2022) argues instead of fractionation, variations in Ne isotope ratios result from mixing between "planetary" (chondritic) and solar components, with Ne in the Earth's interior being primarily solar and atmospheric Ne being primarily planetary. The planetary component consists of dust found on rims of chondrules in both enstatite and carbonaceous chondrites and includes refractory pre-solar grains, such as nano-diamonds, impact-related debris, medium-to-low-temperature phases mainly made of organics, and, in the case of carbonaceous chondrites, hydrated minerals and icy grains. Indeed, the average $^{20}Ne/^{22}Ne$ of carbonaceous chondrites, ≈ 10.7, is far more similar to atmospheric Ne (9.7).

Curiously, the maximum $^{20}Ne/^{22}Ne$ of the MORB reservoir is very similar to that of solar wind-implanted Ne (e.g., Neon B). Taken at face value, this would appear to require three distinct components to explain neon isotopes in the Earth: nebular gas, solar wind-implanted, and a chondritic component rich in ^{20}Ne contained in pre-solar grains. Alternatively, Mukhopadhyay and Parai (2019) point out that if Ne can be subducted into the mantle, Neon in the MORB reservoir may reflect a mixture of primordial nebular Ne of solar composition and subducted atmospheric Ne. This reduces the required primordial Ne components from 3 to 2.

In contrast to Ne, the quite limited Kr isotope data on the mantle are shifted away from both atmospheric and solar toward chondritic Kr (Figure 14.8). Xenon isotope data are somewhat more ambiguous because solar and chondritic Xe isotopic compositions are similar, and mantle Xe is close to atmospheric. However, the light Xe isotope data (Figure 14.10), particularly the Eifel gas data, appear shifted from atmospheric toward chondritic rather than solar. Both atmospheric and mantle $^{38}Ar/^{36}Ar$ also appear to be more similar to chondritic than solar. Hydrogen and carbon isotope ratios are also much closer to chondritic values than solar ones. Thus, there is clear evidence of a chondritic component in the Earth's interior.

Mukhopadhyay and Parai (2019) argue that primordial noble gases in the atmosphere were derived after the Moon-forming impact, but the chondritic noble gases (e.g., Kr and Xe) in Earth's interior were derived before the formation of the Moon. They argue that much of the Earth's initial volatile inventory was lost during the Moon-forming impact and replaced by the subsequently accreted late-stage veneer, supplied by gas-rich carbonaceous chondrites.

At this point, a disclaimer should be added that while heterogeneous accretion can explain much of the variation we see in noble gas isotope ratios, this does not mean that mass fractionation during hydromagmatic escape did not occur at all. Indeed, while some aspects of terrestrial noble gas isotopic compositions can be explained by mixing of solar, chondritic, and radiogenic/nucleogenic components, they cannot be entirely explained this way and mass-dependent fractionation also appears to have been involved.

14.9.2 The Xenon Paradoxes

The abundance and isotopic composition of terrestrial xenon poses several problems that cannot be explained by the processes we have discussed thus far. First, while the Earth's depletion in noble gases relative to solar increases as a function of mass from Kr to He, Xe is more depleted in the Earth than Kr (Figure 14.1), with Xe/Kr ratios a factor of ≈ 20 lower than chondritic. Second, Xe isotope ratios appear to have experienced more mass-dependent fractionation, 20–30‰/u relative to chondritic or solar compositions compared to ≈ 6‰/u for Kr and Ar. Third, even after accounting for mass-dependent fractionation and fissogenic contributions, terrestrial Xe shows a deficit in the heaviest Xe isotopes (Figure 14.18).

Let us consider the last problem first. The deficit in heavy Xe isotopes led Pepin (2000) to hypothesize the existence of a hypothetical "U–Xe" component (where U is derived from

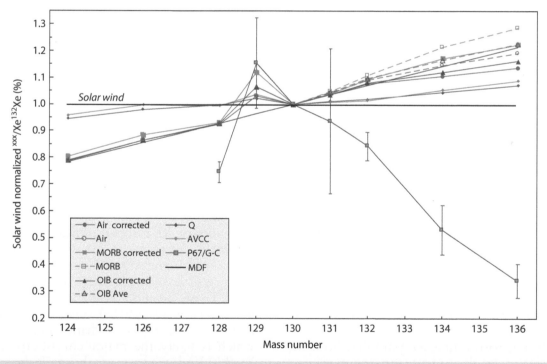

Figure 14.18 Solar wind-normalized abundances of xenon isotopes in terrestrial and Solar System materials with ^{132}Xe $\equiv$ 1. MORB line is the average of Pepin and Porcelli's (2006) average of the data of Kunz et al. (1998), which excludes the least precise, less reliable dada; OIB line is the average of Iceland and Galápagos data of Péron et al. (2021), Q and AVCC are from Pepin and Porcelli (2006), and P67/G-C is from Marty et al. (2017). "Corrected" air, MORB and OIB are approximate (a constant Pu/U ratio was used for all) corrections for ^{238}U and ^{244}Pu fissogenic production of ^{131}Xe through ^{136}Xe. A 2.6‰/μ mass fractionation pattern is shown for comparison.

the German word *Ur*, meaning primordial in this case). Some had speculated the U is a cometary component (Dauphas, 2003; Halliday, 2013). Comets originated in the outer regions of the Solar System where temperatures are cold enough for water, methane, and other gases to form ices and trap heavy noble gases. Above 30 K, argon and krypton are trapped more efficiently than xenon, leading to fractionated noble gas patterns. Most comets appear to be strongly enriched in ^{2}H compared to the Earth, the δ^2H of Halley's Comet, an Oort Cloud comet, has a D/H ratio 13 times that of the Earth (Eberhardt et al., 1995) and Comet P67/G-C has a D/H ratio approximately three times that of the Earth. Comets are similarly enriched in ^{15}N, with δ^{15}N ranging from $\approx$140 to >800‰ (Füri and Marty, 2015). While comets are an unlikely source of the Earth's water and nitrogen, cometary ices are so enriched in noble gases that even if they contribute only a small fraction of the mass of the Earth, cometary material could explain the noble gas abundance pattern of the Earth, as well as the peculiar features of noble gas isotopic compositions (Owen et al., 1992; Dauphas, 2003; Halliday, 2013).

The speculation of this cometary component has now been confirmed. Figure 14.17 compares the Xe isotope patterns of terrestrial and Solar System materials. The pattern of Comet P67/Churyumov–Gerasimenko is distinctive, first in its very strong enrichment in ^{129}Xe, suggesting the comet is iodine-rich, and second in its depletion in ^{134}Xe and ^{136}Xe, reflecting a deficit of r-process nuclides. Based on the analysis of Comet P67/G-C, the U component of Xe hypothesized by Pepin appears to be cometary. Marty et al. (2017) found that atmospheric Xe can be explained by the addition of 22 ± 5% of cometary Xe mixed with the chondritic Q component. With this we have evidence of a cometary component in the Earth in addition to solar and planetary components.

Turning to the first Xe paradox, both the stronger mass-dependent fractionation and the overall terrestrial Xe deficit suggest the Earth has preferentially lost Xe, which is surprising as it is the heaviest stable noble gas, and we expect the lightest gases to be preferentially lost and fractionated. Xenon is far too heavy to be lost through Jean's escape, through which atoms are thermally accelerated to

escape velocity. There has long been a suspicion that noble gases could be lost through hydrodynamic escape, in which gases can be dragged by hydrogen that has been accelerated to high-velocity and drags other gases along with it out of the top of the atmosphere at the poles (e.g., Sasaki and Nakazawa (1988). But this process should also affect lighter noble gases, as well as Xe.

A clue comes from the observation that Xe has a lower-ionization energy than noble gases, as well as gases, such as H_2, CO_2, CH_4, N_2, and H_2O that comprised the Earth's early atmosphere (e.g., Marty, 2012). In a model developed by Hébrard and Marty (2014), strong extreme ultraviolet radiation of the young Sun resulted in ionization of atmospheric Xe, as well as dissociation of H- and C-bearing molecules, such as methane high in the atmosphere ($\approx$100 km). The latter then recombine to form a haze consisting of hydrocarbon aerosols. Experiments have demonstrated that Xe ions are readily trapped in such organics with heavy Xe isotopes preferentially trapped and consequently the lighter isotopes are preferentially lost by hydrodynamic escape. Further fractionation can occur as xenon ions are carried along with H ions (and He) in polar winds channeled by open planetary magnetic field lines. Being more easily ionized than hydrogen, Xe would be dragged by the strong coulomb interactions, whereas the other noble gases would be quickly neutralized. The model predicts that loss should have been strongest in the early Earth when methane was abundant in the atmosphere and the Sun's extreme UV flux was greater.

Studies of Xe trapped in fluid inclusions of ancient rocks over the last dozen or so years have confirmed that the Xe in the early atmosphere was less fractionated than the present one. These results have been summarized by Avice et al. (2018) and are shown in Figure 14.19. The chondritic component Q, which is likely the major contributor to the Earth's Xe inventory, is about 5‰/u fractionated relative to the solar wind while the present atmosphere is about 36‰/ enriched in heavy isotopes. The oldest sample from the 3.48 Ga North Pole outcrop in Australia is 15‰/u fractionated. Most of the increase in fractionation occurred in the Archean, and by about 2 Ga, ancient atmospheric Xe

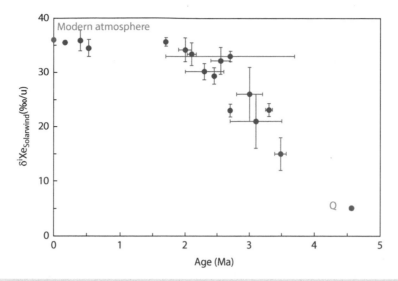

Figure 14.19 Fractionation of atmospheric Xe relative to the solar wind trapped in fluid inclusions in ancient supracrustal rocks as a function of age. The isotopic composition of the Q component of carbonaceous chondrites indicated initial fractionation in the solar nebula was limited to ≈5‰. Fractionation of terrestrial atmospheric xenon increases rapidly during the Archean, reflecting a strong flux of ionizing extreme UV radiation from the young Sun with the present atmospheric composition being obtained in the Proterozoic. Data from Holland et al. (2013) and the compilation of Avice et al. (2018), originally normalized to modern atmospheric, have been renormalized to the solar wind.

is indistinguishable from modern atmospheric Xe. This is consistent with a decreasing solar extreme UV over time, as well as decrease in atmospheric methane, particularly around the GOE at 2.4 Ga.

14.9.3 Degassing of the Earth's interior

While progress is certainly being made, the details of the Earth's formation and how it obtained its inventory of noble gases and other volatiles are still being worked out. Noble gases point to at least three sources: (1) nebular gas with solar isotopic compositions, required by Ne isotopes in the plume reservoir, (2) chondritic (so-called "planetary") noble gases, required by all noble gases, and (3) cometary noble gases, required both by Xe isotopic compositions and the overall abundance of terrestrial noble gases. As we noted earlier, most of the Earth's noble gas inventory is now in the atmosphere, but this may not have always been the case. The energy released by the Moon-forming impact almost certainly resulted in the loss of much and perhaps all of the Earths' primitive atmosphere. Subsequent addition of

a late accretionary veneer could have renewed the atmosphere, but noble gas isotope ratios indicate that much of the present atmosphere has been produced by degassing of the Earth's interior. This process continues, of course, as magmas rising to the surface releases noble gases in volcanic eruptions. Let us briefly consider this in just a bit more detail.

Radiogenic noble gases provide the key constraints on mantle degassing and atmospheric evolution because of the absence of radioactive progenitors in the atmosphere. That is to say that that excess ^{129}Xe in the atmosphere (Table 14.3; Figure 14.18) and radiogenic ^{40}Ar (which constitutes >99% atmospheric Ar and ≈0.9% of the total atmosphere) must have been produced within the solid Earth because neither I nor K is present in the atmosphere. We can infer from this that much of the Earth's atmosphere has been produced by degassing of the Earth's interior.

The next question is: when did this degassing occur? Because $^{129}Xe/^{130}Xe$ ratios are higher in the mantle than the atmosphere, we can conclude that the Earth's interior, particularly the upper mantle source of MORB, must have experienced extensive early degassing, raising the I/Xe ratio, before ^{129}I became effectively extinct at around 100 Ma. Degassing occurring later than that would have resulted in identical mantle and atmospheric ^{129}Xe. Fissogenic Xe provides similar but looser constraints because of the longer half-life of ^{244}Pu. On the other hand, early degassing cannot explain the ^{40}Ar in the atmosphere. Because of the long half-life of ^{40}K, ≈ 1.4 billion years, only a small fraction of the Earth's ^{40}Ar would have been produced in the first 100 Ma.

These observations lead to a two-stage degassing model originally proposed by Sarda et al. (1985) and Allègre et al. (1987). To spare ourselves some of the math in those papers, we will follow the simpler exposition of Moreira (2013). The number of moles of a stable noble gas isotope as a function of time, $S(t)$, in the mantle undergoing degassing is expressed as:

$$\frac{S(t)}{S_0} = Ae^{-\alpha t} + (1 - A)e^{-\beta t} \qquad (14.6)$$

where S_0 is the original amount, and α and β are two distinct degassing constants in year^{-1}, and A describes the proportion of degassing for the two stages. For a radiogenic isotope, F, such as ^{40}Ar or ^{129}Xe, we need to also consider radioactive ingrowth, which is simply $dF/dt = \lambda P(t)$ where P is the amount of the parent and λ is its decay constant. The change in mantle concentration becomes:

$$\frac{dF}{dt} = \frac{\alpha Ae^{-\alpha t} + (1-A)\beta e^{-\beta t}}{Ae^{-\alpha t} + (1-A)e^{-\beta t}} F + \lambda P \qquad (14.7)$$

Allègre et al. (1987) used values of $A = 10^{-3}$, $\alpha = 1.8 \times 10^{-9}$, and $\beta = 2.8 \times 10^{-7}$ to produce the degassing curves shown in Figure 14.20. The model produces rapid degassing in the first few tens of millions of years consistent with outgassing of a magma ocean in the aftermath of the Moon-forming impact. This early degassing releases much of the radiogenic Xe and non-radiogenic noble gases, such as ^{36}Ar, but little of the ^{4}He and ^{40}Ar, which have not yet been produced. The latter are then released at a decreasing rate over time as the Earth cools and mantle convection slows.

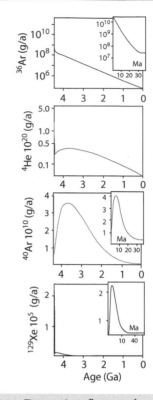

Figure 14.20 Degassing fluxes of noble gases based on a two-stage degassing model, such as Equations (14.6) and (14.7). Much of the unradiogenic gases as well as the radiogenic and fissogenic isotopes of Xe are released to the atmosphere in the first few tens of millions of years, but ^{40}Ar and ^{4}He are released slowly over time. Source: Allègre et al. (1987)/with permission of Elsevier.

Degassing has, of course, affected the upper mantle MORB reservoir much more extensively than the plume reservoir, likely located at the core–mantle boundary. The latter at least partially escaped the early catastrophic degassing. Mukhopadhyay and Parai (2019) suggest a couple of possible reasons. First, the mantle may not have melted entirely after the Moon-forming impact, preserving noble gas inventory in the deep mantle. Alternatively, if crystallization of the magma ocean began in the mid-mantle as some experimental work suggests (Mosenfelder et al., 2009), this could have isolated relatively undegassed material in the deepest mantle (e.g., Coltice et al., 2011).

REFERENCES

Albarède, F. 2009. Volatile accretion history of the terrestrial planets and dynamic implications. *Nature* **461**: 1227–33.

Aldrich, L.T. and Nier, A.O. 1946. The Abundance of ^{3}He in atmospheric and well helium. *Physical Review* **70**: 983–4. doi: 10.1103/PhysRev.70.983.2.

Allègre, C.J., Hofmann, A. and O'Nions, K. 1996. The argon constraints on mantle structure. *Geophysical Research Letters* **23**: 3555–7. doi: 10.1029/96gl03373.

Allègre, C.J., Staudacher, T. and Sarda, P. 1987. Rare gas systematics: formation of the atmosphere, evolution and structure of the Earth's mantle. *Earth and Planetary Science Letters* **81**: 127–50.

Arevalo, R., Jr., McDonough, W.F. and Luong, M. 2009. The K/U ratio of the silicate Earth: insights into mantle composition, structure and thermal evolution. *Earth and Planetary Science Letters* **278**: 361–9. doi: 10.1016/j.epsl.2008.12.023.

Armytage, R.M.G., Jephcoat, A.P., Bouhifd, M.A., et al. 2013. Metal-silicate partitioning of iodine at high pressures and temperatures: implications for the Earth's core and ^{129}Xe budgets. *Earth and Planetary Science Letters* **373**: 140–9. doi: 10.1016/j.epsl.2013.04.031.

Avice, G., Marty, B., Burgess, R., et al. 2018. Evolution of atmospheric xenon and other noble gases inferred from Archean to Paleoproterozoic rocks. *Geochimica et Cosmochimica Acta* **232**: 82–100. doi: https://doi.org/10.1016/j.gca.2018.04.018.

Ballentine, C.J. and Burnard, P.G. 2002. Production, release and transport of noble gases in the continental crust, in Porcelli, D., Ballentine, C. and Weiler, R. (eds). *Noble Gases*. Washington, DC: Mineralogical Society of America, 481–538 pp.

Ballentine, C.J. and Holland, G. 2008. What CO_2 well gases tell us about the origin of noble gases in the mantle and their relationship to the atmosphere. *Philosophical Transactions of the Royal Society A* **366**: 4183–203. doi: 10.1098/rsta.2008.0150.

Ballentine, C.J., Burgess, R. and Marty, B. 2002. Tracing fluid origin, transport and interaction in the crust, in Porcelli, D., Ballentine, C. and Weiler, R. (eds). *Noble Gases*. Washington, DC: Mineralogical Society of America, 539–614 pp.

Ballentine, C.J., Marty, B., Sherwood Lollar, B., et al. 2005. Neon isotopes constrain convection and volatile origin in the Earth's mantle. *Nature* **433**: 33–8. doi: 10.1038/nature03182.

Barlow, M.J., Swinyard, B.M., Owen, P.J., et al. 2013. Detection of a noble gas molecular ion, ^{36}ArH$^+$, in the Crab Nebula. *Science* **342**: 1343–5. doi: 10.1126/science.1243582.

Basu, A.R., Poreda, R.J., Renne, P.R., et al. 1995. High-^{3}He plume origin and temporal-spatial evolution of the Siberian Flood Basalts. *Science* **269**: 822–5. doi: 10.2307/2888490.

Basu, A.R., Renne, P.R., DasGupta, D.K., et al. 1993. Early and late alkali igneous pulses and a high-^{3}He plume origin for the Deccan Flood Basalts. *Science* **261**: 902–6. doi: 10.2307/2882121

Bekaert, D.V., Broadley, M.W. and Marty, B. 2020. The origin and fate of volatile elements on Earth revisited in light of noble gas data obtained from comet 67P/Churyumov-Gerasimenko. *Scientific Reports* **10**: 5796. doi: 10.1038/s41598-020-62650-3.

Bianchi, D., Sarmiento, J.L., Gnanadesikan, A., et al. 2010. Low helium flux from the mantle inferred from simulations of oceanic helium isotope data. *Earth and Planetary Science Letters* **297**: 379–86. doi: 10.1016/j.epsl.2010.06.037.

Boschi, L., Becker, T.W. and Steinberger, B. 2007. Mantle plumes: dynamic models and seismic images. *Geochemistry, Geophysics, Geosystems* **8**. doi: https://doi.org/10.1029/2007GC001733.

Bouhifd, M.A., Jephcoat, A.P., Heber, V.S., et al. 2013. Helium in Earth's early core. *Nature Geoscience* **6**: 982–6. doi: 10.1038/ngeo1959.

Broadley, M.W., Barry, P.H., Bekaert, D.V., et al. 2020. Identification of chondritic krypton and xenon in Yellowstone gases and the timing of terrestrial volatile accretion. *Proceedings of the National Academy of Sciences* **117**: 13997–4004. doi: 10.1073/pnas.2003907117.

Burnard, P., Graham, D. and Turner, G. 1997. Vesicle-specific noble gas analyses of "popping rock": implications for primordial noble gases in Earth. *Science* **276**: 568–71. doi: 10.1126/science.276.5312.568

Busemann, H., Baur, H. and Wieler, R. 2000. Primordial noble gases in "phase Q" in carbonaceous and ordinary chondrites studied by closed-system stepped etching. *Meteoritics and Planetary Science* **35**: 949–73. doi: 10.1111/j.1945-5100.2000.tb01485.x.

Butler, W.A., Jeffery, P.M., Reynolds, J.H., et al. 1963. Isotopic variations in terrestrial xenon. *Journal of Geophysical Research* **68**: 3283–91.

Caffee, M.W., Hudson, G.B., Velsko, C., et al. 1999. Primordial noble gases from earth's mantle: identification of a primitive volatile component. *Science* **285**: 2115–8. doi: 10.1126/science.285.5436.2115.

Caracausi, A., Avice, G., Burnard, P.G., et al. 2016. Chondritic xenon in the Earth's mantle. *Nature* **533**: 82–5. doi: 10.1038/nature17434.

Clarke, W.B., Beg, M.A. and Craig, H. 1969. Excess 3He in the sea: evidence for terrestrial primodal helium. *Earth and Planetary Science Letters* **6**: 213–20. doi: https://doi.org/10.1016/0012-821X(69)90093-4.

Coltice, N., Moreira, M., Hernlund, J., et al. 2011. Crystallization of a basal magma ocean recorded by helium and neon. *Earth and Planetary Science Letters* 308: 193–9. doi: https://doi.org/10.1016/j.epsl.2011.05.045.

Craig, H., Clarke, W.B. and Beg, M.A. 1975. Excess ^{3}He in deep water on the East Pacific Rise. *Earth and Planetary Science Letters* 26: 125–32. doi: 10.1016/0012-821X(75)90079-5.

Damon, P.E. and Kulp, J.L. 1958. Inert gases and the evolution of the atmosphere. *Geochimica et Cosmochimica Acta* 13: 280–92. doi: 10.1016/0016-7037(58)90030-9.

Dauphas, N. 2003. The dual origin of the terrestrial atmosphere. *Icarus* 165: 326–39. doi: 10.1016/S0019-1035(03)00198-2.

Dygert, N., Jackson, C.R.M., Hesse, M.A., et al. 2018. Plate tectonic cycling modulates Earth's ^{3}He/^{22}Ne ratio. *Earth and Planetary Science Letters* 498: 309–21. doi: https://doi.org/10.1016/j.epsl.2018.06.044.

Eberhardt, P., Reber, M., Krankowsky, D., et al. 1995. The D/H and ^{18}O/^{16}O ratios in water from comet P/Halley. *Astronomy and Astrophysics* 302: 301–16.

Füri, E. and Marty, B. 2015. Nitrogen isotope variations in the Solar System. *Nature Geoscience* 8: 515–22. doi: 10.1038/ngeo2451.

Gale, A., Dalton, C.A., Langmuir, C.H., et al. 2013. The mean composition of ocean ridge basalts. *Geochemistry, Geophysics, Geosystems* 13: 489–518. doi: 10.1029/2012gc004334.

Graham, D.W., Michael, P.J. and Shea, T. 2016. Extreme incompatibility of helium during mantle melting: evidence from undegassed mid-ocean ridge basalts. *Earth and Planetary Science Letters* 454: 192–202. doi: https://doi.org/10.1016/j.epsl.2016.09.016.

Halliday, A.N. 2013. The origins of volatiles in the terrestrial planets. *Geochimica et Cosmochimica Acta* 105: 146–71. doi: 10.1016/j.gca.2012.11.015.

Hanyu, T., Tatsumi, Y. and Kimura, J.-I. 2011. Constraints on the origin of the HIMU reservoir from He–Ne–Ar isotope systematics. *Earth and Planetary Science Letters* 307: 377–86. doi: 10.1016/j.epsl.2011.05.012.

Heber, V.S., Baur, H., Bochsler, P., et al. 2012. Isotopic mass fractionation of solar wind: evidence from fast and slow solar wind collected by the *Genesis* mission. *The Astrophysical Journal* 759: 121. doi: 10.1088/0004-637x/759/2/121.

Hébrard, E. and Marty, B. 2014. Coupled noble gas–hydrocarbon evolution of the early Earth atmosphere upon solar UV irradiation. *Earth and Planetary Science Letters* 385: 40–8. doi: https://doi.org/10.1016/j.epsl.2013.10.022.

Hilton, D.R., Fischer, T.P. and Marty, B. 2002. Noble gases and volatile recycling at subduction zones, in Porcelli, D., Ballentine, C. and Weiler, R. (eds). *Noble Gases*. Washington, DC: Mineralogical Society of America, 319–70 pp.

Holland, G. and Ballentine, C.J. 2006. Seawater subduction controls the heavy noble gas composition of the mantle. *Nature* 441: 186–91. doi: http://dx.doi.org/10.1038/nature04761.

Holland, G., Cassidy, M. and Ballentine, C.J. 2009. Meteorite Kr in Earth's mantle suggests a late accretionary source for the atmosphere. *Science* 326: 1522–5. doi: 10.1126/science.1179518.

Holland, G., Lollar, B.S., Li, L., et al. 2013. Deep fracture fluids isolated in the crust since the Precambrian era. *Nature* 497: 357–60. doi: 10.1038/nature12127

Honda, M. and McDougall, I. 1998. Primordial helium and neon in the Earth—A speculation on early degassing. *Geophysical Research Letters* 25: 1951–4. doi: 10.1029/98gl01329.

Honda, M. and Woodhead, J.D. 2005. A primordial solar-neon enriched component in the source of EM-I-type ocean island basalts from the Pitcairn Seamounts, Polynesia. *Earth and Planetary Science Letters* 236: 597–612. doi: 10.1016/j.epsl.2005.05.038.

Horton, F., Curtice, J., Farley, K.A., et al. 2021. Primordial neon in high-^{3}He/^{4}He Baffin Island olivines. *Earth and Planetary Science Letters* 558: 116762. doi: https://doi.org/10.1016/j.epsl.2021.116762.

Hunten, D.M., Pepin, R.O. and Owen, T.C. (1988). Planetary atmospheres, in Kerridge, J.F. and Matthews, M.S. (eds). *Meteorites and the Early Solar System*, Tuscon: University of Arizona Press.

Jackson, C.R.M., Bennett, N.R., Du, Z., et al. 2018. Early episodes of high-pressure core formation preserved in plume mantle. *Nature* 553: 491–5. doi: 10.1038/nature25446.

Jackson, M.G., Konter, J.G. and Becker, T.W. 2017. Primordial helium entrained by the hottest mantle plumes. *Nature* 542: 340–3. doi: 10.1038/nature21023

Jackson, M.G., Blichert-Toft, J., Halldórsson, S.A., et al. 2020. Ancient helium and tungsten isotopic signatures preserved in mantle domains least modified by crustal recycling. *Proceedings of the National Academy of Sciences* 117: 30993–31001. doi: 10.1073/pnas.2009663117.

Jackson, M.G., Kurz, M.D. and Hart, S.R. 2009. Helium and neon isotopes in phenocrysts from Samoan lavas: evidence for heterogeneity in the terrestrial high ^{3}He/^{4}He mantle. *Earth and Planetary Science Letters* 287: 519–28. doi: 10.1016/j.epsl.2009.08.039.

Jenkins, W.J. 1987. ^{3}H and ^{3}He in the Beta Triangle: observations of gyre ventilation and oxygen utilization rates. *Journal of Physical Oceanography* 17: 763–83.

Kellogg, L.H. and Wasserburg, G.J. 1990. The role of plumes in mantle helium fluxes. *Earth and Planetary Science Letters* 99: 276–89. doi: 10.1016/0012-821X(90)90116-F.

King, S.D. and Adam, C. 2014. Hotspot swells revisited. *Physics of the Earth and Planetary Interiors* **235**: 66–83. doi: 10.1016/j.pepi.2014.07.006.

Kipfer, R., Aeschbach-Hertig, W., Peeters, F., et al. 2002. Noble gases in lakes and ground waters. *Reviews in Mineralogy and Geochemistry* **47**: 615–700. doi: 10.2138/rmg.2002.47.14.

Kunz, J., Staudacher, T. and Allègre, C.J. 1998. Plutonium-fission xenon found in Earth's mantle. *Science* **280**: 877–80. doi: 10.1126/science.280.5365.877.

Kurz, M.D., Curtice, J., Fornari, D., et al. 2009. Primitive neon from the center of the Gálapagos hotspot. *Earth and Planetary Science Letters* **286**: 23–34. doi: 10.1016/j.epsl.2009.06.008.

Lupton, J.E., Arculus, R.J., Greene, R.R., et al. 2009. Helium isotope variations in seafloor basalts from the Northwest Lau Backarc Basin: mapping the influence of the Samoan hotspot. *Geophysical Research Letters* **36**: L17313. doi: 10.1029/2009GL039468.

Lyubetskaya, T. and Korenaga, J. 2007. Chemical composition of Earth's primitive mantle and its variance: 1. Method and results. *Journal of Geophysical Research* **112**: B03211. doi: 10.1029/2005jb004223.

Mabry, J., Lan, T., Burnard, P., et al. 2013. High-precision helium isotope measurements in air. *Journal of Analytical Atomic Spectrometry* **28**: 1903–10. doi: 10.1039/C3JA50155H.

Madureira, P., Moreira, M., Mata, J., et al. 2005. Primitive neon isotopes in Terceira Island (Azores archipelago). *Earth and Planetary Science Letters* **233**: 429–40. doi: 10.1016/j.epsl.2005.02.030.

Mark, D.F., Stuart, F.M. and de Podesta, M. 2011. New high-precision measurements of the isotopic composition of atmospheric argon. *Geochimica et Cosmochimica Acta* **75**: 7494–501. doi: 10.1016/j.gca.2011.09.042.

Marty, B. 2012. The origins and concentrations of water, carbon, nitrogen and noble gases on Earth. *Earth and Planetary Science Letters* **313**: 56–66. doi: 10.1016/j.epsl.2011.10.040.

Marty, B. 2020. Origins and early evolution of the atmosphere and the oceans. *Geochemical Perspectives* **9**: 135–313. doi: 10.7185/geochempersp.9.2.

Marty, B. 2022. Meteoritic noble gas constraints on the origin of terrestrial volatiles. *Icarus* **381**: 115020. doi: https://doi.org/10.1016/j.icarus.2022.115020.

Marty, B., Altwegg, K., Balsiger, H., et al. 2017. Xenon isotopes in 67P/Churyumov-Gerasimenko show that comets contributed to Earth's atmosphere. *Science* **356**: 1069–72. doi: 10.1126/science.aal3496.

Matsuda, J., Sudo, M., Ozima, M., et al. 1993. Noble gas partitioning between metal and silicate under high pressures. *Science* **259**: 788–90. doi: 10.1126/science.259.5096.788.

McDonough, W.F. and Sun, S.-S. 1995. The composition of the Earth. *Chemical Geology* **120**: 223–53.

Meshik, A., Hohenberg, C., Pravdivtseva, O., et al. 2014. Heavy noble gases in solar wind delivered by *Genesis* mission. *Geochimica et Cosmochimica Acta* **127**: 326–47. doi: https://doi.org/10.1016/j.gca.2013.11.030.

Moreira, M. 2013. Noble gas constraints on the origin and evolution of earth's volatiles. *Geochemical Perspectives* **2**: 229–403. doi: 10.7185/geochempersp.2.2.

Moreira, M., Kunz, J. and Allègre, C. 1998. Rare gas systematics in popping rock: isotopic and elemental compositions in the upper mantle. *Science* **279**: 1178–81. doi: 10.1126/science.279.5354.1178.

Mosenfelder, J.L., Asimow, P.D., Frost, D.J., et al. 2009. The $MgSiO_3$ system at high pressure: Thermodynamic properties of perovskite, postperovskite, and melt from global inversion of shock and static compression data. *Journal of Geophysical Research: Solid Earth* **114**. doi: https://doi.org/10.1029/2008JB005900.

Mukhopadhyay, S. 2012. Early differentiation and volatile accretion recorded in deep-mantle neon and xenon. *Nature* **486**: 101–4. doi: 10.1038/nature11141.

Mukhopadhyay, S. and Parai, R. 2019. Noble gases: a record of Earth's evolution and mantle dynamics. *Annual Review of Earth and Planetary Sciences* **47**: 389–419. doi: 10.1146/annurev-earth-053018-060238.

O'Nions, R.K. and Oxburgh, E.R. 1983. Heat and helium in the Earth *Nature* **306**: 429–31.

O'Nions, R.K. and Oxburgh, E.R. 1988. Helium, volatile fluxes and the development of continental crust. *Earth and Planetary Science Letters* **90**: 331–47.

Olson, P.L. and Sharp, Z.D. 2022. Primordial helium-3 exchange between Earth's core and mantle. *Geochemistry, Geophysics, Geosystems* **23**: e2021GC009985. doi: https://doi.org/10.1029/2021GC009985.

Ott, U. 2014. Planetary and pre-solar noble gases in meteorites. *Geochemistry* **74**: 519–44. doi: https://doi.org/10.1016/j.chemer.2014.01.003.

Owen, T., Bar-Nun, A. and Kleinfeld, I. 1992. Possible cometary origin of heavy noble gases in the atmospheres of Venus, Earth, and Mars. *Nature* **358**: 43–6.

Ozima, M. and Podosek, F.A. 2002. *Noble Gas Geochemistry*. Cambridge: Cambridge University Press.

Ozima, M. and Zashu, S. 1983. Primitive helium in diamonds. *Science* **219**: 1067–8. doi: 10.1126/science.219.4588.1067.

Ozima, M., Wieler, R., Marty, B., et al. 1998. Comparative studies of solar, Q-gases and terrestrial noble gases, and implications on the evolution of the solar nebula. *Geochimica et Cosmochimica Acta* **62**: 301–4. doi: 10.1016/S0016-7037(97)00339-6.

Palme, H. and O'Neill, H.S.C. 2003. Cosmochemical estimates of mantle composition, in Heinrich, D.H. and Karl, K.T. (eds). *Treatise of Geochemistry: The Mantle and Core*. Amsterdam: Elsevier, 1: 38.

Parai, R. and Mukhopadhyay, S. 2015. The evolution of MORB and plume mantle volatile budgets: constraints from fission Xe isotopes in Southwest Indian Ridge basalts. *Geochemistry, Geophysics, Geosystems* 16: 719–35. doi: 10.1002/2014GC005566.

Parai, R., Mukhopadhyay, S. and Standish, J.J. 2012. Heterogeneous upper mantle Ne, Ar and Xe isotopic compositions and a possible DUPAL noble gas signature recorded in basalts from the Southwest Indian Ridge. *Earth and Planetary Science Letters* 359–60: 227–39. doi: 10.1016/j.epsl.2012.10.017.

Pepin, R.O. 1991. On the origin and early evolution of terrestrial planet atmospheres and meteoritic volatiles. *Icarus* 92: 2–79. doi: 10.1016/0019-1035(91)90036-S.

Pepin, R.O. (2000). On the isotopic composition of primordial xenon in terrestrial planet atmospheres, in Benz, W., Kallenbach, R. and Lugmair, G.W. (eds). *From Dust to Terrestrial Planets*. Dordrecht: Springer Netherlands, 371–95 pp.

Pepin, R.O. 2003. On noble gas processing in the solar accretion disk. *Space Science Reviews* 106: 211–30. doi: 10.1023/a:1024693822280.

Pepin, R.O. 2006. Atmospheres on the terrestrial planets: clues to origin and evolution. *Earth and Planetary Science Letters* 252: 1–14. doi: 10.1016/j.epsl.2006.09.014.

Pepin, R.O. and Porcelli, D. 2002. Origin of noble gases in the terrestrial planets. *Reviews in Mineralogy and Geochemistry* 47: 191–246. doi: 10.2138/rmg.2002.47.7.

Pepin, R.O. and Porcelli, D. 2006. Xenon isotope systematics, giant impacts, and mantle degassing on the early Earth. *Earth and Planetary Science Letters* 250: 470–85. doi: 10.1016/j.epsl.2006.08.014.

Péron, S., Moreira, M., Colin, A., et al. 2016. Neon isotopic composition of the mantle constrained by single vesicle analyses. *Earth and Planetary Science Letters* 449: 145–54. doi: https://doi.org/10.1016/j.epsl.2016.05.052.

Péron, S. and Moreira, M. 2018. Onset of volatile recycling into the mantle determined by xenon anomalies. *Geochemical Perspectives Letters* 9: 21–5. doi: http://dx.doi.org/10.7185/geochemlet.1833.

Péron, S., Moreira, M.A., Kurz, M.D., et al. 2019. Noble gas systematics in new popping rocks from the Mid-Atlantic Ridge (14°N): Evidence for small-scale upper mantle heterogeneities. *Earth and Planetary Science Letters* 519: 70–82. doi: https://doi.org/10.1016/j.epsl.2019.04.037.

Péron, S., Mukhopadhyay, S., Kurz, M.D., et al. 2021. Deep-mantle krypton reveals Earth's early accretion of carbonaceous matter. *Nature* 600: 462–7. doi: 10.1038/s41586-021-04092-z.

Pető, M.K., Mukhopadhyay, S. and Kelley, K.A. 2013. Heterogeneities from the first 100 million years recorded in deep mantle noble gases from the Northern Lau Back-arc Basin. *Earth and Planetary Science Letters* 369–70: 13–23. doi: 10.1016/j.epsl.2013.02.012.

Porcelli, D. and Ballentine, C.J. 2002. Models for distribution of terrestrial noble gases and evolution of the atmosphere. *Reviews in Mineralogy and Geochemistry* 47: 411–80. doi: 10.2138/rmg.2002.47.11.

Porcelli, D. and Elliott, T. 2008. The evolution of He Isotopes in the convecting mantle and the preservation of high ^{3}He/^{4}He ratios. *Earth and Planetary Science Letters* 269: 175–85. doi: 10.1016/j.epsl.2008.02.002.

Porcelli, D. and Halliday, A.N. 2001. The core as a possible source of mantle helium. *Earth and Planetary Science Letters* 192: 45–56. doi: 10.1016/S0012-821X(01)00418-6.

Porcelli, D. and Wasserburg, G.J. 1995. Mass transfer of helium, neon, argon, and xenon through a steady-state upper mantle. *Geochimica et Cosmochimica Acta* 59: 4921–37. doi: 10.1016/0016-7037(95)00336-3.

Porcelli, D., Ballentine, C. and Weiler, R. (eds) 2002. *Noble Gases in Geochemistry and Cosmochemistry*. Washington: Mineralogical Society of America.

Poreda, R.J. and Farley, K.A. 1992. Rare gases in Samoan xenoliths. *Earth and Planetary Science Letters* 113: 129–44.

Raquin, A., Moreira, M.A. and Guillon, F. 2008. He, Ne and Ar systematics in single vesicles: mantle isotopic ratios and origin of the air component in basaltic glasses. *Earth and Planetary Science Letters* 274: 142–150. doi: http://dx.doi.org/10.1016/j.epsl.2008.07.007.

Raquin, A. and Moreira, M. 2009. Atmospheric ^{38}Ar/^{36}Ar in the mantle: implications for the nature of the terrestrial parent bodies. *Earth and Planetary Science Letters* 287: 551–8. doi: 10.1016/j.epsl.2009.09.003.

Roth, A.S.G., Liebske, C., Maden, C., et al. 2019. The primordial He budget of the Earth set by percolative core formation in planetesimals. *Geochemical Perspectives Letters* 9: 26–31. doi: http://dx.doi.org/10.7185/geochemlet.1901.

Rubin, M., Altwegg, K., Balsiger, H., et al. 2018. Krypton isotopes and noble gas abundances in the coma of comet 67P/Churyumov-Gerasimenko. *Science Advances* 4: eaar6297. doi: 10.1126/sciadv.aar6297.

Rudnick, R.L. and Gao, S. 2014. Composition of the continental crust, in Holland, H.D. and Turekian, K.K. (eds). *Treatise on Geochemistry*, 2nd ed. Oxford: Elsevier, 1: 51.

Salters, V.J.M. and Stracke, A. 2004. Composition of the depleted mantle. *Geochemistry Geophysics Geosystems* 5: Q05B07. doi: 10.1029/2003GC000597.

Sarda, P. 2004. Surface noble gas recycling to the terrestrial mantle. *Earth and Planetary Science Letters* 228: 49–63.

Sarda, P., Staudacher, T. and Allègre, C.J. 1985. ^{40}Ar/^{36}Ar in MORB glasses: constraints on atmosphere and mantle evolution. *Earth and Planetary Science Letters* **72**: 357–75. doi: https://doi.org/10.1016/0012-821X(85)90058-5.

Sasaki, S. and Nakazawa, K. 1988. Origin of isotopic fractionation of terrestrial Xe: hydrodynamic fractionation during escape of the primordial H$_2$–He atmosphere. *Earth and Planetary Science Letters* **89**: 323–34. doi: https://doi.org/10.1016/0012-821X(88)90120-3.

Schlosser, P. and Winckler, G. 2002. Noble gases in ocean waters and sediments. *Reviews in Mineralogy and Geochemistry* **47**: 701–30. doi: 10.2138/rmg.2002.47.15.

Smith, G.P., Wiens, D.A., Fischer, K.M., et al. 2001. A Complex Pattern of Mantle Flow in the Lau Backarc. *Science* **292**: 713–6. doi: 10.1126/science.1058763.

Starkey, N.A., Stuart, F.M., Ellam, R.M., et al. 2009. Helium isotopes in early Iceland plume picrites: constraints on the composition of high ^{3}He/^{4}He mantle. *Earth and Planetary Science Letters* **277**: 91–100. doi: 10.1016/j.epsl.2008.10.007.

Staudacher, T. and Allègre, C.J. 1982. Terrestrial xenology. *Earth and Planetary Science Letters* **60**: 389–406. doi: 10.1016/0012-821X(82)90075-9.

Sumino, H., Burgess, R., Mizukami, T., et al. 2010. Seawater-derived noble gases and halogens preserved in exhumed mantle wedge peridotite. *Earth and Planetary Science Letters* **294**: 163–72. doi: https://doi.org/10.1016/j.epsl.2010.03.029.

Torgersen, T. and Clarke, W.B. 1985. Helium accumulation in groundwater, I: an evaluation of sources and the continental flux of crustal ^{4}He in the Great Artesian Basin, Australia. *Geochimica et Cosmochimica Acta* **49**: 1211–8. doi: 10.1016/0016-7037(85)90011-0.

Trieloff, M., Kunz, J. and Allègre, C.J. 2002. Noble gas systematics of the Réunion mantle plume source and the origin of primordial noble gases in Earth's mantle. *Earth and Planetary Science Letters* **200**: 297–313. doi: 10.1016/S0012-821X(02)00639-8.

Trieloff, M., Kunz, J., Clague, D.A., et al. 2000. The nature of pristine noble gases in mantle plumes. *Science* **288**: 1036–8. doi: 10.1126/science.288.5468.1036.

Tucker, J.M., Mukhopadhyay, S. and Schilling, J.-G. 2012. The heavy noble gas composition of the depleted MORB mantle (DMM) and its implications for the preservation of heterogeneities in the mantle. *Earth and Planetary Science Letters* **355–6**: 244–54. doi 10.1016/j.epsl.2012.08.025.

Valbracht, P.J., Honda, M., Matsumoto, T., et al. 1996. Helium, neon and argon isotope systematics in Kerguelen ultramafic xenoliths: implications for mantle source signatures. *Earth and Planetary Science Letters* **138**: 29–38. doi: 10.1016/0012-821X(95)00226-3.

Valbracht, P.J., Staudacher, T., Malahoff, A., et al. 1997. Noble gas systematics of deep rift zone glasses from Loihi Seamount, *Hawaii. Earth and Planetary Science Letter* **150**: 399–411. doi: 10.1016/S0012-821X(97)00094-0.

Wang, K., Lu, X., Liu, X., et al. 2022. Partitioning of noble gases (He, Ne, Ar, Kr, Xe) during Earth's core segregation: a possible core reservoir for primordial noble gases. *Geochimica et Cosmochimica Acta* **321**: 329–42. doi: https://doi.org/10.1016/j.gca.2022.01.009.

White, W.M. 2020. *Geochemistry*, 2nd ed. Oxford: Wiley Blackwell.

White, W.M. and Klein, E.M. 2014. 4.13 - Composition of the oceanic crust, in Holland, H.D. and Turekian, K.K. (eds). *Treatise on Geochemistry*, 2nd ed. Oxford: Elsevier, 457–96 pp.

Wieler, R. 2002. Noble gases in the Solar System. *Reviews in Mineralogy and Geochemistry* **47**: 21–70. doi: 10.2138/rmg.2002.47.2.

Williams, C.D. and Mukhopadhyay, S. 2019. Capture of nebular gases during Earth's accretion is preserved in deep-mantle neon. *Nature* **565**: 78–81. doi: 10.1038/s41586-018-0771-1.

Workman, R. and Hart, S.R. 2005. Major and trace element composition of the depleted MORB mantle (DMM). *Earth Planetary Science Letters* **231**: 53–72. doi: 10.1016/j.epsl.2004.12.005.

Yang, J. and Anders, E. 1982. Sorption of noble gases by solids, with reference to meteorites. II. Chromite and carbon. *Geochimica et Cosmochimica Acta* **46**: 861–75. doi: 10.1016/0016-7037(82)90043-6.

Yokochi, R. and Marty, B. 2004. A determination of the neon isotopic composition of the deep mantle. *Earth and Planetary Science Letters* **225**: 77–88. doi: 10.1016/j.epsl.2004.06.010.

Zadnik, M.G., Smith, C.B. and Begemann, F. 1987. Crushing a terrestrial diamond: ^{3}He/^{4}He higher than solar. *Meteoritics* **22**: 540–1.

PROBLEMS

1. Assume that the atmosphere is steady state and that the isotopic composition of He lost from the top of the atmosphere equals that entering it from the solid Earth. Using the ^{3}He/^{4}He of the atmosphere given in Table 14.2, the He flux and isotopic from the mantle through the oceans given in Section 14.3.2, and assuming that the flux from the continents has an isotopic

composition of $R/R_A = 0.01$ and that from the mantle through the oceans an R/R_A of 8.2, calculate the total flux to the atmosphere. What are the fractions from the crust and oceans?

2. Repeat the calculation of total flux in Problem 1 but assume that ^3He preferentially escapes from the top of the atmosphere such that the ratio of $^3He/^4He$ escaping is 10% higher than the atmospheric concentration (i.e., R/R_A escaping = 1.1).

3. Using the flux you calculated in Problem 1, the atmospheric concentration of 5.2 ppmv, an atmospheric mass of 5.1×10^{21} g, and a molecular weight of 28.8 for atmospheric gas, what is the residence time of ^4He in the atmosphere?

4. Assuming a silicate Earth concentration of U and Th to be 16 and 64 ppb, respectively, and a mass of Earth = 5.97×10^{27} g what is the current production rate of ^4He in the Earth? How does this compare with the rate of ^4He loss to the atmosphere you calculated in Problem 1?

5. The Earth loses heat at a rate of 44 TW, about 16 TW of which is radiogenic, the remainder representing the loss of primordial heat. Assuming the $^3He/^4He$ ratio of the atmosphere represents the ratio of which these isotopes are lost from the solid Earth and that all ^4He is radiogenic and all ^3He is primordial, how do the rates of loss of radiogenic He and heat compare?

Index

Please note that page references to Figures will be followed by the letter '*f*', to Tables by the letter '*t*'

Isotope Geochemistry, Second Edition. William M. White.
© 2023 John Wiley & Sons Ltd. Published 2023 by John Wiley & Sons Ltd.
Companion Website: www.wiley.com/go/white/isotopegeochem2